D0151792

ELECTRONIC DEVICES AND CIRCUITS
Discrete and Integrated

Denton J. Dailey
Butler County Community College

Upper Saddle River, New Jersey
Columbus, Ohio

Library of Congress Cataloging in Publication Data
Dailey, Denton J.
Electronic devices and circuits: discrete and intregrated / Denton J. Dailey.
p. cm.
ISBN 0-13-081110-6
1. Transistors. 2. Digital electronics. 3. Operational amplifiers. 4. Analog electronic systems. 5. Electric circuit analysis—Data processing. I. Title.

TK7871.9.D33 2001
621.3815′28—dc21 99-086432

Vice President and Publisher: Dave Garza
Acquisitions Editor: Scott J. Sambucci
Production Editor: Stephen C. Robb
Production Supervision: Holly Henjum, Clarinda Publication Services
Design Coordinator: Karrie M. Converse-Jones
Cover Designer: Jeff Vanik
Production Manager: Pat Tonneman
Marketing Manager: Ben Leonard

This book was set in American Garamond and Officina by The Clarinda Company. It was printed and bound by Von Hoffmann Press. The cover was printed by Phoenix Color Corp.

Credits: pp. 774–802, Data sheets copyright by ON Semiconductor; pp. 803–807, copyright by Intersil Corporation and used with permission.

OrCAD® PSpice® is a registered trademark of Cadence® Design Systems.

Copyright © 2001 by Prentice-Hall, Inc., Upper Saddle River, New Jersey 07458. All rights reserved. Printed in the United States of America. This publication is protected by Copyright and permission should be obtained from the publisher prior to any prohibited reproduction, storage in a retrieval system, or transmission in any form of by any means, electronic, mechanical, photocopying, recording, or likewise. For information regarding permission(s), write to: Rights and Permissions Department.

10 9 8 7 6 5 4 3 2 1
ISBN 0-13-081110-6

PREFACE

As stated in the title, this book is about the study of electronic devices and circuits. There is an excellent balance of coverage between discrete devices and integrated circuits (ICs), making this book suitable for use in courses that cover either or both of these areas. In general, there is more than enough material covered here for a two-course sequence covering discrete devices, amplifiers, oscillators, and linear ICs. This book is primarily intended for use by students in two- and four-year electronics and electrical engineering technology programs.

Prerequisites for this text are basic knowledge of dc and ac circuit analysis techniques, including Ohm's law, Kirchhoff's laws, the superposition theorem, phasor algebra, and some trigonometry. The use of some calculus is unavoidable, especially when discussing differentiators and integrators; however, no formal calculus background is assumed. When necessary, the basic techniques and applications of differentiation and integration are presented in the text and they are explained in the most straightforward manner possible. It has been the author's experience that more often than not, even students who have never been exposed to calculus appreciate the insight that a brief encounter with derivatives and integrals provides. On the other hand, the book has been written so that the more mathematically advanced discussions can be omitted without loss of continuity. For example, the unit step function is a topic that has traditionally been ignored in devices and IC books. This is included here because it is such an important concept in later studies, and it is interesting and rather easy to understand. It is the author's opinion that students' classroom experience with these analytical tools is equivalent to use of equipment such as spectrum analyzers and logic analyzers in the lab. However, this topic can be omitted without loss of continuity. This flexibility allows the book to be used in a wide variety of programs. This also makes the book more useful as a reference for further study.

The emphasis of this book is on device behavior and modeling. Because of the inherent nonlinearity of electronic devices, their study requires the student to think at a somewhat deeper level of abstraction, as compared to dc and ac circuit analysis. The presentation style used here should make this transition easier. Wherever possible, the emphasis is on the development of analysis equations using the basics: Ohm's law, Kirchhoff's laws, and the superposition theorem, of which Thevenin's theorem is a direct extension. Also, whenever possible, several alternative explanations of various topics are presented. This book is definitely not about the memorization of formulas, although some formulas are used so often that memorization is automatic. Because of the time constraints that instructors must deal with and because of practical book cost and space considerations, it is not possible to develop *every* topic in detail from first principles. However, this approach is taken as often as possible. The instructor may wish to deemphasize or even omit certain topics or sections that are less important, based on the emphasis of his or her program. For example, the coverage of zener regulators in Chapter 2 could be deemphasized, or perhaps deferred until later.

Each chapter ends with a section on computer-aided circuit analysis. PSpice was chosen because it is the most widely used circuit analysis program in the world. A fully functional evaluation version is also available free from OrCAD. This is available on CD, or it can be downloaded from the OrCAD website at www.orcad.com. The examples covered in these sections serve several purposes. First, they are primarily designed to reinforce, and are directly related to, the material covered in each associated chapter. Second, the computer analysis examples sometimes serve to demonstrate the limitations of the manual analysis approximations that are presented. Last, the computer-aided analysis sections allow easy access to more advanced topics such as frequency response in polar and rectangular forms, spectral analysis, transient response, and analog behavioral modeling. Computer-aided analysis was placed in separate sections within each chapter so that it is very easy to find and can more easily be excluded, if so desired.

The author welcomes suggestions for future revisions of this book. Recommendations regarding topic sequencing, errors and omissions, and material that should be deleted or inserted will be appreciated. Please feel free to send your comments to me at djdailey@altavista.com.

What's in this Book?

Chapters 1 and 2 cover diodes and diode applications. Chapter 1 concentrates on the use of diode modeling and the various levels of approximation that are applied in describing diode behavior. The concept of dynamic resistance is also introduced here. Significant coverage is given to zener diodes, photodiodes, and light-emitting diodes as well. Chapter 2 emphasizes rectifier and wave-shaping circuits.

Chapter 3 introduces the bipolar junction transistor (BJT). The emphasis here is on the behavioral characteristics of BJTs rather than on device physics. The basic transistor parameters are covered, including α, β, leakage currents, breakdown voltages, and hybrid parameters. The Early voltage and base-spreading effects are also covered here.

Chapter 4 covers all aspects of bipolar transistor biasing. This is presented strictly from a dc analysis standpoint. In the author's experience, coverage of signal-related topics at this stage tends to confuse many students. The objective here is to make the student comfortable using the basic circuit reduction techniques that allow complicated biasing arrangements to be redrawn in simpler, more familiar forms. Biasing arrangements that are more applicable to linear IC designs, including the current mirror, are also presented here. The dc load line is also covered in this chapter.

Chapter 5 covers BJT small-signal amplifiers using *r* parameters. The basic common-emitter, common-collector, and common-base configurations are presented. Emphasis is placed on the development of the ac-equivalent circuit. Other topics in this chapter include ac load lines, output compliance, decibels, and the Darlington configuration.

Field-effect transistors are covered in Chapters 6 and 7. The basic FET parameters and biasing arrangements are covered in Chapter 6, as well as a few applications such as analog switches, voltage-controlled resistance, and constant-current diodes. Chapter 7 covers the analysis of FET small-signal amplifiers. The relative advantages of FETs and BJTs are discussed.

Differential amplifiers are the subject of Chapter 8. This chapter covers both the biasing and small-signal characteristics of differential amplifiers. The concepts of differential gain, common-mode gain, and common-mode rejection are presented here. This chapter provides a solid foundation for the later study of operational amplifiers.

Multiple-stage amplifiers are covered in Chapter 9. This chapter serves to tie together much of the material covered in previous chapters, including class A BJT and FET amplifiers and differential amplifiers. Coverage of capacitively coupled and direct-coupled stages is also presented, along with the derivation of decibel gain.

Chapter 10 covers power amplifiers and amplifier classifications (A, B, AB, and C). This is a direct extension of the previous chapter on multiple-stage amplifiers. Thermal analysis, overcurrent protection, and some tuned amplifier theory are presented in this chapter.

Chapter 11 introduces the concept of negative feedback and its effects on amplifier characteristics. Some basic filter terminology, the concept of the Bode plot, the Miller effect, and transistor frequency limitations are covered as well. The emphasis here is on high-frequency operation, and coverage of the cascode amplifier follows directly from this perspective.

Operational amplifiers are introduced in Chapter 12. The op-amp is first analyzed as an ideal gain block. Specifications of a sampling of commercial op-amps are presented and discussed as well. Nonideal characteristics of real op-amps, such as small-signal bandwidth, slew rate and power bandwidth, CMRR, PSRR, and offset errors, are examined. Offset compensation techniques are discussed.

Chapter 13 covers a variety of linear op-amp applications, which is a logical continuation of the previous chapter. Basic circuits and applications are analyzed for differential amplifiers, instrumentation amplifiers, V/I and I/V converters, bridged amplifiers, differentiators, and integrators. Brief introductions to differentiation and integration are presented, as well as the concept of the unit step function. These mathematical concepts are discussed primarily in the context of commonly occurring functions, including constant, linear, parabolic, sinusoidal, and exponential functions. Where possible, alternative analysis approaches are presented, such as the use of phasor algebra versus calculus when determining the response of an integrator or differentiator to a sinusoidal input.

Active filters are covered in Chapter 14. This linear application is so extensive that it warrants a chapter of its own. The equal-component, Sallen-Key active filter circuits are emphasized in the coverage of low-pass and high-pass filters. A variety of other filters, including IGMF, biquadratic, and state-variable structures, are presented. Several applications for the all-pass response and a quick introduction to switched capacitor filters are presented.

Chapter 15 covers a variety of nonlinear op-amp applications. Detailed coverage is given to comparators, log and antilog amplifiers, and precision rectifier circuits. Many applications are presented as well, including signal compression, analog multiplication, signal linearization, and digital signal conditioning. These circuits are examined from a transfer characteristic standpoint as well as from a strictly electronic behavior standpoint.

Digital-to-analog and analog-to-digital conversion circuits are covered in Chapter 16. Concepts relating to sample-and-hold, track-and-hold, undersampling, and oversampling are presented as well. Various DAC and ADC error specifications are discussed, as well as applications, including the use of ADCs and DACs in servomechanisms and arbitrary function generators.

Chapter 17 covers oscillator and tuned amplifier circuits. These are closely related topics in many respects. There is a heavy emphasis on transformer coupling, since it is still very commonly employed in high-frequency oscillator and tuned amplifier circuit design. A review of transformer operation and resonant circuit theory is presented. Some additional transistor performance specifications are also introduced, including noise figures. The classical sinusoidal oscillator topologies are presented, including Hartley, Colpitts, and Clapp circuits. A sampling of untuned sinusoidal oscillator circuits is presented as well, including the Wein bridge and phase-shift oscillators.

Chapter 18 covers a variety of important linear integrated circuits. Devices included are the 555 timer, linear and switching regulators, operational transconductance amplifiers, balanced modulators, and phase-locked loops. Many applications are presented, including regulated single-polarity and bipolar power supplies, AM and FM modulators and demodulators, servomechanisms, and phase comparators.

Chapter 19 covers several miscellaneous devices, including SCRs, triacs, unijunction transistors, and tunnel diodes.

Suggestions for Alternative Chapter Sequencing

Although there is necessarily a great dependence of later chapters on earlier chapters, there is also room for quite a bit of flexibility in the sequence of topics covered. For example, you may wish to cover Chapter 17, Tuned Amplifiers and Oscillators, before covering op-amps. Likewise, it is possible to cover Chapter 19, Thyristors and Other Miscellaneous Devices, at any point following Chapter 4. It is even possible to cover Chapter 8, Differential Amplifiers, sometime after covering op-amps.

Acknowledgments

The author thanks the following persons for their efforts in reviewing and providing suggestions during the writing of this book: Jeff Bigelow, Oklahoma Christian University; Tony Hearn, Community College of Philadelphia; Robert Martin; Rich Parrett, ITT; Robert Powell, Oakland Community College; Ken Simpson, Stark State College of Technology.

In addition, the author thanks the people at Motorola, ON Semiconductor, Texas Instruments, Intersil, and OrCAD for their invaluable support.

Denton J. Dailey

CONTENTS

CHAPTER 1

The PN Junction Diode

BEHAVIORAL OBJECTIVES

On completion of this chapter the student should be able to:

- Explain the use of dopants in semiconductor device manufacture.
- Name the commonly used semiconductor materials.
- Choose the appropriate behavioral model of a diode, given the application in which it is being used.
- Analyze diode circuits using Kirchhoff's laws and Thevenin's theorem.
- Explain the difference between dynamic resistance and static (ohmic) resistance.
- Interpret diode specifications as found on device data sheets.
- Use computer simulation to analyze the behavior of circuits containing diodes.
- Analyze circuits containing zener diodes.

INTRODUCTION

This chapter presents an introduction to semiconductor theory in general, and diode operation and characteristics in particular. Diode circuit models and approximations are discussed, as well as diode thermal characteristics. In addition to the standard general-purpose diode, zener diode operating principles are also introduced in this chapter. Common zener diode applications are covered more extensively in Chapter 2.

The major emphasis of this chapter is on the characteristics and behavior of the standard pn junction diode, and they are the most important aspects of diode theory for the technician and engineering technologist to understand. The physical theory of semiconductors requires the use of advanced mathematics and physics that are beyond the scope of this book. If you are planning to be involved in the design of diodes and transistors then fundamental device physics will be extremely important to you. This book is intended for persons that need to understand the behavioral characteristics of semiconductor devices and how they are used in various applications, therefore we cover device physics considerations in a rather qualitative manner, in order to provide some insight into the mechanics of the pn junction.

1.1 SEMICONDUCTOR BASICS

As the name implies, a semiconductor is a material whose conductive characteristics lie somewhere between those of good conductors, such as metals like gold and copper, and insulating materials such as ceramic and plastics. Currently, the most commonly used semiconductor material is silicon (Si). Semiconductor devices are also made using germanium (Ge) and various other materials, including gallium arsenide (GaAs). In this text, we concentrate mainly on silicon devices and discuss the other devices to a lesser extent.

The Bohr Model of the Atom

Danish physicist Neils Bohr postulated that atoms are composed of a nucleus of protons and neutrons that is orbited by electrons. Quantum mechanical arguments dictate that electron orbits can only exist at specific discrete distances from the nucleus of the atom. These allowed orbits are called shells. Electrons of relatively lower energy occupy shells close to the nucleus, while higher energy electrons occupy more distant shells. The outermost shell that is occupied by electrons in a given atom is called the *valence band,* and the electrons located there are called *valence electrons.* The valence electrons are really the only constituents of the atom that are significant in the context of this book.

If we lump the nucleus and all but the valence electrons together, we can represent an atom of silicon, for example, as shown in Fig. 1.1. Although we will not really need it very often, this is the model of the atom we will use in this book.

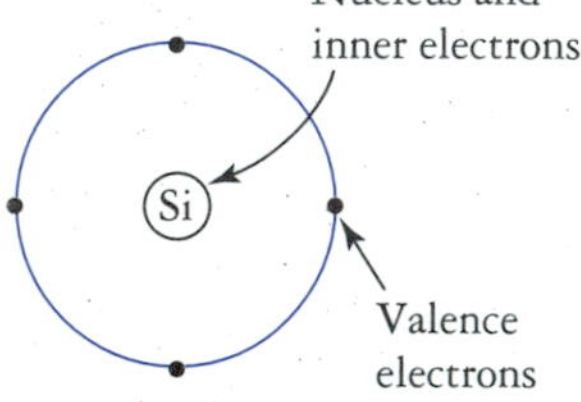

FIGURE 1.1
Bohr model of a silicon atom.

To maintain electrical neutrality, an atom will tend to have the same number of orbiting electrons as it has protons in the nucleus. Atoms that have too many or too few electrons are called *ions.* One of the characteristics of all atoms is an affinity for a full number of electrons in the outermost orbital. This tendency determines the structural characteristics of a given material and is important in the following discussion.

Intrinsic Silicon

Silicon is a Group IV element. That is, four valence electrons are associated with each atom of silicon. In general, the atomic structure of a given element can be found in the *periodic table.* Groups are determined by the valence electron configurations of the various elements.

One of the important atomic characteristics of silicon is that its valence band is completely filled when it contains eight electrons. Because of this, molten silicon forms a crystalline structure, called a *lattice,* when it solidifies. In particular, silicon atoms tend arrange themselves into what is called a face-centered cubic lattice. Figure 1.2 shows a highly idealized two-dimensional representation of the silicon crystal. It turns out that this crystal structure causes the four nearest neighbors of a given silicon atom to share valence electrons, in effect allowing each atom to act as though it has eight valence electrons, without altering electrical neutrality. This type of arrangement, in which valence electrons are shared among neighboring atoms, is called *covalent bonding.*

Note
Other types of bonds are ionic and metallic bonds.

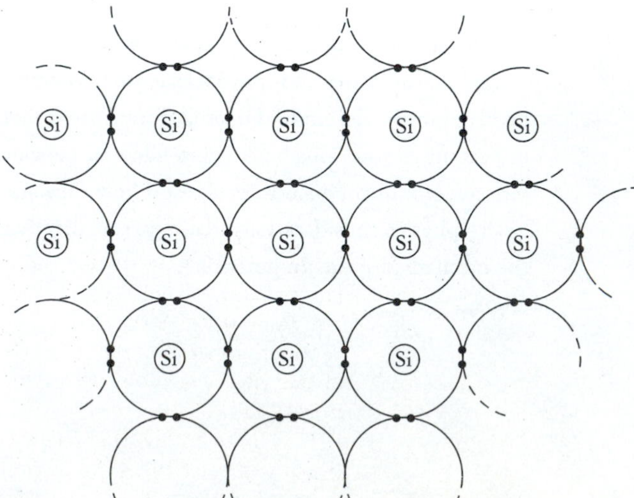

FIGURE 1.2
Two-dimensional cross section of Si atoms in a crystal lattice.

Note
Silicon can also assume a non-crystalline form, called *amorphous* silicon.

A crystal of pure silicon is called *intrinsic silicon.* At normal temperatures, intrinsic silicon is a very poor conductor. This is true because nearly all of the electrons in the silicon crystal are committed to the covalent bonding process. If current is to flow through the crystal, enough external energy must be supplied to pull electrons free from the valence band. This energy is substantial, and to obtain appreciable current flow, very high voltages are required. The conductivity of silicon is affected by temperature, however. Unless the silicon is at absolute zero, thermal energy will be present, and every once in a while, an electron will be stimulated with enough energy to be kicked up from the valence band into what is called the *conduction band.*

Holes and Electrons

When thermal energy releases an electron from the valence band, that electron is free to propagate through the lattice and act as a current carrier. The valence band is now short one electron, leaving what is called a *hole.* This is illustrated from an energy standpoint in Fig. 1.3.

FIGURE 1.3
A photon of the correct energy can raise an electron from the valance band to the conduction band.

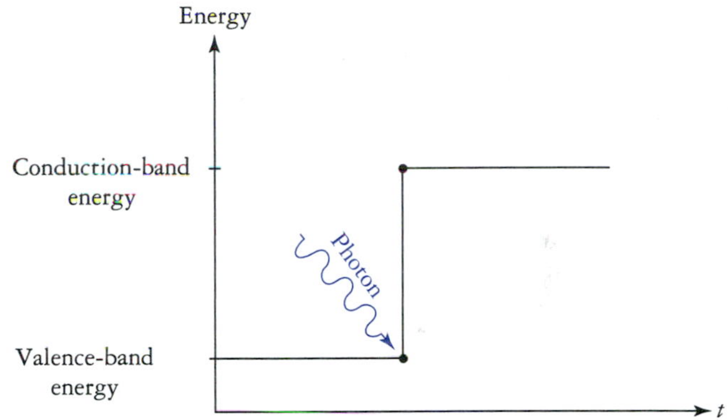

Holes behave just like particles with the same mass and opposite charge of an electron. A voltage applied to the silicon crystal will cause the thermally generated electron to drift toward the positive end, whereas in the valence band, the hole drifts toward the negative end of the crystal. Thus, physically speaking, hole flow is the movement of charge carriers in the valence band. Keep in mind that in the typical semiconductor device, literally millions of electron–hole pairs are being created this way at any instant, at room temperature. Even so, intrinsic silicon is a poor conductor. As a consequence of the electron–hole mechanism, silicon has a negative temperature coefficient (NTC), as does germanium. This means that as temperature increases, the specific resistance of silicon decreases. This fact is put to use in the construction of temperature-sensitive resistors called *thermistors.*

Extrinsic Silicon

To produce a useful electronic device, impurities, called *dopants,* are added to silicon. Recall that silicon is a Group IV element. The most commonly used dopants are the Group V elements arsenic, phosphorus, and antimony, and the Group III elements boron, gallium, and aluminum. Group V elements have five valence electrons, and are called *pentavalent* dopants. Group III elements have three valence electrons and so are called *trivalent* dopants.

N-Type Material

If a small amount of a Group V element, say, arsenic, is present in the silicon (or germanium) lattice, a structure like that shown in Fig. 1.4 is produced. Because arsenic is electronically similar to silicon, arsenic atoms readily join the lattice, but in effect, each arsenic position has an extra electron that is not committed to the covalent bond. This electron is loosely bound and is easily boosted into the conduction band (without creating an associated hole), where it can act as a current carrier. Because of this apparent excess of electrons, silicon that is doped with pentavalent impurities is called *n-type silicon.*

Undoubtedly, thermal energy will disrupt some of the electrons that belong in the valence band, with each event producing a free electron and a hole. Since the pentavalent dopant also provides a large number of conduction-band electrons, but not holes, in n-type material, electrons are

FIGURE 1.4
Arsenic is an n-type dopant. The fifth valence electron requires little energy to move to the conduction band.

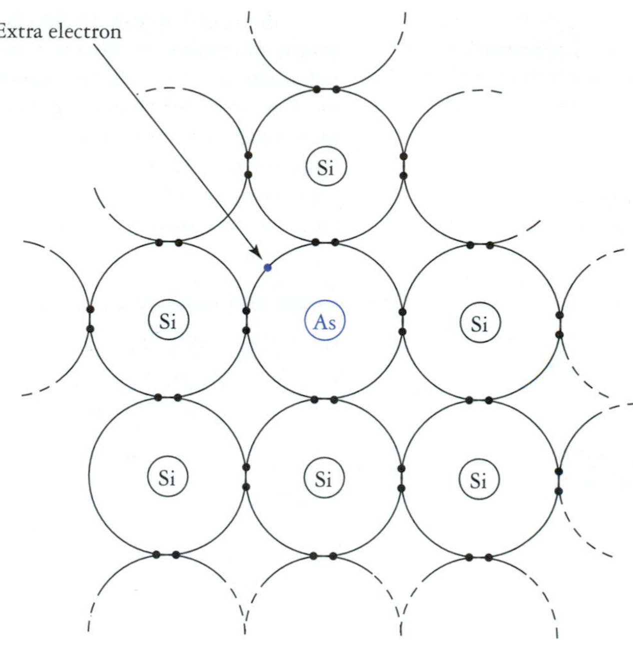

said to be the *majority carriers.* Because relatively fewer holes are present, they are said to be *minority carriers* when they exist in n-type material.

P-Type Material

If small amounts of a trivalent material, such as boron, are added to the silicon lattice we get the structure shown in Fig. 1.5. In this case, the valence band associated with each boron atom has room for one electron. Recall that such a vacancy in the valence band is called a hole. These holes are free to move through the lattice under the influence of an applied voltage, and thus they act as current carriers. Trivalent dopants do not, however, cause the creation of valence-band electrons. The term *p-type silicon* is used because of the positive charge associated with a hole.

As was the case with n-type material, thermally generated electron–hole pairs are also created in p-type material. In this case, the holes are the majority carriers, and thermally generated conduction-band electrons are the minority carriers.

Dopant Concentrations

Although it is not crucial information at this level of study, it is interesting to look at some of the numbers associated with the preceding discussions.

FIGURE 1.5
Boron is a p-type dopant. The hole in the valence band is free to move, acting as a positive charge carrier.

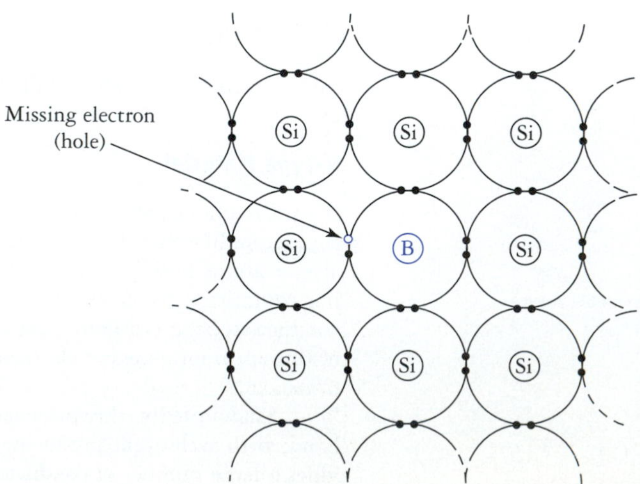

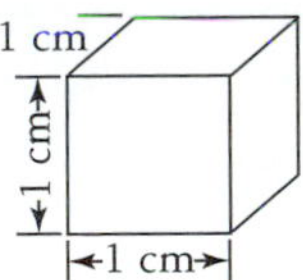

FIGURE 1.6
A 1-cm³ piece of silicon.

A 1-cm³ crystal of silicon (shown to scale in Fig. 1.6) contains approximately 10^{22} silicon atoms. This is called the atom density of silicon. For intrinsic silicon at 27°C (approximately room temperature), there will be approximately 10^{10} thermally generated electron–hole pairs available for conduction at any time. This certainly sounds like a large number of electron–hole pairs. But close examination of the numbers indicates that on the average, for every single thermally generated pair, 10^{12} covalent bonds are still intact.

Typical doping concentrations used for semiconductor devices are in the range of 10^{15} to 10^{16} atom/cm³. Again, these sound like very heavy concentrations of impurities, but comparison of these figures with the atom density of silicon shows that silicon atoms still outnumber impurity atoms from a 1 million-to-one to around a 10 million-to-one ratio. However, even this rather low concentration of impurities increases the conductivity of extrinsic silicon to on the order of 100,000 to 1,000,000 times that of intrinsic silicon. Thus it is reasonable to expect that many aspects of semiconductor device behavior are extremely sensitive to dopant concentration and distribution.

1.2 DIODE FUNDAMENTALS: THE PN JUNCTION

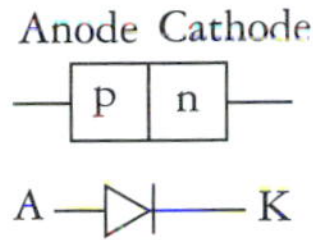

FIGURE 1.7
Representative structure and schematic symbol for a pn junction diode.

Functionally speaking, a diode is a two-terminal device that tends to allow current to flow in one direction, while blocking current flow in the other. In other words, a diode is analogous to a one-way valve, such as a reed valve that would be found on a two-stroke engine. This section lays the foundation for an understanding of how this one-way conduction occurs.

Taken separately, n- and p-type silicon are not too useful. However, if a single crystal of silicon is produced with one side being n-type doped and the other side p-type doped, a truly useful device called a *pn junction diode,* which we'll just call a diode, is produced. Figure 1.7 shows the structure and schematic symbol for a diode. The term *diode* is a contraction of *di,* meaning two, and *electrode.* The p side of the diode is called the *anode* and the n side is called the *cathode.* The letter K is normally used to denote the cathode. On schematic diagrams, diodes are usually denoted D_n, or occasionally CR_n.

The Depletion Region

When a pn junction is initially formed, we have a structure much like that shown in Fig. 1.8(a). The plus signs represent holes associated with trivalent dopants, and the minus signs represent uncommitted electrons associated with pentavalent dopants. Because the n- and p-type regions have different concentrations of holes and electrons, there is a natural tendency for holes to move across to the n side and for electrons to move over to the p side of the junction. This charge carrier movement is called *diffusion.* This diffusion causes recombination of holes and electrons on each side of the junction. Each electron that leaves the n side and recombines with a hole on the p side creates two ions: a negative ion on the p side and a positive ion on the n side.

Recall that ions are electrically charged atoms. Diffusion continues until the electrostatic field created by the junction ions cancels the forces driving the diffusion process. The portion of the junction where hole–electron recombination has created ions is called the *depletion region.* The term *depletion region* is derived from the fact that because of electron–hole recombination, there are no current carriers left in this region, thus the region is depleted of current carriers. Another term given for this region of the pn junction is the *space-charge region.*

Note
Saturation current approximately doubles for every 10°C rise in temperature.

Bias Conditions

The preceding discussion described the behavior of the pn junction with no external voltage applied to the diode. When an external voltage is applied to a diode, the diode is said to be *biased.* Basically, two bias conditions can occur: reverse bias and forward bias. Figure 1.9 illustrates the concept of biasing.

Reverse Bias

The diode in Fig. 1.9(a) is under reverse bias. Under reverse bias, the depletion region widens such that the electric field produced by the ions just cancels the applied reverse bias voltage. Under these conditions, a very small leakage current, the *saturation current* I_S, will flow. Unless a diode is at absolute zero, thermally generated electron–hole pairs will be produced in the depletion region, and these charge carriers constitute the saturation current. As a general rule, saturation current doubles for every 10°C rise in junction temperature. Saturation current is also directly proportional to junction area, and so is sometimes called the *scale current.*

FIGURE 1.8
(a) Hole and electron distributions at formation of the junction.
(b) Diffusion of holes and electrons causes formation of a nonconducting depletion region.

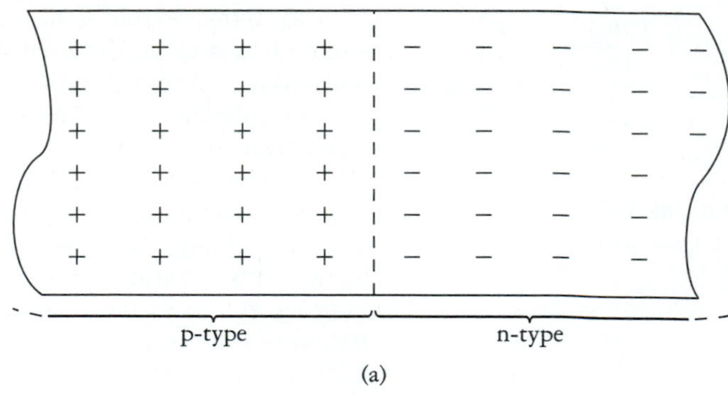

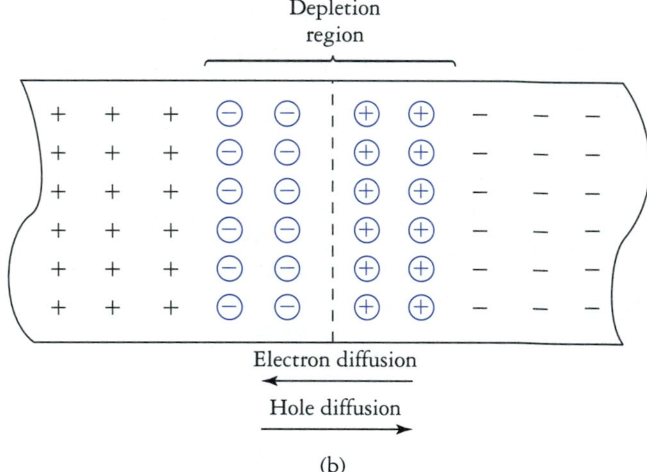

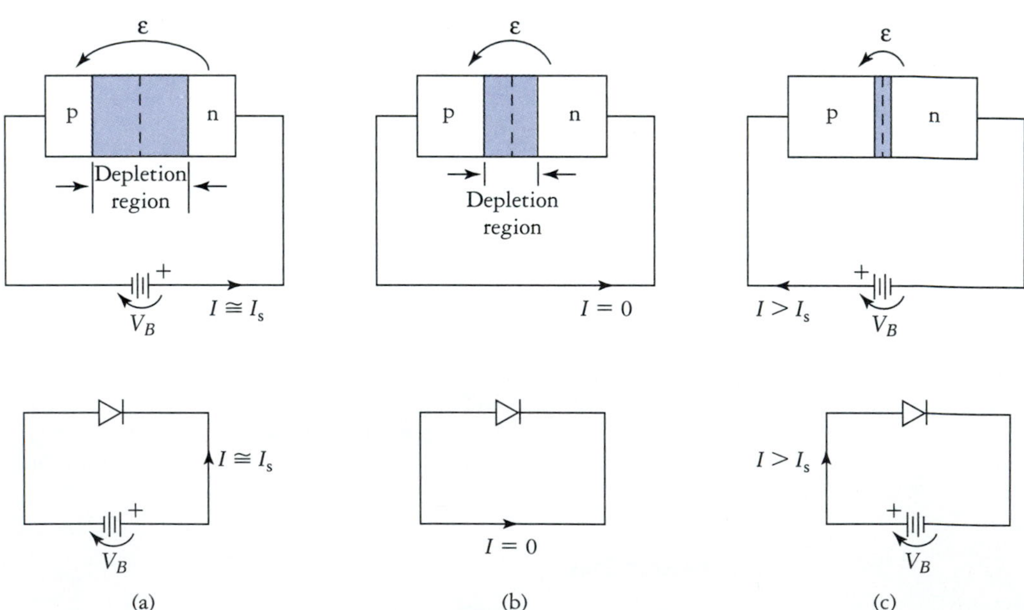

FIGURE 1.9
Biasing a diode. (a) Reverse bias causes the depletion region to widen. (b) No bias. (c) Forward bias causes current to flow across the junction.

For a typical low-power silicon diode, saturation current is on the order of a few nanoamps or less at 25°C. A germanium device of similar construction will normally exhibit a saturation current of several microamps or more.

In addition to thermally generated saturation current, a second component, caused by *surface leakage* is also present. Surface leakage occurs because the silicon atoms located at the sides of the crystal have no neighboring atoms to bond with outside the lattice. Thus, uncommitted electrons are present on the surface of the depletion region to act as current carriers. In silicon devices, surface leakage may be several times larger than the saturation current, at normal operating temperatures. In addition, surface leakage is relatively constant, being nearly independent of temperature and bias voltage. This has some consequences that are examined later.

Forward Bias

Note
For silicon diodes, barrier potential $V\phi \cong 0.7$ V. For germanium diodes, $V\phi \cong 0.3$ V.

When an external voltage is applied to a diode as shown in Fig. 1.9(c), the diode is said to be forward biased. Let us assume that the voltage source in Fig. 1.9(c) is slowly increased from 0 V. As the applied voltage increases, the depletion region shrinks somewhat in width; however, the energy required for charge carriers to cross the depletion region decreases exponentially. For silicon diodes, appreciable current will flow through when the bias is about 0.7 V. This voltage is called the *barrier potential* of the diode. For germanium diodes, the approximate forward voltage drop required for good conduction (the barrier potential) is about 0.3 V. The relationship between forward current and the voltage drop across the diode is nonlinear (exponential). Because of this nonlinearity, Ohm's law does not apply to diodes, except under certain conditions.

Diode Transconductance

A plot of device current as a function of voltage is called a *transconductance curve.* A typical diode transconductance curve is shown in Fig. 1.10. This characteristic is described by *Shockley's equation:*

$$I_D = I_S(e^{V_D/\eta V_T} - 1) \quad (1.1)$$

Note
A graph of current as a function of voltage is a transconductance curve.

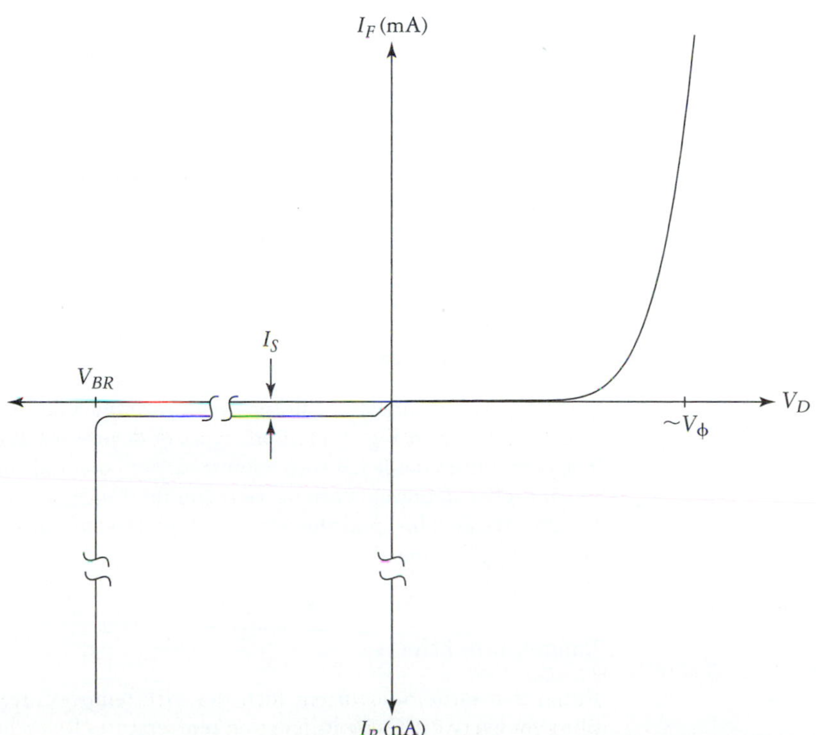

FIGURE 1.10
The diode transconductance curve.

where I_D is the current through the diode (positive for forward bias and negative for reverse bias), I_S is the diode saturation current, V_D is the voltage drop (the bias) across the diode, V_T is the thermal equivalent voltage (approximately 26 mV at room temperature), and η is called the *emission coefficient.* The thermal voltage equivalent is itself given by

Note
$k = 1.38 \times 10^{-23}$J/°K
$q = 1.6 \times 10^{-19}$C
0°K = −273°C

$$V_T = \frac{kT}{q} \tag{1.2}$$

where k is Boltzmann's constant, T is the absolute temperature of the junction, and q is the electron charge. The thermal voltage equivalent is obviously temperature dependent, with a positive temperature coefficient. Again, at 25°C,

$$V_T \cong 26 \text{ mV} \tag{1.3}$$

The emission coefficient η is determined by diode construction and also varies with diode current. For silicon diodes, at relatively low currents (from I_S to around 10 μA) or so, $\eta \cong 2$. At higher currents, $\eta \cong 1$. For germanium devices $\eta \cong 1$ for all current levels.

Fortunately, for nearly all practical analysis and design situations, Shockley's equation need not be used. It is presented here primarily for the sake of completeness and because these parameters are commonly used in computer analysis models.

Forward Bias Revisited

The first (upper-right) quadrant of the diode transconductance graph (Fig. 1.10) contains the forward bias region of operation. The point labeled V_ϕ is the approximate barrier potential of the diode (0.7 V for Si, 0.3 V for Ge). Typically, the forward current I_F will be several milliamps or so at V_ϕ volts.

Reverse Bias Revisited

The third quadrant represents reverse bias operation. Here, the current flow through the diode climbs rapidly to a relatively constant, small level, which is the saturation current. In Fig. 1.10, the I_R current scale is expanded to more readily show the saturation current.

As greater and greater reverse bias is applied to the diode, the depletion region gets wider. There is a limit to this growth, at which point the electrostatic field of the depletion region is strong enough to pull electrons from the valence band. This point is the *reverse breakdown voltage* V_{BR} of the diode. Once reverse breakdown begins, the freed electrons move rapidly through the lattice, causing other electrons to be freed and so on in chain-reaction fashion. This process of reverse breakdown is called *impact ionization* or, more commonly, *avalanche breakdown.* Typical reverse breakdown voltages for standard diodes may range from the tens of volts range to several thousand volts.

Reverse breakdown itself is not necessarily damaging to the diode. It is the combination of high reverse current and high reverse voltage that causes diode power dissipation to become appreciable. Unless the reverse breakdown current is limited by external resistance, the diode will be destroyed very rapidly.

Germanium Devices

For comparison purposes, the typical transconductance curves for similar germanium and silicon diodes is shown in Fig. 1.11. Both types of devices are governed by Shockley's equation, Eq. (1.1). The germanium diode has such a lower barrier potential compared to silicon because of germanium's much higher saturation current. Germanium devices are also much more temperature sensitive than silicon devices. This, plus the relatively high leakage current, limits germanium devices to a few specialized applications.

Temperature Effects

Recall that saturation current increases with temperature, with the value of I_S approximately doubling for every 10°C rise in junction temperature. Reference to Eq. (1.2) indicates that V_T increases in direct proportion to junction temperature. Thus we see that these two parameters I_S and V_T have opposite temperature dependencies in Shockley's equation. Because the increase in I_S is exponential,

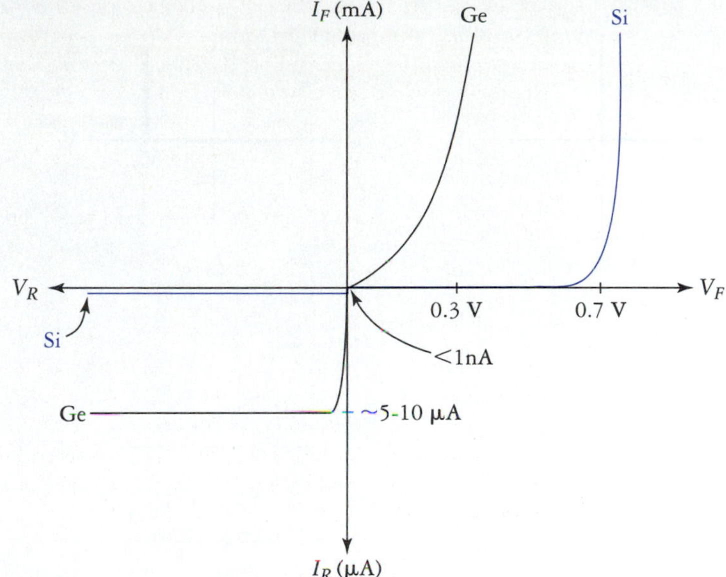

FIGURE 1.11
Comparison of typical germanium and silicon pn junction diode transconductance curves.

it will dominate, causing the diode to have a net negative temperature coefficient. This is an important consideration in many semiconductor device applications. In terms of diode transconductance, an increase in temperature causes the barrier potential to decrease by about 2 mV per degree Celsius. This is shown graphically in Fig. 1.12.

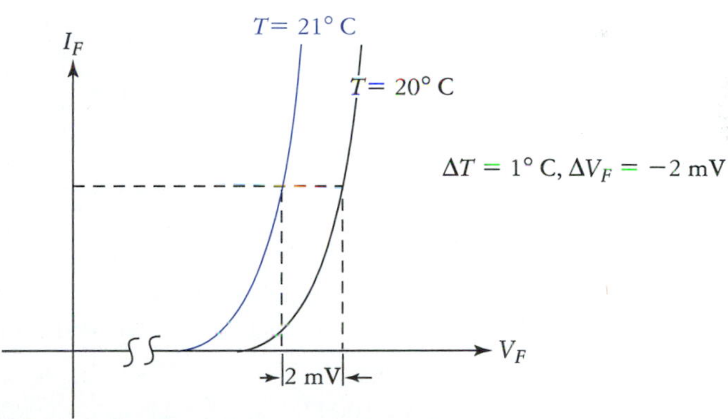

FIGURE 1.12
Increasing temperature effectively shifts the transconductance curve to the left.

1.3 DIODE MODELING AND APPROXIMATIONS

For most applications, the most important characteristic of the diode is its ability to allow current to flow in one direction, while blocking it in the other. The ideal diode would thus be a perfect conductor under forward bias and a perfect insulator under reverse bias. The transconductance curve for such an ideal diode is shown in Fig. 1.13(a). As surprising as it may seem at this point, this approximation of diode behavior is acceptably accurate in many situations.

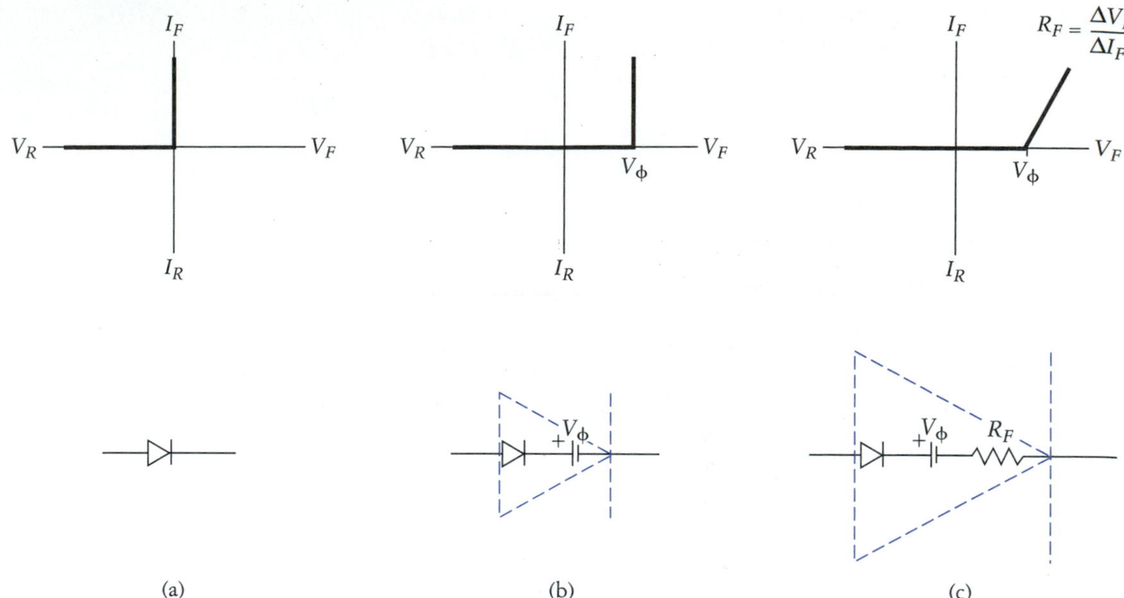

FIGURE 1.13
Diode approximations. (a) Ideal diode. (b) Ideal diode with barrier potential. (c) Ideal diode with barrier potential and linear forward resistance.

EXAMPLE 1.1

Refer to Fig. 1.14. Assuming that the diode is ideal, determine I_D given $R_S = 100\ \Omega$ and $V_B = 5$ V.

FIGURE 1.14
Simple series circuit.

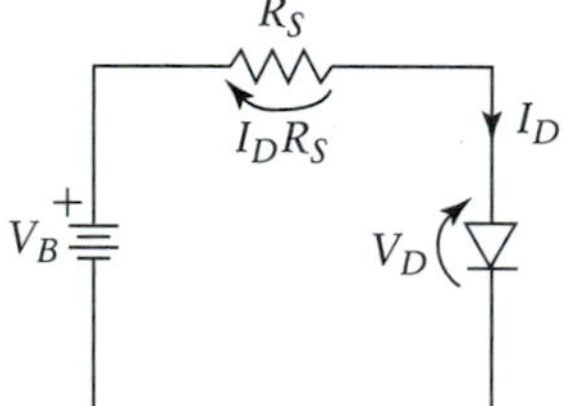

Solution The diode is forward biased (the anode is more positive than the cathode), and since this is an ideal diode, it acts like a perfect conductor. The current is simply

$$\begin{aligned} I_D &= V_B / R_S \\ &= 5\ \text{V} / 100\ \Omega \\ &= 50\ \text{mA} \end{aligned}$$

If the ideal diode approximation is not accurate enough, a more realistic model includes the approximate barrier potential of the real diode in series with the ideal diode. This is shown in Fig. 1.13(b). If we use this model for the diode in Fig. 1.14, the circuit is analyzed as follows. First we write the KVL equation

$$0 = V_B - I_D R_S - V_D$$

Now we solve for I_D, giving

$$I_D = \frac{V_B - V_D}{R_S}$$

Assuming that a typical silicon diode is used, then $V_D = V_\phi = 0.7$ V. Let us assume for illustrative purposes that $V_B = 5$ V and $R_S = 100\ \Omega$. This gives us

$$\begin{aligned} I_D &= \frac{V_B - V_D}{R_S} \\ &= \frac{5\text{ V} - 0.7\text{ V}}{100\ \Omega} \\ &= 43\text{ mA} \end{aligned}$$

This is a more accurate approximation of the actual value of I_D that would exist in a real circuit.

Increased accuracy may be obtained if diode internal resistance is accounted for, as shown in Fig. 1.13(c). Here, R_F is the average resistance of the diode transconductance curve with a bias voltage greater than or equal to V_ϕ dropped across the diode. This resistance is assumed to be linear (constant), and is often called the diode *bulk resistance.* The value of R_F can be obtained from the actual transconductance curve, as shown in Fig. 1.15, but this is usually not necessary. For typical low-power silicon and germanium diodes, R_F will be around 2 to 5 Ω. For higher power devices, R_F will usually be around 1 Ω or less.

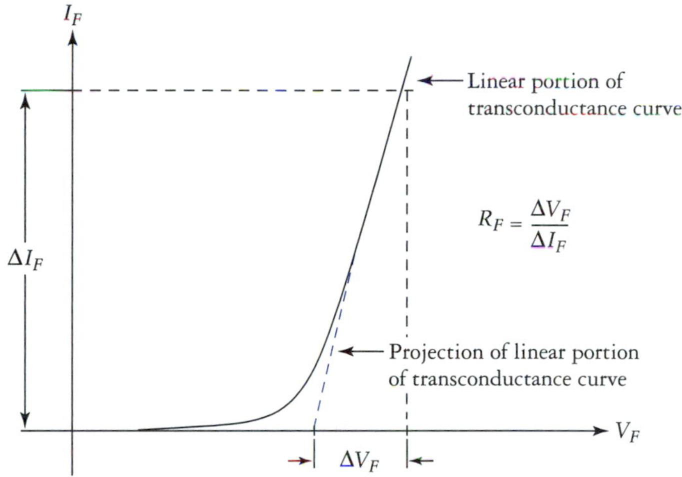

FIGURE 1.15
Determination of the average forward resistance of a diode.

Let us assume that the diode in Fig. 1.14 is a silicon unit, modeled by the device in Fig. 1.13(c), with $R_F = 5\ \Omega$. As in the previous calculations, assume also that $V_B = 5$ V and $R_S = 100\ \Omega$. This equivalent circuit is shown in Fig. 1.16 to emphasize the diode model being applied. Writing the KVL equation for this circuit we get

$$0 = V_B - I_D R_S - I_D R_F - V_\phi$$

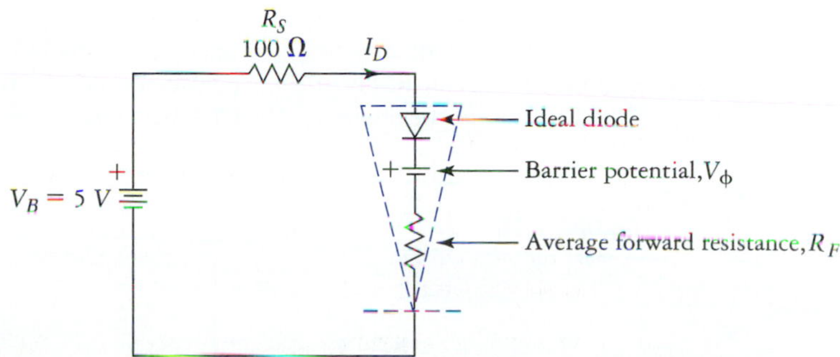

FIGURE 1.16
Series circuit with diode replaced by behavioral model.

Solving for I_D gives

$$I_D = \frac{V_B - V_\phi}{R_S + R_F} = \frac{5\text{ V} - 0.7\text{ V}}{100\ \Omega + 5\ \Omega} = 40.95\text{ mA}$$

It is clear that little accuracy was gained by accounting for bulk resistance in this example. For most of the types of applications covered in this text, we need only account for the diode barrier potential to obtain acceptably accurate results. The bulk resistance of the diode is more important in applications where the diode is required to carry very large currents. For example, an industrial motor control circuit may require diodes to carry hundreds of amps, in which case a small bulk resistance will result in a very large voltage drop. In such a situation, bulk resistance is very important.

EXAMPLE 1.2

Refer to Fig. 1.17. Determine the value of R_L that will produce $I_L = 5$ mA. Account for typical silicon diode barrier potential. Assume that bulk resistance is negligible ($R_F = 0\ \Omega$) and justify this assumption using the results of the analysis.

FIGURE 1.17
Circuit for Example 1.2.

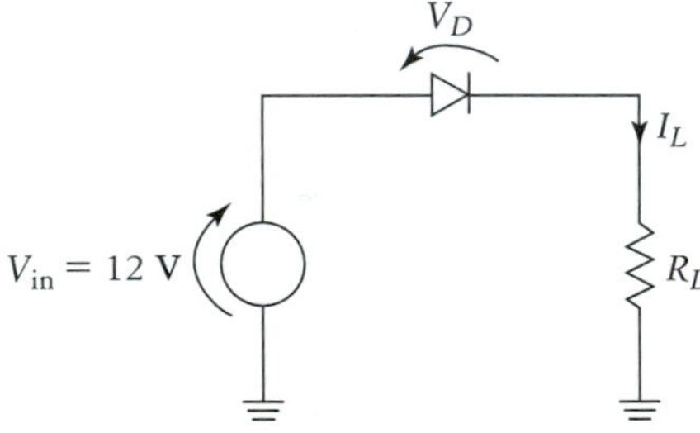

Solution The application of KVL results in

$$0 = V_{in} - V_D - I_L R_L$$

Solving for R_L gives us

$$R_L = \frac{12\text{ V} - 0.7\text{ V}}{5\text{ mA}} = 2.26\text{ k}\Omega$$

Because R_L is so large (2.26 kΩ) compared to the typical value of R_D (less than 5 Ω), it is clear that bulk resistance can be ignored in this case.

Quite often, a diode may be one of many elements in a more complex network than those just presented. The circuit in Fig. 1.18 is one such example. Let us assume that we need to determine the current through the diode I_D. The trick here (and in many other situations too) is to reduce the cir-

FIGURE 1.18
A more complex circuit.

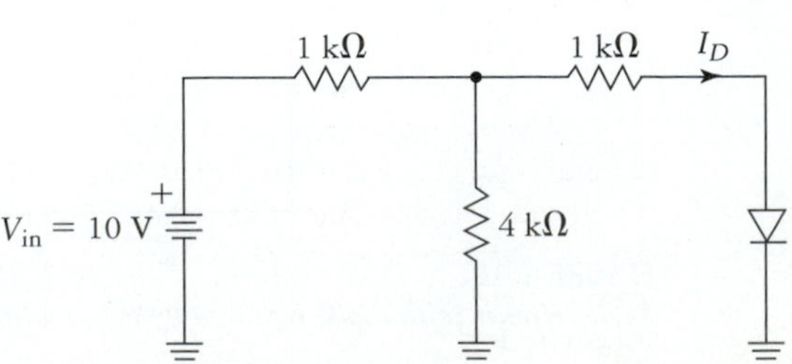

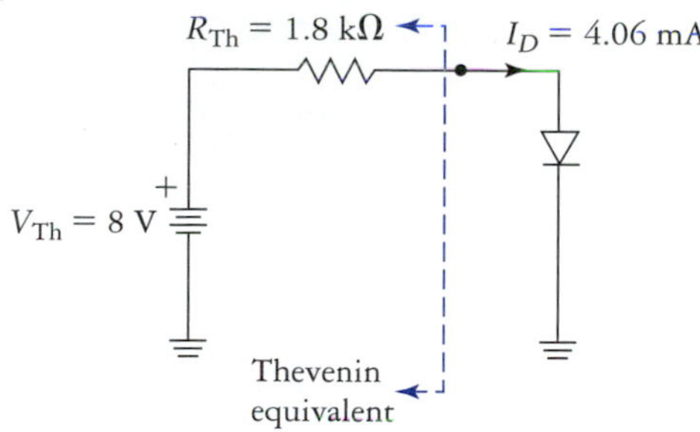

FIGURE 1.19
Thevenin equivalent of Fig. 1.18 allows I_D to be easily determined.

cuit to a form that we already know how to handle. This is done via the application of Thevenin's theorem. The circuit is transformed into a Thevenin equivalent circuit as shown in Fig. 1.19. The diode is forward biased, and the current is now easily determined to be

$$
\begin{aligned}
I_D &= \frac{V_{Th} - V_{\phi}}{R_{Th}} \\
&= \frac{8\text{ V} - 0.7\text{ V}}{1.8\text{ k}\Omega} \\
&= 4.06\text{ mA}
\end{aligned}
$$

Graphical Analysis of Diode Circuits

Up until now, the diode has been modeled in what is termed a *piecewise-linear* manner. That is, the diode transconductance curve is approximated by straight-line segments of various slopes connected end to end. This is an extremely useful technique for creating various composite functions, and it is used extensively in circuit analysis applications. However, if the actual transconductance curve for a given diode is available, then a graphical determination of its current and terminal voltage, called the diode operating point, can be made quite easily.

To predict the exact values of V_F and I_F mathematically, first we need to know all of the diode parameters for Shockley's equation. With these values known, we must simultaneously solve Shockley's equation and the KVL equation for the circuit. This must be done either graphically or using iterative techniques, as there is no closed-form solution to this particular problem. We use the graphical technique.

Refer to the circuit in Fig. 1.20. Let us assume that the diode being used is a silicon device with the transconductance curve of Fig. 1.21. The KVL equation for this circuit is

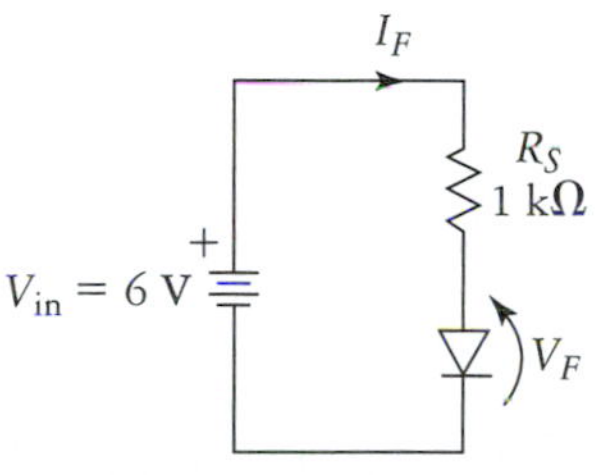

FIGURE 1.20
Simple series circuit.

$$0 = V_{in} - I_F R_S - V_F$$

No matter what else happens, the circuit as a whole must obey this relationship, and the diode must conform to the transconductance curve (Shockley's equation). Since we have the transconductance curve on hand, we must superimpose a plot of the KVL equation on top of it. Since the KVL equation is linear, we need only determine two convenient points and then connect them with a straight line. Solving the KVL equation for I_F gives us

$$I_F = \frac{V_{in} - V_F}{R_S}$$

Now, since V_{in} and R_S are given, we substitute convenient values of V_F into the equation and find the corresponding I_F values. These values can be placed in a table as shown in Fig. 1.21. Here, V_F was set to 0 and 1.4 V, because these are the limits of the graph, and the resulting I_F values were within the range of the current axis given. The line drawn between these two points is called the *load line,* and it intersects the transconductance curve at what is called the *operating point* or *Q point* of the diode. In this case, the coordinates of the Q point are approximately 0.7 V and 5.3 mA. This is in agreement with the results that are obtained using the model of Fig. 1.13(b) with a barrier potential of 0.7 V. The Q in Q point means *quiescent,* or no-signal conditions.

Because one or the other of the piecewise-linear approximations will usually work very well, the graphical analysis is usually not worth the effort, except in cases where the diode barrier potential is very close to the value of the applied voltage. However, graphical load-line construction is a useful way to visualize Q-point position changes with device parameter variations, and so it tends to give you a better feel for the operation of the circuit. Load-line analysis is used later in the context of transistor circuit analysis.

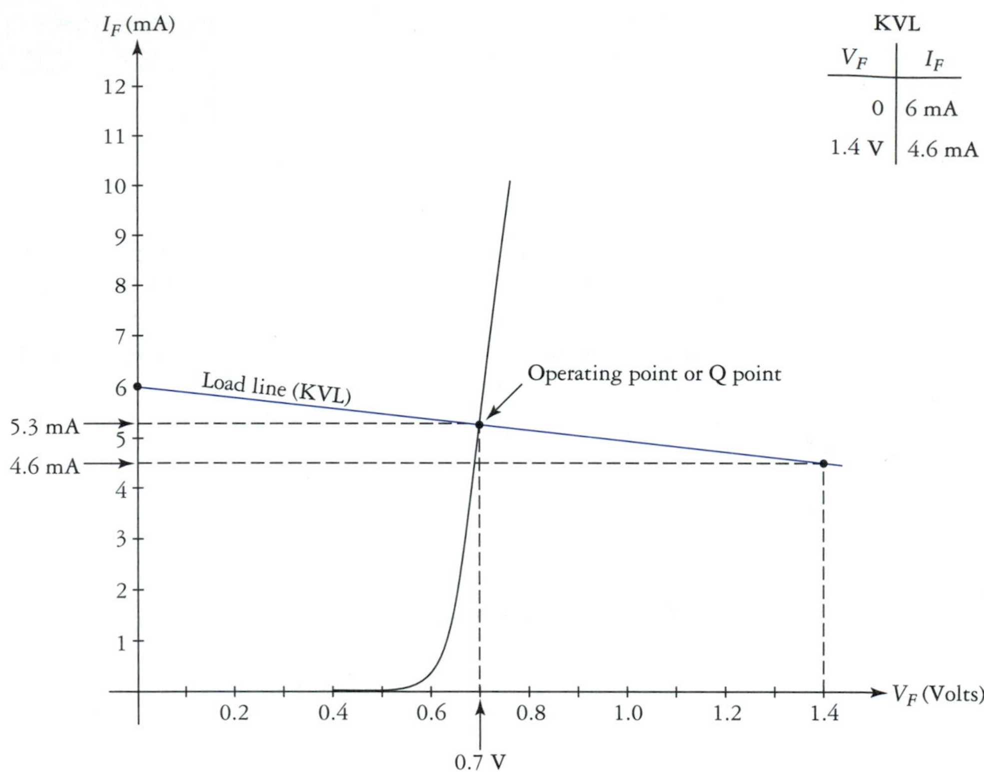

FIGURE 1.21
Location of diode operating (Q) point using a load line.

1.4 DYNAMIC RESISTANCE

We have stated several times previously that the diode is a nonlinear device. A linear device is one that has a constant resistance, and thus if we were to plot device current as a function of terminal voltage, $i(v)$, we would obtain a straight line. Likewise, a plot of $v(i)$ would produce a straight line. Ideally, resistors, capacitors, inductors, independent sources, and linear-dependent sources are all linear devices. Most of the circuit analysis techniques that we use so routinely are useful only on linear circuits. Modeling of the diode using piecewise-linear functions allows us to use those analysis methods that we all know so well.

The diode is fundamentally a nonlinear device and in certain circumstances this must be taken into account. A rather common occurrence in which this is the case is shown in Fig. 1.22, where an independent voltage source is supplying a sinusoidal signal of 1 $V_{\text{p-p}}$, riding on a dc offset of 4 V.

Note
Dynamic resistance is very important in the analysis of transistor amplifier circuits as well.

To determine the total voltage drop across the diode in Fig. 1.22, we must use the principle of superposition. This time we must begin by determining the dc component of the diode current. The dc equivalent circuit is drawn in Fig. 1.23(a). In this step, we use whichever diode approximation will yield satisfactory results. Let us use the ideal diode with a barrier potential model. It seems to be a pretty safe assumption that we won't have very large currents flowing in this circuit, so let's ignore bulk resistance for now. Routine analysis gives us a dc current of

$$I_F' = \frac{4\text{ V} - 0.7\text{ V}}{680\ \Omega}$$

$$= 4.85\text{ mA}$$

The dc component of v_{in} is treated as a biasing voltage. That is, it is a dc level that is used to set the position of the Q point. For this case, the Q-point coordinates are 0.7 V and 4.85 mA.

To determine the ac component of the current, we must determine the slope of the transconductance curve at the Q point. The reciprocal of the slope is called the *dynamic* or *incremental resistance* of the diode, and is designated as r_F. Mathematically, r_F is determined by differentiating V_D with respect to I_D in Shockley's equation. This gives us the diode dynamic resistance equation

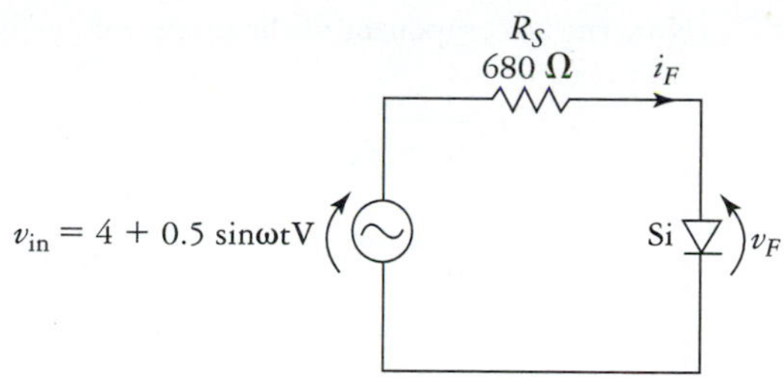

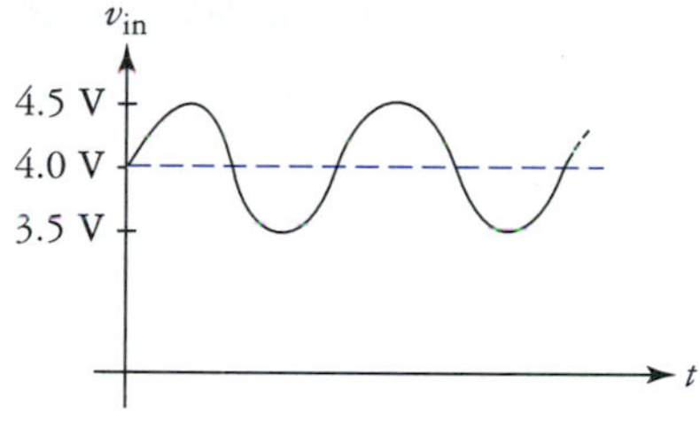

FIGURE 1.22
Diode circuit with dc and ac voltages applied simultaneously.

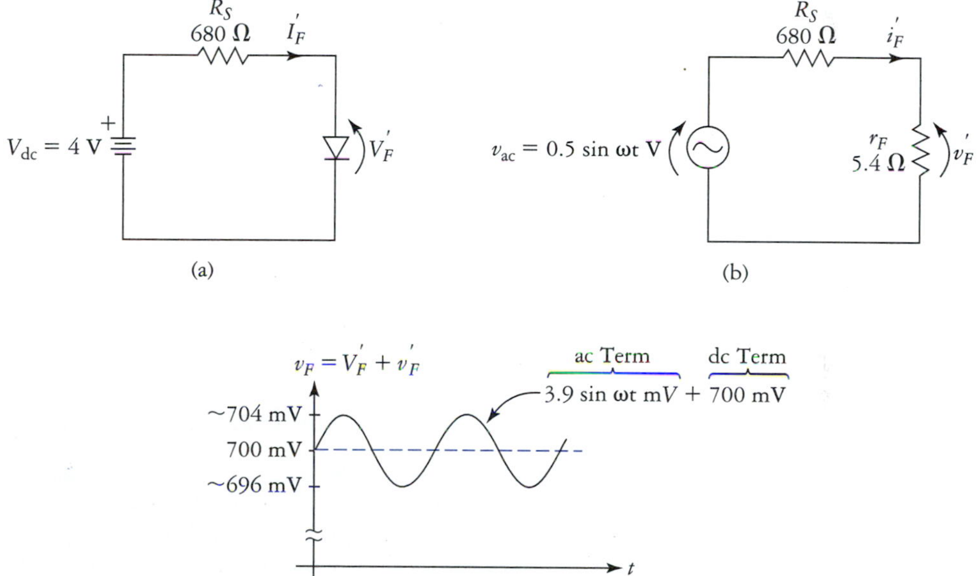

FIGURE 1.23
Superposition is used for the analysis. (a) The dc source is used to find the Q point. (b) The ac signal sees the dynamic resistance of the diode in a voltage divider. (c) The resulting diode voltage waveform.

$$r_F = \frac{\eta V_T}{I_F} \tag{1.4}$$

where V_T is the thermal voltage equivalent (26 mV at room temperature, from Eq. 1.3), η is the emission coefficient, and I_F is the dc forward current.

Using Eq. (1.4), and assuming that $\eta = 1$ (a good approximation if $I_{dc} \geq 1$ mA), the dynamic resistance of the diode is

$$r_F = \frac{26 \text{ mV}}{4.85 \text{ mA}}$$
$$= 5.4\ \Omega$$

Now the ac component of the diode voltage is found using the voltage-divider equation.

$$v_F' = v_{ac}\frac{r_F}{r_F + R_S}$$

$$= (0.5 \sin \omega t \text{ V}) \frac{5.4\ \Omega}{5.4\ \Omega + 680\ \Omega}$$

$$= 3.9 \sin \omega t \text{ mV}$$

This is quite a small ac voltage, but it is to be expected, considering that the diode transconductance curve is very steep once it is well into conduction, and the series resistance $R_S = 680\ \Omega$ is relatively large. The net diode voltage is the instantaneous sum of the dc and ac components.

$$V_F = 700 + 3.9 \sin \omega t \text{ mV}$$

This signal is shown in Fig. 1.23(c).

The dynamic resistance of a forward-biased pn junction is a strong function of Q-point location, as shown in Fig. 1.24. The dashed lines represent tangents to the diode's transconductance curve, and it is obvious that the slope and hence r_F vary significantly with diode bias.

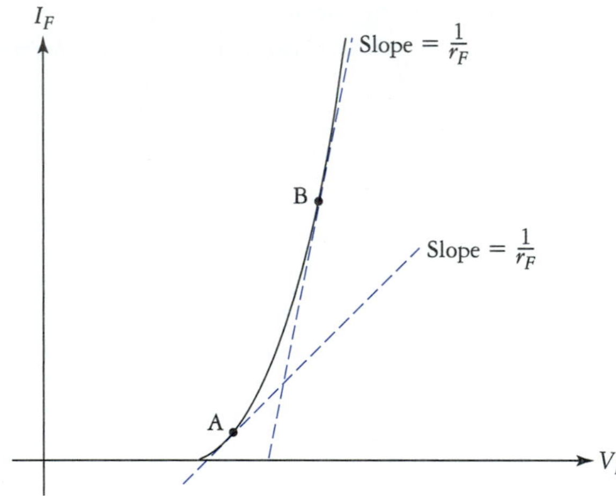

FIGURE 1.24
Dynamic resistance changes as a function of diode current.

Dynamic Resistance and Nonlinearity

A few words of caution are in order here. You may have wondered how it was possible to use the superposition principle to determine v_F in Fig. 1.22 when superposition is only valid in linear networks. There is not really a simple answer to this question, but it was tacitly assumed that the ac voltage present caused very little deviation of the Q point from its original position. This may or may not be the case. We must investigate further to see if this is indeed a valid assumption. But first, let us take a quick look at the effects of diode nonlinearity on a sinusoidal signal.

Note
A p-n junction behaves in a linear manner under small-signal conditions.

The transconductance curve of Fig. 1.25 shows the location of a Q point and the locations to which it is forced when a sinusoidal forward voltage is applied to the diode. In this particular illustration, the diode current is altered significantly on signal peaks. However, because the curve gets steeper to the right of the Q point, positive-going input signal swings cause I_F to increase more rapidly for a given voltage excursion. The net effect is an asymmetrical current waveform. The resulting forward current is shown to the left of the figure. Notice that the waveform is not sinusoidal anymore. This type of distortion is covered in greater detail later. For now, let us say that if this distortion is significant, then in general superposition cannot be reliably used to predict circuit behavior.

If the movement of the Q point is restricted to a small portion of the transconductance curve, then the curve may be approximated as being linear, with a constant slope (the reciprocal of dynamic resistance, r_F) given by the tangent to the curve at the Q point. When such conditions exist, the circuit is said to be operating under *small-signal conditions,* where the effects of nonlinearity are negligible. This was the assumption that was made in the analysis of Fig. 1.22. As a general rule, if the ac component of the input signal causes a change in forward current of less than 10% of that of the Q point, then small-signal conditions prevail, and the dynamic resistance of the pn junction may be considered essentially constant.

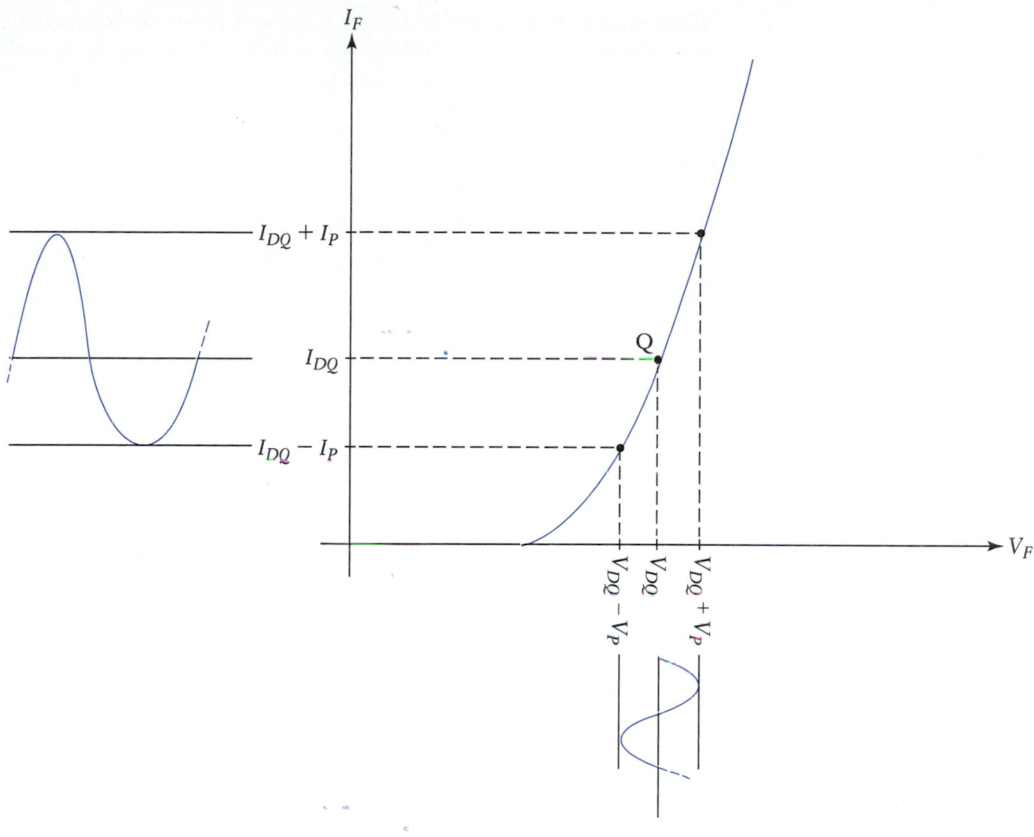

FIGURE 1.25
Symmetrical changes in V_F *result in asymmetrical changes in* I_F. *This is a form of nonlinear distortion.*

EXAMPLE 1.3

Determine the maximum percentage of deviation in I_F and r_F for the circuit in Fig. 1.22.

Solution The Q point has already been established as $V_F = 0.7$ V and $I_F = 4.85$ mA. We again use the ideal diode/barrier potential model of the diode. The maximum change in forward current occurs on positive peaks of the ac signal, because the slope of the transconductance curve increases with increasing v_F. The forward current at the positive peak is

$$I_{F(\text{pk})} = \frac{V_{\text{in(pk)}} - V_\phi}{R_S}$$

$$= \frac{4.5\text{ V} - 0.7\text{ V}}{680\ \Omega}$$

$$= 5.58\text{ mA}$$

This gives us a dynamic resistance of

$$r_F = \frac{26\text{ mV}}{5.58\text{ mA}}$$

$$= 4.7\ \Omega$$

The percent change in I_F relative to the Q point is

$\Delta I_F = +15.1\%$

The percent change in r_F is

$\Delta r_F = -13.7\%$

These results indicate that the diode is somewhat out of the small-signal regime of operation. However, the results determined earlier are accurate enough for most applications.

1.5 DIODE SPECIFICATIONS

Diodes are generally the simplest of electronic devices from the standpoint of operation, applications, and user specifications. The specifications that are most important for a given diode depend on the intended purpose of that device. A given diode will usually fall into one of the following general categories: (1) power or rectifier diodes, (2) switching diodes, (3) small-signal or general-purpose diodes, and (4) special-purpose diodes. For all practical purposes, diodes from any of the first three categories could have been used in the examples up to this point, therefore parameters that are of primary concern in applying these devices will be the main focus of this section. Specifications relating to particular special-purpose diodes will be presented later, as the need arises.

In this section, the 1N400X series diodes are cited as device examples. The 1N400X series (1N4001 through 1N4007) is a very popular family of silicon rectifier diodes. The last digit of the 1N400X designation specifies the reverse breakdown voltage for that particular diode type. For example, the 1N4001 has a reverse breakdown voltage specification of 50 V, while that of the 1N4007 is 1000 V. Aside from different reverse breakdown voltage ratings, the various members of the 1N400X family have essentially the same characteristics in other respects, such as leakage current, forward current ratings, and temperature characteristics.

Reverse Bias Parameters

For most applications, two parameters are important when a diode is operated under reverse bias conditions: reverse breakdown voltage and reverse leakage current. Normally, the reverse breakdown voltage is of primary concern, because it defines a fundamental operating limitation of a particular diode. Reverse leakage current is the second parameter that can be considered, although with modern silicon devices, this current is usually small enough to be considered negligible.

Reverse Breakdown Voltage

The reverse breakdown voltage of a diode can be specified in different ways, depending on the conditions under which the reverse voltage is applied. For example, the 1N4002, a rectifier diode produced by several different manufacturers, has the following reverse breakdown voltage specifications.

Peak repetitive reverse voltage $V_{RRM} = 100$ V

Working peak reverse voltage $V_{RWM} = 100$V

DC blocking voltage $V_R = 100$ V

Nonrepetetive peak reverse voltage $V_{RSM} = 120$ V

For all practical purposes, the first two terms mean the same thing. The variables V_{RRM} and V_{RWM} refer to the maximum peak reverse voltage that the diode can withstand, as applied from an ac voltage source. Typically the 60-Hz ac line frequency is assumed to be used. The term V_R is the maximum continuous dc reverse bias that the diode is designed to withstand.

Another important reverse breakdown specification for the 1N4002 is the nonrepetetive peak reverse voltage. The peak reverse voltage, V_{RSM}, can be applied to the diode over one cycle of a 60-Hz sinusoid. This voltage will generally be slightly higher than the peak repetitive or working reverse voltage ratings.

In general, a diode is selected for a given application such that its reverse breakdown voltage rating is at least 10% greater than the maximum expected reverse bias that would be produced in the circuit. Often a greater safety margin with V_R rated at over twice the expected reverse bias is used.

Given a particular diode geometry, reverse breakdown voltage is dependent primarily on diode doping concentrations. Generally speaking, the lower the density of dopants, the greater the reverse breakdown voltage will be.

Reverse Leakage Current

Recall that diode reverse current consists of two major components: the thermally generated saturation current and the surface leakage current. Because of the strong temperature dependence of the saturation current, reverse leakage current is usually specified at several different temperatures, most often at 25°C and at 100°C.

The 1N400X series of devices has the following reverse leakage current specifications at rated reverse voltage:

$$I_R\big|_{T_J=25°\text{C}} = 0.05\ \mu\text{A (typ.)} \qquad 10\ \mu\text{A (max.)}$$
$$I_R\big|_{T_J=100°\text{C}} = 1.0\ \mu\text{A (typ.)} \qquad 50\ \mu\text{A (max.)}$$

where T_J is the diode junction temperature. The typical values of reverse current are, as the specification implies, the values exhibited by a typical 1N400X diode. Of course, the typical device is a fictitious device and a real 1N400X diode may have a reverse current far lower than the typical value, or as high as the maximum value. A device with a leakage current that is greater than $I_{R(\text{max})}$ is out of specification and would normally be discarded.

As is to be expected, reverse leakage current increases greatly with temperature. In designs where high junction temperatures are expected, the 100°C reverse current specifications would be used. However, in many applications, even the worst-case maximum leakage current for a diode may be negligibly small.

Forward Bias Parameters

Most often, diodes are represented by the piecewise-linear models presented earlier. Often, certain general approximations are made as to forward voltage drop and forward resistance. For example, for silicon diodes, it is assumed that $V_F = 0.7$ V, regardless of the value of forward current I_F. If a more accurate approximation of the forward voltage drop is required, or if forward resistance must be accounted for, then the diode specifications must be consulted.

In higher power applications, it is usually necessary to know the current-handling capabilities of a given diode. This information is also presented in the diode specifications.

Forward Current

The 1N400X series of diodes have the following forward current specifications:

$$I_o\big|_{T_J=75°\text{C}} = 1\ \text{A}$$
$$I_{\text{FSM}} = 30\ \text{A (for 1 cycle of 60 Hz)}$$

The parameter I_0 is the average rectified forward current with a single-phase resistive load at a frequency of 60 Hz, at a junction temperature of 75°C. Basically, this specification says that a 1N400X diode can carry an average forward current of 1 A when used as a rectifier. Rectification is covered in Chapter 2. As usual, in a conservative design, the average forward current would be limited to somewhat less than 1 A.

Diodes that are used in power supply applications (Chapter 2) are often subjected to transient forward currents that are many times greater than the average long-term current. Normally, the surge current-handling capability of the typical diode is also many times greater than the maximum average or continuous forward current rating as well. The maximum forward surge current I_{FSM} for the 1N400X diodes is 30 A for one cycle of 60 Hz. Thus, for reliability, a given circuit should be designed such that surge current is kept well below the I_{FSM} rating of the diode(s) in use.

Forward Voltage

The primary forward voltage specification for the 1N400X series of diodes is:

$$v_F\big|_{I_F=I_o} = 0.93\ \text{V (typ.)} \qquad 1.1\ \text{V (max.)}$$

The parameter v_F is the instantaneous forward voltage drop at rated current ($I_{\text{pk}} = 1$ A) and standard temperature ($T_j = 25$°C). Recall that the typical forward voltage of a silicon diode was assumed to be 0.7 V. This value is based solely on Shockley's equation for the typical silicon diode. The existence of bulk resistance effects and the fact that the diode is operated at maximum forward current are the main reasons why the v_F values given here are higher than 0.7 V.

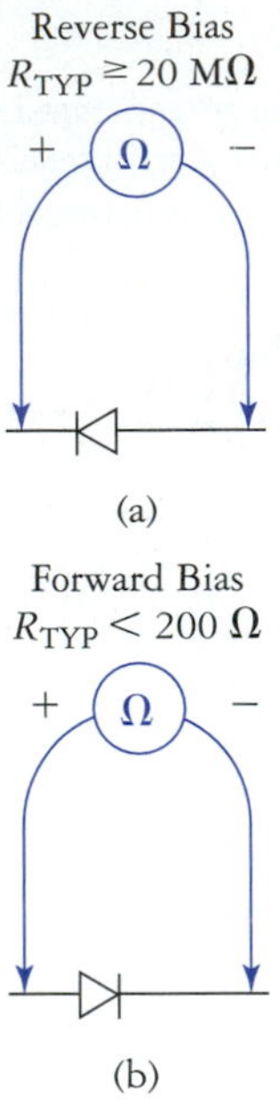

Reverse Bias
$R_{TYP} \geq 20\ M\Omega$
+ Ω −

(a)

Forward Bias
$R_{TYP} < 200\ \Omega$
+ Ω −

(b)

FIGURE 1.26
Testing a diode with an ohmmeter.

A Simple Diode Test

Most general-purpose, rectifier, and switching diodes can be checked quickly and easily using an ohmmeter. The procedure is to check the diode for high resistance by connecting the negative lead of the ohmmeter to the anode of the diode, while the positive lead is connected to the cathode. This is shown in Fig. 1.26(a). The reverse resistance of the typical silicon diode should be well over 20 MΩ (the meter will probably read infinite resistance). Germanium devices may read as low as 1 MΩ in this test.

The diode is checked for forward bias operation by connecting the positive lead of the ohmmeter to the anode and the negative lead to the cathode. The resistance indicated here will vary significantly, however, it will usually be a few hundred ohms or less.

Nearly all analog ohmmeters will forward bias a diode. However, many multimeters (DMMs) do not produce sufficient voltage to forward bias a silicon diode when set to measure resistance. Such instruments normally have a special diode test function. Most DMMs with this capability operate by forcing a test current (1 mA typically) through the diode. The DMM leads are connected to the diode as previously described, and under reverse bias, the diode should be an open circuit, causing an overrange indication. In the forward bias mode, the DMM will normally read the forward voltage drop across the diode. This feature is useful for determining whether a diode is a silicon or germanium device, or whether the apparent barrier potential is abnormally high.

Multiple Diode Connections

Occasions might arise when it is necessary to connect several diodes in series or parallel. Depending on the application, this may be a very simple operation or somewhat more complex.

Parallel Connection

Unless we are examining certain integrated circuit biasing techniques (covered much later), the main reason for connecting diodes in parallel is to increase the forward current-handling capability relative to a single device. Again, it is important to realize that significant parameter variation occurs from one diode to another, even if both diodes are of the same type. This is especially true of saturation current, which greatly influences diode behavior. It is not unusual for saturation current to vary over a 10-to-1 ratio from one diode to another (of the same type). Integrated circuit (IC) fabrication techniques overcome this problem, but in high-power applications, such IC solutions may not be feasible.

If two diodes are parallel connected to increase current handling, series resistors should be used, as shown in Fig. 1.27. Such resistors are called *swamping resistors,* because they swamp out the differences in forward resistance variation between the diodes. Typical swamping resistor values range from around 5 Ω to less than 0.1 Ω. Swamping resistors reduce circuit efficiency, but are necessary to prevent the diode with higher I_S from carrying a significantly larger portion of the total current.

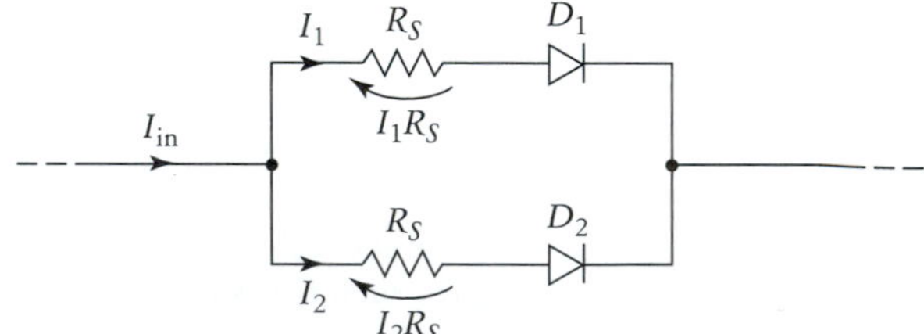

FIGURE 1.27
Swamping resistors must be used to prevent current hogging when diodes are connected in parallel.

Series Connection

There are two common reasons for connecting two or more diodes in series. We might do this in order to obtain a forward voltage that is a multiple of normal diode barrier potential. If the diodes are not going to be subjected to very high reverse bias ($V_D < V_{RM}$) then it is generally acceptable to simply connect the diodes as shown in Fig. 1.28(a). The second common reason that diodes are connected in series is to increase the effective reverse breakdown voltage to a greater value than is available with a single device. For example, two identical series-connected diodes have twice the reverse breakdown voltage of a single device. In an application like this, it is a good idea to connect resistors in parallel with the diodes as shown in Fig. 1.28(b). Large differences between diode saturation currents cause

Forward-biased diodes used to increase V_ϕ in low-power applications.

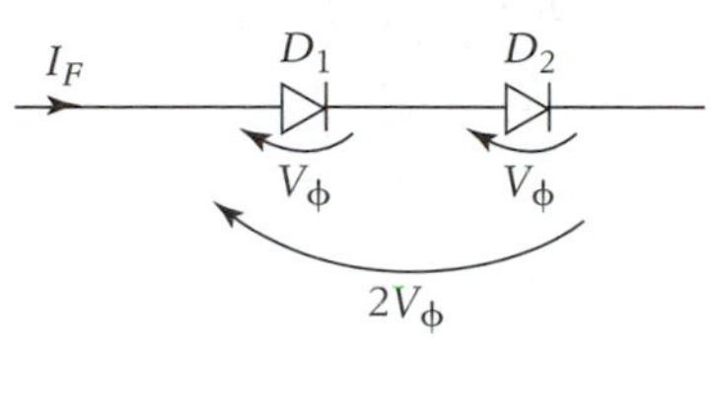

(a)

Reverse-biased diodes used to increase V_{BR} require external resistors.

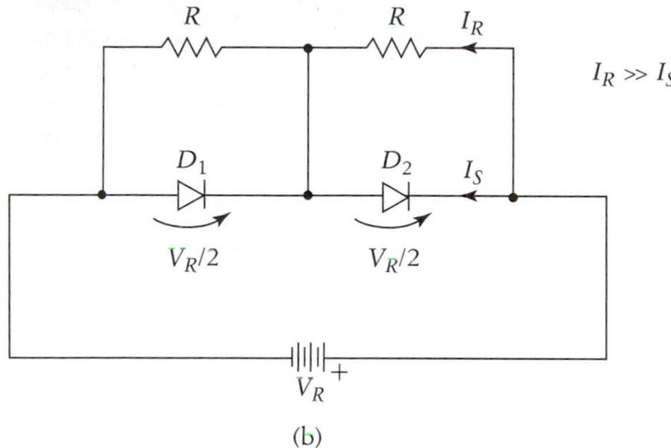

(b)

FIGURE 1.28
Diodes connected in series. (a) For low-voltage applications. (b) For high-voltage applications, parallel high-value resistors help equalize voltage drops across diodes.

large differences between the diodes' effective reverse resistances. This means that one of the diodes will probably end up dropping most of the applied reverse bias, possibly exceeding its specifications. The parallel resistances are chosen to be small compared to the reverse resistance of the diodes, but large enough that the resulting resistor current is negligible. Thus, these resistors swamp-out variations in I_S from diode to diode. Typical values for these resistors are $R \geq 10\ \text{k}\Omega$.

1.6 ZENER DIODES

Structurally, and in many other respects, the zener diode is quite similar to the general-purpose diode discussed earlier. Unlike a standard diode, however, the zener diode is designed specifically to operate under reverse breakdown conditions, and as such has a well-defined reverse breakdown voltage, called the zener voltage V_Z. The schematic symbol for a zener diode is shown in Fig. 1.29.

A K

FIGURE 1.29
Schematic symbol for a zener diode.

Zener Characteristics

Note
Zener diodes are designed to operate in reverse breakdown.

The transconductance curve for a typical zener diode is shown in Fig. 1.30. Under forward bias conditions, a zener diode behaves virtually the same as a general-purpose diode made of the same material (silicon, for example). Indeed, the reverse bias characteristics of the zener diode are much the same as those of any other general-purpose diode, as shown in Fig. 1.30. However, zener diodes are specifically designed to be operated under reverse breakdown conditions. Thus, zener diodes are designed to have a very accurate and predictable reverse breakdown voltage. Typically, the reverse breakdown voltage of a zener diode will be rather low when compared to that of a general-purpose diode, although there are exceptions.

Notice that once a zener diode is well into reverse breakdown that the transconductance curve is nearly vertical. As a consequence, the voltage drop across the zener diode is nearly constant over a wide range of current. This is a very important characteristic that is taken advantage of to produce constant voltage sources.

Zener Voltage

The *zener voltage* V_Z is the *nominal* reverse breakdown voltage drop across the zener diode. Because this is a nominal value, the actual value of V_Z may be slightly higher or lower. The value of V_Z is also somewhat dependent on the current through the diode and on junction temperature. In the context of this book, V_Z is assumed to be a positive voltage, that is,

$$V_Z = V_{KA} \tag{1.5a}$$

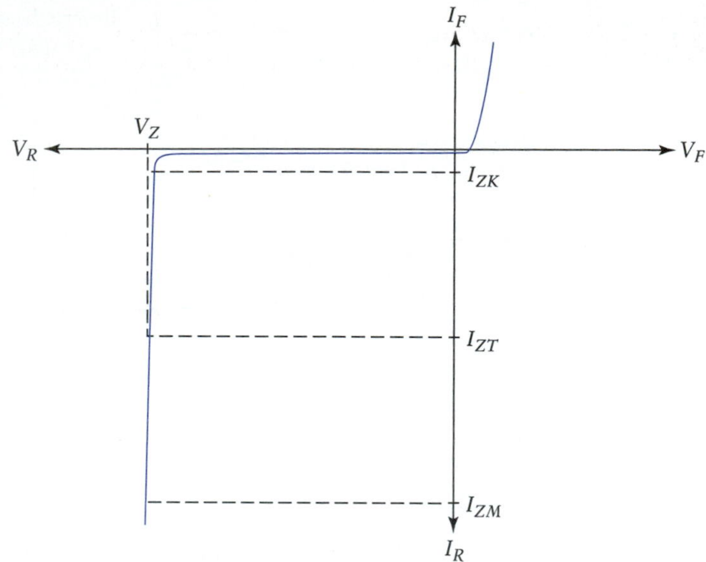

FIGURE 1.30
Typical zener diode transconductance curve.

Zener Test Current

The nominal zener voltage is that voltage drop V_{KA} that is typically produced when $I_Z = I_{ZT}$. That is, the manufacturer tests random samples of a particular type of zener diode at some particular current that is designated as the *zener test current,* I_{ZT}. On the average, a given zener diode will exhibit a voltage drop that is close to the nominal value if $I_Z = I_{ZT}$.

Armed with an understanding of the significance of the zener test current, Eq. (1.5a) can be rewritten more meaningfully as

$$V_{KA}\big|_{I_Z = I_{ZT}} = V_Z \tag{1.5b}$$

Zener Knee Current

The *zener knee current* is the minimum reverse current that allows operation on the near-vertical portion of the zener reverse breakdown characteristic. In a practical circuit design, the minimum zener current would be kept somewhat higher than I_{ZK}, to ensure that V_Z remains relatively constant. Typically, I_{ZK} is less than or equal to 1 mA. If I_{ZK} is not known or specified for a particular zener diode, the following relationship can be used to get a conservative estimate:

$$I_{ZK} \cong 0.05 I_{ZM} \tag{1.6}$$

Maximum Zener Current and Power

The maximum current that may be carried by a given zener diode depends on both the zener voltage and the maximum power dissipation capability of the device. The power dissipation of a zener diode is given by the general equation

$$P_Z = I_Z V_Z \tag{1.7}$$

Because zener voltage remains relatively constant regardless of current, the maximum zener power dissipation rating is given by

$$P_{ZM} \cong I_{ZM} V_Z \tag{1.8}$$

Thus, knowing either I_{ZM} or P_{ZM} we can find the other value.

When designing a circuit that contains zener diodes, the maximum zener power dissipation must be limited to less than P_{ZM}.

EXAMPLE 1.4

Refer to Fig. 1.31. Determine the required value of R_S such that $I_Z = 0.5I_{ZM}$. Also determine the power dissipation of the zener diode and R_S, based on this resistance value.

FIGURE 1.31
Circuit for Example 1.4.

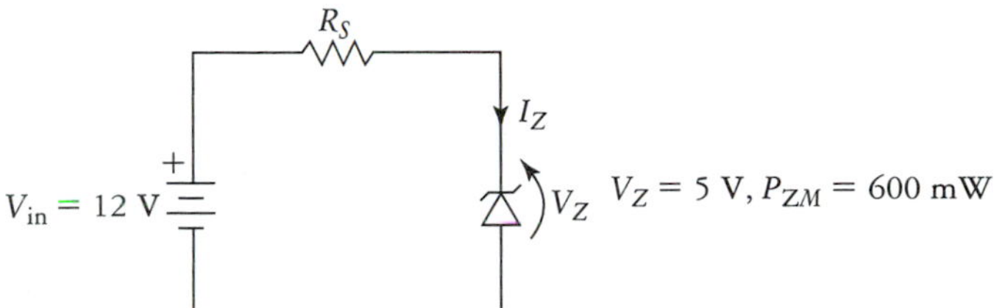

Solution First we must determine I_{ZM}. Solving Eq. (1.8), we obtain

$$I_{ZM} \cong \frac{P_{ZM}}{V_Z} = \frac{600 \text{ mW}}{5 \text{ V}} = 120 \text{ mA}$$

The desired current is $I_Z = I_{ZM}/2 = 60$ mA.

We now write the KVL equation for the circuit, giving us

$$0 = V_{in} - I_Z R_S - V_Z$$

Solving for R_S yields

$$R_S = \frac{V_{in} - V_Z}{I_{ZM}} = \frac{12 \text{ V} - 5 \text{ V}}{60 \text{ mA}} = 116.7 \ \Omega$$

The power dissipation of the resistor is

$$P_{RS} = I_Z R_S = (60 \text{ mA})^2 (116.7 \ \Omega) = 420 \text{ mW}$$

The power dissipation of the zener diode should be 300 mW ($P_{ZM}/2$), because the diode is operating at a current $I_{ZM}/2$. As a check, however, explicit calculation reveals

$$P_Z = I_Z V_Z = (60 \text{ mA})(5 \text{ V}) = 300 \text{ mW}$$

Given that the zener diode is operating well below its maximum power dissipation limit, and well above its zener knee current, the nearest standard value for R_S (120 Ω in this case) would be used. Power dissipation considerations require that the series resistor be rated at 0.5 W or higher.

Zener Impedance

The reciprocal of the slope of the zener transconductance curve between I_{ZK} and I_{ZM} is relatively constant and is called the *zener impedance* Z_Z. Because this portion of the curve is approximately linear, we can define the zener impedance as follows:

$$Z_Z = \frac{\Delta V_Z}{\Delta I_Z} \tag{1.9}$$

The following sampling of 1-W zener diode specs typifies the relationship between the various zener parameters.

Diode Type	V_Z	I_{ZT}	I_{ZK}	Z_Z
1N4728	3.3 V	76 mA	1 mA	10 Ω
1N4740	10 V	25 mA	0.25 mA	7 Ω
1N4751	30 V	8.5 mA	0.25 mA	40 Ω
1N4764	100 V	2.5 mA	0.25 mA	350 Ω

Temperature Characteristics

Recall that the forward-biased pn junction has a negative temperature coefficient. As a result, the effective value of the barrier potential decreases with temperature. The temperature coefficient (TC) of the zener voltage may be *either* positive or negative, depending on the zener voltage. TC is usually expressed as a percent of the nominal zener voltage per Celsius degree (%/°C). To determine the change in zener voltage given a change in temperature, the following formula is applied:

$$\Delta V_Z = (V_Z)(\text{TC})(\Delta T) \tag{1.10}$$

where ΔT is the change in temperature referred to 25°C.

Generally speaking, zener diodes with $V_Z < 4.5$ V have a negative temperature coefficient, while zeners with $V_Z > 4.5$ V have a positive TC. Zener diodes are available at various voltage ratings with near-zero temperature coefficients. Such devices are made by connecting zeners with equal but opposite TCs in series, thus producing a very low net TC.

EXAMPLE 1.5

The 1N5235 has $V_Z = 6.80$ V at $T = 25$°C,, and TC = 0.05%/°C. Determine V_Z at (a) T = 0°C and (b) $T = 100$°C.

Solution

a. Application of Eq. (1.10) gives us

$$\begin{aligned}\Delta V_Z &= (V_Z)(\text{TC})(\Delta T)\\ &= (6.80)(0.0005)(-25)\\ &= -0.085 \text{ V}\end{aligned}$$

The net zener voltage is found as follows

$$\begin{aligned}V_Z &= V_Z + \Delta V_Z\\ &= 6.80 \text{ V} - 0.085 \text{ V}\\ &= 6.72 \text{ V}\end{aligned}$$

b. The zener voltage at T = 100°C is similarly found to be

$$\begin{aligned}V_Z &= V_Z + \Delta V_Z\\ &= 6.80 \text{ V} + 0.26 \text{ V}\\ &= 7.06 \text{ V}\end{aligned}$$

Reverse Breakdown Mechanisms

Two different mechanisms can cause reverse breakdown of a pn junction: avalanche breakdown and zener breakdown. Avalanche breakdown, as discussed earlier, is a high-field effect that occurs when the electrostatic field strength (V/m) associated with the pn junction is strong enough to pull electrons out of the valence band within the depletion region. Such electrons accelerate rapidly and cause other depletion-region electrons to do the same (impact ionization).

The avalanche breakdown mechanism has a positive TC that tends to increase in proportion to V_Z. Avalanche breakdown begins to occur at zener voltages of around 4 V, and becomes the dominant breakdown mechanism at about 5 V. Although all zener diodes are heavily doped in comparison to standard diodes, the doping level is decreased as zener voltage increases. Zener diodes with $V_Z < 4$ V operate via *zener breakdown.* The zener breakdown mechanism is a depletion-region ionization phenomenon that has a negative temperature coefficient.

A third reverse bias conduction mechanism that occurs at very high doping concentrations is called *tunneling.* There is no analog to the quantum tunneling effect in classical physics. Because of the very high doping concentrations used in the fabrication of some diodes, the depletion region is extremely thin, even under significant reverse bias. The extreme narrowness of the depletion region, combined with the intense electrostatic field existing about this region (around 10^6 V/cm), cause the uncertainty principle to become a significant factor in the behavior of electrons in the vicinity of the depletion region.

The uncertainty principle roughly states that the more accurately one knows the momentum of an electron, the less accurately one can define its position in space, and vice versa. Thus, an electron that is held at a rather precise energy level, such as occurs in a crystal lattice, has its location in space smeared out over a significant, but still small distance. The uncertainty principle is only significant over extremely small distances, hence the need for a narrow depletion region. In the tunneling effect, the depletion region is so thin that there is a relatively high probability that even though a given electron does not possess enough energy to cross this region, it can cease to exist as a valence-band electron on the p side and "reappear" on the n side as a conduction-band electron.

Once across the depletion region, this electron is swept out of the lattice by the powerful electrostatic field generated by the applied bias voltage. Meanwhile, an electron is resupplied to the p side by the bias voltage source to make up for the electron that has tunneled away. Tunnel diodes and their applications are covered in Chapter 19.

1.7 OTHER DIODES

This chapter has covered a lot of material, but a brief look at a few other types of diodes is useful.

Light-Emitting Diodes

The *bandgap* is the difference between the energy levels of the majority carriers on the p (holes) and n (electrons) sides of a diode. The larger the bandgap, the higher the effective barrier potential of the pn junction. For silicon diodes, the bandgap is $E_G = 1.12$ eV* (electron-volts), which produces $V_\phi = 0.7$ V. At this low bandgap energy, when charge carriers cross the forward-biased junction, the silicon crystal lattice simply increases in vibrational energy. In other words, the diode warms up.

If we construct a pn junction using the proper materials, such as gallium and arsenic (an alloy called *gallium arsenide,* GaAs) and use the proper dopants, a much larger bandgap can be formed. If the bandgap is large enough, charge carriers crossing the forward-biased depletion region will emit photons of light energy. The larger the bandgap, the shorter the wavelength of the emitted photon. Lower bandgap light-emitting diodes (LEDs) emit infrared radiation, while those with higher bandgap energy emit visible light. A comparison between the bandgaps of a normal silicon diode and a typical LED is shown in Fig. 1.32. The symbol for an LED is shown in Fig. 1.33. The arrows indicate emitted light.

Note
Large bandgap junctions emit photons. Smaller bandgap junctions emit phonons, which are vibrational energy rather than light energy.

LEDs offer several advantages over other light sources such as incandescent lamps. LEDs are extremely rugged. LEDs also have extremely long operating lives, typically exceeding 100,000 hours. LEDs are relatively inexpensive, with per unit prices of less than $0.25 being common. One final good characteristic of LEDs is switching speed. The typical LED can turn on and off in a few microseconds. Special laser diodes can switch even faster (in the low nanosecond range). This makes LEDs and laser diodes ideal for use in high-bandwidth fiber optic data communications systems.

It is interesting to compare the transconductance curves of different types of diodes. Figure 1.34 shows the forward bias characteristics of typical LEDs of various colors and a standard Si diode. Infrared (IR) LEDs tend to have lower barrier potential and they are more efficient than LEDs that produce shorter wavelength light. LEDs typically operate at forward currents ranging from around 10 mA for low-power LEDs to several hundred mA for higher power devices. For example, the Motorola MLED930, an infrared LED, has a rated forward voltage of $V_F = 1.25$ V at $I_F = 50$ mA, with a typical radiant power of 650 μW.

Blue LEDs are a relatively recent invention. Constructed of silicon carbide (SiC), they have the highest barrier potential and are less efficient at converting electrical energy into optical energy than other longer wavelength LEDs. Blue LEDs are also much more expensive than other visible LEDs.

*The electron-volt, eV, is the amount of energy received by an electron falling through a potential of 1 V. This corresponds to the absorption or emission of a photon with a wavelength of 1240 nm.

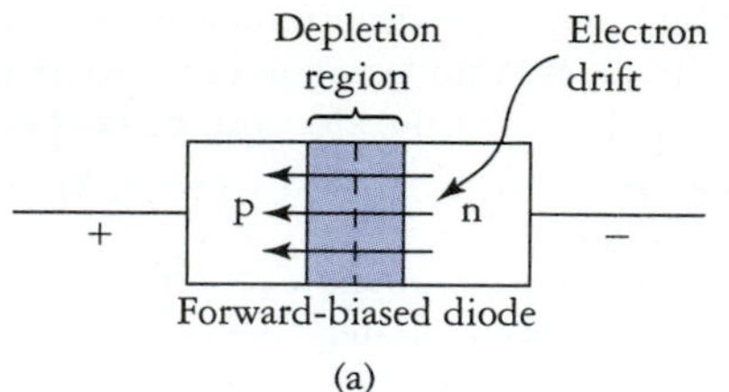

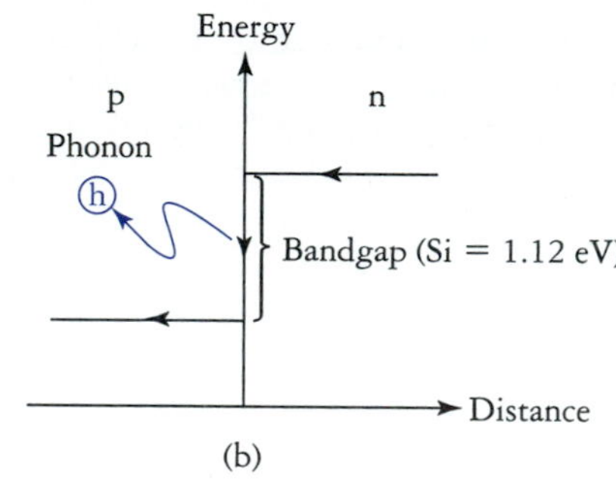

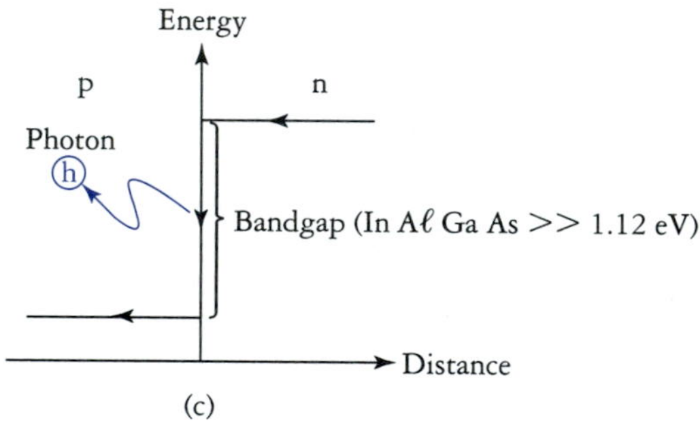

FIGURE 1.32

Bandgap voltage determines the barrier potential of a pn junction. (a) Forward-biased pn junction. (b) Bandgap for normal silicon pn junction. Releases a phonon that simply increases the temperature of the junction via increased vibrational energy. (c) Bandgap for a LED is sufficient to release a photon of electromagnetic energy.

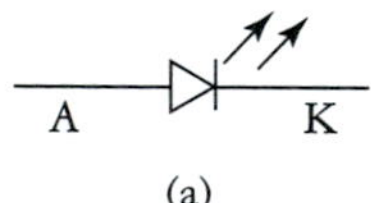

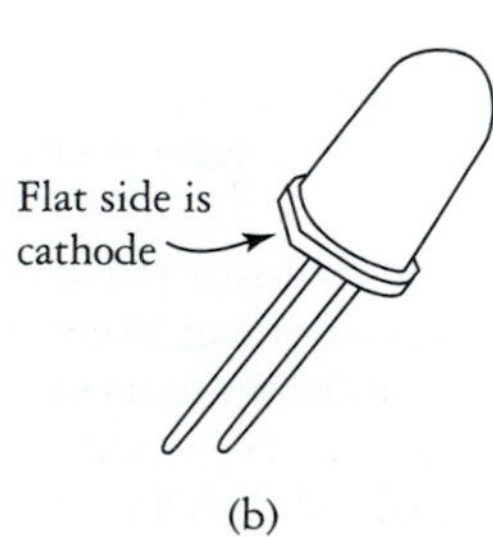

FIGURE 1.33

(a) Schematic symbol for an LED. (b) Typical LED package.

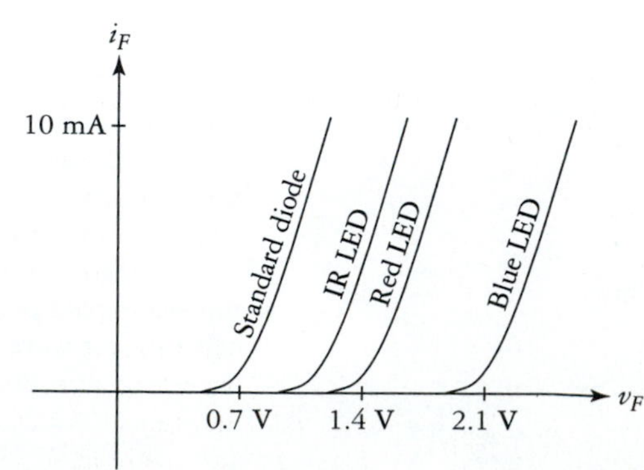

FIGURE 1.34

Comparison of typical LED forward transconductance curves.

LED Circuit Analysis

Circuits that contain LEDs are analyzed using the same techniques that were presented earlier. Most often an LED will be found in a simple series circuit like that shown in Fig. 1.35. To analyze this circuit, we simply apply KVL and come up with the following familiar equation

$$0 = V_{\text{in}} - I_1 R_1 - V_\phi$$

We solve this equation as necessary in a given situation.

FIGURE 1.35
Series circuit containing an LED.

EXAMPLE 1.6

The diode used in Fig. 1.35 is a red LED, which at rated brightness has $V_\phi = 1.5$ V, $I_F = 15$ mA. Given $V_{\text{in}} = 5$ V, determine the closest standard value for R_1 that should be used, the actual diode current, and the power dissipation of the LED and the resistor.

Solution We solve the KVL equation for R_1, giving us

$$\begin{aligned} R_1 &= \frac{V_{\text{in}} - V_\phi}{I_1} \\ &= \frac{5\text{ V} - 1.5\text{ V}}{15\text{ mA}} \\ &= 233.3\ \Omega \text{ (use } 220\ \Omega) \end{aligned}$$

The actual current flow is

$$\begin{aligned} I_F &= \frac{V_{\text{in}} - V_\phi}{R_1} \\ &= \frac{5\text{ V} - 1.5\text{ V}}{220\ \Omega} \\ &= 16\text{ mA} \end{aligned}$$

This slight increase in current will result in a slightly brighter LED, and will generally cause no harm. The power dissipation of the LED and R_1 may be found in several different ways:

$$\begin{aligned} P_{\text{LED}} &= I_1 V_\phi \\ &= (16\text{ mA})(1.5\text{ V}) \\ &= 24\text{ mW} \end{aligned}$$

$$\begin{aligned} P_{R1} &= I_1^2 R_1 \\ &= (16\text{ mA})^2(220\ \Omega) \\ &= 56.3\text{ mW} \end{aligned}$$

Note
LEDS have low reverse breakdown voltage compared to normal diodes.

Like normal diodes, LEDs block current flow when reverse biased. They are almost never used for that purpose, however, because they have very low reverse breakdown voltage, often less than 5 V, and they are easily destroyed when reverse breakdown occurs.

Laser Diodes

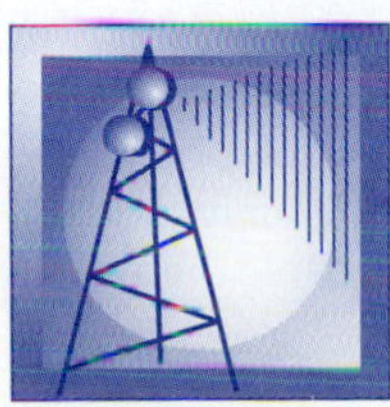

As mentioned earlier, another type of light-emitting diode is the *laser diode.* Laser diodes are structurally more complex than normal LEDs and they emit *coherent* radiation when forward current exceeds a specific threshold value. The term *coherent* means that the emitted photons are all in phase and of the same wavelength or color. Laser diodes have the same schematic symbol as standard LEDs, and they have similar transconductance characteristics, although laser diodes do operate at much higher currents.

Figure 1.36(a) compares the light power emitted by a typical laser diode and LED. At current levels greater than I_{Th}, the laser diode power output increases rapidly and becomes coherent. Laser diodes are also much more *monochromatic* than normal LEDs. This means that the light output of a laser diode concentrates more of its light energy over a smaller range of colors or wavelengths. This is illustrated in Fig. 1.36(b). Wavelength is symbolized by a lowercase lambda, λ. Just to get an idea of

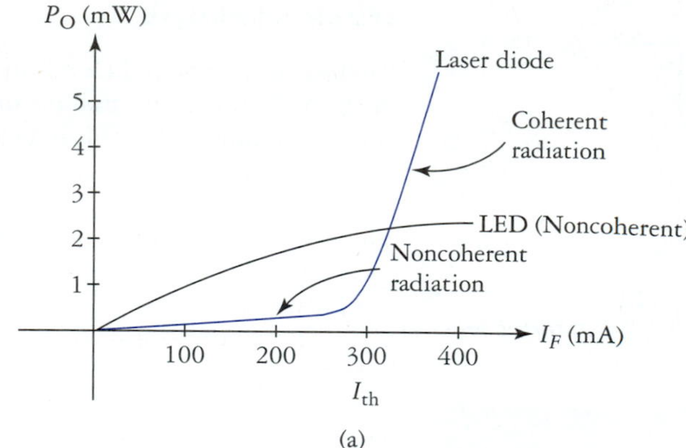

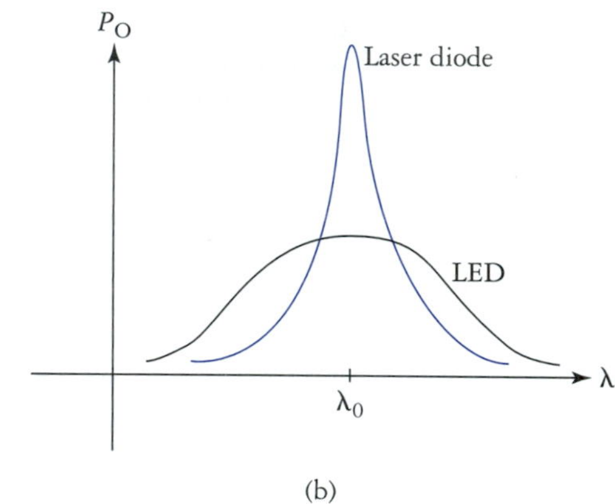

FIGURE 1.36
(a) Power output of a laser diode versus a standard LED. (b) Laser diodes are more monochromatic than normal LEDs.

the scale of the wavelengths involved here, light in the red portion of the spectrum has a wavelength of around 660 nm (nanometers), while a typical infrared wavelength is 850 nm.

Laser diodes are available in *continuous-mode* and *pulse-mode* varieties. The continuous-mode devices are normally used in low-power (a few milliwatts) applications such as laser pointers, barcode scanners, and fiber optic data communications systems. The OLX8000 is an example of a laser diode that is designed for continuous operation. This device has a threshold current of 100 mA, with a rated power output of 3 mW.

Pulse-mode laser diodes are capable of much higher power output, ranging up to hundreds of watts. The light pulses are of very short duration, usually less than 10 μs, once every millisecond or so. Thus, the duty cycle (the ratio of active laser time to the period of operation) of these diodes must be kept very low, somewhere around 1% or less. Pulse-mode laser diodes are often used in military applications, such as target illumination. An example of a pulse-mode laser diode is the model LD-68 from Laser Diode Laboratories. This device is designed to produce a peak emission power of 20 W at a wavelength of 904 nm. The typical threshold current for the LD-68 is 18 A, with a peak forward current rating of 75 A. This particular laser diode has a response time of less than 0.5 ns.

Laser diodes are even more easily damaged by reverse breakdown than LEDs and they are easily damaged by excess power dissipation and forward current. Because of these characteristics, sophisticated protective circuitry is employed in laser diode drive circuits. By the time you finish this book you will easily understand the operation of such sophisticated circuits.

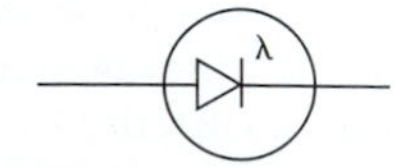

FIGURE 1.37
Two common symbols for the photodiode.

Photodiodes

It is reasonable to assume that since pn junctions can generate light that they should also be sensitive to received light as well. Photodiodes are constructed such that the pn junction may be exposed to external light through a clear window or lens. The schematic symbols commonly used for photodiodes are shown in Fig. 1.37. Several different kinds of photodiode are available; however, we take a quick look at just the standard pn junction device in this section.

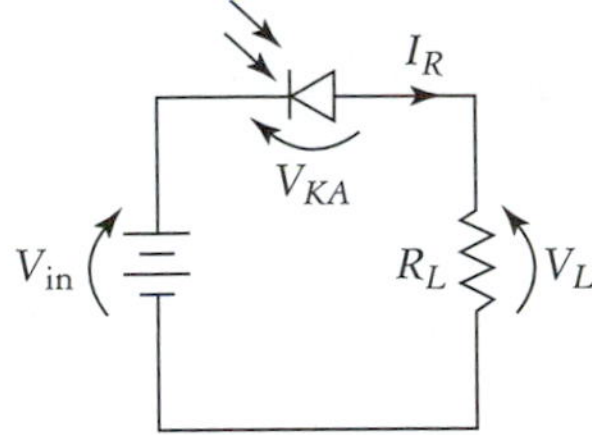

FIGURE 1.38
Photodiode used in the photoconductive mode.

Photoconductive Mode

In the photoconductive mode, the photodiode is operated under reverse bias, as shown in Fig. 1.38. If the diode is not illuminated, a very small *dark* current I_{dark} will flow. This dark current is typically a few nanoamps or so. Because the dark current is so small, nearly all of the applied bias voltage is dropped by the photodiode, and the output voltage is approximately zero.

When the photodiode is illuminated, the light entering the depletion region causes the generation of electrons and holes, which are then free to act as charge carriers through the diode. The amount of current that flows is dependent on the intensity of the intercepted light and, surprisingly, is nearly independent of the applied bias voltage. To see why this is so, examine Shockley's equation, Eq. (1.1). Note that since reverse bias is present $V_{AK} < 0$ V and the exponential term (e^{-x}) approaches zero rapidly, leaving only the saturation current term. The saturation current increases in proportion to the intensity of the received light. Typical curves that would be produced at various light intensity levels *(H)* are shown in Fig. 1.39.

FIGURE 1.39
Family of reverse current versus reverse bias curves for a typical photodiode.

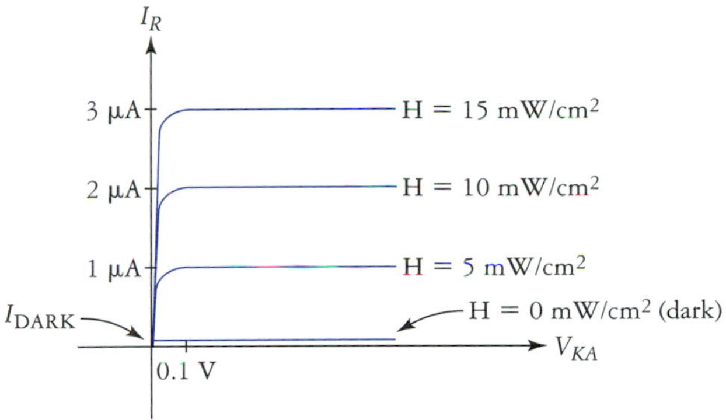

If we vary the intensity of the light received by the photodiode in proportion to an information signal, an electrical signal with the same amplitude variation will be generated by the receiver circuit. This is the idea behind devices such as CD players, optical data communication systems, seismographic instruments, and other devices.

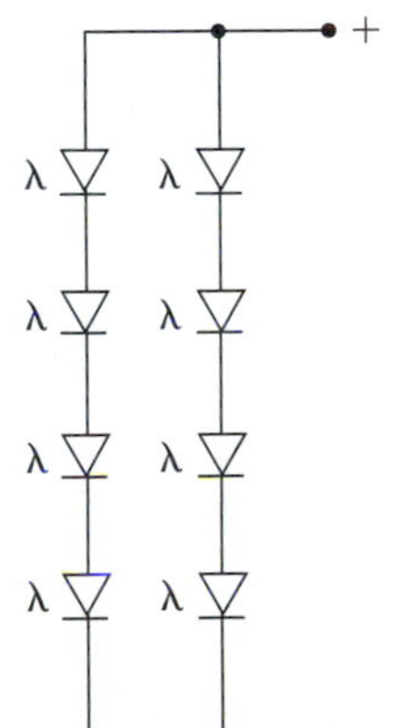

FIGURE 1.40
Connecting photovoltaic cells in series and parallel to increase power output.

Photovoltaic Mode

If no external bias is applied to a pn junction and it is exposed to light radiation of the correct wavelength, the diode will generate a significant voltage across its terminals. This is called the *photovoltaic mode* of operation. In this mode, the photodiode can be used as an energy source. Diodes that have small junction area (typically the size of an LED) produce very low power even under very bright light conditions. Even so, such devices are still usable for light detection and communication applications.

Very large junction devices called *photovoltaic cells* are available with output power levels into the hundreds of watts. These are commonly referred to as *solar cells.* Photovoltaic cells are used extensively in satellite and spacecraft applications, and increasingly in terrestrial applications as well. Photovoltaic cells are often connected in arrays as shown in Fig. 1.40 in order to increase terminal voltage and current sourcing capability.

Schottky Diodes

Two main factors determine how fast a diode can be switched on and off by a time-varying bias voltage: *junction capacitance* C_j and minority carrier stored charge. These factors together determine the *reverse recovery time* T_{rr} of a diode. A common high-voltage diode, the 1N625 has $T_{rr(max)} = 1000$ ns. The practical implication here is that this diode cannot be turned off—that is, driven into reverse bias—in less than 1000 ns. Such a diode would not be suitable for use in digital switching applications. The very popular 1N4148 pn junction diode is listed by National Semiconductor as a *computer diode* because it is designed to switch very quickly. The 1N4148 has $T_{rr(max)} = 4$ ns, which makes it a good choice for use in digital circuit applications. Figure 1.41 shows how reverse recovery time affects the current flow in a simple series circuit. Ideally, we expect the diode to block all reverse current, however, it takes the diode a finite amount of time to respond to the applied reverse bias, as the junction capacitance is charged and carriers are swept out of the depletion region.

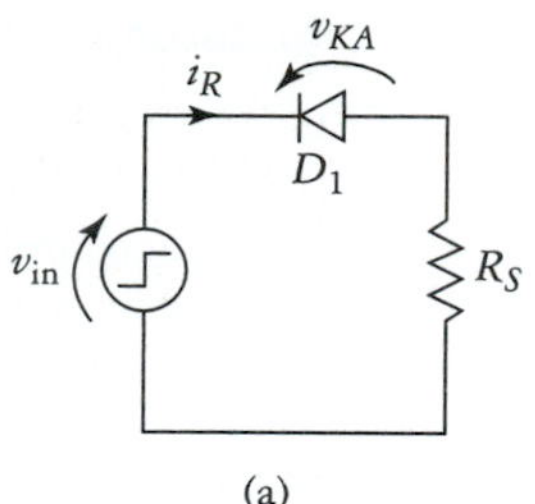

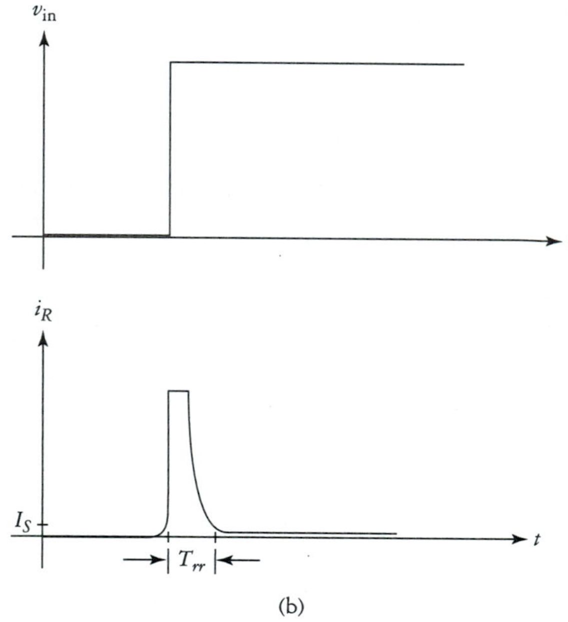

FIGURE 1.41
Diode reverse recovery time. (a) Test circuit. (b) Waveforms for input voltage and diode reverse current.

Junction capacitance is proportional to junction cross-sectional area and depletion region width. High current diodes have large junction areas and so they tend to have high C_j as well. Even if we could eliminate junction capacitance somehow, the injection of minority carriers into the depletion region would slow the diode switching action. The *Schottky diode* eliminates the minority carrier injection problem by using a metal/n-type semiconductor (m-s) junction to form the diode. We won't go into the physics of the m-s junction here, but basically because the majority carriers in metal and n-type silicon are electrons, there is no minority carrier (hole) injection to worry about with a Schottky diode. Thus, the switching speed of a Schottky diode will be faster than for a normal pn junction diode of similar area. The schematic symbol for a Schottky diode is shown in Fig. 1.42(a).

In addition to fast switching speeds, Schottky diodes have a low barrier potential of $V_\Phi = 0.4$ V. Schottky diodes have much larger leakage current than similarly rated silicon diodes. Figure 1.42(b) compares the transconductance curves for typical silicon and Schottky diodes. You may be familiar with the Schottky family of TTL (transistor-transistor logic) devices. Schottky diodes are used in these devices in order to speed switching times, yet at the same time reduce operating current (and power) levels.

Varactor Diodes

The last type of diode we examine in this chapter is the *varactor diode.* As discussed previously, all diodes exhibit junction capacitance. The n- and p-type regions act as the plates of an equivalent capacitor, while the depletion region acts as the dielectric. Junction capacitance varies as a function of the reverse bias applied to the diode. Small reverse bias results in a narrow depletion region and high C_j, while higher values of reverse bias widen the depletion region, thus reducing C_j. This property allows us to create voltage-controlled capacitors, which we call varactor diodes. Varactor diodes are silicon pn junction devices designed specifically to have well-controlled, and usually rather large, junction capacitance characteristics. Common schematic symbols for varactor diodes are shown in Fig. 1.43.

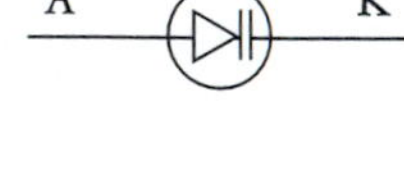

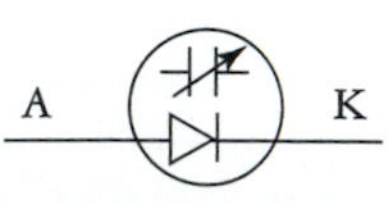

FIGURE 1.43
Common symbols for the varactor diode.

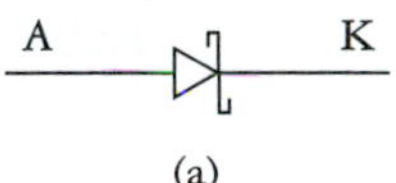

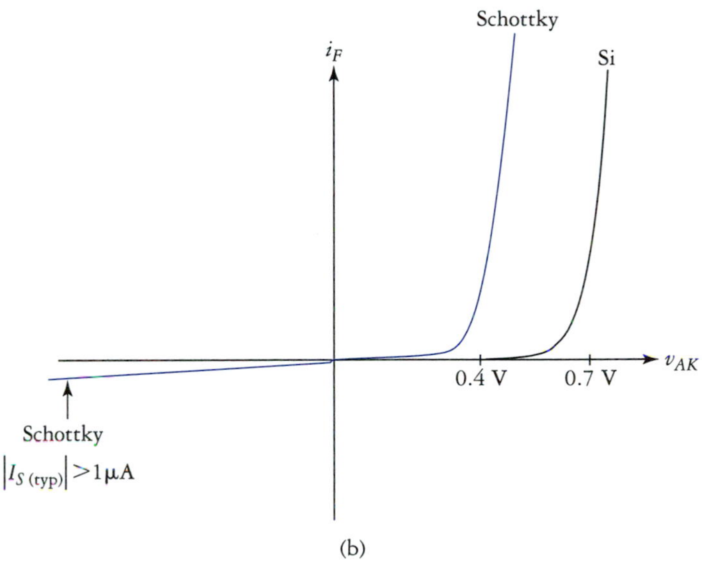

FIGURE 1.42
(a) Schottky diode symbol. (b) Comparison of Schottky diode characteristics with those of a standard Si pn junction diode.

Typical junction capacitance as a function of reverse bias voltage for a varactor diode is shown in Fig. 1.44. The equation describing this relationship is

$$C_j = \frac{C_0}{\sqrt{2\,V_R + 1}} \tag{1.11}$$

where C_0 is the junction capacitance with no bias applied and V_R is the magnitude of the reverse bias.

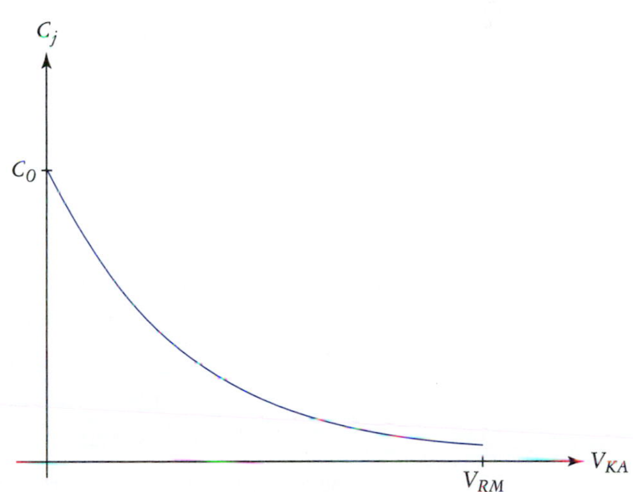

FIGURE 1.44
Variation of junction capacitance with bias voltage.

Varactor diodes are used in electronic tuning circuits like those found in television and radio circuits. Figure 1.45 uses a varactor diode as a variable capacitor in order to adjust the resonant frequency of the resonant circuit. In resonance, the parallel section (C_1, D_1, and L) exhibits very high impedance, allowing most of the ac signal to pass across to v_0. At frequencies above and below resonance, the voltage drop across the tuned circuit decreases, reducing v_0. This forms a bandpass filter, which is useful in many communications applications. Here we've got something we can sink our

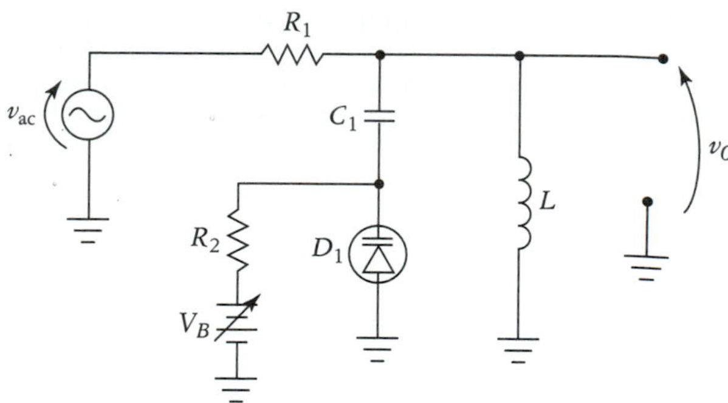

FIGURE 1.45
Electrically tuned resonant circuit.

teeth into, so let's get a little more quantitative again! Recall that resonance occurs when $|X_C| = |X_L|$, which gives us

$$f_0 = \frac{1}{2\pi\sqrt{LC_{eq}}} \tag{1.12}$$

The diode is used as a capacitor, thus the equivalent total capacitance is found to be

$$C_{eq} = \frac{C_1 C_j}{C_1 + C_j} \tag{1.13}$$

The value of R_2 is normally greater than 10 kΩ so as not to lower the Q of the resonant circuit. We must also restrict the amplitude of the ac signal so that the diode does not become forward biased, and so that the signal-induced variation of C_j is negligible as well. Typically, the signal voltage will be kept below 100 $mV_{p\text{-}p}$. You may wish to review the discussion on nonlinearity and small-signal conditions in Section 1.2.

EXAMPLE 1.7

The circuit in Fig. 1.44 has the following values: $0\text{ V} < V_B < 10\text{ V}$, $C_1 = 50$ pF, and $L = 100$ μH. The varactor diode has $C_0 = 50$ pF. Determine the minimum and maximum limits of f_0.

Solution The minimum frequency occurs at $V_B = 0$ V, where $C_j = C_0 = 50$ pF; therefore,

$$\begin{aligned} C_{eq(max)} &= \frac{C_1 C_0}{C_1 + C_0} \\ &= \frac{50\text{ pF} \times 50\text{ pF}}{50\text{ pF} + 50\text{ pF}} \\ &= 25\text{ pF} \end{aligned}$$

$$\begin{aligned} f_{0(min)} &= \frac{1}{2\pi\sqrt{LC_{eq(max)}}} \\ &= \frac{1}{2\pi\sqrt{100\text{ μH} \times 25\text{ pF}}} \\ &= 3.18\text{ MHz} \end{aligned}$$

The maximum resonant frequency occurs when $V_B = 10$ V, which forces C_j to its minimum. Here we obtain

$$\begin{aligned} C_j &= \frac{C_0}{\sqrt{1 + 2V_R}} \\ &= \frac{50\text{ pF}}{\sqrt{1 + 20}} \\ &= 10.9\text{ pF} \end{aligned}$$

Now we find the equivalent capacitance

$$C_{\text{eq(min)}} = \frac{C_1 \times C_{\text{eq(min)}}}{C_1 + C_{\text{eq(min)}}} = \frac{50 \text{ pF} \times 10.9 \text{ pF}}{50 \text{ pF} + 10.9 \text{ pF}} = 8.9 \text{ pF}$$

The maximum resonant frequency is

$$f_{0(\text{max})} = \frac{1}{2\pi\sqrt{LC_{\text{eq(min)}}}} = \frac{1}{2\pi\sqrt{100\ \mu\text{H} \times 8.9 \text{ pF}}} = 5.33 \text{ MHz}$$

1.8 COMPUTER-AIDED ANALYSIS APPLICATIONS

The component library in the evaluation version of PSpice contains models for several diodes, including the following:

1N4148	Switching diode
1N4002	Rectifier diode
1N750	Zener diode
MBD101	Diode
MV2201	Varactor diode

We begin using PSpice here to verify the results of some of the examples that were worked out earlier.

Operating Point Analysis

Note
In PSpice version 8, you can click on the current button (I) on the toolbar instead of using an ammeter.

The circuit of Example 1.2 is now simulated. Recall that we calculated the value of R_S (2.26 kΩ) that should produce a load current of approximately 5 mA. The circuit is drawn with the schematic editor in Fig. 1.46. The 1N4148 diode was chosen because it is extremely common, and this is a low-current application, suitable for such a diode. Only the default operating point analysis was performed. PSpice always performs an operating point analysis in order to establish quiescent or no-signal voltage and current levels in a circuit. Comparing the results of the simulation with those of Example 1.2, we find excellent agreement. This helps give us faith in our techniques.

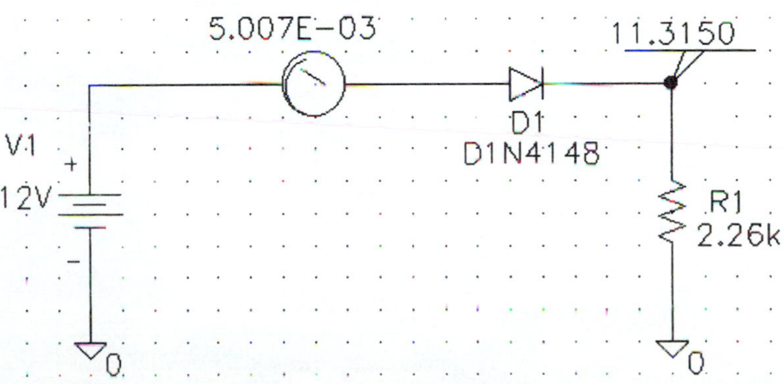

FIGURE 1.46
Schematic capture version of a diode circuit.

The circuit of Fig. 1.18 required us to use Thevenin's theorem in order to find the diode current. This was simulated and the results are shown in Fig. 1.47. Notice that the forward voltage dropped by the diode is approximately 0.67 V and the current is 4.07 mA. In the original example we used $V_\phi = 0.7$ V and obtained $I_D = 4.06$ mA.

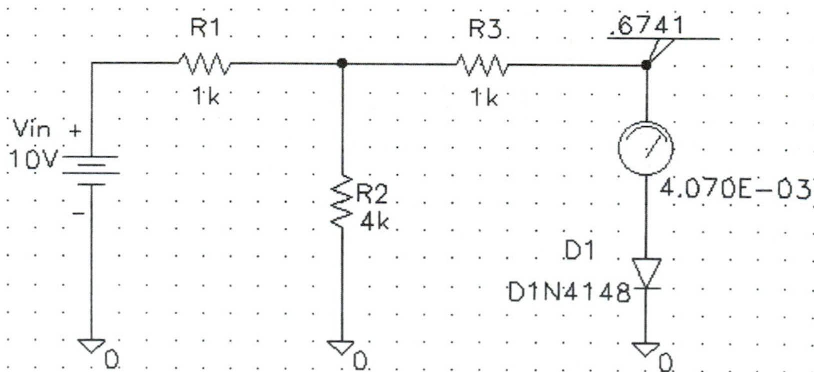

FIGURE 1.47
Circuit that required Thevenin's theorem for analysis.

Plotting the Transconductance Curve

Let's take a look at the transconductance curve of the 1N4148 diode. There are several ways to do this; however, this time we simply draw the schematic as shown in the corner of Fig. 1.48. Notice that the dc voltage source is set to 0 V. This is fine because we are going to sweep the value of V1 from 0 to 0.75V in 0.1-V increments, as shown in the analysis setup boxes. *DC Sweep* allows us to vary the values of voltage and current sources, and it allows us to vary component parameters as well. We will use this feature many times in later chapters.

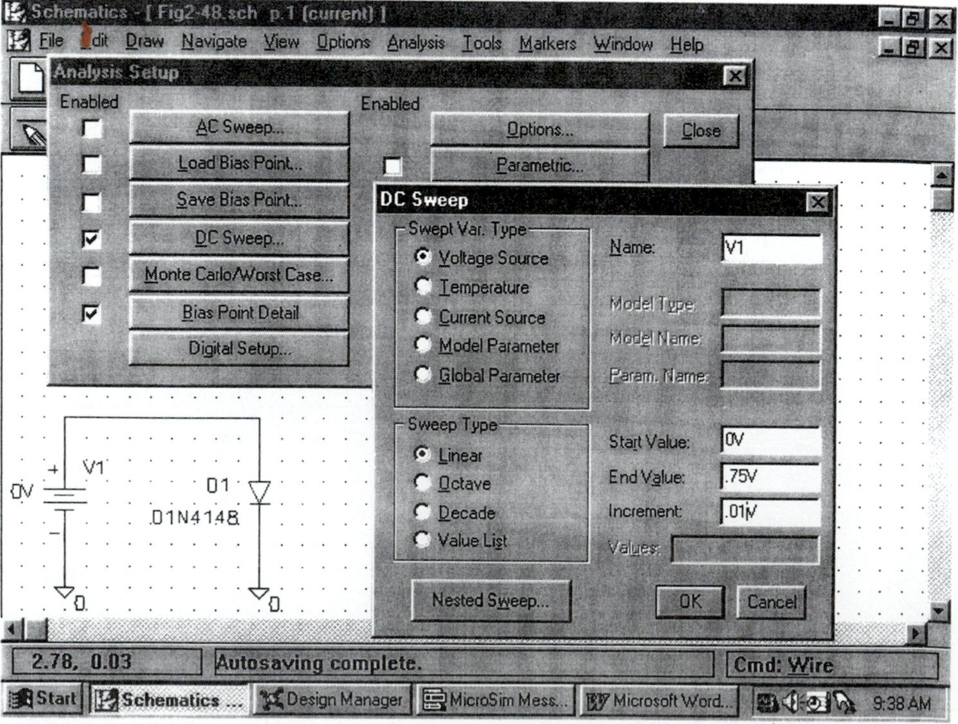

FIGURE 1.48
Setting up a dc sweep of the dc input voltage.

The results of the analysis are shown in Fig. 1.49, where the diode current I(D1) is plotted as a function of diode voltage (actually the applied voltage) V_V1. A cursor was placed at V_V1 = 0.7 V. The cursor data box indicates that the corresponding diode current is 6.69 mA.

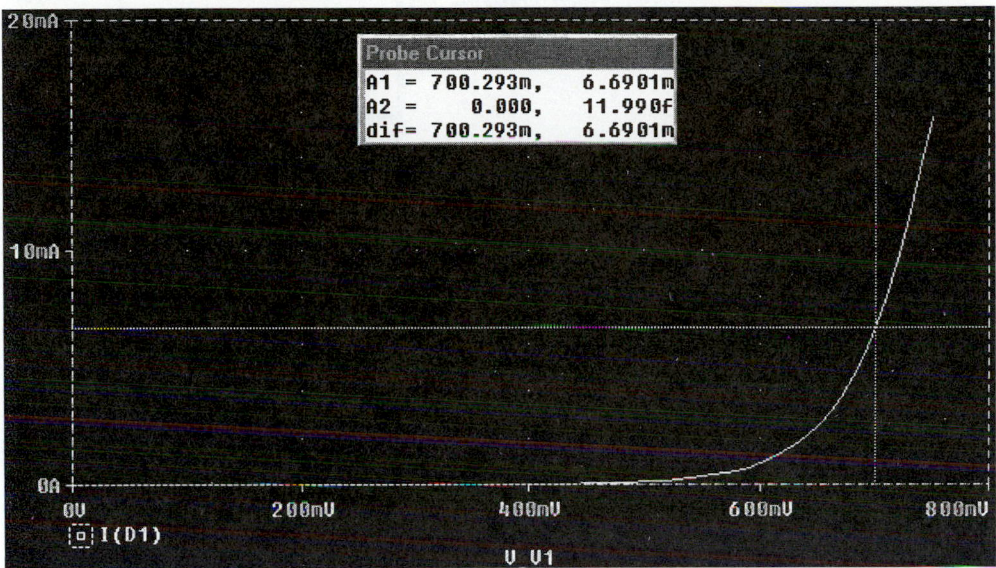

FIGURE 1.49
Forward transconductance curve generated by the dc sweep.

We would not connect a diode directly across a voltage source like this in actual practice, because it is too easy to destroy either the diode or the source by accident. A series resistor would be used to limit current to a safe value. However, this is one of the wonderful things about computer simulation: You don't have to worry about exploding components!

Small-Signal Considerations

As a final example of computer-aided analysis, let us take a look at the circuit used in Fig. 1.22. The circuit is redrawn for simulation in Fig. 1.50. The sinusoidal source was named Vin and its attributes set as shown in the figure. The DC and AC attributes need not be set, because these are used for frequency response studies, which we are not using now. The attributes shown produce a 1-kHz, 0.5-V_{pk} sine wave with a 4-V dc offset, as required.

A transient analysis was performed with a final time of 5 ms and a step increment of 10 μs. Plotting the voltage drop across the diode V(D1:1) produced the plot in Fig. 1.51. Placing cursors at the minimum and maximum peaks of the waveform shows v'_F = 15 $mV_{p\text{-}p}$. Our previous calculations indicated that we should have 7.8 $mV_{p\text{-}p}$. The main cause of discrepancy here is due to our neglecting of bulk resistance in our calculations. It is left as an exercise to show that if we include R_B = 5 Ω in the calculations, the results are in much better agreement. Also note that the average dc voltage across the diode is about 0.683 V (average the min and max values of the signal). This is fairly close to our 0.7-V barrier potential approximation.

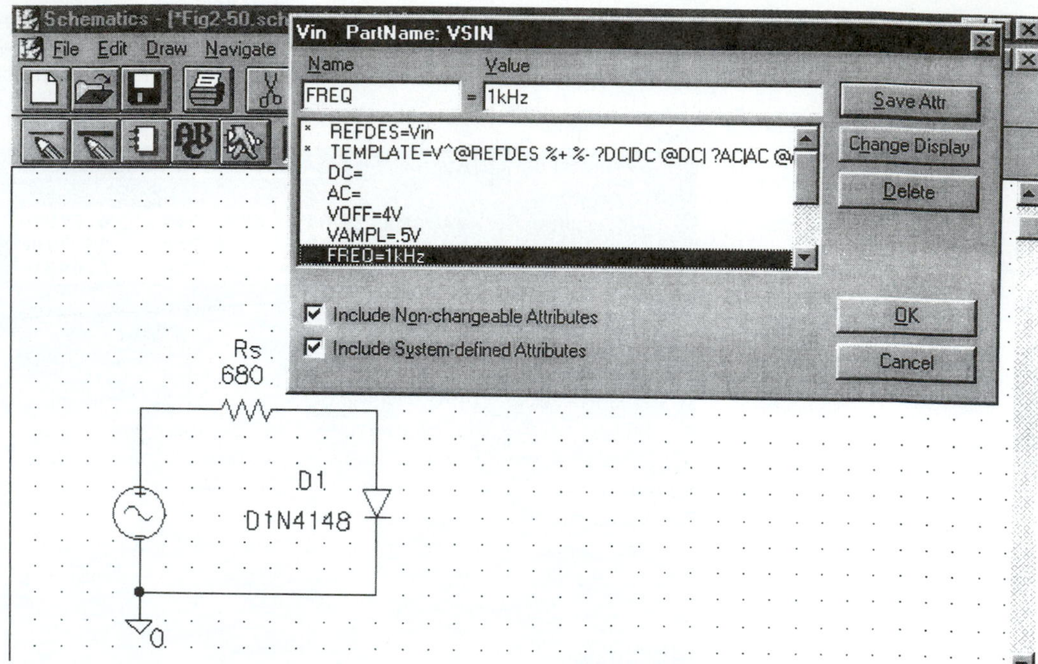

FIGURE 1.50
Small-signal analysis circuit with VSIN atttribute box. Note dc offset of 4 V.

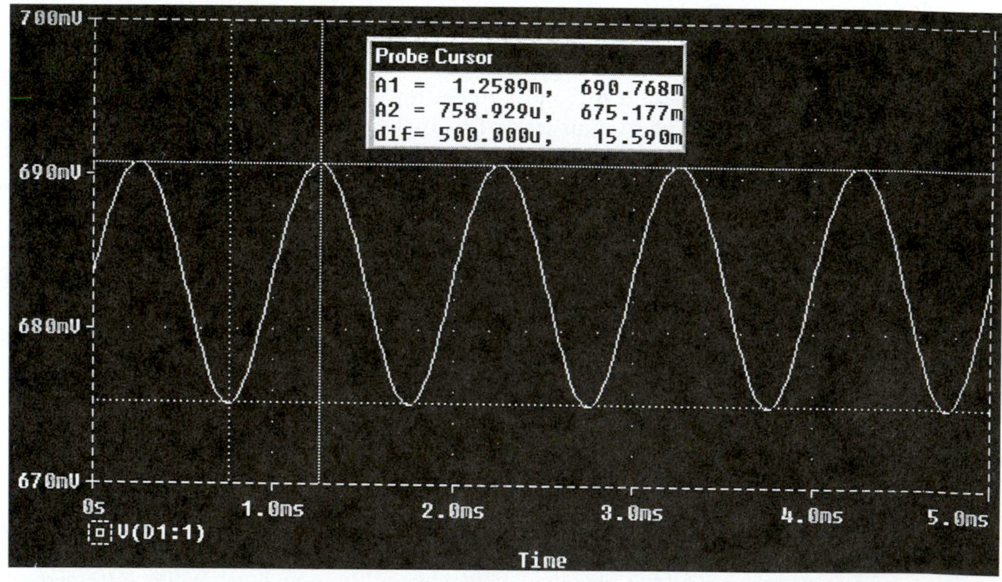

FIGURE 1.51
Voltage waveform produced at the anode of the diode.

■ CHAPTER REVIEW

Semiconductor devices are most commonly made of silicon, which belongs to Group IV of the periodic table. Group III dopants are used to produce p-type material. In p-type material, the majority carriers are called holes. A hole is equivalent to a positively charged particle that flows through the valence band. Group V dopants are used to produce n-type material, in which conduction-band electrons constitute the majority carriers. A metallurgical junction of p- and n-type material produces what is called a depletion region. This region tends to expand under reverse bias, blocking current flow, while allowing current to flow easily under forward bias. Thermally produced electron–hole pairs and surface leakage allow some current to flow across the depletion region, regardless of whether a pn junction is forward or reverse biased.

Diodes are nonlinear devices that tend to allow current to flow relatively unimpeded when forward biased, while tending to block current flow when reverse biased. Currently, silicon is the primary semiconductor material from which diodes are fabricated. The typical barrier potential of the silicon diode is approximately 0.7 V, while that for germanium devices is around 0.3 V. Schottky diodes rely on a metal–semiconductor junction for operation and have fast switching speeds and lower barrier potential than silicon devices. Most of the characteristics of the diode are temperature dependent.

Diodes are usually modeled as linear devices over some specific range of operation. The more complex exponential model of the diode, described by Shockley's equation, is used most often in computer-aided design and analysis situations.

A sufficiently high reverse bias will cause a diode to break down and conduct a relatively high reverse current. For normal diode applications, this situation is to be avoided. However, special zener diodes are designed to operate under reverse breakdown. Zener diodes are generally used to produce very stable voltage drops over a wide range of currents.

Light-emitting diodes are wide bandgap diodes with a pn junction that emits light energy when forward biased. LEDs can produce light ranging from infrared to blue in color. Laser diodes are close relatives to LEDs; however, laser diodes produce much more intense light. Laser diodes are easily damaged and must be handled carefully.

Varactor diodes utilize the variation of depletion region width in order to act as a variable capacitor. This allows varactor diodes to be used in electronic tuning applications.

DESIGN AND ANALYSIS PROBLEMS

1.1. Refer to Fig. 1.52. Determine the value of R_S such that $I_D = 10$ mA. Assume $v_{in} = 15$ V, and the diode is a silicon device. Ignore bulk resistance effects.

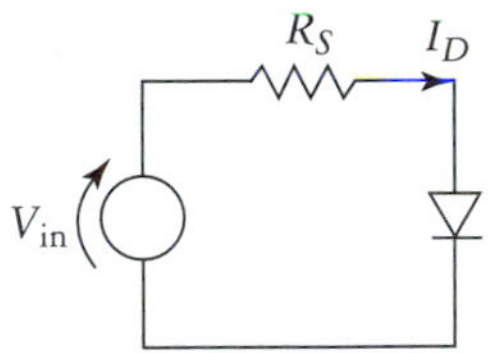

FIGURE 1.52
Circuit for Problem 1.1.

1.2. Refer to Fig. 1.52. Given $R_S = 330\ \Omega$, and the diode is silicon, determine v_{in} such that $I_D = 5$ mA. Assume the bulk resistance is negligible.

1.3. A certain silicon diode has $I_S = 0.2$ nA. Use Shockley's equation to determine I_F given $V_F = 0.65$ V, for emission coefficients of $\eta = 1$ and $\eta = 2$. Assume constant temperature with $V_T = 26$ mV.

1.4. A certain silicon diode has $I_S = 0.5$ nA and $\eta = 1$. Given $V_T = 26$ mV and $I_F = 10$ mA, determine V_F.

1.5. Assume that the diode in Fig. 1.52 has the transconductance curve of Fig. 1.53. Given $v_{in} = 3$ V and $R_S = 150\ \Omega$ use the graphical load-line analysis technique to determine the values of I_F and V_F.

1.6. Repeat Problem 1.5 for $R_S = 200\ \Omega$.

1.7. Assume the diode in Fig. 1.52 has the transconductance curve of Fig. 1.54. Given $V_{in} = 1.5$ V and $R_S = 100\ \Omega$, use the load-line analysis technique to determine I_F and V_F.

1.8. Repeat Problem 1.7 using $V_{in} = 2V$.

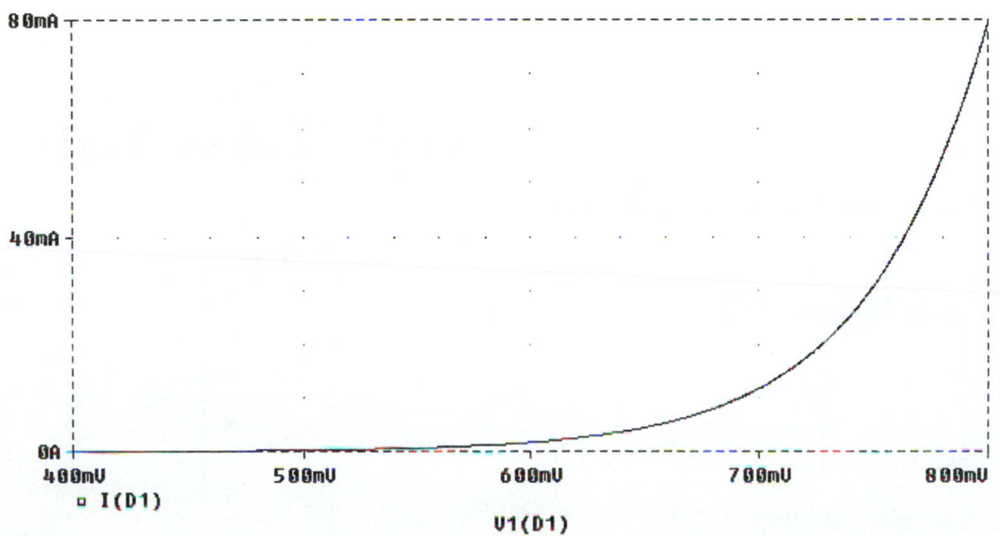

FIGURE 1.53
Diode transconductance curve.

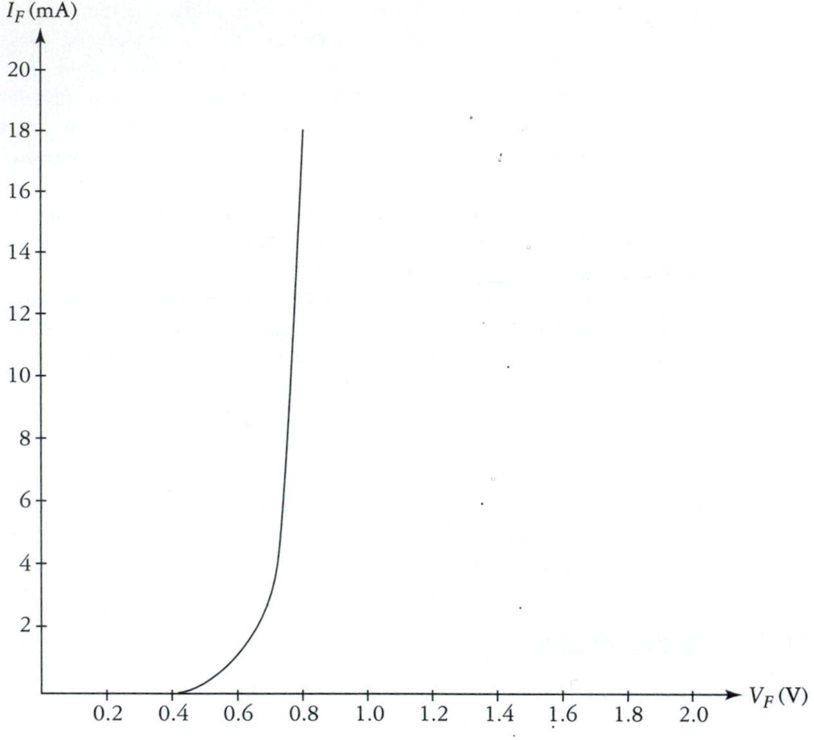

FIGURE 1.54
Diode transconductance curve.

1.9. Use Thevenin's theorem to determine the current for the diode in Fig. 1.55. Assume a silicon device with $R_F = 0\ \Omega$ is used.

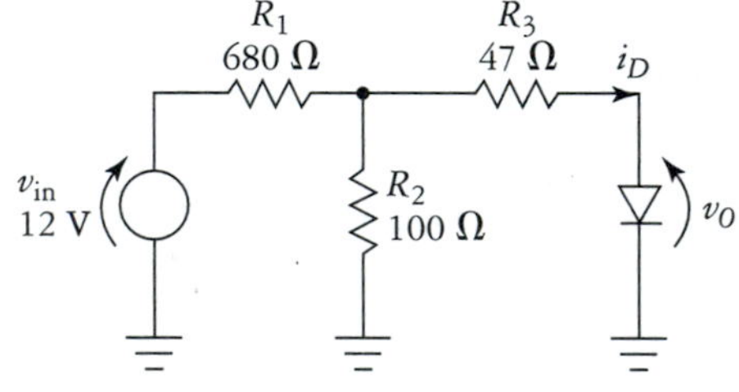

FIGURE 1.55
Circuit for Problem 1.9.

1.10. Use Thevenin's theorem to determine the current for the diode in Fig. 1.56. Assume a silicon device with $R_F = 0\ \Omega$ is used.

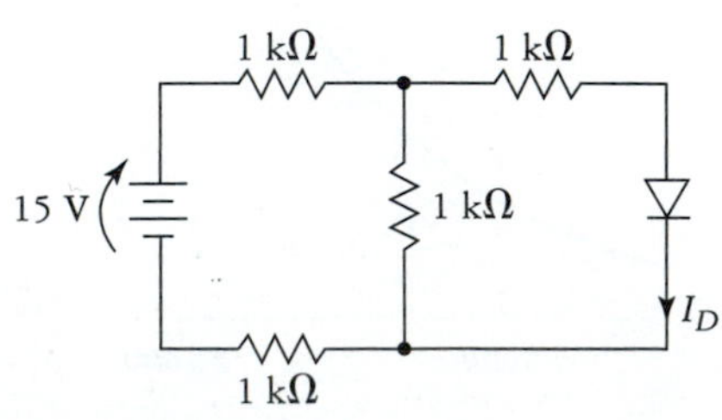

FIGURE 1.56
Circuit for Problem 1.10.

1.11. Refer to Fig. 1.57. Assuming a silicon diode with $R_F = 0\ \Omega$, determine I_D.

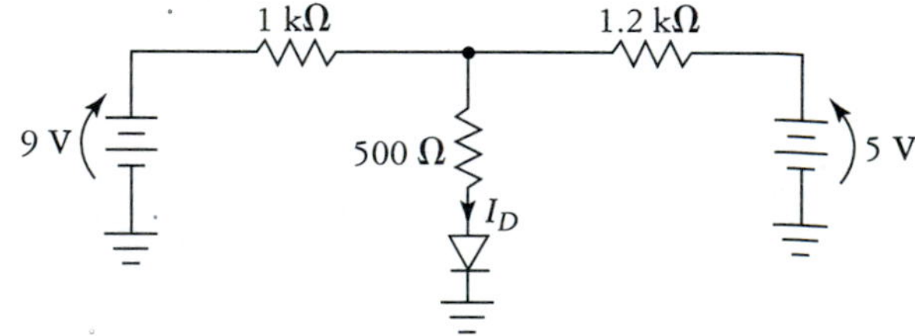

FIGURE 1.57
Circuit for Problem 1.11.

1.12. Refer to Fig. 1.58. Assuming a silicon diode with $R_F = 0\ \Omega$, determine I_D.

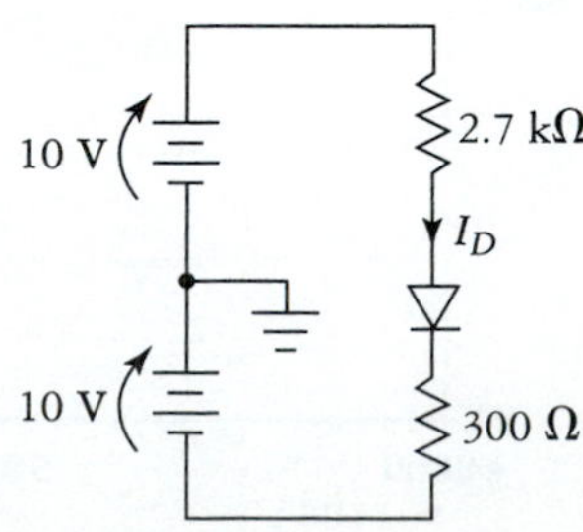

FIGURE 1.58
Circuit for Problem 1.12.

1.13. Refer to Fig. 1.59. Assuming silicon diodes with $R_F = 0\ \Omega$ are used, determine I_D.

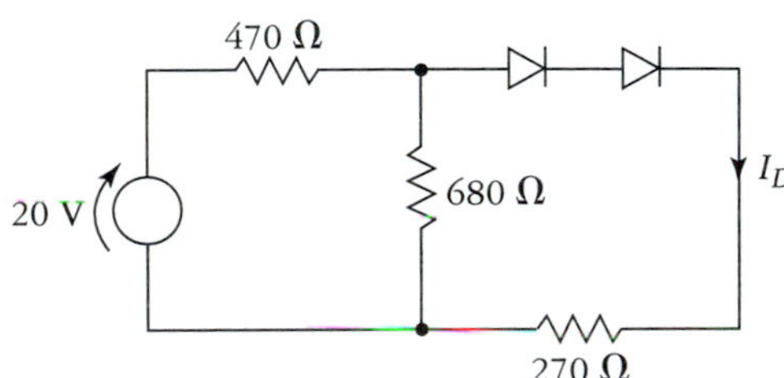

FIGURE 1.59
Circuit for Problems 1.13 and 1.14.

1.14. Refer to Fig. 1.59. Determine I_D if germanium diodes are used. Ignore bulk resistance.

1.15. Refer to Fig. 1.60. Assuming a silicon diode is used, determine the following quantities:

(a) The diode Q point (dc operating conditions).

(b) The dynamic resistance of the diode.

(c) The complete instantaneous expressions for i_D and v_D.

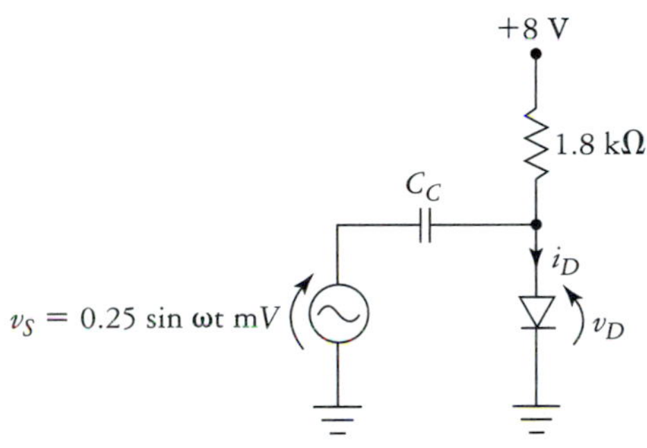

FIGURE 1.60
Circuit for Problem 1.15.

1.16. Refer to Fig. 1.61. Determine the following for the Si diode:

(a) The diode Q point.

(b) The dynamic resistance of the diode.

(c) The instantaneous expressions for i_D and v_D.

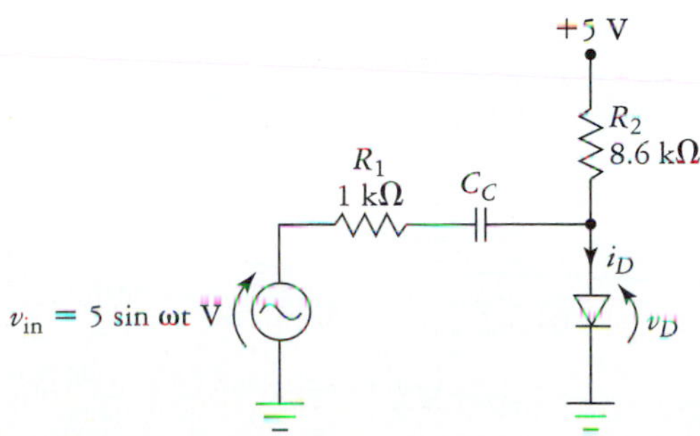

FIGURE 1.61
Circuit for Problem 1.16.

1.17. Refer to Fig. 1.62. Determine the value of R_S that will result in $I_S = 3$ mA. Assume silicon diodes with $R_F = 0\ \Omega$.

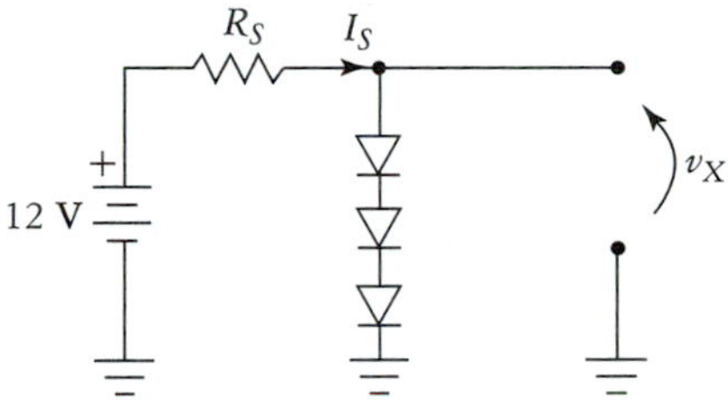

FIGURE 1.62
Circuit for Problems 1.17 and 1.18.

1.18. Refer to Fig. 1.62. Determine the approximate value of v_x if (a) the three diodes are silicon devices; (b) the diodes are germanium devices.

1.19. Refer to Fig. 1.63. Assume that $V_{in} = 15$ V, $V_Z = 8.2$ V, $I_{ZK} = 1$ mA, and $P_{ZM} = 600$ mW. Determine R_S such that (a) $I_Z = 0.9I_{ZM}$; (b) $P_Z = 0.5P_{ZM}$.

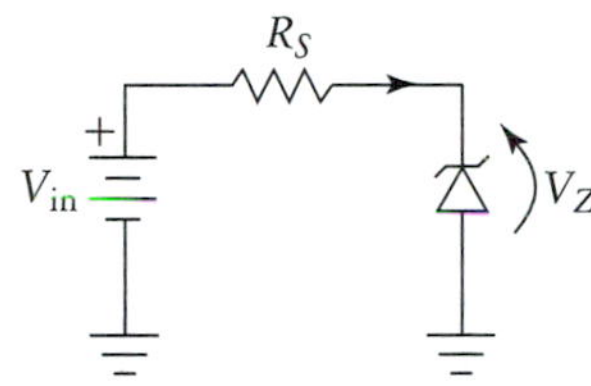

FIGURE 1.63
Circuit for Problems 1.19 and 1.20.

1.20. Repeat Problem 1.19 for $V_Z = 6.8$ V.

1.21. Refer to Fig. 1.64. Given $V_Z = 5.0$ V, $I_{ZM} = 160$ mA, $I_{ZK} = 1$ mA, and $R_L = 150\ \Omega$, determine I_L, I_Z, I_S, and the power dissipation of the zener diode.

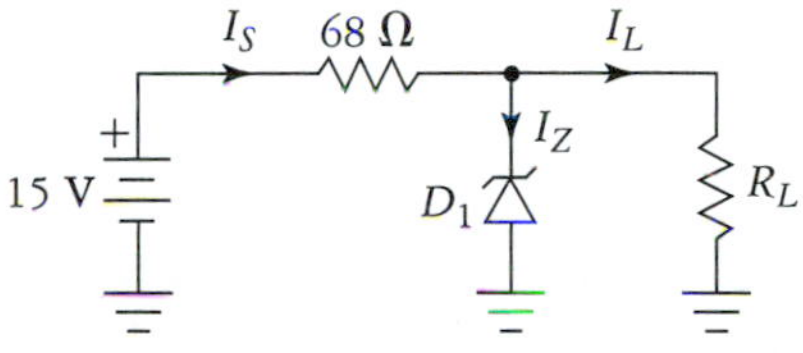

FIGURE 1.64
Circuit for Problems 1.21 and 1.22.

1.22. Refer to Fig. 1.64. Given $V_Z = 6.2$ V, $P_{ZM} = 800$ mW, and $I_{ZK} =$ mA, determine the required values for R_L such that (a) $I_Z = 5I_{ZK}$; (b) $P_Z = 250$ mW.

1.23. Refer to Fig. 1.65. A high-output IR LED is used with $P_0 = 50$ mW at $I_F = 100$mA and $V_F = 1.8$ V. Determine:

(a) The required value of R_1.

(b) The power dissipation of R_1, the LED, and the total power dissipation.

(c) The percent efficiency of the LED.

1.24. The LED in Fig. 1.65 has $I_F = 25$ mA and $V_F = 1.2$ V. Given $R_1 = 120\ \Omega$. Determine the value V_B required for these ratings.

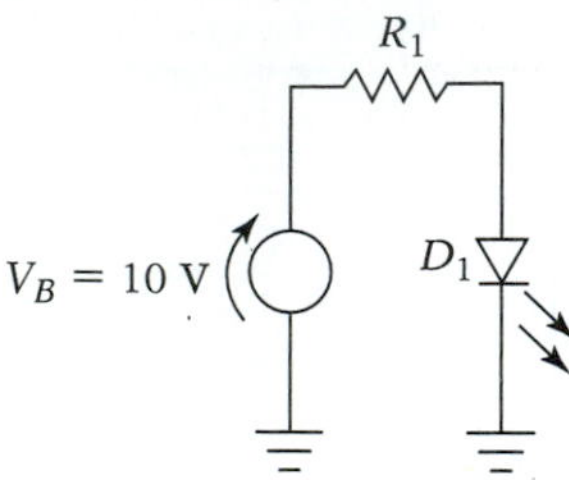

FIGURE 1.65
Circuit for Problems 1.23 and 1.24.

1.25. Refer to Fig. 1.66. The device in the dashed box is an optoisolator. Optoisolators are used to provide optical coupling of signals while electrically isolating the input and output ports. Assume that the optoisolator is designed such that $I_{photo}/I_{LED} = 10\ \mu A/1$ mA. Determine the output voltage V_o.

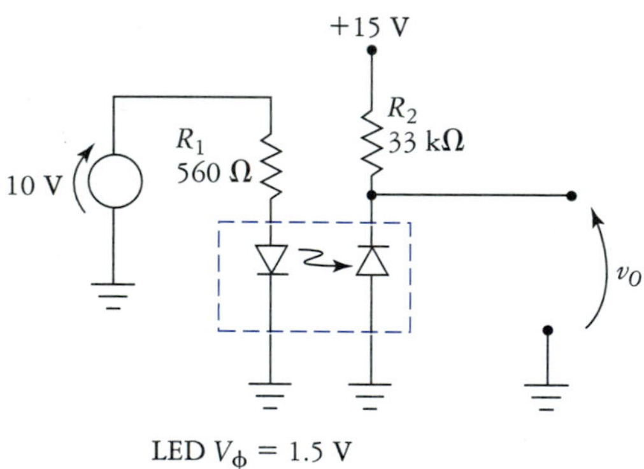

FIGURE 1.66
Circuit for Problems 1.25 and 1.26.

1.26. Using the specifications from Problem 1.25, determine the value of R_1 that will produce $V_o = 5.0$ V.

1.27. Refer to Fig. 1.45. Given $C_1 = 0.001\ \mu F$, $L = 47\ \mu H$, the varactor has $C_0 = 15$ pF and V_B can vary from 0 to 15 V, determine min and max values of C_j and f_0.

1.28. Using the data of Problem 1.27, determine V_B such that $f_0 = 7$ MHz.

TROUBLESHOOTING PROBLEMS

1.29. The circuit in Fig. 1.67 is designed to prevent v_o from exceeding +5.7 or −0.7 V. Sketch the v_o waveform that would be produced if v_{in} was a 10 $V_{p\text{-}p}$ sine wave.

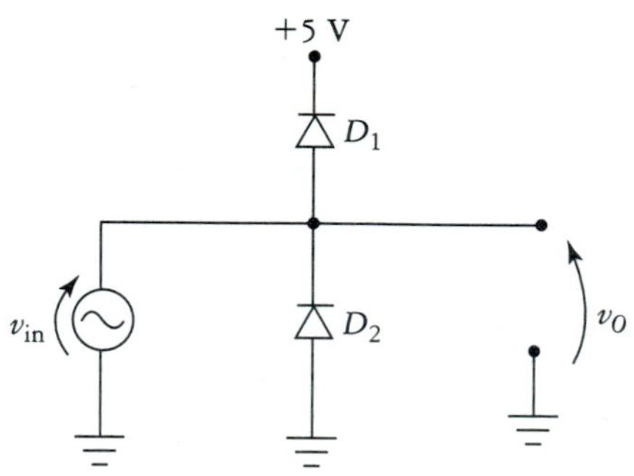

FIGURE 1.67
Circuit for Problem 1.29.

1.30. Repeat Problem 1.29, assuming that D_1 was an open circuit.

1.31. Repeat Problem 1.29, assuming that D_1 is good and D_2 is an open circuit.

1.32. Refer to Fig. 1.18. Suppose that that V_{in} measures 10.0 V and the voltage drop across the 4-kΩ resistor measures 10.0 V as well. What are some likely causes for this occurrence?

1.33. A check of a 1N4001 diode with an analog ohmmeter yields a resistance of 200 Ω in the forward direction, and a resistance of 10 kV in the reverse bias direction. Is this diode most likely good or no good? Explain your answer.

COMPUTER PROBLEMS

1.34. PSpice allows us to compare circuit operation at varying temperatures. Use PSpice to create the circuit of Fig. 1.68. From the Analysis setup menu, select the Temperature button and enter the values shown. Note that 27°C is the default simulation temperature. Now set up the DC Sweep as shown in Fig. 1.69 and run the simulation. You should get a window like that shown in Fig. 1.70. Click OK and Probe will run. A plot I(D1) vs V_V1 should result in a display like that in Fig. 1.71. Use the cursor to compare the diode currents at $V_F = 0.7$ V.

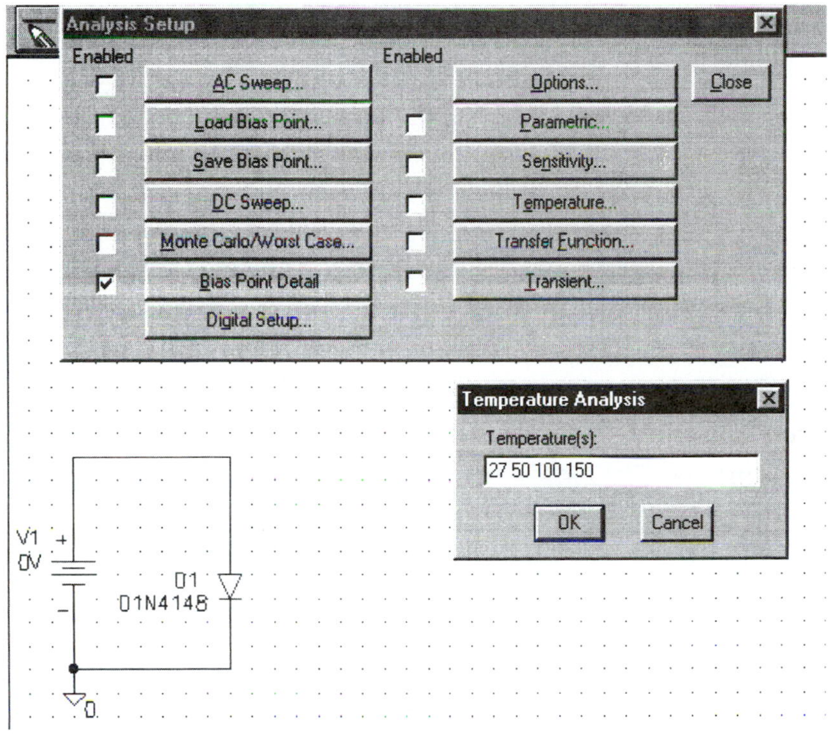

FIGURE 1.68
Diode circuit analyzed at various user-entered temperatures.

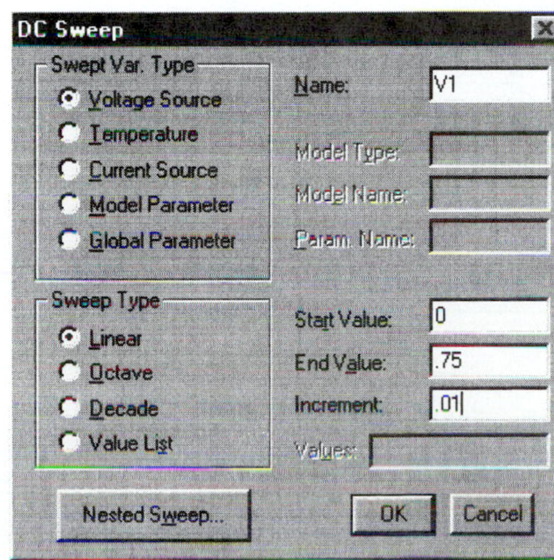

FIGURE 1.69
DC sweep attributes.

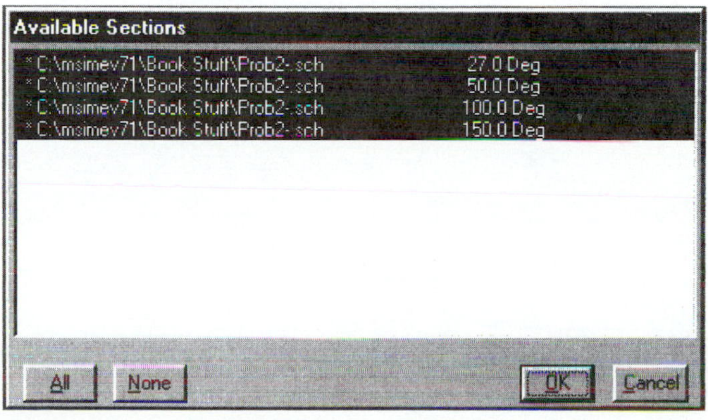

FIGURE 1.70
The temperature analysis sections box allows us to choose data to be plotted.

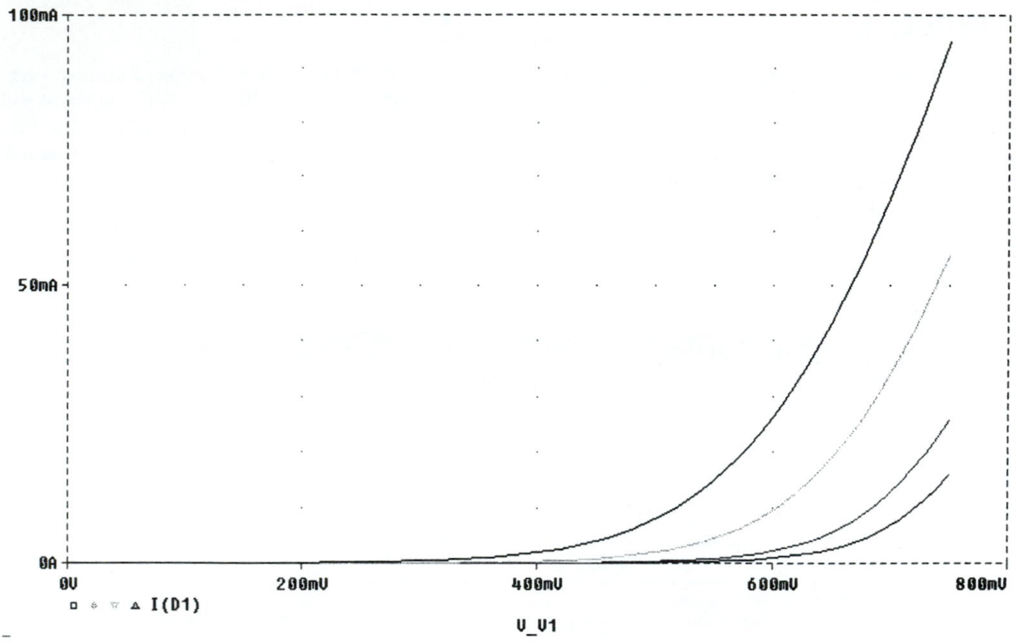

FIGURE 1.71
Diode current versus forward voltage at various temperatures.

1.35. You can find out what parameter values PSpice uses for various device models by running PARTS. Run PARTS (click Start and look under Programs and find the Microsim listing and double click Parts. Under File click Open and select Lib then Eval.lib. Now click on Part then Get. You now have access to the device models. Scroll down through the list until you find the 1N4148 diode. Double click on it and the parameter list will appear. Print out and examine the values listed. You can create your own device models using PARTS, but for now don't change or add anything. Compare the 1N4148 parameters with those of the 1N4002. Try to get a feel for what these parameters mean.

CHAPTER 2

Diode Applications

BEHAVIORAL OBJECTIVES

On completion of this chapter, the student should be able to:

- Describe the characteristics of half-wave and full-wave rectifiers.
- Determine the current levels that should occur in rectifier circuits.
- Determine the effect of filter capacitance and loading on rectifier circuits.
- Explain the significance of peak inverse voltage.
- Design basic signal shaping circuits using diodes.
- Describe the operation of clamping circuits.
- Design and analyze basic zener-regulated power supplies.
- Simulate diode circuits using a computer.

INTRODUCTION

Chapter 1 introduced the fundamental operating principles of the pn junction diode. This chapter expands on that material and also presents the diode from a functional circuit element perspective. Common applications of diodes including rectifier circuits, power supplies, wave-shaping circuits, and simple voltage regulator circuits are presented. Piecewise-linear approximations of the diode are used exclusively in this chapter and, as before, the use of PSpice in diode circuit analysis is also presented.

2.1 HALF-WAVE RECTIFIER

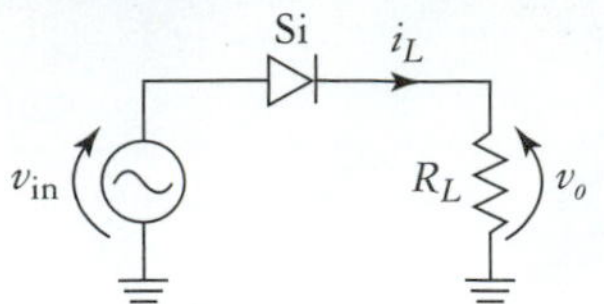

FIGURE 2.1
Basic half-wave rectifier.

Possibly the most common application for the diode is as a rectifier. The term *rectification* literally means correction or adjustment. In the context of diode applications, rectification means the conversion of ac into pulsating dc, which can be thought of as a correction of sorts. The simplest rectifier circuit is the *half-wave rectifier,* as shown in Fig. 2.1. The basic circuit topology is identical to that studied in Chapter 1.

Half-Wave Rectifier Load Voltage and Current

The half-wave rectifier works as follows. Let us assume that v_{in} is a sinusoidal voltage crossing zero, in the positive-going direction. As the upper terminal of the source increases in voltage, very little current flows until v_{in} exceeds the barrier potential V_ϕ of the diode. Once v_{in} has increased beyond V_ϕ, the load current varies approximately in direct proportion to v_{in}. Because of the existence of the barrier potential, the load voltage will be at least V_ϕ volts less than of the peak input voltage. This is illustrated in the center waveform of Fig. 2.2. Note, however, that as the input voltage amplitude is increased, the difference between peak output voltage and peak input voltage, due to barrier potential loss, tends to become insignificant.

On negative-going half-cycles of v_{in}, the diode acts essentially as an open circuit, reducing i_L and v_o to zero. Unless a silicon diode is operating in reverse breakdown, or at extraordinarily high temperatures, reverse leakage current will be negligible in the typical rectifier circuit. Recall that in some applications Schottky diodes are used to provide fast switching speeds and lower barrier potential loss. These diodes will have significantly higher reverse leakage current, which might not be negligible. For now though, we concentrate on silicon pn junction diodes.

Let us designate the maximum output of the half-wave rectifier as $V_{o(\max)}$, and again, let us designate the peak amplitude of the input voltage V_P. If the diode is modeled as an ideal device then

Note
Most often, Eq. 2.1b will be used to determine the maximum output voltage from the half-wave rectifier.

$$V_{o(\max)} = V_P \tag{2.1a}$$

If the barrier potential of the diode is the only significant factor, then, as was assumed in Fig. 2.2,

$$V_{o(\max)} = V_P - V_\phi \tag{2.1b}$$

If greater accuracy is desired, the bulk resistance of the diode could also be accounted for. For peak currents of several amps or more, this may well be necessary. In such cases, the maximum output voltage would be given by

$$V_{o(\max)} = (V_P - V_\phi)\frac{R_L}{R_L + R_F} \tag{2.1c}$$

The following example illustrates one possible situation.

EXAMPLE 2.1

The circuit of Fig. 2.1 has the following specifications: $v_{in} = 40 \sin 377t$ V, $R_L = 10\ \Omega$, and the diode is a silicon device with $R_F = 1.5\ \Omega$. Determine the peak load current and the peak output (load) voltage.

Solution The input voltage is 80 $V_{p\text{-}p}$, 60 Hz, as might be obtained from the residential ac line. Accounting only for barrier potential, the peak output voltage and load current are given by

$$\begin{aligned} V_{o(\max)} &= V_P - V_\phi \\ &= 40\text{ V} - 0.7\text{ V} \\ &= 39.3\text{ V} \end{aligned}$$

The maximum load current is

$$\begin{aligned} I_{L(\max)} &= V_{o(\max)}/R_L \\ &= 39.3\text{ V}/10\ \Omega \\ &= 3.93\text{ A} \end{aligned}$$

Accounting for both the barrier potential and the bulk resistance of the diode, we obtain

$$V_{o(\max)} = (V_P - V_\phi)\frac{R_L}{R_L + R_F}$$

$$= (40\text{ V} - 0.7\text{ V})\frac{10\ \Omega}{10\ \Omega + 1.5\ \Omega}$$

$$= 34.2\text{ V}$$

The maximum load current is found the same way as before.

$$I_{L(\max)} = V_{o(\max)}/R_L$$
$$= 34.2\text{ V}/10\ \Omega$$
$$= 3.42\text{ A}$$

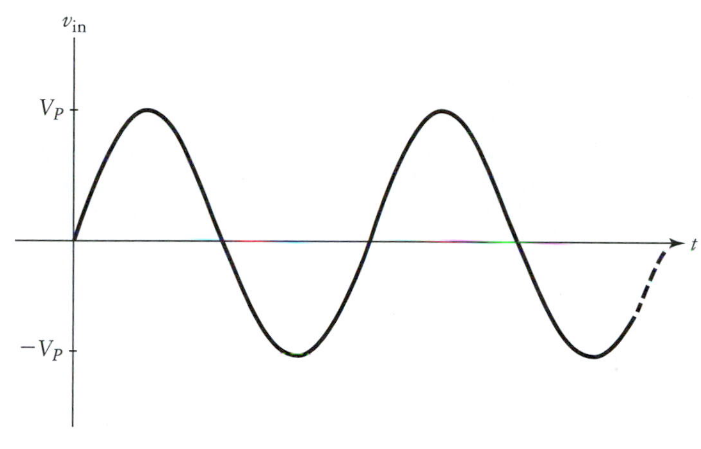

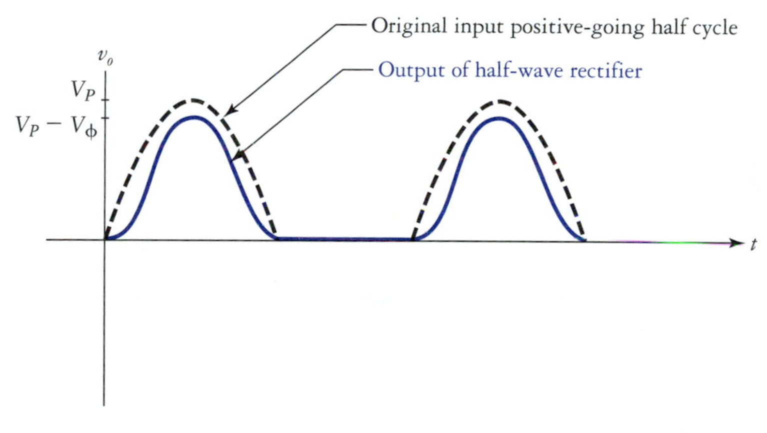

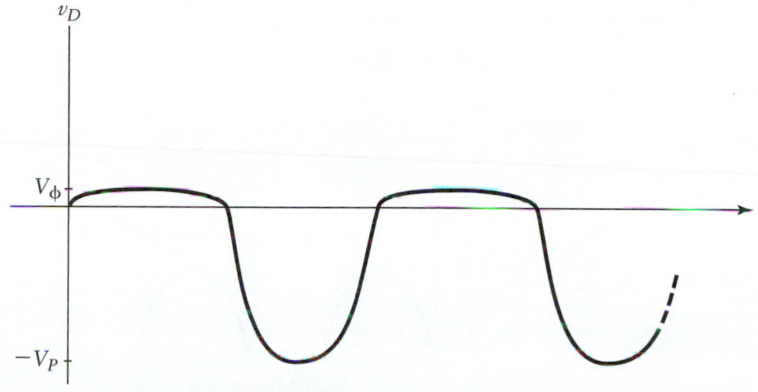

FIGURE 2.2
Rectifier input, output, and diode voltage waveforms.

The differences between the results obtained in the preceding example may not be significant enough to be very meaningful in many applications. As a general rule, if the load resistance is at least 10 times greater than the bulk resistance of the diode, the bulk resistance can be ignored. Of course, loading of the ac voltage source (assumed to be ideal here) could also reduce the maximum output voltage as well. Modification of Eq. (2.1c) to account for this situation is left as an exercise for the reader.

Half-Wave Rectifier Diode Current and Voltage

In a given rectifier application, the diode being used must be able to withstand the forward current and reverse voltage levels to which it is subjected. If we use the load resistor value (10 Ω) and input voltage (40 V_{pk}) as specified in Example 2.1, the diode in Fig. 2.1 must be able to withstand a peak reverse voltage of 40 V, and a peak forward current of possibly 3.93 A. In practice, a diode with $V_{BR} = 50$ V or $V_{BR} = 100$ V, with $I_o = 5$ A would most likely be used.

Peak Inverse Voltage

Note
V_{BR} is a manufacturer's specification, while PIV is determined by the circuit values.

Recall that the maximum reverse voltage a diode is specified to withstand is commonly designated as V_{BR}. The main point here is that V_{BR} is a *device-dependent* parameter. In this text, the maximum *circuit-dependent* reverse voltage that will be applied to a diode will be called the *peak inverse voltage* or *PIV.* A circuit-dependent parameter is a voltage or current that is determined by the conditions existing in the circuit of which the diode (or some other device of interest) is an element. Some authors consider PIV and V_{BR} to mean the same thing.

Half-Wave Rectifier Average Voltage and Current

Note
V_{av} would be read by a dc voltmeter connected across the load.

Up to this point, only peak voltage and current values have been determined. It is useful, however, to examine the half-wave rectifier from a different perspective. Assuming that the diode used in a half-wave rectifier is ideal, the load voltage will appear as shown in Fig. 2.3. The period T of the rectified voltage is the same as that of the original ac source driving the circuit. The average or mean value of the output voltage V_{av} is found by integrating v_o over one cycle, T, of the signal. The derivation of the average voltage is presented in Appendix 3, the result of which is

$$V_{av} = \frac{V_P}{\pi} \tag{2.2a}$$

or, equivalently,

$$V_{av} = 0.318V_P \tag{2.2b}$$

Because the load current and load voltage waveforms have the same general shape, as shown in Fig. 2.3, the average and peak current values are related in the same manner as the voltages, thus

$$I_{av} = \frac{I_P}{\pi} \tag{2.3a}$$

and

$$I_{av} = 0.318V_P \tag{2.3b}$$

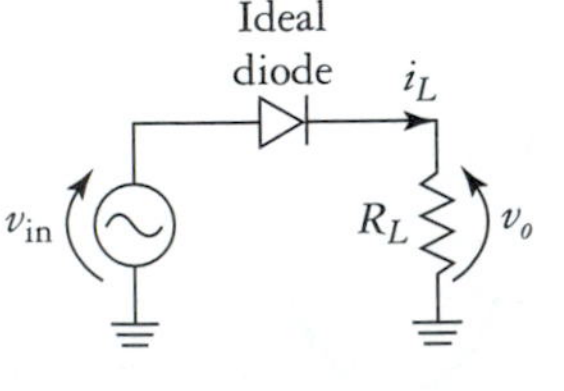

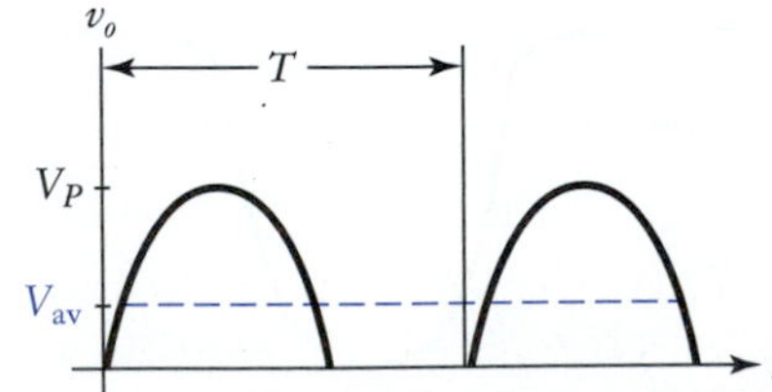

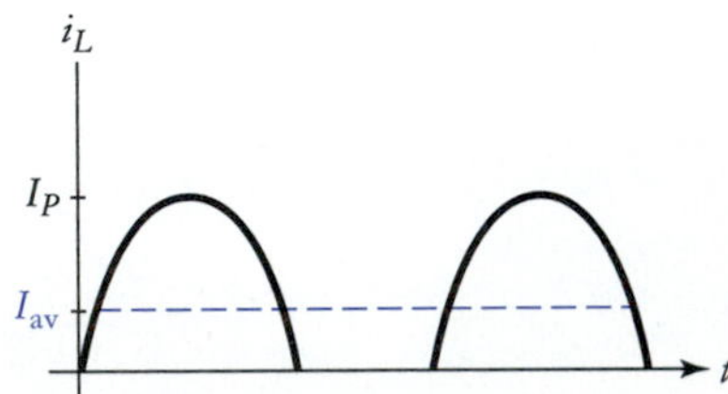

FIGURE 2.3
Average load voltage and current for the half-wave rectifier.

EXAMPLE 2.2

Refer to Fig. 2.4. Transformer T_1 has $n_{pri}/n_{sec} = 8$, and $R_L = 80\ \Omega$. The diode is silicon with $R_F = 1\ \Omega$. Determine the values of $V_{L(av)}$ and $I_{L(av)}$.

FIGURE 2.4
Circuit for Example 2.2.

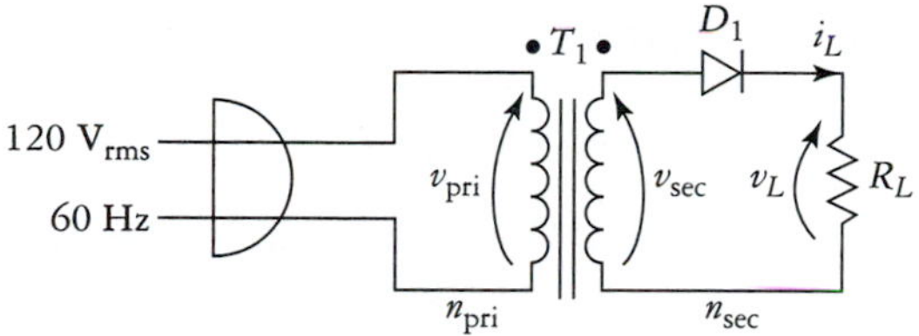

Solution The secondary voltage is found using the turns ratio as follows:

$$\begin{aligned} v_{sec} &= v_{pri}\frac{n_{sec}}{n_{pri}} \\ &= 120\ \text{V}_{rms}/8 \\ &= 16\ \text{V}_{rms} \end{aligned}$$

The peak amplitude of a sinusoid is related to the rms value by the relationship

$$V_P = \sqrt{2}\text{V}_{rms} \tag{2.4}$$

From which we obtain

$$\begin{aligned} V_{sec(pk)} &= (1.414)(16\ \text{V}) \\ &= 22.6\ \text{V} \end{aligned}$$

The peak load voltage is

$$\begin{aligned} V_{L(pk)} &= V_{sec(pk)} - V_\phi \\ &= 22.6\ \text{V} - 0.7\ \text{V} \\ &= 21.9\ \text{V} \end{aligned}$$

The average load current is found via Eq. (2.2):

$$\begin{aligned} V_{L(av)} &= 0.318V_{L(pk)} \\ &= 6.96\ \text{V} \end{aligned}$$

The peak load current is found to be

$$\begin{aligned} I_{L(pk)} &= V_{L(pk)}/R_L \\ &= 21.9\ \text{V}/80\ \Omega \\ &= 273.8\ \text{mA} \end{aligned}$$

from which we find the average load current to be

$$\begin{aligned} I_{L(av)} &= 0.318I_{L(pk)} \\ &= 87.1\ \text{mA} \end{aligned}$$

If we know the average load voltage or current, the remaining value can be determined using the relationship

$$V_{av} = I_{av}R_L \tag{2.5}$$

Conduction Angle

Although simple, the half-wave rectifier has one major drawback: It is relatively inefficient. This should be fairly obvious based on the waveforms shown in Fig. 2.3, where we find that more than 50% of the potential time that is available for delivering energy to the load is wasted during every cycle. The number of electrical degrees, relative to one cycle of v_{in}, over which the diode is conducting current is called the *conduction angle* of the diode. For a half-wave rectifier, the conduction angle is normally slightly less than 180°. To obtain higher efficiency, a higher conduction angle is required. This problem is addressed by the full-wave rectifier.

2.2 FULL-WAVE RECTIFIERS

The *full-wave rectifier* overcomes the inefficiency of the half-wave rectifier by allowing energy to be delivered to the load (again in the form of pulsating dc) over nearly the entire period of the ac input waveform. That is, the conduction angle of the full-wave rectifier is nearly 360°. Full-wave rectification can be accomplished in two ways: via bridge and center-tapped full-wave rectifiers. The bridge rectifier is discussed first.

Bridge Rectifier

The basic bridge rectifier is shown in Fig. 2.5. The key idea behind the bridge rectifier, or any full-wave rectifier for that matter, is to cause the load current to flow the same direction regardless of the polarity of the voltage applied to the bridge.

Note
In signal processing and wave-shaping applications, the full-wave rectifier is commonly called an *absolute value* circuit.

FIGURE 2.5
Bridge rectifier circuit.

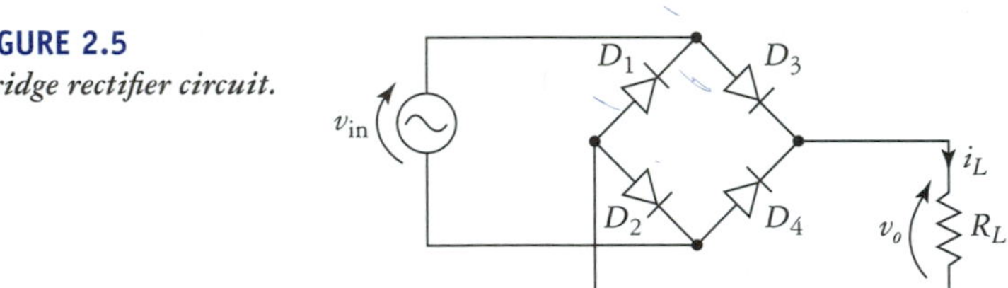

The operation of the bridge rectifier is illustrated in Fig. 2.6. During the time interval over which v_{in} is positive, diodes D_3 and D_2 are forward biased, while D_1 and D_4 are reverse biased, as shown in Fig. 2.6(a). Tracing the load current through the circuit indicates that v_o will be a positive voltage, as shown by the voltage-sensing arrow.

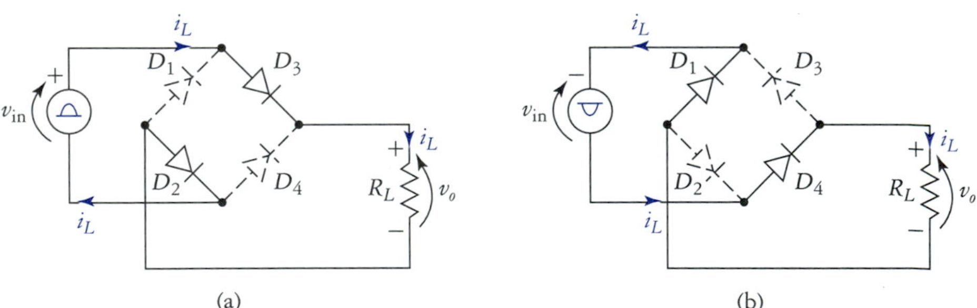

FIGURE 2.6
Operation of the bridge rectifier. (a) Positive and (b) negative half-cycles of v_{in}.

During the negative half-cycle of v_{in}, D_1 and D_4 are forward biased, while D_3 and D_2 are now reverse biased. Tracing the direction of conventional current flow shows that the direction of the load current is the same as during the positive half-cycle. Thus, the load voltage will also be of the same polarity as during the previous half-cycle.

The input and output waveforms for the bridge rectifier are shown in Fig. 2.7. Because two diodes must be forward biased in the bridge rectifier during each half-cycle, the peak output voltage will be an additional barrier potential that is lower than that for a half-wave rectifier used with the same ac source. Also, the bulk resistance of the two diodes is additive as well, which inevitably causes some additional reduction in peak output voltage. Another minor disadvantage to the bridge rectifier is its increased complexity compared to a half-wave rectifier. However, the advantage gained in the increased efficiency of the full-wave bridge more than makes up for these minor problems.

FIGURE 2.7
Full-wave bridge (a) input and (b) output voltage waveforms.

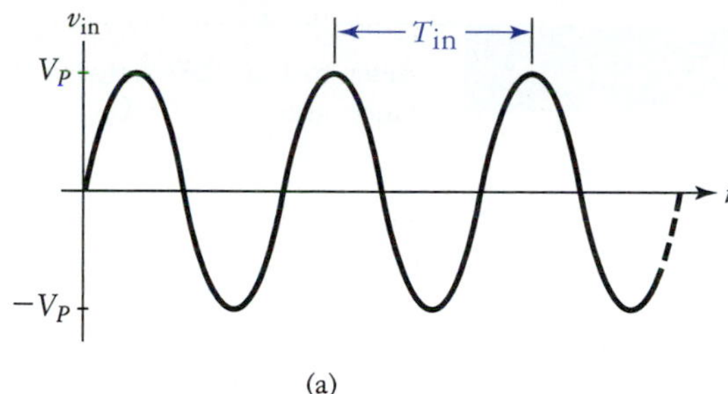

(a)

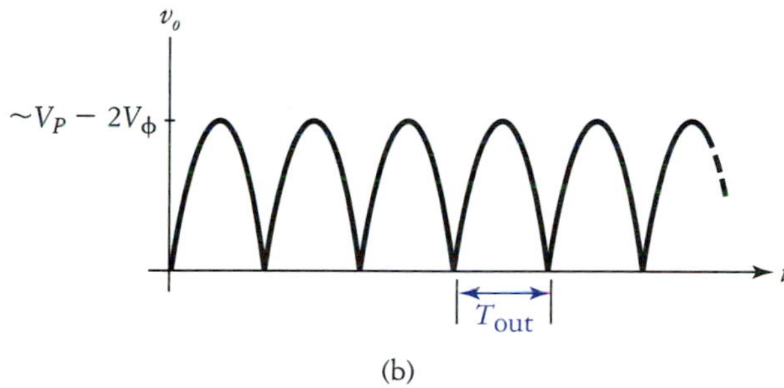

(b)

Full-Wave Rectifier Ripple Frequency

One of the more interesting and useful characteristics of the full-wave rectifier is the fact that the fundamental frequency of the output waveform is twice that of the input. In reference to Fig. 2.7, the period of v_{in} is T_{in}. The period of a signal is defined as the time interval between identical points on that signal. The period of the full-wave rectified output waveform is T_{out}. This is also called the *ripple period* T_{rip}. The input and output periods are related by

$$T_{out} = T_{in}/2 \tag{2.6}$$

Frequency is just the reciprocal of the period, thus

$$f_{out} = 2f_{in} \tag{2.7}$$

Bridge Rectifier Load Voltage and Current

If we assume that the diodes in the bridge rectifier are ideal, then the maximum amplitude of the rectified output voltage is given by

Note
Most of the time we will use Eq. 2.8b to determine the maximum voltage produced at the output of the FW bridge.

$$V_{o(\max)} = V_p \tag{2.8a}$$

where V_p is the peak input voltage. This approximation may generally be used if $V_p > 20V_\Phi$, and if the bulk resistance of the diodes is negligible. If barrier potential is to be accounted for, then

$$V_{o(\max)} = V_p - 2V_\Phi \tag{2.8b}$$

Finally, if barrier potential and bulk resistance R_F are both to be modeled, then we use

$$V_{o(\max)} = (V_p - 2V_\Phi)\frac{R_L}{R_L + 2R_F} \tag{2.8c}$$

In most lower current applications (I_L of only a few amps or so), Eq. (2.6b) provides sufficient accuracy.

The average dc value of the full-wave rectified signal is found by integrating the output waveform over one cycle. Because of the symmetry between the half- and full-wave rectified waveforms, the dc voltage is easily found to be related to the peak voltage by expanding on previous work:

$$V_{av} = \frac{2V_p}{\pi} \tag{2.9a}$$

or, alternatively,

$$V_{av} = 0.637V_{pk} \tag{2.9b}$$

EXAMPLE 2.3

Refer to Fig. 2.8. Assume that $n_p/n_s = 18/5$, the diodes are silicon with negligible bulk resistance, and $R_L = 50\ \Omega$. Determine $V_{o(av)}$ and $I_{L(av)}$.

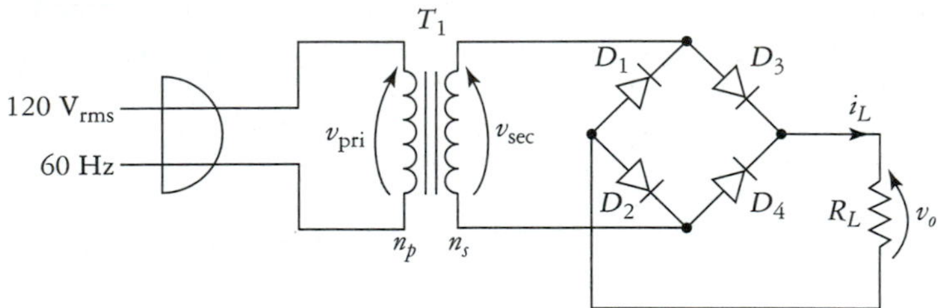

FIGURE 2.8
Circuit for Example 2.3.

Solution Let us start by determining the peak primary voltage:

$$\begin{aligned} V_{pri(pk)} &= 1.414V_{rms} \\ &= (1.414)(120\text{ V}) \\ &= 170\text{ V} \end{aligned}$$

Now the secondary peak voltage is

$$\begin{aligned} V_{sec(pk)} &= (n_s V_{pri(pk)})/n_p \\ &= 850/18 \\ &= 47.2\text{ V} \end{aligned}$$

The peak output voltage is found using Eq. (2.8b):

$$\begin{aligned} V_{o(pk)} &= V_{sec(pk)} - 2V_\phi \\ &= 47.2\text{ V} - 1.4\text{ V} \\ &= 45.8\text{ V} \end{aligned}$$

The average output voltage is now found to be

$$\begin{aligned} V_{o(av)} &= 0.637V_{o(pk)} \\ &= 29.2\text{ V} \end{aligned}$$

And finally, the average load current is

$$\begin{aligned} I_{L(av)} &= V_{o(av)}/R_L \\ &= 29.2\text{ V}/50\ \Omega \\ &= 584\text{ mA} \end{aligned}$$

Bridge Rectifier Diode Voltage and Current Levels

Recall that at any given time, two diodes in the bridge will be reverse biased, while the remaining two diodes will be forward biased. Considering the reverse-biased diodes first, if we examine Fig. 2.6, we find that both reverse-biased diodes will be subjected to a peak inverse voltage equal to the peak amplitude of the output voltage. If we neglect losses associated with the forward-biased diodes in the bridge, then the reverse-biased diodes are subject to the peak secondary voltage $V_{sec(pk)}$, and

$$\text{PIV} = V_{sec(pk)} \tag{2.10}$$

The forward-biased diodes will have a peak forward current given by

$$I_{F(max)} = V_{o(max)}/R_L \tag{2.11}$$

The load on the bridge rectifier is carrying current for nearly 360° of each cycle of v_{in}, but a given diode will only conduct current for approximately half of any cycle of v_{in}, thus

$$I_{F(av)} = I_{L(av)}/2 \tag{2.12}$$

So, the average current carried by any given diode in the full-wave bridge is essentially the same as for a diode in an equivalent half-wave rectifier circuit.

Center-Tap Full-Wave Rectifier

An alternative to the bridge rectifier is the *center-tap rectifier,* shown in Fig. 2.9(a). If we assume that v_{sec} is crossing zero in the positive-going direction, then the instantaneous polarity of the secondary winding is as shown in Fig. 2.9(b). This forward biases D_1 and reverse biases D_2, and the load current flows through the upper half of the secondary, as shown. On negative half-cycles of v_{sec}, the conditions illustrated in Fig. 2.9(c) prevail, and the lower half of the secondary winding supplies the load current. Because only half of the available secondary winding is used on any given half-cycle, the maximum output voltage is approximately half of that which could be obtained using a bridge rectifier and the same transformer.

FIGURE 2.9
The center-tap full-wave rectifier. (a) Basic circuit. (b) Operation for positive half-cycle. (c) Operation for negative half-cycle.

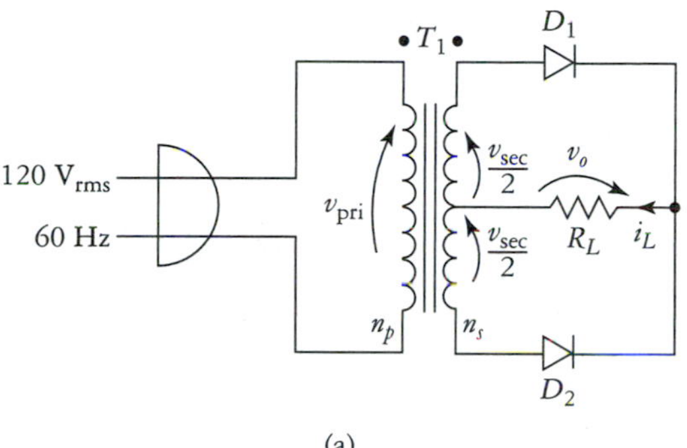

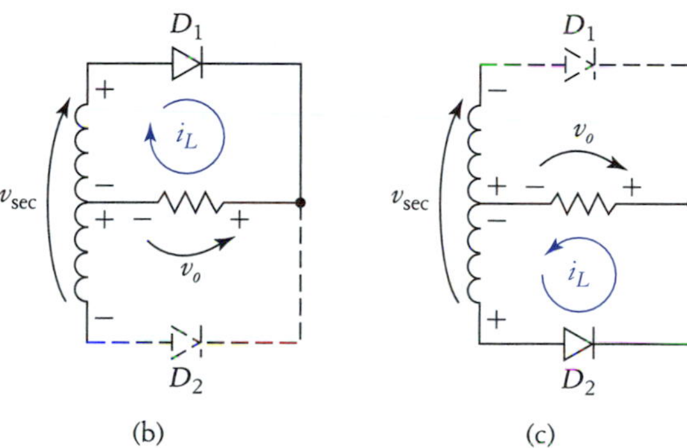

Center-Tap Rectifier Load Voltage and Current

The waveform produced at the output of the center-tap rectifier is just like that produced by the bridge rectifier; therefore, the load voltage and current equations that apply to the two circuits are similar as well. The peak secondary voltage used in the center-tap rectifier output voltage calculation is, of course, one-half of the normal peak secondary voltage. Also, we need to account for only one barrier potential drop and one diode bulk resistance loss on any given half-cycle of v_{in}. For the ideal diode, the maximum output voltage is

Note
Most often we will use Eq. 2.13b to determine the maximum voltage at the output of the rectifier.

$$V_{o(\max)} = \frac{V_{\sec(\text{pk})}}{2} \tag{2.13a}$$

Accounting for barrier potential this is modified to form

$$V_{o(\max)} = \frac{V_{\sec(\text{pk})}}{2} - V_\phi \tag{2.13b}$$

and if bulk resistance is significant as well as barrier potential, then we obtain

$$V_{o(\max)} = \left(\frac{V_{\sec(\text{pk})}}{2} - V_\phi\right)\frac{R_L}{R_L + R_F} \tag{2.13c}$$

The average output voltage is still related to the peak voltage by Eq. (2.9).

Center-Tap Rectifier Diode Current and Voltage

The diodes in the center-tap rectifier are subjected to a peak inverse voltage that is approximately equal to the maximum voltage across the entire secondary of the transformer. This is shown in Fig. 2.10, where we assume that the secondary is at peak voltage with the polarity as shown. Diode D_1 is forward biased while D_2 is reverse biased. Neglecting losses associated with D_1, and applying Kirchoff's voltage law (KVL) around the bottom loop shows that indeed

$$\text{PIV} = V_{\text{sec(pk)}} \tag{2.14}$$

The maximum forward current through a given diode is rather easily found to be

$$I_{F(\text{max})} = \frac{V_{\text{sec(pk)}}}{2R_L} \tag{2.15}$$

The average diode current is again given by

$$I_{F(\text{av})} = \frac{I_{L(\text{av})}}{2} \tag{2.16}$$

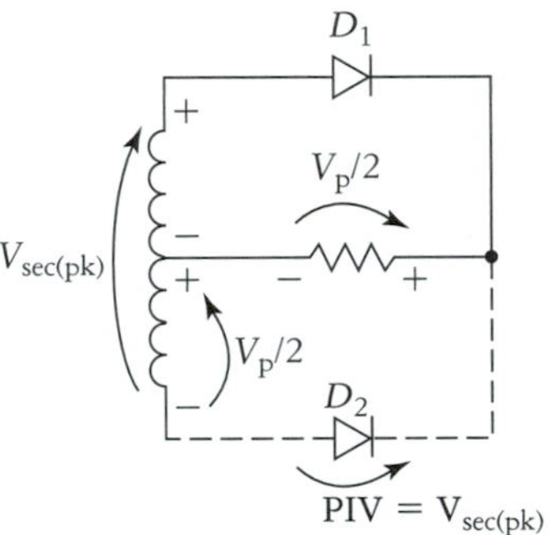

FIGURE 2.10
Determination of diode peak inverse voltage.

2.3 AN ALTERNATE PERSPECTIVE ON RECTIFICATION*

Thus far, we have examined the process of rectification from the time-domain perspective. That is, we determine the effect of a given rectifier circuit on a signal as a function of time, as was done in Fig. 2.7. The time-domain representation is generally used when we are first taught to think of signals and circuit action, primarily because the oscilloscope displays signals in this manner. Some might also argue that the time-domain representation of signals is more intuitively meaningful to most people as well. In any case, an alternative to the time-domain approach is to look at signals in the frequency domain. The time- and frequency-domain representations for a sine function are shown in Fig. 2.11. To completely describe the signal in the frequency domain, a second plot of the phase of the signal versus frequency is also required, but that is not really an important issue in this discussion.

Note
Circuit or device nonlinearities can change the frequency content of a signal. This cannot happen in a linear circuit.

In terms of the *frequency domain,* a sinusoid is just about as simple as a function can get. It consists of just one frequency component with some particular peak amplitude (and a phase angle). The only effect that a linear operation, or in our case a linear circuit, can have on a sinusoidal signal is to change the amplitude and phase of the signal. The frequency content of the signal cannot be changed. Any nonlinear circuit that processes a sinusoidal input signal will produce a response (an output signal) with some deviation from the original sinusoidal shape. This distortion is inherent in any nonlinear process. In the case of a half- or full-wave rectifier, the nonlinearity and hence signal-shape distortion are extreme. The point is, however, that any waveform that is not exactly sinusoidal in shape will contain frequency components that do not exist in a sinusoidal signal. Thus, one way of viewing rectification is as a nonlinear process in which an input signal is distorted such that more useful frequency components than originally existed in the input signal are produced in the output.

In performing rectification, we wish to translate as much input signal energy as possible down to zero frequency, or dc, at the output of the rectifier. Figure 2.12 illustrates the frequency-domain ef-

*This section can be skipped without loss of continuity.

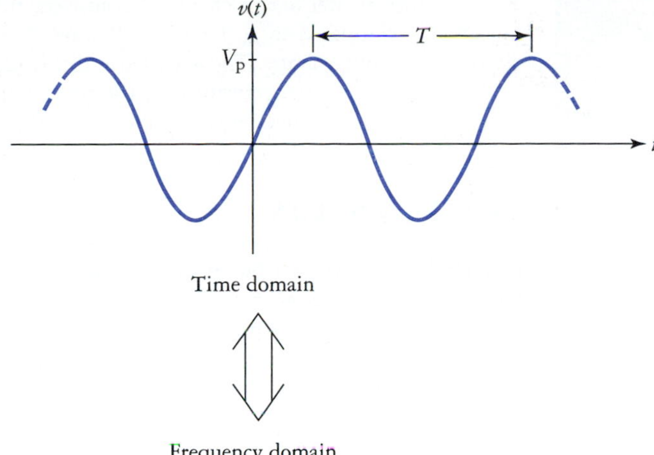

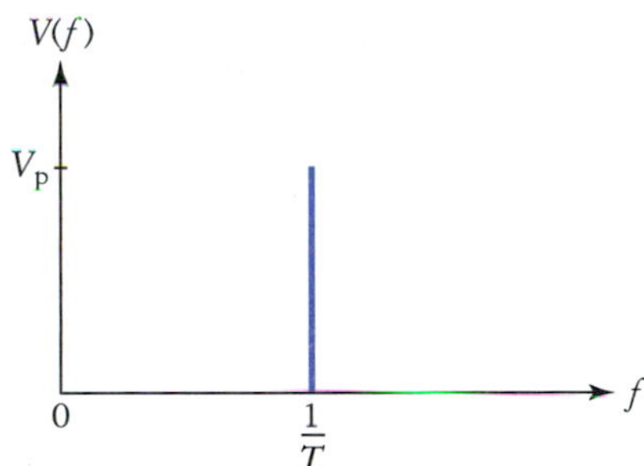

FIGURE 2.11
Time- and frequency-domain representations of a sine voltage.

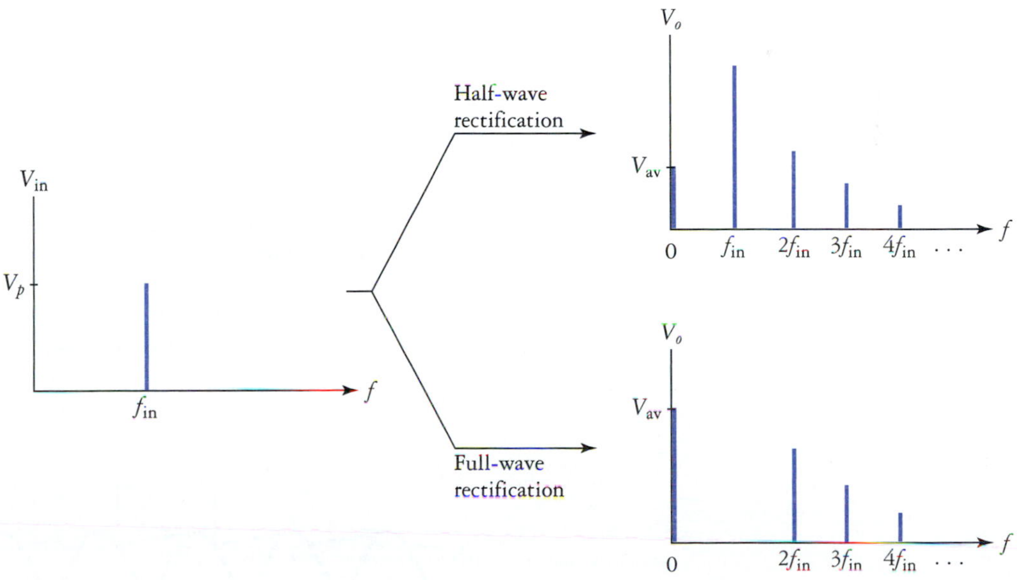

FIGURE 2.12
Rectification of a sine wave generates harmonics (integer multiples) of the fundamental input frequency.

fects of half- and full-wave rectification on a sinusoid. The actual amplitudes of the harmonics are not important, but notice that the average dc value of the full-wave rectified signal is twice as great as that for the half-wave rectifier. This occurs because energy that was present at $f = f_{in}$ in the output of the half-wave rectifier is shifted down to dc at the output of the full-wave rectifier. All of the higher harmonics ($2f_{in}$, $3f_{in}$, etc.) represent wasted energy, but, fortunately, the energy contained in these harmonics drops off rapidly with frequency.

As shown later in this chapter, because the full-wave rectifier produces output harmonics starting at $2f_{in}$, compared to f_{in} for the half-wave rectifier, it is easier to filter them out, producing a cleaner dc output voltage. The actual amplitude of each output frequency component is determined by performing a Fourier transform on the time-domain output of the rectifier. The Fourier transform is an advanced mathematical technique that maps functions of time into the frequency domain.

2.4 FILTERING AND POWER SUPPLIES

Electrical energy is distributed most often in alternating current form. Most electronic equipment, however, requires a dc energy source for proper operation, thus unless such equipment is battery powered, a power supply is required. The term *power supply* is somewhat of a misnomer. It is actually more accurate to describe a power supply as an energy supply (recall that power is the rate of energy transfer). Semantics aside, the basic function of a power supply is to convert an ac voltage into a dc voltage. Additionally, the dc produced usually must be relatively clean, or of constant amplitude. Rectifiers are used to convert ac into pulsating dc, while filters smooth out the pulses, providing a relatively constant dc voltage level.

Capacitor Filter

Filtering for dc power supplies is most often provided by simply using a capacitor, as shown in Fig. 2.13. Typical values for power supply filter capacitors range from a few hundred microfarads to over 50,000 μF. Because such large capacitances are used, power supply filter capacitors are usually polarized electrolytic units.

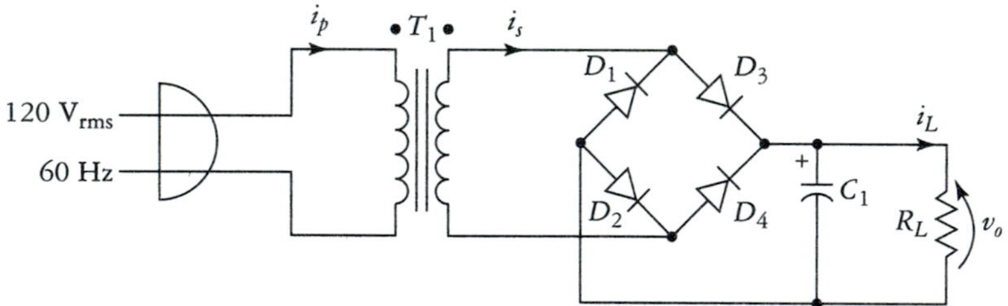

FIGURE 2.13
Simple capacitively filtered dc power supply.

The operation of the capacitive filter is fairly straightforward. Using the circuit in Fig. 2.13 as an example, the output voltage would appear as shown in Fig. 2.14. Here's how the waveform is produced. Assuming that the capacitor is initially uncharged, as the output of the rectifier goes from 0 V to V_p, the capacitor charges and the output voltage increases to V_p as well. The assumption here is that the equivalent series resistance of T_1 and the bridge is approximately 0 Ω, thus the capacitor charges rapidly.

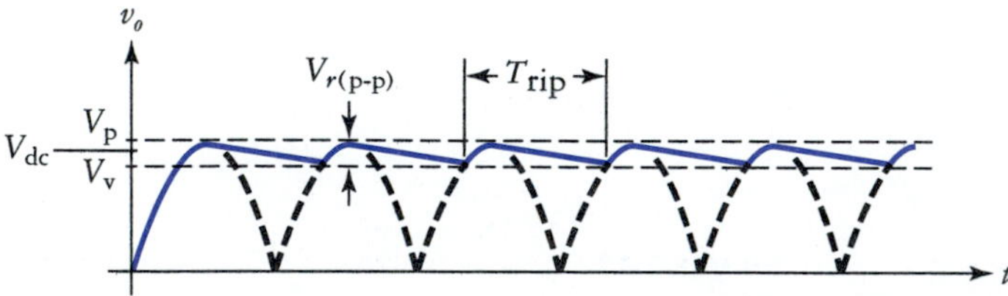

FIGURE 2.14
Output ripple voltage.

Once the output of the rectifier passes the peak and begins dropping, the charge on the filter capacitor reverse biases all of the diodes in the rectifier. Thus the capacitor begins to discharge through R_L. This causes the capacitor voltage to decrease as shown in Fig. 2.14. The larger the value of C_1, the smaller the voltage decrease will be for a given load resistance. Meanwhile, the output of the rectifier has dropped to zero and is increasing toward V_p again. Once this voltage exceeds the capacitor voltage, the capacitor charges up to V_p again and the cycle repeats. This periodic variation in the output voltage is called *ripple voltage.*

If we look at the operation of the filter capacitor from the time-domain perspective, we see that it supplies energy to the load during those times when the output of the rectifier is relatively low. Based on this view of operation, it is desirable for the R_LC_1 time constant to be very long compared to the period of the ripple voltage. For a given load resistance and ac line frequency, it takes half as much capacitance to obtain a given RC time constant for a full-wave rectifier, as it does for a half-wave rectifier.

From the frequency-domain perspective, the function of the capacitor is to provide a low-impedance path to ground for the higher harmonics that exist in the rectifier output. Because X_C is inversely proportional to frequency, the higher the frequency of the ripple voltage components, the more effectively a given value of capacitance filters them out. Thus, the advantage of the full-wave rectifier is that at the lowest ripple frequency component, $f_{rip} = 2f_{in}$, as opposed to the half-wave rectifier where $f_{rip} = f_{in}$.

Ripple Factor

In reference to Fig. 2.14, the peak-to-peak ripple voltage is given by

$$V_{r(p\text{-}p)} = V_p - V_v \tag{2.17}$$

If we zoom in on the upper portion of Fig. 2.14, the ripple voltage appears as shown in Fig. 2.15. In a reasonably loaded power supply, the ripple voltage will be small relative to the average output voltage. If this is true, then the ripple voltage will be approximately triangular in shape, as shown by the dashed-line segments in Fig. 2.15. The average dc value of v_o is then given by

$$V_{dc} = \frac{V_p + V_v}{2} \tag{2.18}$$

This is the dc load voltage that would be measured with a voltmeter.

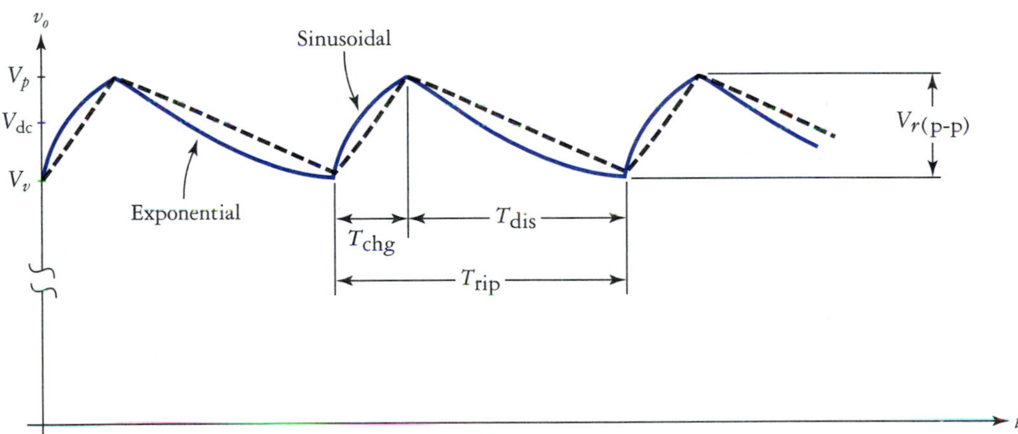

FIGURE 2.15
Enlarged output ripple voltage.

The performance of a power supply may be quantified in several different ways. The ripple factor is one such parameter. The ripple factor of a power supply is the ratio of rms ripple voltage to dc output voltage. That is,

$$r = \frac{V_{r(rms)}}{V_{dc}} \tag{2.19a}$$

Multiplying Eq. (2.19a) by 100 gives the percent of ripple in the output:

$$\%r = \frac{V_{r(rms)}}{V_{dc}} 100 \tag{2.19b}$$

The rms and peak-to-peak voltages of a triangular waveform are related by

$$V_{r(\text{rms})} = \frac{V_{r(\text{p-p})}}{2\sqrt{3}} \tag{2.20}$$

It is very easy to measure peak-to-peak ripple voltage on an oscilloscope. This measurement, plus a dc voltage measurement, allows the ripple factor of a power supply to be determined experimentally under load conditions. For design and analysis purposes, the relationships between dc voltage, ripple voltage, filter capacitance, and load resistance can be expressed in equation form. Referring to Fig. 2.15, if the ripple voltage is *reasonably* small, as it would be in a power supply that is not overloaded, then $T_{\text{dis}} = T_{\text{rip}}$, $V_{\text{dc}} = V_p$, and the linearized approximation of v_r is acceptable. As a general rule, if $r < 0.05$ then the ripple is small enough to make these approximations valid. This being the case, the following expressions apply:

$$V_{r(\text{p-p})} \cong \frac{V_p}{f_{\text{rip}} R_L C_1} \tag{2.21}$$

and

$$r \cong \frac{1}{2\sqrt{3}\, R_L C_1 f_{\text{rip}}} \tag{2.22}$$

where $f_{\text{rip}} = 1/T_{\text{rip}}$.

When used for design purposes, these equations yield conservative results. That is, the ripple voltage and ripple factor that would be present in a real circuit will be somewhat better (lower) than that predicted by the equations.

EXAMPLE 2.4

Refer to Fig. 2.13. The transformer is operated from the 60-Hz ac line and has $V_{\text{sec(pk)}} = 15$ V. The diodes are silicon with $R_F = 0.2\ \Omega$. Determine the required value for C_1 such that $r = 1\%$ with a 25-Ω load and the corresponding rms ripple voltage.

Solution The ratio of load resistance to diode bulk resistance is 125 : 1, therefore the bulk resistance is negligible. Accounting only for barrier potential losses, the maximum output voltage is found to be

$$V_{\text{dc}} \cong V_{o(\text{max})} = 13.6 \text{ V}$$

We now solve Eq. (2.19b) for $V_{r(\text{rms})}$, giving us

$$\begin{aligned} V_{r(\text{rms})} &= 0.01 r\, V_{\text{dc}} \\ &= (0.01)(1\%)(13.6 \text{ V}) \\ &= 136 \text{ mV} \end{aligned}$$

Solving Eq. (2.22) for C_1 yields

$$\begin{aligned} C_1 &= \frac{1}{2\sqrt{3}\, R_L r f_{\text{rip}}} \\ &= \frac{1}{2\sqrt{3} \times 25 \times 0.01 \times 120} \\ &= 9622\ \mu\text{F} \end{aligned}$$

Filter Effects on Diode Current and Voltage

Aside from reducing the power supply output ripple voltage, the filter capacitor also causes rectifier diode current and voltage values to vary in comparison to the unfiltered case.

Peak Inverse Voltage

Consider the filtered half-wave rectifier shown in Fig. 2.16. Unfiltered, the PIV experienced by the diode is simply V_p of the ac source. The filter capacitor effectively doubles the PIV applied to the diode, thus

$$\text{PIV} = 2V_p \tag{2.23}$$

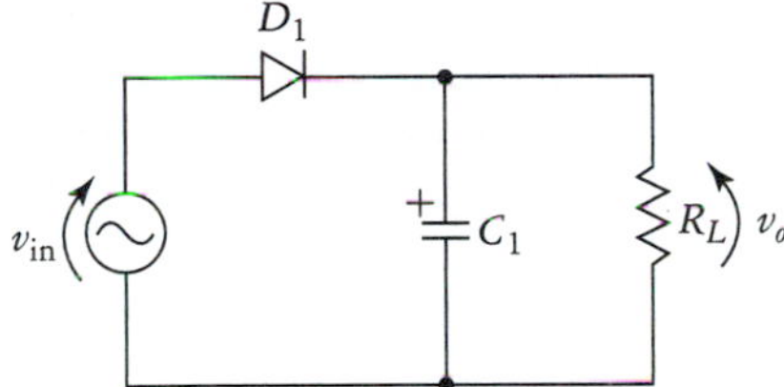

FIGURE 2.16
Basic capacitively filtered half-wave rectifier.

To understand why this doubling of PIV occurs, assume that C_1 has charged to V_p on the positive swing of v_{in}, and now the (ideal) diode is reverse biased. At $t = t_0$, v_{in} is at its negative peak value. At this instant of time, the input voltage source and the charged filter capacitor are equivalent to two series-aiding voltage sources, each of V_P volts reverse biasing the diode. This is shown in Fig. 2.17.

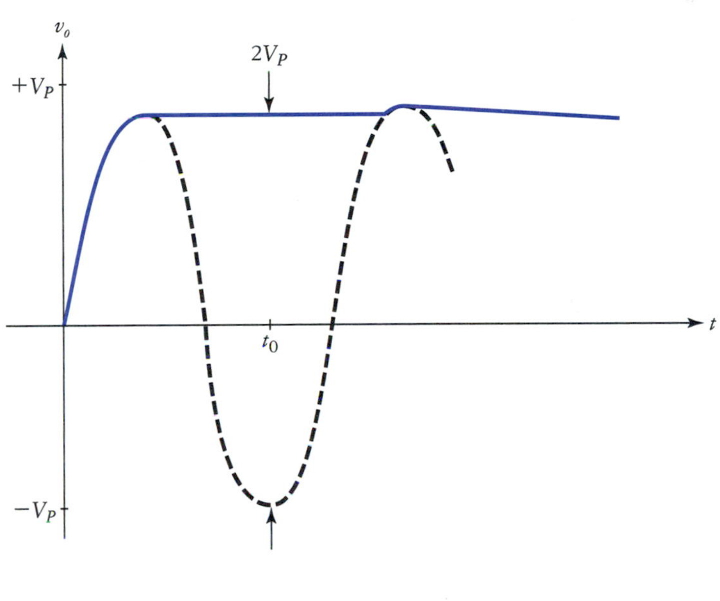

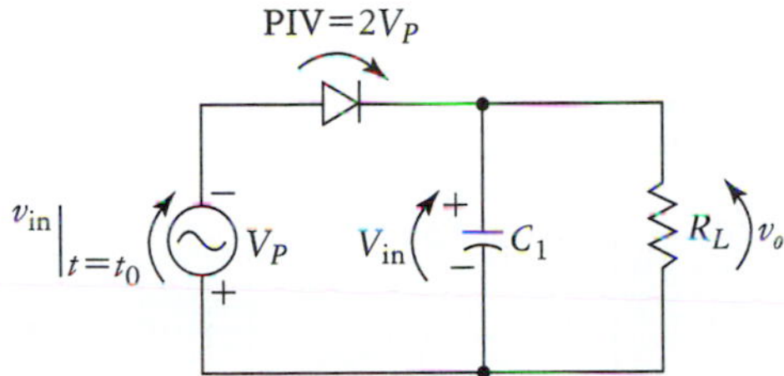

FIGURE 2.17
Output voltage waveform and the circuit showing how capacitor and source are series-aiding on negative half-cycles of v_{in}.

Equation (2.23) also applies to capacitively filtered power supplies using the full-wave bridge rectifier and the center-tap full-wave rectifier. In the case of the center-tap full-wave rectifier, the PIV applied to either diode is twice the peak voltage for half of the secondary.

Filter Capacitor Charging Current

The addition of a filter capacitor can cause surprisingly high repetitive currents to flow through the rectifier diodes. Consider the output voltage waveform shown in Fig. 2.18. For low values of ripple voltage, a desirable situation, the average dc value of the voltage is approximately equal to the maximum output voltage. Now, a given load resistance will draw an average current given simply by

$$I_{dc} = \frac{V_{dc}}{R_L} \tag{2.24}$$

Over each rectifier output cycle, the load will absorb energy from the filter capacitor for T_{dis} seconds. This corresponds to a charge loss of

$$Q_{out} = I_{dc}\, T_{dis} \tag{2.25}$$

The charge lost to the load must be made up during the time T_{chg}, and the areas under the charge and discharge current curves in Fig. 2.18 are equal. Thus we have

$$Q_{in} = I_{chg}\, T_{chg} \tag{2.26}$$

and

$$Q_{out} = Q_{in} \tag{2.27}$$

Solving for I_{chg} yields

$$I_{chg} = I_{dc} \frac{T_{dis}}{T_{chg}} \tag{2.28}$$

Let's look at the significance of Eq. (2.28). Low ripple voltage is obviously very desirable. However, consider that as ripple voltage decreases (due to increasing filter capacitance), T_{dis} approaches T_{rip} and

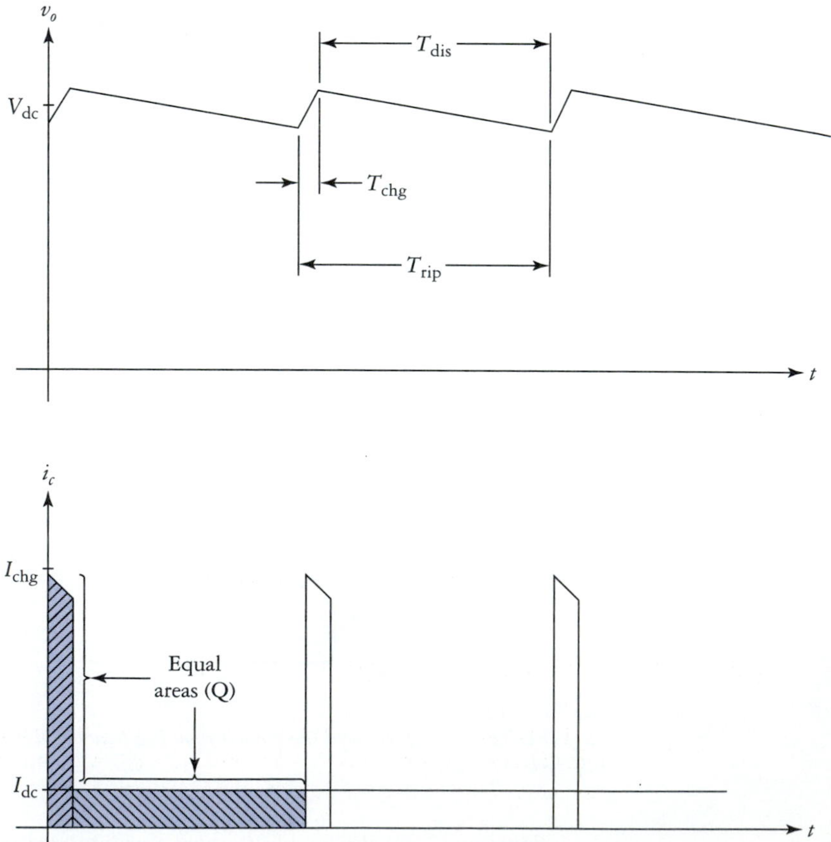

FIGURE 2.18
Charging current waveform for the filter capacitor.

T_{chg} approaches zero. Examination of Eq. (2.28) reveals that when ripple is very low, the peak charging current and hence the peak rectifier current can increase significantly.

Equation (2.28) is difficult to apply to a theoretical pencil-and-paper analysis because of the need to determine either T_{dis} or T_{chg}. However, Eq. (2.28) can be used to calculate peak charging current based on easily obtained oscilloscope readings, so it is more useful as an empirical analysis equation.

Surge Current

Up to this point, we have tacitly assumed that the power supply is turned on at the instant the line voltage is crossing 0 V. If this is the case, then the filter capacitor can charge somewhat gradually during the first quarter-cycle of the sinusoid. However, in real life, one might turn on the power supply at the instant the line voltage is at its maximum value. In such a case, the uncharged filter capacitor acts like a short circuit, as the transformer attempts to charge the capacitor instantaneously. Chances are that the transformer would survive such a momentary current surge, but the rectifier diodes in conduction at the time might not. The peak current drawn under these conditions is limited mainly by the resistance of the transformer and the bulk resistance of the diodes.

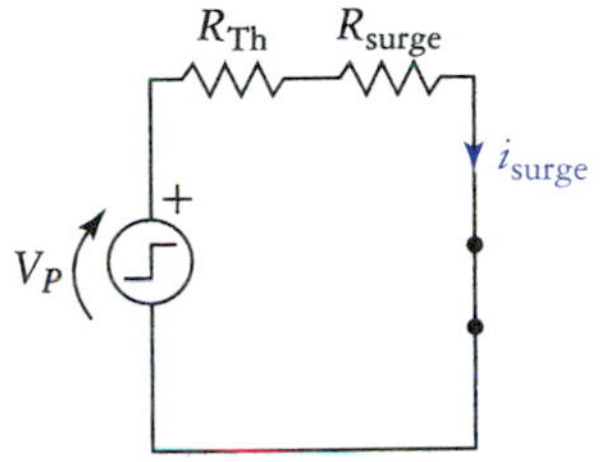

To limit the peak surge current, sometimes a resistance is placed in series with the secondary. Accounting for the Thevenin resistance of the secondary circuit, assuming that a surge-limiting resistor is used, and assuming that power is applied at the peak of the input voltage, the secondary may be modeled as shown in Fig. 2.19. Routine inspection of the circuit reveals that

FIGURE 2.19
Surge current. On application of power, the uncharged capacitor acts like a short circuit.

$$I_{surge(pk)} = \frac{V_p}{R_{Th} + R_{surge}} \tag{2.29}$$

The surge-limiting resistor is chosen such that the peak rectifier current is within the maximum current limits set by the specifications for the diode(s) used in the circuit.

A surge resistor may be placed in the primary side of the transformer rather than in series with the secondary. This is shown in Fig. 2.20(a). A primary-side surge resistor is equivalent to a secondary series resistance given by

$$R_s = \frac{R_P}{(n_p/n_s)^2} \tag{2.30}$$

Proof of this equivalence is left as an exercise at the end of the chapter.

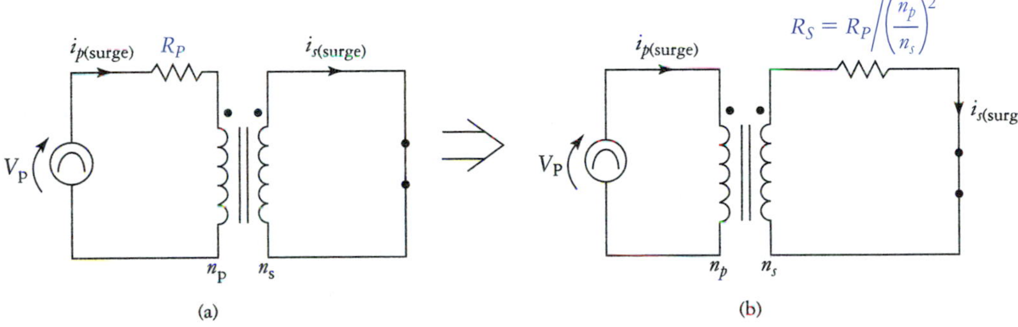

FIGURE 2.20
Surge-limiting resistor located in primary circuit.

2.5 ALTERNATIVE FILTER TECHNIQUES

Peak rectifier current considerations prevent ripple reduction from being achieved by simply using huge filter capacitor values. Several other ripple reduction techniques, however, can be used. This section covers two of them: *RC* filters and *LC* filters.

RC Filters

The addition of a low-pass *RC* filter section to a power supply using basic capacitive filtering results in the circuit of Fig. 2.21. The dc voltage component of v_1 is assumed to be approximately equal to

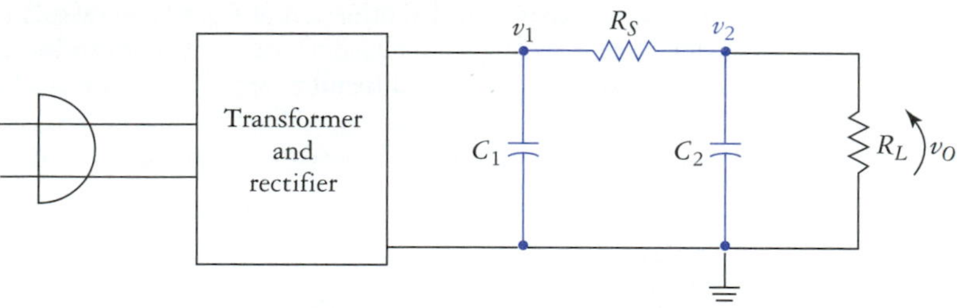

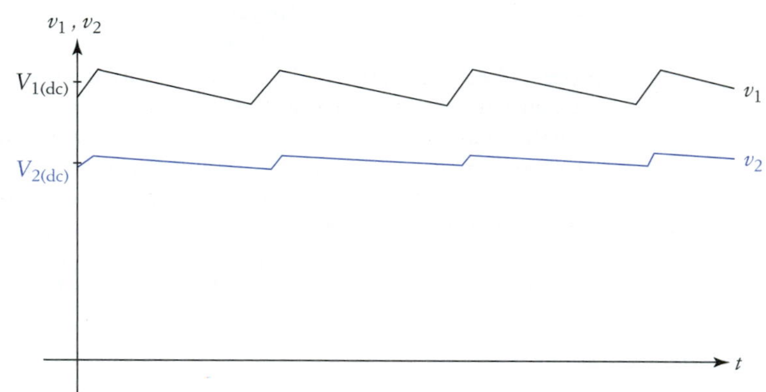

FIGURE 2.21
RC *pi filter.*

Note
The filter in Fig. 2.21 is usually called a *pi-filter* because of its resemblance to the Greek letter π.

the peak rectifier output voltage $[V_{1(dc)} = V_{1(pk)}]$ and the ripple voltage is determined as was done previously using Eq. (2.21), except that the effective load on C_1 is given by

$$R_{eq} = R_S + R_L \tag{2.31}$$

Thus

$$V_{r1(p-p)} \cong \frac{V_{1(dc)}}{f_{rip}\, C_1\, (R_S + R_L)} \tag{2.32}$$

At the output, node v_2, the dc voltage is determined by application of the voltage divider equation

$$V_{2(dc)} = V_{1(dc)} \frac{R_L}{R_L + R_S} \tag{2.33}$$

and finally, the output ripple voltage is approximately

$$V_{r2(p-p)} = V_{r1(p-p)} \frac{1}{\left[1 + (2\pi f_{rip}\, R_L\, C_2)\right]^{1/2}} \tag{2.34}$$

The trade-offs for reduced ripple obtained with the *RC* filter are decreased dc output voltage, increased susceptibility to loading and possibly significant power loss by series resistor R_S. Because of these problems and the superior performance of other filter techniques, at least in power supply applications, the *RC* filter is rarely used.

LC Filters

Losses associated with the series resistor of the *RC* pi filter can be reduced significantly by using an *LC*-type pi filter as shown in Fig. 2.22. The basic principle of operation of the *LC* filter is the same as that of the *RC* filter, except that higher performance is achieved with the same number of circuit elements. To be pragmatic, however, the reader should understand that even the *LC* filter is rarely used in linear power supply design today. One reason for this is that the inductor required for such a filter will generally be rather large, heavy, and expensive. Possibly of even more importance in the decline

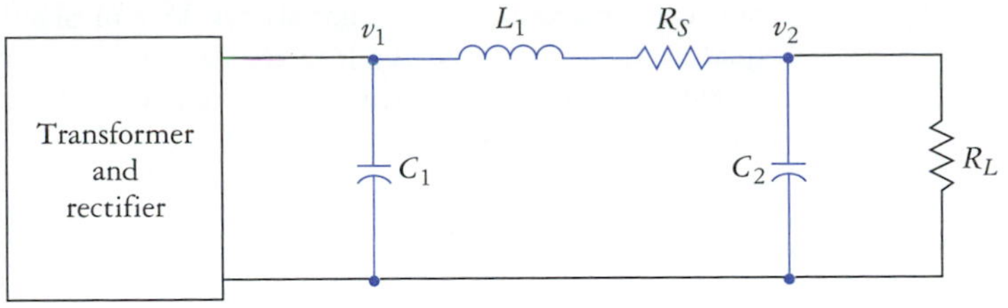

FIGURE 2.22
LC pi filter.

in popularity of the *LC*-filtered power supply is the availability of low-cost, high-performance electronic voltage regulators and the increasing popularity of switching regulators (covered later). However, because *LC* filters are occasionally used in power supplies and also in many other applications, it does not do any harm to take a look at their operation.

Analysis of the *LC* pi filter is performed in two parts. The dc output voltage is found via application of the voltage divider equation, using the series resistance of the inductor and the load resistor:

$$V_{2(\text{dc})} = V_{1(\text{dc})} \frac{R_L}{R_L + R_S} \tag{2.35}$$

In the ac ripple voltage analysis, a few simplifying assumptions are made: We assume that $|X_L| \gg |X_C|$, $|X_L| \gg |R_S|$, and $|R_L| \gg |X_C|$. In a well-designed power supply, these approximations are valid. A conservative estimate (an overestimation) of the v_1 ripple voltage $V_{r1(\text{p-p})}$ is found using Eqs. (2.21) and (2.31). The ripple voltage present at the load is then found using

$$V_{r2(\text{p-p})} = V_{r1(\text{p-p})} \frac{1}{(2\pi f_{\text{rip}})^2 L_1 C_2} \tag{2.36}$$

2.6 BIPOLAR POWER SUPPLIES

A *bipolar power supply* is a power supply that provides one or more output voltages that are negative with respect to ground, in addition to the positive output voltage(s). Most modern electronic equipment requires a source of bipolar dc for proper operation.

The circuit in Fig. 2.23 is typical of most simple bipolar power supply designs. The center tap of the secondary winding serves as the reference point. In Fig. 2.23, the reference is physically

Note
Bipolar power supplies are used extensively in systems that contain linear integrated circuits.

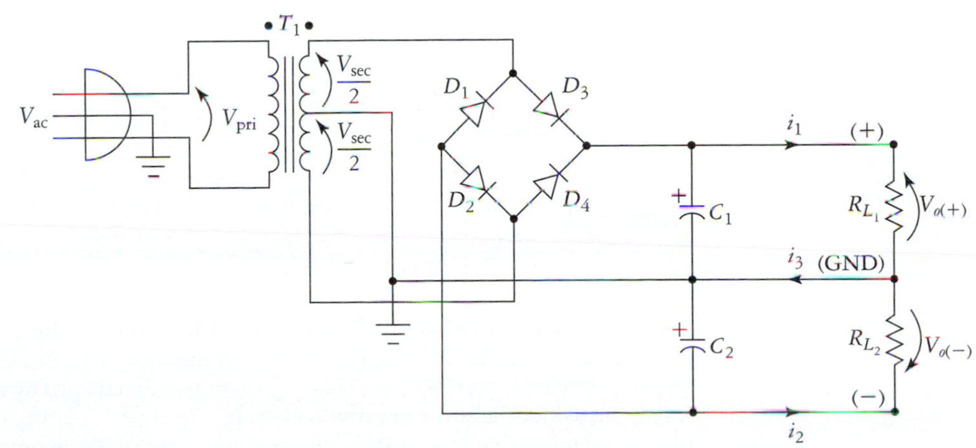

FIGURE 2.23
Basic bipolar power supply.

connected to ground. This is not always done in practice. Most often, filter capacitors C_1 and C_2 are equal in value, although the load resistances are not necessarily equal. Assuming negligible ripple voltage and accounting for diode barrier potential losses, the output voltages are given by the following familiar equations:

$$V_{o(+)} = \frac{V_{\text{sec}}}{2} - 2V_{\Phi} \tag{2.37a}$$

$$V_{o(-)} = -\left(\frac{V_{\text{sec}}}{2} - 2V_{\Phi}\right) \tag{2.37b}$$

RC and *LC* filters can be used in bipolar power supply designs, but as was the case with previously studied power supplies, electronic voltage regulators are preferred.

2.7 VOLTAGE MULTIPLIERS

High-voltage power supplies can be designed using basically the same circuits presented thus far in this chapter. That is, a transformer is used to provide a decrease or increase in the ac line voltage, which is then rectified and filtered. The use of a transformer in the design of a power supply provides several advantages: electrical isolation from the ac line, high efficiency, and virtually unlimited turns ratio options. The disadvantages of transformers are their relatively high cost and weight.

Half-Wave Voltage Doubler

It is possible to step up, rectify, and filter an ac voltage without using a transformer. A circuit that does this is called a *voltage multiplier.* The output of a voltage multiplier is an approximate integer multiple of the peak input voltage. A half-wave voltage doubler is shown in Fig. 2.24(a).

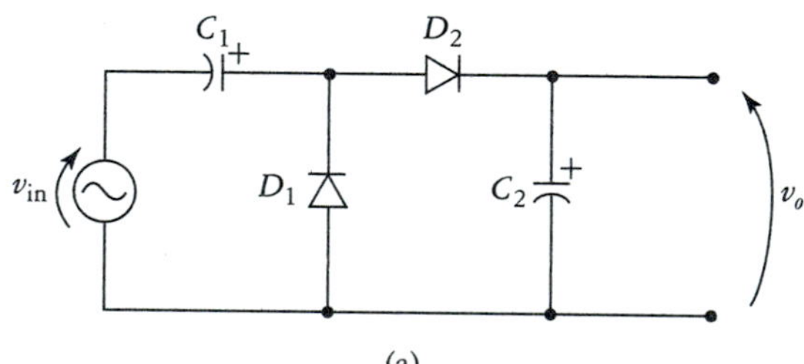

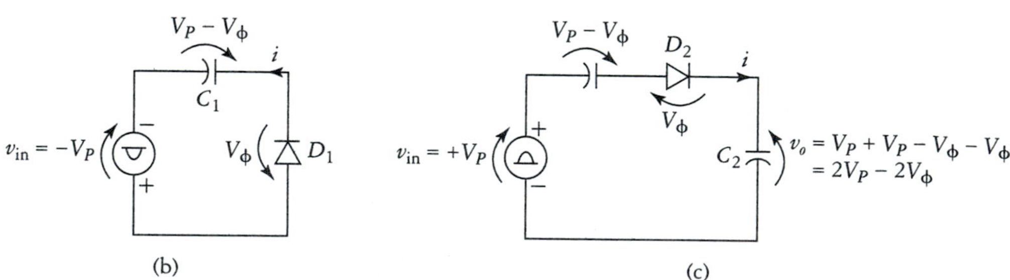

FIGURE 2.24
Half-wave voltage doubler. (a) Circuit. (b) Negative-going input voltage. (c) Positive-going input voltage.

We begin our analysis of the half-wave voltage doubler by assuming that v_{in} is at its negative peak. As shown in Fig. 2.24(b), C_1 has charged through diode D_1. Once the input voltage begins increasing, diode D_1 reverse biases, causing C_1 to remain charged at nearly peak voltage. Diode D_2 is reverse biased for the entire negative half-cycle, effectively isolating C_2 from the input voltage source. Now, on the positive peak of the input voltage, capacitor C_1 is series-aiding with the ac source, forward biasing diode D_2 and causing C_2 to charge to nearly double the peak input voltage. For negligible ripple conditions, the dc output voltage is

$$V_o = 2V_p - 2V_{\Phi} \tag{2.38}$$

The key point in the operation of the half-wave voltage doubler is that the output capacitor (C_2 in this case) is recharged on every other half-cycle of the input waveform. Thus the ripple voltage has the same fundamental frequency as the ac input voltage. To achieve the same ripple voltage as a simple half-wave rectifier/filter driven by the same ac source, C_1 and C_2 must be twice as large as normal. The diodes in the half-wave voltage doubler will be subjected to a peak inverse voltage of

$$\text{PIV} = 2V_p \tag{2.39}$$

Full-Wave Voltage Doubler

The full-wave voltage doubler, shown in Fig. 2.25(a), offers better performance than the half-wave doubler, for a given load current and filter capacitor size. The term *full-wave* comes from the fact that the output receives energy directly on each half-cycle of v_{in}.

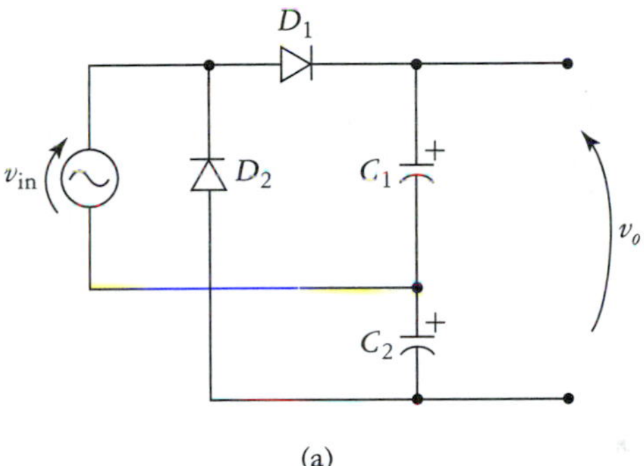

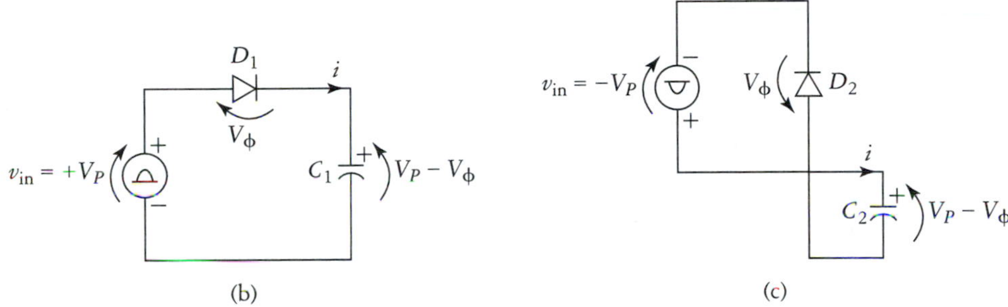

FIGURE 2.25
Full-wave voltage doubler.

For analysis purposes, assume that v_{in} is at the positive peak. The circuit can be modeled as shown in Fig. 2.25(b), where forward-biased diode D_1 provides a low-impedance charging path for C_1. On the negative half-cycle of v_{in}, D_2 is forward biased allowing capacitor C_2 to charge as shown in Fig. 2.25(c). It is apparent from Fig. 2.25(a) that C_1 and C_2 are connected in a series-aiding configuration giving us

$$V_o = 2V_p - 2V_\phi \tag{2.40}$$

The diodes are subject to

$$\text{PIV} = 2V_p \tag{2.41}$$

Higher Order Voltage Multipliers

Half-wave voltage doublers can be cascaded to obtain greater multiples of the peak ac input voltage rather easily. The general structure for such a voltage multiplier is shown in Fig. 2.26. Additional sections can be added to the right as necessary to achieve the desired multiple of the peak input voltage.

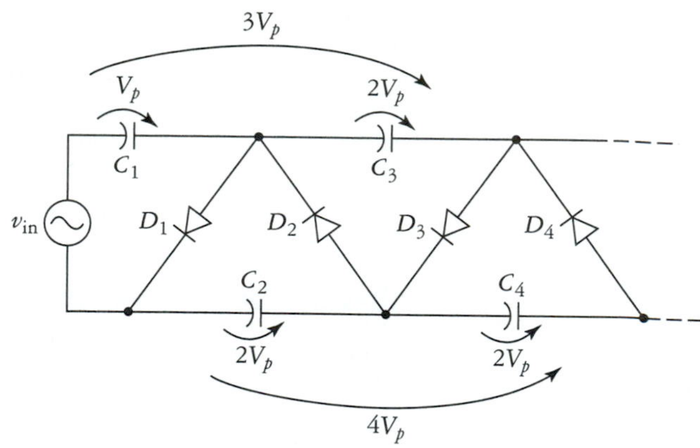

FIGURE 2.26
Generalized voltage multiplier.

It would seem that one could make voltage multipliers of nearly any order; however, a few pitfalls must be taken into consideration. First of all, as the input voltage is multiplied by successively greater degree, the capacitor and diode voltage ratings must be increased as well; a voltage quadrupler requires capacitors that operate at $4V_p$ and diodes that can withstand PIV $= 4V_p$ and so on. A second problem is that as the multiplication factor is increased given a constant load current, proportionately larger capacitance is needed. For example, if a half-wave voltage doubler requires 100 μF capacitors, a voltage quadrupler will generally require 400-μF capacitors for similar performance characteristics. For these reasons, voltage multipliers are usually used in very low-current applications.

2.8 THE CLAMPER

A clamper is a circuit that is designed to shift a time-varying input voltage waveform such that the mean value of that waveform is approximately V_p. That is, the clamper will shift a signal either above or below ground. A positive clamper is shown in Fig. 2.27.

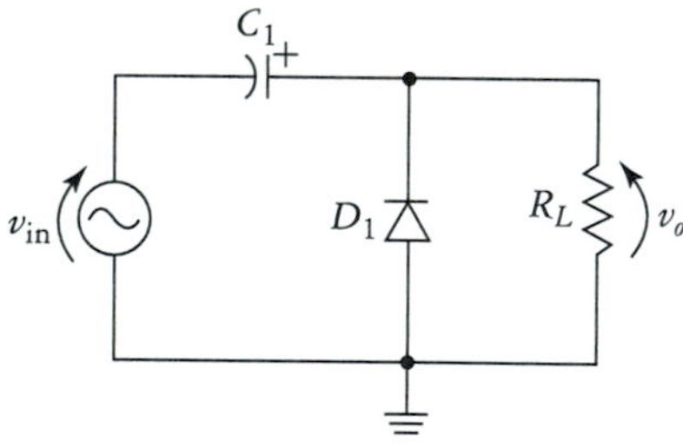

FIGURE 2.27
Positive clamper.

Note
We will encounter the clamper topology again when we cover class-c amplifiers in Chapter 10.

In order for the clamper to work effectively, the following inequality must be satisfied:

$$R_L C_1 \geq 10T \tag{2.42}$$

where T is the period of the ac input voltage. Assuming that this is the case, and v_{in} is now negative going, capacitor C_1 charges to V_p during the first quarter-cycle. As v_{in} crosses zero and begins going positive, capacitor C_1 acts as a dc voltage source of $V_p - V_\phi$ volts in series with the ac source. Thus we see that if the RC time constant is long compared to the period of the input signal, then the capacitor voltage will remain nearly constant between charging times at the negative peaks. The output voltage waveform will appear as shown in Fig. 2.28. The clamper effectively creates a dc offset in a time-varying signal that did not originally contain a dc component. Again, nonlinearity affects the frequency-domain representation of a signal. The greater the peak input voltage is relative to the forward drop across the diode, the more ideally the output voltage is clamped above (or below) ground.

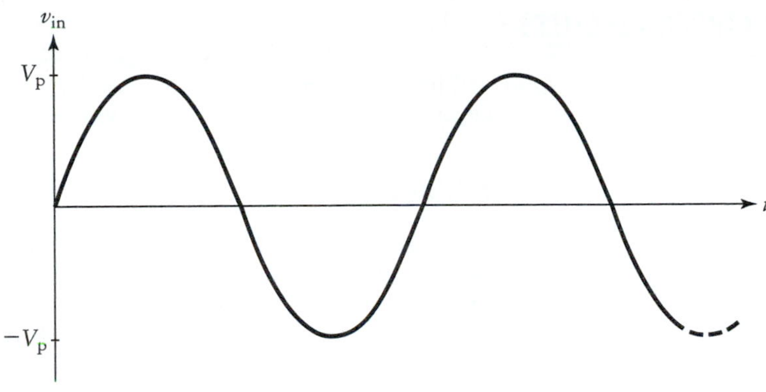

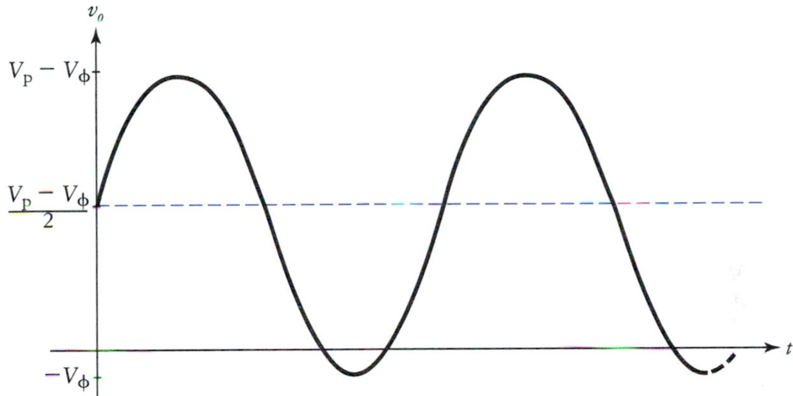

FIGURE 2.28

Positive clamper input and output voltages.

EXAMPLE 2.5

The signal applied to the circuit in Fig. 2.27 is $v_{in} = 5 \sin 1000t$ V. Determine the required value for C_1 such that the circuit will function with $R_L = 1\ \text{k}\Omega$, and write the equation for the resulting output voltage. Assume that a silicon diode is used.

Solution First we must determine the period of the input signal:

$$\omega = 1000 \text{ rad/s and}$$

$$f = \frac{\omega}{2\pi}$$

$$= 159.2 \text{ Hz and } T = 6.28 \text{ ms}$$

The RC time constant must conform to Eq. (2.42); thus,

$$C_1 \geq \frac{10T}{R_L}$$

$$\geq 62.8\ \mu\text{F}$$

The expression for the output voltage is

$$v_o = 4.3 + 5 \sin 1000t \text{ V}$$

2.9 CLIPPER CIRCUITS

Clippers fall into the general catagory of *wave-shaping circuits.* The function of a clipper is to limit the amplitude of a signal to some particular maximum positive or negative value.

Simple Diode Clippers

A simple clipper is shown in Fig. 2.29, along with typical input and output waveforms. On positive half-cycles of v_{in}, the diode acts as an open circuit because it is reverse biased and because $v_o = v_{in}$. On negative half-cycles of the input waveform, the diode becomes forward biased, and limits the output voltage to $v_o = V_\phi$. Thus the negative-going portion of the signal is clipped off. This nonlinear function can be expressed mathematically as

$$v_o = \begin{cases} v_{in} & \text{for } v_{in} > -V_\phi \\ -V_\phi & \text{for } v_{in} \leq V_\phi \end{cases} \tag{2.43}$$

FIGURE 2.29
Negative clipper with typical input and output waveforms.

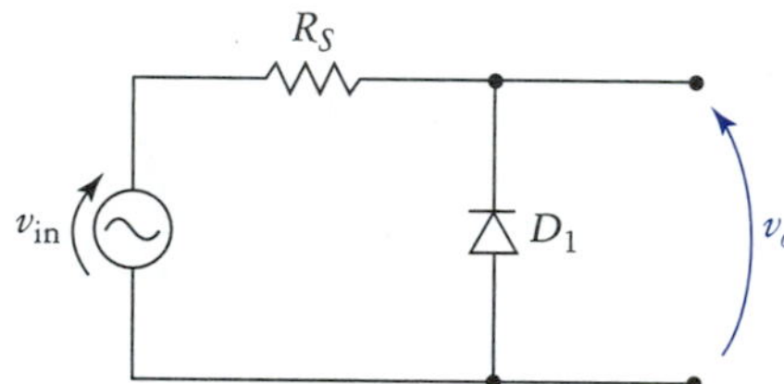

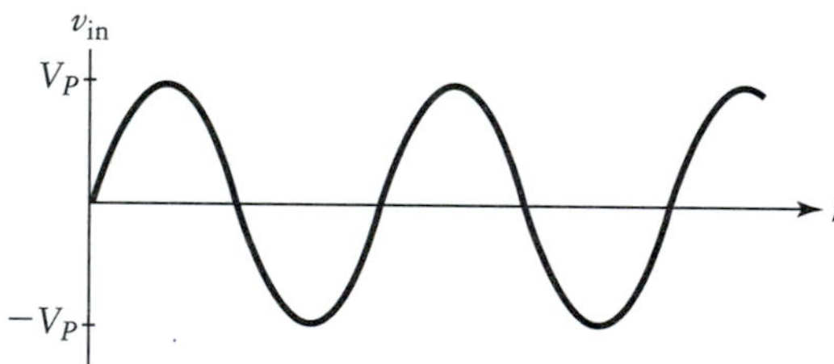

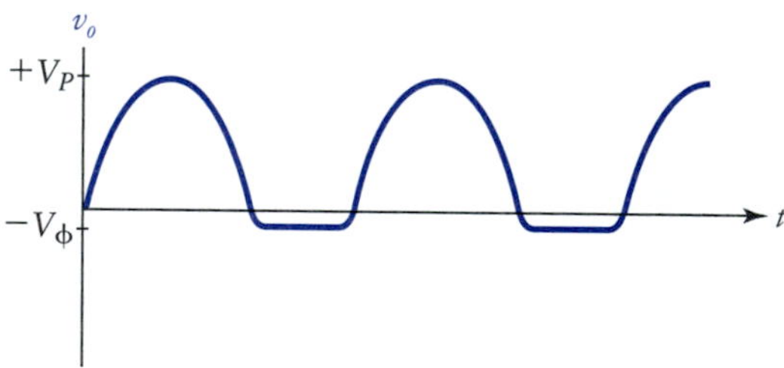

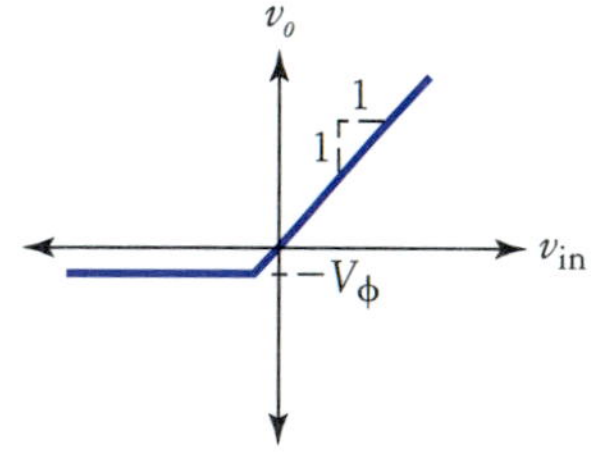

FIGURE 2.30
Transfer characteristics for the negative clipper.

Transfer Characteristic Curves

Sometimes the operation of a nonlinear circuit can be understood more easily if the transfer characteristics of the circuit are plotted. A transfer characteristic curve is a plot of the output of a circuit as a function of the input. For the clipper of Fig. 2.29, we get the transfer characteristic of Fig. 2.30. Notice that the slope of the characteristic is 1 in the first quadrant for this particular circuit. Transfer characteristic curves are useful in many other applications in addition to clipper circuit analysis.

Additional Clipper Variations

Clippers can be constructed using a variety of diode configurations. For example, the circuit in Fig. 2.31(a) uses back-to-back zener diodes to limit both positive and negative peak voltages. The zener voltages do not necessarily have to be the same, and the output voltage expression for this circuit can be expressed in a form similar to that of Eq. (2.42). This is left as an exercise for the reader. The transfer characteristics for this circuit are shown in Fig. 2.31(b).

EXAMPLE 2.6

Refer to Fig. 2.31. Given $v_{in} = 15 \sin \omega t$ V, $V_{Z1} = 4.2$ V, and $V_{Z2} = 6.8$ V, sketch the output voltage waveform.

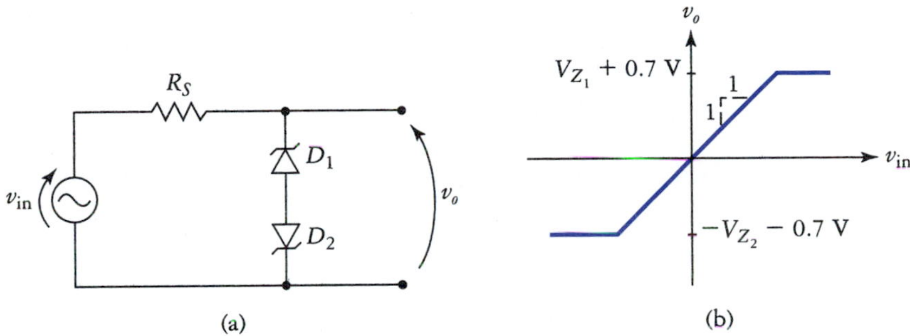

FIGURE 2.31
(a) A symmetrical clipper using Zener diodes. (b) Circuit transfer characteristic.

Solution The output voltage is limited to peak levels $V_{Z_1} + 0.7\text{ V} = 4.9$ V and $-(V_{Z_2} + 0.7\text{ V}) = -7.5$ V. Between these limits, the output voltage is essentially an exact replica of the input voltage, as shown in Fig. 2.32.

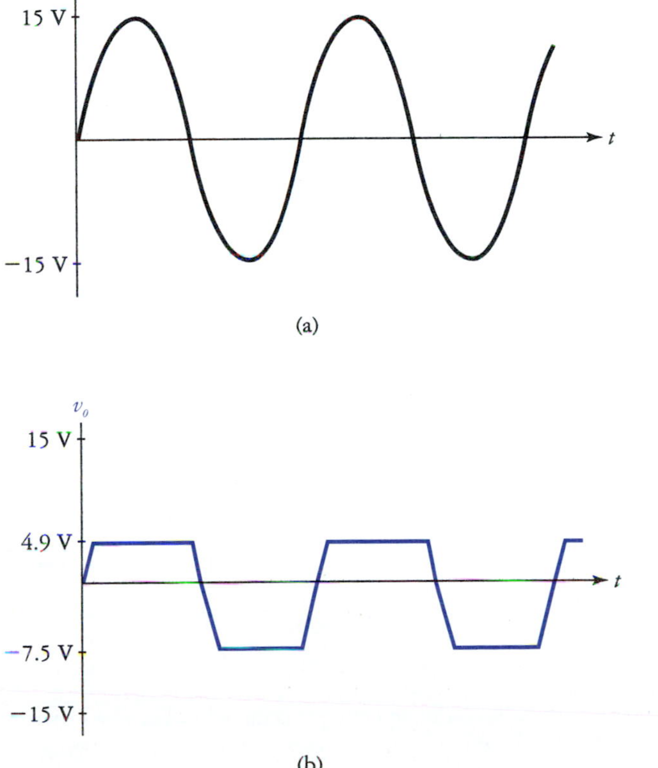

FIGURE 2.32
(a) Input and (b) output waveforms for Example 2.6.

In the preceding example, the load resistance was infinite, and so the output voltage exactly tracked the input voltage between clipping limits. A finite load resistance can change this relationship. Consider the clipper in Fig. 2.33. This circuit clips off most of the positive portion of the input

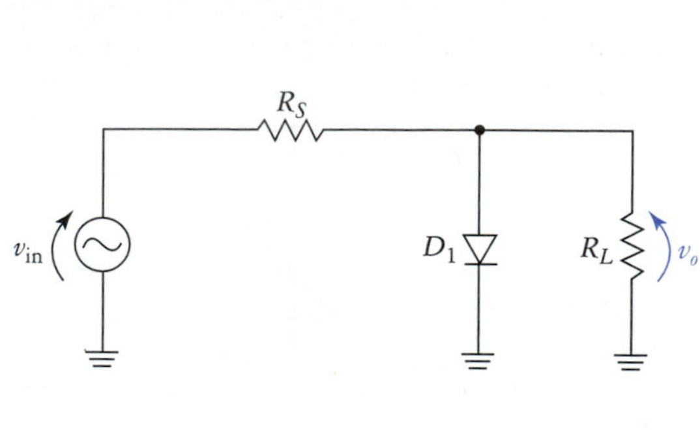

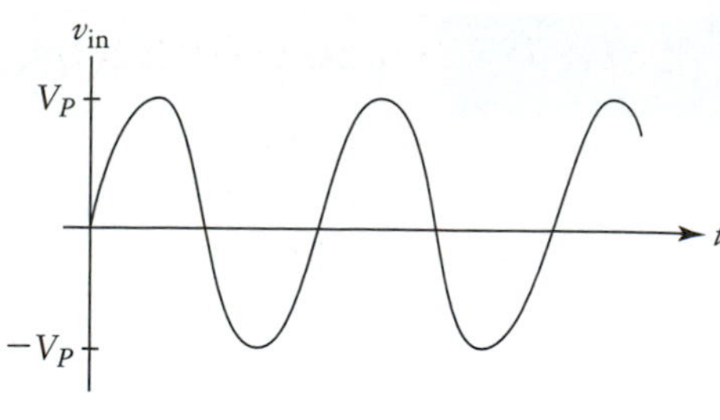

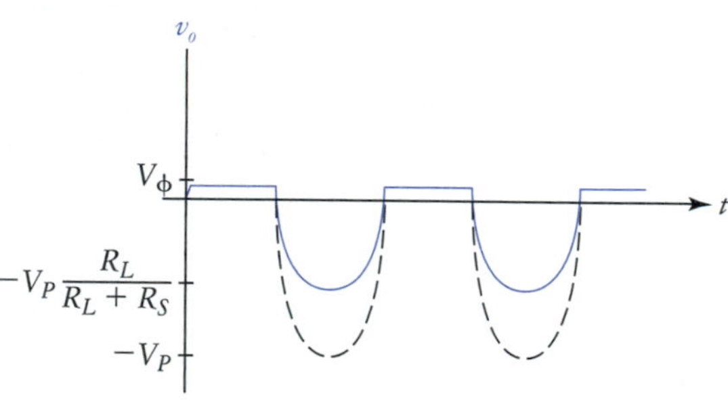

FIGURE 2.33
Positive clipper with a finite load resistance.

signal, while passing the negative portion. The negative peak amplitude of the output is found, however, by application of the voltage divider equation:

$$-V_{o(\text{pk})} = -V_P \frac{R_L}{R_L + R_S} \tag{2.44}$$

The determination of the transfer characteristics for this circuit are left as an exercise for the reader.

Biased Clipper Circuits

A wide variety of clipping limits can be achieved by cascading various rectifier and zener diodes. However, once such a clipper is built, the clipping points cannot be altered without physically adding and/or removing diodes from the circuit. The biased clipper solves this problem.

Note
The bias voltages in a biased clipper are used to effectively increase the barrier potential of the diodes to a desired value.

FIGURE 2.34
Biased clipper circuit.

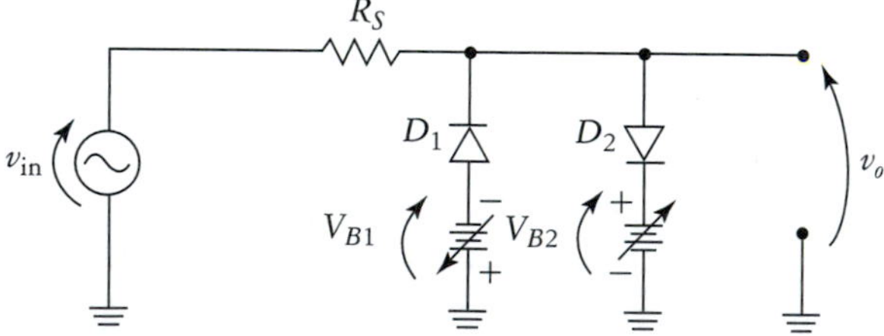

A biased clipper is shown in Fig. 2.34. Biasing voltage sources V_{B_1} and V_{B_2} are adjusted to produce the desired negative and positive clipping levels, respectively. Although the circuit may look rather formidable, it is easy to understand if one simply thinks of the diodes as having adjustable barrier potentials. For example, if we assume that silicon diodes are used, $V_{B1} = 2$ V and $V_{B2} = 5$ V, then the negative clipping voltage is -2.7 V and the positive clipping level is 5.7 V. In equation form, using silicon diodes, the output voltage expression is

$$v_o = \begin{cases} 5.7\text{ V} & \text{for } v_{in} > 5.7\text{ V} \\ v_{in} & \text{for } -2.7\text{V} < v_{in} < 5.7\text{ V} \\ -2.7\text{ V} & \text{for } v_{in} < -2.7\text{V} \end{cases} \tag{2.45}$$

2.10 ZENER-REGULATED POWER SUPPLIES

The final diode application covered in this chapter is that of the zener diode as a voltage regulator. A voltage regulator is a circuit that maintains a constant (or nearly constant) output voltage over a wide variety of load and input voltage ranges. Before we study the zener regulator, however, a few concepts must be introduced.

Regulation Fundamentals

A black box representation for a power supply and a plot of its output voltage as a function of load current are shown in Fig. 2.35. The full-load voltage V_{FL} is the load voltage that is produced when the maximum allowable current I_{FL} is drawn from the supply. The no-load voltage V_{NL} is the supply output voltage that is produced when the load is an open circuit. If the power supply was ideal, the output voltage would remain constant regardless of the load current.

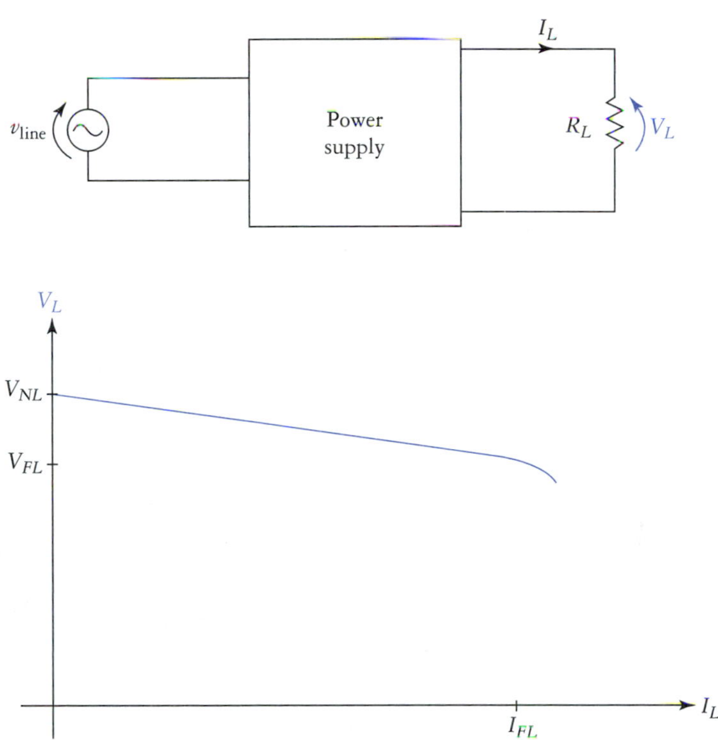

FIGURE 2.35
Supply output voltage variation from a no-load to a full-load situation.

Load Regulation

Load regulation is a measure of the sensitivity of a power supply to changes in load current. Mathematically, load regulation is usually expressed as a percentage, and is defined as

$$\mathrm{VR} = \frac{V_{NL} - V_{FL}}{V_{FL}} \times 100\% \tag{2.46}$$

where V_{FL} is determined by measurement. The value of I_{FL} is predefined and is determined by circuit performance and safety limitations. The smaller VR, the better.

Line Regulation

Changes in the line voltage will also affect a power supply's output voltage. The sensitivity of a power supply to line variations is often expressed in terms of percent change in output voltage per volt of change of the line. That is,

$$LR = \frac{\Delta V_o / V_o}{\Delta V_{in}} \times 100\% \tag{2.47}$$

The concepts of load and line regulation are demonstrated in the following example.

EXAMPLE 2.7

The power supply in Fig. 2.13 has $v_{in} = 120\ V_{rms}$, $V_{NL} = 12$ V, and $V_{FL} = 11$ V. Determine (a) the load regulation of the circuit and (b) the line regulation if V_{in} should change to 105 V_{rms}.

Solution

a. The load regulation is found simply by applying Eq. (2.46):

$$VR = \frac{12\ V - 11\ V}{11\ V} \times 100\%$$
$$= 9.1\%$$

b. The line regulation requires a little more work to determine. Neglecting diode losses, the turns ratio of the transformer is found as follows:

$$\frac{n_p}{n_s} = \frac{V_{pri(pk)}}{V_{NL}}$$
$$= \frac{170}{12}$$
$$= 14.2$$

We determined the peak input voltage, because the capacitor filter produces a dc output voltage that is approximately equal to the peak secondary voltage. Now we determine the *new* peak input and output voltages:

$$V_{pri(pk)} = 1.414 \times 105\ V_{rms}$$
$$= 148.5\ V$$

The resulting output voltage is approximately

$$V_{NL}\Big|_{V_{in} = 105\ V_{rms}} = \frac{148.5\ V}{14.2}$$
$$= 10.5\ V$$

from which $\Delta V_o = 1.5$ V. We now apply Eq. (2.47), which gives us

$$LR = \frac{\Delta V_o / V_o}{\Delta V_{in}} \times 100\%$$
$$= \frac{1.5/12}{15} \times 100\%$$
$$= 0.833\%$$

The load regulation of a power supply can be improved by reducing the equivalent internal resistance of the power supply circuitry. This can be accomplished by brute force, primarily by using a larger transformer, with a heavier secondary winding, by using rectifier diodes with lower bulk resistance, and by using more filter capacitance. Modifications of this type, however, can be rather expensive.

The line regulation specification is somewhat difficult to interpret, because it is dependent primarily on the ratio of output voltage change to input voltage change, which is dictated by the desired output voltage and the available input voltage. In other words, we can't improve line regulation by using "better" parts in the power supply circuit. To improve line and load regulation characteristics, we must use more sophisticated means.

Basic Zener Regulator Operation

Power supply load and line regulation can be improved dramatically with the use of a zener diode regulator. All of the dc power supply circuits covered earlier in this chapter are amenable to zener regulation as shown in Fig. 2.36. A brief analysis of the regulator is now presented.

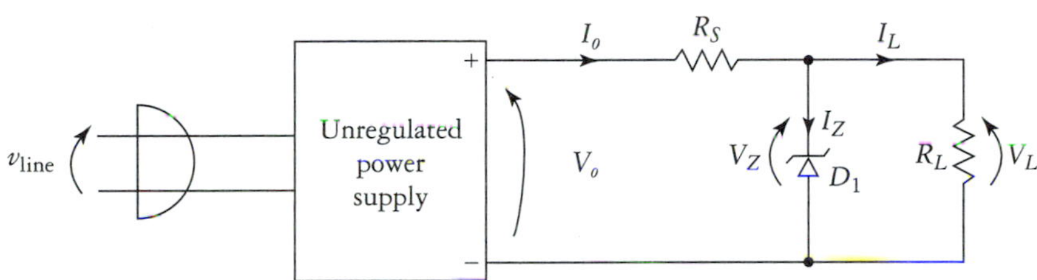

FIGURE 2.36
Basic zener-regulated power supply.

For proper operation, the unregulated supply output voltage V_o must be greater than the zener breakdown voltage V_Z:

$$V_o > V_Z \tag{2.48}$$

We also require that

$$I_Z > I_{Zk} \tag{2.49}$$

If these inequalities are satisfied, then we can proceed. Since the load resistance is in parallel with the zener diode,

$$V_L = V_Z \tag{2.50}$$

The load current is found using Ohm's law:

$$I_L = V_Z/R_L \tag{2.51}$$

If we assume that inequalities (2.48) and (2.49) are still satisfied, the unregulated supply output current I_o is found as follows: First we apply KVL, which gives us

$$0 = V_o - I_oR_S - V_Z$$

Solving for I_o,

$$I_o = \frac{V_o - V_Z}{R_S} \tag{2.52}$$

Note
The zener regulator draws constant current from the unregulated supply, regardless of the load.

Now, we can determine I_Z via application of Kirchoff's current law:

$$I_Z = I_o - I_L \tag{2.53}$$

Derivation of Eq. (2.51) and (2.52) was necessary, because the zener diode is not a linear device, and therefore it does not obey Ohm's law. Remember, however, that even nonlinear devices must satisfy Kirchoff's current law (KCL) and KVL, thus we obtain Eq. (2.53). Also, as long as the zener diode is operating in reverse breakdown, I_o remains constant, regardless of the load conditions.

Close examination of Eq. (2.53) reveals that as the load current increases, the zener current decreases. This means that there is a maximum (full) load (or minimum resistance) that can be applied to the circuit. This maximum load is that load which causes I_Z to drop down to I_{ZK}. To allow some margin for error, in a design situation, we would set a minimum limit for zener current $I_{Z(FL)}$ of about $5I_{Zk}$.

$$I_{Z(FL)} = 5I_{Zk} \tag{2.54}$$

Note
Keep in mind that Equations 2.54–2.56 are *suggested* design equations.

Sometimes, I_{Zk} is not specified, so as a general rule of thumb, we can specify the full-load zener current as a fraction of the maximum allowable zener current I_{ZM}:

$$I_{Z(FL)} = 0.02 I_{ZM} \tag{2.55}$$

Equation (2.53) also tells us that the zener current will be highest under no-load conditions. Thus, the zener diode maximum current specification I_{ZM} must be greater than the no-load zener current $I_{Z(NL)}$. Normally, about a 10% safety margin is acceptable; thus,

$$I_{Z(NL)} = 0.9 I_{ZM} \tag{2.56}$$

As a design rule, Eq. (2.56) can be interpreted as limiting the zener power dissipation to 90% of the maximum allowable (P_{ZM}). Recall that $P_Z = V_Z I_Z$.

It is important for the reader to realize that Eqs. (2.54) through (2.56) are *not* analysis equations, they are suggested design equations; they are not carved in stone. The special requirements existing in any given situation may well dictate that some other design goals be met.

EXAMPLE 2.8

The circuit in Fig. 2.36 has $V_o = 15$ V, $R_S = 50\ \Omega$, $R_L = 100\ \Omega$, $V_Z = 8.2$ V, and $I_{ZK} = 1$ mA. Determine I_Z, P_Z, P_{RS}, $P_{Z(NL)}$, and the minimum recommended load resistance.

Solution We first determine the load current and the output current:

$$\begin{aligned} I_L &= V_Z/R_L \\ &= 8.2\text{ V}/100\ \Omega \\ &= 82\text{ mA} \end{aligned}$$

$$\begin{aligned} I_o &= \frac{V_o - V_Z}{R_S} \\ &= \frac{15\text{ V} - 8.2\text{ V}}{50\ \Omega} \\ &= 136\text{ mA} \end{aligned}$$

$$\begin{aligned} I_Z &= I_o - I_L \\ &= 136\text{ mA} - 82\text{ mA} \\ &= 54\text{ mA} \end{aligned}$$

The zener power is

$$\begin{aligned} P_Z &= V_Z I_Z \\ &= 8.2\text{ V} \times 54\text{ mA} \\ &= 442.8\text{ mW} \end{aligned}$$

The power dissipation of R_S is

$$\begin{aligned} P_{RS} &= I_o R_S \\ &= 136\text{ mA} \times 50\ \Omega \\ &= 924.8\text{ mW} \end{aligned}$$

The no-load zener power dissipation is found by first determining the no-load zener current, which can be shown to be

$$I_{Z(NL)} = I_o$$

Therefore,

$$\begin{aligned} P_{Z(NL)} &= I_o V_Z \\ &= 136\text{ mA} \times 8.2\text{ V} \\ &= 1.12\text{ W} \end{aligned}$$

To find the minimum load resistance, we begin with application of Eq. (2.54):

$$\begin{aligned} I_{Z(FL)} &= 5 I_{ZK} \\ &= 5\text{ mA} \end{aligned}$$

Recall that I_o remains constant, thus we now solve Eq. (2.53) for I_L, which gives us

$$I_{L(\text{max})} = I_o - I_{Z(FL)} = 136 \text{ mA} - 5 \text{ mA} = 129 \text{ mA}$$

Finally, the minimum load resistance is found using Ohm's law:

$$R_{L(\text{min})} = V_Z / I_{L(\text{max})} = 8.2 \text{ V}/129 \text{ mA} = 63.6 \ \Omega$$

Zener Impedance Effects

The very low dynamic resistance of the zener diode is responsible for the superior load and line regulation characteristics of the zener-regulated supply. We now quantify these relationships, beginning with the development of an equation to predict line voltage regulation.

Line Regulation

Broadly speaking, *line regulation* is a measure of a power supply's ability to maintain a constant load voltage under varying line-voltage conditions. In reverse breakdown, the zener diode may be modeled as shown in Fig. 2.37. For the sake of simplicity, let us assume that the load current is negligible. The load voltage is

$$V_L = V_Z' + I_Z Z_Z \tag{2.57}$$

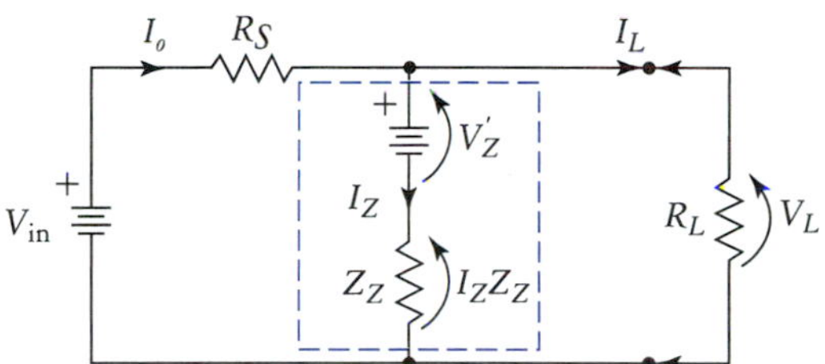

FIGURE 2.37
Equivalent circuit.

A change in input voltage will cause a change in zener current

$$\Delta I_Z = \frac{\Delta V_{in}}{R_S + Z_Z} \tag{2.58}$$

The resulting change in load voltage is

$$\Delta V_L = \Delta I_Z Z_Z \tag{2.59}$$

Substituting Eq. (2.58) into (2.59) gives us

$$\Delta V_L = \frac{\Delta V_{in} Z_Z}{R_S + Z_Z} \tag{2.60}$$

Thus, the expected output voltage is given by Eq. (2.57) and the change in output voltage is given by Eq. (2.60). If we insert the values found using these equations into the line regulation equation, Eq. (2.47), along with the line-voltage change that started the whole output variation, we obtain the line regulation.

Load Regulation

The development of an analytical load regulation relationship is quite similar to that for line regulation. Again, we refer to Fig. 2.37. Under no-load conditions, the load voltage is given by

$$V_{NL} = V_Z' + I_{Z(NL)} Z_Z \tag{2.61}$$

Similarly, under full-load conditions,

$$V_{FL} = V_Z' + I_{Z(FL)} Z_Z \tag{2.62}$$

The numerator of the voltage-regulation equation, Eq. (2.46), is $V_{NL} - V_{FL}$. Substituting Eqs. (2.61) and (2.62) into (2.46) gives us the following equation:

$$\mathrm{VR} = \frac{Z_Z\,(I_{Z(NL)} - I_{Z(FL)})}{V'_Z + I_{Z(FL)}\,Z_Z} \tag{2.63}$$

To apply Eq. (2.63) sensibly, it is necessary to know the minimum and maximum zener currents, which, although not extremely difficult to determine in themselves, require further knowledge of permissible load conditions and so on. With this in mind, the following approximation, expressed in terms of more readily determined parameters, will normally provide sufficient accuracy in predicting the load regulation characteristics of a zener regulator:

$$\mathrm{VR} = \frac{Z_Z\,V_{\mathrm{in}} - Z_Z\,V_Z}{V_Z\,R_S} \tag{2.64}$$

where V_Z is the nominal zener voltage.

Ripple Reduction

Low zener impedance also provides a significant reduction in ripple voltage at the load. If we apply superposition, for ac ripple voltage determination, the power supply can be represented as shown in Fig. 2.38. The input voltage source v_r represents the ripple voltage that appears across the filter capacitor, while v_r is the ripple voltage that is dropped across the zener diode. Once again the voltage divider equation relates the two voltages such that

$$v_r = v'_r\,\frac{Z_Z}{Z_Z + R_S} \tag{2.65}$$

Zener impedance is usually quite low relative to R_S, so the load ripple voltage will also be proportionately smaller. Thus, for a given desired overall ripple factor, the use of a zener diode regulator allows a smaller filter capacitor to be used. This in turn reduces the peak rectifier current (see Eq. (2.28), reducing stress on the rectifier diodes.

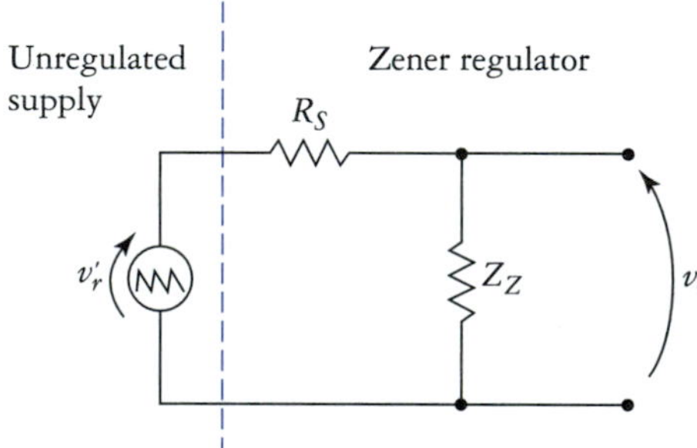

FIGURE 2.38
Ripple attenuation is caused by voltage divider action of the Zener diode.

EXAMPLE 2.9

Refer to Fig. 2.39. This circuit has the following specifications: $V_{\mathrm{sec}} = 12\ \mathrm{V_{rms}}$, $C_1 = 2200\ \mu\mathrm{F}$, $R_S = 220\ \Omega$, $V_Z = 6.2\ \mathrm{V}$, $Z_Z = 4\ \Omega$, $I_{ZM} = 65$ mA, and diodes D_1 through D_4 are silicon with negligible bulk resistance. Determine (a) the p-p ripple voltage across C_1 (V_r), (b) the p-p ripple voltage at R_L, (c) the output ripple factor, (d) the minimum permissible load resistance, (e) the percent load voltage regulation, and (f) the percent line regulation.

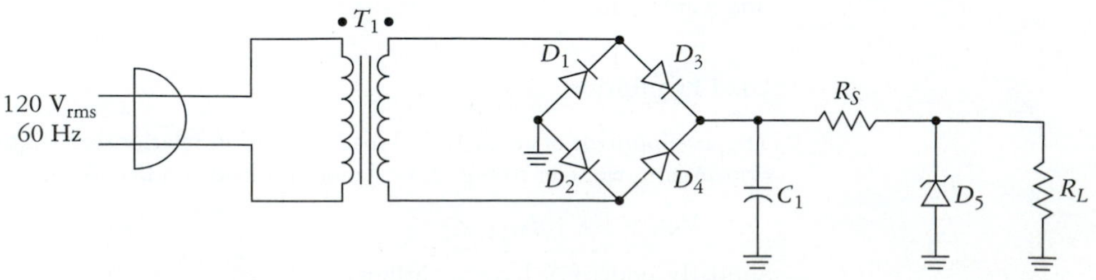

FIGURE 2.39
Circuit for Example 2.9.

Solution The peak filter capacitor voltage is found to be $V_p = 15.6$ V. Now we may begin the requested analysis.

a. To find the p-p ripple across C_1, we apply Eq. (2.21), where in the worst case (the load is a short circuit), the load resistance seen by the filter capacitor is R_S; thus,

$$\begin{aligned} V'_{r(\text{p-p})} &= \frac{V_p}{f_{\text{rip}}\, R_S\, C_1} \\ &= \frac{15.6\text{ V}}{(120\text{ Hz})(220\ \Omega)(2200\ \mu\text{F})} \\ &= 269\text{ mV}_{\text{p-p}} \end{aligned}$$

b. The output peak-to-peak ripple voltage is found using Eq. (2.65):

$$\begin{aligned} V_{r(\text{p-p})} &= V'_{r(\text{p-p})} \frac{Z_Z}{Z_Z + R_S} \\ &= 269\text{ mV} \frac{7}{220 + 7} \\ &= 8.3\text{ mV}_{\text{p-p}} \end{aligned}$$

c. To determine the ripple factor, we must return to the defining equation, Eq. (2.19). Because the ripple voltage is so small, we make the approximation $V_{\text{dc}} = V_Z$, which will be the case in almost every similar situation.

$$\begin{aligned} r &= \frac{V_{r(\text{rms})}}{V_{\text{dc}}} \cong \frac{V_{r(\text{rms})}}{V_Z} \\ &= \frac{2.39\text{ mV}}{6.2\text{ V}} \\ &= 385 \times 10^{-6} \end{aligned}$$

d. To determine the load regulation, we use Eq. (2.64) and multiply by 100 to obtain a percentage.

$$\begin{aligned} \text{VR} &= \frac{(Z_Z V_{\text{in}}) - (Z_Z V_Z)}{V_Z R_S} \times 100\% \\ &= \frac{(7 \times 15.6) - (7 \times 6.2)}{6.2 \times 220} \times 100\% \\ &= 4.82\% \end{aligned}$$

e. To find the line regulation, let us assume that the line voltage drops from 120 to 100 V_{rms}, thus $\Delta V_{\text{line}} = 20\ \text{V}_{\text{rms}} = 28.3\ \text{V}_{\text{p-p}}$. The corresponding change in filter capacitor voltage is $\Delta V_p = 28.3/10 = 2.83$ V, where 10 is the turns ratio of the transformer. For notational convenience, the absolute magnitude of the voltage changes is used here, as the line regulation percentage is stated as a positive number. The corresponding change in load voltage is now found using Eq. (2.60).

$$\begin{aligned} \Delta V_L &= \frac{\Delta V_{\text{in}}\, Z_Z}{R_S + Z_Z} \\ &= \frac{2.83 \times 7}{220 + 7} \\ &= 87.3\text{ mV} \end{aligned}$$

We now invoke Eq. (2.47), which yields

$$\begin{aligned} \text{LR} &= \frac{\Delta V_o/V_o}{\Delta V_{\text{in}}} \times 100\% \\ &= \frac{87.3\text{ mV}/6.2\text{ V}}{20\text{ V}} \times 100\% \\ &= 0.07\% \end{aligned}$$

2.11 COMPUTER-AIDED ANALYSIS APPLICATIONS

This section presents PSpice analysis runs for three types of circuits that were presented in this chapter: a filtered full-wave bridge rectifier, a positive clamper, and a diode clipper.

Filtered Full-Wave Rectifier

The schematic editor version of the filtered full-wave rectifier is shown in Fig. 2.40. Notice that the bridge rectifier must be drawn with the diodes placed horizontally (or vertically, if so desired). The input voltage is provided by a VSIN device set for 10 V_{pk} at 60 Hz. The analysis was set up for Transient Analysis, with a final time of 40 ms, which allows about five cycles of v_{in} to be simulated.

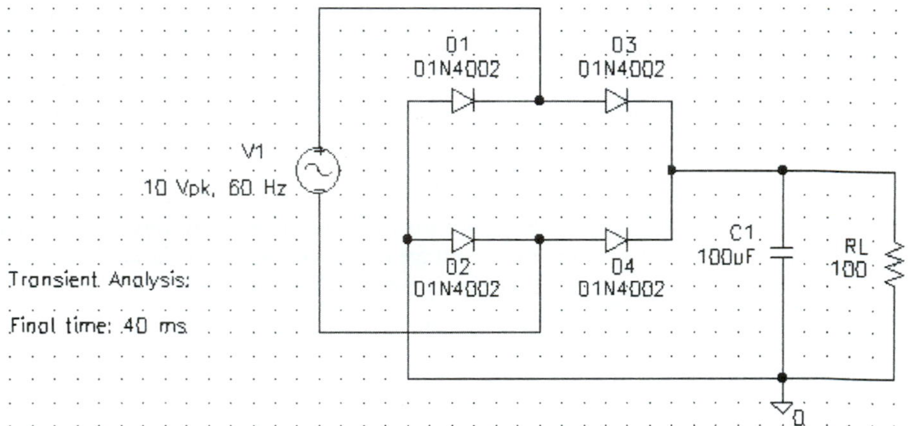

FIGURE 2.40
Schematic editor full-wave rectifier.

The pertinent results of the simulation are shown in Fig. 2.41. Notice that the input sine wave is severely distorted and the large amount of ripple voltage. This occurs because the filter capacitor is too small to be effective given the 100-Ω load. This was done intentionally, to emphasize the ripple voltage content of the output. This amount of ripple would be considered excessive for

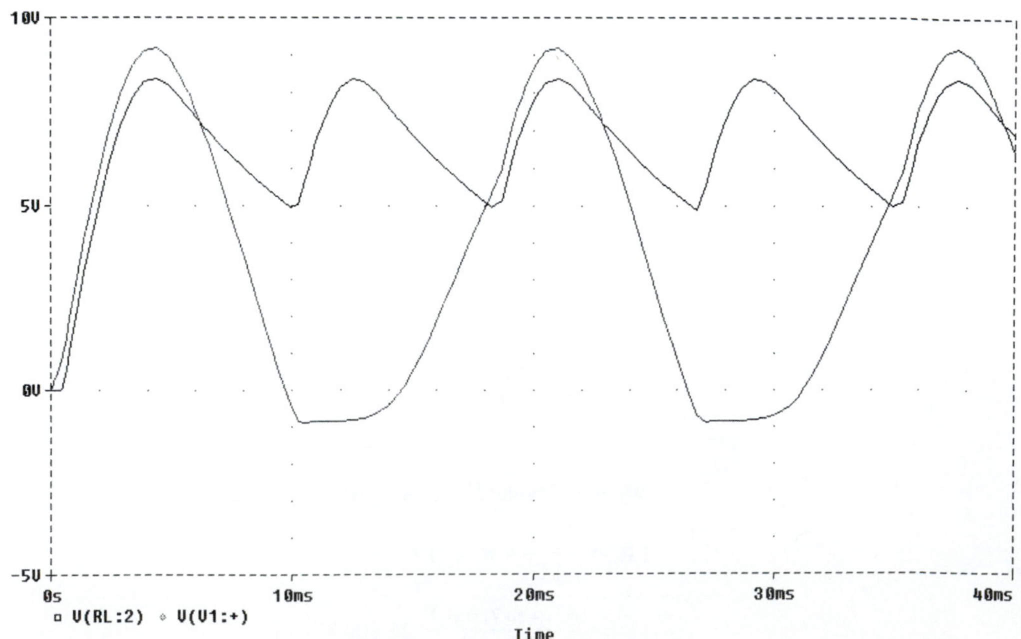

FIGURE 2.41
Resulting input and output voltage waveforms.

most applications. Increasing the size of C_1 would be recommended to make this circuit function more acceptably. Try increasing the capacitor to 1000 or 10,000 μF and note the decrease in ripple voltage.

The Clamper

The clamper to be investigated is shown in Fig. 2.42. The input voltage in this case is a 10-V peak, 1-kHz sine wave. A transient analysis from $t = 0$ to 5 ms allows the response over five cycles of the input signal to be plotted. The maximum step size here is 500 μs, which will give us a plot that contains at least 100 points.

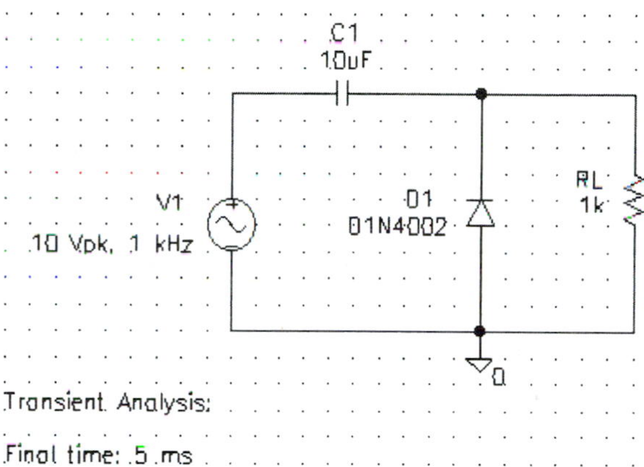

FIGURE 2.42
Positive clamper circuit.

The clamper output voltage waveform is shown in Fig. 2.43. Notice that it takes one cycle of the input voltage for the circuit to settle into its steady-state response. The analysis equations developed in this chapter apply to the steady-state mode of operation. Detailed transient analysis is beyond the scope of this book.

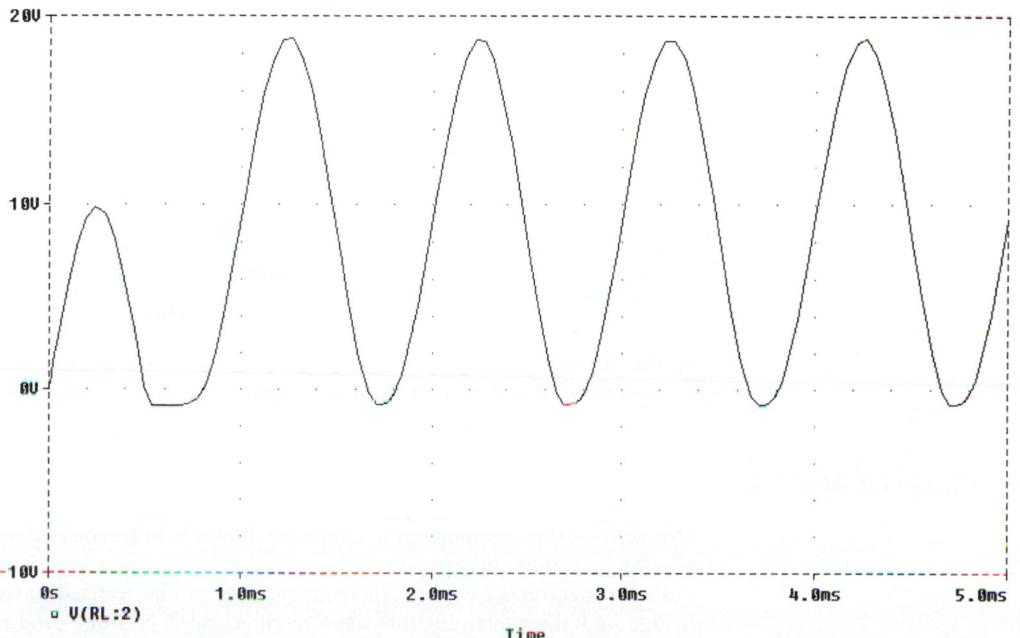

FIGURE 2.43
Positive clamper output voltage waveform.

Asymmetrical Diode Clipper

The last circuit example presented here is the clipper of Fig. 2.44. The input voltage and transient analysis specifications are the same as those used in the preceding clamper circuit. The resulting PROBE plot of the output voltage is shown in Fig. 2.45. Cursors were placed at the minimum and maximum points of the output waveform. The Probe Cursor box shows min and max values of −0.716 and 1.42 V, respectively. These agree closely with our estimates of −0.7 and 1.4 V. Notice that the peaks of the output signal are rounded off. This happens because the diodes turn on gradually, rather than abruptly, as we often approximate their behavior.

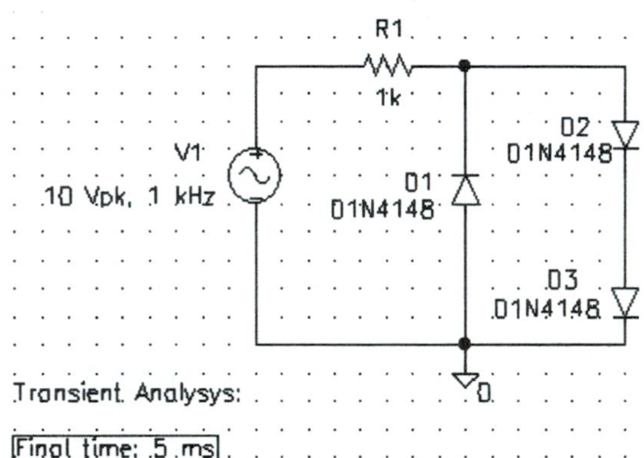

FIGURE 2.44
Asymmetrical clipper.

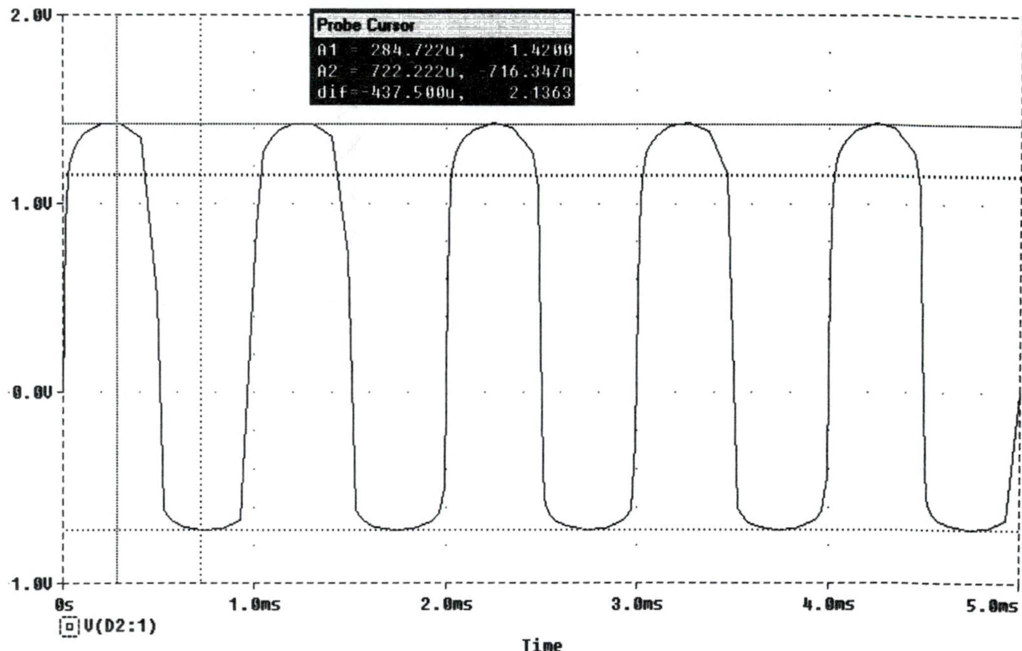

FIGURE 2.45
Asymmetrical clipper output voltage waveform.

■ CHAPTER REVIEW

One of the most common applications for diodes is in rectifier circuits. The function of a rectifier is to convert ac into pulsating dc. Half-wave rectifiers are very simple, but are also relatively inefficient. Full-wave rectifiers are more efficient, and they can be realized in two different forms: the full-wave bridge and the center-tap full-wave rectifier. Both versions produce output voltage pulsations at twice the input line frequency.

Most often, after rectification, filtering is necessary to smooth the dc amplitude variations and provide a constant dc voltage level. Capacitor filters are frequently used to accomplish this task. Less commonly, one might also find *RC* or *LC* filters used in power supply designs as well. Generally speaking, the greater the amount of filter capacitance, the lower the ripple content of the output voltage will be. The use of a large filter capacitor may cause rather high surge currents to flow, however.

A variation of the filtered power supply is the *voltage multiplier.* The output of a voltage multiplier is a dc voltage that is approximately an integer multiple of the peak ac input voltage. Voltage multipliers are usually very easily loaded, so they are commonly used in low-current applications. Unlike standard transformer-coupled power supply designs, voltage multipliers do not provide ac line isolation, which is a disadvantage.

A clamper is a circuit composed of diodes and capacitors that is designed to shift a signal either above or below ground potential. These circuits are sometimes called *dc restorers.*

Clippers are examples of wave-shaping circuits. The function of a clipper is to limit the maximum amplitude of a signal to within some specific range. Zener diodes and general-purpose diodes are both used in the design of clipper circuits. A biased clipper includes one or more dc sources in series with the clipper diodes to provide adjustable clipping limits. Clippers are sometimes called *limiters.*

Very stable, low-ripple power supply designs may be achieved through the use of zener diodes as voltage regulators. The nearly vertical reverse-breakdown transconductance characteristics of the zener diode result in a zener terminal voltage drop that is nearly constant over a wide range of reverse current. Thus, although the load on a zener-regulated power supply may vary significantly, the resulting load voltage will remain constant, yielding very good load regulation characteristics. Zener diode regulators also increase the line regulation and ripple reduction performance characteristics of the power supply as well.

DESIGN AND ANALYSIS PROBLEMS

Note: Unless otherwise stated, assume that all diodes are silicon with 0.7-V barrier potential and negligible bulk resistance.

2.1. Refer to Fig. 2.1. Assume $R_L = 100\ \Omega$, $v_{in} = 5 \sin \omega t$ V. Determine $V_{o(av)}$ and $I_{o(av)}$.

2.2. Again, refer to Fig. 2.1. Given $R_L = 100\ \Omega$ and $v_{in} = 2 + (5 \sin \omega t)$ V, sketch the v_o waveform.

2.3. Refer to Fig. 2.4. Determine the required turns ratio n_p/n_s for T_1 such that $V_{L(av)} = 6$ V for a 50-Ω load.

2.4. Repeat Problem 2.3 such that $I_{L(av)} = 100$ mA for a 50-Ω load.

2.5. Refer to Fig. 2.4. Assuming that $n_p/n_s = 20$, determine the rms value of the secondary voltage and the PIV applied to the diode.

2.6. The voltage applied to the input of Fig. 2.5 is the triangular waveform of Fig. 2.46. Sketch the resulting output voltage. Label all pertinent voltages.

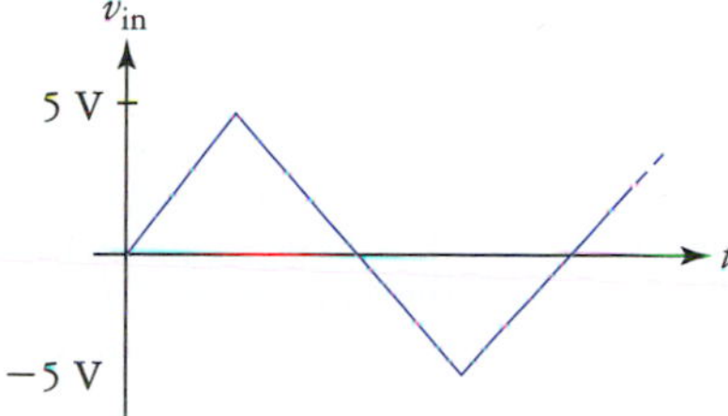

FIGURE 2.46
Waveform for Problem 2.6.

2.7. Refer to Fig. 2.5. Given $v_{in} = 12 \sin 377t$ V, determine the frequency of the resulting ripple voltage.

2.8. Given the circuit of Fig. 2.5 with $V_{in} = 12\ V_{rms}$, 60 Hz, and $R_L = 150\ \Omega$, determine $I_{L(pk)}$, $I_{L(av)}$, $V_{L(pk)}$, $V_{L(av)}$, PIV, and the average diode current $I_{D(av)}$.

2.9. Refer to Fig. 2.13. Given $n_p/n_s = 12$, $C = 4700\ \mu$F, and $R_L = 100\ \Omega$, determine the ripple factor (%r), the peak-to-peak ripple voltage, and the PIV.

2.10. Refer to Fig. 2.13. The transformer has a turns ratio of 15:1 and $R_L = 300\ \Omega$. Determine what filter capacitor value will result in $r = 1\%$.

2.11. Refer to Fig. 2.13. The transformer has a turns ratio of 10:1 with a filter capacitor of 1000 μF. Determine what value of R_L will result in $r = 0.5\%$.

2.12. Refer to Fig. 2.13. Assume $n_p/n_s = 18$. Given a secondary current $i_s = 500 \sin 2\pi 60t$ mA, determine the equation for the primary current.

2.13. Refer to Fig. 2.27. Given $v_{in} = 10 \sin 500t$ V and $R_L = 400\ \Omega$, determine the minimum value required for C_1, and write the equation for the resulting load voltage and load current.

2.14. Refer to Fig. 2.27. Given $v_{in} = 12 \sin 1000t$ V and $C = 220\ \mu$F, determine the minimum usable load resistance and the load voltage equation.

2.15. Sketch the output voltage waveform for Fig. 2.47. Label all pertinent voltages.

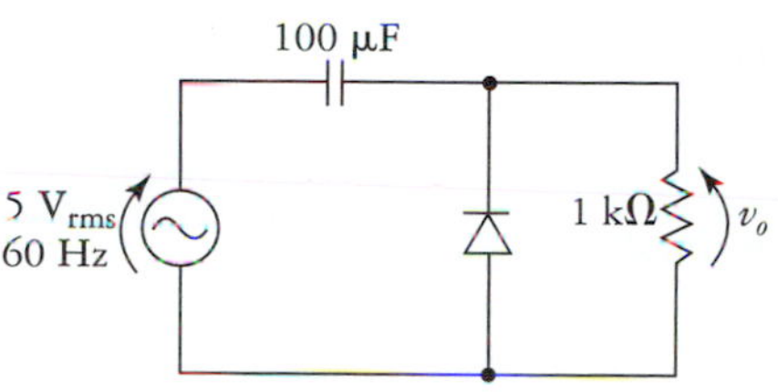

FIGURE 2.47
Circuit for Problems 2.15 and 2.16.

2.16. Assume that the input voltage to Fig. 2.47 is as shown in Fig. 2.48. Sketch the resulting output voltage waveform. Label all pertinent voltages.

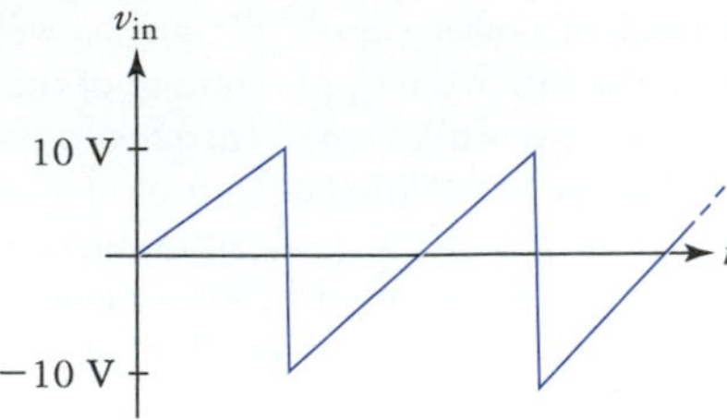

FIGURE 2.48
Waveform for Problem 2.16.

2.17. Refer to Fig. 2.20(a). Assuming that the secondary resistance is negligible, $V_P = 170$ V, $n_p/n_s = 15$, and $R_p = 200\ \Omega$, determine the secondary surge current.

2.18. Refer to Fig. 2.20(a). The secondary resistance is negligible and $n_p/n_s = 12$. Determine the value for R_p that will limit the secondary surge current to 5 A, given $V_P = 170$ V.

2.19. Refer to Fig. 2.21. The circuit has the following values: $f_{rip} = 120$ Hz, $V_{1(dc)} = 12$ V, $R_S = 22\ \Omega$, $C_1 = C_2 = 1000\ \mu$F, and $R_L = 200\ \Omega$. Determine the overall output ripple factor and the peak-to-peak ripple voltage amplitude.

2.20. Repeat Problem 2.19 for a 100-Ω load resistor.

2.21. Refer to Fig. 2.22. Given $f_{rip} = 120$ Hz, $R_S = 2\ \Omega$, $V_{1(dc)} = 10$ V, $R_L = 100\ \Omega$, $L_1 = 2$ H, and $C_1 = C_2 = 2200\ \mu$F, determine $V_{L(dc)}$, $V_{r1(p\text{-}p)}$, $V_{r2(p\text{-}p)}$, and the overall percent ripple factor.

2.22. Refer to Fig. 2.23. Assuming that ripple voltage is negligible, $V_{o(+)} = 12$ V, $V_{o(-)} = -12$ V, $R_{L1} = 100\ \Omega$, and $R_{L2} = 150\ \Omega$, determine I_1, I_2, and I_3.

2.23. Refer to Fig. 2.24(a). Given $V_{in} = 50\ V_{rms}$, determine V_o and PIV.

2.24. Refer to Fig. 2.27. Given $v_{in} = 10 \sin 377\,t$ V and $R_L = 500\ \Omega$, determine the minimum value required for C_1, and write the expression for the resulting v_o.

2.25. The clamper circuit of Fig. 2.27 has $C_1 = 220\ \mu$F and $R_L = 100\ \Omega$. Determine the minimum frequency of v_{in} with which the circuit will function properly.

2.26. Refer to Fig. 2.49. Sketch the output voltage that would be produced given $v_{in} = 9 \sin \omega t$ V. Label all pertinent voltage levels.

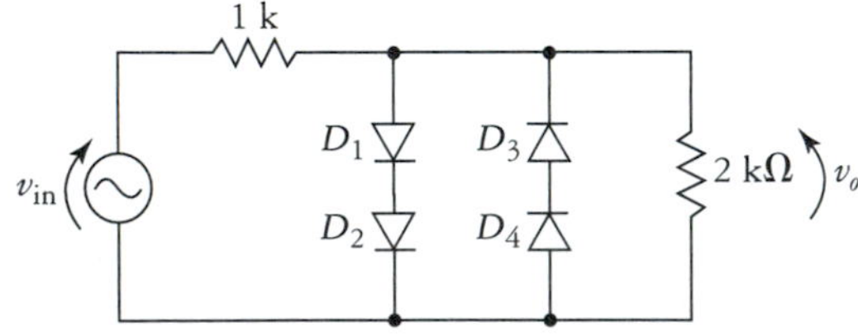

FIGURE 2.49
Circuit for Problems 2.26 and 2.27.

2.27. Sketch the $v_o - v_{in}$ transfer characteristic curve for the circuit of Fig. 2.49.

2.28. Design the biased clipper circuit using silicon rectifier diodes that produces an output given by

$$v_o \cong \begin{cases} 2.1\text{ V} & \text{for } v_{in} > 4.2\text{ V} \\ v_{in}/2 & \text{for } -4.2\text{ V} < V_{in} < 4.2\text{ V} \\ -2.1\text{ V} & \text{for } V_{in} < 4.2\text{ V} \end{cases}$$

2.29. Sketch the $v_o - v_{in}$ transfer characteristic curve for Problem 2.28. Label all pertinent voltages and determine the slopes of the various portions of the plot.

2.30. Sketch the $v_o - v_{in}$ transfer characteristics for the circuit in Fig. 2.50.

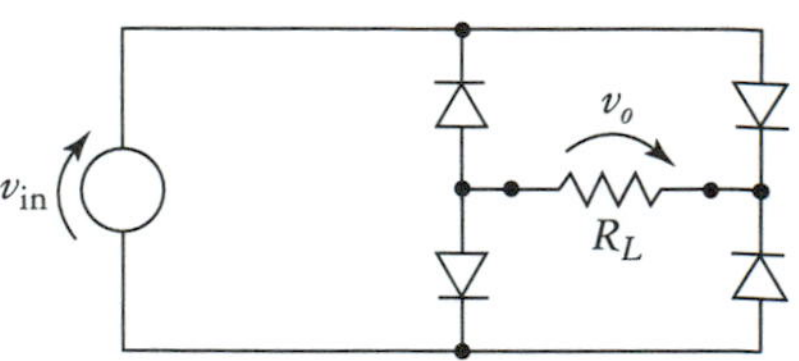

FIGURE 2.50
Circuit for Problem 2.30.

2.31. Refer to Fig. 2.36. Given $V_o = 15$ V, $V_Z = 10$ V, $P_{ZM} = 1$ W, $Z_Z = 8\ \Omega$, and the ripple voltage present in the unregulated voltage is 100 $mV_{p\text{-}p}$, determine R_S, $R_{L(min)}$, P_{Rs}, the overall ripple factor, and the percent load regulation, VR, for $V_{FL} = 9.3$ V and $V_{NL} = 10$ V.

2.32. Refer to Fig. 2.51. Given $V_{sec} = 24\ V_{rms}$, $V_Z = 6.2$ V, $Z_Z = 5\ \Omega$, $I_{ZM} = 200$ mA, $I_{Zk} = 2$ mA, and $C_1 = C_2 = 4700\ \mu$F, determine R_1, R_2, and $R_{L(min)}$. Assume that ripple voltage is negligible.

2.33. Based on the relationships $n_p/n_s = i_s/i_p = v_p/v_s$ and $p_{in} = p_{out}$, prove that R_p and R_s are related by Eq. (2.30).

2.34. Express the graphical transfer characteristics of Fig. 2.31(b) in mathematical form.

2.35. Refer to Fig. 2.33. Sketch the transfer characteristics for this circuit. Write the equation for the slope, and define the min and max values in terms of resistors R_L, R_S, and V_p.

■ TROUBLESHOOTING PROBLEMS

2.36. The circuit of Fig. 2.5 is constructed in the lab. A dc voltmeter indicates $V_{in} = 15.1\ V_{rms}$, while the indicated voltage drop across R_L is 6.31 V. Is the circuit functioning correctly? If not, what is the most likely cause of the problem?

2.37. Refer to Fig. 2.9. Under normal conditions, $V_{sec} = 24\ V_{rms}$. Suppose the upper half of the T_1 secondary winding were to become open circuited. What dc output voltage would be measured?

2.38. A certain power supply tends to burn up rectifier diodes occasionally when first turned on, even under no-load conditions. What is the likely cause for this? List some possible cures for this problem.

2.39. The power supply of Fig. 2.23 is designed properly, such that $V_{o(+)} = 6$ V and $V_{o(-)} = -6$ V. As a quick check of operation, 6-V lamps are connected to the output terminals as loads. With both lamps connected, the supply appears to function normally, lighting both lamps. However, whenever either lamp is disconnected from the supply, the lamp that is still connected goes out. What would cause this problem?

2.40. The circuit of Fig. 2.31 is driven by a 20-$V_{p\text{-}p}$ sinusoidal voltage. Both zener diodes are rated for $V_Z = 5$ V. The resulting output voltage waveform is shown in Fig. 2.52. What is the most likely defect in the circuit?

2.41. Suppse the circuit of Fig. 2.39 is prototyped, with $C_1 = 2200\ \mu$F, $R_S = 100\ \Omega$, $V_Z = 5.0$ V, $I_{Zk} = 1$ mA, and $P_{ZM} = 800$ mW. Power is applied to the circuit, and the no-load output voltage is measured to be 0.81 V. What is the most likely cause of this problem?

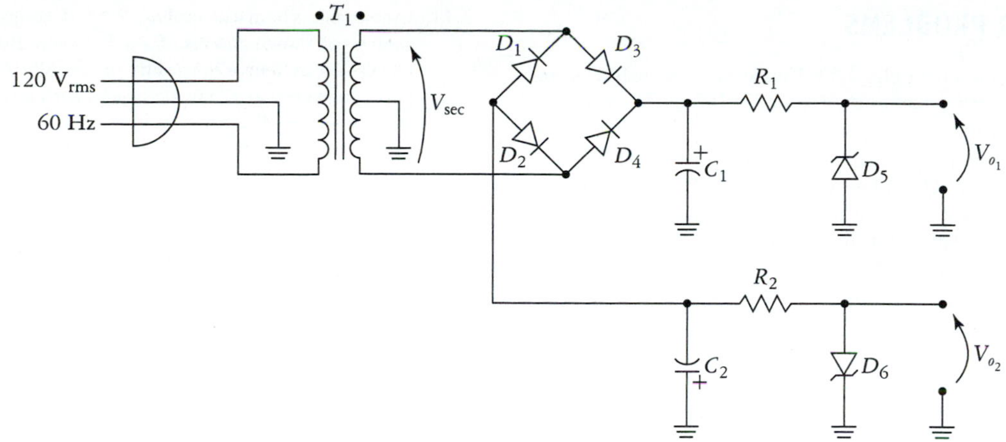

FIGURE 2.51
Circuit for Problem 2.32.

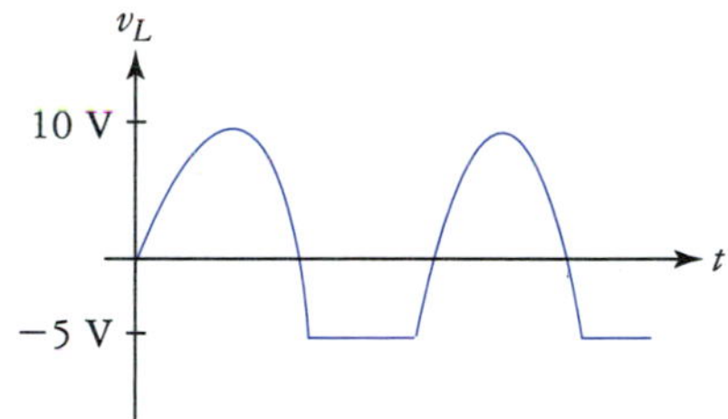

FIGURE 2.52
Waveform for Problem 2.40.

COMPUTER PROBLEMS

2.42. Simulate the circuit in Fig. 2.53. Run a transient analysis from 0 to 200 ms with a maximum step size of 100 μs. Use PROBE to plot v_{in}, v_1, and v_2.

2.43. Draw the schematic for a voltage tripler with $V_{in} = 120\ V_{rms}$, 60 Hz, $R_L = 10\ k\Omega$, and all capacitors are 100 μF. Simulate the circuit using a transient analysis that runs from 0 to 200 ms in 20-μs steps. Plot the resulting load voltage using PROBE.

2.44. Create the schematic in Fig. 2.54. Configure for a transient analysis as shown and also for a dc sweep. Even though there is no explicit dc source in the circuit, the VSIN source (V1) has dc offset characteristics. The dc sweep, with attributes set as listed, will cause the offset to vary from −20 to 20 V in 0.1-V increments. As usual, the transient analysis will allow you to observe v_o as a function of time. The dc sweep will allow you to plot the output voltage V(D1:2) as a function of the input voltage V1. Generate this plot and determine the slope of the resulting transfer characteristic curve.

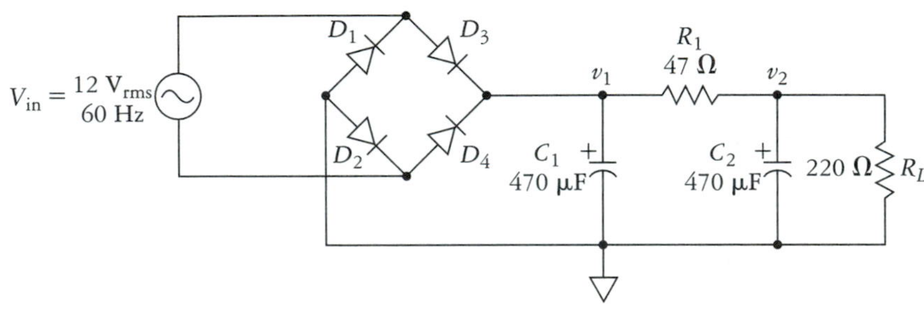

FIGURE 2.53
Circuit for Problem 2.42.

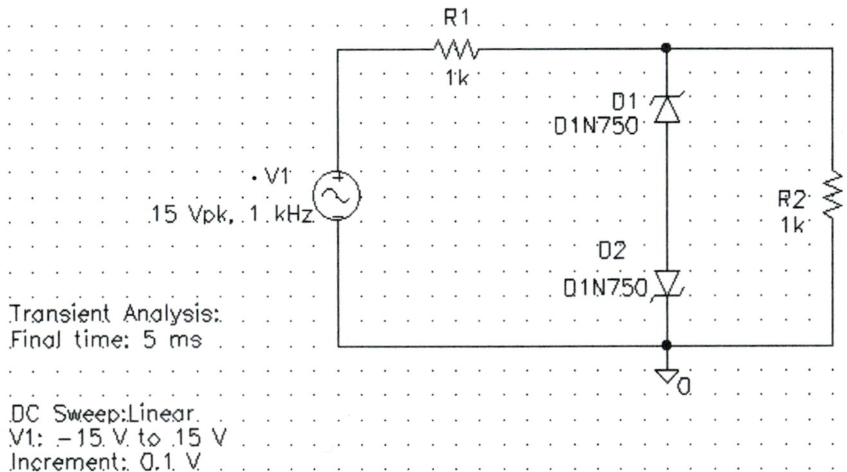

FIGURE 2.54
Circuit for Problem 2.44.

CHAPTER 3

The Bipolar Junction Transistor

BEHAVIORAL OBJECTIVES

On completion of this chapter, the student should be able to:

- Describe the construction of a bipolar junction transistor.
- Explain the terms *diffusion current* and *drift current.*
- Describe the behavior of the BJT in the saturation, cutoff, and active regions.
- Model the BJT as a dependent source (ICIS).
- Define the relationships between collector, base, and emitter currents.
- Describe the effects of collector and emitter leakage currents on the BJT.
- Define the transistor parameters β and α.
- Determine transistor parameters using device characteristic curves.
- Generate transistor characteristic curves using the computer.

INTRODUCTION

Developed by the team of William Shockley, Walter Brattain, and John Bardeen at Bell Laboratories in 1947, the *bipolar junction transistor* (BJT) is certainly one of the most important and far-reaching inventions of the 20th century. The purpose of this chapter is to familiarize you with the basic terminology, notation, and operating principles of the BJT. The plan is to cover transistors rigorously enough to provide a solid foundation for future studies and to develop a good intuitive circuit sense. The ability to visualize the transistor as a dependent source is an excellent way to get a good feel for transistor circuit operation.

Few devices, however, seem to defy allowing an intuitive understanding of their fundamental operation as much as the bipolar transistor does. From a physical standpoint, the bipolar transistor seems to behave in strange ways. This is not to say that transistor circuits are that difficult to analyze. If you can analyze circuits containing dependent sources, you can analyze transistor circuits too. It's just that at the physical level, we must take certain aspects of transistor behavior on faith, because no simple, meaningful explanation is available.

As a final introductory note, most often, when someone says the word *transistor,* it is the bipolar junction transistor that is being discussed. A second class of transistors, called field-effect transistors, also exists, and this can sometimes cause confusion as to which type of device is meant. In this text, when the term *transistor* is used alone, the assumption is that we are referring to a BJT.

3.1 BJT STRUCTURE

Bipolar transistors come in two fundamental varieties; *npn* and *pnp.* The terms npn and pnp are derived from the representative physical structure of the two types of BJTs as shown in Fig. 3.1. The three sections of the transistors are regions of a single, continuous (homogeneous) silicon crystal that have been doped to form two pn junctions. The vertical lines separating the regions represent abrupt changes in doping impurities.

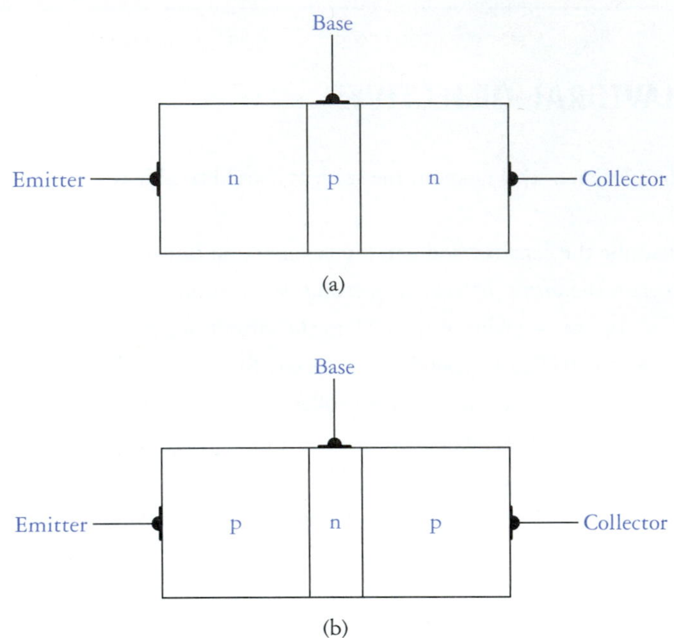

FIGURE 3.1
Cross section of (a) npn and (b) pnp bipolar junction transistors.

Although the transistor structures of Fig. 3.1 are useful for illustrative purposes, a more representative transistor geometry, which can be used in the fabrication of transistors in integrated circuits as well as discrete devices, is shown in Fig. 3.2. Many other transistor geometries are used as well.

FIGURE 3.2
One possible transistor structure.

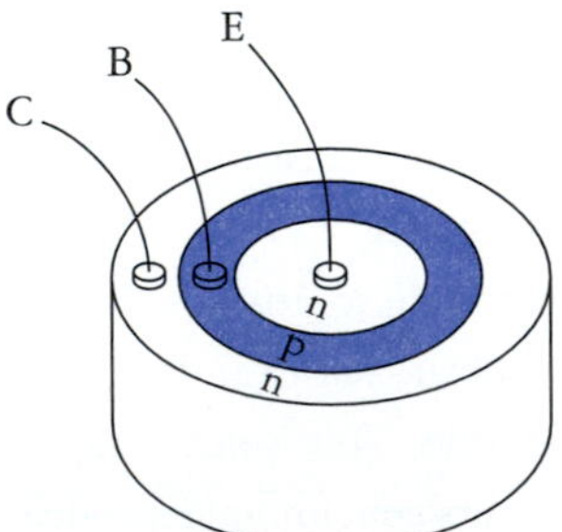

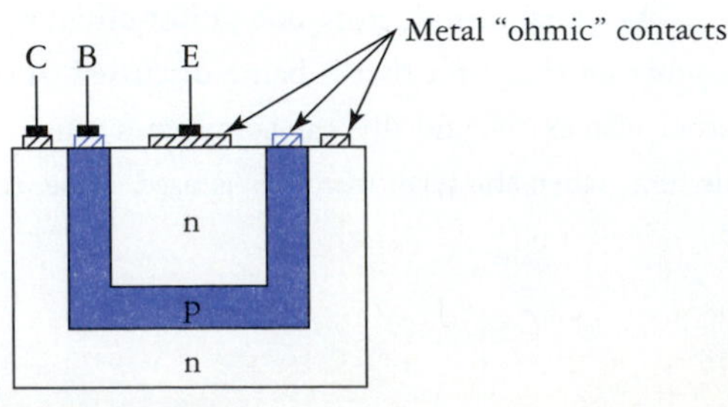

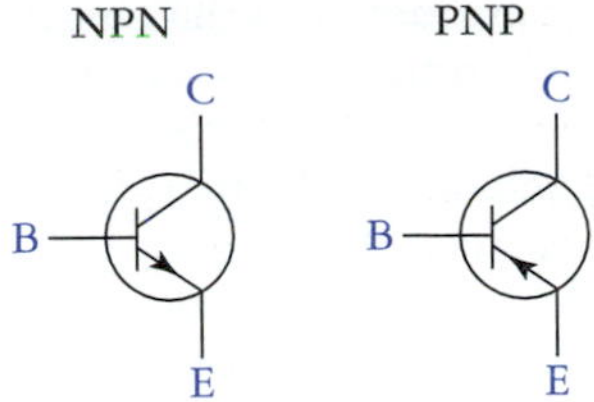

FIGURE 3.3
Schematic symbols for (a) npn and (b) pnp BJTs.

Schematic Symbols

The schematic symbols used to represent npn and pnp bipolar junction transistors are shown in Fig. 3.3. The three terminals of the transistors are called the *collector, base,* and *emitter.* Each terminal is connected to one of the three differently doped regions of the silicon crystal that actually forms the device.

Doping Concentrations

The illustrations in Fig. 3.1 might seem to indicate that bipolar transistors are symmetrical devices, and thus the end terminals (the emitter and collector) are interchangeable. This is not the case. The collector region of the transistor is generally doped at about the same concentration as the typical rectifier diode (around 1 part in 10^6). The base region is normally doped relatively lightly (around 1 part in 10^7 or 10^8) and is also made very thin, as well. A brief explanation of why this is so is presented shortly. The emitter region is heavily doped, much like a zener diode would be. In any case, because of the differences in doping concentrations, the emitter and collector terminals are not interchangeable.

3.2 BASIC TRANSISTOR ACTION

One of the most common configurations in which the transistor is used is shown in Fig. 3.4. In this case, an npn transistor is used; however, a pnp unit could be used in the upcoming discussions, provided the polarities of the two voltage sources are reversed. Notice that conventional current flow is

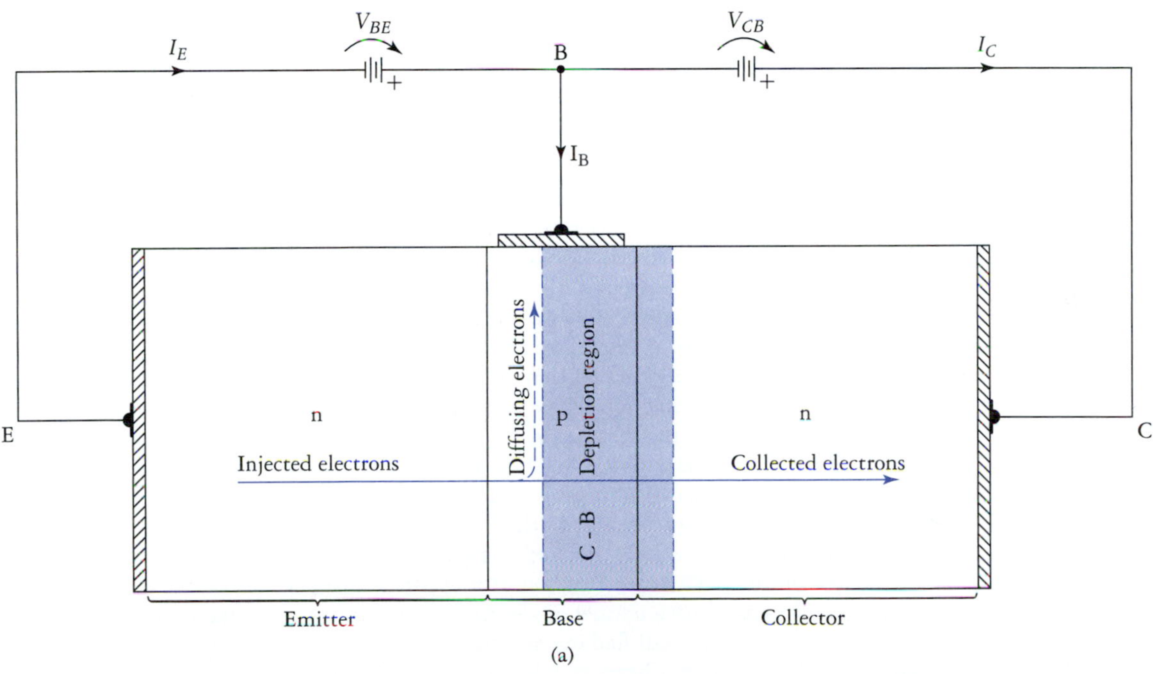

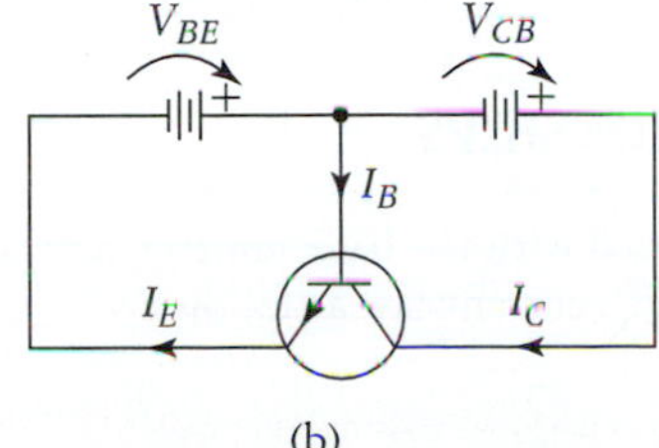

FIGURE 3.4
Biasing up a transistor. (a) Internal view. (b) Circuit diagram.

used external to the transistor, while electron flow is used within the crystal structure of the transistor itself. This is one of the few times when it is necessary to resort to the electron flow viewpoint for purposes of clarity, because both hole and electron flow occur within the transistor, and the two are distinctly different conduction mechanisms. The application of external voltages to the transistor is called *biasing,* thus, the two voltage sources shown are called *bias voltage* sources. Based on the junction structure shown, we see that the collector–base junction is reverse biased, whereas the emitter–base junction is forward biased. Under these conditions, the transistor is said to be operating in the *active region.*

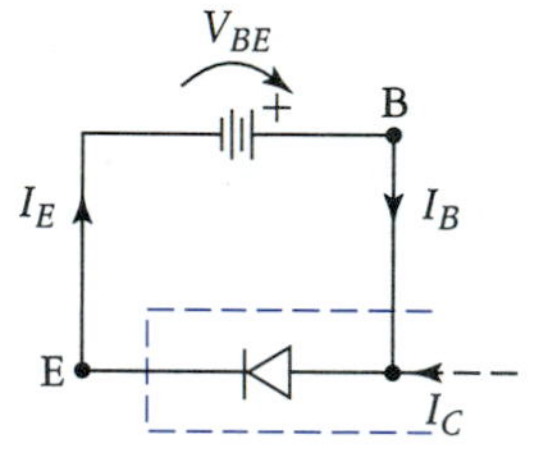

FIGURE 3.5
The base–emitter junction is forward biased.

Base–Emitter Loop

Let us assume that the transistor of Fig. 3.4 is a silicon device. Based on our knowledge of pn junctions, it would appear that if the emitter–base junction is forward biased, the external voltage V_{BE} should be around 0.7 V, and emitter current I_E should be rather large. If we model the E-B junction as a diode, as shown in Fig. 3.5, these assumptions are readily apparent. What is not apparent, however, is the fact that the base current I_B will normally be much smaller than the emitter current. To understand why this occurs, we must examine the collector–base loop.

Collector–Base Loop

Referring back to Fig. 3.4, we see that source V_{CB} reverse biases the C-B junction, causing the depletion region to widen. Because the base is lightly doped, and very thin as well, the depletion region extends deeply into the base region. Now, based on our past experience with diodes we would normally expect collector current I_C to be quite low under these reverse-biased conditions. However, because the emitter and collector both share the base, an unusual event takes place, in which most of the electrons that are injected from the emitter into the base flow right through the C-B depletion region into the collector, rather than out the base terminal. Thus, the base current is relatively small. Typically, the collector current is around 100 times larger than the base current.

Injection Current

In Fig. 3.4, electrons supplied by the V_{BE} source cross from the n-type emitter into the p-type base. These electrons are majority carriers in n-type material but minority carriers in p-type material. It turns out that when minority carriers are injected into a depletion region (as the electrons injected into the base are), those electrons will propagate through that depletion region rather easily. So, forward biasing the E-B junction allows electrons to be injected into the C-B depletion region, where they are easily swept into the collector region by the electric field present across the C-B junction. The net effect is that a large proportion of the electrons that are injected from the emitter into the base will cross the C-B depletion region and travel out the collector terminal.

In order for an electron that is injected from the emitter into the base to flow out the base terminal, it must recombine with a hole in the base. Because the base is lightly doped and very thin, the probability of recombination occurring is rather small (typically about 1 electron in 100 will recombine with a hole this way), so most of the injected emitter current flows out the collector terminal. This is why the base is made so thin and is doped rather lightly.

Based on the preceding discussion, it is apparent that the collector–base junction of the transistor does not behave like a normal diode. At least this is true when the transistor is in the active region of operation. As we will find out shortly, when a transistor is in the active region, the collector terminal of the transistor behaves as a dependent current source.

3.3 TRANSISTOR CURRENT RELATIONSHIPS

The currents associated with the base, emitter, and collector of the transistor are always related to one another by Kirchhoff's current law as follows:

$$I_E = I_C + I_B \tag{3.1}$$

This relationship is determined by direct inspection of Fig. 3.4. Because Kirchhoff's laws are so general, Eq. (3.1) is often written using the notation for instantaneous current (lowercase italic letters), however, since we are dealing strictly with dc conditions now, uppercase italic letters are used for emphasis.

Note
Beta is one of the most important transistor parameters.

There are also other relationships between these three currents that are dependent on transistor geometry, doping concentration, temperature, mode of operation, and other parameters. The dc beta, β_{dc}, of a transistor is defined as the ratio of dc collector current to dc base current. That is

$$\beta_{dc} = \frac{I_C}{I_B} \tag{3.2}$$

Another important relationship is the ratio of dc collector current to dc emitter current. This is called the dc alpha, α_{dc}. In equation form α_{dc} is defined as

$$\alpha_{dc} = \frac{I_C}{I_E} \tag{3.3}$$

Notice that both β_{dc} and α_{dc} are current ratios; they are dimensionless numbers.

It turns out that β_{dc} and α_{dc} are related to each other in a very specific manner as well. For example, if we know the value of β_{dc} for a given transistor, the value of α_{dc} may be determined as follows: We start with Eq. (3.2):

$$\beta_{dc} = \frac{I_C}{I_B}$$

Solving for I_C, we obtain

$$I_C = \beta_{dc} I_B$$

Now, substituting into Eq. (3.3) gives us

$$\alpha_{dc} = \frac{\beta_{dc} I_B}{\beta_{dc} I_B + I_B}$$

Factoring I_B produces

$$\alpha_{dc} = \frac{\beta_{dc} I_B}{I_B(\beta_{dc} + 1)}$$

Now we cancel the I_B factors, producing

$$\alpha_{dc} = \frac{\beta_{dc}}{\beta_{dc} + 1} \tag{3.4}$$

In a similar manner, it can be shown that β_{dc} may be determined from α_{dc} using

$$\beta_{dc} = \frac{\alpha_{dc}}{1 - \alpha_{dc}} \tag{3.5}$$

The derivation of Eq. (3.5) is left as an exercise for the student.

EXAMPLE 3.1

The transistor of Fig. 3.4(b) has $I_B = 20\ \mu\text{A}$ and $I_E = 1.02$ mA. Determine I_C, β_{dc}, and α_{dc}.

Solution In this example, it is easiest to find I_C using KCL and Eq. (3.1). Solving for I_C, we get

$$\begin{aligned} I_C &= I_E - I_B \\ &= 1.02\text{ mA} - 20\ \mu\text{A} \\ &= 1.00\text{ mA} \end{aligned}$$

We can now determine the β_{dc} of the transistor by direct application of Eq. (3.2):

$$\begin{aligned} \beta_{dc} &= \frac{I_C}{I_B} \\ &= \frac{1\text{ mA}}{20\ \mu\text{A}} \\ &= 50 \end{aligned}$$

Similarly, application of Eq. (3.3) gives us the α_{dc} value:

$$\alpha_{dc} = \frac{I_C}{I_E} = \frac{1 \text{ mA}}{1.02 \text{ mA}} = 0.98$$

As a check, we can also determine α_{dc} from β_{dc} using Eq. (3.5):

$$\alpha_{dc} = \frac{\beta_{dc}}{\beta_{dc} + 1} = \frac{50}{51} = 0.98$$

More on β_{dc}

In the previous example, the transistor had $\beta_{dc} = 50$. Depending on the type of transistor in use, this may be considered a fairly typical value, a rather low value, or even a relatively high value. For the typical low-power, general-purpose transistor (the 2N3904 and 2N2222 are good examples), a β_{dc} of 50 is not too unusual, although it is somewhat on the low side. For a power transistor that is designed to handle several amps (the 2N3055 is a good example of such a device), a typical β_{dc} value of 30 would not be unusual, so a β_{dc} of 50 is somewhat on the high side for this type of device. The point here is that expected β_{dc} values depend to a large extent on the particular type of transistor being used.

Note
β_{dc} may vary significantly from one transistor to another of the same type.

We stated earlier that β_{dc} (and α_{dc} as well) depends on several different factors. As it turns out, β_{dc} is very sensitive to manufacturing process variations that are themselves rather difficult to control. Thus, for a given type of transistor, the 2N3904, for example, β_{dc} variations from one device to the next may vary over a 3-to-1 range. For example, if two 2N3904s are selected at random, one may have $\beta_{dc} = 50$, while the other may very well have $\beta_{dc} = 200$ or higher. Tests of good, brand-new 2N3055 transistors performed by the author have revealed β_{dc} variations from as low as 20 to over 300. That's more than a 12-to-1 spread! Thus, it may be somewhat unfortunate, but we have little control over β_{dc}, and we must sometimes either hand-pick transistors for specific β_{dc} values (not too practical most of the time), or learn to live with wild unit-to-unit β_{dc} variations. Later we will find that living with β_{dc} variation is not as bad as it appears right now.

Aside from manufacturing variations, the beta of a transistor is also dependent on temperature and collector current. This is shown in Fig. 3.6. Typical values for general-purpose, low-power transistors are shown in black, while those values that are more representative for high-power transistors are shown in color.

The curves of Fig. 3.6 indicate that β_{dc} increases with temperature and, initially, with collector current. This behavior, along with the inherent negative temperature coefficient of the pn junction, can lead to a phenomenon known as *thermal runaway.*

Thermal Runaway

Assuming that a BJT is biased up as shown in Fig. 3.4, if we monitor the collector current, we will find that as the transistor warms up, the collector current will increase. Normally, the increases in current and temperature are small enough that equilibrium is quickly reached, and the BJT stabilizes. Such stability can be achieved using various circuit design techniques. Let us assume for now that such techniques are not used.

Recall that pn junctions have negative temperature coefficients. That is, as temperature increases, the effective barrier potential of a pn junction will decrease, and saturation current will increase, allowing an increase in current flow across the junction. In Fig. 3.4, this effect allows the V_{BE} source to increase the base current, and hence the collector current as well. At the same time, the β_{dc} of the transistor increases as well, further increasing I_C. This increase in I_C causes an increase in temperature, which in turn increases β_{dc}, which in turn increases temperature, I_B, I_C, and so on. Thus, a self-perpetuating cycle of increasing temperature and collector current is started. This action often continues until the transistor self-destructs (sometimes in a spectacular manner).

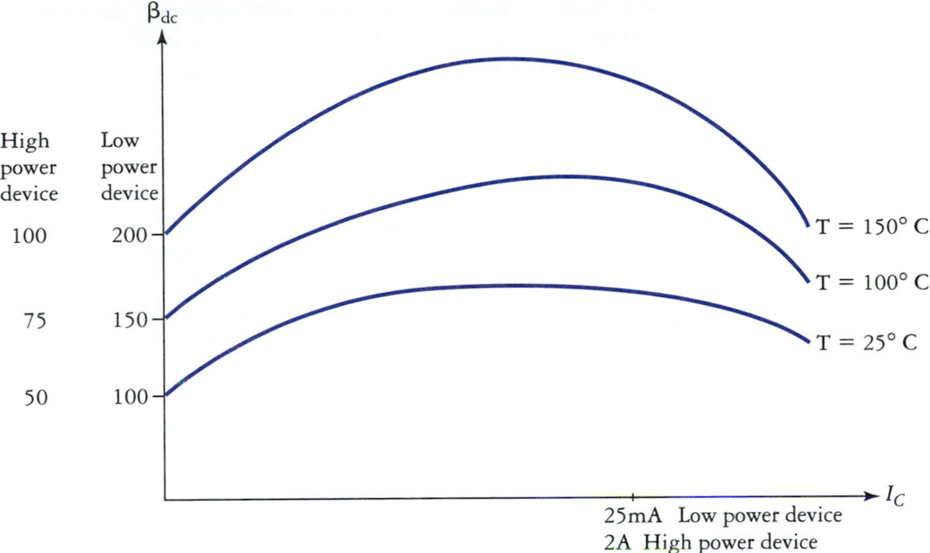

FIGURE 3.6
The β of a BJT varies with temperature and collector current.

More on α_{dc}

Unlike beta, alpha is a transistor parameter that is normally quite well behaved and predictable. This is fortunate, because it will allow us to make some very useful approximations that will help to simplify the task of transistor circuit analysis. The next example should help demonstrate this point.

EXAMPLE 3.2

The 2N2222 is an npn silicon general-purpose transistor for which device-to-device beta variations from 50 to 300 (a 6 : 1 spread) are not uncommon. Determine the corresponding α_{dc} values.

Solution We simply apply Eq. (3.4) for each beta value given, which yields

$$\begin{aligned}\alpha_{min} &= \frac{\beta_{min}}{\beta_{min} + 1}\\ &= \frac{50}{51}\\ &= 0.980\end{aligned}$$

and

$$\begin{aligned}\alpha_{max} &= \frac{\beta_{max}}{\beta_{max} + 1}\\ &= \frac{300}{301}\\ &= 0.997\end{aligned}$$

Note
Most of the time we will approximate alpha using Eq. 3.6.

The point to be made in Example 3.2 is that even though β_{dc} may vary tremendously from device to device, or even for a particular transistor with changing environmental and operating conditions, α_{dc} remains relatively constant, with an approximate value of unity (1). Since this is such a significant point to remember, it will be repeated: Even though beta may vary wildly, it is usually large enough that α_{dc} is practically equal to unity. Even a transistor with $\beta_{dc} = 20$ (generally considered a low β) has $\alpha_{dc} = 0.952$, which is still rather close to 1. Thus, we may often make the following approximation:

$$\alpha_{dc} \cong 1 \tag{3.6}$$

Of course, this approximation cannot be used all of the time, so if you are in doubt, use the actual α_{dc} for the transistor under study, if this information is available to you (in the real world, this may not be available very often).

Transconductance*

Most often in this text, we will treat the bipolar transistor as a current-controlled device. That is, we assume that either the base current controls the collector current ($i_C = \beta i_B$), or the emitter current controls the collector current ($i_C = \alpha i_E$). Occasionally, it is more advantageous to view the bipolar transistor as a voltage-controlled device.

Note

Transconductance is a very important parameter in the study of field-effect transistors, which are introduced in Chapter 6.

When the B-E junction of the transistor is forward biased (as in Fig. 3.4), the collector current is found to be related to the b-e bias voltage by Shockley's equation. That is,

$$I_C = \alpha I_{ES} e^{V_{BE}/\eta V_T} \tag{3.7}$$

where I_{ES} is the saturation current associated with the emitter–base junction. A typical plot of Eq. (3.7) for a low-power npn transistor is shown in Fig. 3.7. The instantaneous slope of the curve in this plot has the units of Siemens or mhos (amps/volt), and is called the *transconductance* g_m of the transistor. The defining equation for transconductance is

$$g_m = \left.\frac{\Delta I_C}{\Delta V_{BE}}\right|_{V_{CE}=\text{const}} \tag{3.8}$$

This equation would be used to determine the value of g_m experimentally.

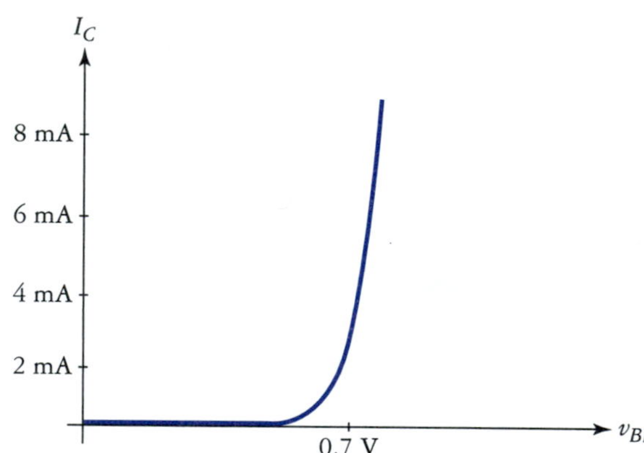

FIGURE 3.7
Transconductance curve for a typical npn transistor.

We can determine the transconductance of a transistor based on circuit-dependent conditions by differentiating i_C with respect to v_{BE} in Eq. (3.7). Assuming nominal room-temperature operation, with the emission coefficient $\eta = 1$, this results in

$$g_m = \frac{I_C}{\alpha \times 26\text{ mV}} \tag{3.9a}$$

Note

Recall that

$V_T = \dfrac{kT}{q} \cong 26\text{ mV}$

at nominal room temperature, T = 25°C.

As stated before, alpha is normally very close to unity, so as a general rule we will use the following approximation:

$$g_m \cong \frac{I_C}{26\text{ mV}} \tag{3.9b}$$

Although transconductance will not be used very extensively in the analysis of bipolar transistors in this text, it does provide an alternative point of view of transistor circuit behavior. Also, transconductance is one of the main device parameters used in the study of field-effect transistors.

*This section can be skipped without loss of continuity.

3.4 REGIONS OF OPERATION

A transistor can be biased into four distinct modes, or regions, of operation. A given region of operation is determined by the polarities of the bias voltages V_{BE} and V_{CB}, as summarized in Table 3.1.

Note
The active region of operation is most important in the study of amplifier circuits.

TABLE 3.1

Region of Operation	C-B Bias	E-B Bias
Active	Reverse	Forward
Saturation	Forward	Forward
Cutoff	Reverse	None/Rev.
Inverse Active*	Forward	Reverse

*Rarely used.

Again, using npn transistors for illustrative purposes, the various modes of operation are shown from a circuit-oriented point of view in Fig. 3.8. In each case, the transistor is assumed to be a silicon device. Because the inverse-active mode of operation is rarely used, discussion of this mode is deferred until later.

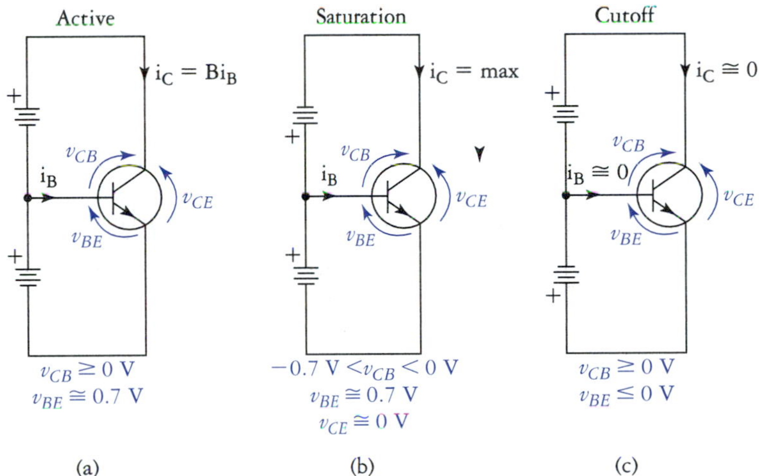

FIGURE 3.8
The three regions of operation: (a) active, (b) saturation, and (c) cutoff.

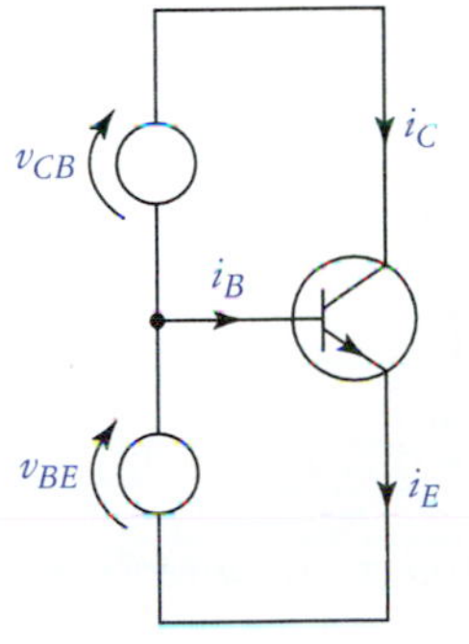

FIGURE 3.9
Common-base biasing configuration.

If we replace the dc sources in Fig. 3.8 with more generalized voltage sources, we obtain the circuit of Fig. 3.9, in which the transistor can assume any possible operating region of operation. Because both voltage sources are connected at the base of the transistor, this particular biasing arrangement is called the *common-base* configuration. More will be said about this later.

Active Region

The *active region* of operation has been the main focus of the discussion up to this point. Active-region operation is central to the realization of amplifiers and many other types of circuits. A great deal of useful information about the transistor in the circuit of Fig. 3.9 can be obtained if we adjust v_{BE} such that i_E is held constant and we plot i_C as a function of v_{CB}. Performing this procedure will generate a set of curves like those shown in Fig. 3.10. These curves are called the *common-base characteristic curves* for the transistor under test. The active region encompasses most of the first (upper-right) quadrant. The area below the lowest i_C curve (the curve for which $i_E = 0$) is the cutoff region. The portions of the curves that lie to the left of the vertical axis comprise the saturation region.

Notice that when the transistor is in the active region, the collector current curves are nearly horizontal. The plot of output current versus terminal voltage for an ideal constant current source would be a horizontal line. Thus the collector terminal of the transistor approximates the behavior of

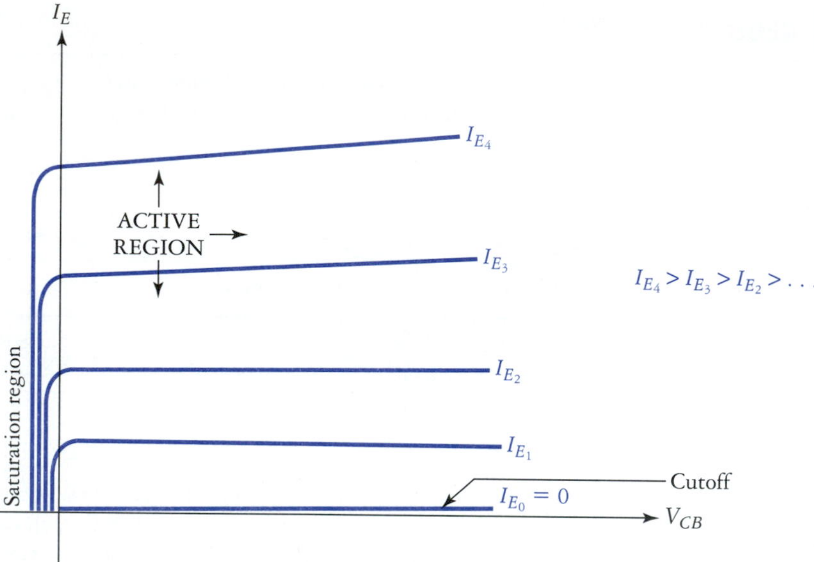

FIGURE 3.10
Common-base family of emitter current curves.

a dependent current source. That is, the collector current can be thought of as being dependent on the emitter current, with the α of the transistor being the proportionality constant relating the two. Of course, it is also possible to relate the collector current to the base–emitter voltage via transconductance. In both cases, however, the base–emitter junction can be represented as a diode, resulting in the transistor models shown in Fig. 3.11. The current-controlled current source model is shown in Fig. 3.11(a); the voltage-controlled current source (transconductance) model is shown in Fig. 3.11(b).

Note
We will use the current-controlled current source model in our study of BJTs.

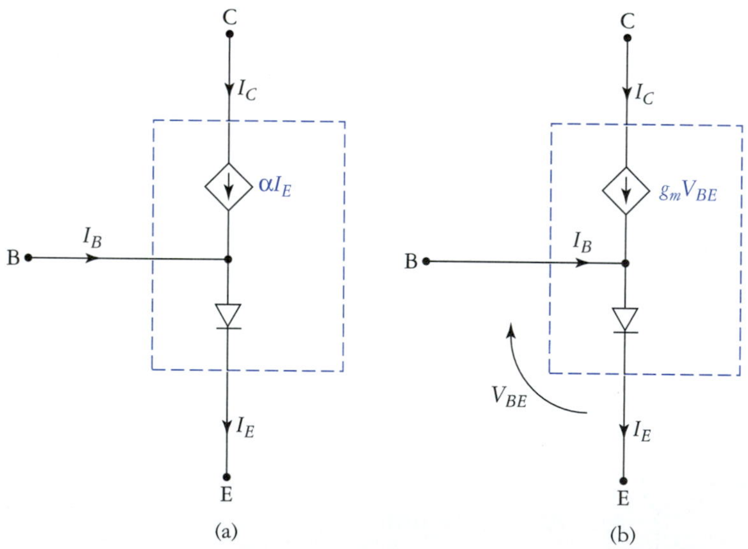

FIGURE 3.11
Linear dependent source models for the common-base-biased BJT in the active region. (a) Current-controlled current source. (b) Voltage-controlled current source.

Cutoff Region

Note
In cutoff, the transistor behaves like an open-circuit with $I_C \cong 0$ and $V_{CE} = \text{max}$.

When a transistor is in cutoff, the collector current is for most practical purposes reduced to zero. This is accomplished by reducing the control quantity (i_E, i_B, or v_{BE}) to zero. In Fig. 3.9, if we reduce v_{BE} to zero, then i_B and i_E are both reduced to zero as well. Collector current can be thought of as being dependent on any one of these three quantities, and so ideally, i_C should be zero regardless of the value of v_{CB}. Thus, in cutoff, the ideal transistor behaves like an open switch when viewed from the collector to the emitter. This is illustrated in Fig. 3.12.

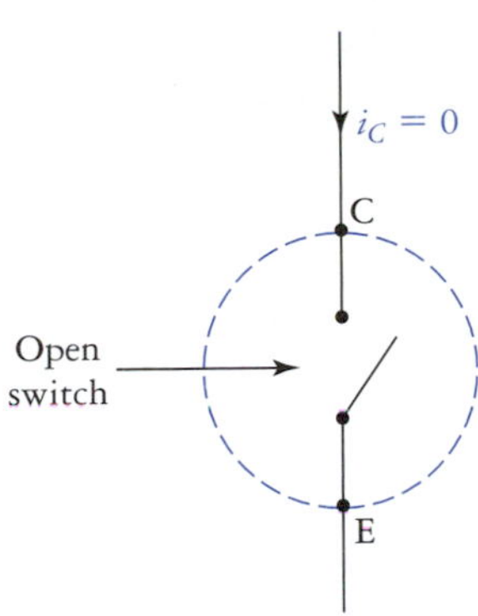

FIGURE 3.12
Behavioral model of the ideal transistor in the cutoff region.

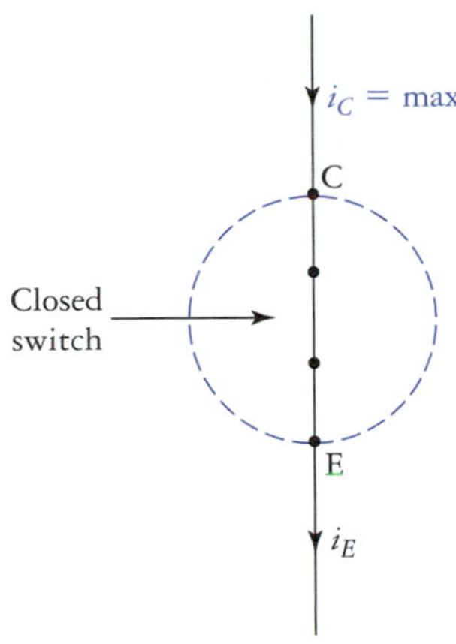

FIGURE 3.13
Model of the ideal transistor in saturation.

Saturation Region

Note
In saturation, the transistor acts like a short circuit with $V_{CE} \cong 0$ and $I_C = \text{max}$.

In the *saturation region,* both the b-e and c-b junctions are forward biased. The barrier potentials of the two pn junctions tend to cancel each other, causing the collector to behave as if it was shorted to the emitter. In this region, very small increases in v_{CB} result in large increases in i_C. A simple model of the saturated transistor is simply that of a dead-short between collector and emitter, as shown in Fig. 3.13.

3.5 BASIC TRANSISTOR BIASING CONFIGURATIONS

The term *biasing* refers to the application of voltages (and sometimes currents) to a transistor in order to achieve some particular operating conditions. There are actually quite a few different ways to bias up a transistor. However, at the fundamental level, any biasing scheme for a single transistor can be considered a common-base, common-emitter, or common-collector topology.

Common-Base Biasing

The discussions of transistor operation up to this point have been presented primarily in terms of the *common-base* (CB) configuration. The CB configuration is very useful for gaining a basic understanding of transistor operation, but it is not the most commonly occurring biasing arrangement in practice. Because CB biasing has been covered sufficiently already, we move onward without further hesitation.

Common-Emitter Biasing

Note
Common-emitter biasing is the most commonly used biasing arrangement.

Common-emitter (CE) biasing is illustrated in its simplest form in Fig. 3.14. The reason for the use of the term *common-emitter* should be apparent, because both external sources each share a node in common with the emitter of the transistor. Notice that a variable independent current source has been used to provide base current to the transistor. This was done here because when we use β as our proportionality constant, the control quantity is base current (I_B controls I_C). Keep in mind that we could have also used the transconductance approach, in which case V_{BE} controls I_C. In such a case, then, it is more appropriate to apply a voltage to forward bias the B-E junction. Notice also that the external voltage source is connected directly in parallel with the collector and emitter terminals of the transistor. This causes $V_{CE} = V_{CC}$.The external source was given a different identity than V_{CE} because in practical circuits $V_{CE} \neq V_{CC}$, except under special conditions. V_{CC} is the standard name applied to this biasing voltage in most cases.

If we were to set I_B at fixed values and then continuously increase V_{CE} (by increasing the supply voltage V_{CC}) and plot I_C as a function of I_B, a set of collector characteristic curves like those shown in Fig. 3.14(b) would be produced. The exact same set of curves would be produced if we used g_m as the control parameter, except the various I_B values would be replaced by equivalent V_{BE} values. As

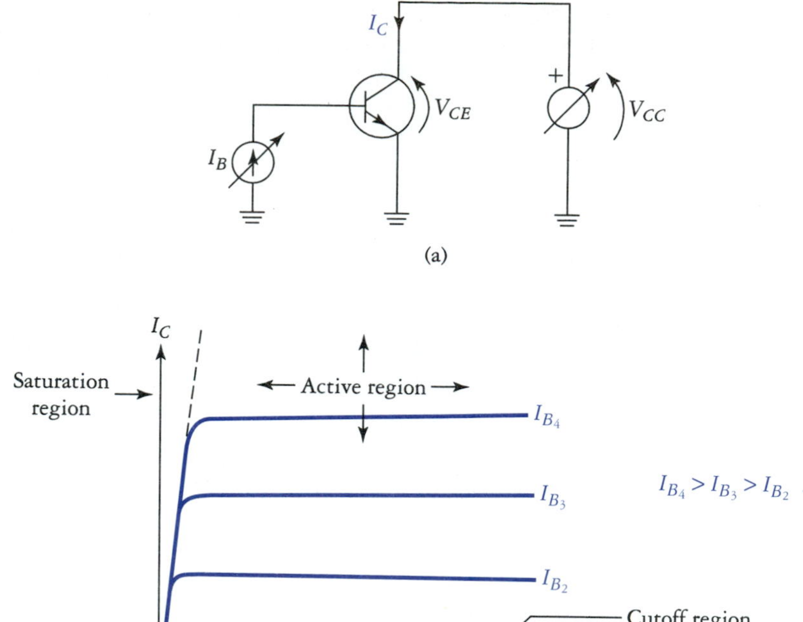

FIGURE 3.14
Common-emitter biasing. (a) The basic circuit. (b) Family of collector characteristic curves.

was the case with the CB configuration, there are three distinct regions of operation: saturation, active, and cutoff.

Active Region

In the active region, the collector terminal of the transistor acts like a dependent current source, with the control variable being either the base current or the base–emitter voltage drop, depending on the analysis approach chosen. Recall that β relates I_C and I_B via Eq. (3.2). You should also recall that typically occurring β values will result in $I_C \gg I_B$. This serves to emphasize a very important point, specifically: In the common-emitter configuration, a small base current controls a large collector current. This is the magic of the transistor in a nutshell. That's what makes solid-state amplifiers and logic gates and oscillators and radios and thousands of other devices possible.

Note
We will use the current-controlled current-source model in our study of BJTs.

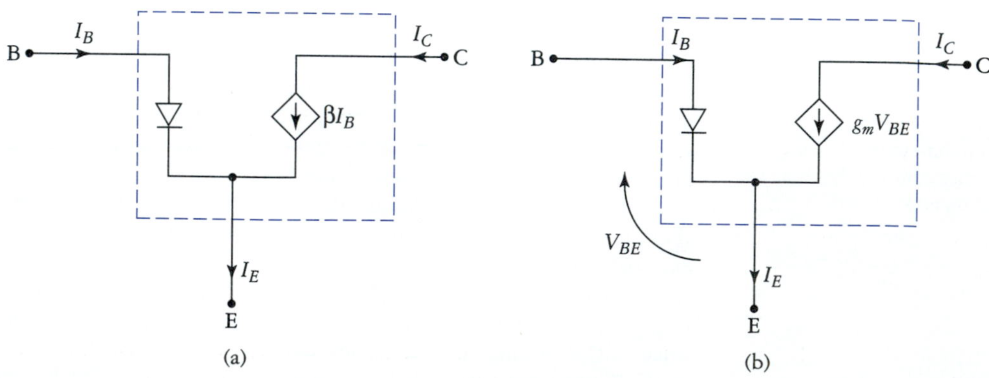

FIGURE 3.15
A useful active-region model of the npn transistor when common-emitter biasing is used. (a) ICIS-type model. (b) Transconductance type.

Models for the nearly ideal npn transistor in the common-emitter bias configuration are shown in Fig. 3.15. Figure 3.15(a) is the ICIS-type model, while Fig. 3.15(b) shows the transconductance model. Again, we normally use the ICIS-type model for the BJT throughout this text.

Saturation Region

Now, back to the collector curves of Fig. 3.14(b). Notice that the saturation region begins to the right of the vertical axis and forms a line of constant slope from which the relatively flat collector curves emanate. The value of V_{CE} that corresponds to the point of intersection of a given active-region I_C curve and this saturation line is called the *collector-to-emitter saturation voltage* $V_{CE(\text{sat})}$. As seen in the graph of Fig. 3.14(b), $V_{CE(\text{sat})}$ increases with collector current. However, for the typical low-power transistor, operating with a collector current level on the order of tens of milliamps or less, $V_{CE(\text{sat})}$ is normally only on the order of a few tenths of a volt. As an approximate rule of thumb,

$$V_{CE(\text{sat})} = 0.2 \text{ V} \tag{3.10}$$

To make life even simpler, it is often possible to neglect this voltage drop altogether. For the time being, however, we assume that $V_{CE(\text{sat})}$ is significant.

Cutoff Region

In the CE configuration, *cutoff* is achieved by reducing I_B to zero (or $V_{BE} = 0$ V for the transconductance approach). Ideally, such conditions would reduce I_C to zero, making the transistor again act like an open circuit from collector to emitter. Of course, some leakage current will always exist, which makes the transistor behave somewhat less than ideally. The effects of such leakage currents are discussed later.

Common-Collector Biasing

The third basic biasing arrangement is *common-collector* (CC) biasing. This is also sometimes called emitter-follower (EF) biasing, for reasons discussed later when amplifier circuits are analyzed. For all practical purposes, CC biasing is essentially the same as CE biasing. A plot of I_E curves for various increments of I_B, as a function of V_{CE}, forms the set of emitter curves for the CC configuration. This is shown in Fig. 3.16. Basically, the only significant difference between the CC curves and the CE curves is that we are now looking at I_E as a function of I_B and V_{CE}, instead of I_C. Remember that α_{dc} is normally almost equal to 1. As a consequence, I_E and I_C are also approximately equal to each other ($I_C = \alpha_{dc} I_E$), and the CC curves are nearly indistinguishable from those of the CE configuration.

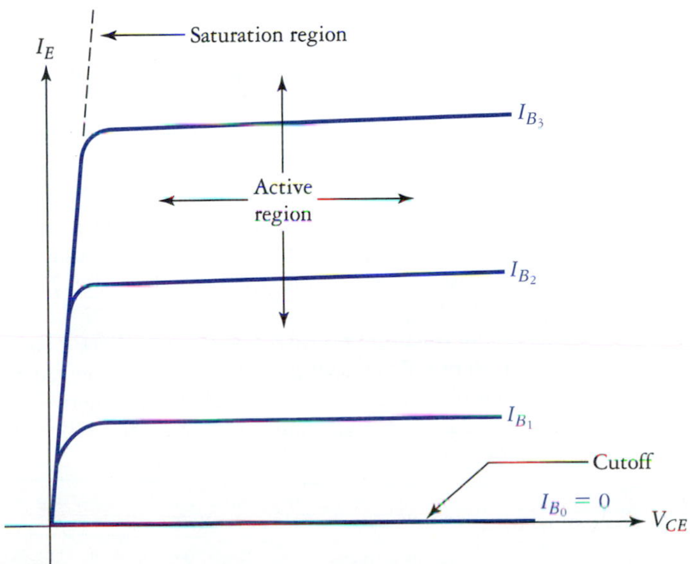

FIGURE 3.16
Family of emitter current curves for the common-collector biasing arrangement.

3.6 TRANSISTOR LEAKAGE CURRENTS

Up to this point, the effects of the various leakage and saturation currents associated with the transistor have been largely ignored. It turns out that in many situations, these leakage currents can indeed be disregarded. But, sometimes they cannot be ignored, and so we must have an understanding of their causes and effects in order to cope with them effectively.

Collector Leakage Current

Collector leakage current is specified primarily in two different manners as either *collector-to-base current with emitter open* (I_{CBO}) or *collector-to-emitter current with base open* (I_{CEO}). Let us begin with I_{CBO}.

I_{CBO}

Consider the circuit in Fig. 3.17(a), in which voltage source V_{CC} is reverse biasing the c-b junction and the emitter terminal is open circuited. Under these conditions, the c-b junction behaves like a typical reverse-biased diode as shown in Fig. 3.17(b). Thus we see that I_{CBO} is equivalent to the saturation current of the equivalent C-B diode and, as such, is strongly temperature dependent.

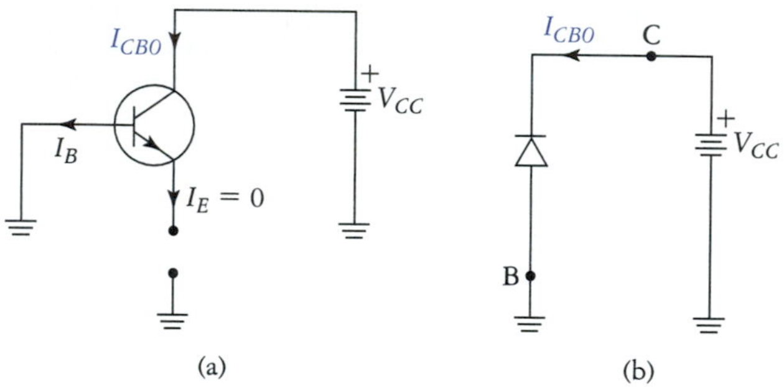

FIGURE 3.17
Collector-to-base leakage current. (a) Test circuit. (b) Diode model.

I_{CBO} is really most applicable to the common-base biasing situation, where it represents the collector leakage current in the cutoff region. This is the value of I_C for the $I_E = 0$ curve (the lowest curve) in Fig. 3.10. From an analytical view, the effect of I_{CBO} is to create a (usually) slight error in the common-base ICIS relationship of the model in Fig. 3.11(a). Thus,

> Note
> Recall that leakage currents are very temperature-dependent, approximately doubling for every 10°C rise in temperature.

$$I_C = \alpha_{dc} I_E + I_{CBO} \tag{3.11}$$

Fortunately, for a typical general-purpose silicon transistor, I_{CBO} will be on the order of a few nanoamps or less. This is usually small enough to ignore. For example, the 2N2222 has a maximum I_{CBO} of 10 nA at 25°C. The typical unit will more than likely have an I_{CBO} of around 1 nA.

Power transistors, having to carry much higher currents than general-purpose transistors, have correspondingly greater junction areas and likewise greater I_{CBO}. For example, the 2N4877 is an npn medium-power transistor, designed to carry a continuous collector current of 4 A. The maximum I_{CBO} for the 2N4877 is 100 μA at 25°C. We must keep things in perspective, however, and consider that most often a power transistor like the 2N4877 will be used in an application in which $\alpha_{dc} I_E \gg I_{CBO}$.

Germanium transistors will have collector leakage currents that are many times higher than for equivalent silicon devices, and so these leakage currents may be of much greater significance. Again, good fortune is on our side though, because germanium transistors are encountered very rarely today. In any case, the total collector current of a CB biased germanium transistor will also be given by Eq. (3.11).

I_{CEO}

If we apply V_{CC}, ground the emitter terminal, and open the base, as shown in Fig. 3.18(a), the resulting collector current is called I_{CEO}. The diode-equivalent circuit of Fig. 3.18(b) shows that the C-B junction is reverse biased. This is equivalent to the common-emitter biasing arrangement of

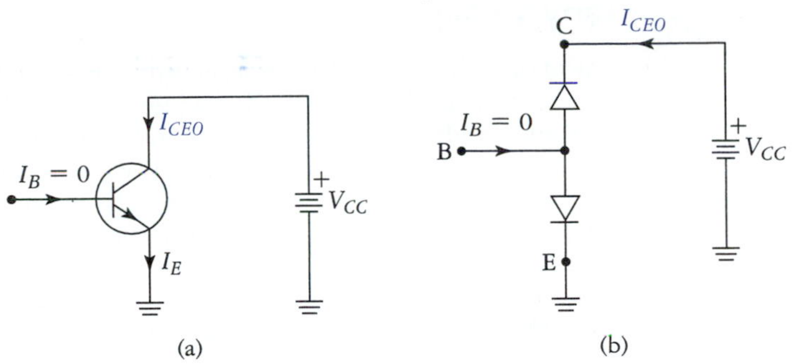

FIGURE 3.18
Collector-to-emitter leakage current. (a) Test circuit. (b) Diode model.

Fig. 3.14, with $I_B = 0$. Now because $I_B = 0$, and C-B junction leakage current that is created must flow through the B-E junction and out the emitter terminal, and none can flow out of the base.

At first glance it appears that $I_{CEO} = I_{CBO}$, after all, it's the saturation current associated with the reverse-biased C-B junction that causes collector leakage current to begin with. This is partly true, however, it's *where* that leakage current goes that is important here. Here's what happens in Fig. 3.18: The collector leakage current crosses the base into the emitter. As far as the transistor is concerned, this leakage current is *equivalent* to a small base current $I_B = I_{CBO}$ being injected by an external source. Now, this effective base current is increased in the collector by a factor of beta, as is expected in the common-emitter configuration; thus,

$$I_{CEO} = \beta_{dc} I_{CBO} \tag{3.12}$$

Recall that β_{dc} may be quite large, and so for a given transistor, leakage current is potentially much more of a problem in a CE biasing arrangement than for common-base operation.

For a given CE-biased transistor, the total collector current is given by

Note
Most often, leakage currents I_{CBO} and I_{CEO} are negligible.

$$I_C = \beta_{dc} I_B + I_{CEO} \tag{3.13}$$

or equivalently

$$I_C = \beta_{dc}(I_B + I_{CBO}) \tag{3.14}$$

The leakage current of the typical modern silicon transistor is normally small enough that even the effects of I_{CEO} may be ignored at relatively low junction temperatures. In some applications, however, high temperatures are unavoidable. In these cases it may be necessary to account for leakage currents in order to prevent maximum device current and/or power handling limits from being exceeded. The problem of thermal runaway is also aggravated by the presence of leakage currents, because I_{CBO} and I_{CEO} both have positive temperature coefficients.

EXAMPLE 3.3

The 2N2222 has the following published characteristics at 25°C:

$$\beta_{dc(typ)} = 100, \quad \beta_{dc(max)} = 300, \quad I_{CBO(max)} = 0.01\ \mu A$$

Determine the worst-case value of I_{CEO} at $T = 25°C$.

Solution We simply multiply the maximum I_{CBO} value by the maximum beta, giving us

$$\begin{aligned} I_{CEO(max)} &= (300)(0.01\ \mu A) \\ &= 3\ \mu A \end{aligned}$$

3.7 ADDITIONAL TRANSISTOR PARAMETERS

In the active region of operation, the typical silicon bipolar transistor can be modeled with good accuracy as a current-controlled current source, as in Figs. 3.11(a) and 3.15(a), or as a voltage-controlled current source as in Figs. 3.11(b) and 3.15(b). These models are very simple, however, and do not account for many characteristics of the transistor that are less than ideal. In this section we examine a few more characteristics of the transistor that have been taken for granted thus far.

Early Voltage

The ideal current source (either dependent or independent) has infinite internal resistance. Of course, there is no way to physically realize an ideal source of any kind, but the transistor sometimes comes close enough. In the active region, the collector terminal of the transistor is modeled as a dependent current source because on the scales over which operation normally occurs, the collector curves are relatively horizontal. The emphasis here is on the term *relative.* Let's use the common-emitter collector curves for illustrative purposes. If we project the more horizontal portions of this set of curves backward, it is found that these projections all intersect the V_{CE} axis at one point called the *Early voltage.* This is illustrated in Fig. 3.19. The Early voltage is designated V_A, and is named after J. M. Early, the first investigator of this phenomenon.

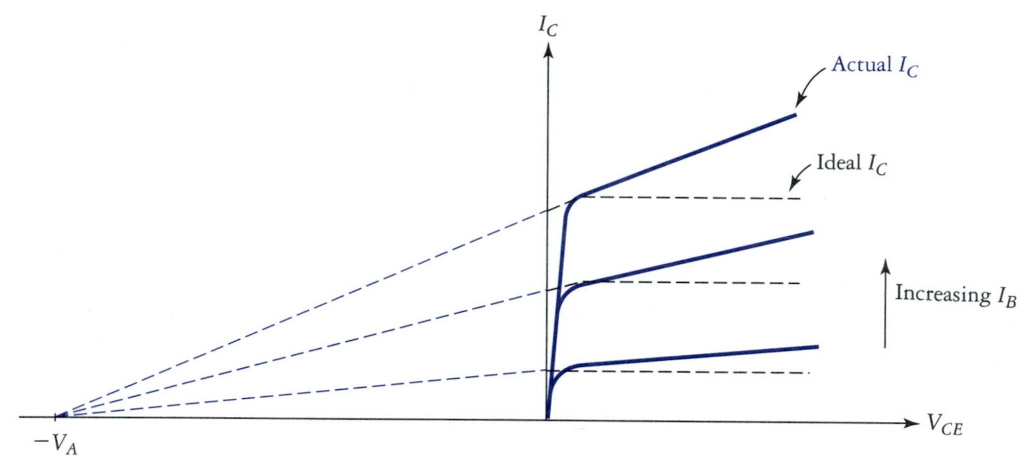

FIGURE 3.19
The Early effect causes I_C (and therefore β) to increase with V_{CB}.

The Early voltage is a constant for a given transistor, normally expressed as a magnitude (positive number). Typical values of V_A range from 50 to 300 V. Because V_A is a constant it is easily shown that as the active-region base current increases, effective collector resistance decreases. In Fig. 3.19, this is obvious because the I_C curves tend to become more vertical as we go up from cutoff. If we know the value of the Early voltage for a given transistor, then we can determine the actual collector current using the following equation:

$$I'_C = I_C\left(\frac{V_{CE}}{V_A} + 1\right) \tag{3.15}$$

where I_C is the ideal expected value of collector current.

Knowledge of the value of the Early voltage also allows us to determine the effective internal resistance or conductance of the collector current source. The conductance (usually referred to as g_o) for a fixed base current in the active region is the slope of the I_C curve. By definition, this is

Note
It is standard practice to use lower-case italic letters to represent transistor parameters that are a function of I_C or V_{CE}. These are dynamic parameters.

$$g_o = \left.\frac{\Delta I_C}{\Delta V_{CE}}\right|_{I_B = \text{const}} \tag{3.16}$$

The reciprocal of the slope is the collector current source internal resistance, r_o, which by definition in equation form is

$$r_o = \left.\frac{\Delta V_{CE}}{\Delta I_C}\right|_{I_B = \text{const}} \tag{3.17}$$

We can express these two equations in terms of the Early voltage as follows:

$$g_o = \left.\frac{I_C}{V_A}\right|_{V_{CE} = 0\text{ V}} \tag{3.18}$$

$$r_o = \left.\frac{V_A}{I_C}\right|_{V_{CE} = 0\text{ V}} \tag{3.19}$$

The addition of r_o to the ICIS transistor models is shown in Fig. 3.20.

Note
Sometimes we use r_{CE} in place of r_O.

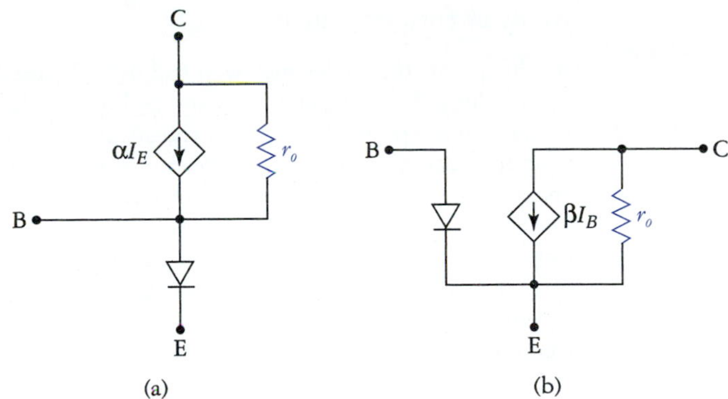

FIGURE 3.20
Transistor active-region bias models that account for finite collector current source resistance. (a) CB model. (b) CE model.

EXAMPLE 3.4

The circuit in Fig. 3.21 has the following parameters: $I_B = 50\ \mu A$ and $V_{CC} = 15$ V. The transistor used is a 2N2222, which has typical values of $\beta_{dc} = 100$ and $V_A = 200$ V. Determine the actual collector current and the effective collector internal resistance.

FIGURE 3.21
Circuit for Example 3.4.

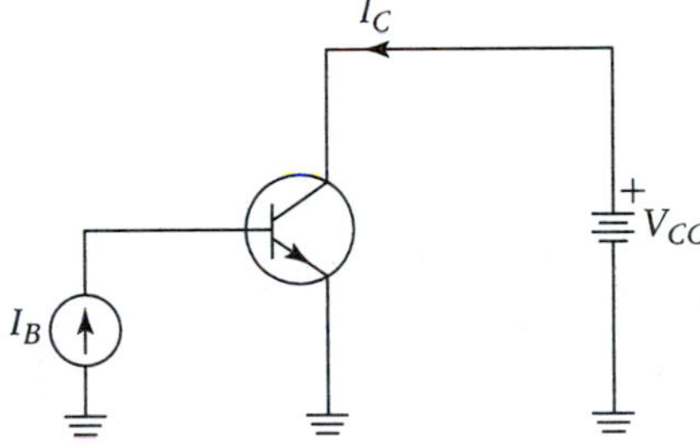

Solution The ideal collector current is

$$\begin{aligned} I_C &= \beta_{dc} I_B \\ &= (100)(50\ \mu A) \\ &= 5\ mA \end{aligned}$$

Substituting into Eq. (3.15), we get

$$\begin{aligned} I'_C &= I_C\left(\frac{V_{CE}}{V_A} + 1\right) \\ &= 5\ mA\left(\frac{15}{200} + 1\right) \\ &= 5.375\ mA \end{aligned}$$

Thus the collector current is about 7.5% higher than expected.

The effective collector current source internal resistance is found via direct application of Eq. (3.19):

$$\begin{aligned} r_o &= \left.\frac{V_A}{I_C}\right|_{V_{CE}=0\ V} \\ &= \frac{200\ V}{5\ mA} \\ &= 40\ k\Omega \end{aligned}$$

Words of Encouragement

At this point, the reader may well feel overwhelmed with transistor characteristics, parameters, and terminology. There is good news and bad news. The bad news is that around 40 or so different parameters can be specified for a given transistor. But, there are several bits of good news. First, we do not have to concern ourselves with most of these parameters; we need only discuss a few more in this chapter. Secondly, most often, when we are analyzing or even designing a given transistor circuit, the effects of the Early voltage and other higher order effects are usually ignored. This normally will cause less than a 5% error in a given situation.

In all honesty, if we were to account for even a small fraction of all of the transistor parameters, the analysis of even the simplest of circuits would become nearly intractable. This is the sort of extremely detailed analysis that computers do best. We humans are more adept at viewing circuits and systems from a holistic perspective.

Base-Spreading Effects

The physical mechanism that causes the Early effect is related to the relatively large width of the c-b junction depletion region compared to the physical width of the base. Refer to the silicon npn transistor structure of Fig. 3.22. This is the common-emitter bias configuration (the topology for which CE characteristic curves are generated). To get a good grasp on the characteristics of this transistor, we would plot I_C versus V_{CE} for fixed values of base current to produce a set of collector curves. By inspection, we see that the b-e junction is forward biased, and for a silicon transistor, such a junction will exhibit a barrier potential of about 0.7 V. Thus, given a fixed base current, as V_{CC} is increased, V_{CB} will increase, while V_{BE} will remain nearly constant. At lower values of V_{CB}, the c-b depletion region extends slightly into the base region. Recall that a small proportion of the electrons that are injected from the emitter will recombine with holes in the base, forming the base current. The ratio of holes that cross into the collector to those that recombine in the base is β_{dc}.

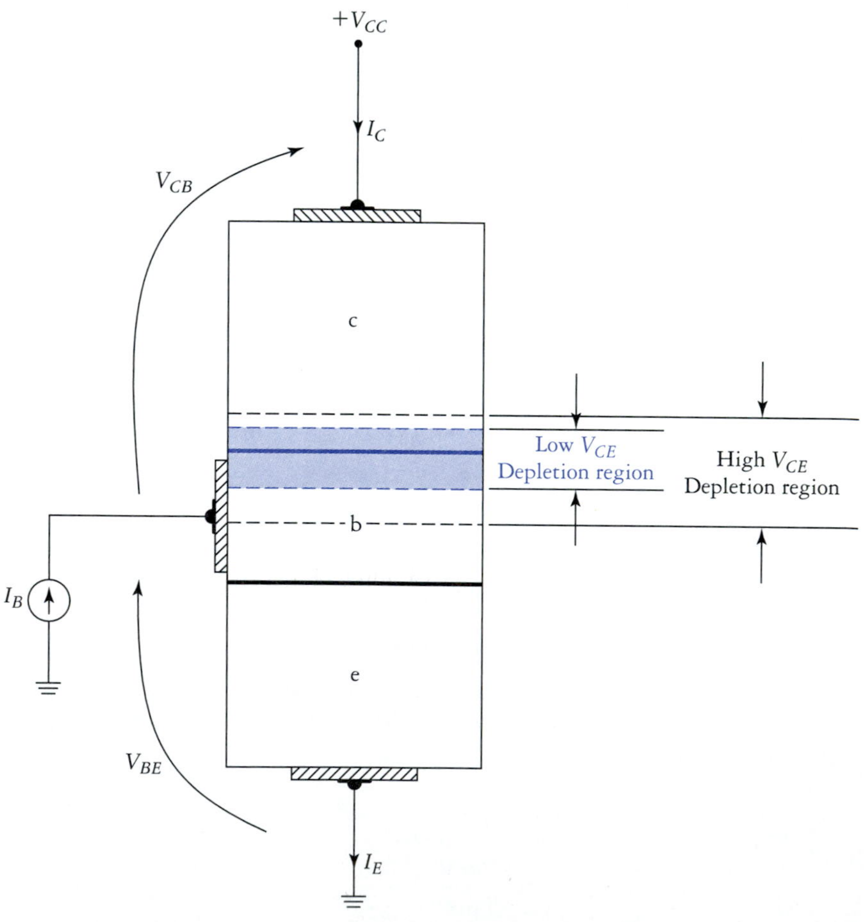

FIGURE 3.22
Base-spreading effect.

If we further increase V_{CB}, while keeping I_B constant, the depth of the c-b depletion region in the base makes the base region appear significantly thinner than it really is. Now the injected electrons from the emitter have a shorter distance to travel to reach the depletion region, and there are fewer holes with which to recombine along the way. The net effect is an increase in β_{dc} with V_{CB}. This effect is called *base spreading* and it accounts for the nonzero slope of the collector curves in the active region.

Base spreading also increases the effective ohmic resistance of the base region. This is called *base-spreading resistance.* Notice in Fig. 3.22 that at higher levels of V_{CB} the external base connection has a smaller area of contact with the part of the base that still has majority carriers. Thus as V_{CB} increases, the base resistance increases as well. Base-spreading resistance r_{bs} is accounted for in the ICIS models of the transistor as shown n Fig. 3.23. Typically r_{bs} will range from around 5 to 20 Ω, and we disregard its effect whenever possible. Making the base physically thicker reduces base-spreading resistance, but at the expense of lower beta. Transistors with very thin base regions usually have relatively high betas, at the expense of greater base-spreading resistance and reduced breakdown voltage.

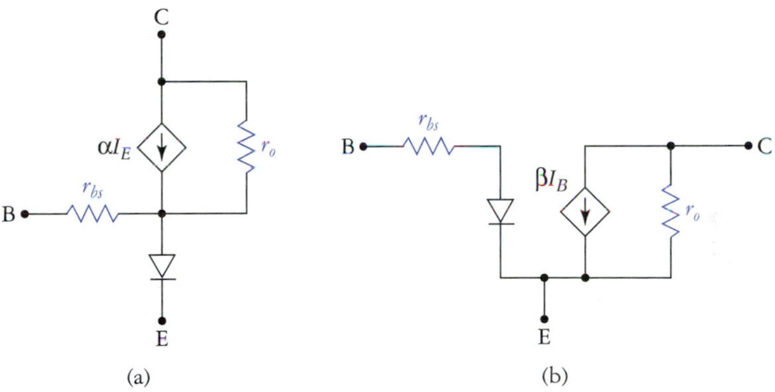

FIGURE 3.23
Addition of base-spreading resistance to active-region models for (a) BC and (b) CE configurations.

A Word on Notation

The collector current source internal resistance and the base-spreading resistance are denoted using lowercase italic letters because their values are dependent on transistor current and voltage conditions. That is, these parameters are nonlinear in the most general sense, and the values that we assign are assumed to be relatively constant in any given circuit. Such dynamic (nonlinear) parameters will normally be represented in this manner, as was the case with the forward resistance of the pn junction, r_F, that was covered in Chapter 1.

Breakdown Voltage

A given transistor is capable of operating over a limited range of voltage and current values. Because a transistor can be configured in many different ways, there are also many different ways in which breakdown voltage limitations can be specified.

BV_{CEO}

One of the more useful breakdown voltage specifications is the *collector-to-emitter breakdown voltage with the base open circuited,* which is designated BV_{CEO}. This specification applies to the common-emitter (and for all practical purposes to the common-collector) biasing configuration. A typical plot of collector curves illustrating BV_{CEO} is shown in Fig. 3.24. The breakdown voltage will be the greatest for the lowest curve ($I_B = 0$) in the plot, which defines BV_{CEO}. This transistor parameter is usually one that is included in even the most incomplete of specification sheets, because it defines the maximum voltage that the transistor can withstand. Notice that as I_B increases, the breakdown voltage decreases. As we will learn in a later chapter, this does not usually present a problem in practical applications.

FIGURE 3.24
Collector-to-emitter breakdown voltage.

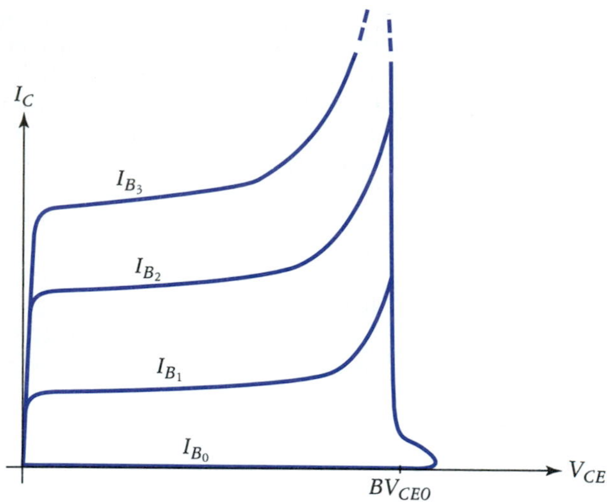

Two major mechanisms account for transistor breakdown: *avalanche* and *reach-through.* In avalanche breakdown, the electrostatic field across the transistor crystal becomes strong enough to pull electrons out of valence in the c-b depletion region, which in turn results in impact ionization, which increases the avalanche effect, and so on in a chain reaction manner.

Reach-through, or *punch-through* as it is sometimes called, occurs whenever the reverse bias applied to the C-B junction causes the depletion region to extend through the base to the emitter. If this occurs, the transistor acts almost like a short circuit from collector to emitter. If a transistor is made with a very thin base region, say, in order to gain high beta, breakdown voltage is sacrificed, and reach-through becomes a problem much sooner. This is why high-voltage transistors usually have rather low betas.

Breakdown of the transistor is usually a destructive occurrence, but in special circumstances it can be caused intentionally without harm to the device. Typical values of BV_{CEO} range from 20 to over 1000 V for some high-voltage devices. As an example, the 2N2222 has a minimum guaranteed $BV_{CEO} = 30$ V.

BV_{CBO}

If the emitter terminal of a transistor is open circuited ($I_E = 0$), and we increase V_{CB}, the transistor will break down at BV_{CBO} volts. This mode of breakdown is important when common-base biasing is used. A typical set of CB collector curves showing BV_{CBO} is given in Fig. 3.25. Increasing the collector current has virtually no effect on breakdown voltage in this case, unlike the situation for the CE configuration curves.

BV_{CBO} is normally much higher than BV_{CEO} for a given transistor. Again using the 2N2222 as an example, the minimum guaranteed value of BV_{CBO} is 60 V. This is twice as high as the minimum BV_{CEO}.

FIGURE 3.25
Collector-to-base breakdown voltage.

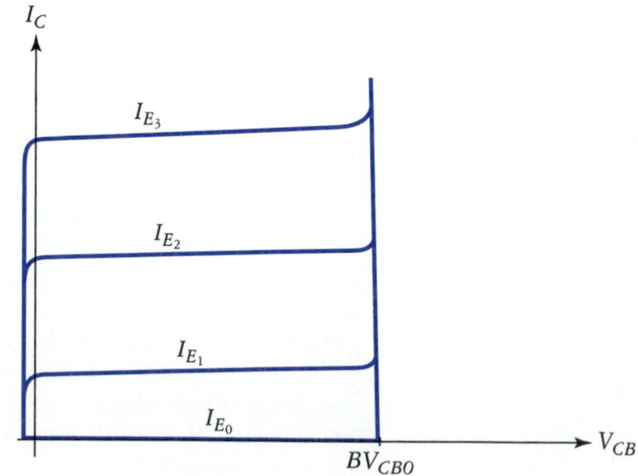

BV_{CER} and BV_{CEX}

We now cover two additional breakdown voltages briefly. If the base of the transistor is connected to the emitter through a resistor and V_{CE} is increased until breakdown occurs, then we have BV_{CER}. This is illustrated in Fig. 3.26. As opposed to breakdown with the base open circuited (BV_{CEO}), there will usually be a very large decrease in V_{CE} once BV_{CER} is exceeded. The value to which V_{CE} drops is sometimes called the *collector-to-emitter breakdown sustaining voltage* $BV_{CE(sus)}$.

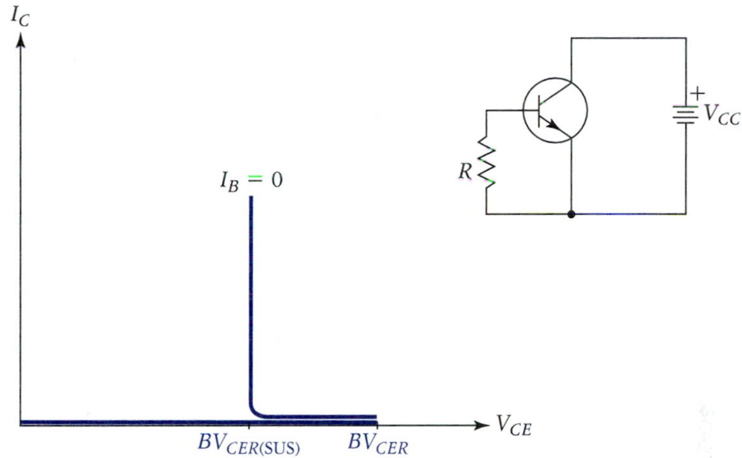

FIGURE 3.26
Other breakdown voltage specifications.

If some special base conditions are to be maintained when collector-to-emitter breakdown voltage is to be determined, then that breakdown voltage is called BV_{CEX}. The subscript X means that base conditions are specified elsewhere in the data sheet.

3.8 SIGNAL-RELATED PARAMETERS

Thus far we have only looked at transistor characteristics from a dc standpoint. The ultimate goal of many transistor circuit designs is to process time-varying signals. These are the applications for which transistor circuit analysis really starts to get interesting as well.

AC Beta and Alpha

As might be suspected, many transistor parameters are frequency sensitive to some degree. Of those examined thus far, beta, alpha, r_o (and its reciprocal g_o), r_{bs} and g_m may vary significantly with frequency. At this point, we concern ourselves only with beta and alpha.

The ac beta is defined as the ratio of the change in collector current to an incremental change in base current. Translated into equation form this is

$$\beta_{ac} = \left.\frac{dI_C}{dI_B}\right|_{V_{CE}=\text{const}} \tag{3.20}$$

If the changes are kept small enough that the transistor behaves linearly, then we can approximate Eq. (3.20) as

$$\beta_{ac} \cong \left.\frac{\Delta I_C}{\Delta I_B}\right|_{V_{CE}=\text{const}} \tag{3.21}$$

EXAMPLE 3.5

A certain npn transistor was placed in a curve tracer and the common-emitter characteristics of Fig. 3.27 resulted. The transistor is intended to be used with a collector current of approximately 1 mA, with V_{CE} approximately 6 V. Determine the values of β_{dc} and β_{ac}.

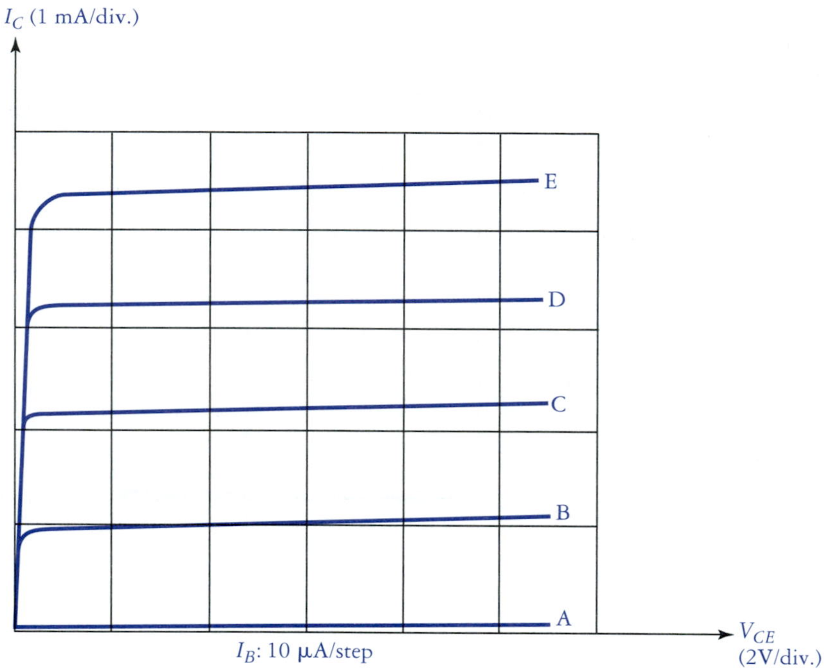

FIGURE 3.27
Curves for Example 3.5.

Solution To determine the dc beta, we use the collector curve that is closest to the expected operational value of I_C, with V_{CE} close to its expected operational value as well. In this case, the curve marked B is used ($I_B = 10\ \mu A$) with $V_{CE} = 6$ V. Thus,

$$\beta_{dc} = \left.\frac{I_C}{I_B}\right|_{V_{CE}=6\text{ V}} = \frac{1\text{ mA}}{10\ \mu\text{A}} = 100$$

The ac beta will be determined using the collector curves above and below curve B (curves A and C), at $V_{CE} = 6$ V. Curves A and C are used because if collector current is alternately increased and decreased, the alternations will apparently be centered about curve B. It is purely coincidental that curve A corresponds to cutoff in this example. First, let us determine the change in I_C from curve A to curve C:

$$\Delta I_C = I_{C(C)} - I_{C(A)} = 2.3\text{ mA} - 0.1\text{ mA} = 2.2\text{ mA}$$

The base current steps are given as 10-μA increments, thus,

$$\Delta I_B = I_{B(C)} - I_{B(A)} = 20\ \mu\text{A} - 0\ \mu\text{A} = 20\ \mu\text{A}$$

Now, the ac beta is determined:

$$\beta_{ac} = \left.\frac{\Delta I_C}{\Delta I_B}\right|_{V_{CE}=6\text{ V}}$$
$$= \frac{2.2\text{ mA}}{20\ \mu\text{A}}$$
$$= 110$$

We see from the preceding example that the ac and dc betas for a given transistor may differ from each other. Fortunately, most of the time the difference is not significant, and so we can write

$$\beta_{ac} \cong \beta_{dc} \tag{3.22}$$

Whenever we are assuming that the equality of Eq. (3.22) holds, or if a small difference in ac and dc beta values is not significant, then beta will be written with no subscript (β).

Note
Unless otherwise noted, we will assume that $\beta_{ac} = \beta_{dc} = \beta$, and $\alpha_{ac} = \alpha_{dc} = \alpha$.

Some texts will subscript beta with the letter *F*, as in β_F. This is done to differentiate forward beta (the beta we've been dealing with all along) from reverse beta (β_R). Recall that there is a region of operation called the inverse-active region. This mode of operation occurs when the collector and emitter terminals are reversed. The CE current gain of the transistor under these conditions is the reverse beta β_R. Reverse beta is normally very low ($\beta_R \cong 1$), so the inverse-active region is rarely used intentionally.

Because beta has separate ac and dc values, it follows that alpha comes in these two flavors as well. Recall that rather drastic changes in beta tend to produce relatively slight, if not completely insignificant, changes in alpha. This characteristic holds true for ac beta and alpha as well. Because of this correlation, differences between α_{dc} and α_{ac} will almost certainly be negligible, and we can write

$$\alpha_{ac} \cong \alpha_{dc} \tag{3.23}$$

Unless otherwise stated, from this point on, we assume that Eq. (3.23) holds, and we use the simpler notation α.

EXAMPLE 3.6

Determine α_{dc} and α_{ac} for the transistor of Example 3.5. Assume the same conditions prevail as before.

Solution

$$\alpha_{dc} = \frac{\beta_{dc}}{\beta_{dc} + 1}$$
$$= \frac{100}{101}$$
$$= 0.990$$

$$\alpha_{dc} = \frac{\beta_{ac}}{\beta_{ac} + 1}$$
$$= \frac{110}{111}$$
$$= 0.991$$

Dynamic Emitter Resistance

When used as a linear amplifier, the transistor is normally required to operate in the active region. If, under normal conditions, the transistor will remain in the active region, the E-B junction will remain forward biased at all times. It proves useful under such circumstances to model the base–emitter junction not as a diode, but as a dynamic resistance. This resistance will be denoted either as r_e or r_π, depending on the exact nature of the transistor model in use.

The *dynamic emitter resistance* is defined as

$$r_e = \frac{dV_{BE}}{dI_E} \tag{3.24a}$$

If the change in emitter current is kept relatively small, then we can define the dynamic emitter resistance by

$$r_e \cong \frac{\Delta V_{BE}}{\Delta I_E} \tag{3.24b}$$

Equation (3.24b) is useful for experimental determination of r_e; however, we can also determine the value analytically, starting with the following form of Shockley's equation:

$$I_E \cong I_{ES}\, e^{V_{BE}/\eta V_T} \tag{3.25}$$

Now, differentiating V_{BE} with respect to I_E as was indicated in Eq. (3.24a), we obtain

$$r_e = \frac{\eta V_T}{I_E} \tag{3.26a}$$

Note
Dynamic emitter resistance is a very important parameter in the analysis of amplifiers.

At nominal room temperature (25°C), $V_T = 26$ mV, and in most applications, $\eta = 1$ and $\alpha \cong 1$ (which means $I_C \cong I_E$) so we can write

$$r_e \cong \left.\frac{26\text{ mV}}{I_C}\right|_{T=25^\circ\text{C}} \tag{3.26b}$$

If you covered the optional discussion on transconductance (Section 3.3), you might notice that there is a close relationship between r_e and g_m. For convenience, the equation for g_m derived in Section 3.3 is rewritten here:

$$g_m \cong \frac{I_C}{V_T} = \frac{I_C}{26\text{ mV}} \tag{3.9b}$$

Now, because $I_C \cong I_E$ (recall that $I_C = \alpha I_E$, and $\alpha \cong 1$), to a very close approximation we have

$$g_m \cong \frac{1}{r_e} \tag{3.27}$$

The active-region model that incorporates r_E is shown in Fig. 3.28. Recall that this model is normally associated with the common-base biasing configuration.

When the common-emitter biasing configuration is used, the active-region model of Fig. 3.29 is usually most appropriate. The emitter–base dynamic resistance in this case is given by

$$r_\pi = (\beta + 1)\, r_e \tag{3.28a}$$

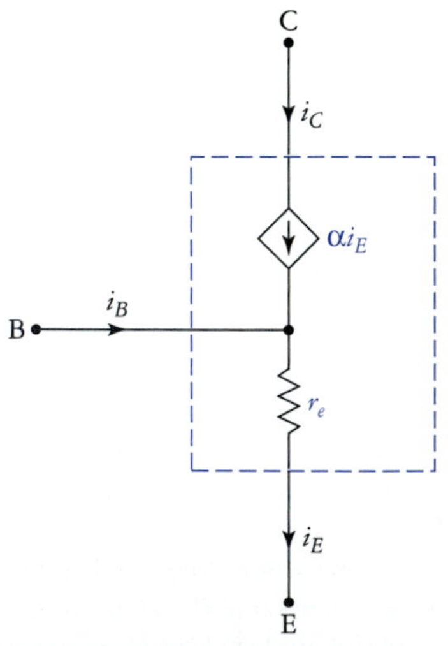

FIGURE 3.28
Small-signal, active-region model for the CB configuration.

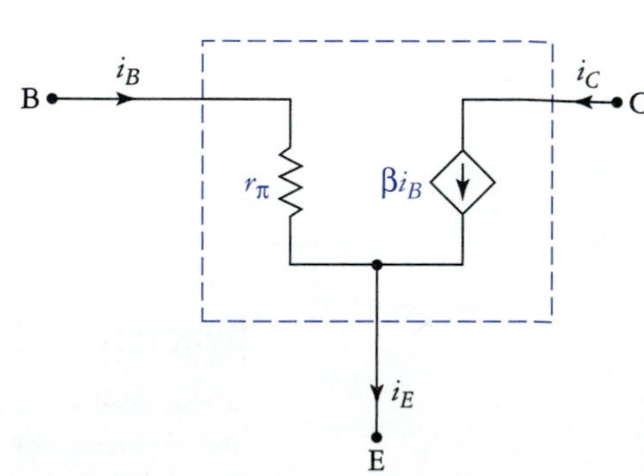

FIGURE 3.29
Small-signal active-region model for the CE configuration.

We know that β is fairly unpredictable; however, one thing that we can be reasonably sure of is that β is usually relatively large. This allows us to simplify Eq. (3.28a) slightly, giving us

$$r_\pi \cong \beta r_e \tag{3.28b}$$

Equations (3.28a) and (3.28b) are given here without further justification. Proof of this relationship is given in Chapter 5.

Before leaving this section, a final comment on notation is in order. Recall that dynamic resistances and conductances are identified with lowercase italics (r_o, r_{bs}, r_e, g_m, and g_o). Dynamic parameters are normally *internal* characteristics of an active device such as a transistor (they are also usually nonlinear). There are exceptions, however, and sometimes it is convenient to denote quantities that are not really *dynamic* with lowercase characters. To prevent any future confusion, whenever a given dynamic parameter is *internal* to a transistor, that parameter will be identified by the use of lowercase italic characters in its subscript (r_e, for example).

3.9 HYBRID PARAMETERS

The transistor parameters we have used so far are usually called *r* parameters. An alternative set of device parameters called *hybrid parameters* (or *h* parameters) is sometimes used as well. Hybrid parameters are derived from a two-port black box model of the transistor. We won't get into the details of the derivation of the hybrid parameters in this text. The relationships between some of the *r* parameters and related *h* parameters is summarized in the equations below. Notice that the dc-related *h* parameters use uppercase subscripts, while the ac-related *h* parameter subscripts are lowercase. The subscript *f* refers to a forward parameter, while *r* is a reverse parameter. Hybrid parameters are dependent on the biasing (or amplifier) configuration used. That is, there are different *h* parameters for common-emitter, common-collector, and common-base configurations. We use *r* parameters throughout this book because they are easier to interpret and apply than *h* parameters:

$$\beta_{dc} = h_{FE}$$

$$\beta_{ac} = h_{fe}$$

$$r_\pi = h_{IE}$$

$$r_e = h_{IE}/(h_{FE} + 1)$$

$$r_o = 1/h_{OE}$$

There are other families of parameters that could be used as well. These include *impedance* (*z* parameters), *conductance* (*g* parameters), *admittance* (*y* parameters), and *scattering* (*s* parameters). Most of these parameters are used in specialized applications. For example, scattering parameters are used to characterize high-frequency (typically VHF, UHF, and microwave) devices and networks.

3.10 TRANSISTOR PACKAGES AND CLASSIFICATION SCHEMES

Thus far, the transistor has been examined from a rather abstract and theoretical point of view. In this section the various physical packages in which transistors are available, the classifications into which transistors are typically divided, and the characteristics of those types of transistors are introduced.

Package Types

There are probably hundreds of different transistor package styles in use today. Of course, some styles have gained wide acceptance and are more commonly encountered than others. Figure 3.30 illustrates several different transistor packages.

Most manufacturers follow standard lead designation schemes (often called the device *pinout*). For example, the TO-3 power transistor package terminals are nearly always identified as shown in Fig. 3.31. Most other packages also have standard pinouts as well. There are, however, always exceptions to the rule, and wiring a transistor into a circuit incorrectly can lead to (sometimes spectacular) device failure and other circuit damage. Beacuse of this, it is best to consult the data sheet for a given transistor in order to positively identify the terminals of a given transistor.

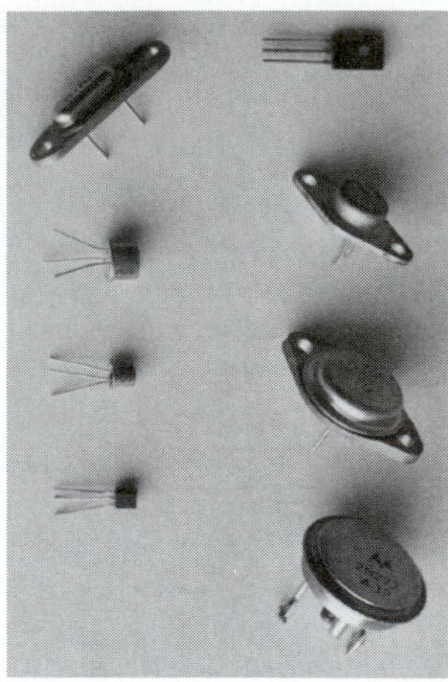

FIGURE 3.30
Assorted transistor packages.

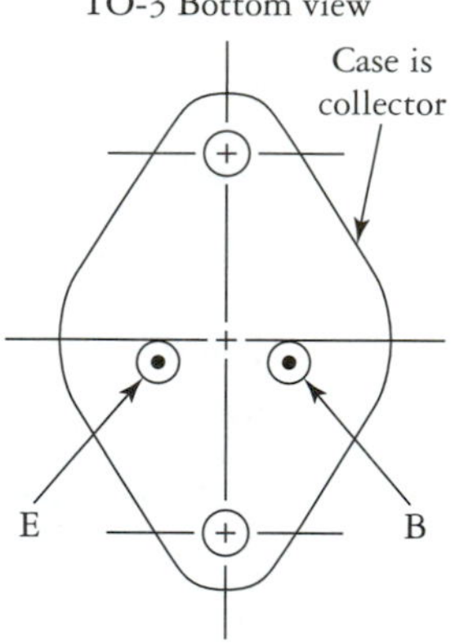

FIGURE 3.31
TO-3 power transistor package, bottom view.

General-Purpose Transistors

General-purpose transistors are normally designed to operate in low-power amplifier and switching applications. Low power, however, is a relative term. If a transistor is designed such that its maximum power dissipation rating is less than 0.5 W or so, then that transistor is generally considered to be a low-power device. A few examples of low-power general-purpose devices are the 2N3904, 2N3906, the 2N2222 and the 2N2222A.

The 2N3904 is an npn silicon device, while the 2N3906 is a pnp silicon device with similar physical characteristics (silicon chip dimensions and geometric features). The 2N3904 and 2N3906 are complementary devices; that is, they are similar in (nearly) all respects, except that the 2N3904 is an npn device, while the 2N3906 is pnp. Both devices are rated for a maximum power dissipation of 350 mW. The 2N2222 and 2N2222A are both npn silicon devices that are very similar in many respects, but they are different enough to warrant distinct designations. Table 3.2 lists a few of the basic device parameters for the 2N2222, 2N2222A, 2N3904, 2N3906, and 2N4877 transistors. Keep in mind that Table 3.2 is not a comprehensive list of device specifications, but rather is intended to provide a general comparison of a few typical transistors.

TABLE 3.2
Basic Device Parameters for Selected Transistors

	2N2222	2N2222A	2N3904	2N3906	2N4877
Device Type	NPN Silicon G.P./Switch	NPN Silicon G.P./Switch	NPN Silicon G.P.	PNP Silicon G.P.	NPN Silicon Power
$I_{C(\max)}$	800 mA	800 mA	200 mA	200 mA	4 A
BV_{CEO}	30 V	40 V	40 V	40 V	60 V
BV_{CBO}	60 V	75 V	60 V	40 V	70 V
$P_{D(\max)}$	800 mW	400 mW	350 mW	350 mW	10 W
$I_{CBO(\max)}$ ($T = 25°C$)	10 nA	10 nA	—	—	100 μA
$V_{CE(\text{sat})}$	0.4 V ($I_C = 150$ mA)	0.3 V ($I_C = 150$ mA)	0.3 V ($I_C = 50$ mA)	0.4 V ($I_C = 50$ mA)	1.0 V ($I_C = 4$ A)
f_T	250 MHz	300 MHz	300 MHz	250 MHz	4 MHz
$\beta_{dc(\min)}$	50 ($I_C = 1$ mA, $V_{CE} = 10$ V)	50 ($I_C = 1$ mA, $V_{CE} = 10$ V)	70 ($I_C = 1$ mA, $V_{CE} = 1$ V)	80 ($I_C = 1$ mA, $V_{CE} = 1$ V)	30 ($I_C = 1$ A, $V_{CE} = 2$ V)

Switching Transistors

In a switching application, the transistor is generally driven either into saturation or cutoff, with very little time spent in the active region. Thus, the transistor is used like a switch; it is either on (conducting, in saturation) or it is off (open circuited, in cutoff). Switching transistors are designed to have relatively fast turn-on and turn-off time characteristics (often less than 1 ns). Often, a plain, general-purpose transistor can be used effectively in a switching application. Likewise, switching transistors are often used in linear applications as well. Thus, the differences between general-purpose and switching transistors (of similar power ratings) are often very slight. The 2N2222 is classified as a switching transistor, in addition to being designed for use as a linear amplifier. As often as not, the design of the circuit in which a given transistor is used will determine whether or not that device performs adequately.

Power Transistors

Power transistors are designed to dissipate substantially more power than general-purpose devices. Most often, the increase in power dissipation is due to an increase in device collector current-handling capacity. To continuously carry large currents, power transistors necessarily have large junction areas. The actual silicon (or possibly germanium) chip that forms a given transistor is called a *die* (the plural is *dice*). Typical cross-sectional dimensions for a power transistor die are around 1/4 in.^{2} (0.6 cm). The dimensions for the typical low-power transistor die are about 1/50 in.^{2} (0.05 cm), or roughly the same cross-sectional area as a grain of table salt. Figure 3.32 illustrates the relative sizes of typical power and general-purpose transistor dice. Both chips are drawn about three times larger than actual size for clarity.

Note
The base and emitter contacts of the power transistor chip in Fig. 3.32 are said to be "interdigitated."

FIGURE 3.32
Comparison of relative sizes of typical power transistor and low-power transistor chips.

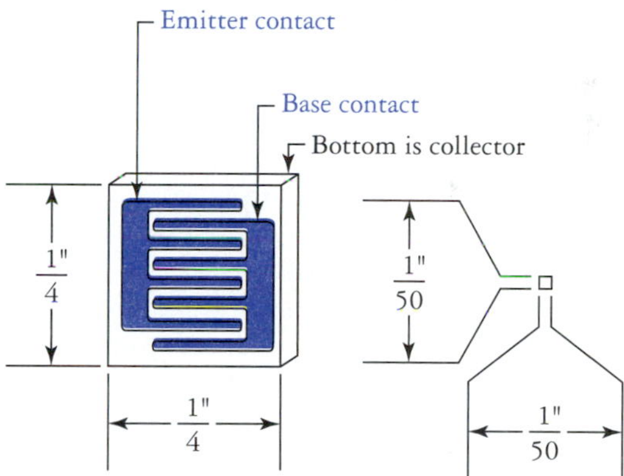

The relatively large junction area associated with power transistor geometry has some effect on other device parameters. For example, power transistors usually have poor high-frequency characteristics when compared to low-power devices. This is primarily caused by the larger capacitances associated with the power transistors' greater junction area. This capacitance can be reduced by making the base region physically thicker, hence increasing the useful frequency range of the transistor. A thicker base region also tends to increase breakdown voltage characteristics as well, which is a positive side effect. On the other hand, however, increased base thickness also tends to cause a decrease in beta, which is usually undesirable. Trade-offs between these various characteristics must be made based on the application in which a given device is to be used.

3.11 COMPUTER-AIDED ANALYSIS APPLICATIONS

Transistors can be modeled many different ways. We like to use the simplest behavioral model that we can get away with when analyzing a circuit visually or with pencil and paper. When performing computer-aided analysis such as with PSpice, the most desirable approach is to use the transistor models available in the device library. The PSpice evaluation version device library Eval.lib contains models for the following devices, as classified according to the *National Semiconductor Discrete Devices Handbook,* 1989 (National Semiconductor Corp, Santa Clara, CA):

2N2222	Silicon, npn, general-purpose amplifier/switch
2N2907A	Silicon, pnp, general-purpose amplifier/switch
2N3904	Silicon, npn, general-purpose amplifier/switch
2N3906	Silicon, pnp, general-purpose amplifier/switch

Additional data for some of these devices is presented in Appendix 2.

Generating Collector Characteristic Curves: Nested Sweep

We begin by generating a family of collector characteristic curves (I_C versus V_{CE}) for the 2N3904 transistor, as modeled in the Eval.lib of PSpice. The circuit used to generate the curves is shown in Fig. 3.33. A dc current source is used to drive the b-e junction, while a dc voltage source provides collector biasing.

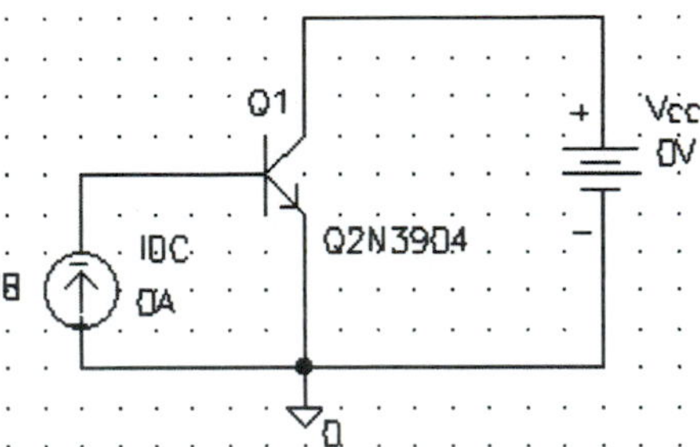

FIGURE 3.33
Circuit for generating collector characteristic curves.

Sweeping the value of a single parameter such as a voltage, current, or resistance, for example, will allow us to generate a single characteristic curve. We can generate a *family* of characteristic curves using what is called a *nested sweep.* A nested sweep works as follows: The main sweep variable is held constant while the nested sweep variable is swept across its range, producing a set of data points. The nested variable is then reset, and the main variable is then incremented. The nested variable is once again swept through its range. This process repeats until the main sweep variable has incremented to its final value.

In this example, the voltage produced by the *Vcc* source is the main sweep variable (*x*-axis variable), and IB is the nested variable. Creating the circuit and clicking on Analysis, Setup, DC Sweep, gives us the main sweep attribute box. This is set up as shown in Fig. 3.34. *Vcc* is swept from 0 to 5 V in 0.01-V steps.

Clicking on the Nested Sweep button produces the nested sweep attribute selection box. These parameters are set up as shown in Fig. 3.35. Here we are sweeping the base current from 0 to 100 μA in 10-μA increments.

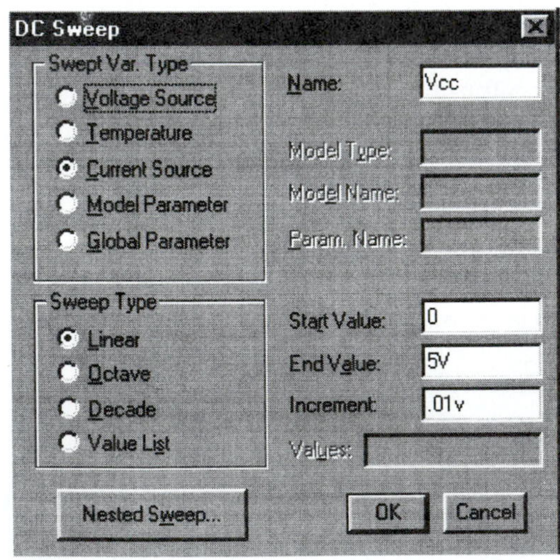

FIGURE 3.34
Setup of dc sweep attributes.

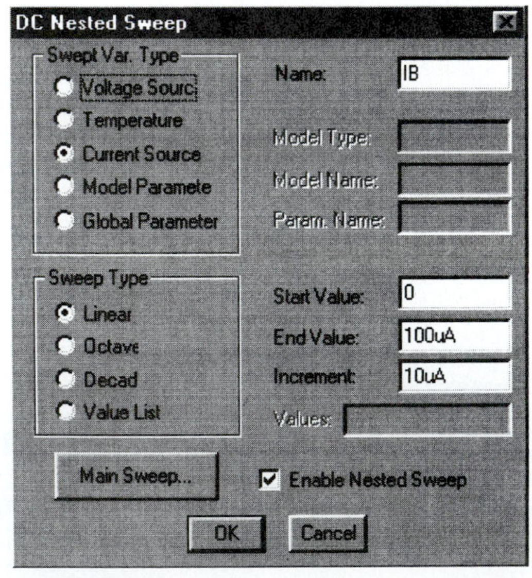

FIGURE 3.35
Nested sweep attributes.

Running the simulation and plotting the collector current IC(Q1) produces the display shown in Fig. 3.36. Close examination of the curves reveals that when $V_{CE} = 3$ V, if $I_B = 60$ μA then $I_C = 10$ mA. Thus $\beta_{dc} = 167$ at this operating point.

If we change the limits of the Vcc sweep, we can look more closely at the saturation characteristics of the transistor. For example, setting the upper limit of the sweep to 0.5 V (and decreasing the

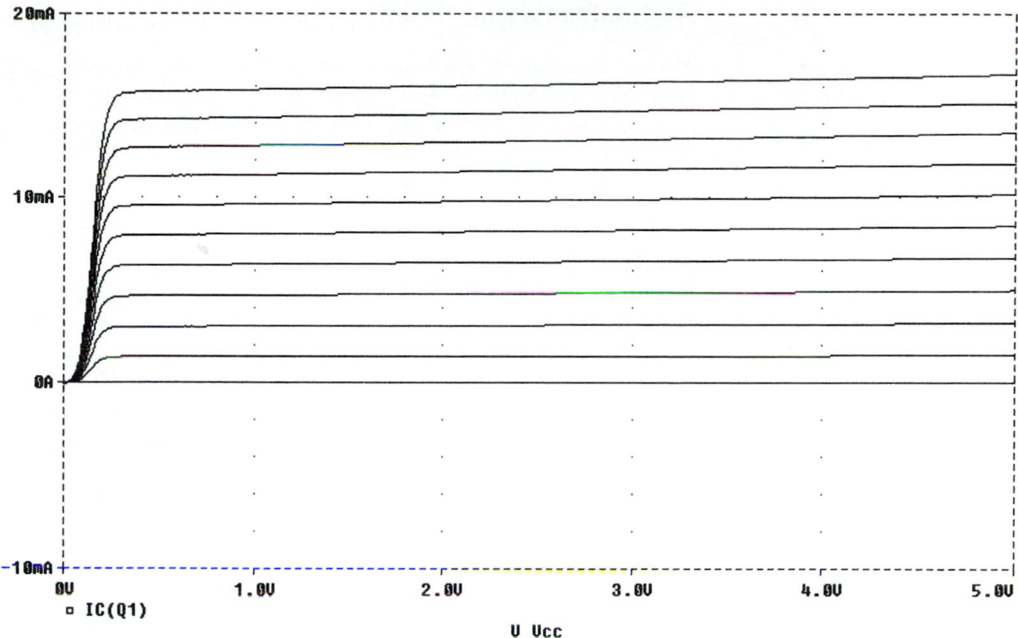

FIGURE 3.36
Resulting plot of collector curves.

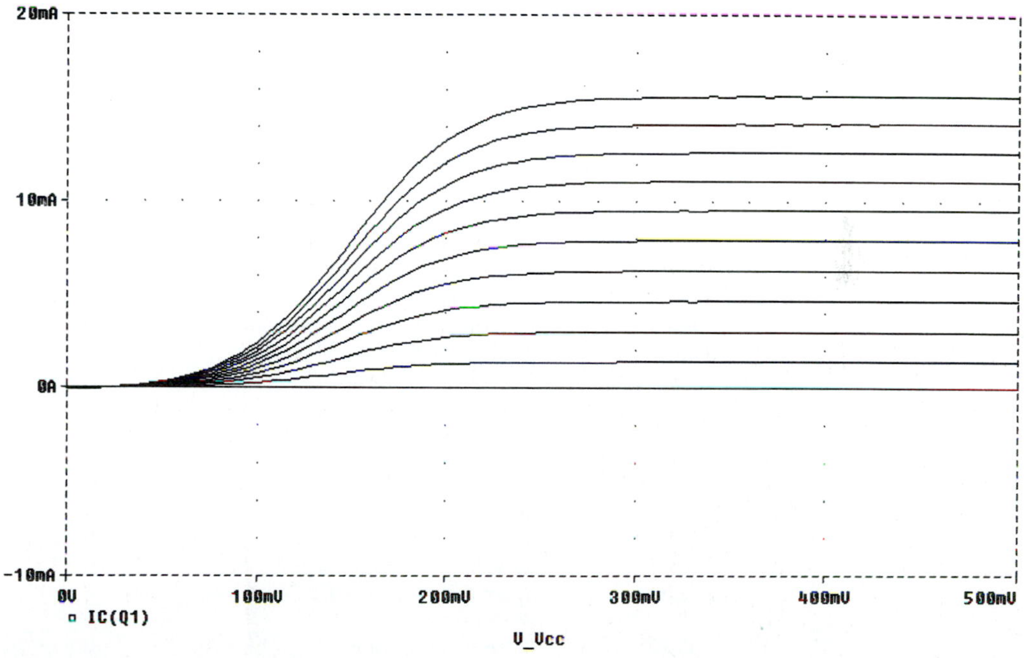

FIGURE 3.37
Zooming in on the saturation characteristics of the transistor.

increment size to 0.005 V) produces the plot in Fig. 3.37. From this plot we can see that the transistor enters the active region for V_{CE} somewhere between 200 and 300 mV for all collector currents less than about 15 mA.

Increasing the Vcc upper limit to a higher voltage of, say, 50 V would allow us to determine the effective r_{CE} value for the transistor, by using the cursors to measure ΔI_C and ΔV_{CE}. This is left as an exercise for the reader.

About Transistor Models

We can approximate the behavior of a transistor by using the various dependent source models that were presented in this chapter. By the way, the transistor models that we have been using are simplified variations of what is called the *Ebers-Moll* model, named after the inventors of this transistor modeling technique. The low-frequency Ebers-Moll model for an npn BJT is shown in Fig. 3.38. As you can see, this model is nonlinear and fairly complex.

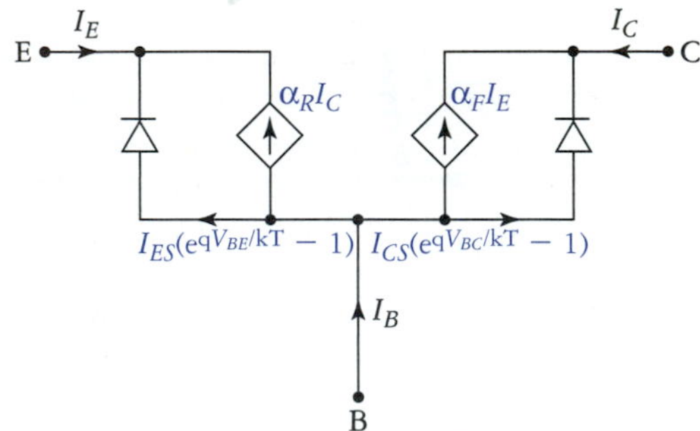

FIGURE 3.38
Low-frequency Ebers-Moll BJT model.

PSpice uses what is called the *Gummell-Poon* model of the transistor. The Gummell-Poon model is an enhanced version of the Ebers-Moll model, with a total of 40 specific device parameters included in the PSpice implementation. These parameters are listed in Table 3.3, along with their default values and values that are typical for a general-purpose, low-power, silicon BJT.

TABLE 3.3
Device Parameters for PSpice Implementation

P	Description	Default	Typical	Unit
IS	Transport saturation current	10^{-16}	10^{-15}	A
BF	Forward beta	100	100	
BR	Reverse beta	1	1	
NF	Forward emission coefficient	1	1	
NR	Reverse emission coefficient	1	1	
ISE	B-E saturation current	0	10^{-13}	A
ISC	C-B saturation current	0	10^{-13}	A
IKF	High-current forward beta corner	∞	0.05	A
IKR	High-current reverse beta corner	∞	0.05	A
NE	B-E leakage coefficient	1.5	1.5	
NC	C-B leakage coefficient	2	2	
VAF	Forward Early voltage	∞	200	V
VAR	Reverse Early voltage	∞	200	V
RC	Collector bulk resistance	0	10	Ω
RE	Emitter bulk resistance	0	1	Ω
RB	Base bulk resistance	0	100	Ω
RBM	Min, base R at high current	0	10	Ω
IRB	I_B for RB $= f(I_B) =$ RB/2	∞	0.05	A
TF	Forward transit time	0	0.1	ns
TR	Reverse transit time	0	10	ns
VJC	B-C barrier potential	0.75	0.65	V
MJC	B-C grading coefficient	0.33	0.5	
CJS	C-to-substrate capacitance	0	2	pF
VJS	Substrate-junction barrier potential	0.75	0.65	V
MJS	Substrate-junction exponent	0	0.5	
XCJC	C-B capacitance fraction to B	1	1	
FC	Fwd. bias depletion cap. coefficient	0.5	0.5	
XTB	Beta temperature coefficient	0	.5	
XTI	Saturation I temp. coefficient	3	3	
EG	Bandgap	1.11	1.11	eV
KF	Flicker noise coefficient	0	0.5	
AF	Flicker noise exponent	1	1	

CHAPTER REVIEW

Bipolar junction transistors (BJTs) are three-terminal devices that are available in npn and pnp forms. The terminals of the transistor are called the base, emitter, and collector. The transistor will normally be operated in one of three important regions of operation: the cutoff region, the saturation region or the active region. In cutoff, the transistor behaves as an open circuit from collector to emitter. In saturation, the transistor acts like a very low resistance between the collector and emitter terminals. In the active region, the collector terminal behaves like a dependent current source. There are three basic biasing arrangements for the BJT: common emitter (CE), common base (CB), and common collector or emitter follower (CC or EF).

The transistor terminal currents are related to each other by the parameters beta (β), alpha (α), and alternatively, transconductance (g_m). Beta is usually rather large (often $\beta > 100$) and unpredictable, whereas alpha is relatively stable, with a value of about 1. Beta, alpha, and transconductance are all temperature sensitive and frequency dependent. Transconductance differs from both beta and alpha in that it is strongly dependent on collector current. Because of the highly temperature-sensitive characteristics of the BJT, thermal runaway can occur. Thermal runaway is a self-driving process whereby the BJT heats up until it destroys itself.

Various leakage currents also exist within the BJT, the main ones being I_{CBO} and I_{CEO}. The common-emitter configuration is most sensitive to leakage current effects, while the common-base configuration is the least sensitive. For silicon BJTs these leakage currents are usually small enough to ignore. Germanium devices will have much higher leakage currents for a given device size. Other nonideal characteristics of the BJT cause the collector to have a finite active-region equivalent resistance. This resistance decreases with increasing base current. The Early voltage is used to quantify this active-region resistance. Base-spreading resistance also plays a part in causing the active-region collector characteristics to exhibit finite dynamic resistance. Increasing the reverse bias of the C-B junction while maintaining constant base current increases beta and the base-spreading resistance. The net effect is an increase in collector current with increasing C-B reverse bias.

Transistor breakdown voltage specifications must be considered in all design situations. Transistors generally break down via the avalanche mechanism or by a phenomenon called reach-through or punch-through. Breakdown can cause permanent damage to the transistor and to other devices in that circuit.

Transistors are often classified based on the application(s) to which they are best suited. Three broad categories that are commonly applied are general-purpose low-power, switching, and power transistors.

DESIGN AND ANALYSIS PROBLEMS

Note: Unless otherwise noted, all transistors are assumed to be silicon devices, leakage currents (I_{CBO} and I_{CEO}) are negligible, $T = 25°C$, and $\eta = 1$. Unless otherwise noted, assume $\beta_{ac} = \beta_{dc} = \beta$.

3.1. Refer to Fig. 3.39. Based on the voltage measurements given, determine the mode of operation (saturation, cutoff, or active) for each transistor. (*Hint:* Application of Kirchhoff's voltage law gives $V_{CE} = V_{CB} + V_{BE}$.)

3.2. A certain transistor has $I_B = 12\ \mu A$ and $I_C = 1.5$ mA. Determine β and α.

3.3. A certain transistor has $\beta = 220$ and $I_E = 8$ mA. Determine I_C and I_B for this device.

3.4. Refer to Fig. 3.40. Determine the values of β, α, and any unknown current for each transistor shown.

3.5. A certain transistor has $\beta = 150$ and $I_{CBO} = 20$ nA. Determine I_{CEO} and the actual collector current, given $I_B = 5\ \mu A$.

3.6. A certain transistor has $\beta = 100$ and $I_{CEO} = 5\ \mu A$. Determine I_{CBO} for this device.

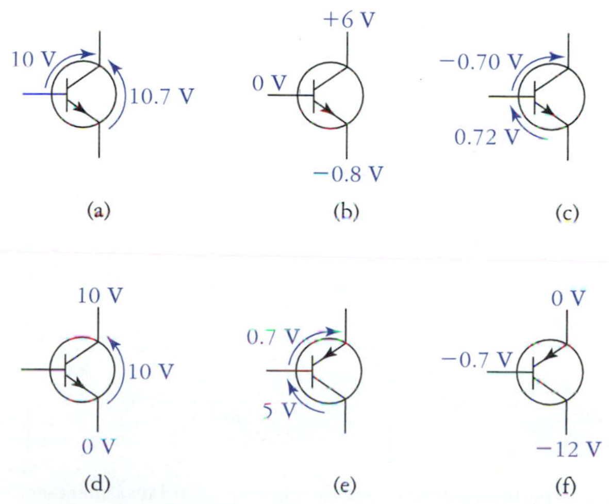

FIGURE 3.39
Transistors for Problem 3.1.

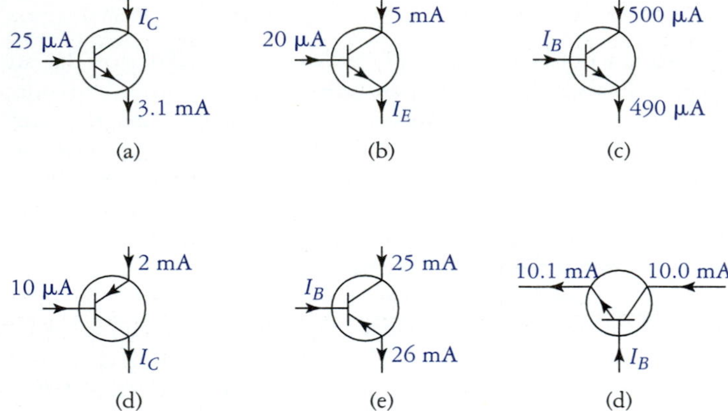

FIGURE 3.40
Transistors for Problem 3.4.

3.7. A certain npn transistor is represented by the model in Fig. 3.41. Determine i_B, i_C, and the dynamic resistance of the emitter region (r_e).

3.8. (Optional) Draw the schematic for the transconductance model of the device in Fig. 3.41.

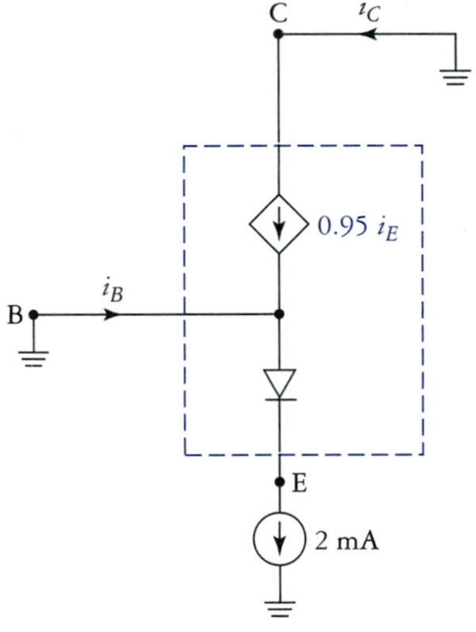

FIGURE 3.41
Transistor model for Problem 3.7.

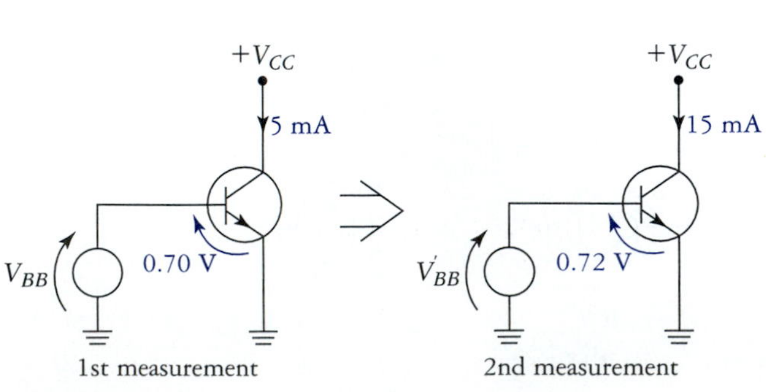

FIGURE 3.42
Circuit for Problem 3.9.

3.9. (Optional) The circuit of Fig. 3.42 was used to characterize an npn transistor, producing the current and voltage values shown. Determine the transconductance of the transistor.

3.10. Experimental analysis of a certain transistor produced the transconductance curve shown in Fig. 3.43. Determine the approximate values of g_m and r_e for this device.

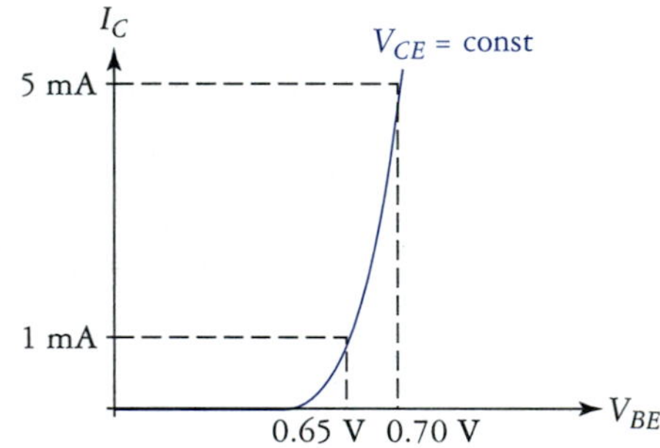

FIGURE 3.43
Transconductance curve for Problem 3.10.

3.11. The analysis of a certain transistor produced the collector characteristic curve of Fig. 3.44. Determine the Early voltage V_A, the output resistance r_o, and assuming that $V_{CE} = 5$ V, determine β and α.

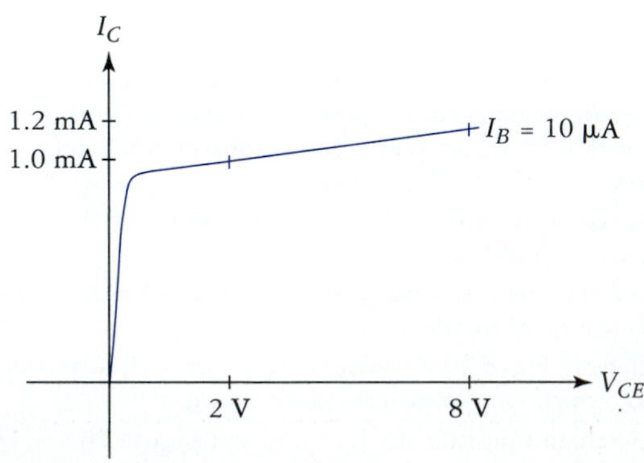

FIGURE 3.44
Collector curve for Problem 3.11.

3.12. Refer to Fig. 3.45. Determine the values for I_{C_1} through I_{C_4}, and the minimum and maximum β values for each corresponding base current. Also, determine: $\beta|_{V_{CE}=10\text{ V}}$ $r_o|_{I_B=5\ \mu\text{A}}$, and $r_o|_{I_B=15\ \mu\text{A}}$. Assume that $V_A = 100$ V.

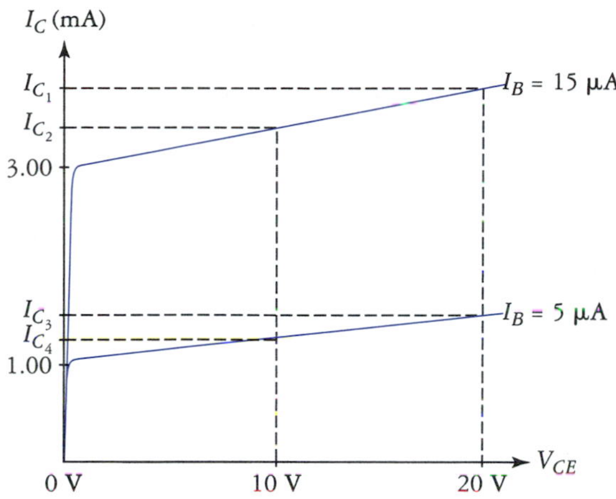

FIGURE 3.45
Collector curves for Problem 3.12.

3.13. Refer to Fig. 3.46. Assume that $\beta = 50$ and $i_B = 40$ μA. Determine r_π and draw the equivalent common-emitter (r_e) model.

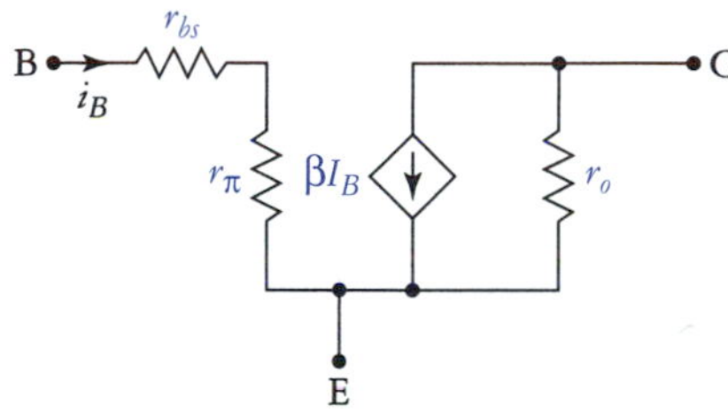

FIGURE 3.46
Model for Problem 3.13.

3.14. Using the linearized collector characteristic curve sketched in Fig. 3.47 for reference purposes, derive Eq. (3.18). (*Hint:* The equation for a linear function is $y = mx + b$, where m is the slope of the function, x is the independent variable, b is the y intercept, and y is the dependent variable.)

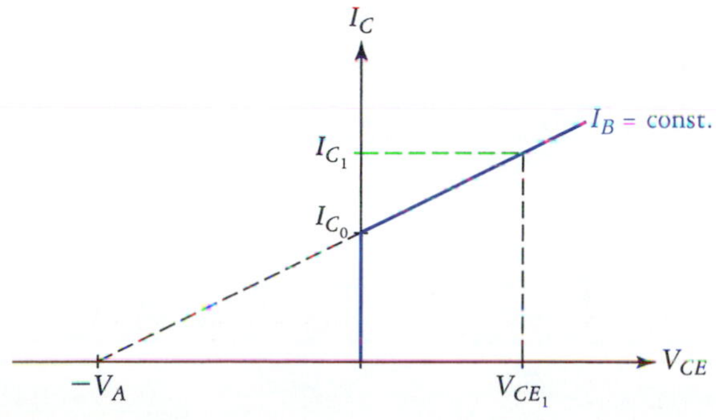

FIGURE 3.47
Collector curve for Problem 3.14.

3.15. Refer to Fig. 3.48. The curve tracer was set up for CE biasing with 2 V/div horizontal sensitivity, 1 mA/div vertical sensitivity, and 10 μA/step base current increments. Determine β_{dc} at $V_{CE} = 10$ V for each of the four curves. Also determine the β_{ac} at $V_{CE} = 10$ V using the I_B = 10-μA and 20-μA curves, and β_{ac} (at $V_{CE} = 10$ V), using the I_B = 20-μA and 30-μA curves.

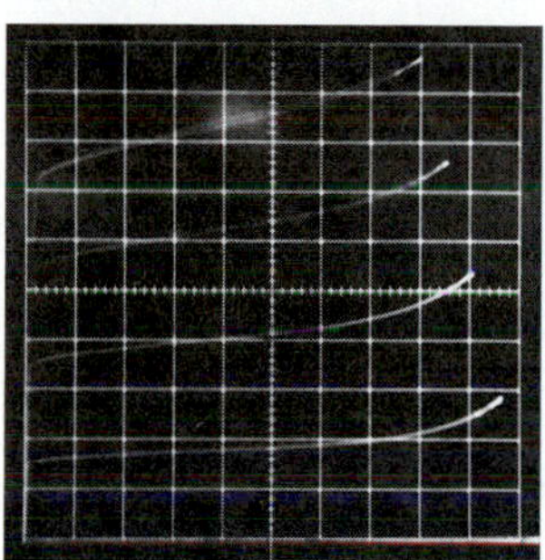

FIGURE 3.48
Collector curves for Problems 3.15 and 3.16.

3.16. Using the I_B = 20-μA collector curve of Fig. 3.48, determine the approximate values for r_o and V_A. Do not use the portion of the curve that curves upward significantly ($V_{CE} > 12$ V), because nonlinearity is too great for meaningful measurements.

3.17. Repeat Problem 3.16 using the I_B = 10-μA collector curve.

■ TROUBLESHOOTING PROBLEMS

3.18. A general-purpose npn silicon transistor is biased up (CE) to provide $I_C = 5$ mA, based on the assumption that $\beta = 100$. Measurements taken show that $I_C = 10$ mA. What is the most likely cause for this collector current discrepancy?

3.19. A general-purpose npn silicon transistor is biased up in the common-base configuration to provide $I_C = 5$ mA. Measurements yield $I_C = 10$ mA. Assuming that the transistor is not defective, what is the most likely cause for this discrepancy?

3.20. Based on the voltage measurements given in Fig. 3.49, what is most likely wrong with the transistor?

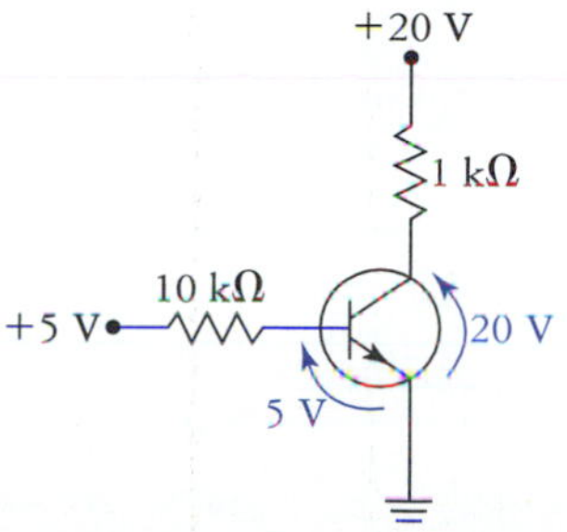

+20 V
1 kΩ
10 kΩ
+5 V
20 V
5 V

FIGURE 3.49
Circuit for Problem 3.20.

3.21. The two junctions of a transistor can be tested (out of circuit) with an ohmmeter just like normal diodes. Assuming that the transistors in Fig. 3.50 are not defective, indicate the resistance that each meter, acting independently, should read (high or low resistance).

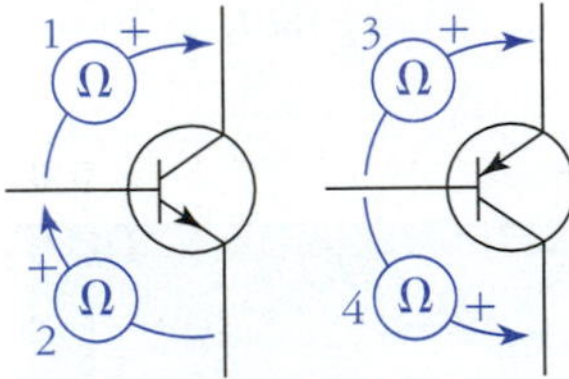

FIGURE 3.50
Transistors for Problem 3.21.

■ COMPUTER PROBLEMS

3.22. Use PSpice to generate a family of collector curves for the 2N2222 transistor. Sweep V_{CE} from 0 to 10 V in 0.1-V increments and I_B from 0 to 100 μA in 20-μA increments.

3.23. Simulate the circuit in Fig. 3.51 using a dc sweep such that i_C is plotted as a function of v_{BE}. Sweep v_{BB} from 0 to 2 V in 0.02-V increments. The transconductance is $g_m = 40$ mS. Use the 1N4148 diode model.

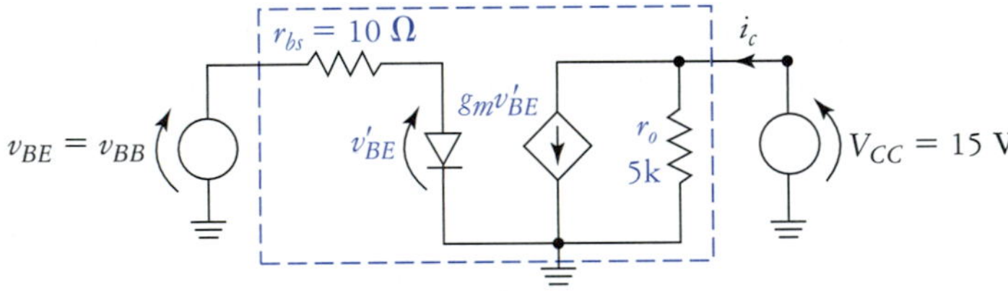

FIGURE 3.51
Circuit for Problem 3.23.

3.24. Repeat the analysis of Problem 3.23 with a 1Ω resistor in place of the 10Ω resistor and compare the resulting curve with that of Problem 3.23.

3.25. The effects of temperature on transistor behavior can also be simulated using PSpice. Configure the simulation for the circuit in Fig. 3.52 such that plots of I_C versus V_{CE} at temperatures of 0, 50, and 100°C are produced.

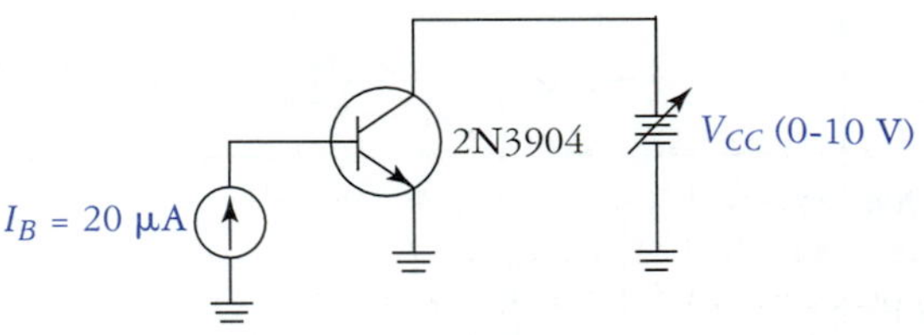

FIGURE 3.52
Circuit for Problem 3.25.

CHAPTER 4

Bipolar Transistor Biasing

BEHAVIORAL OBJECTIVES

On completion of this chapter, the student should be able to:

- Explain the purpose of transistor biasing.
- Determine the Q point for fixed-base bias transistor circuits.
- Apply Thevenin's theorem to the analysis of transistor biasing circuits.
- Determine the power dissipation of a transistor.
- Explain the purpose of an emitter-feedback resistor.
- Determine the Q point of a transistor using collector-feedback bias.
- Apply KVL and KCL in the analysis of transistor biasing circuits.
- Sketch the dc load line for a transistor.
- Perform computer-aided analysis of transistor biasing circuits.

INTRODUCTION

In order for a transistor to perform some useful function, in an amplifier, for example, it is generally necessary to bias up that transistor for a particular collector current and collector-to-emitter voltage drop. One application might require the transistor to be biased into saturation, while another may require that the transistor be biased into cutoff. In the case of a linear amplifier, it is most likely that we would want that transistor to be biased to operate within the active region. Most often in this text, we will be concerned primarily with active-region operation, although after completion of this chapter, you should be able to analyze or design for any of these cases.

One of the most sought-after design goals is a circuit that has a stable Q point. This stability is achieved through the use of appropriate biasing techniques. This chapter will familiarize you with the techniques used to achieve this goal.

4.1 BASE BIASING

Except for the special case of no biasing (which, perhaps surprisingly, is sometimes useful), *base biasing* is the simplest transistor biasing technique. In fact, the preceding chapter just about completely covered this topic already. However, here we briefly reexamine this topic from a slightly different perspective to get things started.

The fundamental form of the base-biasing arrangement is shown in the familiar circuit of Fig. 4.1, which is a common-emitter circuit arrangement. In most practical schematic diagrams, voltage sources, such as the V_{BB} source in the figure, are rarely shown explicitly, as is the case here. Instead, the supply-voltage value is usually just written beside a given node that happens to be connected to the power supply. This is the case for the upper end of the collector resistor in the figure. The circuit is redrawn at the right, however, to emphasize the B-E and C-E loops.

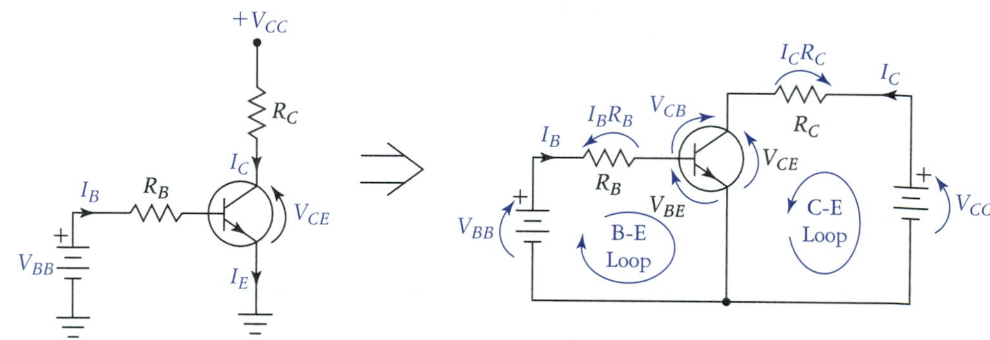

FIGURE 4.1
Fixed-base-biasing arrangement, common-emitter configuration.

The analysis of Fig. 4.1 begins with determination of the base current. We apply KVL around the B-E loop, yielding

$$0 = V_{BB} - I_B R_B - V_{BE}$$

Solving for I_B gives us

$$I_B = \frac{V_{BB} - V_{BE}}{R_B} \tag{4.1}$$

Note
Note the strong dependence of I_C and V_{CE} on β. This is a major disadvantage of base biasing.

The collector current is determined using the fundamental relationship

$$I_C = \beta I_B \tag{4.2}$$

The collector-to-emitter voltage is now determined. Applying KVL around the C-E loop gives us

$$0 = V_{CC} - I_C R_C - V_{CE}$$

which is easily solved to produce

$$V_{CE} = V_{CC} - I_C R_C \tag{4.3}$$

Transistor voltage and current levels that are set by biasing circuit conditions are often called *quiescent* values, which are indicated with a Q appended to the subscript (V_{CEQ} and I_{CQ}, for example). The term *quiescent* means quiet, or no-signal, conditions, where only the presence of biasing voltages (or currents) determines circuit behavior.

Note that application of KVL from the collector to the emitter produces the following relationship:

$$V_{CE} = V_{CB} + V_{BE} \tag{4.4}$$

which holds for any biasing configuration or region of operation.

One point should be emphasized that is not apparent at this time. That is, the analysis of the base-biased transistor requires that we find the base current in order to determine the remaining current and voltage levels. The value of the base current in itself, however, is not important. It is the collector current that we wish to set to some particular value; we let the base current assume whatever value is necessary to achieve this goal. In other words, setting I_C to some value is a rather major design goal, while setting a value for I_B is not (we do normally prefer I_B to be small, but this is more a function of the transistor than the biasing arrangement).

EXAMPLE 4.1

Refer to Fig. 4.1. The transistor has $\beta = 100$, $V_{BB} = 5$ V, and $V_{CC} = 20$ V. Determine values for R_B and R_C such that $I_{CQ} = 5$ mA and $V_{CEQ} = 10$ V.

Solution The analysis of the base-biased transistor starts with the determination of I_B. From this, the value of I_C is then determined. In the base-bias design process, I_B can be determined, but it is found by "working backwards" from the desired value of I_C. Thus we have

$$I_B = \frac{I_B}{\beta} = 50\ \mu\text{A}$$

We now refer to either Eq. (4.1), or the original KVL equation from which it was derived, and simply solve for R_B, giving us

$$\begin{aligned} R_B &= \frac{V_{BB} - V_{BE}}{I_B} \\ &= \frac{5\text{ V} - 0.7\text{ V}}{50\ \mu\text{A}} \\ &= 86\text{ k}\Omega \end{aligned}$$

The required value for R_C is found by solving either Eq. (4.3), or the corresponding C-E KVL equation, producing

$$\begin{aligned} R_C &= \frac{V_{CC} - V_{CEQ}}{I_{CQ}} \\ &= \frac{20\text{ V} - 10\text{ V}}{5\text{ mA}} \\ &= 2\text{k}\Omega \end{aligned}$$

Example 4.1 illustrates the relative ease with which base-biased circuits can be designed. This is the main advantage of the simple base-biasing arrangement.

DC Load Line

The load-line analysis technique that was used in the analysis of diode circuits proves to be useful in the graphical analysis of transistor circuits as well. For the common-emitter biased transistor, the quiescent values of I_C and V_{CE} determine the operating or Q point of the device. In reference to Fig. 4.1, these two parameters are associated with the C-E loop of the circuit. Variations in I_C and V_{CE} are related via the KVL equation for the C-E loop. If we plot every possible combination of I_C and V_{CE} on the collector curves for the transistor in the common-emitter circuit, a straight line is produced, as shown in Fig. 4.2. This is the dc load line for the transistor. Mathematically, the load line is defined as the graph of the set of all possible device operating points. The specific operating point to which a transistor is biased up is called the Q point.

The limits of the load line occur when the transistor reaches either saturation or cutoff. The maximum possible collector current is given by

$$I_{C(\text{sat})} = \frac{V_{CC} - V_{CE(\text{sat})}}{R_C} \tag{4.5}$$

Recall that typically, $V_{CE(\text{sat})}$ is only a few tenths of a volt, which is normally negligible. Thus we are usually justified in using

$$I_{C(\text{sat})} \cong \frac{V_{CC}}{R_C} \tag{4.6}$$

In cutoff, the collector current is $I_C = I_{CEO} \cong 0$, therefore

$$V_{CE(\text{cut})} \cong V_{CC} \tag{4.7}$$

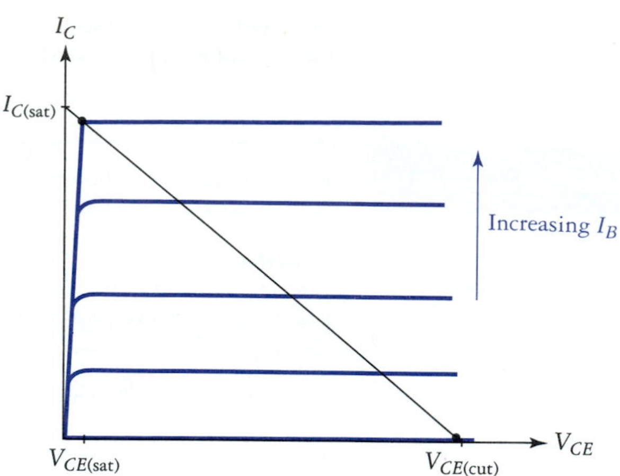

FIGURE 4.2
DC load-line superimposed on collector characteristic curves.

Most often, when a transistor load line is drawn, the device collector current curves are not shown. Such a load line with an arbitrarily located Q point is shown in Fig. 4.3. The collector curves are usually omitted because they are dependent on β, which varies widely between devices, whereas the load line and its limits are essentially independent of transistor parameters. We would need a set of such curves for each transistor used, which is impractical. Even without the collector curves being present, however, the load line provides a good overall picture of circuit limitations.

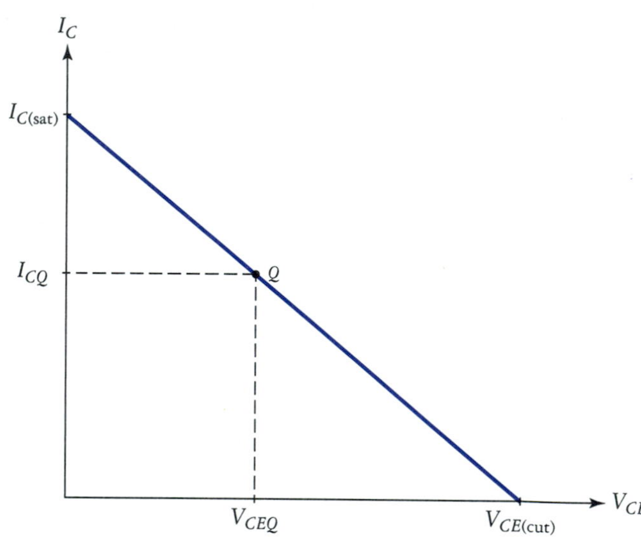

FIGURE 4.3
Important points on the dc load line.

EXAMPLE 4.2

Sketch the dc load line for the circuit in Fig. 4.4. Assume that $V_{CE(sat)} = 0$ V.

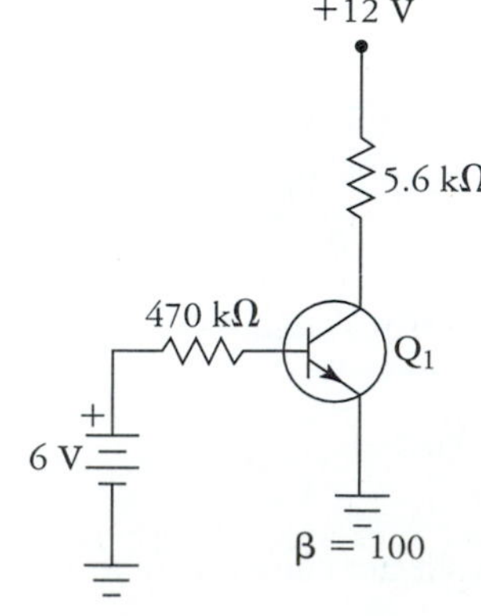

FIGURE 4.4
Circuit for Example 4.2.

Solution First, determine the base current, which by the way can be called the quiescent base current, because it is a bias-determined value:

$$I_{BQ} = \frac{V_{BB} - V_{BEQ}}{R_B}$$
$$= \frac{6\text{ V} - 0.7\text{ V}}{R_B}$$
$$= 11.3\ \mu\text{A}$$

The (quiescent) collector current is

$$I_{CQ} = \beta I_{BQ}$$
$$= (100)(11.3\ \mu\text{A})$$
$$= 1.13\text{ mA}$$

The quiescent collector-to-emitter voltage drop is

$$V_{CEQ} = V_{CC} - I_{CQ}R_C$$
$$= 12\text{ V} - 6.33\text{ V}$$
$$= 5.67\text{ V}$$

The collector saturation current is

$$I_{C(\text{sat})} \cong \frac{V_{CC}}{R_C}$$
$$= \frac{12\text{ V}}{5.6\text{ k}\Omega}$$
$$= 2.14\text{ mA}$$

The cutoff voltage is simply

$$V_{CE(\text{cut})} \cong V_{CC} = 12\text{ V}$$

The resulting dc load line is shown in Fig. 4.5.

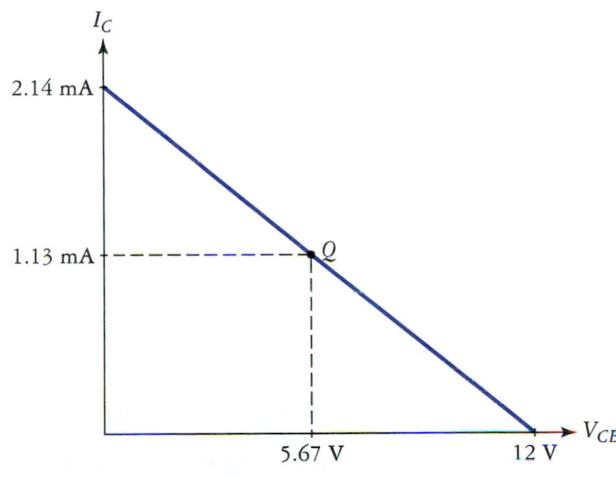

FIGURE 4.5
DC load line for Fig. 4.4.

Please notice in Fig. 4.4 that the transistor is labeled Q_1. It is standard practice to label transistors in this manner on schematic diagrams. Do not confuse this with the notation for the Q point, which has an altogether different meaning.

Base-Biasing Disadvantages

Although simple, there are severe disadvantages associated with base biasing as well, as the next example demonstrates.

EXAMPLE 4.3

The circuit of Fig. 4.4 is constructed using a transistor type that has a possible beta spread from 50 to 200. Determine the effects of these extreme variations on the Q point.

Solution Direct applications of Eqs (4.1), (4.2), and (4.3) yields the following results:

Case 1: $\beta = 50$

$$I_{BQ} = \frac{V_{BB} - V_{BEQ}}{R_B}$$
$$= 11.3\ \mu\text{A}$$

$$I_{CQ} = \beta I_{BQ}$$
$$= (50)(11.3\ \mu\text{A})$$
$$= 565\ \mu\text{A}$$

$$V_{CEQ} = V_{CC} - I_{CQ}R_C$$
$$= 12\text{ V} - 3.16\text{ V}$$
$$= 8.84\text{ V}$$

Case 2: $\beta = 200$

$$I_{BQ} = 11.3\ \mu\text{A}$$
$$I_{CQ} = \beta I_{BQ}$$
$$= (200)(11.3\ \mu\text{A})$$
$$= 2.26\text{ mA}$$

Stop here. Look back at the dc load line for this circuit in Fig. 4.5. The maximum possible collector current is $I_{C(\text{sat})} = 2.14$ mA; thus, for $\beta = 200$, the transistor is in saturation, and $I_{CQ} = I_{C(\text{sat})} = 2.14$ mA . This also means that $V_{CEQ} = V_{CE(\text{sat})} = 0$ V.

There are two lessons to be learned from this example. First, we must *think* about our results as we go along, and not just blindly continue along when absurd answers are produced. The analysis method we are using assumes that the transistor is in the active region at the outset. If this is truly the case, then the analysis will result in consistent, meaningful answers. When the transistor is in cutoff or saturation, we normally obtain results that are impossible (usually they violate Kirchhoff's laws). Secondly, we see that transistor selection is often very critical to the proper operation of the base-biased transistor circuit. The use of a transistor with beta that is too high or low can result in a useless circuit.

Notice in Example 4.3 that the base current remains constant regardless of β, while I_C and V_{CE} vary drastically. If we were designing the circuit to hold a stable value of I_B this would be fine, but the point is that we don't really care what value I_B assumes. Our design goals require I_C and V_{CE} to have specific values.

The main problem with base biasing is that it is totally dependent on device beta, which we know is notoriously unpredictable. We could hand-pick transistors and sort them according to beta, but this would be labor intensive and very costly on a large scale. Given these problems, it might seem that base biasing is completely impractical. This is not always the case, as will be shown shortly.

Base-Biasing Variations

Often, two separate voltage sources are not available to bias up the transistor. The circuit in Fig. 4.6(a) is an example of this situation, where there is no separate V_{BB} supply. It is possible to redraw this circuit in an electrically identical form, as shown in Fig. 4.6(b). Now it is obvious that previously developed analysis methods can be readily used here.

Note
Bipolar power supplies are used very often in electronic equipment, especially when operational amplifiers are part of the design.

Another possibility is shown in Fig. 4.7(a). In this circuit, the emitter is not connected to ground, but rather to the negative supply rail.* Since the negative supply is connected to the emitter, it is usually called $-V_{EE}$. The circuit may be redrawn in more familiar form as shown in Fig. 4.7(b), however, for the purpose of determining the transistor currents. Notice that the equivalent base-biasing voltage is equal in magnitude to the negative supply rail voltage.

*The term *supply rail* is one of the most commonly used terms for the power supply terminals in a circuit. In Fig. 4.7(a), V_{CC} and V_{EE} are the supply rails.

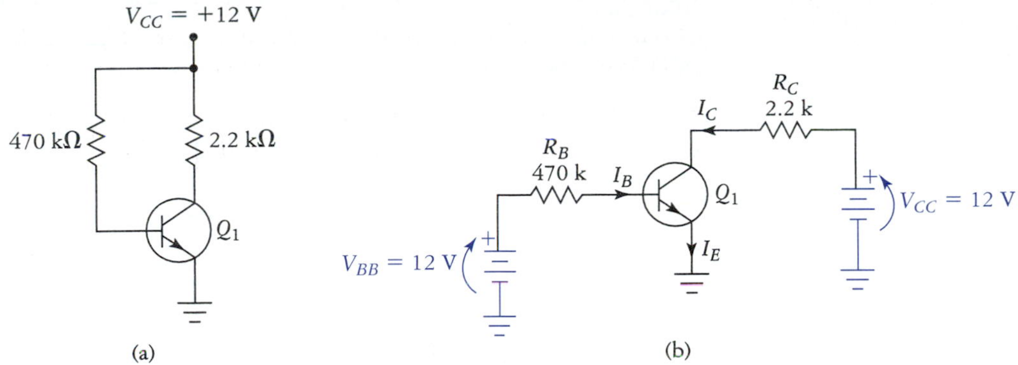

FIGURE 4.6
Fixed-base biasing with a single power supply. (a) Actual circuit. (b) Equivalent circuit redrawn with separate bias voltage sources.

Note
Redraw circuits to make analysis easier.

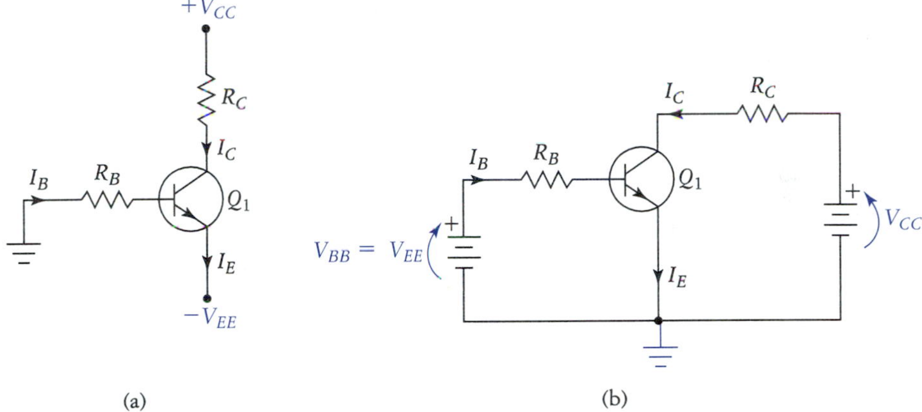

FIGURE 4.7
Fixed-base biasing with a bipolar power supply. (a) Actual circuit. (b) Equivalent circuit redrawn to emphasize biasing arrangement. This equivalent is only used to determine quiescent base and collector currents.

EXAMPLE 4.4

The circuit in Fig. 4.7(a) has $\beta = 100$, $R_B = 220$ kΩ, $R_C = 1$ kΩ, and $V_S = 10$ V. (*Note:* V_S represents the supply voltage in more compact form here. That is, $V_S = 10$ V means $V_{CC} = V_{EE} = 10$ V.) Determine I_C, V_{CE}, V_C, and V_B.

Solution Based on the equivalence of Figs. 4.7(a) and (b), we find the base current using

$$\begin{aligned} I_B &= \frac{V_{EE} - V_{BE}}{R_B} \\ &= \frac{10\text{ V} - 0.7\text{ V}}{220\text{ k}\Omega} \\ &= 42.3\ \mu\text{A} \end{aligned}$$

The collector current is

$$\begin{aligned} I_C &= \beta I_B \\ &= (100)(42.3\ \mu\text{A}) \\ &= 4.23\text{ mA} \end{aligned}$$

To find V_{CE}, we do not use the circuit of Fig. 4.7(b) because it looks too much like the emitter is at ground potential, when in fact it is not. Rather, the circuit is redrawn in Fig. 4.8 to emphasize its true structure.

FIGURE 4.8
Circuit of Fig. 4.7(a) redrawn with explicit bias voltage sources.

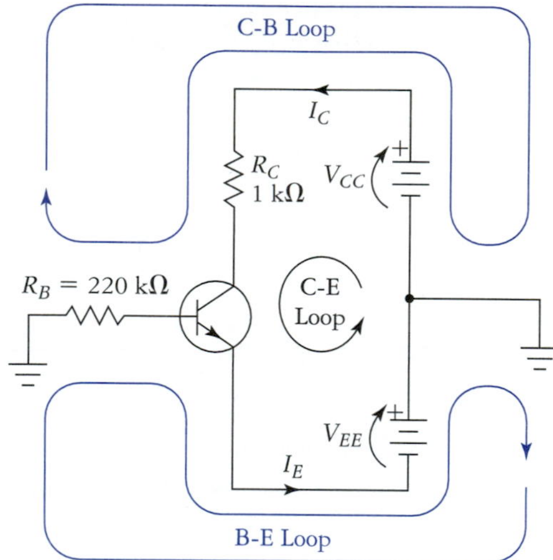

Since we now know I_C, we simply apply KVL around the C-E loop of the circuit giving us

$$0 = V_{CC} + V_{EE} - I_C R_C - V_{CE}$$

Solving for V_{CE}, we obtain

$$\begin{aligned} V_{CE} &= V_{CC} + V_{EE} - I_C R_C \\ &= 10\text{ V} + 10\text{ V} - 4.23\text{ V} \\ &= 15.77\text{ V} \end{aligned}$$

We next find V_C, which is the collector voltage, with respect to ground. Application of KVL gives us

$$\begin{aligned} V_C &= V_{CC} - I_C R_C \\ &= 10\text{ V} - 4.23\text{ V} \\ &= 5.77\text{ V} \end{aligned}$$

The base voltage V_B can be found is several ways. For example,

$$\begin{aligned} V_B &= -I_B R_B \\ &= -(42.3\ \mu\text{A})(220\text{ k}\Omega) \\ &= -9.3\text{ V} \end{aligned}$$

Alternatively, we also can use

$$\begin{aligned} V_B &= -V_{EE} + V_{BE} \\ &= -10\text{ V} + 0.7\text{ V} \\ &= -9.3\text{ V} \end{aligned}$$

It is also possible to apply KVL around the C-B loop to determine V_B as well.

A dc load line for the circuit of Fig. 4.7 can also be generated. The limits of the load line are given by

$$I_{C(\text{sat})} \cong \frac{V_{CC} + V_{EE}}{R_C} \tag{4.8}$$

and

$$V_{CE(\text{cut})} = V_{CC} + V_{EE} \tag{4.9}$$

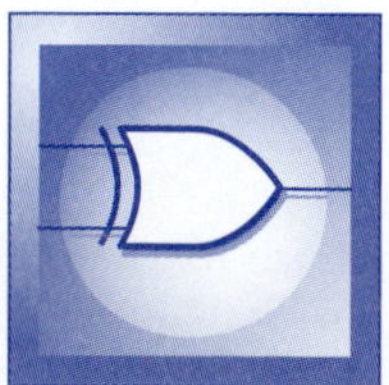

The load line in Fig. 4.9 is based on the values used in Example 4.4. Notice that for an npn transistor $V_{CE} \geq 0$, regardless of the supply voltage distribution.

Saturated Switching Applications

Although limited in usefulness, base biasing is acceptable in some applications. One such application is in the design of saturated switching circuits. When a transistor is used as a saturated switch, it is

FIGURE 4.9
DC load line for Fig. 4.7.

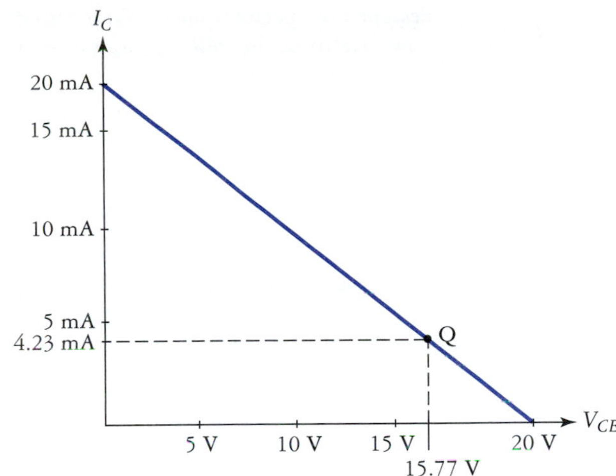

operated in either saturation or in cutoff. The transistor is in the active region only when it is switching from one extreme to the other, which is normally a very short period of time. Saturated switches are typically used to enable the low-power output lines of digital integrated circuits to drive higher power devices such as relays, incandescent lamps, and light-emitting diodes.

The circuit in Fig. 4.10 allows relay K_1 to be controlled through the application of a TTL-compatible logic signal. TTL (transistor-transistor logic) refers to a family of digital logic devices guaranteed to meet the following worst-case specifications:

$V_{OH(\text{min})} = 2.4$ V (minimum output voltage for logic high)

$V_{OL(\text{max})} = 0.4$ V (maximum output voltage for logic low)

$I_{OH(\text{min})} = 600\ \mu$A (minimum output current source capability)

There are many other TTL parameters, but these three are useful to us in this application. The circuit in Fig. 4.10 is to operate such that when the output of the TTL circuit is high, the transistor saturates, causing the relay to close its normally open (NO) contacts. When the output of the TTL circuit is low, the transistor is to be in cutoff, de-energizing the relay.

Let us assume that the relay in Fig. 4.10 has the following specifications; $R_{dc} = 300\ \Omega$ (the dc resistance of the relay winding) and $V_{coil} = 12$ V (the dc voltage required to activate relay). Obviously, the output of a TTL gate could not energize the relay in the logic-high state. Since the relay is

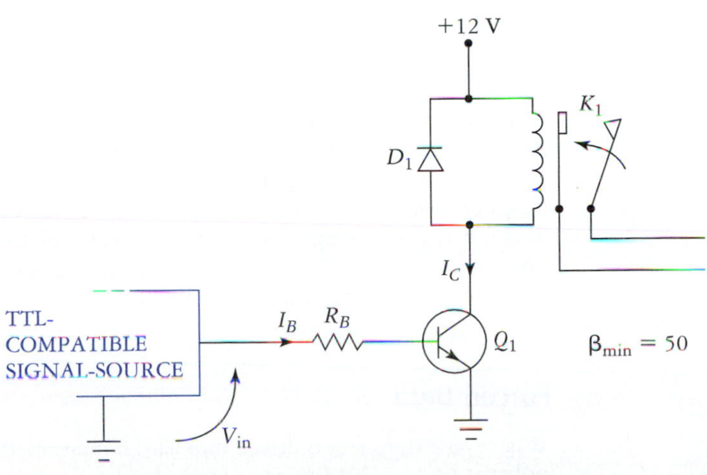

FIGURE 4.10
BJT used as a saturated switch to interface a relay to a TTL-compatible source.

designed to operate from 12 V, no current-limiting resistance is required in the collector circuit. We now determine the collector saturation current. Let us assume that $V_{CE(\text{sat})} = 0$ V

$$I_{C(\text{sat})} = \frac{V_{CC}}{R_{\text{dc}}} = \frac{12\text{ V}}{300\ \Omega} = 40\text{ mA}$$

Note
There are usually many different "correct" solutions to a given problem. Don't be afraid to try different approaches.

At this point, the design process can take any number of different paths. There is no single "right" design approach, but the one taken here is fairly representative of most of them. Let us say that we chose a transistor that has a possible beta spread of 100 to 300 to use in this circuit. We use the worst-case value ($\beta = 100$) in all successive calculations. The maximum base current necessary for driving the transistor into saturation is

$$I_{B(\text{sat})} = \frac{I_{C(\text{sat})}}{\beta_{\text{min}}} = \frac{40\text{ mA}}{100} = 400\ \mu\text{A}$$

This is well within the current-source capabilities of TTL devices. If possible, we would prefer to *overdrive* the transistor, by guaranteeing that $I_{B(\text{high})} \gg I_{B(\text{sat})}$. This produces what is called *hard saturation.* As a general rule of thumb, to obtain hard saturation, we set the base current according to the following formula:

$$I_{B(\text{hard sat})} = \frac{2\ I_{C(\text{sat})}}{\beta_{\text{min}}} \tag{4.10}$$

We cannot afford the luxury of such hard saturation in this case, because of the $I_{OH(\text{min})} = 600\ \mu$A limit, and so we must compromise. Thus, we settle for $I_{B(\text{sat})} = 400\ \mu$A, which leaves at least 200 μA of drive capability in reserve at the TTL output.

We determine the required base resistor value now, based on the preceding discussion:

$$R_B = \frac{V_{OH(\text{min})} - V_{BE}}{I_{B(\text{sat})}} = \frac{2.4\text{ V} - 0.7\text{ V}}{400\ \mu\text{A}} = 4.25\text{ k}\Omega$$

This is not a standard resistor value, so in practice, we would use the next lower standard value (3.9 kΩ) to ensure saturation.

It is readily apparent that a logic-low output from the TTL circuit will cause the transistor to be in cutoff, because $V_{OL(\text{max})} < V_{BE(\text{on})}$ (note that $V_{BE(\text{on})} \cong V_{\phi}$). The last element of the design to consider here is the inclusion of diode D_1. This is called a *freewheeling diode,* and its function is to short-out high-voltage transients that may be generated by the inductance of the coil when the transistor switches off. Consider that when Q_1 is saturated, a powerful magnetic field is created around the relay coil. When the transistor switches off, collector current drops to zero, and the magnetic field collapses, returning energy to the circuit. The polarity of the resulting induced coil voltage forward biases D_1, which provides a low-impedance discharge path for the resulting coil current. Without the diode, it is possible that several hundred volts could be dropped across the transistor, which could cause device failure. Freewheeling diodes are almost always used to prevent device damage when transistors are used to switch inductive loads.

Forced Beta

Whenever a transistor is driven into hard saturation, the effective beta of the transistor decreases. Recall that beta can be defined as the ratio of collector current to base current. In saturation, we reach the point where collector current can no longer increase, while base current can increase. With this in mind, we can define forced beta as follows:

$$\beta_{\text{forced}} = \left.\frac{I_{C(\text{sat})}}{I_B}\right|_{I_B > I_{B(\text{sat})}} \tag{4.11}$$

Although it is not too important to our studies right now, the concept of forced beta does explain some characteristics of amplifier circuits under overdrive conditions, which are covered in detail later.

4.2 EMITTER FEEDBACK

The concept of *negative feedback* is extremely important and useful in virtually all areas of electronics. *Emitter feedback,* which is an example of negative feedback, is realized through the addition of an external resistance in series with the emitter terminal of the transistor. Using the base-bias arrangement as a starting point, the circuit of Fig. 4.11(a) results. Assuming active-region operation, the dependent source model of Fig. 4.11(b) represents this circuit.

FIGURE 4.11
The addition of an emitter-feedback resistor. (a) Biasing arrangement using two sources. (b) Active-region model of the transistor is helpful in understanding the operation of this biasing modification.

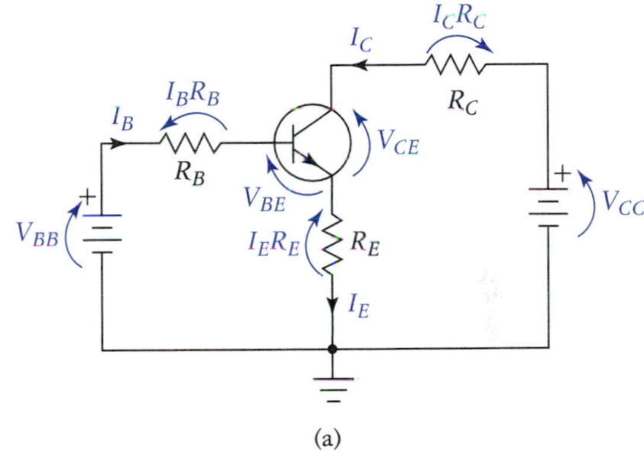

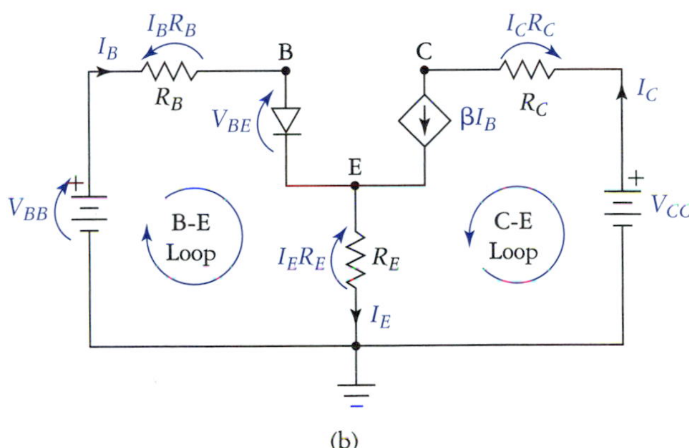

Note
This is one of the most important topics in the book. Please read this section carefully.

The analysis of Fig. 4.11 begins with determination of the base current. Applying KVL around the B-E loop gives us

$$0 = V_{BB} - I_BR_B - V_{BE} - I_ER_E$$

All variables in this expression are found by inspection except for the currents, but as it stands now, there is little we can do to determine their values. The trick is to define the currents such that common factors are produced. Recall that

$$I_E = I_C + I_B$$

which is equivalent to

$$\begin{aligned} I_E &= \beta I_B + I_B \\ &= I_B(\beta + 1) \end{aligned}$$

Substituting back into the KVL equation yields

$$0 = V_{BB} - I_BR_B - V_{BE} - I_B(\beta + 1)$$

Now, we solve for I_B, which gives us one of the most important results in transistor circuit analysis:

$$I_B = \frac{V_{BB} - V_{BE}}{R_B + R_E(\beta + 1)} \tag{4.12}$$

The physical significance of this relationship may not be obvious, but it must be stressed: In terms of base current, any resistance in the emitter is effectively increased by a factor of $(\beta + 1)$.

It is frequently possible to simplify Eq. (4.12) somewhat. Recall that although β is rather unpredictable, it is also usually large as well, thus $\beta \cong \beta + 1$. This is equivalent to saying $I_E \cong I_C$, which we know is generally true because $I_C = \alpha I_E$ and $\alpha \cong 1$. So, armed with this knowledge, we can write

$$I_B \cong \frac{V_{BB} - V_{BE}}{R_B + \beta R_E} \tag{4.13}$$

Note
As usual, we make the approximations $\beta + 1 \cong \beta$ and $\alpha \cong 1$.

Another useful form for this equation can also be written. Multiplying both sides of Eq. (4.13) by beta produces

$$I_C \cong \frac{V_{BB} - V_{BE}}{R_E + \dfrac{R_B}{\beta}} \tag{4.14}$$

Thus, in an analysis situation, we really do not need to determine the base current in order to find the collector current.

The addition of the emitter resistance also affects the saturation current of the transistor. The resulting saturation current equation (assuming $V_{CE(sat)} \cong 0$ V) is

$$I_{C(sat)} = \frac{V_{CC}}{R_C + R_E} \tag{4.15}$$

There is no significant change in the cutoff voltage.

EXAMPLE 4.5

Refer to Fig. 4.12. Assume that $\beta = 100$. Sketch the dc load line for this circuit.

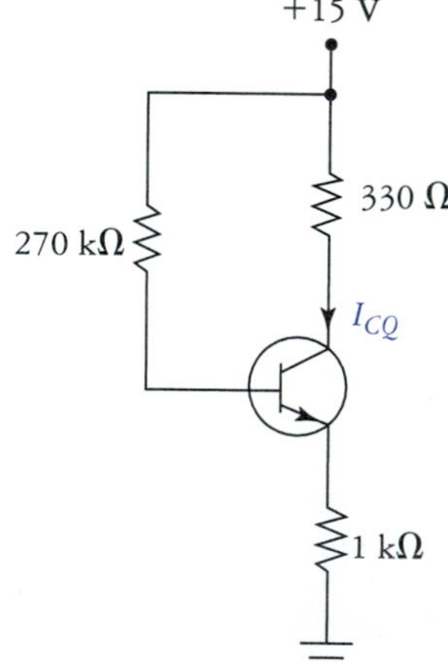

FIGURE 4.12
Circuit for Example 4.5.

Solution For a change of pace, let's determine the limits of the dc load line first this time.

$$\begin{aligned} I_{C(sat)} &= \frac{V_{CC}}{R_C + R_E} \\ &= \frac{15 \text{ V}}{1330\ \Omega} \\ &= 11.29 \text{ mA} \end{aligned}$$

By inspection, we have

$$V_{CE(\text{cut})} = V_{CC} = 15\text{ V}$$

The collector current is found through direct application of Eq. (4.14), where $V_{BB} = V_{CC}$:

$$\begin{aligned} I_{CQ} &= \frac{V_{BB} - V_{BE}}{R_E + \dfrac{R_B}{\beta}} \\ &= \frac{14.3\text{ V}}{3700\ \Omega} \\ &= 3.86\text{ mA} \end{aligned}$$

Then V_{CEQ} is found by applying KVL around the C-E loop as usual:

$$V_{CEQ} = V_{CC} - I_{CQ}R_C - I_{EQ}R_E$$

Keeping in mind that we are assuming $I_{EQ} \cong I_{CQ}$ allows us to use the simpler approximation as follows:

$$\begin{aligned} V_{CEQ} &= V_{CC} - I_{CQ}(R_C + R_E) \\ &= 15\text{ V} - (3.86\text{ mA})(1330\ \Omega) \\ &= 9.87\text{ V} \end{aligned} \tag{4.16}$$

The resulting load line is shown in Fig. 4.13.

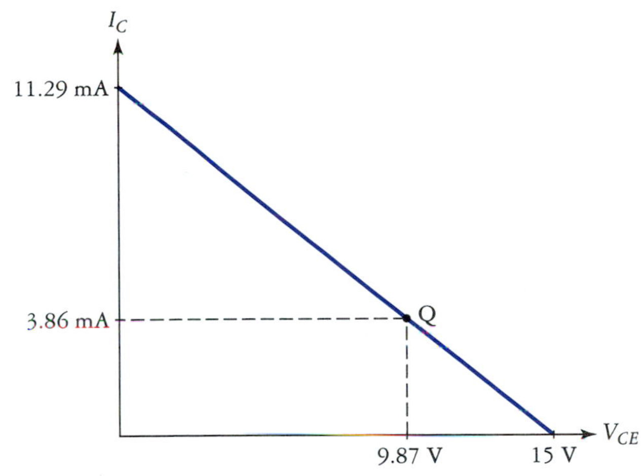

FIGURE 4.13
DC load line for Fig. 4.12.

Effects of Emitter Feedback on Q-Point Stability

So far, our discussion of emitter feedback has been somewhat abstract and mathematical. To gain a good intuitive understanding of the effects of emitter feedback, let us reexamine Eq. (4.14). Notice that if

$$R_E \gg \frac{R_B}{\beta}$$

then fluctuations in beta will have a small effect on the overall value of the denominator.

In physical terms, the stabilizing effects of emitter feedback can be explained as follows: Assume that we have designed the circuit for some specific Q point, based on some particular value of beta. Now, suppose we use a transistor that has a beta that is twice as high as expected. The increased beta causes an increase in collector current. However, increasing I_C also increases the voltage drop across R_E. The polarity of this $I_E R_E$ voltage drop tends to oppose the bias that is applied to the base, thus reducing I_C somewhat. Equilibrium is determined by I_{BQ} and I_{CQ}, as found using Eqs. (4.13) and (4.14). The net result is a smaller shift in I_C than would occur if the emitter resistance was not present. The use of a transistor with lower than expected beta results in similar action, where the resulting decrease in I_C is again less than without emitter feedback.

To minimize the effects of β variation on the Q point, we should make R_B as small as practical. For reasons that will become clear later, in a practical application, the emitter resistance is usually limited to a few thousand ohms at the most. The typical range of β for a general-purpose transistor is from around 50 to 300 or so. Connecting R_B directly to V_{CC}, the base resistance cannot usually be reduced to much less than 100 kΩ. This means that we are somewhat limited in our ability to stabilize the Q point at this point. We are, however, moving in the right direction.

EXAMPLE 4.6

Refer to Fig. 4.12. Determine the percent change in I_{CQ} from that value determined in Example 4.5 if a transistor with $\beta = 200$ is used.

Solution Previous results have given us (for $\beta = 100$)

$$I_{CQ(100)} = 3.86 \text{ mA}$$

Assuming that $\beta = 200$, we obtain

$$I_{CQ(200)} = \frac{V_{CC} - V_{BE}}{R_E + \dfrac{R_B}{\beta}} = \frac{14.3 \text{ V}}{2.35 \text{ k}\Omega} = 6.09 \text{ mA}$$

The percent change in beta is $\Delta\beta = 100\%$. The percent change in collector current is

$$\Delta I_{CQ} = \frac{I_{CQ(200)} - I_{CQ(100)}}{I_{CQ(100)}} \times 100\% = \frac{2.23 \text{ mA}}{3.86 \text{ mA}} \times 100\% = 57.7\%$$

Without emitter feedback, the 100% increase in β would have resulted in a 100% increase in I_{CQ} as well.

The beta sensitivity of the circuit in Fig. 4.12 can be further improved if we increase the value of R_E. If we also decrease the value of R_B by the appropriate amount, we can maintain the same design value of I_{CQ}, while further enhancing stability. The next example illustrates this concept.

EXAMPLE 4.7

The circuit of Fig. 4.12 was redesigned as shown in Fig. 4.14. Determine the values of I_{CQ} for $\beta = 100$ and 200, and the percent increase in I_C that results.

FIGURE 4.14
Circuit for Example 4.7.

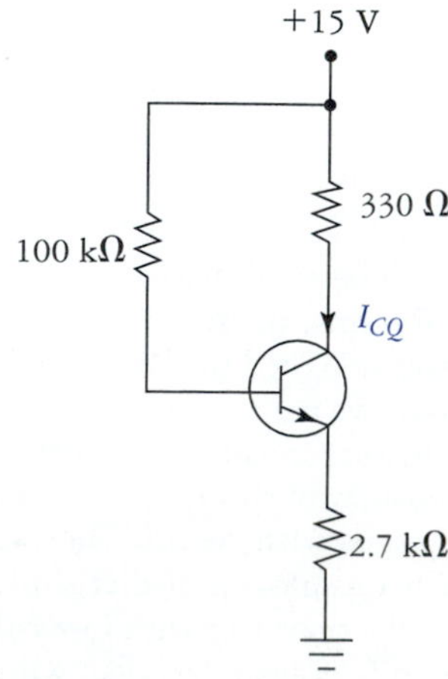

Solution

$$I_{CQ(100)} = \frac{V_{CC} - V_{BEQ}}{R_E + \dfrac{R_B}{\beta}} = \frac{14.3\text{ V}}{3700\ \Omega} = 3.86\text{ mA}$$

Similarly, we find

$$I_{CQ(200)} = 4.47\text{ mA}$$

The percent increase in β is again $\Delta\beta = 100\%$, while the corresponding increase in collector current is now $\Delta I_{CQ} = 15.8\%$.

The circuit of Fig. 4.14 is much more immune to the effects of transistor parameter variation than that of Fig. 4.12; however, there is still much room for improvement. Note also that although the redesigned circuit of Example 4.7 has the same value of I_{CQ} as a design goal, the Q-point location has changed significantly. This occurs because V_{CEQ} has changed from the original design. Given the fixed collector resistance and I_{CQ}, increasing R_E has moved the Q point much closer toward saturation. This is one of the big disadvantages of this biasing arrangement; generally, we must trade flexibility in Q-point location for increased Q-point stability. The design of electronic circuits is almost always dominated by trade-offs and compromises.

4.3 COLLECTOR FEEDBACK

An alternative to the use of emitter feedback is *collector feedback.* Figure 4.15(a) shows the basic circuit, while the dependent-source model is shown in Fig. 4.15(b). Intuitively, the circuit works as follows: Assuming that we are at some Q point, let's say that for some reason β increases. The increase in β tends to cause an increase in I_{CQ}. However, an increase in I_{CQ} causes an increase in the voltage drop across R_C. This voltage drop is series-opposing V_{CC}, which causes a decrease in the bias applied to the B-E junction. As was the case with emitter feedback, the total change in collector current is less than without negative feedback.

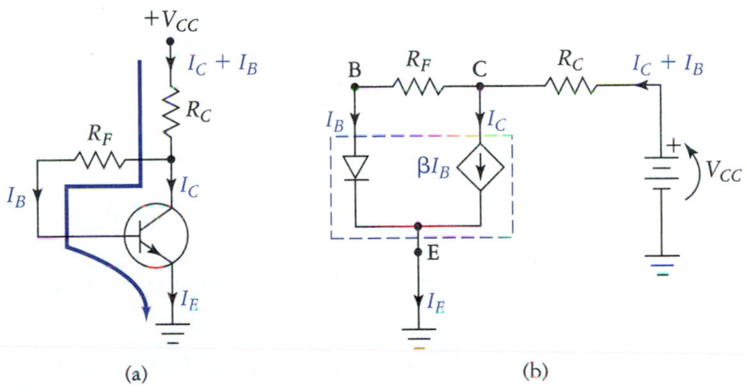

FIGURE 4.15
Collector-feedback biasing. (a) Typical circuit. (b) Equivalent circuit using dependent-source model of the transistor.

We now proceed with the quantitative analysis of the collector-feedback biasing arrangement. Writing the KVL equation for the indicated path (loop) in Fig. 4.15(a), we obtain

$$0 = V_{CC} - R_C(I_C + I_B) - I_B R_F - V_{BE}$$

As usual, we prefer as many common factors as possible, so let us define $I_C = \beta I_B$, giving us

$$0 = V_{CC} - R_C(\beta I_B + I_B) - I_B R_F - V_{BE}$$

Solving for I_B, we obtain

$$I_B = \frac{V_{CC} - V_{BE}}{R_C + \beta R_C + R_F} \tag{4.17a}$$

or, equivalently,

$$I_B = \frac{V_{CC} - V_{BE}}{R_C(\beta + 1) + R_F} \tag{4.17b}$$

Although it may sometimes be useful to know the value of I_B, it is more important to know the value of I_{CQ}. Multiplying both sides of Eq. (4.17b) by β yields

Note
We would have derived Eq. 4.19 directly if we assumed $\beta + 1 = \beta$ in Eq. 4.17b. Try it.

$$I_{CQ} = \frac{V_{CC} - V_{BE}}{R_C + \dfrac{R_C}{\beta} + \dfrac{R_F}{\beta}} \tag{4.18}$$

This equation can be simplified somewhat if we make the now-familiar assumption that $(R_C/\beta) + R_C \cong R_C$, yielding

$$I_{CQ} \cong \frac{V_{CC} - V_{BE}}{R_C + \dfrac{R_F}{\beta}} \tag{4.19}$$

EXAMPLE 4.8

The circuit of Fig. 4.15 has the following specifications: $V_{CC} = 10$ V, $R_C = 2.2$ kΩ, $R_F = 220$ kΩ, and $\beta = 100$. Sketch the dc load line for the circuit.

Solution Let us begin by finding the collector current:

$$I_{CQ} = \frac{V_{CC} - V_{BE}}{R_C + \dfrac{R_F}{\beta}}$$

$$= \frac{9.3 \text{ V}}{4.4 \text{ k}\Omega}$$

$$= 2.11 \text{ mA}$$

We find V_{CEQ} via KVL (assuming that $I_{CQ} + I_B = I_{CQ}$):

$$0 = V_{CC} - I_{CQ} R_C - V_{CEQ}$$

Solving for V_{CEQ},

$$\begin{aligned} V_{CEQ} &= V_{CC} - I_C R_C \\ &= 10 \text{ V} - (2.11 \text{ mA})(2.2 \text{ k}\Omega) \\ &= 5.36 \text{ V} \end{aligned}$$

The feedback resistor R_F has no effect on the saturation or cutoff values of the circuit; therefore,

$$I_{C(\text{sat})} \cong \frac{V_{CC}}{R_C}$$

$$= \frac{10 \text{ V}}{2.2 \text{ k}\Omega}$$

$$= 4.55 \text{ mA}$$

and

$$V_{CE(\text{cut})} \cong V_{CC} = 10 \text{ V}$$

The dc load line is shown in Fig. 4.16.

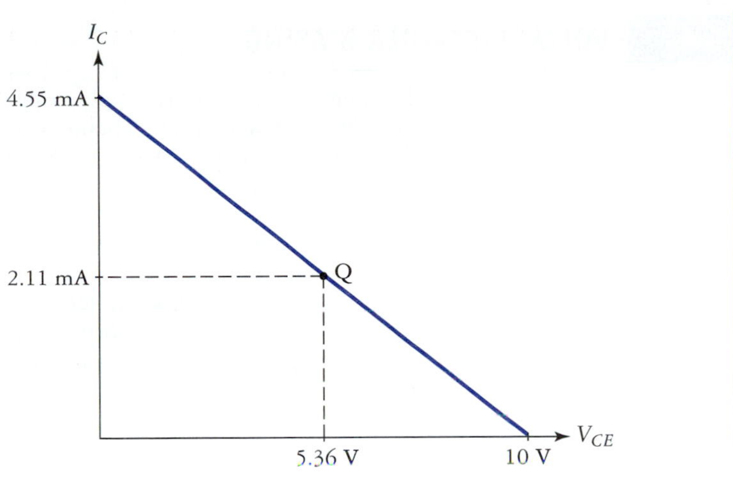

FIGURE 4.16
Load line for Example 4.8.

Collector Feedback versus Emitter Feedback

In terms of Q-point stability, collector-feedback performance is approximately the same as that of emitter feedback when $R_E/R_B = R_C/R_F$. The advantage of collector feedback is that it is easier to design the circuit (in the common-emitter configuration, at least) for a particular Q point because we are only working with two resistors R_C and R_F. When using emitter feedback, we must juggle three resistor values around (unless $R_C = 0$, which is a special case) in order to obtain some desired Q point. Thus, thinking strictly in terms of biasing, the collector-feedback circuit is easier to design. In fact, if the design goal is to center the Q point on the dc load line, then the following formula can be used, given a known collector resistance (this is the usual case):

Note
Keep in mind that Eq. 4.20 is a design equation, not an analysis equation.

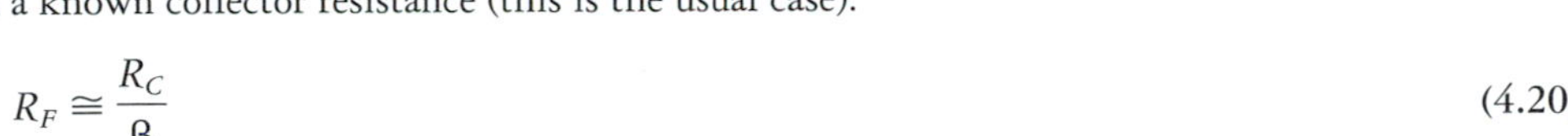

$$R_F \cong \frac{R_C}{\beta} \tag{4.20}$$

This design rule was used in determining the resistor values for Example 4.8, but it must be remembered that Eq. (4.20) is a suggested *design* equation. It is not an *analysis* equation.

Although Eq. (4.20) is useful and interesting, it is shown later that having the Q point centered on the dc load line is not always desirable, so things are not quite as simple as they might seem.

All things being equal, a circuit employing emitter feedback will be easier for a signal source to drive than an equivalent circuit with collector feedback. This is so because the effects of emitter feedback make the effective resistance (looking into the base of the transistor) higher. Detailed proof of this statement must be deferred until later; however, it should be intuitively plausible now, based on our work up to this point.

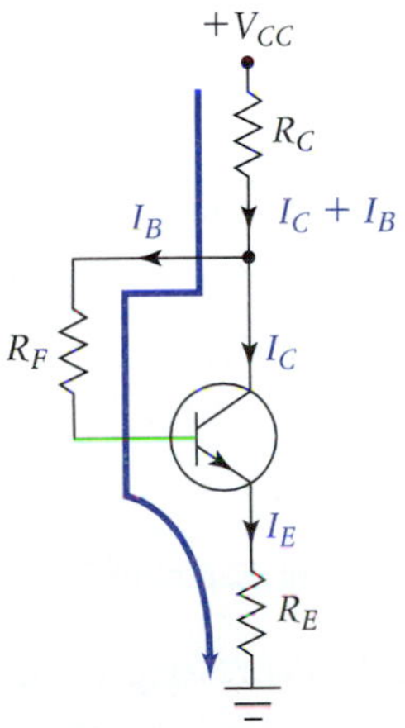

FIGURE 4.17
A circuit combining emitter feedback and collector feedback.

Composite Emitter–Collector Feedback

It is possible to combine both emitter feedback and collector feedback simultaneously. The resulting circuit is shown in Fig. 4.17. The quiescent collector current for this circuit can be approximated by

$$I_{CQ} \cong \frac{V_{CC} - V_{BE}}{R_C + R_E + \dfrac{R_E}{\beta}} \tag{4.21}$$

The derivation of Eq. (4.21) is left as an exercise for the reader. Routine circuit analysis gives the saturation current as

$$I_{C(\text{sat})} \cong \frac{V_{CC}}{R_C + R_E} \tag{4.22}$$

The value of $V_{CE(\text{cut})}$ is again given by $V_{CE(\text{cut})} \cong V_{CC}$.

4.4 VOLTAGE-DIVIDER BIASING

If a single-polarity power supply is all that is available, the technique of *voltage-divider biasing* provides an effective means for obtaining a stable Q point. Recall from the analysis of the basic emitter-feedback circuit that for good Q-point stability we require $R_B \ll \beta R_E$. It happens that voltage-divider biasing makes this a realistic design goal.

The basic voltage-divider biasing arrangement is shown in Fig. 4.18(a). At first glance, this circuit may appear rather ominous and complex looking. However, application of Thevenin's theorem provides us with a convenient method of analysis. Converting the voltage divider composed of R_1 and R_2 and the V_{CC} source into their Thevenin equivalent, splitting V_{CC} between the left- and right-hand halves of the circuit produces the equivalent circuit of Fig. 4.18(b). Thus, the voltage-divider biasing circuit can be decomposed into the more familiar and easily analyzed emitter-feedback circuit.

Note
Thevenin's theorem allows us to redraw circuits in simpler, more familiar forms.

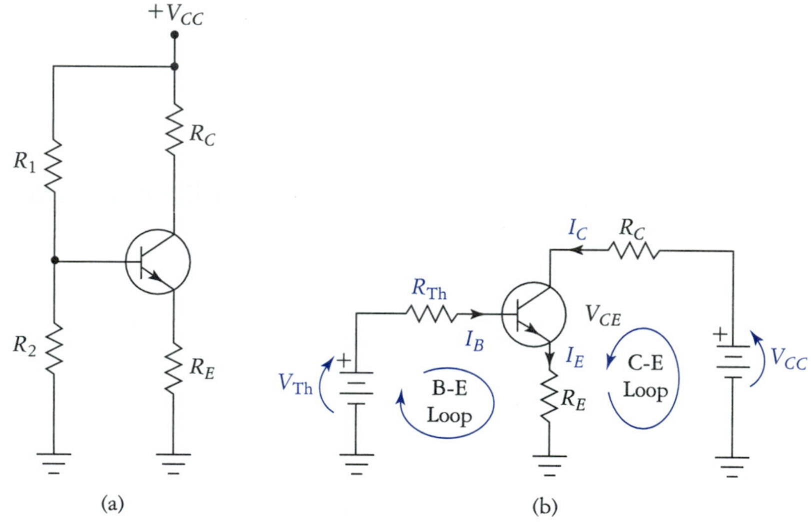

FIGURE 4.18
Voltage-divider biasing. (a) Typical circuit. (b) Circuit redrawn to facilitate analysis after base network is Thevenized.

The following analysis equations are determined for the Thevenin equivalent circuit:

$$R_{Th} = R_1 \parallel R_2 \tag{4.23}$$

and

$$V_{Th} = V_{CC}\frac{R_2}{R_1 + R_2} \tag{4.24}$$

The remaining voltages and currents are determined as usual through the use of Kirchhoff's laws, Ohm's law, and the basic transistor device relationships.

The best way to gain an appreciation of the usefulness of the voltage-divider biasing arrangement is through a thorough analysis. The following example illustrates.

EXAMPLE 4.9

Consider the circuit shown in Fig. 4.19. Assuming that $\beta = 100$, determine the Thevenin equivalent values I_{CQ}, V_{CEQ}, V_{CQ}, V_{EQ}, V_{BQ}, $I_{C(sat)}$, and $V_{CE(cut)}$.

Solution The Thevenin equivalent circuit is shown in Fig. 4.20. The Thevenin resistance and voltage values are determined as follows:

$$\begin{aligned} R_{Th} &= R_1 \parallel R_2 \\ &= 18\text{ k}\Omega \parallel 5.6\text{ k}\Omega \\ &= 4.27\text{ k}\Omega \end{aligned}$$

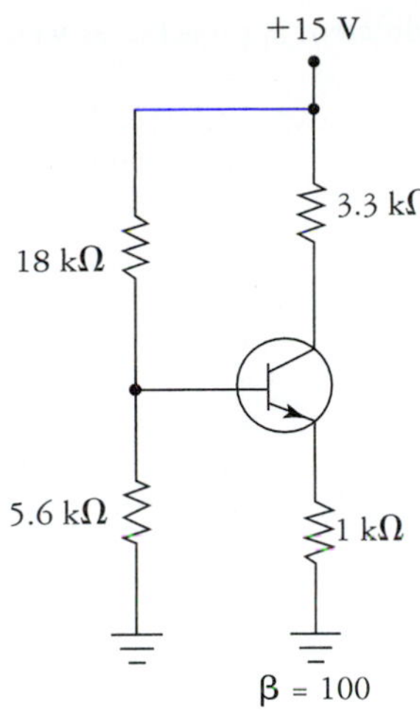

FIGURE 4.19
Circuit for Example 4.9.

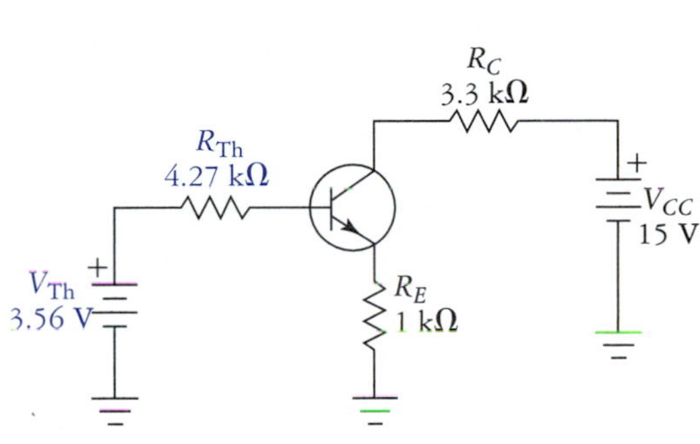

FIGURE 4.20
Thevenized equivalent circuit for Fig. 4.19.

$$\begin{aligned} V_{Th} &= V_{CC}\frac{R_2}{R_1 + R_2} \\ &= 15\text{ V}\frac{5.6\text{ k}\Omega}{5.6\text{ k}\Omega + 18\text{ k}\Omega} \\ &= 3.56\text{ V} \end{aligned}$$

We determine the collector current using a slightly modified form of Eq. (4.14):

$$I_{CQ} = \frac{V_{Th} - V_{BE}}{R_E + \dfrac{R_{Th}}{\beta}} \tag{4.25}$$

$$\begin{aligned} &= \frac{3.56\text{ V} - 0.7\text{ V}}{1\text{ k}\Omega + \dfrac{4.27\text{ k}\Omega}{100}} \\ &= 2.74\text{ mA} \end{aligned}$$

The quiescent collector-to-emitter voltage is found directly from Eq. (4.16), which is repeated here for convenience:

$$V_{CEQ} = V_{CC} - I_{CQ}(R_C + R_E) \tag{4.16}$$

which gives us

$$\begin{aligned} V_{CEQ} &= 15\text{ V} - (2.74\text{ mA})(4.3\text{ k}\Omega) \\ &= 15\text{ V} - 11.78\text{ V} \\ &= 3.22\text{ V} \end{aligned}$$

The various single-subscripted voltages are determined as follows:

$$\begin{aligned} V_{CQ} &= V_{CC} - I_C R_C \\ &= 15\text{ V} - (2.74\text{ mA})(3.3\text{ k}\Omega) \\ &= 15\text{ V} - 9.04\text{ V} \\ &= 5.96\text{ V} \\ V_{EQ} &\cong I_C R_E \\ &= (2.74\text{ mA})(1\text{ k}\Omega) \\ &= 2.74\text{ V} \end{aligned}$$

As a check, let us find V_{EQ} using

$$\begin{aligned} V_{EQ} &= V_{CC} - I_C R_C - V_{CEQ} \\ &= 15\text{ V} - 9.04\text{ V} - 3.22\text{ V} \\ &= 2.74\text{ V} \qquad \text{(OK)} \end{aligned}$$

We shall find the base voltage using two different approaches as well, just to emphasize the self-consistency of the solutions.

$$\begin{aligned} V_{BQ} &= V_{EQ} + V_{BEQ} \\ &= 2.74 \text{ V} + 0.7 \text{ V} \\ &= 3.44 \text{ V} \end{aligned}$$

We need I_B for the next check, so

$$\begin{aligned} I_B &= I_C / \beta \\ &= 27.4\ \mu\text{A} \end{aligned}$$

Thus,

$$\begin{aligned} V_{BQ} &= V_{\text{Th}} - I_B R_{\text{Th}} \\ &= 3.56 \text{ V} - (27.4\ \mu\text{A})(4.27 \text{ k}\Omega) \\ &= 3.56 \text{ V} - 0.12 \text{ V} \\ &= 3.44 \text{ V} \qquad \text{(OK)} \end{aligned}$$

Finally, the cutoff and saturation levels are easily determined.

$$V_{CE(\text{cut})} \cong V_{CC} = 15 \text{ V}$$

$$\begin{aligned} I_{C(\text{sat})} &\cong \frac{V_{CC}}{R_C + R_E} \\ &= \frac{15 \text{ V}}{4.3 \text{ k}\Omega} \\ &= 3.49 \text{ mA} \end{aligned}$$

Voltage-Divider Biasing Stability

The stability of the typical voltage-divider biased circuit is normally quite good. Using the circuit that was analyzed in Example 4.9 as a starting point, we found, for $\beta = 100$: $I_{CQ(100)} = 2.74$ mA and $V_{CEQ(100)} = 3.22$ V. Suppose that we used a transistor that had $\beta = 50$ in this circuit. It is left for the reader to verify that $I_{CQ(50)} = 2.63$ mA and $V_{CEQ(50)} = 3.69$ V. Thus, for a 50% decrease in β, we obtained only a 4% decrease in collector current.

The circuit in Fig. 4.19 is biased relatively close to saturation. It would appear that the use of a transistor with higher than expected beta might cause saturation to occur. Let us assume this time that the circuit is constructed using a transistor with $\beta = 300$. This is a 300% increase from the value assumed in the analysis of Example 4.9. Reanalyzing the circuit yields the following interesting results: $I_{CQ(300)} = 2.82$ mA and $V_{CEQ(300)} = 2.87$ V. The change in collector current is only 2.9%. This is a very stable circuit.

The circuit of Fig. 4.19 is so stable because, over a reasonable range of expected beta variation, we have satisfied the inequality

$$R_E \gg \frac{R_{\text{Th}}}{\beta} \tag{4.26}$$

As a general rule of thumb, when $R_E \geq 100\ R_{\text{Th}}/\beta$ then the circuit will be virtually completely independent of normal beta variations. In such cases, the quiescent collector current of the voltage-divider biased transistor of Fig. 4.18(a) is

$$I_{CQ} \cong \frac{V_{\text{Th}} - V_{BE}}{R_E} \tag{4.27}$$

If we can use Eq. (4.27), then we do not have to resort to the determination of the Thevenin resistance and voltage of the base-resistor network. Basically, Eq. (4.27) assumes that $V_{BQ} = V_{\text{Th}}$ and $I_{BQ} = 0$. If the inequality of (4.26) holds, then these assumptions introduce little error.

EXAMPLE 4.10

Refer to the circuit of Fig. 4.19. Determine the required value for R_C such that $V_{CEQ} = V_{CC}/2$. Sketch the resulting dc load line.

Solution Changing the value of the collector resistor has no effect on the B-E loop of the circuit. Therefore, $I_{CQ} = 2.74$ mA, as determined earlier. The new value of R_C is found by solving the KVL equation of the C-E loop for R_C as follows:

$$0 = V_{CC} - I_{CQ} R_C - V_{CEQ} - I_C R_E$$

$$R_C = \frac{V_{CC} - V_{CEQ} - I_{CQ} R_E}{I_{CQ}}$$

$$= \frac{15\text{ V} - 7.5\text{ V} - 2.74\text{ V}}{2.74\text{ mA}}$$

$$= \frac{4.76\text{ V}}{2.74\text{ mA}}$$

$$= 1.74\text{ k}\Omega$$

In practice, we would probably use the next-closest standard value (1.8 kΩ). The new collector saturation current is

$$I_{C(\text{sat})} \cong \frac{V_{CC}}{R_C + R_E}$$

$$= \frac{15\text{ V}}{1\text{ k}\Omega + 1.8\text{ k}\Omega}$$

$$= 5.36\text{ mA}$$

and, of course, $V_{CE(\text{cut})} = V_{CC} = 15$ V. The load line is shown in Fig. 4.21.

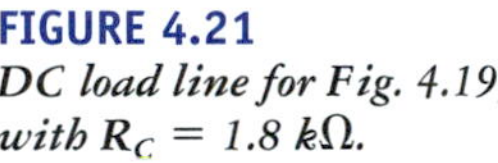

FIGURE 4.21
DC load line for Fig. 4.19, with $R_C = 1.8$ kΩ.

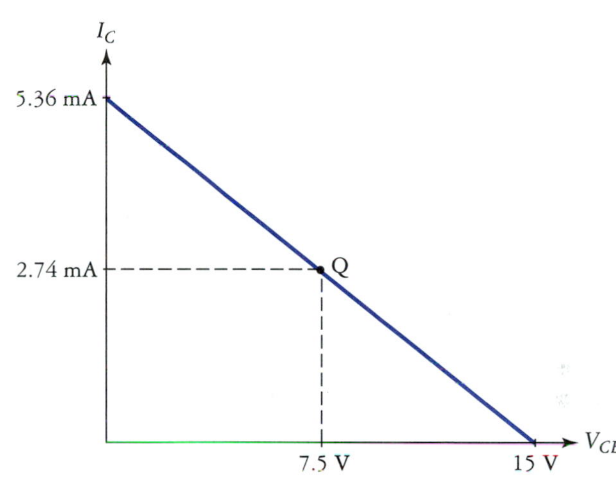

Notice from the preceding example that centering the Q point on the V_{CE} axis also results in operation at the center of the I_C axis.

Some Voltage-Divider Bias Design Guidelines*

Generally speaking, the greater the number of components there are in a given circuit, the greater the number of combinations of values that will produce a good design. It bears repeating that there is really no single right way to design most circuits. usually, a somewhat vague set of goals exists, and as long as they are achieved in a reasonable time and at a reasonable cost the design is a success. Normally, the application will dictate the design; that is, form follows function.

When designing voltage-divider biased transistor circuits, a few general rules can be followed that can make the design process proceed a little more smoothly. Considering designs with low-power transistors, I_{CQ} might range from a few hundred microamps to tens of milliamps, while the

*This section can be skipped without loss of continuity, or it can be covered later if time permits.

supply voltage may range from ±5 V to possibly ±25 V or more. Suppose we have the following specifications to work with (we shall also stay with npn transistors for now):

$$V_{CC} = 12\text{ V}$$
$$I_{CQ} = 1\text{ mA}$$
$$V_{CEQ} = 6\text{ V}$$
$$R_C > 0\ \Omega$$
$$\beta_{\min} \geq 50$$

The specification $R_C > 0\ \Omega$ is included because not all circuits need or use a collector resistor. Many types of amplifiers do, however, require a collector resistor and as far as we know this is the case here. We begin by choosing the emitter voltage such that $V_E \geq 2\ V_{BE}$. This ensures that slight variations in V_{BE} will not significantly affect the emitter current. We have 6 V to work with here ($V_{CC} - V_{CEQ} = 6$ V), so somewhat arbitrarily, let's shoot for $V_E \cong 3V_{BE} = 2.1$ V. This leaves about 3.9 V to drop across the collector resistor. Now, since I_C is to be 1 mA, we write

$$\begin{aligned} R_E &= V_E/I_E \\ &= 2.1\text{ V}/1\text{ mA} \\ &= 2.1\text{ k}\Omega \end{aligned}$$

This is not a standard value, but we'll decide how to handle this situation in a moment.

The collector resistor value can now be determined. Recall that we require the product $I_C R_C = 3.9$ V. This works out well, because $I_C = 1$ mA, which gives us $R_C = 3.9$ kΩ, which is a standard resistor value.

The base voltage is now determined. Because V_E has been decided, we know that the base voltage is

$$\begin{aligned} V_B &= V_E + V_{BE} \\ &= 2.1\text{ V} + 0.7\text{ V} \\ &= 2.8\text{ V} \end{aligned}$$

We now must design the voltage-divider network. Refer back to the basic circuit of Fig. 4.18(a), for reference purposes. To obtain a stable Q point, we require

$$R_{\text{Th}} \ll \beta_{\min} R_E$$

It is certainly clear that $R_2 > R_{\text{Th}}$, so if we choose $R_2 \ll \beta_{\min} R_E$ then a stable Q point will be achieved. Using $R_E = 2.1$ kΩ, we get

$$\begin{aligned} R_2 &\ll \beta_{\min} R_E \\ &\ll (100)(2.1\text{ k}\Omega) \\ &\ll 60\text{ k}\Omega \end{aligned}$$

It may be tempting to use a very low value for R_2, but we don't want to draw excessive current from the power supply just to heat up biasing resistors. (We compromise again: A very slight decrease in bias stability is traded for low-power dissipation.) A reasonable value for R_2 might be anything from 1000 Ω up to around 22 kΩ. Again, rather arbitrarily, let's choose $R_2 = 3.3$ kΩ and see what happens. The circuit is essentially β independent, so we simply solve Eq. (4.24) for R_1, which yields

$$\begin{aligned} R_1 &= R_2\left(\frac{V_{CC}}{V_{\text{Th}}} - 1\right) \\ &= 3.3\text{ k}\Omega\left(\frac{12\text{ V}}{2.8\text{ V}} - 1\right) \\ R_1 &= 10.8\text{ k}\Omega \end{aligned}$$

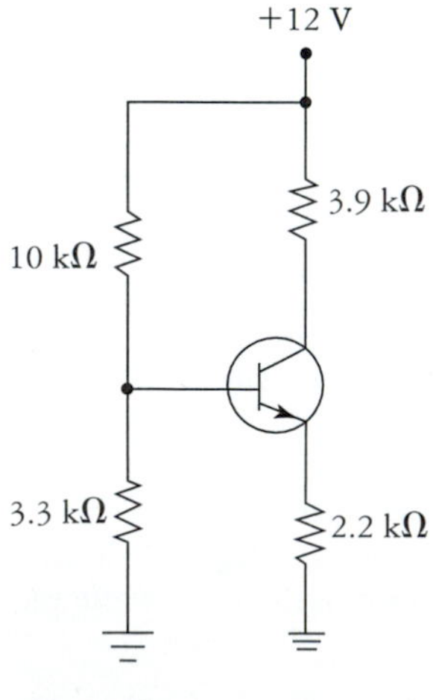

This is not a standard value, but if we use the next lowest standard value here (10 kΩ) and then round R_E up to the next higher standard value (2.2 kΩ), perhaps our compromises will balance. If the resulting design is too far off the mark, we must go back and modify a resistor value or two.

The completed circuit is shown in Fig. 4.22. A detailed analysis, using the worst-case β = 50, produced the following results:

$$I_{CQ} = 1.01\text{ mA} \qquad \text{and} \qquad V_{CEQ} = 5.84\text{ V}$$

FIGURE 4.22
The completed circuit design.

All things considered, these are very close to the desired values (we're in error by less than 3%). In fact, if we are going to use 5% tolerance (or worse) resistors to build the circuit, there is no point in even trying to come closer to the design goals.

You must keep in mind that every design situation is different; some may be trivial and some may be incredibly difficult. Having a holistic or "big picture" view of the design requirements and the basic building blocks and their limitations (and, of course, expertise with the basic circuit analysis techniques) helps make the design process more efficient. This comes with time and experience. Sometimes it is best to follow strict procedures, while at other times it is better to go "by the seat of your pants." Most of the time, a given design will simply require a modification of some standard well-understood circuit. Hopefully, this book will provide you with many such examples.

4.5 EMITTER BIAS

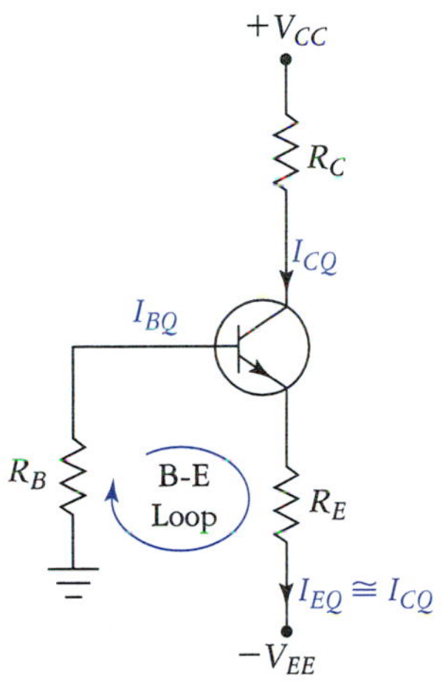

FIGURE 4.23
Basic emitter biasing using a bipolar power supply.

While voltage-divider biasing provides very good stability in single-polarity power supply-type applications, if a bipolar power supply is available, then *emitter biasing* is often preferred. The basic emitter-biased npn transistor circuit is shown in Fig. 4.23.

Here's the idea behind emitter biasing, concentrating on npn transistors: To locate the Q point in the active region, we must forward bias the B-E junction and reverse bias the C-B junction. Previously, we forward biased the B-E junction by connecting the base to a positive voltage via an appropriate resistance, through a voltage divider, or we connected it to the collector, while leaving the emitter grounded. An equivalent forward bias of the B-E junction can be produced by connecting the base to ground and connecting the emitter to a negative voltage ($-V_{EE}$). Incredibly good bias stability is achieved if the base is shorted directly to ground, but in most practical circuits, a series resistance R_B must be used, as shown in Fig. 4.23.

You may have noticed a similarity between the emitter-biased circuit of Fig. 4.23 and the base-biased circuit of Fig. 4.7(a). The addition of the emitter resistor will stabilize the Q point by allowing a smaller R_B to be used to obtain a given I_{CQ}, with only a slight increase in complexity. To be sure there is no confusion, the B-E loop of the emitter-bias circuit is redrawn in Fig. 4.24. We use this circuit only to determine the emitter (or base) current. As usual, for most transistors $\alpha \cong 1$, so we make the simplifying assumption that $I_C = I_E$. Using KVL and working around the B-E loop, we find

$$0 = V_{EE} - I_B R_B - V_{BE} - I_C R_E$$

Defining I_B in terms of I_C and β produces

$$0 = V_{EE} - \frac{I_C R_B}{\beta} - V_{BE} - I_C R_E$$

Solving for I_C yields

$$I_{CQ} = \frac{V_{EE} - V_{BE}}{R_E + \dfrac{R_B}{\beta}} \tag{4.28}$$

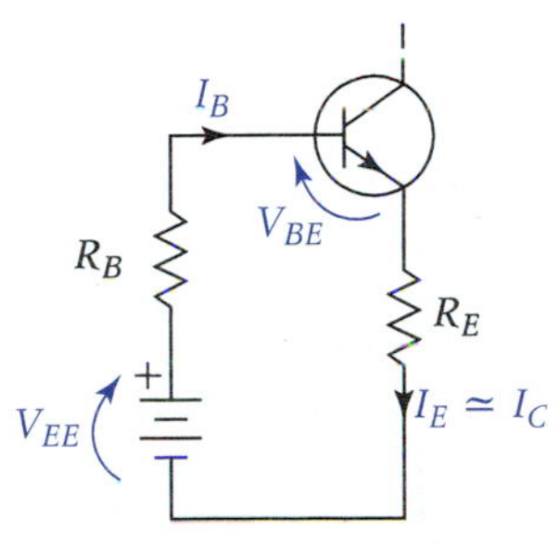

FIGURE 4.24
Base-emitter loop redrawn to facilitate determination of quiescent collector current.

Once the collector current or the emitter current is known, it is simply a matter of using Ohm's law and Kirchhoff's laws on the original circuit in order to complete the analysis. Also, if the inequality of (4.26) is met (substituting R_B for R_{Th}, of course), then we can use a simpler approximation:

$$I_{CQ} \cong \frac{V_{EE} - V_{BE}}{R_E} \tag{4.29}$$

The saturation current for Fig. 4.23 is found by treating the transistor as if it were shorted from collector to emitter. This results in the following equation:

$$I_{C(\text{sat})} \cong \frac{V_{CC} + V_{EE}}{R_C + R_E} \tag{4.30}$$

In cutoff, the collector current will normally be in the nanoamp range, and as a result, we find that

$$V_{CE(\text{cut})} \cong V_{CC} + V_{EE} \tag{4.31}$$

EXAMPLE 4.11

Refer to Fig. 4.23. Given $V_{CC} = V_{EE} = 12$ V, $R_C = 1.2$ kΩ, $R_E = 2.7$ kΩ, $R_B = 1$ kΩ, and $\beta = 100$, sketch the dc load line for the circuit and determine V_{CQ}, V_{BQ}, and V_{EQ}. Use Eq. (4.29) to determine the value of I_{CQ}.

Solution

$$I_{CQ} \cong \frac{V_{EE} - V_{BE}}{R_E} = \frac{12\text{ V} - 0.7\text{ V}}{2.7\text{ k}\Omega} = 4.19\text{ mA}$$

$$\begin{aligned} V_{CEQ} &= V_{CC} + V_{EE} - I_C(R_C + R_E) \\ &= 12\text{ V} + 12\text{ V} - (4.19\text{ mA})(1.2\text{ k}\Omega + 2.7\text{ k}\Omega) \\ &= 12\text{ V} + 12\text{ V} - (4.19\text{ mA})(3.9\text{ k}\Omega) \\ &= 7.66\text{ V} \end{aligned}$$

$$I_{C(\text{sat})} = \frac{V_{CC} + V_{EE}}{R_C + R_E} = \frac{12\text{ V} + 12\text{ V}}{1.2\text{ k}\Omega + 2.7\text{ k}\Omega} = \frac{24\text{ V}}{3.9\text{ k}\Omega} = 6.15\text{ mA}$$

$$\begin{aligned} V_{CE(\text{cut})} &= V_{CC} + V_{EE} \\ &= 24\text{ V} \end{aligned}$$

The dc load line is shown in Fig. 4.25.

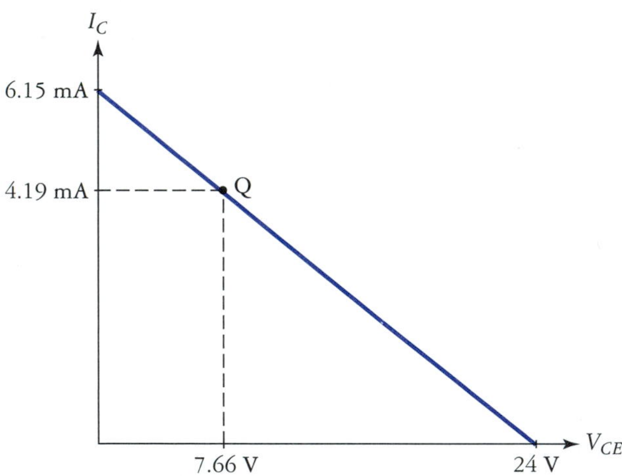

FIGURE 4.25
DC load line for Example 4.11.

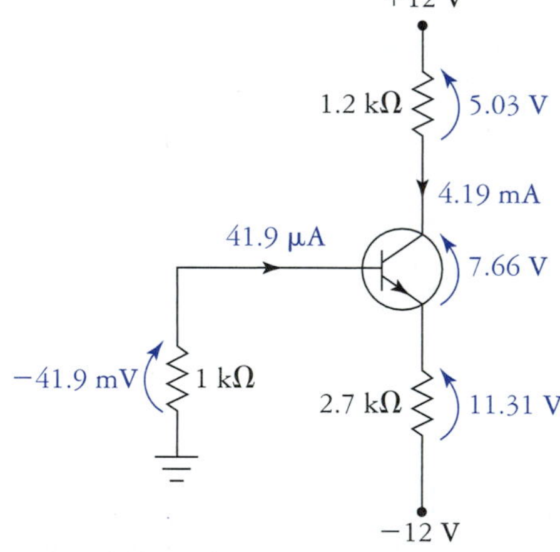

FIGURE 4.26
Circuit redrawn with all currents and voltages determined.

To more clearly show how the collector, base, and emitter voltages are determined, all resistor voltage drops and V_{CE} have been entered on the circuit in Fig. 4.26. The collector voltage is given by

$$\begin{aligned} V_{CQ} &= V_{CC} - I_{CQ} R_C \\ &= 12\text{ V} - (4.19\text{ mA})(1.2\text{ k}\Omega) \\ &= 12\text{ V} - 5.03\text{ V} \\ &= 6.97\text{ V} \end{aligned}$$

The emitter voltage is

$$\begin{aligned} V_{EQ} &= -V_{EE} + I_E R_E \qquad \text{(Remember, we have assumed } I_E \cong I_C.\text{)} \\ &= -12 \text{ V} + (4.19 \text{ mA})(2.7 \text{ k}\Omega) \\ &= -12 \text{ V} + 11.31 \text{ V} \\ &= -0.69 \text{ V} \end{aligned}$$

As a consistency check, we can use the following relationship:

$$V_{CEQ} = V_{CQ} - V_{EQ}$$

Thus, we find

$$V_{CQ} - V_{EQ} = 6.97 \text{ V} - (-0.69 \text{ V}) = 7.66 \text{ V} \qquad \text{(OK)}$$

There are several ways to determine V_{BQ}. Because we have already found V_{EQ}, the easiest way is to use

$$\begin{aligned} V_{BQ} &= V_{EQ} + V_{BEQ} \\ &= -0.69 \text{ V} + 0.7 \text{ V} \\ &= 10 \text{ mV} \end{aligned}$$

However, if we look at Fig. 4.26, we see that we should get $V_{BQ} = -I_B R_B = -41.9$ mV. Which answer is correct? Actually, neither is exactly correct, but because there is some base current flowing, the base *has* to be slightly below ground potential, so -49.1 mV is most likely the more accurate answer. The discrepancy is caused by our assumption that $I_C \cong I_E$, and by the use of Eq. (4.29), which effectively assumes that $I_B = 0$. The important thing to realize is that the errors are quite small. Resistor tolerance errors will normally cause greater errors in a practical circuit.

Memorization and Formulas

We have developed quite a few analysis equations up to this point, and as the next section will show, many more equations are yet to come. There are really far too many equations to commit to memory in the short span of time that you will no doubt spend covering this chapter. With this in mind, the following advice is offered: Memorize only the most fundamental of relationships, such as $I_C = \beta I_B$, $I_C \cong I_E$, and perhaps Eq. (4.14). All of the other relationships are best determined using basic circuit analysis techniques. Even if you keep a list of the formulas and equations that have been developed thus far, there is always the chance that a slight circuit design variation can make a given analysis equation useless.

4.6 ADDITIONAL BIASING VARIATIONS

Many different bias circuit variations are possible. However, most of these different circuits can be grouped under one of the biasing arrangements covered thus far: base-biased, collector-feedback biased, voltage-divider-biased, or emitter-biased. Of course, there are various hybrid combinations of these circuits in use as well, the combination of emitter feedback and collector feedback, for example. In any case, the most important circuit parameter to determine is the emitter current or, equivalently, the collector current. Once I_C is found, the rest of the analysis falls right into place. The key to finding I_C or I_E in an unusual or unfamiliar biasing arrangement is usually to redraw the circuit such that it looks like one of the circuits that we can easily analyze.

EXAMPLE 4.12

Determine I_{CQ}, V_{CEQ}, V_{CQ}, V_{BQ}, $I_{C(\text{sat})}$, and $V_{CE(\text{cut})}$ for the circuit in Fig. 4.27.

Solution This circuit is a combination of voltage-divider biasing and emitter biasing. The best way to start this analysis is by Thevenizing the voltage divider. This produces $V_{\text{Th}} = 1.13$ V and $R_{\text{Th}} = 2.03$ kΩ. Now we can redraw the B-E loop of the circuit to emphasize its true structure.

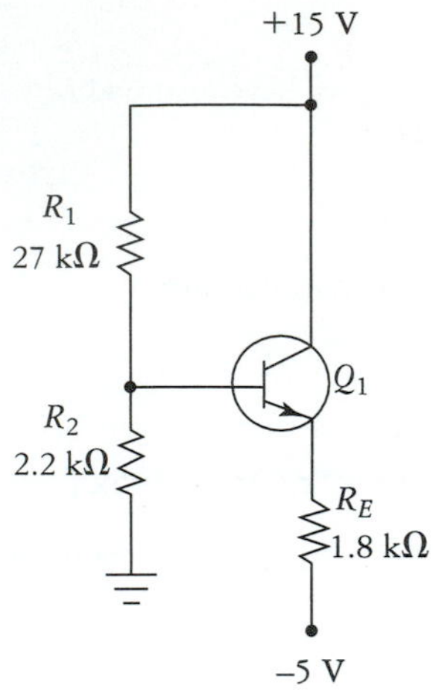

FIGURE 4.27
Circuit for Example 4.12.

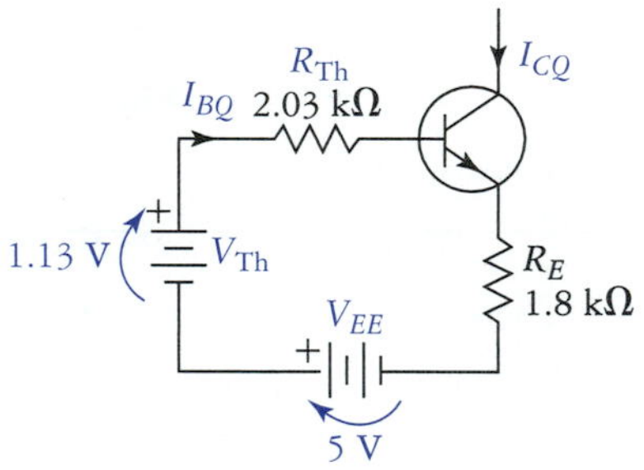

FIGURE 4.28
Equivalent circuit for B-E loop of Fig. 4.27.

This results in the circuit of Fig. 4.28, which is easily analyzed for I_{CQ}. Let us assume we have a typical transistor with $\beta = 100$.

$$I_{CQ} = \frac{V_{EE} + V_{\text{Th}} - V_{BEQ}}{R_E + \dfrac{R_{\text{Th}}}{\beta}}$$

$$= \frac{5\text{ V} - 1.13\text{ V} - 0.7\text{ V}}{1.8\text{ k}\Omega + 20.3\ \Omega}$$

$$= 2.98\text{ mA}$$

We now return to the original circuit of Fig. 4.27. V_{CEQ} is found in the usual manner to be

$$\begin{aligned} V_{CEQ} &= V_{CC} + V_{EE} - I_C R_E \qquad \text{(Note that } R_C = 0\ \Omega.) \\ &= 15\text{ V} + 5\text{ V} - (2.98\text{ mA})(1.8\text{ k}\Omega) \\ &= 15\text{ V} + 5\text{ V} - 5.36\text{ V} \\ &= 14.64\text{ V} \end{aligned}$$

The quiescent collector voltage is found directly by inspection:

$$V_{CQ} = V_{CC} = 15\text{ V}$$

It is easiest to determine V_{BQ} by using KVL as follows:

$$\begin{aligned} V_{BQ} &= -V_{EE} + I_{CQ} R_E + V_{BEQ} \\ &= -5\text{ V} + (2.98\text{ mA})(1.8\text{ k}\Omega) + 0.7\text{ V} \\ &= -5\text{ V} + 5.36\text{ V} + 0.7\text{ V} \\ &= 1.06\text{ V} \end{aligned}$$

The saturation current is now determined:

$$I_{C(\text{sat})} = \frac{V_{CC} + V_{EE}}{R_E}$$

$$= \frac{15\text{ V} + 5\text{ V}}{1.8\text{ k}\Omega}$$

$$= 11.11\text{ mA}$$

Finally, we find

$$\begin{aligned} V_{CE(\text{cut})} &= V_{CC} + V_{EE} \\ &= 15\text{ V} + 5\text{ V} \\ &= 20\text{ V} \end{aligned}$$

EXAMPLE 4.13

Refer to Fig. 4.29. Assuming $\beta = 100$, determine V_{CEQ}, I_{CQ}, V_{CQ}, V_{BQ}, V_{EQ}, and $I_{C(sat)}$.

Solution As in the last example, we start by Thevenizing the voltage-divider network. Before doing this, however, it is critical to understand that it is the voltage drop across R_2 alone that provides forward bias for the B-E junction. The 15-V V_{EE} source is outside of the loop as shown in Fig. 4.29:

$$\begin{aligned} V_{\text{Th}} &= V_{EE}\frac{R_2}{R_1 + R_2} \\ &= 15\text{ V}\frac{2.2\text{ k}\Omega}{2.2\text{ k}\Omega + 10\text{ k}\Omega} \\ &= 2.70\text{ V} \end{aligned}$$

$$\begin{aligned} R_{\text{Th}} &= R_1 \parallel R_2 \\ &= 10\text{ k}\Omega \parallel 2.2\text{ k}\Omega \\ &= 1.8\text{ k}\Omega \end{aligned}$$

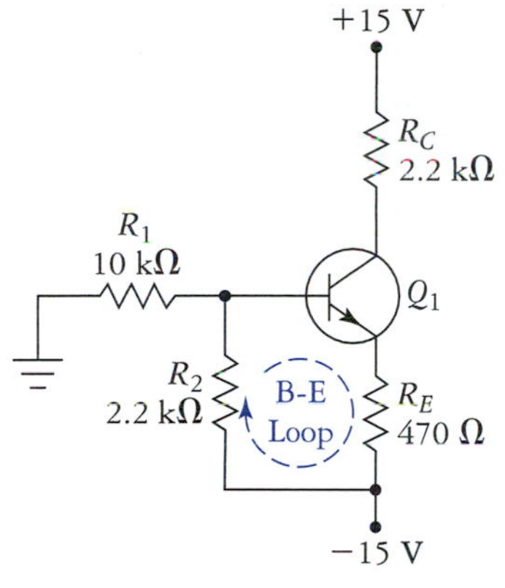

FIGURE 4.29
Circuit for Example 4.13.

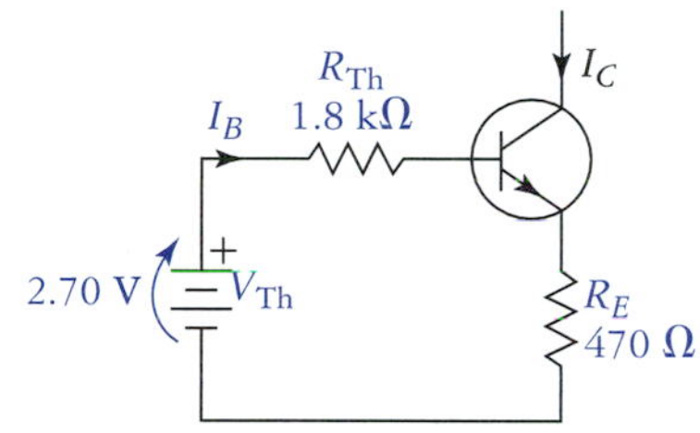

FIGURE 4.30
Emitter-base loop, Thevenized and redrawn for determination of I_{CQ}.

The circuit is now redrawn in Fig. 4.30, for calculation of I_{CQ}. This is virtually identical to the circuit obtained in Fig. 4.28, except as stated before, V_{EE} is outside of the closed B-E loop.

$$\begin{aligned} I_{CQ} &= \frac{V_{\text{Th}} - V_{BEQ}}{R_E + \dfrac{R_{\text{Th}}}{\beta}} \\ &= \frac{2.7\text{ V} - 0.7\text{ V}}{470\ \Omega + 18\ \Omega} \\ &= 4.10\text{ mA} \end{aligned}$$

$$\begin{aligned} V_{CEQ} &= V_{CC} + V_{EE} - I_{CQ}(R_C + R_E) \\ &= 15\text{ V} + 15\text{ V} - 4.10\text{ mA}(2.2\text{ k}\Omega + 470\ \Omega) \\ &= 15\text{ V} + 15\text{ V} - 4.10\text{ mA}(2.67\text{ k}\Omega) \\ &= 15\text{ V} + 15\text{ V} - 10.95\text{ V} \\ &= 19.05\text{ V} \end{aligned}$$

$$\begin{aligned} V_{CQ} &= V_{CC} - I_{CQ}R_C \\ &= 15\text{ V} - (4.10\text{ mA})(2.2\text{ k}\Omega) \\ &= 15\text{ V} - 9.02\text{ V} \\ &= 5.98\text{ V} \end{aligned}$$

$$\begin{aligned} V_{EQ} &= V_{CC} - I_{CQ}R_C - V_{CEQ} \\ &= 15\text{ V} - 9.02\text{ V} - 19.05\text{ V} \\ &= -13.07\text{ V} \end{aligned}$$

$$\begin{aligned} V_{BQ} &= V_{EQ} + V_{BEQ} \\ &= -13.07 \text{ V} + 0.7 \text{ V} \\ &= -12.37 \text{ V} \end{aligned}$$

$$\begin{aligned} I_{C(\text{sat})} &= \frac{V_{CC} + V_{EE}}{R_C + R_E} \\ &= \frac{15 \text{ V} + 15 \text{ V}}{2.2 \text{ k}\Omega + 470 \ \Omega} \\ &= 11.24 \text{ mA} \end{aligned}$$

You are encouraged to confirm the results of the preceding example using alternative applications of KVL.

Sometimes transistors are drawn at rather odd orientations in schematic diagrams. An example of such a drawing is shown in Fig. 4.31. At first glance this circuit appears quite unusual. Undoubtedly this configuration will become more familiar to you as you gain experience. Until then, it is usually helpful to attempt to redraw such a circuit in a form that is more familiar. Study of Fig. 4.31 reveals that we have a simple voltage-divider bias circuit. Nodes that are at the same potential can be connected without affecting circuit operation. Making this connection and flipping the transistor symbol around produces Fig. 4.32. The analysis of this circuit is almost second nature at this point.

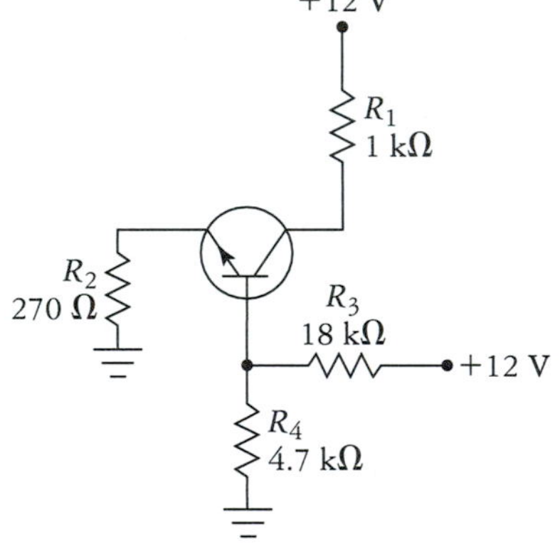

FIGURE 4.31
Unconventionally drawn circuit.

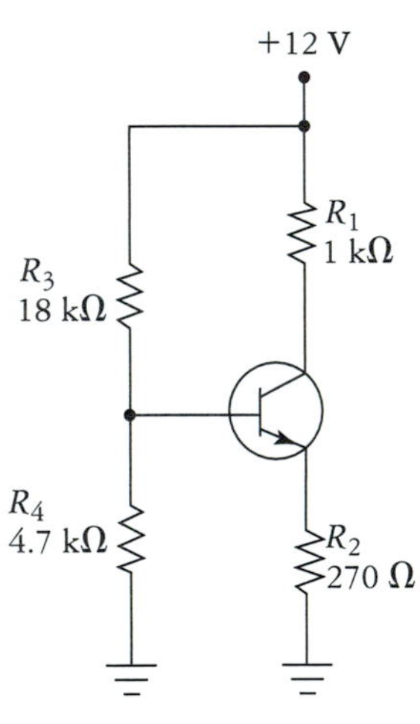

FIGURE 4.32
Redrawing Fig. 4.31 produces this familiar-looking circuit.

We finish this section with the analysis of a circuit that is designed around a pnp silicon transistor. We have concentrated on npn devices thus far, because they are used much more frequently, and there is a far greater variety of npn transistors available from which to choose. Sometimes though, pnp transistors must be used.

One of the more confusing aspects of the analysis of circuits containing pnp transistors is a result of the way such circuits are normally drawn. It is standard drafting practice to draw pnp transistor symbols upside down, relative to npn conventions. When this is done, all quiescent current flow (conventional) is from the top of the schematic to the bottom. Figure 4.33 illustrates this concept. Notice that based on the directions of the various sensing arrows that I_{CQ} is positive, while V_{CEQ} will be negative. It is important to realize that this voltage drop could be made a positive value simply by reversing the sense of the arrow (we would then measure V_{EC}). The following example should help clarify any confusion regarding the pnp transistor.

EXAMPLE 4.14

Refer to Fig. 4.33. Determine the following quantities: I_{CQ}, V_{CEQ}, V_{CQ}, V_{EQ}, V_{BQ}, $I_{C(\text{sat})}$, and $V_{CE(\text{cut})}$. Assume $\beta = 100$.

Solution Inspection of the circuit reveals that it is the voltage drop across R_1 that is forward biasing the B-E junction of the transistor. This drop is the Thevenin biasing voltage. To better visualize exactly how the collector current is determined, the circuit is redrawn in Fig. 4.34 in Thevenin form and flipped over, so that it appears more familiar. Working counterclockwise around the B-E loop we obtain the following KVL expression:

$$0 = V_{\text{Th}} - I_{CQ} R_E - V_{EB} - I_{BQ} R_{\text{Th}}$$

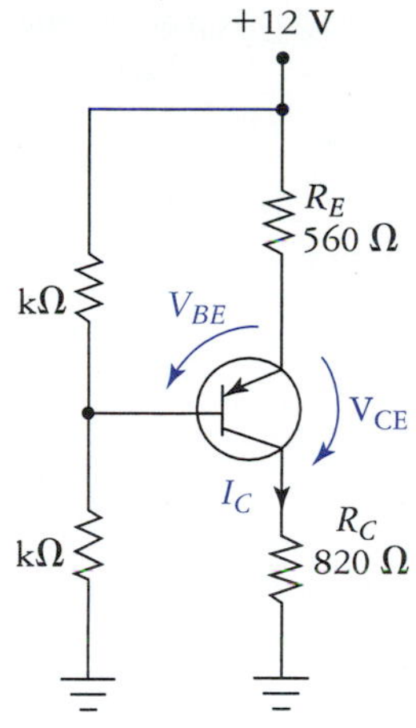

FIGURE 4.33
Circuit for Example 4.14.

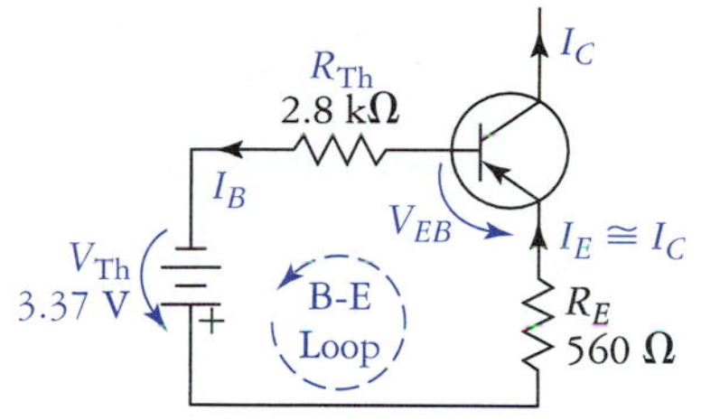

FIGURE 4.34
Circuit of Fig. 4.33 redrawn in Thevenized form for analysis.

We now make the familiar substitution $I_{BQ} = I_{CQ}/\beta$, and solve for I_{CQ}, which produces

$$I_{CQ} = \frac{V_{\mathrm{Th}} - V_{EB}}{R_E + \dfrac{R_{\mathrm{Th}}}{\beta}}$$

This is exactly the same relationship as was derived for npn transistors. Note that I_{CQ} will be positive, based on the direction of the current-sensing arrows used.

We now determine the values of V_{Th} and R_{Th}. The supply voltage is not denoted V_{CC} here, but is instead called V_S. Technically, the supply voltage would be called V_{EE} in this circuit, simply because it is the emitter supply voltage. It is common practice, however, to refer to the positive supply rail as V_{CC} and the negative supply rail (if there is one) as V_{EE}. This is done as a matter of convenience, because both npn and pnp devices may be used in a given circuit.

$$\begin{aligned} V_{\mathrm{Th}} &= V_S \frac{R_1}{R_1 + R_2} \\ &= 12\ \mathrm{V} \frac{3.9\ \mathrm{k}\Omega}{3.9\ \mathrm{k}\Omega + 10\ \mathrm{k}\Omega} \\ &= 3.37\ \mathrm{V} \end{aligned}$$

$$\begin{aligned} R_{\mathrm{Th}} &= R_1 \parallel R_2 \\ &= 3.9\ \mathrm{k}\Omega \parallel 10\ \mathrm{k}\Omega \\ &= 2.8\ \mathrm{k}\Omega \end{aligned}$$

We now substitute the various resistance and voltage values into the I_{CQ} equation, which yields

$$\begin{aligned} I_{CQ} &\cong \frac{V_{\mathrm{Th}} - V_{EB}}{R_E + \dfrac{R_{\mathrm{Th}}}{\beta}} \\ &= \frac{3.37\ \mathrm{V} - 0.7\ \mathrm{V}}{560\ \Omega + 28\ \Omega} \\ &= 4.54\ \mathrm{mA} \end{aligned}$$

The circuit of Fig. 4.34 is only used to determine the collector current. For the remainder of the analysis, we must refer back to the original circuit. V_{CEQ} is found via application of KVL from ground up to V_S, which gives us

$$\begin{aligned} V_{CEQ} &= -V_S + I_C(R_E + R_C) \\ &= -12\ \mathrm{V} + 4.54\ \mathrm{mA}(560\ \Omega + 820\ \Omega) \\ &= -12\ \mathrm{V} + (4.54\ \mathrm{mA})(1380\ \Omega) \\ &= -12\ \mathrm{V} + 6.27\ \mathrm{V} \\ &= -5.73\ \mathrm{V} \end{aligned}$$

The quiescent collector voltage is simply the voltage drop across the collector resistor:

$$\begin{aligned} V_{CQ} &= I_C R_C \\ &= (4.54 \text{ mA})(820 \ \Omega) \\ &= 3.72 \text{ V} \end{aligned}$$

The emitter voltage is lower than the positive supply rail by $I_C R_E$ volts. Thus,

$$\begin{aligned} V_{EQ} &= V_S - I_C R_E \\ &= 12 \text{ V} - (4.54 \text{ mA})(560 \ \Omega) \\ &= 9.46 \text{ V} \end{aligned}$$

Now the base voltage is determined:

$$\begin{aligned} V_{BQ} &= V_{EQ} + V_{BEQ} \\ &= 9.46 \text{ V} + 0.7 \text{ V} \\ &= 10.16 \text{ V} \end{aligned}$$

The collector saturation current is easily determined to be

$$\begin{aligned} I_{C(\text{sat})} &= \frac{V_S}{R_C + R_E} \\ &= \frac{12 \text{ V}}{820 \ \Omega + 560 \ \Omega} \\ &= 8.7 \text{ mA} \end{aligned}$$

Finally, the collector-to-emitter cutoff voltage is

$$V_{CE(\text{cut})} = -V_S = -12 \text{ V}$$

PNP Load Line

Before we leave pnp transistors again, a few points concerning the load line should be mentioned. In all npn transistor load-line examples worked thus far, the load line has been located in the first quadrant, which is entirely correct. Suppose, however, we were to plot the load line for the circuit in Fig. 4.33. Based on the values determined in Example 4.14, this load line would be located in the second quadrant. This is shown by the dashed line in Fig. 4.35.

FIGURE 4.35
Relative locations of npn and pnp load lines.

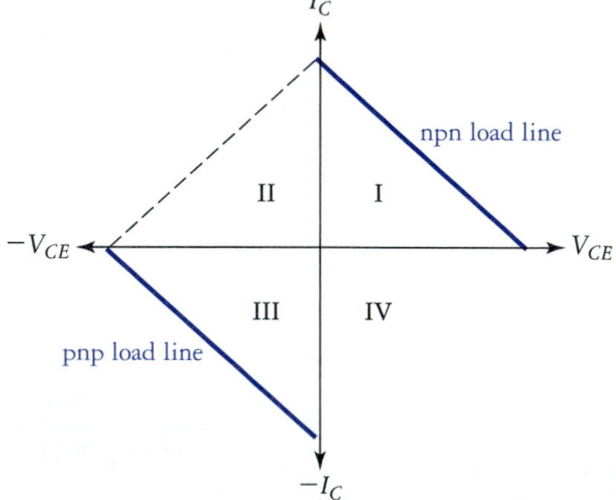

If we are going to stay strictly with passive-sign convention (PSC), then technically the pnp load line should be located in the third quadrant. This is so because the product of I_{CQ} and V_{CE} is device power dissipation, which should be positive, as the transistor will dissipate power. Thus, to conform to PSC, either the collector current, as determined in Example 4.14, should be negative, or we can use V_{EC}, which will be a positive value. The problem is that we do not use V_{EC} on the horizontal axis of the load line; we use V_{CE} regardless of transistor type. So, here we have the point of this brief discussion; that is, when plotting the load line for a pnp transistor, we normally assign negative polarity to the collector current, even though it may be positive with respect to conventional current-sensing arrows in the circuit.

4.7 INTEGRATED CIRCUIT-BIASING TECHNIQUES

Up to this point it has been assumed that we have been working with discrete components. A discrete component is a component (a transistor, for example) that is manufactured or fabricated as a single unit unto itself. When several components that comprise a circuit are fabricated as a single unit (usually from a single silicon substrate), those components are said to be integrated. We call this an *integrated circuit* or *IC*. Many advantages are associated with integrated circuitry, such as small size, light weight, increased reliability, and so on. Though there are other more subtle characteristics, circuit miniaturization is the most widely recognized advantage of IC technology. A typical IC chip (or die) measuring 1/4 in.^{2} may contain the equivalent of more than 500,000 transistors. Two silicon wafers are shown in Fig. 4.36. The individual IC dice or chips are cut from these wafers and mounted in packages similar to those shown below each. The tops have been removed from these packages to show the mounted dice.

Note
IC fabrication techniques are now being used to create miniature mechanical and electromechanical devices such as motors and gas turbines!

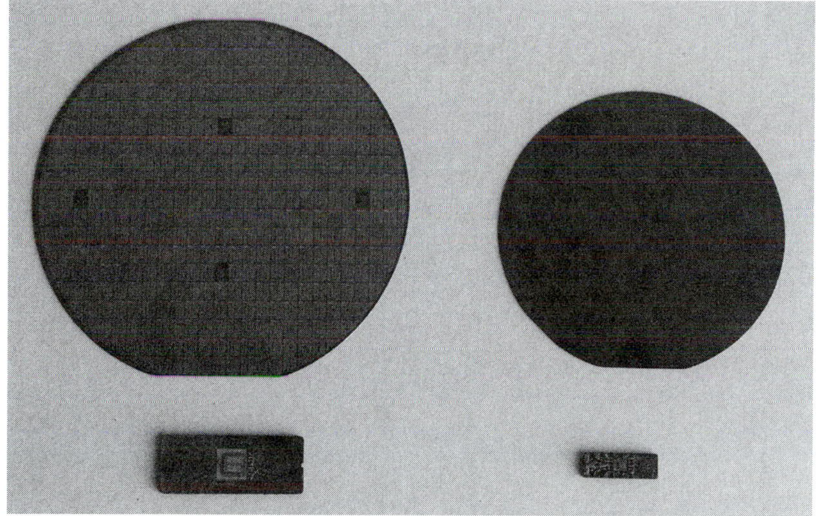

FIGURE 4.36
IC wafers.

ICs are often divided into several different categories. Some are functional classifications (digital, linear, telecommunications, etc.), whereas others are structural classifications (monolithic, hybrid, MOS, bipolar, etc.). Although much of this is beyond the scope of this chapter (detailed discussion of these topics is deferred until later), it does provide some useful background information.

Monolithic ICs

When a circuit is fabricated using IC techniques such that all components are formed from or within a single silicon crystal, that IC is said to be *monolithic*. The term is derived from the greek *mono*, meaning "one," and *lithos*, meaning "stone."

To the circuit designer, one of the main advantages of monolithic ICs is that component parameters can be matched very closely. For example, while the absolute value of β may be very difficult to control, the betas of all similar transistors on a monolithic IC can easily be matched to within 1%. Similar matching is achieved with saturation current, barrier potential, and many other important device parameters. These parameters also tend to exhibit closely matched temperature-tracking characteristics as well. These properties make possible some interesting options for the IC designer.

Current Mirror

A circuit called a *current mirror* is shown in Fig. 4.37. The key to understanding the operation of this circuit is the assumption that the transistor and the diode have very closely matched transconductance curves. If this is the case, then the collector current I_C will be approximately equal to I_1. Thus, the name current *mirror*; the collector current mirrors current I_1.

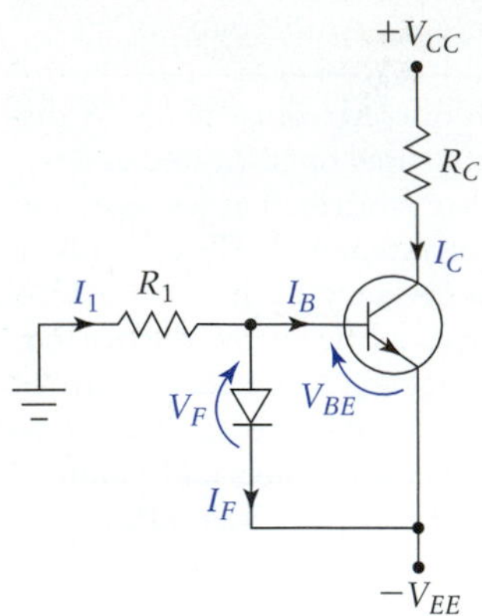

FIGURE 4.37
Current-mirror biasing.

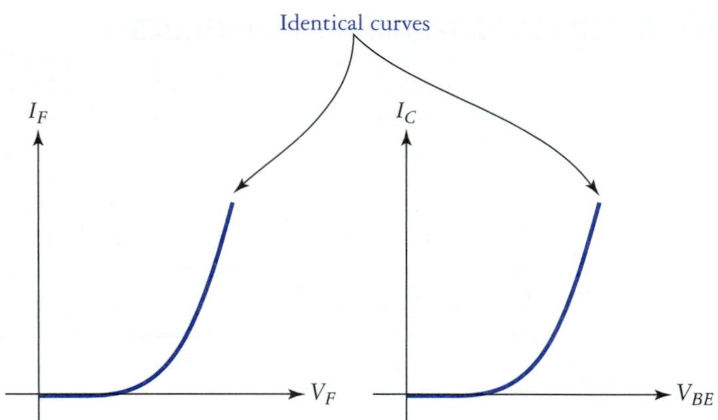

FIGURE 4.38
Current-mirror biasing relies on closely matched device transconductance characteristics.

Here's how the current mirror works. The diode is directly in parallel with the B-E junction of the transistor, thus both pn junctions have the same voltage drop at all times. Because both devices have identical transconductance curves, as shown in Fig. 4.38, then by definition, both devices are operating at the same current level and we have

$$I_C = I_F \tag{4.32}$$

The transistor base current is assumed to be negligibly small compared to I_F (because $I_B = I_C/\beta$ and β is usually large), so we can write

$$I_F \cong I_1 \tag{4.33}$$

Assuming that the base current is negligible is equivalent to assuming that the base represents an open circuit, which it nearly is, compared to the forward-biased diode. Now we can determine current I_1 by

$$I_C = I_1 \cong \frac{V_{EE} - V_{BE}}{R_1} \tag{4.34}$$

The current mirror is used to establish very stable current levels in IC designs.

Diode-Connected Transistors

Note
Discrete transistors can be connected to form diodes as well.

Because of the nature of the IC design process, bipolar transistors are just as easy (if not easier) to fabricate as diodes. Because of this, rather than actually using a diode to forward bias the current mirror, a transistor configured as a diode will normally be used. Simply shorting the collector to the base converts a transistor into a diode, as shown in Fig. 4.39. Since Q_1 is equivalent to a diode, the preceding analysis and Eq. (4.34) still apply.

FIGURE 4.39
Using a diode-connected transistor to implement current-mirror biasing.

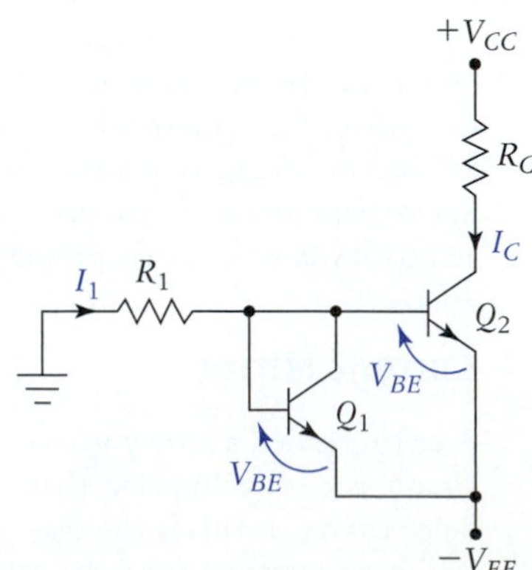

EXAMPLE 4.15

Refer to Fig. 4.39. Given $V_S = \pm 12$ V, choose R_1 and R_C such that $I_C = 10$ mA and $V_{CE} = 14$ V. Assume $V_T = 26$ mV and $\eta = 1$.

Solution We solve Eq. (4.34) for R_1 and substitute the given values yielding

$$R_1 = \frac{V_{EE} - V_{BE}}{I_C} = \frac{11.3\text{ V}}{10\text{ mA}} = 1.13\text{ k}\Omega$$

The collector resistor value is found as usual by using KVL:

$$R_C = \frac{V_{CC} + V_{EE} - V_{CE}}{I_C} = \frac{10\text{ V}}{10\text{ mA}} = 1\text{ k}\Omega$$

Device Scaling Effects*

The previous analysis of the current mirror was predicated on the assumption that the devices used in the circuit are matched. To obtain such matching, transistors of equal junction areas would be used, as shown in the upper part of Fig. 4.40. With all other parameters and environmental conditions being equal, the transistor emitter saturation current I_{ES} will be directly proportional to the B-E junction area. The collector current is determined by Shockley's equation, which is repeated here for convenience:

$$I_C \cong I_{ES}\, e^{V_{BE}\eta V_T}$$

It is clear that I_C is directly proportional to I_{ES}, which is itself directly proportional to junction area. Because of this relationship, the emitter saturation current is often called the *scale current.*

Let us assume that we have fabricated two transistors (Q_3 and Q_4 in Fig. 4.40) on a silicon chip. The E-B junction area of Q_3 is greater than that of Q_4 by a factor of 2. The differences in junction area will cause Q_3 to carry twice as much collector current as Q_4 at a given value of V_{BE}.

Note
In IC design, device parameter ratios are more important than absolute values.

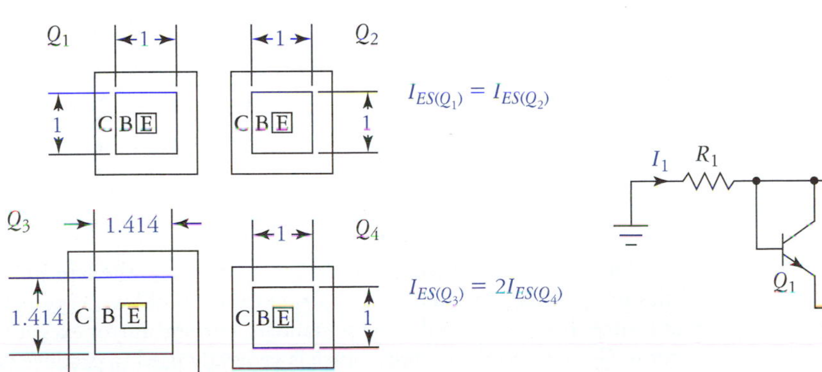

FIGURE 4.40
Changing the relative junction area changes the saturation current proportionately, allowing scaling of mirror current.

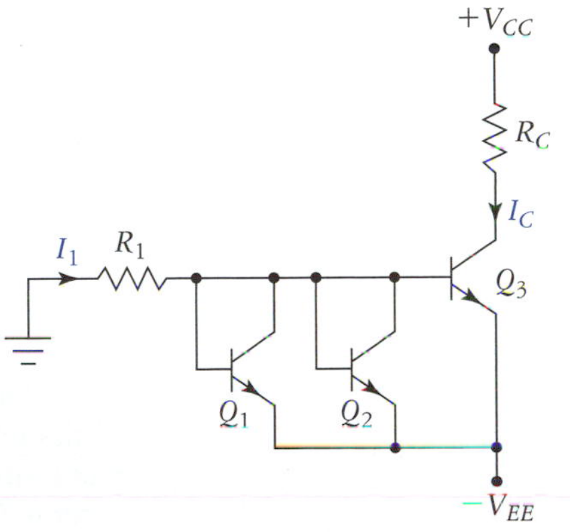

FIGURE 4.41
Connecting of mirror biasing diodes in parallel also allows scaling of mirror current.

*This section can be skipped without loss of continuity, if as is usually the case, time is short.

Using the current-mirror circuit with the diode-connected transistor labeled Q_1 and the normal transistor Q_2, let us define the ratio

$$k = I_{ES(2)}/I_{ES(1)} \tag{4.35}$$

In terms of junction areas this is

$$k = A_2/A_1 \tag{4.36}$$

The current-mirror equation is modified slightly to take this parameter into account as follows:

$$I_C = k\frac{V_{EE} - V_{BE}}{R_1} \tag{4.37a}$$

$$= kI_1 \tag{4.37b}$$

Thus we see that the IC designer can manipulate transistor junction areas, as well as resistor values, to set specific current levels. This is important, because it shows that the ratios of various parameters (which can be accurately controlled) can be used to determine device operating points, rather than the absolute values of these parameters (which are difficult to control).

It is possible to simulate the use of a transistor with twice-normal emitter saturation current by connecting two identical devices in parallel. This is shown in Fig. 4.41, where Q_1 and Q_2 are connected as diodes in parallel. It is left for the reader to show that

$$I_C = \frac{V_{CC} - V_{BE}}{2R_1} \tag{4.38}$$

Widlar Current Source

Although monolithic ICs lend themselves naturally to the fabrication of transistors, resistors too can be produced. Unfortunately, such resistors are limited to a range of from around 10 Ω to around 20 kΩ. This range is sufficient for many applications; however, there are times when this is not the case. For example, suppose that we require a current mirror to operate from $V_S = 15$ V, with $I_C = 50$ μA. Solving Eq. (4.34) for R_1 and assuming that $V_{BE} = 0.7$ V, yields $R_1 = 286$ kΩ. Such a large resistor value would be completely impractical to realize in monolithic IC form.

FIGURE 4.42
The Widlar current source.

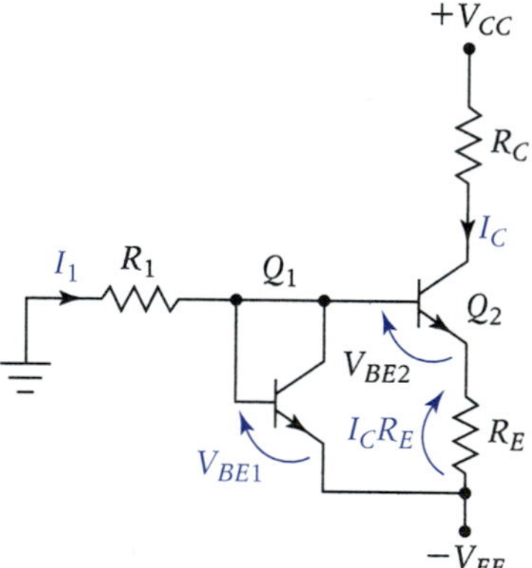

The *Widlar current source* (named for its inventor, the late Robert Widlar, a pioneer of modern IC design) provides an interesting solution to this problem. Figure 4.42 shows the basic Widlar current source. It is quite similar to the standard current mirror except for the addition of an emitter resistor to Q_2. The Widlar current source is generally used to produce constant current levels in the microamp range and lower. A somewhat involved derivation produces the following relationship for matched transistors:

$$I_C = \frac{\eta V_T}{R_E}\ln\left(\frac{I_1}{I_C}\right) \tag{4.39}$$

Equation (4.39) cannot be solved for I_C directly. To use Eq. (4.39) in an analysis application, we must take a first guess at I_C and then refine this guess iteratively. The process is described in the flowchart of Fig. 4.43. Generally speaking, the collector current will be quite small relative to I_1, but a first

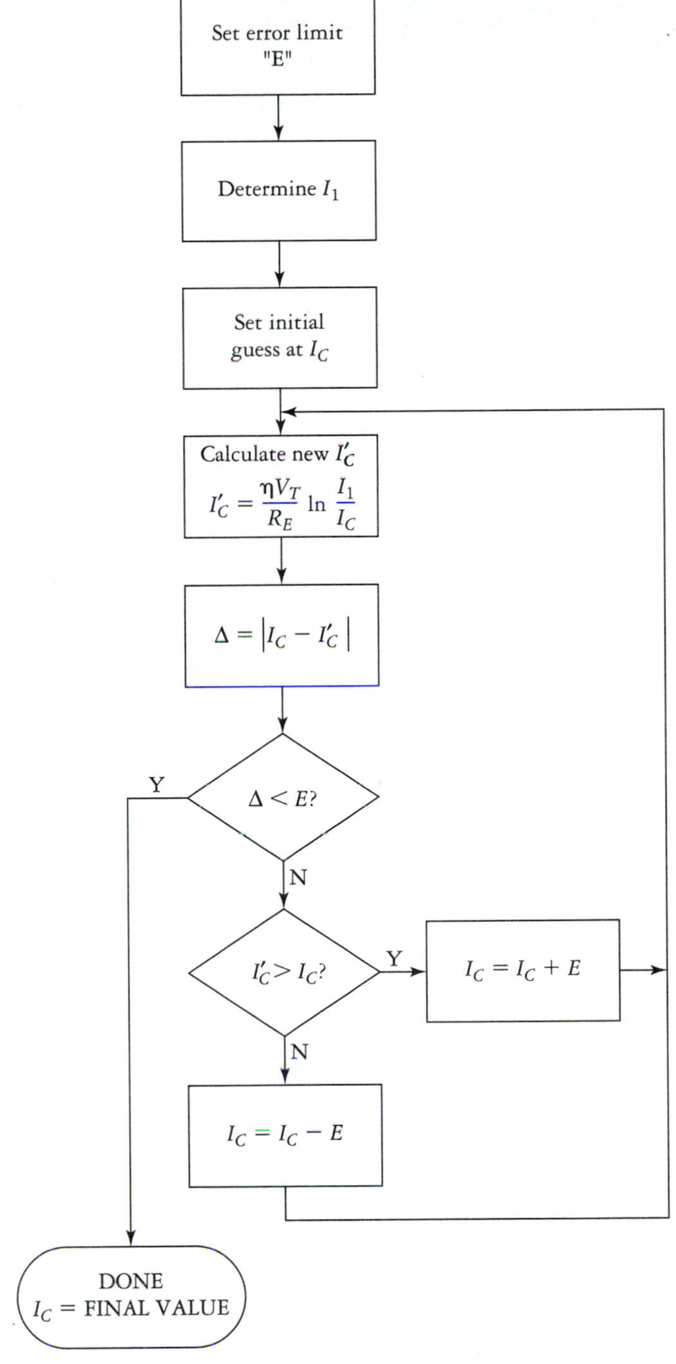

FIGURE 4.43
Flowchart for approximating the Widlar current source Q point.

guess of $I_C = I_1$ is a reasonable starting point. An error limit of 1 μA usually will provide sufficient accuracy. This task is best done via computer program, however, because several hundred iterations may be necessary for the procedure to converge sufficiently close to the final value.

It is actually easier to design a Widlar current source than it is to analyze one (at least just using pencil and paper). Again assuming matched transistors, the following expression can be derived:

$$R_E = \frac{\eta V_T}{I_C} \ln \left(\frac{I_1}{I_C} \right) \tag{4.40}$$

where I_C is the desired collector current and I_1 is set to some convenient value by the designer.

EXAMPLE 4.16

Determine the values required for R_1 and R_E in Fig. 4.42 such that $I_C = 20\ \mu A$ and $I_1 = 3$ mA. Assume that $V_{CC} = 10$ V, $\eta = 1$, and $V_T = 26$ mV.

Solution First we determine R_1:

$$R_1 = \frac{V_{EE} - V_{BE}}{I_1} = \frac{9.3\ \text{V}}{3\ \text{mA}} = 3.1\ \text{k}\Omega$$

R_E is found via direct application of Eq. (4.40):

$$R_E = \frac{\eta V_T}{I_C} \ln\left(\frac{I_1}{I_C}\right) = \frac{26\ \text{mV}}{20\ \mu\text{A}} \ln\left(\frac{3\ \text{mA}}{20\ \mu\text{A}}\right) = 6.51\ \text{k}\Omega$$

4.8 COMPUTER-AIDED ANALYSIS APPLICATIONS

Most of the time a given transistor biasing arrangement will be relatively easy to analyze using pencil, paper, and a calculator. In many situations, only rough estimates of circuit voltage and current levels are necessary. There are the exceptions, of course, such as the Widlar current source, and several other similar circuits that were not covered here, in addition to larger, more complex circuits containing many transistors and other components. It is very often possible to break a larger circuit down into several smaller circuits that can be analyzed individually, without the use of a computer. Still, computer-aided circuit analysis can be very helpful in the analysis of simpler transistor circuits, as is now shown.

Operating-Point Analysis

Recall that another name for the Q point is the operating point. Let's reexamine the collector-feedback bias circuit of Fig. 4.15. The results of the calculations in Example 4.8 were:

$$V_{CEQ} = 5.36\ \text{V} \quad \text{and} \quad I_{CQ} = 2.11\ \text{mA}$$

The circuit was entered using the schematic editor, with voltage viewpoints and current probes inserted as necessary. This is shown in Fig. 4.44, where we find the following current and voltage levels:

$$V_{CEQ} = 4.336\ \text{V} \quad \text{and} \quad I_{CQ} = 2.558\ \text{mA}$$

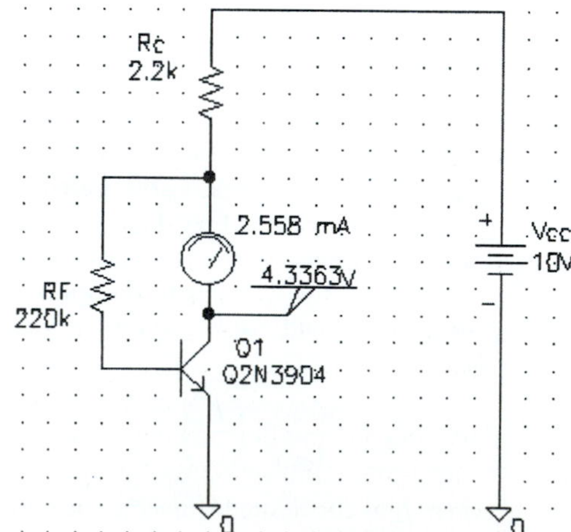

FIGURE 4.44
Collector-feedback biasing.

The main reason for the discrepancies between the two sets of results is caused by differences between the beta values assumed in the example and in the PSpice transistor model. This is the problem with attempting to simulate beta-dependent circuits; the simulation results are only as good as the parameters used in the transistor model. If you were to build the circuit using a 2N3904, chances are that the actual current and voltage values would be closer to the PSpice simulation values.

Let us now use PSpice to simulate a circuit that should be much more insensitive to transistor parameter variation. Using the results of Example 4.9, we have

$$I_{CQ} = 2.74 \text{ mA} \qquad \text{and} \qquad V_{CEQ} = 3.22 \text{ V}$$

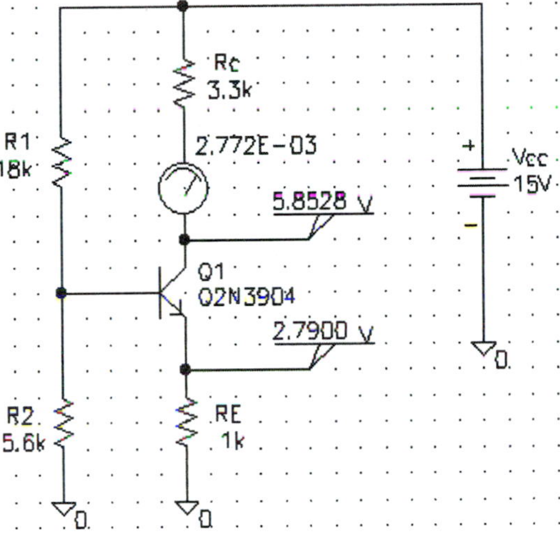

FIGURE 4.45
Voltage-divider biasing.

The circuit of Example 4.9 was simulated, and the results are shown in Fig. 4.45, where we find

$$I_{CQ} = 2.77 \text{ mA}$$

$$\begin{aligned} V_{CEQ} &= V_{CQ} - V_{EQ} \\ &= 5.85 \text{ V} - 2.79 \text{ V} \\ &= 3.06 \text{ V} \end{aligned}$$

The agreement between the simulation and our approximate analysis is good.

CHAPTER REVIEW

Many applications require that a given transistor be biased up into the active region for correct operation. Base biasing is the simplest technique available, but it is very sensitive to device parameter variations. Base biasing is used frequently in saturated switch applications, however. In such applications the transistor is operated either in cutoff or saturation, with little time spent in the active region.

Various types of negative feedback are used to stabilize the transistor's Q point. Emitter feedback and collector feedback are two commonly used feedback techniques. However, in order for either type of feedback to be effective, the resistance "seen" by the base terminal should be minimized. This is accomplished by several different means. Emitter biasing is one such method, in which the emitter terminal of the transistor is connected to a separate bias voltage source (V_{EE}). Another alternative is voltage-divider biasing, whereby the V_{CC} supply voltage is reduced via voltage division, and the Thevenin resistance of the divider network is made relatively small.

Load lines are used to provide a graphical interpretation of circuit bias conditions and the limits of circuit operation. The intersection of the quiescent values of V_{CE} and I_C is called the Q point or the operating point of the circuit.

Integrated circuit design techniques rely primarily on device parameter matching for stability. Two circuits that are commonly used in IC designs are the current mirror and the Widlar current source. Both circuits are used to produce stable current levels. The Widlar current source is most often used to produce very low current levels without having to resort to the use of high-value resistors.

DESIGN AND ANALYSIS PROBLEMS

Note: Unless specified otherwise, assume all transistors are silicon with $\beta = 100$, $\alpha = 1$ and $V_{CE(sat)} \cong 0$ V.

4.1. Refer to Fig. 4.46. Determine V_{BB} such that the Q point is centered on the dc load line.

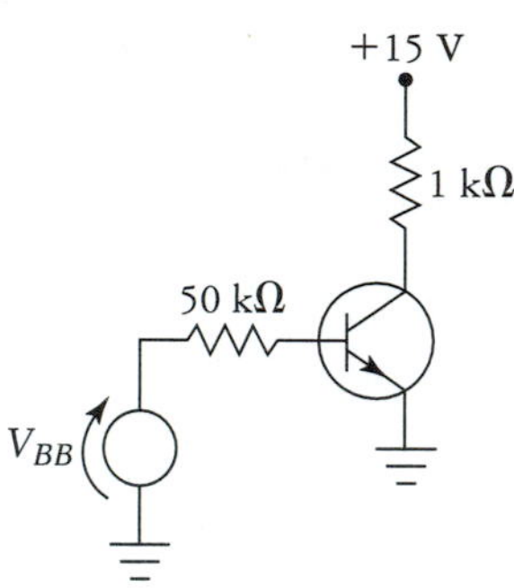

FIGURE 4.46
Circuit for Problems 4.1 and 4.2.

4.2. Refer to Fig. 4.46. Determine the value of V_{BB} required to cause $I_{CQ} = I_{C(sat)}$. Call it $V_{BB(sat)}$.

4.3. Refer to Fig. 4.47. Determine the resistor values such that $I_{C(sat)} = 6$mA and $I_{CQ} = 4$ mA.

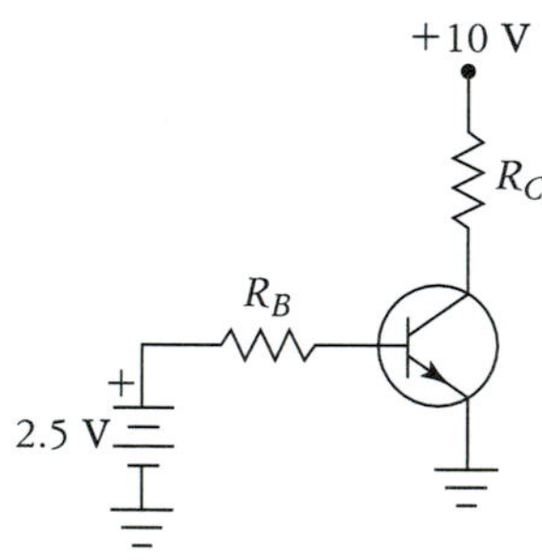

FIGURE 4.47
Circuit for Problem 4.3.

4.4. Sketch the dc load line for the circuit in Fig. 4.48 given $V_{CC} = 20$ V, $R_C = 1$ kΩ, $R_E = 1$ kΩ, and $R_B = 330$ kΩ.

FIGURE 4.48
Circuit for Problem 4.4.

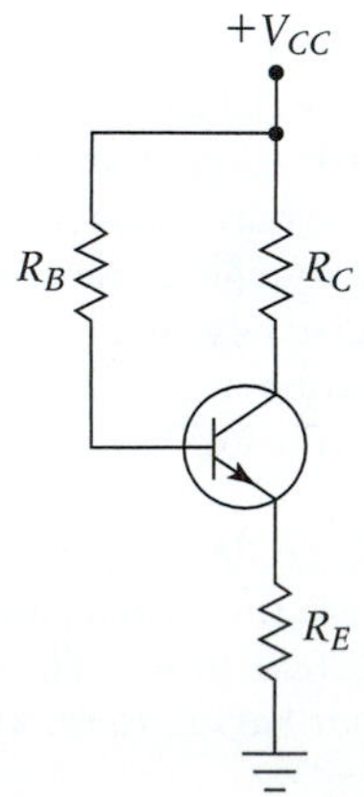

4.5. Refer to Fig. 4.49. Assuming $V_{BB} = 5$ V, determine R_B such that $I_{CQ} = 5$ mA.

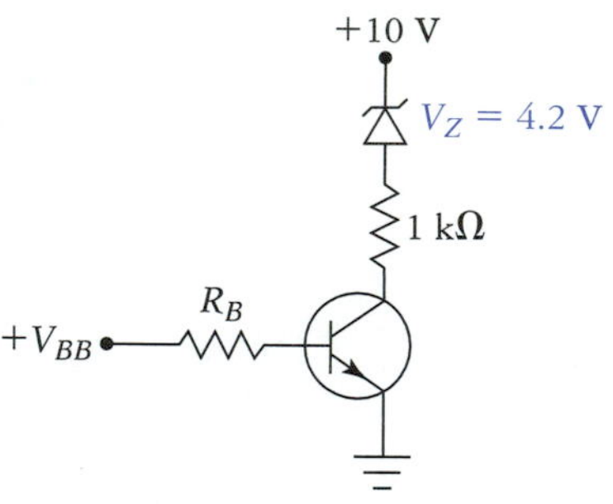

FIGURE 4.49
Circuit for Problems 4.5 and 4.6.

4.6. Determine $I_{C(sat)}$ for Fig. 4.49, and the value of V_{BB} that is necessary to cause saturation, given $R_B = 47$ kΩ.

4.7. Sketch the dc load line for the circuit in Fig. 4.50, given $R_1 = 470$ kΩ and $R_2 = 3.3$ kΩ. Also determine V_{CQ}, V_{BQ}, and V_{EQ}.

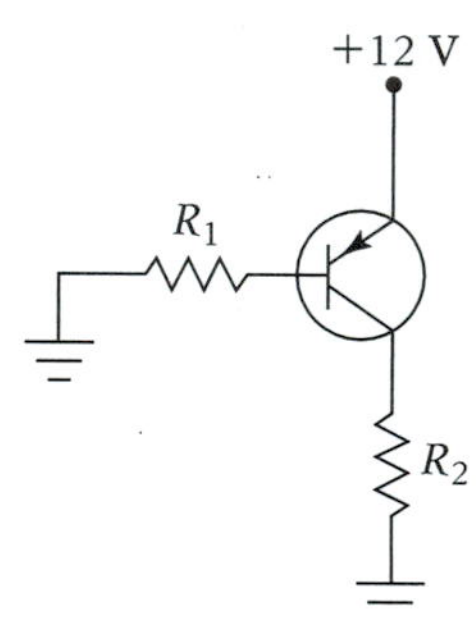

FIGURE 4.50
Circuit for Problems 4.7 and 4.8.

4.8. Refer to Fig. 4.50. Assuming that $R_2 = 1$ kΩ, determine R_1 such that $V_{CEQ} = 6$ V.

4.9. Determine I_{CQ}, V_{CEQ}, V_{EQ}, V_{CQ}, and V_{CB} for the circuit in Fig. 4.51.

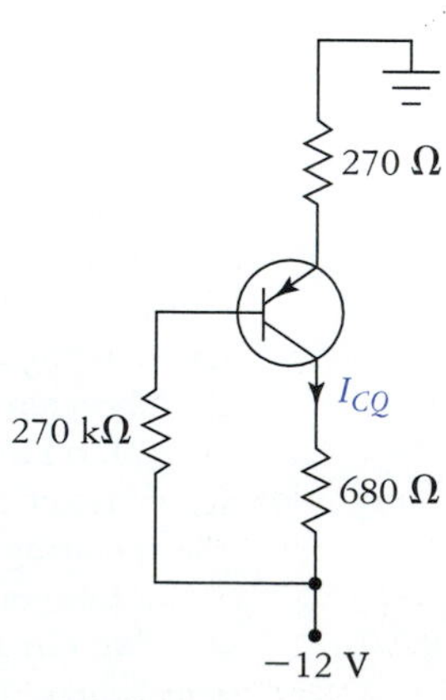

FIGURE 4.51
Circuit for Problem 4.9.

4.10. Determine V_{CEQ}, I_{CQ}, V_{EQ}, V_{CQ}, V_{BQ}, and V_{CB} for the circuit in Fig. 4.52.

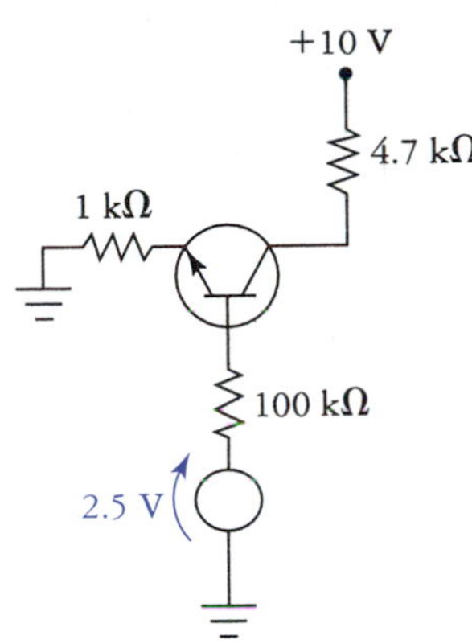

FIGURE 4.52
Circuit for Problems 4.10 and 4.11.

4.11. Refer to Fig. 4.52. Determine the value required for V_{CC} such that $V_{CEQ} = 8$ V.

4.12. Refer to Fig. 4.53. Assume that $R_C = 1.5$ kΩ and $R_F = 220$ kΩ. Sketch the dc load line for this circuit, and determine V_{CB}.

4.13. Refer to Fig. 4.53. Assuming that $\beta = 150$ and $R_F = 270$ kΩ, determine the approximate value of R_C that results in a centered Q point on the dc load line.

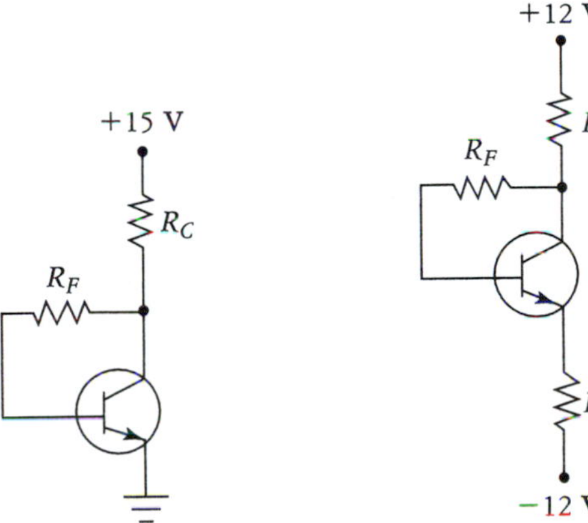

FIGURE 4.53
Circuit for Problems 4.12 and 4.13.

FIGURE 4.54
Circuit for Problems 4.14 and 4.15.

4.14. Refer to Fig. 4.54. Given $R_C = 1.5$ kΩ, $R_E = 500$ Ω, and $R_F = 220$ kΩ, determine I_{CQ}, V_{CEQ}, $I_{C(sat)}$, and $V_{CE(cut)}$.

4.15. Refer to Fig. 4.54. Given $R_C = R_E = 1$ kΩ, choose R_F such that the Q point is centered on the dc load line.

4.16. Given the conditions of Problem 4.15, determine the values of V_{CQ}, V_{EQ}, and V_{BQ}.

4.17. Refer to Fig. 4.55. Determine I_{CQ}, V_{CEQ}, V_{CQ}, and V_{BQ}.

4.18. Sketch the dc load line for the circuit in Fig. 4.56, given $R_1 = 27$ kΩ, $R_2 = 5.6$ kΩ, $R_C = 10$ kΩ, and $R_E = 1.8$ kΩ.

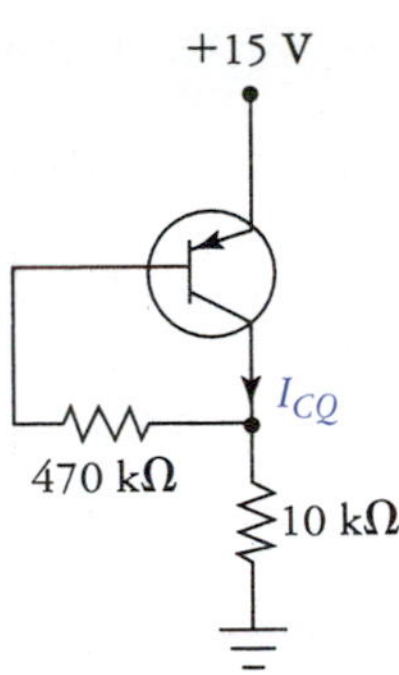

FIGURE 4.55
Circuit for Problem 4.17.

4.19. Refer to Fig. 4.56. Given $R_1 = 15$ kΩ, $R_2 = 4.7$ kΩ, $R_E = 860$ Ω, and $R_C = 1.5$ kΩ, determine V_{CEQ}, I_{CQ}, V_{EQ}, V_{CQ}, and V_{BQ}.

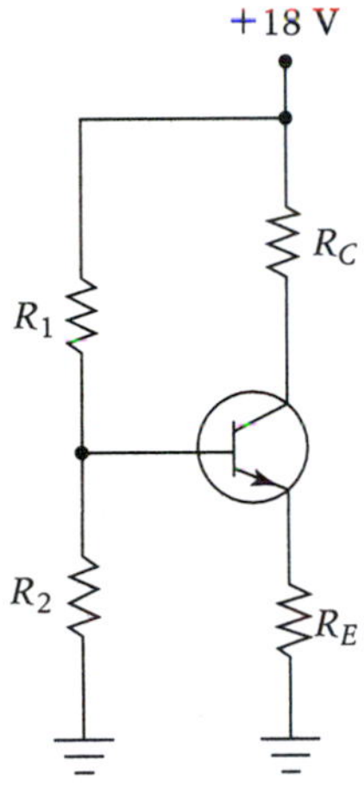

FIGURE 4.56
Circuit for Problems 4.18 and 4.19.

4.20. Refer to Fig. 4.57. Given $R_1 = 18$ kΩ, $R_2 = 12$ kΩ, $R_3 = 2.7$ kΩ, and $V_{CC} = 15$ V, determine V_{CEQ}, I_{CQ}, V_{CQ}, V_{BQ}, V_{EQ}, and V_{CB}.

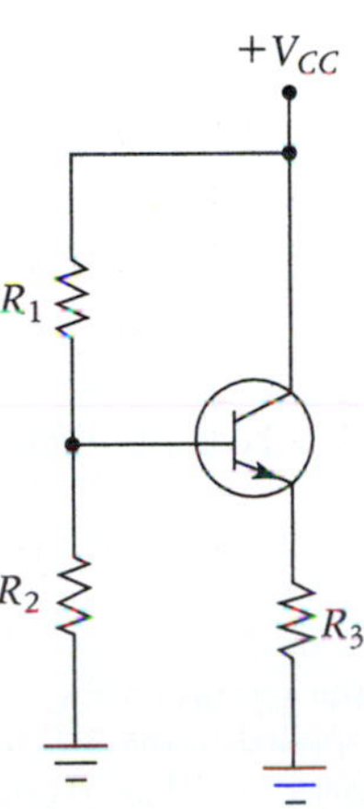

FIGURE 4.57
Circuit for Problem 4.20.

4.21. Determine V_{CEQ}, I_{CQ}, V_{CQ}, V_{BQ}, V_{EQ}, and V_{CB} for the circuit in Fig. 4.58.

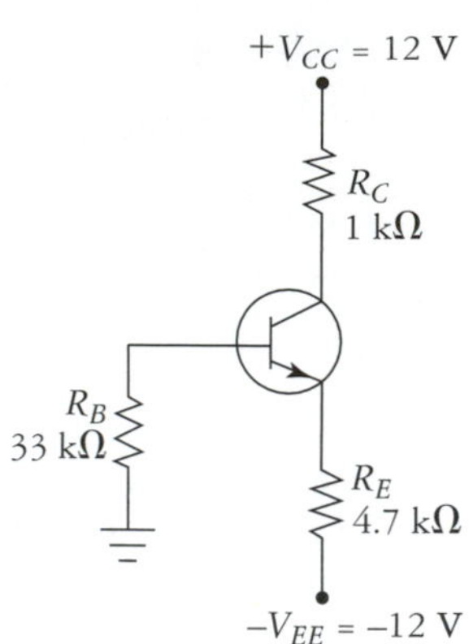

FIGURE 4.58
Circuit for Problem 4.21.

4.22. Determine V_{CEQ}, I_{CQ}, V_{CQ}, V_{BQ}, V_{EQ}, and V_{CB} for the circuit in Fig. 4.59.

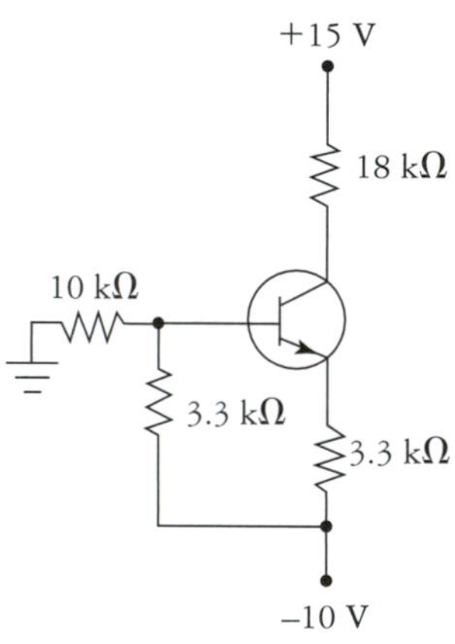

FIGURE 4.59
Circuit for Problem 4.22.

4.23. Determine V_{CEQ}, I_{CQ}, V_{CQ}, V_{BQ}, V_{EQ}, and V_{CB} for the circuit in Fig. 4.60.

4.24. Refer to Fig. 4.61. Determine the value of V_x that will result in $V_{CEQ} = V_{CC}/2$.

4.25. The addition of a diode to the voltage divider in Fig. 4.62 helps to reduce the temperature sensitivity of the circuit by causing the bias applied to the B-E junction to track temperature-induced variations of V_{BE}. Assuming that the diode and the transistor have matched transconductance curves, derive an expression for I_{CQ} for this circuit.

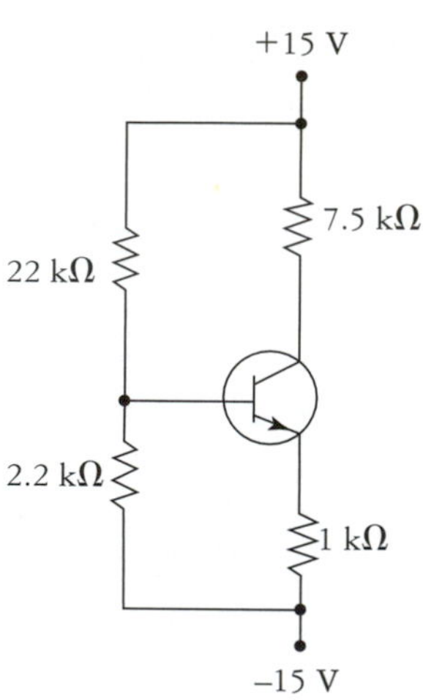

FIGURE 4.60
Circuit for Problem 4.23.

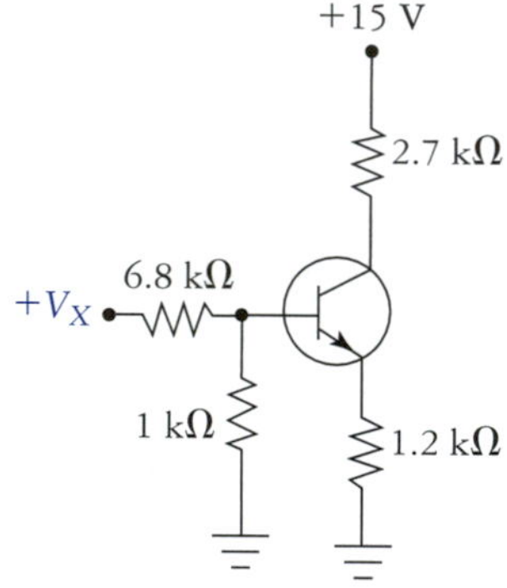

FIGURE 4.61
Circuit for Problem 4.24.

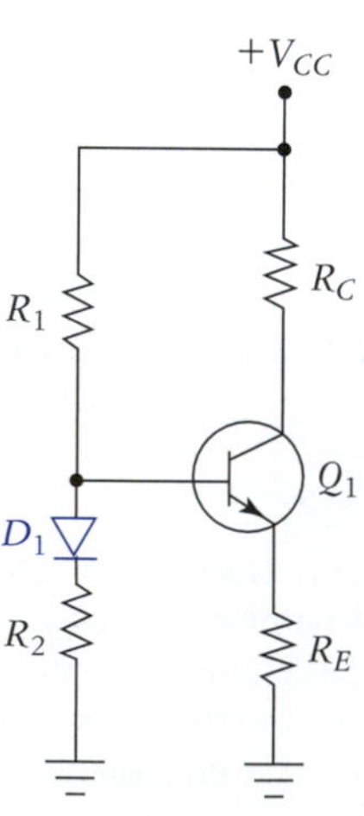

FIGURE 4.62
Circuit for Problem 4.25.

4.26. Refer to Fig. 4.63. Determine R_1 and R_C such that $I_C = 15$ mA and $V_{CEQ} = 10$ V. Assume that $V_{CC} = V_{EE} = 12$ V.

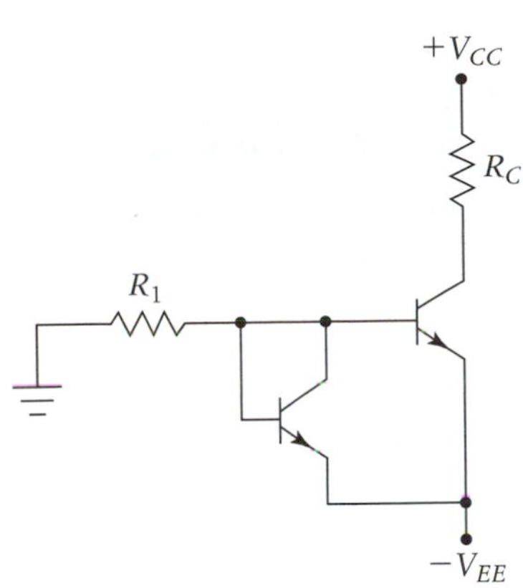

FIGURE 4.63
Circuit for Problem 4.26.

4.27. Refer to Fig. 4.64. Choose R_1 and R_2 such that $I_C = 5$ mA and $V_{C(Q_2)} = 0$ V.

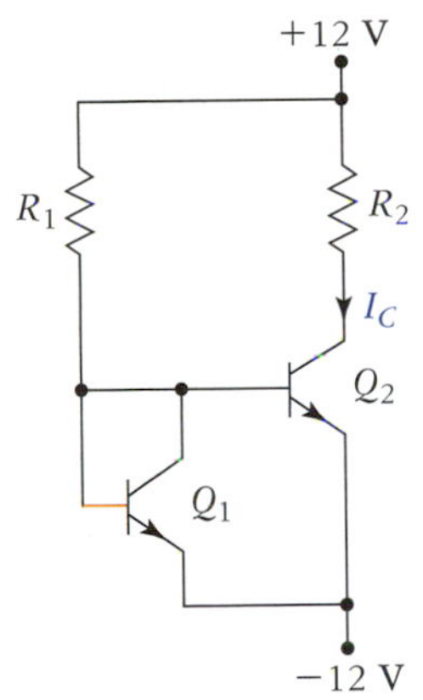

FIGURE 4.64
Circuit for Problem 4.27.

4.28. Refer to Fig. 4.65. Assuming the transistors are matched determine I_L and V_{CE3}. *Note:* The bases of Q_2 and Q_3 are connected. Assume that the base currents of Q_2 and Q_3 are negligibly small.

4.29. Derive Eq. (4.21).

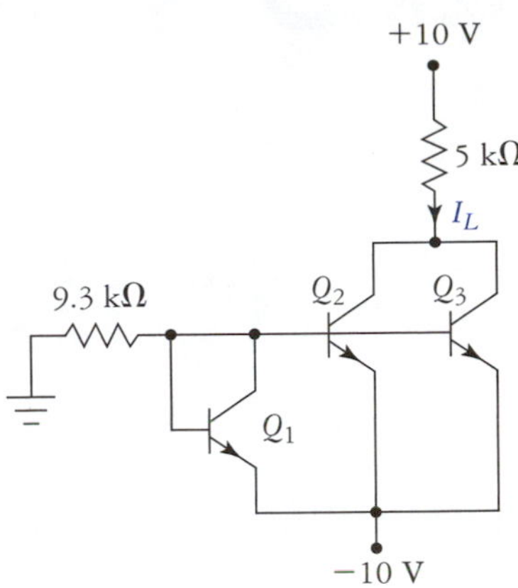

FIGURE 4.65
Circuit for Problem 4.28.

■ TROUBLESHOOTING PROBLEMS

4.30. Refer to Fig. 4.46 and assume that $V_{BB} = 5$ V. A voltage measurement reveals $V_{BE} = 5$ V. What value would you expect to measure for V_{CE}?

4.31. Suppose the zener diode in Fig. 4.49 were to become open circuited. If V_{BB} is set such that $I_B = 50\ \mu$A, then what should be measured for I_E? Explain your answer.

4.32. Refer to Fig. 4.22. Suppose the 3.3-kΩ resistor were to become disconnected from the base of the transistor. What effect would this have on the transistor?

4.33. The circuit of Fig. 4.59 is constructed and a voltmeter measures $V_C = -6$ V. Is the transistor in (or very close to) cutoff or saturation, or is the transistor well within the active region?

4.34. If the transistor in Fig. 4.60 were defective such that internally the B-E junction was shorted out, what (approximately) would a voltmeter read at the collector?

4.35. Generally speaking, if the ambient temperature increases, toward what limit will the Q point of a bipolar transistor circuit drift?

■ COMPUTER PROBLEMS

4.36. Using a high-level language such as BASIC or C, write a program to implement the flowchart of Fig. 4.43. Run the program using the Widlar current source of Fig. 4.42 for $R_1 = 28.6$ kΩ, $R_E = 1$ kΩ, $R_C = 0$ Ω, and $V_S = 15$ V.

4.37. Using the Widlar current source and component values specified in Problem 4.36, determine the operating point of the circuit using PSpice. Use the 2N3904 transistor.

4.38. Use PSpice to plot the response of I_{CQ} for the circuit in Fig. 4.12 to a temperature sweep from 0 to 100°C in increments of 25°C.

4.39. Use PSpice to determine the operating point of the circuit in Fig. 4.65. Use 2N2222 transistors in each position.

4.40. Simulate the circuit of Fig. 4.33 using the 2N3906.

CHAPTER 5

Small-Signal BJT Amplifiers

BEHAVIORAL OBJECTIVES

On completion of this chapter, the student should be able to:

- Describe the three basic transistor amplifier configurations.
- Draw dc and ac equivalent circuits for basic transistor amplifier circuits.
- Perform small-signal analysis of basic transistor amplifier circuits.
- Construct dc and ac load lines for amplifiers.
- Determine the output compliance of a common-emitter amplifier.
- Explain the operation of the Darlington transistor configuration.
- Describe the effects of swamping and emitter bypassing on amplifier gain, input resistance, and stability.

INTRODUCTION

In the previous chapter, we explained that the purpose of biasing is to prepare a transistor to perform some useful function. Many functions can be performed by transistors, but one that is very common and interesting is that of a linear amplifier. The majority of this chapter is devoted to the analysis of the BJT in small-signal linear amplifier applications. In addition, the basic concepts and principles that apply to amplifiers in general are also introduced.

5.1 FUNDAMENTAL AMPLIFIER CHARACTERISTICS

From a functional standpoint, amplifiers fall into the general category of signal processing circuits. Any system that alters some parameter associated with a signal performs signal processing. Filters fall into the signal processing category as well. Some signal processing systems are nonlinear. The clipper circuit of Chapter 2, for example, could be considered a nonlinear signal processing circuit. Nonlinear systems change a signal in a fundamental way, which also alters the frequency content of that signal.

The discussion here is limited to *linear amplifiers.* A linear amplifier is one that produces an output that is a replica of the input, although it is larger (scaled) in amplitude and possibly shifted in phase.

Power Gain

In the most general sense, a useful amplifier is a functional block that exhibits power gain. A network exhibits *power gain* if the power delivered to the load being driven by the network is greater than the power that is absorbed from the original signal source whose output is being amplified. In equation form, we define power gain as

$$A_p = \frac{P_o}{P_{\text{in}}} \tag{5.1}$$

The symbol A_p is used to stress the fact that the power gain is a form of amplification.

Voltage Gain

An amplifier may be thought of as being a "black box," like that shown in Figure 5.1(a), which is also called a *two-port network*. A less generalized symbol that is more commonly used to represent an amplifier is shown in Fig. 5.2.

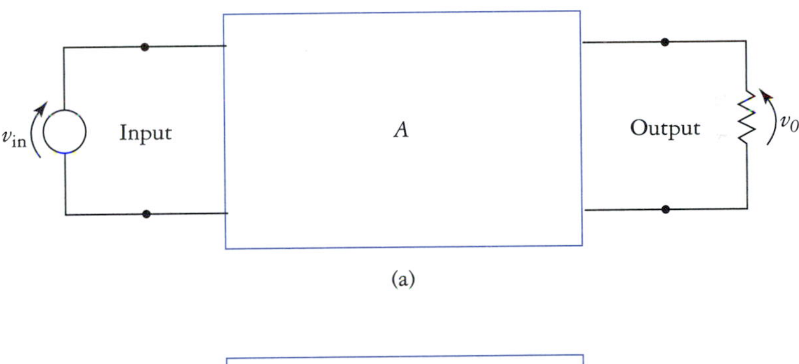

(a)

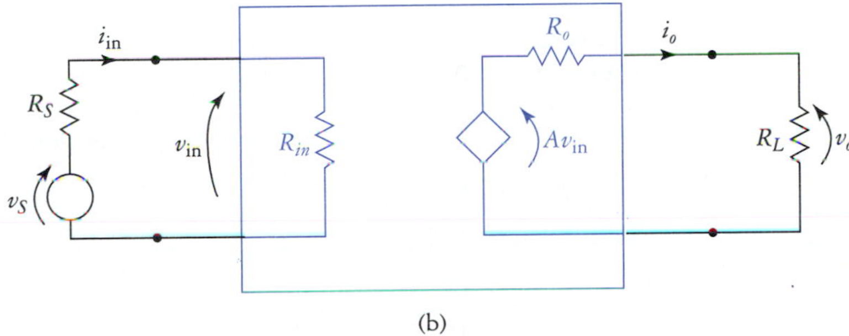

(b)

FIGURE 5.1
Four-terminal black box representation of an amplifier.

FIGURE 5.2
Simplified amplifier symbol (single-ended input and output).

There are four possible ways to model an amplifier: as a voltage-controlled voltage source (VCVS), voltage-controlled current source (VCIS), current-controlled voltage source (ICVS), or a current-controlled current source (ICIS). Most of the time, we shall use the VCVS model, as shown in Fig. 5.1(b). The VCVS is commonly referred to as a voltage amplifier. The proportionality or scaling constant that relates the output voltage to the input voltage is called the *voltage gain* A_v. Voltage gain is defined as

Note
Equation 5.2 is the fundamental definition of voltage gain. It is used to determine A_v experimentally.

$$A_v = \frac{v_o}{v_{\text{in}}} \tag{5.2a}$$

Notice that the gain of the dependent source in Fig. 5.1(b) is denoted simply as A. This was done to provide a distinction between the gain of the dependent source and the overall voltage gain of the system, A_v. The difference is rather subtle, but important. Consider that if R_L is, say, 10 Ω and R_o is 100 Ω then v_o will be less than one-tenth as large as Av_{in}. We may apply the voltage-divider equation to determine the overall voltage gain as follows:

Note
The ideal voltage amplifier has $R_o \to 0\Omega$ and $R_{\text{in}} \to \infty\Omega$.

$$A_v = Av_{\text{in}}\frac{R_L}{R_L + R_o} \tag{5.2a}$$

Equation (5.3a) accounts for loading of the output of the amplifier, therefore A_v is sometimes referred to as the *loaded voltage gain.* It is normally very desirable to make R_o as small as possible (preferably zero). When this is the case, then $A_v = A$.

It is also usually desirable to make R_{in} as large as possible, relative to R_S, so that the source is very lightly loaded. When this is the case, $v_{\text{in}} = v_S$.

Current Gain

Because the amplifier in Fig. 5.1(b) is modeled as a VCVS (a voltage amplifier), we normally assume that $v_o > v_{\text{in}}$. It is also possible that $i_o > i_{\text{in}}$. Whether this is in fact true or not, the *current gain* of any amplifier is defined as

$$A_i = \frac{i_o}{i_{\text{in}}} \tag{5.3}$$

Relationships between Power, Voltage, and Current Gain

It can easily be shown that the *power gain* of any amplifier is given by the product of A_v and A_i. Consider that $p_o = i_o v_o$ and $p_{\text{in}} = i_{\text{in}} v_{\text{in}}$. Substituting these relations into Eq. (5.1) gives us

$$A_p = \frac{i_o v_o}{i_{\text{in}} v_{\text{in}}}$$

which reduces to

$$A_p = A_v A_i \tag{5.4}$$

In practical terms, it is usually easy to measure voltage gain; we simply use an oscilloscope or voltmeter to read the input and output voltages. Current gain is more difficult to measure, and power gain is more difficult yet. However, power gain can be expressed in terms of voltage gain, load resistance, and amplifier input resistance, which are generally easy to determine. We begin with the following equations, which are determined by inspection of Fig. 5.1(b):

$$p_o = v_o^2/R_L, \qquad p_{\text{in}} = v_{\text{in}}^2/R_{\text{in}}$$

Substituting into Eq. (5.1), we get

$$A_p = \frac{v_o^2/R_L}{v_{\text{in}}^2/R_{\text{in}}} = \frac{v_o^2}{v_{\text{in}}^2} \times \frac{R_{\text{in}}}{R_L}$$

which finally reduces to

$$A_p = A_v^2 \frac{R_{\text{in}}}{R_L} \tag{5.5}$$

Note
Equations 5.5 and 5.6 are very general. They apply to all types of amplifiers.

It is possible to define the current gain of an amplifier in terms of A_v, R_{in}, and R_L as well. The relationship is

$$A_i = A_v \frac{R_{in}}{R_L} \tag{5.6}$$

The derivation of Eq. (5.6) is straightforward and is left as an exercise at the end of the chapter.

5.2 BASIC BJT AMPLIFIER CONFIGURATIONS

Three different amplifier configurations are possible using a single transistor. These configurations, shown in Fig. 5.3, are the *common-emitter* (CE), the *common-collector* (CC) [frequently called the *emitter-follower* (EF)], and the *common-base* (CB) configurations. These circuits are shown without any biasing components in order to emphasize the basic structures. You may recall that these configurations were introduced in Chapter 3, in the context of device transfer characteristics.

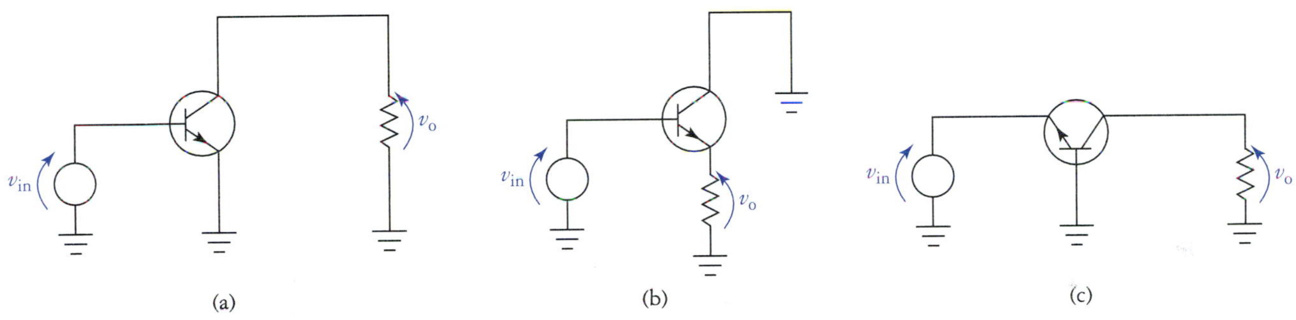

FIGURE 5.3
Three basic BJT amplifier configurations: (a) common-emitter, (b) emitter-follower or common-collector, and (c) common-base.

Each amplifier configuration has its own particular characteristics that make one more suitable than another in a given application. We next examine each configuration individually.

5.3 COMMON-EMITTER CONFIGURATION

A very simple CE amplifier is shown in Fig. 5.4. Notice that the input source and the load resistor have been *capacitively coupled* to the transistor. This was done for several reasons: First, the capacitors prevent the input source and load resistor from disturbing the transistor bias conditions. Second, the capacitors prevent any dc levels that may be present at the transistor terminals from causing dc currents to flow through the input source or the load resistor. The capacitors are chosen such that $|X_C|$ is low enough to be negligible over the range of frequencies for which the amplifier is designed to operate. This circuit, therefore, cannot be used to process dc signals. We will concentrate primarily on the analysis of such capacitively coupled amplifiers in this chapter.

FIGURE 5.4
Simple capacitively coupled common-emitter amplifier.

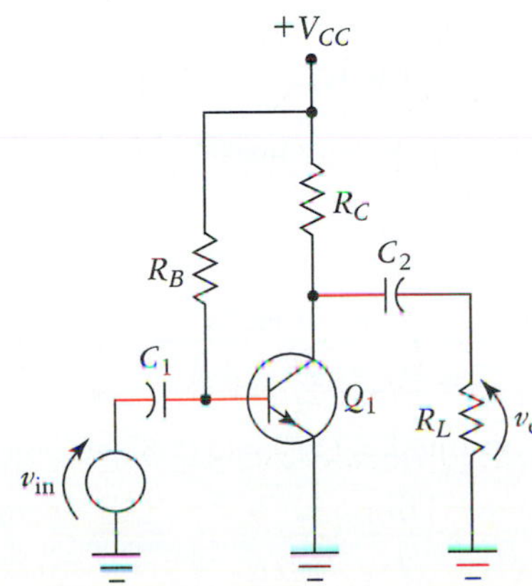

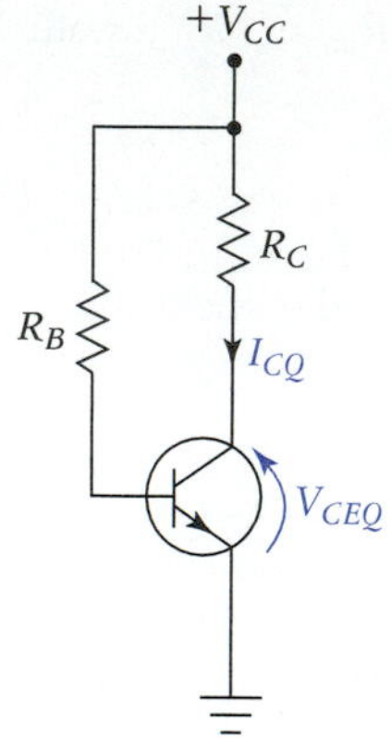

FIGURE 5.5
DC-equivalent circuit of Fig. 5.4.

Analysis of Fig. 5.4 is performed in two parts. First, we must determine the operating point of the transistor. Recall that the operating point (or Q point) is found under no-signal conditions, thus we assume that $v_{in} = 0$ V. We also assume that all coupling capacitors have been charged completely and so may be treated as open circuits under these conditions. It is helpful to redraw the circuit based on these assumptions, producing what is called the *dc-equivalent circuit.* The dc-equivalent circuit for Fig. 5.4 is shown in Fig. 5.5. It is obvious that we have a simple base-biased circuit, for which we could easily determine the Q point.

To determine the effect of an ac signal on the circuit, we must derive the ac-equivalent circuit. This is done by killing all dc sources and replacing all coupling capacitors with short circuits. Recall that the coupling capacitors would have been chosen such that $|X_C| \cong 0\ \Omega$ over the frequency range of interest. Killing the dc power supply makes the V_{CC} node an ac ground point. This produces the ac-equivalent circuit of Fig. 5.6. It is often convenient to combine the load and collector resistors into one equivalent resistance R'_C.

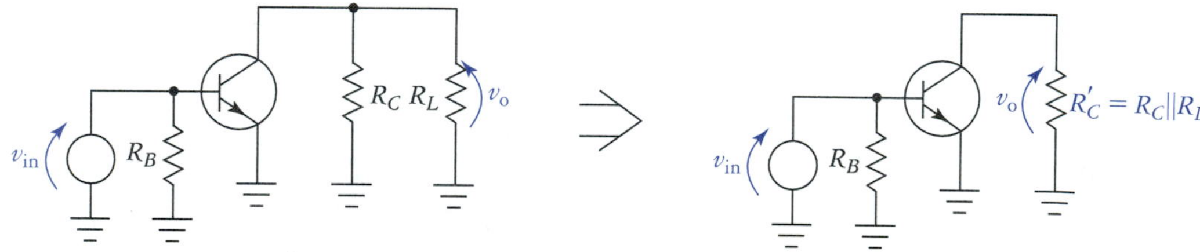

FIGURE 5.6
AC-equivalent circuit of Fig. 5.4.

Because the transistor is connected in the common-emitter configuration, the circuit can be modeled as shown in Fig. 5.7. Resistance r_π is given by Eq. (3.28a), which is repeated here for convenience.

$$r_\pi = r_e(\beta + 1) \tag{3. 28a}$$

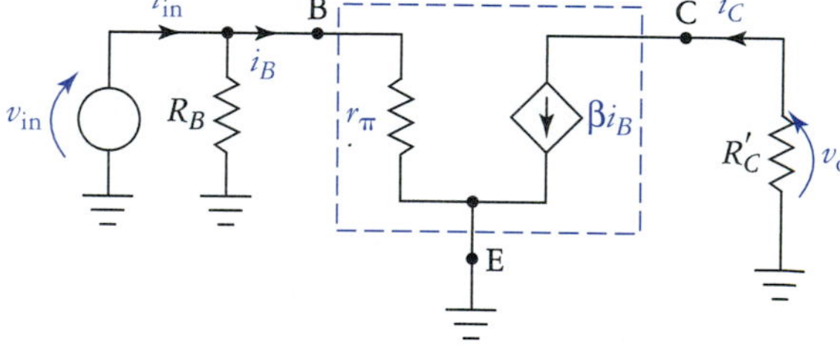

FIGURE 5.7
AC equivalent with small-signal, active-region model of the transistor in place.

Note
Remember, at room temperature (T = 25¡C) we have $V_T \cong 26$ mV.

Recall also from Chapter 3 that

$$r_e \cong \frac{V_T}{I_{CQ}} \tag{3. 26b}$$

Common-Emitter Voltage Gain

We now derive an expression for A_v for Fig. 5.7. The ac base current is found using Ohm's law:

$$i_B = \frac{v_{in}}{r_\pi} \tag{5.7}$$

and the collector current is

$$i_C = \beta i_B \tag{5.8}$$

Substituting Eq. (5.7) into (5.8), we obtain

$$i_C = \frac{\beta v_{in}}{r_\pi} \tag{5.9}$$

The ac output voltage is found using Ohm's law to be

$$v_o = -i_C R'_C \tag{5.10}$$

Notice that the negative sign is necessary because the output voltage sensing arrow is pointing in the same direction as the collector current. Thus, v_o is negative with respect to ground. Substituting Eq. (5.9) into (5.10), we get

$$v_o = -\frac{v_{\text{in}}\beta R'_C}{r_\pi} \tag{5.11}$$

Dividing both sides by v_{in} gives us

$$A_v = \frac{v_o}{v_{\text{in}}} = \frac{\beta R'_C}{r_\pi} \tag{5.12}$$

It is instructive at this point to substitute Eq. (3.28a) into this equation, which yields

$$A_v = -\frac{\beta R'_C}{(\beta + 1)r_e} \tag{5.13a}$$

Separating the fraction into two factors, we have

$$A_v = -\frac{\beta}{\beta + 1}\frac{R'_C}{r_e} \tag{5.13b}$$

If you have a good memory for such things, you will recognize that $\beta/(\beta + 1) = \alpha$, so we obtain

$$A_v = -\alpha\frac{R'_C}{r_e} \tag{5.14}$$

Note
We could also obtain Eq. 5.15 by assuming $r_\pi \cong \beta r_e$ in Eq. 3.23a.

Finally, since for most transistors $\alpha \cong 1$, we can write

$$A_v \cong -\frac{R'_C}{r_e} \tag{5.15}$$

Common-Emitter Input Resistance

By definition, the *input resistance* of any amplifier is given by

$$R_{\text{in}} = \frac{v_{\text{in}}}{i_{\text{in}}} \tag{5.16}$$

Equation (5.16) is a very general definition of input resistance that is best used experimentally. We are really after an expression that gives us R_{in} in terms of circuit component values. Fortunately, this is easy to do. For the CE amplifier of Fig. 5.4, we again refer to the active-region model in Fig. 5.7, where we find

$$R_{\text{in}} = R'_B \parallel r_\pi \tag{5.17}$$

Common-Emitter Phase Relationships

Let's take a moment here and review the significance of phase inversion. Observe in Fig. 5.7 that if v_{in} is positive going that i_B will flow into node B. This causes i_C to flow into node C. Because we are measuring the output voltage with respect to ground, v_o must be a negative-going voltage. If we assume that v_{in} is negative going, then we find that v_o is positive going. This characteristic is called *phase inversion.* Phase inversion is equivalent to a 180° phase shift.

In less abstract terms, phase inversion in the CE amplifier occurs because whenever v_{in} goes positive, the forward bias of the B-E junction is increased. This should cause a proportional increase in I_C (the transistor is driven toward saturation). When this happens, V_{CE} must decrease. This decrease in V_{CE} causes the coupling capacitor to begin to discharge into the collector, which in turn causes current flow from ground up through the load resistor. Under quiescent conditions the collector coupling capacitor is charged to V_{CQ} volts. This is how the load resistor receives energy on negative-going output swings.

When V_{in} is negative going, the bias on the B-E junction is reduced, driving the transistor toward cutoff, reducing I_C. Now V_{CE} increases and the coupling capacitor begins charging. This charging causes current flow through the load resistor to ground, producing a positive-going output voltage swing.

Common-Emitter Current and Power Gain

The easiest way to determine the current gain of the CE amplifier is to use Eq. (5.6). Equation (5.5) provides the easiest way with which to predict the power gain.

It is important to realize that the current gain of the CE amplifier will be negative, as was the case with the voltage gain. Remember, this just means that the load current is 180° out of phase with the input current. The power gain of the amplifier will be positive, because it is the product of two negative quantities $-A_v$ and $-A_i$.

Common-Emitter Amplifier Voltages

A base-biased CE amplifier is shown in Fig. 5.8. If we were to apply a sinusoidal input voltage and look at each node with an oscilloscope, waveforms like those shown in the boxes would be seen. Notice that the coupling capacitors prevent the dc bias voltages from appearing in the input and output signals.

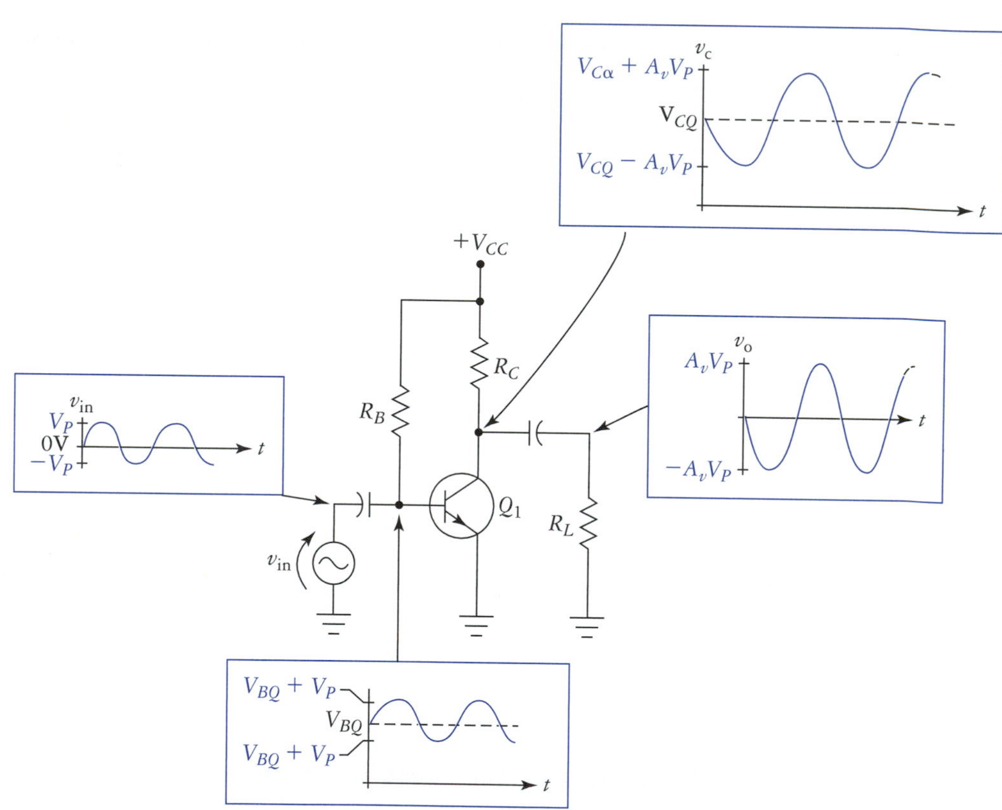

FIGURE 5.8
Voltage waveforms that would be present at various points in the circuit.

EXAMPLE 5.1

The circuit in Fig. 5.4 has $V_{CC} = 12$ V, $R_B = 470$ kΩ, $R_C = 2.7$ kΩ, $R_L = 2.7$ kΩ, and $\beta = 100$. Determine I_{CQ}, V_{CEQ}, A_v, R_{in}, A_i, and A_p.

Solution

Routine analysis gives us

$$I_{CQ} = 2.4 \text{ mA} \qquad \text{and} \qquad V_{CEQ} = 5.52 \text{ V}$$

Perhaps the most useful quantity to determine now is the dynamic emitter resistance as follows:

$$r_e \cong \frac{V_T}{I_{CQ}} = \frac{26 \text{ mV}}{2.4 \text{ mA}} = 10.8\ \Omega$$

The voltage gain is

$$A_v \cong -\frac{R'_C}{r_e} = -\frac{2.7 \text{ k}\Omega \parallel 2.7 \text{ k}\Omega}{10.8\ \Omega} = -125$$

The input resistance is now determined.

$$R_{\text{in}} = R_B \parallel r_\pi = 470 \text{ k}\Omega \parallel 1.08 \text{ k}\Omega \cong 1.08 \text{ k}\Omega$$

The current gain is found as follows:

$$A_i = A_v \frac{R_{\text{in}}}{R_L} = -125 \frac{1.08 \text{ k}\Omega}{2.7 \text{ k}\Omega} = -50$$

And finally, the power gain is

$$A_p = A_v A_i = (-125)(-50) = 6250$$

Note
For the capacitively-coupled CE amplifiers, $A_i < \beta$.

The results of Example 5.1 are significant. One of the more important characteristics of the common-emitter amplifier is the fact that it is capable of significant voltage gain and current gain. The voltage gain is determined primarily by the load resistance, and can easily exceed 200 (actually, -200). The maximum possible current gain magnitude for the CE amplifier will always be less than β. In Example 5.1, for example, we found $|A_i| = 50 = \beta/2$. As you will soon discover, this is really a rather high current gain for this type of circuit. Naturally, since power gain is the product of A_v and A_i, it will tend to be large for the common-emitter amplifier.

5.4 EFFECTS OF EMITTER FEEDBACK ON THE COMMON-EMITTER AMPLIFIER

The circuit of Fig. 5.4 was introduced first because it is simple, and it provides fairly respectable gain values. Of course, we also know that there are problems associated with simple base biasing as well. In particular, this circuit has a very beta-dependent Q point. The gain parameters of the amplifier are also strongly dependent on the Q point, and so they too will vary widely with beta.

Emitter feedback was shown to stabilize the Q-point location. We now show that emitter feedback also tends to stabilize amplifier gain parameters as well. Consider the circuit in Fig. 5.9. This is the voltage-divider biased circuit, configured as a common-emitter amplifier.

The dc- and ac-equivalent circuits for Fig. 5.9 are shown in Fig. 5.10. The dc equivalent circuit is easily analyzed using the techniques developed in Chapter 4. The only significant difference between the ac-equivalent circuits in Figs. 5.10 and 5.6 is the addition of an external ac-equivalent emitter resistance R_e. The notation R_e is used in the ac-equivalent circuit, because the effective external ac emitter resistance is not necessarily the same as the external dc resistance R_E. This is an important distinction that will be used very often in the upcoming discussions.

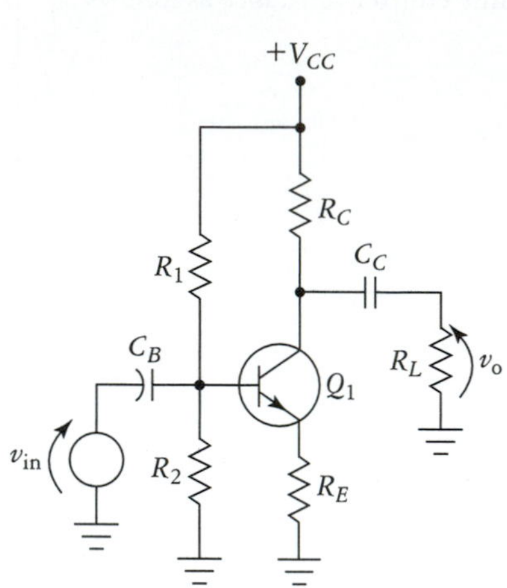

FIGURE 5.9
Voltage-divider biased, capacitively coupled CE amplifier.

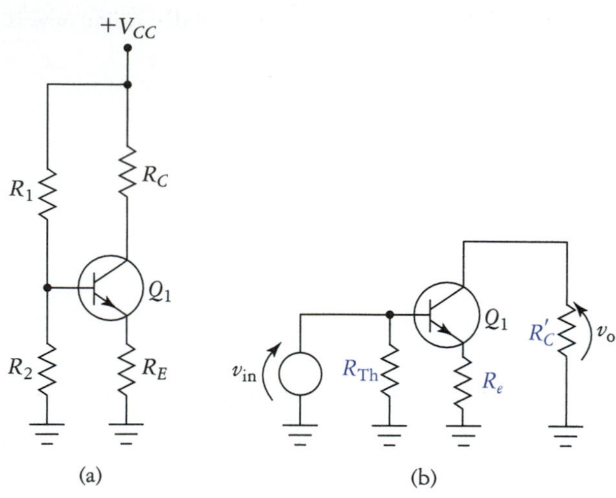

FIGURE 5.10
(a) DC-equivalent circuit. (b) AC-equivalent circuit.

The ac-equivalent circuit of Fig. 5.10 can be modeled as shown in Fig. 5.11. The emitter-feedback resistor complicates the analysis slightly. So, to simplify things a little from the outset, let us assume that $i_E = i_C$, which is nearly always a safe bet.

Note
Remember the following notation:
R'_C = AC equivalent collector resistance
R_C = DC equivalent collector resistance
R_e = AC equivalent emitter resistance
R_E = DC equivalent emitter resistance

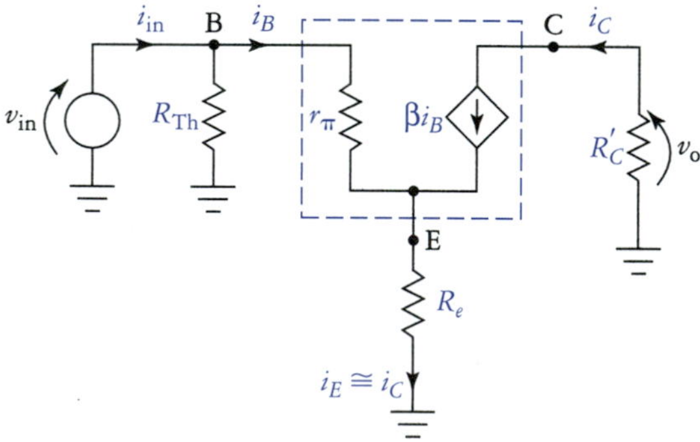

FIGURE 5.11
AC-equivalent circuit using small-signal model of the BJT.

Voltage Gain

What we want to do right now is define i_C in terms of v_{in} and the other easily controlled circuit parameters. We start by applying KVL clockwise from v_{in}, around the B-E loop:

$$0 = v_{in} - i_B r_\pi - i_C R_e \tag{5.18}$$

Substitute i_C/β for i_B, which gives us

$$0 = v_{in} - \frac{i_C}{\beta} r_\pi - i_C R_e \tag{5.19}$$

Now, if we rewrite the second term in this equation we have

$$0 = v_{in} - i_C \frac{r_\pi}{\beta} - i_C R_e \tag{5.20}$$

If you are really alert, you will notice that $r_\pi/\beta \cong r_e$, so we can now write

$$0 = v_{in} - i_C r_e - i_C R_e \tag{5.21}$$

Solving for i_C gives us

$$i_C = \frac{v_{in}}{r_e + R_e} \tag{5.22}$$

We know that $v_o = -i_C R_C$, so now we substitute Eq. (5.22) into this equation and we get

Note
The voltage gain predicted using Eq. 5.24 is sometimes called *swamped voltage gain* because R_e swamps variations in r_e.

$$v_o = -\frac{v_{\text{in}} R'_C}{R_e + r_e} \tag{5.23}$$

Finally, dividing both sides by v_{in} gives us

$$A_v = \frac{v_o}{v_{\text{in}}} = -\frac{R'_C}{R_e + r_e} \tag{5.24}$$

Input Resistance

The input resistance of the common-emitter amplifier with emitter feedback is easily determined to be

$$R_{\text{in}} \cong (r_\pi + \beta R_e) \parallel R'_B \tag{5.25a}$$

or, equivalently,

$$R_{\text{in}} = \cong \beta(r_e + R_e) \parallel R'_B \tag{5.25b}$$

The derivation of these expressions is left as an exercise for the reader. Note for the sake of generality that the external ac resistance between the base terminal and ground has been called R'_B in order to distinguish it from the external dc base resistance. The two are not necessarily the same.

Power Gain and Current Gain

Power gain and current gain for the emitter-feedback circuit are given by Eqs. (5.5) and (5.6), respectively. In fact, Eqs. (5.5) and (5.6) are valid for any amplifier.

Output Resistance

The output resistance of the CE amplifier is the resistance that the load "sees" when it looks at the output of the amplifier. Ideally, the collector-dependent source has infinite internal resistance, $r_o \rightarrow \infty$, and so the output resistance is approximately equal to the collector resistor. That is,

$$R_o \cong R_C \tag{5.26}$$

Equation (5.26) is usually a valid approximation for commonly used values of R_C, so we use it throughout the remainder of this chapter. As a point of general information, however, the effects of finite internal collector resistance (r_{CE} or r_o) on output resistance should be discussed.

Recall that the collector characteristic curves become less horizontal as collector current is increased. This is equivalent to a reduction in r_o with collector current for the transistor. If this resistance is not at least 10 times larger than R_C, then the overall amplifier output resistance should be more accurately determined using

Note
Refer back to page 98 for a review of active-region r_o determination.

$$R_o = R_C \parallel r_o \tag{5.27}$$

The use of emitter feedback causes an increase in r_o that is given by

$$r'_o = r_o\left(\frac{r_\pi \parallel R_e}{r_e} + 1\right) \tag{5.28}$$

Equations (5.27) and (5.28) will not really be crucial in this chapter, but they do point out some additional effects of emitter feedback and nonideal transistor behavior.

Common-Emitter Amplifier Variations

There are many different common-emitter biasing variations. Ultimately though, they all reduce to an ac-equivalent circuit like that of Fig. 5.10(b) (or Fig. 5.11, if you prefer). The best way to gain an understanding of these circuits is to jump right in and do some analysis.

EXAMPLE 5.2

Refer to Fig. 5.12. Determine I_{CQ}, V_{CEQ}, $I_{C(\text{sat})}$, $V_{CE(\text{cut})}$, A_v, A_i, A_p, and R_{in}. Assume that $\beta = 100$.

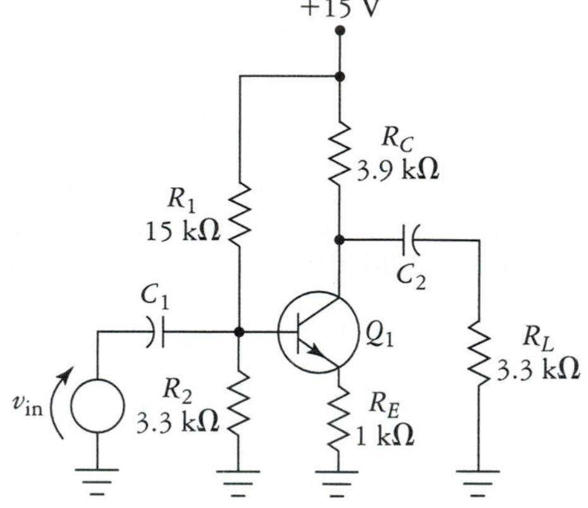

FIGURE 5.12
Circuit for Example 5.2.

Solution The results of the dc analysis are

$$I_{CQ} = 2\text{ mA},\quad V_{CEQ} = 5.2\text{ V},\quad I_{C(\text{sat})} = 3\text{ mA},\quad V_{CE(\text{cut})} = 15\text{ V}$$

It is also useful to record the Thevenin resistance of the voltage-divider network, $R_{\text{Th}} = 2.7\text{ k}\Omega$. The ac-equivalent circuit is shown in Fig. 5.13.

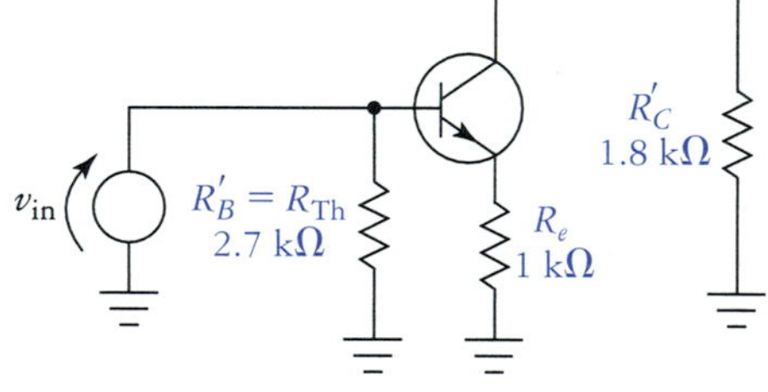

FIGURE 5.13
AC-equivalent circuit.

For future reference, we will now determine r_e, r_π, and R'_C.

$$r_e \cong \frac{V_T}{I_{CQ}} = \frac{26\text{ mV}}{2\text{ mA}} = 13\ \Omega$$

$$r_\pi \cong \beta r_e = (100)(13\ \Omega) = 1.3\text{ k}\Omega$$

$$R'_C = R_C \parallel R_L = 3.9\text{ k}\Omega \parallel 3.3\text{ k}\Omega = 1.8\text{ k}\Omega$$

The voltage gain is

$$\begin{aligned} A_v &= -\frac{R'_C}{R_e + r_e} \\ &= -\frac{1.8\text{ k}\Omega}{1\text{ k}\Omega + 13\ \Omega} \\ &= -1.78 \end{aligned}$$

It is convenient to determine the input resistance at this time. We use Eq. (5.25a):

$$R_{\text{in}} = R'_B \parallel (r_\pi + \beta R_e)$$

In this example, $R'_B = R_{\text{Th}}$ thus

$$\begin{aligned} R_{\text{in}} &= R_{\text{Th}} \parallel (r_\pi + \beta R_e) \\ &= 2.7\text{ k}\Omega \parallel (1.3\text{ k}\Omega + 100\text{ k}\Omega) \\ &= 2.7\text{ k}\Omega \parallel 101.3\text{ k}\Omega \\ &= 2.6\text{ k}\Omega \end{aligned}$$

The current gain is now found:

$$A_i = A_v \frac{R_{\text{in}}}{R_L}$$
$$= -1.78 \frac{1.8}{3.3}$$
$$= -1.40$$

and the power gain is

$$A_p = A_v A_i$$
$$= (-1.78)(-1.40)$$
$$= 2.49$$

Note
Emitter feedback stabilizes the amplifier at the expense of gain.

The results of the preceding example are quite interesting. In all honesty, the amplifier in this example is not very impressive in terms of gain. There is a trade-off between amplifier stability and gain. Generally, the more emitter feedback we use, the more stable the circuit is. Unfortunately, the emitter feedback causes a decrease in voltage gain as well. Reducing the value of the emitter resistor will decrease the amount of negative feedback and increase the voltage gain. This is not practical though because it will move the Q point.

Looking more closely at the voltage gain equation, Eq. (5.24), it is easy to understand why emitter feedback increases gain stability. Without emitter feedback, the R_e term is zero. Thus the gain of the amplifier is completely dependent on the dynamic emitter resistance r_e. This can be a problem, because r_e is very temperature dependent. And, if we get down to even finer details, we should multiply r_e in the A_v equation by the emission coefficient η of the transistor. The main point to be made here is that the CE amplifier without some sort of negative feedback is fairly unpredictable in terms of gain (and input resistance as well). However, we have seen that too much emitter feedback is not necessarily a good thing either.

Emitter Bypassing

The task at hand now is to increase the voltage gain of the CE amplifier with emitter feedback, without disturbing the Q point. This is done by bypassing the emitter resistor with a capacitor. If we connect a capacitor in parallel with the emitter resistor in Fig. 5.12, the emitter is effectively grounded in the ac-equivalent circuit. The bypass capacitor has no effect, however, on the dc analysis of the amplifier. The following example illustrates the effect of emitter bypassing.

EXAMPLE 5.3

Determine A_v, A_i, A_p, and R_{in} for the circuit in Fig. 5.14.

FIGURE 5.14
Circuit for Example 5.3.

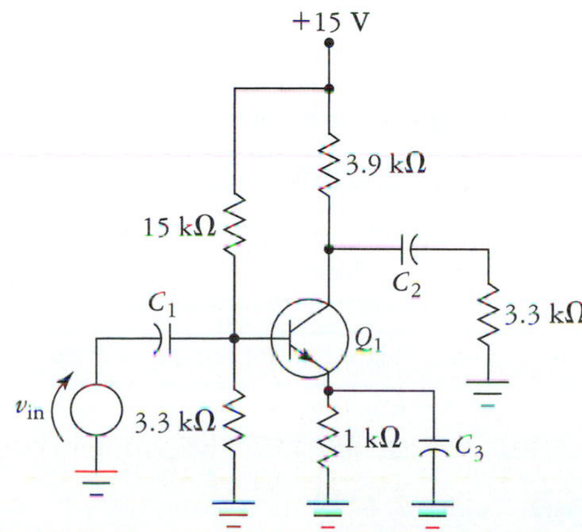

FIGURE 5.15
AC-equivalent circuit.

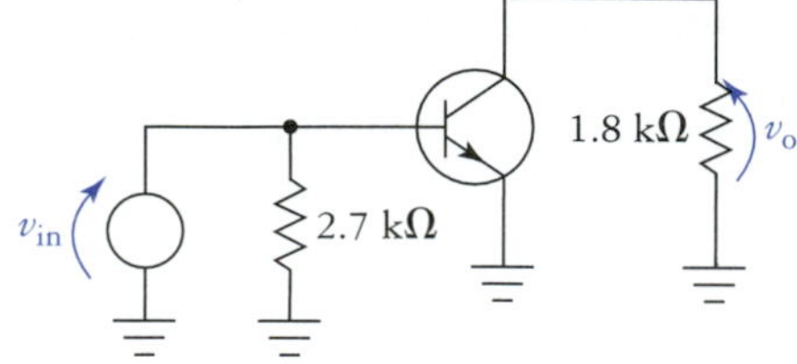

Solution

This circuit is identical to Fig. 5.12, except for the presence of the emitter bypass capacitor. Because the bypass capacitor has no effect on the Q point, we can use the dc analysis results from Example 5.2. The ac-equivalent circuit for this modification is shown in Fig. 5.15.The input resistance and the various gain values are

$$\begin{aligned} R_{\text{in}} &= r_\pi \parallel R'_B \\ &= 1.3\ \text{k}\Omega \parallel 2.7\ \text{k}\Omega \\ &= 878\ \Omega \end{aligned}$$

$$\begin{aligned} A_v &= -\frac{R'_C}{r_e} \\ &= -\frac{1.8\ \text{k}\Omega}{13\ \Omega} \\ &= -138.5 \end{aligned}$$

$$\begin{aligned} A_i &= A_v \frac{R_{\text{in}}}{R_L} \\ &= -138.5 \frac{878\ \Omega}{3.3\ \text{k}\Omega} \\ &= -36.8 \end{aligned}$$

$$\begin{aligned} A_p &= A_v A_i \\ &= (-138.5)(-36.8) \\ &= 5097 \end{aligned}$$

The addition of the emitter bypass capacitor has increased the gain of the circuit tremendously. The power gain has increased by a factor of more than 2000 compared to the circuit with heavy emitter feedback. The Q point of the circuit is still rock-steady though, because the bypass capacitor has no effect on the biasing conditions.

The voltage gain of Fig. 5.14 is relatively high, but because there is no ac emitter feedback at all, the gain is still relatively unpredictable. We can strike a compromise between gain stability and gain magnitude by *partially bypassing* the emitter. This is shown in the next example.

EXAMPLE 5.4

Refer to Fig. 5.16. Determine I_{CQ}, V_{CEQ}, $I_{C(\text{sat})}$, $V_{CE(\text{cut})}$, A_v, A_i, A_p, and R_{in}. Assume that $\beta = 100$.

Solution The dc-equivalent circuit is shown in Fig. 5.17. Notice that the two emitter resistors are combined into a single series resistance. The dc analysis of this circuit yields

$$I_{CQ} = 3.9\ \text{mA}, \quad V_{CEQ} = 7.23\ \text{V}, \quad I_{C(\text{sat})} = 5.58\ \text{mA}, \quad V_{CE(\text{cut})} = 24\ \text{V}$$

The ac-equivalent circuit is shown in Fig. 5.18. The node between the 50-Ω and the 2.75-kΩ resistor in the original circuit is an ac ground point, as are the V_{CC} and V_{EE} nodes. The useful ac analysis parameters are now found:

$$\begin{aligned} r_e &= \frac{V_T}{I_{CQ}} \\ &= \frac{26\ \text{mV}}{3.9\ \text{mA}} \\ &= 6.7\ \Omega \end{aligned}$$

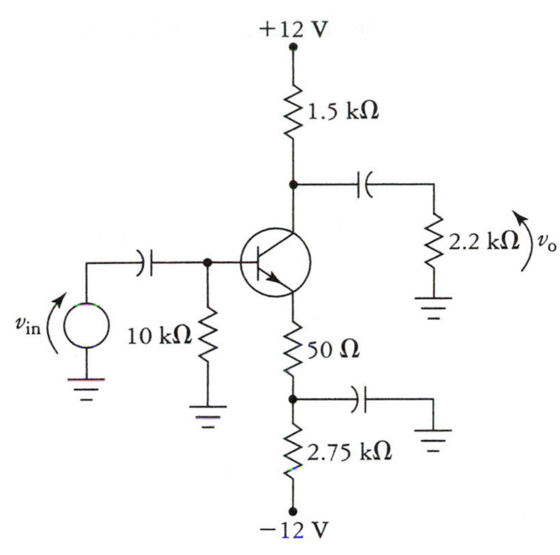

FIGURE 5.16
Circuit for Example 5.4.

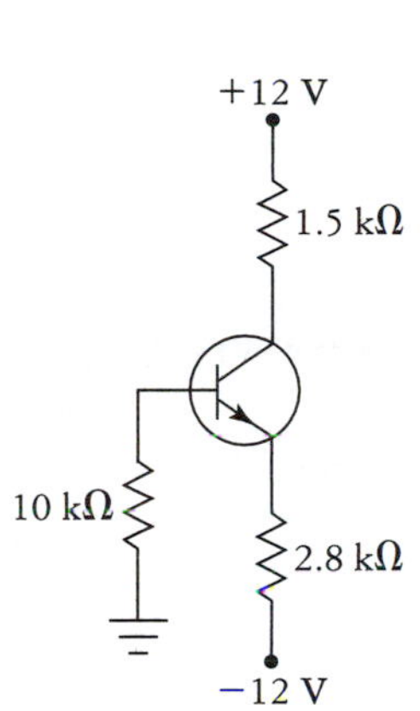

FIGURE 5.17
DC-equivalent circuit.

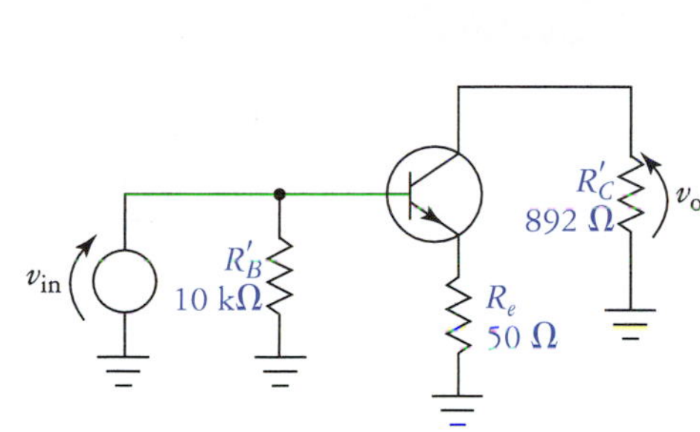

FIGURE 5.18
AC-equivalent circuit.

$$\begin{aligned} r_\pi &= \beta r_e \\ &= (100)(6.7\ \Omega) \\ &= 670\ \Omega \end{aligned}$$

$$\begin{aligned} R'_C &= R_C \parallel R_L \\ &= 1.5\ \text{k}\Omega \parallel 2.2\ \text{k}\Omega \\ &= 892\ \Omega \end{aligned}$$

$$\begin{aligned} R_e &= 50\ \Omega \\ R'_B &= 10\ \text{k}\Omega \end{aligned}$$

Let's start by finding the input resistance first this time:

$$\begin{aligned} R_{\text{in}} &= R'_B \parallel \beta(r_e + R_e) \\ &= 10\ \text{k}\Omega \parallel 100\ (6.7\ \Omega + 50\ \Omega) \\ &= 10\ \text{k}\Omega \parallel 5.67\ \text{k}\Omega \\ &= 3.6\ \text{k}\Omega \end{aligned}$$

The various gains are now determined:

$$\begin{aligned} A_v &= -\frac{R'_C}{R_e + r_e} \\ &= -\frac{892\ \Omega}{50\ \Omega + 6.7\ \Omega} \\ &= -15.7 \end{aligned}$$

$$\begin{aligned} A_i &= A_V \frac{R_{\text{in}}}{R_L} \\ &= -15.7\,\frac{3.6}{2.2} \\ &= -25.7 \end{aligned}$$

$$\begin{aligned} A_p &= A_V A_i \\ &= (-15.7)(-25.7) \\ &= 403.5 \end{aligned}$$

The preceding example has shown that it is possible to achieve a relatively stable circuit without completely throwing away the gain. The 50-Ω resistor in the ac-equivalent circuit is large in comparison to r_e, and so it swamps normal variations in r_e that may be caused by temperature changes and device parameter drift over time. Because of this, the emitter resistor is often called a *swamping resistor.*

EXAMPLE 5.5

Refer to Fig. 5.16. Given $v_{in} = 100 \sin \omega t$ mV, write the expressions and sketch the waveforms for v_C and v_o.

Solution The circuit was analyzed previously. Using those results, we obtain

$$\begin{aligned} v_o &= A_v v_{in} \\ &= (-15.7)(100 \sin \omega t \text{ mV}) \\ &= -1.57 \sin \omega t \text{ V} \end{aligned}$$

The collector voltage is the same as the output voltage except that there is a dc offset of V_{CQ} volts present as well. Going back to the dc-equivalent circuit of Fig. 5.17, and using $I_{CQ} = 3.9$ mA, we obtain

$$\begin{aligned} V_{CQ} &= V_{CC} - I_C R_C \\ &= 12 \text{ V} - (3.9 \text{ mA})(5 \text{ k}\Omega) \\ &= 12 \text{ V} - 5.85 \text{ V} \\ &= 6.15 \text{ V} \end{aligned}$$

The net collector voltage is

$$\begin{aligned} v_C &= V_{CQ} + A_v v_{in} \\ &= 6.15 - 1.57 \sin \omega t \text{ V} \end{aligned}$$

The waveforms are shown in Fig. 5.19.

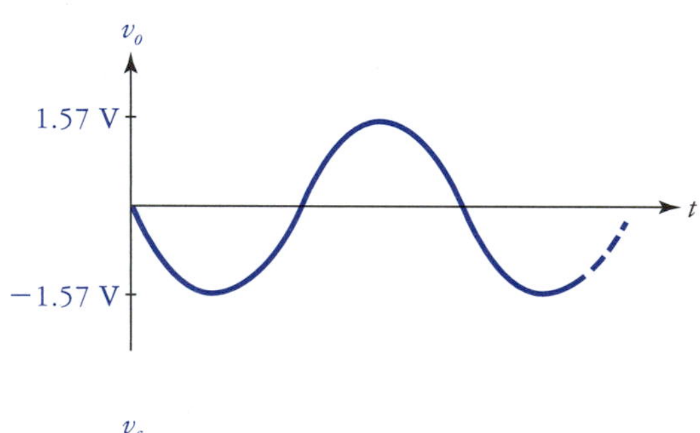

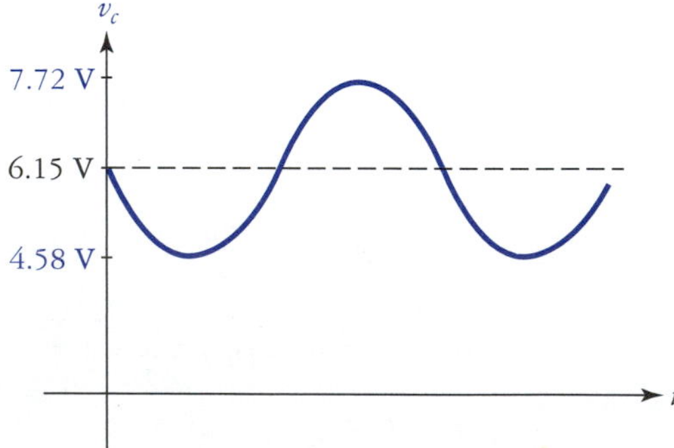

FIGURE 5.19
Output and collector voltage waveforms.

Collector Feedback

Recall that *collector feedback* is sometimes used to obtain a stable Q point. The basic common-emitter amplifier with collector feedback is shown in Fig. 5.20.

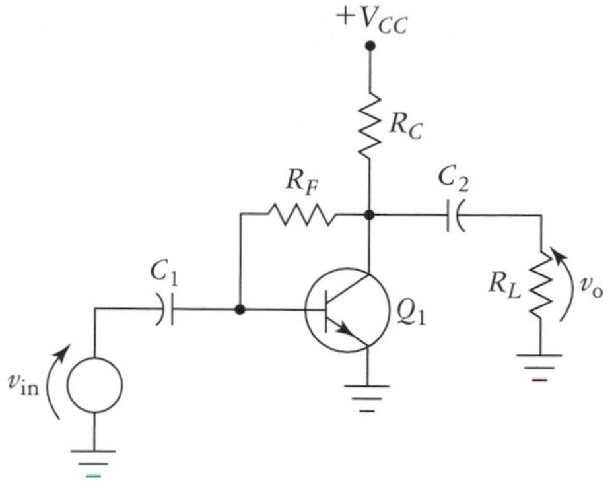

FIGURE 5.20
Common-emitter amplifier using collector-feedback biasing.

The fact that resistor R_F is connected from base to collector complicates the ac analysis of this circuit to some degree. The dependent-source ac model for the collector-feedback circuit is shown in Fig. 5.21. It can be shown that the voltage gain of this circuit is given by

Note
Note that $R_e = 0\Omega$ in the circuit of Fig. 5.20.

$$A_v = -\frac{\dfrac{R_F}{r_e} - 1}{\dfrac{R_F}{R'_C} + 1} \tag{5.29}$$

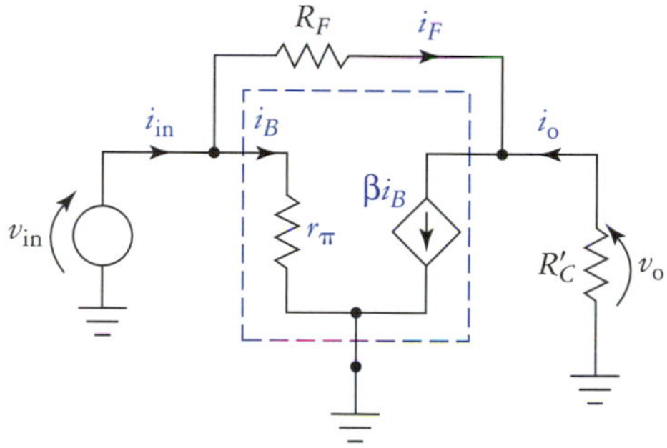

FIGURE 5.21
Small-signal ac-equivalent circuit using dependent-source BJT model.

The input resistance is

$$R_{in} = \beta r_e \left\| \frac{R_F}{|A_v| + 1} \right. \tag{5.30}$$

Finally, the output resistance is

$$R_o = R_C \left\| \left[R_F \left(\frac{|A_v|}{|A_v| + 1} \right) \right] \right. \tag{5.31}$$

As you can see, the collector-feedback resistor, while helping to stabilize the Q point, also makes the ac analysis more cumbersome. There is a trick that really simplifies the ac analysis, however. Consider the circuit in Fig. 5.22(a). The feedback resistor has been split into two parts with the center node ac bypassed to ground. Because of this modification, I_{CQ} is still given by Eq. (4.18), where $R_F = R_1 + R_2$. In the ac-equivalent circuit, however, resistors R_1 and R_2 simply appear in parallel with the input and output, respectively. The voltage gain is given by Eq. (5.15).

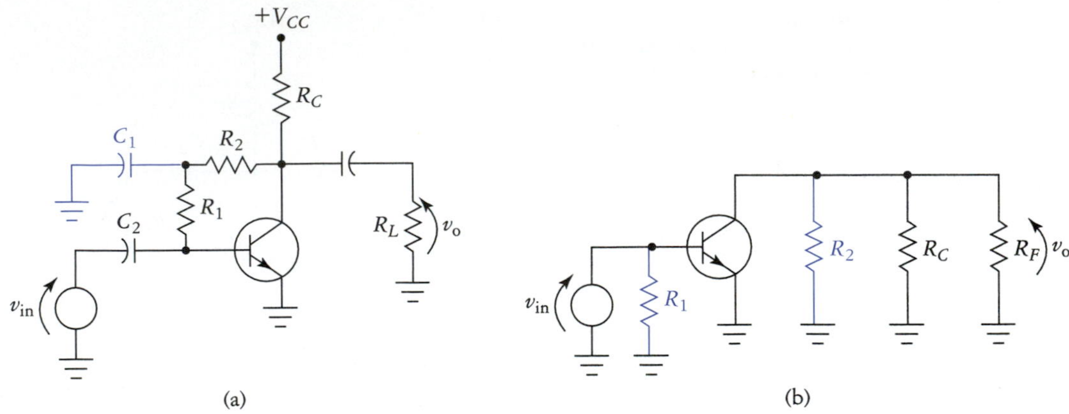

FIGURE 5.22
Simplifying the analysis of the collector-feedback amplifier. (a) Capacitor C_1 removes ac negative feedback. (b) The ac-equivalent circuit.

5.5 EMITTER FOLLOWER

An emitter-follower (also called a common-collector) amplifier, using voltage-divider biasing is shown in Fig. 5.23. The dc analysis of this circuit is routine at this point, so we shall jump right into the ac analysis.

FIGURE 5.23
Emitter-follower amplifier.

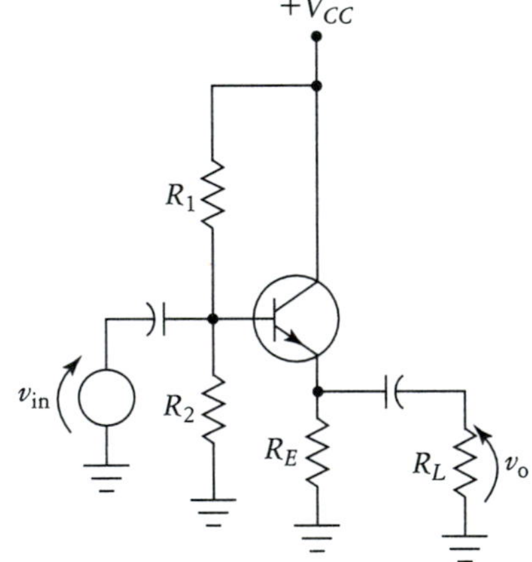

Voltage Gain

The ac equivalent for the emitter follower is shown in Fig. 5.24(a). The dependent-source model is shown in 5.24(b). Note that R_e is the parallel combination of R_E and R_L, and R_B is the Thevenin resistance of the voltage-divider network. The voltage gain equation for this circuit is derived as follows: Applying KVL to the B-E loop, we have

$$0 = v_{in} - i_E r_e - i_E R_e$$

We now make the simplifying assumption that $i_E = i_C$, giving us

$$0 = v_{in} - i_C r_e - i_C R_e$$

Solving for i_C yields

$$i_C = \frac{v_{in}}{r_e + R_e} \tag{5.32}$$

The output voltage is given by

$$v_o = i_C R_e \tag{5.33}$$

Substituting Eq. (5.32) into (5.33), we obtain

Note
The emitter-follower has:
$A_v < 1$
$A_i < \beta$
$A_p < \beta$

$$v_o = \frac{v_{\text{in}}}{r_e + R_e} \tag{5.34}$$

Dividing both sides by v_{in} gives us

$$A_v = \frac{v_o}{v_{\text{in}}} = \frac{R_e}{R_e + r_e} \tag{5.35}$$

Inspection of Eq. (5.35) reveals that the voltage gain of the emitter follower will always be less than 1 (unity), and there is no phase inversion. This type of amplifier is often called a voltage follower because the output voltage is nearly identical to the input.

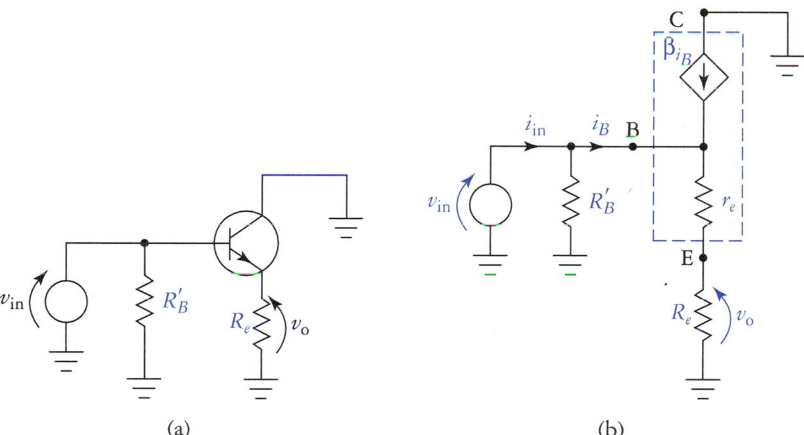

FIGURE 5.24
(a) AC-equivalent circuit. (b) Equivalent circuit with small-signal dependent-source model.

Input Resistance

It is left as an exercise for the reader to prove that

$$R_{\text{in}} = R'_B \parallel \beta(r_e + R_e) \tag{5.36}$$

Output Resistance

The output resistance of the emitter follower is the lowest of the three basic amplifier configurations. This is in sharp contrast to the common-emitter amplifier, where output resistance is typically on the order of several thousand ohms. Low output resistance is a very desirable voltage-amplifier characteristic. It allows the amplifier to drive low-resistance loads very efficiently. The output resistance of the emitter follower is given by

$$R_o = \left(r_e + \frac{R_S}{\beta}\right) \parallel R_E \tag{5.37}$$

If the internal resistance of the signal source that is driving the emitter follower has a low resistance, then we can write

$$R_o \cong r_e \parallel R_E \tag{5.38}$$

EXAMPLE 5.6

Refer to Fig. 5.25. Determine I_{CQ}, V_{CEQ}, A_v, A_i, A_p, R_{in}, and R_o. Assume that $\beta = 100$.

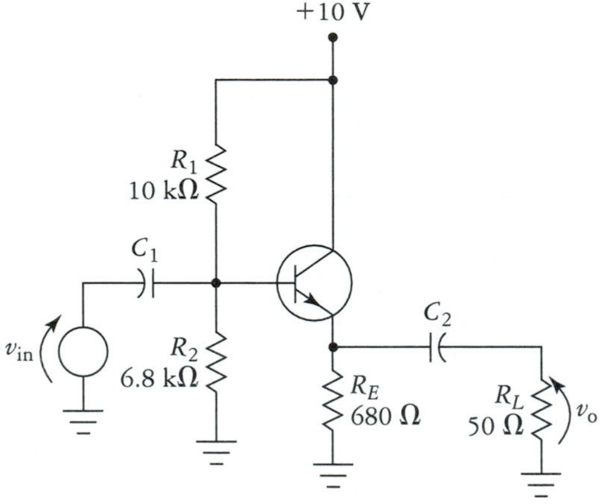

FIGURE 5.25
Circuit for Example 5.6.

Solution

Basic dc analysis gives us $I_{CQ} = 4.7$ mA and $V_{CEQ} = 6.8$ V. The following quantities are required:

$$\begin{aligned} r_e &= V_T/I_{CQ} \\ &= 26\text{ mV}/4.7\text{ mA} \\ &= 5.5\ \Omega \end{aligned}$$

$$\begin{aligned} R_e &= R_E \parallel R_L \\ &= 680\ \Omega \parallel 50\ \Omega \\ &= 46.6\ \Omega \end{aligned}$$

The remaining quantities are now determined:

$$\begin{aligned} A_v &= \frac{R_e}{R_e + r_e} \\ &= \frac{46.6\ \Omega}{46.6\ \Omega + 5.5\ \Omega} \\ &= 0.89 \end{aligned}$$

$$\begin{aligned} R_{in} &= R'_B \parallel \beta(R_e + r_e) \\ &= 4\text{ k}\Omega \parallel 100(46.6\ \Omega + 5.5\ \Omega) \\ &= 2.26\text{ k}\Omega \end{aligned}$$

$$\begin{aligned} A_i &= A_v \frac{R_{in}}{R_L} \\ &= 0.89\,\frac{2.26\text{ k}\Omega}{50\ \Omega} \\ &= 40.2 \end{aligned}$$

$$\begin{aligned} A_p &= A_v A_i \\ &= (0.89)(40.2) \\ &= 35.8 \end{aligned}$$

$$\begin{aligned} R_o &= r_e \parallel R_E \\ &= 5.5\ \Omega \parallel 680\ \Omega \\ &\cong 5.5\ \Omega \end{aligned}$$

Uses of the Emitter Follower

You may be wondering what good an amplifier with such low voltage gain is. As mentioned earlier, emitter followers are used to allow low-resistance loads to be driven more easily by higher resistance sources. Thus, the emitter follower may be thought of as matching impedance between a low-resistance load and a higher resistance source. In some cases, this job is done with a transformer. The emitter follower is often preferable though, because in low-power applications it will usually be less expensive, less sensitive to noise, and it has the advantage of power gain.

Emitter-Follower Stability

One problem that sometimes occurs with the emitter follower is oscillation. For reasons that will be discussed later, from an ac signal standpoint, amplifiers tend to become more unstable as negative feedback is increased. It can be shown that the emitter follower has a large amount of negative feedback, and so these circuits tend to oscillate at high frequency (often at over 100 MHz), especially when breadboarded in the lab.

Note
An unstable emitter-follower amplifier may oscillate at a frequency in the hundreds of MHz.

There are several ways to help eliminate and prevent oscillation. First, the use of short power supply and component leads reduces stray inductance and capacitance. This is not usually a problem in commercial circuits, because printed-circuit techniques are normally used, but it is often a problem when such circuits are prototyped. Bypassing the power supply with a capacitor that is located physically close to the transistor helps prevent oscillation as well. This is often referred to as *decoupling* the power supply. A value in the range from 0. 1 μF to around 10 μF will usually work well.

In addition to capacitive bypassing, sometimes a low-value collector resistor will be used to prevent oscillation. The collector resistor is not necessary for operation of the emitter-follower amplifier since the output is taken from the emitter, but this resistor will decrease the tendency toward oscillation with very little performance or cost penalty.

5.6 COMMON-BASE AMPLIFIER

The final BJT configuration is the common-base amplifier. A common-base amplifier using voltage-divider biasing is shown in Fig. 5.26(a). The dc-equivalent circuit is shown (drawn in more familiar form) in Fig. 5.26(b). This circuit is easily analyzed using the techniques covered previously.

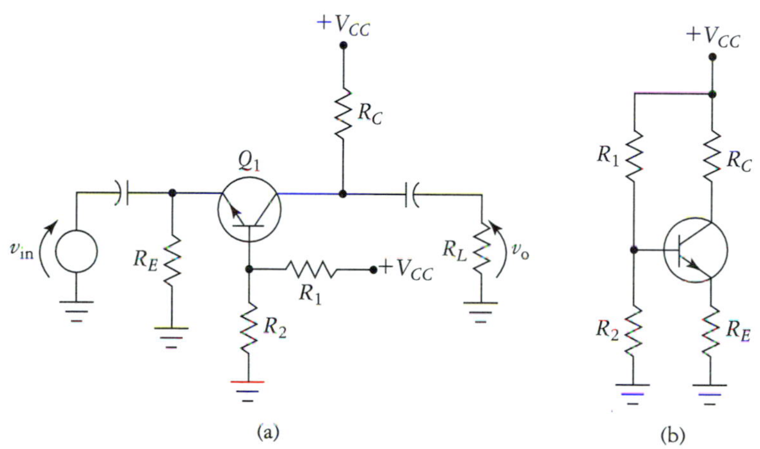

FIGURE 5.26
Common-base amplifier. (a) CB amp using voltage-divider biasing. (b) DC-equivalent circuit redrawn in a more familiar manner.

Voltage Gain

The ac-equivalent of the common-base amplifier is shown in Fig. 5.27(a). Using the dependent-source model of Fig. 5.27(b), the following KVL equation can be written for the E-B loop:

$$0 = v_{in} - i_E r_e - i_B R_B'$$

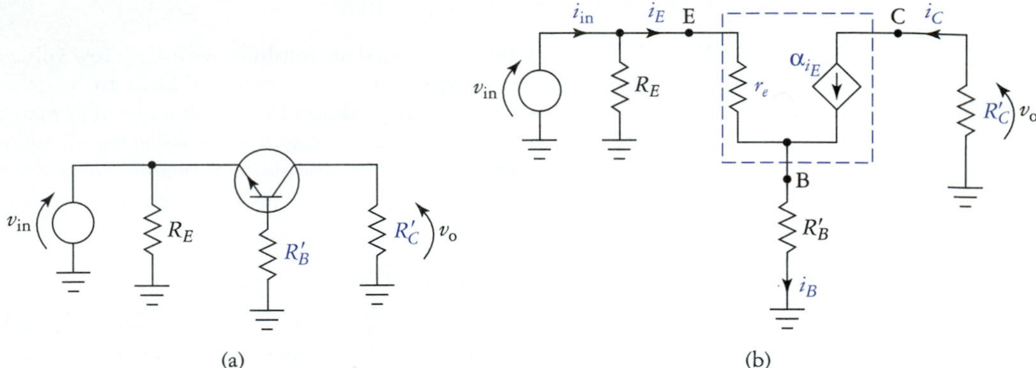

FIGURE 5.27
CB amplifier of Fig. 5.25. AC-equivalent circuit using small-signal dependent-source model.

Using the approximation $i_E = i_C$, we have

$$0 = v_{in} - i_C r_e - i_B R'_B$$

Defining i_B in terms of i_C produces

$$0 = v_{in} - i_C r_e - \frac{i_C}{\beta} R'_B$$

We can rewrite this as

$$0 = v_{in} - i_C r_e - i_C \frac{R'_B}{\beta}$$

Solving for i_C yields

$$i_C = \frac{v_{in}}{r_e + \dfrac{R'_B}{\beta}} \tag{5.39}$$

The output voltage is given by

$$v_o = i_C R'_C \tag{5.40}$$

So, if we substitute Eq. (5.39) into (5.40) and divide by v_{in} we get

$$A_v = \frac{v_o}{v_{in}} = \frac{R'_C}{r_e + \dfrac{R'_B}{\beta}} \tag{5.41}$$

Base Bypassing

Often, a bypass capacitor is used to place the base of the transistor at ac ground potential. This modification is shown in Fig. 5.28. The main effect of base bypassing is an increase in voltage gain. The gain expression for this case is

$$A_v = \frac{R'_C}{r_e} \tag{5.42}$$

This is the same equation as is obtained for the common-emitter amplifier except there is no phase inversion, so the common-base amplifier is capable of providing rather high voltage gain.

Input and Output Resistance

The input resistance of the common-base amplifier is normally the lowest of the three basic amplifier configurations. This can present a problem, because many signal sources cannot drive such a

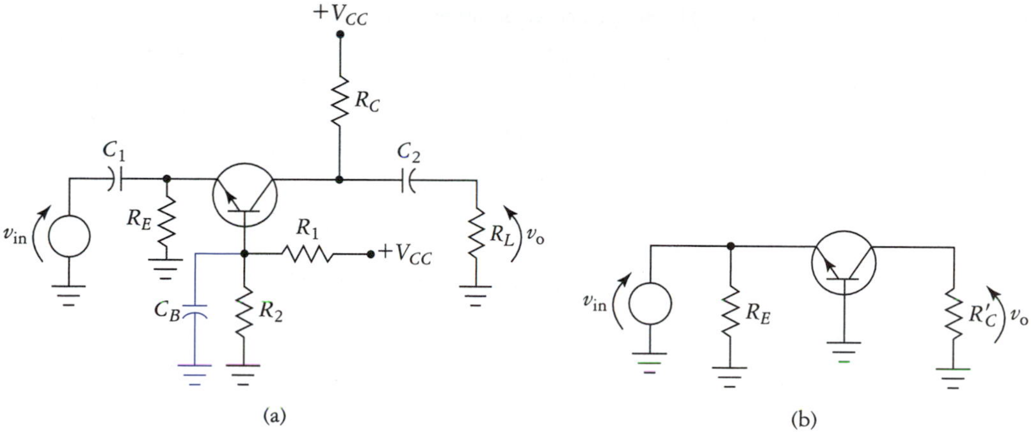

FIGURE 5.28
CB amplifier with base bypassing. (a) Typical circuit. (b) AC-equivalent circuit.

low resistance effectively. Thus, the usefulness of the CB amplifier is limited. The input resistance equation is

$$R_{\text{in}} = R_E \| \left(r_e + \frac{R'_B}{\beta} \right) \quad (5.43)$$

Note
For the common-base amplifier, $A_i < 1$.

If the base is bypassed to ground, then we have

$$R_{\text{in}} \cong R_E \parallel r_e \quad (5.44)$$

Notice the similarities between the CB input resistance equations and the output resistance equations for the emitter follower.

The output resistance of the common-base amplifier is the same as that for the common-emitter amplifier, which is approximated by

$$R_o \cong R_C \quad (5.45)$$

Current Gain

Because we can rather easily determine the voltage gain and the input resistance of the common-base amplifier, we simply use the universal relationship of Eq. (5.6) to determine the current gain. However, the current gain of the common-base amplifier is always less than unity.

EXAMPLE 5.7

Refer to Fig. 5.28. Assume $R_1 = 22\text{ k}\Omega$, $R_2 = 6.8\text{ k}\Omega$, $R_E = 470\ \Omega$, $R_C = 1\text{ k}\Omega$, $R_L = 1\text{ k}\Omega$, $V_{CC} = 15\text{V}$, and $\beta = 100$. Determine I_{CQ}, V_{CEQ}, A_v, A_i, A_p, and R_{in}.

Solution

As usual, we first determine the Q point of the transistor, which is $I_{CQ} = 5.44$ mA and $V_{CEQ} = 7.0$ V. Other necessary values are

$$r_e = \frac{V_T}{I_{CQ}} = \frac{26\text{ mV}}{5.44\text{ mA}} = 4.8\ \Omega$$

$$R'_C = R_C \parallel R_L = 1\text{ k}\Omega \parallel 1\text{ k}\Omega = 500\ \Omega$$

The ac quantities of interest are

$$A_v = \frac{R'_C}{r_e}$$
$$= \frac{500\ \Omega}{4.8\ \Omega}$$
$$= 104.2$$

$$R_{\text{in}} = R_E \parallel r_e$$
$$= 470\ \Omega \parallel 4.8\ \Omega$$
$$\cong 4.8\ \Omega$$

$$A_i = A_v \frac{R_{\text{in}}}{R_L}$$
$$= 104.2 \frac{4.8\ \Omega}{1000\ \Omega}$$
$$= 0.50$$

$$A_p = A_v A_i$$
$$= (104.2)(0.50)$$
$$= 52.1$$

The common-base amplifier has the advantage of superior high-frequency performance relative to the common-emitter and emitter-follower configurations, and so it is used most often in radio-frequency (RF) applications. Knowledge of the common-base configuration is also useful in the analysis of the differential amplifier. This important subject is covered in a later chapter.

5.7 DARLINGTON PAIR

Recall that it is very desirable for a voltage amplifier to have very high input resistance. Of course, the common-base amplifier does not meet this requirement at all, but what about the common-emitter and the emitter-follower amplifiers? The CE and EF circuits that have been analyzed up to now have had input resistances ranging from a few hundred to a few thousand ohms. This may not be sufficiently high enough for many applications. One way to achieve high input resistance is to use a *Darlington pair.*

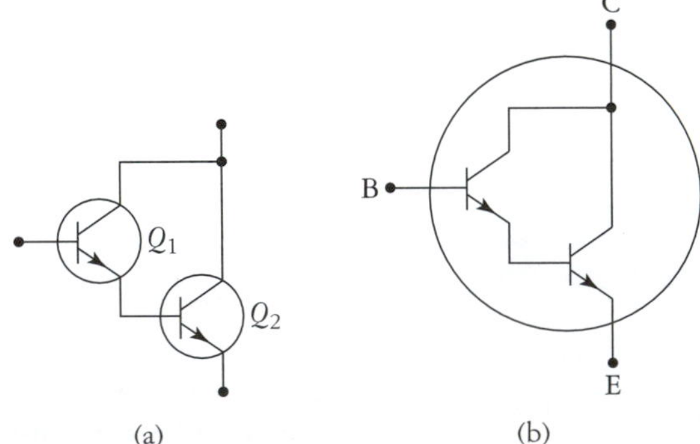

FIGURE 5.29
The Darlington pair. (a) Using discrete transistors. (b) Single three-terminal equivalent Darlington transistor.

Two transistors in the Darlington configuration are shown in Fig. 5.29(a). Normally, the Darlington pair is treated as if it were a single three-terminal device. Darlington transistors are available as single devices, and are usually represented as shown in Fig. 5.29(b).

Darlington Pair DC Analysis

Almost any biasing arrangement will reduce to the form shown in Fig. 5.30 or something even simpler. Our objective here is to derive the expression for total collector current I_{CQ}. This is somewhat lengthy, but important as well. We begin by application of KVL around the B-E loop:

$$0 = V_{BB} - I_{B_1}R_B - 2V_{BE} - I_{E_2}R_E \tag{5.46}$$

To simplify the algebra, let us assume that $\beta_1 = \beta_2 = \beta$ and $\alpha = 1$. Now, let us define I_{E_2} in terms of I_{B_1}. By inspection we find $I_{E_2} = \beta I_{B_2}$ and $I_{B_2} = I_{E_1} = \beta I_{B_1}$. Combining these equations we obtain

$$I_{E_2} = \beta(\beta I_{B_1})$$

or more compactly

$$I_{E_2} = \beta^2 I_{B_1} \tag{5.47}$$

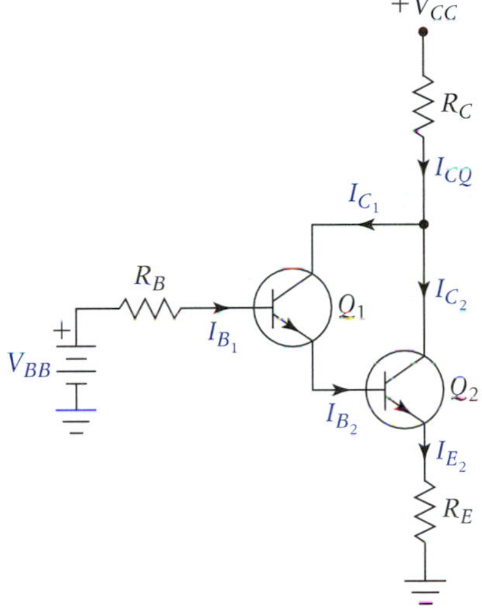

FIGURE 5.30
Base-biased Darlington pair.

Substituting Eq. (5.47) into (5.46) gives us

$$0 = V_{BB} - I_{B_1}R_B - 2V_{BE} - \beta^2 I_{B_1}R_E$$

Solving for I_{B_1} produces the familiar-looking equation

$$I_{B_1} = \frac{V_{BB} - 2V_{BE}}{R_B + \beta^2 R_E} \tag{5.48}$$

Finally, multiplying both sides by β^2 gives us

$$I_{CQ} = \frac{V_{BB} - 2V_{BE}}{\dfrac{R_B}{\beta^2} + R_E} \tag{5.49}$$

This equation is very similar to the equation for the collector current of the basic emitter-feedback circuit [Eq. (4.14)].

The preceding analysis indicates that the Darlington pair may be thought of as being a single device with twice-normal barrier potential and extremely high beta. Consider that if both transistors in the pair have $\beta = 100$, then the total beta is $\beta_T = \beta^2 = 100^2 = 10{,}000$. As a result of this high beta, the external base-biasing resistance can be made quite large without seriously affecting Q-point stability.

It will prove useful in the next discussion for us to derive the relationship between r_{e_1} and r_{e_2}. By inspection, we have

$$I_{C_2} \cong I_{CQ} \tag{5.50}$$

The dynamic emitter resistance equations are

$$r_{e_1} = \frac{V_T}{I_{C_1}} \quad \text{and} \quad r_{e_2} = \frac{V_T}{I_{C_2}}$$

Inspection of the circuit also reveals

$$I_{C_1} = \frac{I_{C_2}}{\beta} \cong \frac{I_{CQ}}{\beta} \tag{5.51}$$

which means that we can write

$$r_{e_1} \cong \beta r_{e_2} \tag{5.52}$$

This relationship will be used in the derivation of the voltage gain of the Darlington pair.

Darlington Pair AC Analysis

Most of the time, the Darlington pair will be used in the common-emitter configuration. The ac-equivalent circuit for this configuration is shown in Fig. 5.31. A slightly different transistor model has been used here in order to make the analysis a bit easier to relate to physically. We begin by writing the KVL equation for the B-E loop. Again, we assume that $\beta_1 = \beta_2 = \beta$ and $\alpha = 1$.

$$0 = v_{\text{in}} - \beta i_{B_1} r_{e_1} - \beta i_{B_2} r_{e_2} - i_{E_2} R_e$$

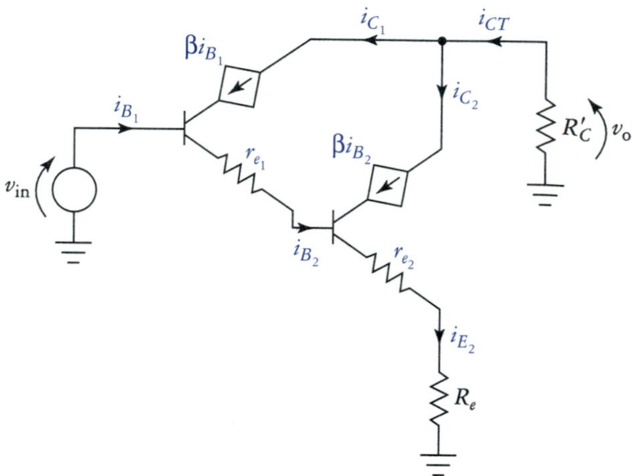

FIGURE 5.31
Small-signal dependent-source models comprising the Darlington pair.

Defining i_{B_2} and i_{E_2} in terms of i_{B_1} produces

$$0 = v_{\text{in}} - \beta i_{B_1} r_{e_1} - \beta^2 i_{B_1} r_{e_2} - \beta^2 i_{B_1} R_e$$

Substituting Eq. (5.52) into this expression yields

$$0 = v_{\text{in}} - \beta^2 i_{B_1} r_{e_2} - \beta^2 i_{B_1} r_{e_2} - \beta^2 i_{B_1} R_e$$

Solving for i_B we obtain

$$i_B = \frac{v_{\text{in}}}{\beta^2 r_{e_2} + \beta^2 r_{e_2} + \beta^2 R_e}$$

$$= \frac{v_{\text{in}}}{2\beta^2 r_e + \beta^2 R_e} \tag{5.53}$$

The total collector current is obtained by multiplying both sides by β^2, giving us

$$i_C = \frac{v_{\text{in}}}{2r_{e_2} + R_e}$$

and since $v_o = -i_C R'_C$,

$$v_o = -\frac{v_{\text{in}} R'_C}{2r_{e_2} + R_e}$$

At long last, we divide both sides by v_{in} and obtain

$$A_v = \frac{v_o}{v_{\text{in}}} = -\frac{R'_C}{2r_{e_2} + R_e} \tag{5.54}$$

For emphasis, we repeat that

$$r_{e_2} = \frac{V_T}{I_{CQ}} = \frac{26 \text{ mV}}{I_{CQ}}$$

Note
The most notable characteristics of the Darlington pair are its high input resistance and high beta.

where I_{CQ} is the total collector current $I_{C_1} + I_{C_2}$.

Note that if complete emitter bypassing is used, Eq. (5.54) reduces to

$$A_v = -\frac{R'_C}{2r_{e_2}} \tag{5.55}$$

The input resistance of any amplifier is given by

$$R_{in} = \frac{v_{in}}{i_{in}}$$

Using Eq. (5.53), and the ac-equivalent circuit, it is relatively easy to show that

$$R_{in} = R'_B \parallel \beta^2(R_e + 2r_{e_2}) \tag{5.56}$$

EXAMPLE 5.8

Refer to Fig. 5.32. Determine I_{CQ}, V_{CEQ}, A_v, A_i, A_p, and R_{in}. Assume that $\beta_1 = \beta_2 = 100$. For the dc analysis, treat the Darlington pair as if it were a single three-terminal device, as labeled in the figure. That is, we assume $V_{CEQ} = V_{CEQ(2)}$.

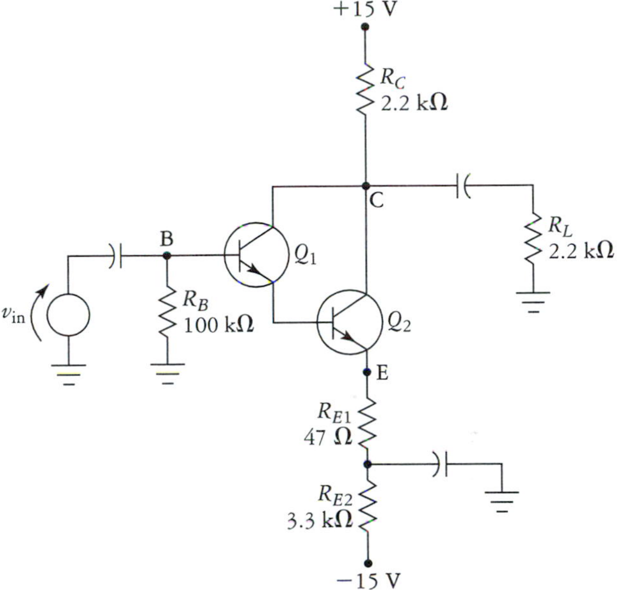

FIGURE 5.32
Partially bypassed Darlington, CE amplifier for Example 5.8.

Solution This is an emitter-biased circuit. The collector current is

$$\begin{aligned} I_{CQ} &= \frac{V_{EE} - 2V_{BE}}{R_{E_1} + R_{E_2} + \dfrac{R_B}{\beta^2}} \\ &= \frac{15\text{ V} - 1.4\text{ V}}{47\ \Omega + 3.3\text{ k}\Omega + 10\ \Omega} \\ &= 4\text{ mA} \end{aligned}$$

We determine V_{CEQ} using KVL and Ohm's law working from V_{CC} down to V_{EE}:

$$\begin{aligned} V_{CEQ} &= V_{CC} + V_{EE} - I_{CQ}(R_C + R_{E_1} + R_{E_2}) \\ &= 15\text{ V} + 15\text{ V} - 4\text{ mA}(2.2\text{ k}\Omega + 47\ \Omega + 3.3\text{ k}\Omega) \\ &= 30\text{ V} - 22.2\text{ V} \\ &= 7.8\text{ V} \end{aligned}$$

The ac analysis requires us to determine the following:

$$r_e = r_{e_2} = \frac{V_T}{I_{CQ}} = \frac{26\text{ mV}}{4\text{ mA}} = 6.5\ \Omega$$

$$
\begin{aligned}
R'_C &= R_C \parallel R_L \\
&= 2.2\ \text{k}\Omega \parallel 2.2\ \text{k}\Omega \\
&= 1.1\ \text{k}\Omega \\
R_e &= R_{E_1} = 47\ \Omega \\
R'_B &= R_B = 100\ \text{k}\Omega
\end{aligned}
$$

We now determine the gain and input resistance values:

$$
\begin{aligned}
A_v &= -\frac{R'_C}{R_e + 2r_e} \\
&= -\frac{1.1\ \text{k}\Omega}{47\ \text{k}\Omega + 13\ \Omega} \\
&= -18.3
\end{aligned}
$$

$$
\begin{aligned}
R_{\text{in}} &= R'_B \parallel \beta^2(R_e + 2r_e) \\
&= 100\ \text{k}\Omega \parallel 600\ \text{k}\Omega \\
&= 85.7\ \text{k}\Omega
\end{aligned}
$$

$$
\begin{aligned}
A_i &= A_v \frac{R_{\text{in}}}{R_L} \\
&= -18.3 \frac{85.7\ \text{k}\Omega}{2.2\ \text{k}\Omega} \\
&= -713
\end{aligned}
$$

$$
\begin{aligned}
A_p &= A_v A_i \\
&= (-18.3)(-713) \\
&= 13{,}048
\end{aligned}
$$

The Darlington pair is capable of voltage gain on the same order of magnitude as the standard common-emitter amplifier. It is in the areas of input resistance and power gain, however, where the Darlington pair exhibits superior performance.

5.8 OVERVIEW OF SMALL-SIGNAL AMPLIFIERS

We have spent a lot of time in this chapter learning how to analyze the various types of bipolar transistor amplifiers, and what a few of their more important characteristics are. Frankly, that is a lot of material to cover, and so this is a good time to put things into perspective and tie together any loose ends.

Amplifier Configuration Summary

Table 5.1 summarizes the general characteristics of the three basic amplifier configurations and the ac analysis equations that apply to them. Because collector feedback is used rather infrequently, it has not been included here.

Distortion

Note
Distortion is very low under small-signal conditions.

The amplifiers that we have studied thus far are classified as *small-signal amplifiers.* As discussed before, the term *small-signal* is somewhat difficult to define. Small-signal operation means that the amplifier is behaving in a linear manner and the output is not significantly distorted. Very often this requires the amplifier output voltage to be small, compared to the maximum possible output.

It is easy to visualize the main source of transistor nonlinearity by examining the base-emitter transconductance curve. When a transistor is biased up, say, as a common-emitter amplifier, we are setting up some particular values for I_{CQ} and V_{BEQ}. Now, when a time-varying voltage is applied to the base of the transistor, it causes variations in V_{BE}, which in turn result in variations in I_C. This is illustrated in Fig. 5.33. If the variation in V_{BE} is kept relatively small, then the transconductance curve is nearly linear, producing an undistorted collector current waveform. It can be shown that if

TABLE 5.1
General Characteristics of the Three Basic Amplifier Configurations

	Common Emitter	Emitter Follower	Common Base
A_v	$-\dfrac{R'_C}{r_e + R_e}$	$\dfrac{R_e}{R_e + r_e}$ $(A_v < 1)$	$\dfrac{R'_C}{r_e + R'_B/\beta}$
R_{in}	$R'_B \parallel \beta(r_e + R_e)$	$R'_B \parallel \beta(r_e + R_e)$	$\left(r_e + \dfrac{R_S}{\beta}\right) \parallel R_E$
R_o	R_C	$\left(r_e + \dfrac{R_S}{\beta}\right) \parallel R_E$	R_C
$A_i = A_v \dfrac{R_{\text{in}}}{R_L}$	$A_i < \beta$	$A_i < \beta$	$A_i < 1$
$A_p = A_v A_i$	High	Medium	Medium
$\Delta\phi$	180°	0°	0°

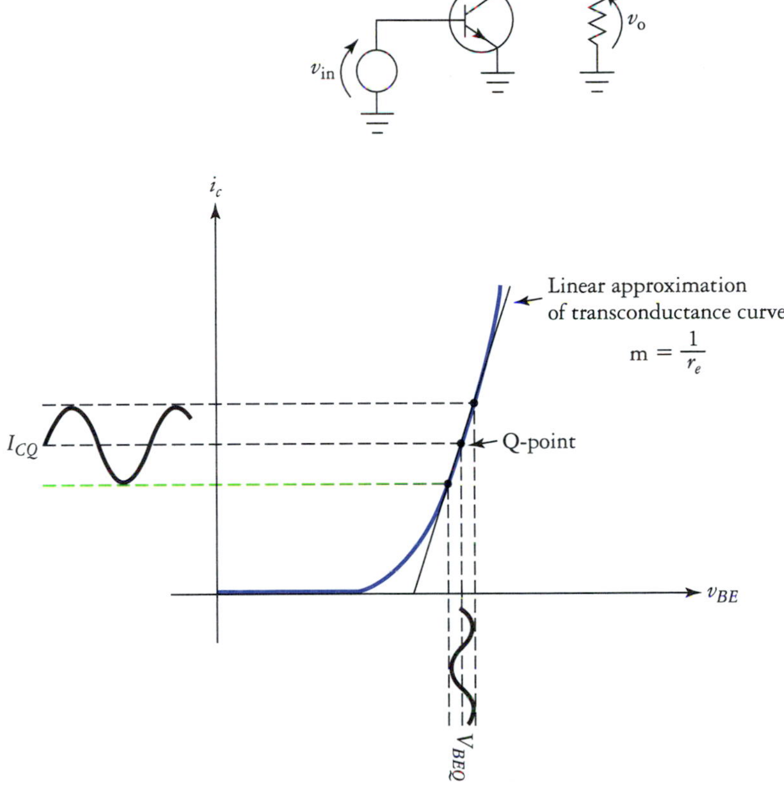

FIGURE 5.33
A small signal does not produce significant distortion because nonlinearity is small for small Q-point movement.

$\Delta V_{BE} \ll V_T$, then the amplification will be linear. A more practical rule is if $\Delta I_{CQ} < 10\%$ then small-signal conditions apply, and the amplifier will behave linearly.

Note
Harmonics are integer multiples of the input frequency.

If a large signal is applied to the amplifier, then distortion of the collector current waveform becomes pronounced. This is apparent in Fig. 5.34. When V_{BE} increases, the curve becomes steeper, causing a large increase in collector current. On negative-going excursions of the input, the corresponding change in collector current is less. Because of this nonlinearity, the collector current waveform is stretched upon increase and squashed upon decrease. This type of distortion produces even and odd harmonics of the input frequency to be generated in the output signal. Such distortion is particularly unpleasant when the signal being amplified is music.

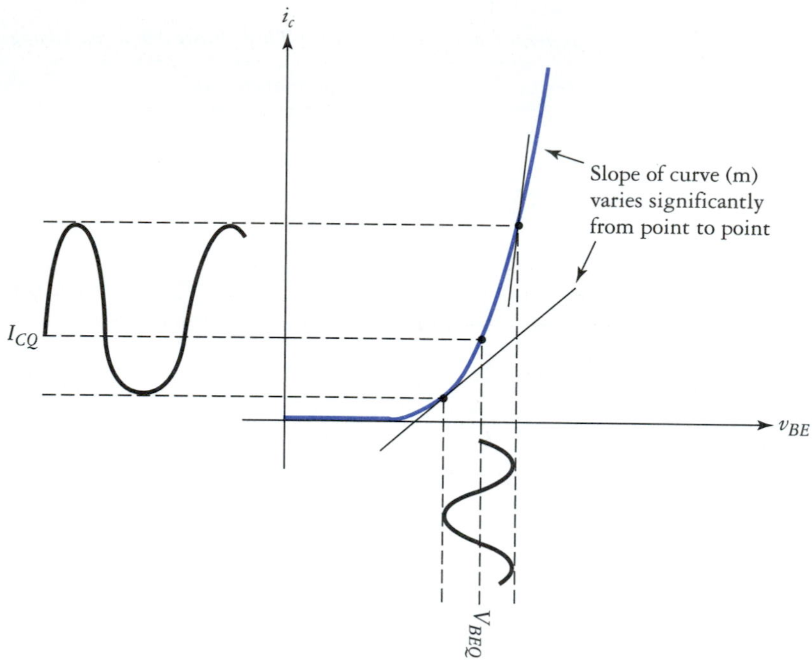

FIGURE 5.34
Large movement of the Q point produces distortion due to nonlinearity of the transconductance curve of the BJT.

Negative feedback (emitter feedback or collector feedback) decreases distortion. An intuitive understanding of why this happens can be had if we look at the voltage gain expression for the common-emitter amplifier. Repeated here for convenience, this is

$$A_v = -\frac{R'_C}{r_e + R_e}$$

Here's one way to think of how distortion is produced: If emitter bypassing is used, then the R_e term reduces to zero and the gain is dependent on r_e alone. Now, when an input signal increases I_C, the value of r_e decreases ($r_e \cong 26 \text{ mV}/I_{CQ}$). This causes A_v to increase. On negative-going input swings, I_C decreases and r_e increases, causing A_v to decrease as well.

When emitter feedback is used, the variations in r_e tend to be swamped, because R_e remains constant. The larger R_e is relative to r_e, the less the gain will vary with the input signal. Hence, R_e linearizes the gain of the circuit. You can think of amplifier nonlinearity as a variation in gain that is a function of the input signal. If you are interested, a detailed discussion of BJT distortion is presented in Appendix 3. For now, it is sufficient that you realize that negative feedback helps reduce distortion.

5.9 THE AC LOAD LINE AND OUTPUT VOLTAGE COMPLIANCE

The dc load line gives us a good idea of how a transistor will behave in a circuit if that circuit is modeled by the dc equivalent. However, under signal conditions, the ac-equivalent circuit is usually quite different from the dc-equivalent circuit. An ac load line is used to obtain a graphical representation of transistor operation under these conditions.

Common-Emitter AC Load-Line Construction

The characteristics of the ac load line of an amplifier are dependent on the component values in the ac-equivalent circuit and by the quiescent operating conditions of the circuit. The quiescent conditions affect the ac load line in that it will always intersect the dc load line at the Q point. The limits of the ac load line for a common-emitter amplifier are given by

$$i_{C(\text{sat})} = I_{CQ} + \frac{V_{CEQ}}{R'_C + R_e} \tag{5.57}$$

and

$$v_{CE(\text{cut})} = V_{CEQ} + I_{CQ}(R'_C + R_e) \tag{5.58}$$

For capacitively coupled amplifiers, the following relationships also hold: $i_{C(\text{sat})} \geq I_{C(\text{sat})}$ and $v_{CE(\text{cut})} \leq V_{CE(\text{cut})}$. If the emitter of the CE amplifier is bypassed, then the R_e terms drop out of Eqs. (5.57) and (5.58). Let us use this information to produce an ac load line for the circuit in Fig. 5.14.

The dc analysis of this circuit was performed in Example 5.2, where we found $I_{CQ} = 2$ mA, $V_{CEQ} = 5.2$ V, $I_{C(\text{sat})} = 3$ mA and $V_{CE(\text{cut})} = 15$ V. The ac analysis of this circuit was performed in Example 5.3, where we obtained $R'_C = 1.8$ kΩ and $R_e = 0$. Substituting the appropriate values into Eqs. (5.57) and (5.58), we obtain

$$\begin{aligned} i_{C(\text{sat})} &= I_{CQ} + \frac{V_{CEQ}}{R'_C} \\ &= 2\text{ mA} + \frac{5.2\text{ V}}{1.8\text{ k}\Omega} \\ &= 4.9\text{ mA} \end{aligned}$$

$$\begin{aligned} v_{CE(\text{cut})} &= V_{CEQ} + I_{CQ}R'_C \\ &= 5.2\text{ V} + (2\text{ mA})(1.8\text{ k}\Omega) \\ &= 8.6\text{ V} \end{aligned}$$

The dc and ac load lines are shown in Fig. 5.35.

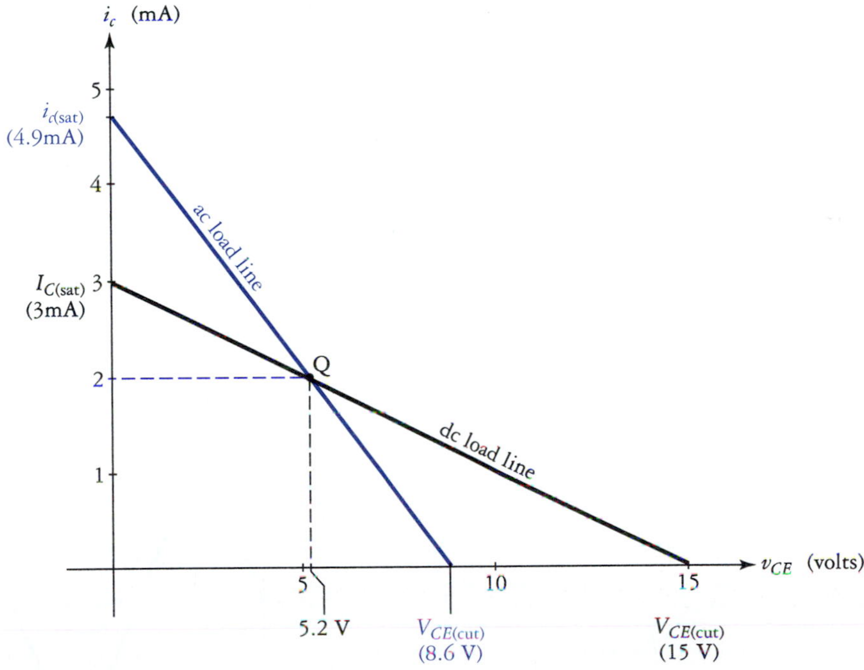

FIGURE 5.35
AC and dc load lines.

The limits of the ac load line tell us the maximum values of v_{CE} and i_C to which the transistor will be subjected at maximum signal levels. Another view of the ac load line is shown in Fig. 5.36. It is clear in this illustration that positive-going excursions of v_{in} cause an increase in i_C and movement of the operating point toward saturation. Negative-going swings of v_{in} cause a decrease in i_C, and drive the transistor toward cutoff. The effective Q point moves along the ac load line.

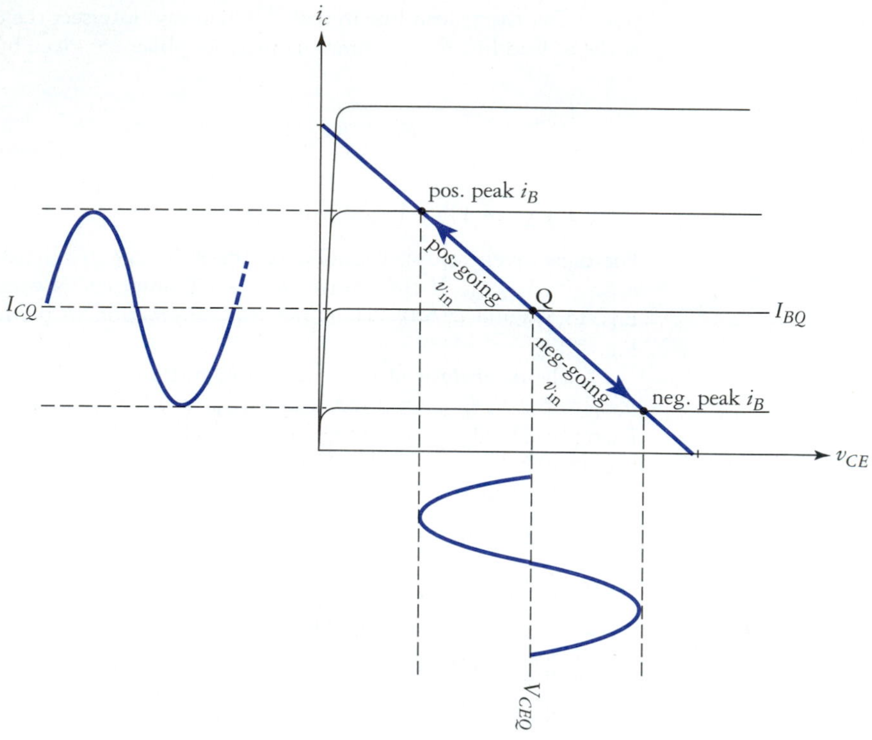

FIGURE 5.36
AC load line superimposed over collector curves.

FIGURE 5.37
Amplifier clipping. (a) Transfer characteristics for a linear amplifier that exhibited clipping. (b) Clipped output voltage in the time domain.

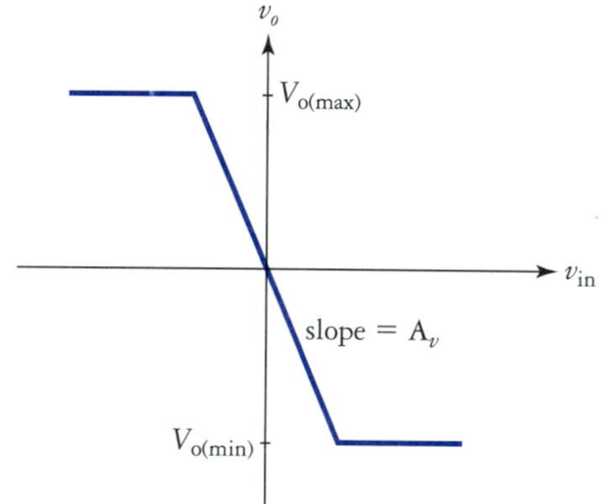

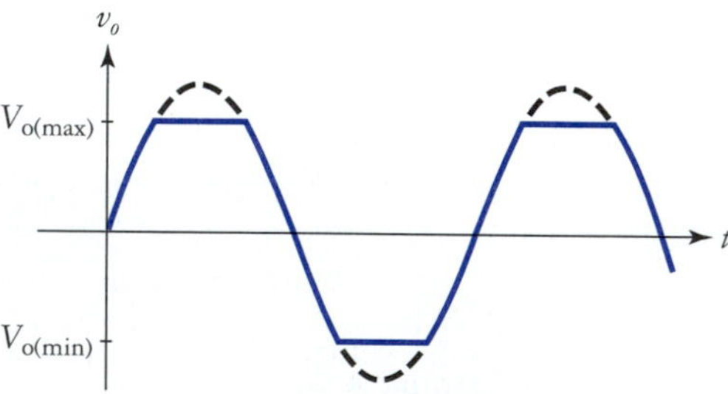

Output Voltage Compliance

Output voltage compliance is the term applied to describe the amplifier maximum output voltage range. In small-signal applications, this is often not an important consideration, but it does follow logically from the ac load-line concept. For a common-emitter amplifier with capacitive coupling, the maximum positive output voltage peaks are given by

$$V_{o(\max)} = I_{CQ}R'_C \tag{5.59}$$

The most negative (minimum) value that v_o can attain is

$$V_{o(\min)} = -V_{CEQ} + \frac{V_{CEQ}R_e}{R_e + R'_C} \tag{5.60}$$

Notice that the negative signal swing limit is affected by the external ac emitter resistance R_e. This occurs because as the transistor is driven toward saturation, V_{CE} drops to nearly zero volts, but at the same time, the increasing emitter current causes an increase in the voltage drop across R_e. The output voltage is the instantaneous sum of V_{CE} and $i_E R_e$.

If complete emitter bypassing is used, then the minimum output voltage limit is

$$V_{o(\min)} = -V_{CEQ} \tag{5.61}$$

Emitter bypassing has no effect on the maximum positive output voltage of the CE amplifier.

Application of an input signal that is too large will result in clipping of the output. When clipping occurs, further increases of the input signal cause no change in the output voltage. If we assume that a given inverting amplifier (a CE amplifier, for example) is linear over the maximum output voltage range, and that clipping occurs symmetrically, then the transfer characteristics of that amplifier will appear as shown in Fig. 5.37(a). An output signal that exceeds the compliance of the amplifier will appear as shown in Fig. 5.37(b).

EXAMPLE 5.9

Determine the output voltage compliance for the circuits of Figs. 5.12 and 5.14.

Solution In terms of dc analysis, both circuits are identical. The quiescent values of interest are found in the load-line graph of Fig. 5.35. $I_{CQ} = 2$ mA and $V_{CEQ} = 5.2$ V. No emitter bypassing is used in Fig. 5.12, thus $R'_C = 1.8$ kΩ and $R_e = 1$ kΩ, and we find

$$\begin{aligned} V_{o(\max)} &= I_{CQ}R'_C \\ &= (2 \text{ mA})(1.8 \text{ k}\Omega) \\ &= 3.6 \text{ V} \end{aligned}$$

The minimum negative output voltage is

$$\begin{aligned} V_{o(\min)} &= -V_{CEQ} + \frac{V_{CEQ}R_e}{R_e + R'_C} \\ &= -5.2 \text{ V} + \frac{5.2 \text{ V} \times 1 \text{ k}\Omega}{1.8 \text{ k}\Omega + 1 \text{ k}\Omega} \\ &= -5.2 \text{ V} + 1.86 \text{ V} \\ &= -3.34 \text{ V} \end{aligned}$$

The circuit in Fig. 5.14 has complete emitter bypassing, thus

$$\begin{aligned} V_{o(\max)} &= I_{CQ}R'_C \\ &= (2 \text{ mA})(1.8 \text{ k}\Omega) \\ &= 3.6 \text{ V} \end{aligned}$$

$$\begin{aligned} V_{o(\min)} &= -V_{CEQ} \\ &= -5.2 \text{ V} \end{aligned}$$

In this particular example, we find that the CE amplifier without emitter bypassing has almost symmetrical positive- and negative-output voltage limits (this is not always the case), while the amplifier with emitter bypassing can produce a larger negative output voltage swing.

The useful output voltage range of a linear amplifier is limited to twice the value of the lowest compliance limit. That is, we consider the useful compliance range of the amplifier to be determined by the lesser of $|V_{o(\min)}|$ and $V_{o(\max)}$. Again, for small-signal amplifiers this is not normally a major concern. In large-signal applications, however, it is normally desirable to have symmetrical clipping points.

The maximum output voltage swing available from the common-base amplifier is given by

$$V_{o(\max)} = I_{CQ}R'_C \tag{5.62}$$

and

$$V_{o(\text{min})} = -V_{CEQ} \tag{5.63}$$

Finally, the maximum output voltage that can be produced by the emitter follower is

$$V_{o(\text{max})} = I_{CQ}R_e \tag{5.64}$$

and the minimum output voltage is

$$V_{o(\text{min})} = -V_{CEQ} \tag{5.65}$$

As was the case with the common-emitter amplifier, the clipping points may be asymmetrical.

EXAMPLE 5.10

Refer to Fig. 5.38. Sketch the dc and ac load lines for this circuit and determine the maximum peak-to-peak output voltage that can be produced. Assume $\beta = 100$.

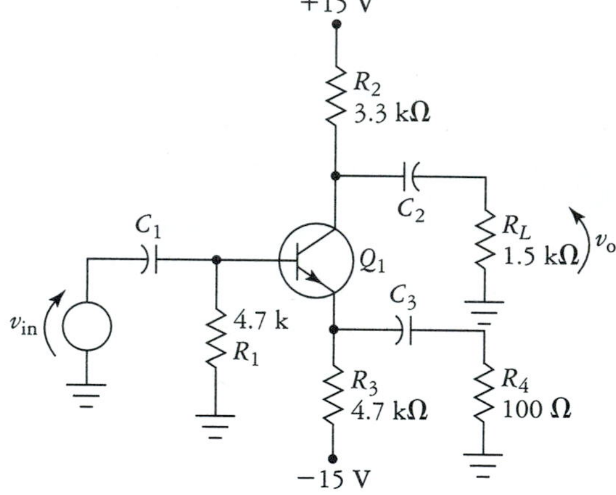

FIGURE 5.38
Circuit for Example 5.10.

Solution

DC analysis shows the Q-point coordinates to be $I_{CQ} = 3$ mA and $V_{CEQ} = 6$ V. The dc load-line limits are

$$\begin{aligned} I_{C(\text{sat})} &= (V_{CC} + V_{EE})/(R_E + R_C) \\ &= 30\text{ V}/8\text{ k}\Omega \\ &= 3.75\text{ mA} \\ V_{CE(\text{cut})} &= V_{CC} + V_{EE} = 30\text{ V} \end{aligned}$$

The ac load-line limits are

$$\begin{aligned} i_{C(\text{sat})} &= I_{CQ} + \frac{V_{CEQ}}{R'_C + R_e} \\ &= 3\text{ mA} + \frac{6\text{ V}}{1.03\text{ k}\Omega + 98\ \Omega} \\ &= 8.3\text{ mA} \end{aligned}$$

$$\begin{aligned} v_{CE(\text{cut})} &= V_{CEQ} + I_{CQ}(R_e + R'_C) \\ &= 6\text{ V} + (3\text{ mA})(98\ \Omega + 1.03\text{ k}\Omega) \\ &= 9.4\text{ V} \end{aligned}$$

The load lines are sketched in Fig. 5.39. The output clipping levels are

$$\begin{aligned} V_{o(\text{max})} &= I_{CQ}R'_C \\ &= (3\text{ mA})(1.03\text{ k}\Omega) \\ &= 3.1\text{ V} \end{aligned}$$

$$V_{o(\min)} = -V_{CEQ} + \frac{V_{CEQ}R_e}{R'_C + R_e}$$
$$= -6\text{ V} + 0.52\text{ V}$$
$$= -5.48\text{ V}$$

The maximum unclipped output voltage is

$$V_{o(\text{p-p})} = 2\,V_{o(\max)}$$
$$= (2)(3.1\text{ V})$$
$$= 6.2\text{ V}_{\text{p-p}}$$

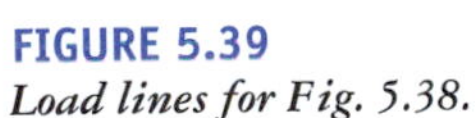

FIGURE 5.39
Load lines for Fig. 5.38.

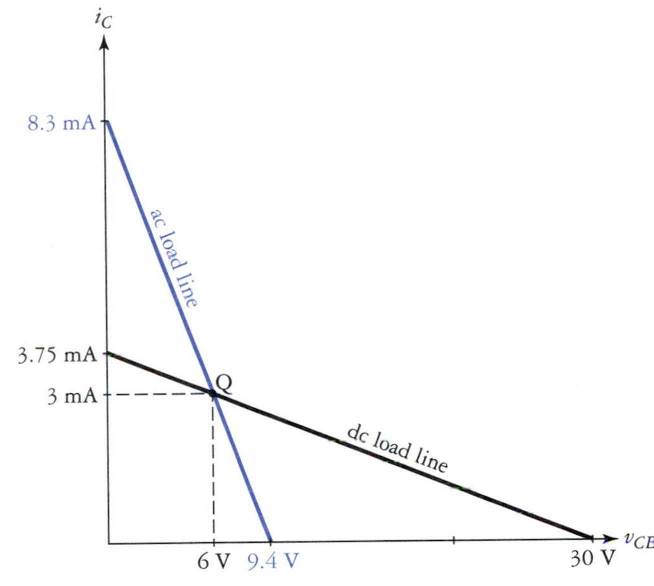

5.10 COMPUTER-AIDED ANALYSIS APPLICATIONS

The analysis equations developed in this chapter are based on the transistor behaving in a linear manner. Under small-signal conditions this is the case. Because of the more accurate modeling of the transistor that is used in circuit simulation programs like PSpice, it is relatively easy to observe the effects of device nonlinearity on amplifier operation.

Voltage Gain and Distortion

Recall that one way to ensure small-signal operation is to limit the variation of v_{BE} to a value such that $v_{BE} \ll 26$ mV. Let's investigate this relationship using PSpice. Without the use of emitter feedback, the simple base-biased common-emitter amplifier can exhibit a high degree of nonlinearity. The circuit that we will use is shown in Fig. 5.40. A quick pencil-and-paper analysis of this circuit assuming $\beta = 150$ (close to what PSpice will use for the 2N3904) gives us a voltage gain of $A_v = -152$. In the ac-equivalent circuit, we find the input source is directly in parallel with the B-E junction, so $\Delta V_{BE} = \Delta V_{\text{in}}$. It seems reasonable to assume we will have small-signal operation if we set $v_{\text{in}} = 5\text{ mV}_{\text{pk}}$. The frequency of the input voltage source is set to 1 kHz.

A transient analysis is set up such that five cycles of the input signal are analyzed. A plot of the output voltage is shown in Fig. 5.41. Notice that the sine wave is inverted, as we would expect from a common-emitter amplifier. The output signal looks fairly clean and undistorted, but placement of cursors on the peaks of the waveform show $V_{o(\max)} \cong 701$ mV and $V_{o(\min)} = 0741$ mV. Given $v_{\text{in}} = 5\text{ mV}_{\text{pk}}$ we find that the voltage gain varies for the positive- and negative-going portions of the input signal. Using the cursor values we have $|A_{v(\min)}| = 140$ and $|A_{v(\max)}| = 140$. Both values are very close to the approximate value calculated by hand, which is very good, considering the simplicity of our approximations.

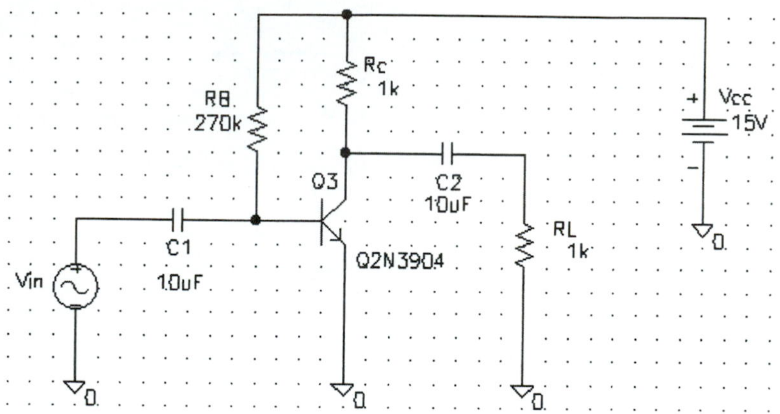

FIGURE 5.40
CE amplifier for computer analysis.

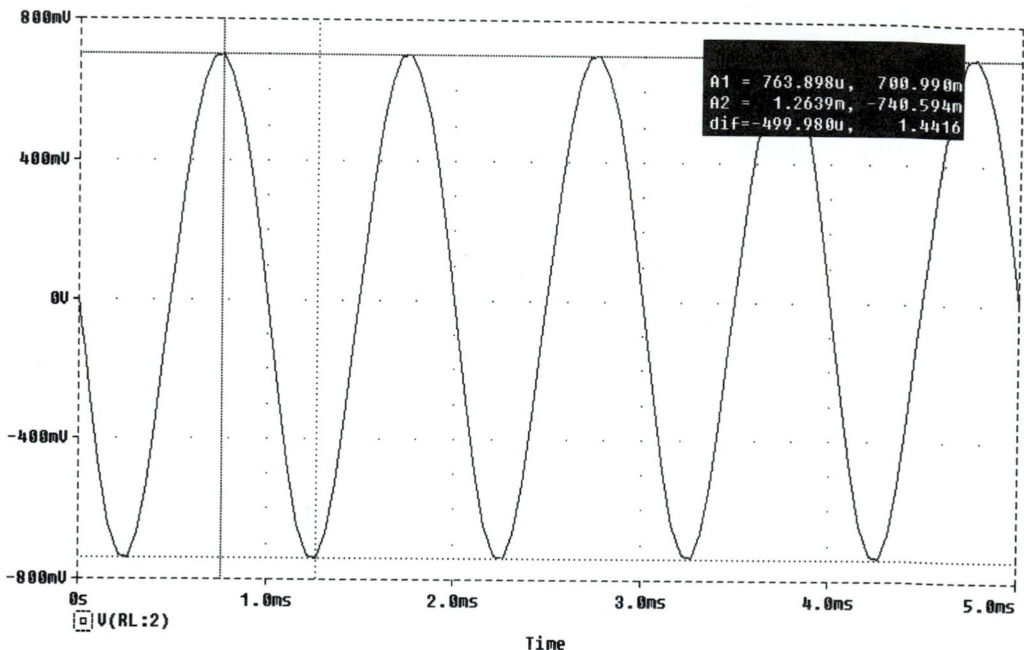

FIGURE 5.41
Output voltage waveform shows low distortion because of small-signal operation.

If we increase the peak amplitude of v_{in} to 20 mV_{pk}, then we certainly should obtain a distorted output signal. This modification is made, and the resulting analysis is plotted in Fig. 5.42. Notice that the upper portion of v_o is rather squashed, while the lower part is stretched out, as expected. The difference in minimum and maximum voltage gain is very significant in this case. The determination of these values is left as an exercise for the reader.

Emitter Feedback

A voltage-divider biased CE amplifier with partial emitter bypassing is shown in Fig. 5.43. Trusting the accuracy of our manual dc analysis techniques, with $\beta = 100$ the Q-point values are approximately $I_{CQ} = 2.8$ mA and $V_{CEQ} = 6.1$ V. The voltage gain is determined to be $A_v = -9.7$.

Set v_{in} to 0.2 V peak at $f = 1$ kHz. This should produce an output voltage of $V_{o(pk)} = 1.94$ V with equal positive and negative amplitudes, assuming that the amplifier behaves in a linear manner.

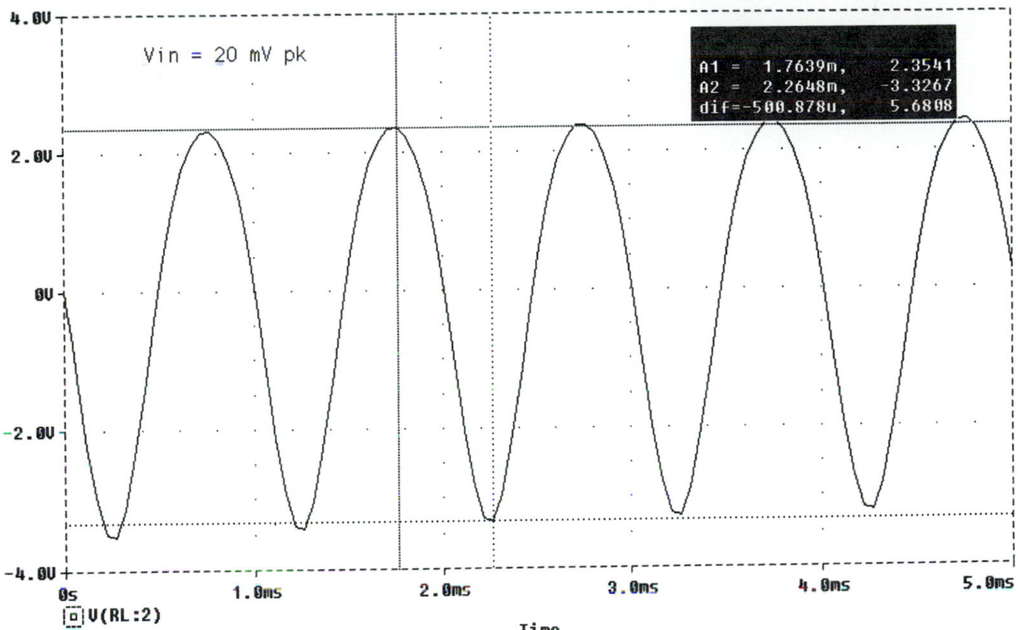

FIGURE 5.42
Large-signal operation produces distorted output.

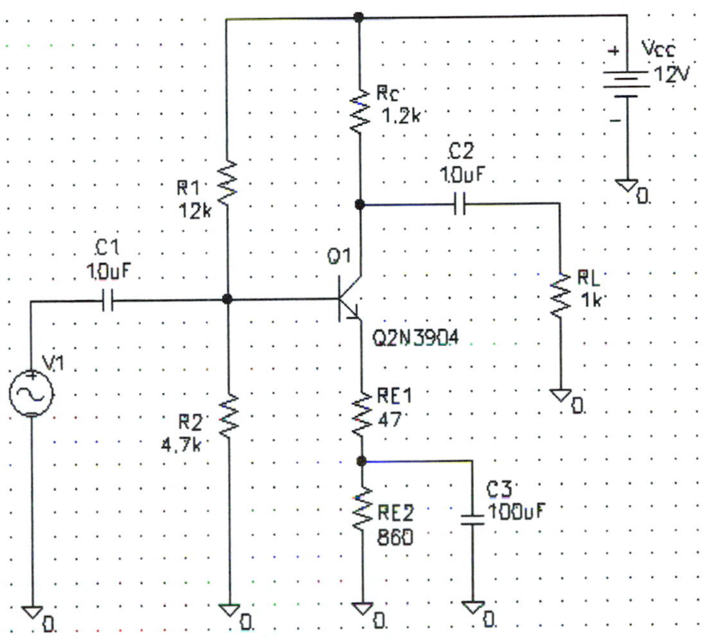

FIGURE 5.43
Partially bypassed CE amplifier.

The result of the analysis run is shown in Fig. 5.44. The smaller waveform is the input voltage. There is noticeable distortion of the output waveform. Using the plot to calculate the average voltage gain on positive- and negative-going half-cycles, we obtain

$$|A_{v(\min)}| = 7.5$$

$$|A_{v(\max)}| = 9.5$$

These results indicate that the average voltage gain in this case is about 14% lower than expected. This is not surprising because the emitter feedback is not very significant in this circuit. The use of a larger R_e would result in more accurate, but lower, voltage gain.

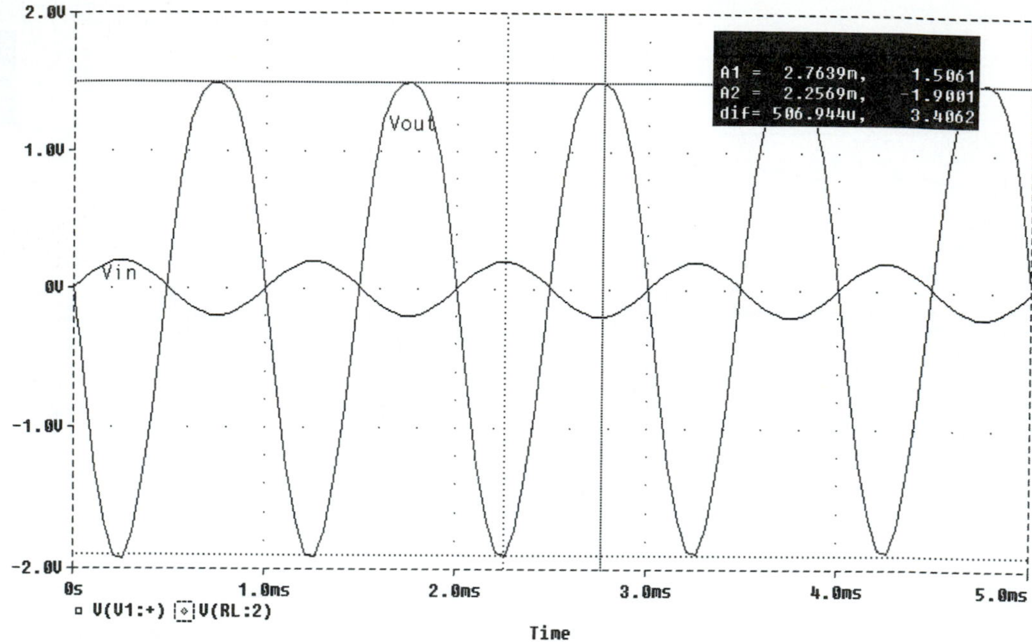

FIGURE 5.44
Input and output voltage waveforms.

Let us now calculate the expected output voltage compliance:

$$V_{o(\max)} = I_{CQ}R_C$$
$$= (2.8\text{ mA})(545\ \Omega)$$
$$= 1.53\text{ V}$$

$$V_{o(\min)} = -V_{CEQ} + \frac{V_{CEQ}R_e}{R_e + R'_C}$$

$$= -6.1\text{ V} + \frac{6.1\text{ V} \times 47\ \Omega}{47\ \Omega + 545\ \Omega}$$
$$= -5.6\text{ V}$$

Changing the peak amplitude of v_{in} from 0. 2 to 1 V overdrives the amplifier heavily, resulting in the output waveform shown in Fig. 5.45. Using the cursor we find

$$V_{o(\max)} = 1.6\text{ V}$$
$$V_{o(\min)} = -4.9\text{ V}$$

Notice that there is a slight positive-going bump in the negative peaks of the waveform in Fig. 5.45. If we had overdriven the amplifier more heavily, then this bump would increase in size as well. What's happening here is that the positive-going portion of the input voltage is forward biasing the C-B junction, causing direct feed-through of the input signal to the load. This is an example of very heavy saturation.

Frequency Response and Input Resistance

Note
The frequency response (AC sweep) analysis maintains small-signal conditions regardless of input amplitude.

Although we have not talked much about frequency limitations in this chapter, PSpice makes it easy for us to take a quick look at this topic now. Let's continue using the circuit of Fig. 5.43. We set the AC attribute of V1 to 1 V, and set up an AC Sweep using 101 points/decade from 10 Hz to 100 MHz. Running the analysis and plotting the output voltage V(RL:2) produces the display of Fig. 5.46.

We know that the circuit clips when the input voltage is 1 V_{pk}; however, PSpice assumes that Q-point-determined small-signal conditions will exist regardless of amplitude levels. PSpice does model large-signal nonlinearities during transient analysis. Inspecting the output voltage response, it appears that the useful frequency range of the amplifier is from around 100 Hz to 50 MHz. The lower

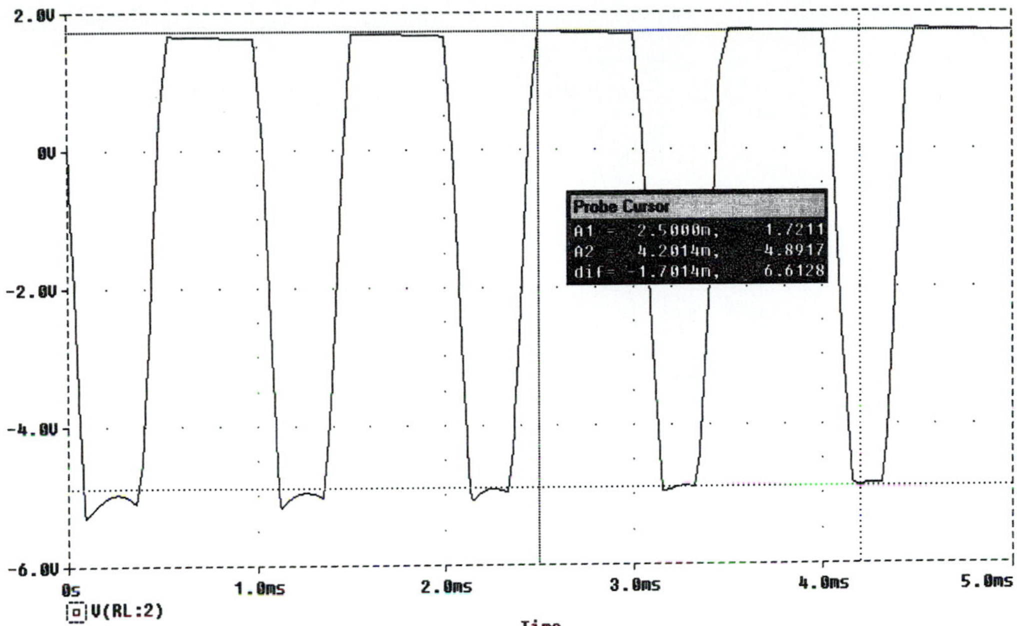

FIGURE 5.45
The amplifier of Fig. 5.43 exhibits asymmetrical clipping.

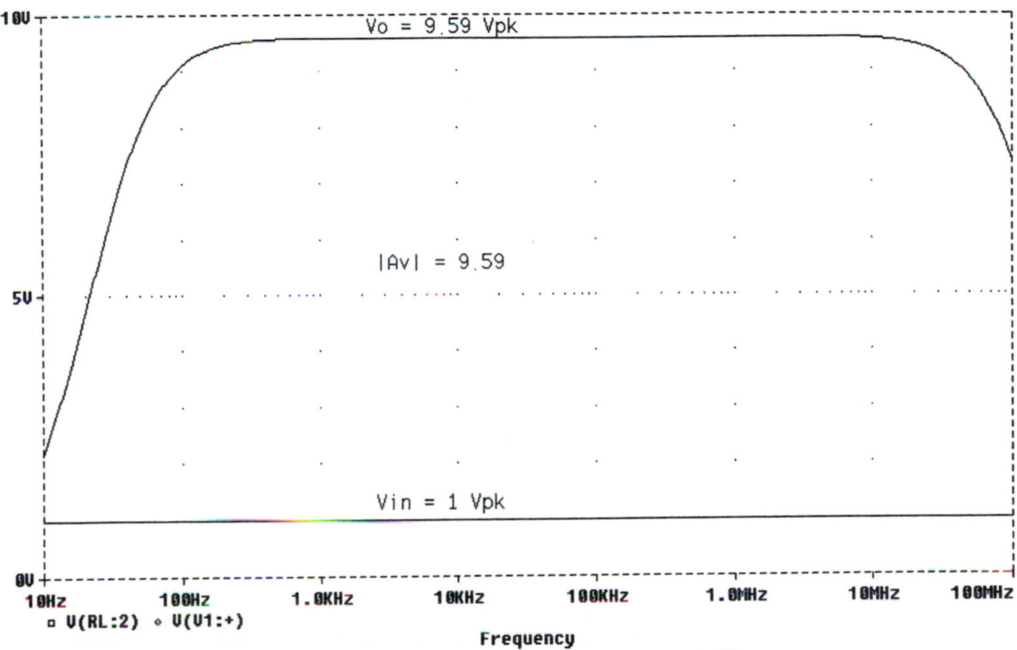

FIGURE 5.46
Frequency response plot for Fig. 5.43.

frequency limit is determined by coupling capacitor reactance. The upper frequency limit is determined by transistor frequency limitations. In the laboratory, however, upper frequency limits are usually determined by parasitic reactance associated with wiring and breadboard layout.

Plotting in Decibels

Often, the dependent-variable (the y-axis variable) of a frequency response plot is expressed in decibels (dB). We can do this in PSpice by adding dB (not case-sensitive) to the trace identifier, as in VdB(RL:2). This produces the graph in Fig. 5.47. The gain in decibels is given by

$$A_{dB} = 20 \log |A_v|$$

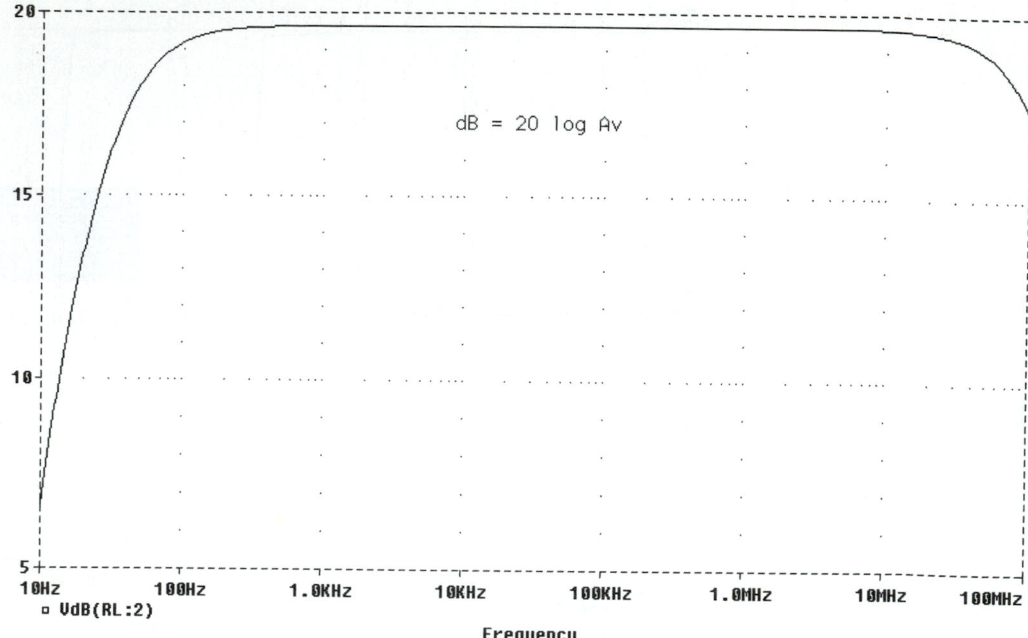

FIGURE 5.47
Frequency response plotted in decibels.

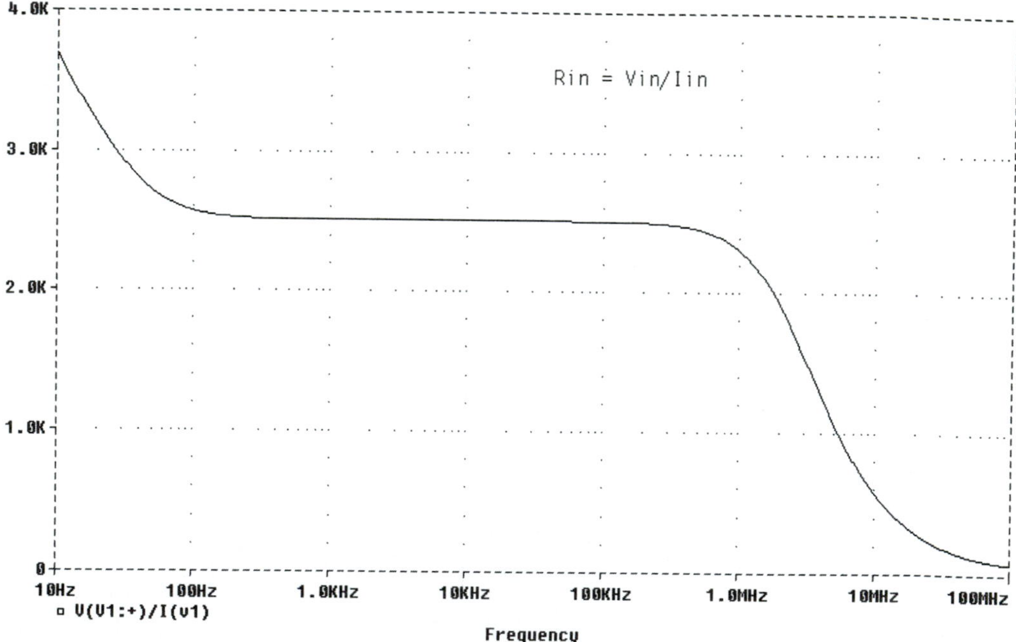

FIGURE 5.48
Variation of input resistance with frequency.

Using the gain from Fig. 5.46 we obtain

$$A_{dB} = 20 \log 9.59$$
$$= 19.6 \text{ dB}$$

This is what we have in the plot of Fig. 5.47, over the useful range of signal frequencies. The frequency limits are determined by where the response drops by 3 dB from the maximum midband value.

Input Resistance

As a final example of PSpice, we determine the input resistance of Fig. 5.43. By definition, input resistance is given by

$$R_{\text{in}} = \frac{V_{\text{in}}}{i_{\text{in}}}$$

We use the AC Sweep analysis to perform this calculation. The results are shown in Fig. 5.48. Notice that PSpice performs mathematical operations entered directly with the trace variable name(s). Because the amplifier parameters are frequency sensitive, the input resistance varies with frequency as well. Over the useful range of input frequencies, the input resistance appears to be approximately 2.5 kΩ.

■ CHAPTER REVIEW

Voltage amplifiers are one of the most common applications for bipolar transistors. There are three basic BJT amplifier configurations: common-emitter, emitter-follower and common-base. The common-emitter amplifier exhibits high voltage and current gain. . The emitter follower has voltage gain less than unity, but does exhibit current gain. The common-base amplifier has current gain less than unity, but exhibits high voltage gain. The common-emitter amplifier has the highest power gain potential for a given transistor, and it also exhibits phase inversion of the output signal, relative to the input.

Emitter bypassing is used to increase the voltage gain of the CE amplifier. This also reduces the input resistance and increases amplifier distortion. The common-base amplifier has very low input resistance. Base bypassing is used to increase the voltage gain of the CB amplifier.

The ac load line is used to graphically represent the instantaneous operating point of the amplifier under signal conditions. The ac load line always intersects the dc load line at the Q point. If the movement of the Q point is relatively small then small-signal conditions prevail, and distortion will be low.

■ DESIGN AND ANALYSIS PROBLEMS

Note: Unless otherwise specified, assume all transistors are silicon with $\beta = 100$ and all coupling and bypass capacitors have $|X_C| \cong 0\ \Omega$.

5.1. Refer to Fig. 5.49. Sketch the dc and ac load lines and determine A_v, A_i, A_p, and R_{in}.

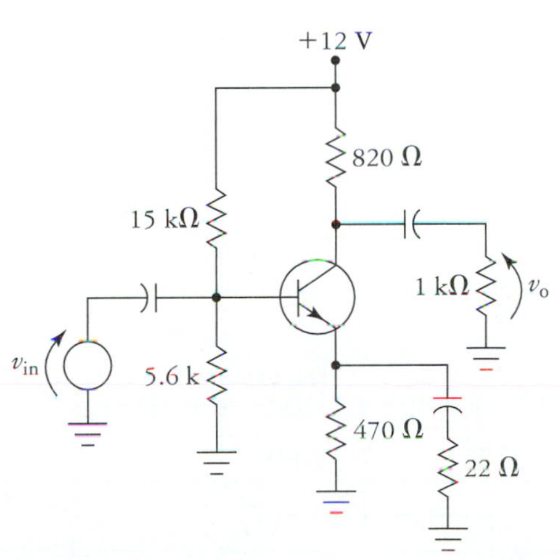

FIGURE 5.49
Circuit for Problems 5.1, 5.2, and 5.3.

5.2. Refer to Fig. 5.49. Determine the peak-to-peak amplitude of v_{in} that will produce the maximum unclipped output voltage.

5.3. Refer to Fig. 5.49. Given $v_{\text{in}} = 25 \sin\omega t$ mV, write the expressions for i_{in}, i_C and i_L.

5.4. Refer to Fig. 5.1(b). Given $v_S = 50$ mV, $R_s = 600\ \Omega$, $R_{\text{in}} = 5$ kΩ, $R_o = 100\ \Omega$, $R_L = 100\ \Omega$ and A = 100, determine A_v, A_i, A_p and v_o.

5.5. Refer to Fig. 5.1. Given the specifications of Problem 5.4, sketch the $v_o - v_{\text{in}}$ transfer characteristic curve for the amplifier.

5.6. Refer to Fig. 5.50. Determine I_{CQ}, V_{CEQ}, A_v, A_i, A_p, and R_{in}.

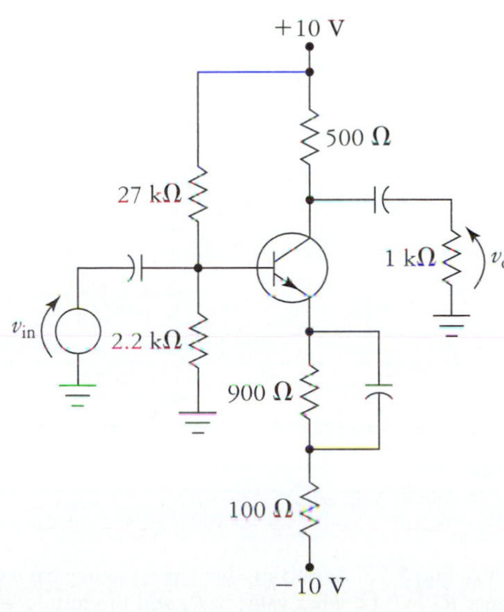

FIGURE 5.50
Circuit for Problems 5.6, 5.7, and 5.8.

5.7. Refer to Fig. 5.50. Determine the positive and negative output voltage clipping levels.

5.8. Refer to Fig. 5.50. Assuming linear operation, if $v_o = 0.5 \sin \omega t$ V, determine v_{in} and v_E.

5.9. Refer to Fig. 5.51. Determine I_{CQ}, V_{CEQ}, R_{in}, A_v, A_i, A_p, $V_{o(max)}$ and $V_{o(min)}$.

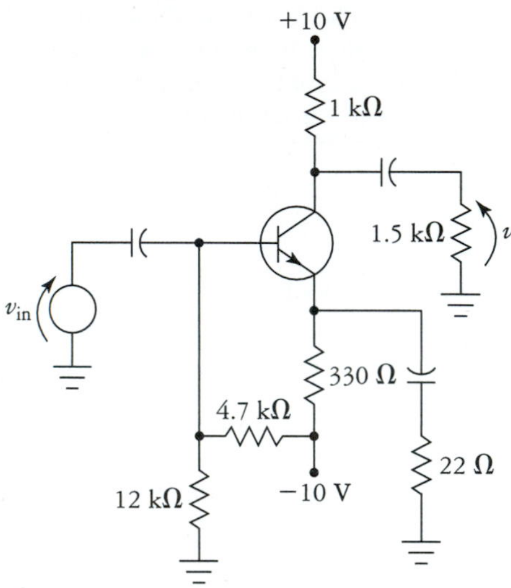

FIGURE 5.51
Circuit for Problems 5.9 and 5.10.

5.10. Refer to Fig. 5.51. Given $v_{in} = 10 \sin \omega t$ mV, determine the expression for the voltage drop across the 22-Ω resistor.

5.11. Refer to Fig. 5.52. Sketch the dc and ac load-lines, and determine A_v, A_i, A_p and R_{in}.

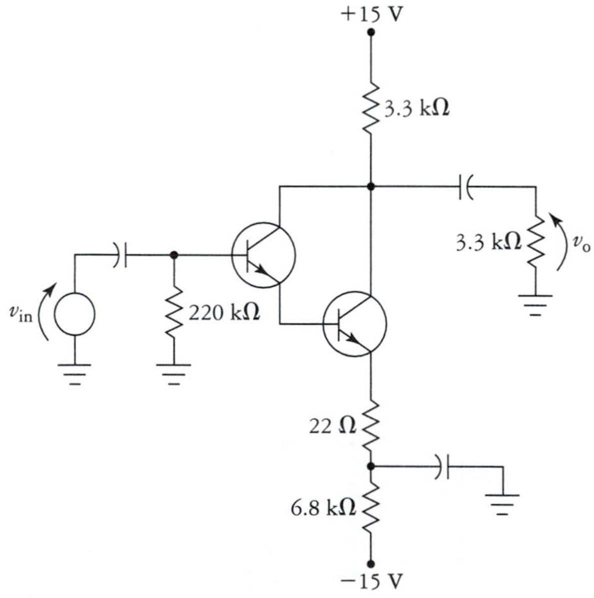

FIGURE 5.52
Circuit for Problems 5.11 and 5.12.

5.12. Refer to Fig. 5.52. Assuming that the v_{in} source has internal resistance $R_S > 0$ Ω, what value of R_S will produce $v_B = v_S/4$?

5.13. Refer to Fig. 5.53. Determine I_{CQ}, V_{CEQ}, A_v, A_i, A_p, R_{in}, $V_{o(max)}$ and $V_{o(min)}$.

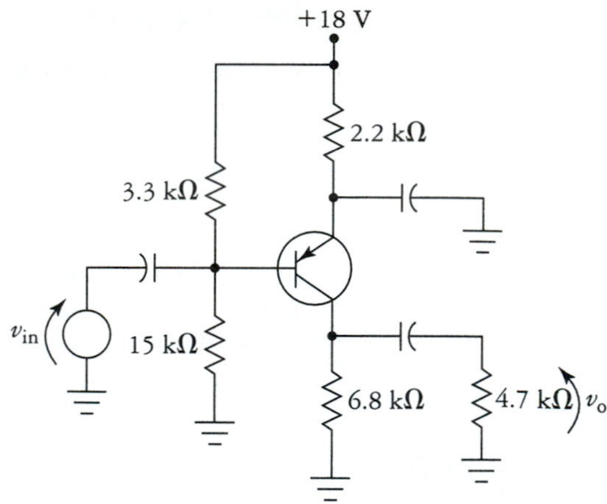

FIGURE 5.53
Circuit for Problems 5.13 and 5.14.

5.14. Repeat the analysis of Fig. 5.53 with the emitter-bypass capacitor removed.

5.15. Refer to Fig. 5.54. Determine I_{CQ}, V_{CEQ}, A_v, A_i, A_p, R_{in} and R_o.

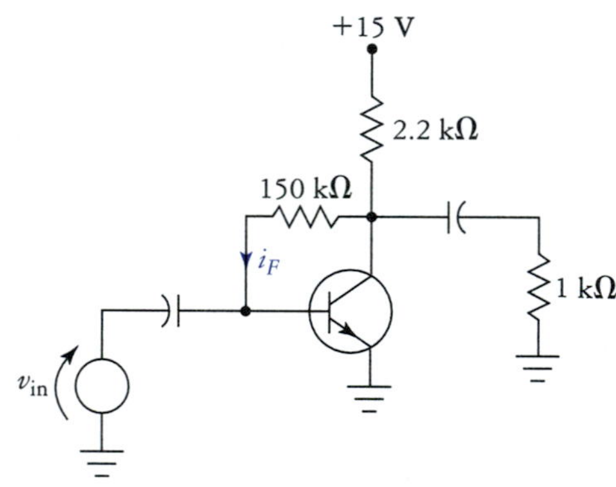

FIGURE 5.54
Circuit for Problems 5.15 and 5.16.

5.16. Refer to Fig. 5.54. Given $v_{in} = 10 \sin \omega t$ mV, determine i_F.

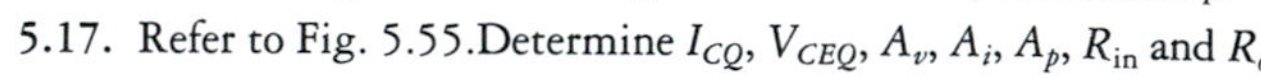

5.17. Refer to Fig. 5.55. Determine I_{CQ}, V_{CEQ}, A_v, A_i, A_p, R_{in} and R_o.

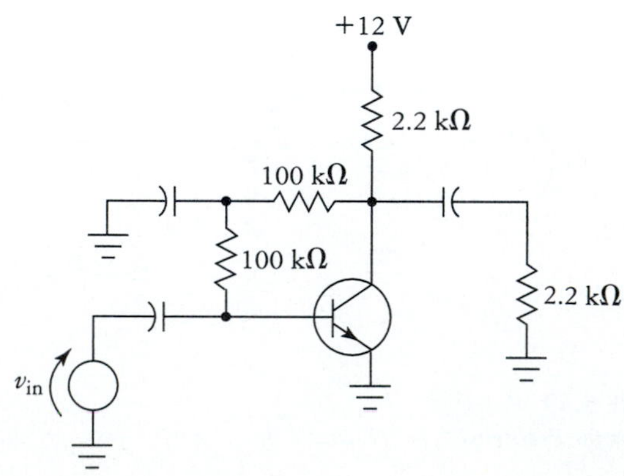

FIGURE 5.55
Circuit for Problem 5.17.

5.18. Refer to Fig. 5.56. Determine I_{CQ}, V_{CEQ}, A_v, A_i, A_p, R_{in} and R_o.

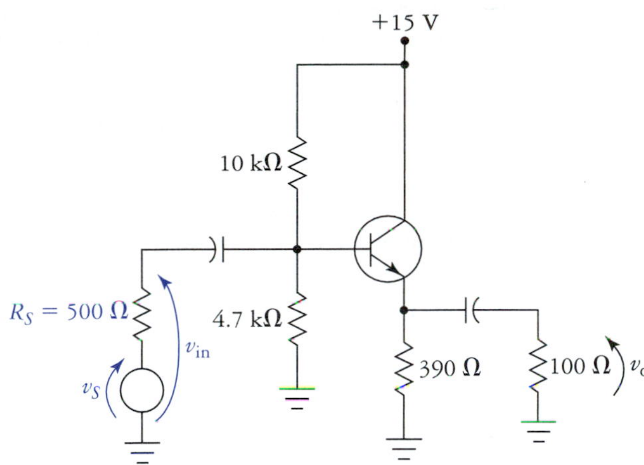

FIGURE 5.56
Circuit for Problems 5.18, 5.19, and 5.20.

5.19. Refer to Fig. 5.56. Given $v_S = 1 \sin \omega t$ V, determine v_o.

5.20. Determine V_o, given $v_S = 0.5\ V_{p\text{-}p}$ for Fig. 5.56.

5.21. Refer to Fig. 5.57. Determine the operating point, A_v, A_i, A_p, R_{in}, R_o.

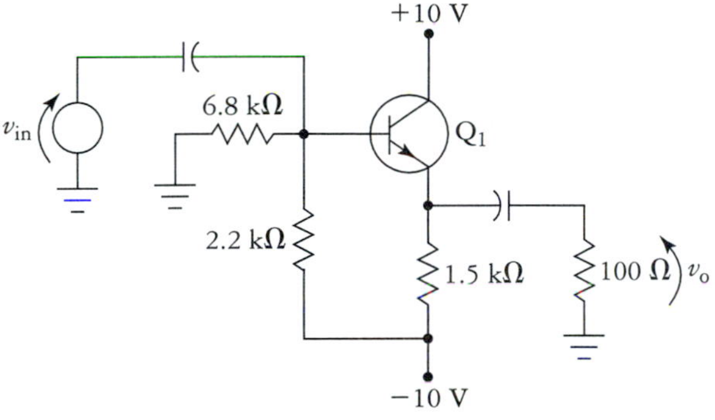

FIGURE 5.57
Circuit for Problem 5.21.

5.22. Refer to Fig. 5.58. Find I_{CQ}, V_{CEQ}, A_v, A_i, A_p, R_{in}, and R_o.

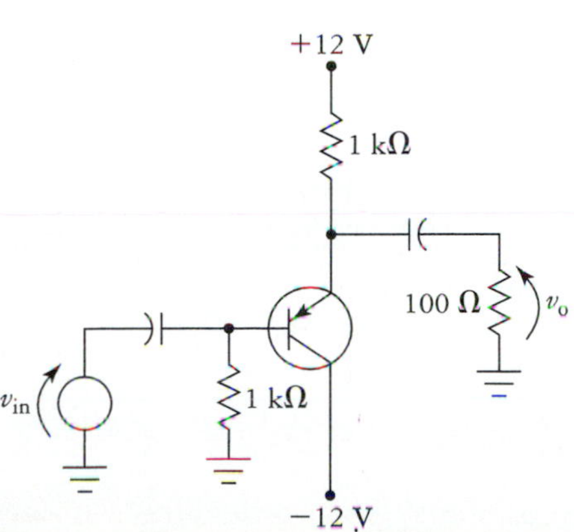

FIGURE 5.58
Circuit for Problem 5.22.

5.23. Refer to Fig. 5.59. Find I_{CQ}, V_{CEQ}, A_v, A_i, A_p, R_{in}, and R_o.

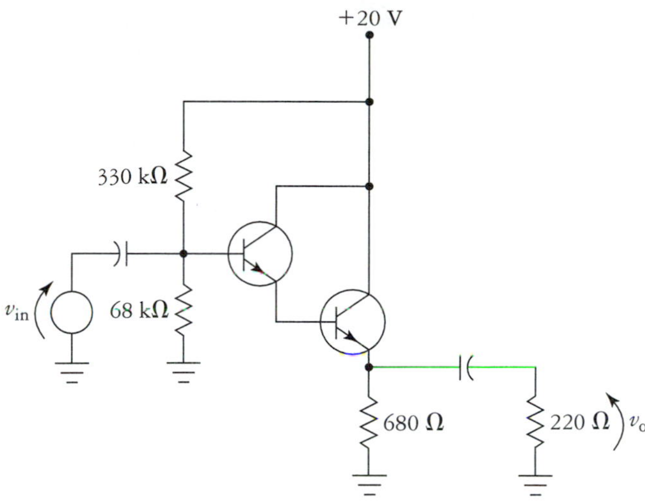

FIGURE 5.59
Circuit for Problem 5.23.

5.24. Refer to Fig. 5.60. Find I_{CQ}, V_{CEQ}, V_{CQ}, A_v, A_i, A_p, and R_{in}.

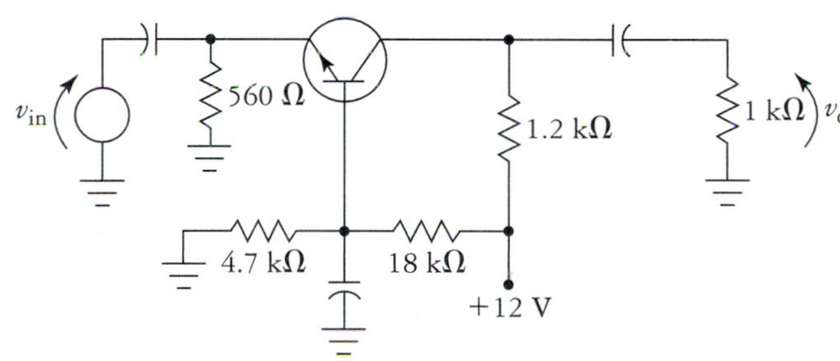

FIGURE 5.60
Circuit for Problem 5.24.

5.25. Refer to Fig. 5.61. Find I_{CQ}, V_{CEQ}, V_{CQ}, A_v, A_i, A_p, and R_{in}.

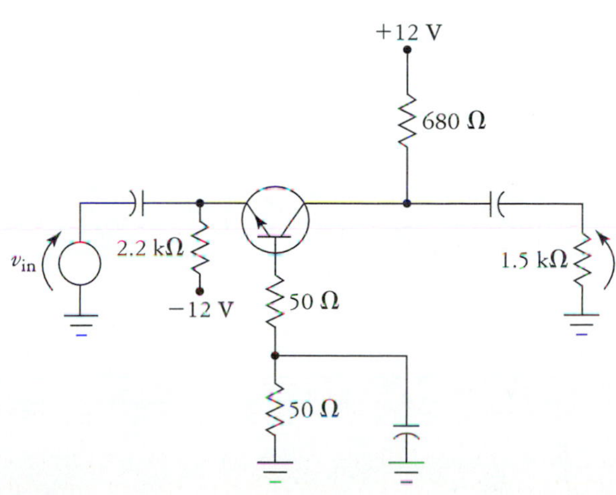

FIGURE 5.61
Circuit for Problem 5.25.

5.26. The relay and TTL-compatible source in Fig. 5.62 are described in Chapter 4, Section 4.1. Determine the value required for R_B, given that each transistor has $\beta = 50$.

5.27. Derive Eq. (5.6) using Fig. 5.1 for reference purposes.

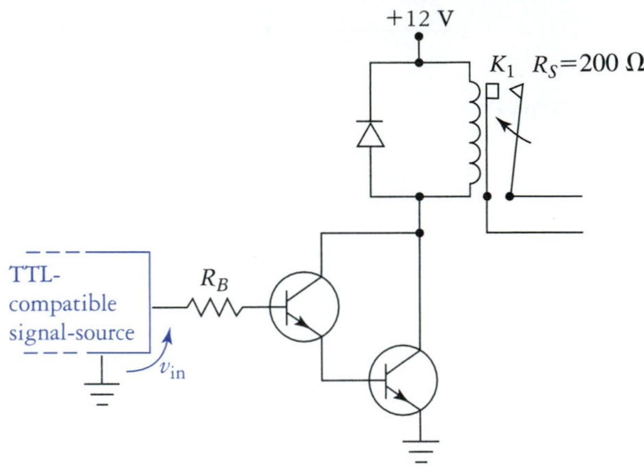

FIGURE 5.62
Circuit for Problem 5.26.

■ TROUBLESHOOTING PROBLEMS

5.28. Refer to Fig. 5.49. Suppose the emitter-bypass capacitor were to become disconnected. What main effect would this have on the circuit?

5.29. Refer to Fig. 5.49. Without actually plugging and chugging on your calculator, what would be the most likely consequence if the emitter-bypass capacitor were to become short circuited?

5.30. An emitter follower is breadboarded in the lab. A digital voltmeter is used to measure V_{CEQ}. The meter reading is not stable, and seems to fluctuate wildly. What is the most likely reason for this behavior?

■ COMPUTER PROBLEMS

5.31. Use PSpice to determine the output voltage compliance for the circuit in Fig. 5.50.

5.32. Using PSpice, perform an operating-point analysis and determine the approximate voltage gain of the amplifier in Fig. 5.32.

5.33. Use PSpice to determine the operating point and v_o waveform for Fig. 5.54. Using $v_{in} = 10 \sin 2\pi 1000t$ mV, plot five cycles of the output voltage waveform.

CHAPTER 6

Field-Effect Transistors

BEHAVIORAL OBJECTIVES

On completion of this chapter, the student should be able to:

- Describe the operational characteristics of junction field-effect transistors (FETs).
- Graphically determine the Q point of a self-biased FET.
- Algebraically determine the Q point of a self-biased FET.
- Determine the voltage, current, and power gain of an FET-based amplifier.
- Describe the advantages and disadvantages of FETs relative to BJTs.
- Explain the differences and similarities between MOSFETs, JFETS, and BJTs.
- Use PSpice to simulate FET-based circuits.

INTRODUCTION

The first commercially available transistors were bipolar devices (pnp germanium devices were actually the most widely available devices, originally). This is one of the reasons why bipolar devices are still so popular. While bipolar transistors were being developed, however, work was also being done on a device called the *field-effect transistor* (FET). From a device physics standpoint, FETs are very different from bipolar transistors. Functionally, however, FETs have many characteristics that are quite similar to those of BJTs. This is especially true in small-signal amplifier applications.

This chapter presents the basic theory of operation of the various types of FETs that are available. FET biasing techniques are presented, as are small-signal amplifier analysis methods. Graphical, algebraic, and computer-assisted methods of analysis are presented.

6.1 THE JUNCTION FET

Junction field-effect transistors (JFETs) are available in two fundamental forms: n channel and p channel. This is analogous to the BJT, which is available in npn and pnp forms. Like npn bipolar transistors, n-channel FETs are more widely available, so we concentrate on those types of devices.

Physical Structure of the JFET

The physical structure of an n-channel JFET is shown in Fig. 6.1. From this illustration, it is apparent how the term *n channel* is derived. The p regions on opposite sides of the n-type body of the JFET restrict current flow from drain to source to a relatively narrow channel. The depletion regions associated with the pn junctions are shown shaded. The main idea behind the operation of the JFET is that we can control the conductivity of the n channel by varying the width of these depletion regions.

FIGURE 6.1
Cross section of an n-channel JFET.

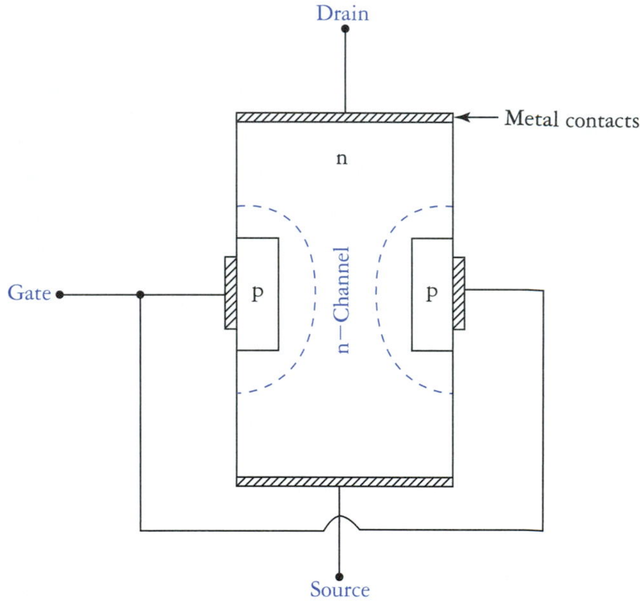

The schematic symbols used to represent JFETs are shown in Fig. 6.2. N-channel JFET symbols are shown in Fig. 6.2(a), and p-channel symbols are shown in Fig. 6.2(b). The left side of the figure shows the symbols for discrete devices. When JFETs are used in integrated circuits, the symbols shown in the left half of the illustration are usually used. A circle drawn around a transistor represents the case of a discrete device.

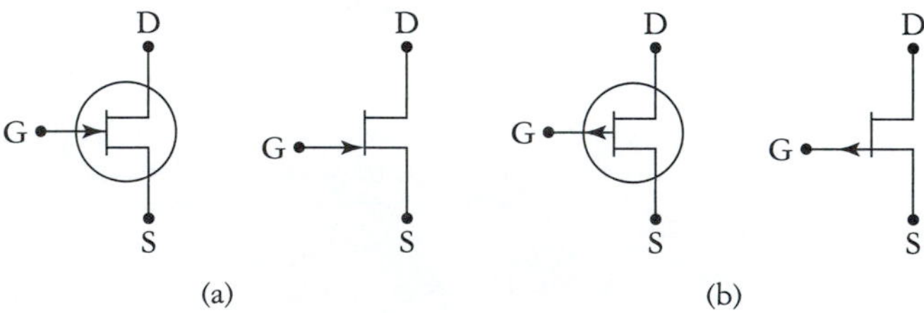

FIGURE 6.2
Schematic symbols. (a) N channel. (b) P channel.

JFET Operation

The operation of the n-channel JFET is represented in Fig. 6.3. Assuming that $V_{DS} = 0$ V and $V_{GS} = 0$ V, the depletion regions are distributed as shown in 6.3(a) where there is still space left in the n channel. Majority carriers (electrons) are available in this area, so if we increase V_{DS}, I_D will increase as well—at least to a point. As V_{DS} increases, I_D increases. This causes a voltage gradient to

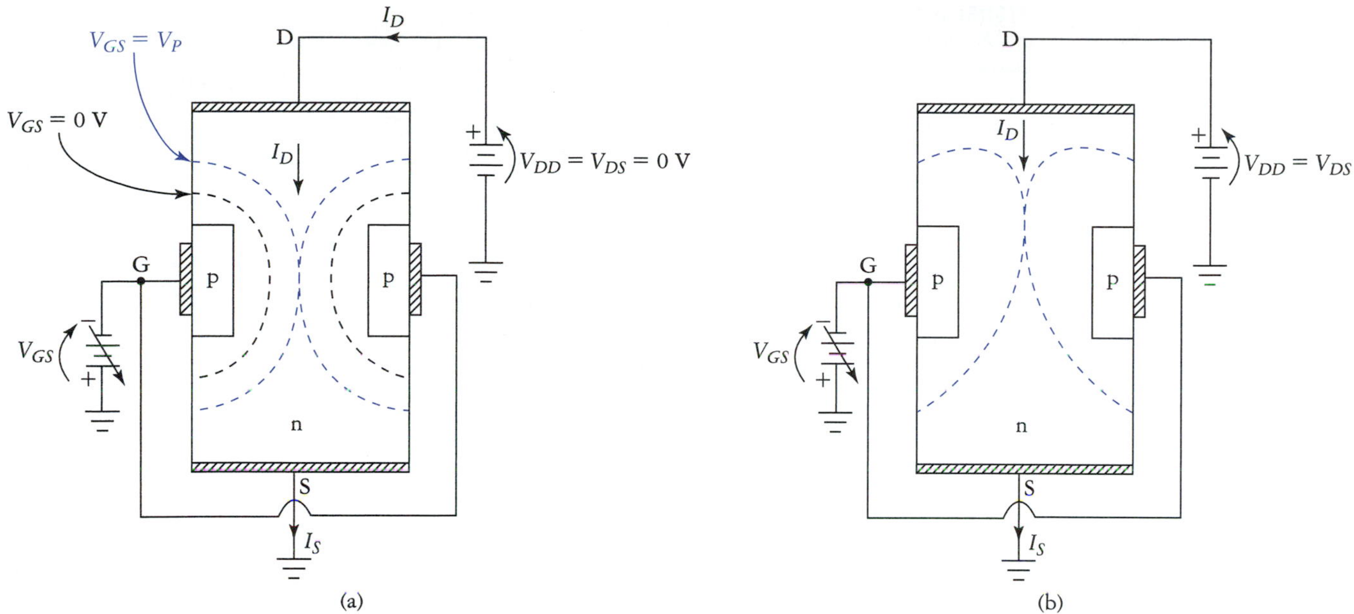

FIGURE 6.3
Effects of bias on JFET depletion regions. (a) V_{GS} varied from 0 to $-V_p$ with $V_{DS} = 0$ V. (b) $V_{GS} > -V_p$, $V_{DS} \geq V_p$.

form across the device, from drain to source. This causes the depletion regions to distort as shown in Fig. 6.3(b). When V_{DS} reaches a particular voltage (we will call this the *pinchoff voltage* V_P), the depletion regions just touch, as shown in 6.3(b). It looks like conduction should cease here, but an equilibrium current flow is reached. Ideally, further increases in V_{DS} do not increase I_D above this value. The FET is said to be *saturated* under these conditions.

JFET Drain Curves

Note
Don't confuse JFET saturation with BJT saturation.

Let us assume that $V_{DS} = 0$ V again. If we make V_{GS} more negative, the depletion regions expand toward each other. Driving V_{GS} further negative forces the depletion regions closer and closer to each other. The value of V_{GS} that just causes the conductive channel to vanish is called $V_{GS(\text{off})}$. This condition is represented by Fig. 6.3(a) where the depletion regions touch. Under these conditions, there are no majority carriers left in the n channel, and ideally $I_D = 0$ regardless of the value of V_{DS}.

If we set $V_{GS(\text{off})} < V_{GS} < 0$ V, then the n channel has an intermediate number of majority carries available, and as V_{DS} is increased, the FET saturates (pinchoff is reached) at a lower value of I_D than was the case when $V_{GS} = 0$ V. If we plot I_D as a function of V_{DS} for various values of V_{GS}, we obtain a family of drain curves like those shown in Fig. 6.4. Notice that when the FET saturates, the drain behaves as a current source. Because the current (I_D) is determined by a voltage (V_{GS}) in the saturation region, the JFET is best modeled as a voltage-controlled current source (VCIS). The dashed curve in Fig. 6.4 represents the set of all points where pinchoff occurs as a function of V_{GS}.

If the JFET is operated with $V_{DS} < V_p$, then the FET is operating in what is called the *ohmic* or *triode* region. In the ohmic region, the JFET behaves like a resistor whose value is dependent on V_{GS}. This voltage-controlled resistance characteristic is not (easily) implemented using BJTs.

Pinchoff Voltage and $V_{GS(\text{off})}$

Let's summarize some of this information before we move on. First, pinchoff voltage is measured from drain to source, and it varies with applied gate-to-source bias. Pinchoff voltage is positive for n-channel devices and we specify V_P at $V_{GS} = 0$ V. The gate-to-source bias required to cause the JFET to turn off is $V_{GS(\text{off})}$, which is a negative number for n-channel devices. It turns out that these two voltages have the same magnitude. That is,

$$|V_{GS(\text{off})}| = V_P$$

It is important to keep in mind that these two voltages are not exactly the same thing. However, because they are the same in magnitude and it is much easier to write V_P or $-V_P$ as necessary, we use this notation in the formulas that are presented in this chapter.

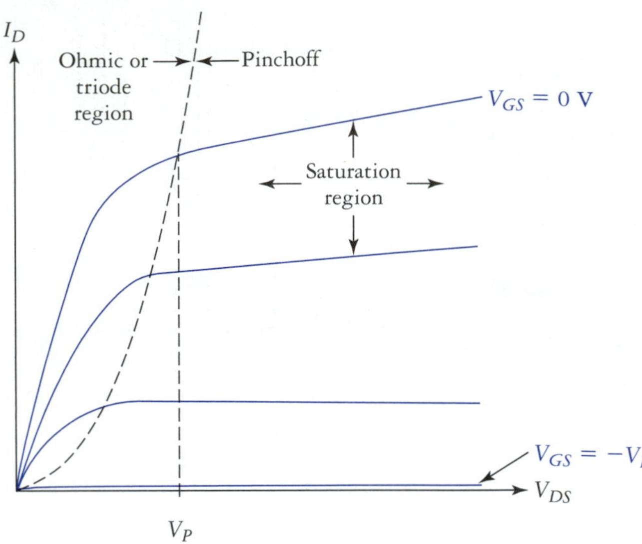

FIGURE 6.4
N-channel JFET family of drain curves.

Transconductance

If we set $V_{DS} > V_p$ and plot I_D as a function of V_{GS} (for an n-channel JFET), we obtain a parabolic *transconductance curve* like that shown in Fig. 6.5. Recall that $-V_p$ is the value of V_{GS} that apparently closes off the conduction channel ($V_{GS(\text{off})}$). I_{DSS} is the value of I_D in saturation (pinchoff), obtained when $V_{GS} = 0$ V. In other words, I_{DSS} is the maximum possible drain current (under normal operating conditions). The JFET transconductance curve is described by Eq. (6.1):

Note
Equation 6.1 is sometimes called *Shockley's equation for FETs*.

$$I_D = I_{DSS}\left(1 - \frac{V_{GS}}{V_{GS(\text{off})}}\right)^2 \tag{6.1}$$

It is important to remember that Eq. (6.1) only applies when the FET is operating in saturation. You must also realize that by saturation, we mean that the FET is behaving like a VCIS. Thus JFET saturation is a *device-dependent* parameter like β for a BJT. Don't confuse this form of saturation with circuit-dependent saturation like we found when determining $I_{C(\text{sat})}$.

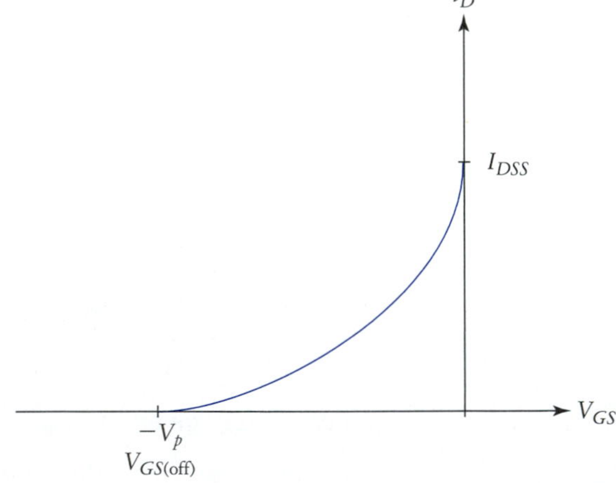

FIGURE 6.5
N-channel JFET transconductance curve.

Gate Leakage Current

Before moving on, it is instructive to take a quick look back at Fig. 6.3. If we redraw the circuit using conventional schematic symbols, we get Fig. 6.6. The source that is driving the gate terminal is simply applying a reverse bias to the JFET pn junctions. This means that the gate current I_G will be very low, consisting mainly of thermally generated charge carriers.

FIGURE 6.6
JFET biasing arrangement.

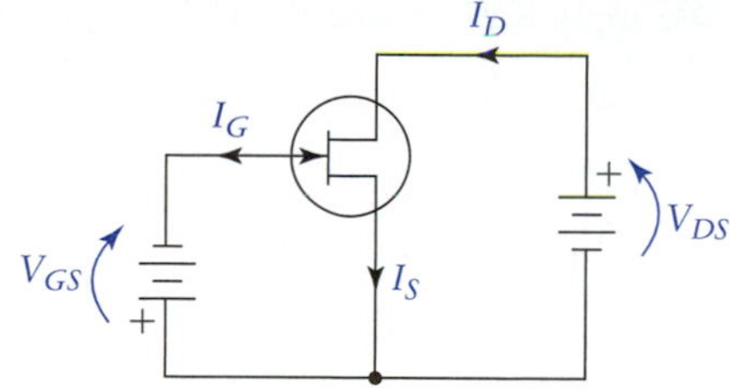

If we set $V_{GS} = 0$ V, with $V_{DS} > V_p$, we get $I_D = I_{DSS}$. Under these conditions the resulting gate leakage current is called I_{GSS}. I_{GSS} usually ranges from a few picoamps to perhaps a few nanoamps. In most situations, I_{GSS} is small enough to ignore. That is,

$$I_{GSS} \cong 0 \text{ A} \tag{6.2}$$

And, because the gate leakage current is essentially independent of V_{GS} (it is thermally dependent), we can write

Note
I_{DSS} is the maximum possible JFET drain current.

$$I_G = I_{GSS} \cong 0 \text{ A} \tag{6.3}$$

With this in mind, if we apply KCL to the JFET in Fig. 6.6, we find that the source and drain currents are equal. That is,

$$I_S = I_D \tag{6.4}$$

Recall the approximation that $I_E \cong I_C$ for BJTs. This approximation was found to be sufficiently accurate for nearly all analysis situations. For FETs, Eq. (6.4) is several orders of magnitude more accurate than the equivalent BJT approximation.

Since we are on the subject of analogies, this is a good time to point out that when operated in the *pinchoff* (or *saturation*) region, the drain, gate, and source of the JFET are analogous to the collector, base, and emitter, respectively, of a BJT.

EXAMPLE 6.1 The JFET in Fig. 6.7 has $I_{DSS} = 15$ mA and $V_p = 4$ V. Determine I_D and V_{DS}.

FIGURE 6.7
Circuit for Example 6.1.

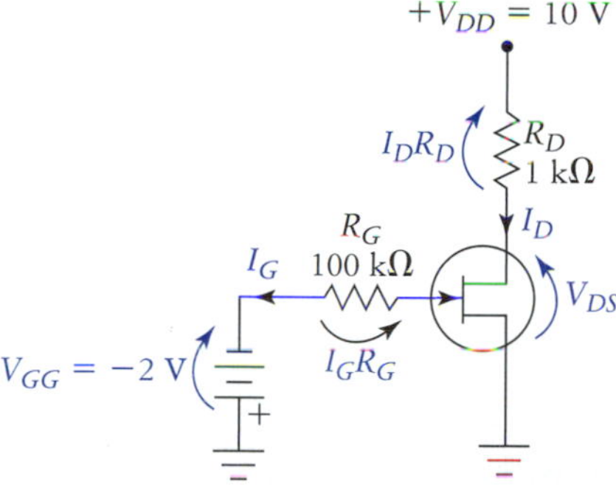

Solution We must determine V_{GS} first. If we write the KVL equation for the G-S loop, and solve for V_{GS}, we obtain

$$V_{GS} = V_{GG} + I_GR_G$$

Now, because $I_G = I_{GSS} = 0$ A, we have

$$V_{GS} = V_{GG} = -2 \text{ V}$$

We now substitute into Eq. (6.1), giving us

$$\begin{aligned} I_D &= I_{DSS}\left(1 - \frac{|V_{GS}|}{V_p}\right)^2 \\ &= 15 \text{ mA}\left(1 - \frac{2 \text{ V}}{4 \text{ V}}\right)^2 \\ &= 3.75 \text{ mA} \end{aligned}$$

We apply KVL around the D-S loop giving

$$0 = V_{DD} - I_D R_D - V_{DS}$$

Solving for V_{DS} we find

$$\begin{aligned} V_{DS} &= V_{DD} - I_D R_D \\ &= 10\text{ V} - (3.75\text{ mA})(1\text{ k}\Omega) \\ &= 6.25\text{ V} \end{aligned}$$

In Example 6.1, we assumed at the outset that the JFET was operating in the pinchoff region. This is standard operating procedure, much like when we assumed active-region operation when analyzing BJT biasing circuits. If we came up with impossible answers, such as a negative value for V_{CE} (using npn BJTs), then we deduced that the transistor is in saturation. Similar situations can occur when analyzing FET circuits as well. The next example demonstrates.

EXAMPLE 6.2

The JFET in Fig. 6.7 has $I_{DSS} = 20$ mA and $V_p = 5$ V. Determine I_D and V_{DS}.

Solution We determine the drain current, where again $V_{GS} = V_{GG} = -2$ V:

$$\begin{aligned} I_D &= I_{DSS}\left(1 - \frac{|V_{GS}|}{V_p}\right)^2 \\ &= 20\text{ mA}\left(1 - \frac{2\text{ V}}{5\text{ V}}\right)^2 \\ &= 7.2\text{ mA} \end{aligned}$$

Now we determine V_{DS}:

$$\begin{aligned} V_{DS} &= V_{DD} - I_D R_D \\ &= 10\text{ V} - (7.2\text{ mA})(1\text{ k}\Omega) \\ &= 2.8\text{ V} \end{aligned}$$

It would appear that the circuit should operate as predicted. After all, we still have a positive V_{DS}. However, there is a problem; specifically $V_{DS} < V_p$. Referring back to Fig. 6.4, we find that this leads to operation in the ohmic region. Thus, the FET is not behaving as modeled by Eq. (6.1), which applies when the JFET acts as a VCIS.

To determine the actual values for I_D and V_{GS} in Example 6.2, we would have to do a graphical dc load-line analysis using the actual drain curves for the FET, or we could develop a transconductance equation that describes the JFET more accurately. We shall do neither here, because we are not that concerned with this mode of operation yet. In linear amplifier applications, the following inequality must be satisfied:

$$V_p < V_{DS} < V_{DD} \tag{6.5}$$

Additional JFET DC Characteristics

Like its bipolar relative, the JFET has certain limitations and nonideal characteristics. A few of these properties are now presented.

Source–Drain Interchangeability

Unlike the collector and emitter terminals of the bipolar transistor, there is no difference between the drain and source terminals of a JFET. Because of this, the JFET will behave the same way regardless of drain and source configuration, although manufacturers designate the terminals to correspond with industry standard practices.

Breakdown Voltage

Using an n-channel JFET as an example, if we apply a large enough value of V_{DS}, then drain curves like those in Fig. 6.8 are generated. Notice that the breakdown voltage of the JFET decreases with decreasing V_{GS}. The maximum breakdown voltage occurs when $V_{GS} = 0$ V. This is designated BV_{DG0}.

FIGURE 6.8
JFET breakdown characteristics.

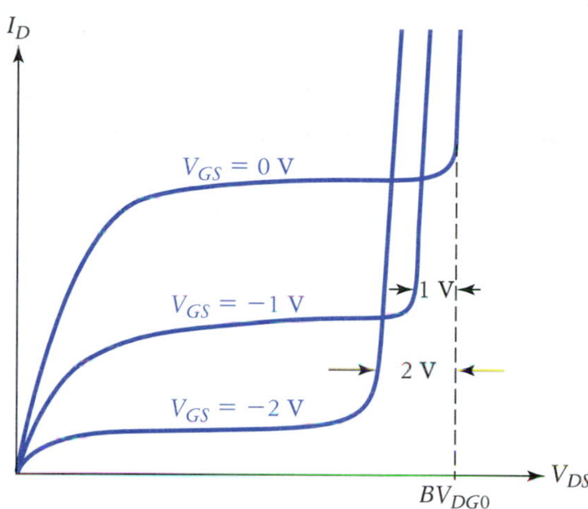

Drain-to-Source Resistance

Like the collector curves for a bipolar transistor, the FET drain curves are not perfectly horizontal. If we project back from the pinchoff region, we find that the drain curves intersect the V_{DS} axis at a common point $-V_A$, which is again called the Early voltage. This is shown in Fig. 6.9. Typical values for V_A range from 50 to 100 V. This is on the same order of magnitude as the typical general-purpose bipolar transistor.

FIGURE 6.9
JFETs have Early voltage characteristics like BJTs.

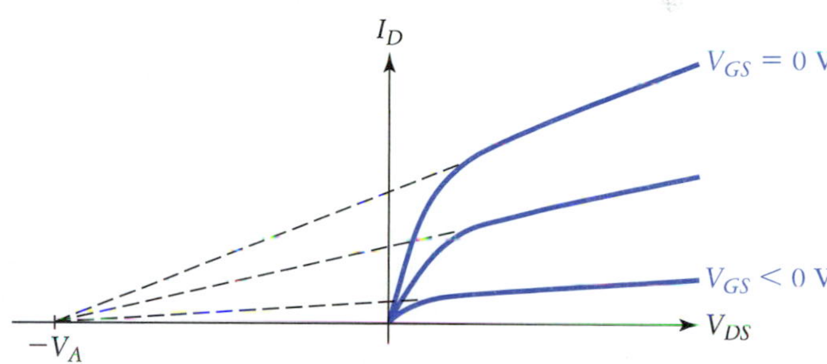

Note
The Early voltage determines the drain-source resistance (r_{DS} or r_o) in saturation.

Thermal Characteristics

Recall that bipolar transistors have negative temperature coefficients. This means that as device temperature increases, so does I_C. Increasing I_C causes an increase in temperature, and so it is possible that the BJT could experience thermal runaway. Because FETs rely only on majority carrier conduction, they have positive temperature coefficients. The positive temperature coefficient characteristic of FETs means they are not subject to thermal runaway. If device temperature increases, I_D tends to decrease. This helps to limit device power dissipation. You may observe the consequences of this characteristic in the laboratory. If a circuit like that of Fig. 6.7 is constructed, you may notice that V_{DS} tends to increase somewhat for a few minutes. What is happening is that as the JFET warms up, the drain current decreases until thermal equilibrium is reached. Because I_D decreases during this time interval, V_{DS} increases.

6.2 JFET BIASING TECHNIQUES

To use a JFET as a linear amplifier, it must first be biased up to some particular operating or Q point. There are several ways to accomplish this task, but as you might have guessed, the simplest method is not usually the best.

A word on notation is in order here. As was the case with bipolar transistor-based circuits, when we bias up a JFET (or any FET for that matter), it is customary to append the letter Q (as in I_{DQ} and V_{DSQ}, for example) to the subscript of the variable in question. This helps to ensure that there is no confusion between quiescent operating values and other values that may be independent of the Q point.

Fixed Bias

Actually, you already know how to analyze the most common form of fixed biasing. It was used in Examples 6.1 and 6.2. In fixed biasing, V_{GS} is simply set to some predetermined value and the drain current flows according to Eq. (6.1). Of course, this assumes operation in the pinchoff region. To emphasize the point, let's try one more example.

EXAMPLE 6.3

Refer to Fig. 6.7. Assume $I_{DSS} = 6$ mA and $V_p = 3$ V. Determine the value of V_{GG} that will produce $I_{DQ} = I_{DSS}/2$.

Solution When a JFET is biased up such that $I_{DQ} = I_{DSS}/2$, we have midpoint biasing, at least in terms of drain current. In this case, we simply solve Eq. (6.1) (with $V_{GS(\text{off})} = -V_p$) for V_{GS} giving us

$$V_{GG} = V_{GSQ} = -V_p\left(1 - \sqrt{\frac{I_{DQ}}{I_{DSS}}}\right)$$

$$= -3\text{ V}\left(1 - \sqrt{\frac{3}{6}}\right)$$

$$= -0.88\text{ V}$$

Note
Like base biasing of BJTs, fixed gate bias is very sensitive to device parameter changes.

The problem with fixed bias is that variations in device parameters have a large effect on Q-point location. The situation is somewhat like the beta-dependency problem we have with fixed-base biasing of BJTs. It turns out that it is difficult for manufacturers to accurately control the absolute values of I_{DSS} and V_p. For example, the 2N3823 is a popular n-channel JFET with a published I_{DSS} spread from 5 to 20 mA. Likewise, V_p may typically vary from around 1 to 4 V from unit to unit as well. This makes fixed bias unsuitable for most linear amplifier applications. Consider the transconductance curves in Fig. 6.10. Using a device with the lower curve results in the Q point being centered on the I_D axis. If a device with the upper curve is used, then the Q point is moved much closer to I_{DSS}. This is usually not the preferred situation.

Self-Bias

Self-bias is particularly well suited for setting up a Q point that is $I_{DQ} = I_{DSS}/x$. That is, the drain current is some particular fraction of I_{DSS}. The basic arrangement for self-bias is shown in Fig. 6.11. A few analysis details are now presented. Because $I_G = 0$ the voltage drop $I_G R_G = 0$ V, and the gate terminal is at 0 V as well. That is,

$$V_{GQ} = 0\text{ V} \tag{6.6}$$

The voltage drop across the source resistor places the source terminal at a more positive potential than the gate, reverse biasing the internal pn junctions. This reverse-biased voltage is given by

$$V_{GSQ} = -I_D R_S \tag{6.7}$$

To determine I_{DQ} it is necessary to solve Eq. (6.7) and (6.1) simultaneously. There are two ways that we can do this: graphically or algebraically. (Actually, there are three ways if we count the use of circuit simulators like PSpice as well.) We look at the graphical method first.

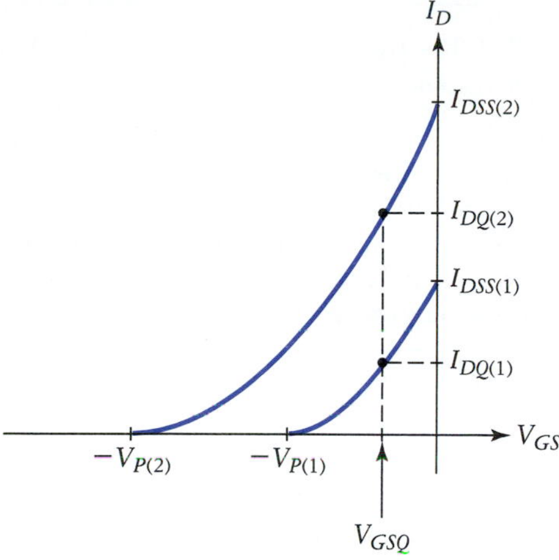

FIGURE 6.10
When fixed, gate voltage source biasing is used, variation of JFET parameters results in large I_{DQ} changes.

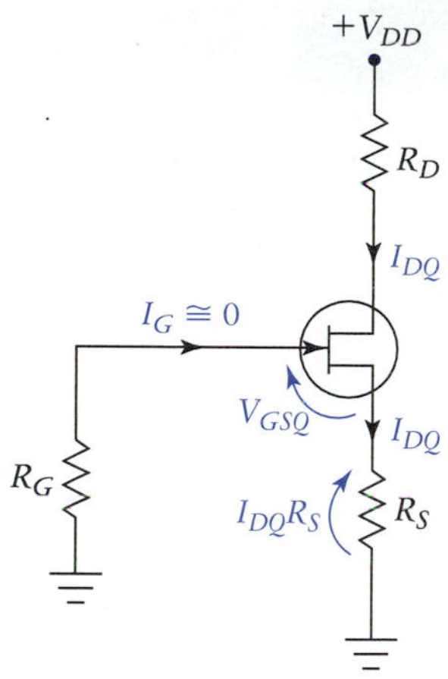

FIGURE 6.11
Source-feedback, or self-bias, arrangement.

Graphical Analysis

Note
Self-bias is also called source-feedback bias.

Graphical analysis of any FET circuit requires a plot of the device transconductance curve. Ideally since all JFETs have parabolic transconductance curves, a normalized parabolic curve can be used. Such a curve is shown in Fig. 6.12. A BASIC program that can be used to generate such curves is presented in Section 6.5. You merely fill in the appropriate values for I_{DSS} and V_p to match your particular JFET. Please keep in mind that graphical analysis is approximate, and that real-life FETs deviate somewhat from the ideal parabolic curve, so extreme precision is not justified here.

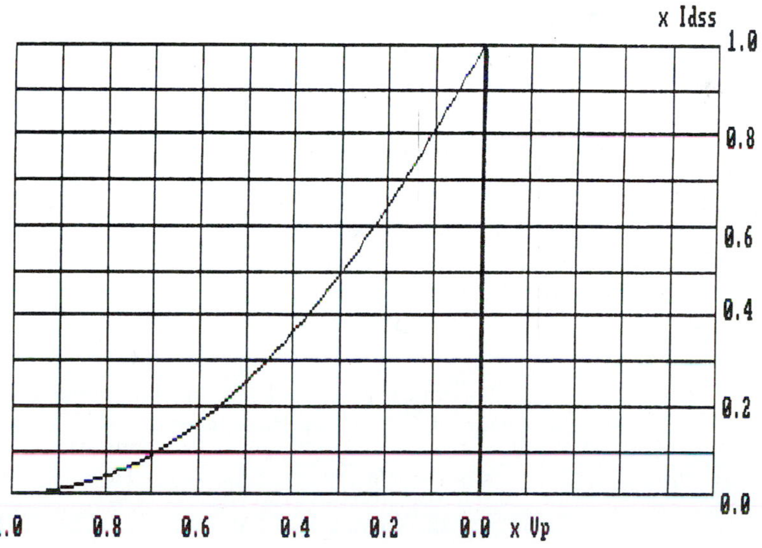

FIGURE 6.12
Normalized n-channel JFET transconductance curve.

To determine the Q point, we must plot the $-I_D R_S$ load line on the same graph as the FET transconductance curve. Since $V_{GS} = -I_D R_S$ is a linear equation, we simply calculate two convenient points and connect them with a straight line. The point of intersection of the R_S load line and the transconductance curve is the Q point of the JFET. An example will help illustrate this concept.

EXAMPLE 6.4

Refer to Fig. 6.11. The JFET has $I_{DSS} = 10$ mA and $V_p = -4$ V, $V_{DD} = 15$ V, $R_S = 100\ \Omega$, and $R_D = 1$ kΩ. Determine I_{DQ} and V_{DSQ}.

Solution

A transconductance curve scaled to match the JFET is shown in Fig. 6.13. We choose two convenient values for I_D and calculate the corresponding voltage drop across R_S. From Eq. (6.7), we know that this voltage drop is $-V_{GS}$. The easiest value to assume first is $I_D = 0$ mA, which causes $V_{GSQ} = 0$ V. This point is located at the origin of the transconductance curve. Arbitrarily, let us assume that $I_D = I_{DSS}$. This gives us the second point $-I_{DSS}R_S = -1$ V. This point is located as shown in Fig. 6.13. Connecting the dots, we find that the I_DR_S line intersects the transconductance curve at about $I_{DQ} = 6.9$ mA.

I_D	V_{GS}
0	0
10 mA	−1 V

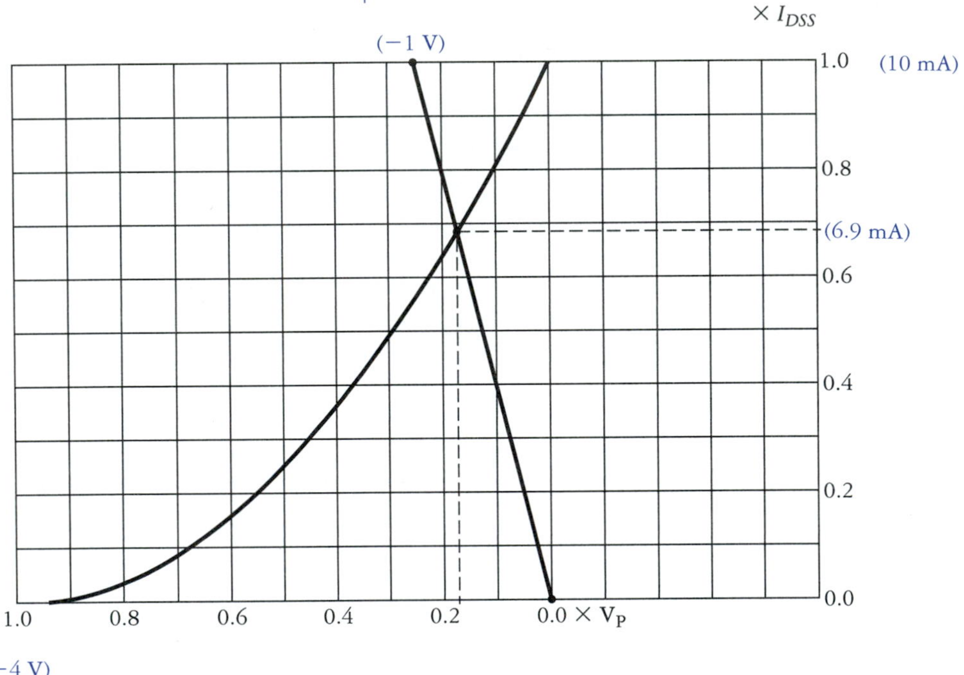

FIGURE 6.13
Transconductance curve for Example 6.4.

The value of V_{GSQ} can be found by dropping down from the Q point to the V_{GS} axis (looks like about −0.7 V). It is perhaps easier to find V_{GSQ} by using Eq. (6.7), which gives us

$$\begin{aligned} V_{GSQ} &= -I_{DQ}\,R_S \\ &= (-6.7\text{ mA})(100\ \Omega) \\ &= -0.67\text{ V} \cong 0.7\text{ V} \end{aligned}$$

We determine V_{DSQ} using KVL:

$$\begin{aligned} V_{DSQ} &= V_{DD} - I_D(R_D + R_S) \\ &= 15\text{ V} - (6.7\text{ mA})(1100\ \Omega) \\ &= 7.6\text{ V} \end{aligned}$$

Since $V_{DD} > V_{DSQ} > V_p$ we can safely assume the FET is operating in the active (saturation) region.

Just to make sure the procedure is clear, let's do one more example.

EXAMPLE 6.5

Refer to Fig. 6.11. The JFET has $I_{DSS} = 5$ mA and $V_p = 3$ V. Given: $R_S = 680\ \Omega$, $R_D = 3.3\ \text{k}\Omega$, $R_G = 1\ \text{M}\Omega$, and $V_{DD} = 12$ V. Determine I_{DQ}, V_{DSQ}, and V_{GSQ}.

Solution We begin again by choosing two arbitrary values of I_D and calculating the corresponding drop across R_S. Let us start with $I_D = 0$ mA, which gives us $V_{GS} = 0$ V. We normally would now set $I_D = I_{DSS}$, but in this case we would obtain a V_{GS} value that is off the graph (−3.4 V). So, we use some other convenient value such as $I_D = 3$ mA, which is $0.6I_{DSS}$. This value produces $V_{GS} = -2.04$ V. The resulting plot is shown in Fig. 6.14 where we find the Q-point coordinates are approximately $I_{DQ} = 1.8$ mA and $V_{GSQ} = -1.2$ V. Kirchhoff's voltage law is used to find V_{DSQ}, producing

$$\begin{aligned} V_{DSQ} &= V_{DD} - I_{DQ}(R_D + R_S) \\ &= 12\text{ V} - (1.8\text{ mA})(3.98\text{ k}\Omega) \\ &= 4.84\text{ V} \end{aligned}$$

I_D	V_{GS}
0	0
3 mA	−2.04 V

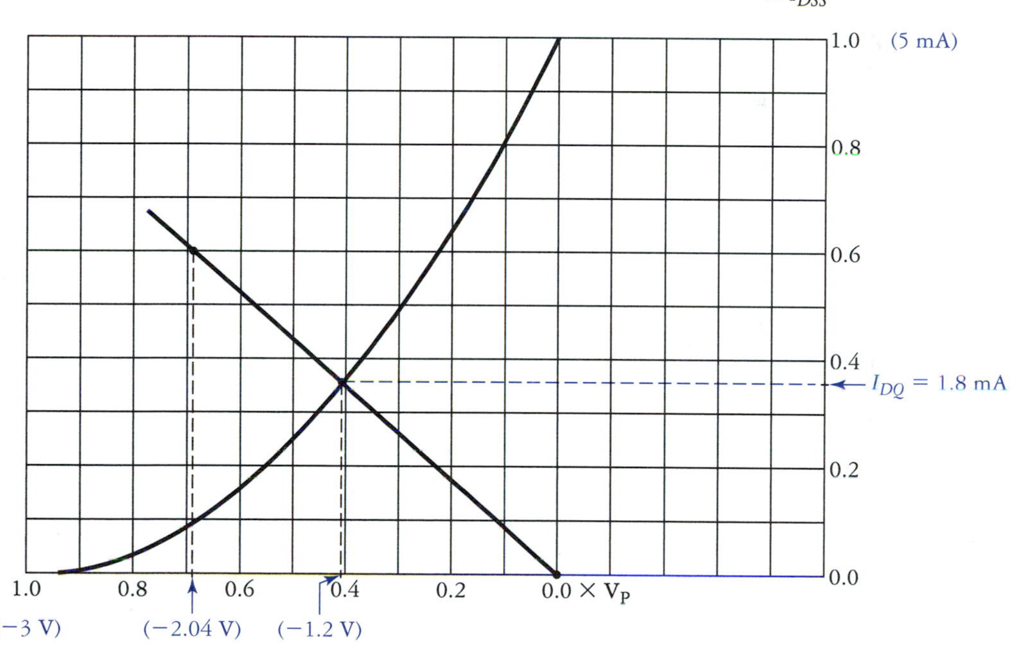

FIGURE 6.14
Transconductance curve for Example 6.5.

A graphical approach may also be used to determine the value of R_S necessary to set a particular drain current. To do this, you first locate the desired drain current on the I_D axis and then project a line to the left (parallel with the x axis) until it intersects the transconductance curve. At the point of intersection, project a line straight down to the V_{GS} axis. This tells us the corresponding V_{GSQ}. The required value of R_S is then found using

$$R_S = \frac{|V_{GSQ}|}{I_{DQ}} \tag{6.8}$$

After R_S is determined, we use KVL and Ohm's law to find the value of R_D that results in the desired V_{DSQ}. Of course, we must make sure we satisfy Eq. (6.4). Generally speaking, the value of R_G is not critical because I_G is so small. A large value of R_G results in higher input resistance, but also higher input noise and temperature sensitivity. Typical values of R_G range from 100 kΩ to around 10 MΩ.

Algebraic Analysis

The algebraic analysis of the self-biased JFET relies on the use of the quadratic formula to determine the Q point. There are always two solutions (or roots) to every quadratic equation. In this case, one of the solutions is not physically possible (an extraneous root) because the transconductance curve is only parabolic for $-V_P < V_{GS}$. The curve flattens once $V_{GS} < -V_p$ (V_p is located at the vertex of the parabola). The formula for I_{DQ} is

$$I_{DQ} = \frac{-B - \sqrt{B^2 - 4AC}}{2A} \tag{6.9}$$

where

$$A = R_S^2, \qquad B = -\left(2R_SV_p + \frac{V_p^2}{I_{DSS}}\right), \qquad \text{and } C = V_p^2$$

EXAMPLE 6.6

Determine I_{DQ} for Examples 6.4 and 6.5 using Eq. (6.9).

Solution In Example 6.4 we have $I_{DSS} = 10$ mA, $V_p = 4$ V, and $R_S = 100\ \Omega$. The parameters for Eq. (6.9) are $A = 10$ k, $B = -2.4$ k, and $C = 16$. Substituting these values into Eq. (6.9) yields

$$I_{DQ} = 6.86 \text{ mA}$$

This is in good agreement with the graphical analysis, which produced $I_{DQ} = 6.9$ mA. Example 6.5 had $I_{DSS} = 5$ mA, $V_p = 3$ V, and $R_S = 680\ \Omega$. The parameters for the quadratic formula are $A = 426.4$ k, $B = -5.88$ k, and $C = 9$. This results in

$$I_{DQ} = 1.75 \text{ mA}$$

Again, we are in excellent agreement with the graphical result of $I_{DQ} = 1.8$ mA.

If you have a programmable calculator, then the algebraic method of determining I_{DQ} is probably most convenient. On the other hand, without a programmable calculator, the algebraic method is rather susceptible to calculator entry errors. The disadvantage of graphical analysis is that you need a transconductance plot to do it. There is really no "best" way to perform this analysis. Use whichever technique you like best.

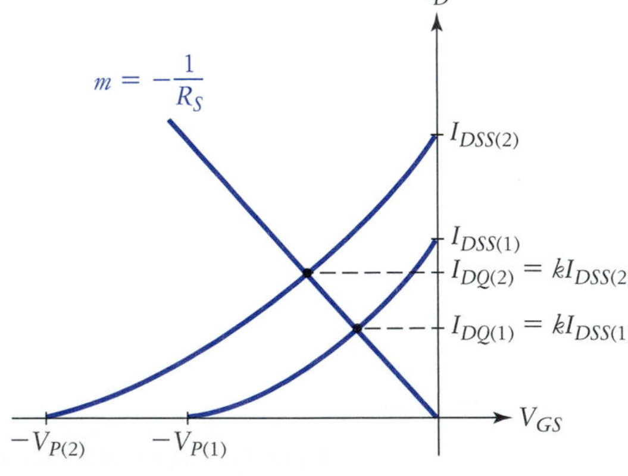

FIGURE 6.15
With source-feedback bias, changes in FET parameters result in constant I_{DQ}/I_{DSS}.

Examination of Fig. 6.15 shows why self-bias is more widely used than fixed bias. If the circuit is designed for midpoint bias ($I_{DQ} = 0.5I_{DSS}$) based on the smaller curve, then even if a JFET with a larger curve is used, we still have midpoint bias in terms of I_{DSS}. From a negative-feedback point of view, what happens is that the increase in I_D caused by using the JFET with greater I_{DSS} and V_p causes a greater voltage drop across R_S, which in turn causes V_{GS} to be more negative, maintaining constant drain current, relative to I_{DSS}.

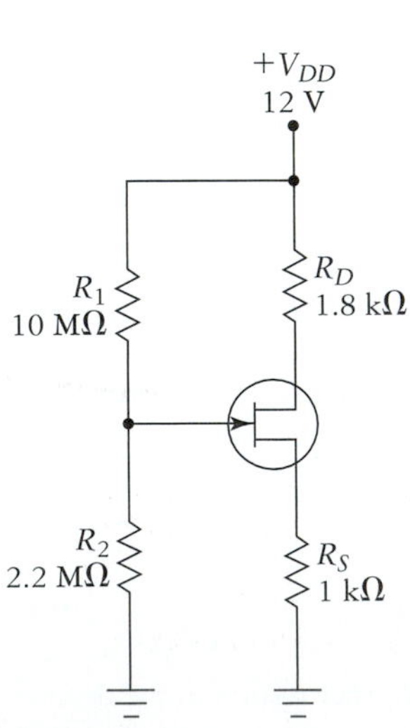

FIGURE 6.16
Voltage-divider biased n-channel JFET.

Voltage-Divider Bias

Another alternative JFET biasing technique is called voltage-divider biasing. This is actually a combination of fixed and self-biasing, as shown in Fig. 6.16. The main effect of the voltage divider is to move the x intercept of the I_DR_S line to the right of the origin, by placing the gate at some fixed voltage level. The gate voltage is given by

$$V_G = V_{DD}\frac{R_2}{R_1 + R_2} \tag{6.10}$$

To determine the operating point of the circuit, we again must solve Eq. (6.1) simultaneously with Eq. (6.7), and we must account for the gate voltage as well. The procedure is very similar to that used for simple self-bias.

Graphical Analysis

To get a better understanding of how voltage-divider biasing works, let us Thevenize the gate circuitry of Fig. 6.16. This gives us Fig. 6.17. Application of KVL around this loop results in

$$V_{GS} = V_{\text{Th}} - I_DR_S \tag{6.11}$$

Notice that since $I_G = 0$,

$$V_G = V_{\text{Th}} \tag{6.12}$$

In order for the circuit to work properly (with an n-channel JFET), Eq. (6.11) requires that $I_DR_S > V_{\text{Th}}$. This ensures that V_{GS} is negative.

The graphical analysis of JFET voltage-divider analysis is performed as shown in the following example.

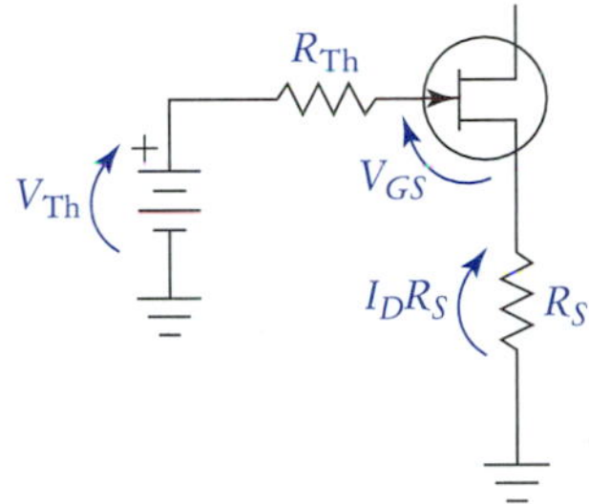

FIGURE 6.17
Thevenized gate network.

EXAMPLE 6.7

Refer to Fig. 6.16. The JFET has $I_{DSS} = 4$ mA and $V_p = 2.5$ V. $R_1 = 10$ MΩ, $R_2 = 2.2$ MΩ, $R_S = 1$ kΩ, and $R_D = 1.8$ kΩ. Determine I_{DQ}, V_{GSQ}, V_{DSQ}, and V_{DQ}.

Solution Begin by finding V_G:

$$\begin{aligned} V_G = V_{\text{Th}} &= 12\text{ V}\frac{2.2}{2.2 + 10} \\ &= 2.16\text{ V} \end{aligned}$$

Using the transconductance curve scaled to match out JFET, we find that V_G is off the graph. This is not a problem, as we can simply draw in a few more divisions to the right (each division is 0.25 V in this example). We now determine two points on the R_S line. Using Eq. (6.11) and choosing $I_D = 0$ mA gives us

$$\begin{aligned} V_{GS} &= V_{\text{Th}} - I_DR_S \\ &= 2.16\text{ V} - 0\text{ V} \\ &= 2.16\text{ V} \end{aligned}$$

We now place this point on the transconductance curve as shown in Fig. 6.18. Choosing $I_D = I_{DSS}$ for the second point yields

$$\begin{aligned} V_{GS} &= V_{\text{Th}} - I_DR_S \\ &= 2.16\text{ V} - (4\text{ mA})(1\text{ k}\Omega) \\ &= 2.16\text{ V} - 4\text{ V} \\ &= -1.84\text{ V} \end{aligned}$$

Plotting this point and connecting the dots gives us $I_{DQ} = 2.6$ mA and $V_{GSQ} = -0.44$ V. To find V_{GSQ} without using the graph, we simply use Eq. (6.11) again.

I_D	V_{GS}
0	2.16 V
4 mA	−1.84 V

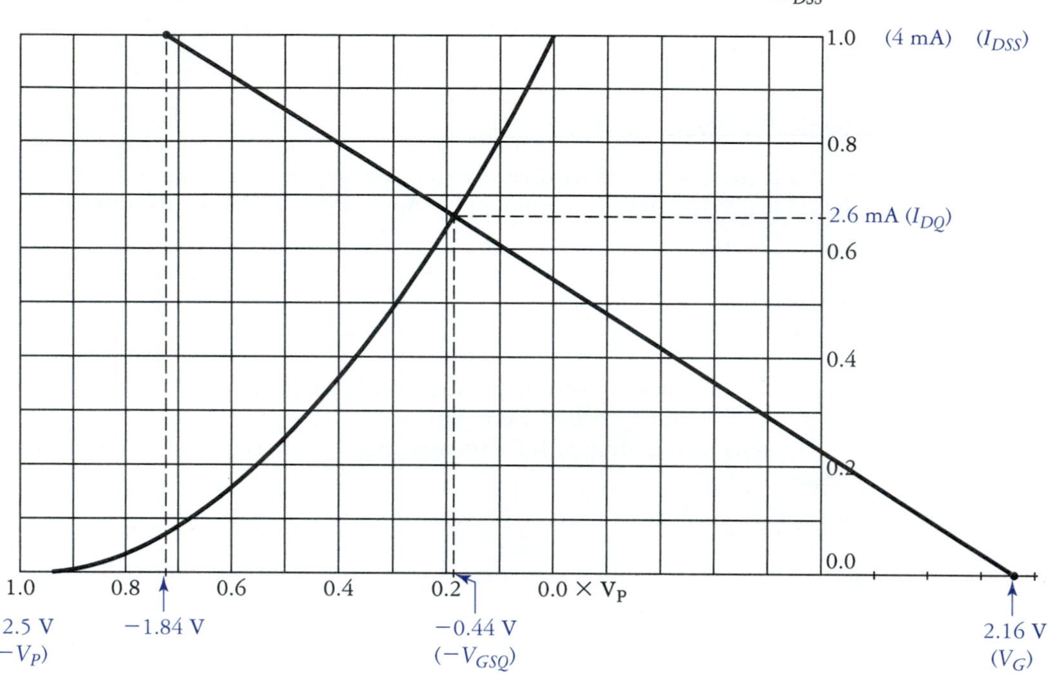

FIGURE 6.18
Transconductance curve for Fig. 6.16.

$$\begin{aligned} V_{GSQ} &= V_{\text{Th}} - I_{DQ}R_S \\ &= 2.16\text{ V} - (2.6\text{ mA})(1\text{ k}\Omega) \\ &= 2.16\text{ V} - 2.6\text{ V} \\ &= -0.44\text{ V} \end{aligned}$$

We now determine V_{DSQ} using KVL:

$$\begin{aligned} V_{DSQ} &= V_{DD} - I_D(R_D + R_S) \\ &= 12\text{ V} - (2.6\text{ mA})(2.8\text{ k}\Omega) \\ &= 12\text{ V} - 7.28\text{ V} \\ &= 4.27\text{ V} \end{aligned}$$

In a similar manner, we find V_{DQ}:

$$\begin{aligned} V_{DQ} &= V_{DD} - I_DR_D \\ &= 12\text{ V} - (2.6\text{ mA})(1.8\text{ k}\Omega) \\ &= 12\text{ V} - 4.68\text{ V} \\ &= 7.32\text{ V} \end{aligned}$$

Algebraic Analysis

Algebraically, we can determine the drain current of Fig. 6.16 by using Eq. (6.9) with the following parameter definitions:

$$A = R_S^2$$

$$B = -\left[2R_S(V_p + |V_G|) + \frac{V_p^2}{I_{DSS}}\right]$$

$$C = (V_p + |V_G|)^2$$

Using this information with the values given in Example 6.7 gives us $A = 1$ M, $B = -10.88$ k, and $C = 21.7$. Substituting these values into Eq. (6.9) yields

$$I_{DQ} = 2.63 \text{ mA}$$

This is in good agreement with the graphical analysis result.

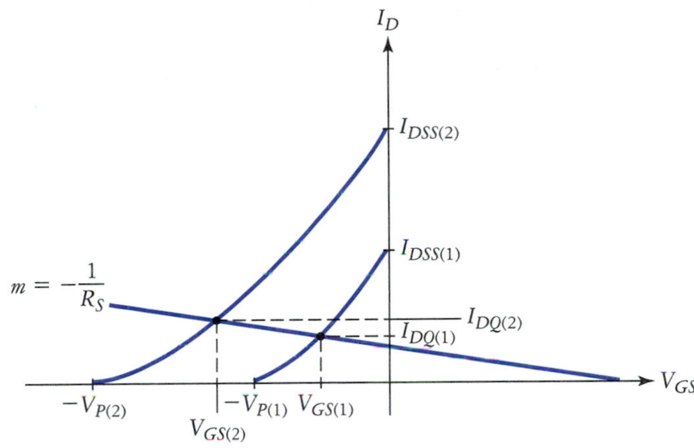

FIGURE 6.19
The gate bias voltage decreases the change in I_{DQ} with changes in JFET parameters.

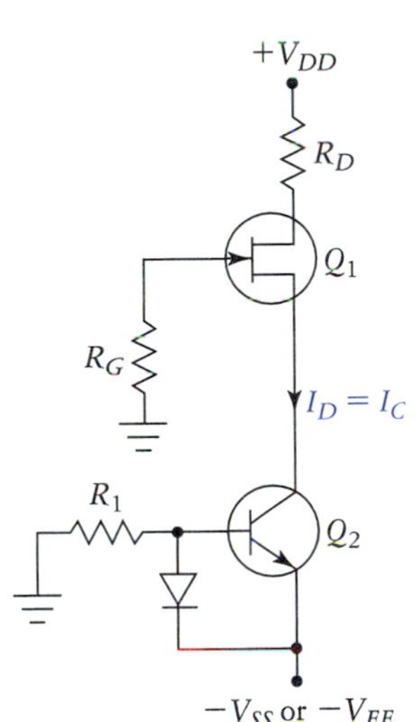

FIGURE 6.20
A current source can be used to establish a constant I_{DQ}.

Voltage-divider biasing is used when a more constant drain current must be established than can be obtained using self-bias alone. Consider the curves in Fig. 6.19. Notice that the large change in transconductance between the two devices results in a rather small change in drain current. If we could move the x-axis intercept of the I_DR_S line out further to the right (increasing V_G) then I_{DQ} becomes even more independent of JFET parameters. As a practical matter, however, V_G would be limited to about $V_{DD}/3$ at the most.

Current-Source Biasing

The most accurate way to set up a fixed value of drain current is to use a current source. For example, a BJT current mirror could be used as shown in Fig. 6.20. The advantage to this approach centers around the fact that the current mirror (and many other BJT-based current sources) can be designed for very accurate operation at nearly any imaginable current level. Notice that the negative supply voltage could be called either V_{EE} (the emitter supply voltage) or V_{SS} (the source supply voltage). If the circuit were composed mostly of FETs, then we would most likely use V_{SS}. In this example, it really does not matter.

The high internal resistance of the BJT collector produces what is effectively a horizontal I_DR_S line. This is shown in Fig. 6.21. We now have a circuit that will produce a fixed drain current that is almost totally independent of JFET parameters.

FIGURE 6.21
Since the internal resistance of a current source is infinite, the R_S line is horizontal.

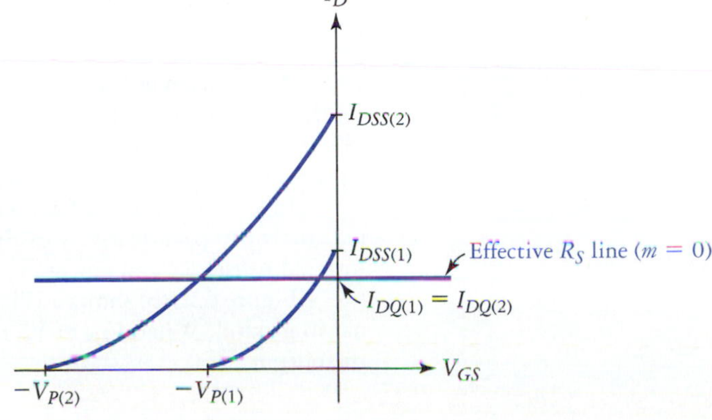

A Few JFET Applications

The unique characteristics of field-effect transistors allow them to be used in many applications for which bipolar devices are not suitable. Two of the more notable applications are use as an analog switch or transmission gate, and as a voltage-controlled resistor.

Analog Switch Applications

If we restrict operation to within about ±50 mV of the origin of the device drain curves, then the JFET is operating in the ohmic region, where it can be modeled as an electronic switch. Consider Fig. 6.22, where we are assuming the use of an n-channel device. When $V_{GS} = 0$ V, the JFET acts as a low resistance from drain to source. If $r_{DS(on)}$ is low compared to the external circuit resistances, then the JFET may be modeled as a closed switch.

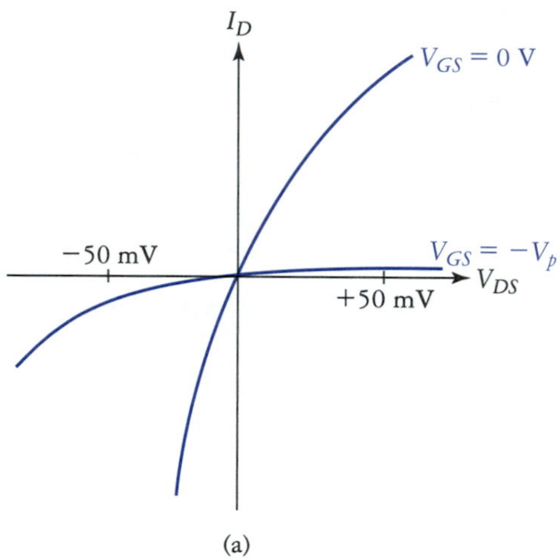

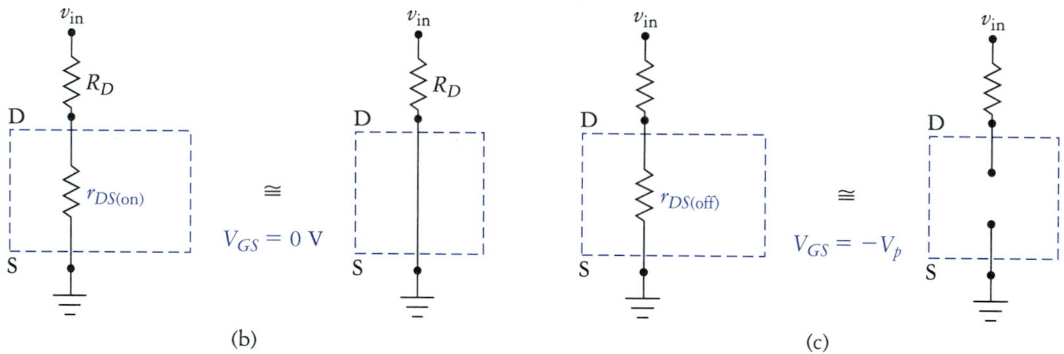

FIGURE 6.22
JFET ohmic-region characteristics. (a) Drain curves. (b) For $V_{GS} = 0$ V the JFET acts as a low resistance. (c) For $V_{GS} = -V_p$ the JFET acts as a high resistance.

Application of $V_{GS} = -V_p$ causes the JFET to turn off and behave like a very high-value resistor. If $r_{DS(off)}$ is large relative to circuit resistances, then the JFET may be modeled as an open switch. We stress, however, that such operation requires V_{DS} to be held to values of much less than V_p. This ensures operation in the ohmic region.

In Fig. 6.23(a), a JFET is configured as a series switch. When $V_{GS} = 0$ V, the FET allows signal current to pass relatively easily to the load. This configuration is best used when the load has low internal resistance and low or no voltage present at its input terminal.

Figure 6.23(b) shows a JFET used as a shunt switch. When $V_{GS} = 0$ V, the FET shunts the signal to ground. When $V_{GS} < V_p$, the FET is an open circuit, and the signal is passed unattenuated to the output.

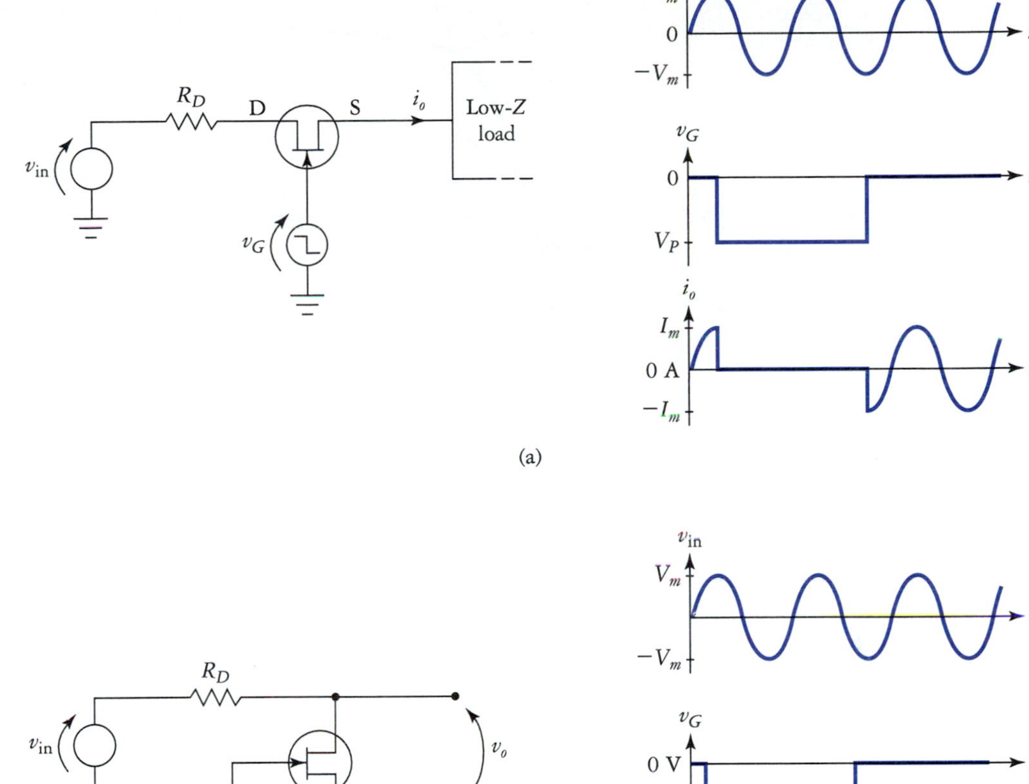

FIGURE 6.23
JFET choppers: (a) Series and (b) shunt configurations.

More sophisticated analog switch circuits are available in integrated circuit form. Many of these devices will operate at higher voltage and current levels than the simple circuits shown here.

Voltage-Controlled Resistors

If we zoom in on the origin of the JFET drain curves, we find that the drain-to-source resistance varies as a function of V_{GS}, as long as operation is maintained in the ohmic region. This is shown in Fig. 6.24. As was the case with the simple analog switch, V_{DS} is severely restricted in amplitude for correct operation.

A JFET voltage-controlled resistor application is shown in Fig. 6.25. In this circuit, which is similar to the shunt switch, the JFET is used with a normal resistor to form a voltage divider. The waveforms in Fig. 6.25(b) illustrate the operation of the circuit. The output of the circuit is given by

$$v_o = v_{in}\frac{r_{DS}}{r_{DS} + R_1} \tag{6.13}$$

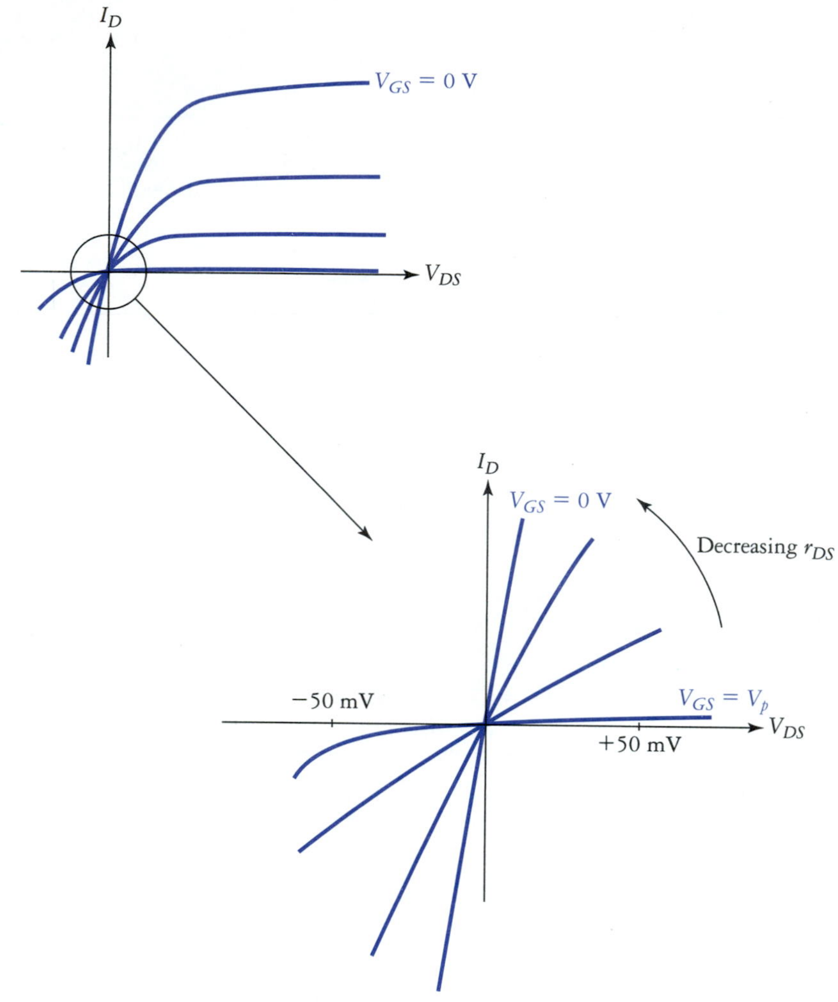

FIGURE 6.24
For small-signal conditions, the JFET can be used as a voltage-controlled resistance.

Constant-Current Diode

If we short the gate and source terminals, the JFET forms a two-terminal device called a *constant-current diode.* Such a device is shown in Fig. 6.26. This device is the complement of the zener diode. The operation of the constant-current diode is simple: As long as V_{AK} (or V_{DS}) is larger than V_p, then the FET will operate in pinchoff, and I_D will be held relatively constant at

$$I_D = I_{DSS} \tag{6.14}$$

The constant-current diode is typically used to establish bias current levels in amplifier circuits.

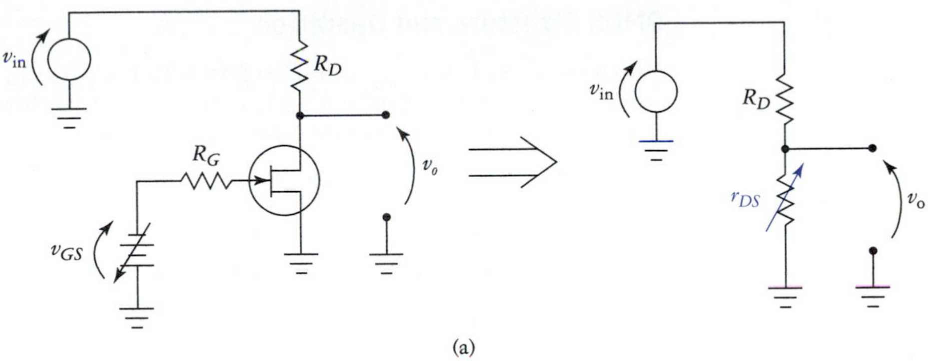

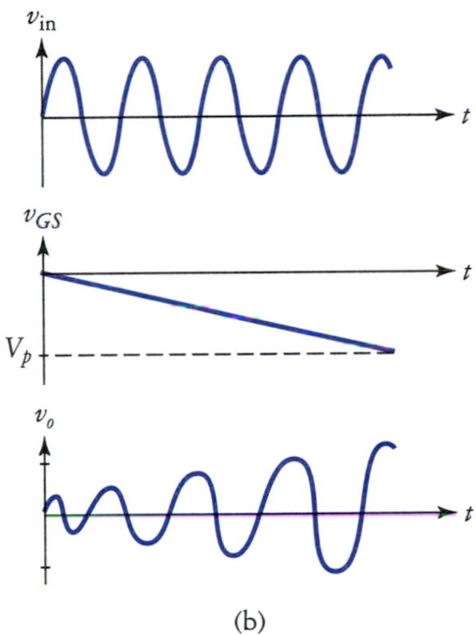

FIGURE 6.25
Voltage-controlled resistance attenuator. (a) Actual circuit and behavioral equivalent. (b) Representative circuit waveforms.

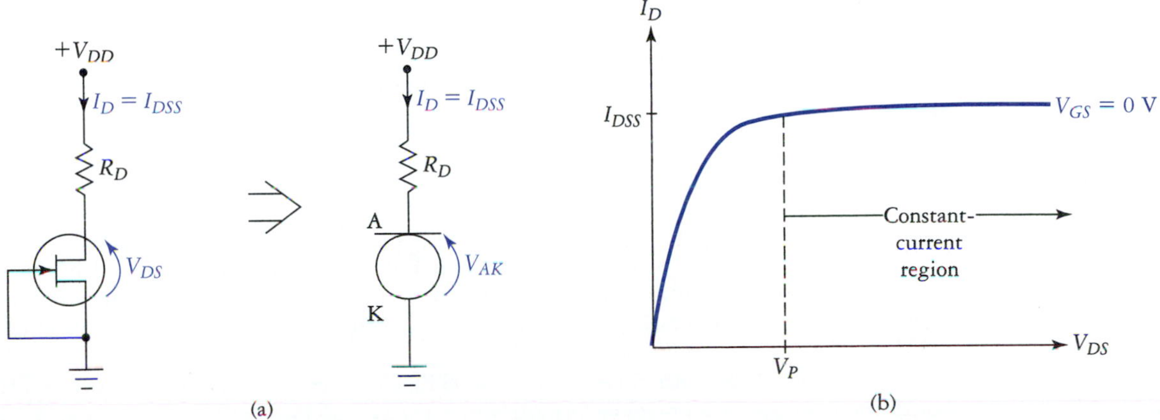

FIGURE 6.26
Constant-current diode. (a) JFET implementation. (b) Schematic symbol. (c) JFET drain curve.

6.3 THE DEPLETION-MODE MOSFET

A second class of field-effect transistors is called *metal-oxide semiconductor* FETs, or *MOSFETs*. There are several varieties of MOSFETs as well. We begin by examining the depletion-mode MOSFET (DMOS).

DMOS Structure and Operation*

The cross-sectional view of an n-channel DMOS FET is shown in Fig. 6.27. The important feature to notice here is that the gate is electrically isolated from the n channel by a thin insulating layer of silicon dioxide. The terminals of the DMOS FET are identical in name and function to those of the JFET.

Note
MOSFETs are very sensitive to static electricity. Handle them with great care.

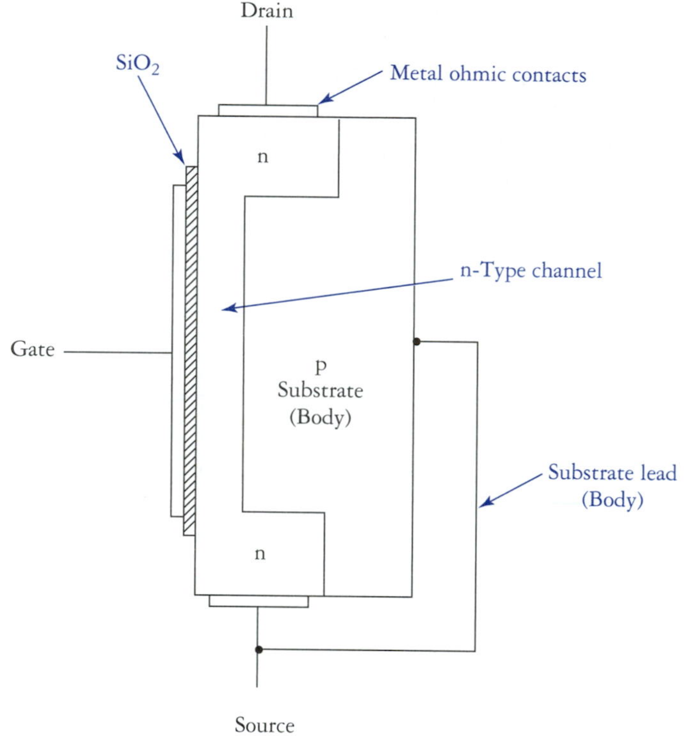

FIGURE 6.27
Cross section of a depletion-mode n-channel MOSFET.

An n-channel DMOS FET is shown with a V_{DS} source and a negative V_{GS} source in Fig. 6.28(a). If $V_{GS} = 0$ V, then I_D increases directly with V_{DS} until the n channel saturates. In saturation, the drain current levels off to a nearly constant value. In this case, because $V_{GS} = 0$ V, $I_{D(\text{sat})} = I_{DSS}$. Saturation current depends mainly on doping density and channel cross-sectional area. Making V_{GS} more negative causes an accumulation of electrons on the gate surface. Now, because the insulating SiO_2 is so thin, a very strong electric field is developed in the n channel. This field forces carriers (electrons) out of the channel, resulting in lower saturation current. A sufficiently negative V_{GS} will cause complete carrier depletion and $I_D = 0$, regardless of V_{DS}; thus, the name depletion-mode MOSFET.

If the gate terminal is driven positively, as in Fig. 6.28(b), then there is a tendency for electrons to be pulled from the substrate into the n channel. This effectively increases the maximum available carrier density in the channel, allowing $I_{D(\text{sat})} > I_{DSS}$. Thus, a positive V_{GS} enhances channel conductivity.

The transconductance curve for the DMOS FET is described by Eq. (6.1), just like a JFET. Because there is no pn junction to be forward biased, the DMOS FET can be operated with V_{GS} of either polarity, whereas the JFET cannot. The transconductance curve of the n-channel DMOS FET is shown in Fig. 6.29(a). When V_{GS} is positive, the FET is said to be operating in the enhancement mode. Negative values of V_{GS} result in operation in the depletion mode. Because of these characteristics, some books refer to the DMOS FET as a D-E MOSFET (depletion-enhancement). MOSFETs are also sometimes called IGFETS (insulated-gate FETs).

If the substrate (or body) terminal of a MOSFET is not connected internally, then the drain and source terminals can be interchanged. However, the terminal to which the substrate (or body) is con-

*Some manufacturers use the DMOS designation to stand for double-diffused MOSFET, where double diffusion is a device fabrication step. Most such MOSFETS are enhancement-mode devices.

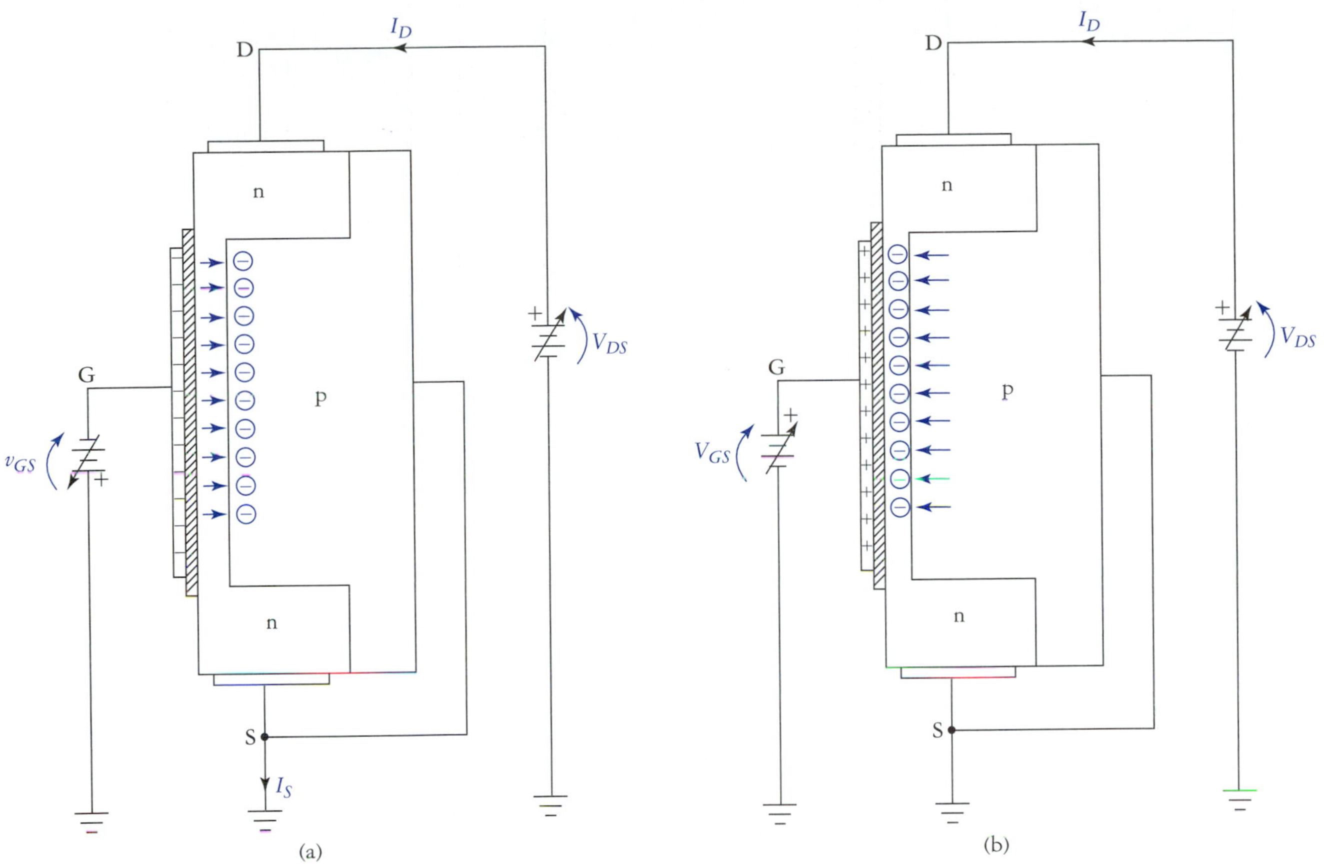

FIGURE 6.28
DMOS operation. (a) Application of negative gate voltage depletes n channel of carriers (electrons). (b) Application of positive gate voltage pulls more carriers into n channel, enhancing conductivity.

nected must always be used as the source. For n-channel devices, the substrate is sometimes connected to the negative supply rail, whereas the substrate terminal of a p-channel device will sometimes be connected to the positive supply rail. In such cases, the source and drain terminals may again be interchanged.

The drain curves for the DMOS FET are shown in Fig. 6.29(b). Notice that drain curves can be produced with $I_D > I_{DSS}$. The value of the pinchoff voltage increases with V_{GS}, as was the case with the JFET.

The schematic symbols used to represent DMOS transistors are shown in Fig. 6.30. Symbols (c) and (f) are used most often in integrated circuit schematic diagrams. Notice that for the integrated circuit (IC) symbols, the arrowhead associated with the source points in the direction of conventional current flow.

Zero Biasing

DMOS transistors are unique in that an especially simple biasing technique can be used with them. Most often, this technique is called *zero biasing.* A zero-biased DMOS FET circuit is shown in Fig. 6.31. As the name implies, in zero biasing, no bias voltage is applied to the FET; that is,

$$V_{GSQ} = 0 \text{ V} \tag{6.15}$$

This can be shown by writing the KVL equation for the G-S loop, and assuming that $I_G = 0$. This is a reasonable assumption because, typically, $I_{GSS} < 0.1$ nA. The gate resistor R_G is necessary in a practical circuit so that a signal may be coupled to the gate terminal. Now, because the gate current is zero, we have

$$I_{DQ} = I_{DSS} \tag{6.16}$$

Thus, the DMOS FET is very easy to bias.

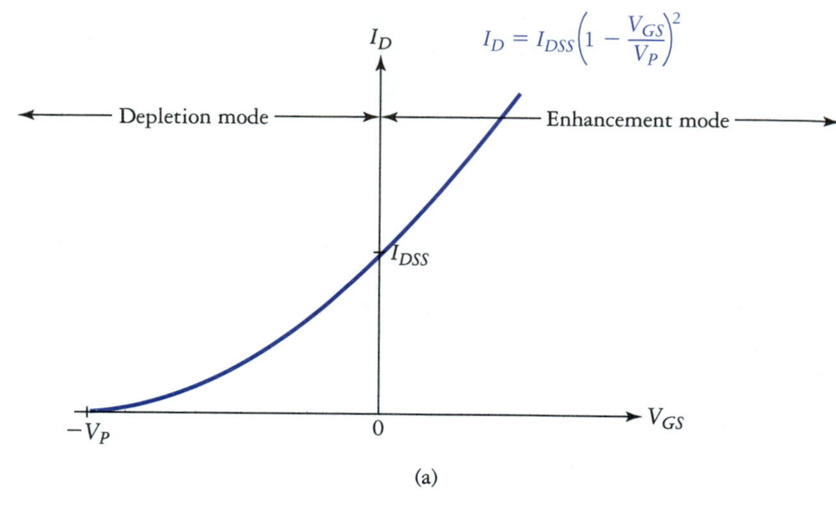

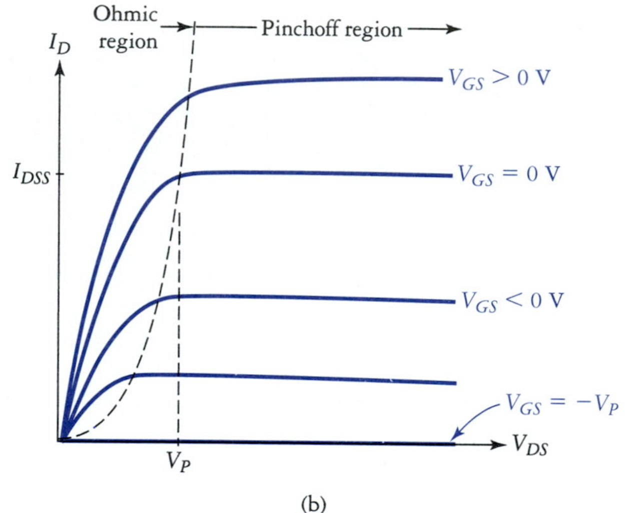

FIGURE 6.29
(a) N-channel DMOS transconductance curve. (b) Family of drain curves.

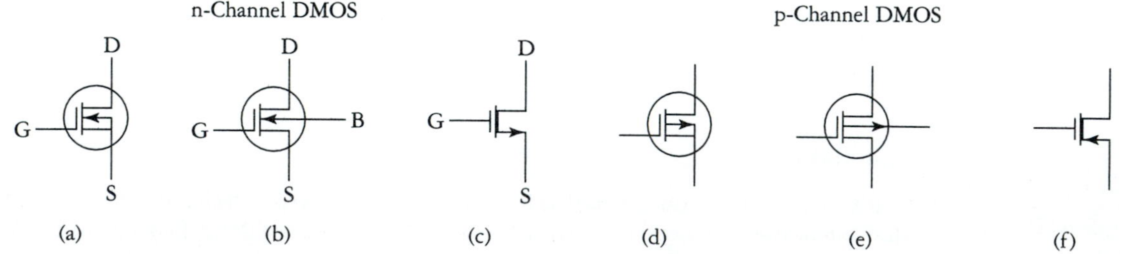

FIGURE 6.30
Various symbols used for DMOS FETs.

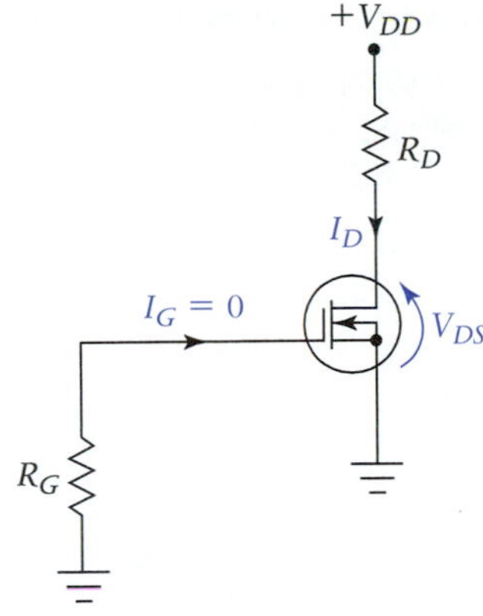

FIGURE 6.31
Zero biasing a DMOS FET.

Biasing Variations

If a source resistor is used, as in Fig. 6.32(a), then the same analysis procedure as that used with JFETs is used (the DMOS device is used strictly in the depletion mode).

The circuit of Fig. 6.32(b) could be used to move the Q point into the first quadrant, if so desired. Because there is no source feedback, however, this arrangement would not be very predictable. Voltage-divider bias combined with source feedback is used in Fig. 6.32. This circuit is analyzed the same way as an equivalent JFET circuit would be, the difference between the two being that there are no restrictions on V_{GS} for the DMOS FET. (The JFET circuit requires $I_{DQ} < I_{DSS}$.)

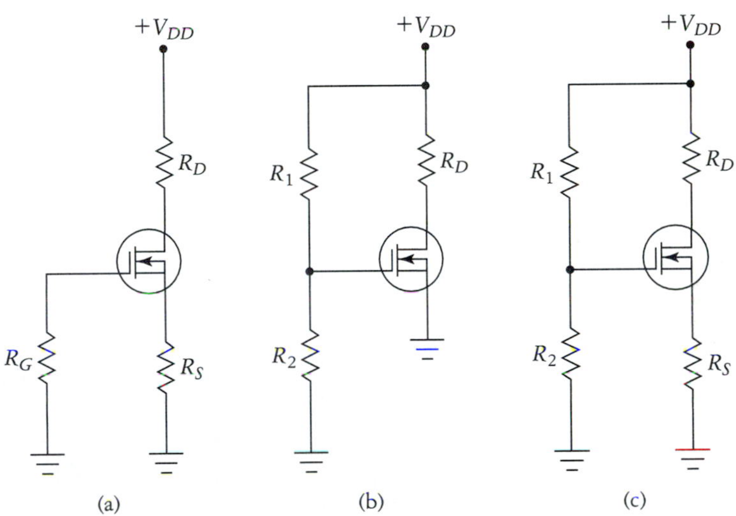

FIGURE 6.32
Other possible DMOS biasing arrangements. (a) Source-feedback bias. (b) Voltage-divider bias. (c) Voltage divider with source feedback.

EXAMPLE 6.8

Refer to Fig. 6.32(c). The DMOS FET has $I_{DSS} = 15$ mA, $-V_p = -4$ V, $R_1 = 10$ MΩ, $R_2 = 1$ MΩ, $R_D = 470$ Ω, $R_S = 100$ Ω, and $V_{CC} = 20$ V. Determine I_{DQ}, V_{DSQ}, V_{GSQ}, V_{DQ}, and V_{GQ}.

Solution As usual, we choose two convenient drain current values and calculate the corresponding values of V_{GS}. The gate voltage is given by Eq. (6.10).

$$V_{GQ} = V_{DD}\frac{R_2}{R_1 + R_2}$$

$$= 20\text{ V}\frac{1}{1 + 10}$$

$$= 1.82\text{ V}$$

Starting with $I_D = 0$, we find V_{GS} to be

$$\begin{aligned} V_{GS} &= V_G - I_D R_S \\ &= 1.82\text{ V} - 0\text{ V} \\ &= 1.82\text{ V} \end{aligned}$$

Setting $I_D = I_{DSS}$, we have

$$\begin{aligned} V_{GS} &= 1.82\text{ V} - (15\text{ mA})(100\ \Omega) \\ &= 1.82\text{ V} - 1.5\text{ V} \\ &= 0.32\text{ V} \end{aligned}$$

The resulting $I_D R_S$ line intersects the transconductance curve at a point where $I_D > I_{DSS}$. Because of this, a transconductance curve with extended V_{GS} and I_D scales was used. This is shown in Fig. 6.33. Plotting the line results in

$$\begin{aligned} I_{DQ} &\cong 1.1 I_{DSS} \\ &= 16.5\text{ mA} \end{aligned}$$

$$\begin{aligned} V_{GSQ} &= V_G - I_{DQ} R_S \\ &= 0.17\text{ V} \end{aligned}$$

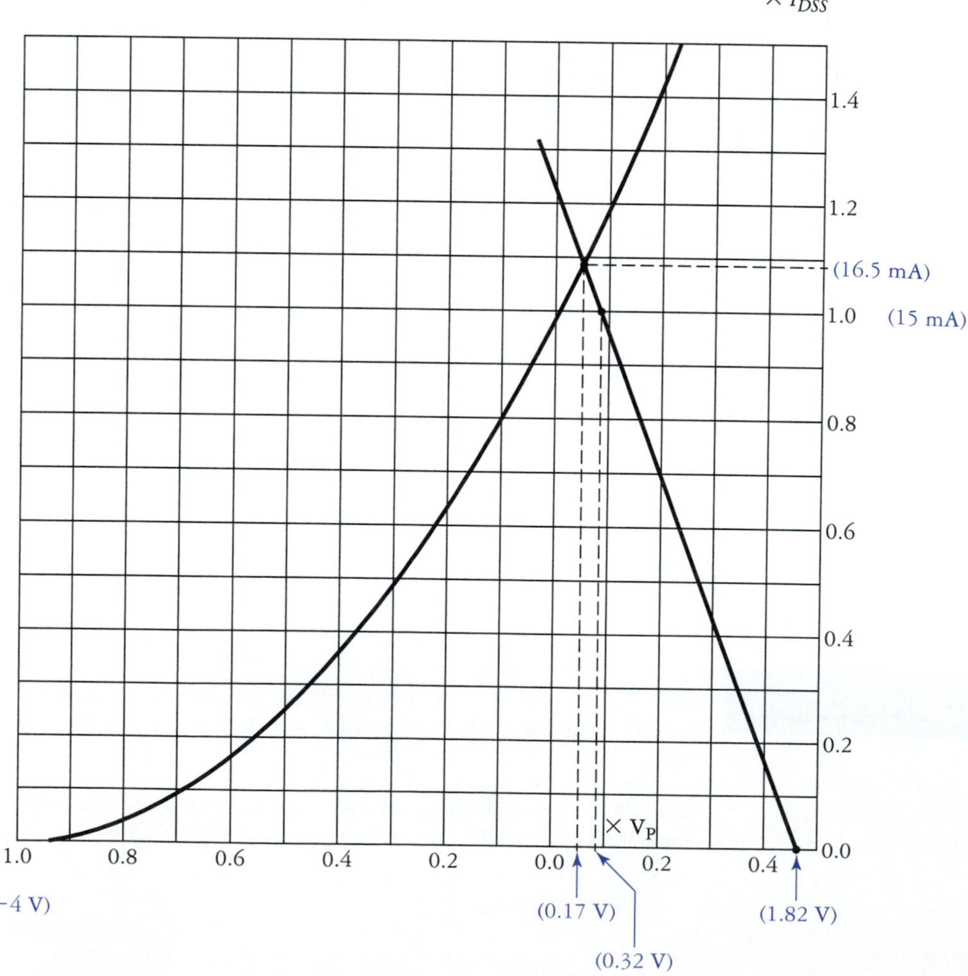

FIGURE 6.33
Transconductance curve for Example 6.8.

All that remains is to determine V_{DQ}:

$$\begin{aligned} V_{DQ} &= V_{DD} - I_{DQ} R_D \\ &= 20\text{ V} - (16.5\text{ mA})(470\ \Omega) \\ &= 20\text{ V} - 7.76\text{ V} \\ &= 12.24\text{ V} \end{aligned}$$

Dual-Gate MOSFETs

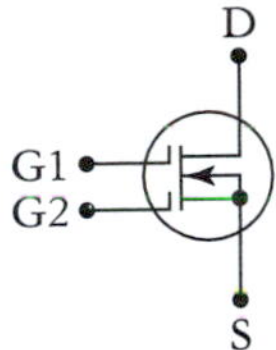

FIGURE 6.34
Dual-gate MOSFET.

DMOS FETs are available with dual-gate terminals as shown in Fig. 6.34. These devices are used most often in radio receiver circuits. A voltage-controlled amplifier can be designed using a dual-gate MOSFET. The input signal is applied to one gate, while a dc control voltage is applied to the other. Varying the dc voltage allows the device Q point to be moved on the transconductance curve. In the next chapter we show that this alters the voltage gain of the FET.

Dual-gate MOSFETs are also used to mix time-varying signals together in a nonlinear manner. Appropriately enough, such circuits are called *mixers*. The output of a mixer will contain frequency components that are not harmonically related to either input frequency. This property proves useful in radio transmission and reception applications.

MOSFET Handling Precautions

MOSFETs are extremely sensitive to damage from static electricity. The reason for this is that the insulating layer of silicon dioxide is typically on the order of a few millionths of a meter thick. The SiO_2 may be considered to be a dielectric between two conductors, the gate and the body of the FET. Even a relatively small potential difference across such a thin dielectric will produce an extremely strong electrostatic field.

Normal potentials of, say, 10 or so volts will not damage the gate dielectric. However, it is possible to develop a charge of several hundred volts just by walking across a tile floor, or by rubbing your arm on a tabletop. Normally, you would not notice the discharge of this potential, but the energy accumulated is more than sufficient to puncture the dielectric of a MOSFET and destroy the device. Thus, some care must be exercised when handling MOSFETs. Often a grounding strap is worn around the wrist to prevent static buildup. The typical resistance of a grounding strap will be several hundred thousand ohms. This is a high enough resistance to eliminate any shock hazard, while at the same time protecting the delicate MOSFETs.

6.4 THE ENHANCEMENT-MODE MOSFET

Enhancement-mode MOSFETs (EMOS) are available as discrete units, but are most widely used in digital integrated circuit designs. We concentrate primarily on linear applications in this section.

EMOS FET Structure and Operation

Structurally, the EMOS FET is similar to a cross between a bipolar transistor and a DMOS FET. If we were to take the DMOS structure of Fig. 6.27 and cut out the n channel and extend the p-type substrate into the space that is left, we would have an n-channel EMOS FET, as shown in Fig. 6.35.

Unlike the JFET and the DMOS FET, the EMOS FET will not conduct current from drain to source unless it is biased on, much like a BJT. Here's how EMOS FETs work: Application of a positive potential to the gate sets up a very strong electric field between itself and the p-type substrate. If the gate voltage is increased sufficiently, the substrate becomes ionized at the surface, creating a thin channel that contains free electrons. The channel is conductive, allowing current to flow between the drain and source. As you might expect, there is a limit to the current density in the channel that when reached causes drain current to level off and remain constant regardless of increases in V_{DS}. The higher V_{GS} is, the deeper the conductive channel, and the higher the saturation current will be. In Fig. 6.35, the conductive channel acts like n-type material, even though it is really p-type silicon. Thus, the conductive channel that is formed is called an *inversion layer*.

Like the other FETs covered so far, EMOS FETs have parabolic transconductance curves. The transconductance curve for an n-channel EMOS FET is shown in Fig. 6.36(a), while a set of typical drain curves is shown in 6.36(b). The transconductance equation for the EMOS FET is

$$I_D = k(V_{GS} - V_{to})^2 \qquad (6.17)$$

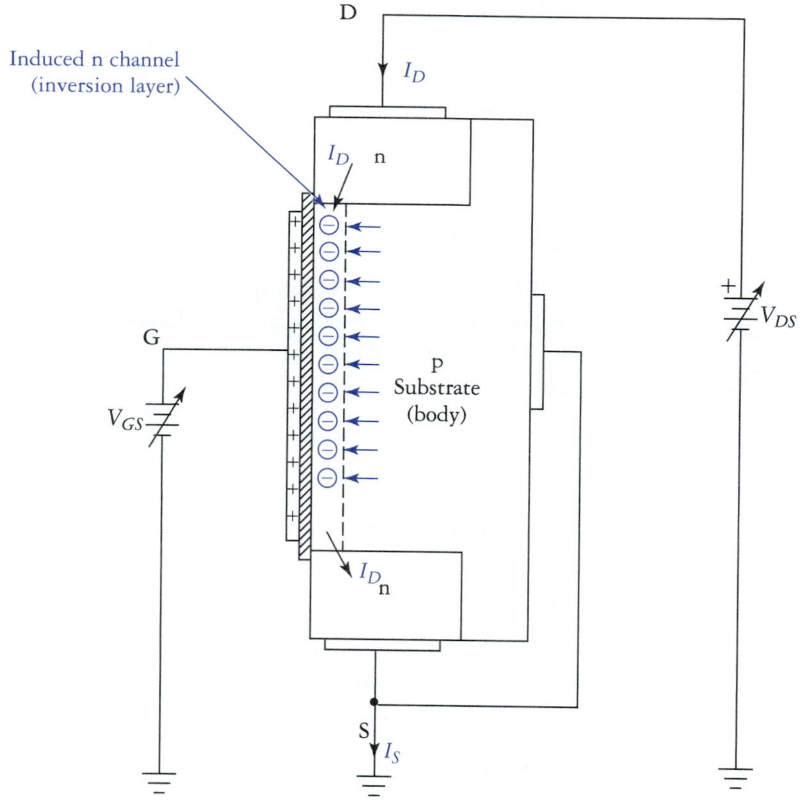

FIGURE 6.35
Structure of an enhancement-mode MOSFET (EMOS).

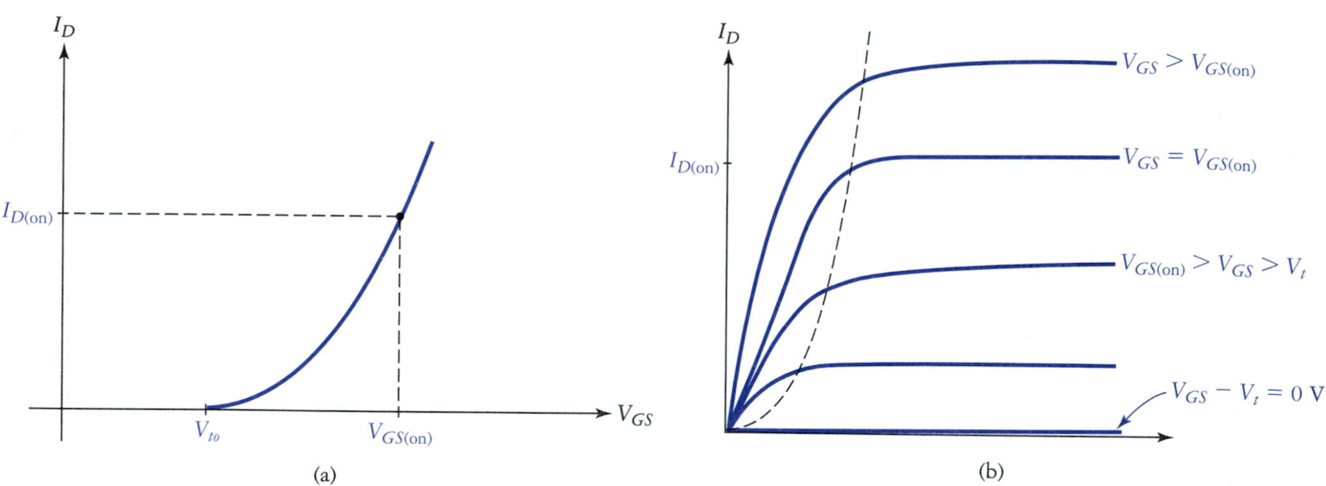

FIGURE 6.36
EMOS (a) transconductance curve and (b) family of drain curves.

where k is a constant that is related to the dimensions of the device structure and V_{to} is the threshold, or turn-on voltage. That is, V_{to} is the gate-to-source voltage that just causes formation of the inversion layer, producing a conductive path between the drain and source.

The schematic symbols used to represent EMOS FETs are shown in Fig. 6.37. Symbols (c) and (f) are used most often in integrated circuit schematic diagrams. In designs using discrete EMOS devices, the substrate or body lead is usually shorted to the source, as shown in (a) and (d). In IC designs, the substrates of the various n-channel MOSFETs are usually tied to the point of lowest potential (this will usually be either ground or V_{SS}). For p-channel MOSFET IC designs, the

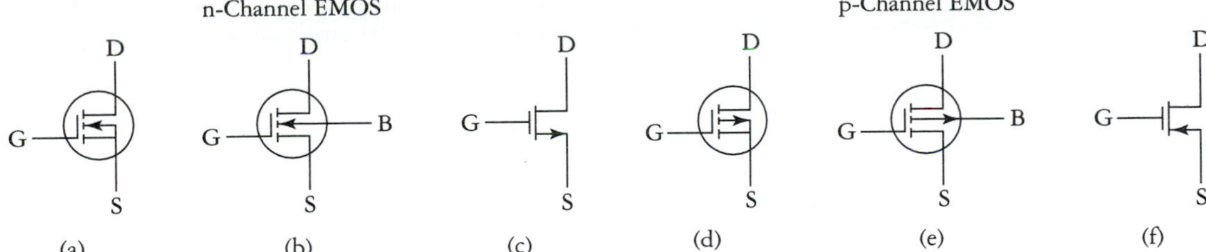

FIGURE 6.37
Various EMOS schematic symbols.

substrate is normally connected to the most positive node, which is often $+V_{DD}$. This is done to prevent the pn junctions between the substrate and the drain and source regions from becoming forward biased.

Enhancement-mode MOSFETs are used extensively in the design of digital ICs. Such ICs that are fabricated using only n-channel devices are called *NMOS* digital ICs, while those fabricated using only p-channel devices are called *PMOS*. Digital ICs that combine both n- and p-channel MOSFETs in their designs are called *CMOS* (complementary MOS) ICs.

EMOS FET Biasing

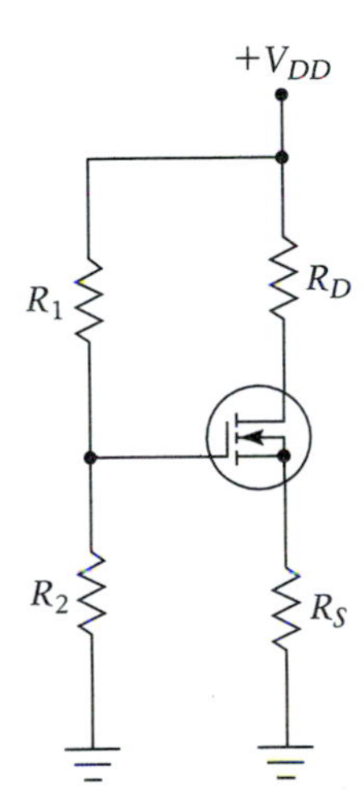

FIGURE 6.38
Voltage-divider biasing with source feedback.

In linear amplifier applications, EMOS FETs are usually either voltage-divider biased or drain-feedback biased. The voltage-divider biasing arrangement is shown in Fig. 6.38. The source resistor serves to stabilize the operating point of the circuit in exactly the same manner as the emitter resistor of an equivalent BJT circuit.

Because the gate terminal presents such a high resistance to the voltage divider, the gate voltage is given by direct application of the voltage-divider equation as follows:

$$V_G = V_{DD}\frac{R_1}{R_1 + R_2} \tag{6.18}$$

As usual, we can determine I_{DQ} either graphically, using the device transconductance curve, or algebraically, using the quadratic formula. The algebraic approach will be used exclusively in this section. We make use of Eq. (6.9) once more, which is repeated here for convenience:

$$I_{DQ} = \frac{-B - \sqrt{B^2 - 4AC}}{2A} \tag{6.9}$$

where

$$A = R_S^2$$

$$B = -\left[2R_S(V_G - V_{to}) + \frac{1}{k}\right]$$

$$C = (V_G - V_{to})^2$$

EXAMPLE 6.9

Refer to Fig. 6.38. For the FET, $V_{to} = 2.5$ V, $I_{D(\text{on})} = 10$ mA, $V_{DS(\text{on})} = 3.5$ V. Given $R_1 = 10$ MΩ, $R_2 = 3.9$ MΩ, $R_S = 150$ Ω, $R_D = 1.2$ kΩ, determine I_{DQ}, V_{GSQ}, V_{DSQ}, and V_{DQ}.

Solution First, we determine the gate voltage:

$$\begin{aligned} V_G &= V_{DD}\frac{R_2}{R_1 + R_2} \\ &= 15\text{ V}\frac{3.9}{3.9 + 10} \\ &= 4.2\text{ V} \end{aligned}$$

Substituting the given circuit parameters into the definitions for A, B, and C yields $A = 22.5$ k, $B = -735$, $C = 2.89$. Substituting these values into the quadratic formula gives us

$$I_{DQ} = 4.57 \text{ mA}$$

The requested circuit voltages are now found:

$$\begin{aligned} V_{GSQ} &= V_G - I_{DQ}R_S \\ &= 4.2 \text{ V} - (4.57 \text{ mA})(150\ \Omega) \\ &= 3.5 \text{ V} \end{aligned}$$

$$\begin{aligned} V_{DSQ} &= V_{DD} - I_D(R_D + R_S) \\ &= 15 \text{ V} - (4.57 \text{ mA})(1.35 \text{ k}\Omega) \\ &= 8.8 \text{ V} \end{aligned}$$

$$\begin{aligned} V_{DQ} &= V_{DD} - I_D R_D \\ &= 15 \text{ V} - (4.57 \text{ mA})(1.2 \text{ k}\Omega) \\ &= 9.5 \text{ V} \end{aligned}$$

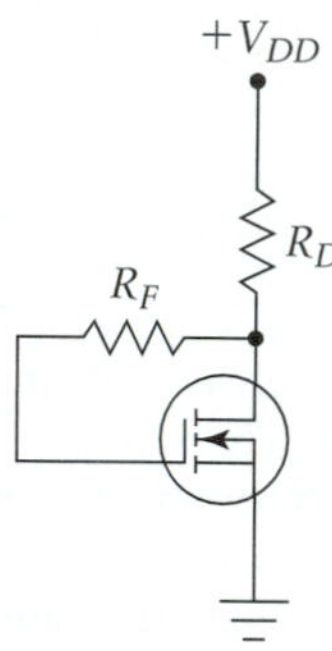

FIGURE 6.39
Drain-feedback biasing.

The basic drain-feedback biasing arrangement is shown in Fig. 6.39. The quiescent drain current is determined algebraically using Eq. (6.9), where

$$A = R_D^2$$

$$B = -\left[2R_D\left(V_{DD} - V_{to}\right) + \frac{1}{k}\right]$$

$$C = (V_{DD} - V_{to})^2$$

The value of R_F is usually not critical, but it will normally be in the range of from a few hundred thousand to several million ohms.

Vertical MOSFET

A very useful variation of the enhancement-mode MOSFET is the vertical MOSFET, or VMOS transistor. The basic principle of operation of the VMOS FET is the same as for the standard

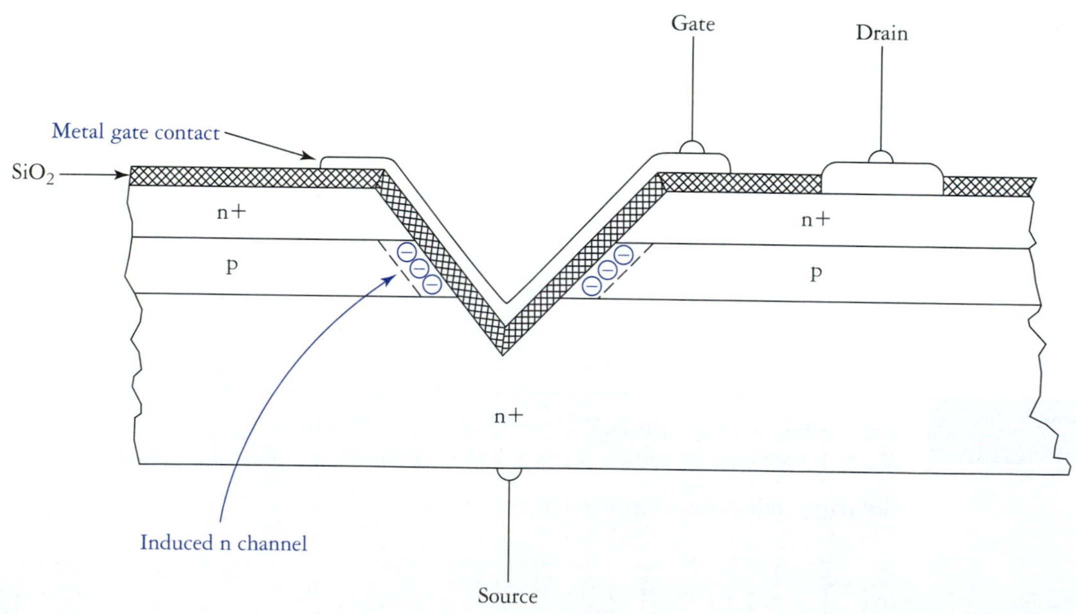

FIGURE 6.40
Cross section of an n-channel vertical MOSFET (VMOS).

EMOS device. The structure of the VMOS FET is much better suited to high-power applications, whereas traditional EMOS structures are typically limited to power dissipation levels in the milliwatt range.

A cross-sectional view of an n-channel VMOS FET is shown in Fig. 6.40. The central p-type region can be made relatively thin, while also extending rather deep (looking into the page). This produces a device that has a large-area inversion layer. This characteristic allows very low resistance ($r_{DS(\text{on})}$) and hence high-current operation is possible.

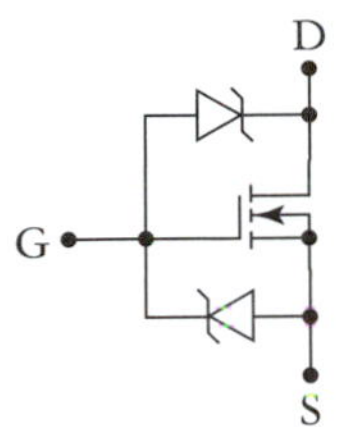

FIGURE 6.41
Zener diodes are often used to provide electrostatic discharge (ESD) protection.

The schematic symbol for the VMOS FET is the same as for a normal EMOS FET. Often, however, such devices are designed with built-in zener diodes as shown in Fig. 6.41. The zener diodes are used to prevent damage caused by static electricity and have no effect on dc operation of the FET.

The fact that VMOS FET has a positive temperature coefficient is a major advantage of the VMOS FET over the bipolar power transistor. Recall that bipolar transistors have negative temperature coefficients, and so are subject to thermal runaway. This is not a problem with field-effect transistors. VMOS devices are also capable of extremely fast switching speeds. A bipolar transistor will switch off relatively slowly once it has been driven into saturation because of stored charge in the form of minority carriers in the base region. FETs do not exhibit this behavior because they rely on majority carriers only.

VMOS FETs can be readily connected in parallel in order to increase overall current-carrying capacity. Recall that in order to do the same with BJTs, emitter swamping resistors must be used to prevent current hogging. Such resistors waste energy and limit overall efficiency. Currently, VMOS FETs have surpassed BJTs in many high-power applications, such as electric motor control and switching power supply designs.

6.5 COMPUTER-AIDED ANALYSIS APPLICATIONS

In addition to representative PSpice applications, this section also presents two BASIC programs that can be used to generate normalized parabolic transconductance curves. These curves are useful for graphical analysis of n-channel JFET and DMOS biasing circuits.

Basic Listings

The program listing in Fig. 6.42 was used to produce the transconductance plot of Fig. 6.12. This program was originally written on an MS-DOS PC clone, using Turbo Basic. If you prefer, you can rewrite the program using another language such as C or C++.

The listing in Fig. 6.43 is a modification of that shown in Fig. 6.42 that extends the graph into the first quadrant. This program was used to create the graph shown in Fig. 6.33.

If you are feeling ambitious, you might want to modify one of the programs such that p-channel transconductance curves are produced. Another possibility would be to modify the program listing in Fig. 6.42 such that the user is prompted to enter the value of R_S and the associated bias line is plotted.

PSpice Applications

The evaluation version of PSpice has the capability of simulating circuits containing JFETs and EMOS FETs. Because of the graphical user interface (GUI) provided by schematic capture, the simulation of such circuits is basically the same as when we used bipolar transistors.

JFET Simulation

PSpice does not allow easy editing of device parameters unless breakout devices are used. Breakout devices are basically user-defined devices, and they are covered later. It is interesting to note, however, that the PSpice JFET model has 13 parameters. Two of these parameters, VTO and BETA, are of primary importance. The parameter VTO is equivalent to $V_{GS(\text{off})}$ for the JFET, and of course it also defines the pinchoff voltage $V_p = -V_{GS(\text{off})}$. VTO is negative for n-channel JFETs

```
rem JFET Normalized Transconductance Curve

screen 2
window (-1.1,-.1)-(.6,1.25)
     for VGS = 0 to -.99 step -0.01
          VGS1 = VGS: VGS2 = VGS -.01
          ID1 = (1 + VGS1)^2
          ID2 = (1 + VGS2)^2
          line (VGS1,ID1)-(VGS2,ID2)
     next VGS

     for x = -1 to .5 step .09999
          line (x,0)-(x,1)
     next x

line (.002,0)-(.002,1)

     for y = 0 to 1 step .0999
          line (-1,y)-(.5,y)
     next y

locate 25,4:  print "1.0";
locate 25,14: print "0.8";
locate 25,23: print "0.6";
locate 25,33: print "0.4";
locate 25,42: print "0.2";
locate 25,51: print "0.0";
locate 25,56: print "x Vp";

locate 4,70:  print "x Idss";
locate 5,77:  print "1.0";
locate 9,77:  print "0.8";
locate 13,77: print "0.6";
locate 16,77: print "0.4";
locate 20,77: print "0.2";
locate 24,77: print "0.0";
```

FIGURE 6.42
QBASIC program for generating JFET transconductance curve.

and positive for p-channel JFETs. The parameter BETA is called the *transconductance coefficient,* and is defined as

$$\text{BETA} = \frac{I_{DSS}}{V_p^2} \tag{6.19}$$

PSpice derives the effective value for I_{DSS} from this device constant.

The evaluation library of PSpice has two JFETs: the 2N3819 and the 2N4393. As is true of all JFETs, these are low-power devices. The 2N3819 has $I_{DSS} = 12$ mA and $V_p = 3$ V, while the 2N4393 has $I_{DSS} = 18$ mA and $V_p = 1.4$ V. These devices are fairly typical of most JFETs that are commonly available.

JFET Transconductance Curve

The circuit in Fig. 6.44 is used to generate the transconductance curve for the 2N3819 JFET. To do this, we set up a DC Sweep of the VGG source from 0 to 3 V in 0.1-V increments. Notice that the negative terminal of VGG is connected to the gate of the JFET. This causes the sweep of V_{GS} to be

```
rem  DMOS Normalized Transconductance Curve

screen 2
window (-1.1,-.1)-(.6,1.5)
     for VGS = 1.5 to -.99 step -0.01
          VGS1 = VGS: VGS2 = VGS -.01
          ID1 = (1 + VGS1)^2
          ID2 = (1 + VGS2)^2
          line (VGS1,ID1)-(VGS2,ID2)
     next VGS

     for x = -1 to .5 step .09999
          line (x,0)-(x,1.5)
     next x

line (.002,0)-(.002,1.5)

     for y = 0 to 1.5 step .0999
          line (-1,y)-(.5,y)
     next y

locate 25,4:  print "1.0";
locate 25,14: print "0.8";
locate 25,23: print "0.6";
locate 25,33: print "0.4";
locate 25,42: print "0.2";
locate 25,51: print "0.0";
locate 25,60: print "0.2";
locate 25,69: print "0.4";
locate 23,56: print "x Vp";

locate 4,70:  print "x Idss";
locate 2,77:  print "1.4";
locate 5,77:  print "1.2";
locate 8,77:  print "1.0";
locate 12,77:  print "0.8";
locate 15,77: print "0.6";
locate 18,77: print "0.4";
locate 21,77: print "0.2";
locate 24,77: print "0.0";
```

FIGURE 6.43
QBASIC program for generating DMOS transconductance curve.

negative, as required. The VDD source is set to 10 V, which is simply a convenient value that is greater than V_p.

Running the simulation produces the plot given in Fig. 6.45. The x-axis variable is changed from the default to the gate voltage V(J1:g) of the FET. The trace that was plotted is the drain current ID(J1).

JFET Drain Curves

We can easily modify the analysis setup for the previous circuit (Fig. 6.44) such that a family of drain characteristic curves is produced. To do this, we again use the nested sweep option. Voltage source Vdd is the main sweep variable (0 to 10 V in 0.1-V increments). The nested sweep variable is voltage

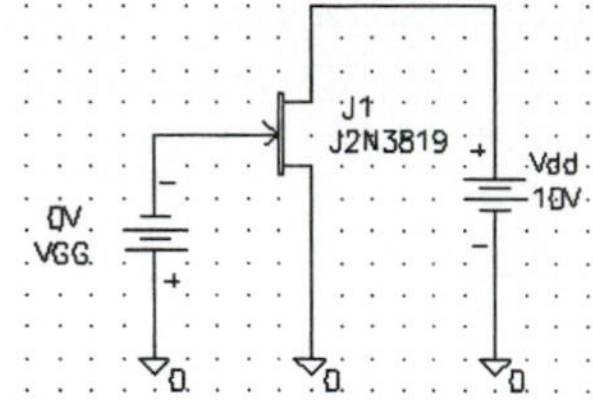

FIGURE 6.44
Circuit used to generate transconductance curve.

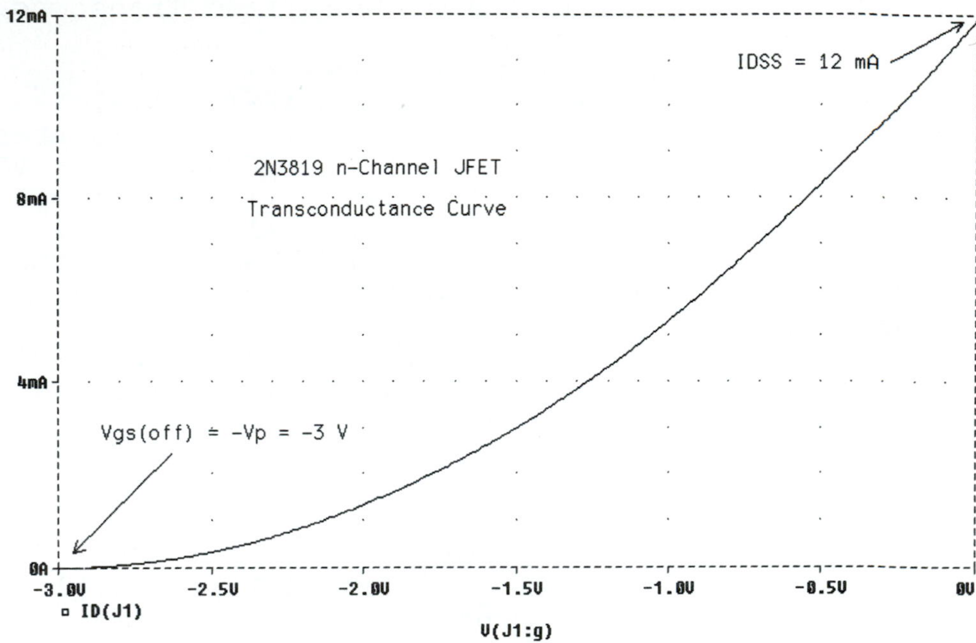

FIGURE 6.45
2N3819 transconductance curve.

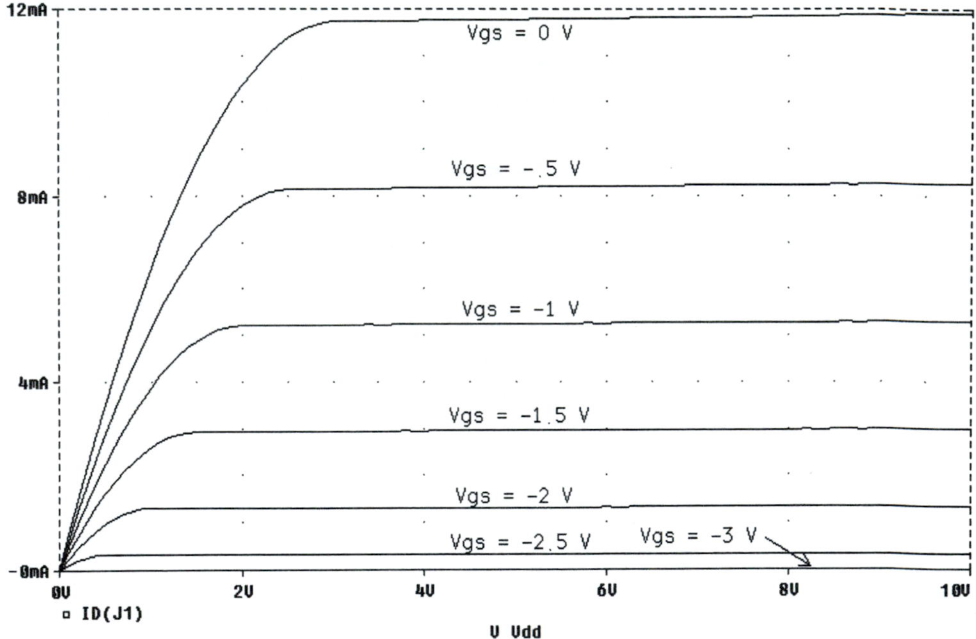

FIGURE 6.46
2N3819 drain curves.

source VGG, which is set for a linear sweep from 0 to 3 V in 0.5-V increments. The resulting family of curves is shown in Fig. 6.46.

JFET Q-Point Analysis

Looking back at Example 6.4 for a moment, we find the results of the graphical analysis were:

$$I_{DQ} = 6.9 \text{ mA}, \qquad V_{DSQ} = 7.6 \text{ V}, \qquad \text{and } V_{GSQ} = -0.67 \text{ V}$$

The hypothetical JFET of the example has $I_{DSS} = 10$ mA and $V_p = 4$ V. These values are close enough to those of the 2N3819 in the evaluation library that a simulation of the circuit using this device should yield similar results. Setting the circuit up and performing the simulation results in the display in Fig. 6.47, where we find

$$I_{DQ} = 6.98 \text{ mA}, \qquad V_{DSQ} = 7.32 \text{ V}, \qquad V_{GSQ} = 0.698 \text{ V}$$

The agreement between these analyses is good, especially considering the very simple model we used.

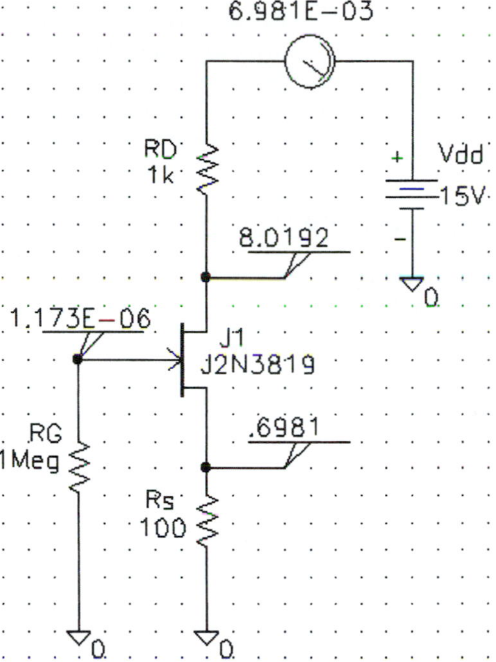

FIGURE 6.47
Self-biased JFET voltages and currents.

EMOS FETs

The PSpice enhancement-mode MOSFET model is quite extensive. Three levels of modeling are available for the EMOS FET. The simplest is level 1, which uses what is called the Shichman and Hodges model. Level 2 is a similar, but more complex theoretical model. Level 3 uses a combination of theoretical and empirical parameters to model the FET. That is, level 3 uses behavioral modeling equations and actual data from test circuits to predict the behavior of the FET under any possible operating condition.

The evaluation library of PSpice has two enhancement-mode MOSFETs: the IRF150, an n-channel device, and the IRF9140, which is a p-channel MOSFET. Both of these are high-power VMOS devices, rated for $I_{D(\text{max})} = 40$ A. It is left as an exercise for the reader to produce transconductance and drain characteristic curves for these devices.

Although many parameters are associated with the MOSFET models, the two main parameters are VTO and KP. VTO is the threshold or turn-on voltage V_{to} of the device. The parameter KP is given by

$$\text{KP} = 2k \tag{6.20}$$

where k is determined from the relationship

$$k = \frac{I_{D(\text{on})}}{(V_{GS(\text{on})} - V_{to})^2} \tag{6.21}$$

Let us assume that we have an EMOS with $V_{to} = 2.5$ V, $I_{D(\text{on})} = 5$ mA, and $V_{GS(\text{on})} = 3.5$ V. Using Eq. (6.21), we find $k = 0.005$, so we would use KP = 0.01 for this transistor.

CHAPTER REVIEW

Field-effect transistors are active devices that behave most like voltage-controlled current sources (VCISs). FETs all come in n-channel and p-channel varieties. JFETs and DMOS FETs are normally on devices, while EMOS FETs are normally off.

FETs have parabolic transconductance curves, and so analysis of these circuits must be performed graphically or algebraically, using the quadratic formula. Unlike bipolar transistors, there are few approximations that can be made in the analysis of biasing circuits using FETs. Thus, the pertinent FET parameters must be known to obtain accurate analysis results.

FETs are used to realize linear amplifiers and nonlinear functions as well. Analog switches, voltage-controlled resistors, and constant-current diodes can be formed using FETs.

JFETs are relatively immune to the effects of routine handling, while MOSFETs are very susceptible to damage caused by static electricity. EMOS FETs are used extensively in the fabrication of digital ICs. In general, the gate current of any FET will be quite small. Usually, it is acceptable to consider gate current to be zero. FETs are positive temperature coefficient devices, and therefore do not suffer from thermal runaway.

DESIGN AND ANALYSIS PROBLEMS

6.1. Refer to Fig. 6.48. Given: I_{DSS} = 14 mA, V_p = 3 V, R_S = 120 Ω, R_D = 860 Ω, and V_{DD} = 15 V. Determine I_{DQ}, V_{DSQ}, V_{GSQ}, and V_{DQ}. Use both graphical and algebraic techniques to find I_{DQ}.

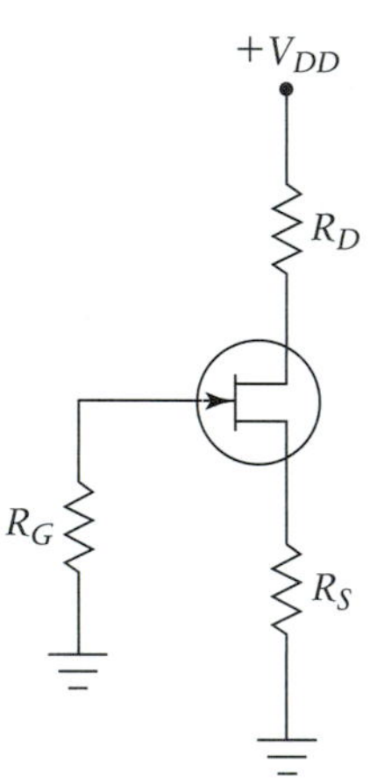

FIGURE 6.48
Circuit for Problems 6.1 to 6.6.

6.2. Refer to Fig. 6.48 Given: I_{DSS} = 8 mA, V_p = 2 V, R_S = 330 Ω, R_D = 1.2 kΩ, and V_{DD} = 15 V. Determine I_{DQ}, V_{DSQ}, V_{GSQ}, and V_{DQ}. Use graphical and algebraic analysis.

6.3. Refer to Fig. 6.48. Given I_{DSS} = 6 mA and V_p = 2 V, choose R_S and R_D such that $I_{DQ} = 0.75 I_{DSS}$ and $V_{DSQ} = V_{DD}/2$. Assume that V_{DD} = 15 V. Use the transconductance plot to determine R_S.

6.4. Refer to Fig. 6.48. Given: I_{DSS} = 20 mA, V_p = 4.2 V, and V_{DD} = 12 V. Determine R_S and R_D such that I_{DQ} = 5 mA and V_{DSQ} = 6 V. Use the transconductance plot to determine R_S.

6.5. Refer to Fig. 6.48. Assume that V_G = 5 mA and R_G = 1 MΩ. Determine the gate leakage current.

6.6. Refer to Fig. 6.48. Given V_G = 2 mV, and R_G = 1 MΩ, determine the gate leakage current.

6.7. Refer to Fig. 6.49. Given: I_{DSS} = 15 mA, V_p = 3.5 V, V_{DD} = 18 V, R_1 = 15 MΩ, R_2 = 2.2 MΩ, R_S = 220 MΩ, and R_D = 220 Ω. Determine I_{DQ}, V_{DSQ}, V_{GSQ}, V_{DQ}, and V_{SQ}. Use both graphical and algebraic analysis.

6.8. Refer to Fig. 6.49. Given: I_{DSS} = 2 mA, V_p = 1.8 V, V_{DD} = 12 V, R_1 = 10 MΩ, R_2 = 1 MΩ, R_S = 1 k Ω, and R_D = 1.5 kΩ. Determine I_{DQ}, V_{DSQ}, V_{GSQ}, V_{DQ} and V_{SQ}. Use both graphical and algebraic analysis.

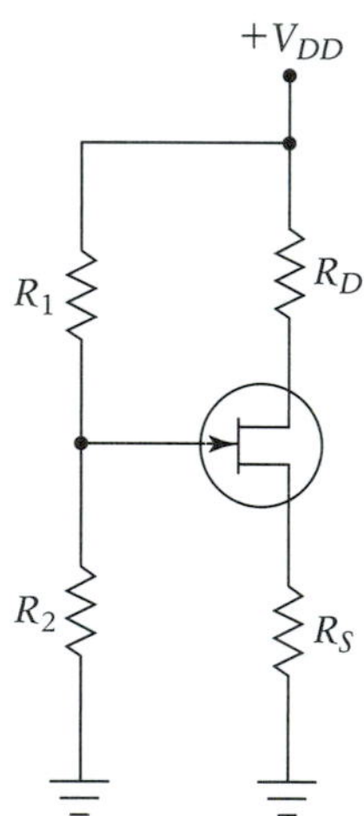

FIGURE 6.49
Circuit for Problems 6.7 to 6.10.

6.9. Refer to Fig. 6.49. Given: I_{DSS} = 4 mA, V_p = 2 V, V_{DD} = 12 V, R_1 = 12 MΩ, and R_2 = 1 MΩ. Determine R_S and R_D such that I_{DQ} = 2 mA and V_{DSQ} = 6 V.

6.10. Refer to Fig. 6.49. Given: I_{DSS} = 8 mA, V_p = 4 V, V_{DD} = 12 V, R_1 = 12 MΩ, and R_2 = 1 MΩ. Determine R_S and R_D such that I_{DQ} = 6 mA and V_{DSQ} = 6 V.

6.11. Refer to Fig. 6.50. Given: I_{DSS} = 5 mA, V_p = 4 V, R_G = 1 MΩ, R_D = 1.2 kΩ, R_S = 1.2 kΩ, V_{DD} = 10 V, and $-V_{SS}$ = −10 V. Determine I_{DQ}, V_{DSQ}, V_{GSQ}, V_{DQ}, and V_{SQ}.

6.12. Refer to Fig. 6.50. Given: I_{DSS} = 3 mA, V_p = 4 V, R_G = 1 MΩ, R_D = 1.5 kΩ, R_S = 560 Ω, V_{DD} = 10 V, and V_{SS} = −10 V. Determine I_{DQ}, V_{DSQ}, V_{GSQ}, V_{DQ}, and V_{SQ}.

6.13. Refer to Fig. 6.51. Given: I_{DSS} = 18 mA, V_p = 4.5 V, R_S = 120 Ω, R_D = 680 Ω, R_G = 10 MΩ, and V_{SS} = 16 V. Determine I_{DQ}, V_{DSQ}, V_{GSQ}, V_{DQ}, and V_{SQ}.

6.14. Refer to Fig. 6.51. Given: I_{DSS} = 18 mA, V_p = 4.5 V, R_S = 220 Ω, R_D = 680 Ω, R_G = 10 MΩ, and V_{SS} = 16 V. Determine I_{DQ}, V_{DSQ}, V_{GSQ}, V_{DQ}, and V_{SQ}.

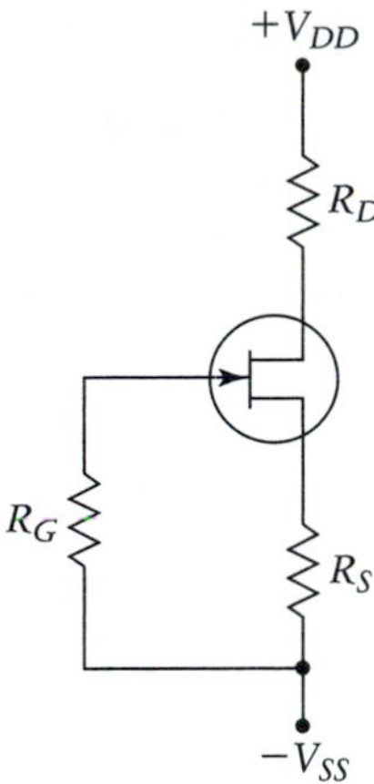

FIGURE 6.50
Circuit for Problems 6.11 and 6.12.

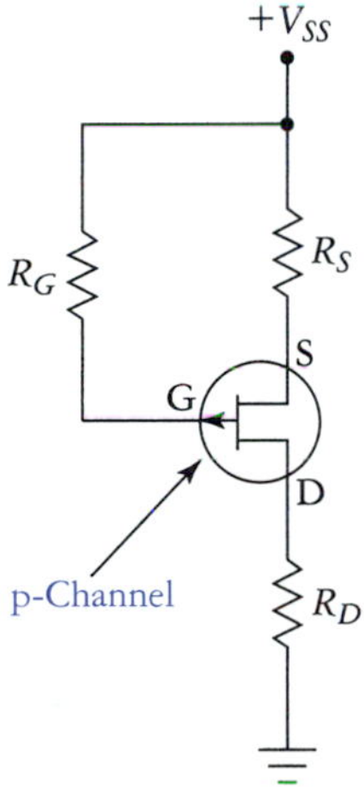

FIGURE 6.51
Circuit for Problems 6.13 and 6.14.

6.15. Refer to Fig. 6.52. Assume that the complementary FETs have $I_{DSS(1)} = I_{DSS(2)} = 4$ mA, and $V_{p(1)} = V_{p(2)} = 3$ V. Determine I_T.

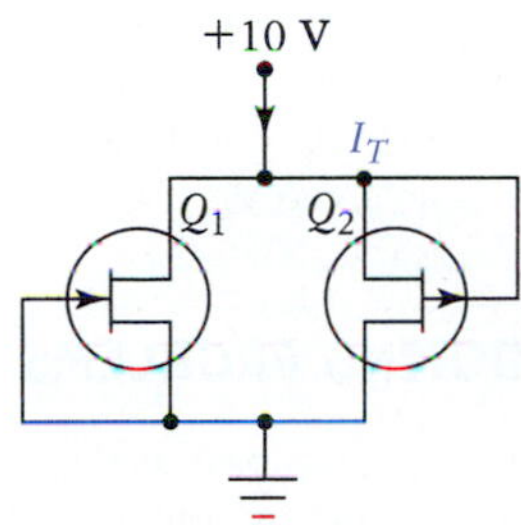

FIGURE 6.52
Circuit for Problem 6.15.

6.16. Refer to Fig. 6.53. Assume that all FETs are matched with $I_{DSS} = 6$ mA and $V_p = 3$ V. Determine I_1, V_{GS1}, V_{GS2}, and V_{GS3}.

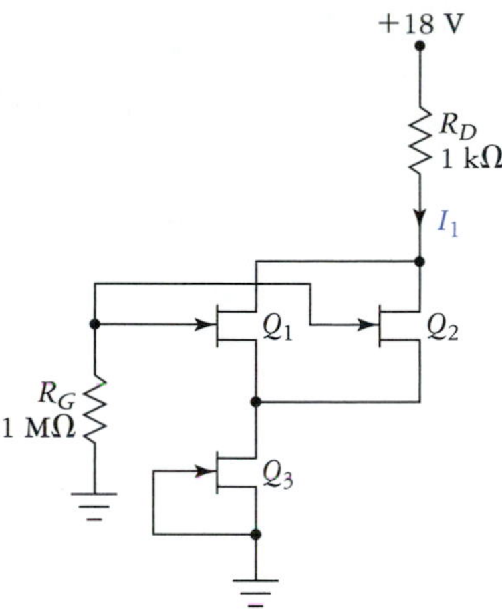

FIGURE 6.53
Circuit for Problem 6.16.

6.17. Refer to Fig. 6.54. Given: $I_{DSS} = 10$ mA, $V_p = 5$ V, $V_{DD} = 20$ V, $R_D = 1$ kΩ, and $R_G = 1$ MΩ, determine I_{DQ}, V_{DSQ}, and V_{GSQ}.

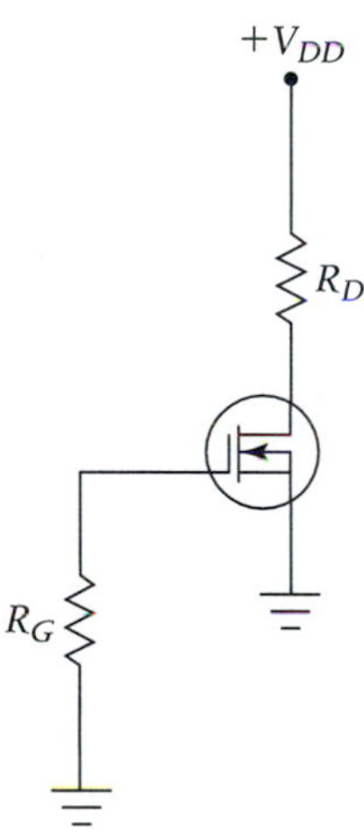

FIGURE 6.54
Circuit for Problems 6.17 to 6.20.

6.18. Refer to Fig. 6.54. Given: $I_{DSS} = 2$ mA, $V_p = 2$ V, $V_{DD} = 12$ V, $R_D = 3.3$ kΩ, and $R_G = 1$ MΩ, determine I_{DQ}, V_{DSQ}, and V_{GSQ}.

6.19. Refer to Fig. 6.54. Given: $I_{DSS} = 5$ mA, $V_p = 1.5$ V, $V_{DD} = 10$ V, choose R_D such that $V_{DSQ} = 5$ V.

6.20. Refer to Fig. 6.54. Given: $I_{DSS} = 12$ mA, $V_p = 3.2$ V, and $V_{DD} = 15$ V, choose R_D such that $V_{DSQ} = 6$ V.

6.21. Refer to Fig. 6.55. Given: $I_{DSS} = 16$ mA, $V_p = 4.5$ V, $R_D = 1$k Ω, $R_S = 270$ Ω, $V_{DD} = 15$ V, and $R_G = 10$ MΩ. Determine I_{DQ}, V_{DSQ}, V_{DQ}, and V_{SQ}.

6.22. Refer to Fig. 6.55. Given: $I_{DSS} = 6$ mA, $V_p = 2.2$ V, $R_D = 470$ Ω, $R_S = 180$ Ω, $V_{DD} = 10$ V, and $R_G = 10$ MΩ. Determine I_{DQ}, V_{DSQ}, V_{DQ}, and V_{SQ}.

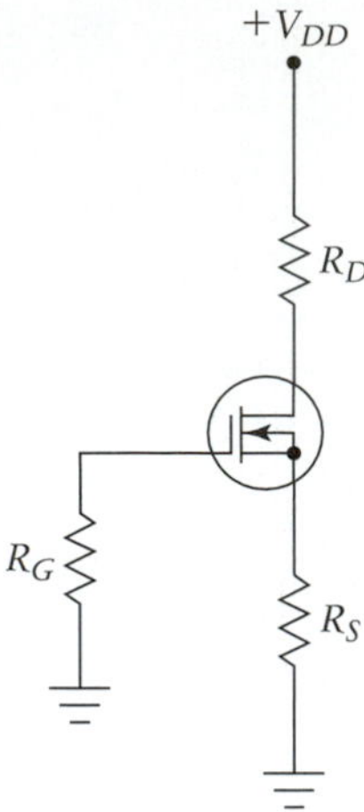

FIGURE 6.55
Circuit for Problems 6.21 and 6.22.

6.23. Refer to Fig. 6.56. Given: $I_{DSS} = 7$ mA, $V_p = 2.2$ V, $R_1 = 5.6$ MΩ, $R_2 = 470$ kΩ, $R_S = 220$ Ω, $R_D = 860$ Ω, and $V_{DD} = 15$ V. Determine I_{DQ}, V_{DSQ}, V_{GSQ}, V_{DQ}, and V_{SQ}.

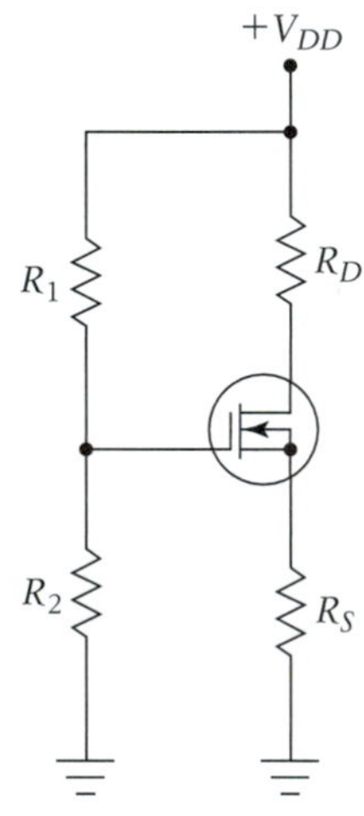

FIGURE 6.56
Circuit for Problems 6.23 and 6.24.

6.24. Refer to Fig. 6.56. Given: $I_{DSS} = 10$ mA, $V_p = 3.6$ V, $R_1 = 5.6$ MΩ, $R_2 = 560$ kΩ, $R_S = 180$ Ω, $R_D = 470$ Ω, and $V_{DD} = 15$ V. Determine I_{DQ}, V_{DSQ}, V_{GSQ}, V_{DQ}, and V_{SQ}.

6.25. Refer to Fig. 6.57. Given: $V_{to} = 2.5$ V, $I_{D(on)} = 10$ mA, $V_{GS(on)} = 4$ V, $R_1 = 10$ MΩ, $R_2 = 3.3$ MΩ, $R_S = 100$ Ω, and $R_D = 2.7$k Ω. Determine I_{DQ}, V_{DSQ}, V_{GSQ}, V_{DQ}, and V_{SQ}. Assume that $V_{DD} = 15$ V.

6.26. Refer to Fig. 6.57. Given: $V_{to} = 2.5$ V, $I_{D(on)} = 10$ mA, $V_{GS(on)} = 4$ V, $R_1 = 5$ MΩ, $R_2 = 2.2$ MΩ, $R_S = 100$ Ω, and $R_D = 680$ Ω. Determine I_{DQ}, V_{DSQ}, V_{GSQ}, V_{DQ}, and V_{SQ}. Assume that $V_{DD} = 15$ V.

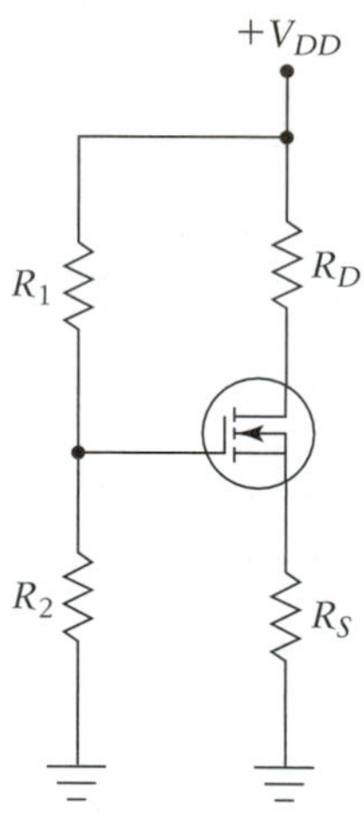

FIGURE 6.57
Circuit for Problems 6.25 and 6.26.

6.27. Refer to Fig. 6.58. Given: $I_{D(on)} = 6$ mA, $V_{GS(on)} = 4$ V, $V_{to} = 3.2$ V, $R_D = 500$ Ω, $R_F = 1$ MΩ, and $V_{DD} = 12$ V. Determine I_{DQ}, V_{DSQ}, and V_{GSQ}.

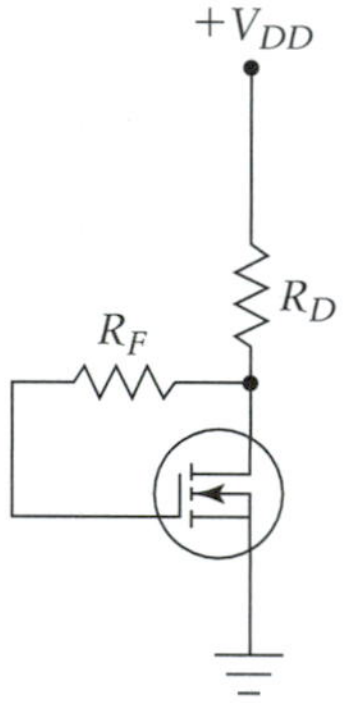

FIGURE 6.58
Circuit for Problems 6.27 and 6.28.

6.28. Refer to Fig. 6.58. Given: $I_{D(on)} = 8$ mA, $V_{GS(on)} = 5$ V, $V_{to} = 4$ V, $R_D = 1000$ Ω, $R_F = 1$ MΩ, and $V_{DD} = 12$ V. Determine I_{DQ}, V_{DSQ}, and V_{GSQ}.

6.29. Refer to Fig. 6.20. Given: $I_{DSS} = 10$ mA, $R_G = 1$ MΩ, $R_D = 1$ kΩ, $R_1 = 2.86$ kΩ, $V_P = 3$ V, and $V_{DD} = V_{SS} = 15$ V. Determine I_D, V_{DS}, V_{CE}, and V_D.

■ TROUBLESHOOTING PROBLEMS

6.30. An n-channel JFET is self-biased, as shown in Fig. 6.11. $R_G = 1$ MΩ, and the gate voltage measures +2 V. Is this likely to be acceptable? What are two causes for such a reading?

6.31. A certain circuit is designed for midpoint bias using a JFET with $I_{DSS} = 5$ mA and $V_p = 2$ V. What value of V_{GS} should be measured in the circuit?

6.32. A JFET with $I_{DSS} = 10$ mA and $V_p = 4$ V is used in the circuit described in Problem 6.31. Would this FET operate at a higher or lower I_{DQ}?

6.33. A DMOS FET is zero biased as shown in Fig. 6.54 with R_G = 10 MΩ, and V_{GS} is measured to be 1.8 mV. Determine the gate leakage current.

COMPUTER PROBLEMS

6.34. Use PSpice to generate a transconductance plot for the 2N4393 n-channel JFET.

6.35. Use PSpice to produce a family of drain curves for the 2N4393 JFET. Sweep V_{DS} from 0 to 10 V in 0.1-V increments. Nest the V_{GG} sweep such that seven curves are produced from $V_{GS} = 0$ to -1.4 V.

6.36. Use PSpice to generate a transconductance plot for the IRF150 n-channel EMOS. This device is in the evaluation library.

6.37. Use PSpice to produce a family of drain curves for the IRF150. Sweep V_{DS} from 0 to 10 V in 0.1-V increments. Nest the V_{GG} sweep such that five curves are produced from $V_{GS} = 0$ to 5 V.

6.38. Use PSpice to perform an operating-point analysis of the circuit in Fig. 6.59, using the IRF150.

6.39. Refer to Fig. 6.60. Determine the α-points for J1 and Q_1.

6.40. Use PSpice to perform an operating-point analysis on the circuit in Fig. 6.53. Use the 2N4393.

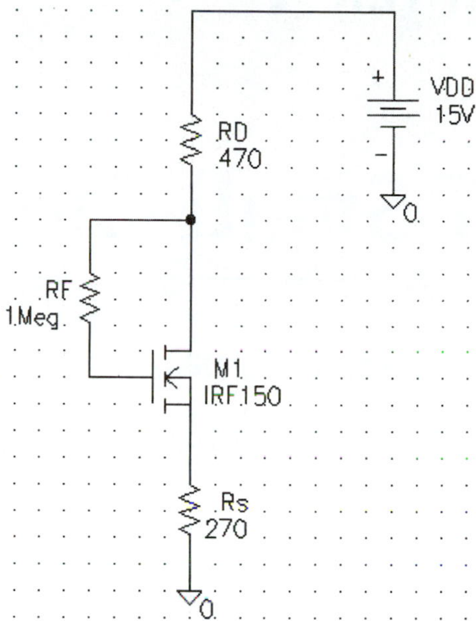

FIGURE 6.59
Circuit for Problem 6.38.

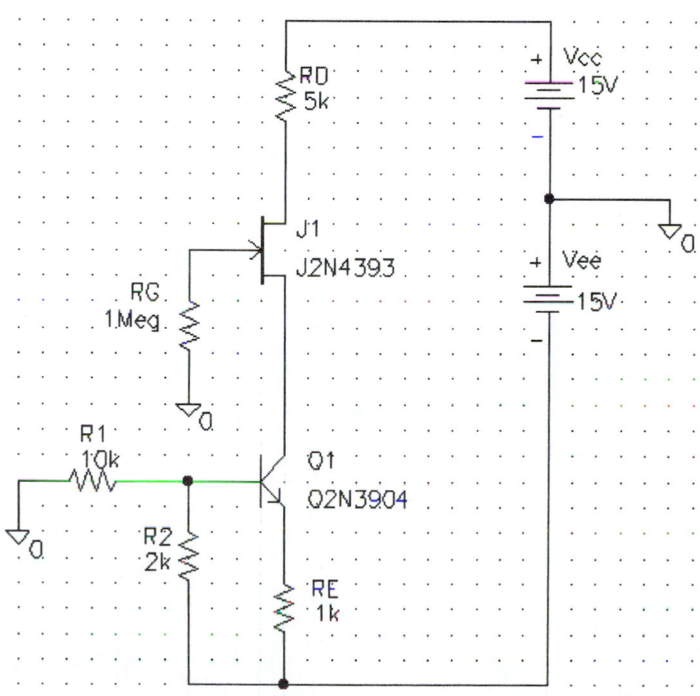

FIGURE 6.60
Circuit for Problem 6.39.

CHAPTER 7

FET Small-Signal Amplifiers

BEHAVIORAL OBJECTIVES

On completion of this chapter, the student should be able to:

- Describe the three basic FET amplifier configurations.
- Draw dc and ac equivalent circuits for FET amplifier circuits.
- Perform small-signal analysis of basic FET amplifier circuits.
- Sketch dc and ac load lines for FET amplifiers.
- Describe the effects of source bypassing.
- Explain the advantages and disadvantages of FET amplifiers relative to bipolar transistor amplifiers.

INTRODUCTION

As was the case with the bipolar transistor, once the field-effect transistor (FET) is biased up it is ready to serve as an amplifier. As we will see, FET amplifiers are capable of extremely high input resistance. This is a significant advantage over bipolar junction transistor (BJT)-based amplifiers in many applications. This chapter introduces the small-signal transconductance model used to represent the various types of FETs. The emphasis of this material is on the determination of voltage, current, and power gain and on input and output resistance.

7.1 BASIC AMPLIFIER CONFIGURATIONS

Three basic amplifier configurations are possible using a single FET: the common-source (CS), source-follower (SF) (or common-drain, CD), and the common-gate (CG) configurations. These are shown in Fig. 7.1, using n-channel JFETs for illustrative purposes. These three configurations are analogous to the bipolar transistor CE, EF, and CC configurations, respectively.

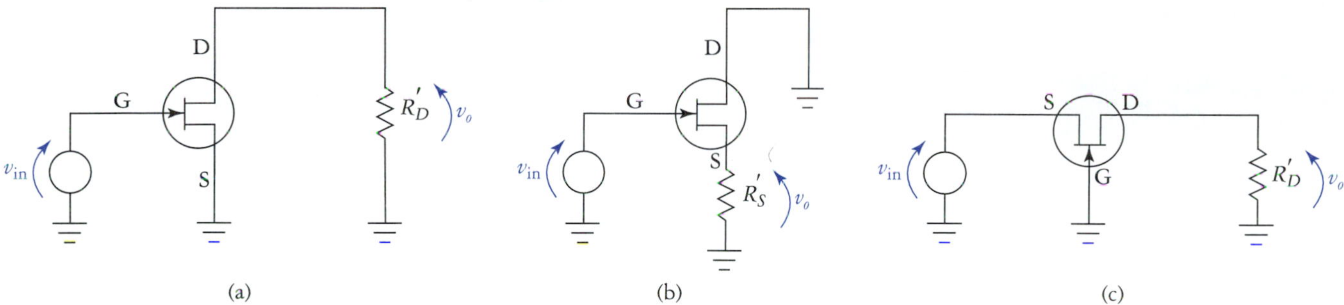

FIGURE 7.1
Basic amplifier configurations. (a) Common source. (b) Common drain or source follower. (c) Common gate.

Many similarities exist between the BJT amplifier configurations and their FET counterparts. For example, like the common-emitter amplifier, the common-drain configuration exhibits phase inversion, current gain, and voltage gain. The source follower, like the emitter follower, always exhibits voltage gain less than unity. Finally, the common-gate configuration has rather low input resistance and less than unity current gain, like the common-base amplifier. Each configuration has its particular strong points; however, the common-source configuration is more widely used than the others.

7.2 SMALL-SIGNAL TRANSCONDUCTANCE

Transconductance is the gain parameter associated with an active device when the output of the device is a current and the input is a voltage. In other words, transconductance is the gain parameter of a voltage-controlled current source (VCIS). In saturation (the active region), the FET is normally best described by the VCIS, so it makes sense that the primary FET gain parameter is transconductance.

The small-signal transconductance of an FET is defined as the instantaneous slope of the i_D versus v_{GS} transconductance curve for that particular device. Mathematically, this is expressed as

$$g_m = \frac{di_D}{dv_{GS}} \tag{7.1}$$

In more physically meaningful terms, transconductance is the proportionality constant associated with the drain terminal current source when the FET is in pinchoff. The general equation for the JFET (and DMOS FET) drain current is

$$i_D = I_{DSS}\left(1 - \frac{|v_{GS}|}{V_p}\right)^2 \tag{7.2}$$

Taking the derivative of i_D with respect to v_{GS} as indicated by Eq. (7.1) yields the transconductance equation

$$g_m = \frac{2I_{DSS}}{V_p}\left(1 - \frac{|v_{GS}|}{V_p}\right) \tag{7.3}$$

If we assume that $v_{GS} = 0$ V, then we have the following important quantity:

$$g_{m0} = \frac{2I_{DSS}}{V_p} \tag{7.4}$$

The parameter g_{m0} is the slope of the transconductance curve where it crosses the vertical axis. The value of g_{m0} (actually a typical value, or range of values) is often given in the data sheets for a given device. It is convenient to combine Eqs. (7.3) and (7.4) into a single expression:

$$g_m = g_{m0}\left(1 - \frac{|v_{GS}|}{V_p}\right) \tag{7.5}$$

Notice that g_{m0} is an intrinsic FET parameter (like β for a BJT), while in general, g_m is dependent on the FET and the drain current I_{DQ} at which it is operating. It is standard practice to denote the transconductance of the device at some particular I_{DQ} as g_{mQ}.

EXAMPLE 7.1

A certain n-channel JFET has $I_{DSS} = 15$ mA, $V_p = 5$ V, and $I_{DQ} = 10$ mA. Determine g_{m0} and g_{mQ}.

Solution Application of Eq. (7.4) gives us

$$g_{m0} = \frac{2I_{DSS}}{V_p} = \frac{30 \text{ mA}}{5 \text{ V}} = 6 \text{ mS}$$

Note that the current SI unit for conductance is the *Siemen,* S. Although obsolete, the older term *mho* $\mho$ is still used very often to denote conductance.

To determine the quiescent transconductance g_{mQ}, we first need to determine V_{GSQ}. Thus, we solve Eq. (7.1), giving

$$V_{GSQ} = V_P\left(1 - \sqrt{\frac{I_{DQ}}{I_{DSS}}}\right) = 5 \text{ V}(1 - \sqrt{0.67}) = -0.92 \text{ V}$$

The quiescent transconductance is now found using Eq. (7.5):

$$g_{mQ} = g_{m0}\left(1 - \frac{|v_{GS}|}{V_P}\right) = 6 \text{ mS}\left(1 - \frac{0.92}{5}\right) = 4.9 \text{ mS}$$

7.3 JFET SMALL-SIGNAL AMPLIFIERS

Field-effect transistors are modeled as voltage-controlled current sources for both dc and ac analysis. JFET amplifier analysis is performed in two parts—dc and ac analyses—just as with similar BJT-based circuits. The dc analysis establishes the operating point of the FET, which as you might have suspected, has a large effect on the ac characteristics of the amplifier.

The ac-equivalent model for an n-channel JFET in the pinchoff region (actually, this model can be used to represent any n-channel FET in the pinchoff region) is shown in Fig. 7.2. Figure 7.2(a) shows the gate-to-source resistance explicitly. At dc and relatively low frequencies (less than a few megahertz or so), this resistance is normally large enough to consider infinite, and so we can usually use the simpler FET model as shown in Fig. 7.2(b).

Common-Source Amplifier

An n-channel JFET CS amplifier is shown in Fig. 7.3. Notice that coupling capacitors are used to prevent dc interaction between the biasing circuitry and the input source and the load.

Note
Since r_{GS} is extremely large, we will use the model of Fig. 7.2b in our work.

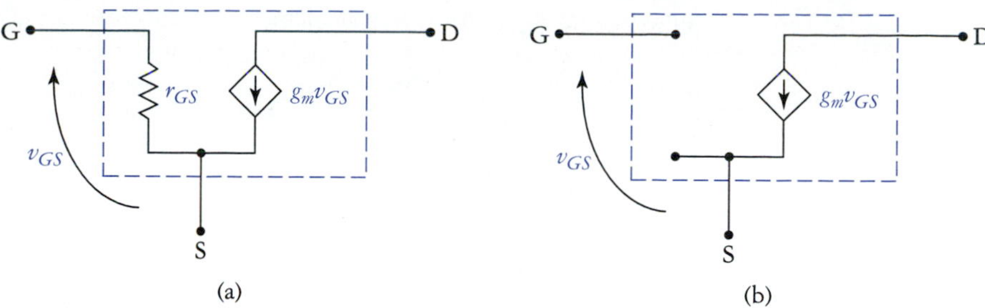

FIGURE 7.2
Small-signal FET models. (a) Finite r_{GS}. (b) Infinite r_{GS}.

FIGURE 7.3
Common-source amplifier with source feedback.

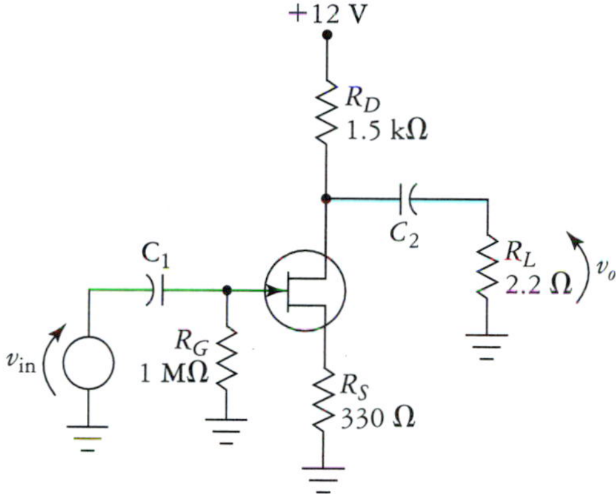

The dc- and ac-equivalent circuits for Fig. 7.3 are shown in Fig. 7.4(a) and (b), respectively. These equivalent circuits are developed in exactly the same manner as those for similar BJT amplifier circuits.

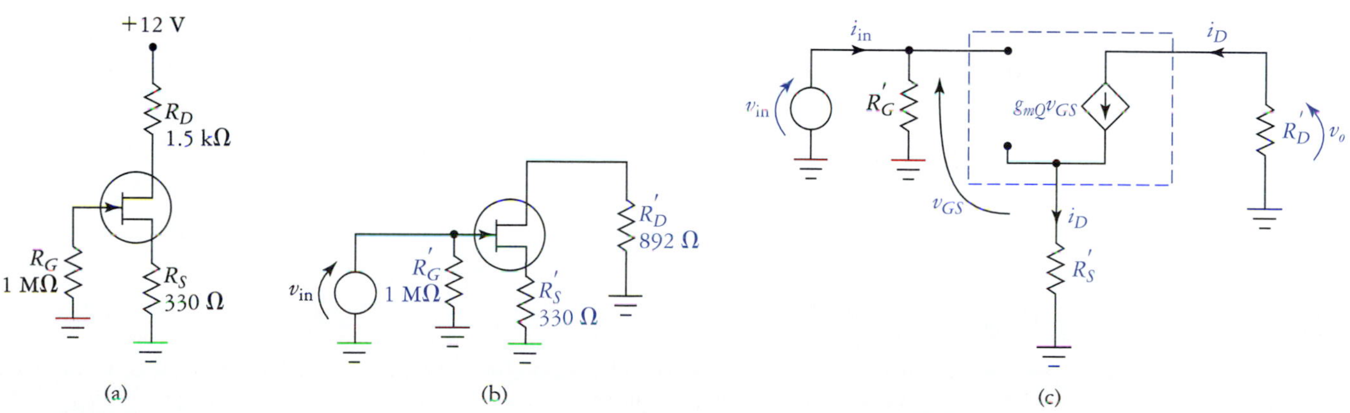

FIGURE 7.4
Common-source amplifier. (a) DC-equivalent circuit. (b) AC-equivalent circuit. (c) AC-equivalent circuit using small-signal dependent-source model.

The circuit of Fig. 7.4(c) uses the simple VCIS model of the FET. A voltage-gain expression is now derived using this circuit. We begin by applying KVL clockwise around the G-S loop:

$$0 = v_{in} - v_{GS} - i_D R_S' \tag{7.6}$$

Inspection of the circuit reveals

$$i_D = g_{mQ}\, v_{GS} \tag{7.7}$$

Substituting Eq. (7.7) into (7.6) yields

$$0 = v_{\text{in}} - v_{GS} - g_{mQ}v_{GS}R'_S \tag{7.8}$$

Solving for v_{GS} produces

$$v_{GS} = \frac{v_{\text{in}}}{1 + g_{mQ}R_S} \tag{7.9}$$

Now, the output voltage is given by

$$v_o = -i_D R'_D \tag{7.10}$$

We substitute Eq. (7.7) into (7.10) giving us

$$v_o = -g_{mQ}v_{GS}R'_D \tag{7.11}$$

Substituting Eq. (7.9) into (7.11) produces

$$v_o = -\frac{g_{mQ}v_{\text{in}}R'_D}{1 + g_{mQ}R'_S} \tag{7.12}$$

Finally, we divide both sides by v_{in}, yielding the voltage-gain equation:

$$A_v = -\frac{g_{mQ}R'_D}{1 + g_{mQ}R'_D} \tag{7.13}$$

Because the input resistance of the FET itself is extremely high ($r_{\text{in}(G)} \cong \infty$), the input resistance of the common-source amplifier is easily found to be

$$R_{\text{in}} \cong R_G \tag{7.14}$$

It is clear that the typical FET CS amplifier will normally have very high input resistance, in comparison to a similar BJT CE amplifier. Recall that the current gain for any amplifier is given by

$$A_i = A_v \frac{R_{\text{in}}}{R_L} \tag{7.15}$$

Likewise, power gain is given by

$$A_p = A_v A_i \tag{7.16}$$

Another amplifier characteristic that is of some importance is output resistance. Assuming that the FET has a high Early voltage, then the drain current source has approximately infinite internal resistance and we have

$$R_o \cong R_D \tag{7.17}$$

EXAMPLE 7.2

Refer to Fig. 7.3. Assuming that $I_{DSS} = 8$ mA and $V_p = 4$ V, determine A_v, A_i, A_p, and R_{in}.

Solution Using the techniques developed in Chapter 6, we find $I_{DQ} = 3.8$ mA. We now determine g_{m0}:

$$g_{m0} = \frac{2I_{DSS}}{V_p} = \frac{16\text{ mA}}{4\text{ V}} = 4\text{ mS}$$

The transconductance at the Q point is

$$\begin{aligned} g_{mQ} &= g_{m0}\left(1 - \frac{|v_{GSQ}|}{V_p}\right) \\ &= 4\text{ mS}\left(1 - \frac{1.25\text{ V}}{4\text{ V}}\right) \\ &= 2.75\text{ mS} \end{aligned}$$

The ac-equivalent drain and source resistances are

$$\begin{aligned} R'_D &= R_D \parallel R_L \\ &= 1.5\text{ k}\Omega \parallel 2.2\text{ k}\Omega \\ &= 892\ \Omega \\ R'_S &= R_S = 330\ \Omega \end{aligned}$$

The voltage gain is

$$\begin{aligned} A_v &= -\frac{g_{mQ} R'_D}{1 + g_{mQ} R'_S} \\ &= -\frac{2.75\text{ mS} \times 892\ \Omega}{1 + (2.75\text{ mS} \times 330\ \Omega} \\ &= -1.3 \end{aligned}$$

The input resistance is $R_{\text{in}} = R_G = 1\text{ M}\Omega$. Thus, we find the current gain to be

$$\begin{aligned} A_i &= A_v \frac{R_{\text{in}}}{R_L} \\ &= -1.3 \frac{1\text{ M}\Omega}{2.2\text{ k}\Omega} \\ &= -591 \end{aligned}$$

and the power gain to be

$$\begin{aligned} A_p &= A_v A_i \\ &= (-1.3)(-591) \\ &= 768.3 \end{aligned}$$

FET amplifiers are capable of producing extremely high current and power-gain figures. Consider that increasing R_G to 10 MΩ in Fig. 7.3 would result in $A_i = -5910$ and $A_p = 7683$.

Note
JFET-based amplifiers usually have much lower voltage gain than similar BJT-based amplifiers.

Source Bypassing

When the source resistor is bypassed with a capacitor, the voltage gain of the common-source amplifier increases as it would for a common-emitter BJT amplifier with emitter bypassing. The ac equivalent for a CS amplifier with complete source bypassing is shown in Fig. 7.5. It is easy to understand why A_v increases without resorting to another long derivation (whew). Complete source bypassing simply makes $R'_S = 0$ in Eq. (7.13), which gives us

$$A_v = -g_{mQ} R'_D \tag{7.18}$$

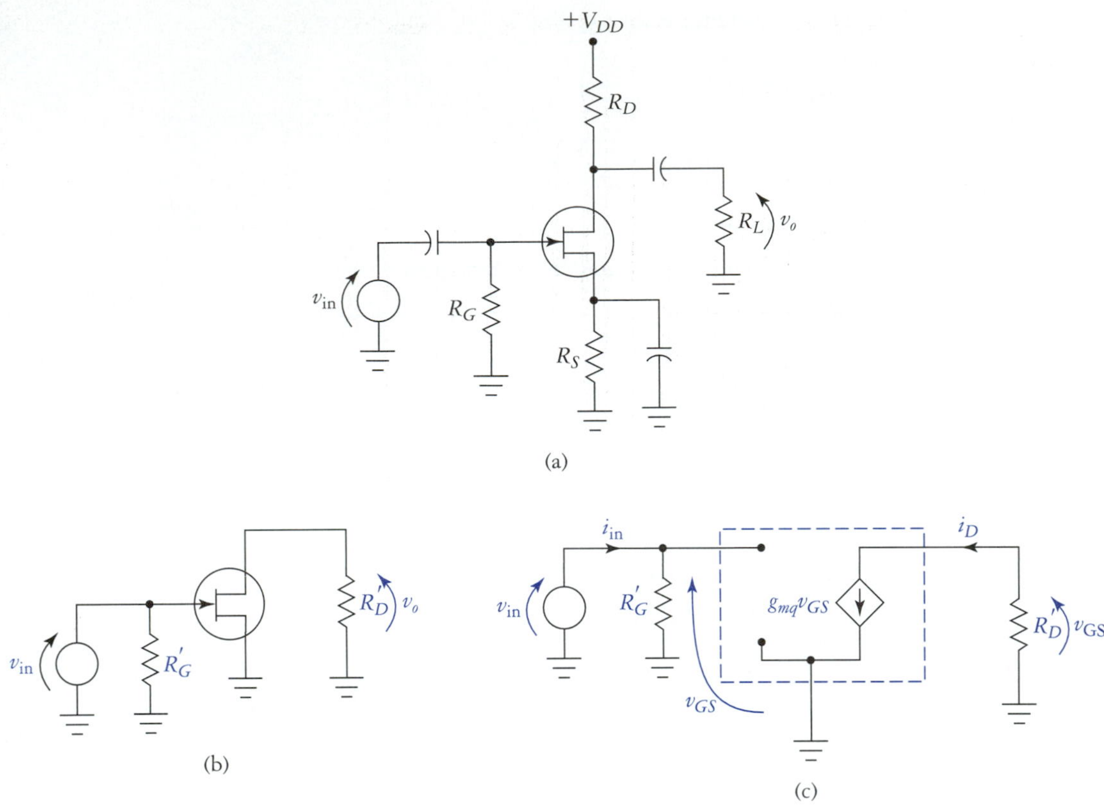

FIGURE 7.5
Common-source amplifier with source bypassing. (a) The circuit. (b) AC-equivalent circuit. (c) AC-equivalent circuit with dependent-source model.

EXAMPLE 7.3

Redetermine the values for A_v, A_i, and A_p for Fig. 7.3 assuming that the source is bypassed to ground. As in Example 7.2, assume that $I_{DSS} = 8$ mA and $V_p = 4$ V.

Solution The source bypass capacitor has no effect on the Q point of the amplifier, so many of the previous results apply. Specifically, $I_{DQ} = 3.8$ mA, $g_{mQ} = 2.75$ mS, and $R'_D = 892\ \Omega$. The ac-equivalent source resistance is $R'_S = 0\ \Omega$. This gives us

$$\begin{aligned} A_v &= (-2.75\text{ mS})(892\ \Omega) \\ &= -2.45 \end{aligned}$$

The current and power gains are

$$\begin{aligned} A_i &= A_v \frac{R_{in}}{R_L} \\ &= -2.45 \frac{1\text{ M}\Omega}{2.2\text{ k}\Omega} \\ &= -1114 \end{aligned}$$

$$\begin{aligned} A_p &= A_v A_i \\ &= (-2.45)(-1114) \\ &= 2729 \end{aligned}$$

The high input resistance of the FET is responsible for the very high current gain of the CS amplifier. This, in turn, allows realization of high power gain as well. We should also point out that unlike emitter bypassing, source bypassing has no direct effect on amplifier input resistance. However, source bypassing eliminates the negative feedback provided by the source resistor, which tends to increase amplifier distortion and decrease the drain-to-source resistance of the FET. Recall that this was also true of BJT emitter bypassing as well.

Source Follower

Note
Like the emitter follower, the source follower always has $A_v < 1$.

A typical source-follower (SF) amplifier is shown in Fig. 7.6. The source follower is analogous to the BJT emitter follower in operation and is used in similar applications. That is, the SF amplifier is used most often to serve an impedance-matching function, allowing a high-resistance source to be coupled effectively to a low-resistance load.

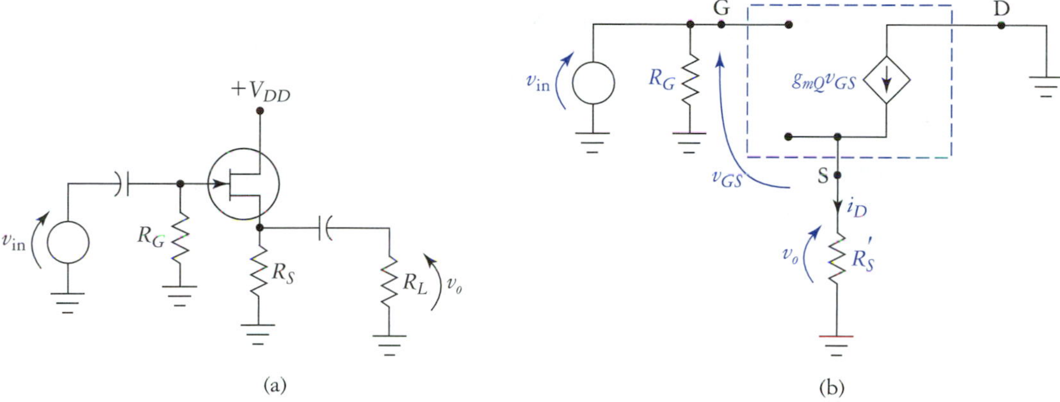

FIGURE 7.6
Source-follower, or common-drain, amplifier. (a) The circuit. (b) Small-signal ac model.

Derivation of the voltage-gain equation for the source follower is based on the dependent-source small-signal model of Fig. 7.6(b). Let us begin by defining the output voltage as

$$v_o = i_D R_S' \tag{7.19}$$

We can define the drain current as

$$i_D = g_{mQ} v_{GS} \tag{7.20}$$

Using these identities and applying KVL clockwise around the G-S loop gives us

$$0 = v_{\text{in}} - v_{GS} - g_{mQ} v_{GS} R_S' \tag{7.21}$$

These expressions are manipulated to form

$$A_v = \frac{g_{mQ} R_S'}{1 + g_{mQ} R_S'} \tag{7.22}$$

Inspection of Eq. (7.22) reveals that the voltage gain of the common-source amplifier will always be somewhat less than unity. The input resistance is simply

$$R_{\text{in}} = R_G \tag{7.23}$$

The output resistance of the source follower is given by

$$R_o = R_S \,\|\, \frac{1}{g_{mQ}} \tag{7.24}$$

EXAMPLE 7.4

Refer to Fig. 7.6. Given $I_{DSS} = 25$ mA, $V_p = 3.5$ V, $R_S = 150\ \Omega$, $R_G = 4.7\ \text{M}\Omega$, $R_L = 100\ \Omega$, and $V_{DD} = 15$ V, determine A_v, A_i, A_p, and R_o.

Solution As usual, we must first determine the Q point and a few preliminary quantities. Thus, we find $I_{DQ} = 9$ mA, $g_{m0} = 14.3$ mS, $V_{GSQ} = -1.35$ V, $g_{mQ} = 8.8$ mS, and $R_S = 60\ \Omega$. The voltage gain is

$$A_v = \frac{g_{mQ}R'_S}{1 + g_{mQ}R'_S} = \frac{0.528}{1.528} = 0.35$$

$$A_i = A_v \frac{R_{\text{in}}}{R_L} = (0.35)(47\text{ k}) = 16.45\text{ k}$$

$$A_p = A_v A_i = (0.35)(16.45\text{ k}) = 5.76\text{ k}$$

The output resistance is

$$R_o = R_S \parallel \frac{1}{g_{mQ}} = 113.6\ \Omega \parallel 150\ \Omega = 64.6\ \Omega$$

Common-Gate Amplifier

A typical common-gate amplifier is shown in Fig. 7.7(a). Although the circuit may look rather odd, with the JFET turned sideways, basic self-biasing is used and the dc analysis is straightforward. Notice that the CG configuration does not require that a gate resistor be used.

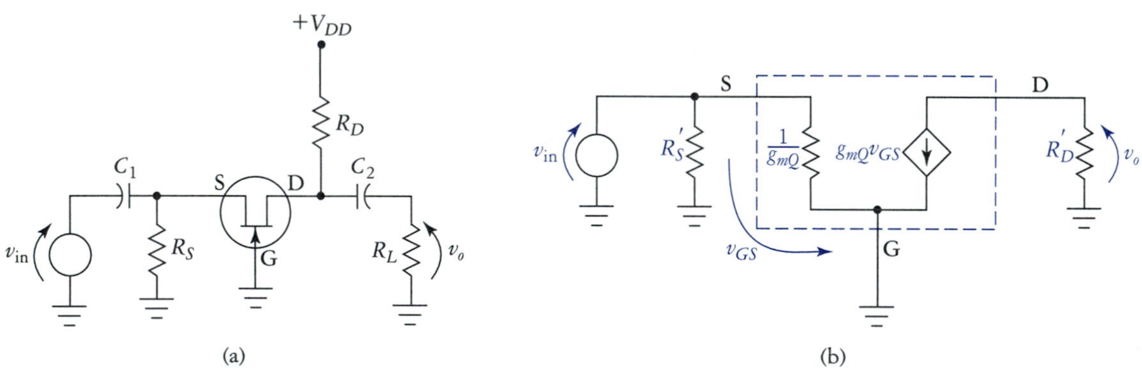

FIGURE 7.7
Common-gate amplifier. (a) Typical circuit. (b) Small-signal ac-equivalent circuit.

The ac-equivalent model for the CG amplifier is shown in Fig. 7.7(b). From this circuit, the following equations are easily derived:

$$A_v = g_{mQ}R'_D \tag{7.25}$$

$$R_{\text{in}} = R_S \parallel \frac{1}{g_{mQ}} \tag{7.26}$$

$$R_o \cong R_D \tag{7.27}$$

Notice that like the BJT common-base amplifier, the CG amplifier is noninverting, has relatively low input resistance, and will *usually* have $A_v > 1$. The following example shows that the current gain of the CG amplifier is always less than unity.

EXAMPLE 7.5

Refer to Fig. 7.7. Given $I_{DSS} = 16$ mA, $g_{m0} = 10$ mS, $R_S = 68\ \Omega$, $R_D = 1\ \text{k}\Omega$, $V_{DD} = 20$ V, and $R_L = 1\ \text{k}\Omega$, determine A_v, A_i, A_p, and R_{in}.

Solution This problem requires that we first determine V_p. Rearranging Eq. (7.4), we have

$$\begin{aligned} V_p &= \frac{2I_{DSS}}{g_{m0}} \\ &= \frac{0.032\ \text{mA}}{10\ \text{mS}} \\ &= 3.2\ \text{V} \end{aligned}$$

The drain current is determined (graphically), and the remaining pertinent dc parameters are found to be $I_{DQ} = 9.9$ mA, $V_{GSQ} = -0.67$ V, and $V_{DSQ} = 9.4$ V. The quiescent transconductance is

$$\begin{aligned} g_{mQ} &= g_{m0}\left(1 - \frac{|V_{GSQ}|}{V_p}\right) \\ &= 10\ \text{mS}\left(1 - \frac{0.67}{3.2}\right) \\ &= 7.9\ \text{mS} \end{aligned}$$

The ac-equivalent drain resistance is

$$\begin{aligned} R'_D &= R_D \parallel R_L \\ &= 1\ \text{k}\Omega \parallel 1\ \text{k}\Omega \\ &= 500\ \Omega \end{aligned}$$

The voltage gain is found next:

$$\begin{aligned} A_v &= g_{mQ}R'_D \\ &= (7.9\ \text{mS})(500\ \Omega) \\ &= 3.95 \end{aligned}$$

The input resistance is

$$\begin{aligned} R_{\text{in}} &= R_S \parallel \frac{1}{g_{mQ}} \\ &= 68\ \Omega \parallel 127\ \Omega \\ &= 44\ \Omega \end{aligned}$$

The current and voltage gain are

$$\begin{aligned} A_i &= A_v \frac{R_{\text{in}}}{R_L} \\ &= (3.95)(0.044) \\ &= 0.17 \end{aligned}$$

$$\begin{aligned} A_p &= A_v A_i \\ &= (3.95)(0.17) \\ &= 0.67 \end{aligned}$$

The amplifier in Example 7.5 has relatively poor performance. This circuit actually absorbs more power from the input source than it delivers to the load, and therefore is really not practical. The low input resistance, high output resistance, and poor current-gain characteristics limit the applicability of the CG amplifier. A later application will show, however, that knowledge of the CG configuration analysis equations is useful.

7.4 DMOS FET SMALL-SIGNAL AMPLIFIERS

Here's a pleasant surprise: JFETs and DMOS FETs share identical ac analysis expressions. Because of this fact, the following two examples constitute our coverage of DMOS FET small-signal amplifiers.

EXAMPLE 7.6

Refer to Fig. 7.8. Given $I_{DSS} = 15$ mA, $V_p = 2$ V, $V_{DD} = 15$ V, $R_G = 10$ MΩ, $R_D = 470$ Ω, and $R_L = 1$ kΩ, determine I_{DQ}, V_{GSQ}, V_{DSQ}, A_v, A_i, R_{in}, and A_p.

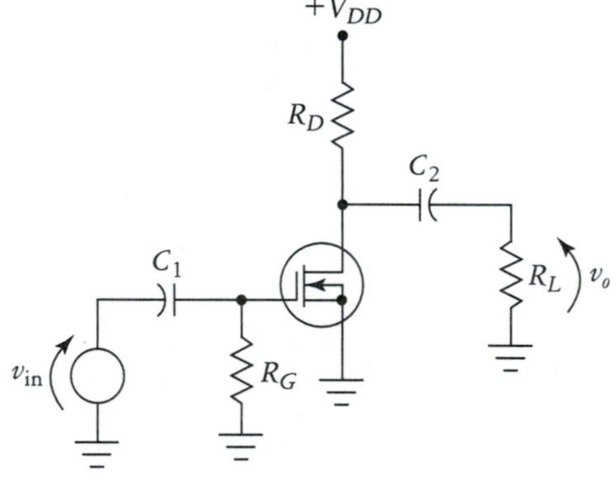

FIGURE 7.8
DMOS common-source amplifier.

Solution Because this FET uses zero biasing, the Q-point parameters are easily found to be

$$g_{mQ} = g_{m0} = \frac{2I_{DSS}}{V_p} = \frac{30 \text{ mA}}{2 \text{ V}} = 15 \text{ mS}$$

$$I_{DQ} = I_{DSS} = 15 \text{ mA}$$

$$V_{GSQ} = 0 \text{ V}$$

$$V_{DSQ} = V_{DD} - I_{DQ}R_D = 15 \text{ V} - (15 \text{ mA})(470 \ \Omega) = 7.95 \text{ V}$$

The small-signal parameters are now easily determined:

$$R_{in} = R_G = 10 \text{ M}\Omega$$

$$R'_D = R_D \parallel R_L = 470 \ \Omega \parallel 1 \text{ k}\Omega = 320 \ \Omega$$

$$A_v = -g_{mQ}R'_D = -(15 \text{ mS})(320 \ \Omega) = -4.8$$

$$A_i = A_v \frac{R_{in}}{R_L} = -4.8 \frac{10 \text{ M}\Omega}{1 \text{ k}\Omega} = -48{,}000$$

$$A_p = A_v A_i = (-4.8)(-48 \text{ k}) = 230.4 \times 10^3$$

As should be expected, the power gain of the CS amplifier is extremely high. If a low-resistance load must be driven effectively, however, a source follower (common drain) amplifier would be a more appropriate choice.

EXAMPLE 7.7

The circuit of Fig. 7.9 uses the same FET as in Example 7.6, with $R_S = 68\ \Omega$, $R_G = 10\ \text{M}\Omega$, $R_L = 47\ \Omega$, and $V_{DD} = 15$ V. Determine I_{DQ}, V_{GSQ}, V_{DSQ}, R_o, A_v, A_i, and A_p.

FIGURE 7.9
DMOS source-follower amplifier.

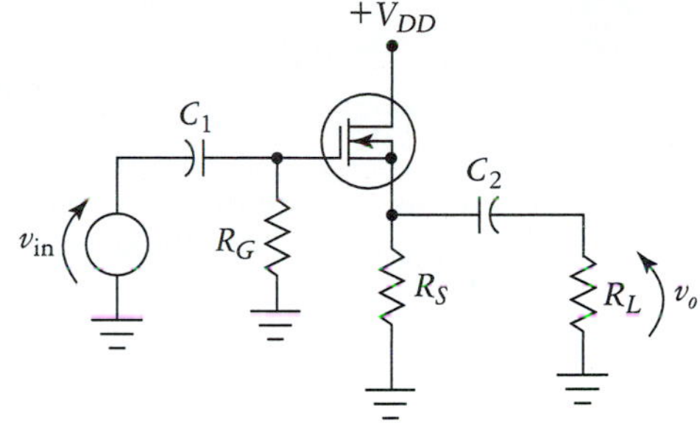

Solution This is a source follower with self-bias. The graphical Q-point analysis was used, and the following quantities determined; $I_{DQ} = 8.1$ mA, $V_{GSQ} = -0.55$ V, $V_{DSQ} = 14.45$ V, $R_S = 28\ \Omega$, $R_{\text{in}} = 10\ \text{M}\Omega$, $g_{m0} = 15$ mS, and $g_{mQ} = 10.9$ mS. Using these values, the various gains are found to be

$$A_v = \frac{g_{mQ}R'_S}{1 + g_{mQ}R'_S} = \frac{0.305}{1.305} = 0.23$$

$$A_i = A_v \frac{R_{\text{in}}}{R_L} = 0.23 \frac{10\ \text{M}\Omega}{47\ \Omega} = 48.9 \times 10^3$$

$$A_p = A_v A_i = (0.23)(48.9\ \text{k}) = 11.2 \times 10^3$$

Finally, the output resistance is

$$R_o = R_S \parallel \frac{1}{g_{mQ}} = 68\ \Omega \parallel \frac{1}{10.9\ \text{mS}} = 39\ \Omega$$

7.5 EMOS FET SMALL-SIGNAL AMPLIFIERS

In general, the analysis of an EMOS-based amplifier is very similar to the analysis of any other FET amplifier; the basic configurations are the same (CS, SF, and CG), and the various gain, input, and output resistance equations are the same as well. Thus, the various formulas that apply to any type of FET amplifier presented in this chapter are summarized in Table 7.1. The main difference between the analysis of the EMOS amplifier compared to DMOS and JFET circuits is in the determination of g_{mQ}. Recall that the transconductance equation for the EMOS FET is

$$I_{DQ} = k\,(V_{GSQ} - V_{to})^2$$

TABLE 7.1
FET Small-Signal Amplifier Analysis Equations

	Common Source	Source Follower	Common Gate
A_v	$-\dfrac{g_{mQ}R'_D}{1 + g_{mQ}R'_S}$	$\dfrac{g_{mQ}R'_S}{1 + g_{mQ}R'_S}$	$g_{mQ}R'_D$
R_{in}	R_G	R_G	$\dfrac{1}{g_{mQ}} \parallel R_S$
R_o	R_D	$\dfrac{1}{g_{mQ}} \parallel R_S$	R_D

where k is determined either from the actual device structural dimensions or empirically, using $I_{D(\text{on})}$ and $V_{GS(\text{on})}$. Recall that the small-signal transconductance of a given FET at some Q point is given by

$$g_{mQ} = \left.\frac{d\,I_{DQ}}{d\,V_{GSQ}}\right|_{V_{DS}=\text{const}}$$

Thus, if we take the derivative of I_{DQ} with respect to V_{GSQ} in the transconductance equation, we obtain

$$g_{mQ} = 2k\,(V_{GSQ} - V_{to}) \tag{7.28}$$

Since we know how to determine the Q point of the EMOS amplifier, and how to determine g_{mQ}, we simply apply the same analysis techniques used for JFET and DMOS amplifiers to those using EMOS devices.

EXAMPLE 7.8

Refer to Fig. 7.10. Given $V_{to} = 3$ V, $I_{D(\text{on})} = 10$ mA, $V_{GS(\text{on})} = 5$ V, $V_{DD} = 18$ V, $R_1 = 15$ MΩ, $R_2 = 4.7$ MΩ, and $R_D = R_L = 1.8$ kΩ, determine I_{DQ}, V_{DSQ}, A_v, A_i, A_p, R_{in}, and R_L.

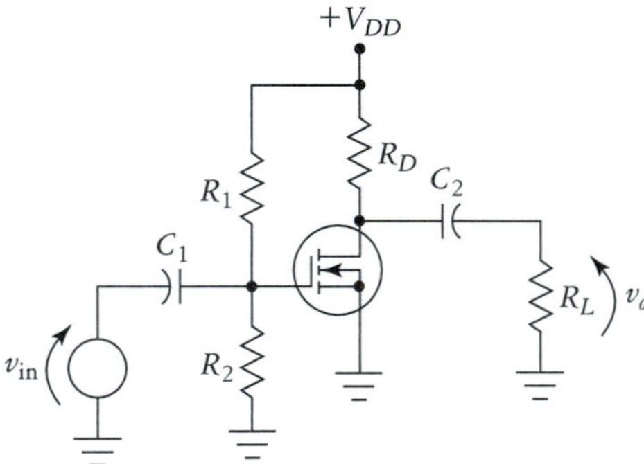

FIGURE 7.10
EMOS common-source amplifier with voltage-divider biasing.

Solution Let us first determine the proportionality constant k. Solving the EMOS transconductance equation, we have

$$\begin{aligned} k &= \frac{I_{D(\text{on})}}{(V_{GS(\text{on})} - V_{to})^2} \\ &= \frac{10 \text{ mA}}{4 \text{ V}^2} \\ &= 0.0025 \text{ mA/V}^2 \end{aligned}$$

Because there is no source resistor, we find by inspection that

$$\begin{aligned} V_{GSQ} = V_G &= V_{DD}\left(\frac{R_2}{R_1 + R_2}\right) \\ &= 18\left(\frac{4.7}{4.7 + 15}\right) \\ &= 4.29 \text{ V} \end{aligned}$$

$$I_{DQ} = k\,(V_{GSQ} - V_{to})^2 = 0.0025(4.29 - 3)^2 = 4.16\text{ mA}$$

$$V_{DSQ} = V_{DD} - I_{DQ}R_D = 18 - (4.16\text{ mA})(1.8\text{ k}\Omega) = 10.5\text{ V}$$

$$R'_D = R_D \parallel R_L = 1.8\text{ k}\Omega \parallel 1.8\text{ k}\Omega = 900\ \Omega$$

$$R_{\text{in}} = R_1 \parallel R_2 = 15\text{ M}\Omega \parallel 4.7\text{ M}\Omega = 3.6\text{ M}\Omega$$

We now apply Eq. (7.28) to determine the transconductance of the FET:

$$g_{mQ} = 2k\,(V_{GSQ} - V_{to}) = 0.005(4.29 - 3) = 6.45\text{ mS}$$

$$A_v = -g_{mQ}R'_D = (-6.45\text{ mS})(900\ \Omega) = -5.8$$

$$A_i = A_v \frac{R_{\text{in}}}{R_L} = (-5.8)(2\text{ k}) = -11.6\text{ k}$$

$$A_p = A_vA_i = (-5.8)(-11.6\text{ k}) = 67.3\text{ k}$$

Note
The effect of R_F on A_v is negligible because R_F is normally very large relative to R_D and R_L.

Drain-feedback biasing is also used with the EMOS FET. A common-source amplifier using this biasing arrangement is shown in Fig. 7.11. Although there is negative feedback from drain to gate, its effect on the voltage gain of the amplifier will be negligible. Thus, the voltage gain of the circuit is given by the familiar equation

$$A_v \cong -g_{mQ}R'_D \tag{7.29}$$

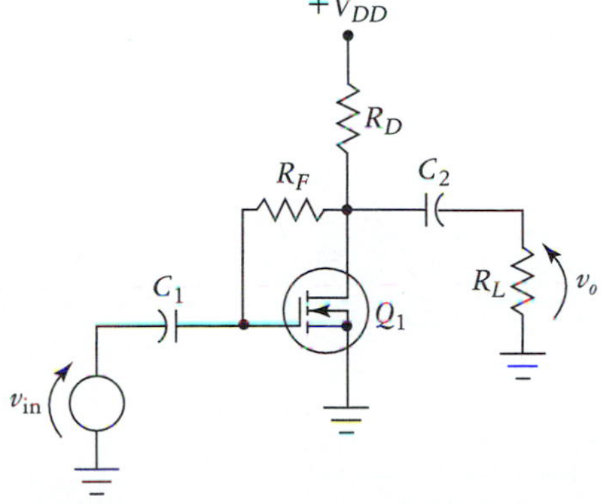

FIGURE 7.11
EMOS common-source amplifier with drain-feedback biasing.

The input resistance of the amplifier is given by

$$R_{\text{in}} = \frac{R_F}{|A_v| + 1} \tag{7.30}$$

Resistor R_F appears to be a lower value than it really is, when referred to the input source. This characteristic is described by *Miller's theorem,* which is discussed in detail in Chapter 9. The output resistance of the amplifier is

$$R_o = R_D \| (|A_v| + 1)R_F \tag{7.31}$$

However, in most cases we will find $R_F >> R_D$, so we can write

$$R_o \cong R_D \tag{7.32}$$

7.6 ACTIVE LOADS

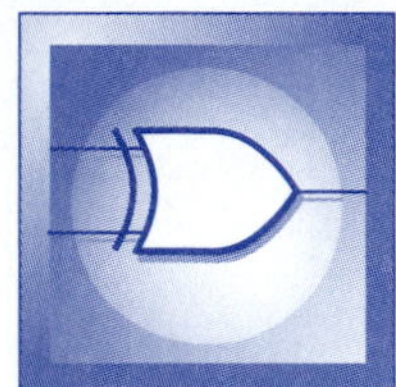

We mentioned earlier that MOSFETs are commonly used in the design of digital integrated circuits (ICs). As it turns out, MOSFETs take up less space (usually called real estate) on a given IC chip than resistors. Because of this, MOSFETs are normally used as active loads in place of resistors in digital IC designs. Currently, such ICs are produced using either NMOS or CMOS technologies. NMOS devices are constructed using n-channel MOSFETs exclusively. CMOS (complementary-MOS) devices are constructed using both n- and p-channel MOSFETs. Because of the simple structure of these FETs, extremely high-density ICs can be fabricated. Microprocessors such as the Intel Pentium, which is commonly used in PCs, are designed using MOS transistors.

Enhancement Loads

When used as an active load, the enhancement-mode MOSFET is connected as shown in Fig. 7.12(a). Once $V_{DS} > V_{to}$, the MOSFET acts as a nonlinear resistance. This is shown in the transconductance plot of Fig. 7.12(b). Because the enhancement-mode active load is used primarily in digital switching applications, the nonlinearity of the effective resistance is not a problem. Digital circuits are nonlinear by their very nature. The drain current is given by the transconductance equation

$$I_D = k(V_{DS} - V_{to})^2 \tag{7.33}$$

Notice that the circuit $V_{GS} = V_{DS}$.

An inverter using the EMOS enhancement load is shown in Fig. 7.13. In this circuit, Q_1 is simply a common-source amplifier, while Q_2 is the active load. When the input signal is low ($V_{\text{in}} = 0$ V), Q_1 is nonconducting, and $V_o \cong V_{DD} - V_{to}$. When the input voltage switches states ($V_{\text{in}} > V_{to}$), Q_1 is driven on, which reduces V_o to about zero volts. Because the resistance of Q_2 varies, switching speed is faster than for a fixed R_D.

Depletion Load

The DMOS transistor is connected as shown in Fig. 7.14(a), and the resulting transconductance characteristic is shown in Fig. 7.14(b). This configuration produces basically the same characteristics as the JFET constant-current diode.

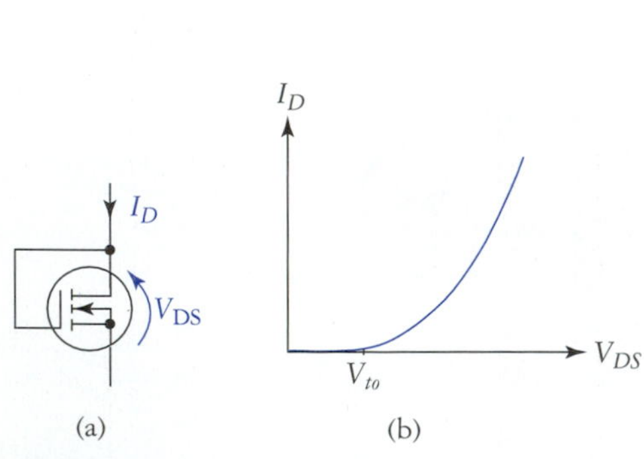

FIGURE 7.12
Using an EMOS FET as a nonlinear resistance. (a) Device connections. (b) Transconductance curve.

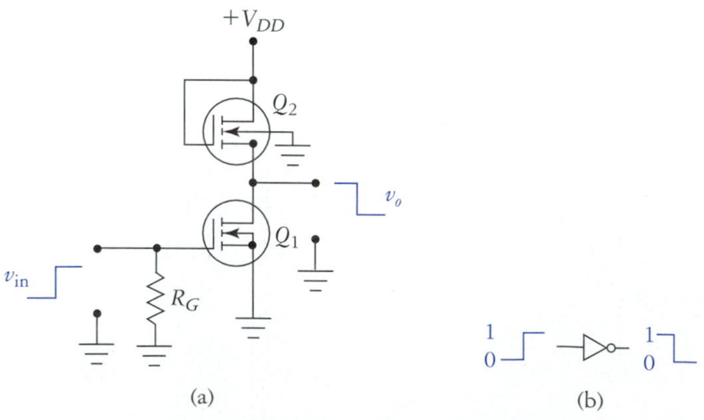

FIGURE 7.13
Enhancement-load connection used to form an inverter.

(a) (b)

FIGURE 7.14
DMOS FET nonlinear resistance. (a) Device connections. (b) Transconductance curve.

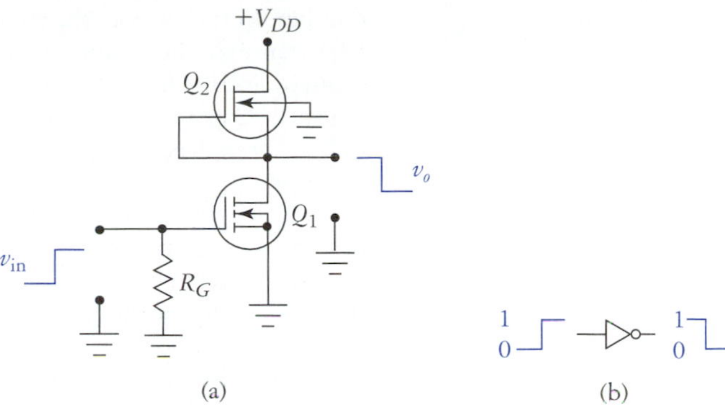

FIGURE 7.15
Using a depletion-load connection to form an inverter.

A depletion-load inverter is shown in Fig. 7.15. Notice that the switching transistor Q_1 is an enhancement-mode device. Normally, the only time that depletion-mode devices are used in digital IC designs is when active loading is implemented. Although it is somewhat more difficult to fabricate EMOS and DMOS devices on the same chip, the depletion-load approach provides superior performance in terms of switching speed.

7.7 ALTERNATIVE FET PARAMETER NOTATION

Several of the symbols used thus far to represent the various FET parameters can be presented in different ways. One of the most important device parameters we use, transconductance, is often denoted as g_{fs}. This notation stands for *forward transconductance in common-source configuration.* This notation allows differentiation between forward and reverse transconductance characteristics (as with forward beta β_F and reverse beta β_R for the bipolar transistor). Reverse transconductance is normally of negligible magnitude and is not necessary for the analyses performed in this text.

Transconductance (g_m or g_{fs}) is represented by a real number. At high frequencies, however, varying phase and amplitude shifts require the gain of the FET to be represented with a complex

TABLE 7.2
FET Small-Signal Parameters

FET Parameter	Description and Interpretation
g_m	General transconductance
g_{m0}	Transconductance, $V_{GS} = 0$ V
g_{fs}	Forward transconductance, $g_{fs} = g_m$, $g_{fs}\|_{V_{GS}=0} = g_{m0}$
y_{fs}	Forward transadmittance, $y = \frac{1}{z}$
$\|y_{fs}\|$	Fwd transadmittance magnitude, $\|y_{fs}\| = g_m$ at low f*
$\mathrm{Re}\{y_{fs}\}$	Real fwd transadmittance, $\mathrm{Re}\{y_{fs}\} = g_m$ at high f†
y_{os}	Output admittance
$\|y_{os}\|$	Output admittance magnitude, $\|y_{os}\| = \frac{1}{r_{DS}}$ at low f*
$\mathrm{Re}\{y_{os}\}$	Real output admittance, $\mathrm{Re}\{y_{os}\} = \frac{1}{r_{DS}}$ at high f†
y_{is}	Input admittance
$\|y_{is}\|$	Input admittance magnitude, $\|y_{is}\| = \frac{1}{r_{DS}}$ at low f*
$\mathrm{Re}\{y_{is}\}$	Real input admittance, $\mathrm{Re}\{y_{is}\} = \frac{1}{r_{GS}}$ at high f*

*Low f is typically 1 kHz.
†High f is typically $\geq$ 100 MHz.

number. In such a case, the transadmittance or *transfer admittance* of the FET is specified. Recall that admittance is the reciprocal of impedance ($Y = 1/Z$). The standard representation for (forward) transadmittance is y_{fs}. The lowercase italic y is used because transadmittance is a dynamic (or instantaneous), Q-point-dependent parameter. Often, a manufacturer will specify only y_{fs}, its magnitude $|y_{fs}|$, or the real-part $\text{Re}\{y_{fs}\}$ for a given FET. Table 7.2 presents the relationships between these various forms.

The drain-to-source resistance of the FET is denoted as r_{DS} in this text. Most often we do not need to use this parameter because usually $R_D \ll r_{DS}$. The complex equivalent of r_{DS} that is most often given in FET data sheets is y_{os}. This is *output conductance in the common-source configuration.* The input resistance of the gate r_{GS} is often expressed as an admittance as well, designated y_{is} in the CS configuration. Table 7.2 summarizes the relationships between these FET admittance, conductance, and resistance parameters as well.

7.8 THE BJT VERSUS THE FET

Being active semiconductor devices, BJTs and FETs share many similarities, especially in terms of applications. A comparison of some of the major functional differences between these classes of devices should be interesting at this point.

Voltage-Gain Characteristics

It has become apparent from the examples presented in this chapter that FET-based amplifiers tend to have rather low voltage gain in comparison with similar BJT-based amplifiers. It is a simple fact that bipolar transistors are generally capable of much higher voltage gains than FETs. It is not unusual for a single-stage CE amplifier to have $|A_v| > 250$, while the FET CS amplifier will rarely exceed $|A_v| > 25$. These relative characteristics can be easily understood if we examine the transconductance curves for the typical BJT and FET. Figure 7.16 shows the transconductance curve for an npn BJT, and the transconductance curve for an n-channel EMOS FET. FETs exhibit parabolic or square-law transconductance behavior, while BJTs produce exponential behavior. The slope of the parabolic transfer curve increases gradually in comparison to the rapidly rising exponential curve of

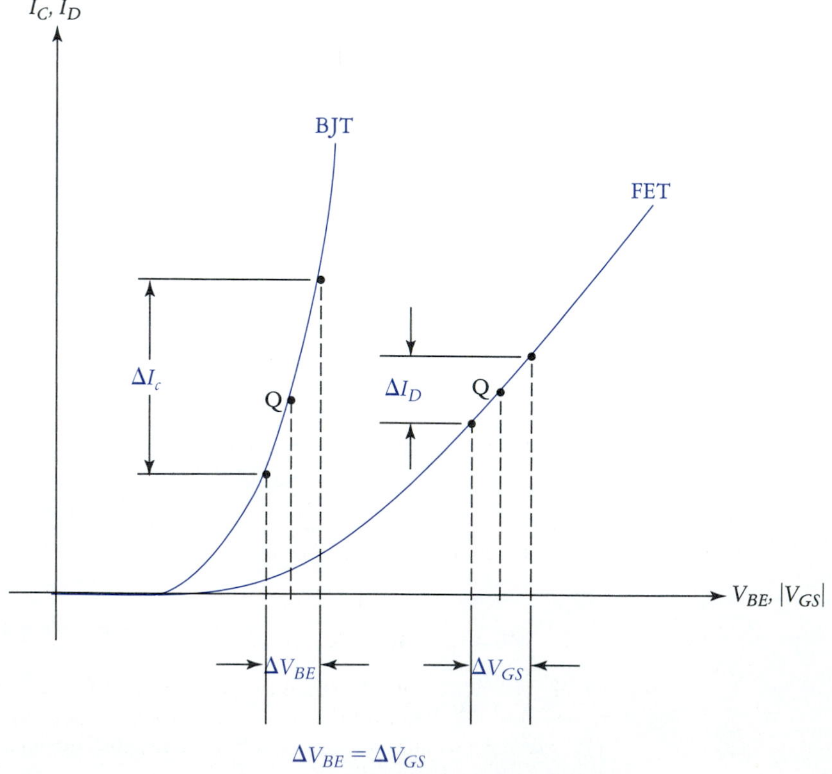

FIGURE 7.16
Comparison of typical transconductance curves of FETs and BJTs.

the typical BJT. Thus, for a given deviation of voltage (v_{BE} or v_{GS}), the BJT will normally experience a much greater change in collector current than the FET will in drain current. This variation in current produces the time-varying output voltage, and so A_v for the BJT will normally be much higher than for an equivalent FET circuit.

Here's a quick numerical example of what this is all driving toward. You may recall that we can define the transconductance of a BJT as

$$g_{m(\mathrm{BJT})} = 1/r_e = I_{CQ}/V_T$$

Now, suppose we are operating a BJT at $I_{CQ} = 5$ mA. This produces

$$g_{m(\mathrm{BJT})}|_{I_{CQ}=5\ \mathrm{mA}} = 192\ \mathrm{mS}$$

The typical 2N3819 n-channel JFET has

$$g_{m0} = 6.3\ \mathrm{mS}$$

Assuming that these are typical devices, the BJT should be capable of about 30 times higher voltage gain than an FET in a given situation. Excellent discussions on this and other related topics can be found in various data books available from companies such as Motorola, Supertex, and Hitachi.

Saturation Characteristics

Bipolar transistors will normally have a lower saturation voltage than FETs with similar power, current, and voltage ratings. The reason for this can be seen in the representative device curves shown in Fig. 7.17, where the collector and drain curves for the typical BJT and FET are shown superimposed on the same graph. The dashed curve represents the FET's pinchoff voltage as a function of v_{GS}.

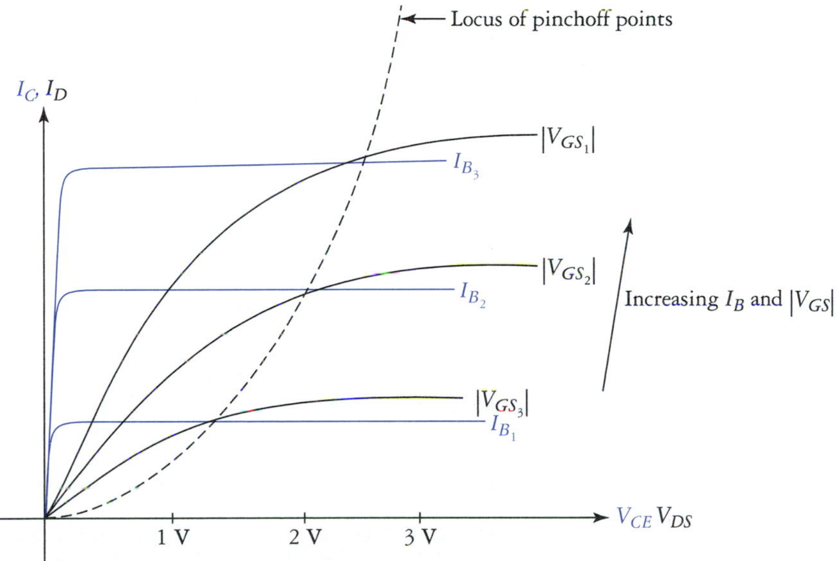

FIGURE 7.17
Comparison of typical low-voltage collector and drain curves.

The main point to be taken from Fig. 7.17 is that the FET will tend to exit the constant-current region sooner than the BJT on signal swings that decrease V_{DS}, as compared to decreasing V_{CE}.

Temperature Characteristics

Recall that FETs have a positive temperature coefficient, while BJTs are negative temperature coefficient devices. This means that destructive thermal runaway cannot normally occur in an FET circuit. This is an important consideration in higher power applications, such as power amplifier design and motor control circuits. Because of the lack of negative feedback, the circuits of Fig. 7.18 would exhibit strong temperature dependency. The curves in Fig. 7.18(c) illustrate the typical temperature-related shift in current for similar FETs and BJTs.

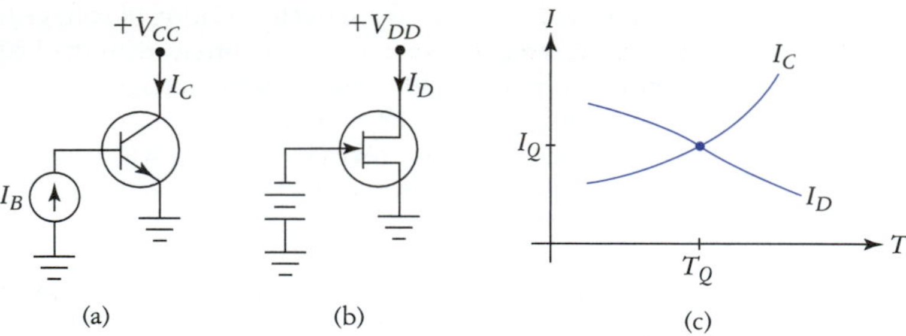

FIGURE 7.18
Temperature sensitivity of BJTs versus FETs. (a) BJT test circuit. (b) FET test circuit. (c) Drain and collector current versus temperature.

Distortion Characteristics

A final point of comparison between FETs and BJTs relates to distortion characteristics. The function of a linear amplifier is to produce a replica of an input signal at its output port. This replica may be inverted and larger in amplitude than the original, but it should be identical to the original signal in all other respects. Distortion is the general term for the deviation of the amplifier's actual output signal from the ideal. Although several types of distortion can occur, in linear applications it seems reasonable that one of the most undesirable types is nonlinear distortion. Such distortion alters the frequency content of a signal. Some degree of nonlinearity will be exhibited by any real amplifier, although it is reduced with the use of negative feedback.

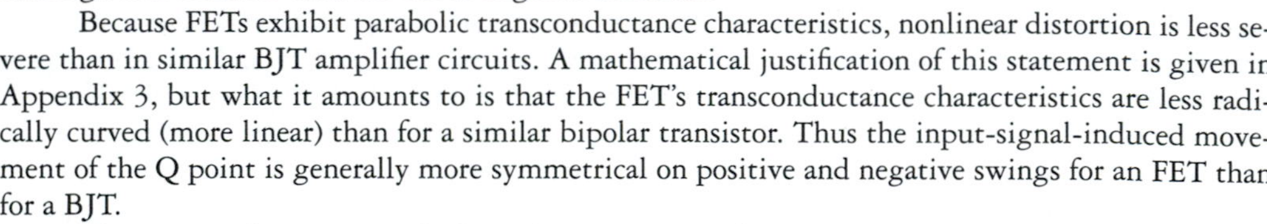

Because FETs exhibit parabolic transconductance characteristics, nonlinear distortion is less severe than in similar BJT amplifier circuits. A mathematical justification of this statement is given in Appendix 3, but what it amounts to is that the FET's transconductance characteristics are less radically curved (more linear) than for a similar bipolar transistor. Thus the input-signal-induced movement of the Q point is generally more symmetrical on positive and negative swings for an FET than for a BJT.

There are applications in which nonlinearity is a useful characteristic, such as in radio and television transmission and reception, and signal synthesis. Both BJTs and FETs are used in these applications; however, the parabolic transconductance of the FETs often makes them the preferred device.

7.9 COMPUTER-AIDED ANALYSIS APPLICATIONS

Small-signal analysis of FET-based amplifiers is very similar to that used in the analysis of similar BJT-based amplifiers. A few representative analysis runs should serve to illustrate this point.

JFET Amplifier

Consider the circuit shown in Fig. 7.19. This is a self-biased n-channel JFET common-source amplifier. The 2N3819 has $I_{DSS} = 11.7$ mA, $V_p = 3$ V, and $g_{m0} = 7.8$ mS. The simulation yields the following Q-point data:

$$I_{DQ} = 6.5 \text{ mA}$$

$$V_{DSQ} = 7.72 \text{ V}$$

$$V_{GSQ} = -0.78 \text{ V}$$

Using this information, we determine

$$g_{mQ} = 5.8 \text{ mS}$$

The value of V_{DSQ} was determined just to make sure that the FET is being operated in the pinchoff region ($V_{DSQ} > V_p$). Since this is the case, we proceed with the analysis.

Analysis of the ac-equivalent circuit gives us

$$\begin{aligned} R'_D &= R_D \parallel R_L \\ &= 1 \text{ k}\Omega \parallel 1 \text{ k}\Omega \\ &= 500 \ \Omega \end{aligned}$$

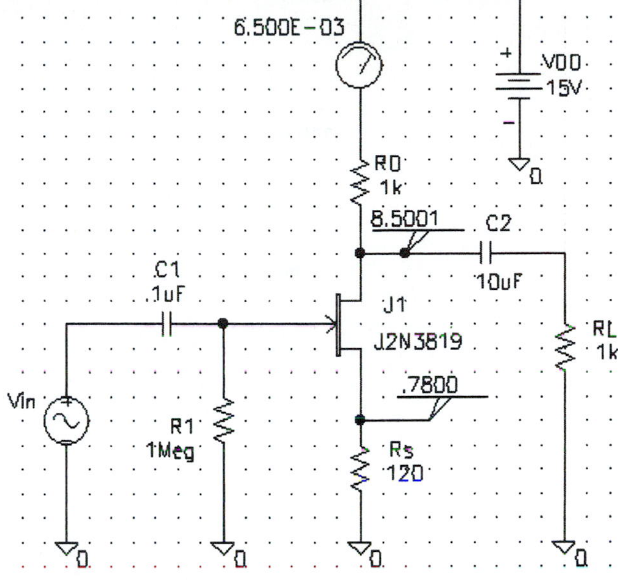

FIGURE 7.19
Self-biased, common-source amplifier, with source feedback.

$$R'_S = R_S = 120\ \Omega$$

$$\begin{aligned} A_v &= -g_{mQ}R_D/(1 + g_{mQ}R_S) \\ &= (-5.8\text{ mS})(500\ \Omega)/(1 + 0.696) \\ &= -1.7 \end{aligned}$$

The input voltage is set to 100 mV_{pk} at 1 kHz and a transient analysis is performed. Plotting the input and output voltages results in the display of Fig. 7.20, where we find the voltage gain to be very close to the hand-calculated value. If a source bypass capacitor is added (10 μF should work), the voltage gain should increase to

$$\begin{aligned} A_v &= -g_{mQ}R'_D \\ &= (-5.8\text{ mS})(500\ \Omega) \\ &= -2.9 \end{aligned}$$

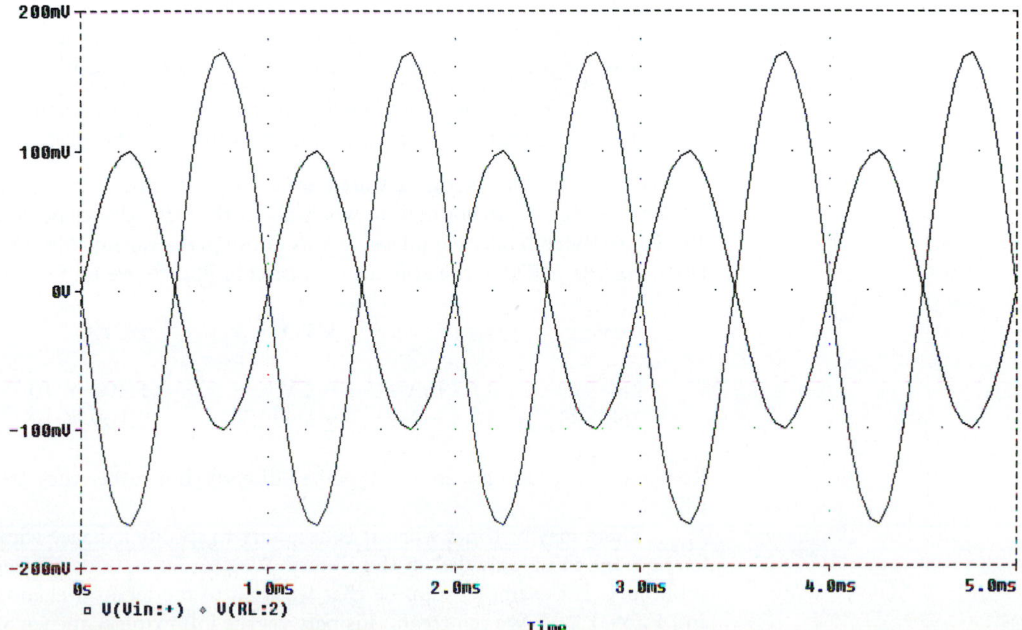

FIGURE 7.20
Input and output waveforms for Fig. 7.19.

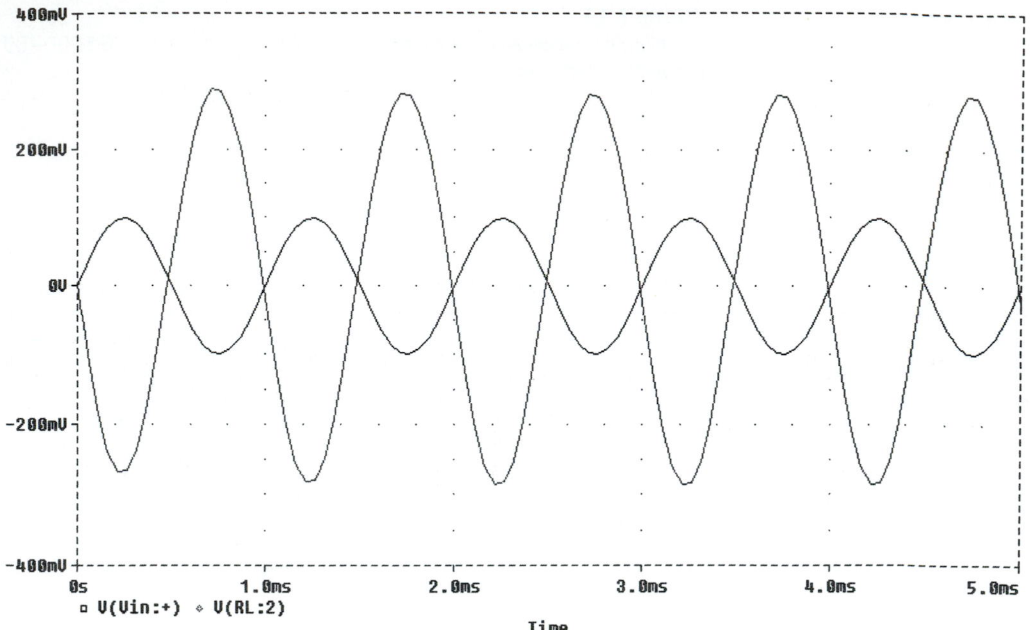

FIGURE 7.21
Bypassing the source in Fig. 7.19 results in an increase in voltage gain.

Making this modification and rerunning the simulation produces the results shown in Fig. 7.21, where we find good agreement with our calculations.

Defining Your Own Device Model

Until now, we have used the device models that are available in the evaluation library. These are excellent for learning the basics of simulation and verifying much of the theoretical information that is presented. Let's look at some of what we have available to work with. The basic FET model parameters used by PSpice were introduced in Chapter 6 and for convenience we restate them here.

Basic JFET Parameters

VTO: Threshold voltage; equivalent to $V_{GS(\text{off})}$. VTO is negative for n-channel JFETs and positive for p-channel devices. Default value: VTO $= -2$ V.

BETA: Transconductance coefficient; related to device parameters by BETA $= I_{DSS}/V_p^2$. This defaults to BETA $= 0.1$ mA/V^2 unless otherwise specified.

Unless the JFET is being operated at extreme temperatures or at relatively high frequencies ($f \gg 1$ MHz), we do not need to worry about the other device parameters. By the way, you can view all of the device model parameters by examining the output file for a given circuit that was simulated. For the JFETs in the evaluation version of PSpice, we find the following basic parameters:

Device	I_{DSS}	VTO ($-V_p$)	BETA	g_{m0}
2N3819	11.7 mA	-3 V	1.304×10^{-3}	7.8 mS
2N4393	18.4 mA	-1.422 V	9.109×10^{-3}	13.6 mS

Remember, I_{DSS} and g_{m0} are not specified directly, but rather they are defined in terms of VTO and BETA.

There may be times when it is necessary to specify a device whose parameters are radically different from those provided in the library. We can define our own device models using the *breakout* sublibrary. For example, suppose that we wish to model an n-channel JFET that has $I_{DSS} = 5$ mA and $V_P = 1.2$ V. We can create this part via the following sequence of steps:

1. In Schematics, choose the part from the sublibrary BREAKOUT.slb (in this case the n-channel JFET JbreakN).

2. Place the part.
3. With the part highlighted in red, click Edit, Model. (Save the schematic now if you are prompted by PSpice.)
4. Click on the Edit Instance Model [Text] command button. This opens the Model Editor window as shown in Fig. 7.22.
5. Change the name of the device to something like MyNJFET and enter the parameters as shown in Fig. 7.23, then click OK.

The new JFET is now part of your schematic. Just to be sure MyNJFET has been entered correctly, a DC Sweep is set up as shown in Fig. 7.24. Notice that the V_{GG} source has its positive terminal connected to the gate of the FET. We compensate by sweeping V_{GG} from 0 to −1.2 V.

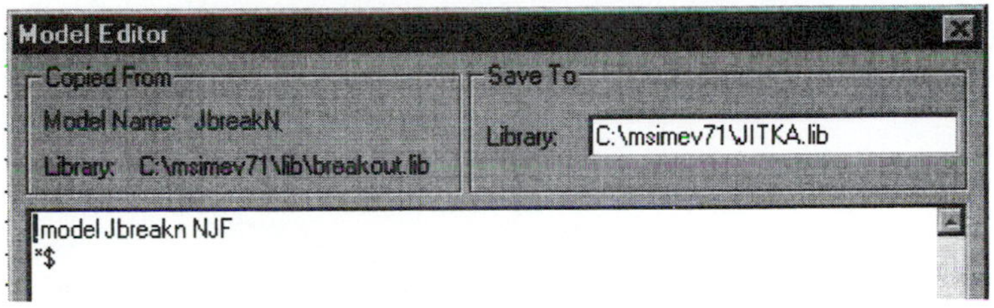

FIGURE 7.22
The Model Editor box.

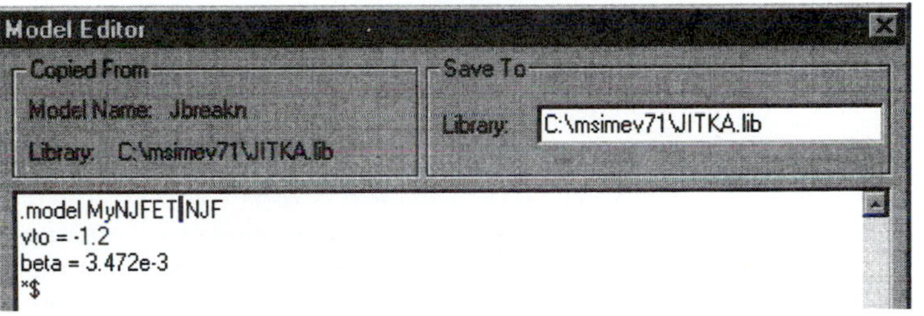

FIGURE 7.23
Modified attributes.

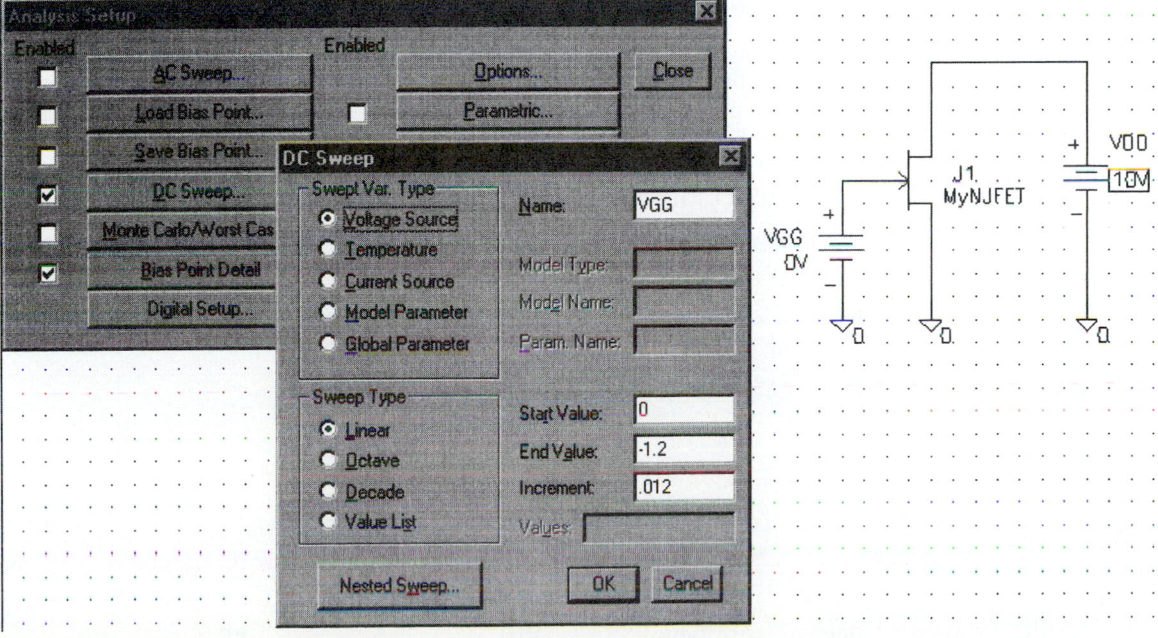

FIGURE 7.24
Setting up the DC Sweep of the gate bias voltage.

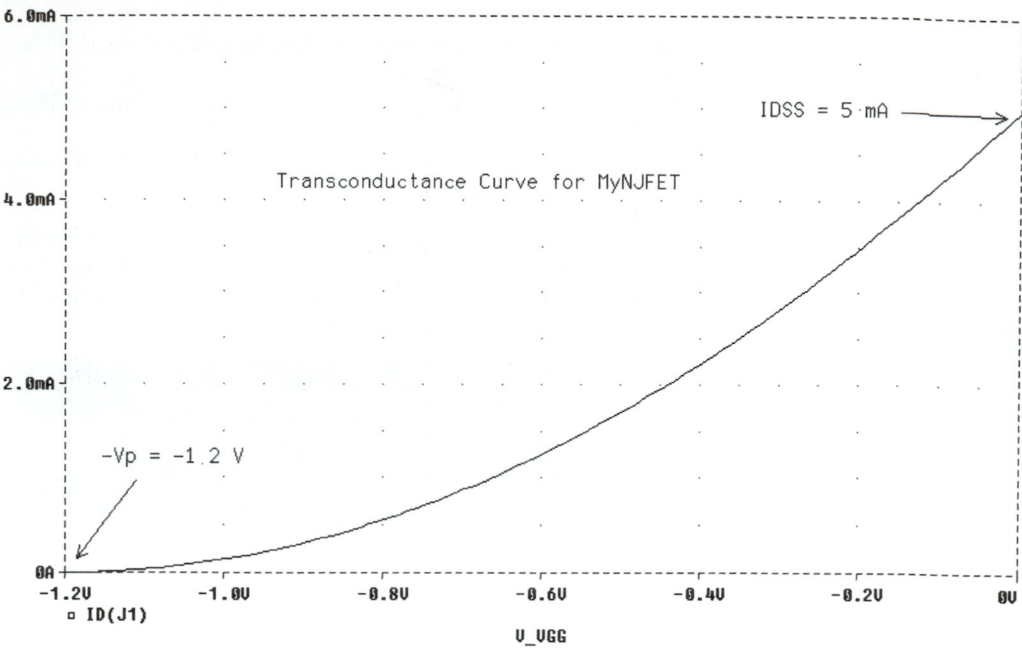

FIGURE 7.25
Transconductance curve.

Running the simulation and plotting the drain current results in the display of Fig. 7.25. This is the transconductance curve that is expected for our new JFET. We must keep in mind that since we only specified the two main device parameters, the remaining ones will be set to default values, which will produce an FET with rather ideal characteristics. This is acceptable for our purposes at this time. This FET model will be used in the problems at the end of this chapter.

The basic steps outlined above can be used to customize models of other devices such as BJTs, resistors, capacitors, and so on. You are encouraged to try this technique with some of the earlier BJT circuits we analyzed.

EMOS FET Amplifiers

PSpice directly models enhancement-mode MOSFETs, but several different model types (levels) are available. Rather than get into the details of all of the levels and their peculiarities, let's concentrate on the two device parameters that are of interest to us at this time, V_{to} (VTO) and KP. These parameters are redefined below for convenience.

Basic MOSFET Parameters

VTO: Threshold voltage; minimum gate-to-source voltage that causes formation of the inversion layer and the onset of drain-to-source conduction.

KP: Transconductance constant; this constant is derived using Eqs. (6.20) and (6.21) from the preceding chapter:

$$\mathrm{KP} = 2k \tag{6.20}$$

$$k = \frac{I_{D(\mathrm{on})}}{(V_{GS(\mathrm{on})} - V_{to})^2} \tag{6.21}$$

The EMOS FETs in the evaluation library have level 3 models and it is difficult to determine how constant KP was determined. However, looking at the output listings and using a probe to generate transconductance curves, the following information can be determined:

Device	Type	VTO	k	KP
IRF150	n-Channel	2.84 V	0.2589	0.5178
IRF9140	p-Channel	−3.67 V	3.5884	7.1767

Again, there will be many occasions in which the characteristics of neither of these devices is close enough to the actual device we wish to model. We will use the EMOS FET breakout device to handle these situations.

EXAMPLE 7.9

The circuit of Fig. 7.26 shows an n-channel EMOS FET configured to provide both common-source and source-follower outputs. Voltage-divider biasing is used, with dc source feedback provided by resistor R_4. The EMOS FET has the following specifications: $V_{to} = 2.5$ V, $I_{D(\text{on})} = 20$ mA, and $V_{GS(\text{on})} = 4$ V. Manually determine the Q-point coordinates and the voltage gain referred to each output (A_{v_1}, A_{v_2}). Verify your results using PSpice.

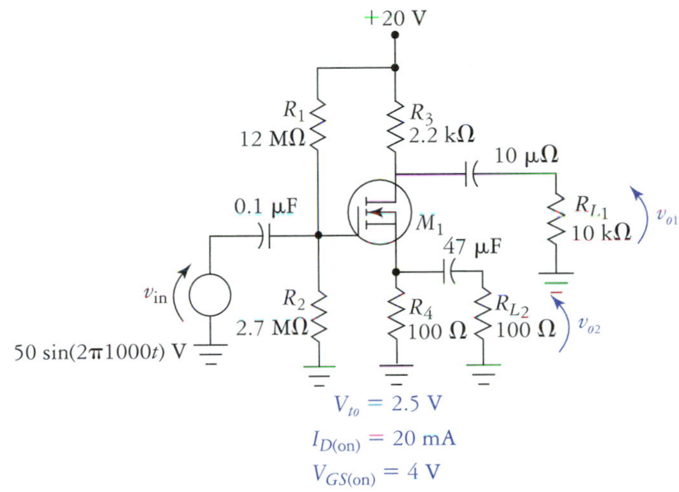

FIGURE 7.26
EMOS FET hybrid amplifier for Example 7.9.

Solution First we determine the device constant k:

$$k = \frac{I_{D(\text{on})}}{(V_{GS(\text{on})} - V_{to})^2}$$
$$= \frac{20 \text{ mA}}{(4 \text{ V} - 2.5 \text{ V})^2}$$
$$= 8.9 \text{ mA/V}^2$$

The gate voltage is found using the voltage divider equation:

$$V_{GQ} = V_{DD}\frac{R_2}{R_1 + R_2}$$
$$= 20 \text{ V}\frac{2.7}{2.7 + 12}$$
$$= 3.7 \text{ V}$$

The algebraic solution for the drain current is

$$I_{DQ} = \frac{-B - \sqrt{B^2 - 4AC}}{2A} \tag{6.9}$$

where

$$A = R_S^2$$

$$B = -(2R_S(V_G - V_{to}) + \frac{1}{k}$$

$$C = (V_G - V_{to})^2$$

These constants evaluate to $A = 10$ k, $B = -352$, and $C = 1.44$. Substituting into Eq. (6.9) yields

$$I_{DQ} = 4.7 \text{ mA}$$

The remaining dc parameters of interest are now determined:

$$\begin{aligned} V_{GSQ} &= V_{GQ} - I_D R_S \\ &= 3.7 \text{ V} - (4.7 \text{ mA})(100 \ \Omega) \\ &= 3.7 \text{ V} - 0.47 \text{ V} \\ &= 3.23 \text{ V} \end{aligned}$$

$$\begin{aligned} V_{DSQ} &= V_{DD} - I_D(R_S + R_D) \\ &= 20 \text{ V} - (4.7 \text{ mA})(100 \ \Omega + 2.2 \text{ k}\Omega) \\ &= 20 \text{ V} - 10.8 \text{ V} \\ &= 9.2 \text{ V} \end{aligned}$$

The operating-point transconductance is determined next:

$$\begin{aligned} g_{mQ} &= 2k\,(V_{GSQ} - V_{to}) \\ &= 0.0178(3.23 - 2.5) \\ &= 13 \text{ mS} \end{aligned}$$

The ac analysis proceeds as follows:

$$\begin{aligned} R'_D &= R_3 \parallel R_5 \\ &= 2.2 \text{ k}\Omega \parallel 10 \text{ k}\Omega \\ &= 1.8 \text{ k}\Omega \end{aligned}$$

$$\begin{aligned} R'_S &= R_4 \parallel R_6 \\ &= 100 \ \Omega \parallel 100 \ \Omega \\ &= 50 \ \Omega \end{aligned}$$

The voltage gain of the common-source output of this amplifier is defined as

$$A_{v(CS)} = \frac{v_{o_1}}{v_{\text{in}}} \tag{A}$$

Similarly, the voltage gain of the source-follower output of this amplifier is defined as

$$A_{v(SF)} = \frac{v_{o_2}}{v_{\text{in}}} \tag{B}$$

Equations (A) and (B) will be used to determine the voltage gain of Fig. 7.26 in conjunction with the PSpice transient analysis results. For manual analysis, the equivalent voltage gains are determined as follows:

$$\begin{aligned} A_{v(CS)} &= -\frac{g_{mQ} R'_D}{1 + g_{mQ} R'_S} \\ &= -\frac{23.4}{1.65} \\ &= -14.2 \end{aligned}$$

$$\begin{aligned} A_{v(SF)} &= \frac{g_{mQ} R'_S}{1 + g_{mQ} R'_S} \\ &= \frac{0.65}{1.65} \\ &= 0.39 \end{aligned}$$

The circuit is drawn with the Schematics editor shown in Fig. 7.27. The breakout device MbreakN was used. This is a level 1 device model (MbreakN3 is a level 3 device and so on). The name was changed to EMOS1, and VTO and KP are set as shown in the schematic. The ac input voltage is a 50-mV peak sine wave with $f = 1$ kHz. The transient analysis runs from $t = 0$ to 3 ms.

Running the analysis we obtain the voltages shown in Fig. 7.28. From these values we obtain the following data. The results of the algebraic analysis are shown in parentheses for comparison. The results are in good agreement.

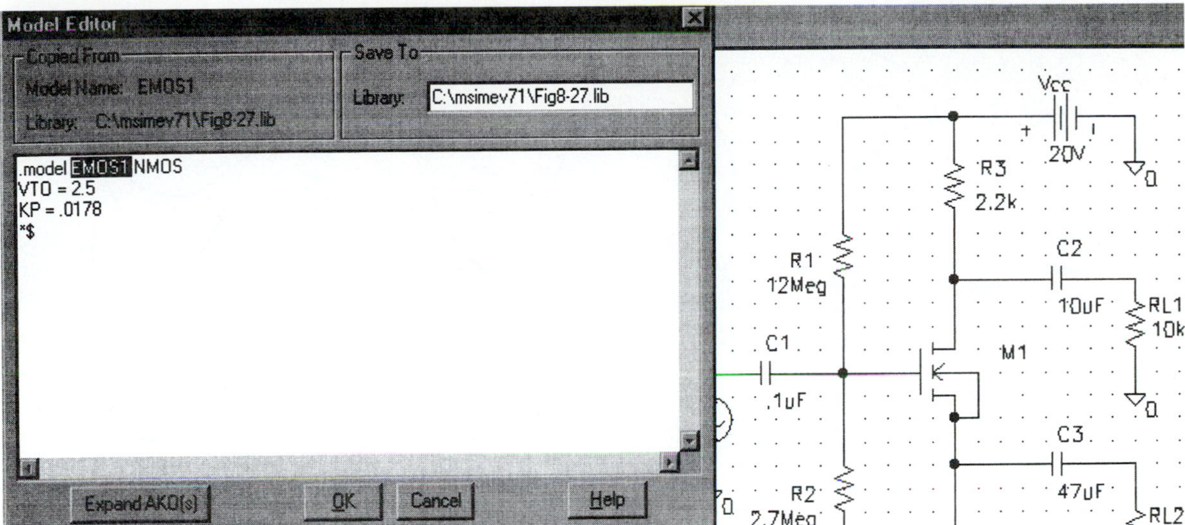

FIGURE 7.27
Setting the EMOS attributes.

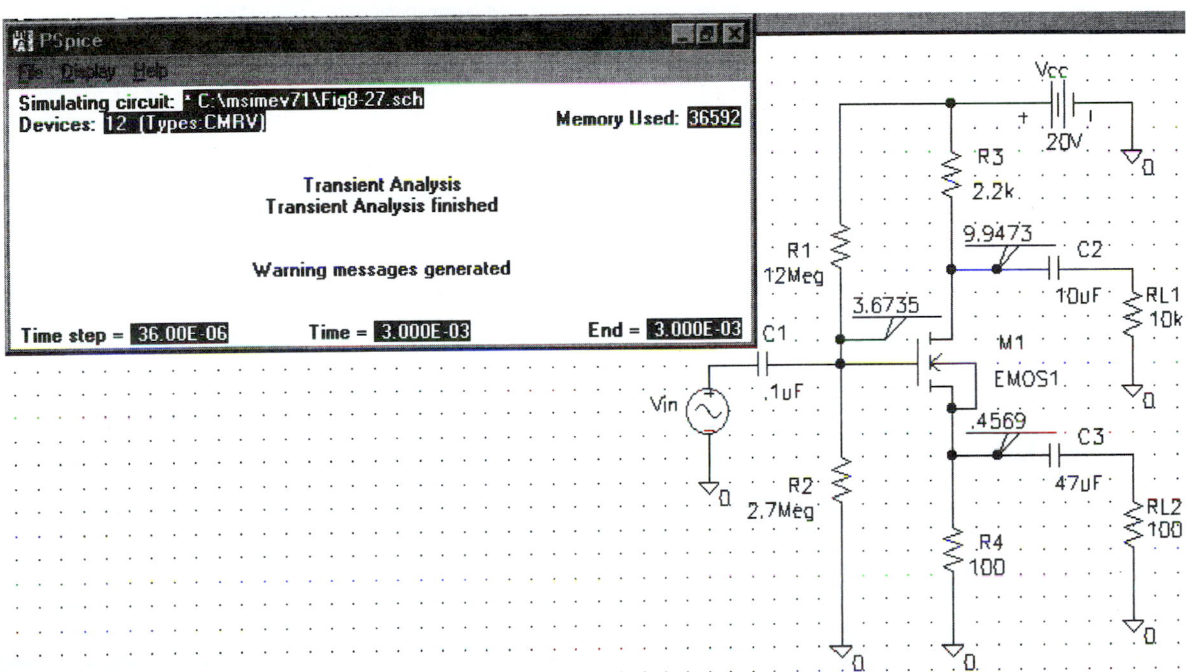

FIGURE 7.28
Warning messages are generated.

$$I_{DQ} = 4.57\text{ mA} \qquad (4.7\text{ mA})$$

$$V_{DSQ} = 9.49\text{ V} \qquad (9.2\text{ V})$$

$$V_{GSQ} = 3.22\text{ V} \qquad (3.23\text{ V})$$

We now determine the voltage gain of the amplifier. Figure 7.29 shows the results of the transient analysis. The CS (v_{o_1}) and SF (v_{o_2}) output signals were plotted in two separate graphs because of the large difference in amplitude between these signals. Reading off the lower graph and using Eq. (A), we find the gain as follows (the hand-calculated gain is in parentheses):

$$A_{v(CS)} = -0.698\text{ V}/0.05\text{ V}$$

$$\cong -14 \qquad (-14.2)$$

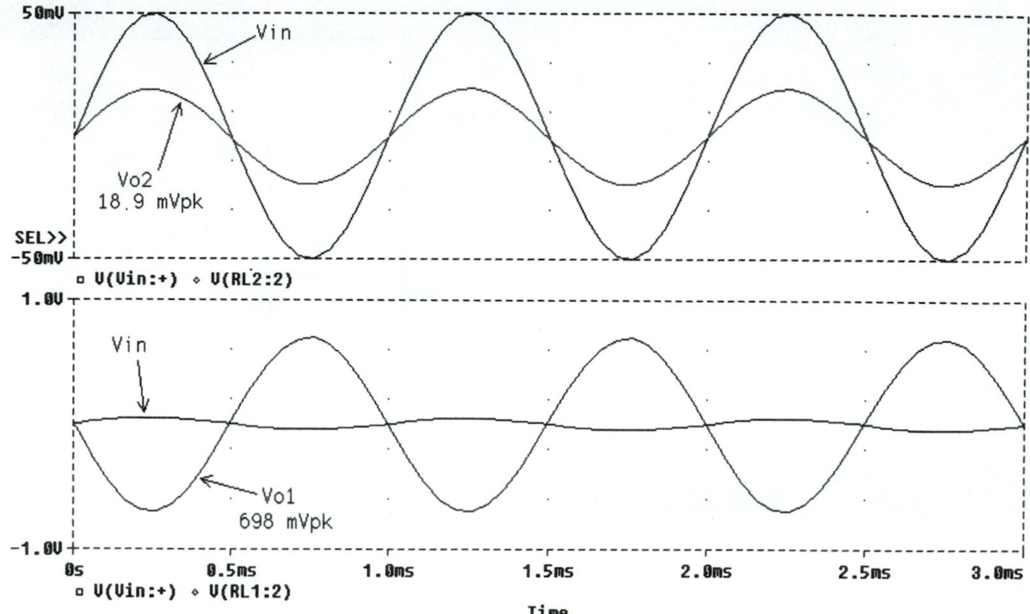

FIGURE 7.29
Input and output voltage waveforms. Note change of scale between upper and lower plots.

The upper plot gives us

$$A_{v(SF)} = 18.9 \text{ mV}/50 \text{ mV}$$
$$\cong 0.38 \qquad (0.39)$$

Again we have good agreement between the PSpice simulation and the theoretical analysis.

CHAPTER REVIEW

Field-effect transistors can be used in practically any application in which a normal bipolar transistor can be used. In fact, FETs are capable of performing some functions that bipolar transistors simply cannot, such as acting as a voltage-controlled resistance or an analog switch, for example. JFETs, depletion-mode MOSFETs, and enhancement-mode MOSFETs are all useful in small-signal amplifier applications, especially when extremely high input resistance is desirable.

Like their BJT counterparts, each FET amplifier configuration exhibits certain inherent characteristics. The common-source amplifier exhibits significant voltage and current gain, high input resistance, moderate output resistance, and phase inversion. The source follower has $A_v < 1$, with high input resistance and relatively low output resistance. The common-gate amplifier is capable of moderate voltage gain and output resistance, but has $A_i < 1$ and suffers from low input resistance. These characteristics limit the usefulness of the CG configuration in small-signal amplifier applications. In general, a bipolar transistor will be capable of producing higher voltage gain than a similar FET; however, FETs are normally capable of much higher current gain and input resistance.

The theoretical transconductance characteristic of all FETs is a parabolic or square-law function. The slope of the transconductance curve at the Q point determines the voltage gain of a given FET when used as a small-signal amplifier. Source bypassing is used to increase voltage gain in CS amplifiers. Unlike emitter bypassing, source bypassing does not significantly reduce amplifier input resistance.

Like small-signal BJT-based amplifiers, FET-based designs will exhibit output signal distortion when operated at high input levels. Unless clipping occurs, the distortion introduced by the FET nonlinearity will be less severe than that produced by a bipolar transistor. In the CS configuration, source feedback reduces distortion at the expense of voltage gain.

DESIGN AND ANALYSIS PROBLEMS

Note: Unless otherwise specified, assume all coupling and bypass capacitors have $X_C \cong 0\ \Omega$ under ac signal conditions.

7.1. Refer to Fig. 7.30. Given: $I_{DSS} = 10$ mA, $V_p = 5$ V, $R_S = 150\ \Omega$, $R_D = 680\ \Omega$, $R_L = 1$ kΩ, $R_G = 1$ MΩ, and $V_{DD} = 15$ V. Determine I_{DQ}, g_{m0}, V_{DSQ}, V_{GSQ}, g_{mQ}, A_v, A_i, A_p, R_{in}, and R_o.

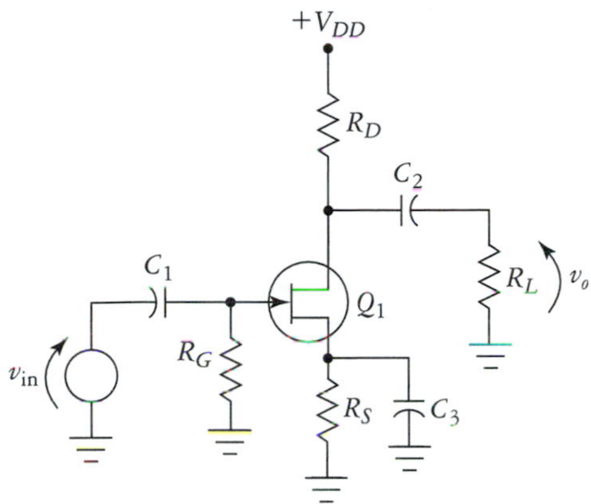

FIGURE 7.30
Circuit for Problems 7.1 to 7.4.

7.2. Refer to Fig. 7.30. Given: $I_{DSS} = 16$ mA, $V_p = 4$ V, $R_S = 100\ \Omega$, $R_D = 680\ \Omega$, $R_L = 1$ kΩ, $R_G = 1$ MΩ, and $V_{DD} = 15$ V. Determine I_{DQ}, g_{m0}, V_{DSQ}, V_{GSQ}, g_{mQ}, A_v, A_i, A_p, R_{in}, and R_o.

7.3. Using the values obtained for Problem 7.1, determine the instantaneous equations for v_o, i_D, and i_L for Fig. 7.30, given $v_{in} = 100 \sin(2\pi 1000t)$ mV.

7.4. Using the values obtained for Problem 7.2, determine the instantaneous equations for v_o, i_D, and i_L for Fig. 7.30, given $v_{in} = 250 \sin(2\pi 1000t)$ mV.

7.5. Refer to Fig. 7.31. Given: $I_{DSS} = 9.6$ mA, $g_{m0} = 6$ mS, $R_1 = 10$ MΩ, $R_2 = 1$ kΩ, $R_3 = 27\ \Omega$, $R_4 = 100\ \Omega$, $R_5 = 1$ kΩ, and $V_{DD} = 15$ V. Determine I_{DQ}, V_p, V_{DSQ}, V_{GSQ}, g_{mQ}, A_v, A_i, A_p, R_{in} and R_o.

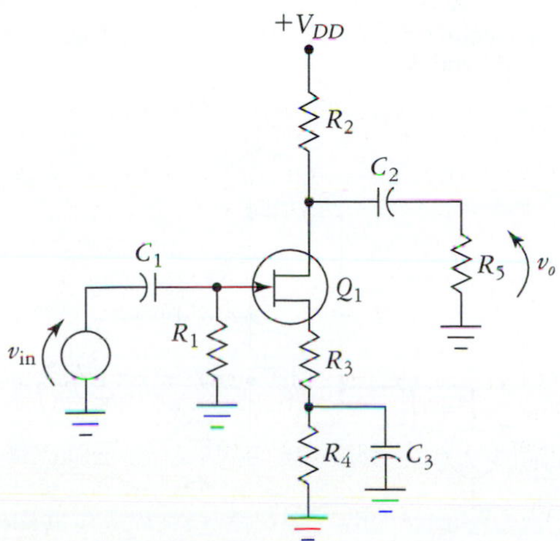

FIGURE 7.31
Circuit for Problems 7.5 and 7.6.

7.6. Refer to Fig. 7.31. Given: $I_{DSS} = 5$ mA, $g_{m0} = 5$ mS, $R_1 = 10$ MΩ, $R_2 = 1.8$ kΩ, $R_3 = 10\ \Omega$, $R_4 = 270\ \Omega$, $R_5 = 2.2$ kΩ, and $V_{DD} = 12$ V. Determine I_{DQ}, V_p, V_{DSQ}, V_{GSQ}, g_{mQ}, A_v, A_i, A_p, R_{in}, and R_o.

7.7. Refer to Fig. 7.32. Given: $V_{DD} = 18$ V, $I_{DSS} = 15$ mA, $V_p = 5$ V, $R_1 = 18$ MΩ, $R_2 = 2.2$ MΩ, $R_3 = 820\ \Omega$, $R_4 = 220\ \Omega$, and $R_L = 4.7$ kΩ. Determine g_{m0}, I_{DQ}, V_{GQ}, V_{GSQ}, V_{DSQ}, R_{in}, R_o, A_v, A_i and A_p.

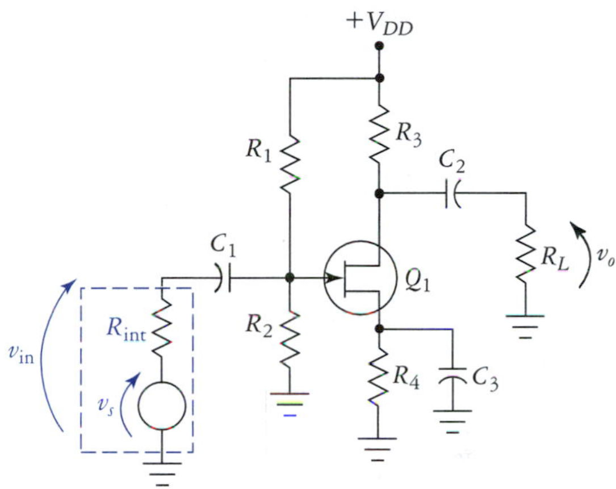

FIGURE 7.32
Circuit for Problems 7.7 to 7.10.

7.8. Refer to Fig. 7.32. Given: $V_{DD} = 15$ V, $I_{DSS} = 5$ mA, $V_p = 2$ V, $R_1 = 15$ MΩ, $R_2 = 1$ MΩ, $R_3 = 500\ \Omega$, $R_4 = 330\ \Omega$, and $R_L = 10$ kΩ. Determine g_{m0}, I_{DQ}, V_{GQ}, V_{GSQ}, V_{DSQ}, R_{in}, R_o, A_v, A_i, and A_p.

7.9. Using the values found in Problem 7.7 (for Fig. 7.32), determine v_o/v_s given $R_S = 100$ kΩ.

7.10. Using the values found in Problem 7.8, determine v_o/v_s given $R_S = 47$ kΩ.

7.11. Refer to Fig. 7.33. Assume $V_{DD} = 15$ V, $R_1 = 4.7$ MΩ, $R_2 = 560\ \Omega$ and $R_3 = 1$ kΩ. For the FETs we have

Q_1: $I_{DSS} = 18$ mA, $V_p = 3$ V

Q_2: $I_{DSS} = 9$ mA, $V_p = 2.5$ V.

Determine $g_{m0(Q_1)}$, $g_{m0(Q_2)}$, I_{DQ}, $V_{GSQ(1)}$, $g_{mQ(1)}$, R_{in}, R_o, A_v, A_i, and A_p.

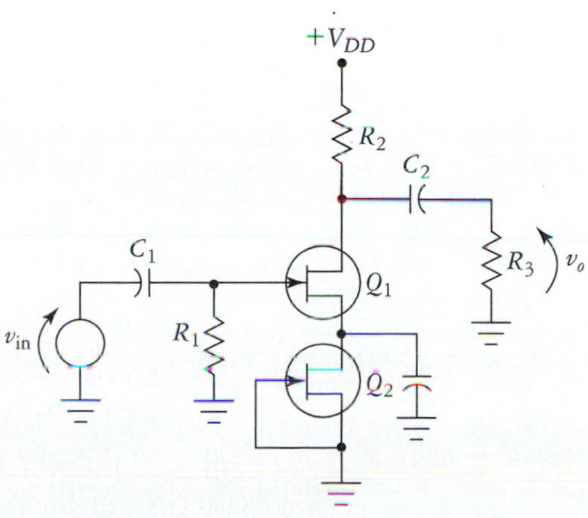

FIGURE 7.33
Circuit for Problems 7.11 and 7.12.

7.12. Refer to Fig. 7.33. Assume $V_{DD} = 15$ V, $R_1 = 1$ MΩ, and $R_2 = R_3 = 2.2$ kΩ. The FET parameters are
Q_1: $I_{DSS} = 10$ mA, $V_p = 2$ V
Q_2: $I_{DSS} = 3$ mA, $V_p = 4$ V.
Determine $g_{m0(Q_1)}$, $g_{m0(Q_2)}$, I_{DQ}, $V_{GSQ(1)}$, $g_{mQ(1)}$, R_{in}, R_o, A_v, A_i, and A_p.

7.13. Refer to Fig. 7.34. Given: $V_{DD} = 15$ V, $R_1 = 1$ MΩ, $R_2 = 100$ Ω, $R_3 = 100$ Ω, $I_{DSS} = 25$ mA, and $V_p = 6$ V. Determine I_{DQ}, V_{GSQ}, g_{m0}, g_{mQ}, A_v, R_{in}, R_o, A_i and A_p.

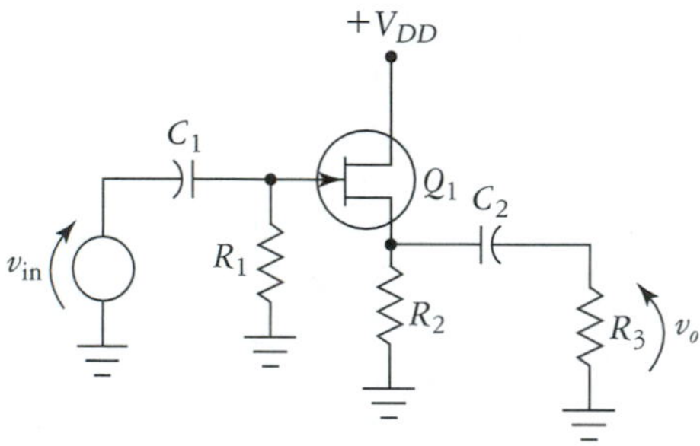

FIGURE 7.34
Circuit for Problems 7.13 and 7.14.

7.14. Refer to Fig. 7.34. Given: $V_{DD} = 12$ V, $R_1 = 1$ MΩ, $R_2 = 50$ Ω, $R_3 = 75$ Ω, $I_{DSS} = 5$ mA, and $V_p = 2.2$ V. Determine I_{DQ}, V_{GSQ}, g_{m0}, g_{mQ}, A_v, R_{in}, R_o, A_i, and A_p.

7.15. Refer to Fig.7.35. Given: $V_{DD} = 15$ V, $R_1 = 180$ Ω, $R_2 = 470$ Ω, $R_3 = 500$ Ω, $I_{DSS} = 15$ mA, and $g_{m0} = 6$ mS. Determine V_p, I_{DQ}, V_{GSQ}, V_{DSQ}, g_{mQ}, R_{in}, R_o, A_v, A_i, and A_p.

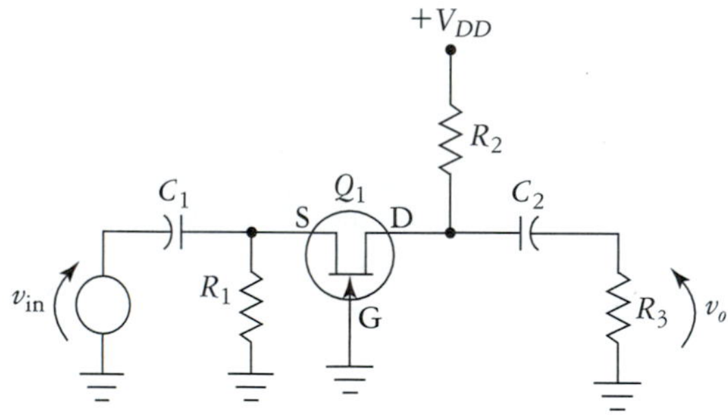

FIGURE 7.35
Circuit for Problems 7.15 and 7.16.

7.16. Refer to Fig. 7.35. Assume $V_{DD} = 15$ V, $R_1 = 100$ Ω, $R_2 = 1.5$ kΩ, $R_3 = 1$ kΩ, $I_{DSS} = 8$ mA, and $g_{m0} = 5$ mS. Determine V_p, I_{DQ}, V_{GSQ}, V_{DSQ}, g_{mQ}, R_{in}, R_o, A_v, A_i, and A_p.

7.17. Refer to Fig. 7.36. Assume $V_{DD} = 12$ V, $R_1 = 1$ MΩ, $R_2 = 220$ Ω, $R_3 = 1$ kΩ, $I_{DSS} = 20$ mA, and $V_p = 4.2$ V. Determine I_{DQ}, V_{DSQ}, g_{mQ}, R_{in}, R_o, A_v, A_i, and A_p.

7.18. Refer to Fig. 7.36. Assume $V_{DD} = 15$ V, $R_1 = 1$ MΩ, $R_2 = 1.8$ kΩ, $R_3 = 1.5$ kΩ, $I_{DSS} = 4$ mA, and $V_p = 1.5$ V. Determine I_{DQ}, V_{DSQ}, g_{mQ}, R_{in}, R_o, A_v, A_i, and A_p.

7.19. Refer to Fig. 7.37. Assume $V_{DD} = 25$ V, $R_1 = 1$ MΩ, $R_2 = 1$ kΩ, $R_3 = 68$ Ω, $R_4 = 10$ kΩ, $R_5 = 50$ Ω, $I_{DSS} = 18$ mA, and $V_p = -3.8$ V. Determine the dc parameters I_{DQ}, V_{GSQ}, V_{DSQ}, and g_{mQ}. v_{o_1} represents the CS configuration output while v_{o_2} is the SF output. Determine the following quantities for each case (CS and SF): A_v, R_{in}, R_o, A_i, and A_p.

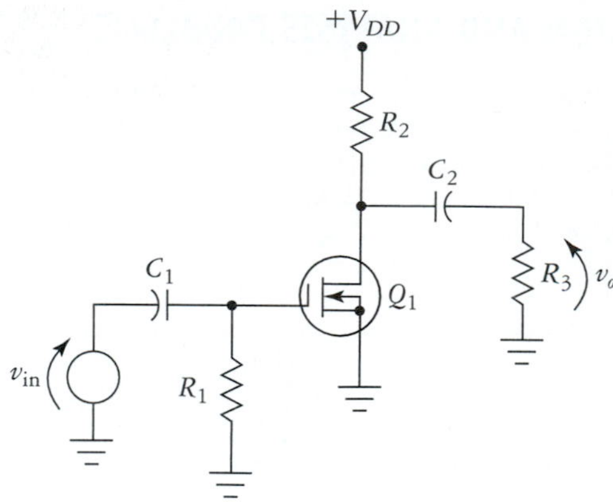

FIGURE 7.36
Circuit for Problems 7.17 and 18.

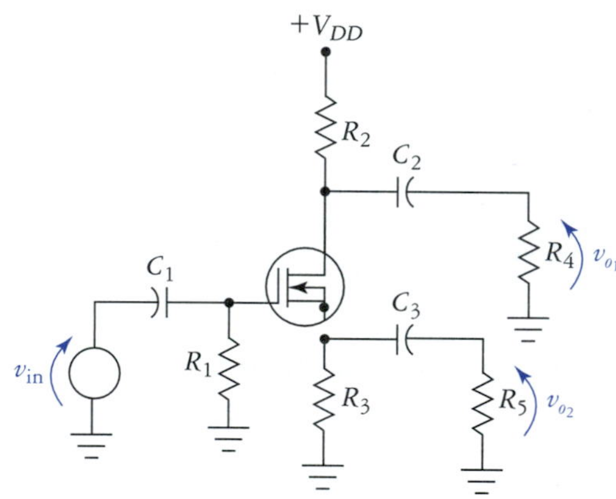

FIGURE 7.37
Circuit for Problems 7.19 and 7.20.

7.20. Refer to Fig. 7.37. Assume $V_{DD} = 18$ V, $R_1 = 5$ MΩ, $R_2 = 1.8$ kΩ, $R_3 = 220$ Ω, $R_4 = 4.7$ kΩ, $R_5 = 25$ Ω, $I_{DSS} = 7$ mA, and $V_p = -2.6$ V. Determine the dc parameters I_{DQ}, V_{GSQ}, V_{DSQ}, and g_{mQ}. v_{o_1} represents the CS configuration output while v_{o_2} is the SF output. Determine the following quantities for each case (CS and SF): A_v, R_{in}, R_o, A_i, and A_p.

7.21. Refer to Fig. 7.38. Given: $V_{DD} = 15$ V, $V_{to} = 3$ V, $I_{D(on)} = 10$ mA, $V_{GS(on)} = 5$ V, $R_1 = 4.7$ MΩ, $R_2 = 3.3$ MΩ, $R_3 = 200$ Ω, and $R_4 = 500$ Ω. Determine V_{GSQ}, I_{DQ}, g_{mQ}, V_{DSQ}, R_{in}, R_o, A_v, A_i, and A_p.

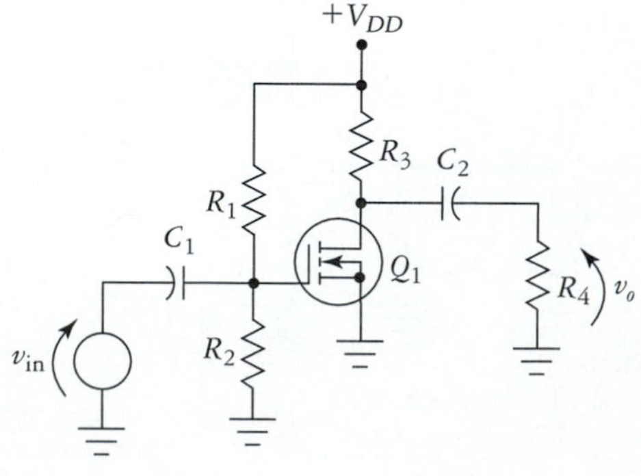

FIGURE 7.38
Circuit for Problems 7.21 to 7.24.

7.22. Refer to Fig. 7.38. Given: $V_{DD} = 20$ V, $V_{to} = 4.5$ V, $I_{D(on)} = 15$ mA, $V_{GS(on)} = 6$ V, $R_1 = 10$ MΩ, $R_2 = 3.9$ MΩ, $R_3 = 1$ kΩ, and $R_4 = 1$ kΩ. Determine V_{GSQ}, I_{DQ}, g_{mQ}, V_{DSQ}, R_{in}, R_o, A_v, A_i, and A_p.

7.23. Refer to Fig. 7.38. Given: $V_{DD} = 22$ V, $V_{to} = 2.5$ V, $V_{GS(on)} = 4.5$ V, $I_{D(on)} = 15$ mA, $R_1 = 12$ MΩ, and $R_4 = 470$ Ω. Determine R_2 and R_3 such that 18 mA $< I_{DQ} <$ 20 mA and 10 V $< V_{DSQ} <$ 12 V. Also determine A_v, R_{in}, R_o, A_i, and A_p for the amplifier. Use standard 5% tolerance resistance values.

7.24. Refer to Fig. 7.38. Given: $V_{DD} = 15$ V, $V_{to} = 3$ V, $V_{GS(on)} = 5$ V, $I_{D(on)} = 20$ mA, $R_2 = 1$ MΩ, and $R_4 = 1$ kΩ. Determine R_1 and R_3 such that 8 mA $< I_{DQ} <$ 12 mA and 6 V $< V_{DSQ} <$ 8 V. Also determine A_v, R_{in}, R_o, A_i, and A_p for the amplifier.

7.25. Refer to Fig. 7.39. Given: $V_{DD} = 15$ V, $R_1 = 10$ MΩ, $R_2 = 4.7$ MΩ, $R_3 = 120$ Ω, $R_4 = 120$ Ω, $V_{to} = 3.8$ V, $I_{D(on)} = 8$ mA, and $V_{GS(on)} = 5$ V. Determine k, I_{DQ}, V_{GQ}, V_{GSQ}, V_{DSQ}, g_{mQ}, R_{in}, R_o, A_v, A_i, and A_p.

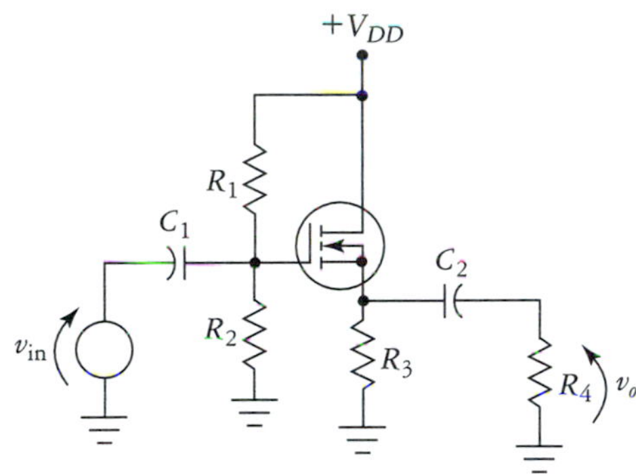

FIGURE 7.39
Circuit for Problems 7.25 and 7.26.

7.26. Refer to Fig. 7.39. Given: $V_{DD} = 12$ V, $R_1 = 470$ kΩ, $R_2 = 470$ kΩ, $R_3 = 220$ Ω, $R_4 = 100$ Ω, $V_{to} = 2.5$ V, $I_{D(on)} = 18$ mA, and $V_{GS(on)} = 5$ V. Determine k, I_{DQ}, V_{GQ}, V_{GSQ}, V_{DSQ}, g_{mQ}, R_{in}, R_o, A_v, A_i, and A_p.

7.27. Refer to Fig. 7.40. Assume $V_{DD} = 12$ V, $V_{to} = 4$ V, $I_{D(on)} = 30$ mA, $V_{GS(on)} = 8$ V, $R_1 = 1$ MΩ, $R_3 = R_2 = 1$ kΩ. Determine I_{DQ}, V_{GSQ}, V_{DSQ}, g_{mQ}, R_{in}, R_o, A_v, A_i, and A_p.

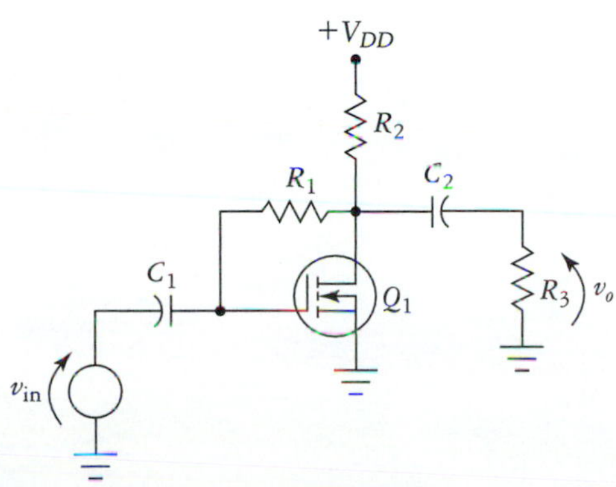

FIGURE 7.40
Circuit for Problems 7.27 and 7.28.

7.28. Refer to Fig. 7.40. Assume $V_{DD} = 15$ V, $V_{to} = 3$ V, $I_{D(on)} = 10$ mA, $V_{GS(on)} = 5$ V, $R_1 = 2.2$ MΩ, and $R_2 = R_3 = 1$ kΩ. Determine I_{DQ}, V_{GSQ}, V_{DSQ}, g_{mQ}, R_{in}, R_o, A_v, A_i, and A_p.

7.29. Refer to Fig. 7.41. This circuit uses an npn bipolar transistor as a current source in order to bias up an EMOS FET for use as a linear amplifier. The BJT has $\beta = 100$ and $V_{BE(on)} = 0.7$ V. The FET has $V_{to} = 4.2$ V, $I_{D(on)} = 15$ mA, and $V_{GS(on)} = 6$ V. Determine I_{DQ}, V_{DSQ}, V_{GSQ}, g_{mQ}, V_{CEQ}, A_v, R_{in}, R_o, A_i and A_p.

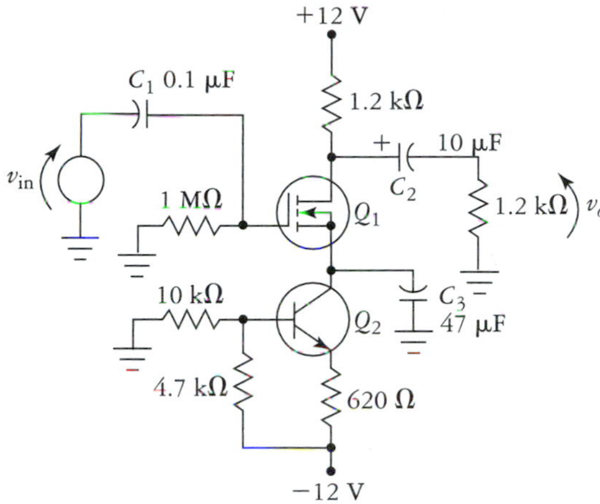

FIGURE 7.41
Circuit for Problem 7.29.

7.30. The circuit shown in Fig. 7.42 uses a feedback technique called *bootstraping* to increase the input resistance as seen by the v_{in} source. Derive an expression for R_{in} in terms of the circuit resistances and A_v.

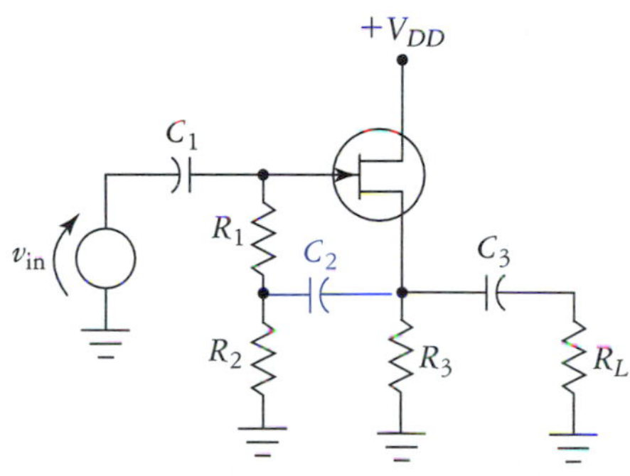

FIGURE 7.42
Circuit for Problem 7.30.

■ TROUBLESHOOTING PROBLEMS

7.31. The circuit of Fig. 7.10 is constructed using $R_1 = 1$ MΩ, $R_2 = 470$ kΩ, $R_D = 100$ Ω, and $V_{DD} = 15$ V. A voltage reading taken at the gate of the FET indicates $V_G = 0.1$ V. State two possible causes for this reading.

7.32. The circuit of Fig. 7.43 is constructed and the voltages shown are obtained using a voltmeter. Assuming that the JFET is good, it should have 5 mA $< I_{DSS} <$ 10 mA, and 1.0 V $< V_p <$ 2.5 V. Explain how one could quickly deduce that the circuit is not functioning properly.

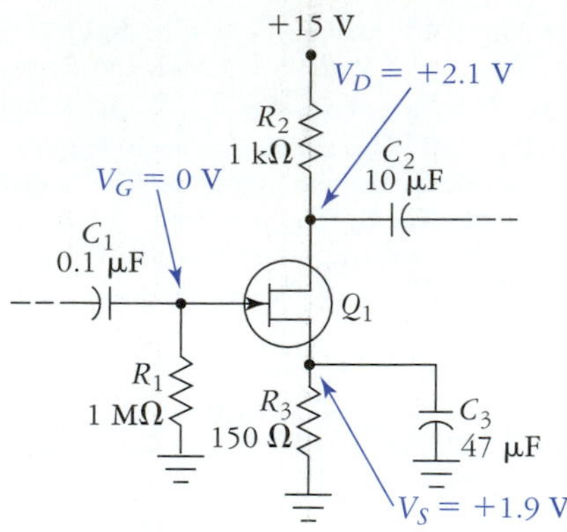

FIGURE 7.43
Circuit for Problem 7.32.

7.33. The voltage waveforms shown in Fig. 7.44 are displayed on an oscilloscope. What is wrong, and what is the most likely cause of the problem?

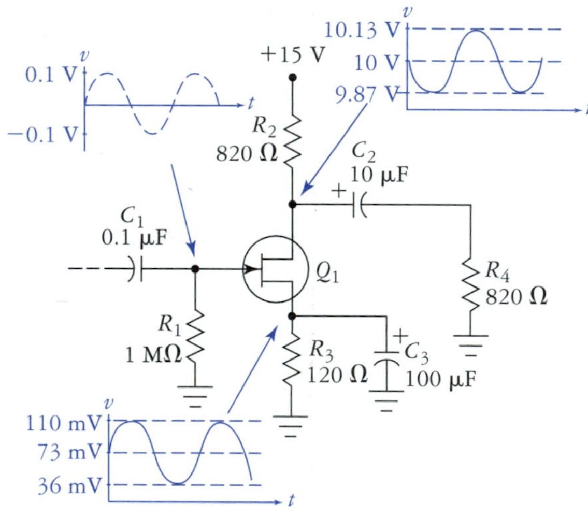

FIGURE 7.44
Circuit for Problem 7.33.

■ COMPUTER PROBLEMS

7.34. Use PSpice to perform an operating-point and transient analysis of Fig. 7.33 using $V_{DD} = 20$ V, $R_1 = 1$ MΩ, $R_2 = 1$ kΩ, $R_3 = 1$ kΩ, $C_1 = 0.1$ μF, $C_2 = 10$ μF, $C_3 = 47$ μF, $I_{DSS(1)} =$ 20 mA, $I_{DSS(2)} = 10$ mA, $V_{p(1)} = 2.5$ V, and $V_{p(2)} = 2$ V. Use breakout devices for the JFETS. Specify v_{in} as a 100-mV$_{\text{pk}}$ sine wave with $f = 1$ kHz. Specify the transient analysis such that three cycles of the output are plotted, with at least 10 data points/cycle.

7.35. Refer to Fig. 7.40. Use a level 1 EMOS FET breakout device with $V_{to} = 5$ V, $I_{D(\text{on})} = 16$ mA, and $V_{GS(\text{on})} = 8$ V. Given $V_{DD} = 25$ V, $R_1 = 1$ MΩ, $R_2 = 1$ kΩ, $R_3 = 1$ kΩ, and $C_1 = 1$ μF, and $C_2 = 10$ μF, use Pspice to determine Q-point and the gain.

7.36. Given the component values of Problem 7.29, use PSpice to determine the operating point and voltage gain for Fig.7.41. Use breakout devices for the EMOS FET (level 1) and the BJT.

7.37. We have seen before that PSpice can be used to determine the input resistance of a circuit as a function of frequency. Using the bootstrapped circuit of Fig. 7.42, $V_{DD} = 15$ V, $R_1 =$ 100 kΩ, $R_2 = 1$ MΩ, $R_3 = 200$ Ω, $R_L = 1$ kΩ, and $C_1 = C_2$ = 10 μF. Use the 2N3819 n-channel JFET from the evaluation library. Sweep the frequency of the input voltage in decades from 10 Hz to 100 MHz, 101 points/decade. Set v_{in} for an ac amplitude of 1 V.

7.38. Using the circuit developed in Problem 7.37, run a transient analysis with $v_{\text{in}} = 100 \sin(2\pi 1000t)$ mV over a 2-ms interval. Plot the input current $[-i(v_{\text{in}})]$ and the input resistance using $v(1)/i(v_{\text{in}})$. You should observe that R_{in} is not a constant for this circuit, but rather it is a function of the amplitude of v_{in}, with R_{in} decreasing to 0 Ω when $v_{\text{in}} = 0$ V. This is not a problem, however, because the input current is also zero at this point (the resistance is actually undefined at these points because we have $R_{\text{in}} = v_{\text{in}}/i_{\text{in}} = 0/0$).

7.39. Repeat the analysis of Problem 7.38 with the feedback capacitor C_2 removed from the circuit. You should find that the input resistance is relatively constant at 1.1 MΩ. There may be some variation at the times when v_{in} is crossing zero, but this is an arithmetic artifact and not an actual reduction of R_{in}.

CHAPTER 8

Differential Amplifiers

BEHAVIORAL OBJECTIVES

On completion of this chapter, the student should be able to:

- Describe the operation of a differential pair.
- Perform a dc Q-point analysis of a differential amplifier.
- Describe the operation of current-source biasing circuits.
- Perform a small-signal ac analysis of a differential amplifier.
- Describe the characteristics of a differential signal.
- Describe the characteristics of a single-ended signal.
- Define common-mode rejection.
- Convert linear gain to decibels and vice versa.
- Use PSpice to analyze differential amplifier circuits.

INTRODUCTION

The differential amplifier is one of the most versatile of all linear electronic circuits. A very large number of the linear integrated circuits that are so widely used today are based on the differential amplifier. In fact, one could reasonably argue that in terms of small-signal linear system analysis and design, the differential pair is the most important of all transistor configurations. This chapter covers the basics of differential amplifier operation, characteristics, and design. Knowledge of the operation of the differential amplifier will allow the reader to gain a deeper understanding of the characteristics and limitations of a large class of linear integrated circuits, operational amplifiers, which are covered later in this text.

8.1 DC ANALYSIS OF THE DIFFERENTIAL AMPLIFIER

Figure 8.1 illustrates a two-transistor configuration that is generally referred to as a *differential pair.* Perhaps the most obvious feature of the differential pair is the connection of the emitters of Q_1 and Q_2 to a single resistor R_{EE}. The significance of this connection will be made clear in the upcoming discussions. Since ultimately, this circuit will be used to amplify signals, it is often simply called a *differential amplifier,* or *diff amp* for short.

Note
Transistors in the differential amplifier should be closely matched. Close matching of device parameters is readily achieved using IC fabrication techniques.

FIGURE 8.1
The basic differential configuration.

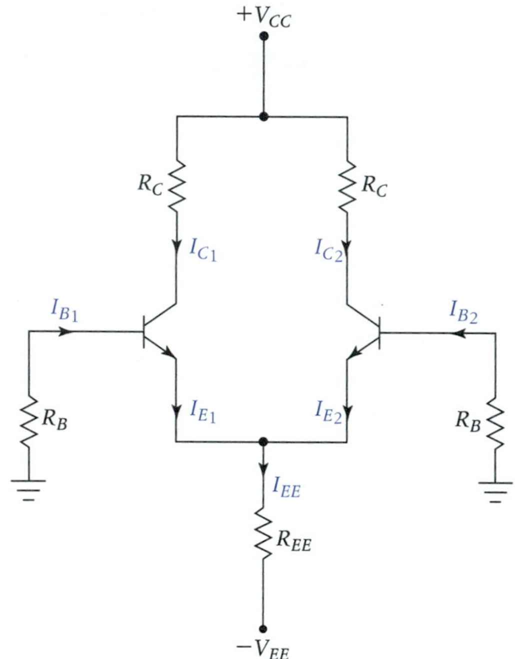

Differential-Pair Current Relationships

The analysis of the differential amplifier in Fig. 8.1 is relatively straightforward, if a few basic assumptions are made at the outset. First, let us assume that Q_1 and Q_2 are a *matched pair.* That is, both transistors have identical transconductance curves, saturation currents, emission coefficients, temperature tracking characteristics, and so on. Let us also assume that both transistors are operating at the same temperature. Resistors will be assumed equal in value if they share the same subscripts. Such close matching of component characteristics is easily achieved using integrated circuit techniques, therefore these are generally valid assumptions for practical differential amplifier circuits.

For dc analysis purposes, the differential amplifier can be modeled by the circuit shown in Fig. 8.2. The emitter current for each of the transistors can be determined by applying Kirchhoff's voltage law (KVL) around loops 1 and 2. Analyzing both loops yields the following two equations:

For Loop 1:

$$0 = V_{EE} - V_{BE} - I_{B_1}R_{B_1} - R_{EE}(I_{E_1} + I_{E_2}) \tag{8.1a}$$

For Loop 2:

$$0 = V_{EE} - V_{BE} - I_{B_2}R_{B_2} - R_{EE}(I_{E_1} + I_{E_2}) \tag{8.1b}$$

Because of the identical characteristics of the components used in each half of the circuit, Eq. (8.1a) and (8.1b) differ only in the subscripts that are used in each I_B term. Remember, the values of the terms in each equation are equal because of the matching of the circuit elements in loops 1 and 2. Because of this matching plus the fact that the two loops are in parallel with each other, we can combine Eqs. (8.1a) and (8.1b) into a single expression:

$$0 = V_{EE} - V_{BE} - I_BR_B - I_ER_{EE} - I_ER_{EE} \tag{8.2}$$

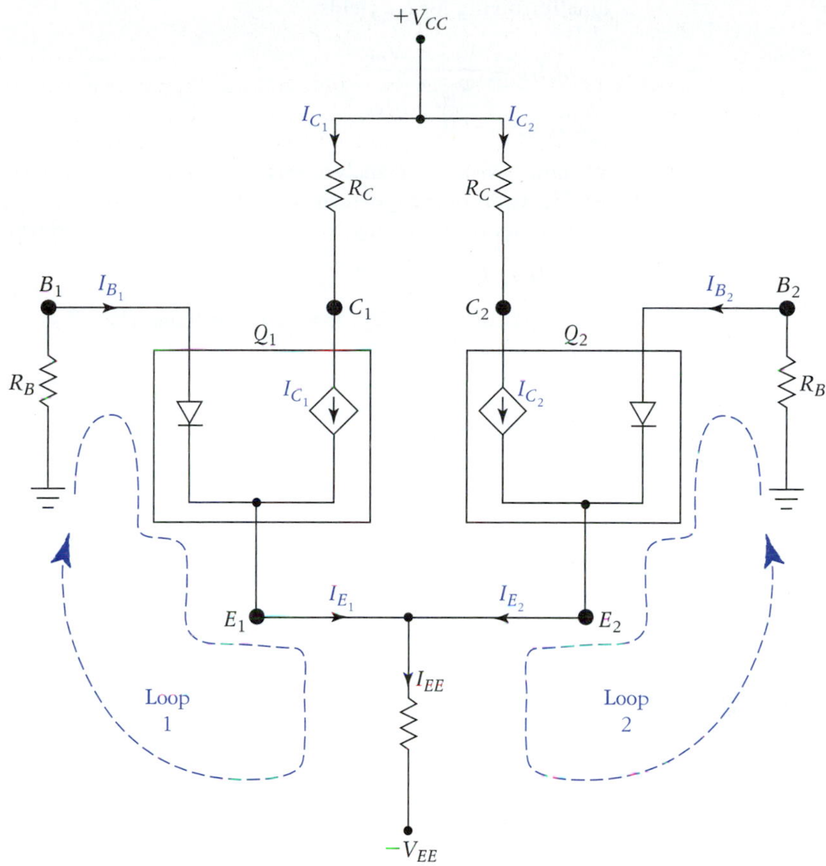

FIGURE 8.2
Differential pair using dependent-source device models.

where

$$I_{E_1} = I_{E_2} = I_E$$
$$I_{B_1} = I_{B_2} = I_B$$
$$R_{B_1} = R_{B_2} = R_B$$

To produce common factors of I_E in each term of Eq. (8.2), we can define I_B in terms of β using the following approximate relation:

$$I_B \cong \frac{I_E}{\beta} \qquad (\text{because } \beta \cong \beta + 1) \tag{8.3}$$

Substituting Eq. (8.3) into (8.2) produces

$$0 = V_{EE} - V_{BE} - \frac{I_E}{\beta} R_B - I_E R_{EE} - I_E R_{EE}$$

Subtracting V_{EE} and V_{BE} from both sides and multiplying by -1 yields

$$V_{EE} - V_{BE} - \frac{I_E}{\beta} R_B - I_E R_{EE} - I_E R_{EE}$$

Factoring the right-hand side of this equation produces

$$V_{EE} - V_{BE} = I_E \left(\frac{R_B}{\beta} + 2R_{EE} \right)$$

Since $I_{EE} = I_E + I_E = I_{E_1} + I_{E_2}$, we can write

$$V_{EE} - V_{BE} = I_{EE} \left(\frac{R_B}{2\beta} + R_{EE} \right)$$

Finally, solving for I_{EE} yields

$$I_{EE} = \frac{V_{EE} - V_{BE}}{\dfrac{R_B}{2\beta} + R_{EE}} \tag{8.4}$$

We now have a useful analysis equation for the differential amplifier. Because both sides of the circuit are identical, under quiescent conditions the current I_{EE} divides equally between the two emitters. This allows the individual emitter currents to be found using

$$I_E = I_{EE}/2 \tag{8.5}$$

In a practical circuit, R_{EE} is very much larger than $R_B/2\beta$. This allows further simplification of Eq. (8.4), producing

$$I_{EE} \cong \frac{V_{EE} - V_{BE}}{R_{EE}} \tag{8.6}$$

Regardless of which I_{EE} equation is used (Eqs. 8.4 or 8.6), the *quiescent* emitter currents for each side of the circuit are assumed to be the same, and equal to $I_{EE}/2$. By the way, we will find out soon that ideally I_{EE} is constant, even when a signal is applied to the circuit. The application of a signal (or signals) causes $I_{E_1} \neq I_{E_2}$, which upsets the balance of the differential pair; however, $I_{EE} = I_{E_1} + I_{E_2}$ will still be true.

EXAMPLE 8.1

Determine the quiescent values of I_{EE}, I_{E_1}, I_{E_2}, I_{C_1}, I_{C_2}, I_{B_1}, and I_{B_2} for the circuit in Fig. 8.3 based on use of Eq. (8.4). Assume that $\beta = 100$ and $V_{BE} = 0.7$ V.

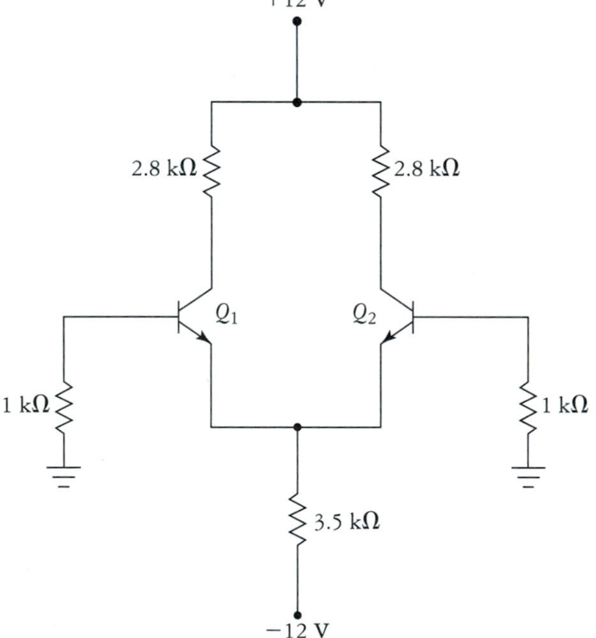

FIGURE 8.3
Circuit for Example 8.1.

Solution

$$\begin{aligned} I_{EE} &= \frac{V_{EE} - V_{BE}}{\dfrac{R_B}{2\beta} + R_{EE}} \\ &= \frac{12\text{ V} - 0.7\text{ V}}{\dfrac{1\text{ k}\Omega}{200} + 3.5\text{ k}\Omega} \\ &= 3.219\text{ mA} \end{aligned}$$

$$I_{E_1} = I_{E_2} = I_{EE}/2$$
$$= 3.219 \text{ mA}/2$$
$$= 1.610 \text{ mA}$$

$$I_{C_1} = I_{C_2} = \alpha I_E \quad [\text{Note: } \alpha = \beta/(\beta + 1)]$$
$$= (0.990)(1.610 \text{ mA})$$
$$= 1.594 \text{ mA}$$

$$I_{B_1} = I_{B_2} = I_E - I_C$$
$$= 1.610 \text{ mA} - 1.594 \text{ mA}$$
$$= 16 \ \mu\text{A}$$

Notice in Example 8.1 that the values of I_C and I_E are very nearly equal to one another, and that I_B is relatively small in comparison. Recall that for typical small-signal transistors, $\alpha \cong 1$ and so $I_C \cong I_E$. In most cases, the analysis of the differential amplifier need not be so exact. To illustrate this point, let us use Eq. (8.6) to determine I_{EE} and I_E for Fig. 8.3. We assume that base current is negligible ($I_B \cong 0$) and $I_C = I_E$.

Note
As usual, most of the time we will assume that $I_E = I_C$.

$$I_{EE} = \frac{V_{EE} - V_{BE}}{R_{EE}}$$
$$= \frac{12 \text{ V} - 0.7 \text{ V}}{3.5 \text{ k}\Omega}$$
$$= 3.23 \text{ mA}$$

$$I_C = I_E = I_{EE}/2$$
$$= 1.62 \text{ mA}$$

Equation (8.6) becomes an even closer approximation for I_{EE} as the ratio of R_{EE}/R_B increases. We show in the next section that the effective value of R_{EE} can be increased to millions of ohms through the use of an active I_{EE} current source. The utility of the more approximate relation of Eq. (8.6) becomes more apparent when one considers that slight variations in transistor V_{BE}, β, η, and other device parameters can easily negate the small amount of increased accuracy provided by Eq. (8.4).

Differential Amplifier Voltage Relationships

Once the emitter currents are known, it is possible to determine the voltage drops across the various components in the circuit via basic application of Kirchhoff's laws and Ohm's law. With equal resistances in both left and right halves of the amplifier, analysis of both sides of the circuit will yield identical values. The following example applies the values of the currents that were determined in Example 8.1.

EXAMPLE 8.2

Determine the following quiescent voltages for Fig. 8.3: V_{CE_1}, V_{CE_2}, V_{C_1}, and V_{C_2}. Use the approximate current values $I_{EE} = 3.23$ mA and $I_C = 1.62$ mA to calculate the voltage drops across the appropriate resistors.

Solution Because both sides of the circuit are identical, we can equate the voltage drops across the transistors and resistors. That is, an analysis of one side yields voltage drops that are applicable to the other side. Using KVL and Ohm's law, and the current values from the preceding analysis, we obtain

$$V_{CE_1} = V_{CE_2} = V_{EE} + V_{CC} - I_C R_C - I_{EE} R_{EE}$$
$$= 12 \text{ V} + 12 \text{ V} - (1.62 \text{ mA})(2.8 \text{ k}\Omega) - (3.23 \text{ mA})(3.5 \text{ k}\Omega)$$
$$= 24 \text{ V} - 4.54 \text{ V} - 11.31 \text{ V}$$
$$= 8.15 \text{ V}$$

$$V_{C_1} = V_{C_2} = V_{CC} - I_C R_C$$
$$= 12 \text{ V} - (1.62 \text{ mA})(2.8 \text{ k}\Omega)$$
$$= 12 \text{ V} - 4.54 \text{ V}$$
$$= 7.46 \text{ V}$$

Many different variations are possible in the design details of differential amplifiers. For example, in some cases one side of the differential pair may not include a collector resistor. When this situation occurs, the voltage drops associated with the differential-pair transistors will not be equal. The next example demonstrates this possibility.

EXAMPLE 8.3

Determine the values of I_{C_1}, I_{C_2}, V_{CE_2}, and V_{CE_2} for the circuit shown in Fig. 8.4. Use Eq. (8.6) to determine circuit current levels.

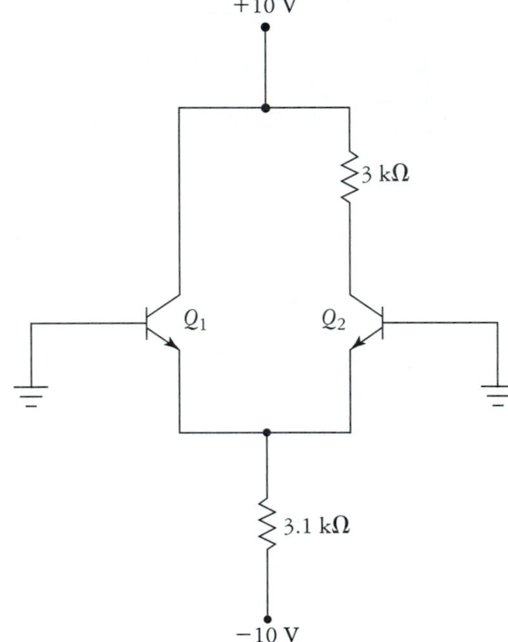

FIGURE 8.4
Circuit for Example 8.3.

Solution From basic transistor analysis theory, we know that unless the transistor is either saturated or in cutoff, I_C is essentially independent of the resistance present between the collector and the power supply. This means that $I_{C_1} = I_{C_2} = I_{EE}/2$; therefore,

$$I_{EE} \cong \frac{V_{EE} - V_{BE}}{R_{EE}}$$
$$= \frac{10\text{ V} - 0.7\text{ V}}{3.1\text{ k}\Omega}$$
$$= 3.0\text{ mA}$$

$$I_{C_1} = I_{C_2} = I_{EE}/2$$
$$= 1.5\text{ mA}$$

Summing voltage drops from the positive supply rail to the negative supply rail across Q_1 yields

$$V_{CE_1} = V_{CC} + V_{EE} - I_{EE}R_{EE}$$
$$= 10\text{ V} + 10\text{ V} - (3\text{ mA})(3.1\text{ k}\Omega)$$
$$= 20\text{ V} - 9.3\text{ V}$$
$$= 10.7\text{ V}$$

Similarly for Q_2 we obtain

$$V_{CE_2} = V_{CC} + V_{EE} - I_{EE}R_{EE} - I_{C_1}R_C$$
$$= 10\text{ V} + 10\text{ V} - (3\text{ mA})(3.1\text{ k}\Omega) - (1.5\text{ mA})(3\text{ k}\Omega)$$
$$= 20\text{ V} - 9.3\text{ V} - 4.5\text{ V}$$
$$= 6.2\text{ V}$$

Although the effects of unequal collector loads will normally be very slight, it is good practice to maintain equal load conditions in the collector circuits of the differential pair. This helps to keep both transistors operating at the same points on their characteristic curves, which results in improved circuit performance. There are times, however, when unbalanced collector loads are unavoidable, producing situations like that of the preceding example.

8.2 ACTIVE CURRENT SOURCES

The emitter currents in the differential amplifier play the primary role in determining the circuit's operating conditions. Because of this, it is advantageous to have a very stable source producing I_{EE}. In other words, it is very desirable that I_{EE} be made independent of the individual emitter currents flowing in the differential pair. These requirements lead us to drive the emitters of the differential pair with an ideal current source. A differential amplifier employing an idealized constant current source for emitter biasing is shown in Fig. 8.5.

FIGURE 8.5
Current-source biasing of a differential pair.

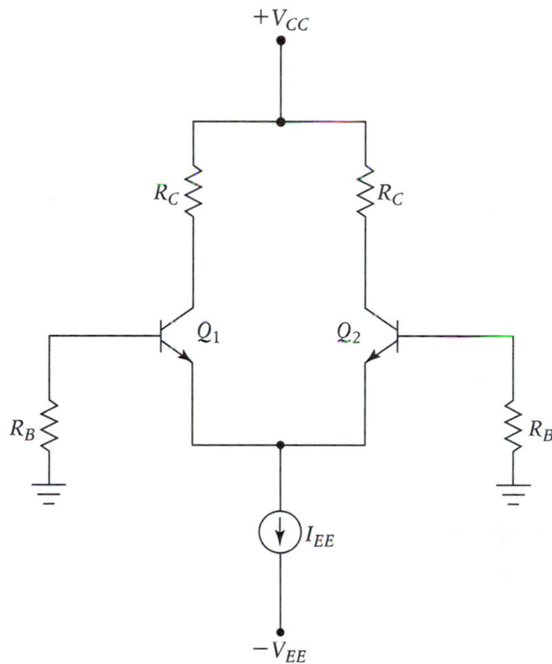

Note
In practice, active current sources are nearly always used to bias differential pairs.

The internal resistance (or impedance) of an ideal current source is infinite. To approximate such an ideal current source, we would need to use an extremely high resistance for R_{EE}. This would invariably result in extremely low quiescent currents in the differential amplifier, which may not be desired in some cases. It also turns out that it is very difficult to fabricate high-value resistors using integrated circuit design techniques as well. It has been shown in Chapter 4, Section 4.7, that the transistor can be used as an active current source. The effective internal resistance of such a circuit can typically range from a few hundred thousand ohms for the basic current mirror to well over 10 MΩ for the Widlar current source. This property makes active current sources ideal for use in differential amplifier biasing applications.

Voltage-Divider Biased Current Source

One type of bipolar transistor current source that may be realized uses voltage-divider biasing, which is illustrated in Fig. 8.6(a). Thevenizing the voltage-divider network of Q_3, the current source can be redrawn as shown in Fig. 8.6(b). The collector current (I_{EE}) is given by

$$I_{EE} = \frac{V_{\text{Th}} - V_{BE}}{\frac{R_{\text{Th}}}{\beta} + R_E} \tag{8.7}$$

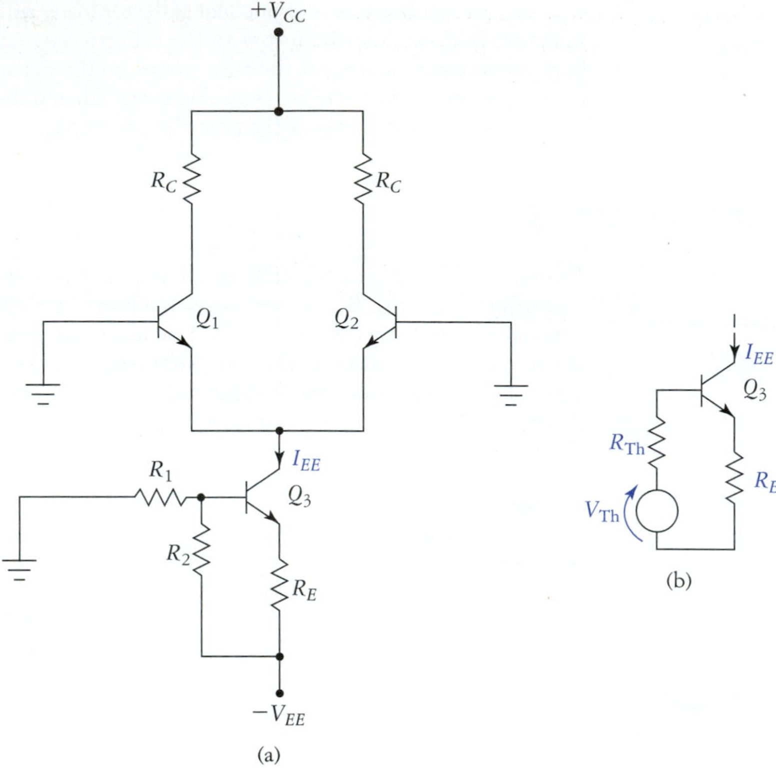

FIGURE 8.6
Analysis of the current source. (a) The actual circuit. (b) Current source redrawn with Thevenized base network.

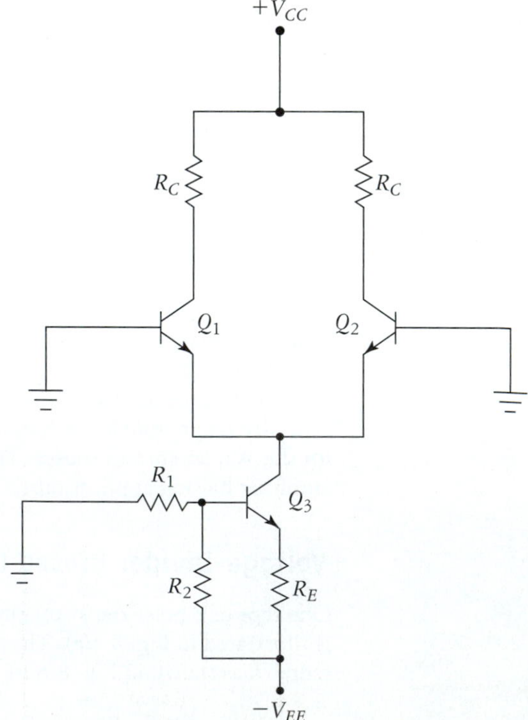

FIGURE 8.7
Typical differential amplifier stage.

where

$$V_{\text{Th}} = V_{EE}\frac{R_2}{R_1 + R_2} \qquad \text{and} \qquad R_{\text{Th}} = R_1 \parallel R_2$$

As usual, if $\beta R_E \gg R_{\text{Th}}$ then Eq. (8.7) can be reduced to

$$I_{EE} \cong \frac{V_{\text{Th}} - V_{BE}}{R_E} \tag{8.8}$$

Several other active current-source designs are much more widely used than the voltage-divider-biased approach of Fig. 8.7 because they are simpler and actually provide superior performance. There are other applications (nonlinear mixers and modulators, for example) in which a voltage signal source would be coupled to the base of the current-source transistor. The voltage-divider biased current source is a good choice in such cases. This capability is sometimes used in differential amplifier applications as well.

EXAMPLE 8.4

Determine I_{EE}, I_{C_1}, I_{C_2}, V_{CE_1}, V_{CE_2}, and V_{CE_3} for the circuit in Fig. 8.7, given $V_{CC} = V_{EE} = 15$ V, $R_1 = 10$ kΩ, $R_2 = 2.2$ kΩ, $R_E = 330$ Ω, and $R_C = 3.3$ kΩ. Assume that $\beta = 100$.

Solution The Thevenin-equivalent values for Q_3 are determined first:

$$\begin{aligned} R_{\text{Th}} &= R_1 \parallel R_2 \\ &= 10\text{ k}\Omega \parallel 2.2\text{ k}\Omega \\ &= 1.8\text{ k}\Omega \end{aligned}$$

$$\begin{aligned} V_{\text{Th}} &= V_{EE}\frac{R_2}{R_1 + R_2} \\ &= 15\text{ V}\frac{2.2\text{ k}\Omega}{10\text{ k}\Omega + 2.2\text{ k}\Omega} \\ &= 2.7\text{ V} \end{aligned}$$

Because $\beta R_E \gg R_{\text{Th}}$, we can closely approximate the emitter biasing current as

$$\begin{aligned} I_{EE} &\cong \frac{V_{\text{Th}} - V_{BE}}{R_E} \\ &= \frac{2.7\text{ V} - 0.7\text{ V}}{330\ \Omega} \\ &= 6\text{ mA} \end{aligned}$$

and

$$\begin{aligned} I_{C_1} = I_{C_2} &= I_{EE}/2 \\ &= 3\text{ mA} \end{aligned}$$

We must now determine either V_{CE_1} or V_{CE_2} (they are equal in this case), without using Q_3 in the KVL loop, because V_{CE_3} is unknown at this time. To solve this problem, we first apply KVL from V_{CC} down through R_C and the C-B junction of Q_1 (or equivalently Q_2) to ground:

$$0 = V_{CC} - I_C R_C - V_{CB}$$

Solving for V_{CB} gives us

$$\begin{aligned} V_{CB_1} = V_{CB_2} &= V_{CC} - I_C R_C \\ &= 15\text{ V} - (3\text{ mA})(3.3\text{ k}\Omega) \\ &= 15\text{ V} - 9.9\text{ V} \\ &= 5.1\text{ V} \end{aligned}$$

Recall from basic theory (Eq. 4.4, to be exact) that

$$V_{CE} = V_{CB} + V_{BE}$$

We now apply this relationship to obtain

$$\begin{aligned} V_{CE_1} = V_{CE_2} &= 5.1\text{ V} + 0.7\text{ V} \\ &= 5.8\text{ V} \end{aligned}$$

Finally, application of KVL and Ohm's law (arbitrarily using the left side of the differential pair) gives us

$$\begin{aligned} V_{CE_3} &= V_{CC} + V_{EE} - I_{C_1}R_C - V_{CE_1} - I_{EE}R_E \\ &= 15\text{ V} + 15\text{ V} - (3\text{ mA})(3.3\text{ k}\Omega) - 5.8\text{ V} - (6\text{ mA})(330\ \Omega) \\ &= 15\text{ V} + 15\text{ V} - 9.9\text{ V} - 5.8\text{ V} - 2\text{ V} \\ &= 12.3\text{ V} \end{aligned}$$

Current Mirror

Current mirrors and other closely related circuits are most frequently used to bias up differential amplifiers. The basic operating principles behind the current mirror were covered in Chapter 4, Section 4.7.

Recall that the current mirror can be implemented using either a matched diode-transistor pair as in Fig. 8.8(a), or using a matched diode-connected transistor as in Fig. 8.8(b). Both approaches are equivalent, with current I_1 being given by

$$I_1 = \frac{V_{EE} - V_{BE}}{R_1} \tag{8.9}$$

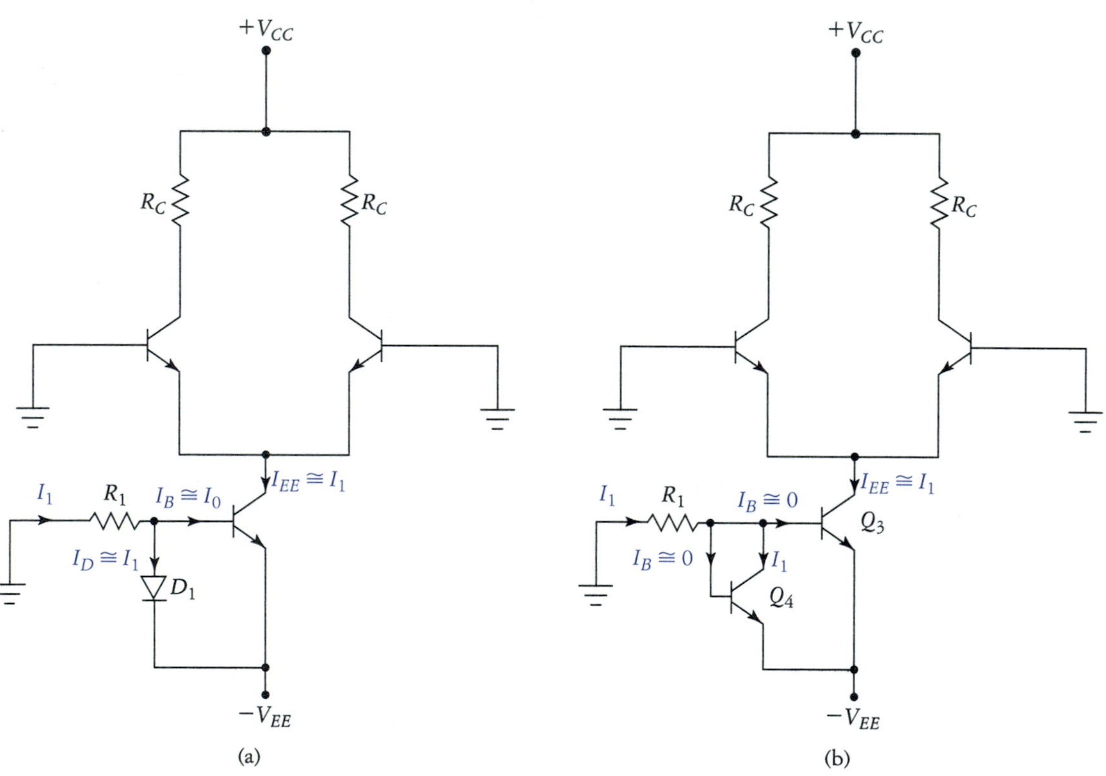

FIGURE 8.8
Current-mirror biasing. (a) Biasing with a diode. (b) Biasing using a diode-connected transistor.

Equation (8.9) is valid because the base of Q_3 presents a very light load on R_1, in comparison to the diode, or diode-connected Q_4. The circuit is called a *current mirror* because the collector current I_{EE} "mirrors" reference current I_1. This occurs because the transconductance curves for pairs D_1–Q_3 (or Q_4–Q_3) are identical, thus I_C for Q_3 must equal I_1 at a given operating point. This allows us to write $I_{EE} = I_1$ and

$$I_{EE} = \frac{V_{EE} - V_{BE}}{R_1} \tag{8.10}$$

EXAMPLE 8.5

Determine I_{EE}, I_{C_1}, I_{C_2}, V_{CE_1}, V_{CE_2}, and V_{CE_3} for the circuit shown in Fig. 8.9.

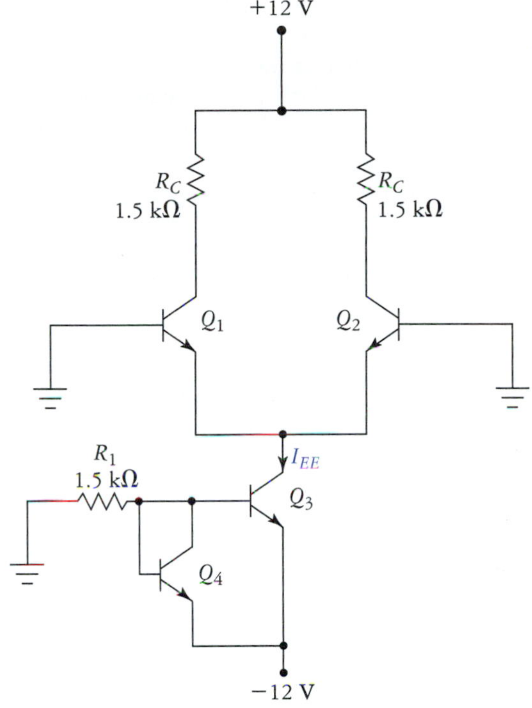

FIGURE 8.9
Circuit for Example 8.5.

Solution The emitter bias current is found using Eq. (8.10):

$$I_{EE} = \frac{V_{EE} - V_{BE}}{R_1}$$

$$= \frac{12\text{ V} - 0.7\text{ V}}{1.5\text{ k}\Omega}$$

$$= 7.53\text{ mA}$$

$$I_{C_1} = I_{C_2} = I_{EE}/2$$
$$= 3.77\text{ mA}$$

To determine the requested voltages, it is necessary to sum voltages from one supply node to the other. As usual, since both sides of the diff amp are identical, we need only analyze one side. Writing the KVL equation for the left half of the circuit we obtain

$$0 = V_{CC} + V_{EE} - I_{C_1}R_{C_1} - V_{CE_1} - V_{CE_3}$$

As was shown in Example 8.4, we must determine V_{CB} for either Q_1 or Q_2. Choosing Q_1 arbitrarily

$$V_{CB_1} = V_{CC} - I_C R_C$$
$$= 12\text{ V} - 5.66\text{ V}$$
$$= 6.34\text{ V}$$

We now determine V_{CE_1} and V_{CE_2}

$$V_{CE_1} = V_{CE_2} = V_{CB} + V_{BE}$$
$$= 6.34\text{ V} + 0.7\text{ V}$$
$$= 7.04\text{ V}$$

Now, solving the original KVL equation for V_{CE_3} we obtain

$$\begin{aligned} V_{CE_3} &= V_{CC} + V_{EE} - I_{C_1}R_{C_1} - V_{CE_1} \\ &= 12\text{ V} + 12\text{ V} - 5.66\text{ V} - 7.04\text{ V} \\ &= 11.3\text{ V} \end{aligned}$$

One of the advantages of the current mirror compared to the voltage-divider biased current source is increased bias stability in terms of temperature. If the circuit of Fig. 8.9 is implemented in integrated circuit form, which is most likely, all devices will operate at nearly the same temperature. Because the transistors are matched, a temperature shift that causes a change in Q_3's parameters will cause identical changes in those of Q_4, thus stabilizing the value of I_{EE}.

Widlar Current Source

It is often desirable to operate the transistors of the differential pair at very low current levels. For example, let us assume that we require $I_{EE} = 20\ \mu\text{A}$ using the circuit of Fig. 8.9. Application of Eq. (8.10) gives us $R_1 = 565\ \text{k}\Omega$. The fact that this is not a standard value is not a problem. Keep in mind that differential amplifiers are nearly always implemented in integrated circuit (IC) form, and almost no resistor values used in IC designs are standard. Higher value resistors, where $R > 50\ \text{k}\Omega$ or so, are extremely difficult to fabricate in IC form, and the tolerances of these high-value resistors is on the order of 20% as well. The Widlar current source is most useful in these situations.

Note
The Widlar current source allows us to obtain small I_{EE} without using high-value resistors.

FIGURE 8.10
Differential pair using a Widlar current source to establish I_{EE}.

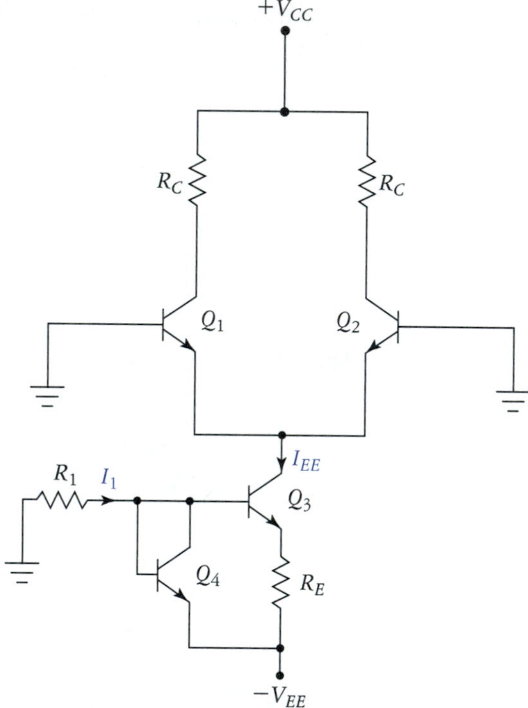

The differential amplifier in Fig. 8.10 uses a Widlar current source to establish I_{EE}. The analysis and design expressions relating to the Widlar source are (from Chapter 4)

$$I_{EE} = \frac{\eta V_T}{R_E} \ln\left(\frac{I_1}{I_{EE}}\right) \tag{8.11}$$

and

$$R_E = \frac{\eta V_T}{I_{EE}} \ln\left(\frac{I_1}{I_{EE}}\right) \tag{8.12}$$

where I_1 is found using Eq. (8.9). For convenience, we normally assume that the emission coefficient $\eta = 1$. The following example illustrates the design of a differential amplifier using the Widlar current source.

EXAMPLE 8.6

Refer to Fig. 8.10. Assume that all transistors have $\beta = 100$ and $V_{CC} = V_{EE} = 15$ V. Design the amplifier such that $I_{C_1} = I_{C_2} = 10\ \mu$A and $I_1 = 1$ mA.

Solution We first determine the required value for R_1. Using Eq. (8.9), we get

$$R_1 = \frac{V_{EE} - V_{BE}}{I_1} = \frac{15\text{ V} - 0.7\text{ V}}{1\text{ mA}} = 14.3\text{ k}\Omega$$

Using $I_{EE} = 2I_C = 20\ \mu$A, we apply Eq. (8.12), giving us

$$R_E = \frac{V_T}{I_{EE}}\ln\left(\frac{I_1}{I_{EE}}\right) = \frac{26\text{ mV}}{20\ \mu\text{A}}\ln\left(\frac{1\text{ mA}}{20\ \mu\text{A}}\right) = 5.09\text{ k}\Omega$$

Both of the resistor values found in the preceding example are readily realized in monolithic IC form.

An interesting and important characteristic of the Widlar current source is the fact that the effective resistance seen looking into the collector terminal r_o is extremely high. For the standard current mirror, r_o will typically be on the order of 100 kΩ. The actual value is determined using I_{EE} and the Early voltage V_A (refer back to Section 3.7) using

$$R_{EE} = r_o = \frac{V_A}{I_{EE}}$$

It can be shown that for the Widlar current source, the output resistance is

$$R_{EE} = r_o = \beta\frac{V_A}{I_{EE}}$$

which means that the Widlar current source more closely approximates the ideal current source.

Supply-Voltage-Independent Current Source

All of the current sources discussed up to this point produce an I_{EE} that is dependent on the value of V_{EE}. Often this does not pose a problem. However, it is sometimes desirable to maintain a relatively constant I_{EE} level in spite of supply-voltage variations. One way to realize such a current source uses a zener diode in its biasing circuit, as shown in Fig. 8.11.

Recall that zener diodes produce very predictable and stable voltage drops over a wide range of current, and they have very low dynamic resistances when operated in reverse breakdown. This means that as long as V_{EE} is sufficiently larger than V_Z such that $I_Z > I_{Zk}$, the bias applied to Q_3 in Fig. 8.11 will remain essentially constant and independent of V_{EE}. Because the dynamic resistance of the zener diode is quite low ($Z_{Z(\text{typ})} < 10\ \Omega$), the Thevenin resistance at the base of Q_3 will be very low, which also contributes to the stability of the circuit.

The two factors that determine I_{EE} for Fig. 8.11 are V_Z and R_E. Assuming that $V_{EE} > V_Z$, the following equation for I_{EE} can be developed by summing voltage drops around loop 1:

Note
The zener test current I_{Z_t} is found on the data sheet for a given zener diode.

$$0 = V_Z - V_{BE} - I_{EE}R_E$$

Solving for I_{EE} yields

$$I_{EE} = \frac{V_Z - V_{BE}}{R_E} \tag{8.13}$$

Notice that the V_{EE} source does not enter into Eq. (8.13), because it is outside of the KVL loop.

For design purposes, we need to ensure that the zener diode carries enough current to operate beyond its zener knee. This requirement can be conveniently met by choosing R_1 such that

$$I_Z = I_{Zt}$$

FIGURE 8.11
Using a zener diode to produce a stable reference voltage for current-source biasing

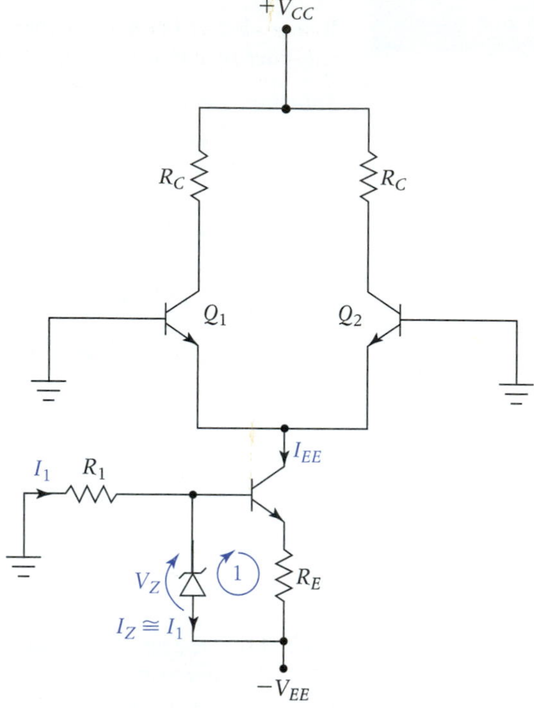

where I_{Zt} is the zener test current, as specified by the manufacturer. For analysis purposes, the zener diode current is found by

$$I_Z = \frac{V_{EE} - V_Z}{R_1} \tag{8.14}$$

EXAMPLE 8.7

Determine the values required for R_1 and R_E in Fig. 8.11 such that $I_{EE} = 1$ mA. Assume that $V_{EE} = V_{CC} = 15$ V, $V_Z = 4.2$ V, and $I_{Zt} = 5$ mA.

Solution R_1 is chosen to produce $I_Z = I_{Zt} = 5$ mA. Solving Eq. (8.14) for R_1 gives us

$$\begin{aligned} R_1 &= \frac{V_{EE} - V_Z}{I_Z} \\ &= \frac{15\text{ V} - 4.2\text{ V}}{5\text{ mA}} \\ &= 2.16\text{ k}\Omega \end{aligned}$$

Now, solving Eq. (8.13) for R_E, we find

$$\begin{aligned} R_E &= \frac{V_Z - V_{BE}}{I_{EE}} \\ &= \frac{4.2\text{ V} - 0.7\text{ V}}{1\text{ mA}} \\ &= 3.5\text{ k}\Omega \end{aligned}$$

Another approach that may be used to bias up a differential pair is shown in Fig. 8.12. In this circuit two series-connected diodes serve to forward bias the B-E junction of Q_3. If we assume that the forward voltages dropped by the diodes are $V_F = V_{BE}$, then applying KVL around loop 1 gives us

$$0 = 2V_{BE} - V_{BE} - I_{EE}R_E$$

Solving for I_{EE} produces

$$I_{EE} = \frac{V_{BE}}{R_E} \tag{8.15}$$

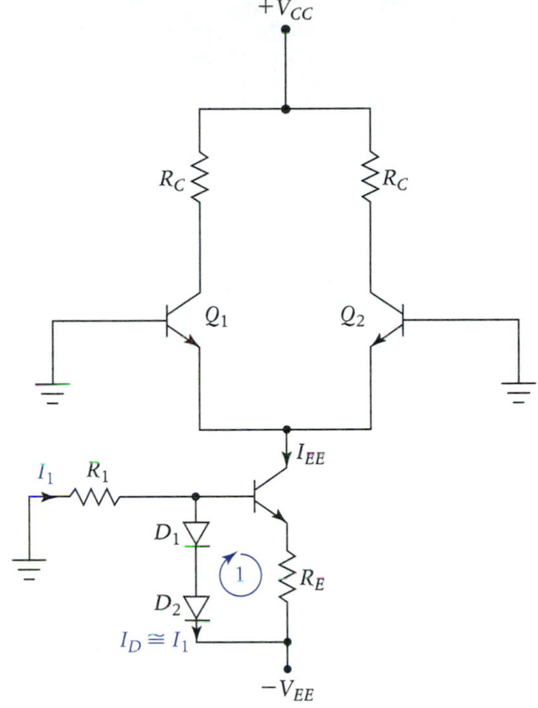

FIGURE 8.12
Series-connected diodes provide a stable reference voltage for current-source biasing.

The current flow through the diodes is easily shown to be

$$I_D = \frac{V_{EE} - 2V_{BE}}{R_1} \tag{8.16}$$

In some designs it is desirable to set $I_D = I_{EE}$ to ensure that the diodes will have nearly the same operating points as Q_3. Again, this helps decrease the temperature sensitivity of the circuit.

EXAMPLE 8.8

Determine the values required for R_1 and R_E in Fig. 8.12 such that $I_{EE} = I_D = 5$ mA. Assume that $V_{CC} = V_{EE} = 15$V.

Solution Manipulation of Eq. (8.15) produces

$$R_E = \frac{V_{BE}}{I_{EE}}$$
$$= \frac{0.7 \text{ V}}{5 \text{ mA}}$$
$$= 140\ \Omega$$

Now, R_1 is selected such that $I_D = 5$ mA. Solving Eq. (8.16) gives us

$$R_1 = \frac{V_{EE} - 2V_{BE}}{I_D}$$
$$= \frac{15 \text{ V} - 1.4 \text{ V}}{5 \text{ mA}}$$
$$= 2.72 \text{ k}\Omega$$

In addition to the bipolar designs just covered, FETs can also be used in differential amplifier biasing arrangements (they may also used to form the differential pair as well). Recall that JFETs and depletion-mode MOSFETs can be easily configured as constant-current sources. Figure 8.13 illustrates how these current sources are set up. In each case, $I_{EE} = I_{DSS}$. Of course, this only applies if $V_{DS} > V_p$ for a given FET. Under quiescent conditions, it can be shown that this inequality is satisfied if $V_{EE} > V_p + V_{BE}$, which is usually very easily met.

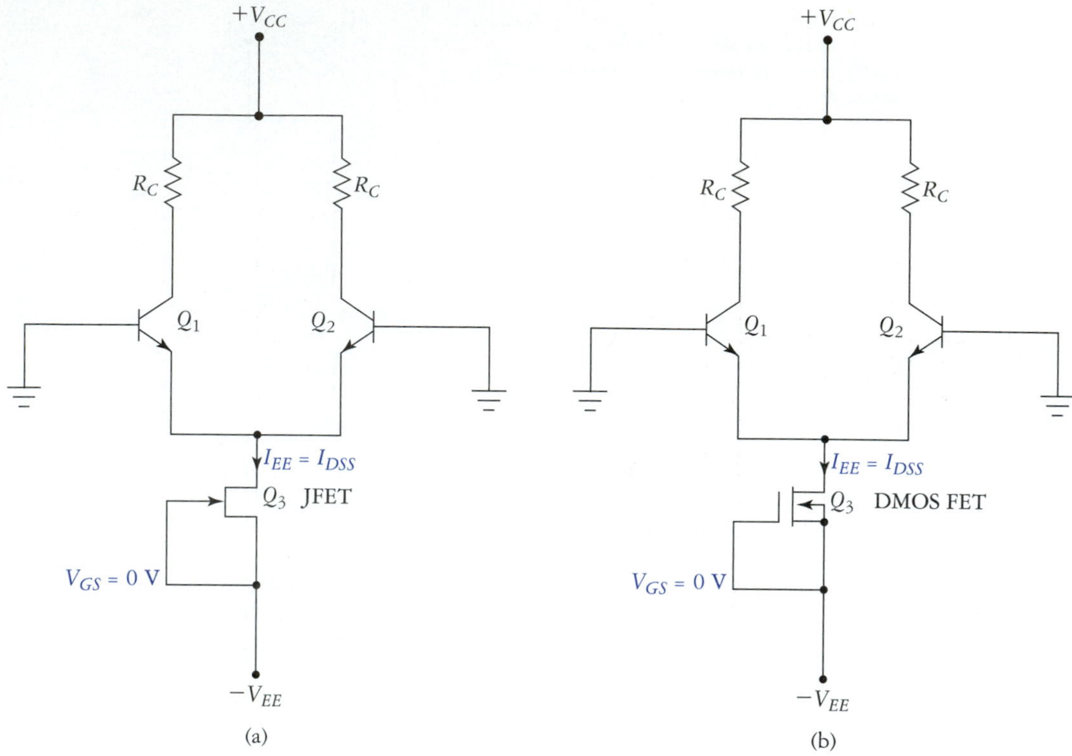

FIGURE 8.13
Current-source biasing variations. (a) Using a JFET as a constant-current diode. (b) A DMOS FET current source.

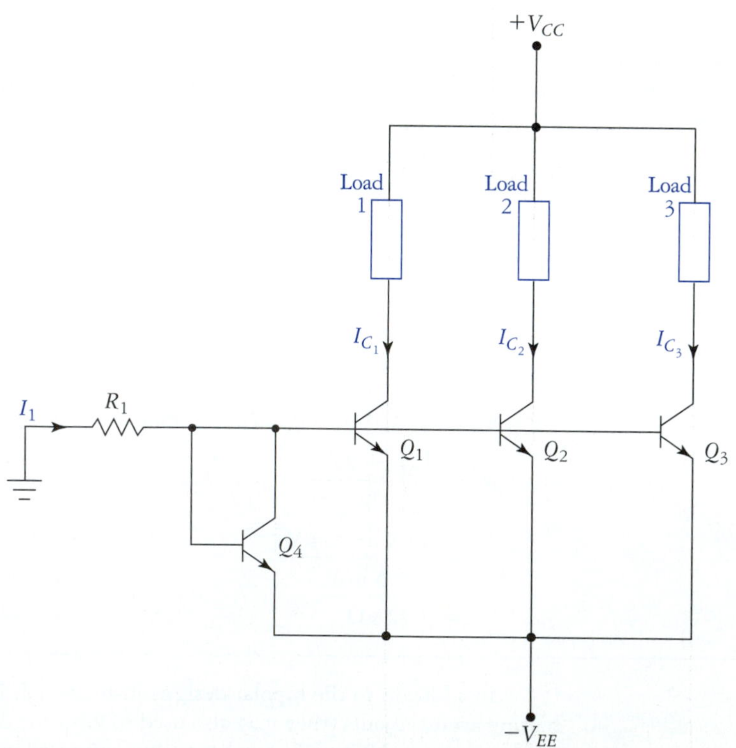

FIGURE 8.14
Using a single reference device to bias multiple current sources.

Additional Active Current-Source Variations

Often, it is necessary to use multiple current sources in a given circuit design. One possible design for such a circuit is the variation of the current mirror shown in Fig. 8.14. Although this circuit looks formidable, it is really quite easy to analyze. First of all, the solid line bisecting the transistors represents a common connection of the bases of Q_1, Q_2, and Q_3 to diode-connected transistor Q_4. Even though the three bases are in parallel, they still present a light load at the junction of R_1 and Q_4, because of beta multiplication of their emitter resistances. Since this is a current-mirror-type circuit, all of the transistors are normally matched (or their saturation currents scaled appropriately) for correct operation. Assuming the transistors are matched we find that

$$I_{C_1} = I_{C_2} = I_{C_3} = I_1 \tag{8.17}$$

where I_1 is found using Eq. (8.9).

8.3 AC ANALYSIS OF THE DIFFERENTIAL AMPLIFIER

There are four differential amplifier configurations: (1) single-ended input, single-ended output; (2) single-ended input differential output; (3) differential input, differential output; and (4) differential input, single-ended output. These configurations are shown in Fig. 8.15. Each configuration has certain characteristics that make it useful in a given situation. All four types are discussed in this section.

Single-Ended-Input, Single-Ended-Output Configuration

Perhaps the easiest of the differential amplifier connections to analyze is that of Fig. 8.16(a), the single-ended-input, single-ended-output configuration. This circuit is redrawn in Figs. 8.16(b) and (c) in its ac-equivalent and dependent-source forms to make the its operation easier to visualize. Inspection of Fig. 8.16(b) reveals that Q_1 is operated as an emitter follower, while Q_2 is in the common-base configuration. In the ac-equivalent circuit, the emitter current source has been replaced with its equivalent internal resistance (ideally), an open circuit. To simplify the upcoming analysis, let us make the reasonable assumption that $i_C = i_E$ for a given transistor.

The input resistance to common-base amplifier Q_2 is given by

$$r_{\text{in}_2} = r_{e_2} \tag{8.18}$$

The input resistance of emitter-follower transistor Q_1 is the resistance seen by the input source, and we have

$$R_{\text{in}} = R_{\text{in}_1} = \beta(r_{e_1} + R_{\text{in}_2}) \tag{8.19}$$

Substituting Eq. (8.18) into (8.19), we obtain

$$R_{\text{in}} = \beta(r_{e_1} + r_{e_2}) \tag{8.20}$$

since $I_{CQ_1} = I_{CQ_2}$, then $r_{e_1} = r_{e_2} = V_T/I_{CQ}$, and we can write

$$R_{\text{in}} = 2\beta r_e \tag{8.21}$$

The ac output voltage of the amplifier, under no-load conditions, is given by

$$v_o = i_{C_2} R_C \tag{8.22}$$

where $i_{C_2} = i_{C_1}$, $i_{C_1} = \beta i_{B_1}$, and

$$i_{B_1} = \frac{v_{\text{in}}}{R_{\text{in}}} = \frac{v_{\text{in}}}{2\beta\, r_e}$$

Now, substituting into Eq. (8.22) gives us

$$v_o = \frac{v_{\text{in}} R_C}{2r_e} \tag{8.23}$$

Dividing by v_{in} gives us the voltage gain of the single-in single-out differential amplifier:

$$A_v = \frac{R_C}{2r_e} \tag{8.24}$$

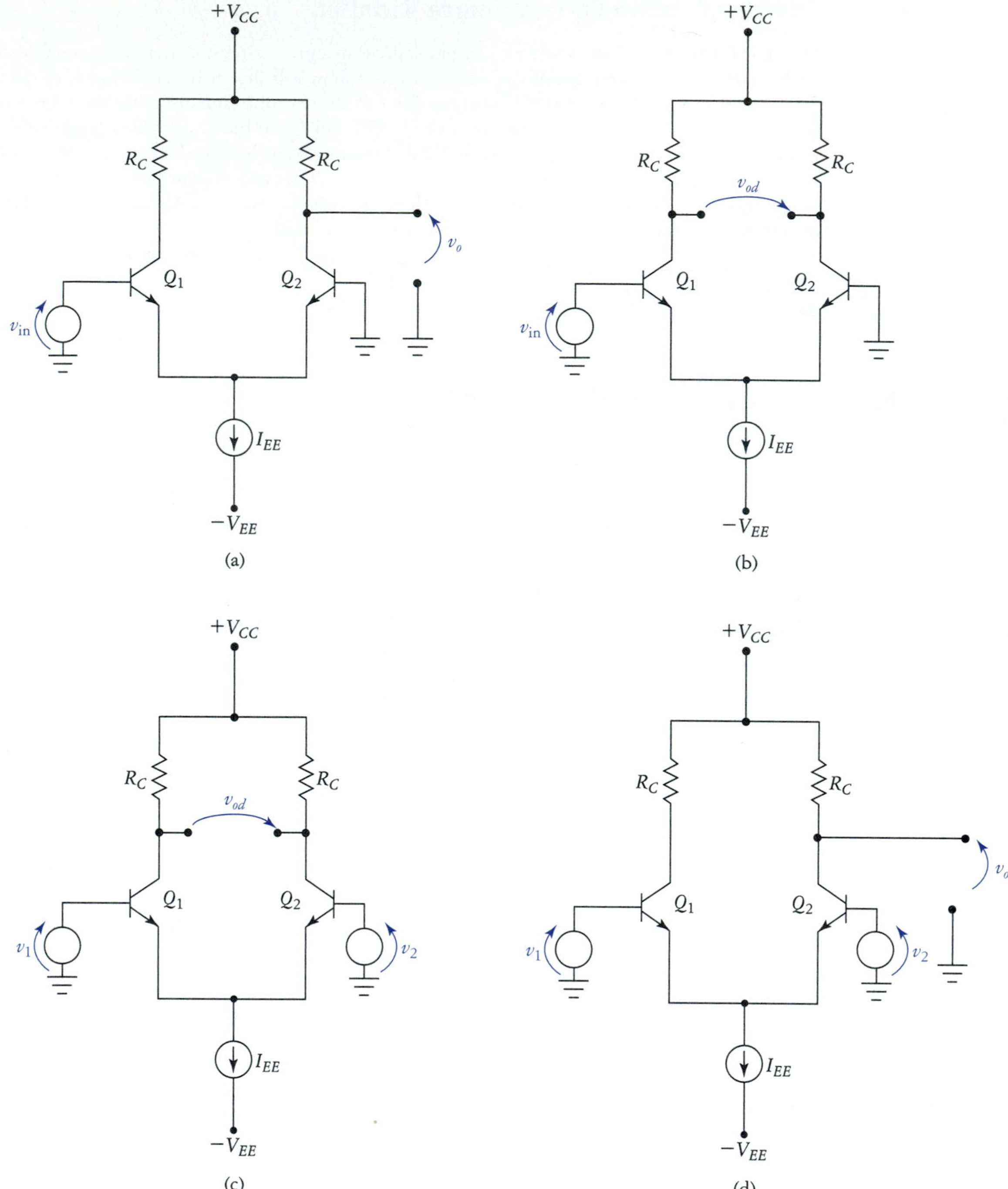

FIGURE 8.15
The four diff amp configurations. (a) Single-ended input, single-ended output. (b) Single-ended input, differential output. (c) Differential input, differential output. (d) Differential input, single-ended output.

The net output voltage is the algebraic sum of Eq. (8.23) and the quiescent collector voltage V_{oQ}. The reader should also keep in mind that when a load is applied to the output of the amplifier, the voltage gain equation must be modified accordingly. Referring to a given load as R_L, Eq. (8.24) would become

$$A_v = \frac{R'_C}{2r_e} \tag{8.25}$$

where $R'_C = R_C \parallel R_L$.

FIGURE 8.16
Analysis of the single-ended-input, single-ended-output configuration. (a) Actual circuit. (b) Small-signal ac equivalent. (c) Small-signal ac equivalent using dependent source models.

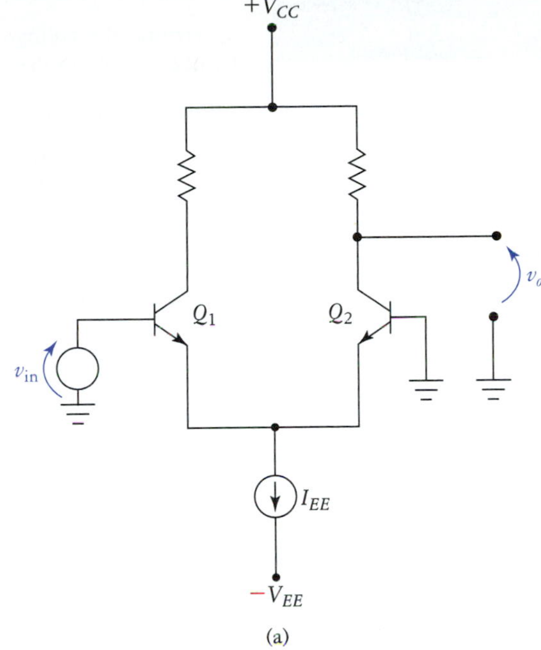

(a)

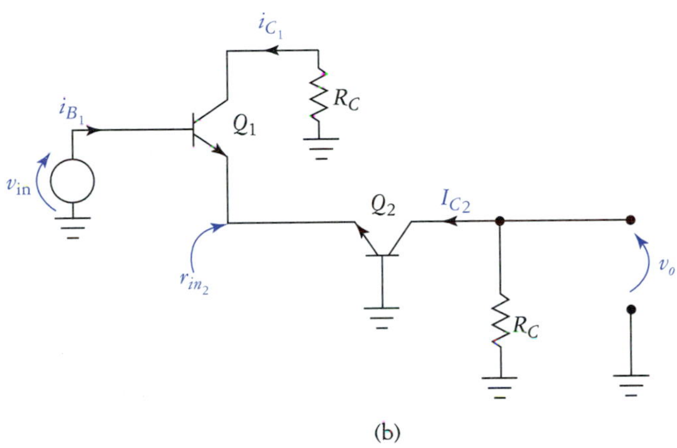

(b)

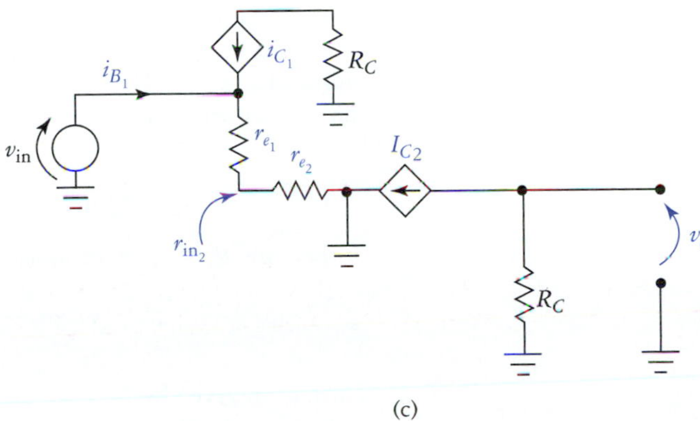

(c)

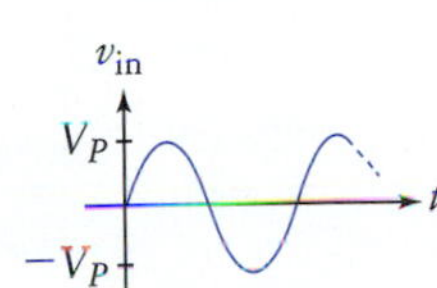

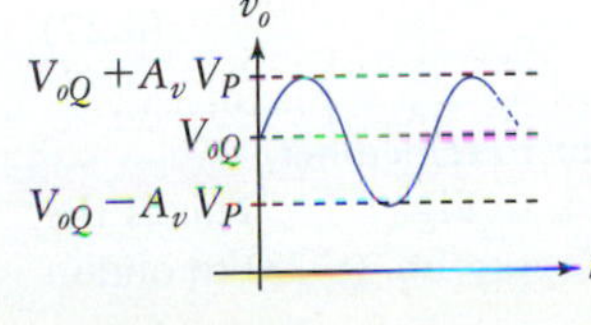

FIGURE 8.17
Input and output signal relationships.

The phase relationships between v_{in} and v_o are shown in Fig. 8.17. This amplifier configuration produces an output signal that is in phase with the input. This should seem plausible considering that neither the common-base nor the emitter-follower configurations, which form the ac-equivalent circuit, produce phase inversion.

The output resistance of Fig. 8.16 is the same as for a common-emitter amplifier, which is approximately

$$R_o \cong R_C \tag{8.26}$$

EXAMPLE 8.9

Determine the voltage gain and input resistance of the amplifier in Fig. 8.16(a) given $R_C = 10\ \text{k}\Omega$, $I_{EE} = 1.5$ mA, and $\beta = 100$.

Solution Under quiescent conditions, the emitter currents are equal, which yields

$$\begin{aligned} I_{C_1} = I_{C_2} &= I_{EE}/2 \\ &= 1.5\ \text{mA}/2 \\ &= 0.75\ \text{mA} \end{aligned}$$

Thus,

$$\begin{aligned} r_e &= V_T/I_C \\ &= 26\ \text{mV}/0.75\ \text{mA} \\ &= 34.7\ \Omega \end{aligned}$$

The input resistance is found via Eq. (8.21):

$$\begin{aligned} r_{\text{in}} &= 2\beta r_e \\ &= (200)(34.7\ \Omega) \\ &= 6.94\ \text{k}\Omega \end{aligned}$$

The voltage gain is

$$\begin{aligned} A_v &= \frac{R_C}{2r_e} \\ &= \frac{10\ \text{k}\Omega}{69.4\ \Omega} \\ &= 144.1 \end{aligned}$$

The single-ended-input, single-ended-output configuration is not commonly used in practice. However, the analysis that was just presented forms a useful starting point for the study of the other three differential amplifier configurations.

Single-Ended-Input, Differential-Output Configuration

The single-ended-input, differential-output configuration is shown in Fig. 8.18. In this circuit, the output voltage is the difference between the collector voltages of the differential pair. The direction of the output voltage-sensing arrow is chosen arbitrarily in this case, such that the differential output voltage is $v_{od} = v_{C_2} - v_{C_1}$. Because the output voltage is not referred to ground, it is assigned the extra subscript *d.* This differential amplifier configuration can be used to drive a floating (ungrounded) load.

Because the differential amplifier behaves linearly (hopefully), the ac analysis of Fig. 8.18(a) is performed in two parts, with superposition giving us our final answers. Let us first determine the response of $v_{C_1}$to v_{in}. The ac-equivalent circuit is shown in Fig. 8.18(b). Transistor Q_1 is operating as a common-emitter amplifier, with its voltage gain (referred to ground) being

$$A_{v_1} = -\frac{R_C}{r_e + R_e}$$

In this case, $R_e = r_{e_2} = r_e$, so we can write

$$A_{v_1} = -\frac{R_C}{2r_e}$$

Thus, v_{C_1} is given by

$$v_{C_1} = -\frac{v_{\text{in}}R_C}{2r_e} \tag{8.27}$$

Notice that v_{C_1} is 180° out of phase with v_{in}. This is a very important point to remember.

To determine the response of the collector of Q_2 to v_{in}, we refer back to Fig. 8.16. This is the model for Fig. 8.18 when we are determining v_{C_2}. Thus, we may simply apply Eq. (8.23) to obtain the expression for v_{C_2}, which is

$$v_{C_2} = \frac{v_{\text{in}}R_C}{2r_e} \tag{8.28}$$

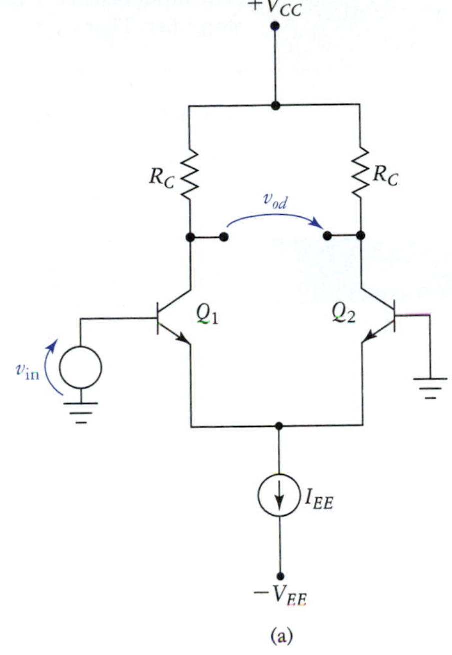

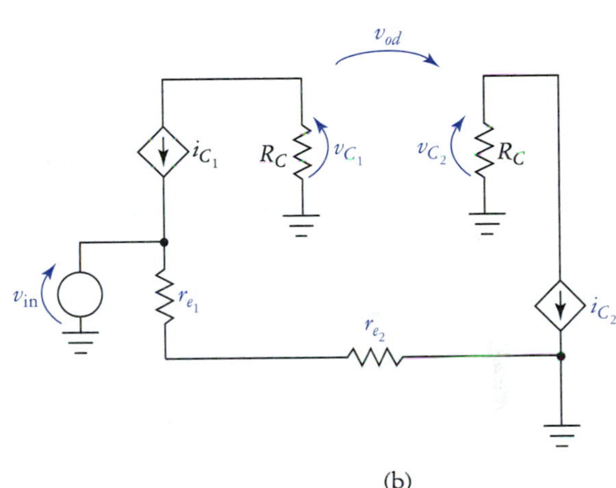

FIGURE 8.18
Analysis of the single-ended-input, differential-output configuration. (a) Basic circuit. (b) Small-signal ac equivalent.

Because v_{od} is the difference between v_{C_2} and v_{C_1} we can write

$$v_{od} = \frac{v_{in}R_C}{2r_e} - \frac{-v_{in}R_C}{2r_e}$$

$$= \frac{v_{in}R_C}{2r_e} + \frac{v_{in}R_C}{2r_e}$$

$$= \frac{2v_{in}R_C}{2r_e} \tag{8.29}$$

which reduces to

$$v_{od} = \frac{v_{in}R_C}{r_e} \tag{8.30}$$

The voltage gain expression is obtained simply by dividing both sides of Eq. (8.30) by v_{in}, which gives us

$$A_v = \frac{V_{od}}{v_{in}} = \frac{R_C}{r_e} \tag{8.31}$$

If the output voltage-sensing arrow is reversed, then the gain of the circuit is

$$A_v = -\frac{R_C}{r_e}$$

The input resistance of the single-in, diff-out amplifier is the same as that of the single-in, single-out amplifier. That is,

$$R_{in} = 2\beta r_e \tag{8.32}$$

The output resistance is defined in the same manner as for the single-in, single-out amplifier, thus

$$R_o \cong R_C \tag{8.33}$$

EXAMPLE 8.10

Determine R_{in} and the peak-to-peak and instantaneous forms of v_{od} for the circuit in Fig. 8.18(a), given $I_{EE} = 2$ mA, $R_C = 6.8$ kΩ, and $v_{in} = 8 \sin(2\pi 1000t)$ mV.

Solution The collector currents for Q_1 and Q_2 are equal, and are determined to be

$$I_C = I_{EE}/2 = 1 \text{ mA}$$

The dynamic resistances of the emitters are found next:

$$\begin{aligned} r_e &= 26 \text{ mV}/I_C \\ &= 26 \text{ mV}/1 \text{ mA} \\ &= 26\ \Omega \end{aligned}$$

Using Eq. (8.31), the differential-output voltage is now calculated. Since v_{in} was given in instantaneous form, its peak value (8 mV) was specified, giving us the output voltage as follows:

$$\begin{aligned} v_{od} &= \frac{v_{in} R_C}{2 r_e} \\ &= \frac{8 \sin(2\pi 1000t) \text{ mV} \times 6.8 \text{ k}\Omega}{2 \times 26\ \Omega} \\ &= 2.1 \sin(2\pi 1000t) \text{ V} \end{aligned}$$

The peak-to-peak amplitude is

$$V_{od} = 4.2 \text{ V}_{\text{P-P}}$$

The input resistance is

$$\begin{aligned} R_{in} &= 2\beta r_e \\ &= (200)(26\ \Omega) \\ &= 5.2 \text{ k}\Omega \end{aligned}$$

Differential-Input, Differential-Output Configuration

The differential-input, differential-output configuration is shown in Fig. 8.19(a). The output of this circuit is proportional to the instantaneous difference between v_2 and v_1. The proportionality constant is the *differential voltage gain.* Because the voltage gain relates to the difference between two input signals, it is called the differential gain A_d. As in the last section, the output voltage-sensing arrow has arbitrarily been directed such that $v_{od} = v_{C_2} - v_{C_1}$.

To derive expressions for v_{od} and A_d, we must determine the change in collector voltages, v_{C_1} and v_{C_2}, in response to each input source acting independently of the other. Superposition is then used to obtain net results. Let us arbitrarily start the analysis assuming v_1 is active, while v_2 is "killed" and replaced with its internal resistance, a short circuit. The equivalent circuit under these conditions is shown in Fig. 8.19(b).

The differential output v_{od_1}, due to v_1, is determined in the same manner as was v_{od} for Fig. 8.18. This gives the equation

$$v_{od_1} = \frac{v_1 R_C}{r_e} \tag{8.34}$$

The differential-output voltage caused by source v_2 acting alone has the same magnitude, but the opposite sign, thus

$$v_{od_2} = -\frac{v_2 R_C}{r_e} \tag{8.35}$$

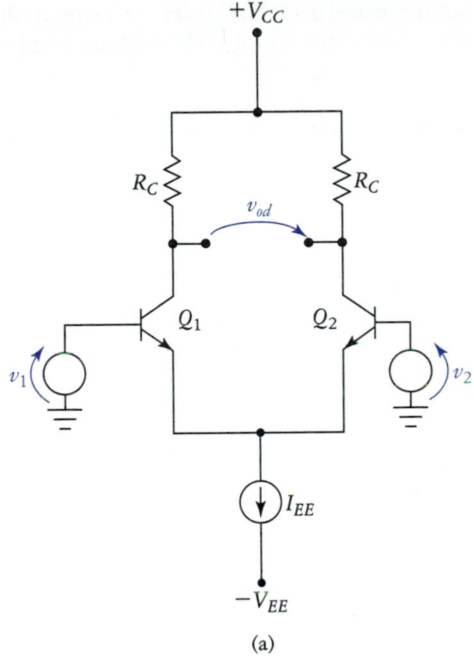

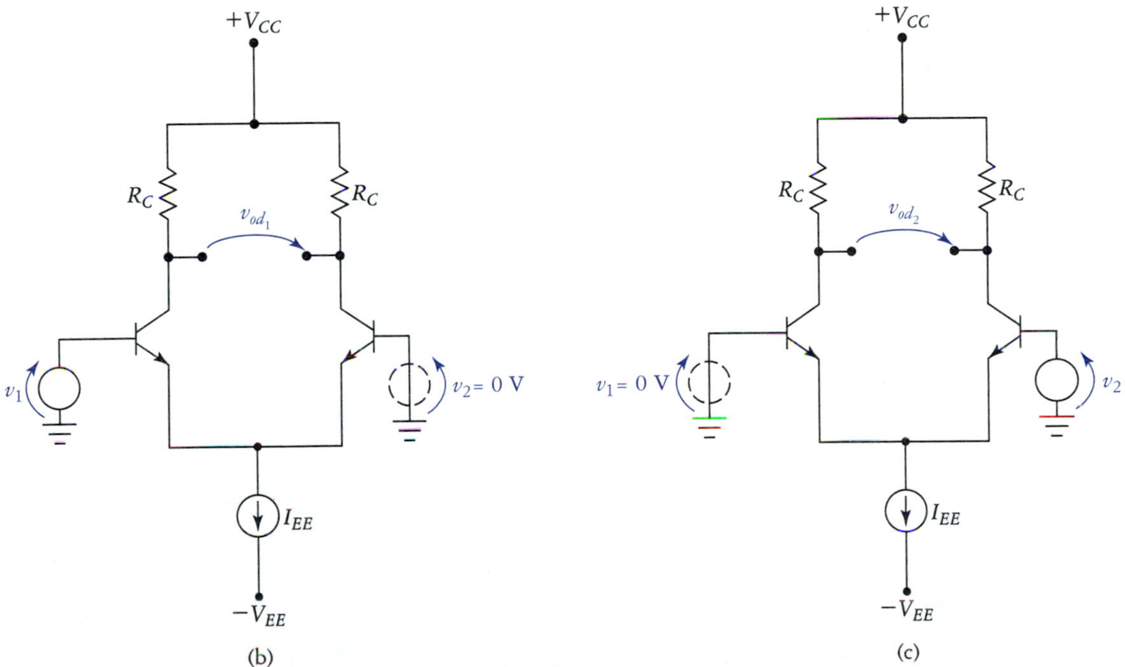

FIGURE 8.19
Analysis of the differential-input, differential-output configuration. (a) Basic circuit. (b) Using superposition with source 1 active. (c) Using superposition with source 2 active.

The net differential output v_{od} is the instantaneous algebraic sum of Eq. (8.34) and (8.35):

$$v_{od} = \frac{v_1 R_C}{r_e} + \frac{-v_2 R_C}{r_e}$$

The factor R_C/r_e, is called the *differential gain.* That is

$$A_d = \frac{R_C}{r_e} \qquad (8.36)$$

For notational convenience, from here on we shall refer to the voltage gain of any diff amp configuration that uses differential inputs as A_d. The differential input voltage is given by

$$v_{id} = v_1 - v_2 \tag{8.37}$$

Combining Eqs. (8.36) and (8.37) we can write

$$v_{od} = A_d v_{id} \tag{8.38}$$

or equivalently

$$v_{od} = \frac{R_C}{r_e}(v_1 - v_2) \tag{8.39}$$

The input resistance of the diff-in, diff-out amplifier is the same as for the previous configurations. That is,

$$R_{in} = 2\beta r_e \tag{8.40}$$

The output resistance is

$$R_o \cong R_C \tag{8.41}$$

EXAMPLE 8.11

Determine I_{EE}, I_{CQ}, V_{CEQ}, V_{CQ}, v_{od}, v_{C_1}, v_{C_2}, and R_{in} for the circuit in Fig. 8.20.

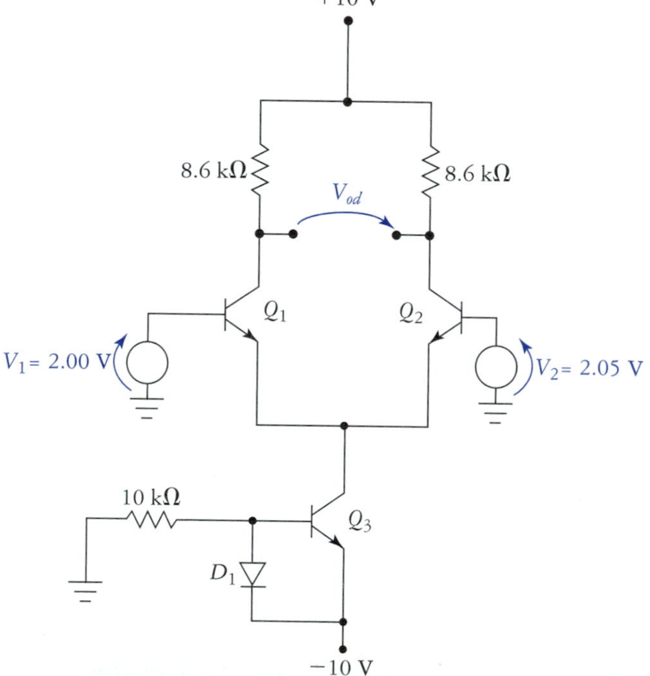

FIGURE 8.20
Circuit for Example 8.11.

Solution Analyzing the current mirror yields

$$I_{EE} = \frac{V_{EE} - V_{BE}}{R_1} = \frac{10\text{ V} - 0.7\text{ V}}{10\text{ k}\Omega}$$
$$= 930\ \mu\text{A}$$

and we have

$$I_{C_1} = I_{C_2} = I_{EE}/2 = 465\ \mu\text{A}$$

The quiescent collector voltages are now determined, applying the standard technique of killing the input voltage sources:

$$V_{CQ_1} = V_{CQ_2} = V_{CC} - I_C R_C$$
$$= 10 \text{ V} - (465 \ \mu\text{A})(8.6 \text{ k}\Omega)$$
$$= 10 \text{ V} - 4 \text{ V}$$
$$= 6 \text{ V}$$

Because the input sources have been killed for the quiescent analysis, both bases are at ground potential and we have $V_{CBQ} = V_{CQ}$, thus V_{CEQ} values are given by

$$V_{CEQ_1} = V_{CEQ_2} = V_{CBQ} + V_{BE}$$
$$= 6 \text{ V} + 0.7 \text{ V}$$
$$= 6.7 \text{ V}$$

Calculating dynamic emitter resistances gives

$$r_{e_1} = r_{e_2} = r_e = V_T / I_{CQ}$$
$$= 26 \text{ mV}/465 \ \mu\text{A}$$
$$= 56 \ \Omega$$

We can now determine the differential-output voltage. The differential gain is given by

$$A_d = R_C / r_e$$
$$= 8.6 \text{ k}\Omega \ / \ 56 \ \Omega$$
$$= 153.6$$

Applying Eq. (8.36), we get

$$v_{od} = A_d(v_1 - v_2)$$
$$= 153.6(2.00 \text{ V} - 2.05 \text{ V})$$
$$= -7.68 \text{ V}$$

It can be shown that the differential output will be divided evenly between the two collectors. That is, V_{C_1} and V_{C_2} will change by equal amounts in opposite directions. In this case, since $v_2 > v_1$, V_{C_2} will decrease by 3.84 V ($v_{od}/2$) and V_{C_1} will increase by 3.84 V. This produces

$$V_{C_1} = V_{CQ_1} + \Delta V_{C_1}$$
$$= 6 \text{ V} + 3.84 \text{ V}$$
$$= 9.84 \text{ V}$$

$$V_{C_2} = V_{CQ_2} + \Delta V_{C_2}$$
$$= 6 \text{ V} + (-3.84 \text{ V})$$
$$= 2.16 \text{ V}$$

Notice that Q_1 is very close to cutoff.

The collector voltages V_{C_1} and V_{C_2} in Example 8.11 could have been found by applying Eq. (8.27) to the amplifier in three steps; first with v_1 acting alone and then for v_2 acting alone, and then superposition is applied to determine the net collector voltages. This method produces impossibly large intermediate collector voltages on its way to the final answer. You may wish to try this analysis method just to be reassured that it will indeed produce the same results as were found in the example. You should also be aware that if the output voltage-sensing arrow were reversed, then the polarity of the output would be reversed as well.

Differential-Input, Single-Ended-Output Configuration

Note
The differential-input single-ended output configuration is the most widely used of the four configurations.

We have saved the best for last. If there is one differential amplifier configuration that you should know about, it is the differential-input, single-ended-output configuration. Such an amplifier is shown in Fig. 8.21. Although the collector of Q_2 is used to provide the output voltage, the collector of Q_1 could have been used just as well. The choice is completely arbitrary at this point. (We prefer to have inputs on the left and outputs on the right, which makes the Q_2 collector the more convenient output terminal.)

The ac analysis of Fig. 8.21 is quite easily performed at this time, because the basic groundwork has already been laid in the analysis of the previous three configurations. By inspection, it can be seen that the output voltage in response to v_{in_1} acting alone will be given by

$$v_{o_1} = v_1 \frac{R_C}{2r_e} \tag{8.42}$$

FIGURE 8.21
The differential-input, single-ended-output configuration.

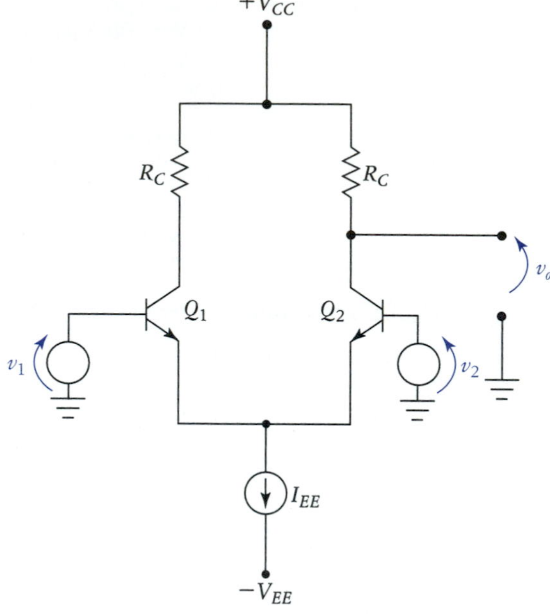

The output voltage that is developed in response to v_2 alone is

$$v_{o_2} = v_2 \frac{-R_C}{2r_e} \tag{8.43}$$

The net output voltage is found using superposition and a little algebraic manipulation:

$$v_o = \frac{R_C}{2r_e}(v_1 - v_2) \tag{8.44}$$

The factor $R_C/2r_e$ is the differential voltage gain of the circuit. That is,

$$A_d = \frac{R_C}{2r_e} \tag{8.45}$$

Again, for consistency, whenever we are working with an amplifier in the differential-input configuration, the voltage gain of that circuit will be called the differential gain A_d. The gain of the single-ended-input configurations is simply A_v.

When applying Eq. (8.44) and (8.23) (the single-ended-output equations) it is important to keep in mind that V_{CQ} is present, and it must be accounted for when the net output voltage is determined. This is not necessary when considering the configurations that produce a differential output because the quiescent collector voltages are equal and thus cancel each other.

EXAMPLE 8.12

Refer to Fig. 8.22. Determine I_{EE}, I_{CQ}, V_{oQ}, and v_o. Sketch the resulting output voltage waveform.

Solution

Current source Q_3 is analyzed first:

$$\begin{aligned} I_{EE} &= V_{BE}/R_4 \\ &= 0.7\text{ V}/1.4\text{ k}\Omega \\ &= 500\ \mu\text{A} \end{aligned}$$

Thus,

$$I_{C_1} = I_{C_2} = I_{EE}/2 = 250\ \mu\text{A}$$

and

$$r_e = 26\text{ mV}/250\ \mu\text{A} = 104\ \Omega$$

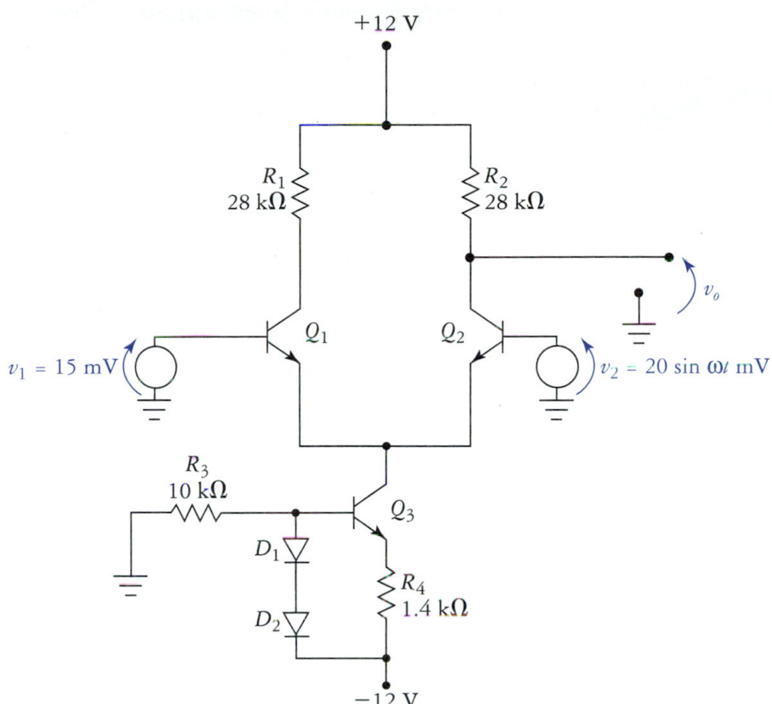

FIGURE 8.22
Circuit for Example 8.12.

Using KVL, the quiescent collector voltage for Q_2, which is also V_{oQ}, is found to be

$$\begin{aligned} V_{oQ} &= V_{CC} - I_C R_C \\ &= 12\text{ V} - (250\ \mu\text{A})(28\text{ k}\Omega) \\ &= 12\text{ V} - 7\text{ V} \\ &= 5\text{ V} \end{aligned}$$

The gain of the amplifier is

$$\begin{aligned} A_d &= R_C/2r_e \\ &= 28\text{ k}\Omega\,/208\ \Omega \\ &= 134.6 \end{aligned}$$

Now, since one of the inputs is a dc voltage and the other is a sinusoid, the differential input voltage is simply

$$v_{id} = 15 - 20\sin\omega t\text{ mV}$$

and the output voltage is

$$\begin{aligned} v_o &= V_{oQ} + A_d v_{id} \\ &= 5\text{ V} + 134.6(15 - 20\sin\omega t\text{ mV}) \\ &= 5 + 2 - 2.7\sin\omega t\text{ V} \\ &= 7 - 2.7\sin\omega t\text{ V} \end{aligned}$$

This voltage is illustrated in Fig. 8.23.

FIGURE 8.23
Output voltage for Fig. 8.22.

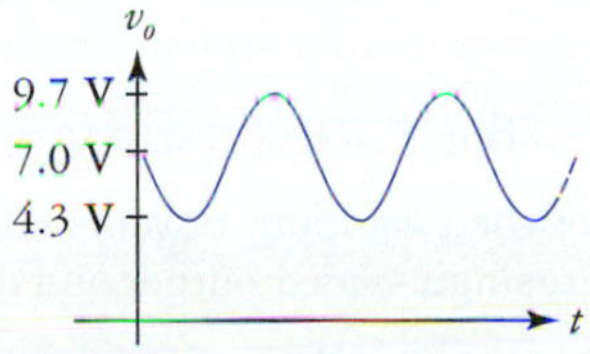

The next example presents a design-oriented problem.

EXAMPLE 8.13

Determine the values of R_C required in Fig. 8.22 to produce $A_d = 100$ and $A_d = 500$.

Solution The dc analysis was performed in the previous example, so we shall use those results that are applicable. Because $A_d = R_C/2r_e$, we can solve for R_C, which gives $R_C = 2r_eA_d$. For $A_d =$ 100 we have

$$\begin{aligned} R_C &= 2 \times 104\ \Omega \times 100 \\ &= 20.8\ \text{k}\Omega \end{aligned} \qquad (1)$$

For $A_d = 500$ we have

$$\begin{aligned} R_C &= 2 \times 500 \times 104 \\ &= 104\ \text{k}\Omega \end{aligned} \qquad (2)$$

Solution (1) is valid, however, solution (2) is not realizable. Consider that in order for Q_2 to operate in its active region, its C-B junction must be reverse biased. If $R_C = 104\ \text{k}\Omega$, then at $I_C = 250$ μA, V_{CQ} calculates to be -14 V. Obviously, the transistor would be heavily saturated because with its base held at 0 V, the lowest value that V_{CQ} could possibly attain (assuming that the transistor is not defective) is about -0.7 V.

The ac analysis equations applicable to the basic differential-pair configurations are summarized in Table 8.1. The voltage gain (or differential gain) is expressed in absolute magnitude because of the many variations in input and output orientations that are possible.

TABLE 8.1
Differential Amplifier AC Characteristics

	Single In, Single Out	Single In, Diff Out	Diff In, Diff Out	Diff In, Single Out
$\lvert A_d \rvert$	$\dfrac{R_C}{2r_e}$	$\dfrac{R_C}{r_e}$	$\dfrac{R_C}{r_e}$	$\dfrac{R_C}{2r_e}$
R_{in}	$2\beta r_e$	$2\beta r_e$	$2\beta r_e$	$2\beta r_e$
R_o	R_C	R_C	R_C	R_C

8.4 DIFFERENTIAL AMPLIFIER VARIATIONS

Differential amplifiers are implemented in different ways in order to better suit a given application. This section examines a few of the more significant variations.

Swamped Differential Amplifier

Recall that the addition of emitter swamping resistors has two important beneficial effects on amplifier characteristics; input resistance is increased and voltage gain is stabilized. On the other hand, the main disadvantage of emitter swamping is a decrease in voltage gain. This disadvantage, however, is often outweighed by the improved input resistance and stability provided by emitter swamping.

Note
Equation 8.47 applies to the single-ended output configurations.

A differential amplifier that uses emitter swamping is shown in Fig. 8.24. Routine analysis of this circuit shows that for all four amplifier configurations.

$$R_{in} = 2\beta(r_e + R_E) \qquad (8.46)$$

The effect of the swamping resistors on voltage gain is seen in the following equations. For single-ended-input, single-ended-output and differential-input, single-ended-output configurations,

$$A_v = A_d = \frac{R_C}{2(r_e + R_E)} \qquad (8.47)$$

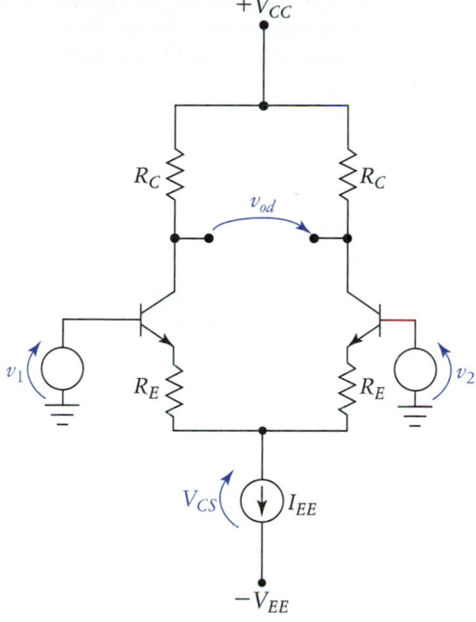

FIGURE 8.24
Differential amplifier using emitter swamping resistors.

Note
Equation 8.48 applies to the differential output configurations.

For single-ended-input, differential-output and differential-input, differential-output configurations,

$$A_v = A_d = \frac{R_C}{r_e + R_E} \tag{8.48}$$

The value of R_E will usually be small enough that its effect on the dc analysis is negligible, as shown in the next example.

EXAMPLE 8.14

Refer to Fig. 8.24. Given $V_{CC} = V_{EE} = 15$ V, $I_{EE} = 2$ mA, and $R_C = 10$ kΩ, determine V_{CEQ}, R_{in}, A_d (diff-in, diff-out), and the voltage drop V_{CS} across the current source, for $R_E = 0\ \Omega$ and for $R_E = 100\ \Omega$. Assume $\beta = 100$.

Solution The collector current is independent of the emitter swamping resistors and is

$$I_{CQ} = I_{EE}/2 = 1 \text{ mA}$$

The value of V_{CEQ} is determined using KVL and Ohm's law. For the C-B junctions we have

$$\begin{aligned} V_{CB} = V_C &= V_{CC} - I_{CQ}R_C \\ &= 15 \text{ V} - (1 \text{ mA})(10 \text{ k}\Omega) \\ &= 5 \text{ V} \end{aligned}$$

which we substitute into

$$\begin{aligned} V_{CE} &= V_{CB} + V_{BE} \\ &= 5 \text{ V} + 0.7 \text{ V} \\ &= 5.7 \text{ V} \end{aligned}$$

This result is valid for both $R_E = 0\ \Omega$ and $R_E = 100\ \Omega$, or any other value of R_E.

The voltage drop across the current source V_{CS} is dependent on the emitter resistor value. Applying KVL, we have

$$V_{CS} = V_{CC} + V_{EE} - I_{CQ}(R_C + R_E) - V_{CEQ}$$

For $R_E = 100\ \Omega$, we find

$$\begin{aligned} V_{CS} &= 15 \text{ V} + 15 \text{ V} - (1 \text{ mA})(10 \text{ k}\Omega + 100) - 5.7 \text{ V} \\ &= 14.2 \text{ V} \end{aligned}$$

For $R_E = 0$,

$$\begin{aligned} V_{CS} &= 15 \text{ V} + 15 \text{ V} - (1 \text{ mA})(10 \text{ k}\Omega) - 5.7 \text{ V} \\ &= 14.3 \text{ V} \end{aligned}$$

Thus we see that for typical R_E values there is virtually no effect on the dc analysis. This is not the case in the ac analysis, however. For $R_E = 100\ \Omega$, we have $R_{in} = 25.2\ k\Omega$ and $A_d = 79.4$. For the case of $R_E = 0\ \Omega$, we have $R_{in} = 5.2\ k\Omega$ and $A_d = 385$.

Darlington- and FET-Based Differential Pairs

To increase the input resistance of the differential amplifier without using swamping resistors, either Darlington pairs or field-effect transistors can be used. A Darlington-type diff amp is shown in Fig. 8.25. The pertinent ac analysis equations for this circuit are as follows: The gain magnitude for both single-ended-output configurations is

$$A_d = \frac{R_C}{4r_e} \tag{8.49}$$

For single-ended-output with emitter swamping we have

$$A_d = \frac{R_C}{4r_e + 2R_E} \tag{8.50}$$

For both differential-output configurations the gain magnitude is given by

$$A_d = \frac{R_C}{2r_e} \tag{8.51}$$

And if emitter swamping resistors are used we have

$$A_d = \frac{R_C}{2r_e + R_E} \tag{8.52}$$

The input resistance for all four basic configurations is approximately

$$R_{in} \cong 4\beta^2 r_e \tag{8.53}$$

and for the swamped-emitter case

$$R_{in} \cong 2\beta^2 R_E + 4\beta^2 r_e \tag{8.54}$$

A JFET-type diff amp is shown in Fig. 8.26. The input resistance of the FET differential amplifier will be the gate-to-source resistance for the particular devices used. Typically this is around $10^{10}\ \Omega$ at frequencies below 1 MHz. Notice that although JFETs are shown here, MOSFETs could be used also.

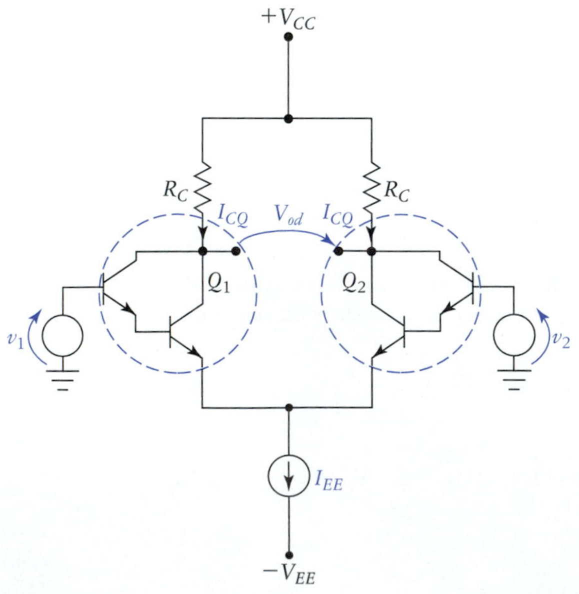

FIGURE 8.25
Darlington transistor diff amp.

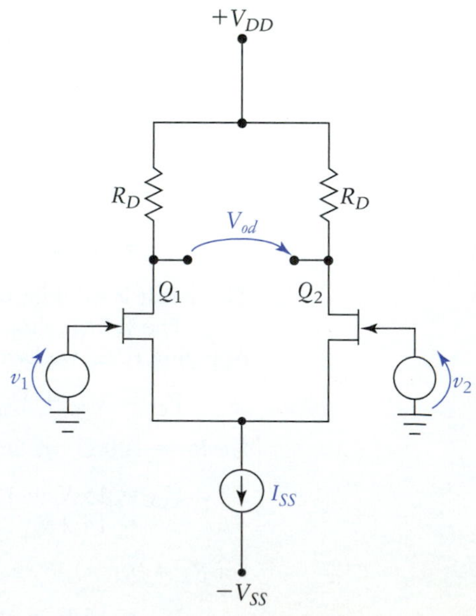

FIGURE 8.26
Differential amplifier using n-channel JFETs.

The choice between FET and BJT usage in differential amplifier designs is determined by the characteristics that are needed in certain applications. For example, using monolithic IC fabrication techniques, bipolar transistors can be matched very closely in V_{BE} characteristics ($\Delta V_{BE} < 1$ mV), whereas the pinchoff voltages of JFETs in a differential pair may easily differ by more than 10 mV. On the other hand, since FETs require nearly zero gate bias current (typically I_{GSS} is around 10^{-12} A, which is usually several orders of magnitude lower than that of I_B for a BJT), the effects of slight inequalities between these bias currents and external gate resistances will be very low. This makes FET input differential amplifiers better suited for operation with very high resistance sources. Other pros and cons are associated with each device type; however, these are two of the more important characteristics to consider.

Active Loading

You may have noticed that it is rather easy to achieve large voltage gain from a differential pair, even though no bypassing capacitors are used. This is a strong advantage of the differential pair; however, it is possible to increase the voltage gain to much higher levels using *active loading.*

Note
Active loading allows very high voltage gain to be realized without the use of high-value resistors.

Active loading is used in the circuit of Fig. 8.27(a). Here's how it works: Transistors Q_1 and Q_2 form the usual differential pair. PNP transistors Q_3 and Q_4 form a current mirror, where Q_3 is connected as a diode. Under quiescent conditions, Q_1 and Q_2 carry the same collector current, and I_{CQ_1} acts as the reference current for the pnp current mirror. Thus we find that under quiescent conditions, all of the transistors naturally carry equal collector currents. The application of a differential-input voltage causes Q_1 and Q_2 to carry different collector currents. Let's assume that $v_{\text{in}_1} > v_{\text{in}_2}$. This tends to increase I_{C_1} and decrease I_{C_2} by equal amounts. An increase in I_{C_1} causes an increase in reference current I_{C_3}. This in turn should cause an increase in I_{C_4}; however, $I_{C_4} = I_{C_2}$, and we know that I_{C_2} is tending to decrease. As a result of these conflicting actions, a very small differential-input voltage will produce a very large output voltage. Using the small-signal model of Fig. 8.27(b), the magnitude of the differential gain of the circuit is found to be

$$A_d = \frac{r'_{CE}}{2r_e} \tag{8.55}$$

where $r'_{CE} = r_{CE_2} \| r_{CE_4}$.

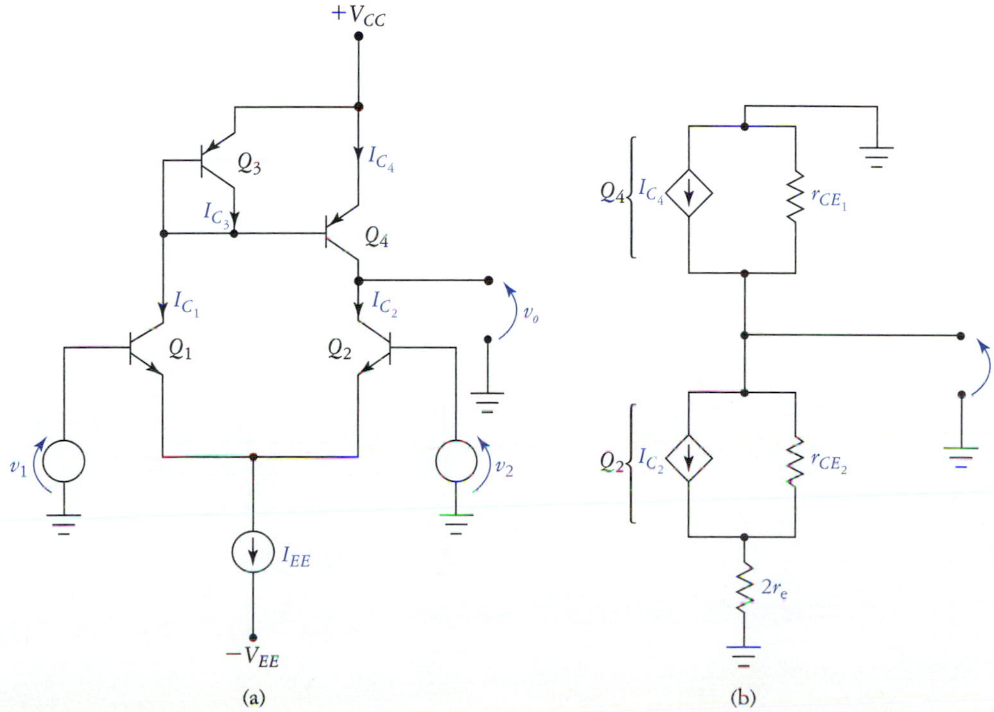

$+V_{CC}$, Q_3, Q_4, I_{C_4}, I_{C_3}, I_{C_1}, I_{C_2}, Q_1, Q_2, v_o, v_1, v_2, I_{EE}, $-V_{EE}$

(a)

Q_4, I_{C_4}, r_{CE_1}, v_o, Q_2, I_{C_2}, r_{CE_2}, $2r_e$

(b)

FIGURE 8.27
Differential amplifier using active loading to achieve high voltage gain. (a) Typical circuit. (b) Small-signal ac equivalent of the right side of the circuit.

To appreciate the effectiveness of active loading, let us assume that in Fig. 8.27(a), $I_{EE} = 2$ mA. This means that all transistors have $I_{CQ} = 1$ mA. Let us assume that we also have $r_{CE_2} = r_{CE_4} =$ 100 kΩ. Under these conditions, $r_e = 26\ \Omega$ and $r_{CE} = 50$ kΩ. Application of Eq. (8.55) gives us

$$A_d = \frac{r_{CE}}{2r_e}$$
$$= \frac{50\ \text{k}\Omega}{52\ \Omega}$$
$$= 962$$

It has been the author's experience that differential gains of several thousand in the laboratory can be obtained using this active-loading technique with integrated transistor arrays.

8.5 DIFFERENTIAL-PAIR ERROR SOURCES

The characteristics of the differential pair have been presented in an idealized manner up to this point. Before leaving this topic, however, a few of the less ideal aspects of the differential pair should be examined.

Common-Mode Response

It is very important to understand that the output of an amplifier with differential inputs is proportional to the difference between the input voltages. This means that if $v_{id} = 0$ V (that is, $v_1 - v_2 =$ 0 V), then ideally, $v_o = 0$ V.* When $v_1 = v_2 \neq 0$ V we have what is called a *common-mode input voltage* v_{icm}. Unfortunately, common-mode inputs do cause an undesirable change in the output of the practical differential amplifier.

A reasonable question that might be asked at this point is, what could cause the amplifier to produce an output v_{ocm} in response to a common-mode input? The primary cause of the common-mode response of the differential amplifier is the finite internal resistance of the I_{EE} emitter current source. Ideally, the internal resistance of a current source is infinite. As we all know, however, this is impossible to attain in practice. A differential amplifier with a common-mode input and nonideal current source is shown in Fig. 8.28.

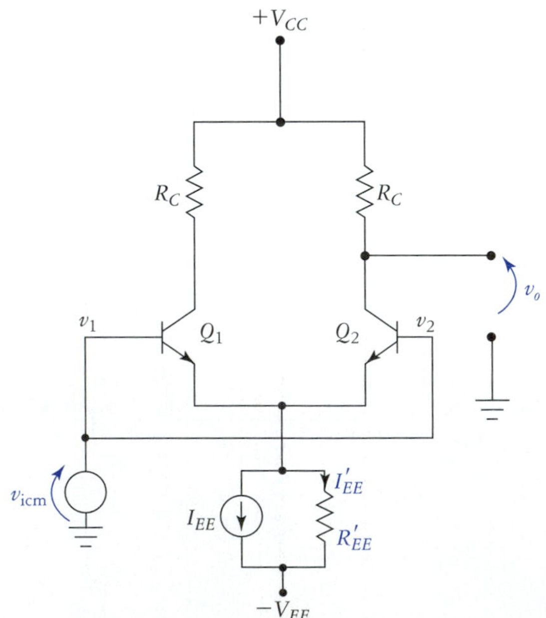

FIGURE 8.28
Nonideal emitter current source reduces common-mode rejection.

*Actually, for single-ended output $v_o = V_{oQ} + A_d v_{id}$, so technically we should write $\Delta v_o = \Delta V_{oQ} = 0$ V if $v_{\text{in}_1} = v_{\text{in}_2}$.

Common-mode input voltage is defined as the average of the two voltages applied to the inputs of the differential amplifier. That is,

$$v_{icm} = \frac{v_1 + v_2}{2} \tag{8.56}$$

The gain of the differential amplifier to common-mode inputs is given by the approximate relationship

$$A_{cm} \cong -\frac{R_C}{2R_{EE}} \tag{8.57}$$

It is desirable to make A_{cm} as small as possible. This is one of the reasons that active current sources are used in differential amplifier circuits. The very high R_{EE} values that occur in active current sources (especially the Widlar current source) make the denominator of Eq. (8.57) very large.

Most often when working with differential amplifier circuits, a parameter called the *common-mode rejection ratio (CMRR)* is specified. CMRR is defined as

$$\text{CMRR} = \left|\frac{A_d}{A_{cm}}\right| \tag{8.58}$$

The larger the value of CMRR, the better. It is not uncommon to have CMRR $> 10^5$ for off-the-shelf IC differential amplifiers.

EXAMPLE 8.15

Refer to Fig. 8.28. Assume that $v_{icm} = 2$ V, $R_{EE} = 100$ kΩ, $R_C = 10$ kΩ, $V_{CC} = 15$ V, and $I_{EE} =$ 2 mA. Determine V_{oQ}, A_d, A_{cm}, V_{ocm}, CMRR, and the net output voltage v_o.

Solution The collector currents are $I_{CQ} = I_{EE}/2 = 1$ mA, therefore $r_e = 26$ Ω. The quiescent output voltage is

$$\begin{aligned} V_{oQ} &= V_{CC} - I_{CQ}R_C \\ &= 15 \text{ V} - (1 \text{ mA})(10 \text{ k}\Omega) \\ &= 5 \text{ V} \end{aligned}$$

The various gain values are

$$\begin{aligned} A_d &= \frac{R_C}{2r_e} \\ &= \frac{10 \text{ k}\Omega}{52\ \Omega} \\ &= 192.3 \end{aligned}$$

$$\begin{aligned} A_{cm} &= -\frac{R_C}{2R_{EE}} \\ &= -\frac{10 \text{ k}\Omega}{200 \text{ k}\Omega} \\ &= -0.05 \end{aligned}$$

The common-mode rejection ratio is

$$\begin{aligned} \text{CMRR} &= \left|\frac{A_d}{A_{cm}}\right| \\ &= \frac{192.9}{0.05} \\ &= 3.85 \times 10^3 \end{aligned}$$

The common-mode output voltage is

$$\begin{aligned} v_{ocm} &= A_{cm}v_{icm} \\ &= (-0.05)(2 \text{ V}) \\ &= -100 \text{ mV} \end{aligned}$$

The net output voltage is

$$\begin{aligned} v_o &= V_{oQ} + v_{ocm} \\ &= 5\text{ V} - 100\text{ mV} \\ &= 4.9\text{ V} \end{aligned}$$

For the circuits that we have worked with so far in this chapter, the minimum (most negative) allowable common-mode input voltage is approximately

$$v_{icm(\text{min})} \cong -V_{EE} \tag{8.59}$$

Common-mode input voltages more negative than this will forward bias C-B junctions in the I_{EE} current-source part of the circuit, possibly causing device destruction.

The maximum allowable common-mode input voltage is approximately

$$v_{icm(\text{max})} \cong V_{CQ} \tag{8.60}$$

The reason for this limitation is clear: Application of an input voltage greater than V_{CQ} causes forward biasing of the collector–base junctions of the differential pair, hence terminating normal transistor action. Exceeding the $v_{icm(\text{max})}$ specifications for any differential amplifier circuit will usually result in the destruction of the circuit.

Decibel Representation of CMRR

Very often, CMRR is expressed in *decibels* (dB). The decibel is a logarithmic expression of a ratio of two quantities. A formal derivation of the decibel is presented in Chapter 9, Section 9.6. However, for now, we can determine the CMRR in decibels using the relationship

$$\text{CMRR}_{\text{dB}} = 20 \log \left| \frac{A_d}{A_{cm}} \right|$$

where log is the common or base-10 logarithm. If you are given CMRR in decibels, you can determine the linear ratio using

$$\text{CMRR} = \log^{-1}\left(\frac{\text{CMRR}_{\text{dB}}}{20}\right)$$

where $\log^{-1}$ is the inverse log or antilogarithm.

Bias Current, Offset Current, and Offset Voltage

In addition to common-mode considerations, practical differential amplifiers have several other nonideal characteristics that may be important in certain applications. These are briefly examined now.

Bias-Current Effects

In most of the preceding examples the base-bias currents were assumed to be extremely small (essentially zero) and the input voltage sources had zero internal resistance as well. In practice, the base-bias currents may be quite small (typically a few nanoamps for bipolar designs), but nevertheless they will produce voltage drops across any base resistance that may be present. The effects of unequal bias currents and/or voltage drops across the base resistances can cause significant error at the output of a differential amplifier.

Consider the differential ampifier of Fig. 8.29(a), where the input source has an internal resistance R_S. When $v_S = 0$ V, we expect $v_{od} = 0$ V and the single-ended output $v_{o(\text{SE})} = V_{oQ}$ as well. However, the flow of Q_1's base current through R_S causes a voltage drop I_BR_S to exist. Now, because the base of Q_2 is connected directly to ground, this amounts to a differential-input voltage $v_{id} = -I_BR_S$. Thus erroneous output voltage responses are produced. Specifically,

$$v_{od} = -I_BR_S\frac{R_C}{r_e} \tag{8.61}$$

and

$$v_{o(\text{SE})} = V_{oQ} + I_BR_S\frac{R_C}{2\,r_e} \tag{8.62}$$

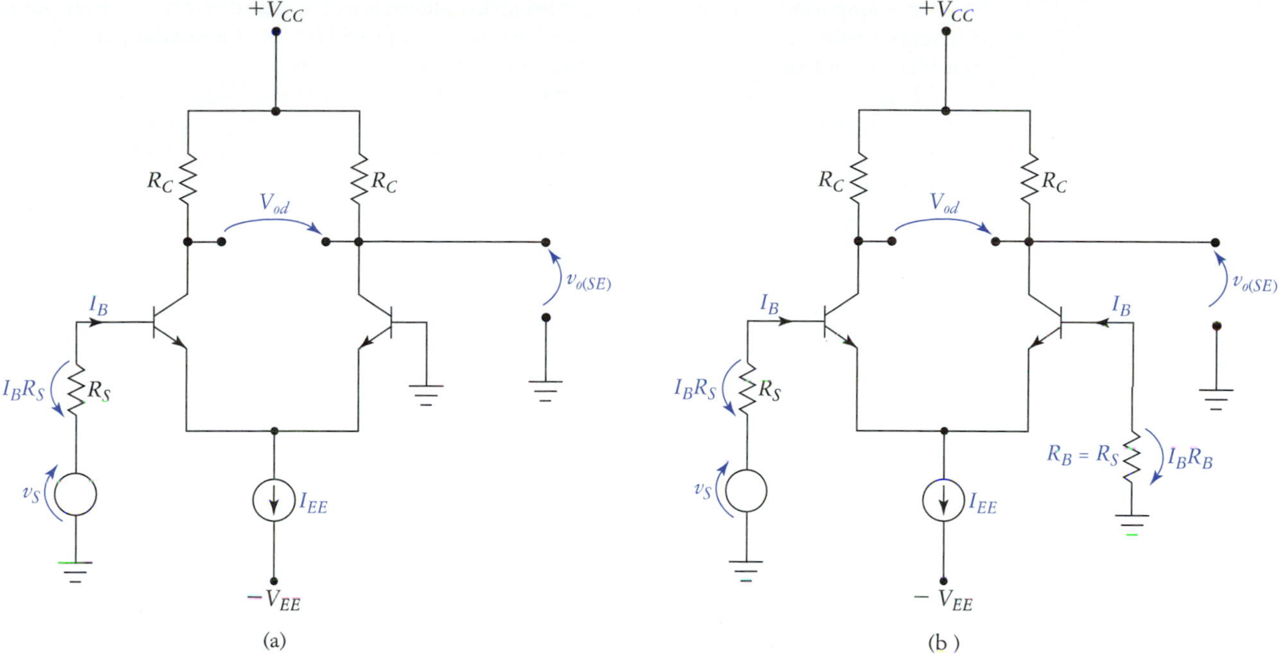

FIGURE 8.29

Bias currents can induce an output error voltage. (a) Bias current causes drop across source resistance. (b) Equalizing resistances of the input terminals cancels bias-current-induced error.

The effects of this bias-current-induced voltage can be reduced significantly if the base of Q_2 is grounded through a resistance $R_B = R_S$, as shown in Fig. 8.29(b). This causes both bases to float at the same potential, reducing the differential-input voltage to zero. Equalizing the base resistances produces a bias-current-induced common-mode input voltage, but typically this does not cause problems because this voltage and A_{cm} are both quite small.

Typical bias currents for BJT-based differential amplifiers range from about 1 nA to around 20 nA at 25°C. FET-based diff amps typically have bias currents of around 1 pA or so.

Offset-Current Effects

The assumption that the base currents of the differential pair are equal depends on having perfectly matched transistors. We know that in practice this is impossible. In particular, slight mismatching between β will cause $I_{B_1} \neq I_{B_2}$. The difference between these currents is called the input offset current I_{io}, which is defined as

$$I_{io} = |I_{B_1} - I_{B_2}| \tag{8.63}$$

The absolute magnitude of the current difference is used because in general, it is not known which base-bias current is larger, thus I_{io} is assumed to be positive.

It is impractical to determine the exact values of the bias currents for most diff amp applications; therefore, the bias current that is specified is usually just the average of the typical individual bias currents. That is,

$$I_B = \frac{I_{B_1} + I_{B_2}}{2} \tag{8.64}$$

Offset-Voltage Effects

Even if the effects of bias and offset currents are accounted for or eliminated, the output of the differential amplifier may still deviate from the expected value. One cause for such an erroneous output is mismatching between V_{BE} drops in the differential pair. This causes the collectors of the differential pair to operate at slightly different quiescent voltages, which is undesirable. Recall that typically, this mismatch is within 1 mV for bipolar IC designs and 10 mV for FET-based designs.

It is important to realize that V_{BE} mismatch-induced errors appear directly at the output of the differential pair, while bias-current-induced errors are amplified by the differential pair. This means that bias-current-induced errors will usually be of greater significance.

It is standard practice to simply lump the effects of bias currents, offset currents, and V_{BE} mismatching together to obtain the input offset voltage V_{io} for a differential pair. The input offset voltage is the hypothetical differential-input voltage that exists when external signal sources are inactive. Assuming a differential-in, single-ended-out configuration, we define V_{io} in equation form as follows:

$$V_{io} = \frac{V_{oo}}{A_d} \tag{8.65}$$

where V_{oo} is the output offset voltage. In other words, V_{oo} is the output voltage that is caused only by the error sources that constitute V_{io}.

8.6 PRACTICAL DIFFERENTIAL AMPLIFIER IMPLEMENTATION

It is impractical to implement a useful differential amplifier using discrete transistors because of parameter matching problems. For this reason, differential amplifiers are usually implemented in monolithic or hybrid IC form, or they are made using transistor arrays. The monolithic form has transistors, diodes, and resistors formed on a single silicon die or chip. All the user of such a device must do is provide the necessary supply voltages and possibly a few external resistors to have a working differential amplifier. This approach is generally the least expensive, while at the same time providing good performance.

FIGURE 8.30
The CA3028A is a monolithic differential amplifier. (a) Internal circuit. (b) Pin configurations.

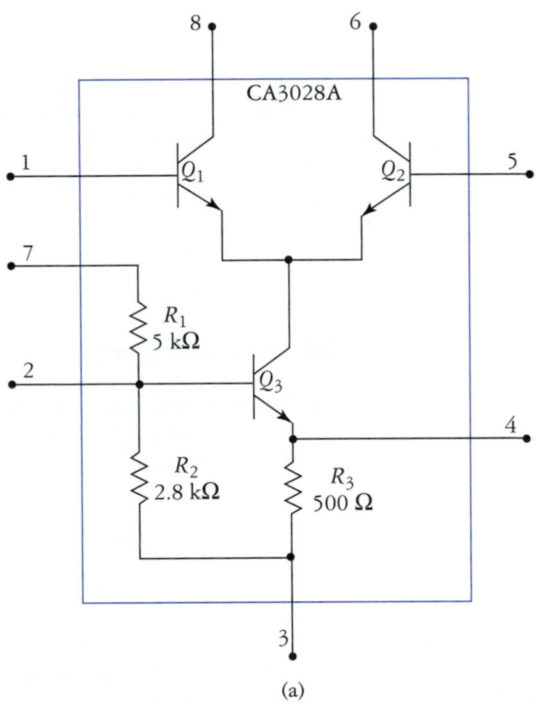

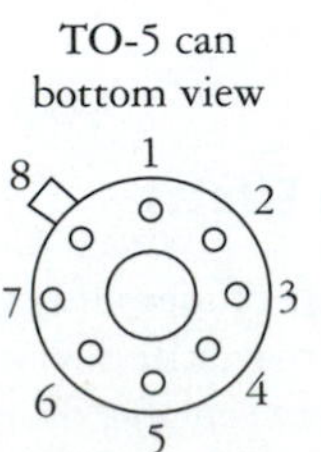

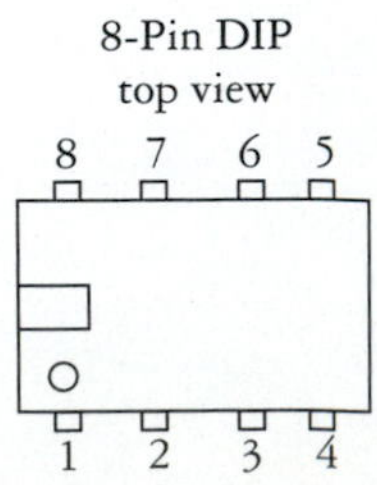

(b)

An example of a monolithic differential amplifier is the CA3028A from RCA. The pin assignments and internal circuitry for this device are shown in Fig. 8.30. Transistors Q_1 and Q_2 form the differential pair, while Q_3 is an active current source. Resistors R_1, R_2, and R_3 are integrated into the device, while the user supplies the appropriate collector resistors for the application at hand.

An example of an integrated transistor array is the LM3086 from National Semiconductor. The pin assignments and internal transistor arrangements for the LM3086 are shown in Fig. 8.31. Transistor pairs Q_1–Q_2, and Q_4–Q_5 are V_{BE} matched to within 5 mV, while the input bias currents of pair Q_1–Q_2 are matched to within 2 μA. Transistor Q_3 is not closely matched to any specific pair, and is for general-purpose use. When using the LM3086, the emitter of Q_5 must be connected to the most negative point in the circuit, relative to the other transistors in the LM3086. This prevents accidental forward biasing of parasitic pn junctions, which could easily destroy the array. Transistor arrays are used in the design of custom and special-purpose differential amplifiers.

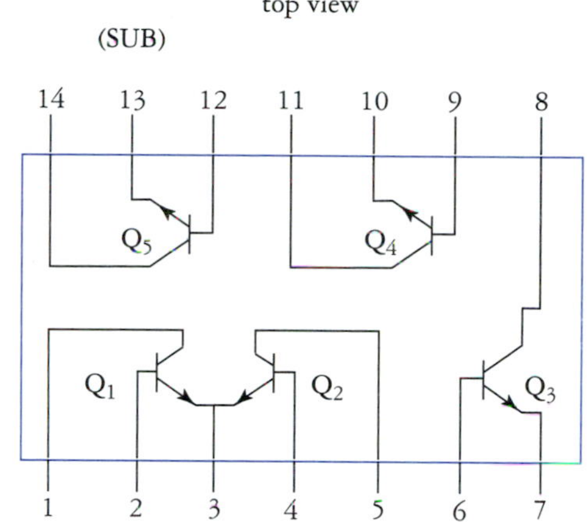

FIGURE 8.31
The LM3086 monolithic transistor array.

Hybrid differential amplifiers generally contain active devices (BJTs and/or FETs) in monolithic form, along with thick- or thin-film resistors, all mounted on a ceramic substrate. Often the film resistors are laser trimmed while the circuit is active in order to reduce offset errors. These devices are normally rather expensive, but they usually have superior performance in terms of gain accuracy, common-mode rejection, offset voltage and current, and temperature-drift characteristics.

8.7 COMPUTER-AIDED ANALYSIS APPLICATIONS

A differential amplifier that is representative of those presented in this chapter is shown in Fig. 8.32. This circuit is referred to extensively throughout the following discussion.

Manual Analysis Results

Assuming that the input signal sources are inactive, and ignoring the slight effect of the 1-kΩ source resistances, pencil-and-paper analysis of this circuit yields the following results:

$$I_{EE} = \frac{V_{EE} - V_{BE}}{R_3} = \frac{15\text{ V} - 0.7\text{ V}}{14.3\text{ k}\Omega} = 1\text{ mA}$$

$$I_{C_1} = I_{C_2} = I_{EE}/2 = 500\ \mu\text{A}$$

$$\begin{aligned} V_{CB_1} = V_{CB_2} &= V_{CC} - I_C R_C \\ &= 15\text{ V} - (500\ \mu\text{A})(20\text{ k}\Omega) \\ &= 5\text{ V} \end{aligned}$$

$$\begin{aligned} V_{CE_1} = V_{CE_2} &= V_{CB} + V_{BE} \\ &= 5\text{ V} + 0.7\text{ V} \\ &= 5.7\text{ V} \end{aligned}$$

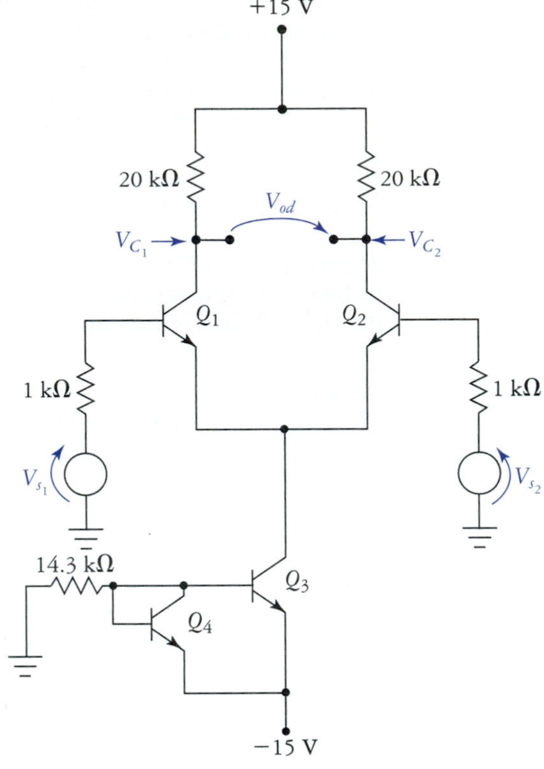

FIGURE 8.32
Typical differential amplifier.

$$V_{C_1} = V_{C_2} = V_{CB} = 5\text{ V}$$

A quick ac analysis yields the following results:

$$|A_d| = \frac{R_C}{r_e} \quad \text{(for differential-mode outputs)}$$

$$= \frac{20\text{ k}\Omega}{52\ \Omega}$$

$$= 385$$

and for single-ended outputs

$$|A_v| = 192.5$$

The input and output resistances of the amplifier are

$$R_{\text{in}} = 2\beta r_e = (200)(52\ \Omega) = 10.4\text{ k}\Omega$$

$$R_o \cong R_C = 20\text{ k}\Omega$$

PSpice Q-Point Analysis Results

The circuit is created using 2N3904 transistors from the evaluation library. Pertinent quiescent voltages and currents are shown by the Viewpoints and Iprobes in Fig. 8.33. In this simulation, $I_{EE} = 1.17$ mA, which is 17% higher than the manual analysis predicted. This is the primary reason for the variation from the calculated values in the other Q-point quantities. We could tweak the circuit easily by adjusting the value of R_3, but we'll work with what we have for now.

It is interesting to examine the BJT parameters that were used by PSpice. The 2N3904 specifications are shown in Fig. 8.34. Notice that the forward beta (BF) is listed as 416.4, which is apparently quite high. The actual beta value that exists at a given Q point (PSpice calls this β_{dc}) will be much lower.

The Q-point-dependent transistor parameters are shown in Fig. 8.35. Here we find that β_{dc} ranges from 129 to 157, depending on the transistor under scrutiny. These data are found in the output file that is generated by the bias-point analysis. A few other interesting things here are worth ex-

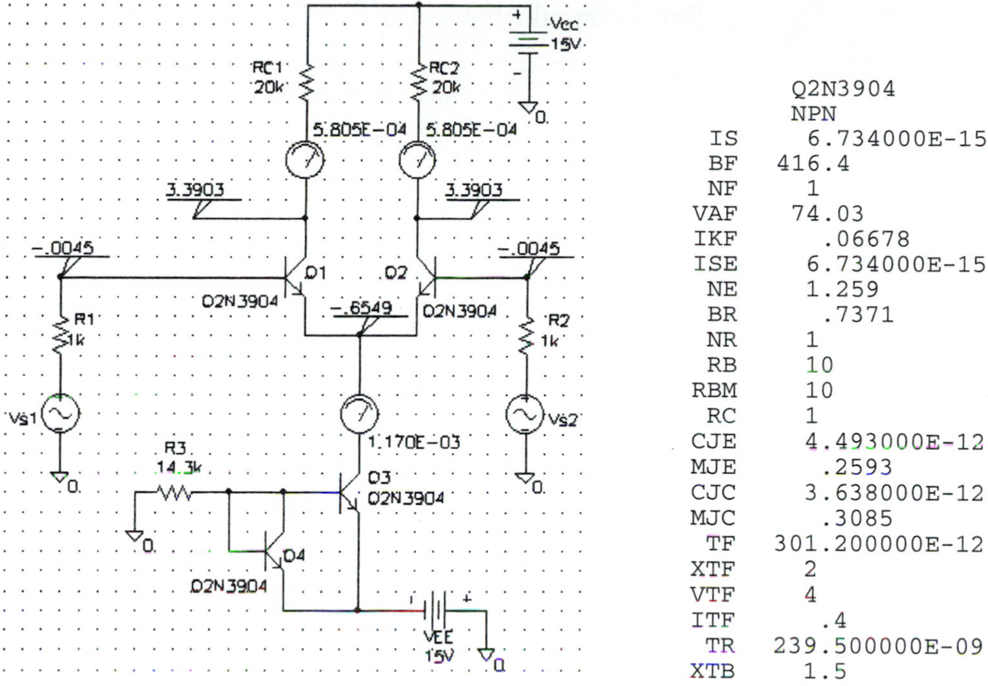

	Q2N3904
	NPN
IS	6.734000E-15
BF	416.4
NF	1
VAF	74.03
IKF	.06678
ISE	6.734000E-15
NE	1.259
BR	.7371
NR	1
RB	10
RBM	10
RC	1
CJE	4.493000E-12
MJE	.2593
CJC	3.638000E-12
MJC	.3085
TF	301.200000E-12
XTF	2
VTF	4
ITF	.4
TR	239.500000E-09
XTB	1.5

FIGURE 8.33
The diff amp showing bias-point information.

FIGURE 8.34
2N3904 transistor model parameters.

amination. Notice that instead of listing r_e, PSpice produces the equivalent transconductance parameter GM ($g_m \cong 1/r_e$). Also notice that the pi-model equivalent resistance RPI (that is, r_π) is given. Recall that $r_\pi = r_e(\beta + 1)$.

So, we know that the simulation run differs somewhat from our more idealized analysis, but the degree of disagreement is not that severe, all things considered. As we will find out in later chapters, most often amplifiers (especially differential amplifiers) are operated with large amounts of negative feedback. This feedback tends to stabilize the various amplifier parameters, or make their absolute values large or small enough to ignore.

**** BIPOLAR JUNCTION TRANSISTORS

NAME	Q_Q3	Q_Q4	Q_Q2	Q_Q1
MODEL	Q2N3904	Q2N309	Q2N3904	Q2N3904
IB	7.45E-06	7.45E-06	4.52E-06	4.52E-06
IC	1.17E-03	9.88E-04	5.80E-04	5.80E-04
VBE	6.65E-01	6.65E-01	6.50E-01	6.50E-01
VBC	-1.37E+01	0.00E+00	-3.39E+00	-3.39E+00
VCE	1.43E+01	6.65E-01	4.05E+00	4.05E+00
BETADC	1.57E+02	1.33E+02	1.29E+02	1.29E+02
GM	4.46E-02	3.76E-02	2.23E-02	2.23E-02
RPI	4.03E+03	4.03E+03	6.69E+03	6.69E+03
RX	1.00E+01	1.00E+01	1.00E+01	1.00E+01
RO	7.50E+04	7.50E+04	1.33E+05	1.33E+05
CBE	1.99E-11	1.78E-11	1.31E-11	1.31E-11
CBC	1.46E-12	3.64E-12	2.15E-12	2.15E-12
CJS	0.00E+00	0.00E+00	0.00E+00	0.00E+00
BETAAC	1.80E+02	1.52E+02	1.49E+02	1.49E+02
CBX	0.00E+00	0.00E+00	0.00E+00	0.00E+00
FT	3.32E+08	2.79E+08	2.32E+08	2.32E+08

FIGURE 8.35
Transistor parameters at the Q point.

Small-Signal Analysis

We can determine the gain of the differential amplifier in Fig. 8.33 in any of several ways. The most obvious way to find A_v (or A_d) at this point is to set one of the input SIN sources, say, v_{S_1}, to a convenient value and examine the resulting collector voltages. The results of this approach are shown in Fig. 8.36, where we find for single-ended output

$$|A_v| = \frac{v_C}{v_{in}} = \frac{1.66 \text{ V}}{10 \text{ mV}} = 166$$

This compares favorably with the value we obtained earlier (192.6). The absolute magnitude was used, because the gain can be considered inverting or noninverting, depending on which collector is used as the output. The gain referred to the differential output configuration is simply $2A_v = 332$. The dc offset in each of the collector output signals is $V_{CQ} = 3.39$ V, which can be seen in Fig. 8.36 as well.

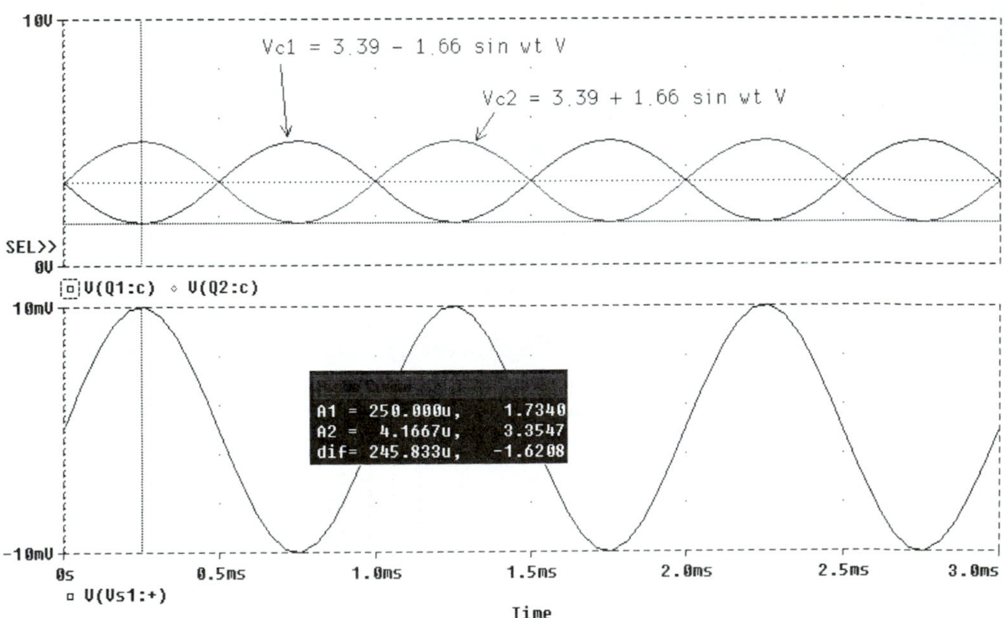

FIGURE 8.36
Diff amp input and output voltage waveforms. Input signal is in lower plot.

Determining A_v with a DC Sweep

Because the differential amplifier is direct coupled (no input or output coupling capacitors are used), we can determine the voltage gain by sweeping the dc value of an input voltage source. For example, sweeping the dc attribute of v_{S_1} from to −50 to 50 mV in 50-μV increments and plotting the Q_2 collector voltage V(Q2:c) yields the display in Fig. 8.37. Placing the cursors at convenient points, we find that the voltage gain is simply the slope of the $\Delta V_C/\Delta V_{S_1}$ transfer characteristic.

$$A_v = \frac{\Delta V_C}{\Delta v_{S_1}} = \frac{6.58 \text{ V}}{40 \text{ mV}} = 164.5$$

This is very close to the value obtained from the transient analysis.

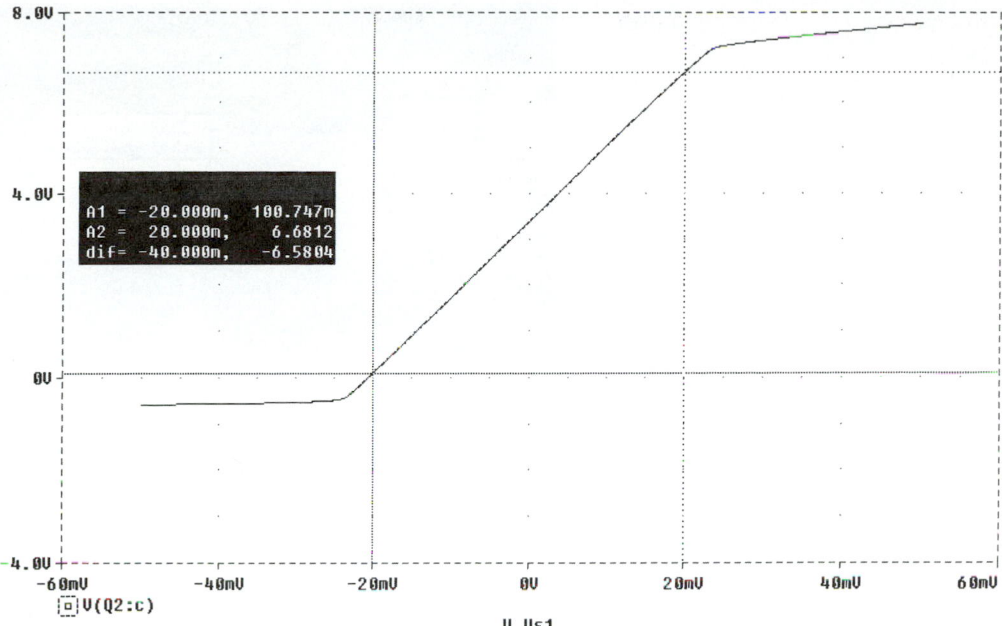

FIGURE 8.37
Plot of amplifier output voltage versus input voltage transfer characteristic. This is obtained by performing a dc sweep of source V_1.

Notice that the transfer characteristic in Fig. 8.37 is very linear for V_C ranging from about 0 to 7 V. This is a good characteristic. Nonlinearity, such as waviness, jumps, and other deviations of the transfer characteristic from a straight line, indicates that the amplifier will produce distortion.

Finally, plotting the transfer characteristics allows us to easily determine the clipping points of the amplifier. If we had plotted the Q_1 collector voltage, we would obtain a similar display, except the slope would be of opposite sign. Why don't you give this a try?

Transfer Function Option

The transfer function is a system-oriented concept that we will use extensively in later chapters. For the time being though, we can think of a transfer function as a fancy name for the mathematical relationship between the output and input of a network. Transfer functions can be very complex, involving systems of nonlinear differential equations, for example, or they may be very simple, involving nothing more than a real number. The voltage gain of an amplifier can be considered a transfer function, and over a limited range (the small-signal range at midband frequencies), A_v is approximately a constant.

A transfer function option can be enabled under the Analysis Setup window of PSpice. The PSpice transfer function analysis determines the ratio

$$\text{TF} = \frac{\text{Output varible}}{\text{Input varible}}$$

where the output and input variables are dc currents and/or voltages associated with the circuit under study. In this case, if we want to use v_{s_1} as the input variable and collector voltage v_{C_2} as the output variable, we set up the transfer function analysis prior to simulation as shown in Fig. 8.38.

Running the simulation and examining the resulting output file we find the information shown in Fig. 8.39. The transfer function also returns the small-signal input resistance and output resistance. Here we find

$$A_v = -168$$

$$R_{\text{in}} = 15.39\ \text{k}\Omega$$

$$R_o = 18.69\ \text{k}\Omega$$

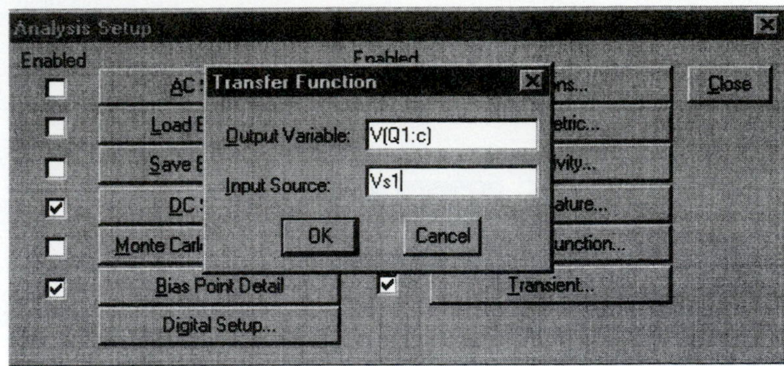

FIGURE 8.38
Setting up the transfer function analysis options.

```
** 07/01/98 23:08:23 ** NT Evaluation PSpice (October 1996)**
***** SMALL-SIGNAL CHARACTERISTICS *************************

      V($N_0013)/V_Vs1 = -1.682E+02

      INPUT RESISTANCE AT V_Vs1 = 1.539E+04

      OUTPUT RESISTANCE AT V($N_0013) = 1.869E+04

************************************************************
```

FIGURE 8.39
Results of the transfer function analysis.

The voltage gain is about what is expected. The input resistance is slightly higher than expected for several reasons: First, PSpice used $\beta = 129$ for Q_1 and Q_2 while we assumed $\beta = 100$. Second, we did not account for the 1-kΩ series resistors like the program does (we would ignore R_{s_1}, because it is considered internal to the v_{s_1} source). Factoring these effects into our calculations would produce much closer agreement. The output resistance is lower than expected, primarily because PSpice accounts for r_o of the output collector terminal. Recall that

$$r_o = V_A/I_{CQ}$$

Examining Figs. 8.34 and 8.35 we see that $V_A = \text{VAF} = 74.03$ V and $I_{CQ_2} = 580\ \mu\text{A}$. A little arithmetic gives us

$$r_o = 127.6\ \text{k}\Omega$$

The collector output resistance is also listed explicitly in the output file, and in Fig. 8.35 we find

$$r_o = \text{RO} = 133\ \text{k}\Omega$$

PSpice has used higher order effects to determine this value, so it differs from our more approximate calculation. In any case, we could account for this finite internal collector resistance to more accurately estimate the amplifier output resistance:

$$\begin{aligned} R_o &= R_C \parallel r_o \\ &= 20\ \text{k}\Omega \parallel 133\ \text{k}\Omega \\ &= 17.4\ \text{k}\Omega \end{aligned}$$

Again, the slight difference between results is due to PSpice modeling higher order effects that we normally ignore.

Common-Mode Gain

We can establish a common-mode input voltage by setting the dc attributes of the input voltage sources to convenient values. For example, setting $V_{s_1} = V_{s_2} = 2$ V produces the values shown in Fig. 8.40. Calculating the common-mode response, we have

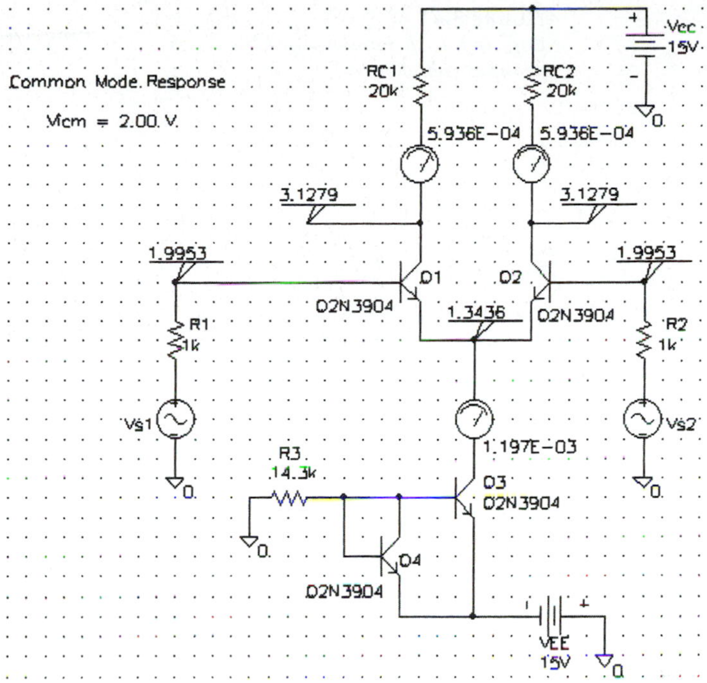

FIGURE 8.40
Common-mode response. The dc attributes of both input sources are set for 2 V. Only steady-state dc response is reflected in this bias-point analysis.

$$\begin{aligned} A_{cm} &= \frac{\Delta V_C}{V_{icm}} \\ &= \frac{3.1279\text{ V} - 3.3309\text{ V}}{2\text{ V}} \\ &= -0.1312 \end{aligned}$$

In practice, we would use the magnitude of the common-mode gain 0.1312. The common-mode rejection ratio is found (using the single-ended-output voltage gain) to be

$$\begin{aligned} \text{CMRR} &= \left|\frac{A_V}{A_{cm}}\right| \\ &= \frac{168}{0.1312} \\ &= 1280 \\ &= 62.14\text{ dB} \end{aligned}$$

Ideally the common-mode input signal would have no effect on the output voltage. The main reason we get a response here is due to the finite internal collector resistance of Q_3. Using the information in Figs. 8.34 and 8.35, we find $V_{AF} = 74.03$ V and $I_{CQ_3} = 1.17$ mA, which gives us

$$\begin{aligned} R_{EE} = r_{o_3} &= V_{AF}/I_{CQ} \\ &\cong 74\text{ V}/1.17\text{ mA} \\ &\cong 63.2\text{ k}\Omega \end{aligned}$$

According to Eq. (8.57), we should have

$$\begin{aligned} A_{cm} &= -\frac{R_C}{2R_{EE}} \\ &= -\frac{20\text{ k}\Omega}{126.4\text{ k}\Omega} \\ &= -0.316 \end{aligned}$$

FIGURE 8.41
Diff amp with a common-mode ac input signal.

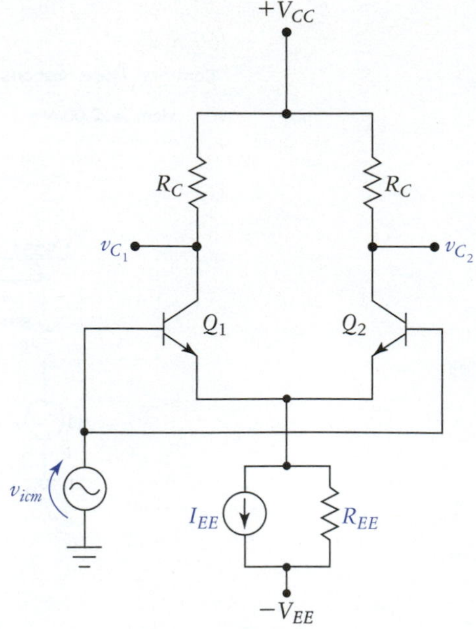

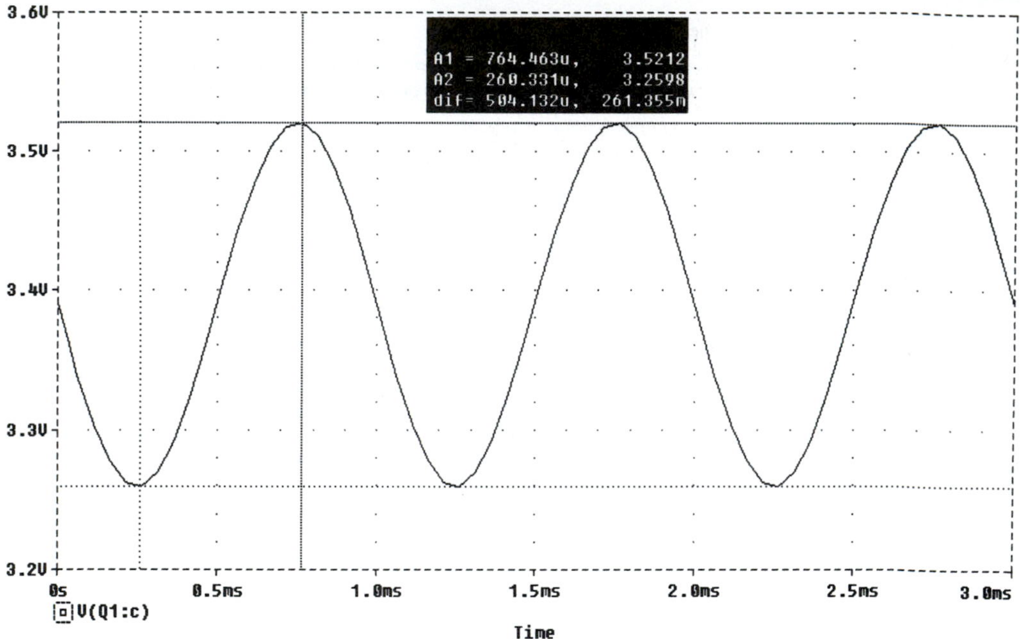

FIGURE 8.42
Common-mode response of Q_1 with $V_{icm} = 2\ V_{p\text{-}p}$ at 1 kHz.

The simulation results gave us $A_{cm} = -0.1312$. The difference is mainly due to our neglect of r_o for Q_1 and Q_2. You may wish to verify this for yourself.

AC Common-Mode Gain

Let us now determine the common-mode gain of the differential amplifier using a slightly different approach than before. An interesting way to do this is to apply a sinusoidal signal to both inputs simultaneously and plot the response at the collector of Q_1 or Q_2. Figure 8.41 illustrates this idea.

Setting the V_P attributes of both of the input signal sources of the circuit in Fig. 8.33 to 1 V_{pk} with $f = 1$ kHz produces $v_{icm} = 2\ V_{p\text{-}p}$. Using the collector of Q_1 as the output terminal, the response is shown in Fig. 8.42. The output signal varies from about 3.26 to 3.52 V, and it is inverted with respect to the input voltage. The common-mode output voltage is therefore

$$v_{ocm} = -(3.52\text{ V} - 3.26\text{ V})$$
$$= -260\text{ mV}_{\text{p-p}}$$

The ac common-mode gain is then found to be

$$A_{cm(\text{ac})} = v_{ocm}/v_{icm}$$
$$= -260\text{ mV}/2\text{ V}$$
$$= -0.13$$

This is approximately the same A_{cm} that we obtained for a dc common-mode input voltage. Common-mode gain will eventually increase with frequency. This is an undesirable but unavoidable characteristic of real differential amplifiers.

CHAPTER REVIEW

The four basic configurations of the differential amplifier are (1) single-ended input, differential output; (2) single-ended input, differential output; (3) differential input, differential output; and (4) differential input, single-ended output. Differential amplifiers must be constructed using components that have closely matched characteristics. These strict device matching requirements are most easily met using monolithic integrated circuit and integrated transistor array implementation. Most often, differential amplifiers are designed to operate from bipolar power supplies. Coupling and bypass capacitors are generally not used in diff amp designs.

Amplifiers that have differential inputs exhibit common-mode rejection. Common-mode rejection is a desirable characteristic. Differential amplifiers often use active current sources to provide emitter bias currents. This results in a very stable Q point and also enhances common-mode rejection characteristics.

To achieve very high differential gain, active loading is often used. Active loads are current-mirror-type circuits, and they can only be used effectively in stages that use the single-ended-output configuration. FETs and Darlington pairs are also used in differential amplifier design, in order to obtain very high input resistance.

DESIGN AND ANALYSIS PROBLEMS

Note: Unless otherwise specified, all transistors are assumed to be matched with $\beta = 100$ and $V_{BE} = 0.7$ V. For notational convenience, A_d will refer to voltage gain with a differential-output configuration; A_v will represent the voltage-gain magnitude for the single-ended-output configuration. Reference to gain with the single-ended-output or differential-output configuration can also be inferred from the context of a given problem. The term A_{cm} refers to common-mode gain for the single-ended-output configuration, and the CMRR is defined as the ratio of the *single-ended-output gain* A_v (which may very well be referred to a differential-input voltage) to the common-mode gain. Note that $A_{cm} = 0$ for the differential-output configuration, with perfectly matched transistors, as we assume to be true here.

8.1. Refer to Fig. 8.43. Given: $V_{CC} = V_{EE} = 15$ V, $R_{EE} = 4.7$ kΩ, and $R_C = 3.3$ kΩ. Determine the approximate values for I_{EE}, I_{C_1}, I_{C_2}, V_{CE_1}, V_{CE_2}, A_d, A_v, R_{in}, A_{cm}, and CMRR. Determine v_{od} given $v_1 = 1.00$ V and $v_2 = 1.05$ V.

8.2. Refer to Fig. 8.43. Given: $V_{CC} = V_{EE} = 15$ V, $R_{EE} = 10$ kΩ, and $R_C = 12$ kΩ. Determine the approximate values for I_{EE}, I_{C_1}, I_{C_2}, V_{CE_1}, V_{CE_2}, A_d, A_v, R_{in}, A_{cm}, and CMRR. Determine v_{od} given $v_1 = 50$ mV and $v_2 = 20$ mV.

8.3. Refer to Fig. 8.43. Assume $V_{CC} = V_{EE} = 15$ V, $R_{EE} = 4.7$ kΩ, and $R_C = 3.3$ kΩ, as in Problem 8.1. Determine ΔV_{C_1} given $V_1 = V_2 = 5$ V.

8.4. Refer to Fig. 8.43. Assume $V_{CC} = V_{EE} = 15$ V, $R_{EE} = 10$ kΩ, and $R_C = 12$ kΩ, as in Problem 8.2. Determine ΔV_{C_1} given $V_1 = V_2 = 5$ V.

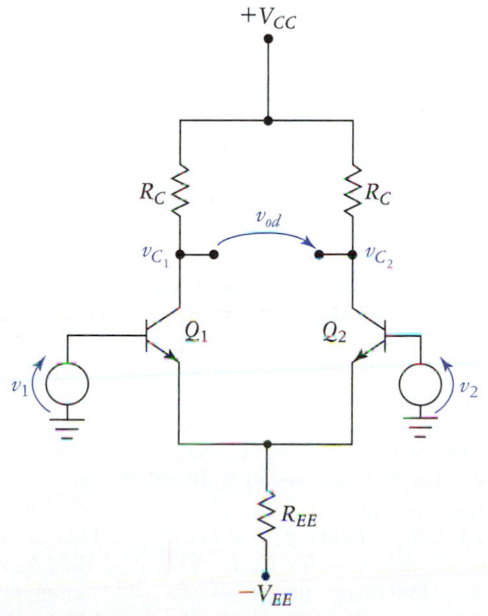

FIGURE 8.43
Circuit for Problems 8.1 to 8.4.

8.5. Refer to Fig. 8.44. Given $V_{CC} = V_{EE} = 15$ V, $R_1 = R_2 = 6$ kΩ, $R_3 = 8.2$ kΩ, $R_4 = 2.2$ kΩ, $R_5 = 920$ Ω, and $V_A = 100$ V for all transistors, determine I_{EE}, I_{C_1}, I_{C_2}, V_{CE_1}, V_{CE_2}, V_{CE_3}, A_d, A_v, R_{in}, A_{cm}, $r_{o(Q_3)}$, and CMRR. Determine v_{od} given $v_1 = 120$ mV and $v_2 = 110$ mV. Disregard the effect of r_o on the differential voltage gain.

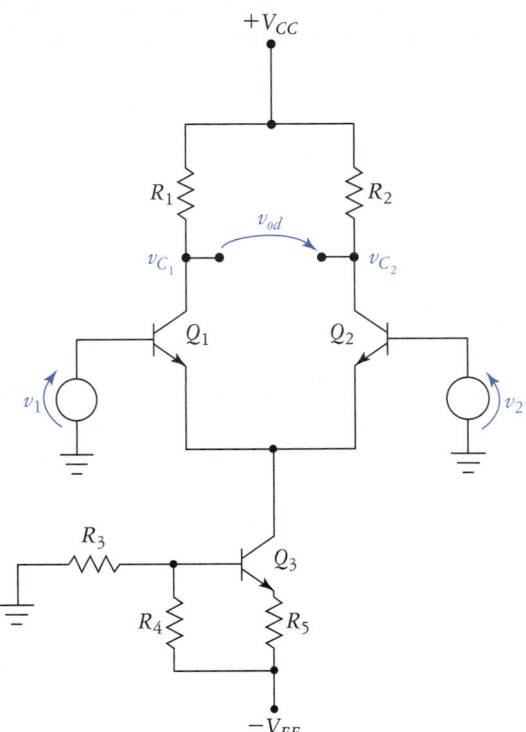

FIGURE 8.44
Circuit for Problems 8.5 to 8.8.

8.6. Refer to Fig. 8.44. Given $V_{CC} = V_{EE} = 12$ V, $R_1 = R_2 = 8.75$ kΩ, $R_3 = 10$ kΩ, $R_4 = 5$ kΩ, $R_5 = 2.7$ kΩ, and $V_A = 100$ V for all transistors, determine I_{EE}, I_{C_1}, I_{C_2}, V_{CE_1}, V_{CE_2}, V_{CE_3}, A_d, A_v, R_{in}, A_{cm}, $r_{o(Q3)}$, and CMRR. Determine v_{od} given $v_1 = 0$ V and $v_2 = 15 \sin(2\pi 1000t)$ mV. Disregard the effect of r_o on differential voltage gain.

8.7. Refer to Fig. 8.44. Assuming $-V_{EE} = -10$ V and $R_5 = 1$ kΩ, determine R_3 and R_4 such that $I_{EE} = 4$ mA and $R_3 \parallel R_4 \ll \beta R_5$. Since this is a design problem, there are many correct answers, just keep in mind that some may be better than others, depending on various factors.

8.8. Refer to Fig. 8.44. Assume $-V_{EE} = -12$ V, $R_4 = 1.5$ kΩ, and $R_5 = 1.5$ kΩ. Determine R_3 such that $I_{EE} = 1$ mA.

8.9. Refer to Fig. 8.45. Given $V_{CC} = V_{EE} = 12$ V, $R_1 = R_2 = 1.2$ kΩ, $R_3 = 15$ kΩ, $R_4 = 5$ kΩ, $R_5 = 520$ Ω, and $V_A = 120$ V for Q_3, determine I_{EE}, I_{C_1}, I_{C_2}, V_{CE_1}, V_{CE_2}, V_{CE_3}, A_d, A_v, R_{in}, and A_{cm}. Determine the maximum and minimum allowable common-mode input voltages, and determine v_{C_1} and v_{C_2} given $v_1 = 10$ mV and $v_2 = 10 \sin(2\pi 1000t)$ mV.

8.10. Refer to Fig. 8.45. Given $V_{CC} = V_{EE} = 15$ V, $R_1 = R_2 = 5$ kΩ, $R_3 = 10$ kΩ, $R_4 = 2$ kΩ, $R_5 = 1.3$ kΩ, and $V_A = 150$ V for Q_3, determine I_{EE}, I_{C_1}, I_{C_2}, V_{CE_1}, V_{CE_2}, V_{CE_3}, A_d, A_v, R_{in}, and A_{cm}. Determine the maximum and minimum allowable common-mode input voltage, and determine v_{C_1} and v_{C_2} given $v_1 = -5$ mV and $v_2 = -8$ mV.

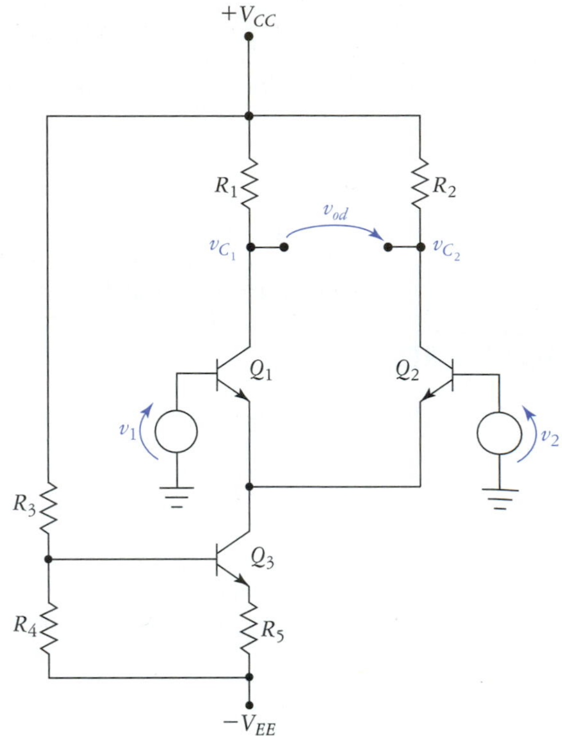

FIGURE 8.45
Circuit for Problems 8.9 and 8.10.

8.11. Refer to Fig. 8.46. Given: $V_{CC} = V_{EE} = 10$ V, $R_1 = 20$ kΩ, $R_2 = 18.6$ kΩ, and $V_A = 100$ V for Q_3. Determine I_{EE}, I_{C_1}, I_{C_2}, V_{CE_1}, V_{CE_2}, V_{CE_3}, V_{oQ}, A_v, R_{in}, and A_{cm}. Determine v_o for $v_1 = 0$ V and $v_2 = 12$ mV.

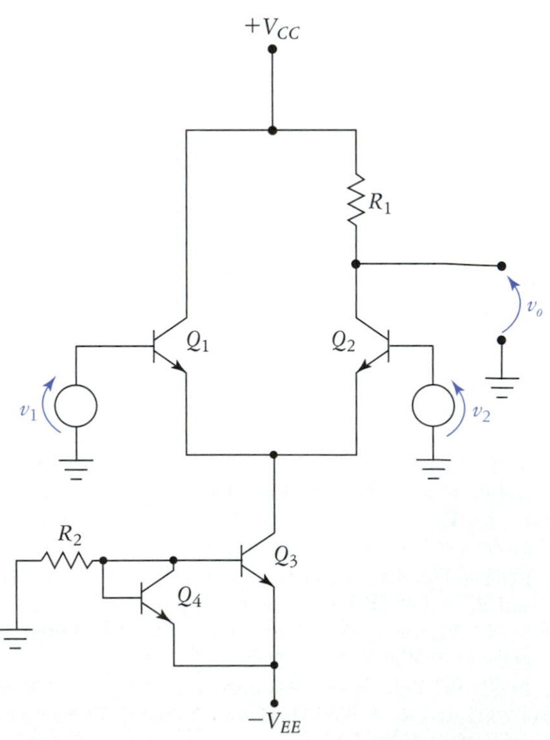

FIGURE 8.46
Circuit for Problems 8.11 and 8.12.

8.12. Refer to Fig. 8.46. Given: $V_{CC} = V_{EE} = 15$ V, $R_1 = 4.7$ kΩ, $R_2 = 5.6$ kΩ, and $V_A = 100$ V for Q_3. Determine I_{EE}, I_{C_1}, I_{C_2}, V_{CE_1}, V_{CE_2}, V_{CE_3}, V_{oQ}, A_v, R_{in}, and A_{cm}. Determine v_o for $v_1 = 6$ mV and $v_2 = 18$ mV.

8.13. Given the conditions of Problem 8.9, determine v_o given $v_1 = v_2 = 4$ V.

8.14. Given the conditions of Problem 8.10, determine v_o given $v_1 = v_2 = -2$ V.

8.15. Refer to Fig. 8.47. Given: $V_{CC} = V_{EE} = 12$ V, $V_Z = 4.2$ V, $R_1 = R_2 = 10$ kΩ, $R_3 = R_4 = 5$ kΩ, $R_5 = 680$ Ω, $R_6 = 1.75$ kΩ, and Q_3 has $r_o = 100$ kΩ. Determine I_{EE}, I_{C_1}, I_{C_2}, V_{CE_1}, V_{CE_2}, V_{CE_3}, V_{icm}, I_Z, A_d, A_v, R_{in}, and A_{cm}.

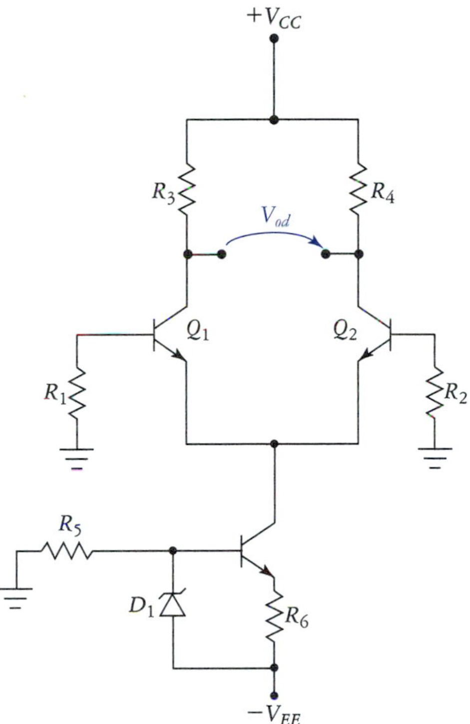

FIGURE 8.47
Circuit for Problems 8.15 to 8.18.

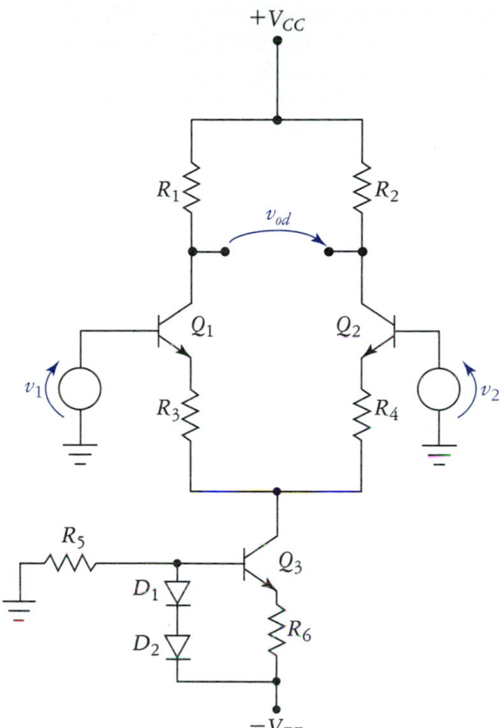

FIGURE 8.48
Circuit for Problems 8.19 and 8.20.

8.16. Refer to Fig. 8.47. Given: $V_{CC} = V_{EE} = 15$ V, $V_Z = 5.6$ V, $R_1 = R_2 = 10$ kΩ, $R_3 = R_4 = 14$ kΩ, $R_5 = 680$ Ω, $R_6 = 11.2$ kΩ, and Q_3 has $r_o = 100$ kΩ. Determine I_{EE}, I_{C_1}, I_{C_2}, V_{CE_1}, V_{CE_2}, V_{CE_3}, V_{icm}, I_Z, A_d, A_v, R_{in}, and A_{cm}.

8.17. Refer to Fig. 8.47. Given $V_{CC} = V_{EE} = 15$ V and $V_Z = 4.8$ V, design the circuit such that $I_{EE} = 6$ mA, $I_Z = I_{Zk} = 12$ mA, and $V_{CE_1} = V_{CE_2} = 7$ V. Assume that $R_1 = R_2 \cong 0$ Ω.

8.18. Refer to Fig. 8.47. Given $V_{CC} = V_{EE} = 18$ V and $V_Z = 6.2$ V, design the circuit such that $I_{EE} = 1.5$ mA, $I_Z = I_{Zk} = 20$ mA, and $V_{CE_1} = V_{CE_2} = 7$ V. Assume that $R_1 = R_2 \cong 0$ Ω.

8.19. Refer to Fig. 8.48. Given: $V_{CC} = V_{EE} = 12$ V, $R_1 = R_2 = 4$ kΩ, $R_3 = R_4 = 100$ Ω, $R_5 = 4.7$ kΩ, and $R_6 = 350$ Ω. Determine I_{EE}, I_{C_1}, I_{C_2}, V_{CE_1}, V_{CE_2}, V_{CE_3}, A_d, A_v, and R_{in}.

8.20. Refer to Fig. 8.48. Given: $V_{CC} = V_{EE} = 15$ V, $R_1 = R_2 = 2.5$ kΩ, $R_3 = R_4 = 50$ Ω, $R_5 = 6.8$ kΩ, and $R_6 = 175$ Ω. Determine I_{EE}, I_{C_1}, I_{C_2}, V_{CE_1}, V_{CE_2}, V_{CE_3}, A_d, A_v, and R_{in}.

8.21. Refer to Fig. 8.49. Given: $V_{CC} = V_{EE} = 15$ V, $R_1 = R_2 = 100$ kΩ, $R_3 = 10$ kΩ, $R_4 = 256$ Ω, $V_T = 26$ mV, $\eta = 1$, and $V_A = 100$ V for Q_3. Determine I_{EE}, I_{C_1}, I_{C_2}, V_{CE_1}, V_{CE_2}, A_d, A_v, R_{in}, A_{cm} and CMRR.

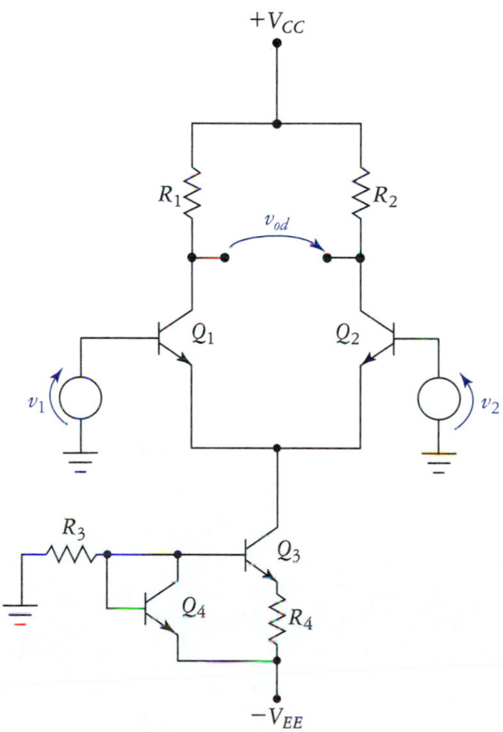

FIGURE 8.49
Circuit for Problems 8.21 to 8.24.

8.22. Refer to Fig. 8.49. Given: $V_{CC} = V_{EE} = 15$ V, $R_1 = R_2 = 20$ kΩ, $R_3 = 4.7$ kΩ, $R_4 = 68$ Ω, $V_T = 26$ mV, $\eta = 1$, and $V_A = 100$ V for Q_3. Determine I_{EE}, I_{C_1}, I_{C_2}, V_{CE_1}, V_{CE_2}, A_d, A_v, R_{in}, A_{cm} and CMRR.

8.23. Refer to Fig. 8.49. Given $V_{CC} = V_{EE} = 12$ V and $i_{R_3} = 1$ mA, determine R_3 and R_4 such that $I_{EE} = 200\ \mu$A.

8.24. Refer to Fig. 8.49. Given $V_{CC} = V_{EE} = 15$ V and $i_{R_3} = 5$ mA, determine R_3 and R_4 such that $I_{EE} = 10\ \mu$A.

8.25. Refer to Fig. 8.50. Given $V_{CC} = V_{EE} = 15$ V, $R_1 = R_2 = 6$ kΩ, $R_3 = 7.2$ kΩ, determine I_{EE}, I_{C_1}, I_{C_2}, V_{CE_1}, V_{CE_2}, V_{CE_3}, A_d, A_v, and R_{in}. Assuming $v_1 = 5$ mV, determine the value of v_2 that will result in $v_{od} = 4$ V.

8.26. Refer to Fig. 8.50. Given: $V_{CC} = V_{EE} = 15$ V, $R_1 = R_2 = 5$ kΩ, $R_3 = 4.7$ kΩ. Determine I_{EE}, I_{C_1}, I_{C_2}, V_{CE_1}, V_{CE_2}, V_{CE_3}, A_d, A_v, and R_{in}. Assuming $v_{\text{in}_1} = -15$ mV, determine the value of v_{in_2} that will produce $v_{od} = 3$ V.

8.27. Refer to Fig. 8.51. All FETs have $V_p = 2.5$ V and $I_{DSS} = 5$ mA. Given $V_{DD} = V_{SS} = 15$ V and $R_1 = R_2 = 2.2$ kΩ, determine I_{SS}, I_{D_1}, I_{D_2}, V_{DS_1}, V_{DS_2}, V_{DS_3}, and g_{mQ}.

8.28. Refer to Fig. 8.51. All FETs have $V_p = 2.0$ V and $I_{DSS} = 8$ mA. Given $V_{DD} = V_{SS} = 15$ V and $R_1 = R_2 = 1.8$ kΩ, determine I_{SS}, I_{D_1}, I_{D_2}, V_{DS_1}, V_{DS_2}, V_{DS_3}, and g_{mQ}.

8.29. Derive the expressions for A_d and R_{in} (Eqs. 8.49 and 8.54) for the single-ended-input, single-ended-output diff amp, implemented using Darlington transistors. Assume that $I_C = I_E$ for the transistors.

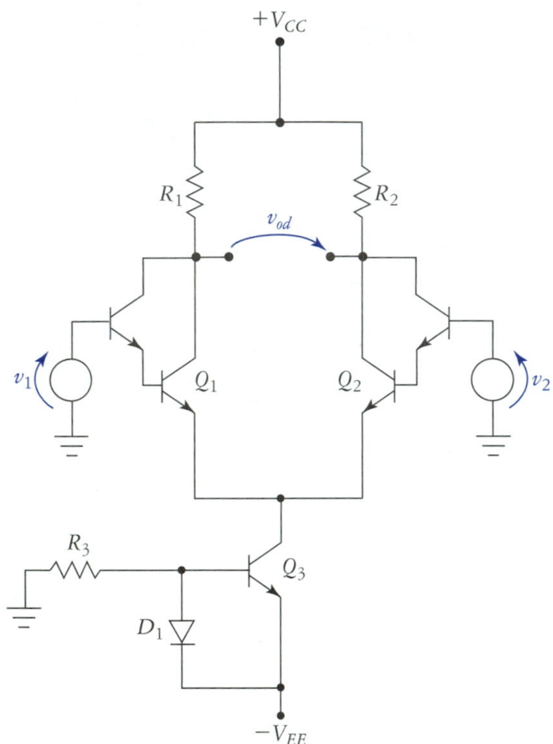

FIGURE 8.50
Circuit for Problems 8.25 and 8.26.

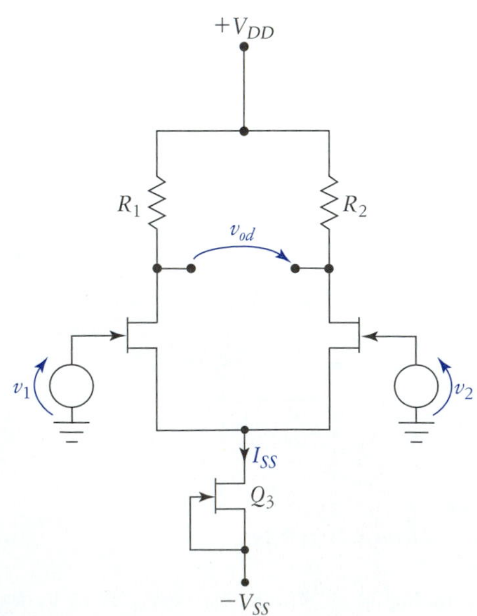

FIGURE 8.51
Circuit for Problems 8.27 and 8.28.

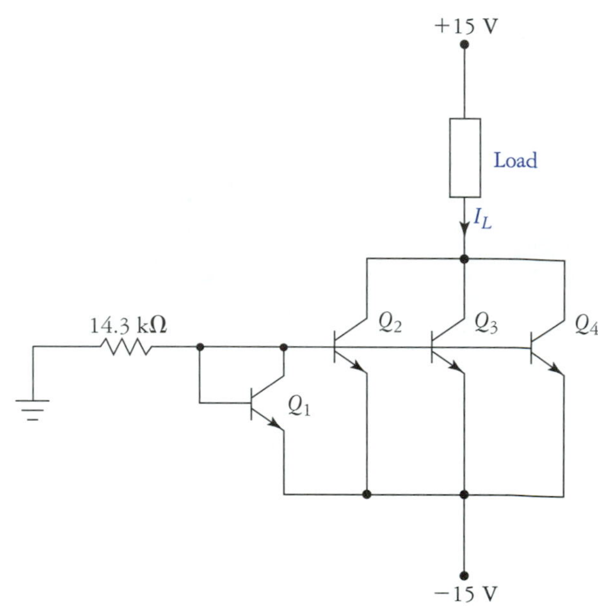

FIGURE 8.52
Circuit for Problem 8.31.

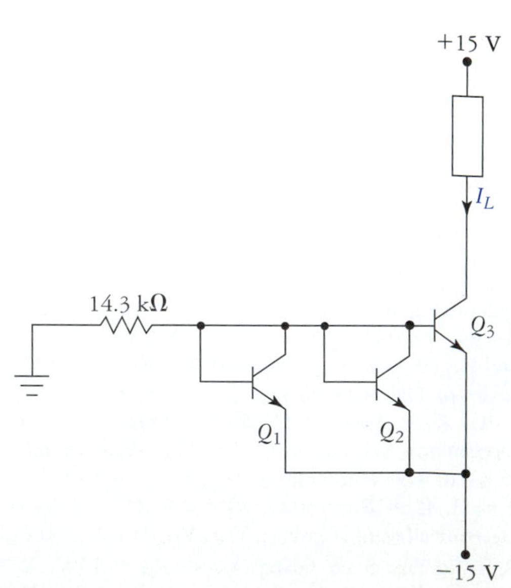

FIGURE 8.53
Circuit for Problem 8.32.

8.30. Using the analysis equations presented in this chapter, mathematically derive the expression CMRR $= \dfrac{2R_{EE}}{r_e}$.

8.31. Determine I_L for the circuit in Fig. 8.52.

8.32. Determine I_L for the circuit in Fig. 8.53.

8.33. Refer to Fig. 8.54. Determine I_{EE}, I_{C_1}, I_{C_2}, V_{CE_1}, V_{CE_2}, V_{CE_3}, V_{C_1}, V_{C_2}, A_d, A_v, and R_{in}.

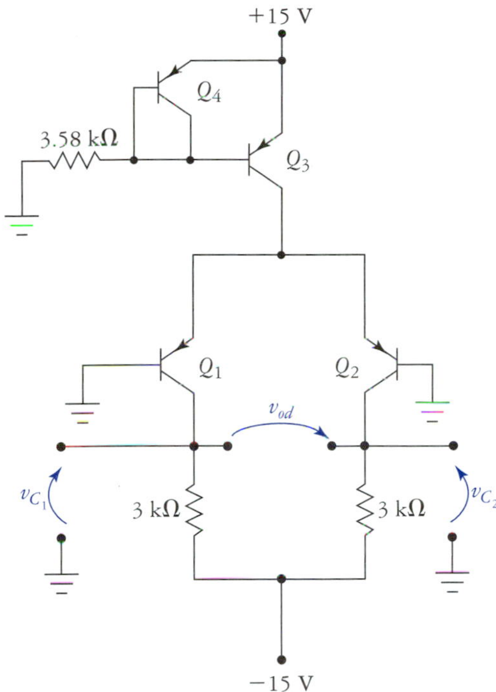

FIGURE 8.54
Circuit for Problem 8.33.

8.34. Derive an expression for the output resistance of the circuit in Fig. 8.27(a).

■ TROUBLESHOOTING PROBLEMS

8.35. Refer to Fig. 8.9. Suppose that Q_4 became an open circuit. Explain how this would affect collector voltages V_{C_1} and V_{C_2}.

8.36. Refer to Fig. 8.20. Both input sources are inactive, and the following voltage measurements are obtained: $V_{B_3} = -9.25$ V, $V_{E_1} = V_{E_2} = 0$ V, $V_{C_1} = V_{C_2} = 9.98$ V. Assuming that Q_3 is defective, the most likely failure mode would appear to be (a) shorted B-E junction, (b) open B-E junction, (c) open C-B junction, or (d) shorted C-B junction.

■ COMPUTER PROBLEMS

8.37. Using PSpice, determine the operating point, A_{cm}, and A_v (referred to the collector of Q_2) for the circuit in Fig. 8.55. Use breakout devices QbreakN and QbreakP with BF = 100 and VAF = 100 V. Follow this procedure:

(a) Place all transistors on the schematic.

(b) Highlight one of the npn transistors.

(c) Choose Edit, Model, and click the Edit Instance Model [Text] button.

(d) Edit the device parameters as shown in Fig. 8.56.

(e) Highlight the remaining npn transistors by holding the Shift key and single clicking on each of them.

(f) Choose Edit, Model, and click the Change Model Reference button.

(g) Enter the new model name (Qntyp, which stands for *typical npn*) for each device as shown in Fig. 8.57.

Repeat this procedure for the pnp devices using the name QPtyp.

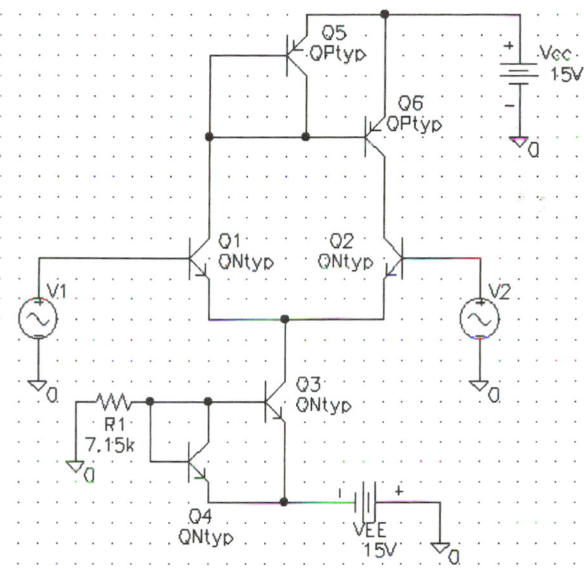

FIGURE 8.55
Circuit for Problem 8.37.

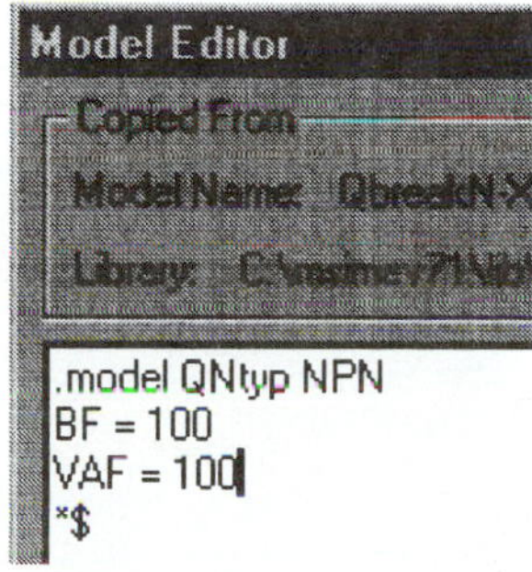

FIGURE 8.56
Model editor box.

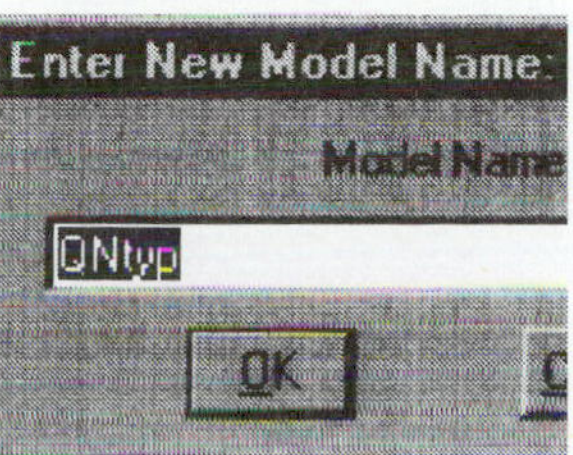

FIGURE 8.57
Changing the device model name.

8.38. Verify the analysis of the circuit in Fig. 8.22 and the output-voltage waveform shown in Fig. 8.23. Use 2N3904 transistors and 1N4148 diodes from Eval.lib.

8.39. Refer to Fig. 8.58. Set up this schematic using QbreakN and QbreakP with VAF = 100 V and BF = 100. Using a SIN voltage source for V1, set DC = 0, AC = 1, VOFF = 0, VAMPL = 10 μV, and FREQ = 1 kHz. Determine V_{oQ} and determine the voltage gain using the output voltage waveform V(Q:9C). Use the Transfer Function option to determine R_{in}, A_d, and R_o. Perform a dc sweep of V1 from −100 to 100 μV in 1-μV increments to determine the clipping points of the output voltage.

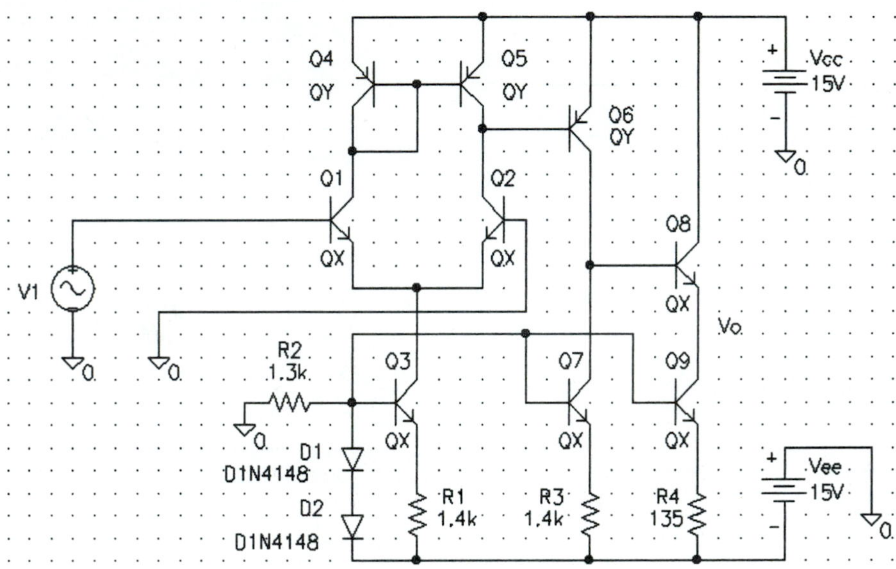

FIGURE 8.58
Circuit for Problem 8.39.

CHAPTER 9

Introduction to Multiple-Stage Amplifiers

BEHAVIORAL OBJECTIVES

On completion of this chapter, the student should be able to:

- Explain why multiple-stage amplifiers are necessary.
- Describe the advantages and disadvantages of direct coupling.
- Analyze direct-coupled differential amplifiers.
- Analyze level-translation amplifier stages.
- Convert between linear and decibel gain specifications.
- Determine amplifier overall performance specifications based on individual stage parameters.
- Use PSpice to analyze multiple-stage amplifiers.

INTRODUCTION

The performance demands of many practical applications simply cannot be met using a single stage of amplification. Most often a complete amplifier will consist of a series of stages, each of which contributes to the overall gain of the system. These circuit configurations are referred to as *multiple-stage amplifiers.* To use very familiar applications as examples, the amplifiers used in stereos, televisions, and other consumer electronic devices are virtually always multiple-stage designs.

So far, we have examined amplifier operation in a rather isolated manner. We have really only examined what are generally considered to be single-stage amplifiers. A single-stage amplifier will usually use a single transistor to provide signal amplification. The Darlington pair might be considered an exception, but it is normally treated as if it were a single device, thus comprising a single stage of amplification. Likewise, the differential pair is also considered to comprise a single stage of amplification, although the circuit may be viewed as an emitter-follower-common-base combination for analysis purposes.

This chapter investigates the operation of amplifiers that consist of several stages, and the various techniques used to couple one stage to another. The concept of decibel gain is also expanded on in this chapter.

9.1 BASIC GAIN RELATIONSHIPS

Recall that the triangular symbol shown in the upper part of Fig. 9.1 represents an amplifier with voltage gain A_v. The specific circuitry comprising the amplifier is unknown; however, for this discussion, let us assume that the amplifier actually consists of three cascaded amplifiers as shown in the lower half of Fig. 9.1. Each of these internal triangles could represent single transistors (FETs or BJTs), or differential pairs, and their associated biasing, coupling, and bypass circuitry.

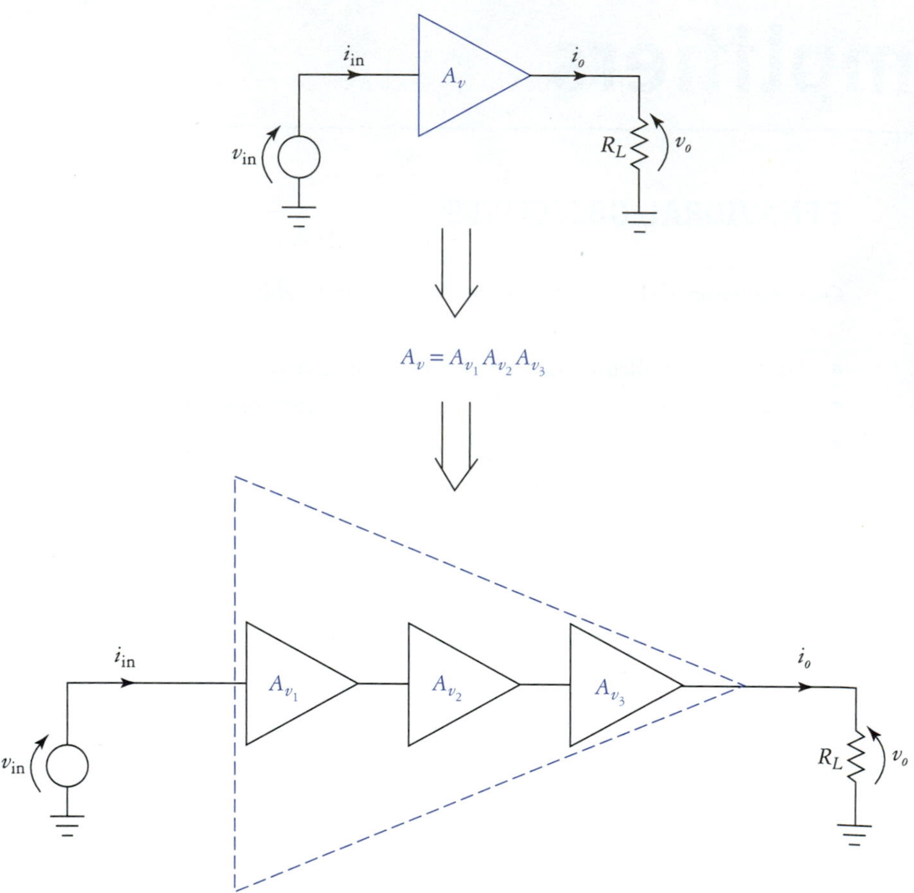

FIGURE 9.1
A typical amplifier may consist of several stages. The overall gain is the product of the individual stage gains.

Let us denote the overall voltage gain of the amplifier in Fig. 9.1 as simply A_v, while the gains of the individual stages are written A_{v_1}, A_{v_2}, and A_{v_3}. The overall voltage gain is simply the product of the individual stage gains. That is,

$$A_v = A_{v_1} A_{v_2} A_{v_3} \tag{9.1}$$

An amplifier may have an arbitrary number n of stages, and so a more general relationship is

$$A_v = A_{v_1} A_{v_2} \cdots A_{v_n} \tag{9.2}$$

Both Eqs (9.1) and (9.2) assume that the individual stage gains are independent of, or already defined in terms of, any loading effects.

Most often we will be dealing with so-called voltage amplifiers or VCVSs. A more detailed, two-port representation of such an amplifier is shown in Fig. 9.2. This VCVS model of the amplifier allows us to more easily determine the loading effects of one stage upon another. We now examine Fig. 9.2 more closely. The input source has an internal resistance R_s, and an unloaded terminal voltage of v_s. Because of loading effects, the actual input voltage applied to the amplifier is given by the voltage-divider equation.

$$v_{in} = v_s \frac{R_{in}}{R_{in} + R_s} \tag{9.3}$$

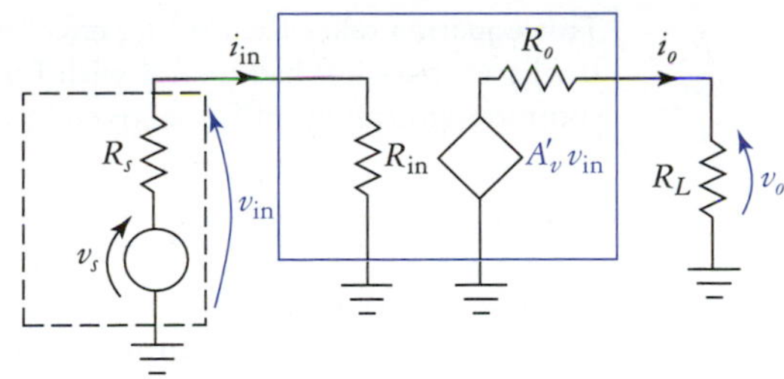

FIGURE 9.2
Behavioral model of a voltage amplifier.

The unloaded voltage gain of the amplifier (gain with $R_L = \infty$) is denoted as A'_v. With a finite-resistance load connected, the output voltage is given by

$$v_O = A'_v v_{\text{in}} \frac{R_L}{R_L + R_O} \tag{9.4}$$

The voltage gain, referred to the unloaded source, is by definition

$$A_v = \frac{v_o}{v_s} \tag{9.5}$$

Solving Eqs (9.3) and (9.4) for v_s and v_o and substituting into Eq. (9.5) gives us the following voltage-gain expression:

$$A_v = A'_v \left(\frac{R_{\text{in}}}{R_{\text{in}} + R_s}\right)\left(\frac{R_L}{R_L + R_o}\right) \tag{9.6}$$

EXAMPLE 9.1

The circuit in Fig. 9.2 has $v_s = 100$ mV, $R_s = 2\ \text{k}\Omega$, $R_{\text{in}} = 8\ \text{k}\Omega$, $A'_v = 150$, $R_o = 1\ \text{k}\Omega$, and $R_L = 3\ \text{k}\Omega$. Determine the overall voltage gain A_v and the current gain A_i.

Solution The voltage gain is found simply by substituting the given values into Eq. (9.6). Thus, we have

$$\begin{aligned} A_v &= (150)\left(\frac{8\text{ k}}{8\text{ k} + 2\text{ k}}\right)\left(\frac{3\text{ k}}{3\text{ k} + 1\text{ k}}\right) \\ &= (150)(0.8)(0.75) \\ &= 90 \end{aligned}$$

The current gain is found as usual by

$$\begin{aligned} A_i &= A_v \frac{R_{\text{in}}}{R_L} \\ &= 90\left(\frac{8\text{ k}}{3\text{ k}}\right) \\ &= 240 \end{aligned}$$

Equation (9.6) is useful, but very often we simply define the voltage gain of the amplifier as

$$A_v = \frac{v_o}{v_{\text{in}}} \tag{9.7}$$

This is equivalent to ignoring the actual value of v_s, or assuming that $v_{\text{in}} = v_s$. In other words, in many cases it doesn't really matter what value v_s is, because it is internal to the loaded source that cannot be measured directly anyway. This does not mean, however, that source loading is not ever important. Heavy loading could damage or destroy some sources, and it is necessary to know R_s in order to consider power transfer characteristics as well.

In preceding chapters we determined the loaded voltage gain of the amplifier at the outset. For example, recall that the (loaded) voltage gain of a common-emitter amplifier is given by

$$A_v = \frac{R'_C}{r_e + R_e} \tag{9.8}$$

This equation takes the loading effect of R_L into account automatically, because $R'_C = R_C \parallel R_L$. This is why we have not had to deal with Eq. (9.6) explicitly before now. The unloaded voltage gain of the common-emitter amplifier is given by the slightly altered equation

$$A_v = -\frac{R_C}{R_e + r_e} \tag{9.9}$$

Using Eq. (9.9), we must explicitly determine the effect of loading to determine the effective voltage gain of the amplifier. Accounting for loading by R_L we use the voltage-divider relationship to obtain

$$A_v = -\left(\frac{R_C}{R_e + r_e}\right)\left(\frac{R_L}{R_L + R_C}\right) \tag{9.10}$$

Equations (9.8) and (9.10) are equivalent expressions, giving identical results for all component values.

We have used Eq. (9.8) in previous work because it is usually easiest to account for loading this way, provided that the load resistance is known at the outset of the analysis. The use of Eq. (9.8) and similar gain equations does tend to disguise the fact that loading is really accounted for. If, however, the load is not specified, then we may determine unloaded gain and account for loading via the voltage-divider equation, as in Eq. (9.10).

EXAMPLE 9.2

Refer to Fig. 9.3. Determine the expressions for v_o/v_{in} and v_o/v_s in terms of the resistances and unloaded voltage gains given.

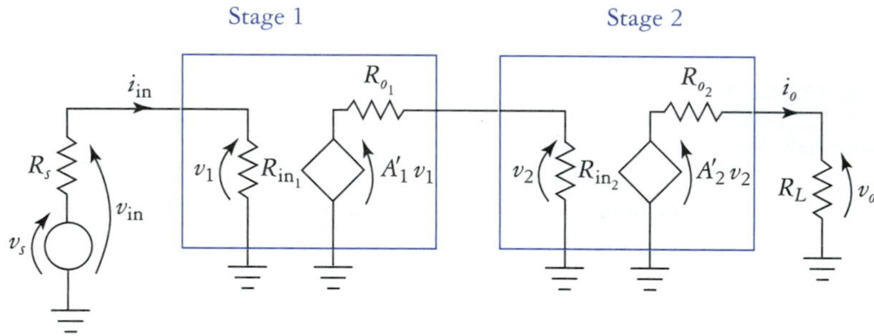

FIGURE 9.3
Two-stage amplifier model for Example 9.2.

Solution It is a little more convenient to start at the output of the amplifier. The output voltage is

$$v_o = A'_2 v_2 \frac{R_L}{R_L + R_{o_2}}$$

with v_2 given by

$$v_2 = A'_1 v_1 \frac{R_{\text{in}_2}}{R_{\text{in}_2} + R_{o_1}}$$

At this point, $v_1 = v_{\text{in}}$, so we may combine expressions to form

$$A_v = \frac{v_o}{v_{\text{in}}} = \left(A'_1 \frac{R_{\text{in}_2}}{R_{\text{in}_2} + R_{o_1}}\right)\left(A'_2 \frac{R_L}{R_L + R_{o_2}}\right)$$

Accounting for the loading effect on the source we have

$$A_v = \frac{v_o}{v_s} = \left(\frac{R_{\text{in}_1}}{R_{\text{in}_1} + R_s}\right)\left(A'_1 \frac{R_{\text{in}_2}}{R_{\text{in}_2} + R_{o_1}}\right)\left(A'_2 \frac{R_L}{R_L + R_{o_2}}\right)$$

Because it is very generalized, the two-port VCVS representation of amplifier operation is most suitable for more theoretical studies. With these concepts behind us, we now examine the actual implementation of multiple-stage amplifiers using transistors.

9.2 CAPACITIVELY COUPLED AMPLIFIERS

Recall that the function of a coupling capacitor is to allow a time-varying signal to be coupled from one node to another, while at the same time preventing the flow of direct current between these nodes. One of the more useful benefits of capacitive coupling is that it allows for relatively simple analysis of multiple-stage amplifier circuits.

Note
We always begin the analysis of an amplifier with the determination of the α-points for each stage.

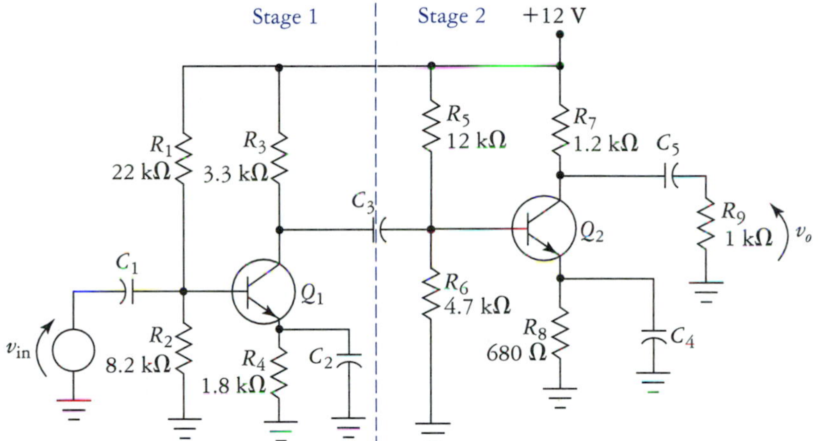

FIGURE 9.4
A two-stage, capacitively coupled amplifier.

The schematic diagram in Fig. 9.4 is that of a two-stage capacitively coupled BJT amplifier. In this case, both stages are common-emitter amplifiers. As usual, the analysis of this circuit begins with the determination of the operating points of the transistors. The dc analysis is performed under the assumption that all capacitors are open circuits. Thus, the dc-equivalent circuit for Fig. 9.4 is as shown in Fig. 9.5, where we have what is effectively two separate voltage-divider-biased transistors. We can determine the Q points for these circuits using the techniques developed in previous chapters. In terms of derived parameters such as voltage gain, Thevenin resistance, and input resistance, numerical subscripts will generally specify the stage with which that given parameter is associated. This notation will be used whenever possible from here on. Also, to simplify notation somewhat, the subscript Q will not be used to indicate quiescent values unless it is for special emphasis. Most often, uppercase italic letters will represent quiescent quantities anyway.

FIGURE 9.5
DC equivalents for the two stages of Fig. 9.4.

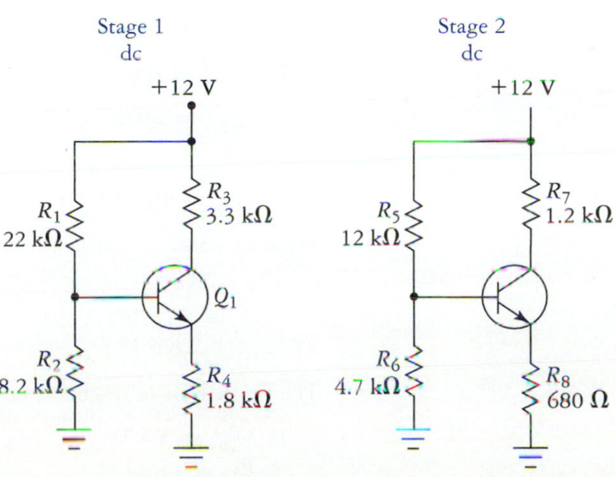

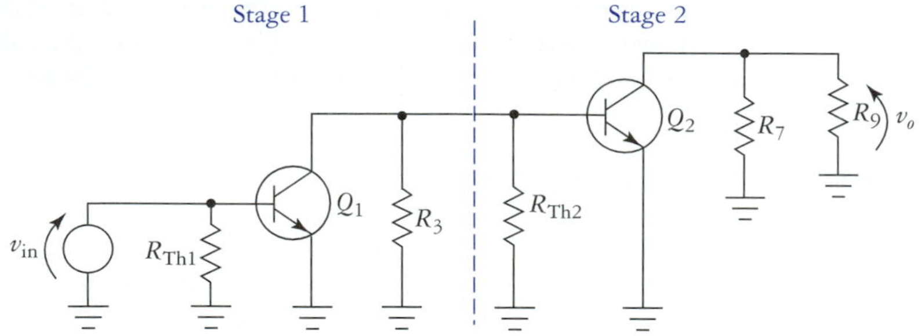

FIGURE 9.6
AC equivalent for the circuit of Fig. 9.4.

The ac-equivalent circuit for Fig. 9.4 is shown in Fig. 9.6. If we wish to account for loading effects as in Eq. (9.8), the ac analysis must be performed starting with the output stage (stage 2) of the amplifier. This is so because we must determine the input resistance of stage 2 in order to define R'_{C_1}. That is,

$$R'_{C_1} = R_3 \parallel R_{\text{in}_2} \tag{9.11}$$

which allows us to use the voltage-gain equation:

$$A_{v_1} = -\frac{R'_{C_1}}{R_{e_1} + r_{e_1}} \tag{9.12}$$

The loading effect of stage 2 on stage 1 is now accounted for.

The overall input resistance of the amplifier is simply the input resistance of the first stage, while the overall output resistance is the output resistance of the last stage (stage 2 in this case). The following example ties all of this information together.

EXAMPLE 9.3

Refer to Fig. 9.4. Perform a dc analysis, determining V_{CE}, I_C, V_C, and V_B for both stages. Determine the individual stage voltage gains, input resistances and output resistances, and the overall gains (A_v, A_i, and A_p), input and output resistances. Assume that all transistors have $\beta = 100$ and $\alpha \cong 1$.

Solution Let us begin the dc analysis arbitrarily with stage 1. Routine analysis gives us

$$V_{\text{Th}_1} = 3.25 \text{ V}$$
$$R_{\text{Th}_1} = 6 \text{ k}\Omega$$

Now, because $\beta R_E >> R_{\text{Th}}$, we can closely approximate the collector current with

$$\begin{aligned} I_{C_1} &\cong \frac{V_{\text{Th}_1} - V_{BE}}{R_4} \\ &= \frac{3.25 \text{ V} - 0.7 \text{ V}}{1.8 \text{ k}\Omega} \\ &= 1.4 \text{ mA} \end{aligned}$$

Application of KVL and Ohm's laws gives us

$$\begin{aligned} V_{CE_1} &= 4.9 \text{ V} \\ V_{C_1} &= 7.4 \text{ V} \\ V_{B_1} &= V_{\text{Th}_1} = 3.3 \text{ V} \end{aligned}$$

Since we know I_{C_1}, we can now determine r_{e_1}:

$$\begin{aligned} r_{e_1} &= V_T/I_{C_1} \\ &\cong 26 \text{ mV}/1.4 \text{ mA} \\ &= 18.6 \ \Omega \end{aligned}$$

The dc analysis of stage 2 is now performed. Routine analysis gives us

$$\begin{aligned} V_{\text{Th}_2} &= 3.4 \text{ V} \\ R_{\text{Th}_2} &= 3.4 \text{ k}\Omega \end{aligned}$$

As was the case with the first stage, the second stage also has $\beta R_E >> R_{\text{Th}}$ so the collector current is approximately

$$I_{C_2} \cong \frac{V_{\text{Th}_2} - V_{BE}}{R_8}$$
$$\cong \frac{3.4\text{ V} - 0.7\text{ V}}{680\ \Omega}$$
$$= 3.9\text{ mA}$$

The remaining dc parameters are easily determined to be

$$V_{CE_2} = 4.7\text{ V}$$
$$V_{C_2} = 7.3\text{ V}$$
$$V_{B_2} \cong V_{\text{Th}_2} = 3.4\text{ V}$$
$$r_{e_2} = 6.7\ \Omega$$

The ac analysis begins with the second stage, where

$$R'_{C_2} = R_7 \parallel R_9$$
$$= 1\text{ k}\Omega \parallel 1.2\text{ k}\Omega$$
$$= 545\ \Omega$$

The voltage gain for stage 2 is

$$A_{v_2} = -\frac{R'_{C_2}}{r_{e_2}}$$
$$= -\frac{545\ \Omega}{6.7\ \Omega}$$
$$= -81.3$$

The input resistance of stage 2 is

$$R_{\text{in}_2} = R_{\text{Th}_2} \parallel r_{\pi_1} \text{ (Remember } r_\pi \cong \beta r_e\text{)}$$
$$= 3.4\text{ k}\Omega \parallel (100)(6.7\ \Omega)$$
$$= 560\ \Omega$$

The output resistance of stage 2 is

$$R_{o_2} \cong R_7 = 1.2\text{ k}\Omega$$

We now begin the ac analysis of the first stage, where the ac collector resistance is

$$R'_{C_1} = R_3 \parallel R_{\text{in}_2}$$
$$= 3.3\text{ k}\Omega \parallel 560\ \Omega$$
$$= 479\ \Omega$$

The voltage gain of stage 1 is now found to be

$$A_{v_1} = -\frac{R'_{C_1}}{r_{e_1}}$$
$$= -\frac{479\ \Omega}{18.6\ \Omega}$$
$$= -25.8$$

The input resistance of stage 1 is

$$R_{\text{in}_1} = R_{\text{Th}_1} \parallel r_{\pi_1}$$
$$= 6\text{ k}\Omega \parallel (100)(18.6\ \Omega)$$
$$= 1.4\ \Omega$$

The output resistance of stage 1 is

$$R_{o_1} \cong R_3 = 3.3\text{ k}\Omega$$

The overall amplifier specifications are

$$R_{in} = R_{in_1} = 1.4\ k\Omega$$

$$R_o = R_{o_2} = 1.2\ k\Omega$$

$$\begin{aligned} A_v &= A_{v_1}A_{v_2} \\ &= (-25.8)(-81.3) \\ &= 2098 \end{aligned}$$

$$\begin{aligned} A_i &= A_v \frac{R_{in}}{R_L} \\ &= 2098 \frac{1.4\ k\Omega}{1\ k\Omega} \\ &= 2937 \end{aligned}$$

$$\begin{aligned} A_p &= A_v A_i \\ &= (2098)(2937) \\ &= 6.16 \times 10^6 \end{aligned}$$

The analysis of the multiple-stage capacitively coupled amplifier is not really any more complicated than that for a single-stage amplifier; there is simply more information to keep track of. This is one of the big advantages of capacitive coupling. The reader should also note that the overall gain of the amplifier in Example 9.3 is noninverting. This should seem reasonable, since both stages invert, and the product of two negative numbers is a positive number.

EXAMPLE 9.4

Based on the analysis presented in Example 9.3, write the gain expression for the amplifier in terms of unloaded stage gains [as in Eq. (9.6)]. Note that the source resistance is not a factor in this example, because $R_s = 0\ \Omega$.

Solution Because we are dealing with unloaded amplifier gains, we can start the ac analysis with the first stage if so desired. This gives us

$$\begin{aligned} A'_{v_1} &= -\frac{R_{C_1}}{r_{e_1}} \\ &= -\frac{3.3\ k\Omega}{18.6\ k\Omega} \\ &= -177.4 \end{aligned}$$

The unloaded voltage gain for stage 2 is

$$\begin{aligned} A'_{v_2} &= -\frac{R_{C_2}}{r_{e_2}} \\ &= -\frac{1.2\ k\Omega}{6.7\ \Omega} \\ &= -179.1 \end{aligned}$$

Generalizing Eq. (9.6), we have

$$\begin{aligned} A_v &= A'_{v_1}\left(\frac{R_{in_2}}{R_{in_2} + R_{o_1}}\right) A'_{v_2}\left(\frac{R_L}{R_L + R_{o_2}}\right) \\ &= (-177.4)\left(\frac{560\ \Omega}{560\ \Omega + 3300\ \Omega}\right)(-179.1)\left(\frac{1\ k\Omega}{1\ k\Omega + 1.2\ k\Omega}\right) \\ &= (-177.4)(0.145)(-179.1)(0.455) \\ &= 2096 \end{aligned}$$

The slight difference between the voltage-gain values determined in the previous examples is due to rounding and is basically insignificant. The point to be taken from these two examples is that there are two ways (at least) to approach the ac analysis of multiple-stage amplifiers. The reader should use whichever approach is most comfortable.

Because capacitive coupling does not really increase the complexity of the analysis procedure very much, a final example is presented before we move on to the next topic.

EXAMPLE 9.5

Refer to Fig. 9.7. This circuit uses a JFET common-source amplifier to obtain high input resistance with moderate voltage gain, while the BJT emitter-follower provides low output resistance. Determine the Q points for each transistor and the overall voltage, current, and power gain. Assume the FET has $I_{DSS} = 8$ mA, $V_p = 2$ V, and $g_{m0} = 8$ mS. The BJT has $\beta = 100$.

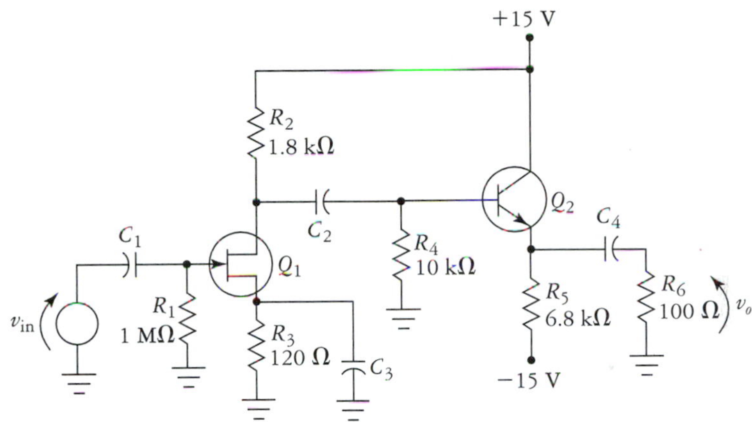

FIGURE 9.7
Two-stage amplifier using an FET and a BJT for Example 9.5.

Solution Beginning the dc analysis with the JFET, we have the equivalent circuit shown in Fig. 9.8. A quick graphical analysis gives us

$$I_{D_1} = 4 \text{ mA}$$

$$\begin{aligned} V_{DS_1} &= V_{DD} - I_{D_1}(R_D + R_S) \\ &= 15 \text{ V} - (4 \text{ mA})(1.92 \text{ k}\Omega) \\ &= 7.3 \text{ V} \end{aligned}$$

$$\begin{aligned} V_{D_1} &= V_{DD} - I_{D_1}R_D \\ &= 15 \text{ V} - (4 \text{ mA})(1.8 \text{ k}\Omega) \\ &= 7.8 \text{ V} \end{aligned}$$

$$\begin{aligned} V_{GS_1} &= I_{D_1}R_S \\ &= (-4 \text{ mA})(120 \ \Omega) \\ &= -0.48 \text{ V} \end{aligned}$$

$$\begin{aligned} g_{mQ} &= g_{m0}\left(1 - \frac{|V_{GSQ}|}{V_p}\right) \\ &= 8 \text{ mS}\left(1 - \frac{0.48 \text{ V}}{2 \text{ V}}\right) \\ &= 6.1 \text{ mS} \end{aligned}$$

The dc analysis of the second stage yields

$$\begin{aligned} I_{C_2} &\cong \frac{V_{EE} - V_{BE}}{R_5} \qquad \text{(Neglecting } R_4 \text{ because } \beta R_5 \gg R_4\text{)} \\ &= \frac{15 \text{ V} - 07. \text{ V}}{6.8 \text{ k}\Omega} \\ &= 2.1 \text{ mA} \end{aligned}$$

$$\begin{aligned} V_{CE_2} &= (V_{EE} + V_{CC}) - I_{C_2}R_5 \\ &= (15\text{ V} + 15\text{ V}) - (2.1\text{ mA})(6.8\text{ k}\Omega) \\ &= 30\text{ V} - 14.3\text{ V} \\ &= 15.7\text{ V} \end{aligned}$$

$$\begin{aligned} r_{e_2} &= V_T/I_{C_2} \\ &= \frac{26\text{ mV}}{2.1\text{ mA}} \\ &= 12.4\ \Omega \end{aligned}$$

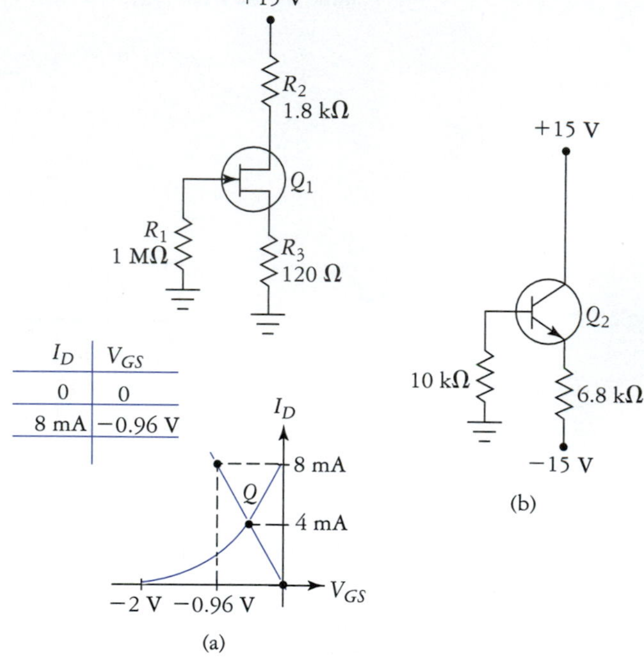

FIGURE 9.8
DC analysis for Fig. 9.7. (a) The JFET stage using graphical determination of Q point. (b) The BJT circuit.

The ac equivalent circuit is shown in Fig. 9.9. We begin the ac analysis with stage 2. The external ac emitter resistance is

$$\begin{aligned} R_e &= R_4 \parallel R_5 \\ &= 6.8\text{ k}\Omega \parallel 100\ \Omega \\ &\cong 100\ \Omega \end{aligned}$$

The loaded voltage gain is

$$\begin{aligned} A_{v_2} &= \frac{R_e}{R_e + r_e} \\ &= \frac{100\ \Omega}{100\ \Omega + 12.4\ \Omega} \\ &= 0.89 \end{aligned}$$

The input resistance is now found:

$$\begin{aligned} R_{\text{in}_2} &= R_3 \parallel \beta(R_e + r_e) \\ &= 10\text{ k}\Omega \parallel (100)(112.4\ \Omega) \\ &= 10\text{ k}\Omega \parallel 11.24\text{ k}\Omega \\ &= 5.3\text{ k}\Omega \end{aligned}$$

To determine the output resistance of the emitter-follower, we must determine the equivalent ac resistance R_B as "seen" looking out of the base of the BJT. From the ac-equivalent circuit we find

$$\begin{aligned} R'_B &= R_2 \parallel R_3 \\ &= 1.8\text{ k}\Omega \parallel 10\text{k}\Omega \\ &= 1.53\text{ k}\Omega \end{aligned}$$

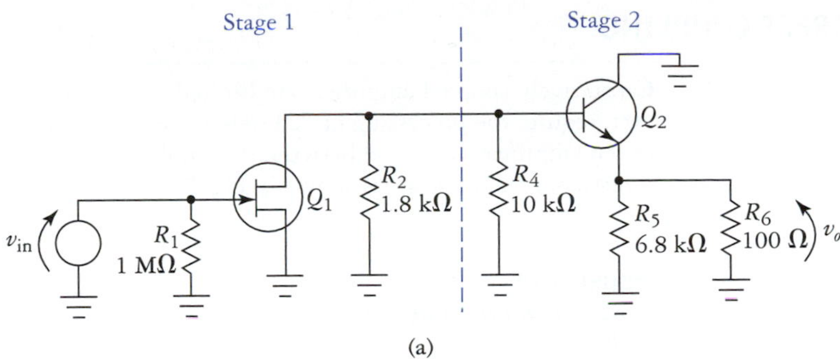

FIGURE 9.9
AC equivalent of the circuit in Fig. 9.7.

Now we can find R_{o_2}:

$$\begin{aligned} R_{o_2} &= R_E \parallel \left(r_e + \frac{R'_B}{\beta} \right) \\ &= 6.8\ \text{k}\Omega \parallel (12.4\ \Omega + 15.3\ \Omega) \\ &= 6.8\ \text{k}\Omega \parallel 27.7\ \Omega \\ &\cong 27.7\ \Omega \end{aligned}$$

The ac analysis of the first stage yields

$$\begin{aligned} R_{\text{in}_1} &= R_1 = 1\ \text{M}\Omega \\ R_{o_1} &= R_2 = 1.8\ \text{k}\Omega \end{aligned}$$

The equivalent ac drain resistance R_D is given by

$$\begin{aligned} R'_D &= R_2 \parallel R_{\text{in}_2} \\ &= 1.8\ \text{k}\Omega \parallel 5.3\ \text{k}\Omega \\ &= 1.3\ \text{k}\Omega \end{aligned}$$

The loaded voltage gain of stage 1 is

$$\begin{aligned} A_{v_1} &= -g_{mQ} R'_D \\ &= -(6.1\ \text{mS})(1.3\ \text{k}\Omega) \\ &= -7.93 \end{aligned}$$

The overall amplifier parameters are

$$\begin{aligned} R_{\text{in}} &= R_{\text{in}_1} = 1\ \text{M}\Omega \\ R_o &= R_{o_2} = 27.7\ \Omega \\ A_v &= A_{v_1} A_{v_2} \\ &= (-7.93)(0.89) \\ &= -7.1 \end{aligned}$$

$$\begin{aligned} A_i &= A_v \frac{R_{\text{in}}}{R_L} \\ &= -7.1 \frac{1\ \text{M}\Omega}{100\ \Omega} \\ &= -71 \times 10^3 \end{aligned}$$

$$\begin{aligned} A_p &= A_v A_i \\ &= (-7.1)(-71\ \text{k}) \\ &= 504.1 \times 10^3 \end{aligned}$$

9.3 DIRECT COUPLING

Capacitively coupled amplifiers are limited to the processing of ac signals. Many applications, however, require the processing of dc levels as well. Such applications normally require *direct coupling* between amplifier stages and between the amplifier and the signal source and the load. An example of a direct-coupled amplifier is shown in Fig. 9.10.

Note
Integrated circuit designs use direct coupling almost exclusively.

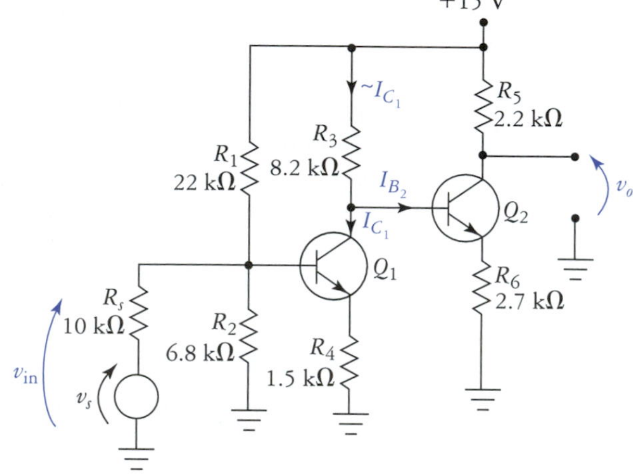

FIGURE 9.10
Direct-coupled two-stage amplifier for Example 9.10.

Figure 9.10 uses basically the same kind of amplifier configurations that we studied in Chapters 5 and 6. An analysis of this circuit will point out a few of the complications associated with this approach to direct-coupled amplifier design.

EXAMPLE 9.6

Refer to Fig. 9.10. Determine the Q points for both transistors, the overall voltage gain, and the expression for v_o given $v_{in} = 50 + 200 \sin \omega t$ mV. Assume that $\beta = 100$.

Solution Because direct coupling is used, the analysis procedure is rather long and tedious. Please be patient. There is a light at the end of the tunnel.

The dc analysis must always be performed first. However, in the case of direct coupling, we must start at the input stage and work toward the output, and in order to keep the mathematics from getting out of hand, a few simplifying assumptions are also required. Thus, we begin by finding the Thevenin voltage and resistance at the base of Q_1. It is important to notice that because of the direct coupling of the input source to the base that the source resistance appears in the voltage-divider equation. Thus, let us define

$$\begin{aligned} R_2' &= R_s \parallel R_2 \\ &= 10\text{ k}\Omega \parallel 6.8\text{ k}\Omega \\ &= 4.05\text{ k}\Omega \end{aligned}$$

Now, the Thevenin voltage at the base of Q_1 is

$$\begin{aligned} V_{\text{Th}_1} &= V_{CC}\frac{R_2'}{R_2' + R_1} \\ &= 15\text{ V}\frac{4.05}{4.05 + 22} \\ &= 2.33\text{ V} \end{aligned}$$

For the purposes of determining bias stability, we determine

$$\begin{aligned} R_{\text{Th}_1} &= R_s \parallel R_1 \parallel R_2 \\ &= 10\text{ k}\Omega \parallel 22\text{ k}\Omega \parallel 6.8\text{ k}\Omega \\ &= 3.4\text{ k}\Omega \end{aligned}$$

From this we find $\beta R_4 >> R_{\text{Th}_1}$ so that

$$I_{C_1} \cong \frac{V_{\text{Th}_1} - V_{BE}}{R_4}$$
$$= \frac{2.33 \text{ V} - 0.7 \text{ V}}{1.5 \text{ k}\Omega}$$
$$= 1.1 \text{ mA}$$

and

$$r_{e_1} = V_T / I_{C_1}$$
$$= 26 \text{ mV}/1.1 \text{ mA}$$
$$= 23.6\ \Omega$$

Now, things get a little tricky. Because there is no coupling capacitor between Q_1 and Q_2, the current through R_3 is somewhat less than I_{C_1}. In fact, it is $I_{R_3} = I_{C_1} - I_{B_2}$. An exact determination of the currents and voltages in this circuit would best be found using nodal or mesh analysis. However, in order to simplify the analysis, let us assume that $I_{B_2} << I_{C_1}$ This inequality will hold for a well-designed circuit, and so KVL gives us

$$V_{CE_1} \cong V_{CC} - I_{C_1}(R_3 + R_4)$$
$$= 15 \text{ V} - 1.1 \text{ mA}(8.2 \text{ k}\Omega + 1.5 \text{ k}\Omega)$$
$$= 15 \text{ V} - 10.67 \text{ V}$$
$$= 4.33 \text{ V}$$

The collector voltage of Q_1 is

$$V_{C_1} = V_{CC} - I_{C_1} R_3$$
$$= 15 \text{ V} - (1.1 \text{ mA})(8.2 \text{ k}\Omega)$$
$$= 6 \text{ V}$$

It is the collector voltage of Q_1 that is biasing up Q_2. For good bias stability of Q_2 we require $\beta R_6 >> R_3$. In this case the inequality is well satisfied (by a ratio of about 33 to 1), thus we have

$$I_{C_2} \cong \frac{V_{C_1} - V_{BE}}{R_6}$$
$$= \frac{6 \text{ V} - 0.7 \text{ V}}{2.7 \text{ k}\Omega}$$
$$= 2 \text{ mA}$$

The dynamic emitter resistance of Q_2 is

$$r_{e_2} = V_T / I_{C_2}$$
$$= 26 \text{ mV} / 2 \text{ mA}$$
$$= 13\ \Omega$$

We can now determine the remaining dc quantities:

$$V_{CE_2} = V_{CC} - I_{C_2}(R_5 + R_6)$$
$$= 15 \text{ V} - 2 \text{ mA}(2.2 \text{ k}\Omega + 2.7 \text{ k}\Omega)$$
$$= 15 \text{ V} - 9.8 \text{ V}$$
$$= 5.2 \text{ V}$$

The quiescent output voltage V_{oQ} is the same as the collector voltage in this case, so

$$V_{oQ} = V_{C_2} = V_{CC} - I_{C_2} R_5$$
$$= 15 \text{ V} - (2 \text{ mA})(2.2 \text{ k}\Omega)$$
$$= 15 \text{ V} - 4.4 \text{ V}$$
$$= 10.6 \text{ V}$$

If a resistive load were connected to the output terminal, we would have to account for the loading effect on V_{0Q} (we could use superposition, mesh, or nodal analysis, but in any case it would be very messy unless $R_L >> R_5$). To keep things somewhat simple, this amplifier is driving an open circuit.

The ac analysis must begin with the output stage and work backward toward the input. The voltage gain of stage 2 is

$$A_{v_2} = -\frac{R'_{C_2}}{r_{e_2} + R_{e_2}}$$
$$= -\frac{2.2\ \text{k}\Omega}{13\ \Omega + 2.7\ \text{k}\Omega}$$
$$= -0.81$$

Note that $R'_{C_2} = R_5$ because $R_L = \infty$. The input resistance of stage 2 is

$$R_{\text{in}_2} \cong \beta(R_6 + r_{e_2})$$
$$= 100(2700\ \Omega + 13\ \Omega)$$
$$= 271\ \text{k}\Omega$$

The output resistance of stage 2 is

$$R_{o_2} \cong R_5 = 2.2\ \text{k}\Omega$$

Moving to stage 1, we find

$$R_{\text{in}_1} = R_{\text{Th}_1} \parallel \beta(R_4 + r_{e_1})$$
$$= 5.2\ \text{k}\Omega \parallel 100(1500\ \Omega + 23.6\ \Omega)$$
$$= 5.1\ \text{k}\Omega$$

The ac collector resistance of Q_1 is

$$R'_{C_1} = R_3 \parallel R_{\text{in}_2}$$
$$= 8.2\ \text{k}\Omega \parallel 271\ \text{k}\Omega$$
$$= 8\ \text{k}\Omega$$

The voltage gain of stage 1 is

$$A_{v_1} = -\frac{R'_{C_1}}{r_{e_1} + R_{e_1}}$$
$$= -\frac{8\ \text{k}\Omega}{23.6\ \Omega + 1.5\ \text{k}\Omega}$$
$$= -5.3$$

Because the input source has a significant internal resistance, we must account for the loading effect of Q_1. Let us designate this loading factor as α. Application of the voltage-divider equation gives us

$$\alpha = \frac{R_{\text{in}}}{R_{\text{in}} + R_s}$$
$$= \frac{5.1\ \text{k}\Omega}{5.1\ \text{k}\Omega + 10\ \text{k}\Omega}$$
$$= 0.338$$

The overall amplifier parameters are

$$R_{\text{in}} = R_{\text{in}_1} = 5.1\ \text{k}\Omega$$
$$R_o = R_{o_2} = 2.2\ \text{k}\Omega$$
$$A_v = \alpha A_{v_1} A_{v_2}$$
$$= (0.338)(-5.3)(-0.81)$$
$$= 1.45$$

The current gain is undefined since $R_L = \infty$.

The output voltage is given by

$$v_o = V_{oQ} + A_v v_{\text{in}}$$
$$= 10.6\ \text{V} + (1.45)(50 + 200 \sin \omega t\ \text{mV})$$
$$= 10.6\ \text{V} + (72.5 + 290 \sin \omega t\ \text{mV})$$
$$\cong 10.67\ \text{V} + 0.29 \sin \omega t\ \text{V}$$

The preceding example warrants further discussion. First, the dc level present in the input signal was amplified and produced at the output. This dc level would shift the Q point of the amplifier to some extent, but the assumption here is that the shift is of little consequence; that is, the dc and ac signals are "small signals." This is close enough to the truth in this case, but things can get very tricky and complicated very quickly. As a general rule, if the maximum shift in I_{CQ} (in any stage) is less than 10% then the small-signal approximation is valid. Large-signal considerations are taken up in the next chapter.

Note
The main advantage of direct coupling is that it allows the processing of DC as well as AC signals.

The reader should also notice that because bypass capacitors are not used the overall voltage gain of the circuit is rather low. Emitter bypass capacitors could indeed be used to increase the ac gain of the circuit. This would also cause the dc gain of the amplifier to differ from the ac gain, a situation we do not usually desire in a dc amplifier. Ideally, the gain of the amplifier should be constant or flat from dc on up to some arbitrarily high frequency. It should also be observed that when direct coupling of CE amplifiers is used, the base voltage of each successive transistor must increase as we go from the input stage to the output stage. This often presents significant problems in designs that require three or more stages. Mixing pnp and npn transistors in the design can often help to eliminate this problem as well.

As far as the input source and the load are concerned, when they are connected to a circuit like that of Fig. 9.10, each must be able to withstand a significant quiescent dc level. This too presents problems in many applications. Finally, direct-coupled amplifiers are much more difficult to troubleshoot than their capacitively coupled cousins. Consider for example that if Q_1 in Fig. 9.10 were to become open circuited at the collector terminal, then the base of Q_2 would be pulled up to near V_{CC} in voltage. This would heavily saturate Q_2, making it look defective as well. This propagation of "bad voltages" not only makes successive stages appear defective, it can also cause a chain reaction of device failures to occur as one stage after another is driven to its operational limits.

Although there are definitely problems associated with direct coupling, it may come as a surprise that it is actually more commonly used than any other coupling technique. This is especially true of integrated circuit designs, partly because capacitors larger than a few picofarads require lots of IC surface area, and surface area is at a premium on IC chips. A number of the problems associated with direct coupling are eliminated through the use of differential pairs, which we examine next.

9.4 MULTIPLE-STAGE DIFFERENTIAL AMPLIFIERS

Figure 9.11 shows a two-stage differential amplifier. The first stage (Q_1, Q_2, Q_3, D_1, D_2, and resistors R_1, R_2, R_3, R_4) is a differential-input, differential-output section, using a two-diode active current source. The second stage is a differential-input, single-ended-output type section. Overall, the circuit would be treated as a differential-input, single-ended-output amplifier. This is the most common

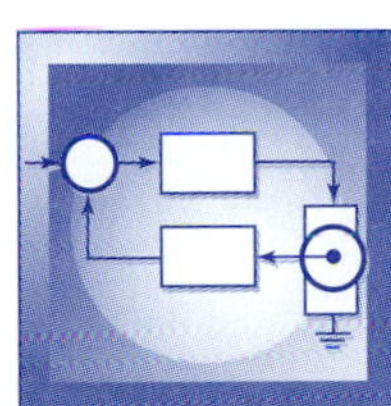

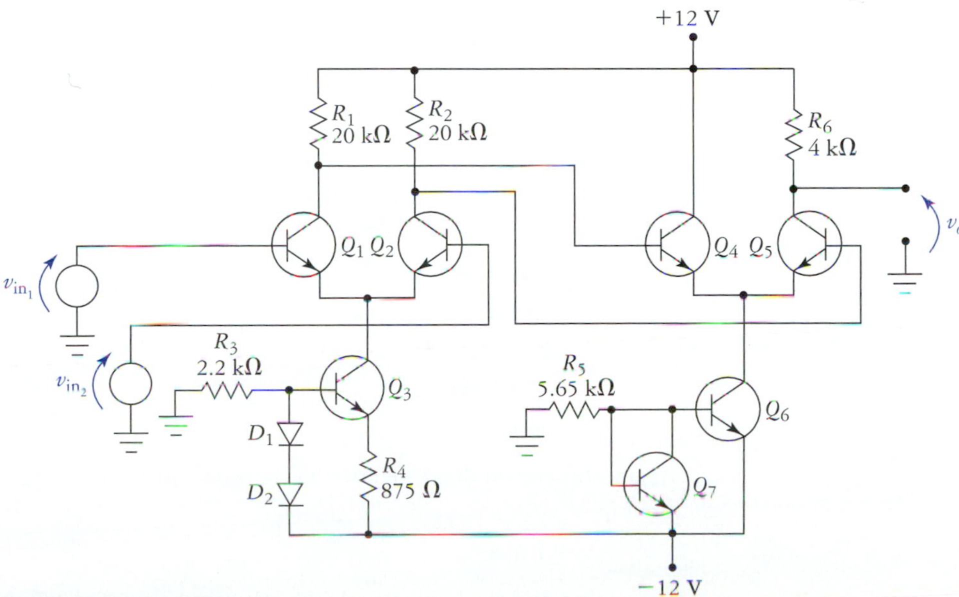

FIGURE 9.11
Two-stage differential amplifier.

configuration for such multiple-stage differential amplifiers. The analysis of this circuit is presented in the next example.

EXAMPLE 9.7

Determine I_{CQ} and V_{CEQ} for each transistor in Fig. 9.11. Also determine V_{oQ}, the loaded differential gain of stage 1, A_{d_1}; the differential gain of stage 2, A_{d_2}; the overall differential gain, A_d; the overall input resistance R_{in} as seen by either source 1 or source 2 (it is the same for both inputs); and the overall output resistance R_o. Assume $\beta = 100$ and $V_{BE} = 0.7$ V.

Solution The dc analysis is performed first. At the outset, we make the assumption that the second stage has little effect on the first stage, as far as the dc analysis goes. This is based on the assumption that the base-bias currents I_{B_4} and I_{B_5} are very small compared to collector currents I_{C_1} and I_{C_2}. Calculating I_{C_3} we have

$$I_{C_3} = \frac{V_{BE}}{R_4} = \frac{0.7\text{ V}}{875\ \Omega} = 800\ \mu\text{A}$$

This gives us

$$I_{C_1} = I_{C_2} = I_{C_3}/2 = 400\ \mu\text{A}$$

Because of the symmetry between both sides of stage 1 we may assume that equal voltage drops will exist around each transistor. The input voltage sources are inactive, which places both bases at ground (0 V) potential, and the loading caused by stage 2 is assumed to be negligible. Application of KVL yields

$$\begin{aligned} V_{CB_1} = V_{CB_2} &= V_{CC} - I_c R_C \\ &= 12\text{ V} - (400\ \mu\text{A})(20\text{ k}\Omega) \\ &= 12\text{ V} - 8\text{ V} \\ &= 4\text{ V} \end{aligned}$$

Now, since $V_{BE} = 0.7$ V, we have

$$\begin{aligned} V_{CE_1} = V_{CE_2} &= V_{CB} + V_{BE} \\ &= 4.7\text{ V} \end{aligned}$$

We find V_{CE_3} applying KVL yielding

$$\begin{aligned} V_{CE_3} &= V_{CC} + V_{EE} - I_{C_1}R_1 - V_{CE_1} - I_{C_3}R_4 \\ &= 12\text{ V} + 12\text{ V} - (400\ \mu\text{A})(20\text{ k}\Omega) - 4.7\text{ V} - (800\ \mu\text{A})(875\ \Omega) \\ &= 12\text{ V} + 12\text{ V} - 8\text{ V} - 4.7\text{ V} - 0.7\text{ V} \\ &= 10.6\text{ V} \end{aligned}$$

The dc analysis of the second stage can now be performed. First, we determine the collector current of active current source Q_6:

$$I_{C_6} = \frac{V_{EE} - V_{BE}}{R_5} = \frac{12\text{ V} - 0.7\text{ V}}{5.65\text{ k}\Omega} = 2\text{ mA}$$

Again, this current divides equally in the differential pair giving

$$I_{C_4} = I_{C_5} = I_{C_6}/2 = 1\text{ mA}$$

Since stage 2 is not symmetrical (the collector resistances differ), Q_4 and Q_5 will have different voltage drops existing across them. It is also very important to note that the bases of these transistors are not at ground potential, since they are directly coupled to the collectors of Q_1 and Q_2.

Based on these factors, and from the analysis of stage 1, we have $V_{B_4} = V_{C_1}$ and $V_{B_5} = V_{C_2}$. Therefore,

$$\begin{aligned} V_{C_1} = V_{C_2} &= V_{CC} - I_{C_1}R_1 \\ &= 12\text{ V} - (400\ \mu\text{A})(20\text{ k}\Omega) \\ &= 4\text{ V} \end{aligned}$$

This produces

$$\begin{aligned} V_{CB_4} &= V_{CC} - V_{B_4} \\ &= 12\text{ V} - 4\text{ V} \\ &= 8\text{ V} \end{aligned}$$

$$\begin{aligned} V_{CB_5} &= V_{CC} - I_{C_5}R_6 - V_{B_5} \\ &= 12\text{ V} - 4\text{ V} - 4\text{ V} \\ &= 4\text{ V} \end{aligned}$$

We always assume that the transistor is biased on, therefore $V_{BE} = 0.7$ V and we can write

$$\begin{aligned} V_{CE_4} &= V_{CB_4} + V_{BE} \\ &= 8\text{ V} + 0.7\text{ V} \\ &= 8.7\text{ V} \end{aligned}$$

$$\begin{aligned} V_{CE_5} &= V_{CB_5} + V_{BE} \\ &= 4\text{ V} + 0.7\text{ V} \\ &= 4.7\text{ V} \end{aligned}$$

We determine V_{CE_6} using KVL:

$$\begin{aligned} V_{CE_6} &= V_{CC} + V_{EE} - V_{CE_4} \\ &= 12\text{ V} + 12\text{ V} + 8.7\text{ V} \\ &= 15.3\text{ V} \end{aligned}$$

Because Q_7 is connected as a diode, we have $V_{CE_7} = 0.7$ V.

The quiescent output voltage is the collector voltage of Q_5, therefore we obtain

$$\begin{aligned} V_{oQ} = V_{C_5} &= V_{CC} - I_{C_5}R_6 \\ &= 12\text{ V} - (1\text{ mA})(4\text{ k}\Omega) \\ &= 12\text{ V} - 4\text{ V} \\ &= 8\text{ V} \end{aligned}$$

This concludes the dc analysis of Fig. 9-11. Since stage 2 is loading stage 1, the ac analysis is best performed starting with stage 2. We can begin by determining the dynamic emitter resistances of Q_4 and Q_5:

$$\begin{aligned} r_{e_4} = r_{e_5} &= V_T/I_{CQ} \\ &= 26\text{ mV}/1\text{ mA} \\ &= 26\ \Omega \end{aligned}$$

Now, because stage 2 is a differential-input, single-ended-output circuit, the voltage gain is

$$\begin{aligned} A_{d_2} &= \frac{R_6}{2\,r_e} \\ &= \frac{4\text{ k}\Omega}{52\ \Omega} \\ &= 76.9 \end{aligned}$$

The input resistance of stage 2 is

$$\begin{aligned} R_{\text{in}_2} &= 2\beta r_e \\ &= (200)(26\ \Omega) \\ &= 5.2\text{ k}\Omega \end{aligned}$$

The ac analysis of stage 2 is complete. A similar process is applied to stage 1:

$$\begin{aligned} r_{e_1} = r_{e_2} &= V_T/I_{CQ} \\ &= 26\text{ mV}/400\ \mu\text{A} \\ &= 65\ \Omega \end{aligned}$$

In determining the differential gain of stage 1, we must take into account the loading effect of stage 2. This means that

$$\begin{aligned} R'_{C_2} = R'_{C_1} &= R_1 \parallel R_{\text{in}_2} \\ &= 20\ \text{k}\Omega \parallel 5.2\ \text{k}\Omega \\ &= 4.13\ \text{k}\Omega \end{aligned}$$

So the differential gain of stage 1 is

$$\begin{aligned} A_{d_1} &= \frac{R'_C}{r_e} \\ &= \frac{4.13\ \text{k}\Omega}{65\ \Omega} \\ &= 63.5 \end{aligned}$$

The input resistance to stage 1 is now determined:

$$\begin{aligned} R_{\text{in}_1} &= 2\beta r_e \\ &= (200)(65\ \Omega) \\ &= 13\ \text{k}\Omega \end{aligned}$$

The overall voltage gain of the amplifier is the product of the individual stage gains.

$$\begin{aligned} A_d &= A_{d_1}A_{d_2} \\ &= (76.9)(63.5) \\ &= 4883 \end{aligned}$$

The overall input resistance is the input resistance of stage 1, thus

$$R_{\text{in}} = R_{\text{in}_1} = 13\ \text{k}\Omega$$

The output resistance is

$$R_o \cong R_6 = 4\ \text{k}\Omega$$

Note
Differential amplifiers are especially well-suited to direct coupling.

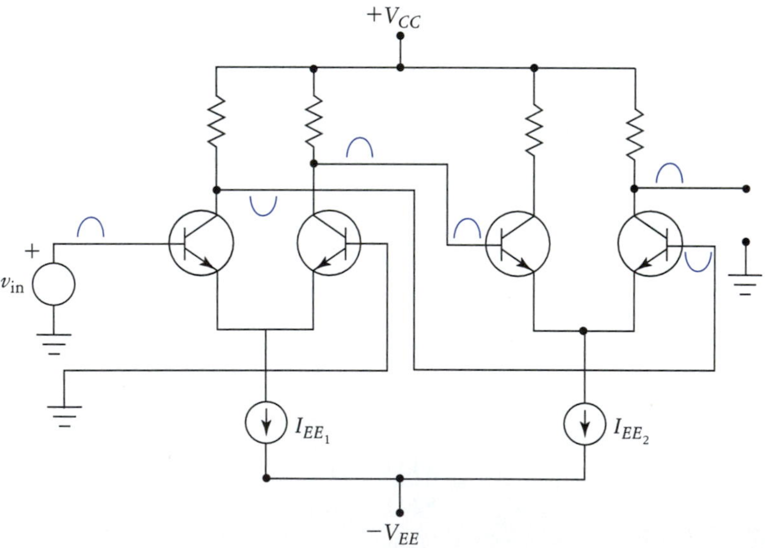

FIGURE 9.12
Phase relationships in the two-stage diff amp.

There are a few more important points to consider concerning Fig. 9.11. Because, overall, this circuit is a differential-input, single-ended-output amplifier, one of the inputs will act as an inverting (−) input and the other will act as a noninverting (+) input. Identifying the gain polarity of the input terminals is very important. One way to do this is outlined in the following steps:

1. Assume that one input is grounded and that a positive voltage exists at the remaining input. The positive half-cycle of a sine wave, a plus sign, or an arrow pointing up are good symbols to use at the active input.

2. Determine the direction in which the collector voltages of the input-stage transistors would be driven in response to the input and draw half-cycles or arrows designating this response.
3. Determine the resulting response from the next stage, and repeat for each successive stage.

These three steps are applied to the simplified two-stage differential amplifier in Fig. 9.12, providing identification of the inverting and noninverting inputs.

9.5 LEVEL-TRANSLATING STAGES

As explained earlier, most differential amplifier designs use direct coupling. This is done primarily to allow the processing of dc voltage levels, as well as ac signals. Because of this dc amplification ability, it is desirable to have the quiescent output V_{oQ} of the final stage of the differential amplifier at zero volts. That is, for $v_{id} = 0$ V, we desire $v_o = V_{oQ} = 0$ V. To meet this requirement, a stage called a *level shifter* or *level translator* is often used.

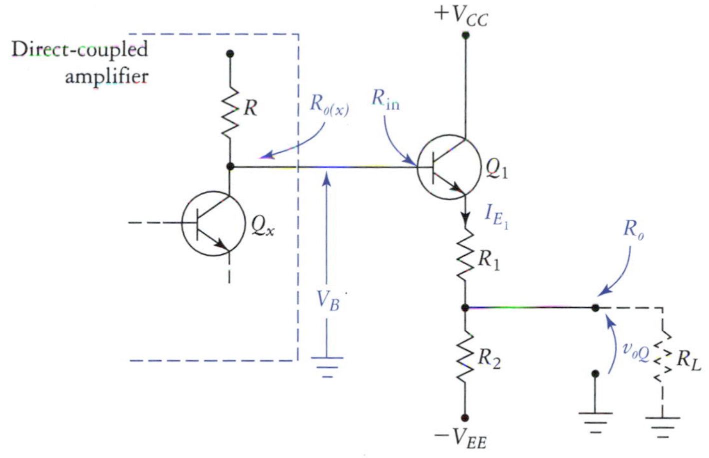

FIGURE 9.13
A level translator circuit is used to remove the output offset voltage while still retaining dc response.

The idea behind a level shifter is to shift or translate one voltage level (typically V_{oQ} of a single-ended-output differential amplifier) to another level (usually 0 V). Most often, level translators are designed using emitter-follower-type circuits like the one shown in Fig. 9.13. Assuming very high r_o for the transistors, the analysis prodeeds as follows.

The emitter current of Q_1 is given by

$$I_E = \frac{V_{EE} + V_B - V_{BE}}{R_E} \tag{9.13}$$

where $R_E = R_1 + R_2$. Application of KVL shows that

$$V_{oQ} = V_{EE} - I_E R_2 \tag{9.14}$$

For design purposes, V_B is given, a desired value of I_E is chosen, and the values required for R_1 and R_2 would be determined as follows. Solving Eq. (9.13) for R_E, we obtain

$$R_E = \frac{V_{EE} + V_B - V_{BE}}{I_E} \tag{9.15}$$

Now, solving Eq. (9.14) for R_2, we obtain

$$R_2 = \frac{V_{EE} - V_{oQ}}{I_E} \tag{9.16a}$$

which if $V_{oQ} = 0$ V can be simplified to

$$R_2 = \frac{V_{EE}}{I_E} \tag{9.16b}$$

With the value of R_2 now known, R_1 can be determined by

$$R_1 = R_E - R_2 \tag{9.17}$$

Under no-load conditions, the voltage gain of the level translator is

$$A_v = \frac{R_2}{R_1 + R_2 + r_e} \tag{9.18}$$

The output resistance is

$$R_o = R_2 \left\| \left(R_1 + r_e + \frac{R_{o(x)}}{\beta} \right) \right. \tag{9.19}$$

Note
The level shifters of Figs. 9.13 and 9.14 are basically emitter followers with $A_v \leq 1$.

and the input resistance is

$$R_{in} = \beta(r_e + R_1 + R_2) \tag{9.20}$$

It is interesting to look at the operation of the level shifter when a load is connected as in the case of the dashed-line resistor R_L in Fig. 9.13. When the circuit is designed for $V_{oQ} = 0$ V (which is the most common case), the load resistance has no effect on the biasing of Q_1. This is because both sides of R_L are at ground potential, and the load current is zero. The connection of a load does, however, affect some of the signal-related parameters. Specifically, we have

$$A_v = \frac{R'_L}{R'_L + R_1 + r_e} \tag{9.21}$$

and

$$R_{in} = \beta(r_e + R_1 + R'_L) \tag{9.22}$$

where in both cases $R'_L = R_L \parallel R_2$.

The output resistance of the circuit is unaffected by the addition of R_L and is given by Eq. (9.19). Also, since the level shifter is basically an emitter-follower, its voltage gain will always be less than unity.

Although the level shifter of Fig. 9.13 will perform adequately in many applications, it is common practice to use an active current source in place of R_2. The circuit in Fig. 9.14 illustrates one such possibility. Even though a current mirror is used in this example, any of the other current sources that were discussed could be used in the design.

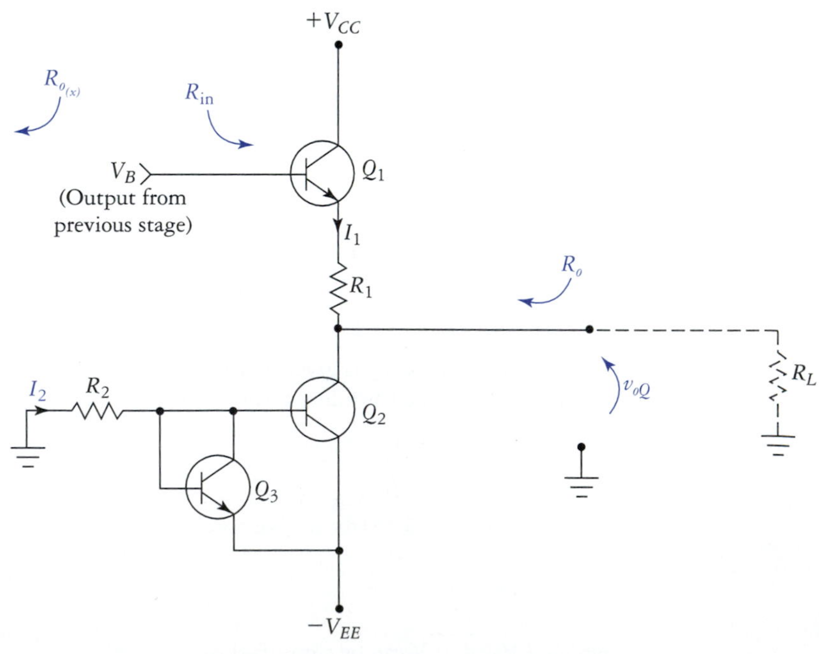

FIGURE 9.14
Level translator using current-mirror biasing.

An analysis of Fig. 9.14 begins with the determination of the collector current of Q_2. This is done using the procedures developed earlier, which may vary depending on the type of current source used. In this case,

$$I_1 = I_{C_2} \cong I_2 = \frac{V_{EE} - V_{BE}}{R_2}$$

From a design perspective, this current source would be set up to produce some desired value for I_1. Now, given a known value for I_1, V_{oQ} is found using

$$V_{oQ} = V_B - V_{BE} - I_1 R_1 \tag{9.23}$$

For design purposes, V_B and V_{BE} are given, I_1 is chosen, and we solve Eq. (9.23) for R_1 giving

$$R_1 = \frac{V_B - V_{oQ} - V_{BE}}{I_1} \tag{9.24}$$

Again, it is most common for us to design for $V_{oQ} = 0$ V, and so for this special case, we may simplify Eq. (9.24) to

$$R_1 = \frac{V_B - V_{BE}}{I_1} \tag{9.25}$$

Assuming active-region collector resistance $r_o \cong \infty$, the no-load small-signal parameters for the level shifter with current-source biasing are

$$A_v \cong 1 \tag{9.26}$$

$$R_o \cong R_1 + r_e + \frac{R_{o(x)}}{\beta} \tag{9.27}$$

$$R_{in} \cong \infty \tag{9.28}$$

When a load is connected to the output of the circuit, the output resistance is unaffected, but the other small-signal parameters are then given by

$$A_v = \frac{R_L}{R_L + R_1 + r_e} \tag{9.29}$$

$$R_{in} = \beta(R_1 + R_L + r_e) \tag{9.30}$$

The following example uses these relationships.

EXAMPLE 9.8

Refer to Fig. 9.14. Given $V_B = 5$ V and $V_{CC} = V_{EE} = 12$ V, determine the resistor values necessary to produce $V_{oQ} = 0$ V with $I_1 = 5$ mA. Determine the voltage gain and input resistance for $R_L = \infty$, $R_L = 10$ kΩ, and $R_L = 1$ kΩ.

Solution Begin by determining R_2:

$$R_2 = \frac{V_{EE} - V_{BE}}{I_1} = \frac{11.3 \text{ V}}{5 \text{ mA}} = 2.26 \text{ k}\Omega$$

We now determine R_1 using Eq. (9.25):

$$R_1 = \frac{V_B - V_{BE}}{I_1} = \frac{5 \text{ V} - 0.7 \text{ V}}{5 \text{ mA}} = 860\ \Omega$$

For the case where $R_L = \infty$, we have $A_v = 1$ and $R_{in} = \infty$. For $R_L = 10\ k\Omega$, we obtain

$$A_v = \frac{R_L}{R_L + R_1 + r_e}$$

$$= \frac{10\ k}{10\ k + 860} \text{ (we may disregard the small value of } r_e \text{ here)}$$

$$\cong 0.92$$

$$R_{in} = \beta(R_1 + R_L + r_e)$$
$$= 100(860\ \Omega + 10\ k\Omega)$$
$$= 1.1\ M\Omega$$

For the case where $R_L = 1\ k\Omega$, we find in a similar manner $A_v \cong 0.54$ and $R_{in} = 186\ k\Omega$.

9.6 DECIBELS

Most of the time the voltage, current, and power gains used so far in this text have been expressed simply as linear ratios of two quantities. We did have a brief encounter with decibels in Chapter 8, which we now expand. To reiterate, the decibel (dB) is a logarithmic measure of the ratio of two quantities. The various gains that we work with are ratios so the decibel may be easily applied in these cases. Decibels are used extensively in filter and communication circuit analysis applications.

Decibel Power Gain

Named in honor of Alexander Graham Bell, the inventor of the telephone, the Bel is a logarithmically related measure of the power gain (or power loss) of an amplifier or some other system. Specifically, the power gain of an amplifier in Bels is given by

$$A_{p(B)} = \log \frac{P_o}{P_{in}} \tag{9.31a}$$

or equivalently

$$A_{p(B)} = \log A_p \tag{9.31b}$$

where we are using the common or base-10 logarithm. As an example of the use of this relationship, assume that a certain amplifier has an overall power gain $A_p = 2 \times 10^6$. The power gain in Bels is then

$$A_{p(B)} = \log(2 \times 10^6)$$
$$= 6.301\ B$$

By taking the logarithm of the linear power gain, very large and very small values are compressed into a more manageable size. It turns out that the Bel produces values that are compressed too much. Because of this, the decibel dB is the more commonly used gain unit. A decibel is one-tenth of a Bel. Symbolically, the relationship between Bels and decibels is written

$$A_{p(dB)} = 10A_{p(B)} \tag{9.32}$$

It is standard practice to simply define the decibel power gain of an amplifier as

$$A_{p(dB)} = 10 \log \frac{P_o}{P_{in}} \tag{9.33a}$$

or equivalently

$$A_{p(dB)} = 10 \log A_p \tag{9.33b}$$

Again using the case where $A_p = 2 \times 10^6$ as an example, the gain of this amplifier in decibels is

$$A_{p(dB)} = 10 \log(2 \times 10^6)$$
$$= 63\ dB$$

Because the dimension of the power gain is given explicitly in the result, it is not really necessary to specify dB in the subscript. That is, it is correct to write

$$A_p = 2 \times 10^6 \qquad \text{or} \qquad A_p = 63 \text{ dB}$$

Decibel Voltage Gain

In a real-world setting, it is much more convenient to measure voltage than power. Because of this, decibel gain is often expressed in terms of voltage ratios. The equivalence between decibel voltage gain and decibel power gain can be derived using the circuit of Fig. 9.15. The power delivered to the amplifier is given by

$$P_{\text{in}} = \frac{v_{\text{in}}^2}{R_{\text{in}}} \tag{9.34}$$

FIGURE 9.15
Amplifier used for derivation of decibel voltage gain.

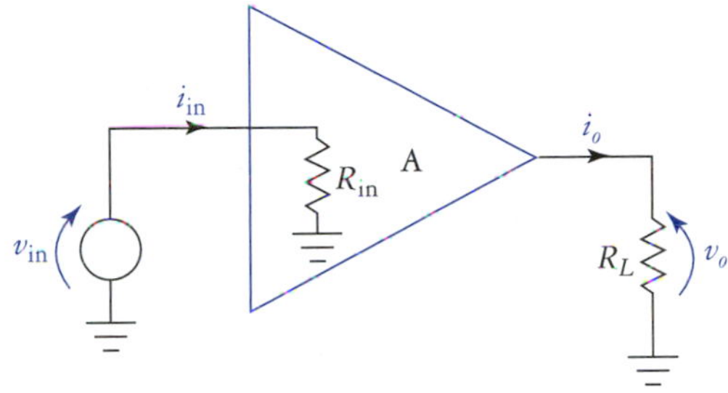

Note
Aside from A_v and A_p, many other parameters that are defined by ratios are also expressed in dB.

The power delivered to the load is given by

$$P_o = \frac{V_o^2}{R_L} \tag{9.35}$$

Substituting these expressions into Eq. (9.33a) produces

$$\begin{aligned} A_{p(\text{dB})} &= 10 \log \frac{P_o}{P_{\text{in}}} \\ &= 10 \log \frac{v_o^2 / R_L}{V_{\text{in}}^2 / R_{\text{in}}} \\ &= 10 \log \left[\frac{V_o^2}{V_{\text{in}}^2} \times \frac{R_L}{R_{\text{in}}} \right] \\ &= 10 \log \left[\left(\frac{V_o}{V_{\text{in}}} \right)^2 \left(\frac{R_L}{R_{\text{in}}} \right) \right] \end{aligned}$$

Although it is often not the case, it is standard practice to assume that $R_{\text{in}} = R_L$. This allows us to write

$$A_{p(\text{dB})} = 10 \log \left(\frac{V_o}{V_{\text{in}}} \right)^2 \tag{9.36}$$

Applying the rule $\log x^y = y \log x$ gives us

$$A_{p(\text{dB})} = 20 \log \left| \frac{V_o}{V_{\text{in}}} \right| \tag{9.37a}$$

or equivalently

$$A_{p(\text{dB})} = 20 \log |A_v| \tag{9.37b}$$

Because the application of Eqs. (9.37) only yields the true decibel power gain when $R_{\text{in}} = R_L$, we use the standard voltage-gain notation to help prevent any future confusion. Thus, we can simply write

$$A_{v(\text{dB})} = 20 \log \left| \frac{V_o}{V_{\text{in}}} \right| \tag{9.38a}$$

or

$$A_{v(\text{dB})} = 20 \log |A_v| \tag{9.38b}$$

The absolute magnitude of the voltage gain is used because even though inverting amplifiers have negative voltage gain, the power gain of such a system is still positive valued.

EXAMPLE 9.9

Refer to Fig. 9.16. Assuming that $R_{\text{in}} = R_L$ for all sections of the circuit, determine the overall linear voltage gain, the overall decibel voltage gain of the circuit, and the individual stage gains in decibels.

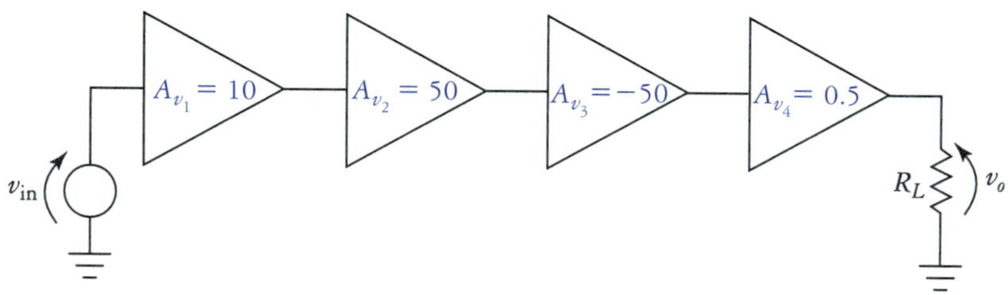

FIGURE 9.16
System for Example 9.9.

Solution The overall voltage gain is the product of the individual stage gains, thus we have

$$\begin{aligned} A_v &= A_{v_1}A_{v_2}A_{v_3}A_{v_4} \\ &= (10)(50)(-50)(0.5) \\ &= -12.5 \text{ k} \end{aligned}$$

The overall voltage gain in decibels is

$$\begin{aligned} A_{v(\text{dB})} &= 20 \log |A_v| \\ &= 20 \log 12{,}500 \\ &= 81.94 \text{ dB} \end{aligned}$$

The individual stage gains are

$$\begin{aligned} A_{v_1(\text{dB})} &= 20 \log |A_{v_1}| \\ &= 20 \log 10 \\ &= 20 \text{ dB} \end{aligned} \qquad \begin{aligned} A_{v_2(\text{dB})} &= 20 \log |A_{v_2}| \\ &= 20 \log 50 \\ &= 33.98 \text{ dB} \end{aligned}$$

$$\begin{aligned} A_{v_3(\text{dB})} &= 20 \log |A_{v_3}| \\ &= 20 \log 50 \\ &= 33.98 \text{ dB} \end{aligned} \qquad \begin{aligned} A_{v_4(\text{dB})} &= 20 \log |A_{v_4}| \\ &= 20 \log 0.5 \\ &= -6.02 \text{ dB} \end{aligned}$$

Notice in the preceding example that a fractional linear gain results in negative decibel gain.

Another useful property of logarithms is written symbolically as

$$\log(x_1 x_2 \cdots x_n) = \log x_1 + \log x_2 + \cdots + \log x_n$$

This allows us to determine the overall decibel gain of a multiple-stage amplifier by simply summing the individual decibel stage gains. This is shown in the next example.

EXAMPLE 9.10

Using the individual decibel stage gains of Example 9.9, determine the overall decibel voltage gain of the amplifier in Fig. 9.16.

Solution We sum the decibel stage gains, giving us

$$\begin{aligned} A_{v(\text{dB})} &= A_{v_1(\text{dB})} + A_{v_2(\text{dB})} + A_{v_3(\text{dB})} + A_{v_4(\text{dB})} \\ &= 20 \text{ dB} + 33.98 \text{ dB} + 33.98 \text{ dB} - 6.02 \text{ dB} \\ &= 81.94 \text{ dB} \end{aligned}$$

This agrees with the previous result.

Whenever you are given the gain of an amplifier expressed in decibels, the linear gain magnitude of that amplifier can be determined (assuming that $R_{\text{in}} = R_L$) by rearranging Eq. (9.38b) to form

$$A_v = \log^{-1}\left(\frac{A_{v(\text{dB})}}{20}\right) \tag{9.39}$$

EXAMPLE 9.11

Refer to Fig. 9.17. Determine the overall decibel gain and linear gain, as well as the individual linear stage gains for this system.

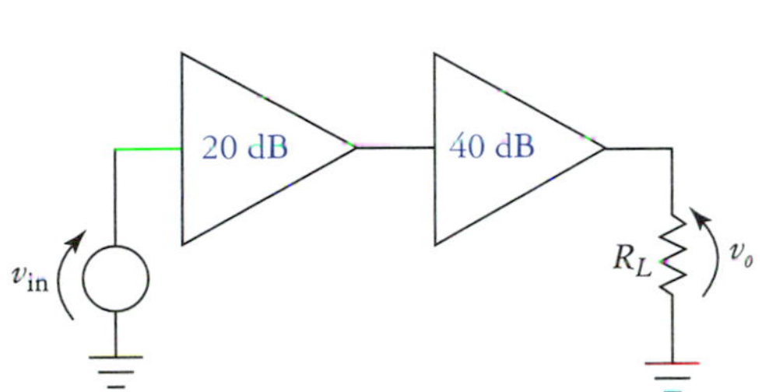

FIGURE 9.17
System for Example 9.11.

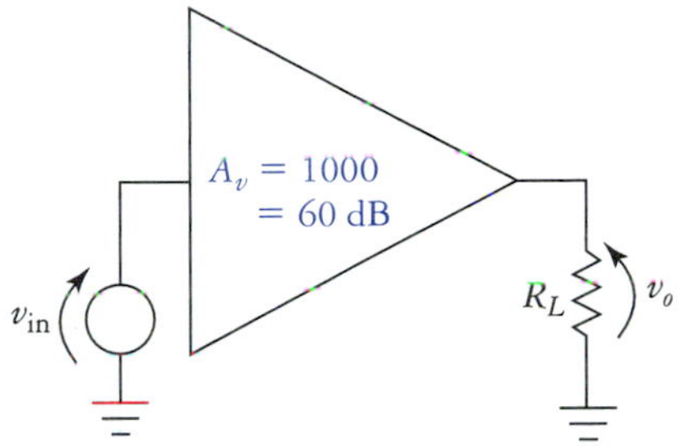

FIGURE 9.18
Equivalent block diagram for Fig. 9.17.

Solution The overall decibel gain is

$$\begin{aligned} A_{v(\text{dB})} &= A_{v_1(\text{dB})} + A_{v_2(\text{dB})} \\ &= 20\text{ dB} + 40\text{ dB} \\ &= 60\text{ dB} \end{aligned}$$

The individual linear stage gains are

$$\begin{aligned} A_{v_1} &= \log^{-1}\left(\frac{A_{v_1(\text{dB})}}{20}\right) \\ &= \log^{-1}\left(\frac{20}{20}\right) \\ &= 10 \end{aligned} \qquad \begin{aligned} A_{v_2} &= \log^{-1}\left(\frac{A_{v_2(\text{dB})}}{20}\right) \\ &= \log^{-1}\left(\frac{40}{20}\right) \\ &= 100 \end{aligned}$$

The overall gain magnitude can be found in two different ways. Using linear stage gains,

$$\begin{aligned} |A_v| &= |A_{v_1}A_{v_2}| \\ &= (10)(100) \\ &= 1000 \end{aligned}$$

Using the overall decibel gain,

$$\begin{aligned} |A_v| &= \log^{-1}\left(\frac{A_{v(\text{dB})}}{20}\right) \\ &= \log^{-1}\left(\frac{60}{20}\right) \\ &= 1000 \end{aligned}$$

An equivalent simplified representation of the two-stage amplifier of Fig. 9.17 is shown in Fig. 9.18.

9.7 COMPUTER-AIDED ANALYSIS APPLICATIONS

In general, at moderate frequencies (typically the audiofrequency range) capacitively coupled amplifier circuits are relatively straightforward to analyze manually. Because of this, we do not examine any examples of such applications in this section. Computer-aided analysis becomes extremely helpful, however, when we are dealing with direct-coupled circuits. This is especially true, for example, when the base current of a given transistor is significant when compared to the collector current of the transistor that is driving it. These situations often require very time-consuming and iterative analysis. This, however is the sort of task for which the computer is ideally suited.

Direct-Coupled CE Amplifier: Q-Point Analysis

The circuit that was analyzed in Example 9.6 was entered and simulated using PSpice. The schematic is shown in Fig. 9.19. 2N3904 transistors from eval.lib were used in this case. The transistor operating-point values are excerpted from the output file and presented in Fig. 9.20.

The results of the PSpice analysis and the manual analysis are summarized here:

Parameter	PSpice	Manual
I_{C_1}	1.08 mA	1.1 mA
I_{C_2}	1.96 mA	2.0mA
V_{CE_1}	4.37 V	4.33 V
V_{CE_2}	5.37 V	5.20 V
V_{oQ}	10.69 V	10.60 V

The agreement between the Q-point analyses is excellent. Let's see if the small-signal analysis results are equally impressive.

Direct-Coupled CE Amplifier: Signal Analysis

As in Example 9.6, the simulation run used an input voltage of 200 mV$_{pk}$ with a dc offset of +50 mV. The overall voltage gain and resulting output voltage were found in the example to be

$$A_v = 1.45$$
$$v_o = 10.67 + 0.29 \sin \omega t \text{ V}$$

The lower waveform of Fig. 9.21 is the collector voltage waveform for Q_1. This is an inverted sinusoid with $V_p = 350$ mV superimposed on an offset of $V_{dc} = V_{CQ_1} \cong 6$ V. Considering only the ac component, voltage gain A_{v_1} is

$$\begin{aligned} A_{v_1} &= v_{C_1(ac)} / v_{s(ac)} \\ &= -350 \text{ mV} / 200 \text{ mV} \\ &= -1.75 \end{aligned}$$

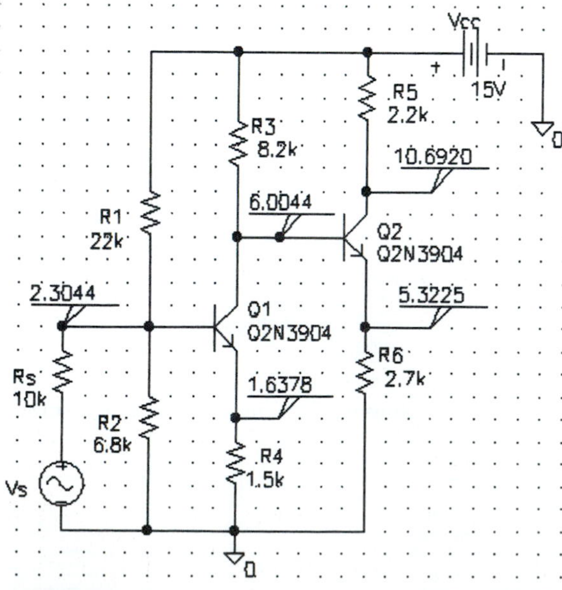

FIGURE 9.19
Q-point analysis of a direct-coupled amplifier.

**** BIPOLAR JUNCTION TRANSISTORS ****

NAME	Q_Q2	Q_Q1
MODEL	Q2N3904	Q2N3904
IB	1.29E−05	7.74E−06
IC	1.96E−03	1.08E−03
VBE	6.82E−01	6.67E−01
VBC	−4.69E+00	−3.70E+00
VCE	5.37E+00	4.37E+00
BETADC	1.52E+02	1.40E+02
GM	7.37E−02	4.13E−02
RPI	2.31E+03	3.88E+03
RX	1.00E+01	1.00E+01
RO	4.02E+04	7.17E+04
CBE	2.87E−11	1.89E−11
CBC	1.97E−12	2.10E−12
CJS	0.00E+00	0.00E+00
BETAAC	1.71E+02	1.60E+02
CBX	0.00E+00	0.00E+00
FT	3.82E+08	3.13E+08

FIGURE 9.20
Transistor model parameters.

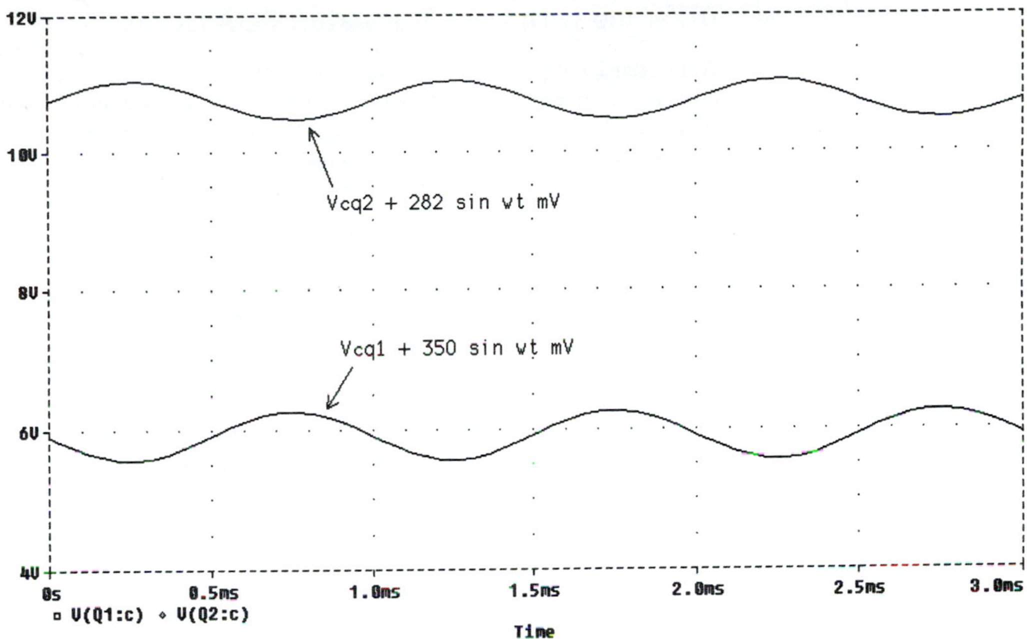

FIGURE 9.21
Collector voltage waveforms for the two stages.

This differs from the stage gain of the example where $A_{v_1} = -5.3$ because the simulator has factored in the effects of source loading. Accounting for loading effects we have

$$
\begin{aligned}
A_{v_1} &= \alpha A_{v_1} \\
&= (0.338)(-5.3) \\
&= -1.79
\end{aligned}
$$

This is in good agreement with the computer-derived result.

The waveform at the top of Fig. 9.21 is the output voltage of the amplifier, a sinusoid of 282 mV_{pk} superimposed on a dc level of $V_{dc} = V_{CQ_2} \cong 10.7$ V. Working with the ac components of V_s and V_o, and using this information to determine the overall voltage gain results in

$$
\begin{aligned}
A_v &= v_{o(ac)}/v_{s(ac)} \\
&= 282 \text{ mV} / 200 \text{ mV} \\
&= 1.41
\end{aligned}
$$

Again, there is reasonably good agreement between the manual and computer-derived values.

The input resistance, output resistance, and overall voltage gain were also found using the transfer function option under the PSpice analysis setup. The input variable is Vs and the output variable is V[C:Q2], which is an alias for V($N_0004). The resulting simulation values are shown in Fig. 9.22. Notice that PSpice has added R_s to the input resistance computation. If we subtract this out we have $R_{in} = 5.07$ kΩ, which is close to the 5.1 kΩ found by hand. The output resistance $R_o \cong 2.2$ kΩ closely matches our approximation, and the voltage gain is close as well.

```
****   SMALL-SIGNAL CHARACTERISTICS   ****

   V($N_0004)/V_Vs   =  1.416E+00

   INPUT RESISTANCE AT V_Vs  =   1.507E+04

   OUTPUT RESISTANCE AT V($N_0004)   =   2.197E+03
```

FIGURE 9.22
Transfer function analysis results.

Diff-Amp with Level Translator: DC Analysis

A differential amplifier with a level-translator output stage is shown in Fig. 9.23. Both stages are biased up using active current sources, which receive a constant base voltage derived from D_1 and D_2. Note that because of the slight loading of Q_2 by the level translator that $V_{C_1} \neq V_{C_2}$. Also, the quiescent output voltage is not zero, as would be desired; however, it is acceptably small. As we will see in the next chapter, this problem can be greatly reduced with the application of negative feedback.

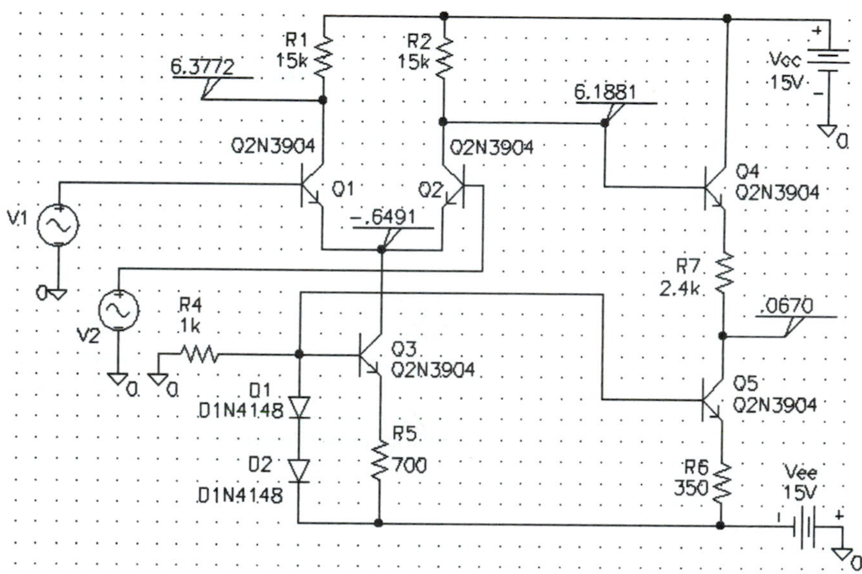

FIGURE 9.23
Q-point analysis of a diff amp with a level translator.

Diff-Amp with Level Translator: Signal Analysis

The amplifier of Fig. 9.23 was simulated using the following options: sine voltage source V1 set to 100 mV_{pk},1 kHz, AC = 1 V, DC = 0 V; source V2 has all attributes set to zero except f = 1 kHz; transient analysis, transfer function analysis (Output = V[Q5:c], Input = V1), dc sweep of V1 from −0.1 V to 0.1 V, and a decade-based ac sweep of V1 from 100 Hz to 10 MHz.

The transient analysis yielded A_d = 148, R_{in} = 1.4 kΩ and R_o = 2.5 kΩ. A comparative pencil-and-paper analysis of the amplifier is left as an exercise for the reader.

The results of the dc sweep are shown in Fig. 9.24, where we find clipping occurs gradually as v_o approaches about 8 V. The negative swing of the output clips a little more abruptly at about −6 V. The amplifier is reasonably linear for $-5 \text{ V} < v_o < 5 \text{ V}$.

Plotting the output voltage results in the display of Fig. 9.25. The gradual clipping of the positive peaks and abrupt clipping of the negative peaks is apparent here. The larger waveform is a plot of v_{in}, scaled by a factor of 148 (the small-signal gain). This is the output that would occur if the amplifier were linear over the range required for the given input signal.

A plot to the amplifier gain as a function of frequency is shown in Fig. 9.26. The output voltage was expressed in decibels [VdB(Q5:c)] and plotted over five decades from 100 Hz to 10 MHz. Since this amplifier is direct coupled, the frequency response extends down to 0 Hz, which cannot be shown on a logarithmic scale (it is infinitely far to the left). The midband gain in decibels is calculated from the linear gain as follows.

$$\begin{aligned} A_{d(\text{mid})} &= 20 \log 148 \\ &= 43.4 \text{ dB} \end{aligned}$$

This agrees with the value found in Fig. 9.26. The highest frequency at which the amplifier provides useful performance is called the *upper corner frequency*. This is the frequency at which the response drops by 3 dB from the midband value. The cursor was used to locate this point, where we find $f_c \cong 2.94$ MHz. We will deal more extensively with decibels and frequency response in later chapters.

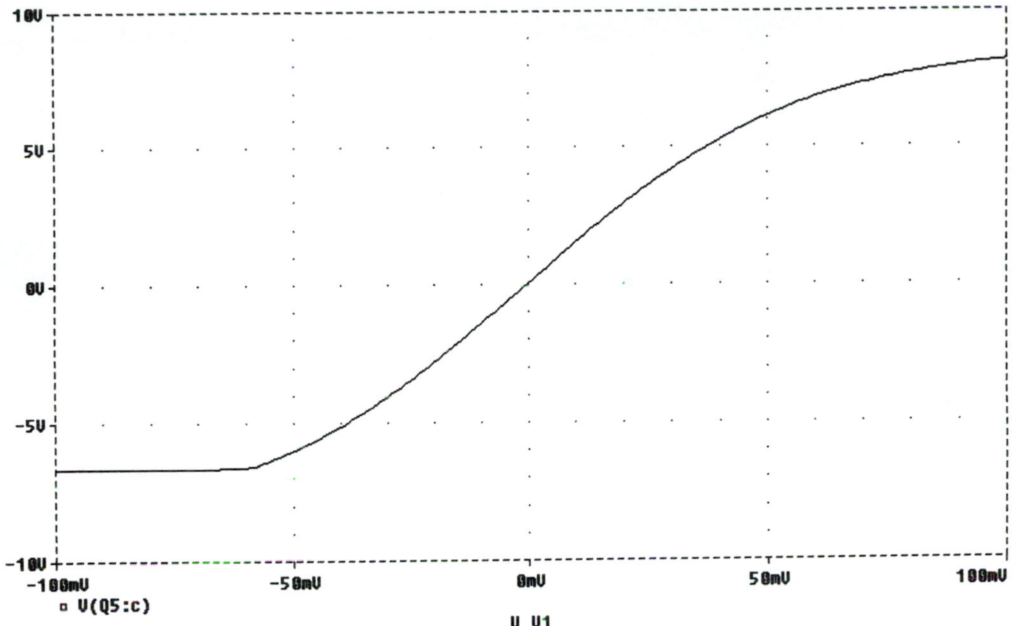

FIGURE 9.24
Transfer characteristic for the amplifier.

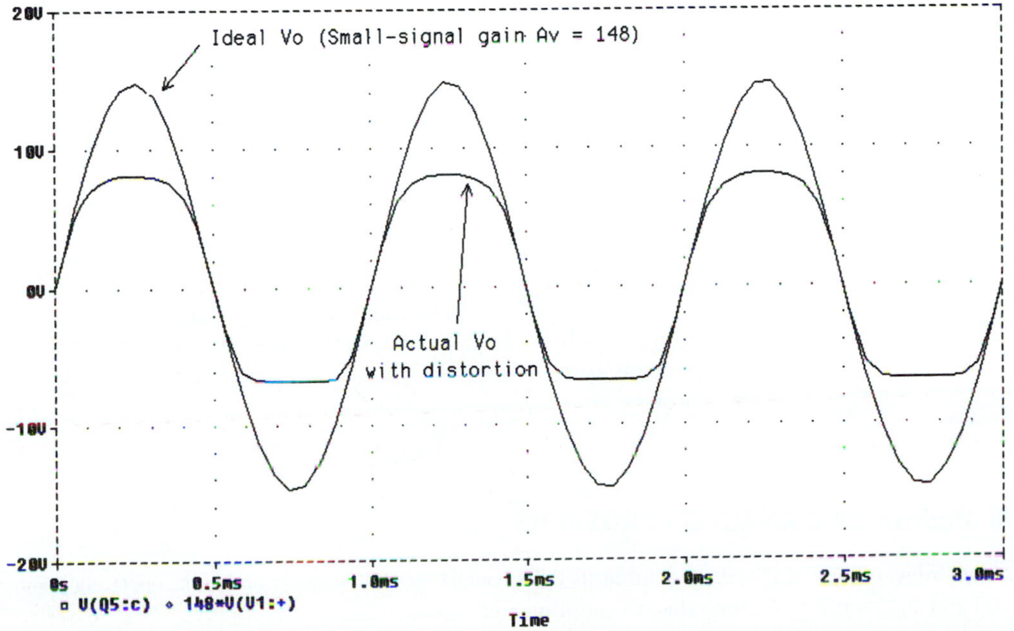

FIGURE 9.25
Comparison of output voltage that would be present if clipping did not occur (large waveform) and actual clipped output voltage.

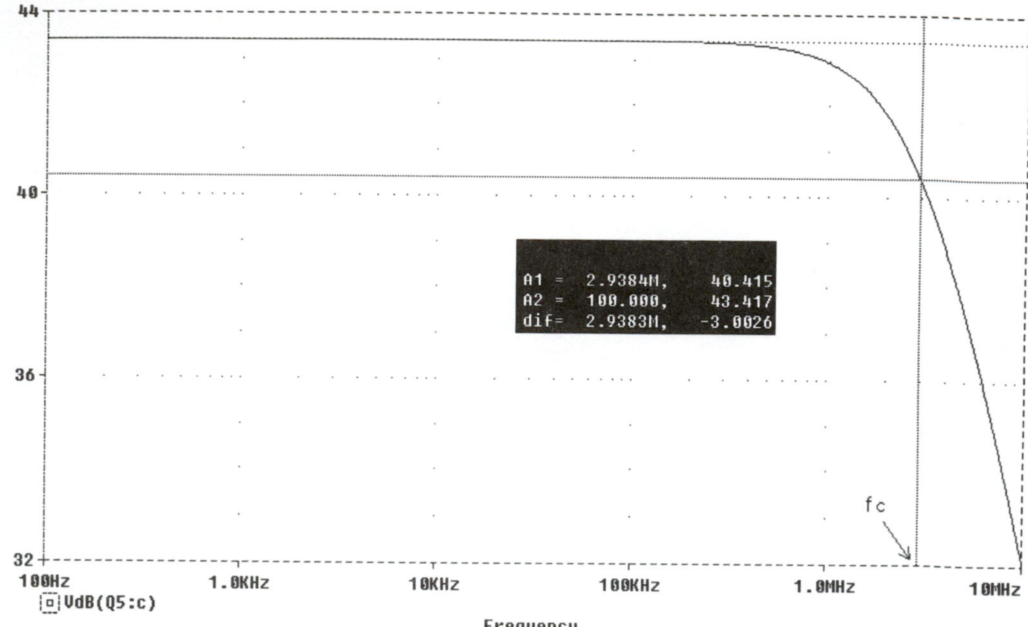

FIGURE 9.26
Frequency response of the amplifier.

CHAPTER REVIEW

More often than not, a multiple-stage amplifier will be required to provide the gain needed in a given application. Multiple-stage amplifiers are constructed using BJTs, FETs, and combinations of both in order to take advantage of the unique features of each class of device.

Capacitive coupling simplifies the design and analysis process because the coupling capacitors prevent dc interaction between successive amplifier stages. The main disadvantage of capacitive coupling is that it does not allow the amplifier to be used to process dc levels. Direct coupling overcomes this limitation through the elimination of coupling (and usually bypass) capacitors. Direct-coupled amplifiers are more difficult to design, analyze, and troubleshoot than similar capacitively coupled circuits. A defective component in one stage of a direct-coupled amplifier can affect other stages as well, possibly causing damage to devices in those other stages.

Differential amplifiers are best suited for use in direct-coupled applications because of their inherent common-mode rejection capability and because high gain can be obtained without the use of bypass capacitors. A level-shifting circuit will normally be used at the output of a direct-coupled amplifier in order to produce a quiescent output voltage of zero volts. These level shifters are usually emitter-follower-type circuits.

The decibel provides a logarithmic measure of the power gain of an amplifier. This allows very large and small gain values to be represented by numbers of a more convenient size. Decibel gains are additive, which is also an advantage in some situations. Because it is easier to measure voltage than power, decibel voltage gains are often used. To simplify the work involved in determining decibel voltage gains, it is standard practice to assume that the input resistance of the amplifier equals the load resistance.

DESIGN AND ANALYSIS PROBLEMS

Note: Unless otherwise specified, assume all BJTs have $\beta = 100$ and $V_{BE} = 0.7$ V. Assume that all coupling and bypass capacitors are ideal.

9.1. Refer to Fig. 9.27. Given $R_1 = 22$ kΩ, $R_2 = 6.8$ kΩ, $R_3 = 8.2$ kΩ, $R_4 = 100$ Ω, $R_5 = 2.1$ kΩ, $R_6 = 18$kΩ, $R_7 = 5.6$ kΩ, $R_8 = 2.7$ kΩ, $R_9 = 1$ kΩ, $R_L = 3.3$ kΩ, and $V_{CC} = 12$ V, determine the operating point of each transistor, the loaded voltage gain of each stage, and the overall values for A_v, $A_{v(\mathrm{dB})}$, R_{in} and R_o.

9.2. Refer to Fig. 9.27. Assume $R_1 = 27$ kΩ, $R_2 = 12$ kΩ, $R_3 = 3.9$ kΩ, $R_4 = 100$ Ω, $R_5 = 2.2$ kΩ, $R_6 = 15$ kΩ, $R_7 = 5.6$ kΩ, $R_8 = 1.5$ kΩ, $R_9 = 860$ Ω, $R_L = 1$ kΩ, and $V_{CC} = 15$ V,

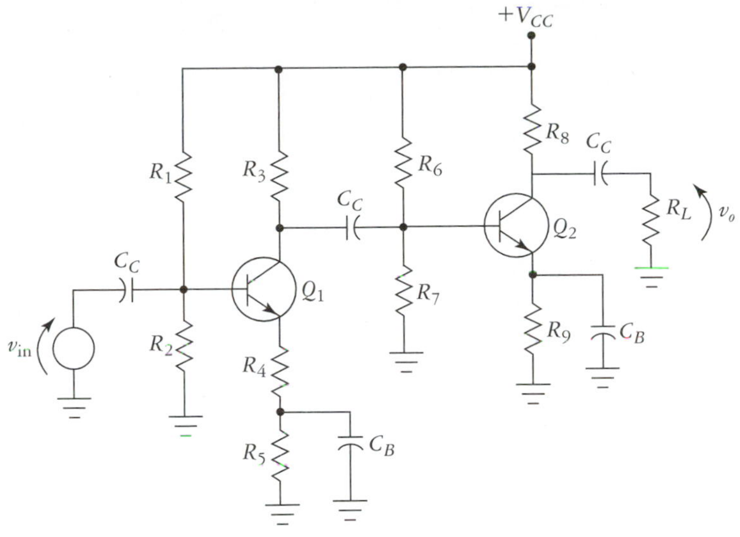

FIGURE 9.27
Circuit for Problems 9.1 to 9.4.

determine the operating point of each transistor, the loaded voltage gain of each stage, and the overall values for A_v, $A_{v(\text{dB})}$, R_{in}, and R_o.

9.3 Given the conditions of Problem 9.1, write the equation for v_o given $v_{\text{in}} = 2 \sin(2\pi 1000t)$ mV.

9.4. Given the conditions of Problem 9.2, write the equation for v_o given $v_{\text{in}} = 2 \sin(2\pi 1000t)$ mV.

9.5. Refer to Fig. 9.28. Assume $R_1 = 10$ kΩ, $R_2 = 500$ Ω, $R_3 = 1.86$ kΩ, $R_4 = 10$ kΩ, $R_5 = 100$Ω, $R_6 = 1$ kΩ, $R_L = 1$ kΩ, and $V_{CC} = V_{EE} = 10$ V. Determine the operating point of each transistor, the input resistance and loaded voltage gain of each stage, and the overall values for A_v, $A_{v(\text{dB})}$, R_{in}, and R_o.

9.6. Refer to Fig. 9.28. Assume $R_1 = 10$ kΩ, $R_2 = 2.2$ kΩ, $R_3 = 3.9$ kΩ, $R_4 = 10$ kΩ, $R_5 = 100$ Ω, $R_6 = 2.2$ kΩ, $R_L = 50$ Ω, and $V_{CC} = V_{EE} = 12$ V. Determine the operating point of each transistor, the input resistance and loaded voltage gain of each stage, and the overall values for A_v, $A_{v(\text{dB})}$, R_{in}, and R_o.

9.7. Refer to Fig. 9.29. Given $R_1 = 165$ kΩ, $R_2 = 33$ kΩ, $R_3 = 1.8$ kΩ, $R_4 = 220$ Ω, $R_5 = 220$ Ω, $R_6 = 22$ kΩ, $R_7 = 2.7$ kΩ, $R_8 = 2.2$ kΩ, $R_9 = 220$ Ω, $R_L = 10$ kΩ, and $V_{CC} = 15$ V, determine the operating point of each transistor, the loaded voltage gain of each stage, and the overall values for A_v, $A_{v(\text{dB})}$, R_{in}, and R_o.

9.8. Refer to Fig. 9.29. Given $R_1 = 470$ kΩ, $R_2 = 150$ kΩ, $R_3 = 3.9$ kΩ, $R_4 = 1.1$ kΩ, $R_5 = 100$ Ω, $R_6 = 12$ kΩ, $R_7 = 3.3$ kΩ, $R_8 = 1.2$ kΩ, $R_9 = 470$ Ω, $R_L = 1$ kΩ, and $V_{CC} = 15$ V, determine the operating point of each transistor, the loaded voltage gain of each stage, and the overall values for A_v, $A_{v(\text{dB})}$, R_{in}, and R_o.

9.9. Refer to Fig. 9.30. For Q_1: $I_{DSS} = 12$ mA and $g_{m0} = 10$ mS. For Q_2: $I_{DSS} = 6$ mA and $g_{m0} = 6$ mS. Given $R_1 = 10$ MΩ, $R_2 = 820$ Ω, $R_4 = 18$ kΩ, $R_5 = 4.7$ kΩ, $R_6 = 3.3$ kΩ, $R_7 = 1.2$ kΩ, $V_{CC} = 15$ V, and $R_L = 100$ Ω, determine the operating point for each transistor, the loaded stage gains, the overall

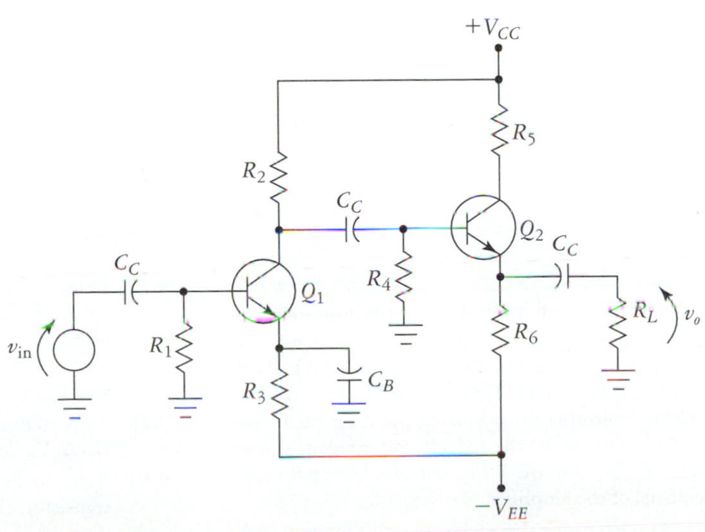

FIGURE 9.28
Circuit for Problems 9.5 and 9.6.

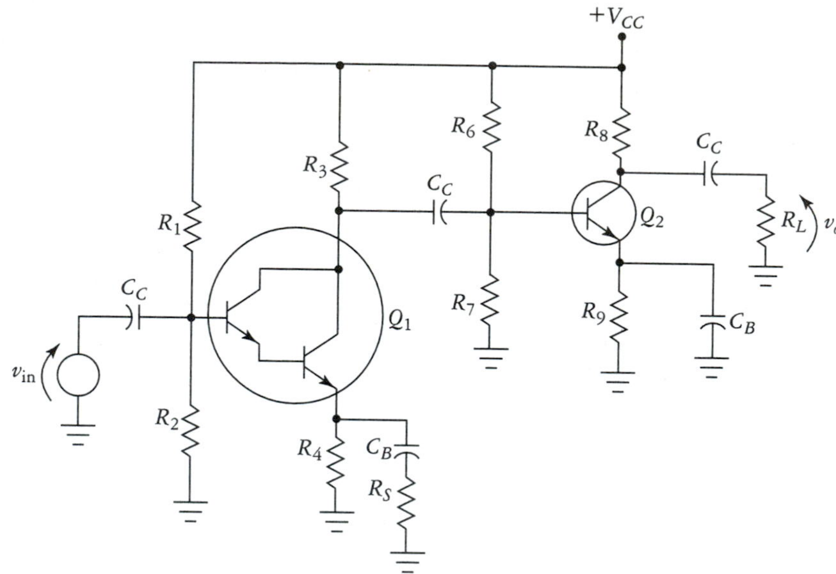

FIGURE 9.29
Circuit for Problems 9.7 and 9.8.

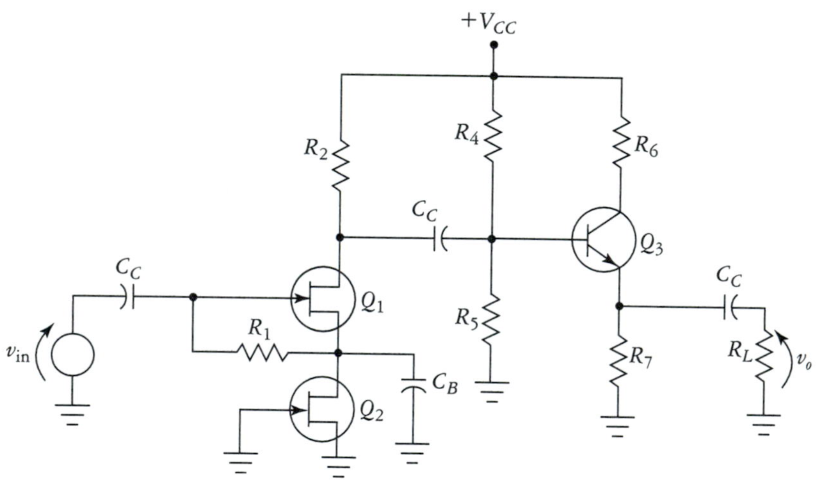

FIGURE 9.30
Circuit for Problems 9.9 and 9.10.

voltage gain (linear and in decibels), R_{in}, R_o, and the actual overall power gain (linear and in decibels) of the amplifier.

9.10. Refer to Fig. 9.30. For Q_1: $I_{DSS} = 8$ mA and $g_{m0} = 8$ mS. For Q_2: $I_{DSS} = 6$ mA and $g_{m0} = 6$ mS. Given $R_1 = 10$ MΩ, $R_2 = 820$ Ω, $R_4 = 18$ kΩ, $R_5 = 10$ kΩ, $R_6 = 100$ Ω, $R_7 = 890$ Ω, and $R_L = 100$ Ω, determine the operating point for each transistor, the loaded stage gains, the overall voltage gain (linear and in decibels), R_{in}, R_o, and the actual overall power gain (linear and in decibels) of the amplifier.

9.11. Refer to Fig. 9.31. Determine the operating point for each transistor, the individual loaded stage gains, and the overall values of A_v, $A_{v(dB)}$, R_{in}, R_o, A_i, A_p, and $A_{p(dB)}$.

9.12. Refer to Fig. 9.32. Determine the operating point for each transistor, the individual loaded stage gains, and the overall values of A_v, $A_{v(dB)}$, R_{in}, R_o, A_i, A_p, and $A_{p(dB)}$. Assume that the FET has $I_{DSS} = 15$ mA and $V_p = -3.5$ V. All BJTs have $\beta = 100$ and $V_{BE} = 0.7$ V.

9.13. Refer to Fig. 9.33. Determine the operating point for each transistor, V_{B_1} and V_{oQ}.

9.14. Refer to Fig. 9.34. Determine the operating points for each transistor, the overall voltage gain A_v, V_{oQ}, and v_o, given $v_{in} = 2 \sin \omega t$ V.

9.15. Refer to Fig. 9.35. Determine the operating points for each transistor and the overall voltage gain of the circuit.

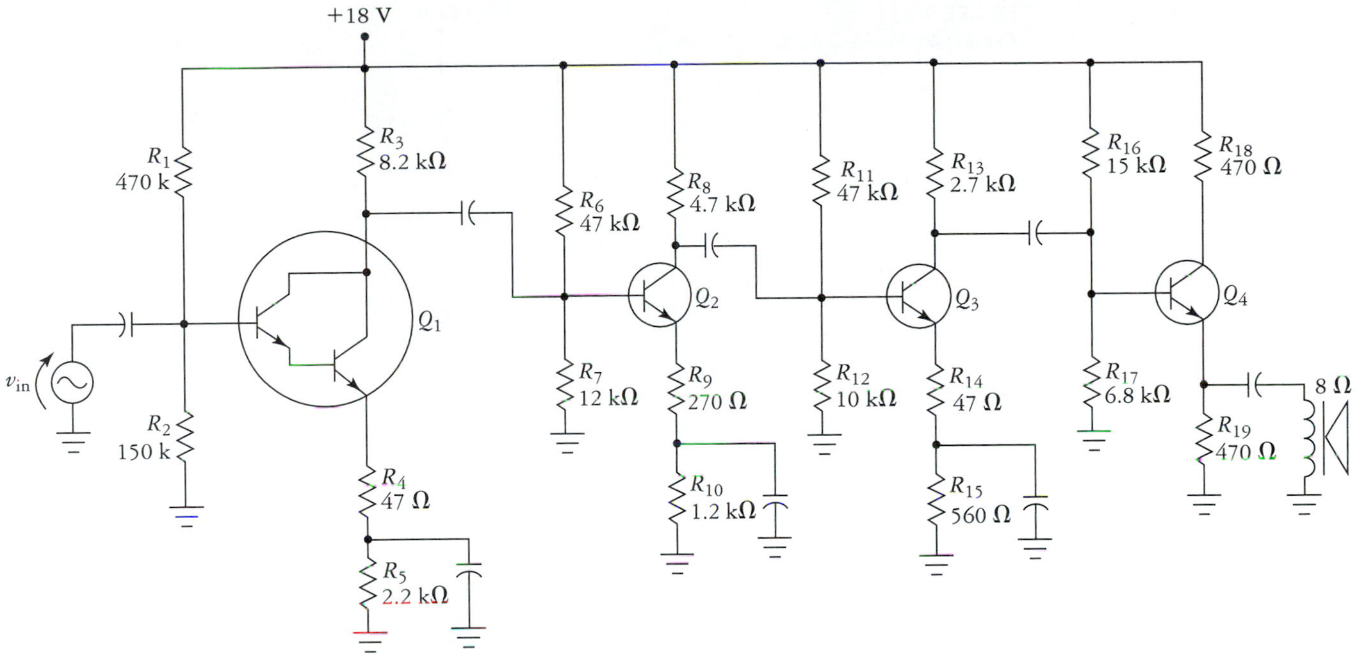

FIGURE 9.31
Circuit for Problem 9.11.

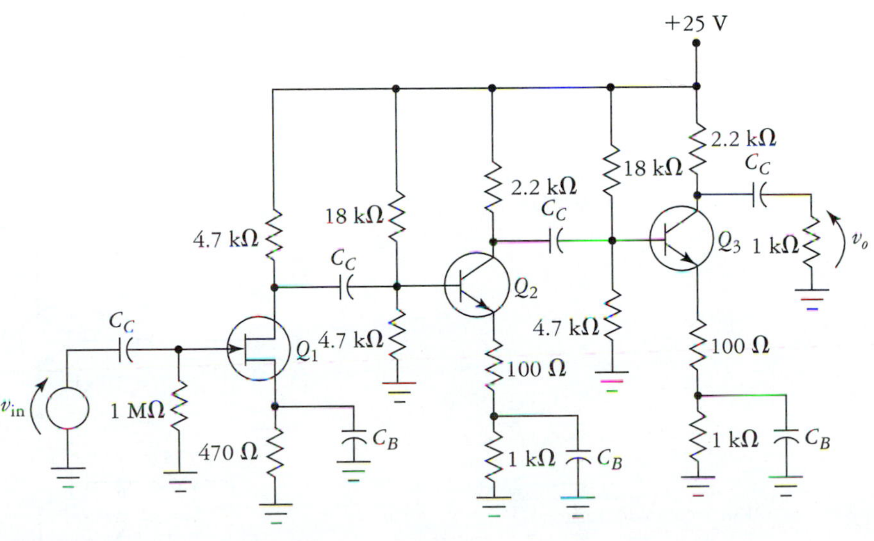

FIGURE 9.32
Circuit for Problem 9.12.

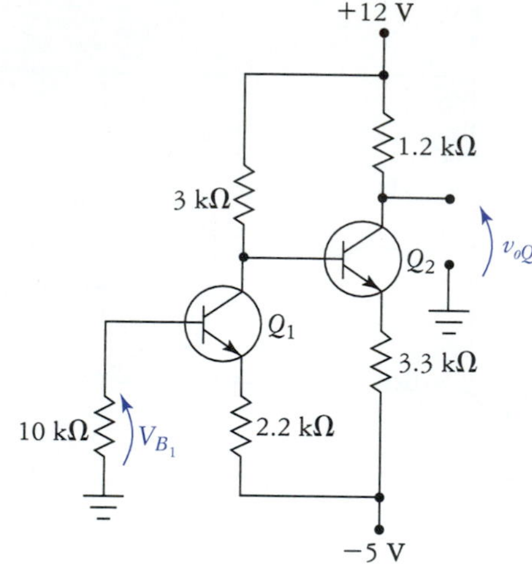

FIGURE 9.33
Circuit for Problem 9.13.

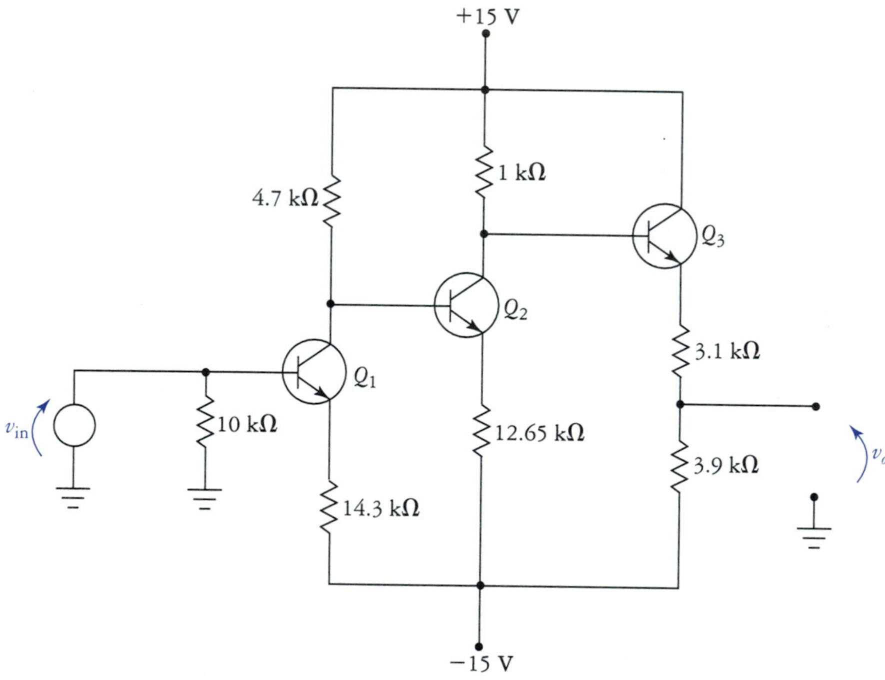

FIGURE 9.34
Circuit for Problem 9.14.

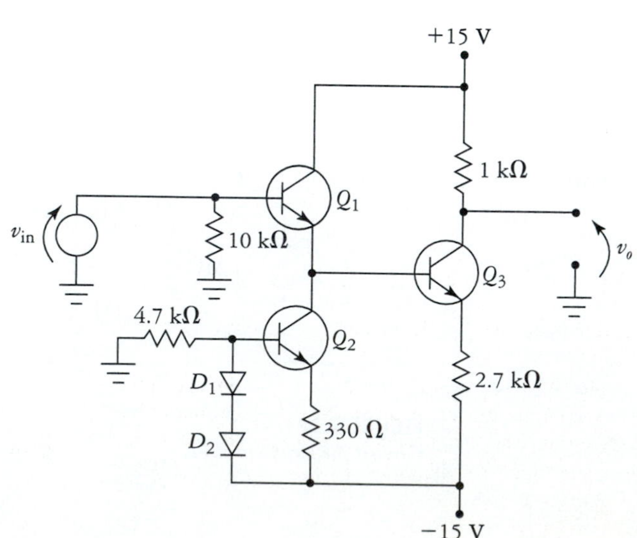

FIGURE 9.35
Circuit for Problem 9.15.

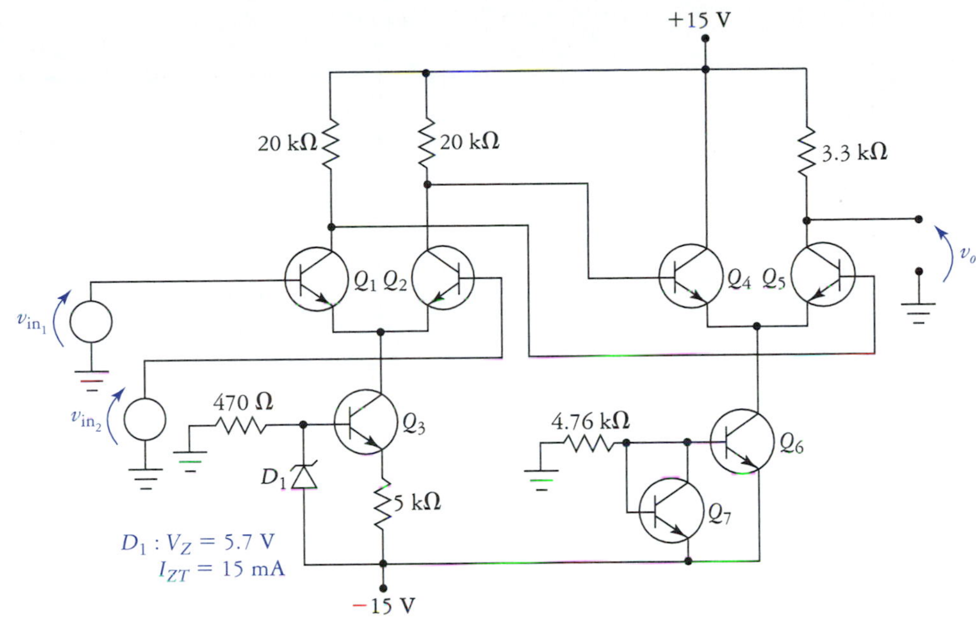

FIGURE 9.36
Circuit for Problem 9.16.

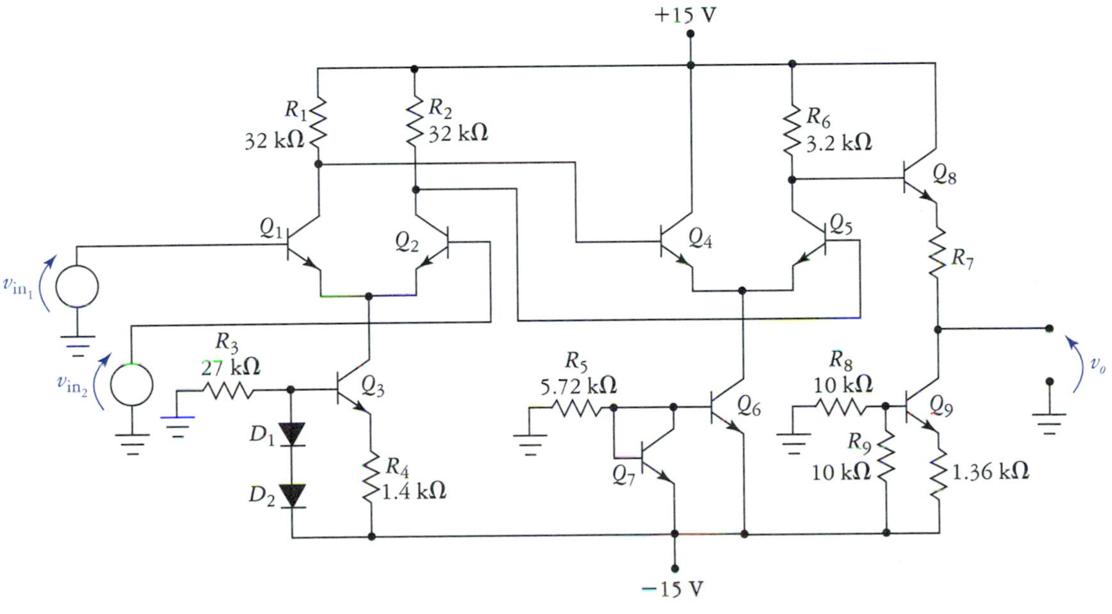

FIGURE 9.37
Circuit for Problems 9.17 and 9.18.

9.16. Refer to Fig. 9.36. Determine the operating points for each transistor, the loaded stage gains, the overall voltage gain A_d (linear and in decibels), R_{in}, R_o, V_{oQ}, and v_o given $v_{in_1} = 0.2 \sin \omega t$ mV, $v_{in_2} = 0$ V.

9.17. Refer to Fig. 9.37. Determine the operating points for each transistor and determine the required value for R_7 such that $V_{oQ} = 0$ V. Determine the overall values of A_d, $A_{d(\text{dB})}$, R_{in}, and R_o.

9.18. In reference to Fig. 9.37, determine v_o for the following input voltages. Assume that CMRR = ∞.

(a) $v_{in_1} = 25$ mV, $v_{in_2} = 22$ mV

(b) $v_{in_1} = 2 \sin 100t$ mV, $v_{in_2} = 2$ mV

(c) $V_{in_1} = 3 \angle 0°$ mV, $V_{in_2} = 3 \angle 30°$ mV

(d) $v_{in_1} = 5 \sin 1000t$ mV, $v_{in_2} = 4 \cos 1000t$ mV

9.19. Refer to Fig. 9.38. Design the level-shifting circuit using npn bipolar transistors. Use a current mirror to establish the circuit Q point.

9.20. Refer to Fig. 9.39. Determine the Q point for each transistor Assume that interstage loading is negligible.

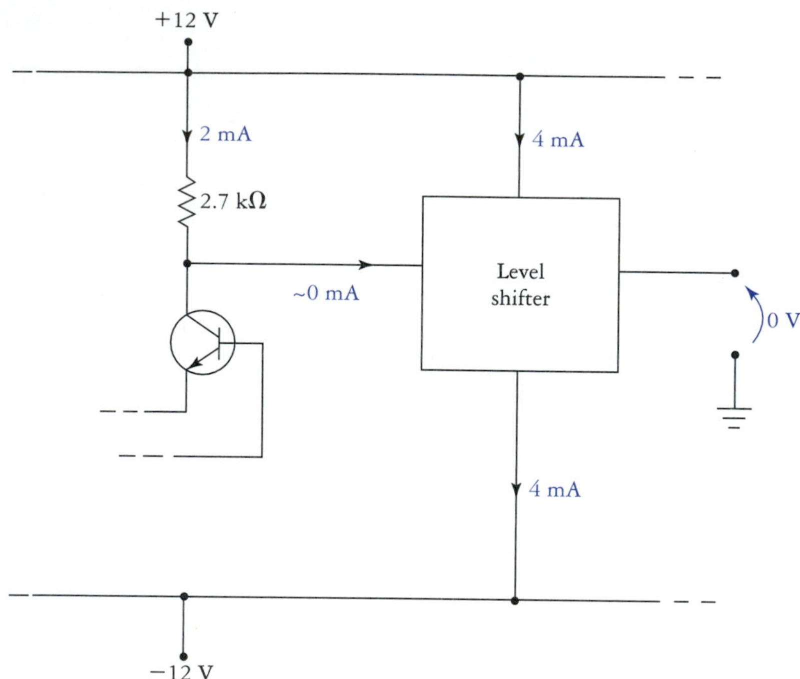

FIGURE 9.38
Circuit for Problem 9.19.

FIGURE 9.39
Circuit for Problem 9.21.

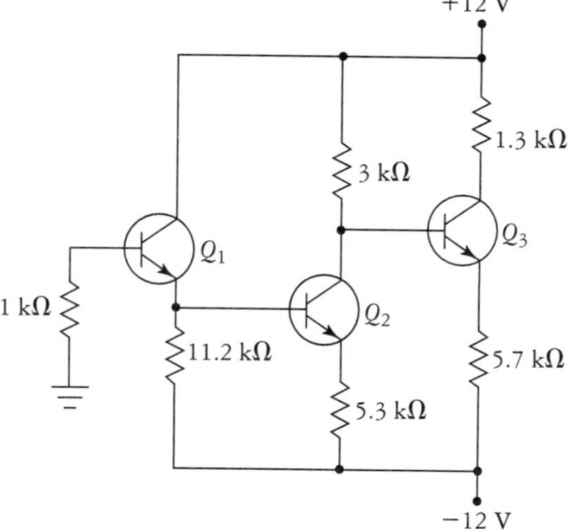

9.21. Refer to Fig. 9.40. Assuming that all transistors have $\beta = 120$ and $V_{BE} = 0.7$ V, determine the Q points for each transistor, the dc voltage that will be present at each terminal of the IC, and the gain v_o/v_s.

9.22. Refer to Fig. 9.41. Assuming that all transistors have $\beta = 150$ and $V_{BE} = 0.7$ V, determine the Q points for each transistor, the dc voltage that will be present at each terminal of the IC, and the gain v_o/v_s.

9.23. Refer to Fig. 9.42. Determine the Q point for each transistor, the zener current I_Z, V_{oQ}, A_d, R_{in}, and R_o. Assume all transistors have $V_A = \infty$.

9.24. Refer to Fig. 9.43. Determine the Q point for each transistor and the current for each diode. Assume all transistors have $V_A = \infty$. Determine V_{oQ}, A_d, R_{in}, and R_o.

TROUBLESHOOTING PROBLEMS

9.25. Refer to Fig. 9.4. Suppose that C_3 became short circuited. Determine the general effect this would have on the operating points of the transistors.

9.26. Assuming that the circuit of Fig. 9.4 is functioning properly, and v_{in} is adjusted such that $V_o = 2\ V_{p\text{-}p}$, what voltage waveform should be observed at the emitter of Q_2 on an oscilloscope?

9.27. Refer to Fig. 9.10. Suppose that an input source with very low internal resistance was connected to the input of the amplifier. Describe the effect this would have on V_{CE_1} and V_{CE_2}.

9.28. Refer to Fig. 9.30. Suppose that Q_2 was replaced with an FET that had the same I_{DSS} as Q_1. Would the amplifier function properly? What effect would this change have on V_{DS} of Q_1? In what region would Q_1 most likely be operating?

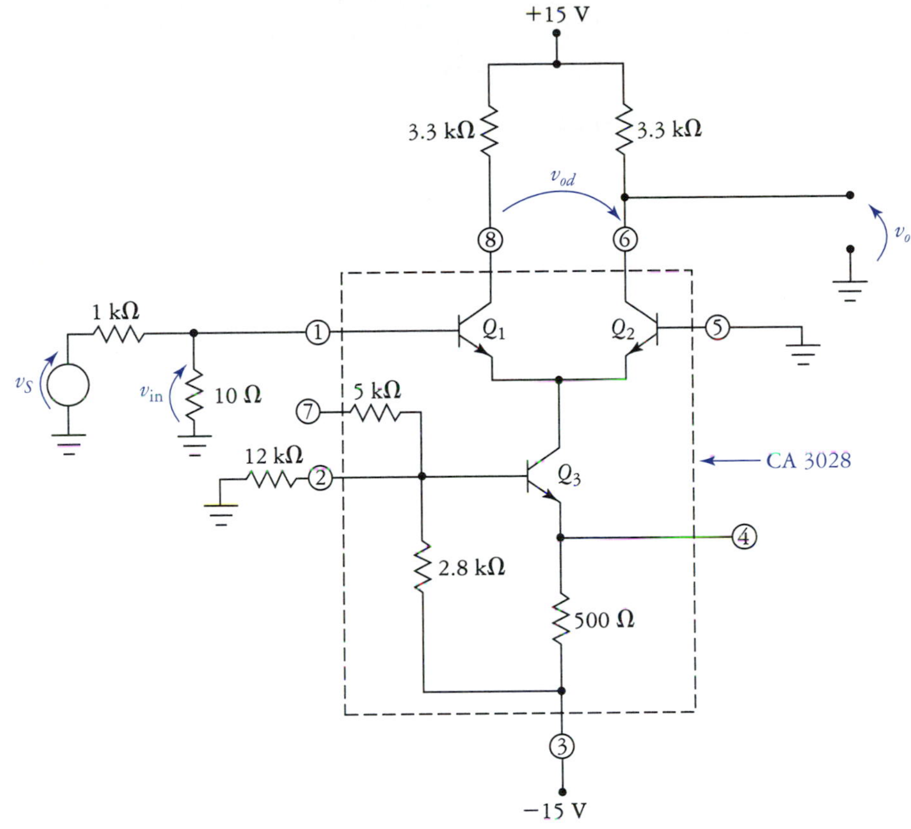

FIGURE 9.40
Circuit for Problem 9.22.

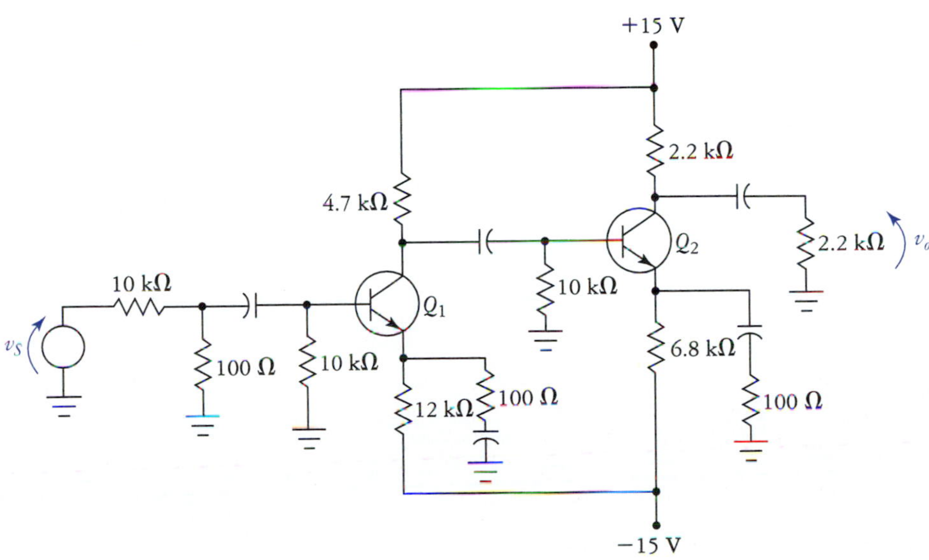

FIGURE 9.41
Circuit for Problem 9.23.

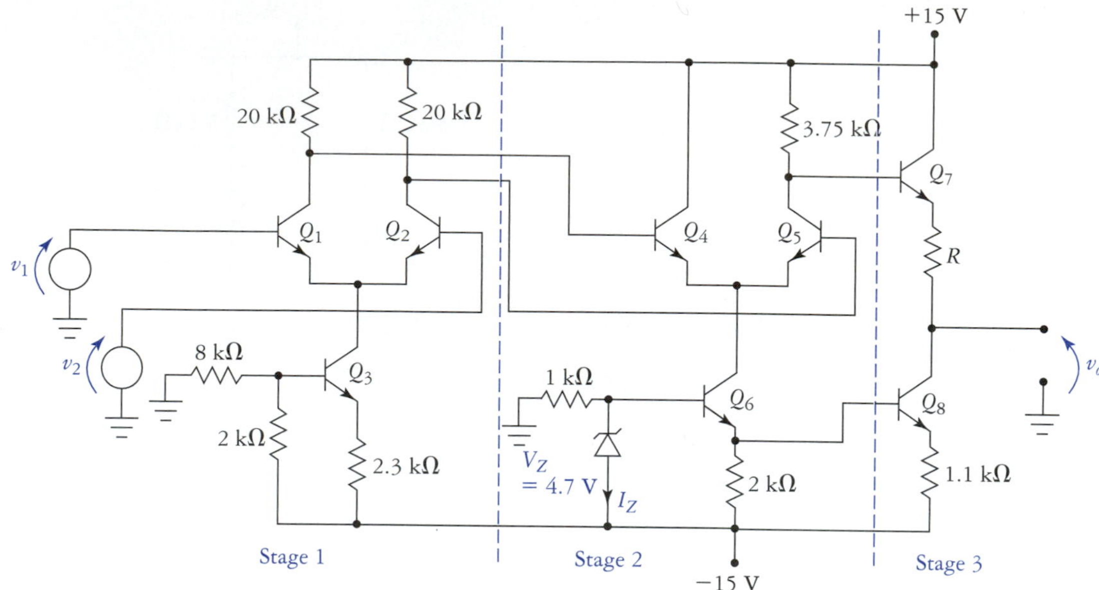

FIGURE 9.42
Circuit for Problem 9.24.

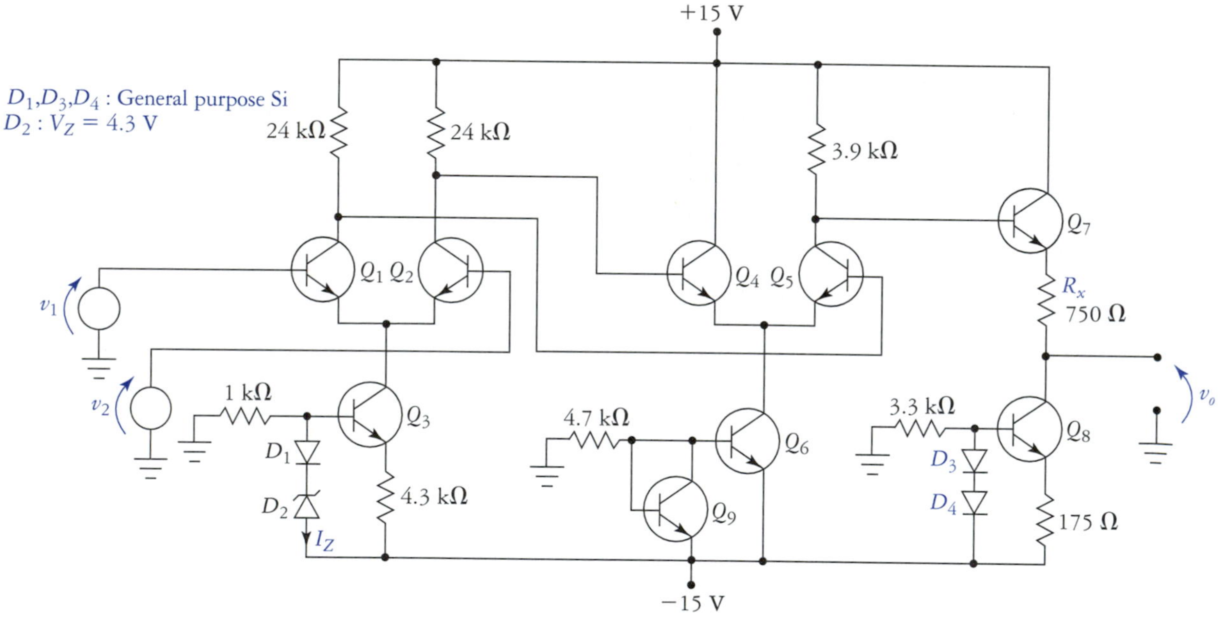

FIGURE 9.43
Circuit for Problem 9.25.

■ COMPUTER PROBLEMS

9.29. Refer to Fig. 9.23. Add a 10-pF capacitor across the C-B junction of transistor Q_4 and simulate the circuit. Does this affect any of the analysis results? If so, which ones are affected and by how much?

9.30. Using PSpice, determine the operating-point information for Fig. 9.35. Use the transfer function option to determine A_v, R_{in}, and R_o for the circuit. Use 2N3904 transistors and 1N4148 diodes.

9.31. Refer to Fig. 9.29. Using the values of Problem 9.8, simulate this circuit and determine the Q points of the transistors and the output voltage, given $v_{in} = 2\ \sin(2\pi 1000t)$ mV. Use 2N3904 transistors in the simulation.

CHAPTER 10

Power Amplifier Fundamentals

BEHAVIORAL OBJECTIVES

On completion of this chapter, the student should be able to:

- Explain the significance of the conduction angle of a transistor.
- Determine the class of operation for which a transistor is biased.
- Discuss the relative efficiency of various classes of amplifiers.
- Explain the function of the complementary-symmetry output stage.
- Identify the quasi-complementary-symmetry configuration.
- Analyze class AB amplifiers.
- Analyze the V_{BE} multiplier circuit.
- Describe crossover distortion and explain how it is reduced.
- Describe the advantages and disadvantages of transformer coupling.

INTRODUCTION

The amplifier circuits discussed to this point are most commonly used in low-power, small-signal applications. Power amplifiers are required in applications where a large amount of energy must be delivered to a load, such as in music reproduction and in control system and servomechanical designs. Because of the special demands placed on power amplifiers, such circuits are normally designed differently than the (small-signal) amplifiers that we have been dealing with so far. This chapter presents a representative sample of the basic power amplifier circuits commonly used today. The fundamental concepts that are of primary concern in power amplifier analysis and design are also presented.

10.1 AMPLIFIER CLASSIFICATIONS

Amplifiers in general can be classified in many different ways. Familiar examples are capacitively coupled and direct-coupled classifications. Amplifiers can also be classified in terms of the frequencies at which they are designed to operate. A few examples of this classification scheme are audio amplifiers, radio-frequency amplifiers, microwave amplifiers, video amplifiers, and wideband amplifiers. We also can differentiate between amplifiers in terms of the input/output variables with which they operate, this gives us the VCVS, VCIS, ICVS, and ICIS designations. An amplifier can also be classified in terms of the power it is designed to deliver to a load. This leads us to the small-signal and power amplifier classifications.

Small-signal amplifiers are designed to amplify rather weak signals (often in the microvolt range or smaller), and they operate at relatively low output power levels (usually $P_o \ll 1$ W). These amplifiers are often quite similar to those that we have looked at previously in this book. Although there is no hard-and-fast rule, recall that a small signal may usually be considered to be a signal that causes a shift of 10% or less in the Q point of an amplifier. Thus, small-signal amplifiers normally are not driven to the limits of their load lines. Power amplifiers (or large-signal amplifiers) are often designed to operate with larger input-signal voltages (V_{in} around 1 $V_{p\text{-}p}$ or so is common), they produce relatively large output power levels, and they are also more likely to be driven to their load-line limits.

A complete amplifier will usually consist of several cascaded stages of amplification. The first few stages are generally small-signal amplifiers, often called *preamplifiers.* These are followed by a driver stage and an output stage, as shown in Fig. 10.1. If we view the overall system as a voltage amplifier, the small-signal stages normally provide the majority of the voltage gain, while the power amplifier stage provides high voltage and current output capability. The driver acts as a transitional stage between the output and the preamplifier sections. Even though it may be considered separately, the driver section is most often an integral part of the power amplifier.

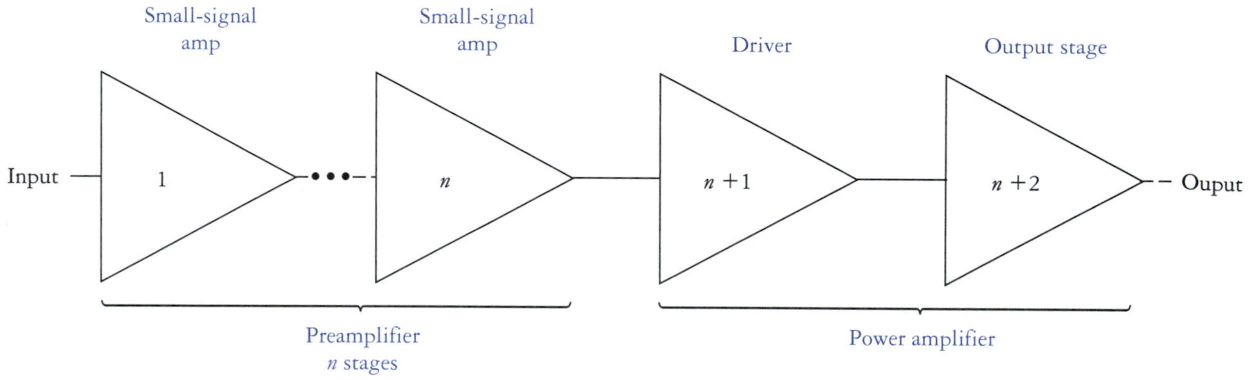

FIGURE 10.1
Stages of a typical high-fidelity audio amplifier.

Conduction Angle

Amplifiers can be described in terms of a parameter called the *conduction angle.* The concept of the conduction angle was introduced in Chapter 2 with regard to rectifier operation. The meaning here is basically the same; the conduction angle of an amplifier is the number of degrees per sine wave input cycle over which the output of the amplifier can continuously vary. All of the amplifier circuits studied in the previous two chapters have had conduction angles of 360°. This means that unless the amplifier is driven into clipping, the output will vary continuously with the input signal.

In terms of conduction angle θ there are four classical amplifier types:

Class A:	$\theta = 360°$
Class B:	$\theta < 180°$ ($\theta \rightarrow 180°$ as signal amplitude increases)
Class AB:	$\theta = 180°$
Class C:	$\theta \ll 180°$

You may have heard of another amplifier category, class D, which is a discrete-time approach to amplification (we are dealing with continuous-time systems here) using a technique called pulse-width modulation. This is a more advanced topic that we will discuss in later chapters. The details concerning the implementation of the various classes of amplifiers and the consequences of the conduction angle variations are presented in the next sections.

10.2 CLASS A AMPLIFIERS

A familiar design that is sometimes used in class A amplifier designs is shown in Fig. 10.2(a). This is a capacitively coupled, common-emitter amplifier with an ac-equivalent circuit as shown in Fig. 10.2(b). Although it is limited in its usefulness as a power amplifier, this is a good place at which to start because the circuit is so familiar.

A typical ac load line for the class A amplifier is shown in Fig. 10.3. Because the class A power amplifier is designed to provide a large output voltage swing, the Q point will be located close to, or preferrably right at the center of, the ac load line as shown. It can be seen in Fig. 10.3 that the output voltage ($v_o = v_{CE}$ for the CE amp with full emitter bypassing) does vary continuously with the input for 360° as required for class A operation.

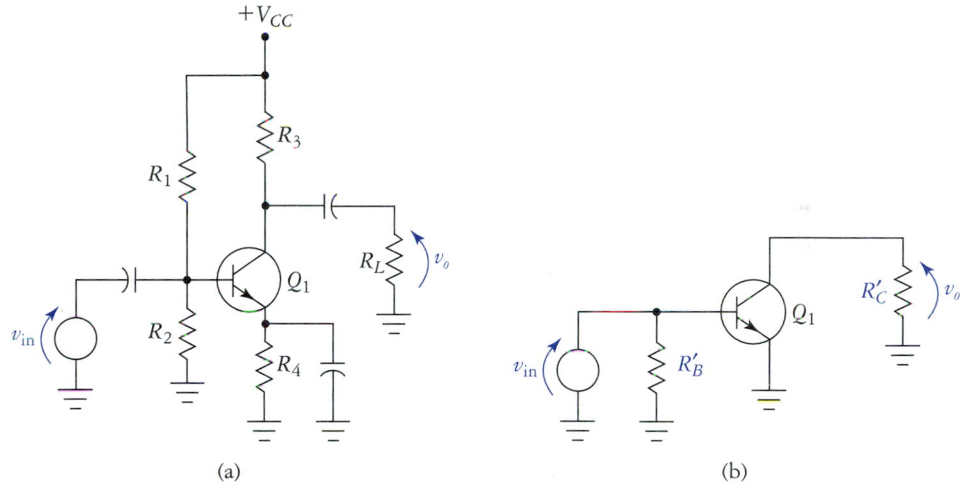

FIGURE 10.2
Class A amplifier. (a) Capacitively coupled, using voltage-divider biasing. (b) AC-equivalent circuit.

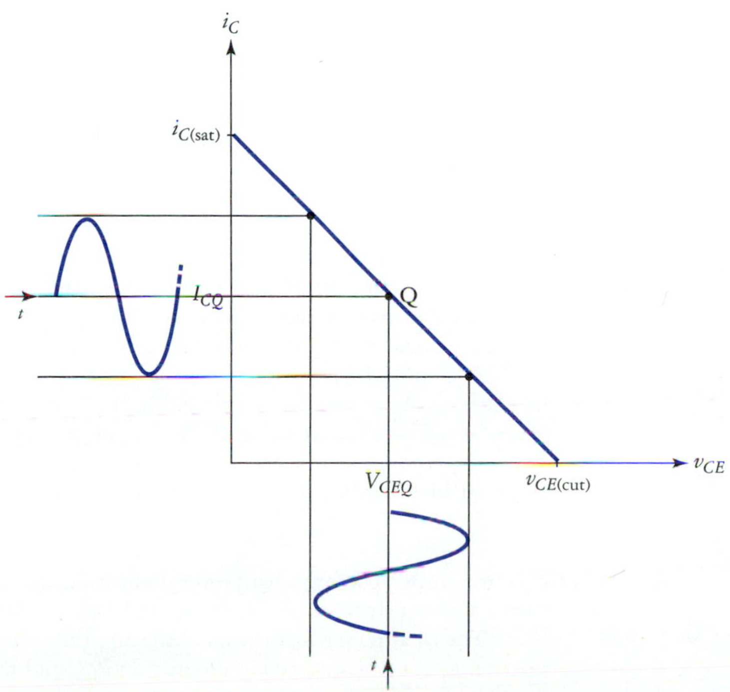

FIGURE 10.3
Effect of a sinusoidal input voltage on Q point.

Amplifier Power Dissipation

The instantaneous power dissipation of the transistor in Fig. 10.2 is given by

$$p_D = i_C v_{CE} \tag{10.1}$$

Under no-signal conditions, the quiescent power dissipation is

$$P_{DQ} = I_{CQ} V_{CEQ} \tag{10.2}$$

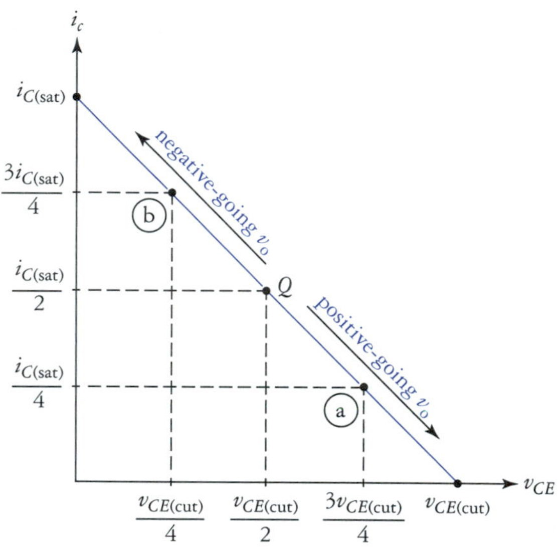

FIGURE 10.4
Q-point movement in relation to input signal polarity.

Let's examine the significance of these relationships. Consider the ac load line of Fig. 10.4. The Q point is centered, and the quiescent power dissipation is P_{DQ} watts. Application of an input signal causes the Q point to move back and forth along the load line. Halfway to cutoff or saturation [points *a* and *b* in Fig. 10.4] the power dissipation is

$$P_{D(a)} = P_{D(b)} = 0.75P_{DQ}$$

Assuming that the transistor behaves nearly ideally, that is $i_{C(cut)} = 0$ A and $v_{CE(sat)} = 0$ V, then at the limits of the load line, we find

$$P_{D(sat)} = P_{D(cut)} = 0 \text{ W}$$

The important conclusion that can be drawn here is that the power dissipation of a class A amplifier is greatest under no-signal conditions, at the Q point. The transistor power dissipation decreases as an input signal drives the transistor toward cutoff or saturation. This leads to the interesting characteristic that class A amplifiers will run hottest when not producing an output signal. When a class A amplifier is being driven to the verge of clipping, the output transistor will, on the average, be dissipating the least amount of power. Most home stereo amplifiers (but not all) behave much differently than this; they usually run hotter as the loudness of the music is increased. This is a good indication that the typical stereo power amplifier is not a class A design.

A graphical representation of the power dissipation of the class A transistor over one cycle of the output voltage signal is shown in Fig. 10.5. Notice that the instantaneous transistor power dissipation varies at twice the rate of the output voltage. Working with a sinusoidal input signal that drives the circuit to the verge of clipping, the average power dissipation of the transistor in a class A amplifier like that in Fig. 10.2 is found by integrating Eq. (10.1) over one cycle, giving us

$$P_{D(ave)}\big|_{v_o = \max} = 0.875P_{DQ} \tag{10.3}$$

It is important to know what power dissipation the transistor is expected to handle so that a device with a sufficiently high power rating can be chosen for the design.

The total power dissipation of the amplifier circuit P_{CC} is the sum of the transistor power dissipation P_D, the biasing power dissipation P_{bias} and the power dissipation of the collector resistance P_{RC}. The dissipation of emitter resistance P_{R_E} is also accounted for, if the emitter is not fully bypassed. The power dissipation of the biasing network should be very low compared to the other terms, and so it may be neglected. This allows us to express the *total power dissipation* of the class A amplifier with a single-polarity power supply as

$$P_{CC} = I_{CQ}V_{CC} \tag{10.4}$$

If a bipolar power supply is used, then we have

$$P_{CC} = I_{CQ}(V_{CC} + V_{EE}) \tag{10.5}$$

FIGURE 10.5
Amplifier output voltage and power dissipation waveforms.

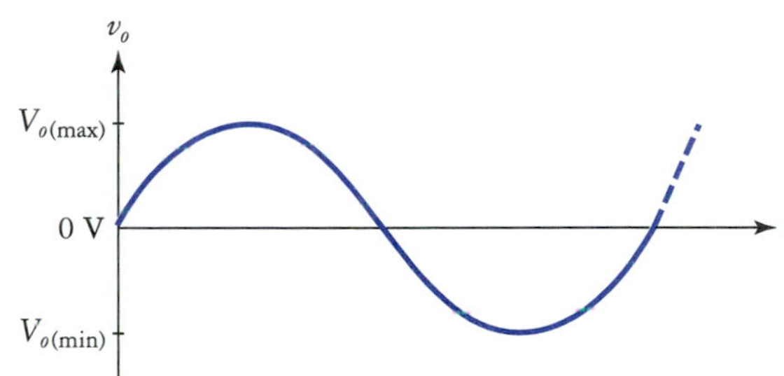

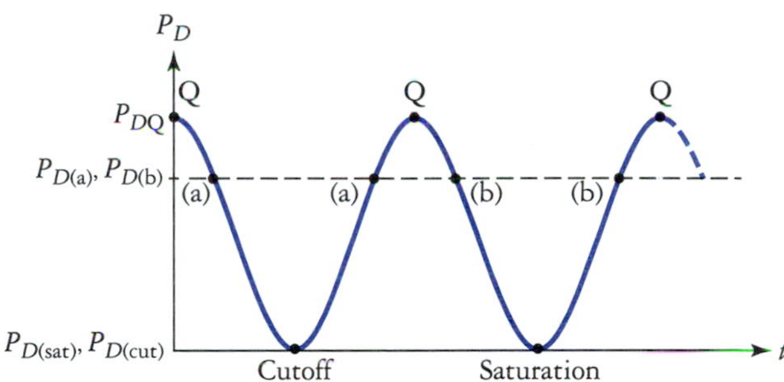

Note
The transistor power dissipation waveform is a $\sin^2$ function, so its frequency is twice that of the output voltage signal.

Load Power Dissipation

The power delivered to the load by an amplifier is also a very important consideration. Using a sinusoidal output signal, the average output power delivered to a resistive load is given by

$$P_{o(\text{ave})} = \frac{V_{o(\text{rms})}^2}{R_L} \tag{10.6}$$

It is sometimes more convenient to work with peak or peak-to-peak output voltage. Using the peak output voltage V_p we can express Eq. (10.4) as

$$P_{o(\text{ave})} = \frac{(0.707V_p)^2}{R_L}$$

which reduces to the more convenient form

$$P_{o(\text{ave})} = \frac{V_p^2}{2R_L} \tag{10.7}$$

Note
For a sine wave, $P_{\text{ave}} = P_{\text{rms}}$

In terms of peak-to-peak output voltage $V_{\text{p-p}}$ we have

$$P_{o(\text{ave})} = \frac{(0.354\ V_{\text{p-p}})^2}{R_L}$$

This also reduces to the more easily used form

$$P_{o(\text{ave})} = \frac{V_{\text{p-p}}^2}{8R_L} \tag{10.8}$$

The greatest *average* output power is produced when the amplifier is producing the maximum (unclipped) output voltage. In terms of the maximum peak-to-peak output voltage this is

$$P_{o(\text{ave})} = \frac{V_{\text{p-p(max)}}^2}{8R_L} \tag{10.9}$$

EXAMPLE 10.1

Refer to Fig. 10.6. Given $\beta = 100$, determine the operating point, sketch the ac and dc load lines, calculate the maximum unclipped output voltage, and determine P_{DQ}, P_{CC}, and $P_{o(max)}$.

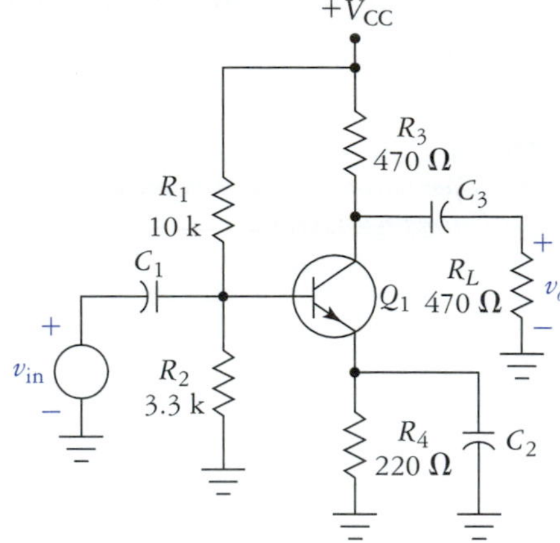

FIGURE 10.6
Circuit for Example 10.1.

Solution The Thevenin voltage and resistance of the base-biasing network are found to be

$$R_{Th} = 10\text{ k}\Omega \parallel 3.3\text{ k}\Omega = 2.5\text{ k}\Omega$$

$$V_{Th} = V_{CC}\frac{R_2}{R_1 + R_2} = 25\text{ V}\frac{3.3\text{ k}\Omega}{10\text{ k}\Omega + 3.3\text{ k}\Omega} = 6.2\text{ V}$$

The operating point is determined next:

$$I_{CQ} \cong \frac{V_{Th} - V_{BE}}{R_E} = \frac{6.2\text{ V} - 0.7\text{ V}}{220\ \Omega} = 25\text{ mA}$$

$$V_{CEQ} = V_{CC} - I_C(R_C + R_E) = 25\text{ V} - 25\text{ mA}(470\ \Omega + 220\ \Omega) = 7.8\text{ V}$$

The limits of the dc load line are

$$V_{CE(cut)} = V_{CC} = 25\text{ V}$$

$$I_{C(sat)} = \frac{V_{CC}}{R_C + R_E} = 36\text{ mA}$$

The limits of the ac load line are found using Eqs. (5.57) and (5.58):

$$v_{CE(cut)} = V_{CEQ} + I_{CQ}R_C' = 7.8\text{ V} + (25\text{ mA})(235\ \Omega) = 13.7\text{ V}$$

$$i_{C(sat)} = I_{CQ} + \frac{V_{CEQ}}{R_C'} = 25\text{ mA} + \frac{7.8\text{ V}}{235\ \Omega} = 58\text{ mA}$$

The load lines are shown in Fig. 10.7.

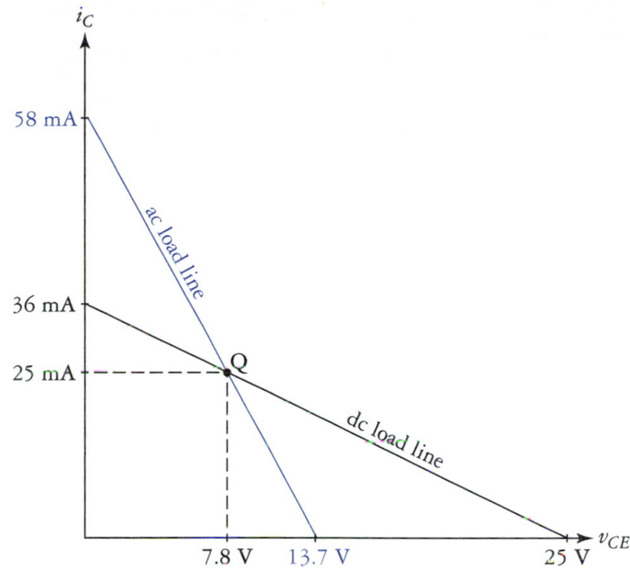

FIGURE 10.7
Load lines for Fig. 10.6.

Clipping of the output signal will occur first when the transistor is driven toward cutoff (a positive-going output voltage swing), which has a maximum amplitude of 5.9 V. The negative-going portion of the output signal could decrease as far as −7.8 V before clipping occurs. Thus, the maximum peak-to-peak unclipped output voltage is set by the positive-going output limit giving us

$$\begin{aligned} V_{\text{p-p(max)}} &= (2)(5.9\text{ V}) \\ &= 11.8\text{ V} \end{aligned}$$

The quiescent power dissipation of the transistor is now determined:

$$\begin{aligned} P_{DQ} &= I_{CQ}V_{CEQ} \\ &= (25\text{ mA})(7.8\text{ V}) \\ &= 195\text{ mW} \end{aligned}$$

The total average power dissipated by the amplifier is

$$\begin{aligned} P_{CC} &= I_{CQ}V_{CC} \\ &= (25\text{ mA})(25\text{ V}) \\ &= 625\text{ mW} \end{aligned}$$

The maximum average power delivered to the load is found to be

$$\begin{aligned} P_{o(\text{max})} &= \frac{V^2_{\text{p-p(max)}}}{8R_L} \\ &= \frac{(139.2\text{ V})^2}{3760\ \Omega} \\ &= 37\text{ mW} \end{aligned}$$

Efficiency of the Capacitively Coupled Class A Amplifier

Efficiency is a major concern in power amplifier design. The efficiency η of an amplifier, expressed as a percentage, is defined by

$$\eta = \frac{P_{o(\text{max})}}{P_{\text{in(dc)}}} \times 100\% \tag{10.10}$$

where $P_{\text{in(dc)}}$ is the average power dissipated by the amplifier [the transistor(s) and associated biasing components] and $P_{o(\text{max})}$ is the average power that is delivered to the load, under maximum output signal conditions. For a single-supply class A amplifier like that of Fig. 10.2, we have

$$P_{\text{in(dc)}} \cong P_{CC} = I_{CQ}V_{CC} \tag{10.11}$$

Note
Although very inefficient, class-A amplifiers are capable of delivering very low distortion and good transient response.

If a bipolar power supply were used, we would have

$$P_{\text{in(dc)}} \cong P_{CC} = I_{CQ}(V_{CC} + V_{EE}) \tag{10.12}$$

Do not confuse Eq. (10.10) with $A_p = p_o/p_{\text{in}}$, which is the power-gain equation. Power gain is the ratio of output power to power delivered to the amplifier by the input signal source. In Eq. (10.10), we are dealing with power dissipation that is caused by energy that is taken from the power supply (V_{CC} and/or V_{EE}) and power dissipated by the load.

As it turns out, class A amplifiers are the least efficient of all of amplifier configurations. The maximum theoretical efficiency of a capacitively coupled class A amplifier is

$$\eta_{\max} = 25\%$$

In practice, the actual efficiency of a given amplifier (class A or otherwise) will often be much less than the theoretical maximum. The next example illustrates this point.

EXAMPLE 10.2

Determine the efficiency of the amplifier shown in Fig. 10.6.

Solution This circuit was analyzed in Example 10.1. The pertinent results of that analysis are presented again for convenience:

$$\begin{aligned} P_{\text{in(dc)}} = P_{CC} &= I_{CQ}V_{CC} \\ &= (25\text{ mA})(25\text{ V}) \\ &= 625\text{ mW} \end{aligned}$$

$$\begin{aligned} P_{o(\max)} &= \frac{V^2_{\text{p-p(max)}}}{8R_L} \\ &= \frac{(11.8\text{ V})^2}{3760\ \Omega} \\ &= 37\text{ mW} \end{aligned}$$

The efficiency is now determined:

$$\begin{aligned} \eta &= \frac{P_{o(\max)}}{P_{\text{in(dc)}}} \times 100\% \\ &= \frac{37\text{ mW}}{625\text{ mW}} \times 100\% \\ &= 5.92\% \end{aligned}$$

Transformer-Coupled Class A Amplifier*

For audio frequency power amplifier applications, direct coupling and capacitive coupling are most commonly used. Less commonly encountered in this area of application (although very popular in tuned radio-frequency amplifier designs) is transformer coupling. A transformer-coupled, class A, common-emitter amplifier is shown in Fig. 10.8. Notice that an iron-core transformer is used to couple the load to the collector. This is typical for audio frequency applications, but note that air and ferrite cores are normally used at higher frequencies.

Advantages and Disadvantages of Transformer Coupling

Various advantages and disadvantages are associated with the use of transformer coupling. In particular, for low-frequency class A power amplifier applications, the primary disadvantages follow:

1. Low-frequency response is limited ($f_{\text{sig}} > 0$ Hz.)
2. Iron-core transformers generally have poor transient and high-frequency response.
3. Iron-core transformers tend to be large and heavy, essentially eliminating the possibility of miniaturization, as in integrated circuit designs.
4. Iron-core transformers are relatively expensive.
5. Without shielding, transformers are relatively susceptible to noise.

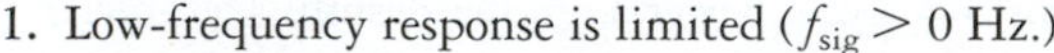

*If time is limited, this material may be skipped without loss of continuity.

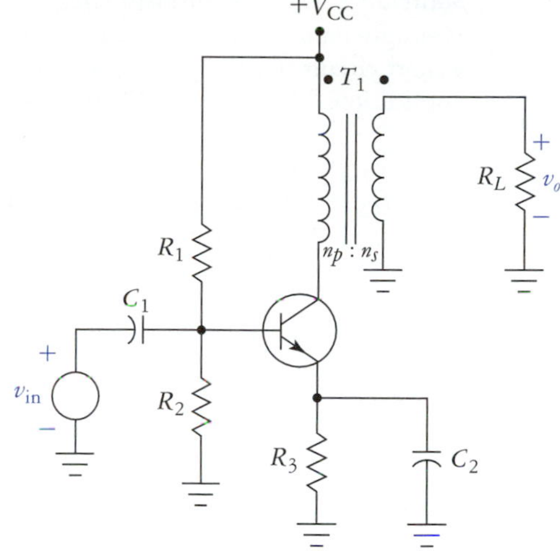

FIGURE 10.8
Transformer-coupled class A amplifier.

Some of the main advantages of transformer coupling are listed here:

1. Impedance matching is easily accomplished.
2. Amplifier efficiency is increased.
3. Phase inversion can easily be produced or eliminated.
4. The load can be electrically isolated from the amplifier, reducing the potential for shock hazards (especially in biomedical electronic devices).

Note
Transformer-coupled class-A amplifiers can approach 50% efficiency.

Modern low-frequency and audio amplifier designs tend to be transformerless, but the increased efficiency and electrical isolation do outweigh the disadvantages in some instances. It can be shown that the theoretical maximum efficiency of a transformer-coupled, class A amplifier is $\eta_{max} = 50\%$. This is twice as good as can be achieved using capacitive coupling alone.

Much of the analysis of a transformer-coupled amplifier, like that shown in Fig. 10.8, is the same as would be performed on a similar capacitively coupled amplifier. As usual, we begin with the determination of the amplifier Q point; then the ac analysis is performed. The details of such an analysis are best shown through an example.

EXAMPLE 10.3

Refer to Fig. 10.9. Determine the operating point, the dc load line, R_{in}, A_v, A_i, A_p, ac load line, the maximum unclipped output voltage swing, and the maximum efficiency. Assume $\beta = 100$, the transformer winding resistances are negligible ($r_p = r_s = 0\ \Omega$), and the coupling capacitors are ideal.

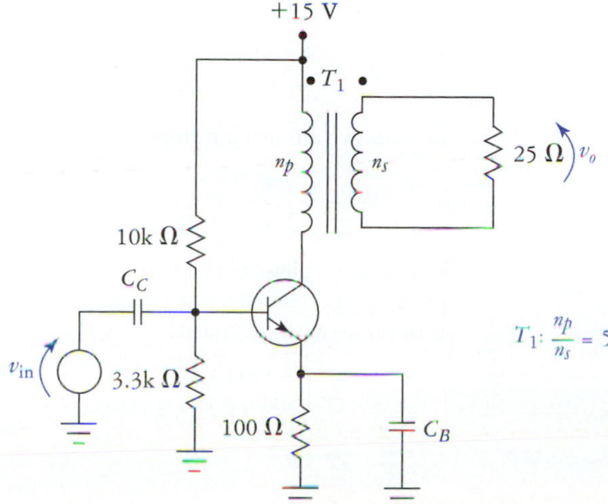

FIGURE 10.9
Circuit for Example 10.3.

Solution The dc-equivalent circuit used for this analysis is shown in Fig. 10.10. The transformer is simply replaced with its primary winding resistance r_p, which is small enough to be considered a short circuit in this application (the dc winding resistance is more important in tuned amplifier applications, the subject of Chapter 16). Routine analysis of the dc equivalent circuit yields

$$V_{Th} = 3.72 \text{ V} \qquad R_{Th} = 2.48 \text{ k}\Omega$$
$$I_{CQ} = 24.1 \text{ mA} \qquad V_{CEQ} = 12.6 \text{ V}$$
$$I_{C(sat)} = 24.1 \text{ mA} \qquad V_{CE(cut)} = 15 \text{ V}$$
$$r_e = 1.1\ \Omega$$

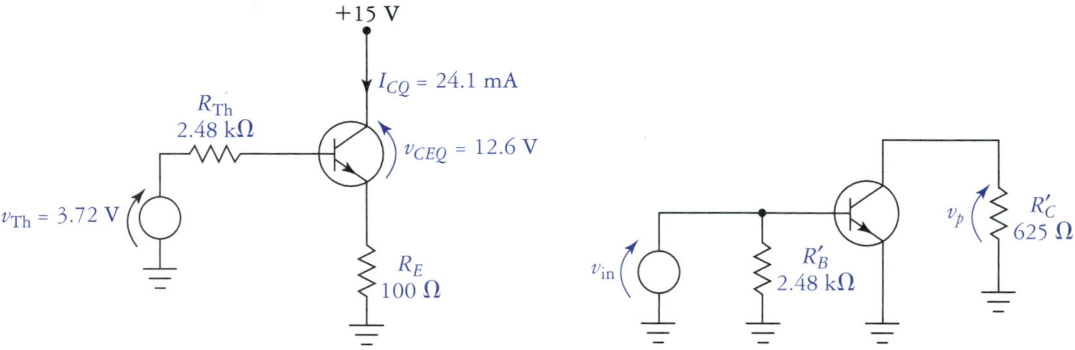

FIGURE 10.10
DC-equivalent circuit with base network Thevenized.

FIGURE 10.11
AC-equivalent circuit.

To form the ac-equivalent circuit, the effective collector resistance must be determined. The reflected load resistance R'_L is determined by

$$\begin{aligned} R'_L &= R_L\left(\frac{n_p}{n_s}\right)^2 \\ &= 25\ \Omega\left(\frac{5}{1}\right)^2 \\ &= 25\ \Omega \times 25 \\ &= 625\ \Omega \end{aligned} \tag{10.13}$$

The ac-equivalent collector resistance is

$$\begin{aligned} R'_C &\cong R'_L \\ &= 625\ \Omega \end{aligned}$$

The ac-equivalent circuit is shown in Fig. 10.11.

Let's determine the voltage gain of the circuit. Up until now, the A_v equations we have used have been derived in terms of the collector voltage and the input voltage. This apparent voltage-gain magnitude will be denoted as A'_v in this discussion; thus

$$A'_v = \frac{R'_C}{r_e}$$

Because we are using a step-down transformer in this application the actual voltage gain is

$$A_v = A'_v\frac{n_p}{n_s} \tag{10.14}$$

The relative phase of the output voltage will depend on the orientation of the phasing dots on the primary and secondary windings of the transformer. The circuit in Fig. 10.9 will produce v_o that is in phase with v_{in}. Substituting the values obtained in the ac-equivalent circuit into Eq. (10.14),

$$\begin{aligned} A_v &= \left(\frac{625\ \Omega}{1.1\ \Omega}\right)\left(\frac{1}{5}\right) \\ &= (568.2)(0.2) \\ &= 114 \end{aligned}$$

The input resistance of the circuit is

$$\begin{aligned} R_{\text{in}} &= R'_B \parallel \beta r_e \\ &= 2.48\ \text{k}\Omega \parallel 110\ \Omega \\ &= 105\ \Omega \end{aligned}$$

The current gain of the amplifier is found as usual, using

$$\begin{aligned} A_i &= A_v \frac{R_{\text{in}}}{R_L} \\ &= 114\left(\frac{105\ \Omega}{25\ \Omega}\right) \\ &= 479 \end{aligned}$$

This is an interesting result. For a capacitively coupled amplifier we must have $A_i < \beta$; however, transformer coupling allows $A_i > \beta$. The power gain is found using the familiar relation

$$\begin{aligned} A_p &= A_v A_i \\ &= (114)(479) \\ &= 54{,}606 \end{aligned}$$

This is based on the assumption that the transformer is lossless, which will normally be a valid approximation.

The limits of the ac load line are now found:

$$\begin{aligned} i_{C(\text{sat})} &= I_{CQ} + \frac{V_{CEQ}}{R'_C} \\ &= 24.1\ \text{mA} + \frac{12.6\ \text{V}}{625\ \Omega} \\ &= 24.1\ \text{mA} + 20.2\ \text{mA} \\ &= 44.3\ \text{mA} \end{aligned}$$

$$\begin{aligned} v_{CE(\text{cut})} &= V_{CEQ} + I_{CQ}R'_C \\ &= 12.6\ \text{V} + 15.1\ \text{V} \\ &= 27.7\ \text{V} \end{aligned}$$

We have generated another interesting result. That is, the peak value of V_{CE} (27.7 V) is greater than V_{CC} (15 V). This can occur because as the Q point moves toward cutoff, v_{CE} increases toward V_{CC}. However, at the same time i_C, and hence the primary current of T_1, is decreasing. The resulting collapse of the magnetic field around the primary winding of the transformer generates a voltage that is series aiding with the V_{CC} source. In this manner it is possible to have $V_{CE} > V_{CC}$. A plot of the resulting dc and ac load lines is shown in Fig. 10.12. Notice that the Q point is close to the center of the ac load line. In many cases, the dc load line of this type of amplifier will be nearly vertical, making scaling of the load-line plot rather difficult.

The maximum output voltage swing for the amplifier is given as twice the value of the smaller of the following two quantities:

$$V_{o(\text{max})} = (v_{CE(\text{cut})} - V_{CEQ})\left(\frac{n_s}{n_p}\right)$$

$$|V_{o(\text{min})}| = V_{CEQ}\left(\frac{n_s}{n_p}\right)$$

For our circuit we have

$$\begin{aligned} V_{o(\text{max})} &= (15.1\ \text{V})(0.2) \\ &= 3.02\ \text{V} \end{aligned}$$

$$\begin{aligned} |V_{o(\text{min})}| &= (12.6\ \text{V})(0.2) \\ &= 2.52\ \text{V} \end{aligned}$$

thus

$$\begin{aligned} V_{\text{p-p(max)}} &= 2|V_{o(\text{min})}| \\ &= (2)(2.52\ \text{V}) \\ &= 5.04\ \text{V}_{\text{p-p}} \end{aligned}$$

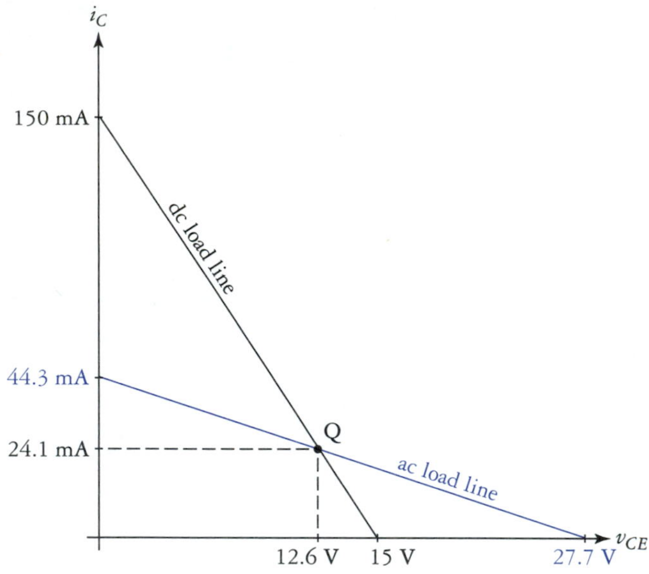

FIGURE 10.12
Load lines for the circuit of Fig. 10.9.

To determine the efficiency of the amplifier, we must determine the maximum load power. For an unclipped output signal we have

$$\begin{aligned} P_{o(\max)} &= \frac{V_{\text{p-p(max)}}^2}{8R_L} \\ &= \frac{(5.04\ \text{V})^2}{8 \times 25\ \Omega} \\ &= 127\ \text{mW} \end{aligned}$$

The power dissipated by the amplifier is

$$\begin{aligned} P_{\text{in(dc)}} = P_{CC} &= I_{CQ}V_{CC} \\ &= (24.1\ \text{mA})(15\ \text{V}) \\ &= 362\ \text{mW} \end{aligned}$$

The efficiency of the amplifier is now found to be

$$\begin{aligned} \eta &= \frac{P_{o(\max)}}{P_{\text{in(dc)}}} \times 100\% \\ &= \frac{127\ \text{mW}}{362\ \text{mW}} \times 100\% \\ &= 35.1\% \end{aligned}$$

This is less than the theoretical maximum efficiency, but much better than could be obtained using capacitive coupling alone.

The preceding example points out some of the more interesting aspects of transformer coupling, and provides some background for the later study of tuned amplifiers. Most applications, however, utilize other power amplifier designs, which are discussed next.

10.3 CLASS B AND CLASS AB AMPLIFIERS: BIASING CONSIDERATIONS

The main disadvantage of the class A amplifier is its relatively low efficiency. It can be shown that a class B amplifier can have an efficiency approaching 79%. This efficiency is achieved at the expense of having a conduction angle that is somewhat less than 180°. Class B amplifiers are normally designed using emitter-followers, as shown in the ac-equivalent circuit of Fig. 10.13(a). This provides another advantage. That is, since power amplifiers are often required to drive low-impedance loads (usually less than 10 Ω), the low output resistance of the emitter-follower configuration provides better power transfer and impedance matching under such conditions.

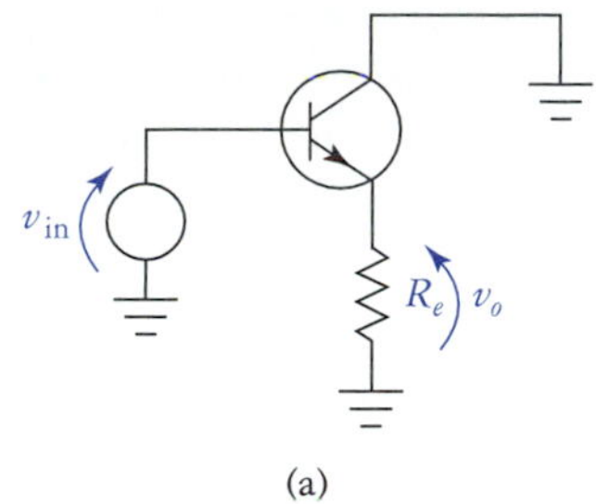

(a)

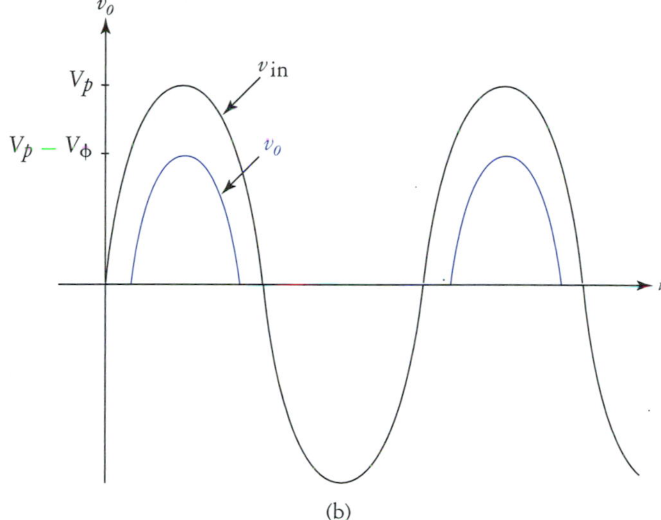

(b)

FIGURE 10.13
Class B amplifier. (a) Possible circuit configuration. (b) Input and output voltage waveforms.

In terms of dc operation, the key characteristic of the class B amplifier is that no bias is applied to the transistor. Using an npn-based circuit as shown in Fig. 10.13(a), the transistor produces an output only for positive-going input swings. During the negative-going portions of v_{in} the transistor is driven into cutoff, which results in $v_o = 0$ V. Representative input and output voltage waveforms are shown in Fig. 10.13(b). Notice that the transistor is not driven into conduction until v_{in} exceeds the barrier potential of the B-E junction. The fact that the class B amplifier is an emitter-follower, plus the barrier potential loss, accounts for the output voltage being smaller than the input.

Complementary-Symmetry Amplifiers

The high distortion that is present in the output signal of Fig. 10.13(b) prevents single-transistor class B amplifiers from being very useful in most applications. A more practical circuit is shown in Fig. 10.14(a). This circuit is commonly referred to as a *class B push-pull* or *complementary-symmetry* amplifier. As the name implies, transistors Q_1 and Q_2 are complementary devices; that is, npn and pnp transistors of similar electrical and temperature characteristics. These transistors are also mounted such that there is good thermal coupling between them as well. The more closely the complementary transistors are matched, the more effective the operation of the amplifier.

The operation of the class B push-pull amplifier is described as follows [refer to Fig. 10.14(a)]: Let us assume that no signal is present yet, and the input and output coupling capacitors have charged to $V_{CC}/2$. Because the transistors are in cutoff (with $I_{CQ} = 0$ A), this charging could take a long time; however, once a signal is applied, the capacitors will charge quickly such that the average voltage drop across them is $V_{CC}/2$. This action is fundamental to the basic operation of Fig. 10.14(a).

Let's assume that v_{in} now begins going positive. This tends to forward bias the B-E junction of Q_1, while at the same time reverse biasing the B-E junction of Q_2. Once v_{in} exceeds the B-E barrier potential of Q_1 (V_ϕ or $V_{BE(on)}$), the output coupling capacitor begins to charge from $V_{CC}/2$ toward V_{CC}. This causes v_o to begin to increase toward $V_{CC}/2$ as well. Note that KVL is satisfied because

$$V_{CC} = v_o + v_C + v_{CE_1}$$

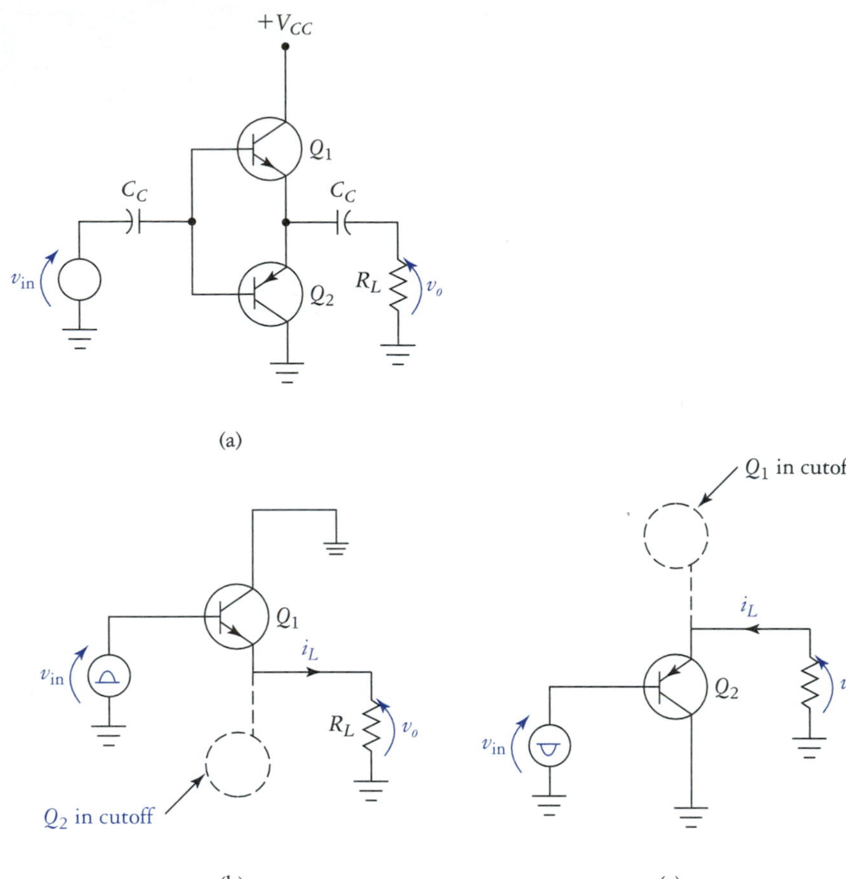

FIGURE 10.14
A practical class B amplifier. (a) Complementary-symmetry, push-pull stage. (b) AC-equivalent circuit for positive-going input voltage. (c) AC-equivalent circuit for negative-going input voltage.

Note
The output coupling capacitor supplies energy to the load on negative-going output signal swings.

where v_{CE_1} is dropping toward 0 V, and the voltage across the coupling capacitor $v_C = V_{CC}/2$. At this time, since Q_2 is in cutoff it is an open circuit, allowing the ac-equivalent circuit of Fig. 10.14(b) to model the circuit during these positive-going excursions.

The circuit behaves in a similar manner during negative-going input signal excursions, except that Q_1 is driven into cutoff and Q_2 is driven into conduction. The discharge of the output coupling capacitor provides energy to the load during this half-cycle. This action is necessary because the V_{CC} supply is isolated from the load by cutoff transistor Q_1 during these negative-going signal excursions. Because of the direction of the load current during C_C discharge, v_o is a negative voltage during this interval. The load line for the npn transistor and the input voltage, collector current, and v_{CE} waveforms are shown in Fig. 10.15.

Assuming that the complementary transistors are matched, the maximum output voltage levels that can be produced by Fig. 10.14 are approximately

$$V_{o(\max)} \cong \pm \left| \frac{V_{cc}}{2} - V_{BE} \right| \tag{10.15}$$

where V_{BE} is the barrier potential loss from base to emitter.

Coupling Capacitor Selection

Although this subject is treated in greater detail later in Chapter 11, a brief discussion of coupling capacitor size requirements is presented here, because of the large capacitor values that are often needed in power amplifier applications. Push-pull amplifiers, and power amplifiers in general, are normally required to drive low-resistance loads. Thus a rather large output coupling capacitor will usually be needed.

Several different approaches can be used in determining the required coupling capacitor size. One method is to choose C_C such that the following inequality is satisfied:

$$X_C|_{f=\min} \leq 0.1R_L \tag{10.16}$$

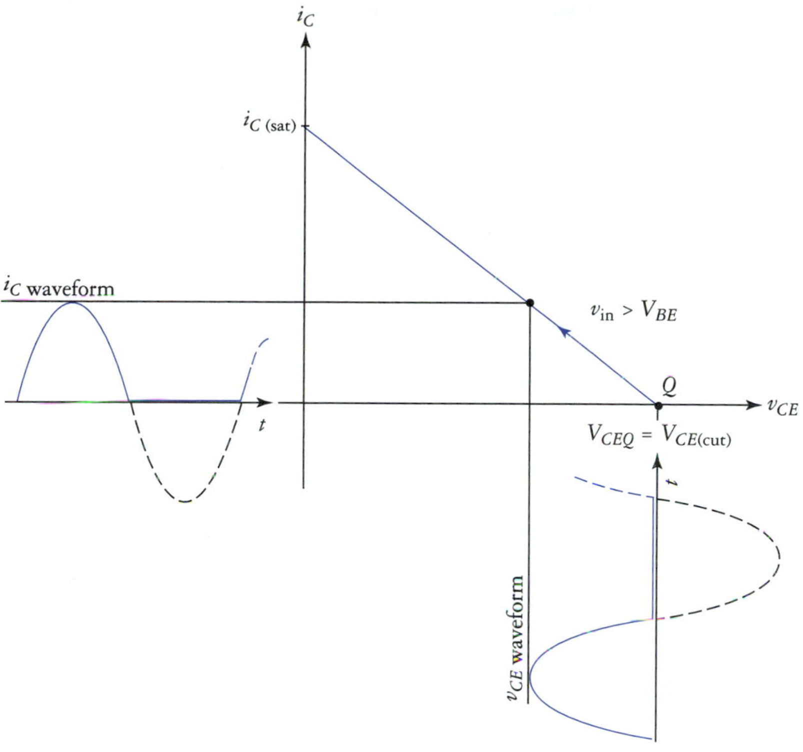

FIGURE 10.15
AC load line for the npn transistor.

where the magnitude of X_C is determined at the lowest frequency to be amplified. If we satisfy Eq. (10.16), then the voltage drop and phase shift caused by the coupling capacitor at the lowest expected signal frequency will be negligibly small.

Equation (10.16) is very conservative. That is, although it ensures that output signals at f_{min} will experience negligible attenuation or phase shift, we may not need such extreme overdesign. Because capacitive coupling networks are really just simple high-pass filters, we can use the corner (or cutoff) frequency equation to set the capacitor value as well. The corner frequency is given by

$$f_C = \frac{1}{2\pi R_T C_C} \tag{10.17}$$

where R_T is the total resistance in the coupling network ($R_T = R_s + R_{\text{in}}$ for input coupling and $R_T = R_o + R_L$ for output coupling). We can use Eq. (10.17) directly to determine the low-frequency limit of a coupling network, or we can solve it for C_C to select the capacitor value that produces a given corner frequency. If input and output corner frequencies are not equal, then the higher of the two dominates, and is considered to set the lower frequency limit of the amplifier.

EXAMPLE 10.4

An audio power amplifier with $R_o = 2\ \Omega$ is to be designed to drive an 8-Ω resistive load with a minimum signal frequency of 20 Hz. Determine the necessary output coupling capacitor value required using both Eqs. (10.16) and (10.17).

Solution We begin by solving Eq. (10.16) for C_C, with $X_C = \dfrac{1}{2\pi f_{\text{min}} C_C}$, producing

$$\begin{aligned} C_C &\geq \frac{10}{2\pi f_{\text{min}} R_L} \\ &\geq \frac{10}{(2\pi)(20\text{ Hz})(8\ \Omega)} \\ &\geq 9947\ \mu\text{F} \end{aligned}$$

In practice, a 10,000-μF electrolytic capacitor would probably be used.

Using Eq. (10.17), we solve for C_C, giving

$$C_C = \frac{1}{2\pi R_T f_C}$$

$$= \frac{1}{(2\pi)(10\ \Omega)(20\ \text{Hz})}$$

$$= 796\ \mu F \quad \text{(use next larger standard value 1000 } \mu\text{F)}$$

As the last example shows, it is not unusual to find very high-value coupling capacitors used in power amplifier designs. This is a disadvantage, because such capacitors are large, bulky and relatively expensive. In addition, large electrolytic capacitors have other less than ideal characteristics associated with them. For example, large electrolytics often behave more inductively than capacitively at higher frequencies, causing poor high-frequency response. Large electrolytic capacitors can also decrease in value as the internal electrolyte dries up over time. These problems can be eliminated through the use of direct coupling, which is covered shortly.

Crossover Distortion

Application of a sinusoidal input signal to the class B complementary-symmetry amplifier produces an output signal that appears much like that shown in Fig. 10.16(a). The flat portions of the output that occur around zero crossings of v_{in} cause a form of distortion called *crossover distortion.* Crossover distortion is worst at low signal levels, decreasing in severity as the amplifier is driven harder.

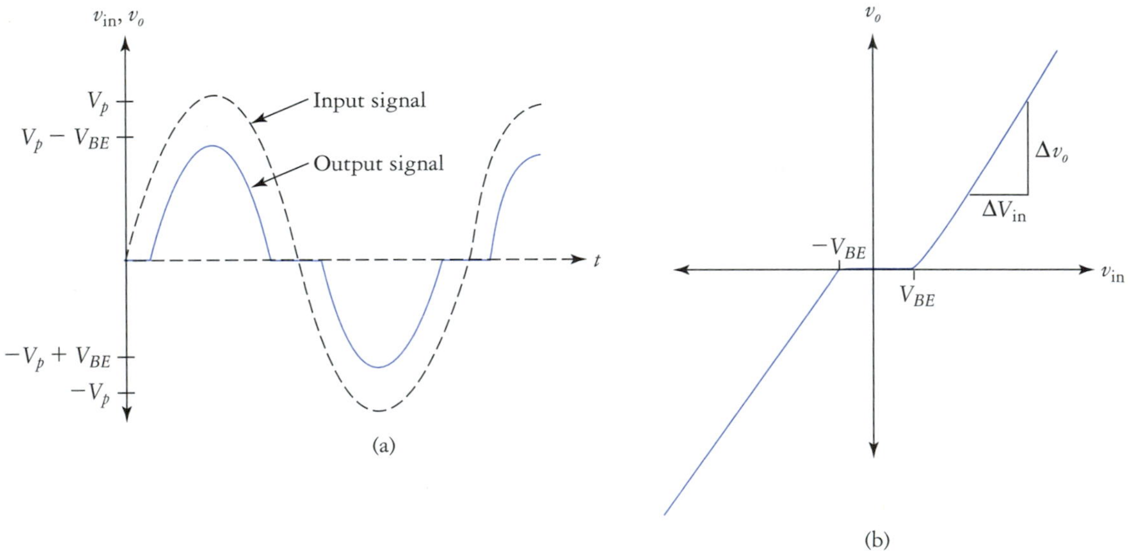

FIGURE 10.16
Crossover distortion. (a) Time-domain view. (b) Transfer characteristic curve.

An alternative way of visualizing crossover distortion is through the examination of the transfer characteristics of the amplifier, as shown in Fig. 10.16(b). The voltage gain of the amplifier is the slope (v_o/v_{in}) of the transfer characteristic curve as determined using the right triangle shown in the figure. Nonlinearities in the transistors show up as deviations from a straight line in the plot. For example, the flat crossover region and the saturation limit portions of the characteristic curve are nonlinearities.

If both transistors are perfectly matched, nonlinearities will be symmetrical, as shown in Fig. 10.17(a). This means that whatever output distortion *is* present will affect the positive- and negative-going portions of the output signal symmetrically as well. A mathematical technique called Fourier analysis shows that for a given input signal frequency f_{sig} this type of symmetrical distortion produces only odd harmonics ($3f_{sig}$, $5f_{sig}$, $7f_{sig}$, etc.)in the output signal. Asymmetrical distortion, as produced by a circuit with a transfer characteristic like that of Fig. 10.17(b), will cause the

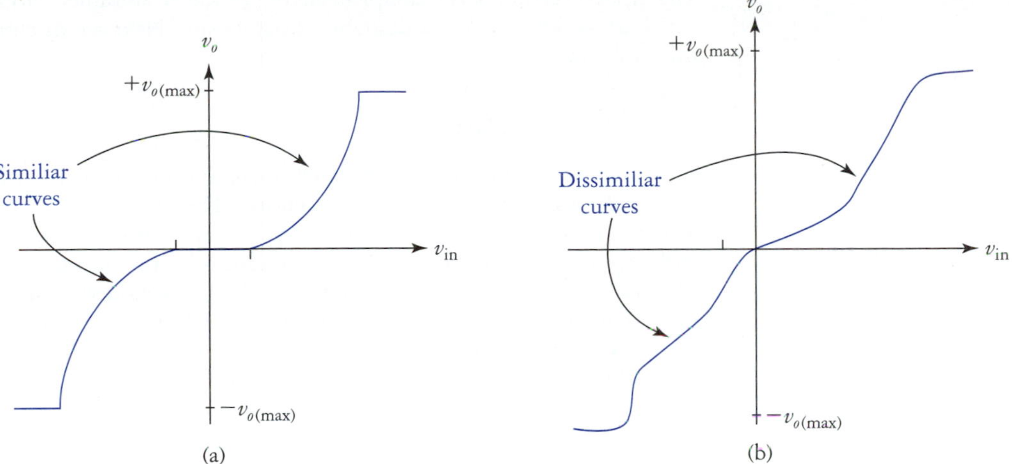

FIGURE 10.17
Additional distortion characteristics. (a) Symmetrical nonlinearities cause output distortion to contain only odd harmonics of a sinusoidal input. (b) Asymmetrical nonlinearities cause even and odd harmonics to be created.

output signal to contain both odd and even harmonics, and possibly even nonharmonically related signal energy. This is an undesirable characteristic, except in special cases where nonlinear behavior is desired.

Bipolar Supply Operation

Complementary-symmetry amplifiers are commonly operated from bipolar power supplies, as shown in Fig. 10.18. The main advantage of using bipolar power supplies with the complementary-symmetry amplifier is that it allows the elimination of coupling capacitors. Assuming that both transistors are matched, under no-signal conditions the transistors will each drop half of the applied supply voltage. Because the supply voltages are usually equal in magnitude ($V_{CC} = V_{EE}$) we have

$$V_{CE_1} = V_{CC} = V_{EC_2} = V_{EE}$$

This also means that the bases and emitters of the transistors are at ground potential $V_{B_1} = V_{B_2} = 0$ V, $V_{E_1} = V_{E_2} = 0$ V). This eliminates the need for coupling capacitors. This is a form of direct coupling, and one of the advantages of this approach is that dc signals can be amplified just as well as time-varying ones.

FIGURE 10.18
Class B, complementary-symmetry amplifier using a bipolar power supply.

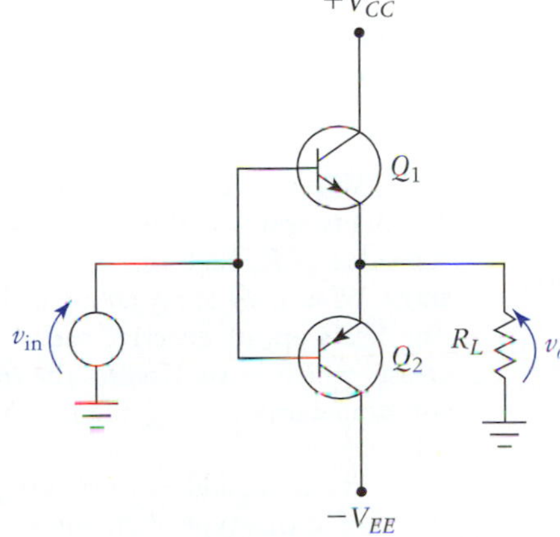

The principle of operation of Fig. 10.18 is exactly the same as that of the circuit of Fig. 10.14. The approximate maximum and minimum output voltages for the circuit of Fig. 10.18 are given by

$$V_{o(\max)} = V_{CC} - V_{BE}$$
$$V_{o(\min)} = -V_{EE} + V_{BE}$$

The bipolar version of the complementary-symmetry amplifier will also exhibit crossover distortion, which as we know is not a desirable characteristic. Next we discuss some methods for eliminating this problem.

Class AB Amplifiers

Crossover distortion can be eliminated or significantly reduced using several different techniques. Class AB biasing provides one such solution. The main idea behind class AB biasing is to provide just enough bias to the complementary-symmetry transistor pair to overcome barrier potential losses without causing a significantly large quiescent collector current to flow. This keeps quiescent power dissipation low while at the same time providing a conduction angle of 360° when using the push-pull configuration. Thus we see how the class AB designation originates: It is not class B, because there is some bias applied to the output transistors, yet class A implies that a transistor is biased into the middle of its active region. In class AB, the transistors are biased just slightly above cutoff.

Note
The small quiescent current that flows through the push-pull pair is often called the *idling current*.

Simple Voltage-Divider Biasing

As shown in Fig. 10.19, a voltage divider can be used to provide class AB biasing. Normally, we set $R_1 = R_2$, while R_3 is chosen to produce a voltage drop of around 1.2 V. In Fig. 10.19(a), two input coupling capacitors are required. If we split R_3 into two possibly equal parts R_{3A} and R_{3B} then we can use a single input coupling capacitor as shown in Fig. 10.19(b).

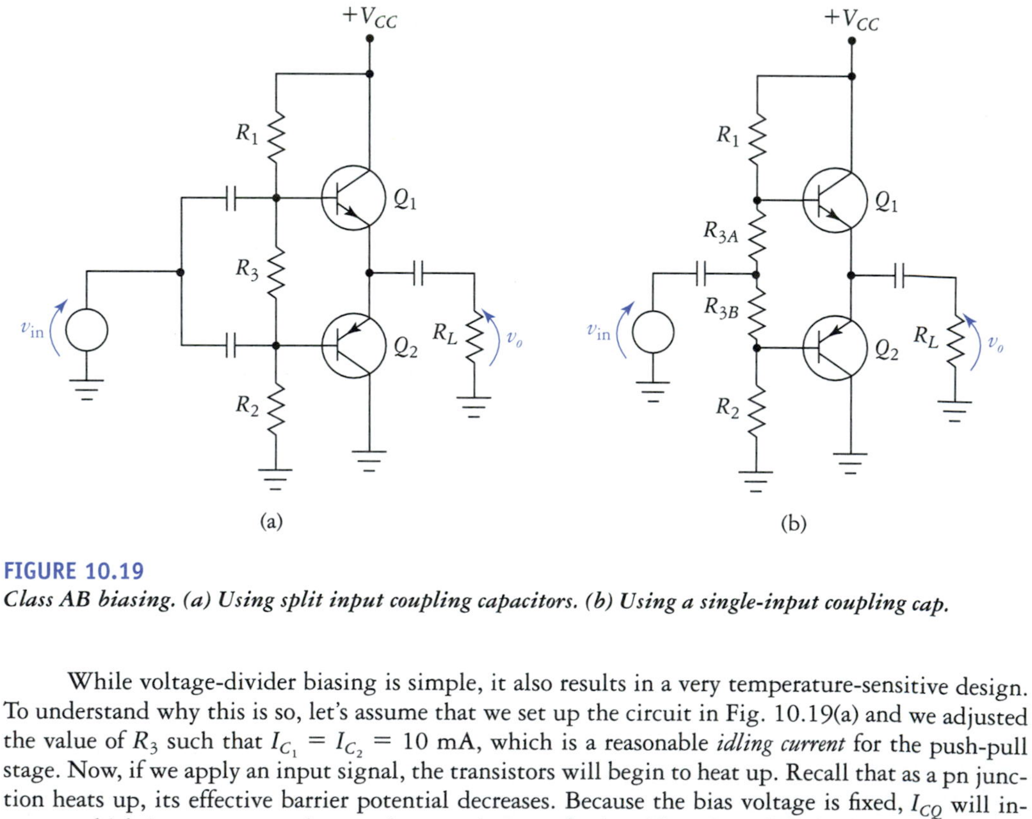

+V_{CC} R_1 Q_1 R_3 v_{in} Q_2 R_L v_o R_2

(a)

+V_{CC} R_1 Q_1 R_{3A} R_{3B} v_{in} Q_2 R_L v_o R_2

(b)

FIGURE 10.19
Class AB biasing. (a) Using split input coupling capacitors. (b) Using a single-input coupling cap.

While voltage-divider biasing is simple, it also results in a very temperature-sensitive design. To understand why this is so, let's assume that we set up the circuit in Fig. 10.19(a) and we adjusted the value of R_3 such that $I_{C_1} = I_{C_2} = 10$ mA, which is a reasonable *idling current* for the push-pull stage. Now, if we apply an input signal, the transistors will begin to heat up. Recall that as a pn junction heats up, its effective barrier potential decreases. Because the bias voltage is fixed, I_{CQ} will increase, which in turn causes the transistors to heat up further. Thus Q_1 and Q_2 both tend to turn on simultaneously causing a large current to flow from V_{CC} to ground. This can easily result in thermal runaway.

We could add emitter swamping resistors to eliminate this problem. However, if they are to be effective in this type of circuit, such resistors would have to be too large to be practical. Rather than dwell on this circuit, we shall move on to a more practical biasing technique.

Current-Mirror Biasing

Application of the current-mirror technique provides a nearly ideal way to obtain class AB biasing. In Fig. 10.20, diodes D_1 and D_2 provide the bias voltage for the push-pull transistors. Let us make

FIGURE 10.20
Current-mirror biasing of a class AB amplifier.

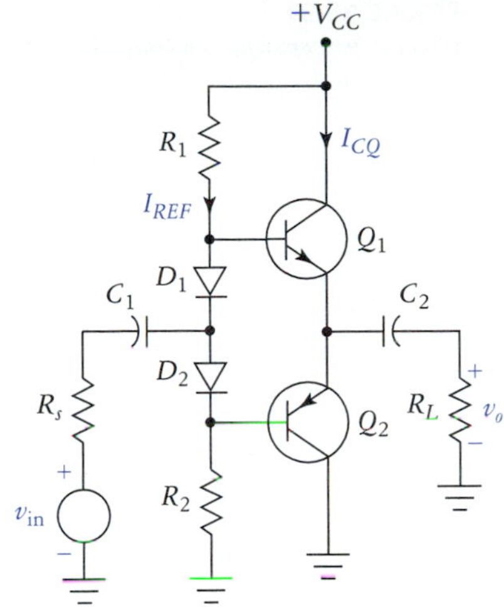

Note
Current mirror biasing provides good temperature stability.

the following assumptions: The transistors and diodes are matched devices, $I_{B_1} = I_{B_2} \cong 0$, and $R_1 = R_2 = R$. Based on previous study of the current mirror, we know that $I_{CQ} \cong I_{\text{REF}}$, so based on the assumptions that were given, we have

$$I_{CQ} \cong I_{\text{REF}} = \frac{V_{CC} - 2V_{BE}}{2R} \tag{10.18}$$

Thus we can set the quiescent (idling) current for the push-pull output stage simply by choosing appropriate values for R_1 and R_2.

If this biasing technique is to work properly, each diode should be in close thermal contact with its respective transistor. In this way, when the transistor heats up, it heats the biasing diode as well. Thus, the shifts in barrier potential and leakage current for the transistor are matched by equal shifts in diode parameters. The net result is a stable, temperature-independent Q point. Keep in mind, however, that Eq. (10.18) is really only accurate when we have well-matched components.

EXAMPLE 10.5

Refer to Fig. 10.20. Assume $V_{CC} = 25$ V and $V_{BE(\text{on})} = 0.7$ V for all devices. Choose R_1 and R_2 such that $I_{CQ} \cong 20$ mA.

Solution We simply solve Eq. (10.18) for R, giving us

$$R = \frac{V_{CC} - 2V_{BE}}{2I_{\text{REF}}}$$
$$= \frac{25\text{ V} - 1.4\text{ V}}{40\text{ mA}}$$
$$= 573\ \Omega$$

If a bipolar power supply is used, as shown in Fig. 10.21, *then* Eq. (10.18) must be modified slightly, giving us

$$I_{CQ} \cong I_{\text{REF}} = \frac{V_{CC} + V_{EE} - 2V_{BE}}{2R} \tag{10.19}$$

Again, notice that using a bipolar power supply allows us to eliminate coupling capacitors. As before, however, careful matching of components is required for this to be accomplished. The use of unmatched devices will result in significant dc voltages being present at the input and output terminals of the circuit. If you were to set up a class AB push-pull amplifier in the lab, you would most likely not have closely enough matched devices to eliminate the coupling capacitors in the basic circuit we are using now. In such a situation, you should generally use large-value resistors for R_1 and R_2 ($R \geq 4700\ \Omega$) in order to be on the safe side in preventing destruction of the transistors. Later, we will examine other techniques using negative feedback that can reduce the undesirable effects of mismatched components.

FIGURE 10.21
Bipolar power supply operation.

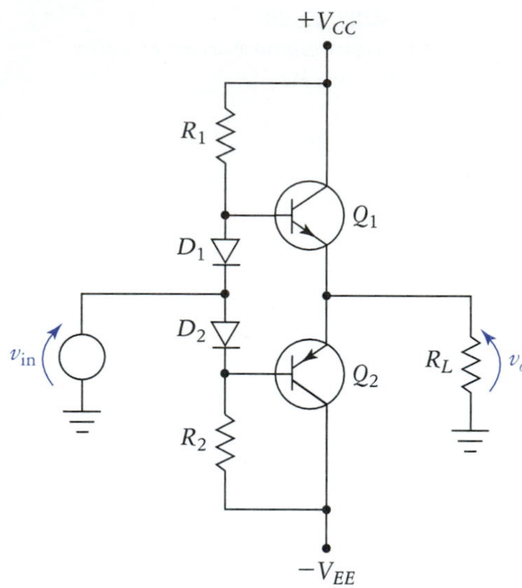

Darlington Pair Complementary-Symmetry Amplifier

Darlington configuration transistors are sometimes used to increase the input resistance of the push-pull stage. The details relating to signal analysis of these amplifiers are presented in the next section. For now, we wish to examine the consequences of such alternative configurations on biasing conditions and circuitry.

A darlington push-pull amplifier is shown in Fig. 10.22. Four diodes are required to provide sufficient bias voltage to overcome the barrier potential of the Darlington transistors used in this circuit. Because the transistors in a given Darlington pair are operating at significantly different collector currents, there is no simple way to predict the quiescent current I_{CQ}. However, I_{CQ} will be about the same order of magnitude as I_{REF}. Making R_1 or R_2 adjustable would allow trimming of I_{CQ} to a desired value. As before, it is necessary for the biasing diodes and the Darlington transistors to have close thermal coupling.

If we replace the diodes with diode-connected matched Darlingtons, then the current mirror equation can be modified to predict I_{CQ}. This modification is shown in Fig. 10.23. Now, under quiescent conditions, the current distribution within each push-pull Darlington and its diode-connected partner will be identical. This allows us to write

$$I_{CQ} \cong I_{REF} = \frac{V_{CC} - 4\,V_{BE}}{2\,R} \tag{10.20}$$

FIGURE 10.22
Darlington-based complementary-symmetry class AB amplifier.

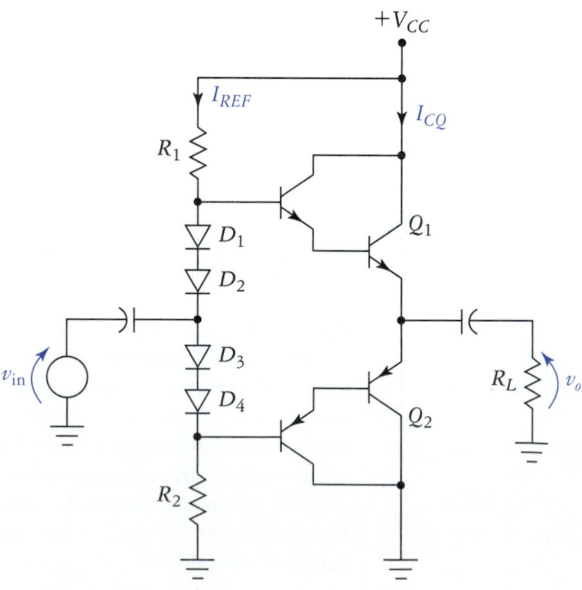

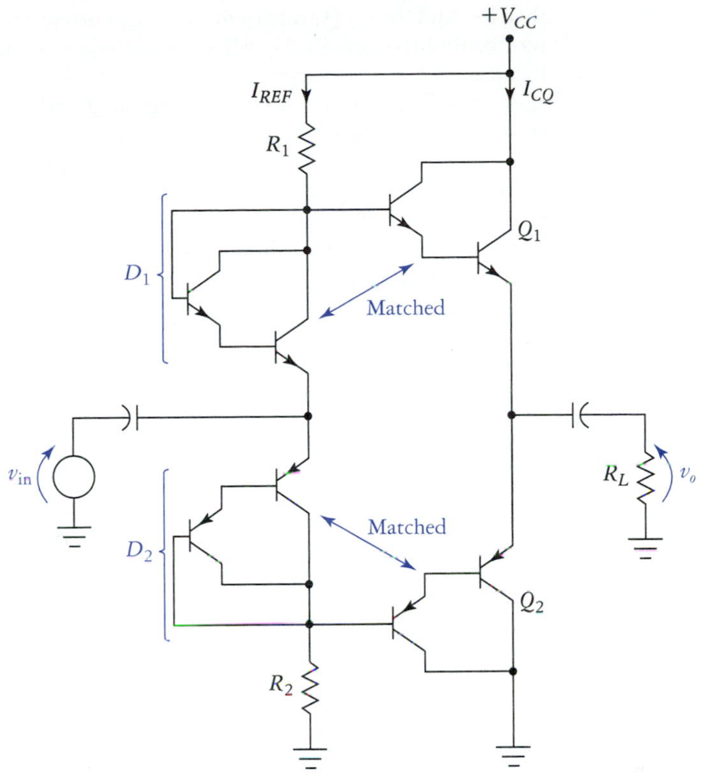

FIGURE 10.23
Darlington transistors can be connected as diodes to provide current-mirror biasing.

If a bipolar power supply were used, the expression would be written

$$I_{CQ} \cong I_{\text{REF}} = \frac{V_{CC} + V_{EE} - 4\,V_{BE}}{2\,R} \tag{10.21}$$

Quasi-Complementary-Symmetry Amplifiers

Quasi-complementary symmetry allows the use of two npn power transistors as shown in Fig. 10.24. The upper pair (Q_1–Q_2) is a standard npn Darlington that handles positive-going output voltage swings. The lower pair (Q_3–Q_4) is a composite connection that behaves like a pnp Darlington except

FIGURE 10.24
Quasi-complementary-symmetry amplifier.

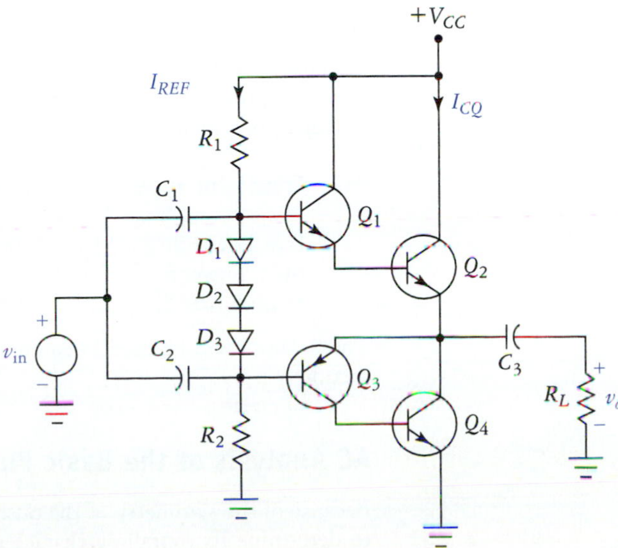

that the npn device Q_4 carries most of the current on negative-going signal swings. Here's the basic idea behind the circuit: A positive-going input voltage turns on pair Q_1–Q_2 while at the same time turning off Q_3, which starves Q_4 of base current, sending it into cutoff. Negative-going input voltages cut off pair Q_1–Q_2, while turning on Q_3, which in turn drives Q_4 into conduction. Thus, the Q_3–Q_4 pair acts much like a pnp Darlington.

There is no simple way to set the exact idling current for the quasi-complementary-symmetry push-pull stage as shown in Fig. 10.24. Fortunately, I_{CQ} is again the same order of magnitude as I_{REF}, so we can tweak R_1 and/or R_2 to obtain a desired current level. Also, the asymmety of the push-pull transistors results in a much more complex large-signal analysis for this type of circuit. We deal with this in the next section.

Finally, you may be wondering why quasi-complementary symmety is even used at all. There are several reasons. First, if you look through a few transistor data books you will notice that there are many more npn devices available than pnp. This is so because in general, an npn device will have slightly better high-frequency performance than an equivalent pnp device (because of hole mobility versus electron mobility). Currently, improved manufacturing processes have reduced these differences significantly. Secondly, in many applications we only need one type of transistor (npn, for example) anyway. For example, unless we are using active loading, or a push-pull stage, we could use npn BJTs exclusively, if we wanted to. Thus, manufacturers simply produce more npn than pnp devices. At the present time, however, true complementary symmetry is very practical and is generally used whenever possible.

10.4 CLASS B AND AB AMPLIFIERS: LARGE-SIGNAL CONSIDERATIONS

Class B and AB amplifiers are used extensively in large-signal (usually power amplifier) applications. Under such conditions, amplifiers tend to behave in very nonlinear ways that can greatly complicate detailed analysis. As stated before, we examine some negative-feedback techniques later that can be used reduce the effects of nonlinearity (and device parameter variation). With this in mind, we shall make liberal use of approximations in this section. However, the basic principles that will be presented are still valid and very important.

Average Emitter Resistance

Many important small-signal transistor amplifier characteristics are dependent on emitter dynamic resistance. The same is true of large-signal amplifiers, except that it is usually preferable to work with the *average dynamic emitter* resistance r'_e. Technically, we determine r'_e using the limits of the Q point for a given transistor as shown in Fig. 10.25. Here, we are assuming class AB biasing, where the Q point swings from cutoff to saturation at the extremes. Notice that even if strict class B biasing were used, as long as $I_{C(sat)}$ did not change, r'_e would have the same value. The average resistance is given by

$$r'_e = \frac{\Delta V_{BE}}{\Delta I_C} \tag{10.22}$$

Because Eq. (10.22) uses the actual transistor transconductance curve, the effects of bulk resistance are automatically factored in as well. The fact that Eq. (10.22) requires the use of a transconductance curve is a problem because we rarely have access to this detailed information. Because of this, the following approximations can be used in most situations:

Transistor Type	Typical Device	Typical r'_e	Typical β
Low Power Si	2N3904	5 Ω	>100
Med. Power Si	2N3053	1 Ω	100
High Power Si	2N3055	0.5 Ω	50

These values will be used in conjunction with large-signal amplifiers through the remainder of this section.

AC Analysis of the Basic Push-Pull Amplifier

Because of the symmetry of the push-pull amplifier, we usually need only analyze half of the circuit to determine its overall ac characteristics. For example, if we determined that the voltage gain of Fig. 10.26 was 0.80 for positive-going signals (the npn transistor is active), then we assume that

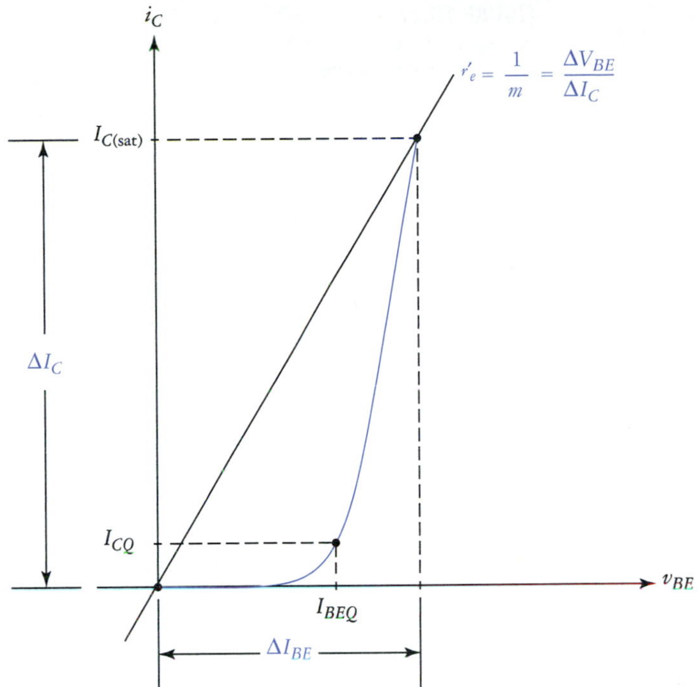

FIGURE 10.25
Defining the average large-signal emitter resistance r'_e.

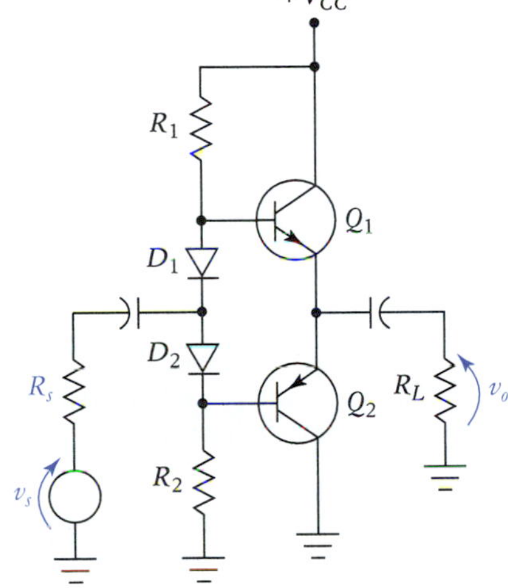

FIGURE 10.26
Typical class B push-pull, complementary-symmetry amplifier.

the voltage gain is 0.80 for negative-going signals as well (when the pnp transistor is active). This reasoning applies to input and output resistance calculations as well.

Analyzing Fig. 10.26 for positive-going input signals, we obtain the ac-equivalent circuit shown in Fig. 10.27. Transistor Q_2 is driven into cutoff and therefore is effectively removed from the circuit. The diodes in the biasing network are forward biased and have negligible forward resistance so they are simply replaced with short circuits. The power supply is a signal ground point, which places R_1 and R_2 in parallel with each other. Routine analysis of this circuit assuming $\alpha \cong 1$ produces the following results:

$$R_{\text{in}} = R_1 \parallel R_2 \parallel [\beta(r'_e + R_L)] \tag{10.23}$$

$$A_v = \frac{v_o}{v_{\text{in}}} = \frac{R_L}{R_L + r'_e} \tag{10.24}$$

$$R_o = r'_e + \frac{R_s \| R_1 \| R_2}{\beta} \tag{10.25}$$

FIGURE 10.27
AC-equivalent circuit of Fig. 10.26 for positive-going input voltage.

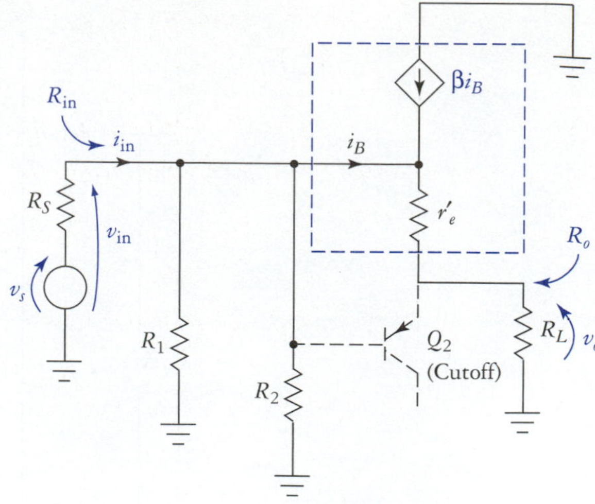

These equations apply to the bipolar power supply version of the amplifier as well, because it has the same ac-equivalent representation.

The simpler class B circuit of Fig. 10.14(a) has even simpler ac analysis equations [draw the ac equivalent for Fig. 10.14(b) to see this], therefore the derivation of these equations is left as an exercise for the reader.

The maximum possible load current that will flow occurs when either Q_1 or Q_2 is saturated, which is the collector saturation current $I_{C(sat)}$. For the single-supply version of the circuit we have

$$I_{L(max)} = I_{C(sat)} \cong \frac{V_{CC}}{2(R_L + r'_e)} \tag{10.26}$$

and for the bipolar power supply version of the circuit with $V_{CC} = V_{EE}$, we have

$$I_{L(max)} = I_{C(sat)} \cong \frac{V_{CC}}{R_L + r'_e} \tag{10.27}$$

We have ignored the transistor saturation voltage (i.e., $V_{CE(sat)} \cong 0$ V). If necessary, this can be accounted for simply by subtracting it from the numerator of the expression.

EXAMPLE 10.6

The circuit in Fig. 10.26 has $V_{CC} = 25$ V, $R_1 = R_2 = 1$ kΩ, $R_L = 8$ Ω, $R_s = 0$ Ω, and the diodes are matched to Q_1 and Q_2, which are high-power transistors with $\beta = 50$ and $r'_e = 0.5$ Ω. Determine I_{CQ}, the transistor power dissipation P_{DQ}, the total quiescent power dissipation of the circuit P_{CC}, $I_{C(sat)}$, $P_{o(max)}$ and $P_{o(rms)}$, A_v, R_{in}, A_i, A_p, and R_o.

Solution First we determine the amplifier quiescent current:

$$\begin{aligned} I_{CQ} \cong I_{REF} &= \frac{V_{CC} - 2\,V_{BE}}{2\,R} \\ &= \frac{25\text{ V} - 1.4\text{ V}}{2\text{ k}\Omega} \\ &= 11.8\text{ mA} \end{aligned}$$

Both transistors dissipate the same power, giving us

$$\begin{aligned} P_{DQ_2} = P_{DQ_1} &= I_{C_1} V_{CE_1} \\ &= I_{C_1} (V_{CC} / 2) \\ &= (11.8\text{ mA})(12.5\text{ V}) \\ &= 147.5\text{ mW} \end{aligned}$$

The total quiescent power dissipation of the transistors is

$$\begin{aligned} P_{DQ} &= P_{DQ_1} + P_{DQ_2} \\ &= 147.5\text{ mW} + 147.5\text{ mW} \\ &= 295\text{ mW} \end{aligned}$$

The biasing network power dissipation can be found as follows:

$$\begin{aligned} P_{\text{bias}} &= V_{CC} I_{\text{bias}} \\ &= (25\text{ V})(11.8\text{ mA}) \\ &= 295\text{ mW} \end{aligned}$$

Thus the total quiescent power dissipation is

$$\begin{aligned} P_{CC} &= P_{DQ} + P_{\text{bias}} \\ &= 295\text{ mW} + 295\text{ mW} \\ &= 590\text{ mW} \end{aligned}$$

The collector saturation current is

$$\begin{aligned} I_{C(\text{sat})} &\cong \frac{V_{CC}}{2\,(R_L + r_e')} \\ &= \frac{25\text{ V}}{17\ \Omega} \\ &= 1.47\text{ A} \end{aligned}$$

One way to find the peak output power is

$$\begin{aligned} P_{o(\max)} &= I_{L(\max)}^2 R_L \\ &= I_{C(\text{sat})}^2 R_L \\ &= (1.47\text{ A})^2 (8\ \Omega) \\ &= 17.29\text{ W} \end{aligned}$$

Assuming that the amplifier is driven to its maximum output limits with a sinusoidal signal, the maximum rms output power is the same as the average output power. Under these conditions, the average (and rms) load power is half the peak power:

$$\begin{aligned} P_{o(\text{rms})} = P_{o(\text{ave})} &= \frac{P_{o(\max)}}{2} \\ &= \frac{17.29\text{ W}}{2} \\ &= 8.65\text{ W}_{\text{rms}} \end{aligned}$$

The voltage gain is found to be

$$\begin{aligned} A_v &= \frac{R_L}{R_L + r_e'} \\ &= \frac{8\ \Omega}{8.5\ \Omega} \\ &= 0.94 \end{aligned}$$

The input resistance is

$$\begin{aligned} R_{\text{in}} &= R_1 \parallel R_2 \parallel \beta(r_e' + R_L) \\ &= 1\text{ k}\Omega \parallel 1\text{ k}\Omega \parallel 50\,(0.5\ \Omega + 8\ \Omega) \\ &= 230\ \Omega \end{aligned}$$

The current gain is found using the universal relationship

$$\begin{aligned} A_i &= A_v \frac{R_{\text{in}}}{R_L} \\ &= 0.94\,\frac{230\ \Omega}{8\ \Omega} \\ &= 28.8 \end{aligned}$$

The power gain is

$$\begin{aligned} A_p &= A_v A_i \\ &= (0.94)(28.8) \\ &= 27.1 \end{aligned}$$

At last we determine the output resistance using Eq. (10.25) with $R_s = 0\ \Omega$, which is simply

$$R_o = r'_e = 0.5\ \Omega$$

Let's run through another example using a bipolar power supply this time.

EXAMPLE 10.7

Refer to Fig. 10.28. Assume $V_{CC} = V_{EE} = 30$ V, $R_1 = R_2 = 2.7$ kΩ, $R_L = 4\ \Omega$, $R_s = 100\ \Omega$, and the diodes are matched to Q_1 and Q_2, which are high-power transistors with $\beta = 50$ and $r'_e = 0.5\ \Omega$. Determine I_{CQ}, the transistor power dissipation P_{DQ}, the total quiescent power dissipation of the circuit P_{CC}, $I_{C(\text{sat})}$, $P_{o(\max)}$, and $P_{o(\text{rms})}$, A_v, R_{in}, A_i, A_p, and R_o.

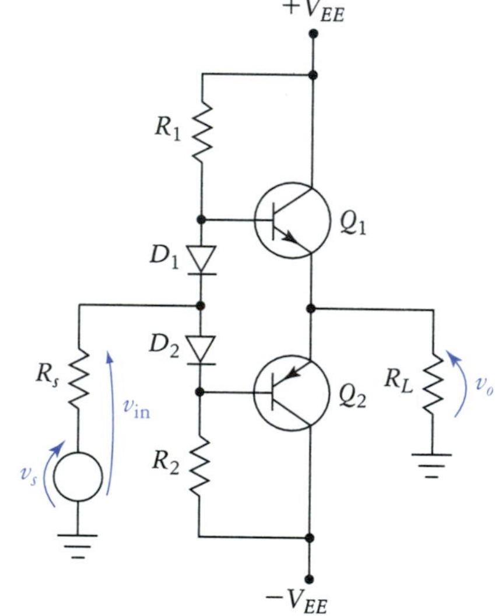

FIGURE 10.28
Amplifier using a bipolar power supply.

Solution As before, we determine the amplifier quiescent current

$$I_{CQ} \cong I_{\text{REF}} = \frac{V_{CC} + V_{EE} - 2V_{BE}}{2R} = \frac{30\text{ V} + 30\text{ V} - 1.4\text{ V}}{5.4\text{ k}\Omega} = 10.9\text{ mA}$$

Both transistors dissipate the same power, giving us

$$P_{DQ_2} = P_{DQ_1} = I_{C_1} V_{CE_1} = I_{C_1} V_{CC} = (10.9\text{ mA})(30\text{ V}) = 327\text{ mW}$$

The total quiescent power dissipation of the transistors is

$$P_{DQ} = P_{DQ_1} + P_{DQ_2} = 327\text{ mW} + 327\text{ mW} = 654\text{ mW}$$

The biasing network power dissipation can be found as follows:

$$P_{\text{bias}} = (V_{CC} + V_{EE})\, I_{\text{bias}} = (60\text{ V})(10.9\text{ mA}) = 654\text{ mW}$$

Thus the total quiescent power dissipation is

$$P_{CC} = P_{DQ} + P_{\text{bias}} = 654\text{ mW} + 654\text{ mW} = 1.308\text{ W}$$

The collector saturation current is

$$I_{C(\text{sat})} \cong \frac{V_{CC}}{R_L + r'_e}$$

$$= \frac{30\text{ V}}{4.5\ \Omega}$$

$$= 6.67\text{ A}$$

The peak output power is

$$\begin{aligned} P_{o(\max)} &= I_{L(\max)} R_L \\ &= I_{C(\text{sat})} R_L \\ &= (6.67\text{ A})^2 (4\ \Omega) \\ &= 178\text{ W} \end{aligned}$$

Again, we assume that the amplifier is driven to its maximum output limits with a sinusoidal signal; therefore, the maximum rms output power is the same as the average output power. Let's use Eq. (10.7) this time, assuming that the output voltage can swing approximately 27 V_{pk} (I cheated because I already know $A_v = 0.89$, and this is $A_v V_{CC}$):

$$P_{o(\text{rms})} = P_{o(\text{ave})} = \frac{V_p^2}{2R_L}$$

$$= \frac{27^2}{8}$$

$$= 91\text{ W}$$

which is slightly different than $P_{o(\max)}/2 = 89$ W, due to rounding errors. Quantities that are squared are greatly affected by rounding.

The voltage gain is found to be

$$A_v = \frac{R_L}{R_L + r'_e}$$

$$= \frac{4\ \Omega}{4.5\ \Omega}$$

$$= 0.89$$

The input resistance is

$$\begin{aligned} R_{\text{in}} &= R_1 \parallel R_2 \parallel \beta(r'_e + R_L) \\ &= 2.7\text{ k}\Omega \parallel 2.7\text{ k}\Omega \parallel 50\,(0.5\ \Omega + 4\ \Omega) \\ &= 193\ \Omega \end{aligned}$$

The current gain is found as usual using

$$A_i = A_v \frac{R_{\text{in}}}{R_L}$$

$$= 0.89 \frac{193\ \Omega}{4\ \Omega}$$

$$= 42.9$$

The power gain is

$$\begin{aligned} A_p &= A_v A_i \\ &= (0.89)(42.9) \\ &= 38.2 \end{aligned}$$

At last we determine the output resistance using Eq. (10.25), producing

$$R_o = r'_e + \frac{R_s \parallel R_1 \parallel R_2}{\beta}$$

$$= 0.5\ \Omega + \frac{93.1\ \Omega}{50}$$

$$= 0.5\ \Omega + 1.9\ \Omega$$

$$= 2.4\ \Omega$$

If used in an audio amplifier application, the circuit of Example 10.7 would produce enough output power to satisfy most listeners. However such high power output could not be achieved in practice using such a simple circuit. In a practical application, we would have to connect several power transistors in parallel in order to keep individual device power dissipation down to an acceptable level. This would require the use of emitter swamping resistors, which in turn decrease A_v and the maximum output voltage compliance. Also, we would need a driver stage that is capable of handling this push-pull output stage. As it turns out, the driver stage is often the most critical and difficult to design as part of a complete power amplifier. We will address this subject soon enough, however.

Emitter Swamping

Very often, emitter swamping resistors are incorporated into push-pull amplifier circuitry, as shown in Fig. 10.29. These resistors can serve several purposes. One function is to provide swamping of variations in r'_e, which produces a more stable and linear voltage gain from the stage. Emitter resistors also limit the maximum collector current that could flow should the output accidentally become shorted to ground. Without emitter resistances, such current would be limited by the transistor that is conducting at that time, which would almost certainly cause destruction of that device. The external resistance can limit such a current to a safe level. Also, if several transistors are placed in parallel, then emitter swamping resistors must be used to prevent current hogging. These are the positive aspects of emitter swamping. On the negative side, swamping resistors reduce amplifier efficiency, increase output resistance, and reduce voltage gain.

FIGURE 10.29
Push-pull amplifier with emitter swamping resistors.

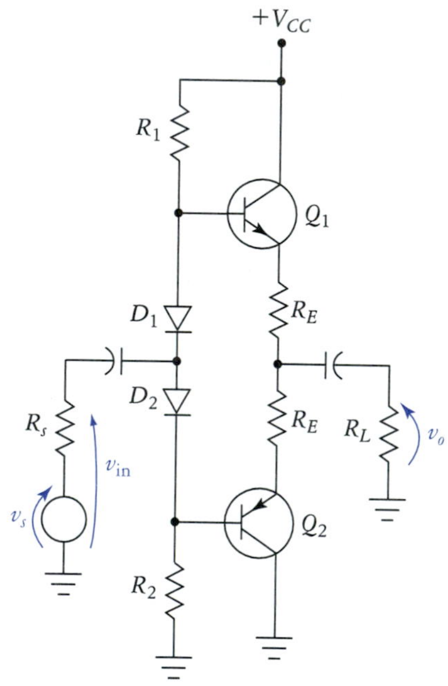

Emitter swamping resistors typically range from around 10 Ω in lower power applications to 0.1 Ω in high-power amplifiers. The addition of emitter swamping resistors can easily be accounted for. The various gain and signal limits for Fig. 10.29 are given by the following equations:

$$A_v = \frac{R_L}{R_L + R_E + r'_e} \tag{10.28}$$

$$R_{\text{in}} = R_1 \parallel R_2 \parallel \beta(r'_e + R_E + R_L) \tag{10.29}$$

$$R_o = r'_e + R_E + \frac{R_1 \parallel R_2 \parallel R_s}{\beta} \tag{10.30}$$

$$I_{L(\max)} = I_{C(\text{sat})} \cong \frac{V_{CC}}{2(R_E + R_L + r'_e)} \tag{10.31}$$

and if a bipolar power supply is used

$$I_{L(\max)} = I_{C(\text{sat})} \cong \frac{V_{CC}}{R_E + R_L + r'_e} \tag{10.32}$$

The addition of emitter swamping resistances also makes the simple current-mirror approximation more inaccurate. Inspection of Fig. 10.29 reveals that the quiescent collector current might best be determined using the Widlar current source equation from Chapter 4 [see Eq. (4.39)]. It is important to realize though, that when we are dealing with power amplifiers, we usually look at circuits from a somewhat different perspective. To put it briefly, it is nice when we can perform reasonably accurate quantitative analysis, but in many cases (especially in power amplifier circuits), this is not practical. Because of this, we must sometimes make less exact approximations, and we realize that ballpark values often result in acceptable answers.

EXAMPLE 10.8

Refer to Fig. 10.29. Assume $V_{CC} = 22$ V, $R_1 = R_2 = 560\ \Omega$, $R_E = 1\ \Omega$, $R_L = 8\ \Omega$, $R_s = 0\ \Omega$, and the diodes are matched to Q_1 and Q_2, which are high-power transistors with $\beta = 50$ and $r'_e = 0.5\ \Omega$. Estimate I_{CQ} and $V_{o(\max)}$, and determine $I_{C(\text{sat})}$, $P_{o(\max)}$, and $P_{o(\text{rms})}$, A_v, R_{in}, A_i, A_p, and R_o. Determine the average power dissipation of the emitter swamping resistors $P_{R_E(\text{ave})}$ under full-power operation.

Solution Even though emitter swamping resistors are used, the quiescent collector current is still approximated by

$$\begin{aligned} I_{CQ} \cong I_{\text{REF}} &= \frac{V_{CC} - 2V_{BE}}{R_1 + R_2} \\ &= \frac{20.6\text{ V}}{1120\ \Omega} \\ &= 18.4\text{ mA} \end{aligned}$$

The collector saturation current is

$$\begin{aligned} I_{C(\text{sat})} &\cong \frac{V_{CC}}{2\,(R_E + R_L + r'_e)} \\ &= \frac{22\text{ V}}{19\ \Omega} \\ &= 1.16\text{ A} \end{aligned}$$

Because this amplifier has a single-polarity supply, the maximum theoretical output voltage is

$$\begin{aligned} V_{o(\max)} &= A_v \frac{V_{CC}}{2} \\ &= \frac{R_L}{R_L + R_E + r'_e}\, 11\text{ V} \\ &= 9.2\text{ V} \end{aligned}$$

The peak output power is

$$\begin{aligned} P_{o(\max)} &= I^2_{L(\max)} R_L \\ &= I^2_{C(\text{sat})} R_L \\ &= (1.16\text{ A})^2\,(8\ \Omega) \\ &= 10.8\text{ W} \end{aligned}$$

We assume that the amplifier is driven to its maximum output limits with a sinusoidal signal, therefore the maximum rms output power is the same as the average output power.

$$\begin{aligned} P_{o(\text{rms})} = P_{o(\text{ave})} &= \frac{P_{o(\max)}}{2} \\ &= \frac{10.8\text{ W}}{2} \\ &= 5.4\text{ W} \end{aligned}$$

The voltage gain was previously found to be

$$A_v = \frac{R_L}{R_L + R_E + r'_e}$$
$$= \frac{8\ \Omega}{8\ \Omega + 1\ \Omega + 0.5\ \Omega}$$
$$= 0.84$$

The input resistance is

$$R_{\text{in}} = R_1 \parallel R_2 \parallel \beta(r'_e + R_E + R_L)$$
$$= 560\ \Omega \parallel 560\ \Omega \parallel 50\,(0.5\ \Omega + 1\ \Omega + 8\ \Omega)$$
$$= 176\ \Omega$$

The current gain is found as usual using

$$A_i = A_v \frac{R_{\text{in}}}{R_L}$$
$$= 0.84\,\frac{176\ \Omega}{8\ \Omega}$$
$$= 18.5$$

The power gain is

$$A_p = A_v A_i$$
$$= (0.84)(18.5)$$
$$= 15.5$$

The output resistance is found using Eq. (10.30) with $R_s = 0\ \Omega$, which gives us

$$R_o = r'_e + R_E$$
$$= 0.5\ \Omega + 1\ \Omega$$
$$= 1.5\ \Omega$$

We can determine the power dissipation of the emitter resistors in many different ways. Using the peak collector current we proceed as follows:

$$P_{R_E(\text{ave})} = \frac{I^2_{C(\text{sat})}\,R_E}{2}$$
$$= \frac{(1.16\ \text{A})^2\,(1\ \Omega)}{2}$$
$$= 0.67\ \text{W}$$

This power dissipation is divided evenly between the two emitter resistors, causing each to dissipate 335 mW at full output.

Class B and AB Efficiency

Efficiency is an important consideration in the design of power amplifiers. For class B amplifiers, the maximum theoretical efficiency is

$$\eta = \pi/4 = 78.5\% \qquad (10.33)$$

For class AB amplifiers the efficiency will be lower because of biasing circuit dissipation. The typical class AB amplifier will operate at around 50% efficiency.

10.5 ALTERNATIVE BIASING AND DRIVER CIRCUITS

In comparison to small-signal amplifiers, complete power amplifier designs tend to be much more complex and there are hundreds of variations in those circuit designs. In this section, we look at a representative sampling of those variations.

Direct-Coupled Drivers

A common push-pull amplifier configuration that uses a direct-coupled driver is shown in Fig. 10.30. This version is shown using a single-polarity power supply, although bipolar supply operation is possible as well. In this circuit, Q_1 is called a *driver.* It serves to establish the quiescent current for push-pull transistors Q_1 and Q_2, while at the same time operating as a class A amplifier.

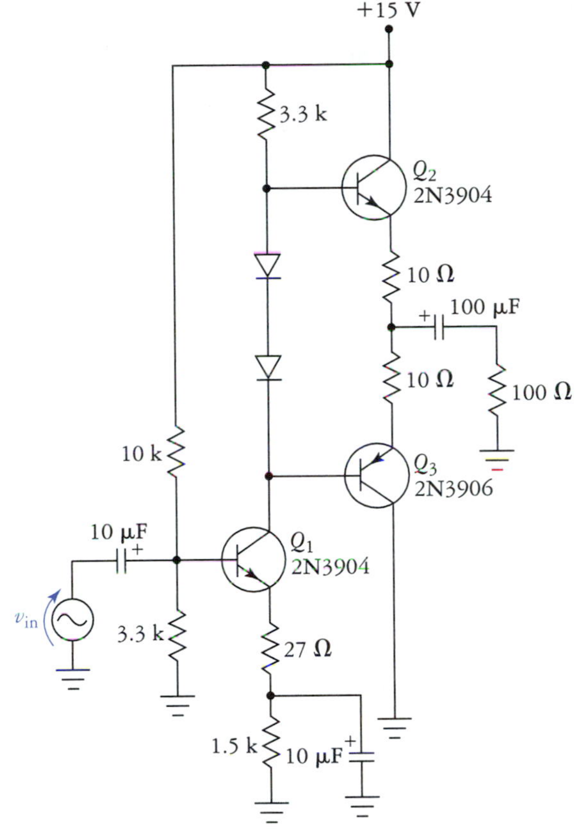

FIGURE 10.30
Push-pull amplifier with direct-coupled driver stage.

DC Analysis

We begin the analysis of this circuit by determining the Q point of Q_1. Assuming $\beta = 100$, we find

$$\begin{aligned} R_{Th} &= 10\ \text{k}\Omega \parallel 3.3\ \text{k}\Omega \\ &= 2.5\ \text{k}\Omega \end{aligned}$$

$$\begin{aligned} V_{Th} &= 15\ \text{V}\frac{3.3\ \text{k}\Omega}{10\ \text{k}\Omega + 3.3\ \text{k}\Omega} \\ &= 3.72\ \text{V} \end{aligned}$$

$$\begin{aligned} I_{C_1} &= \frac{3.72\ \text{V} - 0.7\ \text{V}}{1.5\ \text{k}\Omega + 27\ \Omega + 25\ \Omega} \\ &= 1.9\ \text{mA} \end{aligned}$$

$$\begin{aligned} V_{CE_1} &= V_{CC} - 2V_{BE} - I_{C_1}(3.3\ \text{k}\Omega + 27\ \Omega + 1.5\ \text{k}\Omega) \\ &= 4.4\ \text{V} \end{aligned}$$

Before we go any further, recall that for a capacitively coupled CE class A amplifier the maximum negative-going signal swing is approximately $-V_{CEQ}$. Since this signal drives the emitter-follower push-pull stage (where $A_v < 1$), we find immediately that the output of the amplifier will clip somewhere before reaching -4.4 V (about -4 V). The maximum positive signal swing for the driver is found to be $+5$ V (refer back to Chapter 5).

The dynamic emitter resistance of Q_1 is

$$\begin{aligned} r_{e_1} &= 26\ \text{mV}/1.9\ \text{mA} \\ &= 13.7\ \Omega \end{aligned}$$

Due to the presence of emitter swamping resistors and because 1N4148 diodes do not exactly match 2N3904 and 2N3906 transistors, the best we can do is estimate the quiescent current of the output stage at somewhat less than 1.9 mA. If the output transistors are reasonably matched, they will divide the supply voltage evenly.

AC Analysis

We begin with the output stage (stage 2). Assuming $\beta = 100$ and $r'_e = 5\ \Omega$ we obtain

$$A_{v_2} = \frac{100\ \Omega}{100\ \Omega + 10\ \Omega + 5\ \Omega}$$
$$= 0.87$$

$$R_{\text{in}_2} = 3.3\ \text{k}\Omega \parallel (100\ \Omega + 10\ \Omega + 5\ \Omega)\ 100$$
$$= 2.6\ \text{k}\Omega$$

The first stage is now analyzed, where by inspection we find

$$R'_{C_1} = R_{\text{in}_2}$$

$$A_{v_1} = -\frac{R'_{C_1}}{r_{e_1} + R_{e_1}}$$
$$= \frac{2.6\ \text{k}\Omega}{13.7\ \Omega + 27\ \Omega}$$
$$= -64$$

$$R_{\text{in}_1} = R_{\text{Th}} \parallel \beta(R_{e_1} + r_{e_1})$$
$$= 2.5\ \text{k}\Omega \parallel 100(27\ \Omega + 13.7\ \Omega)$$
$$= 1.5\ \text{k}\Omega$$

The overall specifications are

$$R_{\text{in}} = R_{\text{in}_1} = 1.5\ \text{k}\Omega$$

$$A_v = A_{v_1}A_{v_2}$$
$$= (0.87)(-64)$$
$$= -56$$

$$A_i = A_v\frac{R_{\text{in}}}{R_L}$$
$$= -56\frac{1.5\ \text{k}\Omega}{100\ \Omega}$$
$$= -840$$

$$A_p = A_vA_i$$
$$= (-56)(-840)$$
$$\cong 47000$$

The maximum and minimum output voltage limits are found using the clipping points of the driver stage as follows:

$$V_{o(\text{max})} = A_{v_2}V_{o_1(\text{max})}$$
$$= (0.87)(5\ \text{V})$$
$$= 4.4\ \text{V}$$

$$V_{o(\text{min})} = A_{v_2}V_{o_1(\text{min})}$$
$$= (0.87)(-4\ \text{V})$$
$$= -3.5\ \text{V}$$

Thus the maximum unclipped output voltage is $2 \times 3.5\text{ V} = 7\text{ V}_{\text{p-p}}$. The peak power delivered to the load (without clipping) is

$$P_{o(\max)} = \frac{V_{\text{pk}}^2}{R_L}$$

$$= \frac{3.5\ V^2}{100\ \Omega}$$

$$= 122.5\text{ mW}$$

For a maximum continuous sine wave output the rms and average power are the same:

$$P_{o(\text{rms})} = P_{o(\text{ave})} = 61.25\text{ mW}$$

This is not a very high-power amplifier; it might be sufficient for driving headphones with a portable CD player. Even though it is not too practical, the amplifier we just analyzed serves to illustrate how many of the concepts we covered earlier can be applied to new circuit configurations.

The V_{BE} Multiplier

An interesting circuit that is used often to provide a stable bias voltage is the V_{BE} multiplier. Both single- and bipolar power supply versions are shown in Fig. 10.31, with all pertinent voltages and currents labeled. We will analyze the single-supply version by making the assumption that $I_B \ll I_1$, and the transistor is in the active region. These assumptions will be valid if $\beta > 50$ and certain restrictions on resistor values are met.

Note
V_{BE} multipliers are used in both discrete and integrated circuit amplifier designs.

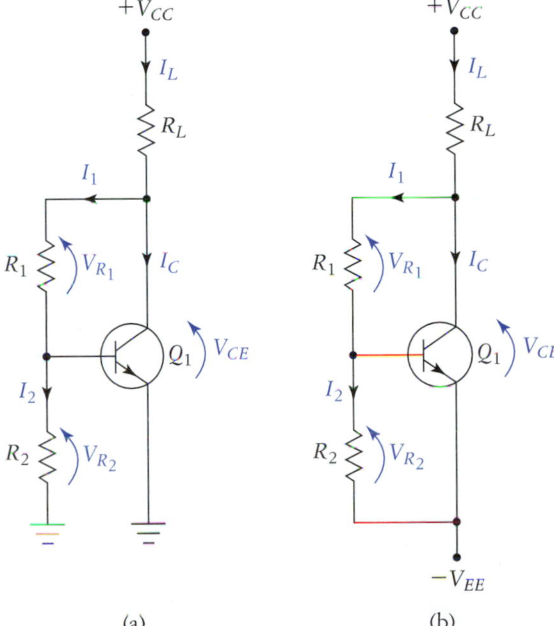

FIGURE 10.31
A V_{BE} multiplier. (a) Single-supply version. (b) Bipolar-supply version.

We begin with direct inspection, which reveals that since R_2 is in parallel with the base–emitter junction of Q_1 we have

$$V_{R_2} = V_{BE}$$

So we can easily determine I_2 using

$$I_2 = \frac{V_{BE}}{R_2}$$

Now, assuming the base current is negligibly small $I_1 = I_2$, which means

$$V_{R_1} = I_2 R_1 = \frac{V_{BE}}{R_2} R_1$$

Applying KVL we find

$$V_{CE} = V_{R_1} + V_{R_2} = \frac{V_{BE}}{R_2} R_1 + V_{BE}$$

This is normally factored to produce

$$V_{CE} = V_{BE}\left(1 + \frac{R_1}{R_2}\right) \tag{10.34}$$

It is clear where the term V_{BE} multiplier comes from. As Eq. (10.34) indicates, the collector-to-emitter voltage of Q_1 is a multiple of V_{BE} that is easily set by choosing two resistor values.

EXAMPLE 10.9

Refer to Fig. 10.31. Assume $R_1 = 3\ \text{k}\Omega$, $R_2 = 1\ \text{k}\Omega$, $R_L = 1\ \text{k}\Omega$, and $V_{CC} = 15$ V.

(a) Determine I_1, I_L, I_C, and V_{CE} (not necessarily in that order).
(b) Determine the minimum supply voltage $V_{CC(\text{min})}$ that can be used.
(c) Determine the maximum allowable load resistance $R_{L(\text{max})}$.
(d) Determine the minimum allowable load resistance $R_{L(\text{min})}$, assuming that the transistor can handle a maximum current of $I_{C(\text{max})} = 500$ mA.

Solution (a) From the derivation that was just presented we have

$$I_1 = I_2 = \frac{V_{BE}}{R_2} = \frac{0.7\ \text{V}}{1\ \text{k}\Omega} = 700\ \mu\text{A}$$

Using Eq. (10.34),

$$V_{CE} = V_{BE}\left(1 + \frac{R_1}{R_2}\right) = 0.7\ \text{V}\left(1 + \frac{3\ \text{k}\Omega}{1\ \text{k}\Omega}\right) = 2.8\ \text{V}$$

Next we apply Ohm's law and KVL to find I_L:

$$\begin{aligned} I_L &= V_{RL}/R_L \\ &= (V_{CC} - V_{CE})/R_L \\ &= (15\ \text{V} - 2.8\ \text{V})/1\ \text{k}\Omega \\ &= 12.2\ \text{mA} \end{aligned}$$

Using KCL allows us to find I_C:

$$\begin{aligned} I_C &= I_L - I_1 \\ &= 12.2\ \text{mA} - 0.7\ \text{mA} \\ &= 11.5\ \text{mA} \end{aligned}$$

(b) We make the reasonable assumption that as the supply voltage changes, there should be little (ideally zero) change in V_{BE}. This means that there will be little change in V_{CE} and I_1. Using KCL a little earlier we found $I_C = I_L - I_1$. But we can define these terms as follows:

$$I_L = \frac{V_{CC} - V_{CE}}{R_L} \qquad I_1 = \frac{V_{BE}}{R_2}$$

Back substituting gives us

$$I_C = \frac{V_{CC} - V_{CE}}{R_L} - \frac{V_{BE}}{R_2} \tag{10.35}$$

The minimum current that we could possibly have is $I_C = 0$, so substituting this into Eq. (10.35) and solving for V_{CC} gives us

$$V_{CC(\text{min})} = \frac{V_{BE}R_L}{R_2} + V_{CE} \tag{10.36}$$

Substituting the given resistor values yields

$$V_{CC(\text{min})} = 3.5 \text{ V}$$

(c) Making the same assumptions as in part (b), we find that V_{R_L} is constant, given by

$$V_{R_L} = V_{CC} - V_{CE}$$

and KCL tells us that $I_L = I_C + I_1$. We assume that I_1 is constant and if V_{R_L} is constant then so is I_{R_L}. Thus it is the collector current that changes in order to compensate for varying load and supply voltage conditions. Again, the ideal minimum collector current is $I_{C(\text{min})} = 0$, therefore we must determine the value of R_L that produces this result. When this occurs $I_L = I_1$ but since $I_1 = I_2$, then

$$\begin{aligned} I_L &= I_2 \\ &= \frac{V_{BE}}{R_2} \end{aligned}$$

Since $I_L = V_{RL}/R_L = (V_{CC} - V_{CE})/R_L$, we have (remember, this applies only when $I_C = 0$)

$$\frac{V_{CC} - V_{CE}}{R_L} = \frac{V_{BE}}{R_2}$$

Solving for $R_{L(\text{max})}$ we get the final equation:

$$R_{L(\text{max})} = \frac{R_2(V_{CC} - V_{CE})}{V_{BE}} \tag{10.37}$$

Substituting the values specified in the problem we get

$$R_{L(\text{max})} = 17.4 \text{ k}\Omega$$

(d) The minimum load resistance is limited by either the current-handling capability of the transistor, the power dissipation limit of the transistor, or possibly by power supply current limitations. In this example, we assume that the transistor current rating is the limiting factor. Applying KCL gives us

$$I_C = I_L - I_1$$

where I_1 is constant given by $I_1 = V_{BE}/R_2$ and as derived previously, $I_L(V_{CC} - V_{CE})/R_L$. Back substituting results in

$$I_C = \frac{V_{CC} - V_{CE}}{R_L} - \frac{V_{BE}}{R_2}$$

Solving this for $R_{L(\text{min})}$ we have

$$R_{L(\text{min})} = \frac{V_{CC} - V_{CE}}{I_{C(\text{max})} + \dfrac{V_{BE}}{R_2}} \tag{10.38}$$

The maximum allowable collector current was given as $I_{C(\text{max})} = 500$ mA, which on substitution into Eq. (10.32) yields

$$R_{L(\text{min})} = 24.4\ \Omega$$

Overcurrent Protection

Power amplifiers are often required to drive low-resistance loads to high current and voltage levels that push the output transistors to near their extreme operating limits. Occasionally the output may be subject to short- or open-circuit conditions. In earlier chapters, we discussed the dangers of open-circuit output conditions; this increases the voltage gain of the amplifier very drastically, which may cause severe overdrive even at low signal levels. Similarly, without precautions being taken, a shorted output will usually result in destruction of the push-pull output transistors and possibly other devices.

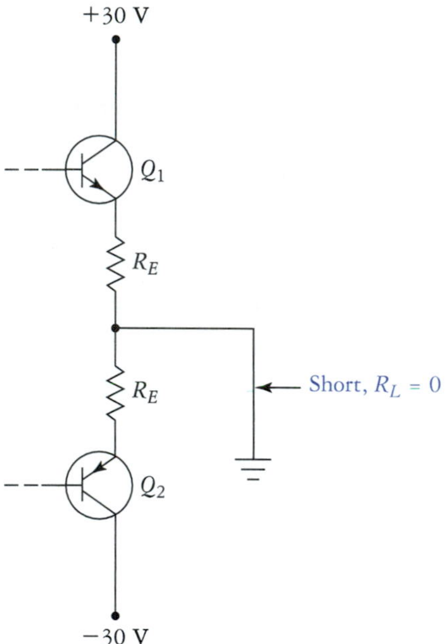

FIGURE 10.32
If the output becomes shorted to ground R_E will limit current.

The addition of emitter swamping resistors adds some degree of overcurrent protection; however, in order to perform this job effectively, too large of a value is generally required. Consider the circuit in Fig. 10.32, where the output of the amplifier has been shorted to ground. Suppose that Q_1 is being driven on (Q_2 is in cutoff). The short-circuit saturation current is given by Eq. (10.32), setting $R_L = 0\ \Omega$, and to make things even simpler, let's estimate $r'_e \cong 0\ \Omega$ as well. This gives us

$$I_{C(\text{short})} \cong \frac{V_{CC}}{R_E}$$

Suppose we wish to limit the short-circuit current in Fig. 10.32 to 1 A. This requires

$$\begin{aligned} R_E &= \frac{V_{CC}}{I_{SC}} \\ &= \frac{30\text{ V}}{1\text{ A}} \\ &= 30\ \Omega \end{aligned}$$

This is most likely far too large to be used in this amplifier; it would dissipate too much power under normal operating conditions.

The solution to the large-value emitter resistor problem is shown in Fig. 10.33. In this circuit, transistors Q_3 and Q_4 have been connected to provide overcurrent protection for the output transistors. Here is how the circuit works. Suppose that the output is shorted to ground and Q_1 is being driven on. As the emitter current of Q_1 increases, the voltage drop across R_E increases as well. However, once the drop across R_E reaches about one V_{BE} drop (about 0.7 V), transistor Q_3 turns on, robbing base current from Q_1. This effectively limits the maximum current for Q_1 (or Q_2) to

$$I_{C(\max)} = \frac{V_{BE}}{R_E} \tag{10.39}$$

Note
This form of overcurrent protection is used in the design of voltage regulators as well.

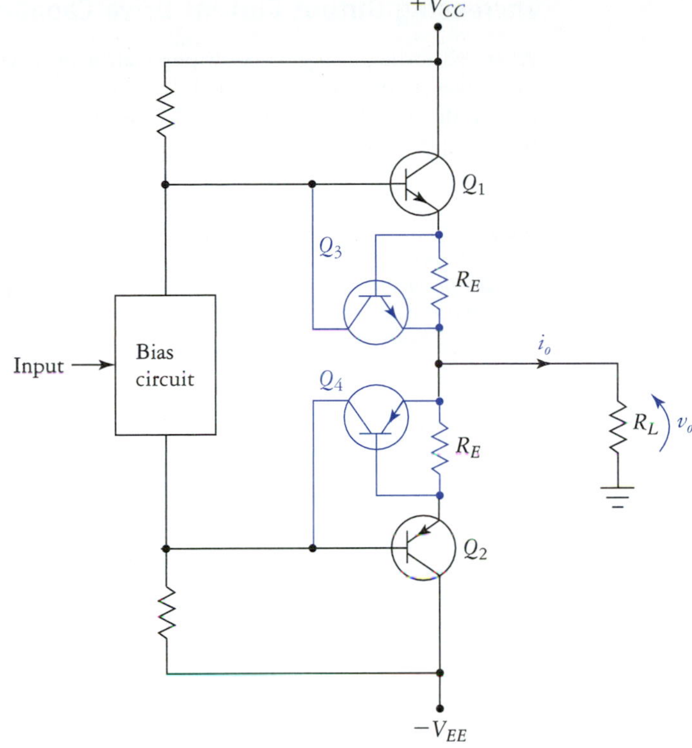

FIGURE 10.33
Short-circuit protection using alternative current-limiting circuitry.

Even though they make the circuit look very complicated, protective transistors Q_3 and Q_4 have no effect on the circuit unless an overcurrent condition exists.

EXAMPLE 10.10

Refer to Fig. 10.33. Given $V_{CC} = V_{EE} = 30$ V, determine the required values for R_E such that $I_{C(\max)} = 1$ A. Assuming that $V_o = 25$ V, determine the value of R_L that will result in current limiting.

Solution To set the maximum collector current, we solve Eq. (10.39), giving us

$$R_E = \frac{V_{BE}}{I_{C(\max)}} = \frac{0.7\text{ V}}{1\text{ A}} = 0.7\ \Omega$$

This is a reasonable value for these resistors, much preferable to the 30 Ω required with simple resistive current limiting.

The load resistance that will induce current limiting at a given output voltage is given by

$$R_L = \frac{V_o}{I_{C(\max)}} = \frac{25\text{ V}}{1\text{ A}} = 25\ \Omega$$

The implication here is that if we are driving a load resistance that is less than 25 Ω (an 8-Ω loudspeaker, for example), the amplifier will current limit the output long before maximum output voltage can be achieved.

Increasing Output Current Drive Capability

As stated earlier, a single npn or pnp transistor may not be able to handle the load placed on it in a given power amplifier application. In such cases transistors are connected in parallel as shown in the push-pull amplifier of Fig. 10.34. Connecting transistors in parallel effectively doubles the current-handling and power dissipation ratings, compared to using a single device.

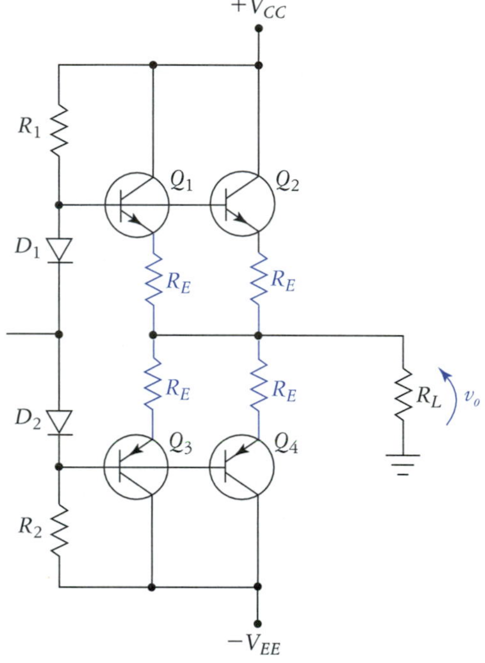

FIGURE 10.34
Parallel-connected output transistors double current-handling capability but require emitter swamping resistors to be used.

When bipolar transistors are connected in parallel, emitter swamping resistors are an absolute necessity. The emitter resistors provide thermal stability and prevent one transistor from hogging all of the load current.

You can think of two parallel connected transistors as a single device with twice the normal current and power rating. The breakdown voltage limits of the pair, however, are the same as for a single device. Also, the effective β of a parallel connected pair is half that of a single transistor. Assuming that all of the parallel connected devices are matched, we have the following relationships where n is the number of transistors connected and the subscript T is the total for the composite device:

$$I_{CT(\max)} = nI_{c(\max)}$$

$$P_{DT(\max)} = nP_{D(\max)}$$

$$\beta_T = \beta/n$$

Connecting transistors in parallel also slows down switching speeds. This may be a concern in high-frequency applications.

10.6 FET POWER AMPLIFIERS

Field-effect transistors, particularly enhancement-mode VMOS devices, are frequently used in the design of power amplifiers. FETs offer the advantages of higher input resistance, faster switching speeds, and lack of second breakdown, compared to BJTs. Additionally, because FETs have a positive temperature coefficient, they can be connected in parallel without the need for efficiency-reducing source swamping resistors. The main disadvantage of high-power FETs is that they have lower transconductance than BJTs.

A VFET push-pull amplifier is shown in Fig. 10.35. The VMOS devices used are the 2SK134 and 2SJ49, which are complementary devices rated for $I_{D(\max)} = 7$ A and $P_{D(\max)} = 100$ W, with typical forward transconductance $g_{fs} = 1$ S. The V_{BE} multiplier would be adjusted to set the quies-

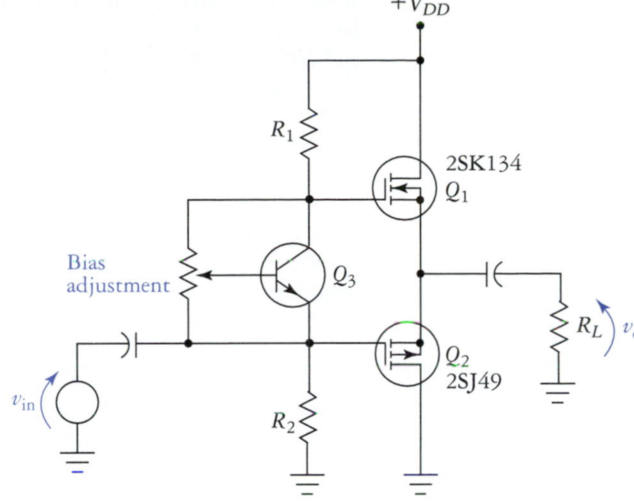

FIGURE 10.35
Class AB, EMOS complementary-symmetry amplifier.

cent current of the FETs to some small level (tens of milliamps) that will eliminate crossover distortion. Because the amplifier is a source follower, as in the case of the emitter-follower, the voltage gain will be less than unity.

10.7 AMPLIFIER THERMAL CONSIDERATIONS

Up to this point we have not really considered the implications of high-power operation in terms of transistor power ratings. Clearly this is an important aspect of power amplifier operation.

Transistor Current Voltage and Power Limitations

Transistors have three primary high-power limiting factors: collector-to-emitter breakdown voltage BV_{CEO}, maximum collector current $I_{C(\max)}$, and maximum power dissipation $P_{D(\max)}$. These three parameters can be interpreted graphically as shown in Fig. 10.36. The device voltage and current limits are the easiest to interpret. Device instantaneous power dissipation is given by

$$p_D = i_C v_{CE}$$

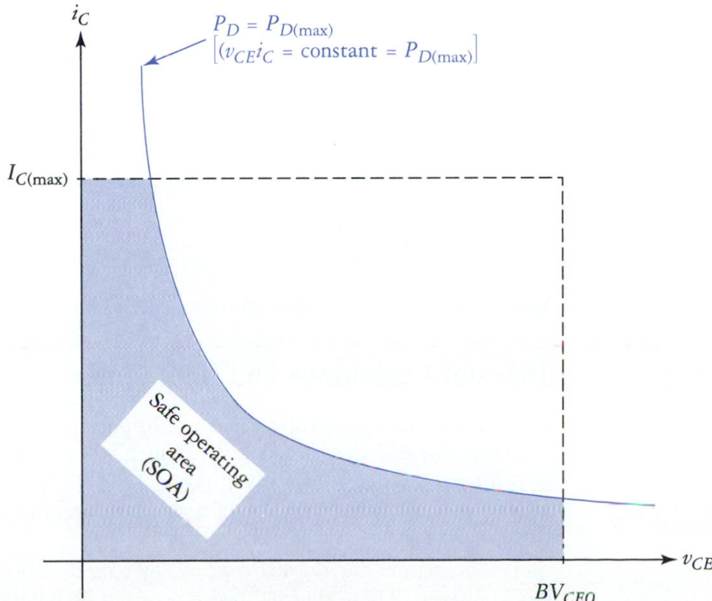

FIGURE 10.36
The maximum power hyperbola of a transistor defines its safe operating area (SOA).

If we hold p_D constant at its maximum value ($p_D = P_{D(max)}$) then the graph of this equation is a *hyperbola,* like that shown in Fig. 10.36. The area bounded by the maximum power hyperbola, $I_{C(max)}$ and BV_{CEO}, is called the *safe operating area* (SOA) of the transistor. This is the shaded portion of Fig. 10.36. We must ensure that the instantaneous operating point of the transistor never leaves this area, or it could be damaged or destroyed.

If we were to plot the ac load line of a poorly designed amplifier and the maximum power hyperbola for the transistor being used, we might obtain a graph like that shown in Fig. 10.37(a). If we did not include the maximum power hyperbola, it would appear that this design is acceptable because neither $I_{C(max)}$ nor BV_{CEO} is exceeded during operation. Because the max power hyperbola is included though, it is obvious that this is not a good design. If the amplifier were class A, the Q point would be outside of the SOA and the transistor would overheat and be destroyed quickly. If a class B or AB design were used, damage to the transistor would be delayed until the operating point was driven out of the SOA. If we select another transistor with appropriate power handling ability, we might obtain a graph similar to that in Fig. 10.37(b). Here, the ac load line is contained within the SOA, so the transistor is not overstressed.

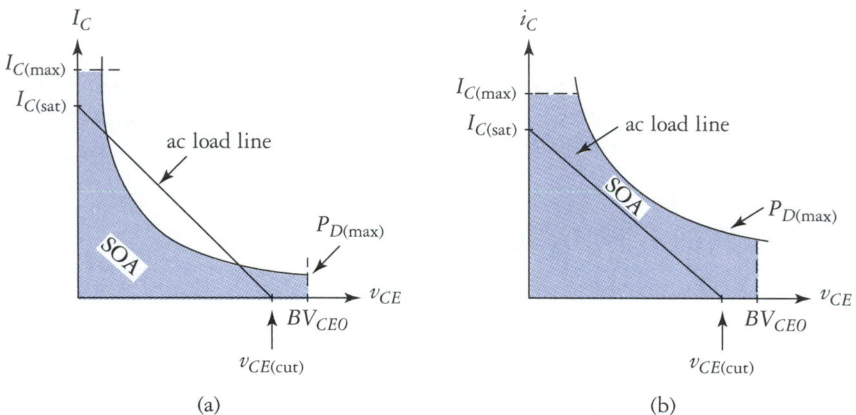

FIGURE 10.37
(a) AC load line exceeds SOA. Not good. (b) AC load line within SOA. Good.

It is clear from Fig. 10.36 that we (normally) cannot operate a transistor at its maximum current and maximum voltage limits simultaneously. For example, let's examine the the very popular 2N3055 npn Si power transistor. According to the Motorola Power Transistor Data Book, revision 7, the 2N3055 (and its pnp complement the MJ2955) has the following ratings:

> **Note**
> In a practical situation, a 2N3055 would be limited to about 15 or 20 W maximum power dissipation, even using a good heat sink.

$$P_{D(max)} = 115 \text{ W}$$

$$I_{C(max)} = 15 \text{ A}$$

$$BV_{CEO} = 60 \text{ V}$$

If we operated this device at its voltage and current limits we would have

$$\begin{aligned} P_D &= (60 \text{ V})(15 \text{ A}) \\ &= 900 \text{ W} \end{aligned}$$

Under these conditions the transistor would rapidly self-destruct in a spectacular manner.

Thermal Resistance and Heat Sinks

Some additional transitor ratings are important in high-power applications. One rating is *thermal resistance from junction to ambient* θ_{JA}, which is measured in °C/W. This parameter tells us how hot the C-B junction will get at a given power level if the transistor is mounted such that the case radiates heat to free air. It is a measure of how well the transistor can conduct heat from the C-B junction to the outside air.

Sometimes the *thermal resistance from junction to case* θ_{JC} and *thermal resistance from case to ambient* θ_{CA} are given. These thermal resistances can be handled mathematically like normal resistances and we have

$$\theta_{JA} = \theta_{JC} + \theta_{CA} \tag{10.40}$$

Often a heat sink is required in order to reduce the thermal resistance from junction to ambient. A heat sink is simply a piece of thermally conductive metal (usually aluminum) that provides better transfer of heat from the device to ambient air. A common heat sink design is shown in Fig. 10.38. A manufacturer will specify the thermal resistance of a given heat sink to ambient θ_{SA}. The actual value of θ_{SA} also depends on the orientation of the heat sink such that convection can occur, and whether forced-air cooling is available and so on.

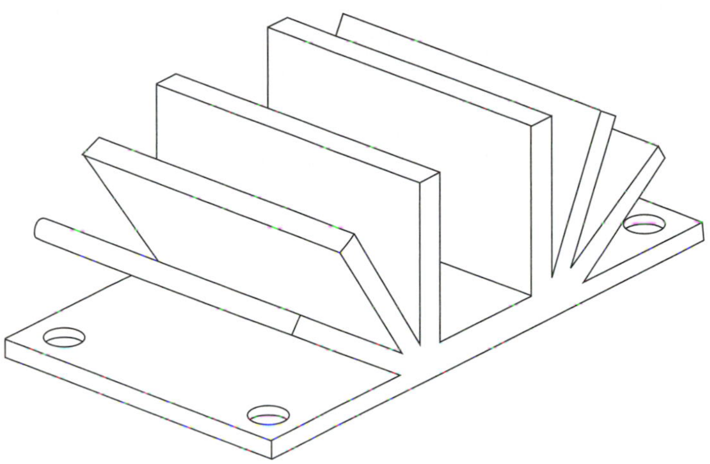

FIGURE 10.38
Typical radial-finned heat sink.

Different transistor case types have different thermal resistance ratings. The two most common power transistor case styles are the TO-3 and TO-220. These cases are shown in Fig. 10.39. The approximate thermal ratings for these cases are given in Table 10.1.

FIGURE 10.39
Common power transistor packages.

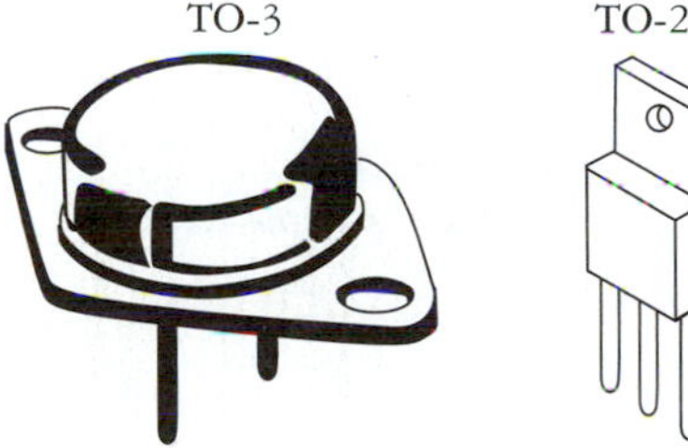

TABLE 10.1
Thermal Ratings for Transistor Packages

Case Style	θ_{JC}	θ_{CA}	θ_{JA}	(°C/W)
TO-3	1.5	33	34.5	
TO-220	4.2	63	58.8	

Given TO-3 and TO-220 devices in free air, the TO-3 transistor is about twice as effective at transferring heat to the surrounding air than the TO-220 device. Mounting these devices to an appropriate heat sink will normally result in about a 3-to-1 advantage in favor of the TO-3 package.

A thermal system consisting of thermal resistance, ambient temperature, and device power dissipation is analogous to an electrical circuit as shown in Fig. 10.40. Using this equivalent circuit, we can determine the junction temperature at any given power level using

$$T_J = T_A + P_D(\theta_{JC} + \theta_{CS} + \theta_{SA}) \tag{10.41}$$

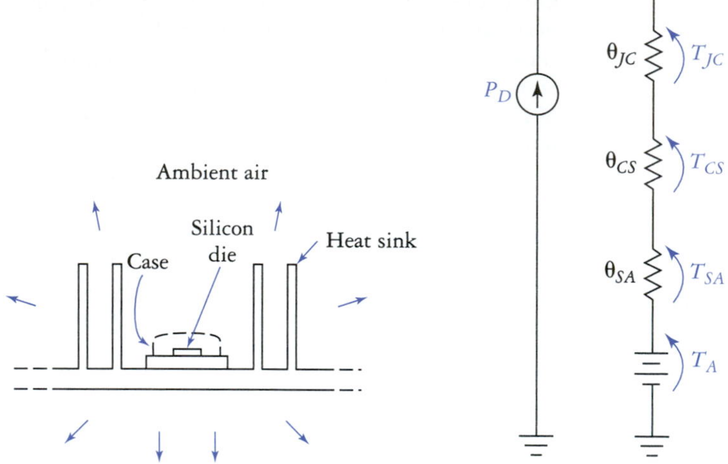

FIGURE 10.40
A transistor mounted on a heat sink can be modeled by an analogous electrical circuit to determine various temperatures.

If no heat sink is used, then the equation is

$$T_J = T_A + P_D\theta_{JA} \tag{10.42}$$

Let's put all of this information to use in a typical situation.

EXAMPLE 10.11

A 2N3055 is biased up as shown in Fig. 10.41. This is the sort of situation that could occur in the design of a regulated power supply for example, which is a topic we cover in more detail in Chapter 18. The biasing circuitry holds the output voltage constant at $V_o = 5.0$ V. The input voltage is constant at $V_{in} = 12$ V. Assume the transistor has a TO-3 case exposed to free air at $T_A = 25$°C without a heat sink. Determine (a) P_D and T_J for the transistor given $I_L = 200$ mA and (b) the maximum allowable load current given $T_{J(max)} = 200$°C.

FIGURE 10.41
Circuit for Example 10.11.

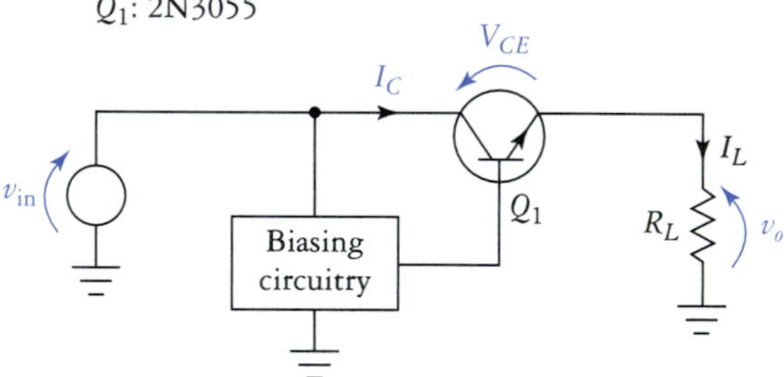

Solution (a) Assuming active region operation, the power dissipation of the transistor is $P_D = I_C V_{CE}$. Adapting this equation to the problem at hand,

$$\begin{aligned}P_D &= I_C(V_{in} - V_o)\\ &= 200 \text{ mA } (12 \text{ V} - 5 \text{ V})\\ &= (200 \text{ mA})(97 \text{ V})\\ &= 1.4 \text{ W}\end{aligned}$$

The transistor junction temperature is found using Eq. (10.42):

$$\begin{aligned}T_J &= T_A + P_D\theta_{JA}\\ &= 25° + (1.4 \text{ W})(34.5°/\text{W})\\ &= 25° + 48.3°\\ &= 73.3°\text{C}\end{aligned}$$

(b) To determine the maximum load current, we first solve Eq. (10.42) for P_D and substitute the given parameter values, thus

$$P_{D(\max)} = \frac{T_{J(\max)} - T_A}{\theta_{JA}} = \frac{200° - 25°}{34.5°/\mathrm{W}} = 5.07\ \mathrm{W}$$

Now we solve the transistor power dissipation equation and back substitute giving us

$$I_{L(\max)} = \frac{P_{D(\max)}}{V_{\text{in}} - V_L} = \frac{5.07\ \mathrm{W}}{12\ \mathrm{V} - 5\ \mathrm{V}} = 724\ \mathrm{mA}$$

Think about the previous example for a moment. The maximum allowable power dissipation and collector current found were far below the published limits for the 2N3055. This is a very common occurrence and it helps explain why even moderately powerful audio amplifiers (50 to 100 W_{rms}) usually require parallel connection of several power transistors and large heat sinks.

Derating Curves

The maximum power dissipation rating for a transistor is *derated* as case (or possibly ambient) temperature is increased above a reference value (typically 25°C). This means that if the case temperature is held at or below 25°C, the transistor can be operated at full-rated power. Case temperatures above 25°C require the maximum device dissipation to be limited below a value obtained from a graph such as that shown in Fig. 10.42. Here we can see that once the case temperature reaches the maximum junction temperature (200°C) the transistor cannot dissipate any additional power.

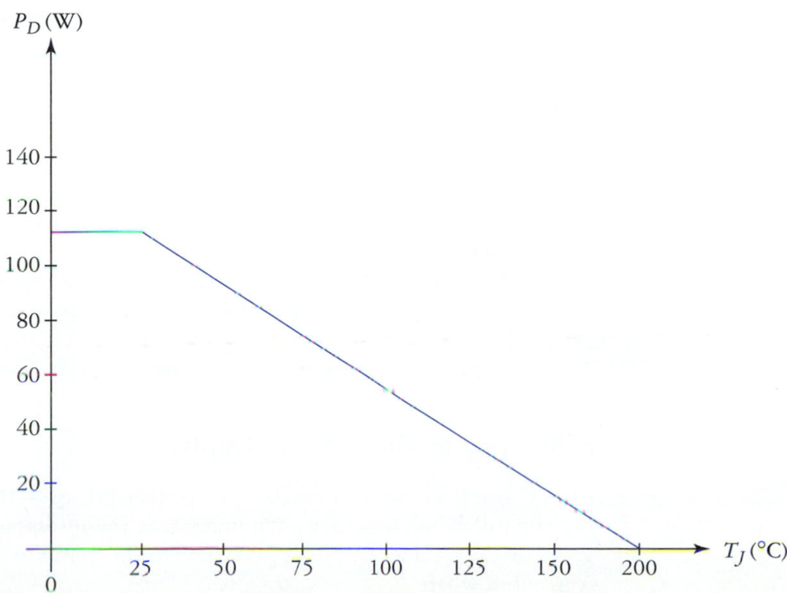

FIGURE 10.42
Transistor power derating curve.

10.8 CLASS C AMPLIFIERS

Class A amplifiers have a conduction angle $\alpha = 360°$, which is achieved by biasing the transistor well into the active region. Class B amplifiers have $\alpha \cong 180°$, which is achieved by using no biasing, thus placing the Q point at cutoff. Class C amplifiers are biased *beyond cutoff* resulting in $\alpha \ll 180°$. This produces a virtual Q-point location as shown in Fig. 10.43. The actual Q point will be located at $v_{CE(\text{cut})}$ because assuming an npn transistor is used it is not possible for I_C to have a negative value (we have seen $v_{CE(\text{cut})} > V_{CC}$ however). This virtual Q-point location is obtained by applying reverse bias to the B-E junction of the transistor, thus the transistor will not go into conduction until the reverse bias is overcome.

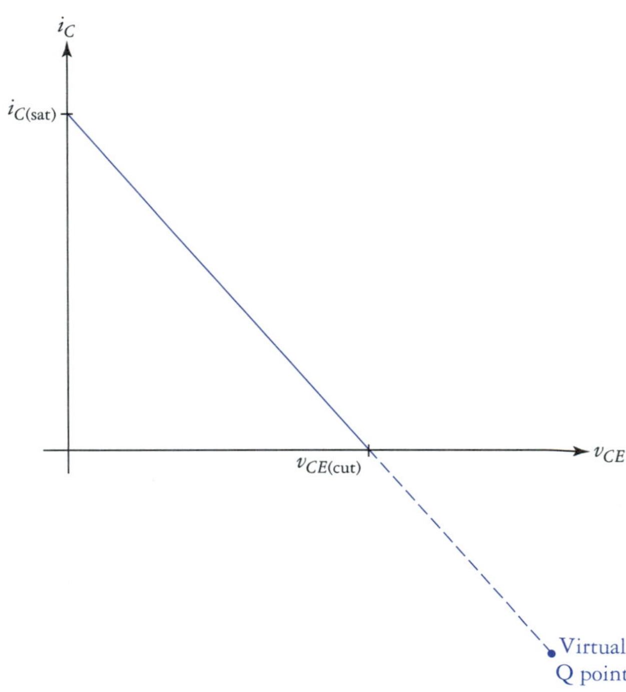

FIGURE 10.43
Load line for the class C amplifier.

It is possible to apply a reverse bias to the B-E junction; however, most often the reverse bias is produced via a clamping action using the circuit shown in Fig. 10.44. The ac equivalent of the circuit is the familiar negative clamper shown in the lower part of the diagram. Representative input and base voltage waveforms are shown as well. The only time the transistor is not in cutoff is during the short interval when the B-E junction is forward biased.

If we plot the collector-to-emitter voltage of the class C amplifier with a sinusoidal input voltage, a graph like that of Fig. 10.45 results. The dashed waveform represents the input voltage. Most of the time the transistor is in cutoff where $V_{CE} = V_{CC}$. The transistor is driven into saturation for a short time T_W that is a complicated function of the amplitude and frequency of v_{in}, R_1, C_1, and the transistor characteristics. The main point is that the transistor has a low conduction angle, typically on the order of $\alpha < 30°$.

If the class C amplifier is to work effectively, we must satisfy the following inequality:

$$R_1C_1 \geq 10T_{\text{in}} \qquad (10.43)$$

This is the same inequality that was presented for the clamper circuit in Chapter 2 [see Eq. 2.42].

Efficiency of the Class C Amplifier

Class C amplifiers are very efficient, with efficiency $\eta \to 100\%$. This can be understood by considering the following argument. We know that the instantaneous power dissipation of a transistor is given by the product $p_D = i_C v_{CE}$. We determine from Fig. 10.45 that most of the time the transistor is in cutoff where

$$\begin{aligned} P_{D(\text{cut})} &= I_{C(\text{cut})}V_{CE(\text{cut})} \\ &\cong (0\text{ mA})(V_{CC}) \\ &= 0\text{ W} \end{aligned}$$

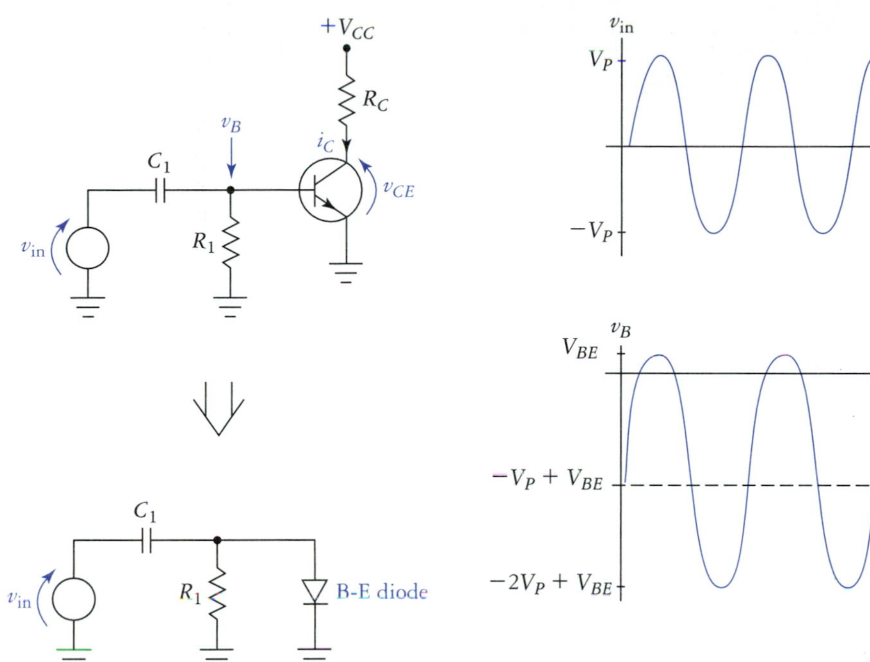

FIGURE 10.44
Class C amplifier and typical voltage waveforms.

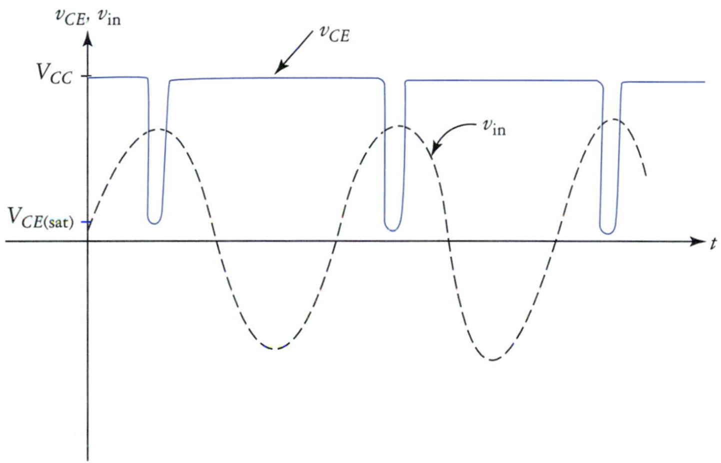

FIGURE 10.45
Class C amplifier output voltage waveform.

When the transistor is in saturation, the power dissipation also very low, assuming that $V_{CE(\text{sat})}$ is low, thus

$$
\begin{aligned}
P_{D(\text{sat})} &= I_{C(\text{sat})}V_{CE(\text{sat})} \\
&\cong (I_{C(\text{sat})})(0\ \text{V}) \\
&= 0\ \text{W}
\end{aligned}
$$

Transistor power dissipation is maximum when $v_{CE} = V_{CC}/2$ and $i_C = I_{C(\text{sat})}/2$. This power dissipation is plotted in relation to v_{CE} in Fig. 10.46, where the peak dissipation is seen to be

$$P_{D(\text{max})} = \frac{I_{C(\text{sat})} - V_{CE(\text{cut})}}{4} \tag{10.44}$$

Notice that the power peaks occur in pairs and that the average power dissipation is very low. This is why the class C amplifier is so efficient.

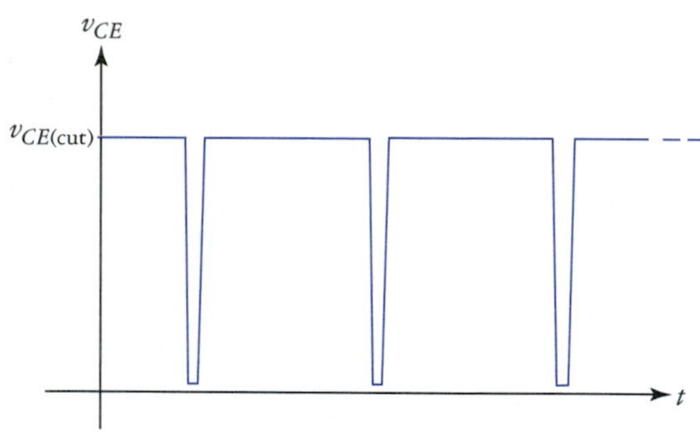

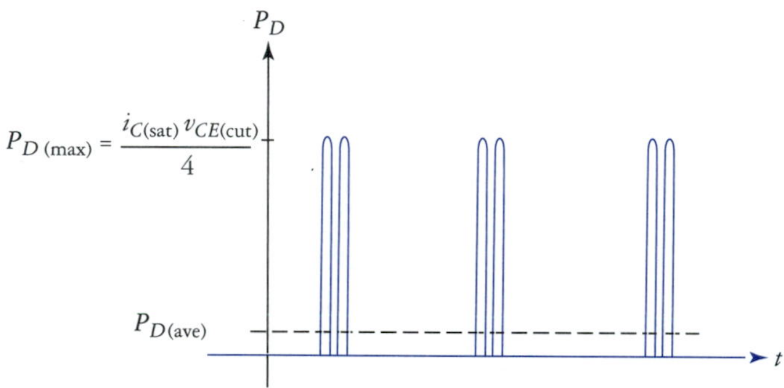

FIGURE 10.46
Class C amplifier output voltage and power dissipation waveforms.

Class C Distortion

The disadvantage of the class C amplifier is the very high degree of distortion that it produces. In practical terms, the distortion of the class C amplifier is so great that looking at the output waveform, about all we can determine about the input signal is its frequency. Information relating to wave shape and amplitude is lost at the output. This means that in its current form, the class C amplifier is not suitable for linear amplifier applications.

Base–Emitter Junction Protection

If we use the circuit of Fig. 10.44, we must ensure that v_{in} is sufficiently low in amplitude that breakdown of the B-E junction does not occur. Recall that the emitter of a BJT is heavily doped, which like a zener diode, results in a relatively low breakdown voltage. The B-E breakdown voltage for the 2N3904, for example, is $BV_{BEO} = 5$ V. Because of this, we may wish to place a diode in series with either the base or the emitter terminals as shown in Fig. 10.47. In both circuits, the diode causes an

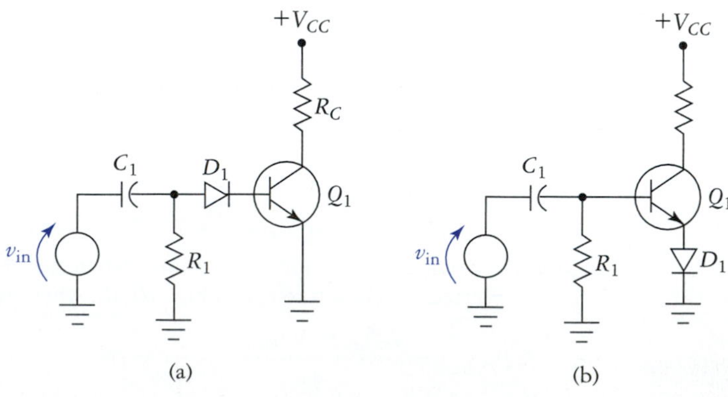

FIGURE 10.47
Protecting the B-E junction.

extra voltage shift of one barrier potential to the base waveform of Fig. 10.44. In the case where the diode is located in the emitter, it also increases $V_{CE(sat)}$ by 0.7 V as well. The purpose of the extra diode, with its comparatively large reverse breakdown voltage, is to protect the B-E junction from damage due to to excessive reverse bias.

Tuned Class C Amplifiers

There is a way in which we can "clean up" the output of a class C amplifier. As shown in Fig. 10.48, placing a parallel resonant network in series with the collector serves to filter out nearly all signal energy, except that at the resonant frequency of the tuned circuit. The parallel resonant circuit is often called a *tank* circuit.

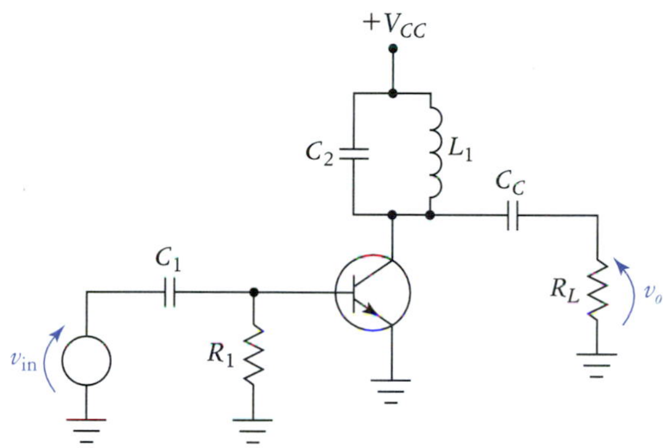

FIGURE 10.48
Tuned class C amplifier.

Assuming the Q of the tank circuit is high ($Q \geq 10$) then the resonant frequency is given by the famous equation

$$f_o = \frac{1}{2\pi\sqrt{L_1C_2}} \tag{10.45}$$

Recall that this equation is obtained by setting $|X_L| = |X_C|$ and solving for f. This means that at resonance X_{L_1} and X_{C_2} are of equal magnitude and opposite phase (they are complex conjugates). Because of the conjugate values of inductive and capacitive reactance, the impedance of the tank circuit will be extremely high at f_o (ideally infinite). The coupling capacitor is chosen to have negligible capacitive reactance at the resonant frequency, and so it does not have a significant effect on the resonant frequency of the amplifier. Using Eq. (10.43), C_1 and R_1 are chosen as before.

Basically, the higher the Q of the tuned circuit, the more clean the output sine wave will be. Factors that affect Q are the series resistance r_s of the inductor, the load resistance, and possibly the transistor active region resistance r_o. At the resonant frequency, the Q of the inductor is given by

$$Q_L\Big|_{f=f_o} = \frac{|X_L|}{r_s} \tag{10.46}$$

and, also at resonance, the equivalent resistance of the tank circuit is given by

$$R_P\Big|_{f=f_o} = r_s(1 + Q_L^2) \tag{10.47}$$

The total effective ac collector resistance is therefore

$$R'_C\Big|_{f=f_o} = R_P \parallel R_L \parallel r_o \tag{10.48}$$

and the overall Q of the resonant circuit is

$$Q = \frac{R'_C}{|X_L|} \tag{10.49}$$

EXAMPLE 10.12

Refer to Fig. 10.48. Given $R_1 = 10\ \text{k}\Omega$, $C_1 = 0.1\ \mu\text{F}$, $C_2 = 0.001\ \mu\text{F}$, $L_1 = 100\ \mu\text{H}$, $r_s = 10\ \Omega$, $C_C = 1\ \mu\text{F}$, $R_L = 10\ \text{k}\Omega$, $r_o = \infty$, and $V_{CC} = 12$ V, determine f_o, R'_C, and Q, and determine whether the inequality of Eq. (10.43) is met if $f_{in} = f_o$.

Solution The resonant frequency is

$$f_o = \frac{1}{2\pi\sqrt{L_1C_2}}$$
$$= \frac{1}{(2\pi)(100\ \mu\text{H} \times 0.001\ \mu\text{F})^{1/2}}$$
$$= 503\ \text{kHz}$$

We now determine the various resonant-frequency-related factors:

$$Q_L = \frac{|X_L|}{r_s}$$
$$= \frac{316\ \Omega}{10\ \Omega}$$
$$= 31.6$$

$$R_P = r_s(1 + Q_L^2)$$
$$= 10\Omega\ (1 + 31.6^2)$$
$$\cong 10\ \text{k}\Omega$$

$$R'_C = R_P \,\|\, R_L \,\|\, r_o$$
$$= 10\ \text{k}\Omega \,\|\, 10\ \text{k}\Omega \,\|\, \infty$$
$$= 5\ \text{k}\Omega$$

$$Q = \frac{R'_C}{|X_L|}$$
$$= \frac{5\ \text{k}\Omega}{316\ \Omega}$$
$$= 15.8$$

At the resonant frequency

$$10\ T_{\text{in}} = 10/f_{\text{in}}$$
$$= 19.9\ \mu\text{s}$$

The RC time constant is

$$R_1C_1 = (10\ \text{k}\Omega)(0.1\ \mu\text{F})$$
$$= 1\ \text{ms}$$

Thus the inequality of Eq. (10.43) is satisfied.

As a point of interest, aside from capacitive coupling, transformer coupling is also used often in class C amplifier designs. We address this topic in Chapter 17, however.

Where Are Class C Amplifiers Used?

You may be wondering what the class C amplifier is good for if it requires a tuned circuit (a bandpass filter, actually) to produce a clean sine wave output, and even then, it is only useful at one frequency or over a very limited range of frequencies. As stated before, class C amplifiers are certainly not useful in the traditional applications we have discussed up until now. However, tuned class C amplifiers are useful in radio-frequency (RF) applications where we are interested in operation at one particular frequency or a very narrow band of frequencies centered around f_o. In these applications, the narrow frequency response characteristics of a tuned amplifier are desirable, plus we have the added benefit of very high efficiency as well. Also, because of the high frequencies involved, it is often possible to use relatively small air-core inductors and transformers in the design of the amplifier.

Frequency Multipliers

Tuned class C amplifiers are sometimes used as so-called *frequency multipliers.* Other more flexible and sophisticated techniques are used to produce frequency multiplication, but the class C amplifier has some uses as well. The basic idea is that if we apply an input signal such that $f_{in} < f_o$, then the output of the tuned class C amplifier will still be a sine wave of f_o Hertz. Thus in a sense we have frequency multiplication. This works best when f_o is an integer multiple of f_{in}. Noninteger multiples produce a distorted output waveform. If the Q of the amplifier is high enough, we can set f_o as high as $4f_{in}$ without having much peak amplitude fluctuation in the output of the amplifier. This is illustrated in Fig. 10.49. One way of thinking about this is to assume that the short-duration collector current pulses cause the tank circuit to resonate or *ring* at f_o. Successive pulses supply energy before much is lost due to finite Q.

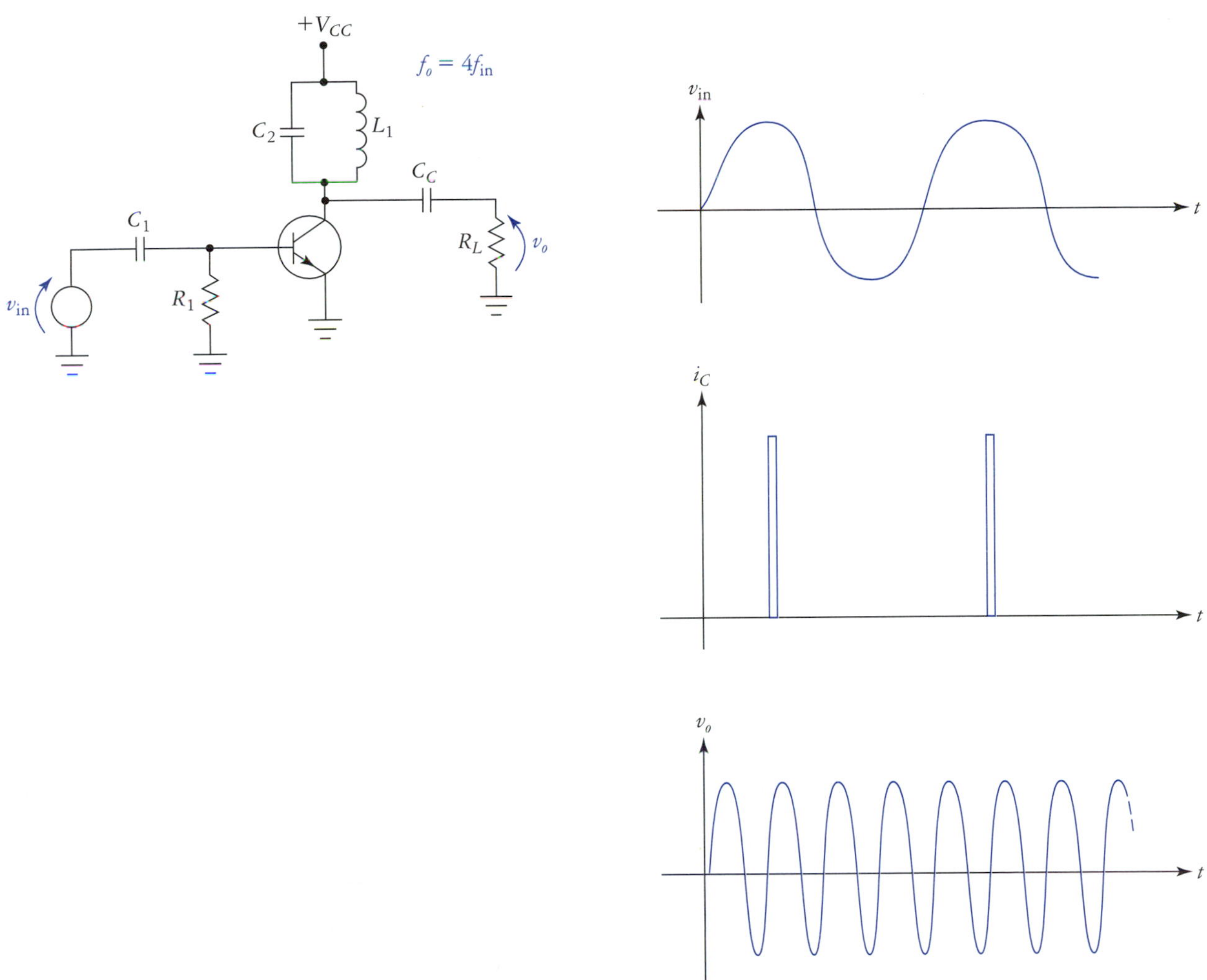

FIGURE 10.49
Tuned class C amplifier and typical voltage waveforms for $f_{in} < f_o$.

An interesting application occurs when we set $f_{in} \ll f_o$. If f_o is an audio frequency, we can use this sort of circuit to simulate a bell or gong, which is simply a mechanical resonant circuit that is mathematically equivalent. The application of a short-duration pulse to the input of the amplifier is equivalent to striking the bell, and the resulting output signal decays exponentially just as the ringing of the bell does. This type of response is called an *exponentially damped sinusoid,* and is shown in Fig. 10.50. The peak amplitude of the sinusoid decays as $e^{-\alpha t}$ where α is called the *damping coefficient.* More is said about α when we discuss active filters. This exponential portion of the response is the *transient response,* where the term *transient* implies a short-lived characteristic. The sinusoidal component of the output is called the *steady-state response* of the system. The combination of transient and steady-state responses is the total response of the system. Phasor algebra can be used to perform steady-state analysis, while differential equations and/or *Laplace transforms* are used to determine transient and total response.

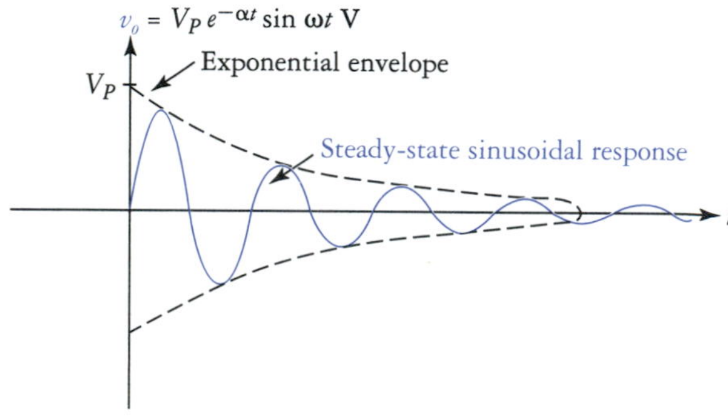

FIGURE 10.50
A short pulse applied to a resonant circuit results in an exponentially decaying "ringing" response.

10.9 COMPUTER-AIDED ANALYSIS APPLICATIONS

In this section we examine the behavior of the class AB and tuned class C amplifiers. Breakout devices are used, and the pulse voltage source is used to generate a pulse train.

Simulation of the Basic Class AB Amplifier

A capacitively coupled class AB push-pull amplifier is shown in Fig. 10.51. The circuit is initially drawn using QbreakN and QbreakP devices from the breakout library of PSpice. The only nondefault parameters specified for these devices are $\beta_F = 50$ and $V_{AF} = 100$ V.

Recall that the model parameters are set by highlighting the desired device, choosing Edit, Model, Edit Instance Model {Text} and inserting the desired parameter values and model name. The Model Editor box is shown in Fig. 10.52. This procedure is done one time. For other devices that are

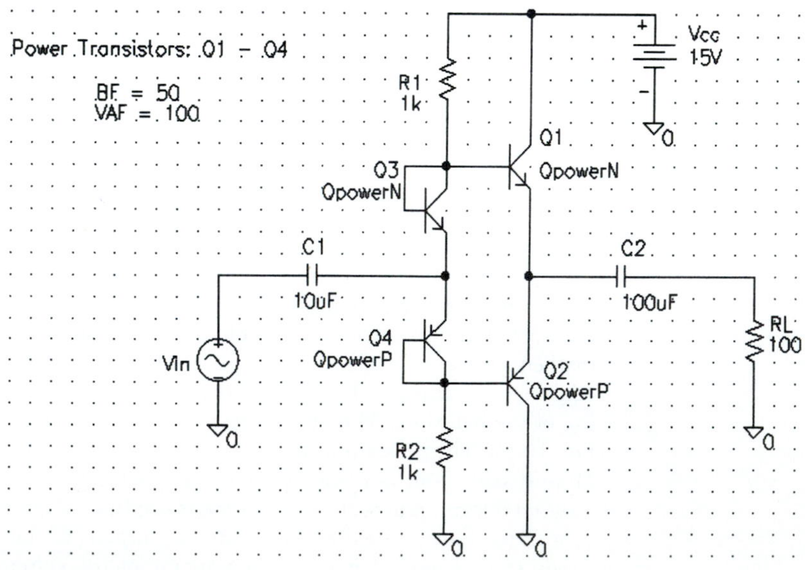

FIGURE 10.51
Class AB amplifier.

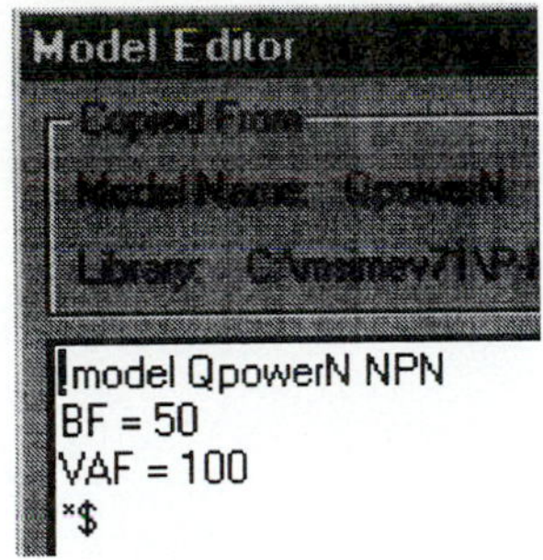

FIGURE 10.52
Editing the transistor model parameters.

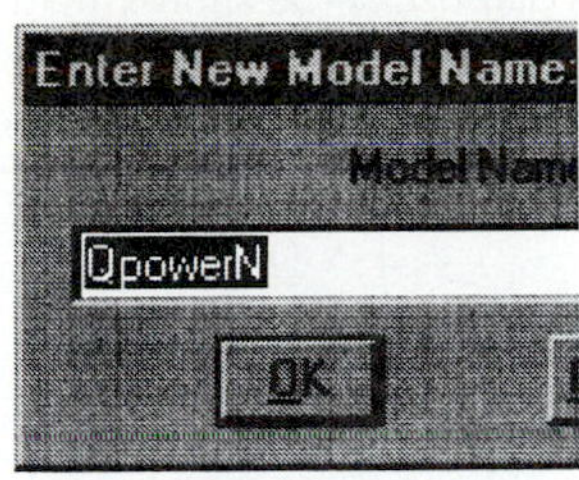

FIGURE 10.53
Assigning a model name.

to use the same model, we change the *model reference* by highlighting the device(s) of interest and choosing Edit, Model, Change Model Reference and inserting the new model name as shown in Fig. 10.53.

Transient Analysis

With the transistor parameters set as required, the sine voltage source is set for DC = 0, AC = 1, Amplitude = 5, Offset = 0, and Freq = 1 kHz. Performing a transient analysis running from 0 to 3 ms and plotting v_{in} and v_o [alias V(Vin:+) and V(RL:2)] results in the plot of Fig. 10.54. Simply taking the ratio of the positive peak amplitudes of these signals produces $A_v = 0.92$. Notice, however, that there is a slight negative dc offset in the output voltage. This causes a slight error in the voltage gain as calculated, but it is not significant here. If we make some reasonable assumptions and hand calculate the voltage gain we obtain

$$A_v = \frac{R_L}{R_L + r'_e}$$

$$= \frac{100\ \Omega}{100.5\ \Omega}$$

$$= 0.995$$

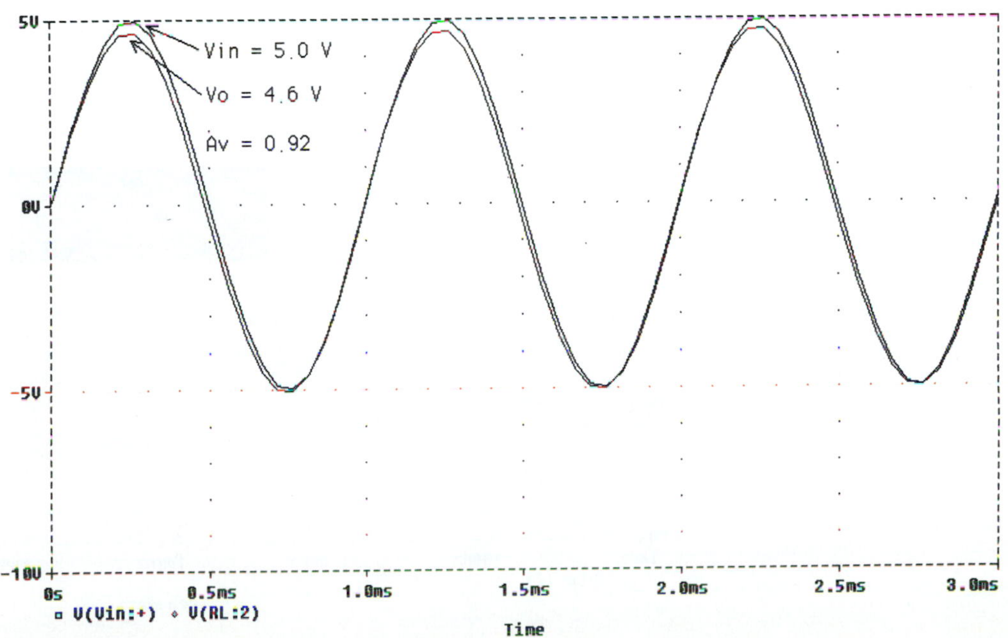

FIGURE 10.54
Inout and output voltage waveforms.

Low-Frequency Response

Let's determine the approximate lower frequency limit of this amplifier. We arbitrarily determine the corner frequency of the input coupling network first. The input resistance of the circuit is

$$\begin{aligned} R_{\text{in}} &\cong R_1 \parallel R_2 \parallel \beta(r_e' + R_L) \\ &= 1\text{ k}\Omega \parallel 1\text{ k}\Omega \parallel 50(0.5\ \Omega + 100\ \Omega) \\ &= 455\ \Omega \end{aligned}$$

which gives us

$$\begin{aligned} f_{C(\text{in})} &= \frac{1}{2\pi R_{\text{in}} C_1} \\ &= \frac{1}{0.0286} \\ &= 35\text{ Hz} \end{aligned}$$

The corner frequency of the output coupling network is

$$\begin{aligned} f_{C(\text{out})} &= \frac{1}{2\pi R_L C_2} \\ &= \frac{1}{0.0623} \\ &= 16\text{ Hz} \end{aligned}$$

Thus we find that the input coupling network dominates the low-frequency response limit; therefore,

$$f_{C(\text{low})} = f_{C(\text{in})} = 35\text{ Hz}$$

Performing a logarithmic (decade) AC Sweep from 10 Hz to 10 MHz and plotting the output voltage in decibels [VdB(RL:2)] produces the display shown in Fig. 10.55. The midband response (where $f_{C(\text{low})} \ll f \ll f_{C(\text{high})}$ is $A_{\text{mid}} \cong -0.2\ dB$, so the lower corner frequency is found by placing the cursor at the frequency where $A_{v(\text{dB})} = -3.2$ dB. This indicates $f_{C(\text{low})} = 40.8$ Hz. Our hand-calculated frequency is acceptably close to this value.

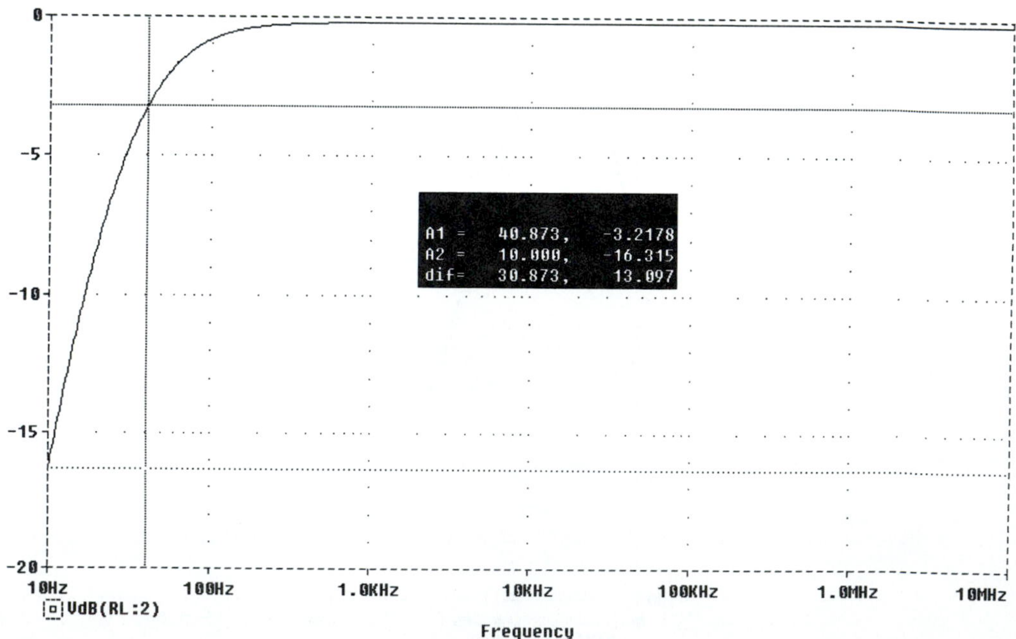

FIGURE 10.55
Lower corner frequency.

Short-Circuited Output

One of the nice things about computer simulations is that we can set up conditions that might result in smoke and flames if we set them up in a lab instead. Let's investigate what happens if we short the output of the amplifier to ground while applying the 5-V_{pk} input signal. To do this, we simply replace R_L with a wire and rerun the simulation. Plotting the collector current of Q_1 produces the plot of Fig. 10.56. The collector current appears to be approaching 400 mA_{pk}. In this case, the main current limiting factor is the reactance of C_2. Increasing the frequency of the input signal or the value of C_2 will result in an increase in peak collector current. You are encouraged to try this (simulation!) on your own.

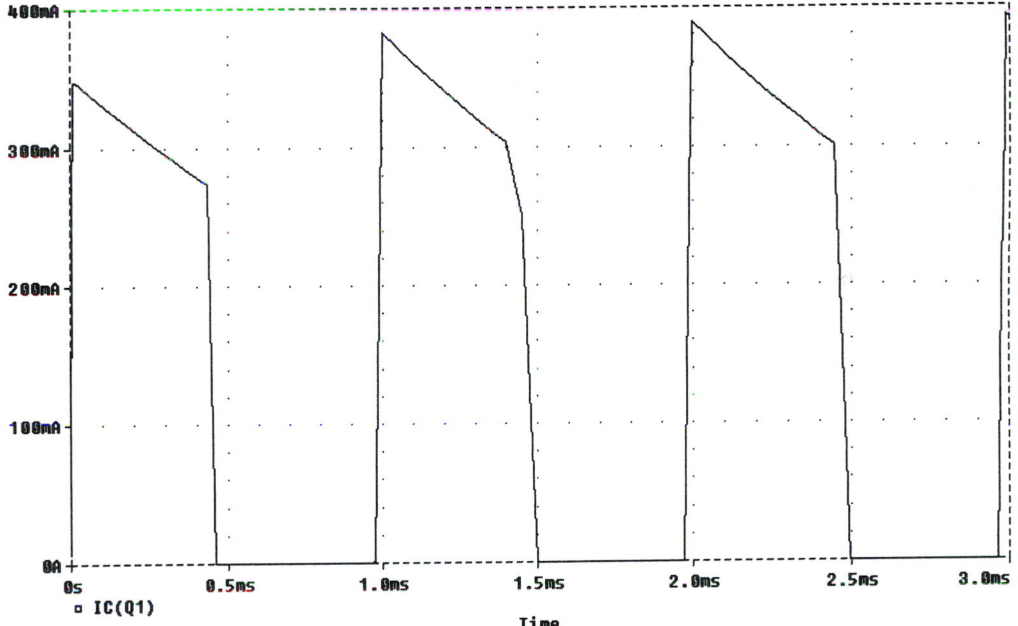

FIGURE 10.56
Collector current under short-circuited output conditions.

Overcurrent Protection

Figure 10.57 is a modification of the previous circuit to include overcurrent protection. Assuming typical transistor behavior, the maximum collector (and load) current is

$$\begin{aligned} I_{L(\max)} &= I_{C(\text{sat})} \\ &= V_{BE}/R_E \\ &= 0.7\ \text{V}/10\ \Omega \\ &= 70\ \text{mA} \end{aligned}$$

This is much lower than the value obtained before, and is presumably a safe maximum limit for our hypothetical output transistors.

Running the simulation and plotting the collector current of Q_1 gives us the waveform shown in Fig. 10.58, where we find $I_{C(\text{sat})} = 80$ mA. This is not too far from the predicted value. It is really very encouraging when such simple approximations produce results that are so accurate. You are encouraged to verify that the current-limiting transistors actually do have negligible effect on the operation of the amplifier when $i_C < i_{C(\text{sat})}$.

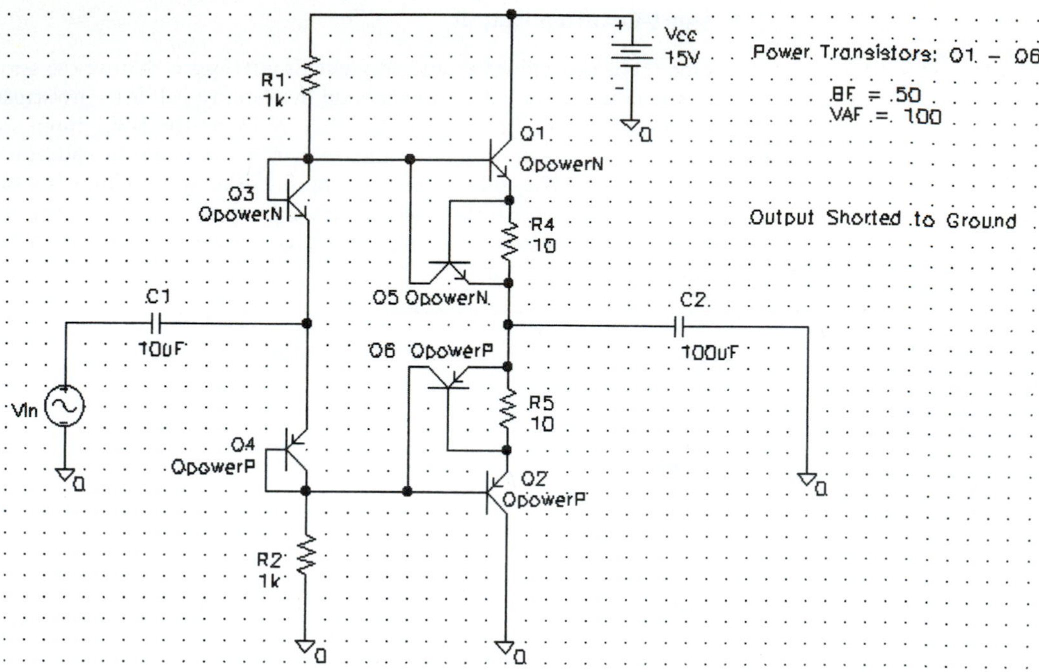

FIGURE 10.57
Implementation of overcurrent protection circuitry.

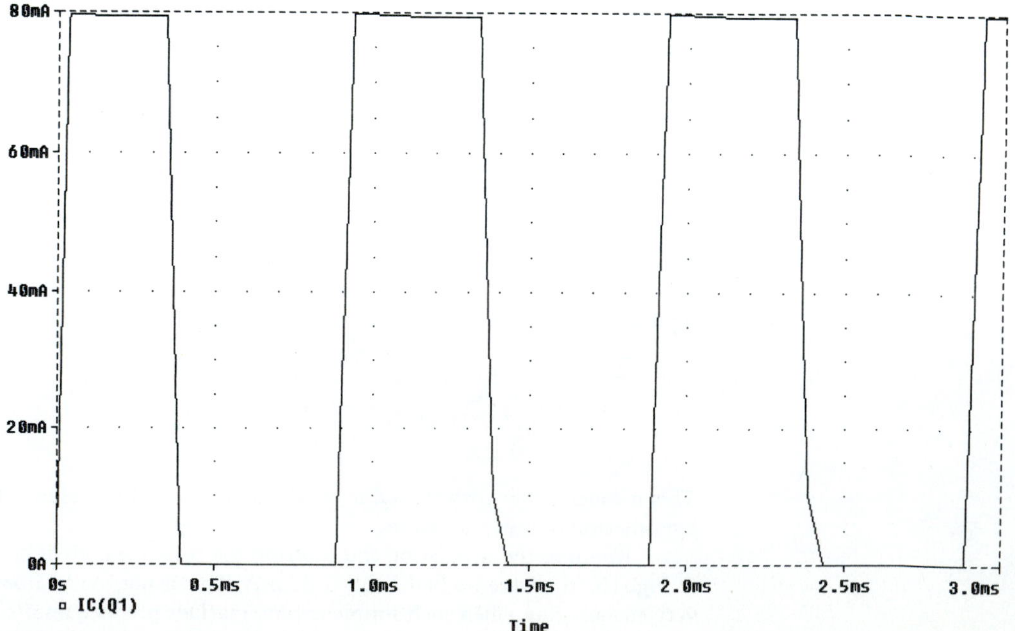

FIGURE 10.58
Resulting short-circuited output current waveform.

Tuned Class C Amplifier

Figure 10.59 is the schematic for a tuned class C amplifier. We hand calculate the resonant frequency and the Q of the tank circuit, assuming that the inductor is ideal ($r_s = 0\ \Omega$ and $Q_L \rightarrow \infty$) and the transistor has $r_o \rightarrow \infty$.

$$f_o = \frac{1}{2\pi\sqrt{L_1C_2}}$$
$$= \frac{1}{(2\pi)(1.53 \times 10^{-4})}$$
$$= 1038 \text{ Hz}$$

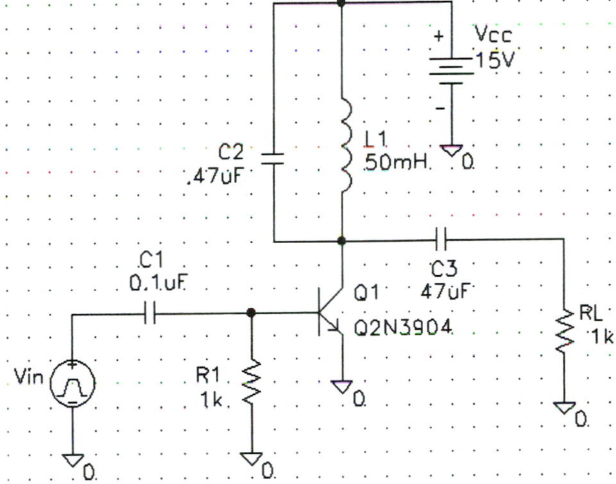

FIGURE 10.59
Tuned class C amplifier.

Because the Q of the inductor is infinitely high, so is the impedance of the tank circuit at resonance. The ac collector resistance is simply $R_C' = R_L = 1\ \text{k}\Omega$, and the reactance of the inductor at resonance is

$$|X_L| = 2\pi f_o L$$
$$= (2\pi)(1038 \text{ Hz})(50 \text{ mH})$$
$$= 326\ \Omega$$

Therefore the Q of the circuit is

$$Q = \frac{R_C'}{X_L}$$
$$= \frac{1\ \text{k}\Omega}{326\ \Omega}$$
$$= 3.07$$

This is a relatively low Q, therefore if we drive the amplifier with a short-duration pulse (an impulse), the output should *ring* at approximately 1038 Hz, while decaying rather quickly. The higher the Q, the more slowly the ringing dies out.

Pulse Source Setup

The input voltage is produced using a Vpulse source. The pulse source is used to produce a single pulse, or a periodic series of pulses (a pulse-train) of user-determined delay, rise time, fall time, pulse width, and period. For this example, we want to produce a pulse train like that shown in Fig. 10.60. These pulses will cause the amplifier to ring, but they occur at long enough intervals that the exponential decay will be very apparent.

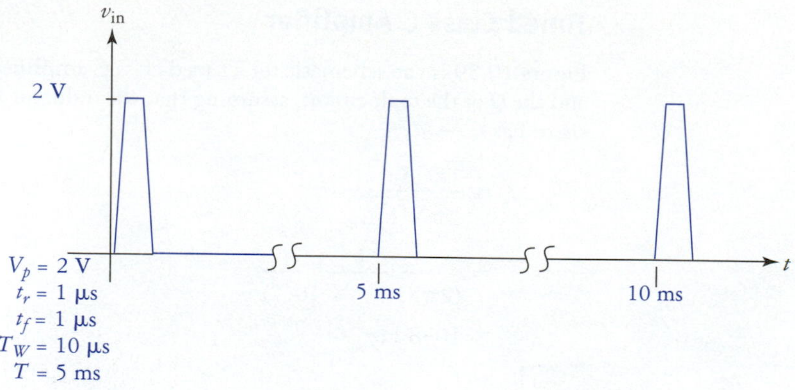

FIGURE 10.60
Pulse train to be used as input voltage.

The pulse generator attributes and values that are required to produce the desired input signal are listed below.

Parameter	Description	Desired Value
DC	DC offset voltage	0
AC	Steady-state signal for AC Sweep	0
V1	Initial value of pulse	0
V2	Final value of pulse	2
TD	Delay to first pulse occurrence	0
TR	Rise time	1 μs
TF	Fall time	1 μs
PW	Pulse width	10 μs
PER	Period of pulse train	5 ms

Simulation Results

Performing a transient analysis from 0 to 15 ms and plotting the output voltage yields the waveform shown in Fig. 10.61. Placing the cursors at the first two peaks allows us to calculate the resonant frequency as follows:

$$\begin{aligned} f_o &= 1/T \\ &= 1/(1.440 \text{ ms} - 449 \ \mu\text{s}) \\ &= 1/991 \ \mu\text{s} \\ &= 1009 \text{ Hz} \end{aligned}$$

A Totally Optional Excursion

Since the information is presented in Fig. 10.61, it is instructive to calculate the damping coefficient α for the circuit. This is important information if we are interested in knowing the total response of the system. We might wish to know this if the circuit is to be used in the guidance system of a cruise missile, or or in a heart pacemaker.

The individual ringing portion of the waveform shown in Fig. 10.61 is described by the equation

$$v = V_P e^{-\alpha t} \sin(2\pi f)t$$

In this case, we are just interested in the exponential part of this equation, which allows us to write

$$V_2 = V_1 e^{-\alpha(\Delta t)}$$

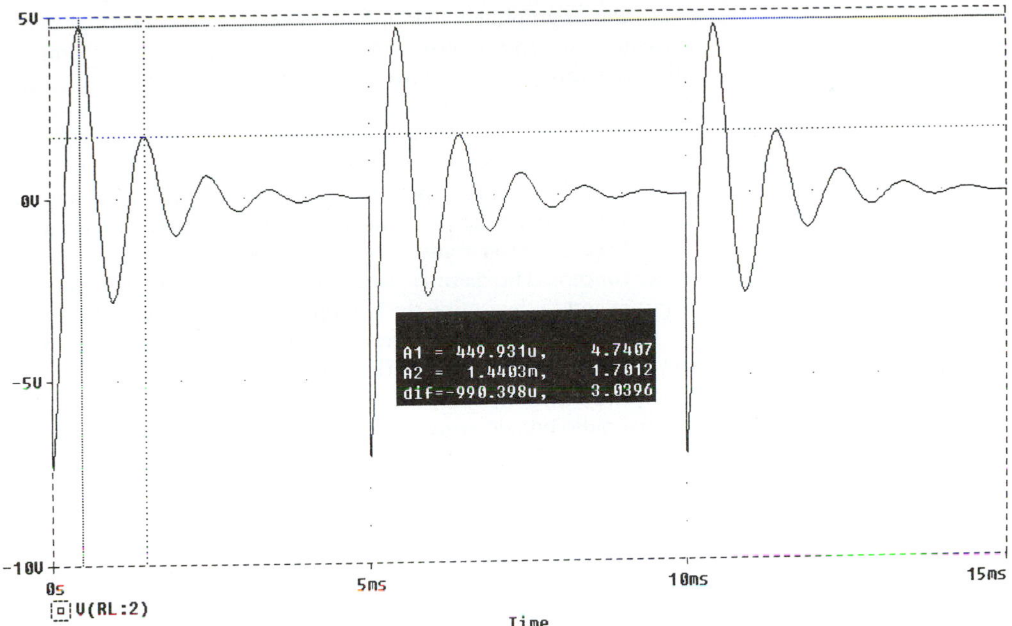

FIGURE 10.61
Resulting output voltage waveform.

where V_1 is the voltage at the first cursor (4.74 V), V_2 is the voltage at the second cursor (1.70 V), e is the base of the natural logarithms (2.718 . . .), Δt is the time interval between V_1 and V_2 (990.4 μs), and α is the damping coefficient. We now solve the exponential equation for α giving us

$$\begin{aligned}\alpha &= -\frac{\ln\left(\frac{V_2}{V_1}\right)}{\Delta t} \\ &= -\frac{\ln\left(\frac{1.70\text{ V}}{4.74\text{ V}}\right)}{990.4\ \mu\text{s}} \\ &= \frac{1.025}{990.4\ \mu\text{s}} \\ &= 1036\text{ s}^{-1}\end{aligned}$$

The damping coefficient is also the reciprocal of the *time constant* of the system, thus we can write

$$\begin{aligned}\tau &= 1/\alpha \\ &= 1/1036 \\ &= 956\ \mu\text{s}\end{aligned}$$

The damping coefficient and time constant give us two equivalent ways of looking at the decay of the system response. Recall from your studies of *RC* circuits that for all practical purposes, the final value of an exponential function (such as during the discharging of a capacitor) is reached after five time constants have elapsed ($t = 5\tau$). For this circuit, $5\tau \cong 4.8$ ms. Examination of Fig. 10.61 reveals that the ringing does indeed decay to nearly zero amplitude 5 ms after the input pulse is applied.

CHAPTER REVIEW

Power amplifiers are designed to deliver relatively large current and voltage levels to a load. Typically the load has low internal resistance. Class A amplifiers are used infrequently as power amplifiers because of their low efficiency and asymmetrical nonlinear distortion characteristics. Class B amplifiers have higher efficiency, but require the use of complementary push-pull configured transistors for practical implementation. Class AB amplifiers eliminate crossover distortion by providing a small forward bias of the push-pull output transistors. Class AB biasing is often accomplished using current-mirror circuits and also using V_{BE} multiplier circuits.

Power amplifiers that operate from a single-polarity power supply require the use of coupling capacitors, which very often must be large. Direct-coupled amplifiers, using bipolar power supplies eliminate this problem. Power transistors can be connected in the Darlington configuration to increase current gain, and they can be connected in parallel in order to increase current- and power-handling capability. Emitter swamping resistors are required to prevent current-hogging when parallel connected transistors are used. VMOS field-effect transistors are sometimes used in push-pull amplifier designs because they have very good high-frequency characteristics. FETs are also used because their positive temperature coefficient property prevents thermal runaway.

Because power transistors are subject to high current and voltage levels, heat dissipation is a major concern. The thermal characteristics of a given transistor are used to predict its operating temperature and to determine the appropriate heat sinking requirements.

Class C amplifiers are very nonlinear, however they are used primarily in high-frequency, narrow-band applications where tuned circuits can be used to filter the output signal as required. Class C amplifiers are very efficient, which makes them attractive in high-power applications, such as commercial radio broadcasting and radar systems.

DESIGN AND ANALYSIS PROBLEMS

Note: Unless otherwise specified, assume all transistors are silicon with $\beta = 100$, $V_{BE(on)} = 0.7$ V, and $r_o = \infty$ and for output and load power calculations assume maximum unclipped output.

10.1. Refer to Fig. 10.62. Determine the Q point, P_{DQ}, $i_{C(sat)}$, $v_{CE(cut)}$, $V_{o(max)}$, $V_{o(min)}$, A_v, $P_{o(max)}$, $P_{o(ave)}$, and the efficiency η of the amplifier. Because this is a large-signal amplifier, use r'_e to determine the voltage gain.

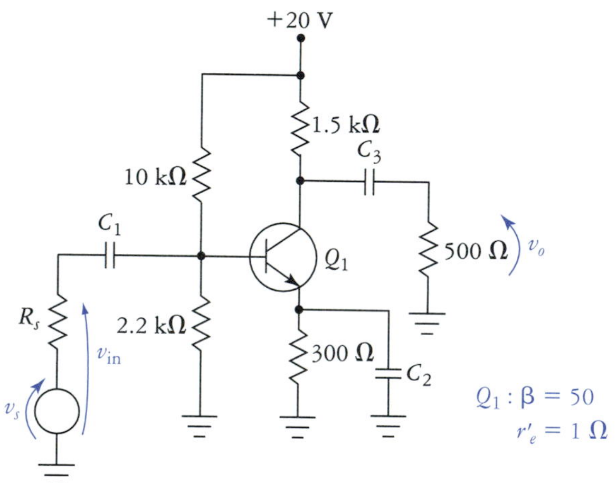

FIGURE 10.62
Circuit for Problems 10.1 to 10.3.

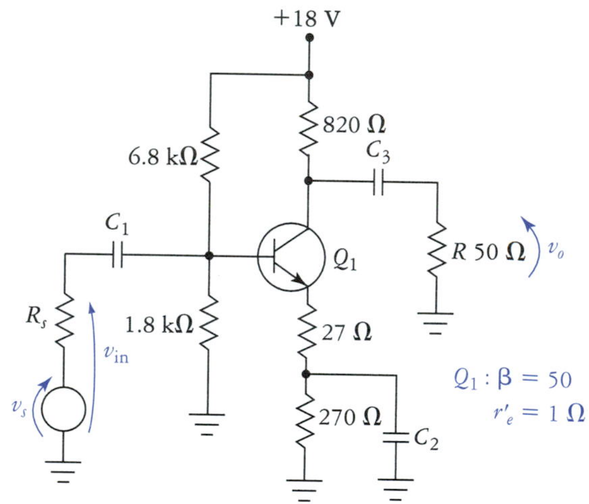

FIGURE 10.63
Circuit for Problem 10.4.

10.2. Assume the transistor in Problem 10.1 has $\theta_{JA} = 100°C/W$. Determine the junction temperature under no-signal conditions if $T_A = 27°C$.

10.3. Assume the transistor in Problem 10.1 has $\theta_{JA} = 75°C/W$. Determine the increase in junction temperature under no-signal conditions.

10.4. Refer to Fig. 10.63. Determine the Q point, P_{DQ}, $i_{C(sat)}$, $v_{CE(cut)}$, $V_{o(max)}$, $V_{o(min)}$, A_v, $P_{L(max)}$, $P_{L(ave)}$, and the efficiency η of the amplifier. Because this is a large-signal amplifier, use r'_e to determine the voltage gain.

10.5. Refer to Fig. 10.64. Using the information given, determine the required values for R_E, R_C, and R_X such that the Q point is centered on the resulting ac load line. Determine the resulting A_v, R_{in}, A_i, and A_p. Use r'_e in place of r_e as required in all calculations. Determine P_{DQ}, $P_{L(max)}$, and $P_{L(ave)}$.

10.6. Refer to Fig. 10.65. Using the information given, determine the required value for R_L to produce an ac load line with a centered Q point. Determine the resulting A_v, R_{in}, A_i, and A_p. Use r'_e in place of r_e as required in all calculations. Determine P_{DQ}, $P_{L(max)}$, and $P_{L(ave)}$.

10.7. A certain transistor is used in a circuit where $P_{D(ave)} = 5$ W. The transistor has $\theta_{JC} = 3°C/W$. The transistor is mounted on a heat sink such that the thermal resistance from case to sink is $\theta_{CS} = 1.5°C/W$. Determine the minimum thermal resistance from heat sink to ambient θ_{SA} such that the temperature rise of the junction is a maximum of 75°C.

10.8. A certain transistor is used in a circuit where $P_{D(ave)} = 15$ W. The transistor has $\theta_{JC} = 1.5°C/W$. The transistor is mounted on a heat sink such that the thermal resistance from case to sink is $\theta_{CS} = 1.0°C/W$. Determine the minimum thermal resistance from heat sink to ambient θ_{SA} such that the temperature rise of the junction is a maximum of 80°C.

10.9. Refer to Fig. 10.66. Assume $V_{CC} = 15$ V. Determine R_1, R_2, and R_3 such that $V_{CE} = 3.5$ V and $I_C = 10$ mA.

10.10. Using the data from Problem 10.9, determine the maximum allowable value of R_3.

10.11. Refer to Fig. 10.66. Assume $R_1 = 2.2$ kΩ, $R_2 = 6.8$ kΩ, and $V_{CC} = 12$ V. The transistor is rated such that $P_{D(max)} = 500$ mW. Determine the minimum value of R_3 that can be used.

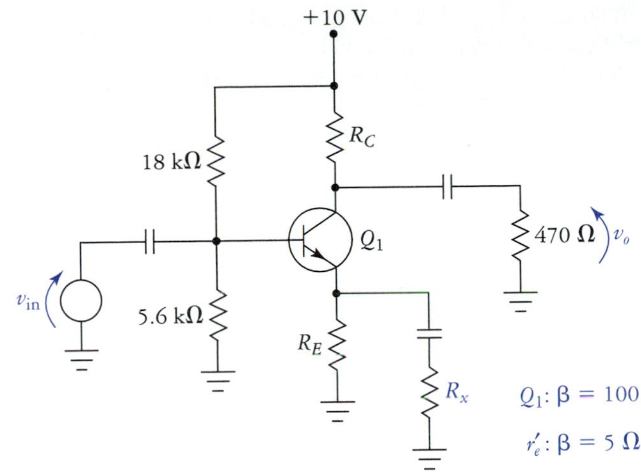

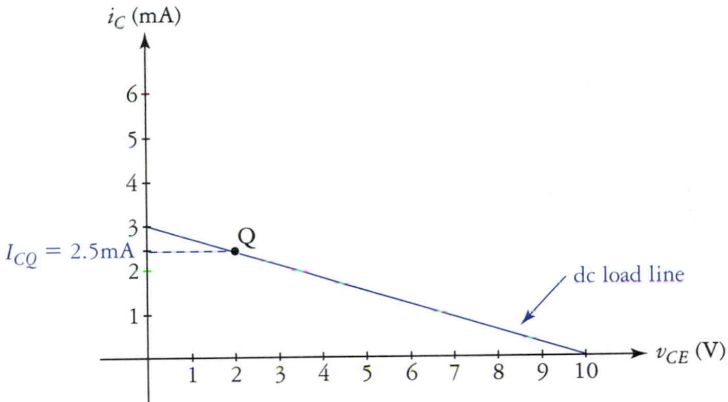

FIGURE 10.64
Circuit for Problem 10.5.

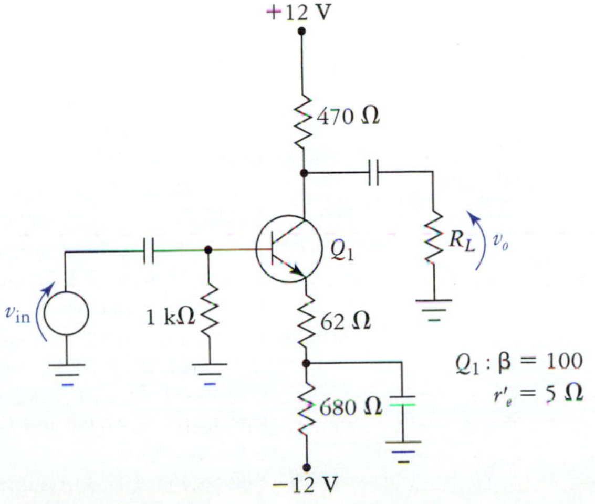

FIGURE 10.65
Circuit for Problem 10.6.

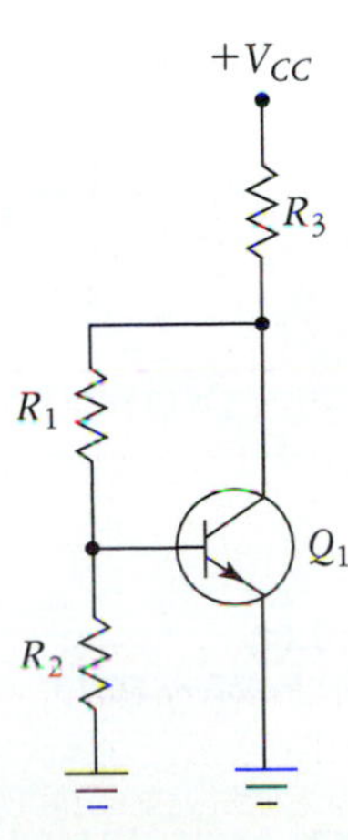

FIGURE 10.66
Circuit for Problems 10.9 to 10.12.

10.12. Refer to Fig. 10.66. Given $V_{CC} = 15$ V and $R_2 = 2$ kΩ, determine R_1, and R_3 such that $V_{CE} = 4.2$ V and $I_C = 10$ mA.

10.13. The circuit in Fig. 10.67 has $V_{CC} = 30$ V and $R_2 = 2.2$ kΩ. Determine R_1 such that $V_{CE_1} = 1.3$ V.

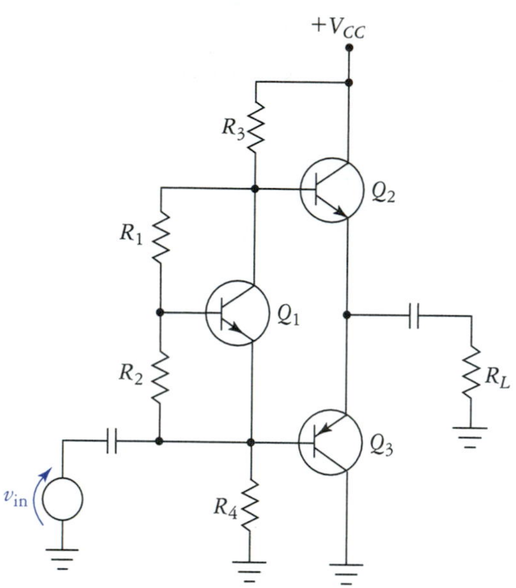

FIGURE 10.67
Circuit for Problems 10.13 and 10.14.

10.14. The circuit in Fig. 10.67 has $V_{CC} = 22$ V and $R_2 = R_3 = R_4 = 1$ kΩ. Determine R_1 and R_4 such that $V_{CE_1} = 1.4$ V. Determine the resulting I_{C_1}.

10.15. Refer to Fig. 10.68. Determine V_{CEQ}, I_{CQ}, $I_{C(\text{sat})}$, $V_{CE(\text{cut})}$, R'_C, R_{in}, A_v (use r'_e), A_i, A_p, $i_{C(\text{sat})}$, $V_{CE(\text{cut})}$, P_{DQ}, $V_{o(\max)}$, $V_{o(\min)}$, $P_{o(\max)}$ and $P_{o(\text{ave})}$.

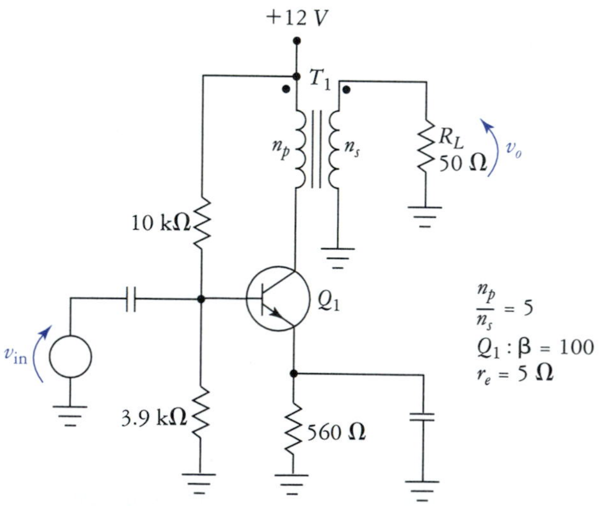

FIGURE 10.68
Circuit for Problem 10.15.

10.16. Refer to Fig. 10.69. Determine V_{CEQ}, I_{CQ}, $I_{C(\text{sat})}$, $V_{CE(\text{cut})}$, R'_C, R_{in}, A_v (use r'_e), A_i, A_p, $i_{C(\text{sat})}$, $v_{CE(\text{cut})}$, P_{DQ}, $V_{o(\max)}$, $V_{o(\min)}$, $P_{o(\max)}$, and $P_{o(\text{ave})}$.

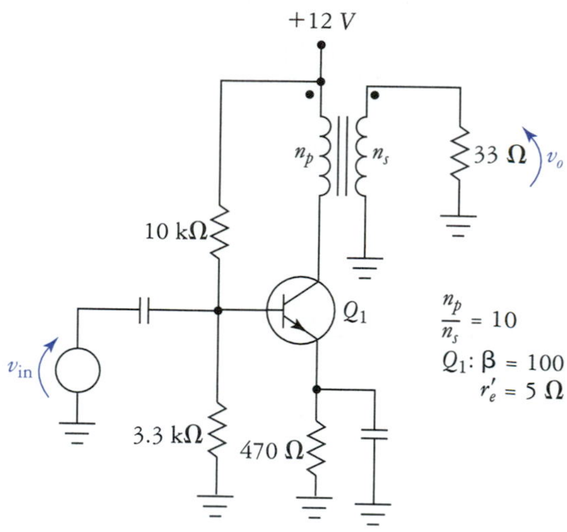

FIGURE 10.69
Circuit for Problem 10.16.

10.17. Refer to Fig. 10.70. Given $R_1 = R_2 = 2.2$ kΩ, $R_L = 8$ Ω, $R_s = 300$ Ω, and $V_{CC} = 25$ V, determine I_{CQ}, $I_{C(\text{sat})}$, R_{in}, R_o, A_v, A_i, A_p, $P_{o(\max)}$, and $P_{o(\text{ave})}$.

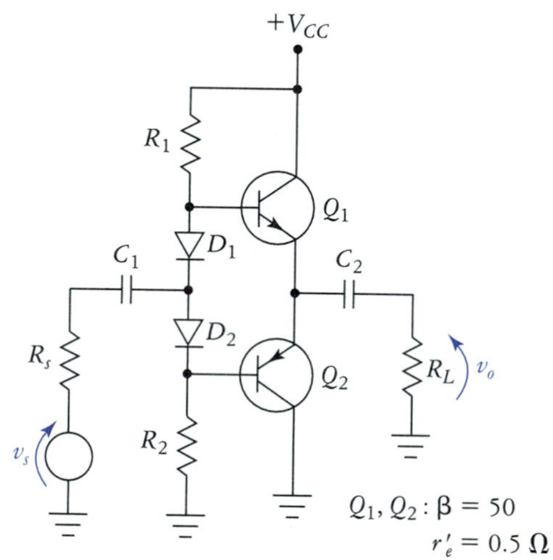

FIGURE 10.70
Circuit for Problem 10.17.

10.18. Refer to Fig. 10.71. Given $R_1 = R_2 = 1.8$ kΩ, $R_L = 4$ Ω, $R_s = 200$ Ω, and $V_{CC} = V_{EE} = 20$ V. Determine I_{CQ}, $I_{C(\text{sat})}$, R_{in}, R_o, A_v, A_i, A_p, $P_{o(\max)}$, and $P_{o(\text{ave})}$.

10.19. Refer to Fig. 10.72. Assume $R_1 = R_2 = 1.2$ kΩ, $R_E = 0.47$ Ω, $R_L = 8$ Ω, $R_S = 200$ Ω, and $V_{CC} = 40$ V. Determine I_{CQ} (approximate), $I_{C(\text{sat})}$, R_{in}, R_o, A_v, A_i, A_p, $P_{o(\max)}$, and $P_{o(\text{ave})}$.

10.20. Refer to Fig. 10.72. Assume $R_1 = R_2 = 2.2$ kΩ, $R_E = 0.22$ Ω, $R_L = 8$ Ω, $R_S = 75$ Ω, and $V_{CC} = 28$ V. Determine I_{CQ} (approximate), $I_{C(\text{sat})}$, R_{in}, R_o, A_v, A_i, A_p, $P_{o(\max)}$, and $P_{o(\text{ave})}$. Assuming $C_1 = 22$ μF and $C_2 = 4700$ μF, determine the $f_{C(\text{in})}$ and $f_{C(\text{out})}$.

10.21. Refer to Fig. 10.73. Assume $R_1 = R_2 = 820$ Ω, $R_E = 0.22$ Ω, $R_L = 8$ Ω, $R_S = 100$ Ω. Determine I_{CQ} (approximate), I_{sc}, R_{in}, R_o, A_v, A_i, A_p, $P_{o(\max)}$, and $P_{o(\text{ave})}$. Assuming $C_1 = 10$ μF and $C_2 = 3300$ μF, determine the $f_{C(\text{in})}$ and $f_{C(\text{out})}$.

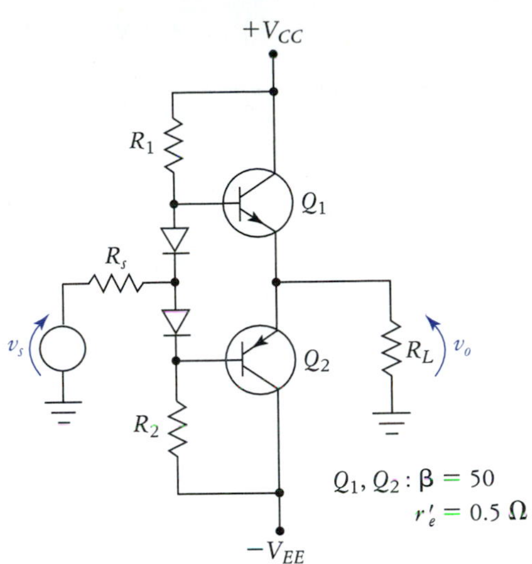

FIGURE 10.71
Circuit for Problem 10.18.

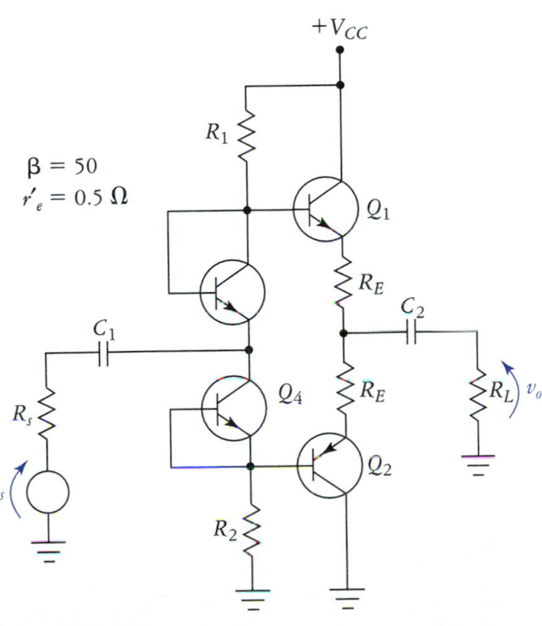

FIGURE 10.72
Circuit for Problems 10.19 and 10.20.

10.22. Refer to Fig. 10.73. Assume $R_1 = R_2 = 1\ \text{k}\Omega$, $R_E = 0.33\ \Omega$, $R_L = 4\ \Omega$, $R_S = 120\ \Omega$. Determine I_{CQ} (approximate), I_{sc}, R_{in}, R_o, A_v, A_i, A_p, $P_{o(\text{max})}$, and $P_{o(\text{ave})}$.

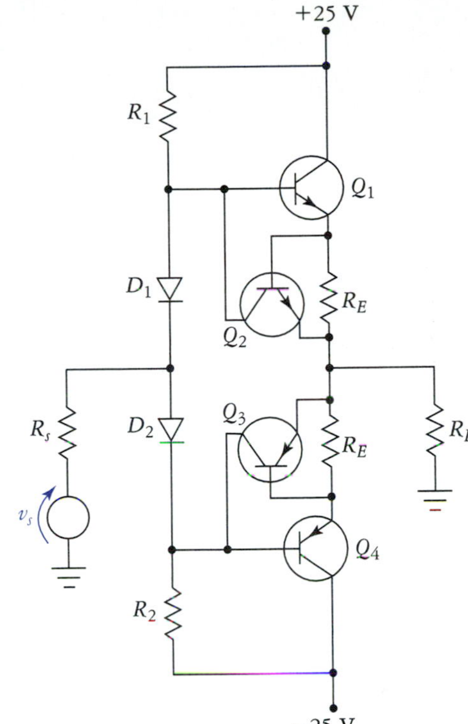

FIGURE 10.73
Circuit for Problems 10.21 and 10.22.

10.23. Refer to Fig. 10.74. Given $R_1 = 3.6\ \text{k}\Omega$, $R_2 = 1.2\ \text{k}\Omega$, $R_3 = R_4 = 1\ \text{k}\Omega$, and $V_{CC} = 20$ V, determine I_{C_1}.

10.24. Refer to Fig. 10.74. Given $R_1 = 1.2\ \text{k}\Omega$, $R_2 = 330\ \Omega$, and $V_{CC} = 20$ V, determine R_3 and R_4 such that $I_{C_1} = 10$ mA.

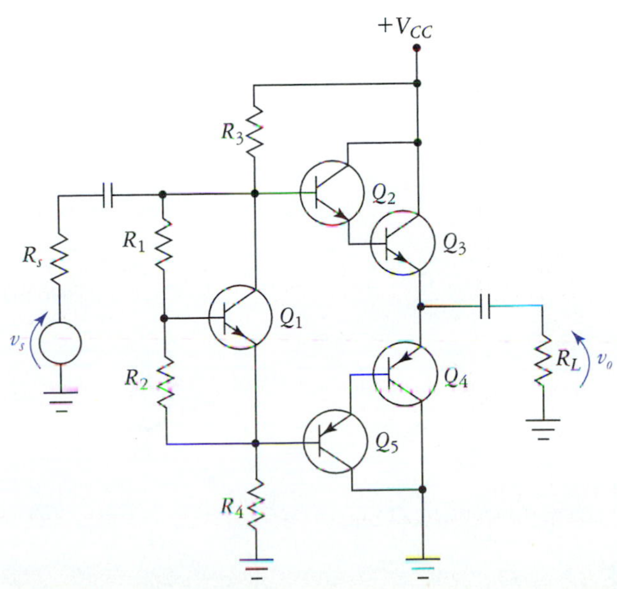

FIGURE 10.74
Circuit for Problems 10.23 and 10.24.

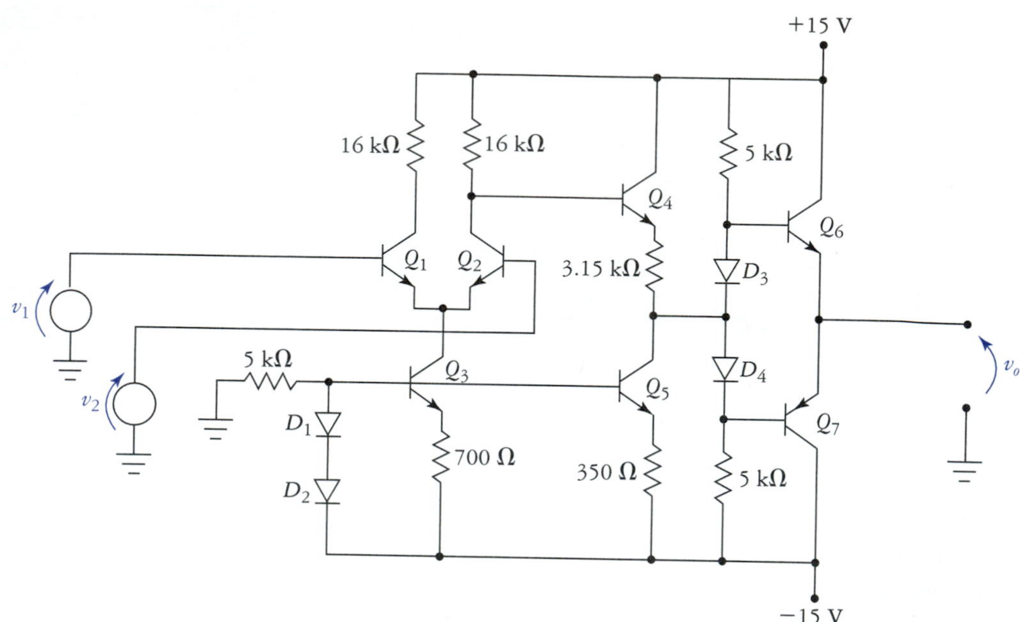

FIGURE 10.75
Circuit for Problem 10.25.

10.25. Refer to Fig. 10.75. Stage 1 is a differential-input, single-ended-output stage operating under small-signal conditions. Stage 2 is a level translator under small-signal conditions as well. Stage 3 is a large-signal amplifier. Assume that all transistors have $\beta = 100$, $V_{BE} = 0.7$ V, and negligible base current is drawn (that is, $I_B \cong 0$). The output transistors have $r'_e = 5\ \Omega$. Determine the Q point for each transistor and determine the overall voltage gain, input resistance, and output resistance. Identify the inverting and noninverting input terminals.

10.26. Refer to Fig. 10.76. For Q_1, $\beta = 10{,}000$. For Q_2 and Q_3, $\beta = 50$ and $r'_e = 1\ \Omega$. Determine the overall voltage gain, input resistance, output resistance, current gain, and power gain.

10.27. A certain amplifier has $R_{in} = 4\ \text{k}\Omega$. Determine the necessary coupling capacitor size that will result in $f_{C(in)} = 50$ Hz.

10.28. A certain amplifier has $R_o = 200\ \Omega$ and $R_L = 50\ \Omega$. Determine the value of the coupling capacitor required to produce $f_{C(out)} = 20$ Hz.

10.29. Refer to Fig. 10.77. Assume $v_{in} = 2\sin(2\pi 500t)$ V and $C_1 = 0.022\ \mu$F. Determine the minimum required value for R_1. Sketch the resulting v_B and v_o waveforms.

10.30. Refer to Fig. 10.77. Assume $v_{in} = 2 \sin 2000t$ V and $R_1 = 3.3$ kΩ. Determine the minimum required value for C_1. Sketch the resulting v_B and v_o waveforms.

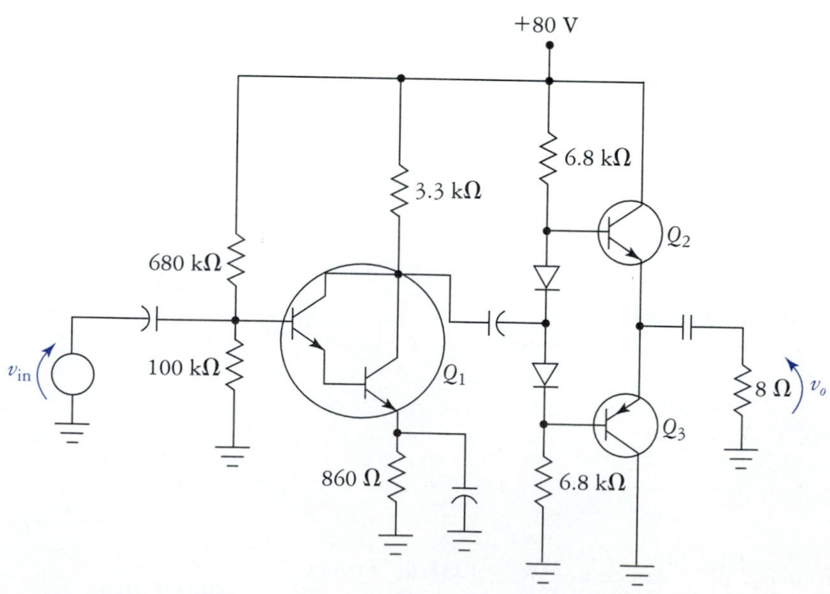

FIGURE 10.76
Circuit for Problem 10.26.

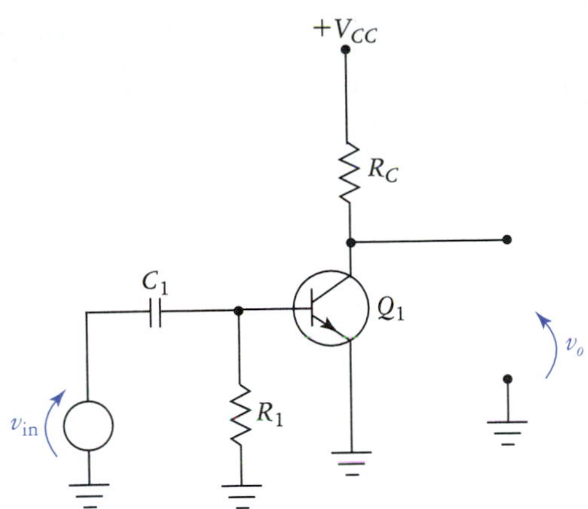

FIGURE 10.77
Circuit for Problems 10.29 and 10.30.

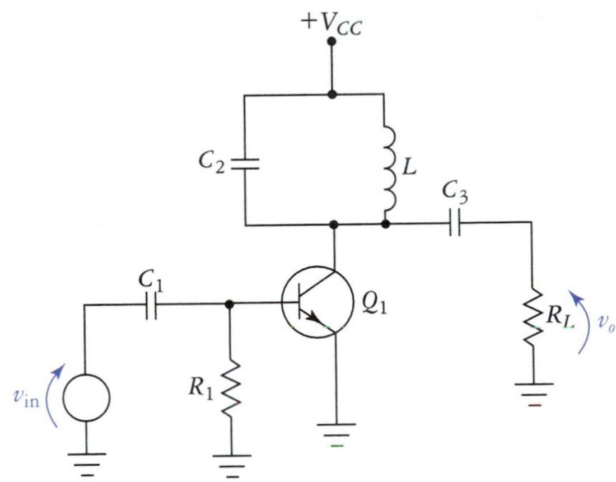

FIGURE 10.79
Circuit for Problems 10.32 to 10.35.

10.31. Assume the circuit in Fig. 10.77 produces the waveform shown in Fig. 10.78. Determine the average power dissipation of the transistor if $R_C = 1\ \text{k}\Omega$.

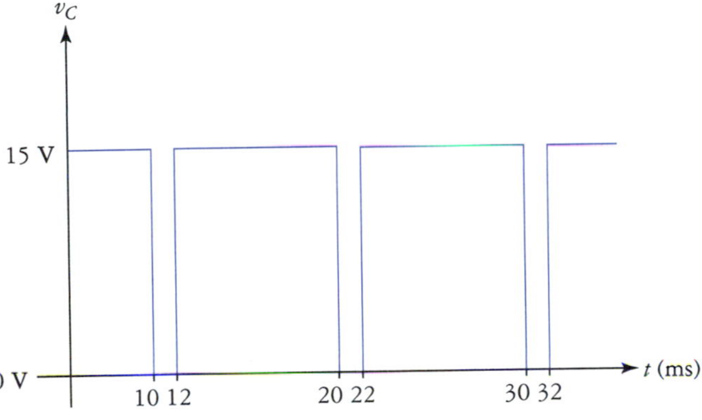

FIGURE 10.78
Waveform for Problem 10.31.

10.32. Refer to Fig. 10.79. Assume C_1 and R_1 have appropriate values for adequate clamping of the input voltage and $C_3 >> C_2$. Given $L_1 = 47\ \mu\text{H}$ with $r_s = 5\ \Omega$, $C_2 = 220$ pF, and $R_L = 1\ \text{k}\Omega$, determine f_o, Q_L, and Q.

10.33. Refer to Fig. 10.79. Assume C_1 and R_1 have appropriate values for adequate clamping of the input voltage and $C_3 >> C_2$. Given $L_1 = 100\ \mu\text{H}$ with $r_s = 12\ \Omega$, $C_2 = 270$ pF and $R_L = 2.2\ \text{k}\Omega$, determine f_o, Q_L, and Q.

10.34. Refer to Fig. 10.79. Assume C_1 and R_1 have appropriate values for adequate clamping of the input voltage and $C_3 >> C_2$. Given $L_1 = 1$ mH with $r_s = 25\ \Omega$, and $R_L = 1.5\ \text{k}\Omega$, determine C_2 such that $f_o = 250$ kHz. Determine the resulting Q_L and Q as well.

10.35. Refer to Fig. 10.79. Assume C_1 and R_1 have appropriate values for adequate clamping of the input voltage and $C_3 >> C_2$. Given $L_1 = 470\ \mu\text{H}$ with $r_s = 15\ \Omega$, determine C_2 such that $f_o = 250$ kHz and determine R_L such that $Q = 10$.

■ TROUBLESHOOTING PROBLEMS

10.36. Refer to Fig. 10.6. Suppose the load resistor became disconnected from the amplifier. What effect would this have on voltage gain and output voltage compliance?

10.37. Refer to Fig. 10.9. If the secondary winding of T_1 were to become short circuited, what effect would this have on I_{CQ} and V_{CEQ}?

10.38. Refer to Fig. 10.80. Assuming a typical class B push-pull amplifier is being used, based on the waveforms shown, what is the most likely defective component in the circuit?

10.39. Refer to Fig. 10.29. Suppose that D_1 became short circuited. What effect would this have if a sinusoidal output signal was expected?

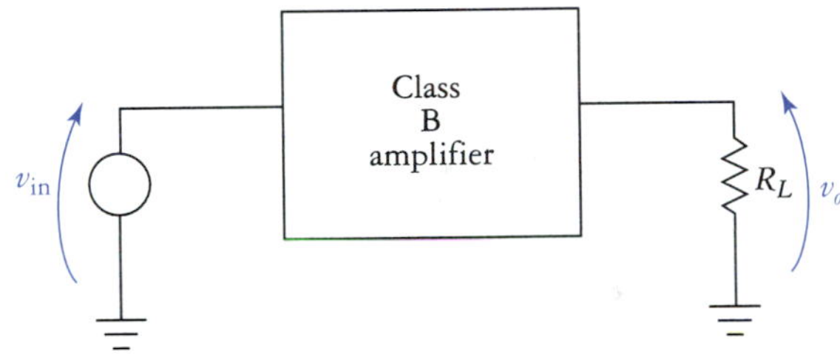

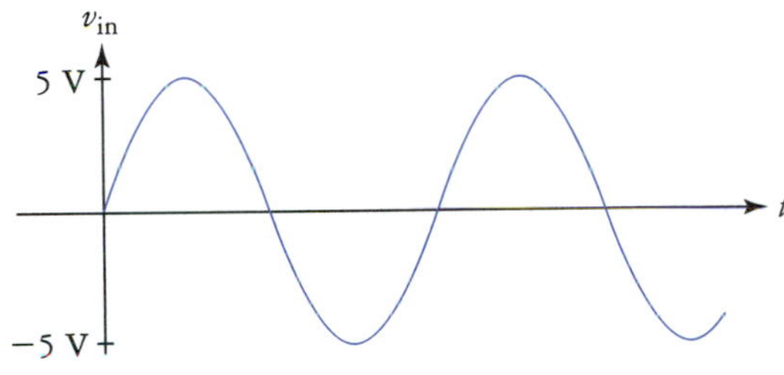

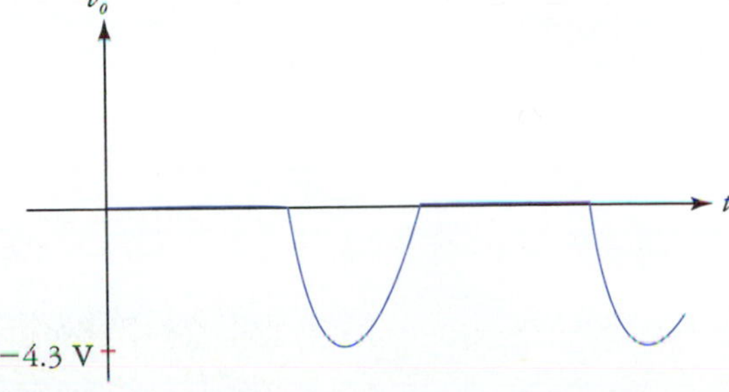

FIGURE 10.80
Waveforms for Problem 10.38.

10.40. Refer to Fig. 10.29. Suppose that D_1 became open circuited. What effect would this have if a sinusoidal output was expected?

■ COMPUTER PROBLEMS

10.41. Simulate the class B amplifier of Fig. 10.14 using 2N3904 and 2N3906 transistors, V_{CC} = 15 V, C_1 = 10 μF, C_2 = 100 μF, R_L = 50 Ω, and v_{in} is a 10-$V_{p\text{-}p}$, 1-kHz sine wave. Determine the voltage gain.

10.42. Simulate the circuit of Fig. 10.75 using breakout transistors with BF = 100 and VAF = 100 V. Determine the voltage gain and maximum unclipped output voltage. Use 1N4148 in all diode locations.

CHAPTER 11

Negative Feedback and Frequency Response

BEHAVIORAL OBJECTIVES

On completion of this chapter, the student should be able to:

- Determine the lower corner frequency for an amplifier.
- Determine the upper corner frequency for an amplifier.
- Explain the significance of beta cutoff and beta transition frequency.
- Explain the use of a Bode plot.
- Describe the advantages of the cascode amplifier.
- Describe the effects of negative feedback on amplifier voltage gain.
- Describe the Miller effect.

INTRODUCTION

This chapter presents additional details regarding frequency response and negative feedback. In previous chapters we looked at frequency response analysis with regards to coupling and bypass capacitor effects. These low-frequency related topics are briefly recapped here. High-frequency performance limitations of transistors and the device parameters that affect them are analyzed in detail.

Negative feedback is a topic that is extremely important from both the component- and system-level perspectives. This chapter introduces the basic terminology and some of the primary derivations and applications of negative feedback in relation to the circuits that have already been covered. Because certain aspects of high-frequency response are greatly affected by negative feedback, high-frequency response is covered after the negative-feedback material. Much of the information presented here is used as a foundation for later chapters of the text, particularly in the study of operational amplifiers and oscillators. This means that although this is a short chapter, the information presented is very important.

11.1 FREQUENCY RESPONSE TERMS AND DEFINITIONS

A given amplifier will operate over a certain range of frequencies called the *midband.* The gain of the amplifier over this range is often called the *midband gain* A_{mid}, a term we have used before. Lower frequency limits are generally determined by coupling and bypass capacitor values. The high-frequency limits of circuit operation are more complex and are usually determined by parasitic capacitances associated with the device or the physical layout of the circuit in which it is used.

Bode Plots

The frequency response of an amplifier will typically appear as shown in Fig. 11.1. Such a plot of gain magnitude versus frequency with logarithmic axes is called a *Bode plot* (pronounced Bo-dee), so named after Hendrick Bode of Harvard University. Notice that since the logarithm on the vertical axis is scaled by 20, this is equivalent to using decibel voltage gain. This is one of the reasons why the decibel is commonly used to represent voltage gain; it is an inherently logarithmic unit.

Note
Frequency response is fundamental to the study of active filters, the subject of Ch. 14.

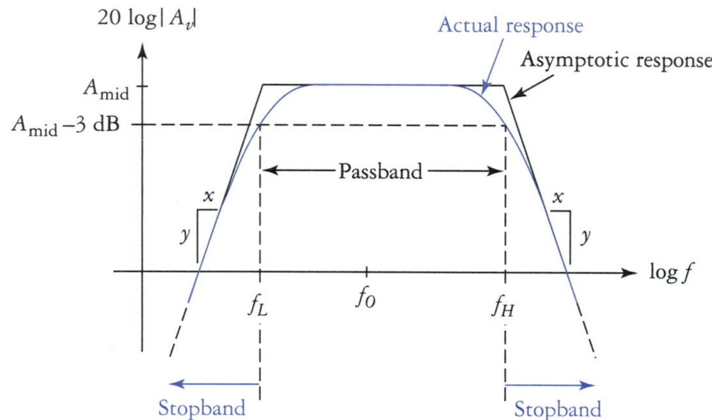

FIGURE 11.1
Important parameters of a Bode plot.

The actual response of the amplifier will be a continuous curve as shown in Fig. 11.1. In many cases, however, the response is approximated using straight-line segments that have the same slope as the actual response at its *asymptotes.* This is also shown in the figure and such a graph is usually called an *asymptotic Bode plot.* You may recall that an asymptote is the value a function approaches as its independent variable approaches some limit. In the case of frequency response plots, the limits are often f_o, f_L, f_H, 0 Hz, and ∞ Hz.

Corner Frequency

A corner frequency of a response plot is the frequency at which power transfer from the amplifier to the load has dropped to 50% of the maximum midband value. Because of this, an amplifier's corner frequency is sometimes called the *half-power point* or *critical frequency.* The term *break frequency* is applied to any point on the asymptotic Bode plot at which the slope changes.

Referring to Fig. 11.1, in terms of the midband gain, the corner frequencies are defined as the frequencies at which the response drops by 3 dB from the midband value. A −3-dB change in A_v corresponds to a halving of power gain, while a +3-dB change in A_v is a doubling of power gain. Remember that the decibel is fundamentally a measure of power gain, although we prefer to work in terms of voltage gain for reasons of convenience. If we do not use decibels, the corner frequency is defined as the frequency at which the response drops to 0.707 times the midband value. That is,

$$A_{corner} = 0.707\, A_{mid} \tag{11.1}$$

We can show this by working with the definition of the decibel

$$A_{v(dB)} = 20 \log A_v$$

Substituting −3 dB for $A_{v(\text{dB})}$ gives us

Note
For a review of the derivation of the decibel, refer to Ch.9, pp. 340–342.

$$-3 \text{ dB} = 20 \log A_v$$

Solving for A_v then gives us

$$A_v = \log^{-1}(-3/20) \cong 0.707$$

Thus a decibel voltage gain decrease of 3 dB is the same as multiplying the linear voltage gain by 0.707.

You may be wondering how this relates to the half-power point. The connection is found when we realize that power gain is proportional to the square of voltage gain (see Section 9.6), so $0.707^2 = 0.5$, which means that if voltage gain decreases to $A_{v(\text{corner})} = 0.707A_{v(\text{mid})}$ then power gain decreases to $A_{p(\text{corner})} = 0.5A_{p(\text{mid})}$.

Bandwidth, Passband, and Stopband

The terms *passband* and *stopband* are usually associated with filter analysis, but an amplifier with frequency limits is basically a filter circuit. The passband is the range of frequencies between lower and upper corner frequencies. This range is also called the *bandwidth* (BW) of the amplifier, which we define mathematically as

$$\text{BW} = f_H - f_L \tag{11.2}$$

In the special case where the amplifier response extends down to dc, the bandwidth is simply

$$\text{BW} = f_H$$

We also sometimes use this as an approximation when the upper corner frequency is very high relative to the lower corner. For example, if f_L = 10 HZ and f_H = 1 MHz, then BW ≅ 1 MHz.

All frequencies outside of the passband are said to lie in the *stopband.* The response curves in Figs. 11.1 and 11.2 actually have two stopbands. Frequencies within the stopband(s) are not amplified efficiently and will generally exhibit large amounts of phase shift compared to those in the passband.

Center Frequency

The center frequency f_o in Fig. 11.1 is the midpoint between upper and lower corner frequencies. Because we are using a logarithmic scale, we must use the *geometric mean* to determine this midpoint. The equation is

$$f_o = \sqrt{f_L f_H} \tag{11.3}$$

EXAMPLE 11.1

Refer to Fig. 11.2. The vertical axis is scaled logarithmically (in decades). Determine the midband gain in decibels, the corner frequency gain in decibels, and linear forms, bandwidth, and f_o.

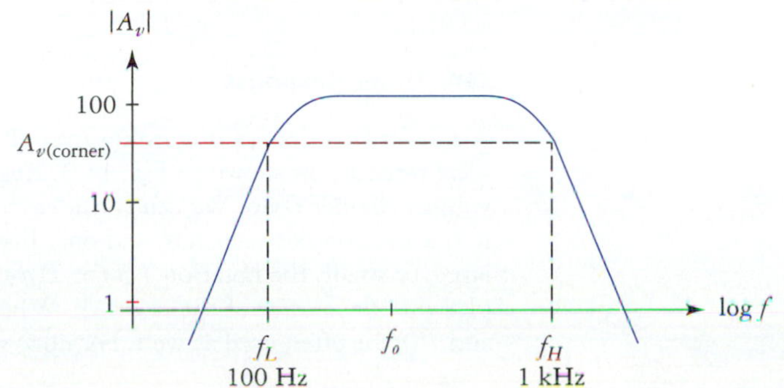

FIGURE 11.2
Bode plot for Example 11.1.

Solution In decibels, the midband gain is

$$\begin{aligned} A_{\text{mid(dB)}} &= 20 \log A_v \\ &= 20 \log(100) \\ &= 40 \text{ dB} \end{aligned}$$

To simplify notation, let's call $A_{v\text{(corner)}}$, the gain at the corner frequencies, A_L or A_H in linear units and $A_{L\text{(dB)}}$ or $A_{H\text{(dB)}}$ in decibels. The decibel gain at either corner frequency is

$$\begin{aligned} A_{L\text{(dB)}} = A_{H\text{(dB)}} &= \text{A}_{\text{mid(dB)}} - 3 \text{ dB} \\ &= 40 \text{ dB} - 3 \text{ dB} \\ &= 37 \text{ dB} \end{aligned}$$

The linear gain at the corner frequencies is

$$\begin{aligned} A_H = A_L &= 0.707 A_{v\text{(mid)}} \\ &= (0.707)(100) \\ &= 70.7 \end{aligned}$$

The bandwidth is

$$\begin{aligned} \text{BW} &= f_H - f_L \\ &= 1 \text{ kHz} - 100 \text{ Hz} \\ &= 900 \text{ Hz} \end{aligned}$$

The center frequency is

$$\begin{aligned} f_o &= (f_L f_H)^{1/2} \\ &= (100 \times 1000)^{1/2} \\ &= 316.2 \text{ Hz} \end{aligned}$$

Notice that the geometric center frequency (geometric mean) is much different from the arithmetic mean (550 Hz).

Rolloff Rate

In the passband, the asymptotic Bode plot is horizontal. That is, the slope, m, equals 0 in the passband. In the stopband(s), the slope of the Bode plot will be constant between break frequencies. Multiple break frequencies are possible as well. The slope of the Bode plot in the stopband is called the *rolloff rate,* or simply *rolloff.* The rolloff rate is usually given in dB/decade or dB/octave. The rolloff rate will normally be a multiple of ± 20 dB/decade or ± 6 dB/octave. In Fig. 11.1, the rolloff rate would be determined using $m = \Delta x/\Delta y$. We discuss rolloff in more detail when we cover active filter circuits.

One last bit of terminology. The sloping portions of the asymptotic Bode plot in the stopbands are called the response *skirts* or *filter skirts.* The shape of the skirt is often important in filter applications.

Response Order

Frequency-dependent networks can be classified in terms of *order.* This terminology is derived from the theory of differential equations used to describe the transient behavior of a circuit. Fortunately, we do not need to resort to using differential equations to understand and apply this information to real-world circuits.

Zero-Order Response

A zero-order response is constant over all frequencies. An example would be a voltage divider using ideal resistors, as shown in Fig. 11.3. Regardless of the frequency, the output is simply scaled by the voltage-divider ratio. We can consider this ratio to be the *gain* of the network, although in this case it is a fraction between zero and one. Because the scaling constant can be real, imaginary, complex, large, or small, the notation $T(f)$ or $T(jf)$ is sometimes used to represent the generalized real or complex *transfer function* of the network. When we are dealing specifically with filters, the notation $H(f)$ and $H(jf)$ is often used as well. For now, we just stick with A_v.

FIGURE 11.3
A purely resistive network (a) has a constant frequency response (b).

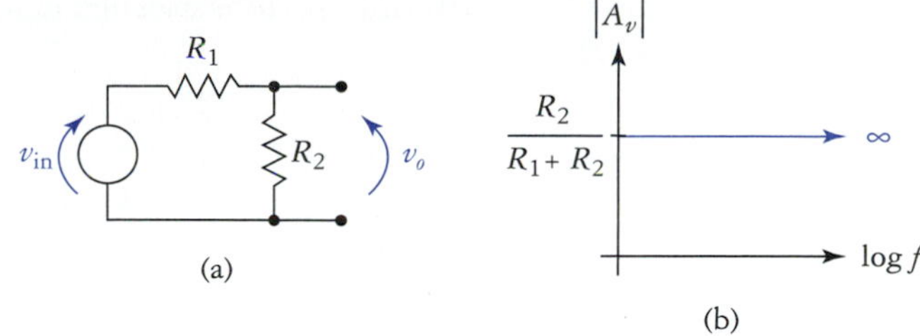

First-Order Response

A first-order network exhibits a first-order response. This may be either a *high-pass* or *low-pass* response. A first-order low-pass network and its frequency response plot are shown in Fig. 11.4. This is really just a frequency-dependent voltage divider. At low frequencies ($f_{in} < f_c$), $|X_C|$ is extremely large and $v_o \cong v_{in}$. At very high frequencies ($f_{in} >> f_c$), $|X_C|$ is very small and $v_o \rightarrow 0$ V. At the corner frequency $|X_C| = R$. The equation is

$$\frac{1}{2\pi f_c C} = R$$

Solving for f_c we have the familiar equation

$$f_c = \frac{1}{2\pi RC} \tag{11.4}$$

The factor 2π converts cyclic frequency (Hz) into radians/second ω, so we can also write

$$\omega_c = \frac{1}{RC} \tag{11.5}$$

Note
In control system applications, a low-pass RC section is often called a *lag network* because the output phase lags the input.

FIGURE 11.4
(a) First-order low-pass RC filter. (b) Frequency response plot.

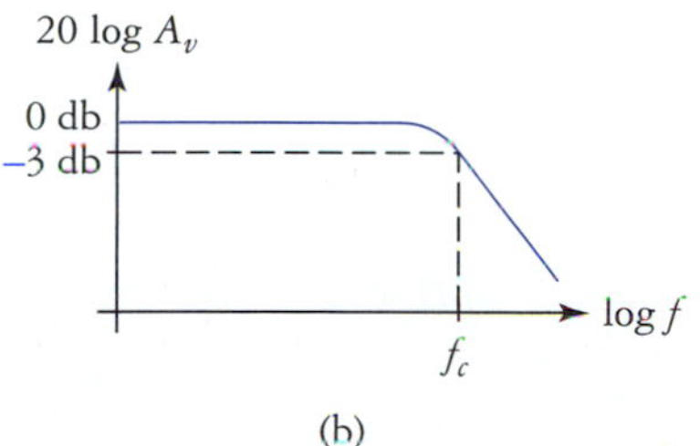

The equation for the response of the low-pass filter as a function of frequency (for steady-state sinusoidal signals) is found by writing the voltage-divider equation out in full phasor form:

$$V_o = V_{in} \frac{X_C}{X_C + R}$$

From the definition of voltage gain (as a function of complex frequency), we can write

$$A(j\omega) = \frac{V_o}{V_{in}} = \frac{X_C}{X_C + R}$$

Substituting the rectangular form for X_C, we obtain

$$A(j\omega) = \frac{\dfrac{1}{j\omega C + R}}{\dfrac{1}{j\omega C} + R}$$

Combining the denominator terms producers

$$A(j\omega) = \frac{\dfrac{1}{j\omega C}}{\dfrac{1}{j\omega C} + \dfrac{j\omega RC}{j\omega C}} = \frac{\dfrac{1}{j\omega C}}{\dfrac{j\omega RC + 1}{j\omega C}}$$

Multiplying by the reciprocal of the denominator we find

$$A(j\omega) = \frac{1}{j\omega C} \times \frac{j\omega C}{j\omega RC + 1}$$

$$= \frac{1}{j\omega RC + 1} \tag{11.6}$$

We now have expressed the gain as a complex function of frequency. Because we are only interested in the magnitude of this function, we apply the Pythagorean theorem, which gives us

$$A(\omega) = |A(j\omega)| = \frac{1}{\sqrt{(\omega RC)^2 + 1}} \tag{11.7}$$

In terms of frequency in hertz, this is

$$A(f) = \frac{1}{\sqrt{(2\pi f RC)^2 + 1}} \tag{11.8}$$

We can use Eqs. (11.7) and (11.8) to determine the value of A_v at any arbitrary frequency. However, we also know that

$$f_c = \frac{1}{2\pi RC} \qquad \text{and} \qquad \omega_C = \frac{1}{RC}$$

So if we substitute these into Eqs. (11.7) and (11.8) we can define the gain magnitude in terms of the input frequency and the corner frequency as

$$A(\omega) = \frac{1}{\sqrt{\left(\dfrac{\omega_{\text{in}}}{\omega_c}\right)^2 + 1}} \tag{11.9}$$

and

$$A(f) = \frac{1}{\sqrt{\left(\dfrac{f_{\text{in}}}{f_c}\right)^2 + 1}} \tag{11.10}$$

A first-order high-pass network and its gain magnitude response curve are shown in Fig. 11.5. The response equations for this configuration are

$$A(\omega) = \frac{1}{\sqrt{\left(\dfrac{\omega_c}{\omega_{\text{in}}}\right)^2 + 1}} \tag{11.11}$$

$$A(f) = \frac{1}{\sqrt{\left(\dfrac{f_c}{f_{\text{in}}}\right)^2 + 1}} \tag{11.12}$$

It is easy to recognize a zero-order network; there are no reactive components (capacitors or inductors) in the circuit. A first-order network has only sources, resistors, and a single equivalent

FIGURE 11.5
(a) First-order high-pass RC filter. (b) Frequency response plot.

Note
Coupling capacitors in multiple-stage amplifiers and emitter bypass capacitors form high-pass networks.

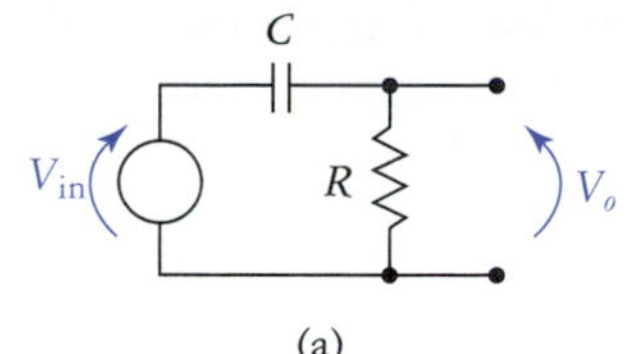

(a)

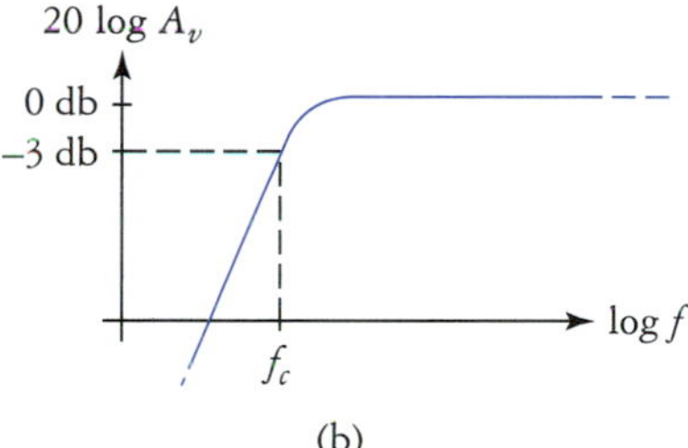

(b)

capacitance, or a single equivalent inductance. If your equivalent circuit contains both capacitors and inductors, or multiple capacitors (or inductors) that cannot be placed directly in series or parallel, then the network is higher than first order. This is not necessarily a bad thing, it is just more complex. For now, we limit ourselves to first-order networks.

As implied by the preceding paragraph, not all circuits are as neat and pretty as those we just covered. If, for example, we have a network like that of Fig. 11.6(a), we still have a first-order low-pass circuit. To determine the corner frequency of the circuit, we must first apply standard circuit reduction techniques. Moving R_2 to the left, we get the circuit shown in Fig. 11.6(b). Now it is easy to

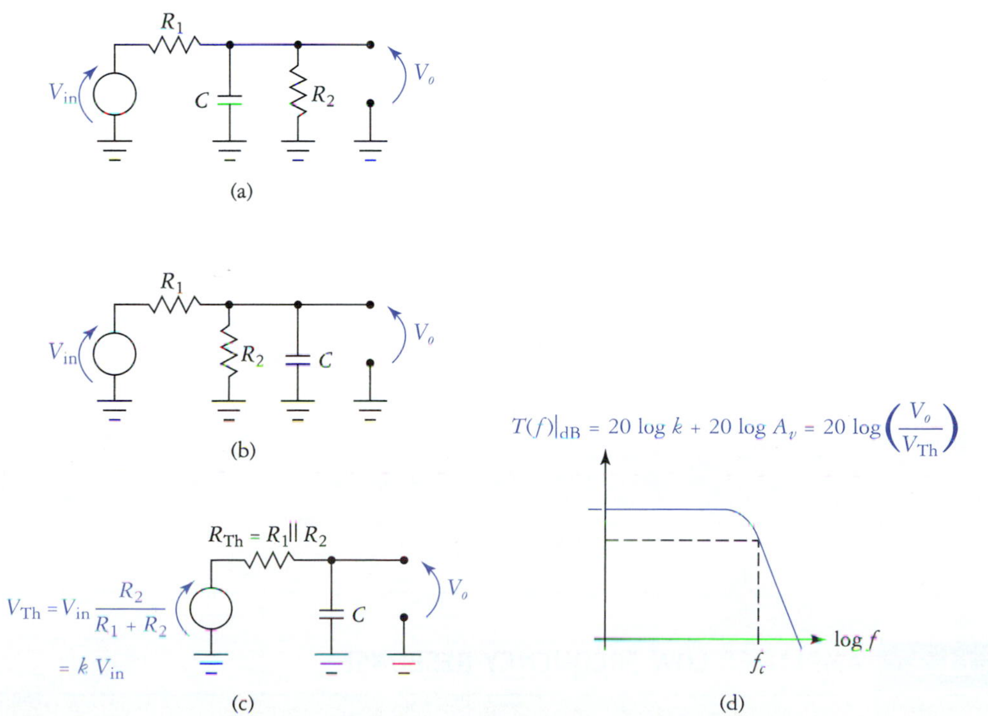

FIGURE 11.6
Circuit reduction. (a) The original circuit. (b) Moving $\mathbf{R_2}$ ***to emphasize the voltage divider. (c) Thevenizing the resistive divider. (d) Resulting frequency response plot.***

see that we apply Thevenin's theorem to produce the equivalent circuit in Fig. 11.6(c). The corner frequency is

$$f_c = \frac{1}{2\pi R_{\text{Th}} C} \tag{11.13}$$

The response curve for this circuit is shown in Fig. 11.6(d). Here, things get a little complicated because the resistors produce a constant reduction in gain over all frequencies. Most of the time we are only interested in the corner frequency when we are dealing with coupling and bypass circuits.

EXAMPLE 11.2

Determine the response type, the corner frequency, and the resulting frequency response magnitude expression for the circuit of Fig. 11.7.

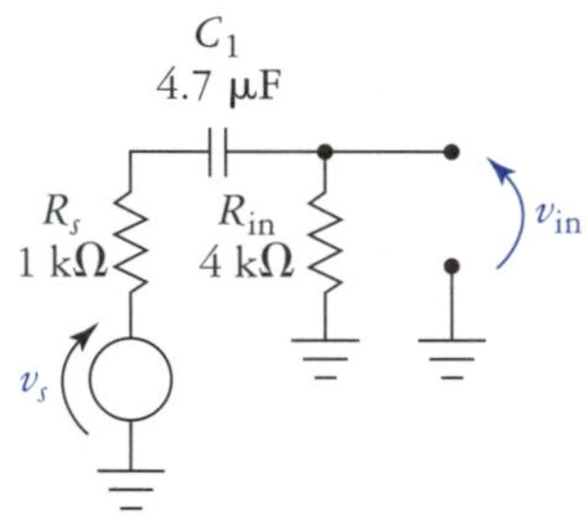

FIGURE 11.7
Circuit for Example 11.2.

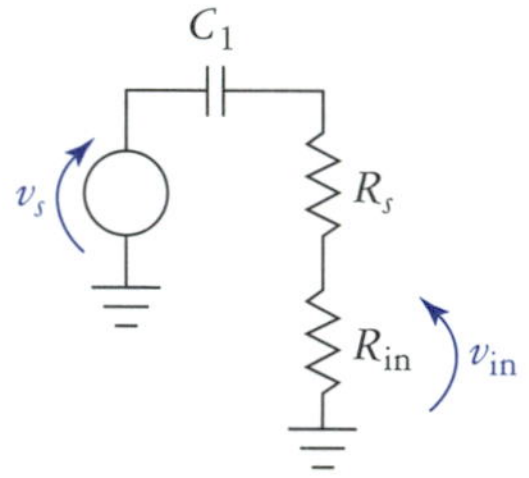

FIGURE 11.8
Simplified network.

Solution This is a simple series circuit which can be rearranged as shown in Fig. 11.8 where we lump the resistances together. Since the capacitor is in series with the source, this is a high-pass network with $R_T = R_s + R_1 = 5\ \text{k}\Omega$. The corner frequency is

$$\begin{aligned} f_c &= \frac{1}{2\pi R_T C_1} \\ &= \frac{1}{(2\pi)(5\ \text{k}\Omega)(4.7\ \mu\text{F})} \\ &= 6.7\ \text{Hz} \end{aligned}$$

This is a high-pass filter with a 6.7-Hz corner frequency. The response equation is found using Eq. (11.12) with k determined from the voltage-divider equation:

$$\begin{aligned} k &= \frac{R_{\text{in}}}{R_{\text{in}} + R_s} \\ &= 0.8 \end{aligned}$$

Thus we have

$$\begin{aligned} A(f) &= \frac{k}{\sqrt{\left(\dfrac{f_c}{f_{\text{in}}}\right)^2 + 1}} \\ &= \frac{0.8}{\sqrt{\left(\dfrac{f_c}{f_{\text{in}}}\right)^2 + 1}} \end{aligned}$$

11.2 AMPLIFIER LOW-FREQUENCY RESPONSE

We know from previous study that if capacitive coupling is used, an amplifier will have a lower frequency limit $f_{c(\text{low})}$ that is greater than dc, that is, $f_{c(\text{low})} > 0$ Hz. (We use the notation f_L rather than $f_{c(\text{low})}$ when possible in order to simplify the notation.) Thus, we find that the low-frequency response of an amplifier is controlled primarily by coupling and bypass capacitor sizes. Direct coupling allows us to achieve low-frequency response extending down to 0 Hz.

Because we have already covered most aspects of low-frequency response, a single example will serve to illustrate the main points.

EXAMPLE 11.3

Refer to the circuit shown in Fig. 11.9. Determine the corner frequency for each coupling and bypass network. Sketch the resulting low-frequency asymptotic Bode plot for the amplifier.

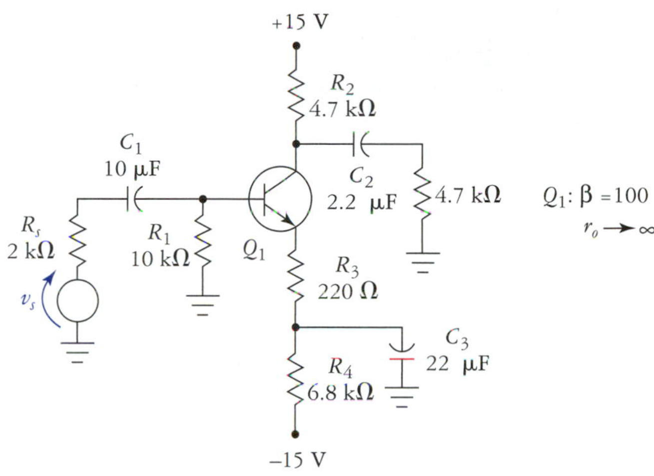

FIGURE 11.9
Circuit for Example 11.3.

Solution As always, we must first analyze the circuit to determine the Q point and other important parameters. The results of this analysis are

$$I_{CQ} \cong 2 \text{ mA}$$
$$V_{CEQ} = 6.6 \text{ V}$$
$$r_e = 13\ \Omega$$
$$R'_C = 2.35 \text{ k}\Omega$$
$$R_e = 220\ \Omega$$
$$R_{\text{in}} = 7 \text{ k}\Omega$$
$$R_o = 4.7 \text{ k}\Omega$$
$$|A_v| = 10 = 20 \text{ dB}$$

Starting with the input coupling network we can use the equivalent circuit of Fig. 11.10. This is a high-pass filter with $R_T = R_s + R_{\text{in}} = 9 \text{ k}\Omega$. Using this we find

$$\begin{aligned} f_{L_1} &= \frac{1}{2\pi R_T C_1} \\ &= \frac{1}{(2\pi)(9 \text{ k}\Omega)(10\ \mu\text{F})} \\ &= 1.8 \text{ Hz} \end{aligned}$$

FIGURE 11.10
Simplified input coupling network.

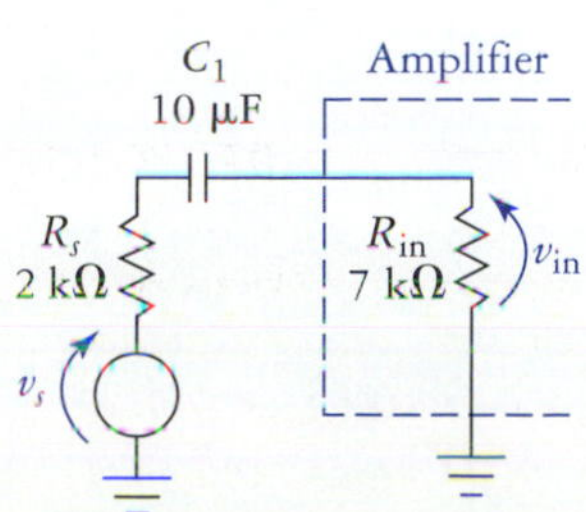

The output coupling network is redrawn in Fig. 11.11. The total resistance is $R_T = R_o + R_L = 9.4\ \text{k}\Omega$, which gives us

$$f_{L_2} = \frac{1}{2\pi R_T C_2}$$

$$= \frac{1}{(2\pi)(9.4\ \text{k}\Omega)(2.2\ \mu\text{F})}$$

$$= 15.4\ \text{Hz}$$

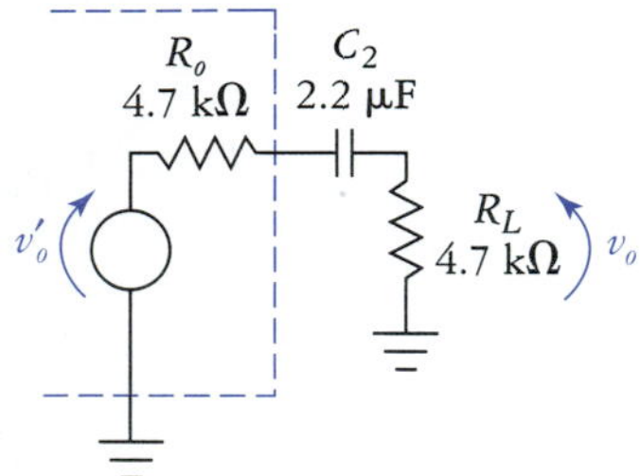

FIGURE 11.11
Simplified output coupling network.

The emitter bypass network is redrawn in Fig. 11.12. Although it is somewhat difficult to visualize, this too is a high-pass filter with the rather complicated total resistance

$$R_T = \left[R_3 + r_e + \frac{R_s \parallel R_1}{\beta}\right] \parallel R_4$$

from which the corner frequency is

$$f_{L_3} = \frac{1}{2\pi R_T C_3}$$

$$= \frac{1}{(2\pi)(241\ \Omega)(22\ \mu\text{F})}$$

$$= 30\ \text{Hz}$$

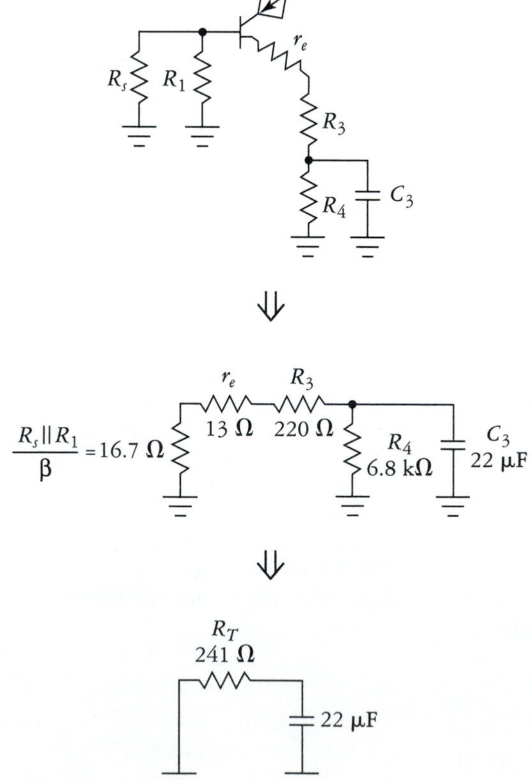

FIGURE 11.12
Analysis of the emitter-bypass network.

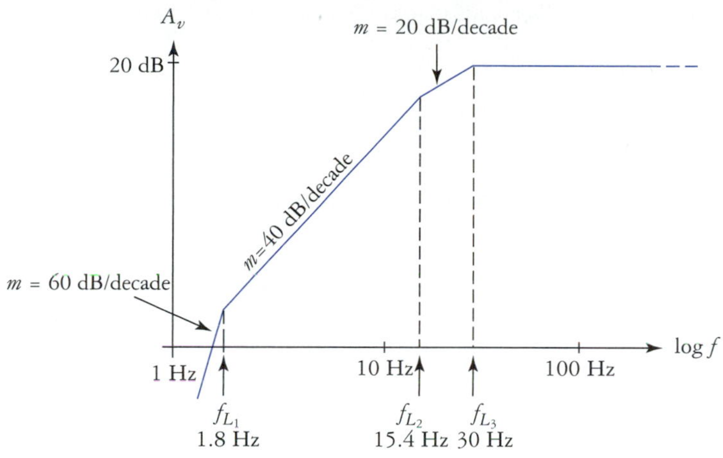

FIGURE 11.13
Resulting low-frequency asmyptotic Bode plot.

The asymptotic Bode plot is shown in Fig. 11.13.

In the previous example, the dominant lower corner frequency was 30 Hz, caused by the emitter-bypass circuit. This sets the overall lower corner frequency of the amplifier.

A Word About Coupling Capacitors and Approximations

In practice, approximations are made whenever possible. This is often done when analyzing coupling and bypass circuits, which are usually overdesigned anyway. Another practical consideration is based on the fact that large coupling and bypass capacitors ($C \geq 1\ \mu$F) are typically aluminum electrolytic units, which have notoriously wide tolerances; they are often as much as 50% greater than the marked value. Such variations swamp the effects of many terms in the corner frequency equations. The main thing to remember is to use common sense in the analysis process, which mainly comes with experience.

11.3 NEGATIVE FEEDBACK

We will delay our study of high-frequency response until we have covered the basics of negative feedback. As stated before, negative feedback is an extremely important concept; it is responsible for improvements in many amplifier characteristics, such as bandwidth, input resistance, output resistance, and linearity. Positive feedback is also encountered, but this is normally used to implement oscillators and latches.

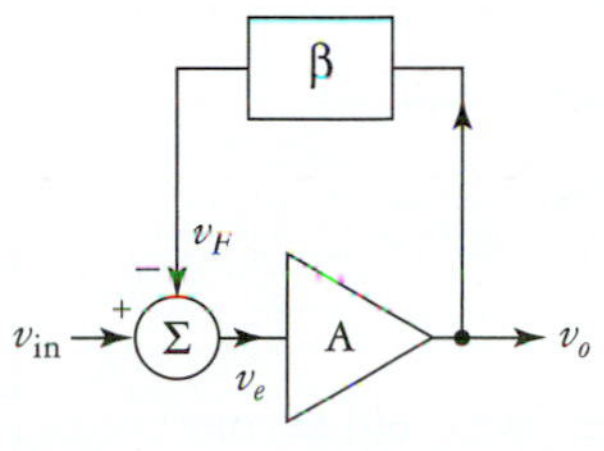

FIGURE 11.14
General amplifier with negative feedback.

A block diagram for an amplifier with feedback is shown in Fig. 11.14. We will deal strictly with voltages and voltage gains in this discussion. The triangle is the basic amplifier with voltage gain A. Sometimes this is referred to as the *open-loop voltage gain* A_{OL}; it would be the gain of the circuit if the feedback loop were opened. The *feedback factor,* β, determines the fraction of the output voltage that is sampled and fed back to the input of the amplifier. The negative sign in the feedback loop indicates that the feedback voltage is the opposite phase of the input voltage, thus we have negative feedback. The circle with sigma Σ inside it represents the summing of the feedback voltage v_f and the input voltage v_{in}. The output of the summation is the *error voltage* v_e. Because we have negative feedback the equation for v_e is

$$v_e = v_{in} - v_f \tag{11.14}$$

When the feedback loop is closed, the overall voltage gain of the amplifier may be designated as the gain with feedback A_f. The equation for A_f is derived next. Inspection of Fig. 11.14 reveals that the output voltage can be given by

$$v_o = A_f v_{in} \tag{11.15}$$

and equivalently

$$v_o = Av_e \tag{11.16}$$

Substituting Eq. (11.14) into (11.16),

$$v_o = A(v_{\text{in}} - v_f) \tag{11.17}$$

We can also define the feedback voltage as

$$v_f = \beta v_o \tag{11.18}$$

Substitute Eq. (11.18) into Eq. (11.17):

$$v_o = A(v_{\text{in}} - \beta v_o) \tag{11.19}$$

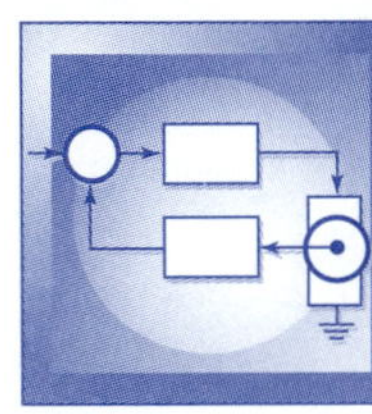

Now, we substitute Eq. (11.15) into (11.19) to obtain

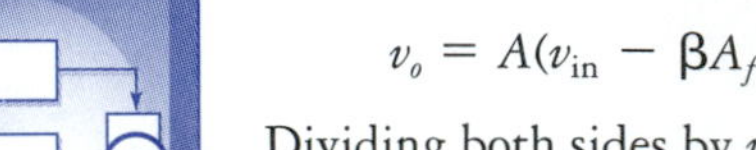

$$v_o = A(v_{\text{in}} - \beta A_f v_{\text{in}})$$

Dividing both sides by v_{in} we have

$$\frac{v_o}{v_{\text{in}}} = A_f = A(1 - \beta A_f)$$

We wish to combine the A_f factors. Distributing A across the right side yields

$$A_f = A - \beta A A_f$$

Adding $\beta A A_f$ to both sides, we get

$$A = A_f + \beta A A_f$$

Factor A_f from the right side

$$A = A_f(1 + A\beta)$$

Finally, we solve for A_f, producing the equation we have been after:

$$A_f = \frac{A}{1 + A\beta} \tag{11.20}$$

This is an important result, which applies to any amplifier that employs feedback. In fact, Eq. (11.20) also applies when we have positive feedback, for which $A\beta < 0$.

EXAMPLE 11.4

Refer to Fig. 11.15. This is a two-stage capacitively coupled amplifier that employs negative feedback via R_f and C_3. Assume that both transistors have $\beta = 100$ and $r_o = \infty$. Assume that $R_s \cong 0\ \Omega$ and that under signal conditions, the amplifier is operating in its midband. Determine the Q points for each stage, the overall open-loop voltage gain, the feedback factor β, and the closed-loop gain with feedback A_f for $R_F = 1$ and 10 kΩ. Also verify that the feedback is indeed negative.

Solution The dc equivalent circuits for both stages are shown in Fig. 11.16(a). Coupling capacitor C_3 prevents the feedback loop from affecting the dc analysis. Both stages are nearly identical, with the difference between dc emitter resistances being negligible. The quiescent values for the two circuits are found using basic analysis techniques to be

$$I_{C_1} \cong I_{C_2} \cong 2\text{ mA}$$

$$V_{CE_1} \cong V_{CE_2} \cong 6.6\text{ V}$$

$$r_{e_1} \cong r_{e_2} \cong 13\ \Omega$$

We perform the ac analysis assuming that the feedback loop is open. We could get more accurate results if we factored in the effects of R_F loading, but since this resistor varies by a factor of 10 (from 1 to 10 kΩ) in this example, this is impractical. We also perform this open-loop ac analysis under no-load conditions. The ac-equivalent circuits for the two stages are shown in Fig. 11.16(b). These circuits were analyzed beginning with stage 2, where we find

$$R'_{C_2} = 4.7\text{ k}\Omega$$

$$R_{e_2} = 0\ \Omega$$

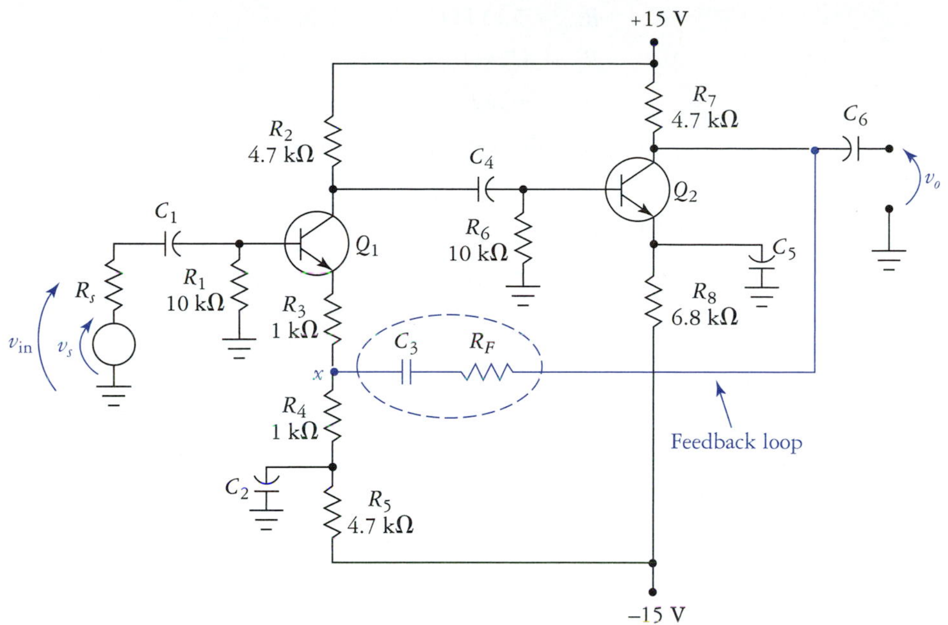

FIGURE 11.15
Two stage amplifier with negative feedback network.

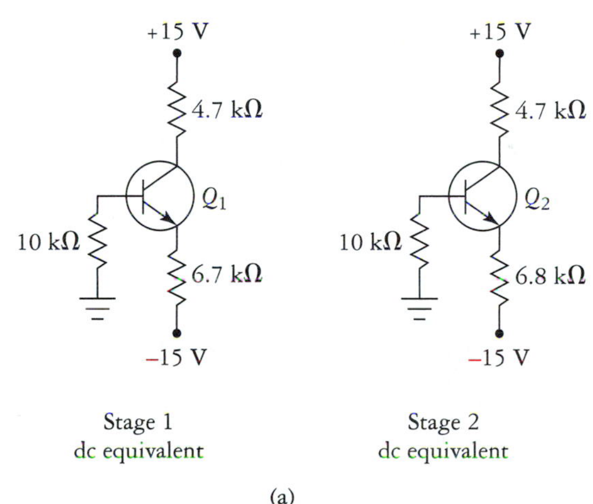

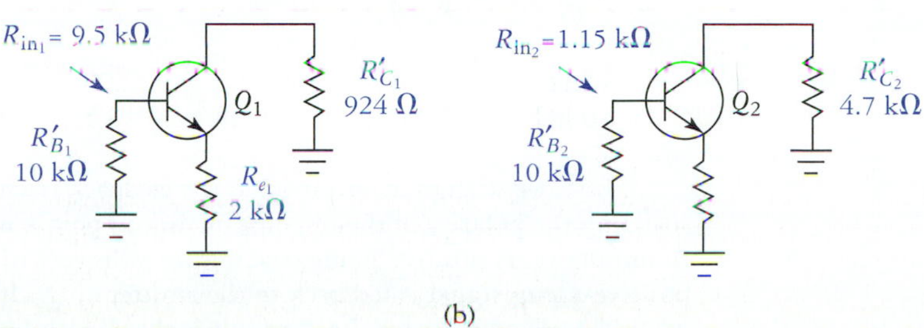

FIGURE 11.16
DC- and AC-equivalent circuits.

$$R_{\text{in}_2} = 1.15\ \text{k}\Omega$$
$$R_{o_2} = 4.7\ \text{k}\Omega$$
$$A_{v_2} = -362$$

Analysis of the first stage gives us

$$R'_{C_1} = R_{C_1} \parallel R_{\text{in}_2} = 924\ \Omega$$
$$R_{e_1} = 0\ \Omega$$
$$R_{\text{in}_1} = 9.5\ \text{k}\Omega$$
$$A_{v_1} = -0.46$$

The overall open-loop characteristics are summarized below and the equivalent circuit using a VCVS is drawn in Fig. 11.17.

$$A = A_{v_1} A_{v_2} = 167$$
$$R_{\text{in}} = R_{\text{in}_1} = 9.5\ \text{k}\Omega$$
$$R_o = R_{o_2} = 4.7\ \text{k}\Omega$$

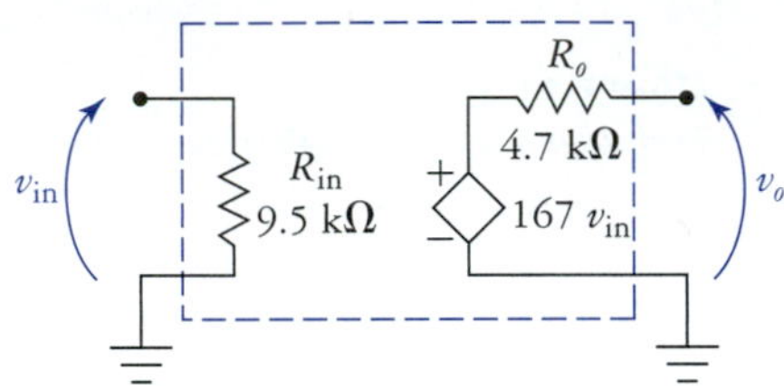

FIGURE 11.17
Overall open-loop model of the amplifier.

The feedback loop is approximately modeled as shown in Fig. 11.18. We determine the feedback factor by using the voltage-divider equation:

$$\beta \cong \frac{R_4 \parallel R_3}{(R_4 \parallel R_3) + R_f}$$

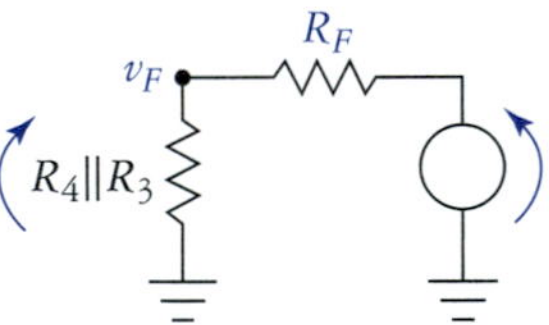

FIGURE 11.18
Simplified model of the feedback loop.

Applying the β expression above, and Equation (11.20) for the two feedback resistor values requested, we obtain the following values:

R_F	$R_4 \| R_3$	β	A_f
1 kΩ	500Ω	0.33	2.98
10 kΩ	500Ω	0.0048	18.5

Last, we wish to determine that the feedback is actually negative. This is done by simply visualizing the polarity of the response at various points within the circuit under signal conditions. Assuming v_{in} is positive going, the output voltage will also be positive going. A portion of this positive-going signal is fed back to the emitter of Q_1. If this were the input terminal, the output would be driven negative. Because this is the opposite polarity of the feedback signal, we do have negative feedback.

Some interesting conclusions can be drawn based on the last example. First, we find that application of negative feedback reduces the gain of the amplifier. While this may not seem desirable, what is happening is that we are trading off high, unpredictable gain for lower, but more stable gain. Suppose for example that the circuit was constructed and the actual open-loop gain turned out to be 100, which would be disappointing, but not too unusual. If that were the case, the following feedback analysis would result.

R_F	$R_4 \| R_3$	β	A_f
1 kΩ	500Ω	0.33	2.94
10 kΩ	500Ω	0.0048	17.2

The open-loop gain decreased by about 40% while the feedback gain decreased by 1% and 4%. Negative feedback made the voltage gain much more stable.

Real-World Amplifiers, Negative Feedback, and Voltage Gain

The previous analysis was fairly complicated. At this point, such detailed work is important in order to reinforce basic concepts and principles. Normally the open-loop gain of a given amplifier like that used in Example 11.4 is made as high as possible. In this way, we can use negative feedback to reduce the gain to the desired value and produce a very stable circuit. In fact, if we let the open-loop gain approach infinity, the feedback gain equation becomes

$$\lim_{A\to\infty} \frac{A}{1 + A\beta} = \frac{1}{\beta} \tag{11.21}$$

Thus, when the open-loop gain is very large, the closed-loop gain is controlled by the feedback resistors, which are very accurate, predictable components. Equation (11.21) is called the *ideal gain equation.*

Because of the implications of Eq. (11.21), most amplifiers that are produced in integrated circuit form are designed to have an extremely high open-loop voltage gain ($A > 200$ k is very common). The user configures a negative-feedback loop in order to produce the desired voltage gain. Increasing the open-loop gain of the circuit in Example 12-4 without limit produces

R_F	$R_4 \| R_3$	β	A_f
1 kΩ	500Ω	0.33	3
10 kΩ	500Ω	0.0408	21

The higher the open-loop gain, the closer the closed-loop gain will be to the limiting values listed.

Thousands of circuit variations are possible, and the previous example was just one possibility. In practice, however, certain amplifier configurations are used more often. One of the more commonly encountered feedback amplifier configurations is examined in the next example.

EXAMPLE 11.5

Refer to Fig. 11.19. This is a three-stage direct-coupled amplifier composed of a diff-in, single-ended-output amplifier, a level translator, and a class B push-pull output stage. The input labeled (+) is the noninverting input; the output voltage is in phase with the voltage applied here. The output voltage is sampled by the voltage divider formed by R_f and R_A, and applied to the inverting input. Determine the open-loop voltage gain and the closed-loop feedback gain given

(a) $R_A = 10$ kΩ and $R_F = 10$ kΩ.

(b) $R_A = 10$ kΩ, $R_F = 100$ kΩ

Estimate the level translator (stage 2) and push-pull output (stage 3) to have voltage gains $A_{v_2} = A_{v_3} = 0.7$.

Solution The primary quiescent values for the circuit are determined assuming the feedback-loop is open:

$$I_{C_1} = I_{C_2} = 515\ \mu\text{A}$$

$$r_{e_1} = r_{e_2} \cong 50\ \Omega$$

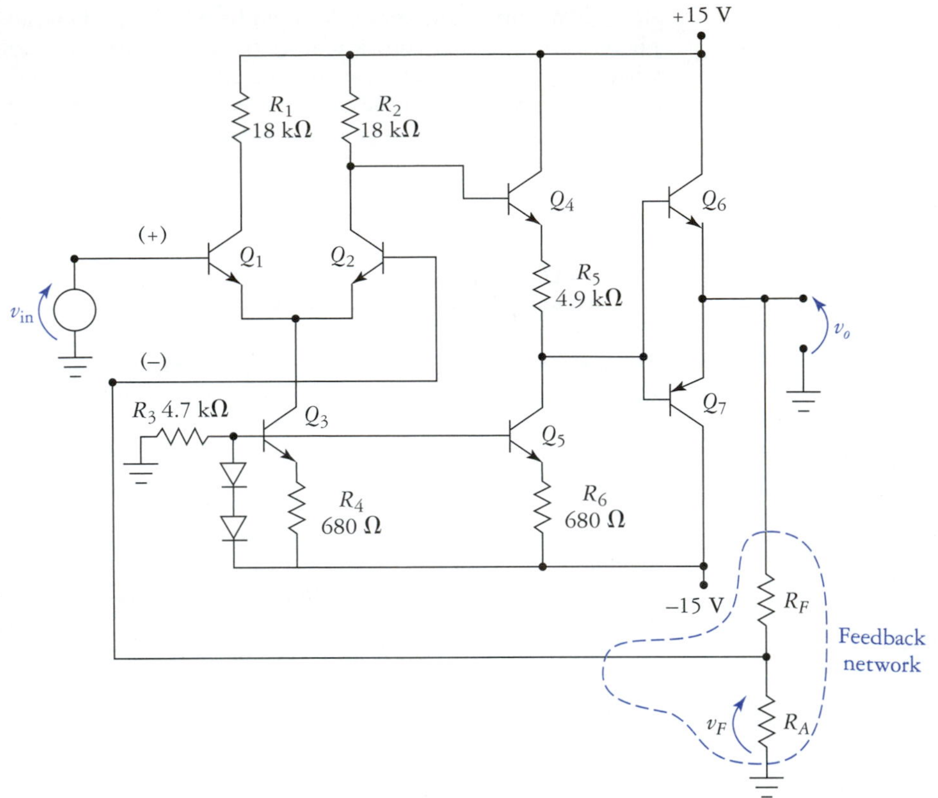

FIGURE 11.19
Circuit for Example 11.5.

$$V_{CE_1} = V_{CE_2} = 6.43\text{ V}$$
$$V_{C_1} = V_{C_2} = 5.73\text{ V}$$
$$I_{C_4} = I_{C_5} = 1.03\text{ mA}$$
$$r_{e_4} = 25\ \Omega$$
$$V_{C_5} \cong 0\text{ V}$$
$$V_{CE_6} = -V_{CE_7} = 15\text{ V}$$
$$V_{oQ} = 0\text{ V}$$

Under no-load conditions, the input resistance of level translator Q_4 (stage 2) is extremely high ($R_{\text{in}_2} \to \infty$), therefore the differential gain of stage 1 is approximately

$$\begin{aligned} A_{d_1} &\cong \frac{R'_C}{2r_e} \\ &= \frac{18\text{ k}\Omega}{50\ \Omega} \\ &= 180 \end{aligned}$$

And the overall open-loop gain of the amplifier is approximately

$$\begin{aligned} A &= A_{d_1}A_2A_3 \\ &= (180)(0.7)(0.7) \\ &= 88.2 \end{aligned}$$

The open-loop input resistance is

$$\begin{aligned} R_{\text{in}} &= 2\beta r_{e_1} \\ &= (200)(50\ \Omega) \\ &= 10\text{ k}\Omega \end{aligned}$$

The open-loop output resistance (assuming Q_6 is active, with $r'_{e_6} = 5\ \Omega$) is given by the formidable expression

$$R_o = r'_{e_6} + \frac{\dfrac{R_2}{\beta} + r_{e_4} + R_5}{\beta}$$
$$= 56\ \Omega$$

Under no-signal conditions, the actual output resistance will be very high (thousands of ohms), because Q_6 and Q_7 are both in cutoff (class B biasing) until one or the other is driven into conduction. The output resistance calculated assumes that the output transistor is in the active region.

The feedback factor is found using the familiar voltage-divider equation

$$\beta = \frac{R_A}{R_A + R_F}$$

The closed-loop feedback gain is given by Eq. (11.20). The feedback resistors specified give us the following results.

R_F	R_A	β	A	A_f
(a) 10 kΩ	10 kΩ	0.5	88.2	1.96
(b) 10 kΩ	100 kΩ	0.0909	88.2	9.78

Increasing the open-loop gain of the amplifier would give us closed-loop voltage gains that approach 2 and 11 in the limit. Options that might be considered are the addition of another differential gain stage or the use of active loading.

The amplifier in Fig. 11.19 is totally direct coupled, and so is the feedback loop. Because of this, in addition to providing gain stability, the negative feedback also tends to stabilize the Q point, forcing the output to zero when $v_{in} = 0$ V.

Negative Feedback and Distortion

You might be wondering about the severity of crossover distortion in the amplifier of Fig. 11.19. Under open-loop conditions, considerable crossover distortion will be present in the output. If we intended to use the circuit open loop, as a linear amplifier, then class AB biasing of the output stage would likely be required.

Under closed-loop conditions, the crossover distortion is reduced drastically. The main idea here is that when we close the feedback loop, the amplifier forces the feedback voltage to track the applied input voltage, forcing the differential input voltage to a small value. Recall the differential amplifier equation

$$v_o = A_d v_{id}$$
$$= A_d(v_1 - v_2)$$

In this case, $A_d = A$ (overall open-loop gain), $v_1 = v_{id}$ and $v_2 = v_f$; therefore,

$$v_o = A\,(v_{in} - v_f)$$

This gives us

$$v_{id} = \frac{v_o}{A}$$

Thus for a given amount of negative feedback, the higher the open-loop gain, the smaller the differential input voltage will be. Because of this, the amplifier forces the output to track small changes in v_{in}, by increasing the drive to the output transistor, compensating for its barrier potential. The effective barrier potential with negative feedback $V_{\phi F}$ is

$$V_{\phi F} = \frac{V_\phi}{1 + A\beta} \qquad (11.22)$$

As the open-loop gain increases, the effective barrier potential decreases. It can also be shown that other amplifier nonlinearities are reduced as well. Commercial IC amplifiers like that of Fig. 11.19 have very high open-loop gain, which in conjunction with negative feedback can produce very low-distortion amplifiers for a low cost. We discuss additional feedback-related issues in Chapter 12.

11.4 AMPLIFIER HIGH-FREQUENCY RESPONSE

The high-frequency response of an amplifier is determined by a combination of many factors, including input and source resistances, stray capacitances associated with the physical layout of the circuit, device parasitic capacitances, and circuit topology. We know, for example, that all things being equal, the common-base configuration will have the highest frequency response for a given transistor. The reason why this is so is explained in this section. Of course, circuit layout can also have a drastic effect on the high-frequency performance of an amplifier. This is one of the reasons why it is so difficult to work with frequencies above 1 MHz or so when using breadboards and point-to-point wiring.

The (usually undesired) capacitances associated with all electronic components are called *parasitic* or *stray capacitances.* At relatively low frequencies, these stray capacitances are usually negligibly small. At higher frequencies, usually beginning around 1 MHz, stray capacitances often become significantly large and it may become necessary to factor their effects into circuit operation. Usually, we can lump capacitances and resistances into single equivalent components and apply standard circuit analysis procedures to determine the critical frequencies. At very high frequencies (typically around 100 MHz and up), we must resort to using *distributed parameters.* Distributed parameter analysis is beyond the scope of this text, but it is used extensively in VHF, UHF, and microwave circuit design.

Transistor Junction Capacitances

The pn junctions within BJTs and JFETs, and the oxide insulating layers within MOSFETs behave like low-value capacitors. Also, the close spacing of transistor leads contributes what is called *interelectrode capacitance.* These capacitances are illustrated in Fig. 11.20. Sometimes alternative names are given to these capacitances on transistor data sheets. For example, BJT data sheets often list C_{ob} and C_{ib}, which relate to Fig. 11.20 as

$$C_{ob} = C_{CB}$$
$$C_{ib} = C_{BE}$$

FET data sheets often list C_{iss}, C_{oss}, and C_{rss}, which are defined as

$$C_{iss} = C_{GD} + C_{GS}$$
$$C_{oss} = C_{GD} + C_{DS}$$
$$C_{rss} = C_{GD}$$

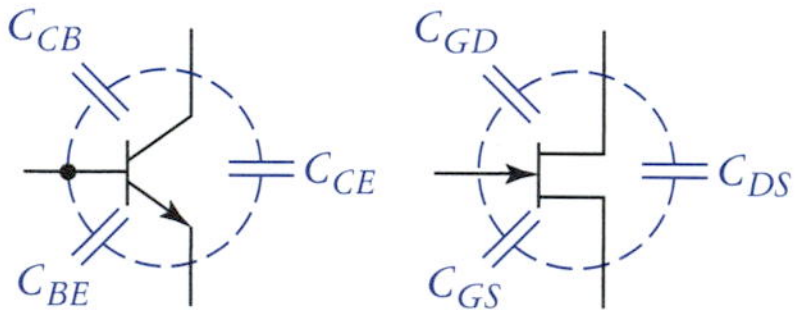

FIGURE 11.20
Transistor junction capacitances.

For bipolar transistors, junction capacitance is a function of transistor junction area, physical chip geometry, bias voltage, and package configuration. Small, low-power transistors tend to have low junction capacitances, while larger area power transistors have higher junction capacitances. Because of this, it is difficult to obtain good high-frequency performance from high-power transistors.

The Miller Effect

In the common-emitter configuration, the most important parasitic capacitance is C_{BC}. Modeling the CE amplifier as a voltage-gain black box, we obtain the circuit shown in Fig. 11.21(a). Because the CE amplifier inverts, the stray capacitance C_{BC} forms a negative-feedback loop. This causes the feedback capacitance to effectively appear much larger at the input and slightly larger at the output. We

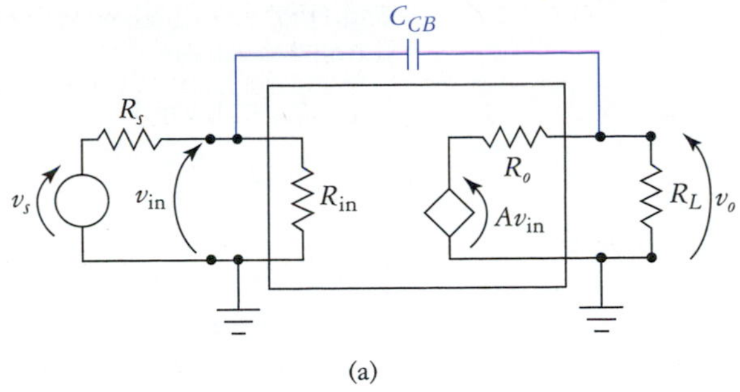

(a)

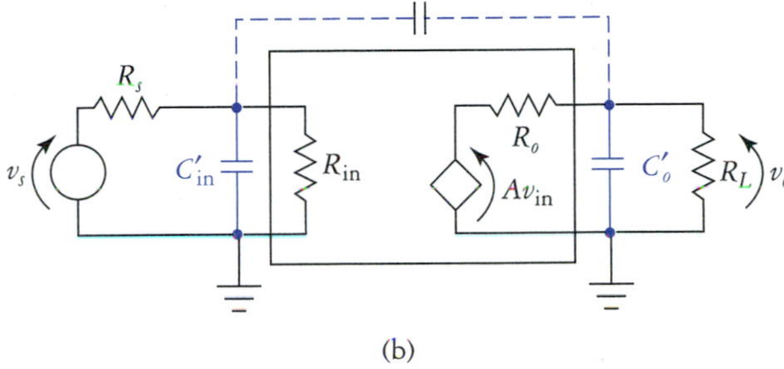

(b)

FIGURE 11.21
C_{CB} can have a major effect on the frequency response of the CE configuration. (a) AC small-signal model. (b) Miller equivalent model.

can model this effect as shown in Fig. 11.21(b). An analysis of the circuit generates the following relationships:

$$C'_{in} = C_{CB}\,(|A_v| + 1) \tag{11.23}$$

$$C'_o = C_{CB}\frac{|A_v| + 1}{|A_v|} \tag{11.24}$$

The same relationships hold for the common-source FET amplifier as well. Because the voltage gain of the typical CE amplifier can be quite large, the associated Miller input capacitance may also be substantial.

Qualitatively, the effect of C_{CB} can be understood as follows. At low frequencies the reactance of C_{CB} is so large that very little signal energy is fed back from collector to base: There is little negative feedback, so the gain of the amplifier is high. As frequency increases, X_{CBC} decreases, allowing heavier negative feedback and decreasing the gain of the amplifier.

EXAMPLE 11.6

A certain person with poor television reception and no access to cable TV intends to use the amplifier in Fig. 11.22 as a booster amplifier between his antenna and his television. The transistor is a 2N3904 with $C_{BC} = 4$ pF, $C_{BE} = 2$ pF, $C_{CE} = 1$ pF, and $\beta \cong 100$. Determine whether this amplifier will perform adequately.

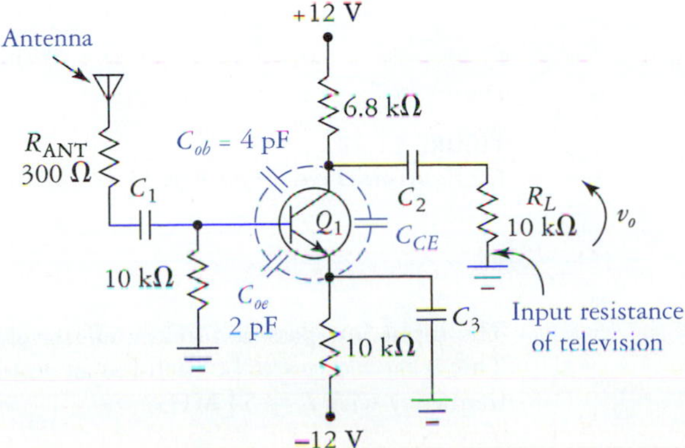

FIGURE 11.22
Circuit for Example 11.6.

Solution As usual, we determine the Q point of the amplifier first, which gives us the following reasonable values:

$$I_{CQ} = 1.1 \text{ mA}$$
$$V_{CEQ} = 5.2 \text{ V}$$
$$r_e = 24\ \Omega$$

Performing our standard small-signal analysis we obtain

$$R'_C = 4 \text{ k}\Omega$$
$$A_v = -167$$
$$R_{\text{in}} = 1.9 \text{ k}\Omega$$
$$R_o = 6.8 \text{ k}\Omega$$

Everything looks good so far. We now apply Miller's theorem, which gives us

$$C'_{\text{in}} = C_{BC}(|A_v| + 1) = (4 \text{ pF})(168) = 672 \text{ pF} \qquad C'_o = C_{BC}\frac{|A_v| + 1}{|A_v|} = (4 \text{ pF})(1.006) \cong 4 \text{ pF}$$

The equivalent circuit appears in Fig. 11.23(a). Reducing the circuit further produces Fig. 11.23(b). The resulting low-pass networks have the following corner frequencies:

$$f_{c(\text{in})} = \frac{1}{2\pi R_{\text{Th}} C_{\text{in}T}} = \frac{1}{(2\pi)(259)(674 \times 10^{-12})} = 912 \text{ kHz} \qquad f_{C(\text{out})} = \frac{1}{2\pi R'_C C_{oT}} = \frac{1}{(2\pi)(4 \text{ k})(5 \times 10^{-12})} = 8 \text{ MHz}$$

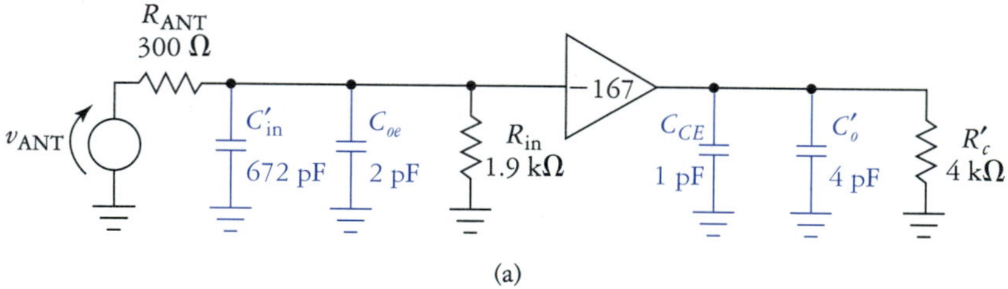

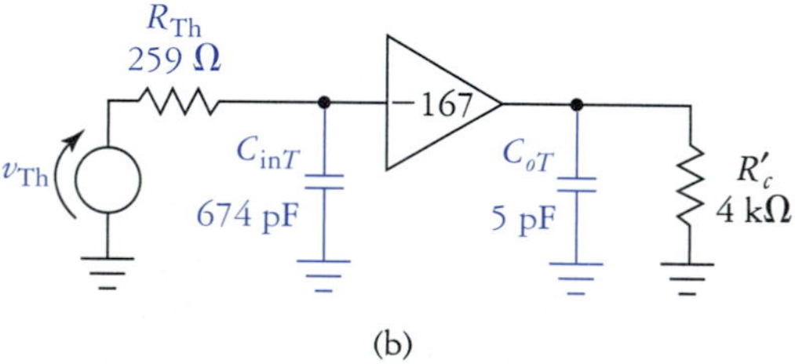

FIGURE 11.23
(a) Equivalent circuit for Fig. 11.22. (b) Simplified equivalent.

The input low-pass dominates, effectively limiting the maximum input frequency to 912 kHz. This is far too low to be useful as an amplifier in this application, since channel 2 has the lowest frequency with $f_o = 54$ MHz.

The common-base and emitter-follower configurations do not exhibit the Miller effect, and so are generally useful at higher frequencies than the common-emitter configuration. Between the emitter-follower and the common-base configurations, however, the common-base configuration has superior high-frequency performance. This is explained next.

Beta and Alpha Cutoff and Transition Frequencies

Aside from stray capacitances and the Miller effect, the high-frequency operation of the transistor is also limited by frequency-related decreases in β and α. A typical plot of β versus frequency is shown in Fig. 11.24. Because the rate of rolloff in β is similar in character to that of a first-order network, a log-log graph is normally used.

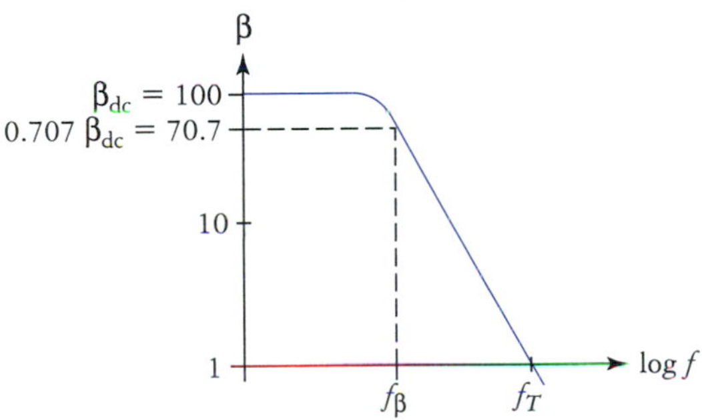

FIGURE 11.24
Beta cutoff and transition frequencies.

Beta Cutoff Frequency

Note
Beta rolls off at -20dB/decade beyond f_β.

The frequency at which beta decreases to 0.707 times the low-frequency (β_{dc}) value is called the *beta cutoff frequency* f_β. This is the frequency at which the power gain of a common-emitter or emitter-follower amplifier will drop to 50% of the midband value. Many times, these amplifiers are still useful beyond f_β, because the voltage gain is not a strong function of β. In the case of the common-emitter amplifier, however, the Miller effect will usually become significant before the beta cutoff frequency is reached. The typical 2N3904 has $f_\beta \cong 10$ MHz, but as we saw in Example 11.6, the Miller effect limited the high-frequency response long before this limit was reached. An emitter-follower might have worked better if we weren't after voltage gain, but we still would have exceeded f_β. It is easy to intuitively understand why β is a primary factor in the power gain of CE and EF amplifiers: Both configurations have i_B as the basic input variable, while i_C and i_E are their respective controlled variables and these are related via $i_E \cong i_C = \beta I_B$.

Beta Transition Frequency

The frequency at which the beta of a transistor decreases to unity ($\beta = 1$) is called the *beta transition frequency* f_T. This is basically the absolute maximum frequency at which the transistor can be used. The transistor is incapable of exhibiting power gain at frequencies greater than f_T. The 2N3904 has $f_{T(\text{min})} \cong 300$ MHz.

Alpha Cutoff Frequency

A typical plot of α versus frequency is shown in Fig. 11.25. Because most often we assume that $\alpha \cong 1$, this graph applies to any typical transistor. The frequency at which α decreases to 0.707 times the low-frequency value is called the *alpha cutoff frequency* f_α. The alpha cutoff frequency will usually be many times higher than the beta cutoff frequency. This can be understood by considering a transistor with $\beta_{dc} = 100$. Recall that

$$\begin{aligned}\alpha &= \frac{\beta}{\beta + 1}\\ &= \frac{100}{101}\\ &= 0.99\\ &\cong 1\end{aligned}$$

Based on these numbers, the beta cutoff frequency occurs when $\beta = 70.7$ and the alpha cutoff frequency occurs when $\alpha = 0.707$. Knowing alpha, we can find the corresponding beta using

$$\begin{aligned}\beta &= \frac{\alpha}{1-\alpha}\\ &= \frac{0.707}{0.293}\\ &= 2.4\end{aligned}$$

Thus, the alpha cutoff frequency does not occur until we have nearly reached the beta transition frequency. Because the difference between f_T and f_α is usually so slight, manufacturers do not often publish f_α. Normally, we make the approximation

$$f_\alpha \cong f_T \tag{11.25}$$

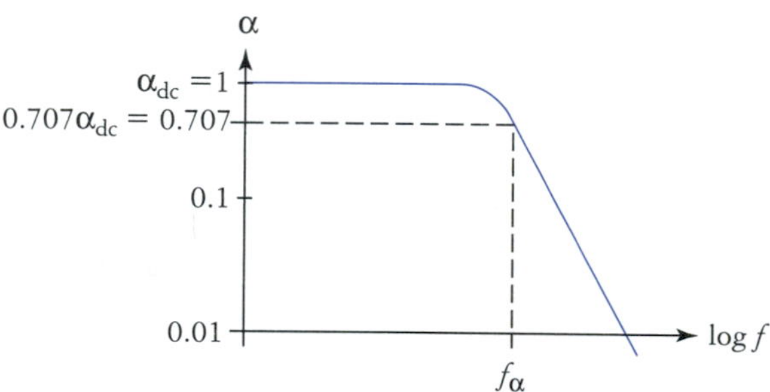

FIGURE 11.25
Alpha cutoff frequency.

Here is what is significant about these results. If we can design an amplifier that depends on α as its primary gain parameter, then we can use the amplifier at much higher frequencies than f_β. This is why the common-base (and common-drain) configuration has superior high-frequency performance. The input variable to the CB amplifier is i_E, while the controlled output variable is i_C, thus fundamentally the gain is controlled by alpha, because $\alpha = i_C/i_E$.

The Cascode Amplifier

Two characteristics of common-base amplifiers that can be undesirable are the inherently low input resistance and less than unity current gain. The cascode amplifier largely eliminates these problems, while maintaining excellent high-frequency response. A cascode amplifier is shown in Fig. 11.26. This is basically a direct-coupled connection of a common-emitter stage driving a common-base stage. Both transistors operate at the same collector current, which is set by Q_1.

The CE–CB connection can most easily be seen in the ac-equivalent circuit of Fig. 11.27(a). The C_{CE} stray capacitances have been neglected, because they are usually very small compared to C_{BC}

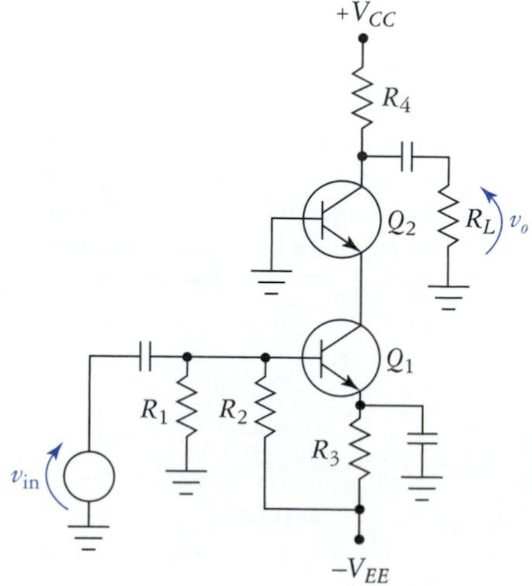

FIGURE 11.26
Typical cascode amplifier configuration.

(a)

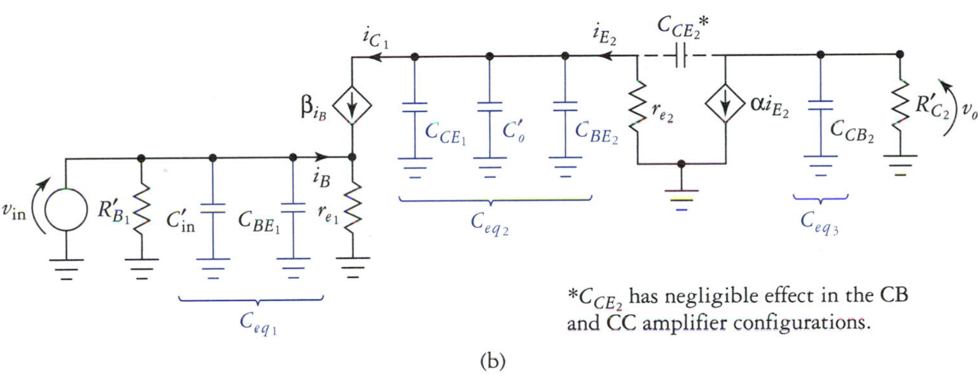

(b)

FIGURE 11.27
(a) AC-equivalent circuit. (b) AC-equivalent circuit using small-signal dependent source models.

and C_{BE}. The small-signal model is shown in Fig. 11.27(b), where the C_{BC} capacitance of Q_1 has been split into its Miller components. The circuit looks pretty hideous, but really reduces easily. As explained earlier, the frequency response of the CB amplifier is inherently high. The reason why the CE stage has good high-frequency response is due to the fact that its voltage gain is held to about unity because of its low equivalent collector resistance. Thus, the Miller effect only doubles the equivalent value of C_{CB}. This is shown as follows: The input resistance of stage 2 is

$$R_{in_2} = r_{e_2}$$

The voltage gain of stage 2 is

$$A_{v_2} = \frac{R'_{C_2}}{r_{e_2}}$$

Now, the voltage gain of stage 1 is

$$A_{v_1} = -\frac{R_{in_2}}{r_{e_1}} = -\frac{r_{e_2}}{r_{e_1}}$$

But since both transistors operate at the same collector current, $r_{e_1} = r_{e_2} = r_e$ and we can write

$$A_{v_1} = -\frac{r_e}{r_e} = -1$$

Thus, the Miller input capacitance is

$$C'_{in} = C_{CB}(|A_v| + 1) = 2C_{CB} \tag{11.26}$$

The Miller output capacitance is

$$C'_o = C_{CB}\frac{|A_v| + 1}{|A_v|} = 2C_{CB} \tag{11.27}$$

The input resistance of stage 1 is

$$R_{\text{in}_1} = \beta r_e \parallel R'_{\beta_1}$$

Overall, the cascode amplifier has the following gain and resistance equations:

$$A_v = -\frac{R'_{C_2}}{r_e} \tag{11.28}$$

$$R_{\text{in}} = R_{\text{in}_1} = \beta r_e \parallel R'_{B_1} \tag{11.29}$$

$$R_o = R_{C_2} \tag{11.30}$$

EXAMPLE 11.7

The circuit in Fig. 11.26 has the following values: $V_{CC} = V_{EE} = 15$ V, $R_1 = 10$ kΩ, $R_2 = 3.3$ kΩ, $R_3 = 620$ Ω, $R_4 = R_L = 2.2$ kΩ, and for both transistors, $\beta = 100$, $C_{CB} = C_{BE} = 2$ pF and $C_{CE} = 1$ pF. Determine the overall voltage gain, current gain, power gain, and the dominant high-frequency corner.

Solution The collector currents are determined first. Thevenizing the bias network of Q_1 we have $R_{\text{Th}} = 2.5$ kΩ and $V_{\text{Th}} = 3.7$ V. Now, the collector currents are

$$I_{C_2} \cong I_{C_1} = V_{EE}\frac{V_{\text{Th}} - V_{BE}}{\dfrac{R_{\text{Th}}}{\beta} + R_E} = 15\text{ V}\frac{3.7\text{ V} - 0.7\text{ V}}{\dfrac{2500\ \Omega}{100} + 680\ \Omega} = 4.3\text{ mA}$$

The collector-to-emitter voltages are determined using KVL, resulting in the following equation:

$$0 = V_{CC} + V_{EE} - I_C R_4 - V_{CE_1} - V_{CE_2} - I_C R_3$$

We have one equation and two unknowns, but this can be solved since we know that $V_{CE} = V_{CB} + V_{BE}$ and $V_{BE} \cong 0.7$ V. Using this information, we obtain $V_{CE_1} = 11.4$ V and $V_{CE_2} = 6.2$ V. The dynamic emitter resistances are

$$r_{e1} = r_{e2} = V_T/I_{CQ} = 26\text{ mV}/4.3\text{ mA} = 6\ \Omega$$

The small-signal collector load on Q_2 is

$$R'_{C_2} = R_4 \parallel R_L = 2.2\text{ k}\Omega \parallel 2.2\text{ k}\Omega = 1.1\text{ k}\Omega$$

Using Eqs. (11.28) through (11.30) gives us

$$A_v = \frac{-R'_{C_2}}{r_e} = \frac{-1.1\text{ k}\Omega}{6\ \Omega} = -183$$

$$R_{\text{in}} = \beta r_e \parallel R'_{B_1} = (100)(6\ \Omega) \parallel 2.5\text{ k}\Omega = 484\ \Omega$$

$$A_i = A_v \frac{R_{\text{in}}}{R_L}$$
$$= -183 \frac{484\ \Omega}{2.2\ \text{k}\Omega}$$
$$\cong -40$$

$$A_p = A_v A_I$$
$$= (-183)(-40)$$
$$= 7230$$

Application of Eqs. (11.26) and (11.27) provide the Miller input and output capacitance for Q_1:

$$C'_{\text{in}} = C'_o = 2C_{BC}$$
$$= (2)(2\ \text{pF})$$
$$= 4\ \text{pF}$$

Using the equivalent circuit of Fig. 11.27(b), we have (the effect of C_{CE_2} can be ignored):

$$C_{\text{eq}_1} = C'_{\text{in}} + C_{BE_1} = 4\ \text{pF} + 2\ \text{pF} = 6\ \text{pF}$$
$$C_{\text{eq}_2} = C_{CE_1} + C'_o + C_{BE_2} = 1\ \text{pF} + 4\ \text{pF} + 2\ \text{pF} = 7\ \text{pF}$$
$$C_{\text{eq}_3} = C_{CB_2} = 2\ \text{pF}$$

The corresponding equivalent resistance are

$$R_{\text{eq}_1} = R_{in_1} = 484\ \Omega \qquad R_{\text{eq}_2} = r_{e_2} = 6\ \Omega \qquad R_{\text{eq}_3} = R'_{C_2} = 1.1\ \text{k}\Omega$$

We can now determine the various corner frequencies:

$$f_{c_1} = \frac{1}{2\pi R_{\text{eq}_1} C_{\text{eq}_1}} = \frac{1}{(2\pi)(484)(6\ \text{pF})} = 54.8\ \text{MHz}$$
$$f_{c_2} = \frac{1}{2\pi R_{\text{eq}_2} C_{\text{eq}_2}} = \frac{1}{(2\pi)(6\ \Omega)(7\ \text{pF})} = 3.8\ \text{GHz}$$
$$f_{c_3} = \frac{1}{2\pi R_{\text{eq}_3} C_{\text{eq}_3}} = \frac{1}{(2\pi)(1.1\ \text{k}\Omega)(2\ \text{pF})} = 72.3\ \text{MHz}$$

The dominant high-frequency corner occurs at $f = 54.8$ MHz. This amplifier would be useful in the application of Example 11.6—for channel 2, at least.

11.5 COMPUTER-AIDED ANALYSIS APPLICATIONS

As you have probably noticed, feedback and high-frequency analysis of amplifiers can become very complex, very quickly. Computer simulations are immensely useful for this type of work. Here we use PSpice to verify (or refute) the results presented in this chapter. Let's concentrate on the circuit of Example 11.5, the three-stage direct-coupled amplifier, which is redrawn with a few slight modifications in Fig. 11.28.

Open-Loop Analysis

The amplifier has an open feedback loop, so the inverting input must be grounded to provide a bias-current path for Q_2. A large-value load resistor ($R_L = 10\ \text{M}\Omega$) has also been placed on the output for convenient lookup of the output voltage. The input voltage source is set as shown in Fig. 11.29. The AC amplitude is set to 1 V in order to make decibel voltage-gain interpretation easier. The simulation is configured for a transient analysis (0 to 3 ms), ac sweep from 10 Hz to 10 MHz in decades. In addition, because direct coupling is used, we can also perform a transfer function analysis using v_{in} and V(RL:2) as the input and output variables, respectively.

Simulation of the circuit and examination of the resulting output file produces the small-signal characteristics listed in Fig. 11.30, along with the results of the previous manual analysis:

Parameter	Hand Analysis	PSpice
A	88.2	88.4
R_{in}	10 kΩ	15 kΩ
R_o	56Ω	49 kΩ

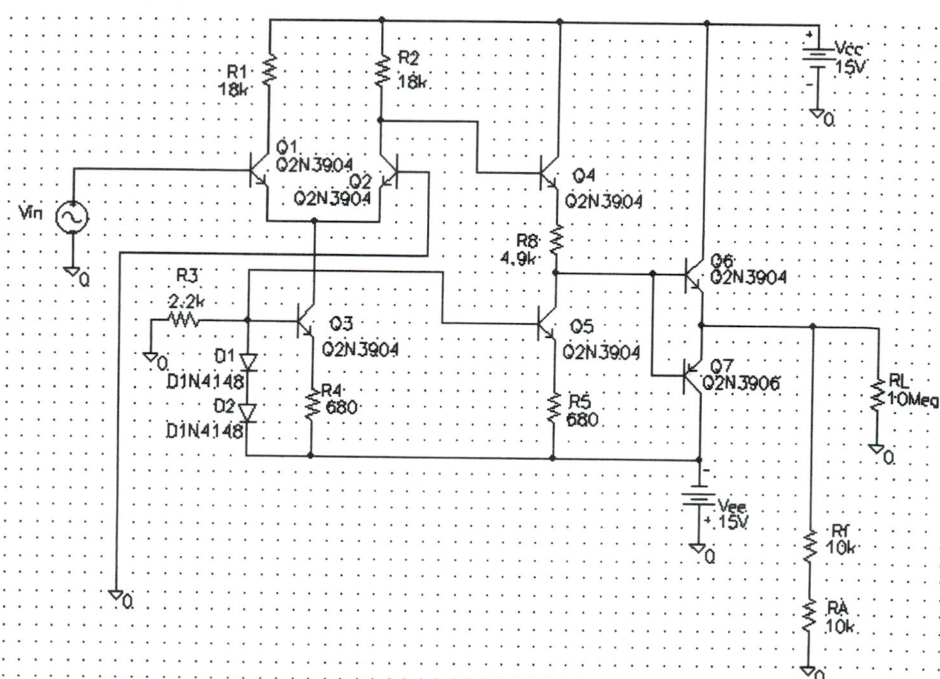

FIGURE 11.28
Circuit for computer analysis.

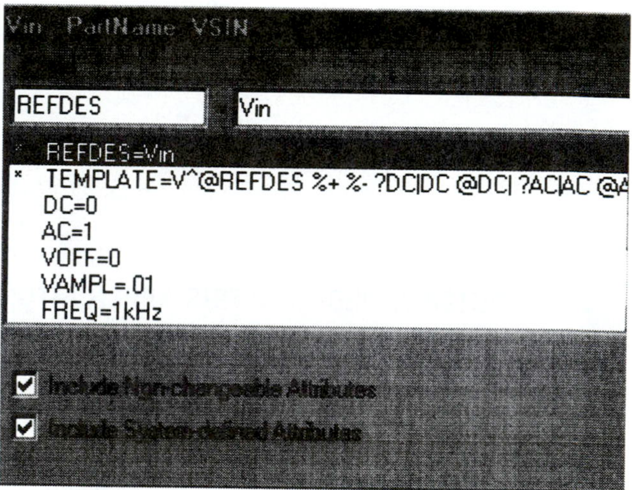
Vin PartName VSIN

REFDES | Vin

REFDES=Vin
* TEMPLATE=V^@REFDES %+ %- ?DC|DC @DC| ?AC|AC @A
DC=0
AC=1
VOFF=0
VAMPL=.01
FREQ=1kHz

Include Non-changeable Attributes
Include System-defined Attributes

FIGURE 11.29
Input voltage source attributes.

FIGURE 11.30
Open-loop transfer function analysis results.

```
******* SMALL-SIGNAL CHARACTERISTICS *******

V($N_0003)/V_Vin = 8.843E+01
INPUT RESISTANCE AT V_Vin = 1.489E+04
OUTPUT RESISTANCE AT V($N_0003) = 4.938E+04

********************************************
```

The main discrepancy is in the output resistance, which PSpice determines as nearly 50 kΩ. This is actually correct, under no-signal conditions, because the output transistors are in cutoff. Our hand-calculated value assumes active-region operation, which will exist when a signal is present. So we see that depending on the context, either result could be considered correct. This is a perfect example of why manual circuit analysis will never become obsolete.

Examination of the transient (time-domain) output voltage with Probe yields the display shown in Fig. 11.31. As predicted, there is a large amount of crossover distortion. We also see that there is a great deal of asymmetry in the output as well; this prevents a meaningful determination of voltage gain using this waveform.

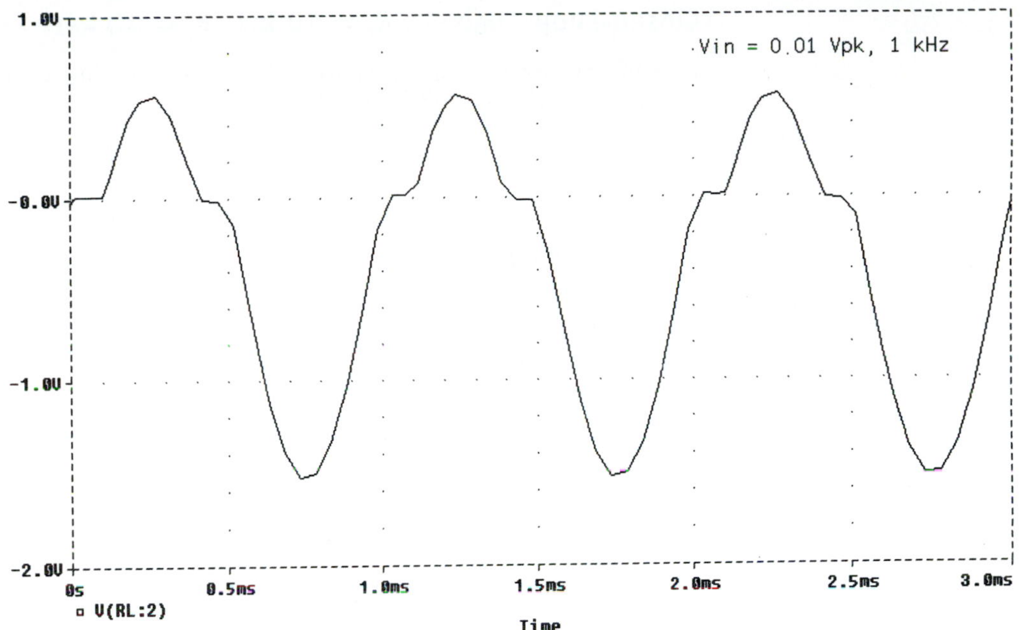

FIGURE 11.31
Open-loop output voltage waveform exhibits severe crossover distortion.

Plotting V_o in decibels, using the ac (frequency-domain) analysis gives the waveform of Fig. 11.32. The midband gain is about 88.1, as we expected; however, there is a large degree of peaking right before the response rolls off. This is a sign of an underdamped system, which is second order or greater. We address these topics more thoroughly in Chapter 14. If we are after relatively constant voltage gain, the amplifier is flat up to about 100 kHz.

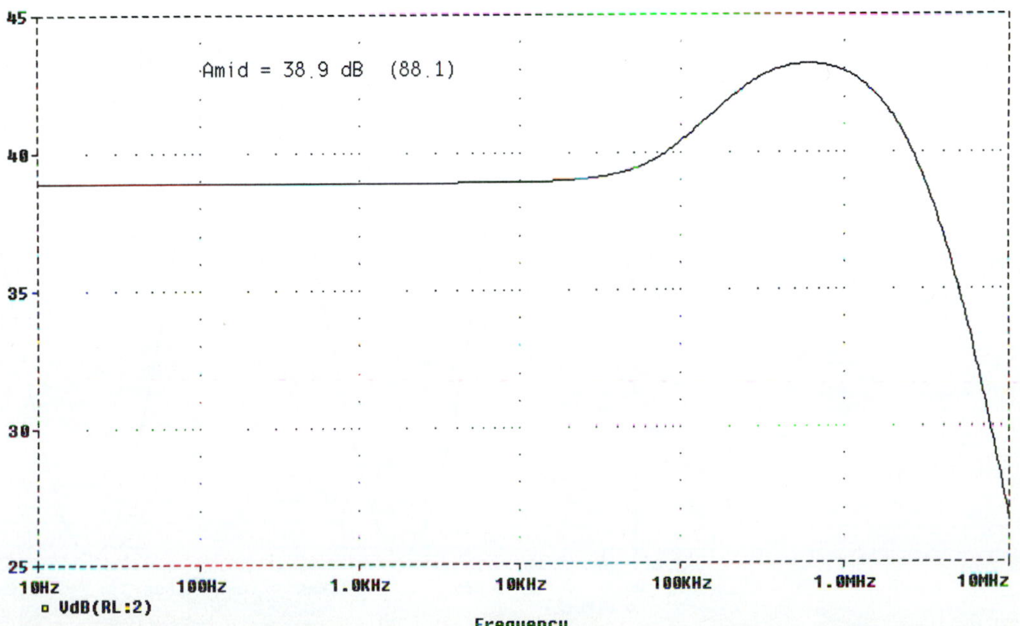

FIGURE 11.32
Open-loop frequency response.

Closed-Loop Analysis: $R_F = 10\ \text{k}\Omega$, $R_A = 10\ \text{k}\Omega$

If we lift the inverting input terminal from ground and connect it to the junction of R_F and R_A, we close the feedback loop and we have $\beta = 0.5$ (Fig. 11.33). The small-signal transfer-function analysis produces these results:

$$A_f = 1.95$$

$$R_{\text{in}F} = 814\ \text{k}\Omega$$

$$R_{oF} = 141\ \Omega$$

FIGURE 11.33
Closed-loop transfer function analysis results, $\beta = 0.5$.

```
******* SMALL-SIGNAL CHARACTERISTICS *******

  V($N_0003)/V_Vin = 1.954E+00
  INPUT RESISTANCE AT V_Vin = 8.143E+05
  OUTPUT RESISTANCE AT V($N_0003) = 1.411E+02

********************************************
```

In the example, we calculated $A_f = 1.96$, which agrees well with the simulation. Notice that the input resistance has increased greatly with negative feedback. In theory, it can be shown that for this circuit

$$R_{\text{in}F} = R_{\text{in}}\,(1 + |A|\beta) \tag{11.31}$$

$$R_{oF} = \frac{R_o}{1 + |A|\beta} \tag{11.32}$$

If we substitute the PSpice-determined values of these parameters into these equations, we obtain $R_{\text{in}F} \cong 678\ \text{k}\Omega$ and $R_{oF} \cong 1\ \text{k}\Omega$. The calculated input and output resistances with feedback differ from the simulation values because of higher order nonlinear effects that are too complex to quantify at this level. The important points to observe for this circuit are that negative feedback increases input resistance and decreases output resistance, which is desirable for a voltage amplifier.

A plot of the output voltage is shown in Fig. 11.34. The crossover distortion and gain asymmetry have been reduced greatly. This is a good side effect of negative feedback.

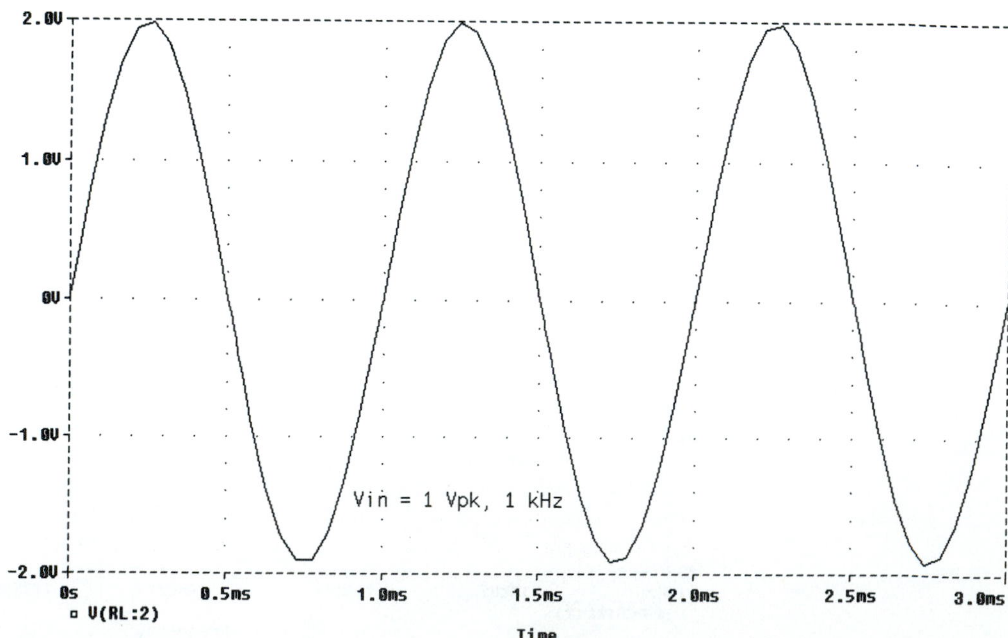

FIGURE 11.34
Negative feedback ($\beta = 0.0909$) greatly reduces crossover distortion.

The frequency response of the amplifier is shown in Fig. 11.35. The gain is essentially constant at 1.95 up to the corner frequency $f_c = 9.4$ MHz. Negative feedback has flattened the gain peak, which is usually desirable.

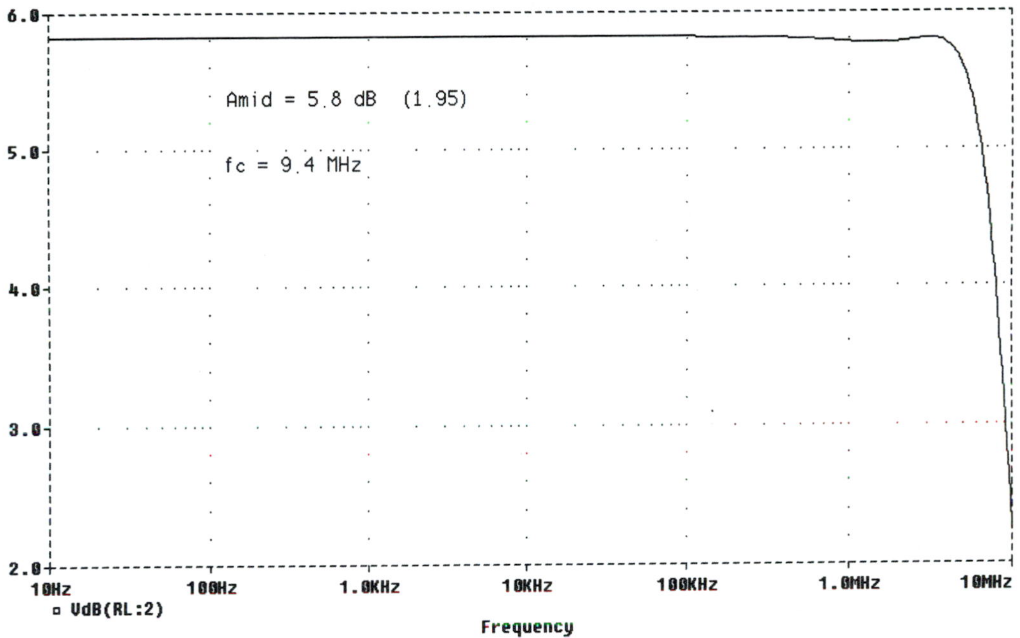

FIGURE 11.35
Closed-loop frequency response for $\beta = 0.5$.

Closed-Loop Analysis: $R_F = 100\ \text{k}\Omega$ $R_A = 10\ \text{k}\Omega$

Changing the value of R_F to 100 kΩ results in $\beta = 0.0909$. The small-signal transfer function results now become (Fig. 11.36):

$$A_f = 19.5$$

$$R_{inF} = 220\ \text{k}\Omega$$

$$R_{oF} = 926\ \Omega$$

Calculating these values using Eq (11.31) and (11.32) produces $R_{inF} = 120\ \text{k}\Omega$ and $R_{oF} = 1.7\ \text{k}\Omega$.

FIGURE 11.36
Closed-loop transfer function analysis results, $\beta = 0.0909$.

```
******* SMALL-SIGNAL CHARACTERISTICS *******

V($N_0003)/V_Vin = 9.829E+00
INPUT RESISTANCE AT V_Vin = 2.200E+05
OUTPUT RESISTANCE AT V($N_0003) = 9.263E+02

**************************************************
```

A plot of the output voltage is shown in Fig. 11.37. Notice, however, that a significant dc offset (0.27 V) is present in the output signal. This is caused by unequal resistances in the bias current paths for Q_1 and Q_2. The bias current of the noninverting input (the base of Q_1) flows through the ideal voltage source v_{in}, and so no voltage drop is produced. The equivalent resistance at the inverting input (the base of Q_2) is

$$\begin{aligned} R_{B(-)} &= R_F \parallel R_A \\ &= 100\ \text{k}\Omega \parallel 10\ \text{k}\Omega \\ &= 9.09\ \text{k}\Omega \end{aligned}$$

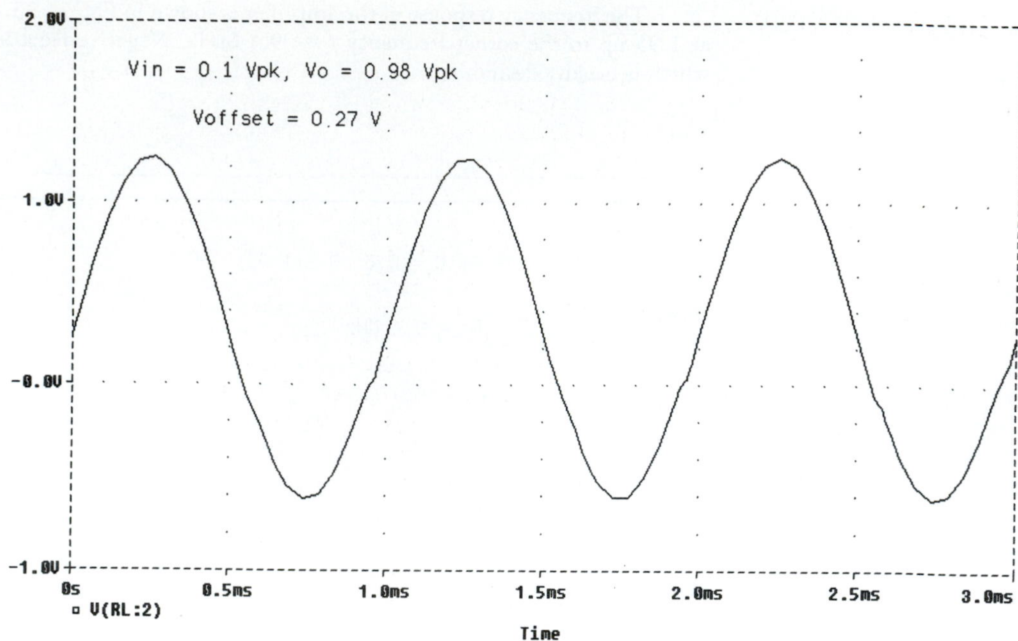

FIGURE 11.37
Output voltage with β = *0.0909. Some crossover distortion is visible.*

The bias current flows through this resistance causing a voltage drop across equivalent resistance $R_{B(-)}$, resulting in a differential input voltage, called an *input offset voltage.* Placing an equal-value resistor in series with the noninverting input should produce an equal voltage drop, cancelling the input differential and hence reducing the output offset voltage as well.

The frequency response of the amplifier for this simulation run is shown in Fig. 11.38. Midband gain is flat at 9.8 up to $f_c = 1$ MHz. Notice the significant reduction in f_c compared to when $A_f \cong 2$. With most amplifier topologies, the use of negative feedback allows us to trade gain for bandwidth. This is explored further in the next chapter as well.

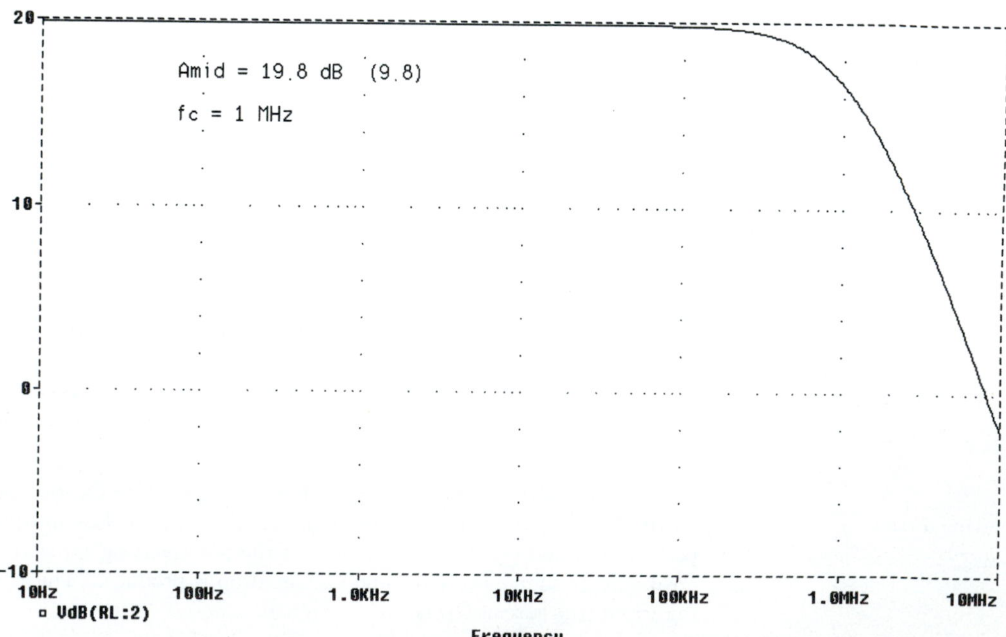

FIGURE 11.38
Frequency response for β = *0.0909.*

CHAPTER REVIEW

The frequency response of a system is often plotted on a log-log graph called a Bode plot. A piecewise-linear approximation of the actual continuous response is called an asymptotic Bode plot. The corner frequency of an amplifier is that frequency at which power gain has decreased to half its mid-band value. The difference between upper and lower corner frequencies is the bandwidth of the amplifier.

Any circuit that uses capacitive (or transformer) coupling will have a lower frequency limit. The low-frequency performance limits of an amplifier can often be determined using first-order high-pass filter analysis equations. All capacitively coupled and bypassed terminals must be analyzed to determine which forms the dominant corner frequency.

Negative feedback is used to reduce overall gain, while at the same time increasing amplifier linearity. In many cases, negative feedback also increases bandwidth. It is also possible for negative feedback to alter the input and output resistance of an amplifier. Miller's theorem quantifies the effect of components in the feedback loop of an amplifier. Common-base and common-emitter amplifiers are immune to the Miller effect, but each has certain weaknesses of its own. Cascode amplifiers are used in high-frequency applications because they are nearly immune to the Miller effect, and provide good all-around gain characteristics.

High-frequency amplifier characteristics are normally determined largely by device parasitic capacitances and stray wiring capacitances. Operation of common-emitter and emitter-follower amplifiers at very high frequencies may be limited by transistor beta cutoff frequency. Common-base amplifiers are fundamentally frequency limited by the alpha cutoff frequency of the transistor.

DESIGN AND ANALYSIS QUESTIONS

Note: Unless otherwise noted, assume all BJTs have $\beta = 100$, $\alpha \cong 1$, $V_\phi = 0.7$ V.

11.1. A certain amplifier has $f_L = 100$ kHz and $f_H = 1$ MHz. Determine the bandwidth BW and center frequency f_o.

11.2. A certain amplifier has BW = 100 kHz and $f_H = 1$ MHz. Determine f_L and f_o.

11.3. A certain amplifier has $A_{\text{mid}} = 60$ dB and $f_H = 5$ MHz. The response rolls off at -20 dB/decade. Determine A at $f = 50$ MHz and $f = 500$ MHz.

11.4. A certain amplifier has $A_{\text{mid}} = 100$ and $f_H = 1$ MHz. The response rolls off at -20 dB/decade. Determine A at $f = 10$ MHz and $f = 100$ MHz.

11.5. A certain amplifier has $A_{\text{mid}} = 80$ dB and $f_H = 200$ Hz. The response rolls off at -6 dB/octave. Determine A at $f = 400$ Hz and $f = 800$ Hz. At what frequency will $A = 0$ dB?

11.6. A certain amplifier has $A_{\text{mid}} = 100$ k and $f_H = 10$ Hz. The response rolls off at -6 dB/octave. Determine A at $f = 20$ Hz and $f = 80$ Hz. At what frequency will $A = 1$?

11.7. Refer to Fig. 11.39. Given $R_1 = 600\ \Omega$, $R_2 = 1.2$ kΩ, $R_3 = 1$ kΩ, and $C_1 = 22\ \mu$F, determine the response type and the corner frequency.

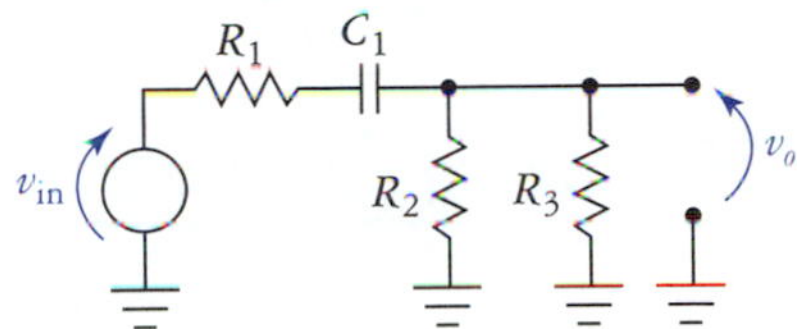

FIGURE 11.39
Circuit for Problems 11.7 to 11.10.

11.8. Refer to Fig. 11.39. Given $R_1 = 50\ \Omega$, $R_2 = 500\ \Omega$, $R_3 = 500\ \Omega$, and $C_1 = 33\ \mu$F, determine the corner frequency.

11.9. Refer to Fig. 11.39. Given $R_1 = 100\Omega$, $R_2 = 10$ kΩ, and $R_3 = 2$ kΩ, determine the value of C_1 such that $f_c = 15$ Hz.

11.10. Refer to Fig. 11.39. Given $R_1 = 100\Omega$, $R_2 = 10$ kΩ, and $C_1 = 1\mu$F, determine the value of R_3 such that $f_c = 100$ Hz.

11.11. Refer to Fig. 11.40. Given $R_1 = 25\ \Omega$, $R_2 = 1$ kΩ, $C_1 = 22\ \mu$F, and $C_2 = 4.7\ \mu$F, determine the response type and the corner frequency.

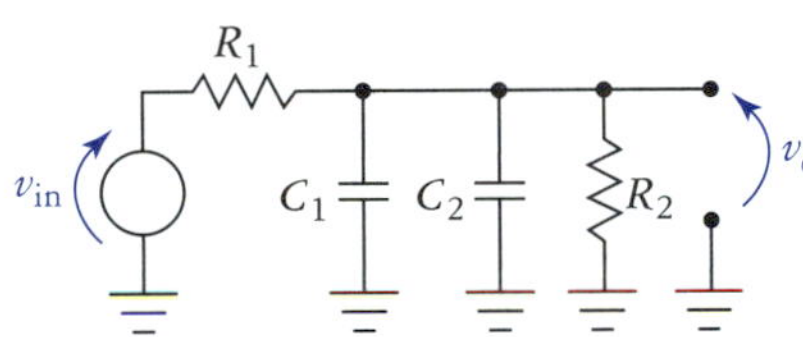

FIGURE 11.40
Circuit for Problems 11.11 to 11.14.

11.12. Refer to Fig. 11.40. Given $R_1 = 2.2$ kΩ, $R_2 = 1$kΩ, $C_1 = 100\ \mu$F, and $C_2 = 100\ \mu$F, determine the corner frequency.

11.13. Refer to Fig. 11.40. Given $R_1 = 1$ kΩ and $R_2 = 10$ kΩ, determine the total capacitance required such that $f_c = 20$ Hz.

11.14. Given the conditions of Problem 11.13, assuming that $C_1 = 4.7\ \mu$F, what value should C_2 have?

11.15. Refer to Fig. 11.41. Assume $V_{CC} = 15$ V, $R_1 = 12$ kΩ, $R_2 = 3.3$ kΩ, $R_3 = 2.7$ kΩ, $R_4 = 1.2$ kΩ, $R_L = 2.7$ kΩ, $C_1 = 2.2\ \mu$F, $C_2 = 47\ \mu$F, and $C_3 = 1\ \mu$F. Determine the Q point, R_{in}, A_v, f_{C_1}, f_{C_2}, and f_{C_3}. Sketch the resulting low-frequency asymptotic Bode plot.

11.16. Refer to Fig. 11.41. Assume $V_{CC} = 15$ V, $R_1 = 15$ kΩ, $R_2 = 3.9$ kΩ, $R_3 = 1.2$ kΩ, $R_4 = 560\ \Omega$, $R_L = 1$ kΩ, $C_1 = 10\ \mu$F, $C_2 = 22\ \mu$F, and $C_3 = 10\ \mu$F. Determine the Q point, R_{in}, A_v, f_{C_1}, f_{C_2}, and f_{C_3}. Sketch the resulting low-frequency asymptotic Bode plot.

11.17. Refer to Fig. 11.42. Assume $V_{DD} = 15$ V, $R_1 = 1$ MΩ, $R_2 = 1$ kΩ, $R_3 = 100\ \Omega$, $R_L = 1$ kΩ, $C_1 = 0.1\ \mu$F, $C_2 = 10\ \mu$F, and $C_3 = 1\ \mu$F. The JFET has $I_{DSS} = 10$ mA and $V_P = 3$ V. Determine the Q point, R_{in}, A_v, f_{C_1}, f_{C_2} and f_{C_3}. Sketch the resulting low-frequency asymptotic Bode plot.

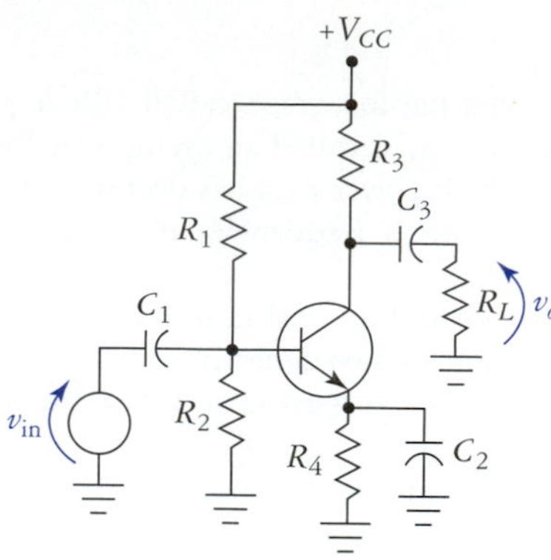

FIGURE 11.41
Circuit for Problems 11.15, 11.16, 11.21, and 11.22.

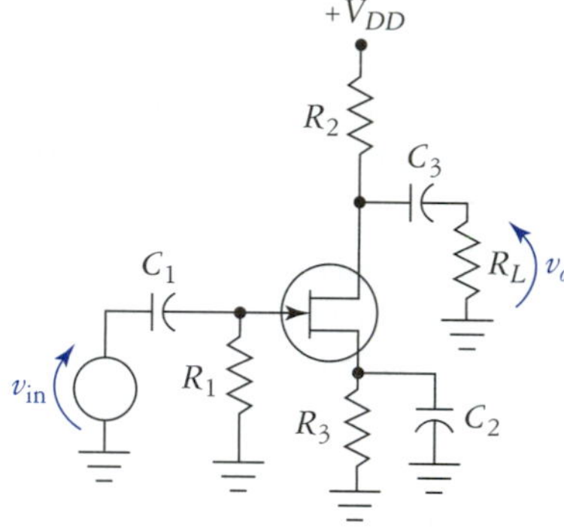

FIGURE 11.42
Circuit for Problems 11.17, 11.18, 11.23, and 11.24.

11.18. Refer to Fig. 11.42. Assume $V_{DD} = 15$ V, $R_1 = 10$ MΩ, $R_2 = 1.2$ kΩ, $R_3 = 180$ Ω, $R_L = 1.2$ kΩ, $C_1 = 0.1$ μF, $C_2 = 22$ μF, and $C_3 = 10$ μF. The JFET has $I_{DSS} = 8$ mA and $V_P = 3$ V. Determine the Q point, R_{in}, A_v, f_{C_1}, f_{C_2} and f_{C_3}. Sketch the resulting low-frequency asymptotic Bode plot.

11.19. Refer to Fig. 11.43. Given $A = -200$ and $C_F = 10$ pF, determine C'_{in} and C'_o.

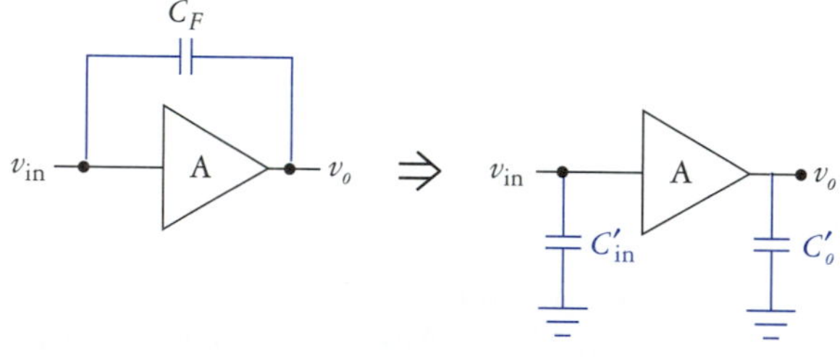

FIGURE 11.43
Circuit for Problems 11.19 and 11.20.

11.20. Refer to Fig. 11.43. Given $A = -100$ k and $C_F = 100$ μF, determine C'_{in} and C'_o.

11.21. Refer to Fig. 11.41. Using the results of Problem 11.15 and assuming that the transistor has $C_{BC} = C_{BE} = 3$ pF and $C_{CE} = 2$ pF, determine the high-frequency equivalent input and output capacitances and the upper corner frequency of the amplifier for $R_S = 100$ Ω.

11.22. Refer to Fig. 11.41. Using the results of Problem 11.16 and assuming that the transistor has $C_{BC} = C_{BE} = 2$ pF and $C_{CE} = 0.5$ pF, determine the high-frequency equivalent input and output capacitances and the upper corner frequency of the amplifier with $R_S = 600$ Ω.

11.23. Refer to Fig. 11.42. Using the results of Problem 11.17 and assuming that the transistor has $C_{GD} = C_{GS} = 2$ pF and $C_{DS} = 1$ pF, determine the upper corner frequency of the amplifier with $R_S = 200$ Ω.

11.24. Refer to Fig. 11.42. Using the results of Problem 11.18 and assuming that the transistor has $C_{GD} = C_{GS} = 1$ pF and $C_{DS} = 0.5$ pF, determine the upper corner frequency of the amplifier for $R_S = 10$ kΩ.

11.25. The circuit in Fig. 11.44 has $V_{CC} = V_{EE} = 15$ V, $R_1 = 10$ kΩ, $R_2 = 2.2$ kΩ, $R_3 = 2.2$ kΩ, $R_L = 100$ Ω, $C_{BC} = 10$ pF, $C_{BE} = 12$ pF, $R_S = 50$ Ω, and $C_{CE} = 5$ pF. Determine the upper corner frequency of the amplifier.

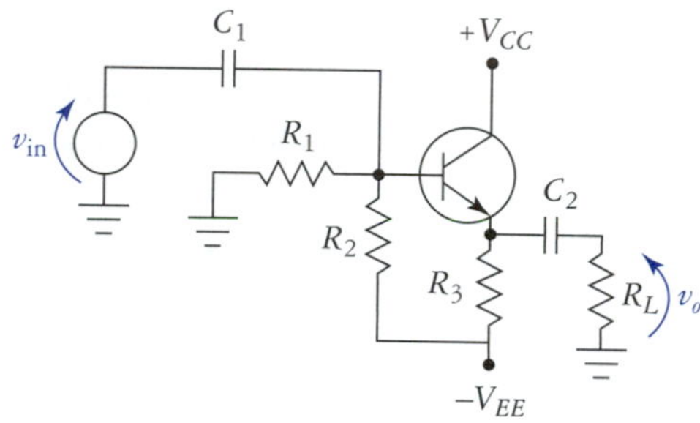

FIGURE 11.44
Circuit for Problems 11.25 and 11.26.

11.26. The circuit in Fig. 11.44 has $V_{CC} = V_{EE} = 15$ V, $R_1 = 4.7$ kΩ, $R_2 = 4.7$ kΩ, $R_3 = 3.3$ kΩ, $R_L = 50$ Ω, $C_{BC} = 5$ pF, $C_{BE} = 6$ pF, $R_S = 25$ Ω, and $C_{CE} = 2$ pF. Determine the upper corner frequency of the amplifier.

11.27. The circuit in Fig. 11.45 has $V_{CC} = V_{EE} = 15$ V, $R_1 = 6.8$ kΩ, $R_2 = 10$ kΩ, $R_3 = 3.9$ kΩ, $R_L = 10$ kΩ, $R_s = 5$ Ω, $C_{BC} = 5$ pF, $C_{BE} = 6$ pF, and $C_{CE} = .2$ pF. Determine the upper corner frequency of the amplifier.

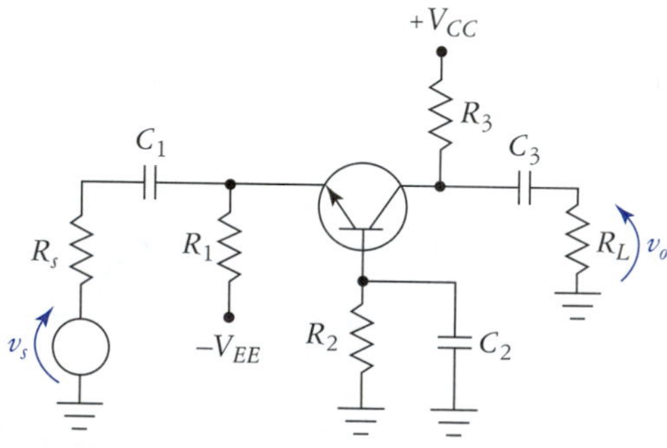

FIGURE 11.45
Circuit for Problems 11.27 and 11.28.

11.28. The circuit in Fig. 11.45 has $V_{CC} = V_{EE} = 15$ V, $R_1 = 3.3$ kΩ, $R_2 = 10$ kΩ, $R_3 = 2.2$ kΩ, $R_L = 2.2$ kΩ, $R_s = 10$ Ω, $C_{BC} = 2$ pF, $C_{BE} = 2$ pF, and $C_{CE} = .1$ pF. Determine the upper corner frequency of the amplifier.

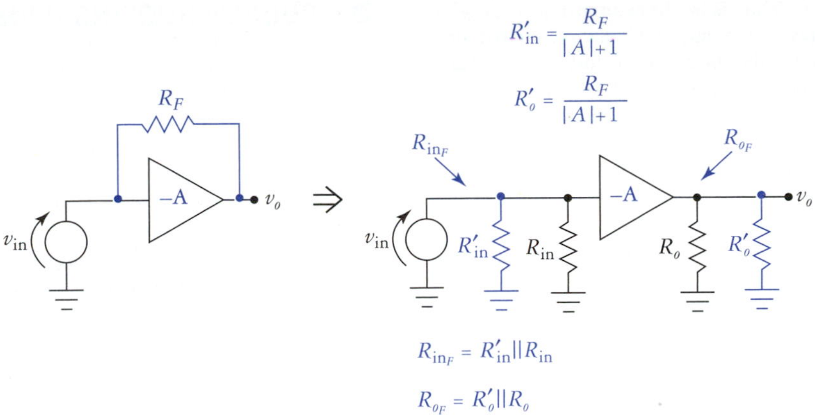

FIGURE 11.46
Circuit for Problem 11.29.

11.29. Miller's theorem also applies to resistance in the feedback loop of an inverting amplifier as shown in Fig. 11.46. Here we find the input resistance with feedback is lower than the open-loop input resistance. That is, $R_{inF} < R_{in}$. Given $R_f = R_{in} = 10\ k\Omega$ determine $|A|$ such that $R'_{in} = 1\ k\Omega$.

11.30. Refer to Fig. 11.47. Use the relationships of Fig. 11.46 to determine equations for R_{in} and R_o. Assume the transistor has $r_o = \infty$.

11.31. Refer to Fig. 11.48. Assume $V_{CC} = V_{EE} = 15$ V, $I_1 = I_2 = 1$ mA, $I_3 = 2$ mA, $R_1 = R_2 = 24\ k\Omega$, $R_3 = 12\ k\Omega$, and $R_4 = 4.15\ k\Omega$. Determine the open-loop differential voltage gain A_d, and determine the closed-loop gain for $R_f = 47\ k\Omega$, $R_A = 1\ k\Omega$, and $R_F = 9\ k\Omega$, $R_1 = 1\ k\Omega$.

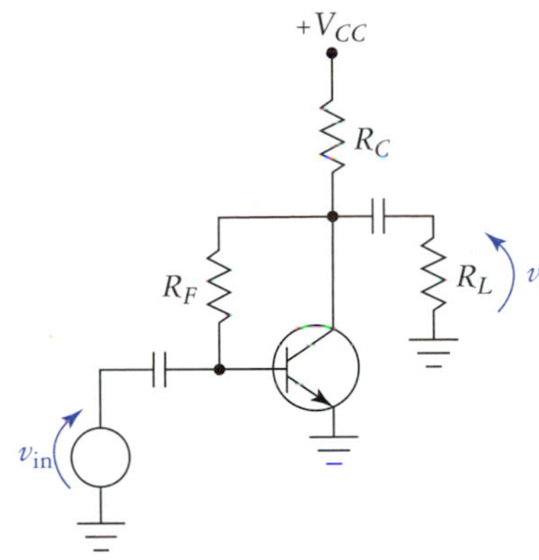

FIGURE 11.47
Circuit for Problem 11.30.

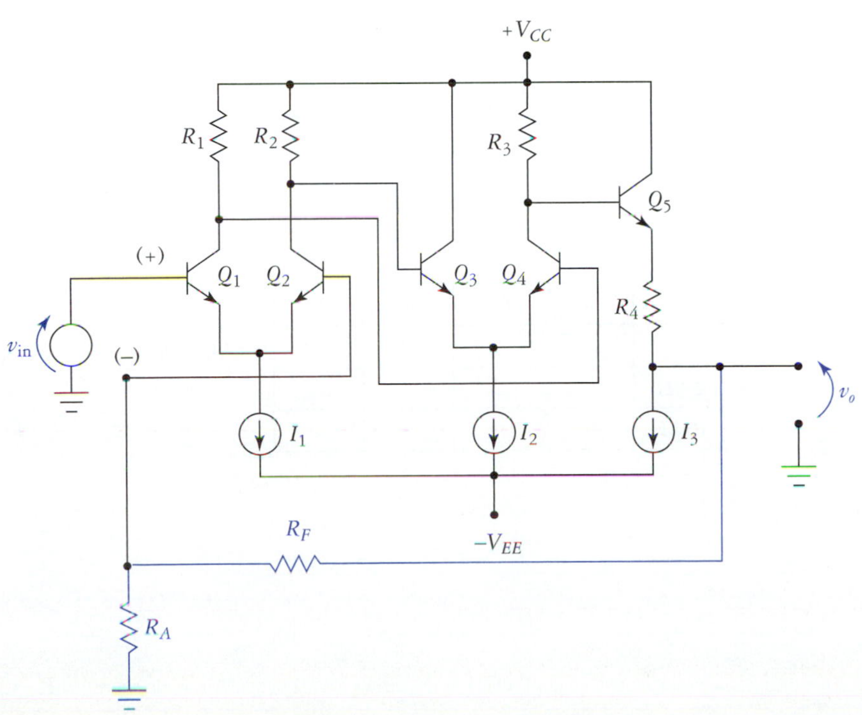

FIGURE 11.48
Circuit for Problem 11.31.

11.32 Refer to Fig. 11.49. This is a differential-input, single-ended-output amplifier, commonly called an operational amplifier or op amp. Internally, the circuit would likely be similar to that of Fig. 11.19 or possibly 11.48, optimized for very high open-loop gain. Assume that the open-loop parameters are A_d = 200 k, R_{in} = 2 MΩ, and R_o = 50 Ω. Determine the closed-loop gain A_F, R_{inF}, and R_{oF} given R_F = 90 kΩ and R_A = 10 kΩ.

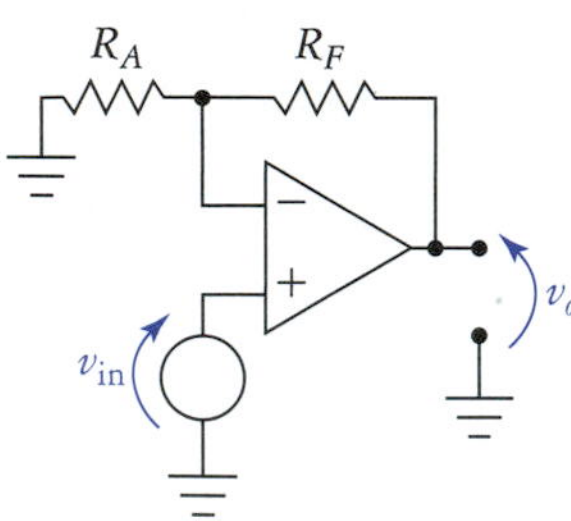

FIGURE 11.49
Circuit for Problem 11.32.

■ TROUBLESHOOTING PROBLEMS

11.33. Refer to Fig. 11.9. Suppose capacitor C_3 becomes open circuited. How would this affect the lower corner frequency of the amplifier?

11.34. Assume the circuit of Fig. 11.14 was operating correctly for a while and then it malfunctioned. The symptom observed (under proper input signal conditions) was severe clipping of the output voltage. What is a likely cause of this problem?

11.35. An emitter-follower is prototyped in the lab. The circuit tends to break into oscillations of around 150 MHz most of the time. Describe several modifications that might help alleviate this problem and explain your rationale for each solution.

■ COMPUTER PROBLEMS

11.36. Simulate the circuit in Fig. 11.50. Using an ac sweep of v_{in}, determine the small-signal open-loop midband gain, input resistance, and the lower and upper corner frequencies. To effectively open the feedback loop, simply set R_F to a very high value, such as 100 MΩ. Repeat these measurements for the R_F = 1 kΩ and 10 kΩ. Compile your results in a table like that shown in the figure.

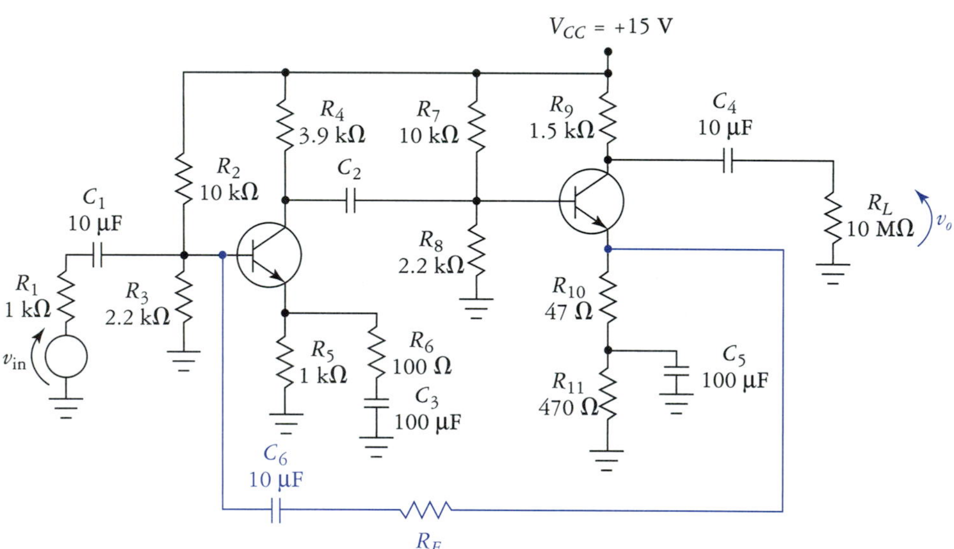

R_F	A	R_{in}	f_L	f_H
∞					← Open-loop
1 kΩ				
10 kΩ				

FIGURE 11.50
Circuit for Problem 11.36.

CHAPTER 12

Operational Amplifiers

BEHAVIORAL OBJECTIVES

On completion of this chapter, the student should be able to:

- Derive the voltage-gain equations for ideal operational amplifiers.
- Determine the input and output resistance of an op amp with negative feedback.
- Determine the closed-loop bandwidth of an op amp.
- Determine the power bandwidth of an op amp.
- Describe the significance of slew-rate limiting.
- Explain the effects of bias currents and offset voltages on the output of an op amp.
- Analyze op amp offset compensation circuitry.

INTRODUCTION

This chapter covers the basic characteristics of operational amplifier circuits employing negative feedback. The effects of various op amp errors are investigated, along with techniques used to reduce these errors. Large-signal aspects of op amp behavior are also studied.

12.1 WHAT IS AN OPERATIONAL AMPLIFIER?

Operational amplifiers, usually just called op amps, are the fundamental integrated circuit (IC) building blocks of analog electronics. Without getting into technical definitions, we will use the term *analog* to mean any application that is not primarily implemented using digital logic devices. The analog functions performed by op amps may be linear or nonlinear, however, and sometimes we combine both analog and digital functions into a single system, making the distinction somewhat blurred.

Note
Op amps are the basic building blocks of analog circuits.

The schematic symbol for an operational amplifier is shown in Fig. 12.1. Originally, vacuum-tube operational amplifiers were developed in the late 1940s for use in analog computer circuits. Typically, these circuits were used to perform mathematical operations such as integration and differentiation, hence, the name operational amplifier. These early op amps were large and expensive, and suffered from poor performance compared to modern versions.

FIGURE 12.1
Op amp schematic symbol.

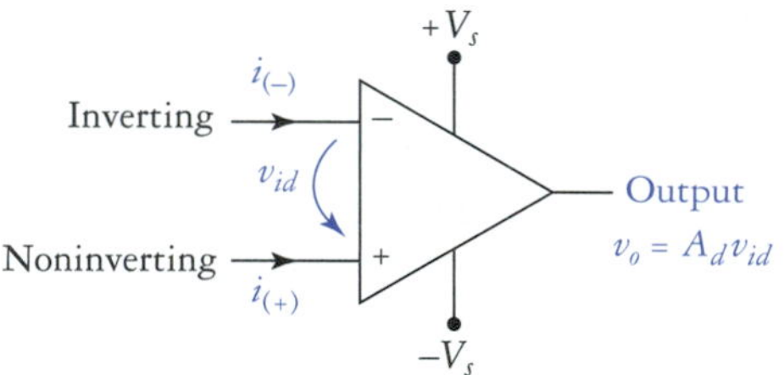

Internally, most operational amplifiers are similar in design to the circuits shown back in Figs. 8.58 and 11.19. All true op amps have differential inputs, and most have single-ended outputs. (A few are available with differential outputs, but these are rare.) Most operational amplifiers are designed to operate from bipolar power supplies as well. The implementation of high-performance op amps requires the use of very closely matched devices. This was made possible by the advent of integrated circuit technology. In addition, IC fabrication techniques have made such high-performance devices very inexpensive. The performance of a typical modern $0.50 IC op amp is as good as that of an op amp made with discrete devices that would cost hundreds of dollars.

12.2 IDEAL VOLTAGE AMPLIFIERS

Most operational amplifiers behave fundamentally like voltage amplifiers. Recall that a voltage amplifier is an amplifier that accepts a voltage as an input and produces a scaled version of that voltage at its output. Because an input voltage causes the generation of the output voltage, a voltage amplifier is a voltage-controlled voltage source (VCVS). We use negative feedback to modify the various VCVS parameters to suit a given application.

The ideal open-loop voltage amplifier (VCVS) would have the following characteristics:

Open-loop voltage gain	$A_{OL} \to \infty$
Bandwidth	$\text{BW} \to \infty$ Hz
Input resistance	$R_{\text{in}} \to \infty\ \Omega$
Output resistance	$R_o \to 0\ \Omega$

Of course, in real life such ideal characteristics cannot be achieved. However, in many practical situations, operational amplifiers can reasonably approximate this performance. Because operational amplifiers have such high open-loop gain, we can use negative feedback to trade some of this gain to obtain improvements in other characteristics.

12.3 SAMPLING OF COMMERCIAL OP AMPS

Ideally, we know what basic open-loop characteristics a voltage amplifier should have. Table 12.1 summarizes a few of the fundamental open-loop characteristics for a few popular op amps. Most of these parameters are very familiar already, and the others will be discussed in greater detail very soon. The frequency response of all of these op amps extends down to 0 Hz because direct coupling is used in each device. The f_c specification is the typical frequency at which the open-loop voltage gain is down by 3 dB. The f_{unity} specification is the typical frequency at which the open-loop voltage gain has dropped to unity (0 dB).

TABLE 12.1
Selected Op Amps and Their Open-Loop Characteristics*

Device	LM741C	LF351	OP-07	LH0003	AD549K
Technology	BJT	BiFET	BJT	Hybrid BJT	BiFET
$A_{OL(typ)}$	200 k	100 k	400 k	40 k	100 k
R_{in}	2 MΩ	10^{12} Ω	8 MΩ	100 kΩ	10^{13}Ω ∥ 1 pF
R_o	50 Ω	30 Ω	60Ω	50 Ω	100 Ω (approx.)
f_c	8 Hz	40 Hz	1 Hz	10 kHz	10 Hz
f_{unity}	1 MHz	4 MHz	1 MHz	30 MHz	1 MHz
SR	0.5 V/μs	13 V/μs	0.3 V/μs	70 V/μs	3 V/μs
I_B	80 nA	50 pA	1.8 nA	400 nA	75 fA
I_{io}	20 nA	25 pA	0.8 nA	20 nA	20 fA
V_{io}	2 mV	5 mV	60 μV	0.4 mV	0.15 mV
I_{sc}	25 mA†	20 mA†	20 mA†	125 mA(max)	20 mA†
CMRR	90 dB	100 dB	110 dB	90 dB	90 dB
Freq. Comp.	Internal	Internal	Internal	External	Internal
Offset Adj.	Yes	Yes	Yes	No	Yes

* All specs are typical values at $T_A = 25°\text{C}$, $V_s = \pm 15$ V unless specified otherwise.
† Continuous short-circuit overcurrent protection (not implemented on LH0003).

All of the devices listed in Table 12.1 except the LH0003 are monolithic ICs. Monolithic ICs have all internal circuitry fabricated on a single silicon chip or *die.* The LH0003 is a *hybrid IC.* Hybrid ICs contain components (usually resistors or power transistor chips) that are mounted along with the main IC chip on a ceramic substrate. For example, thin-film resistors can be deposited and then laser trimmed to very precise values during production (laser trimming may also be done to monolithic ICs like the AD549K, as well). Hybrid ICs require more processing steps and are more difficult to produce in general, and therefore are usually much more expensive than monolithic devices.

As you can see in the table, op amp parameters vary widely from one device type to another. The intended use of a given op amp may require optimization of one parameter at the expense of others. For example, the LM741C is an extremely popular general-purpose op amp that is easy to use, with moderately good characteristics in general. The LH0003 is designed to have superior high-frequency characteristics, consequently many other parameters such as voltage gain are sacrificed to obtain this performance. We will refer to Table 12.1 extensively in upcoming discussions. One thing that should be clear though is that all of the op amps have very high open-loop gain, and most have very low corner frequencies. A large amount of this open-loop gain is traded for increased bandwidth in most applications.

12.4 OP AMP FEEDBACK ANALYSIS

Analysis of circuits containing nonideal op amps tends to be very complex and often is simply not necessary. Fortunately, under conditions for which a given op amp was designed to operate, its parameters will normally be good enough to approximate the op amp as being ideal. Before proceeding, recall the basic relationship for a diff-in, single-ended-output amplifier like that shown in Fig. 12.1, where A_{OL} is the open-loop differential voltage gain:

$$v_o = A_{OL}\, v_{id} \tag{12.1}$$

This is easily rearranged to form

$$v_{id} = \frac{v_o}{A_{OL}} \tag{12.2}$$

Note
We will use rules 1 and 2 often when analyzing op amp circuits. This is called *heuristic*, or *ruled-based* analysis.

When an ideal op amp is operated with negative feedback, two basic rules are followed:

1. **The differential input voltage is forced to zero ($v_{id} = 0$ V).** This occurs because of the assumption that the open-loop gain $A_{OL} \to \infty$. Thus, when the feedback loop is closed, the output forces the differential input to zero. Look at Eq. (12.2). It is clear that for any output voltage, as $A_{OL} \to \infty$ the differential input voltage $v_{id} \to 0$.
2. **No current enters or leaves the input terminals.** This is a consequence of the assumption that $R_{in} \to \infty$. Thus, the currents $i_{(+)}$ and $i_{(-)}$ are zero at all times.

These two simple rules allow us to analyze many seemingly complex op amp circuits with relative ease. Of course, they only apply as long as the op amp is operating normally. Saturation conditions, excessive loading, or operation at frequencies that are too high can invalidate these rules. Fortunately, we have techniques that allow us to determine whether such conditions are present.

The Noninverting Configuration

An op amp is configured for use as a noninverting amplifier in Fig. 12.2. Resistors R_F and R_A comprise the feedback loop for this circuit. Notice that no power supply connections are shown on the schematic. As is done with digital logic diagrams, the power supply connections, decoupling capacitors, and so on are usually shown in a separate diagram. There are exceptions, but these will be handled on a case-by-case basis.

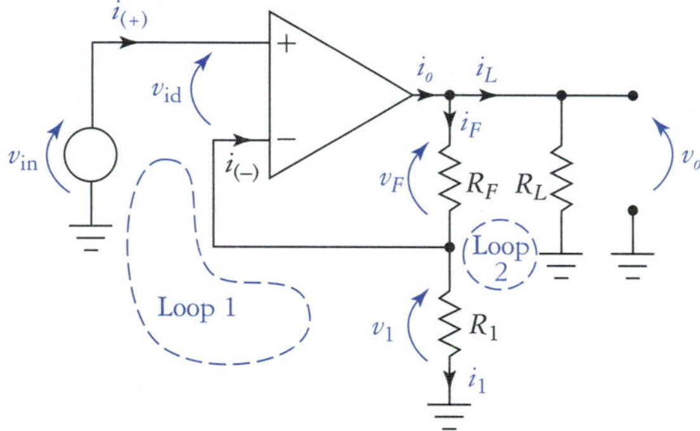

FIGURE 12.2
Noninverting configuration with feedback.

Closed-Loop Gain

Assuming the op amp in Fig. 12.2 is ideal, it is easy to derive the voltage-gain expression. Application of KVL around loop 1 produces

$$0 = v_{in} - v_{id} - v_1$$

However, rule 1 says that $v_{id} = 0$ V, so

$$v_1 = v_{in} \tag{12.3}$$

Applying KVL to loop 2, we have

$$0 = v_1 + v_F - v_o$$

Solving for v_o,

$$\begin{aligned} v_o &= v_1 + v_F \\ &= v_1 + i_F R_F \end{aligned} \tag{12.4}$$

But from Eq. (12.3) we have

$$v_o = v_{in} + i_F R_F \tag{12.5}$$

Rule 2 implies that $i_F = i_1$, and Ohm's law lets us determine i_F from

$$i_F = \frac{v_{in}}{R_1} \tag{12.6}$$

Substituting into Eq. (12.5) we find

$$\begin{aligned} v_o &= \frac{v_{in} R_F}{R_1} + v_{in} \\ &= v_{in}\left(\frac{R_F}{R_1} + 1\right) \end{aligned} \tag{12.7}$$

Dividing both sides by v_{in} produces the closed-loop voltage-gain equation. (We will use A_v to designate closed-loop gain for now.)

$$A_v = \frac{R_F}{R_1} + 1 \tag{12.8}$$

Sometimes this equation is expressed in the equivalent form

$$A_v = \frac{R_F + R_1}{R_1} \tag{12.9}$$

We went through quite a bit of KVL and alegbraic gyrations to arrive at these useful gain expressions. There is another way to derive this expression that is a little quicker. Please recall that the closed-loop gain equation presented in the previous chapter is

$$A_v = A_F = \frac{A_{OL}}{1 + A_{OL}\,\beta} \tag{12.10}$$

You may also recall that as open-loop gain $A \rightarrow \infty$ the closed-loop gain approaches

$$A_v = \frac{1}{\beta} \tag{12.11}$$

Beta is the feedback factor; that is, the portion of the output that is returned to the input. Direct examination of Fig. 12.2 indicates that the feedback network is simply a voltage divider, therefore

$$\beta = \frac{R_1}{R_1 + R_F} \tag{12.12}$$

The voltage gain is $1/\beta$, which is exactly the same as Eq. (12.9). As usual, there are several different ways to arrive at the same result. Also keep in mind that if we wish to determine the actual closed-loop gain (not the ideal approximation) we can use Eq. (12.10). As a practical matter, we almost always use the ideal voltage-gain equation. The next example shows why this is the case.

EXAMPLE 12.1

Refer to Fig. 12.2. Given $R_F = 99\ \text{k}\Omega$, $R_1 = 1\ \text{k}\Omega$, and $v_{\text{in}} = 0.5$ V, determine A_v assuming the op amp is ideal, and for a typical LM741C.

Solution For the ideal op amp we simply use Eq. (12.9), producing

$$\begin{aligned} A_{v(\text{ideal})} &= \frac{R_1 + R_F}{R_1} \\ &= \frac{100\ \text{k}\Omega}{1\ \text{k}\Omega} \\ &= 100 \end{aligned}$$

Note that $\beta = 1/A_{v(\text{ideal})} \cong 0.01$. Consulting Table 12.1, we find $A_{OL} = 200$ k for the typical LM741C. Thus we have

$$\begin{aligned} A_v &= \frac{A_{OL}}{1 + A_{OL}\beta} \\ &= \frac{200\ \text{k}}{1 + (0.01)(200\ \text{k})} \\ &= 99.95 \end{aligned}$$

This slight difference $(-0.005\ \%)$ would certainly be swamped by variations in feedback loop resistance values and A_{OL} variations.

The Voltage Follower

If we connect the output of the noninverting amplifier directly back to the inverting input terminal as shown in Fig. 12.3, then we have 100% feedback. In equation form this means $\beta = 1$. This is obtained by taking Eq. (12.12) and letting $R_1 \rightarrow \infty$. Often, the feedback resistor is simply replaced with a short circuit, but as long as R_F is finite the feedback factor is 1. We will see situations where it is desirable to place a resistance in the feedback loop of this circuit. For the ideal op amp we have

$$\begin{aligned} A_v &= 1/\beta \\ &= 1 \end{aligned}$$

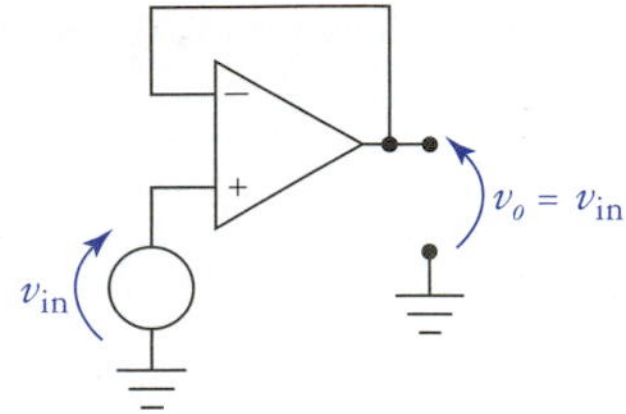

FIGURE 12.3
Unity-gain amplifier (voltage follower).

Because the voltage gain is unity, this circuit is called a *voltage follower,* or sometimes a *unity-gain buffer.*

Closed-Loop Input Resistance

The closed-loop input resistance of the noninverting configuration is extremely high. In fact, for the ideal op amp ($A_{OL} \to \infty$) the input resistance is

$$R_{\text{in}F(\text{ideal})} = \infty \tag{12.13}$$

The actual input resistance for a nonideal op amp with feedback is given by

$$R_{\text{in}F} = R_{\text{in}}(1 + \beta A_{OL}) \tag{12.14}$$

The next example shows how astronomically high closed-loop input resistance can become.

EXAMPLE 12.2

Unity-gain buffers, like that shown in Fig. 12.3, are constructed using typical LM741C and AD549K op amps. Determine the closed-loop input resistance for each amplifier.

Solution The noninverting voltage follower has $\beta = 1$. We use the data from Table 12.1:

$$R_{\text{in}F} = R_{\text{in}}(1 + \beta A_{OL})$$

LM741C	AD549K
$\cong (2\ \text{M}\Omega)(200\ \text{k})$	$\cong (10^{13}\ \Omega)(100\ \text{k})$
$= 400\ \text{G}\Omega$	$= 1 \times 10^{19}\ \Omega$

Because the unity-gain configuration has the highest degree of feedback, it also produces the highest possible input resistance. Increasing the closed-loop gain A_v of the noninverting amplifier will decrease the input resistance, but even so, $R_{\text{in}F}$ will still be extremely large.

Closed-Loop Output Resistance

Negative feedback causes the noninverting amplifier to have very low output resistance. Ideally, the output resistance is

$$R_{oF(\text{ideal})} = 0\ \Omega \tag{12.15}$$

The actual output resistance with feedback is

$$R_{oF} = \frac{R_O}{A_{OL}\beta + 1} \tag{12.16}$$

For op amps used in typical situations, the effective output resistance will usually be a few milliohms.

Some Practical Considerations

There are practical limits to the resistor values that we can use in the feedback network of an op amp. Generally speaking, the lower the bias currents of an op amp, the larger the resistors we can use in the feedback loop. Most often we shall try to keep resistor values below 1 MΩ, to avoid bias-current-induced offset voltage problems.

As far as minimum resistor values are concerned, we must consider the effects of loading on the ouptut of the op amp. A scan through Table 12.1 shows that most op amps have rather low output current drive capability. We do not want to draw 24 mA from the output terminal of a 741 op amp just to drive the feedback loop: We might not be able to drive a reasonable load sufficiently! Generally, we prefer to keep the minimum resistor values around 1 kΩ or higher. We determine these limits quantitatively in later sections.

Operational amplifiers are so popular because they are relatively easy to use. Another example will demonstrate how simple it can sometimes be to design circuits using operational amplifiers.

EXAMPLE 12.3

An audio signal generated by a particular microphone is 10 $mV_{p\text{-}p}$. The microphone has $R_s = 10$ kΩ. Using an LM741C, design a noninverting amplifier that will boost this signal to $V_o \geq 250$ $mV_{p\text{-}p}$. The amplifier should cause minimal loading of the microphone. The amplifier is to drive a 10-kΩ load.

Solution We require an amplifier with

$$A_v = 250\text{ mV} / 10\text{ mV} = 25$$

Using the noninverting configuration will ensure that $R_{inF} >> R_s$ to prevent loading of the microphone. Let us choose $R_1 = 1.5$ kΩ. The ideal gain equation is

$$A_v = \frac{R_1 + R_F}{R_1}$$

Solving for R_F

$$\begin{aligned} R_F &= R_1 (A_v - 1) \\ &= (1.5\text{ k}\Omega)(24) \\ &= 36\text{ k}\Omega \text{ (use 39 k}\Omega \text{ for } A_v = 27) \end{aligned}$$

The amplifier is shown in Fig. 12.4. The input resistance of the amplifier is

$$\begin{aligned} R_{inF} &= R_{in} (1 + \beta A_{OL}) \\ &= 2\text{ M}\Omega[1 + (0.037)(200\text{ k})] \\ &= 14.8\text{ G}\Omega \end{aligned}$$

The microphone will definitely not be loaded by this amplifier. The output resistance of the circuit is

$$\begin{aligned} R_{oF} &= \frac{R_o}{1 + A_{OL}\beta} \\ &= \frac{50\Omega}{1 + (0.037)(200\text{ k})} \\ &= 6.8\text{ m}\Omega \end{aligned}$$

This is a very low output resistance, and the amplifier apparently should easily be able to drive a 10-kΩ load.

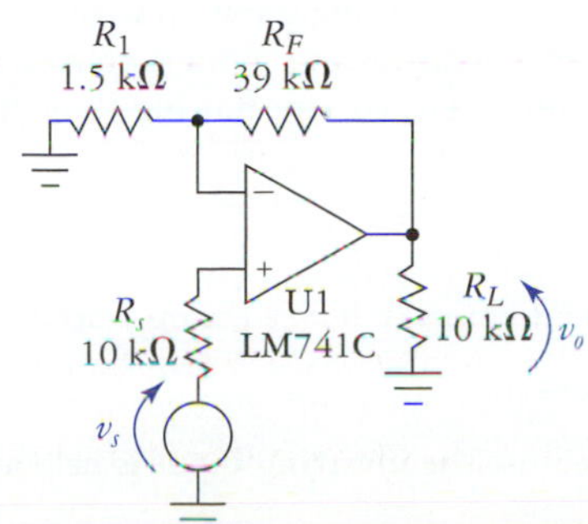

FIGURE 12.4
Circuit for Example 12.3.

The initial choice for R_1 in the previous example happened to cause R_F to come out fairly close to a standard resistor value. Using the next higher standard value also allowed us to meet the inequality requirements of the problem. Sometimes it is necessary to try several different initial resistor values in order to obtain a solution that is close enough to standard values. The details of the problem to be solved will dictate which parameters must be tighly controlled and which can be relaxed. Also, although no power supply requirements were specified, it is standard practice to assume ±15-V supplies are used with most op amp circuits.

The Inverting Configuration

Some applications require the use of an inverting amplifier. The operational amplifier is easily configured to perform this operation. The basic inverting amplifier circuit is shown in Fig. 12.5. Before deriving the ideal voltage-gain equation for the inverting configuration, a few very interesting and important side effects of negative feedback should be discussed.

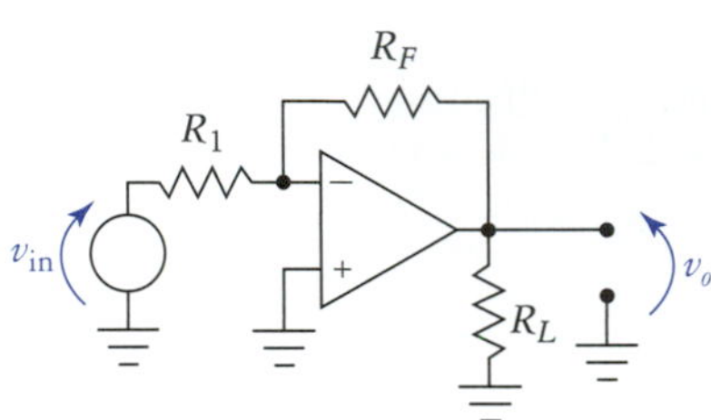

FIGURE 12.5
Inverting configuration with feedback.

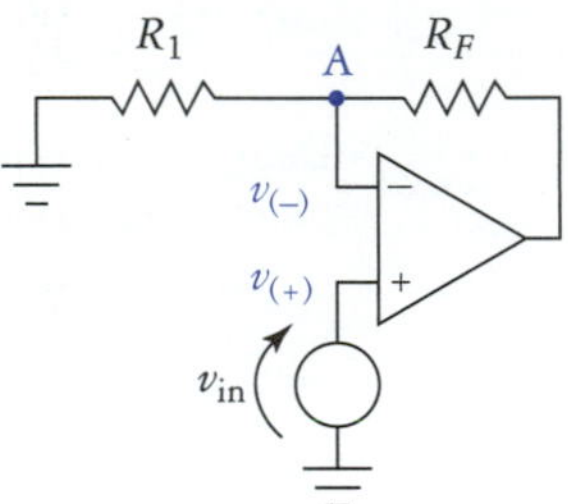

FIGURE 12.6
Node A is a virtual short to the noninverting input.

Note
The virtual ground and virtual short circuit concepts are very helpful for understanding the operation of many op amp circuits.

The Virtual Short Circuit and Virtual Ground

Take a look at Fig. 12.6. Although this is basically a noninverting amplifier, let's think about the voltages present on the op amp input terminals $v_{(+)}$ and $v_{(-)}$. Ideal op amp analysis rule 1 says that the differential input will always be forced to zero volts. Another way to express this is to write $v_{(-)} = v_{(+)}$. In other words, under closed-loop conditions, the inverting input terminal is forced to track the noninverting input. Thus, in Fig. 12.6 (or any of the circuits used in this chapter so far) regardless of what voltage is present at v_{in}, $v_{(-)}$ will exactly match it (ideally) at all times. This is called a *virtual short circuit.* It isn't a true short circuit: the relationship is one way, as $v_{(-)}$ tracks $v_{(+)}$, not vice versa.

If we set $v_{in} = 0$ V in Fig. 12.4, then $v_{(+)} = 0$ V as well. Because of the virtual short-circuit characteristic, the inverting input is forced to ground potential as well. Under these special conditions, the node A is called a *virtual ground.* It is not a true ground, but electrically it behaves as such. For reasons to be covered soon, node A is sometimes referred to as a *summing junction.* Be sure you understand the concepts of the virtual short circuit and the virtual ground. These are extremely useful in the analysis of op-amp-based systems.

Closed-Loop Voltage Gain

The basic inverting configuration with various voltages and currents identified is shown in Fig. 12.7. Because the noninverting input is connected directly to ground, we have a virtual ground at the inverting input. The current flowing through R_1 is

$$i_1 = \frac{v_{in}}{R_1} \tag{12.17}$$

No current enters or leaves the op amp input terminals (rule 2), therefore

$$i_F = i_1 \tag{12.18}$$

Now, because the inverting input is held at ground potential, the output voltage is given by

$$v_o = -i_F R_F \tag{12.19}$$

Substituting Eq. (12.17) into (12.19), we have

$$v_o = -\frac{v_{\text{in}}}{R_1} R_F \tag{12.20}$$

Dividing both sides by v_{in} gives us the ideal voltage-gain equation

$$A_v = -\frac{R_F}{R_1} \tag{12.21}$$

Most of the time the ideal equation will produce sufficiently accurate results. If it is necessary to determine the gain more precisely, we use the general relationship

Note
Remember, for the noninverting configuration, $\beta = \frac{R_1}{R_1 + R_F}$, while for the inverting configuration, $\beta = \frac{R_1}{R_F}$.

$$A_v = A_F = \frac{A_{OL}}{1 + \beta A_{OL}} \tag{12.22}$$

where in the case of the inverting configuration, the feedback factor is

$$\beta = \frac{R_1}{R_F} \tag{12.23}$$

Notice that the ideal voltage gain is again the inverse of the feedback factor ($A_v = 1/\beta$).

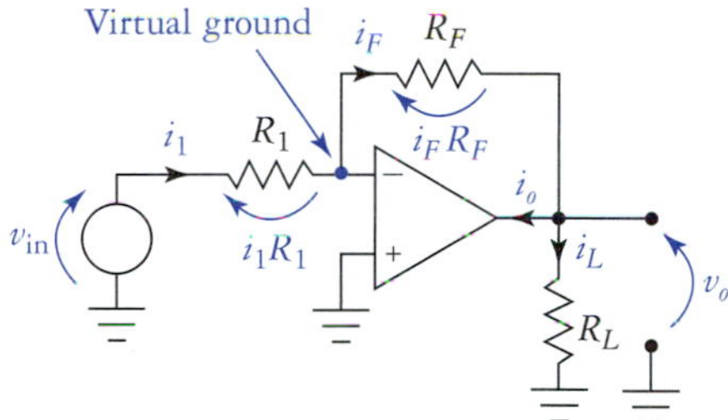

FIGURE 12.7
Special case of the virtual short circuit forces the inverting input to virtual ground.

EXAMPLE 12.4

Refer to Fig. 12.7. Given $V_{\text{in}} = 1.00$ V, $R_1 = 1\ \text{k}\Omega$, $R_F = 10\ \text{k}\Omega$, and $R_L = 1\ \text{k}\Omega$, determine A_v, V_o, I_F and I_o. Assume the op amp is ideal.

Solution The voltage gain is

$$\begin{aligned} A_v &= -R_F/R_1 \\ &= -10\ \text{k}\Omega/1\ \text{k}\Omega \\ &= -10 \end{aligned}$$

The output voltage is now easily found:

$$\begin{aligned} V_o &= A_v V_{\text{in}} \\ &= (-10)(1.00\ \text{V}) \\ &= -10\ \text{V} \end{aligned}$$

The feedback current may be found several ways, but let's use Eq. (12.18), which gives us

$$\begin{aligned} I_F = I_1 &= V_{\text{in}}/R_1 \\ &= 1.00\ \text{V}/1\ \text{k}\Omega \\ &= 1\ \text{mA} \end{aligned}$$

To find the op amp output current, we apply KCL. Because we already assumed that the I_F current-sensing arrow is positive, we must assume that currents entering the output node are positive while those currents leaving the node are negative, if we are to use any equations derived previously using I_F. This produces

$$0 = -I_L - I_o + I_F$$

Solving for I_o

$$I_o = -I_L + I_F$$

We already know the value of I_F, and Ohm's law tells us that $I_L = V_o/R_L$, therefore

$$\begin{aligned} I_o &= -\frac{V_o}{R_L} + I_F \\ &= -\frac{-10\text{ V}}{1\text{ k}\Omega} + 1\text{ mA} \\ &= 10\text{ mA} + 1\text{ mA} \\ &= 11\text{mA} \end{aligned}$$

The positive value of I_o indicates that the output current is flowing in the direction indicated by the arrow.

Closed-Loop Input Resistance

The input resistance of the inverting configuration is relatively low in comparison to the noninverting configuration. In fact, if the op amp is ideal, the input resistance for Fig. 12.7 is

$$R_{\text{in}} = R_1 \tag{12.24}$$

This should seem reasonable because the input source is driving *virtual ground* through resistor R_1. In nearly all practical situations, the ideal relationship is sufficiently accurate.

The actual input resistance of the inverting amplifier using a nonideal op amp is slightly greater than R_1. The equivalent resistance between the inverting input and true ground is

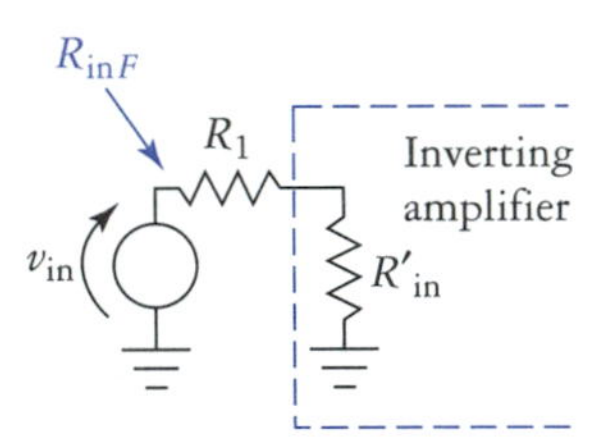

FIGURE 12.8
Modeling input resistance of op amp with feedback.

$$R'_{\text{in}} \cong \frac{R_F}{1 + A_{OL}} \tag{12.25}$$

Including this resistance, the input of the amplifier is equivalent to Fig. 12.8, where

$$R_{\text{in}} = R_1 + R'_{\text{in}} \tag{12.26}$$

The equivalent input resistance is normally very small (a few milliohms), so it can usually be disregarded. For example, if the op amp in Example 12.4 were a typical LM741C, the equivalent resistance would be $R'_{\text{in}} \cong 0.05\ \Omega$. Thus the actual input resistance as seen by the input source would be $1000.05\ \Omega$, which is insignificantly different.

Closed-Loop Output Resistance

Using an ideal op amp, the output resistance of the inverting configuration is zero ohms. Again, we shall use this approximation whenever it is possible because it makes life much easier for us. If necessary, we can determine the actual ouptut resistance for the nonideal case using

$$R_{oF} = \frac{R_o}{1 + \beta A_{OL}} \tag{12.27}$$

where R_o is the open-loop output resistance of the op amp and the feedback factor is

$$\beta = \frac{R_1}{R_1 + R_F} \tag{12.28}$$

Notice that we are using the same feedback factor as for the noninverting amplifier. We used a different feedback factor ($\beta = R_1/R_F$) when determining the closed-loop gain of the inverting amplifier. Sometimes it gets confusing, but thankfully, most often we can simply assume $R_{oF} \cong 0\ \Omega$. If we apply this information to Example 12.3, assuming a typical LM741C is used, we obtain $R_{oF} = 2.7\text{ m}\Omega$, which is quite a low output resistance.

12.5 SMALL-SIGNAL BANDWIDTH

The bandwidth of an open-loop op amp will usually be one of its least ideal characteristics. Recall that the ideal amplifier has infinite bandwidth. Look back at Table 12.1 and you will find that the corner frequency of most op amps is quite low. As we shall soon find out, this is not as big a problem as it might first appear.

The Bode plot for the LM741C is shown in Fig. 12.9. For convenience both decibel and linear open-loop voltage gains are given on the vertical axis. The open-loop corner frequency is about 8 Hz, and the gain rolls off at −20 dB/decade, dropping to 0 dB at $f = 1$ MHz. This is a first-order low-pass response, and many op amps exhibit this behavior under open-loop conditions. In fact, all of the *internally compensated* op amps in Table 12.1 exhibit first-order low-pass open-loop response. The function of internal compensation is to ensure first-order gain rolloff until unity gain is reached. For reasons that we shall investigate later, it turns out that op amps for which A_{OL} rolls off in a first-order manner are easier to keep from oscillating under closed-loop operation. Once the open-loop gain drops below 0 dB, the rolloff rate no longer matters. The LH0003 requires external compensation, which means that we must use a few extra components to tailor its rolloff characteristics to suit our needs. Before we can quantitatively determine the bandwidth of a circuit, we must first define another amplifier parameter.

FIGURE 12.9
Bode plot for the 741 op amp.

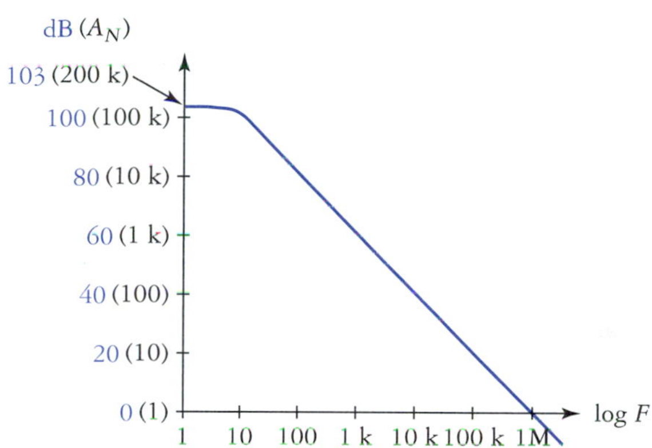

Noise Gain

Note
The noise gain equation (12.29) applies to both inverting and noninverting configurations.

Whenever we are performing calculations to determine closed-loop bandwidth (and offset voltage errors), we use a parameter called *noise gain.* Unlike normal voltage gain, the noise gain of an amplifier is independent of whether it is inverting or noninverting to external signals. The upper portion of Fig. 12.10 shows a noninverting amplifier while in the middle is an inverting amp. If we kill the input voltage sources, the resulting equivalent circuits are identical. These equivalent circuits will respond to disturbances (noise) according to

$$A_N = \frac{R_1 + R_F}{R_1} \tag{12.29}$$

Thus, the noise gain of the noninverting amplifier is the same as its normal voltage gain, while the noise gain of the inverting configuration is always higher than the closed-loop gain magnitude by one.

We next use the noise gain in the determination of amplifier closed-loop bandwidth. Later in this chapter, we use the noise gain to determine how an op amp will respond to various error sources.

The Gain-Bandwidth Product (GBW)

Okay, you might be asking yourself how op amps can be useful at higher frequencies if their open-loop gain begins dropping at such low frequencies. It turns out that for the op amps encountered most often, we can trade gain for bandwidth. For op amps with first-order gain rolloff, the product of the closed-loop bandwidth and the noise gain is a constant given by

$$\text{GBW} = A_N \text{BW} \tag{12.30}$$

Observe in Fig. 12.9 that if we choose any noise gain, say, $A_N = 1$ k, and multiply by the corresponding frequency, in this case 1 kHz, we obtain the product GBW = 1 MHz. Thus, the gain-bandwidth product for the 741 op amp is 1 MHz. The lower the closed-loop gain of an op amp, the wider its bandwidth will be. You can use Eq. (12.30) to find the closed-loop bandwidth of an amplifier if you know the GBW for the op amp being examined (and of course you must determine A_N). In Table 12.1, the frequency at which A_{OL} drops to unity is the same as GBW for all the op amps except the LH0003, which is not internally compensated.

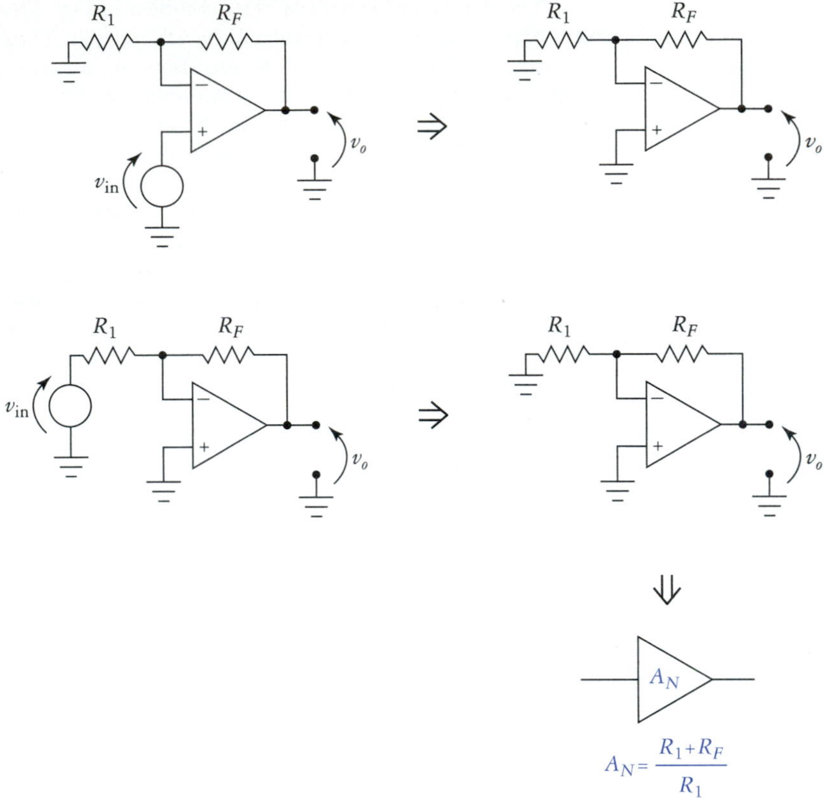

FIGURE 12.10
Noise-gain equivalent circuits.

EXAMPLE 12.5

Using Table 12.1, determine GBW for all op amps except the LH0003.

Solution Because these op amps are compensated for first-order response, each has GBW equal to the unity-gain frequency, thus we have

Device	LM741C	LF351	OP-07	AD549K
GBW	1 MHz	4 MHz	1 MHz	1 MHz

Please keep in mind that the product of noise gain and closed-loop bandwidth is only constant for op amps that have a first-order low-pass rolloff characteristic (−20 dB/decade). Some op amps roll off at a higher rate, while another kind of op amp called a *current-feedback op amp* has bandwidth that is independent of closed-loop gain. For now though, we use only op amps for which GBW is applicable.

EXAMPLE 12.6

Assume both amplifiers in Fig. 12.10 are constructed using LM741C op amps. The noninverting amplifier has $R_F = R_1 = 10\ \text{k}\Omega$. The inverting amplifier has $R_F = 10\ \text{k}\Omega$ and $R_1 = 5\ \text{k}\Omega$. Determine the closed-loop voltage-gain A_v, the noise gain A_N, and the bandwidth for each case.

Solution The gain of the noninverting amplifier is

$$A_v = \frac{R_1 + R_F}{R_1}$$

$$= \frac{10\ \text{k}\Omega + 10\ \text{k}\Omega}{10\ \text{k}\Omega}$$

$$= 2$$

The noise gain of the noninverting amplifier is the same as the normal voltage gain, therefore $A_N = A_v = 2$. The gain-bandwidth product of the 741 op amp is GBW = 1 MHz, therefore

$$\begin{aligned} \text{BW} &= \frac{\text{GBW}}{A_N} \\ &= \frac{1\text{MHz}}{2} \\ &= 500 \text{ kHz} \end{aligned}$$

The inverting amplifier has a closed-loop voltage gain of

$$\begin{aligned} A_v &= -\frac{R_F}{R_1} \\ &= -\frac{10 \text{ k}\Omega}{5 \text{ k}\Omega} \\ &= -2 \end{aligned}$$

The noise gain of the inverting amplifier is

$$\begin{aligned} A_N = \frac{R_1 + R_F}{R_1} &= |A_v| + 1 \\ &= 2 + 1 \\ &= 3 \end{aligned}$$

The closed-loop bandwidth of the inverting amplifier is

$$\begin{aligned} \text{BW} &= \frac{\text{GBW}}{A_N} \\ &= \frac{1\text{MHz}}{3} \\ &= 333.3 \text{ kHz} \end{aligned}$$

This example illustrates how the bandwidth of the inverting and noninverting configurations can differ substantially, even though both amplifiers have the same closed-loop gain magnitude. If you are after the most bandwidth possible for a given gain, the noninverting configuration is the way to go.

Remember that bandwidth is a small-signal characteristic. That is, we have tacitly assumed that the amplifier is operating in a linear manner at all times when a signal is applied. There are conditions for which we need to know more about the frequency-related characteristics of the op amp than the small-signal bandwidth.

Bandwidth of Cascaded Amplifiers

The amplification requirements of many applications cannot be met using a single op amp. For example, assume that 741 op amps are to be used in a design that requires a closed-loop gain of 200 (noninverting), and a bandwidth of at least 40 kHz. A single 741 configured for $A_v = 200$ will have BW = GBW/A_v = 5 kHz. Obviously, a single 741 will not do the job. Realization of this circuit using a single op amp requires GBW = (200)(40 kHz) = 8 MHz.

Many op amps are available with GBW > 8 MHz, but let's see if the design requirements can be met by cascading two very inexpensive 741 op amps. If both stages are set up for a closed-loop gain of approximately 14.14, as shown in Fig. 12.11 the overall gain is 200. For each stage, the bandwidth will be

$$\begin{aligned} \text{BW}_1 = \text{BW}_2 &= \frac{\text{GBW}}{A_N} \\ &= \frac{1 \text{ MHz}}{14.14} \\ &= 70.72 \text{ kHz} \end{aligned}$$

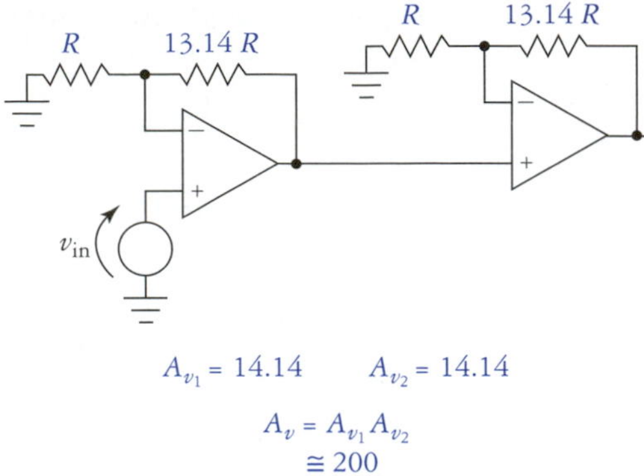

FIGURE 12.11
Distributing gain equally throughout identical amplifier stages results in maximum bandwidth.

This looks promising, but just because the individual stage bandwidths are sufficient does not necessarily mean the total bandwidth BW_T will be sufficient. For identical stages that are cascaded, the overall effective bandwidth is

$$\mathrm{BW}_T = \mathrm{BW}_s(2^{1/n} - 1)^{1/2} \tag{12.31}$$

where BW_s is the closed-loop bandwidth of each of the stages and n is the number of stages cascaded. Applying Eq. (12.31) to the circuit of Fig. 12.11, we find

$$\begin{aligned} BW_T &= 70.72 \text{ kHz } (2^{1/2} - 1)^{1/2} \\ &= 45.5 \text{ kHz} \end{aligned}$$

We have discovered that by cascading two 741 op amps, we exceed the design requirements (40 kHz), but not by as great a margin as might be expected. Whenever multiple (first-order rolloff) stages have identical corner frequencies, the rolloff rate increases according to

$$\text{Rolloff} = 20n \text{ dB/decade} \tag{12.32}$$

where n is the number of stages. It is also possible to use stages with different corner frequencies. As in the case of discrete amplifiers, which were covered earlier, the section with the lowest corner frequency dominates. However, maximum bandwidth will be obtained when the gain is divided equally between identical amplifier stages.

EXAMPLE 12.7

Determine the overall voltage gain and bandwidth of the circuit in Fig. 12.12. Assume op amps U_1–U_3 are LF351s.

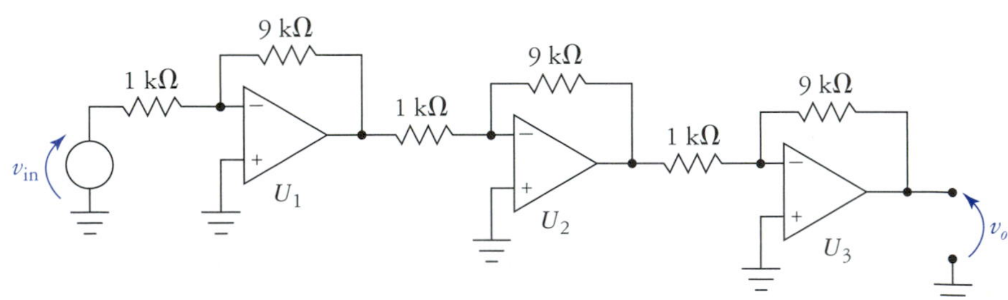

FIGURE 12.12
Circuit for Example 12.7.

Solution The gain stages have identical voltage gain given by

$$\begin{aligned} A_{v_1} = A_{v_2} = A_{v_3} &= -\frac{9\text{ k}}{1\text{ k}} \\ &= -9 \end{aligned}$$

The overall voltage gain is the product of the individual stage gains.

$$\begin{aligned} A_v &= A_{v_1}A_{v_2}A_{v_3} \\ &= (-9)(-9)(-9) \\ &= -729 \end{aligned}$$

To determine the bandwidth of each stage, we must determine the noise gain:

$$\begin{aligned} A_{N_1} = A_{N_2} = A_{N_3} &= \frac{1\text{ k} + 9\text{ k}}{1\text{ k}} \\ &= 10 \end{aligned}$$

The LF351 has GBW = 4 MHz, therefore the bandwidth of each stage is

$$\begin{aligned} \text{BW}_1 = \text{BW}_2 = \text{BW}_3 &= \frac{\text{GBW}}{A_N} \\ &= \frac{4\text{ MHz}}{10} \\ &= 400\text{ kHz} \end{aligned}$$

The overall bandwidth is

$$\begin{aligned} \text{BW}_T &= \text{BW}_s(2^{1/n} - 1)^{1/2} \\ &= 400\text{ kHz}\,(2^{1/4} - 1)^{1/2} \\ &= 174\text{ kHz} \end{aligned}$$

12.6 SLEW RATE AND POWER BANDWIDTH

Bandwidth is a small-signal phenomenon. The slew rate of an op amp is the main parameter that determines frequency-related large-signal characteristics. The slew rate of an amplifier is the maximum possible rate of change of its output voltage. In equation form this is

$$SR = \left.\frac{dv_o}{dt}\right|_{\text{max}} \cong \frac{\Delta v_o}{\Delta t}_{\text{max}} \tag{12.33}$$

Slew rate is normally independent of the closed-loop gain of the op amp. However, the slew rate of some op amps, like the LH0003, can be altered by selection of external frequency compensation components.

Perhaps the best way to intuitively understand the concept of slew rate is to consider the response of an amplifier to a step input, as shown in Fig. 12.13. Ideally, the output of the op amp is a

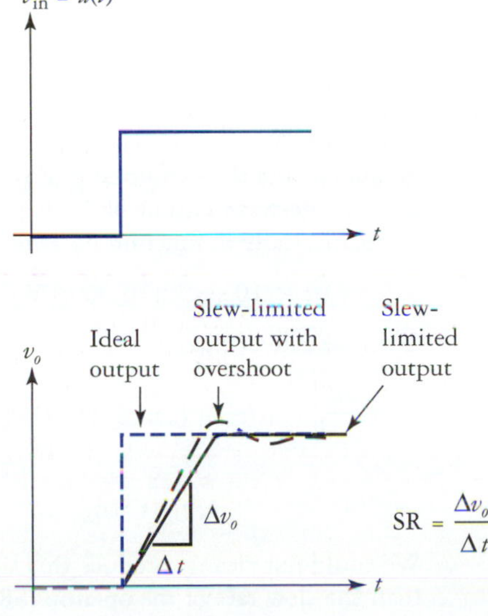

FIGURE 12.13
Effects of slew limiting on large-signal step response.

scaled version of the input, with exactly the same general shape. In practice, the slew-limited output ramps to the final value in a linear manner. In addition, some op amps may also overshoot the final value, but the maximum rate of change of the output is still a constant.

EXAMPLE 12.8

Determine the time required for an LM741C to slew from 0 to 10 V. Repeat this for an LF351 op amp.

Solution The output of the op amp slews linearly from 0 to 10 V, therefore $\Delta v_o = 10$ V. The slew rate for the 741 is SR = 0.5 V/μs = 500 kV/s. These are related by

$$\begin{aligned}\Delta t &= \frac{\Delta v_o}{\text{SR}}\\ &= \frac{10\text{ V}}{1} \times \frac{1\text{ s}}{500\text{ kV}}\\ &= 20\ \mu\text{s}\end{aligned}$$

For the LF351, SR = 13 V/μs, therefore

$$\begin{aligned}\Delta t &= \frac{10\text{ V}}{1} \times \frac{1\text{ s}}{13 \times 10^6\text{ V}}\\ &= 0.769\ \mu\text{s}\end{aligned}$$

Slew Rate and Sine Waves

Note
$\cos(\omega t) = \sin(\omega t + 90°)$

Because they are so commonly encountered, it is important that we be able to relate the slew limitations of an op amp to sinusoidal signals. The following well-known result from basic calculus will be very useful in this section:

$$\frac{d}{dt} V_p \sin(\omega t) = \omega V_p \cos(\omega t) \tag{12.34}$$

Note
The maximum rate of change of a sine or cosine occurs at zero crossing.

This equation says that the instantaneous rate of change (the *derivative*) of a sine function is given by the cosine, scaled by the frequency ω in radians per second. We do not need the complete form of the derivative at this time; all we need to know is the *maximum* rate of change of the sine function, which is

$$\left.\frac{dv}{dt}\right|_{\max} = \omega V_p \tag{12.35}$$

Since we usually prefer to work with cyclic frequency in hertz, we can express this in the alternative form

$$\left.\frac{dv}{dt}\right|_{\max} = 2\pi f V_p \tag{12.36}$$

To understand the usefulness of Eq. (12.35), let's suppose that a particular op amp is expected to produce a sine-wave output of 10 V peak at a frequency of 10 kHz. This output voltage is expressed mathematically in function notation as

$$v_o(t) = 10 \sin(2\pi 10{,}000 t)\text{ V}$$

The max rate of change is

$$\begin{aligned}\left.\frac{d}{dt}\right|_{\max} v_o(t) &= (2\pi 10{,}000\text{ Hz})(10\text{ V})\\ &\cong 628.3\text{ kV/s}\\ &\cong 0.63\text{ V}/\mu\text{s}\end{aligned}$$

We could not cleanly produce this signal with a 741 because the maximum rate of change is greater than the slew rate of the op amp ($\text{SR}_{741} = 0.5$ V/μs). A faster op amp such as the LF351, however, could easily produce this signal.

When the rate of change of a sinusoidal signal greatly exceeds the slew rate of an op amp, the resulting output is distorted as shown in Fig. 12.14. As you can see, the slew-limited output signal tends to become triangular in shape, with slope $m = \pm SR$. If we were to increase the amplitude or frequency of the sine wave $v(t)$, the output of the op amp would decrease because it would have even less time to slew back and forth in response.

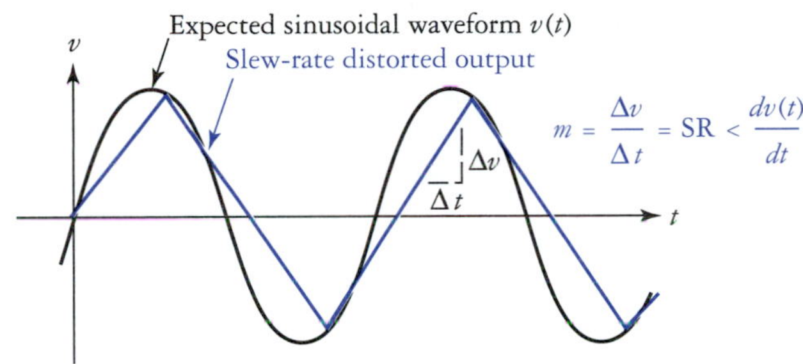

FIGURE 12.14
Effects of slew limiting on a large-signal sinusoid.

Power Bandwidth

Note
Power bandwidth is a large-signal parameter.

The desired output signal amplitude and the slew rate specification of an op amp are used to determine what is called the amplifier's *power bandwidth.* Power bandwidth specifies the maximum frequency at which a sinusoidal output signal can be produced without causing slew limiting to occur. We assume that frequency response extends to dc, hence the term *power bandwidth* is applied.

To determine the power bandwidth BW_p of an amplifier, we specify some full-power output voltage $V_{o(\max)}$ and find the resulting frequency at which the maximum rate of change of V_o just equals the slew rate of the op amp. That is,

$$2\pi f V_{o(\max)} = SR$$

We solve this equation for f, which is BW_p, giving us

$$BW_p = \frac{SR}{2\pi V_{o(\max)}} \tag{12.37}$$

To use Eq. (12.37), you must express the slew rate of the op amp in volts/sec.

EXAMPLE 12.9

Determine the power bandwidth for a 741 op amp with $V_{o(\max)} = 10$ V.

Solution The 741 has SR = 0.5 V/μs = 500 kV/s. Using Eq. (12.37), we find

$$\begin{aligned} BW_p &= \frac{SR}{2\pi V_{o(\max)}} \\ &= \frac{500 \text{ kV/s}}{(2\pi)(10 \text{ V})} \\ &= 7.96 \text{ kHz} \end{aligned}$$

Remember, unlike small-signal bandwidth, the power bandwidth of an op amp is not a function of noise gain. Because of this, in some situations a signal may exceed the slew rate before exceeding the small-signal BW or vice versa.

EXAMPLE 12.10

Refer to Fig. 12.15. The op amp is a 741. Determine the voltage gain, the small-signal bandwidth, and determine whether any of the following input voltages will cause the amplifier to exceed BW or SR.

(a) $v_{in} = 500 \sin(2\pi 20{,}000t)$ mV
(b) $v_{in} = 20 \sin(2\pi 100{,}000t)$ mV

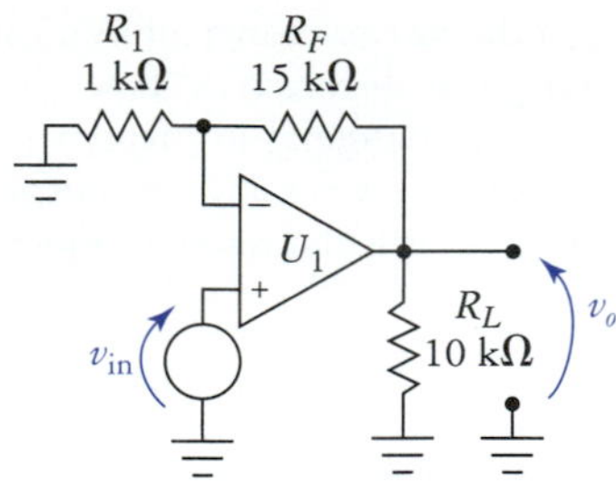

FIGURE 12.15
Circuit for Example 12.10.

Solution

$$A_N = A_v = \frac{R_1 + R_F}{R_1}$$
$$= \frac{1\text{ k}\Omega + 15\text{ k}\Omega}{1\text{ k}\Omega}$$
$$= 16$$

$$\text{BW} = \frac{\text{GBW}}{A_N}$$
$$= \frac{1\text{ MHz}}{16}$$
$$= 62.5\text{ kHz}$$

(a) We have $f <$ BW (20 kHz < 62.5 kHz). This signal does not exceed the small-signal bandwidth of the amplifier. The resulting output voltage should be

$$\begin{aligned} v_o &= A_v v_{\text{in}} \\ &= (16)[0.5 \sin(2\pi 20{,}000t)]\text{ V} \\ &= 8 \sin(2\pi 20{,}000t)\text{ V} \end{aligned}$$

The maximum rate of change of v_o is

$$\begin{aligned} \left.\frac{dv_o}{dt}\right|_{\max} &= (8)(2\pi 20{,}000) \\ &= 1\text{ MV/s} \\ &= 1\text{ V/}\mu\text{s} \end{aligned}$$

This output signal exceeds the slew rate of the typical 741 op amp, which would cause severe distortion.

(b) In this case we have $f >$ BW (100 kHz > 62.5 kHz). The signal exceeds the small-signal bandwidth of the amplifier. Because of this, in order to determine the expected output, we would have to use the first-order response equation Eq. (11.10) as follows:

$$|A_v(f)| = \frac{A_{\text{mid}}}{\sqrt{\left(\frac{f_{\text{in}}}{f_c}\right)^2 + 1}}$$

At the given input frequency the gain of the amplifier is

$$\begin{aligned} \left.A_v(f)\right|_{f=100\text{ kHz}} &= \frac{16}{\sqrt{\left(\frac{100\text{ k}}{62.5\text{ k}}\right)^2 + 1}} \angle\theta \\ &= 8.5 \angle\theta \end{aligned}$$

where θ is a phase shift given by $\theta = -\tan^{-1}(f_{\text{in}}/f_c) = -58°$. The output voltage should therefore be

$$\begin{aligned} V_o &= (8.5 \angle -58°)[20 \sin(2\pi 100{,}000t)\text{ mV}] \\ &= 170 \angle -58°\text{ mV} \\ v_o &= 170 \sin[(2\pi(100{,}000t) - 58°]\text{ mV} \end{aligned}$$

The phase shift has no effect on the rate of change of the output, therefore

$$\left.\frac{dv_o}{dt}\right|_{\max} = (0.170)(2\pi 100{,}000)$$

$$= 0.107\ \text{V}/\mu\text{s}$$

In this case, the amplifier is limited by the small-signal bandwidth, and even if there were no band-limited gain reduction, the slew rate would still not be exceeded.

Distortion and Frequency Limitations

As shown in the previous example, if a signal only exceeds the small-signal bandwidth of an amplifier, the output will be smaller than expected and shifted in phase. It is significant, however, that this distortion will still only contain energy at the original signal frequency(s). If a more complex signal

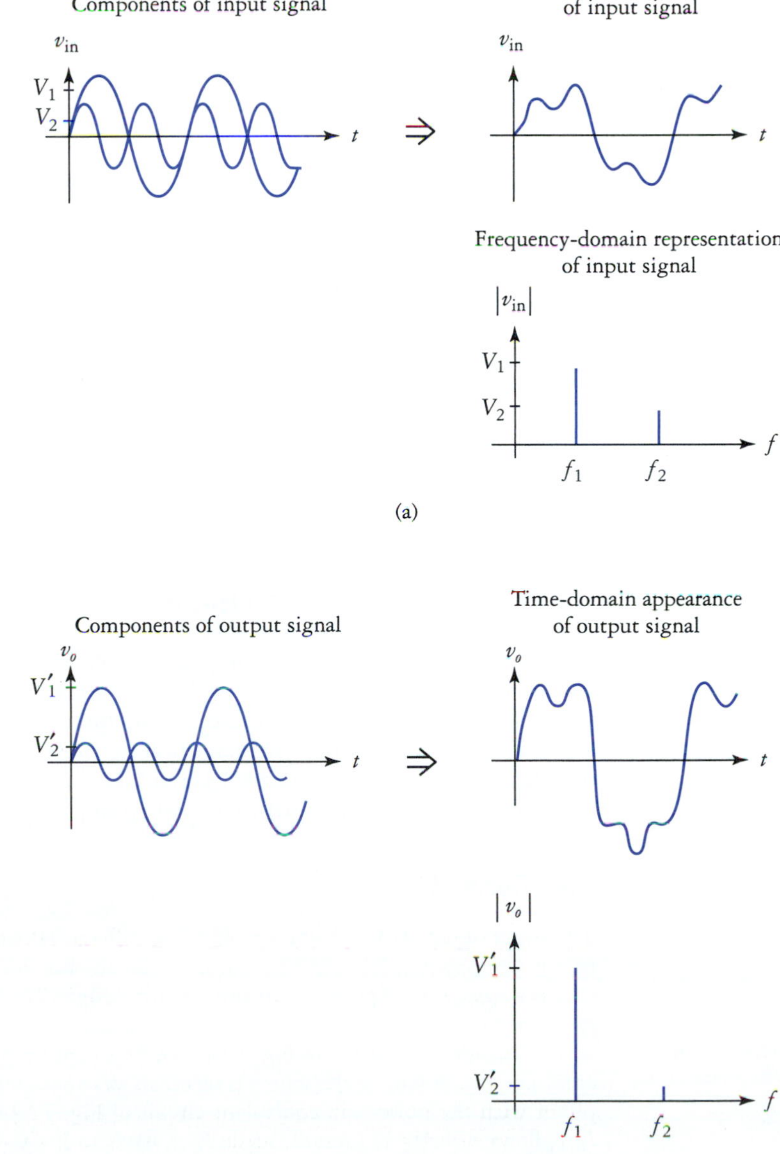

FIGURE 12.16
Effect of bandwidth limiting. (a) Input signal components with net waveform and spectrum. (b) Output of band-limited amplifier exhibits relative amplitude and phase changes.

is applied, such as an audio signal, the relative amplitude and phase of the spectral components (harmonics) will be altered. In audio applications, this type of distortion causes the sound to be *colored.* This is shown in Fig. 12.16. Figure 12.16(a) shows the input signal and its frequency spectrum. Processing this signal in an amplifier with $f_c < f_2$ causes the v_2 component to experience less gain and undesired phase shift, changing the shape of the signal, as shown in Fig. 12.16(b). Because the relative frequency content of the output is the same as the input, however, this kind of distortion is not extremely objectionable, and may be reduced or eliminated using various filtering schemes.

If a signal exceeds the slew rate of the amplifier, a nonlinear distortion occurs, which fundamentally alters the spectral content of the signal. This phenomenon is illustrated in Fig. 12.17. A simple sinusoidal input is used here to keep the frequency-domain illustration uncluttered. Slew distortion causes the output to contain energy at odd multiples of the input signal frequency. In general, these frequency components cannot be removed by filtering, because they are not distinguishable from the desired signal frequencies that happen to be located at odd multiples of f_{in}.

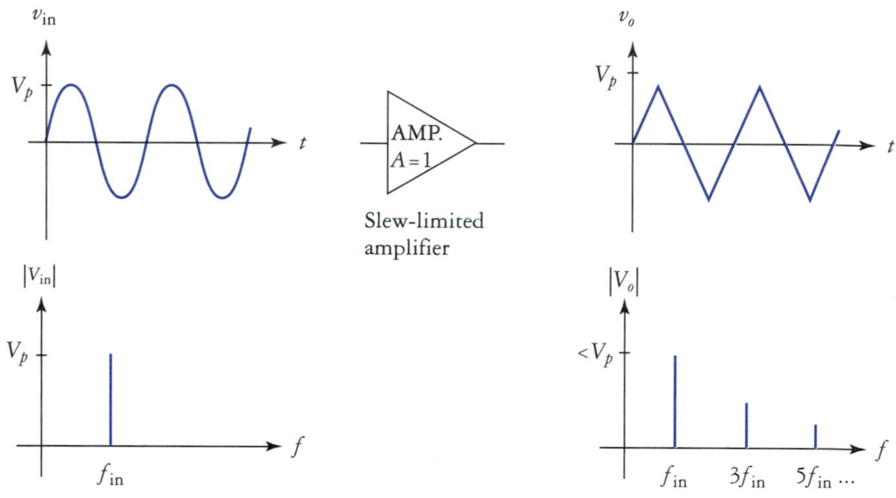

FIGURE 12.17
Exceeding the slew rate of the amplifier creates odd harmonics in the output signal.

12.7 BIAS CURRENTS AND OTHER NONIDEAL CHARACTERISTICS

Because the frequency response of most op amps extends to 0 Hz (dc), they are used extensively in applications where low-level dc and very low-frequency signals are common. An example might be an industrial application where we must measure the output of a strain gauge with an output of a few millivolts. Another example is in biomedical electronics, where electrical activity of the brain is to be monitored. These signals fall below 1 Hz and are in the microvolt range as well. Such applications require detailed knowledge of many nonideal op amp characteristics.

Bias Current

The input stage of the typical op amp is a differential pair. Regardless of whether the differential pair is composed of BJTs or FETs, a path must be provided for each input bias current. Remember this as a basic rule of practical op amp circuit design: *The input terminals require a dc path to ground for bias currents.*

Note
Op amp input terminals require a dc path to ground for bias currents.

Consider the circuit in Fig. 12.18(a). This could represent either an inverting or noninverting amplifier, but as long as the source is inactive, we cannot tell the difference. We can represent the amplifier with the noise-gain equivalent circuit of Fig. 12.18(b). The noninverting input bias current $I_{B(+)}$ flows directly to ground, while $I_{B(-)}$ flows to ground via equivalent resistance R_{eq}. Using the current directions indicated, positive values would indicate that npn BJTs or n-channel FETs are used. Assuming this is the case, and observing that $V_{(+)} = 0$ V, the differential input voltage present is

$$V_{id} = V_{(-)} = -I_{B(-)}R_{eq}$$

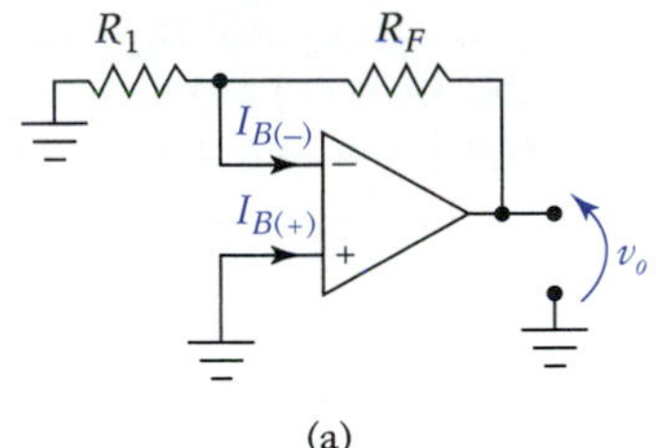

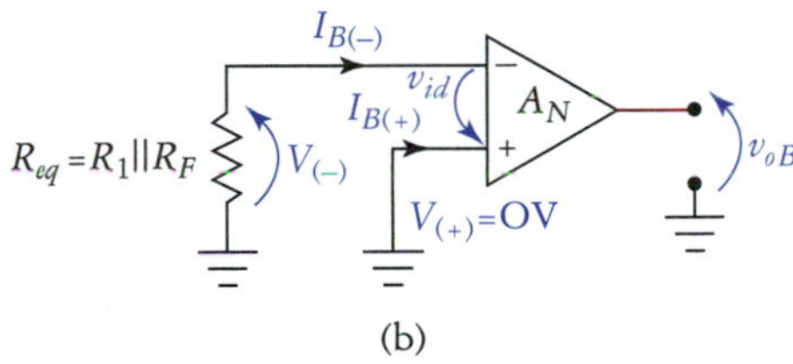

FIGURE 12.18 *Bias current effects. (a) Simple closed-loop amplifier. (b) Noise-gain model of amplifier with Thevenized feedback network.*

The resulting output is an error voltage that is a *bias-current-induced output offset voltage* V_{oB}, which is given by

$$V_{oB} = A_N I_{B(-)} R_{eq} \tag{12.38}$$

This is one of several error sources for which we must sometimes account.

EXAMPLE 12.11

Refer to Fig. 12.18. Given $R_F = 47\ \text{k}\Omega$ and $R_1 = 2.7\ \text{k}\Omega$, determine the output offset voltage that would be produced if the circuit is constructed using a 741 op amp and an LF531 op amp.

Solution In both cases the noise gain is

$$A_N = \frac{R_1 + R_F}{R_1}$$
$$= \frac{2.7\ \text{k} + 47\ \text{k}}{2.7\ \text{k}}$$
$$= 18.4$$

The equivalent inverting input resistance is

$$R_{eq} = R_F \| R_1$$
$$= 47\ \text{k}\Omega \| 2.7\ \text{k}\Omega$$
$$= 2.6\ \text{k}\Omega$$

The resulting output offset voltage is

$$V_{oB} = A_N I_{B(-)} R_{eq}$$

For the 741 we have

$$V_{oB(741)} = (18.4)(80\ \text{nA})(2.6\ \text{k}\Omega)$$
$$= 3.8\ \text{mV}$$

For the LF351 the offset voltage is

$$V_{oB(351)} = (18.4)(50\ \text{pA})(2.6\ \text{k}\Omega)$$
$$= 2.4\ \mu\text{V}$$

In this example we found that compared to the 741, the LF351 output offset voltage caused by bias current is very low. Ideally, we would like this error to be zero. Using lower value resistors reduces R_{eq}, which in turn reduces the effect of bias-current-induced errors. Depending on the application, however, small amounts of output offset voltage may be tolerable.

It is possible to reduce the bias-current-induced offset voltage by equalizing the resistances at the inverting and noninverting inputs. This is shown in Fig. 12.19. Assuming both bias currents are equal, the voltage drops across both input resistances are equal as well, causing

$$V_{id} = V_{(+)} - V_{(-)} = 0 \text{ V}$$

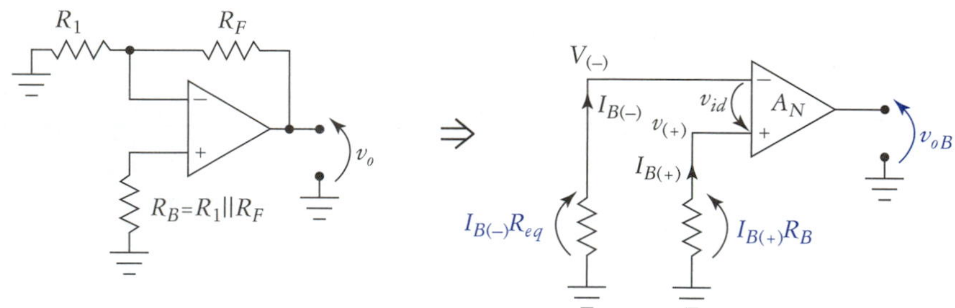

FIGURE 12.19
Compensation resistor eliminates differential input voltage due to bias currents.

Therefore the resulting output offset voltage is zero as well. The input bias current is defined by the following equation:

$$I_B = \frac{I_{B(+)} + I_{B(-)}}{2} \tag{12.39}$$

Equation (12.39) implies that we do not normally differentiate between inverting and noninverting input bias currents; we assume they are equal. If the two bias currents are indeed equal, then Eq. (12.39) is completely accurate.

The addition of the bias equalization resistor has virtually no effect on other amplifier characteristics. High-precision low-level dc amplifiers require the use of low bias current op amps, the bias current compensation resistor, and often other compensation circuitry as well.

Input Offset Current

In the previous discussion we assumed that the bias currents entering the op amp inputs were equal. In reality, these currents will differ from one another. The mismatch of bias currents is called *input offset current* I_{io}. This difference is caused by slight β mismatches for BJTs and by gate leakage current mismatches for FET inputs. These mismatches are random, and it is not possible to predict which input bias current will be larger. Because of this, the input offset current is defined as

$$I_{io} = |I_{B(+)} - I_{B(-)}| \tag{12.40}$$

Input offset current is usually much smaller than the input bias current.

The addition of a bias current compensating resistor will not reduce the output offset voltage error caused by input offset current. In addition, the polarity of the error caused by I_{io} is random. It may be possible to increase or reduce R_B as necessary to null error caused by I_{io}, but this is not always practical.

Input Offset Voltage

Another error source found on op amp spec sheets is *input offset voltage* V_{io}. Input offset voltage is caused by V_{BE} and V_{GS} mismatches in the input-stage differential pair. Like input offset current, the actual polarity of V_{io} is random.

There is no analytical equation for input offset voltage. Usually, V_{io} is simply lumped together with other input error sources and we define the effective input offset voltage as

$$V_{io} = \frac{V_{oo}}{A_N} \tag{12.41}$$

where V_{oo} is the total output offset voltage due to all error sources combined.

Offset Compensation

Many op amps have offset compensation terminals that allow for very easy elimination of output offset voltage. All of the op amps listed in Table 12.1 except the LH0003 have this feature.

Built-In Offset Compensation

Details for the compensation components and their connections vary for each op amp, but most designed with built-in offset compensation terminals are configured similar to the 741, which is shown in Fig. 12.20. If offset compensation is not required, the compensation terminals are simply left open. Offset compensation is often referred to as the *offset-nulling* or *offset-null adjustment.* This type of offset adjustment is effective in nulling output offset voltages on the order of tens of millivolts.

The output of an op amp is nulled by reducing all input signal sources (ac and dc) to zero and then adjusting the offset comp potentiometer for $V_o = 0$ V. In some cases there is interaction between amplifier gain and offset adjustment. If this occurs, the gain and offset adjustments are repeated until convergence occurs.

External Offset Compensation

Many op amps do not have built-in offset null terminals. In these cases external offset null circuitry can be used. The inverting amplifier may be offset nulled using either circuit in Fig. 12.21. The circuit in Fig. 12.21(a) can be used for op amps with input offset voltage specs similar to those for the LM741 (several millivolts). Basically, we are using this circuit to apply a small voltage V_{adj} to the noninverting input that is set such that it just cancels the input offset voltage. As a general rule of thumb, we would choose R_y such that when the potentiometer is at either extreme, $V_{\text{adj}} \cong 2V_{io(\max)}$ for the op amp being compensated. Potentiometer R_x is chosen to be as small as possible (without drawing excessive power supply current) in order to obtain adjustment linearity. Assuming

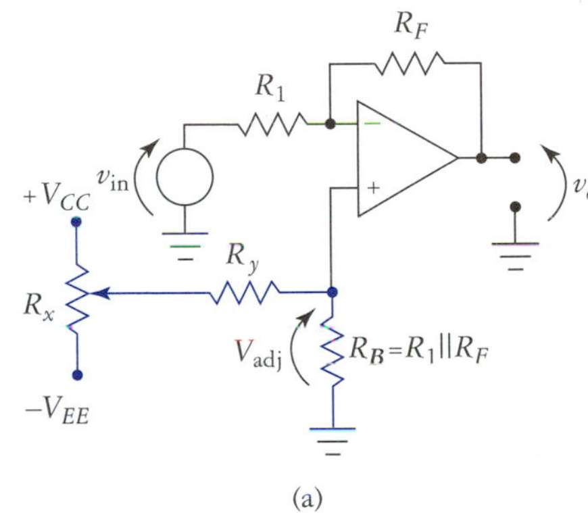

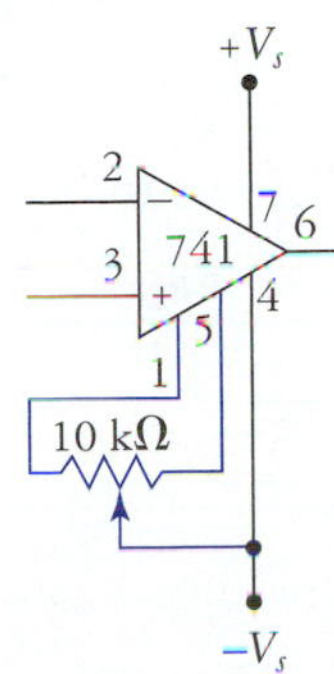

FIGURE 12.20
Offset compensation adjustment of the 741.

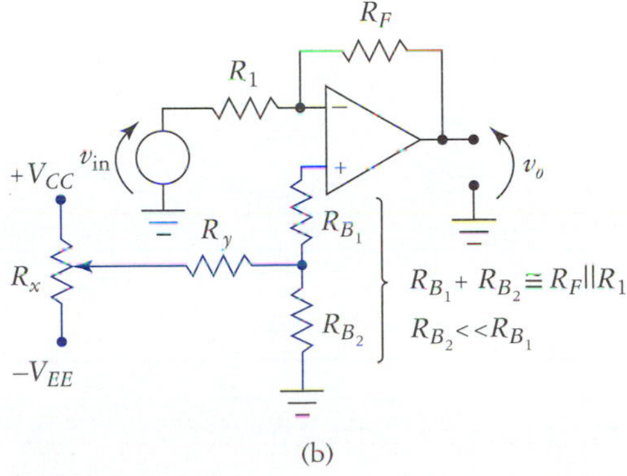

FIGURE 12.21
External offset compensation networks. (a) Inverting amplifier. (b) Modification for op amps with very low V_{io}.

symmetrical supply voltages, a few useful design equations that quantify these suggestions are

$$R_B = R_1 \parallel R_F \tag{12.42}$$

$$R_y = R_B\left(\frac{V_{CC}}{2V_{io(\max)}} - 1\right) \tag{12.43}$$

$$R_x \leq R_y/4 \tag{12.44}$$

If the input offset voltage of the op amp is very low, Eq. (12.43) will produce an impractically large resistance for R_y. If this occurs, then the bias current compensation resistor can be split as shown in Fig. 12.21(b). In this case, we simply choose a convenient value for R_{B_2} and use Eq. (12.43) and (12.44) to find R_x and R_y.

EXAMPLE 12.12

The LM108 is an op amp with $V_{io(\max)} = 2$ mV. This op amp is to set up in the inverting configuration with $R_F = 47$ kΩ, $R_1 = 4.7$ kΩ, and $V_{CC} = V_{EE} = 15$ V. Design the offset compensation circuitry for this amplifier.

Solution The bias current compensation resistor is

$$\begin{aligned} R_B &= R_F \parallel R_1 \\ &= 47\text{ k}\Omega \parallel 4.7\text{ k}\Omega \\ &= 4.3\text{ k}\Omega \end{aligned}$$

If we use Eq. (12.43) directly, we obtain $R_y = 16$ MΩ. This is far too large to be practical (we would like to keep resistances less than 1 MΩ most of the time). Let's spilt the resistor into two parts. Because we will have some variation in bias current anyway, the absolute value of R_B is not critical, so let's choose $R_{B_1} = 3.9$ kΩ and $R_{B_2} = 100$ Ω. Now, applying Eq. (12.43),

$$\begin{aligned} R_y &= R_{B_2}\left(\frac{V_{CC}}{V_{io(\max)}} - 1\right) \\ &= 100\ \Omega\left(\frac{15\text{ V}}{4\text{ mV}} - 1\right) \\ &= 374\text{ k}\Omega \end{aligned}$$

Using the next lower standard value guarantees sufficient adjustment range. This is $R_y = 330$ kΩ. There are many standard-value potentiometers that can be used for R_x. A good value that satisfies inequality (12.44) is $R_x = 10$ kΩ. This potentiometer will have very low power dissipation and will provide very linear adjustment of the offset voltage. The resulting circuit is shown in Fig. 12.22.

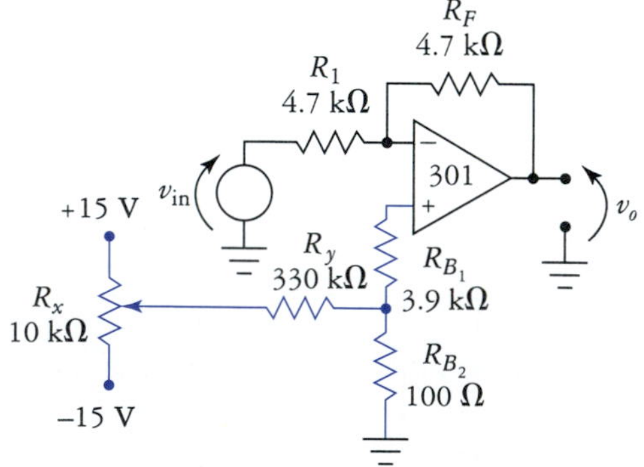

FIGURE 12.22
Circuit for Example 12.12.

As you can see from the previous example, designing an offset null circuit is not that difficult, but it requires a lot of "feel" for the circuit. As usual, this comes with experience.

The offset null circuitry used with the noninverting amplifier is shown in Fig. 12.23(a). In this case, we are applying a small voltage to the inverting side of the amp in order to compensate for input offset voltage. Unfortunately, this compensation scheme alters the closed-loop gain of the amplifier.

The compensation network reduces to the equivalent resistance shown in Fig. 12.23(b). The equivalent resistance will always be less than R_z. The desired accuracy of the voltage gain determines the size of R_z; the smaller R_z is relative to R_1, the smaller the gain error will be. For less than 1% gain error, we use the inequality

$$R_z \leq \frac{R_1}{100} \tag{12.45}$$

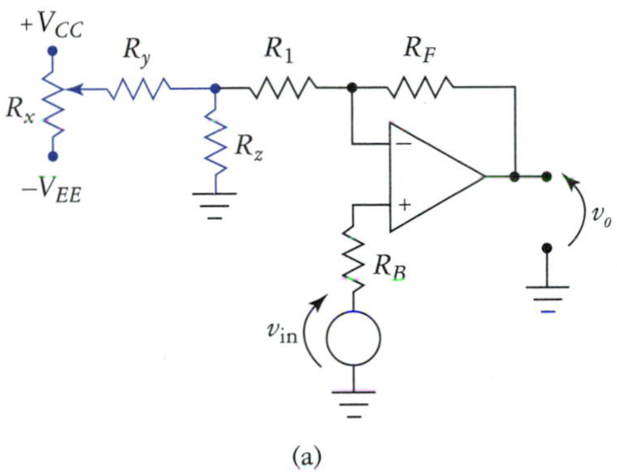

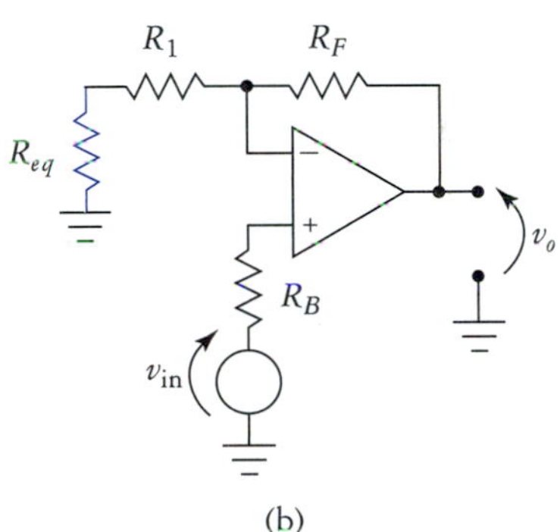

FIGURE 12.23 *Noninverting amplifier offset null. (a) The actual circuit. (b) Circuit with compensation network reduced.*

Resistor R_y is determined using a relationship similar to that used with the previous offset null circuit

$$R_y = R_z\left(\frac{V_{CC}}{V_{io(\max)}} - 1\right) \tag{12.46}$$

Likewise, we choose R_x using

$$R_x \leq R_y/4 \tag{12.47}$$

EXAMPLE 12.13

Recall that the LM108 is an op amp with $V_{io(\max)} = 2$ mV. This op amp is to set up in the noninverting configuration with $R_F = 22\ \text{k}\Omega$, $R_1 = 2.2\ \text{k}\Omega$, and $V_{CC} = V_{EE} = 15$ V. Design the offset compensation circuitry for this amplifier such that the gain error (caused by the compensation circuitry) is 1% or less.

Solution Using inequality (12.45), $R_z \leq 22\ \Omega$. This is a standard value, so we use it directly in Eq. (12.46):

$$\begin{aligned} R_y &= R_z\left(\frac{V_{CC}}{V_{io(\max)}} - 1\right) \\ &= 22\left(\frac{15\text{ V}}{4\text{ mV}} - 1\right) \\ &= 82.5\ \text{k}\Omega \end{aligned}$$

We use the next lower standard value of 82 kΩ. A convenient value for the potentiometer is 10 kΩ. The circuit is shown in Fig. 12.24.

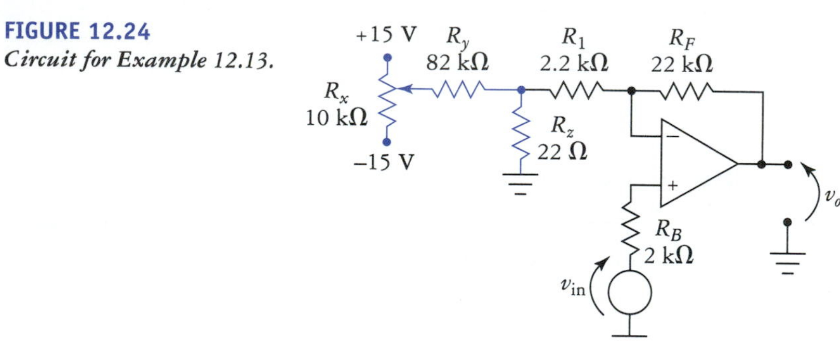

FIGURE 12.24
Circuit for Example 12.13.

12.8 CMRR AND PSRR

We studied the concept of common-mode rejection ratio at length in Chapter 8. This material applies directly here because the input stage of the op amp is a differential amplifier. The power supply rejection ratio (PSRR) has some similarities to CMRR and so is covered here as well.

Common-Mode Rejection Ratio

Common-mode signals are encountered frequently in many op amp applications. Because of this, the ability of the op amp to reject common-mode input voltage is very important. Even very inexpensive op amps like the 741 have very good CMRR characteristics. Recall that CMRR is usually expressed in decibels and is defined as

$$\text{CMRR} = 20 \log \left| \frac{A_d}{A_{cm}} \right| \tag{12.48}$$

So far in this chapter we have only dealt with amplifiers that are single-ended-input type circuits. In such cases, we would use the noise gain A_N in the numerator of Eq. (12.48).

If it is necessary to determine the response of an amplifier to a common-mode input voltage, we must determine A_{cm}. Op amp data sheets only list CMRR, so we must solve Eq.(12.48). This gives us

$$A_{cm} = \frac{A_N}{\log^{-1}\left(\dfrac{\text{CMRR}}{20}\right)} \tag{12.49}$$

The common-mode input voltage is defined as the average of the voltages present at the inverting and noninverting inputs of the op amp. That is,

$$v_{icm} = \frac{v_{(+)} + v_{(-)}}{2} \tag{12.50}$$

Figure 12.25 illustrates schematically how these equations are related. Assuming all other error sources have been nulled, the output voltage consists of two parts: the desired output caused by the differential input voltage and an error voltage due to the presence of the common-mode input voltage. The polarity of the common-mode gain depends on the op amp internal circuitry. For the 741, the common-mode gain is inverting.

Even though we are only dealing with single-ended-input circuits right now, there is a situation in which high CMRR is important. Consider the noninverting amplifier in Fig. 12.26. Ideally negative feedback forces the differential input voltage to zero, which means that $V_1 = V_{in}$, therefore, for the noninverting amplifier

$$V_{icm} = \frac{V_{in} + V_{in}}{2} = v_{in} \tag{12.51}$$

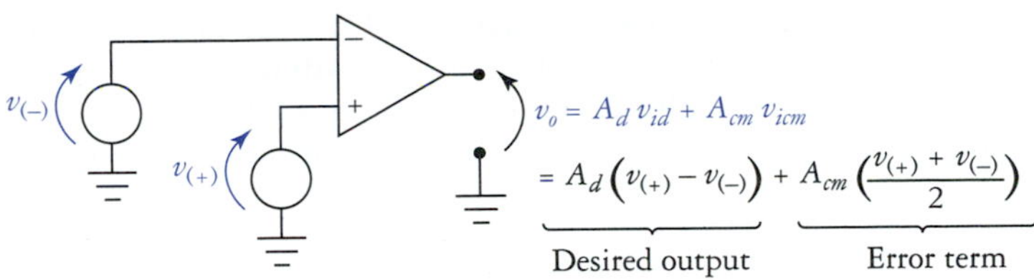

FIGURE 12.25
Output voltage expression accounting for common-mode error.

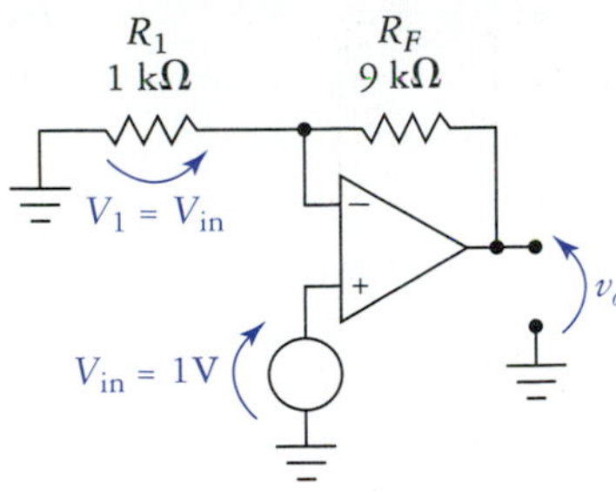

FIGURE 12.26
The noninverting configuration produces a common-mode input voltage.

Thus we see that the noninverting configuration always has a self-induced common-mode input voltage. The inverting configuration does not share this error source because its noninverting input is tied to ground, and the inverting input is held at virtual ground.

EXAMPLE 12.14

Determine the output voltage for the amplifier in Fig. 12.26. Assume a 741 op amp is used with CMRR = 90 dB.

Solution The noise gain and closed-loop gain are the same for the noninverting configuration, thus

$$A_N = A_v = \frac{1\text{ k} + 9\text{ k}}{1\text{ k}}$$
$$= 10$$

The common-mode gain is (remember, A_{cm} is negative for the 741)

$$A_{cm} = \frac{A_N}{\log^{-1}\left(\frac{\text{CMRR}}{20}\right)}$$
$$= \frac{10}{\log^{-1}\left(\frac{90}{20}\right)}$$
$$= -316 \times 10^{-6}$$

The common-mode input voltage is

$$v_{ocm} = V_{in} = 1\text{ V}$$

Therefore,

$$v_{icm} = A_{cm}v_{icm}$$
$$= (-316 \times 10^{-6})(1\text{ V})$$
$$= -316\ \mu\text{V}$$

The desired output voltage is

$$V_{od} = A_v V_{in}$$
$$= (10)(1\text{ V})$$
$$= 10\text{ V}$$

The net output voltage is

$$V_o = V_{od} + v_{ocm}$$
$$= 10\text{ V} - 316\ \mu\text{V}$$
$$= 9.9997\text{ V}$$

The common-mode error in the previous example is quite small and most likely can be ignored. In later applications, however, we will encounter situations where even the 90-dB CMRR of the 741 is not good enough to provide adequate performance.

Note
CMRR rolls off at −20 dB/decade with frequency.

The common-mode rejection ratio specification given in Table 12.1 applies to dc common-mode voltages. CMRR decreases with frequency, so most op amp data sheets will include a Bode plot of $CMRR_{(dB)}$ versus log f. In environments where high-frequency common-mode noise is expected, appropriate shielding and grounding precautions must also be taken to ensure adequate common-mode rejection.

Power Supply Rejection Ratio

One of the reasons why op amps are so popular is because they can usually be operated from a wide variety of power supply voltages. For example, the 741 can be operated from bipolar supplies ranging from ±5 to ±18 V with only slight parameter shifts.

If the power supply voltage does change while an op amp is operating, it will cause a slight change in output voltage. The ratio of the change in supply voltage to change in output voltage is the *supply voltage rejection ratio* (SVRR), usually expressed in decibels. In equation form this is

$$\text{SVRR} = 20 \log \frac{\Delta V_s}{\Delta V_o} \tag{12.52}$$

SVRR is not listed in Table 12.1, however, the typical 741 op amp has SVRR = 96 dB over the specified supply voltage range of ±5 to ±18 V. The supply voltage rejection ratio is important in applications where the op amp power supply is subject to noise pickup.

12.9 COMPUTER-AIDED ANALYSIS APPLICATIONS

The 741 is an extremely popular op amp, and its behavioral model is included in the library of the evaluation version of PSpice. This section demonstrates how the op amp device is used and how to assign global power supply voltages.

Figure 12.27 shows the schematic editor version of a 741-based noninverting amplifier with $A_v = 2$. The particular op amp modeled here is the μA741. The prefix μA is used by Fairchild Instruments, the original developers of the 741 (the LM prefix seen previously is used by National Semiconductor). You can find the μA741 in eval.lib. The op amp is connected to external components just like any other device we have simulated previously. The offset compensation terminals are simply left open in this example.

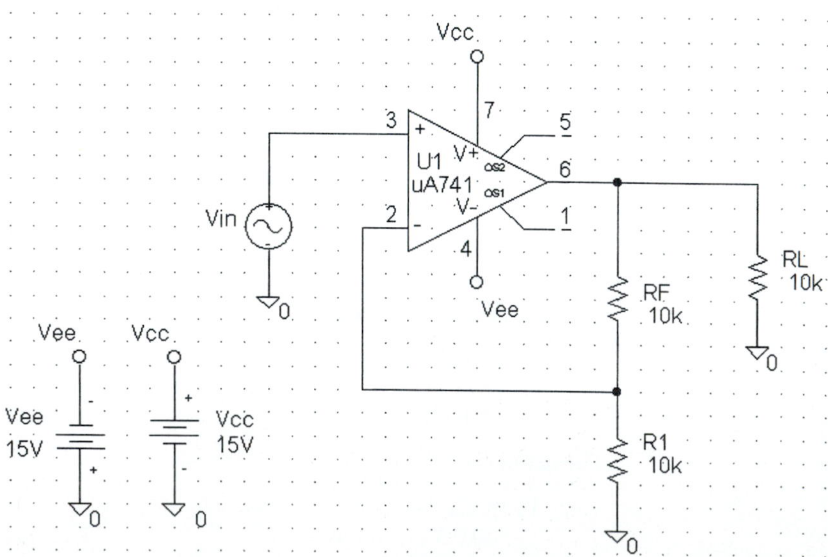

FIGURE 12.27
Noninverting op amp.

Designating Supply Voltages

Note
In PSpice, Release 9 and later, supply voltages are implemented using the VCC_CIRCLE /CAPSYM symbol. Details are given in Appendix 4.

Rather than cluttering up the schematic by directly connecting the supply voltage sources to the op amp, we can place these sources in a convenient location and reference them just like in a hand-drawn schematic. To do this, we use the *bubble* component, which is found in the PORT sublibrary. A bubble is connected to each terminal we wish to connect in this manner. We then double-click on each bubble and assign a name that identifies that particular node. The bubble effectively connects all like-named terminals to a single node. In this case, the nodes are labeled Vcc and Vee. This can be done any time you want to unclutter your schematic.

Closed-Loop Frequency Response

With the component values shown in Fig. 12.27, assuming a typical 741 op amp, we expect a closed-loop small-signal bandwidth of

$$\begin{aligned} BW &= \frac{GBW}{A_N} \\ &= \frac{1\ \text{MHz}}{2} \\ &= 500\ \text{kHz} \end{aligned}$$

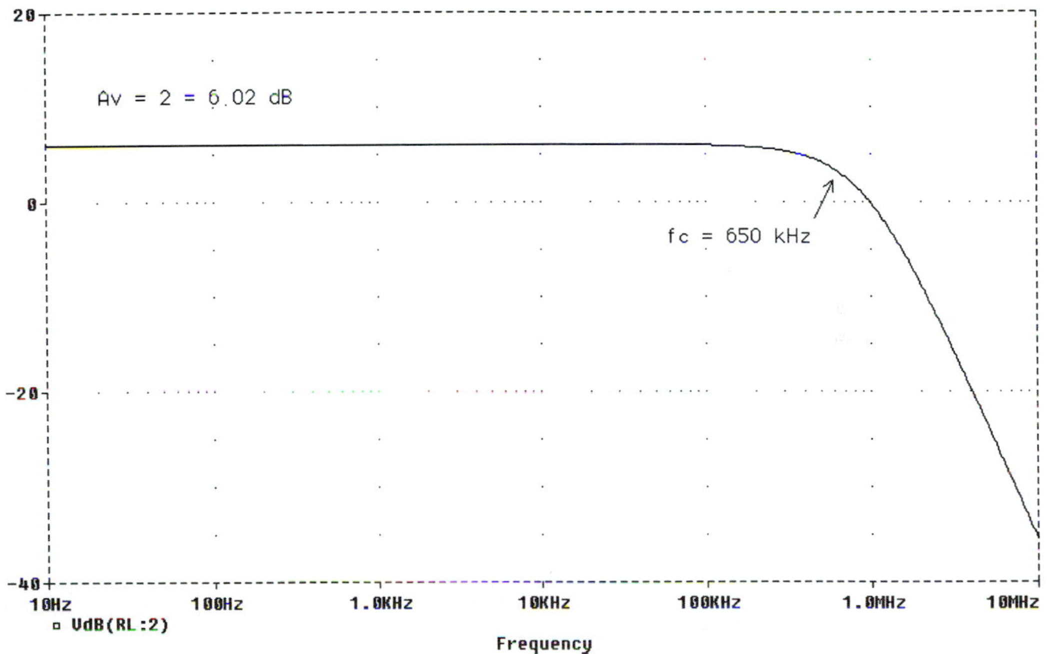

FIGURE 12.28
Frequency response (in decibels) of noninverting op amp.

We set the input sinusoidal voltage source for ac = 1 V and perform a decade ac sweep from 10 Hz to 10 MHz. Plotting the resulting output voltage in decibels produces the Bode plot of Fig. 12.28. The midband voltage gain is 6.02 dB (or 2 if you prefer) and the corner frequency is approximately 650 kHz. The closed-loop voltage gain is very accurate, as we would hope. The closed-loop bandwidth differs from our estimate, but this variance would not be surprising even if we actually built the circuit and measured the bandwidth. The reason is because the closed-loop bandwidth is directly proportional to the op amp's open-loop bandwidth. The absolute value of the open-loop parameters can vary significantly from device to device, with supply voltage, and so on. Closed-loop gain is much less dependent on open-loop gain, as long as $A_{OL} \gg A_{CL}$.

Because the gain of the op amp is really a function of frequency, we have assumed that the gain rolls off in a first-order manner, at −20 dB/decade. Examination of Fig. 12.28 indicates a gain rolloff approaching −40 dB/decade, which is actually a second-order response characteristic. Don't worry. This is actually correct once the gain drops below 0 dB, which occurs very soon on this Bode plot. For gains higher than 0 dB, the response is very close to first order. You are encouraged to verify this for yourself, perhaps by increasing the closed-loop gain to 100 and repeating the ac sweep.

Phase Response

Since the gain of the op amp is a function of frequency, it will also have a phase response that is a function of frequency. In other words, the voltage gain of the amplifier is a complex number. Most of the time we will restrict signal frequencies to the midband, where the gain is basically a real number. When we plot ac voltages, such as V(RL:2) or VdB(RL:2) using Probe, the resulting plot is the *magnitude* of the complex number. At frequencies approaching the limit(s) of the midband, however, as the gain rolls off, its phase angle changes as well. We can observe this phase relationship by inserting P to display the phase angle of the complex number. For example plotting VP(RL:2) gives us the graph in Fig. 12.29 where we see the phase of V_o (with respect to V_{in}) begins to lag significantly at 100 kHz. The combination of the gain magnitude plot and the phase plot tell us everything we need to know about the small-signal frequency response of the amplifier.

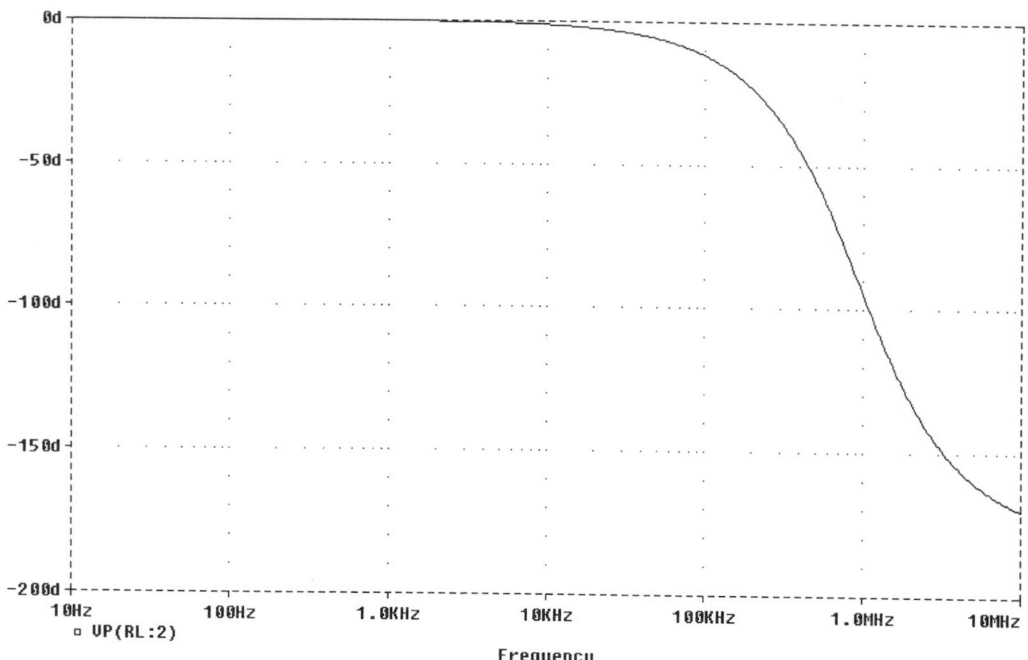

FIGURE 12.29
Phase response of noninverting op amp.

Real and Imaginary Response

The amplitude and phase response plots of a network present its frequency response in *polar form.* Using PSpice it is also possible to present this information in *rectangular form;* that is, we can display the real and imaginary parts of the response, rather than the magnitude and angle. This is shown in Fig. 12.30, where VR(RL:2) is the real part of the output voltage and VI(RL:2) is the imaginary part. As we would expect, the imaginary part is negligibly small until around 100 kHz. Comparing Fig. 12.30 with Fig. 12.28, we find that the imaginary part of V_o reaches its maximum magnitude very close to the corner frequency.

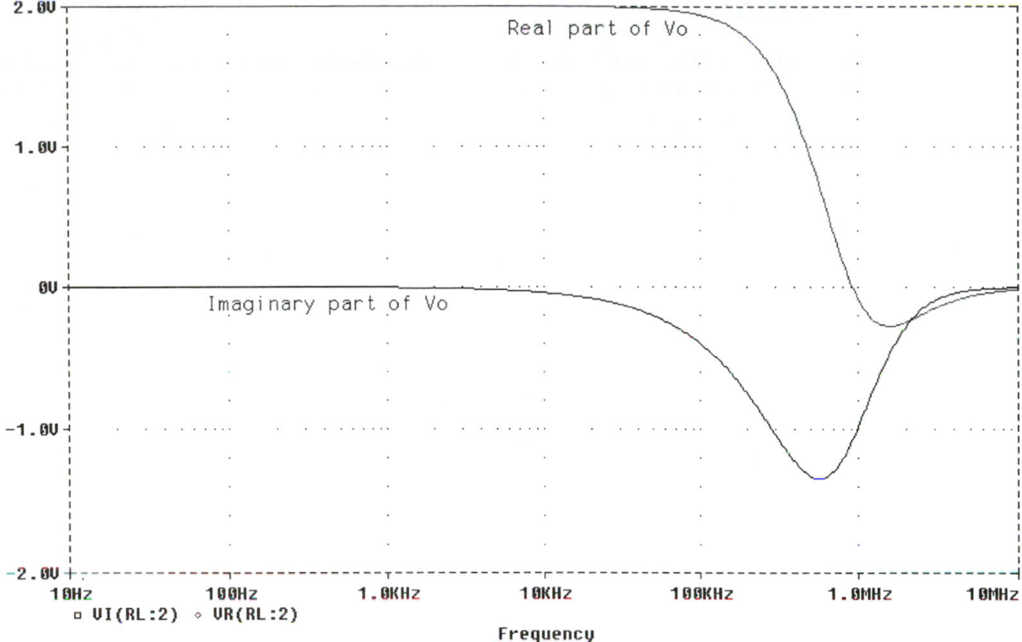

FIGURE 12.30
Frequency response of the op amp in rectangular form (real and imaginary components).

Input Resistance as a Function of Frequency

If we plot V_{in}/I_{in}, a graph of the amplifier input resistance versus frequency results. This is shown in Fig. 12.31. At low frequencies, the input resistance is extremely high, as we would expect. The closed-loop small-signal input resistance begins to decrease more quickly than we might have imagined though. However, in spite of this rapid decrease, R_{inF} is still around 50 MΩ at 10 kHz, approaching the open-loop value of $R_{in} \cong 2$ MΩ at the limit. This is one of the reasons why JFETs, with their intrinsically high input resistance, are used to form the input stage of may op amps.

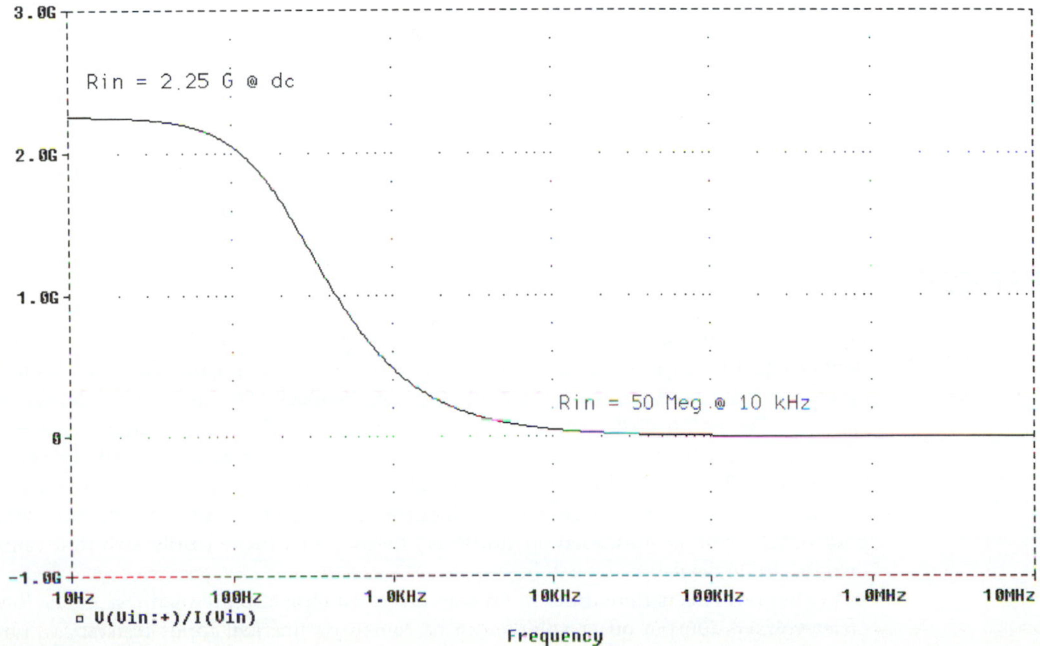

FIGURE 12.31
Input resistance as a function of frequency.

Slew Rate

Slew rate is a large-signal time-domain phenomenon that must be investigated using a transient analysis. As presented earlier, a good working estimate of the slew rate for the 741 is 0.5 V/μs. Setting the frequency of v_{in} to 1 MHz ensures that the maximum rate of change of the expected output greatly exceeds SR for the op amp (6.26 V/μs versus 0.5 V/μs). Plotting the output voltage over a 10-μs time period gives the display in Fig. 12.32. The program accurately simulates start-up transient behavior, which is why the waveform initially has a dc offset, and is drifting toward a mean value of 0 V. Disregarding this transient behavior, we can use the cursors to estimate the slew rate of the op amp to be about 0.4 V/μs, which is fairly close to the estimated value.

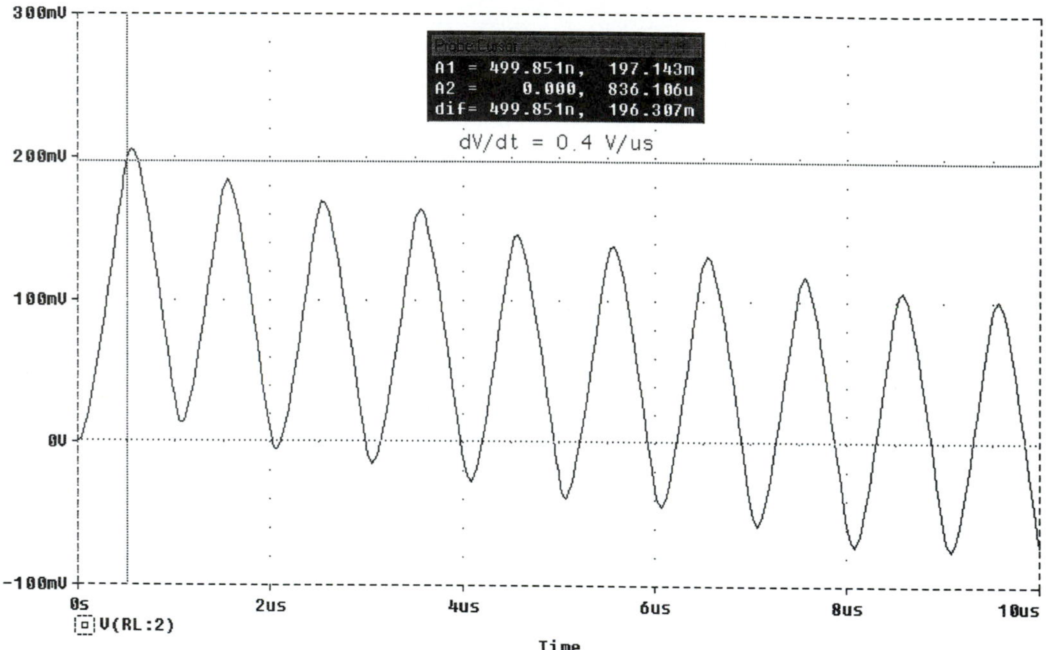

FIGURE 12.32
Slew-limited output voltage.

If you use a digital storage oscilloscope, you can observe transient start-up behavior like that in Fig. 12.22 in the laboratory without too many problems. These kinds of measurements are very difficult to obtain using a normal analog oscilloscope.

■ CHAPTER REVIEW

Operational amplifiers are convenient, easy-to-use gain blocks with frequency response that extends down to dc. Op amps are designed to have very high open-loop gain. The high open-loop gain of the op amp is reduced to a useful level using negative feedback. This also causes changes in other characteristics. The noninverting configuration exhibits very high input resistance, while the input resistance of the inverting configuration is comparatively low. Most op amps have very low open-loop bandwidth. Negative feedback increases bandwidth at the expense of gain. Large-signal frequency response is limited by the power bandwidth of the op amp, which is a function of op amp slew rate and peak output voltage. Cascaded op amps may be used to achieve bandwidth that cannot be obtained using a single device.

Various effects contribute to op amp error, which is usually manifest in the form of an output offset voltage. Output offset voltage can be caused by unequal input resistances, unequal bias currents called input offset current, and by input offset voltages. Some op amps have provisions for offset null adjustment while others require more elaborate external compensation circuitry. The common-mode rejection ratio of most op amps is very high, but decreases with frequency.

DESIGN AND ANALYSIS PROBLEMS

12.1. The amplifier in Fig. 13.2 has A_{OL} = 10 k and β = 0.005. Determine the closed-loop gain of the circuit.

12.2. Based on the conditions stated in Problem 12.1, given v_{in} = 10 sin ωt mV, write the expressions for v_o and v_{id}.

12.3. Refer to Fig. 12.2. Assuming the op amp is ideal, determine the feedback factor required to produce A_v = 80.

12.4. Determine β, A_v, R_{in}, R_o, and BW for the circuit in Fig. 12.33, given A_{OL} = 500, R_{in} = 10 kΩ, R_o = 100 Ω, GBW = 10 MHz, R_F = 50 kΩ, and R_1 = 1 kΩ.

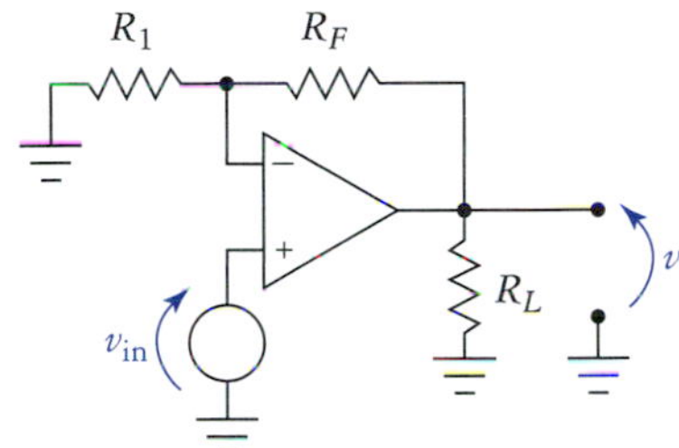

FIGURE 12.33
Circuit for Problems 12.4, 12.5, 12.8, 12.9, 12.12, 12.13, 12.23, 12.24, and 12.29.

12.5. Determine β, A_v, R_{in}, R_o, and BW for the circuit in Fig. 12.33, given A_{OL} = 1000, R_{in} = 20 kΩ, R_o = 20 Ω, GBW = 1 MHz, R_F = 10 kΩ, and R_1 = 2 kΩ.

12.6. Determine β, A_v, R_{in}, R_o, and BW for the circuit in Fig. 12.34, given A_{OL} = 5000, R_{in} = 5 kΩ, R_o = 200 Ω, GBW = 1 MHz, R_F = 10 kΩ, and R_1 = 5 kΩ.

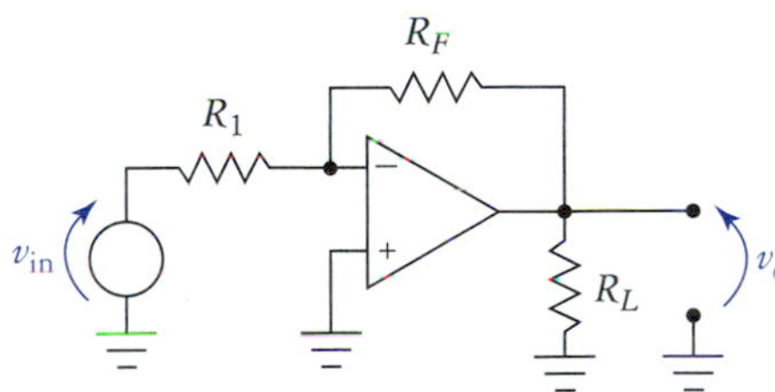

FIGURE 12.34
Circuit for Problems 12.6, 12.7, 12.10, and 12.11.

12.7. Determine β, A_v, R_{in}, R_o, and BW for the circuit in Fig. 12.34, given A_{OL} = 200, R_{in} = 10 kΩ, R_o = 50 Ω, GBW = 4 MHz, R_F = 15 kΩ, and R_1 = 1.5 kΩ.

12.8. Refer to Fig. 12.33. Assuming that the op amp is ideal, determine A_v, i_{R_1}, i_{R_F}, i_L, and v_o, given R_F = 68 kΩ, R_1 = 27 kΩ, R_L = 5 kΩ, and v_{in} = 2 sin (2π 1000t) mV.

12.9. Refer to Fig. 12.33. Assuming that the op amp is ideal, determine A_v, i_{R_1}, i_{R_F}, i_L, and v_o, given; R_F = 12 kΩ, R_1 = 2.2 kΩ, R_L = 1 kΩ, and v_{in} = 2 sin (2π 1000t) V.

12.10. Refer to Fig. 12.34. Assuming that the op amp is ideal, determine A_v, i_{R_1}, i_{R_F}, i_L, and v_o, given R_F = 47 kΩ, R_1 = 10 kΩ, R_L = 5 kΩ, and v_{in} = 2 sin (2π 1000t) mV.

12.11. Refer to Fig. 12.34. Assuming that the op amp is ideal, determine A_v, i_{R_1}, i_{R_F}, i_L, and v_o, given R_F = 4.7 kΩ, R_1 = 1 kΩ, R_L = 5 kΩ, and v_{in} = 2 sin (2π 1000t) V.

12.12. Refer to Fig. 12.33. Given R_F = 330 kΩ, what value should R_1 have to result in A_v = 20? Assume the op amp is ideal.

12.13. Refer to Fig. 12.33. If the op amp has GBW = 1 MHz, what gain is required (via negative feedback) to produce f_C = 10 kHz? Assume that response if flat from dc to f_C.

12.14. A certain noninverting amplifier has BW = 20 kHz when the closed-loop gain is −100. Determine GBW for the op amp. At what frequency will this op amp exhibit A_{OL} = 0 dB? Assume first-order response.

12.15. Determine the overall voltage gain and bandwidth for the circuit in Fig. 12.35. Assume the op amps are AD549Ks with R_F = 90 kΩ and R_1 = 10 kΩ.

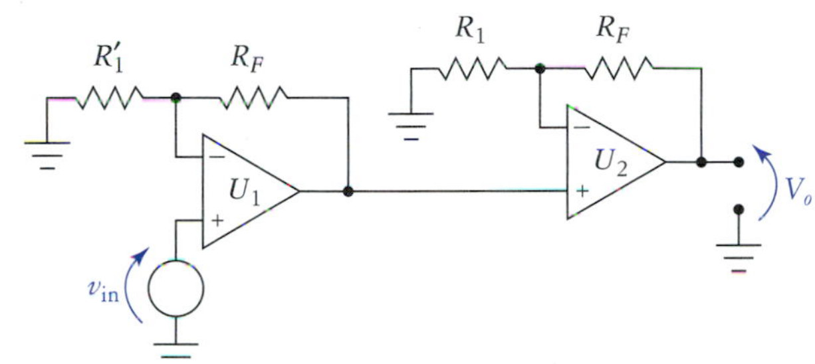

FIGURE 12.35
Circuit for Problems 12.15 to 12.18, 12.21, and 12.22.

12.16. Determine the overall voltage gain and bandwidth for the circuit in Fig. 12.35. Assume the op amps are LF351s with R_F = 90 kΩ and R_1 = 10 kΩ.

12.17. Refer to Fig. 12.35. Determine the value of v_{in} such that v_o = 5 sin(2π 1000t) V.

12.18. Refer to Fig. 12.35. Determine the value of v_{in} such that v_o = 12 sin(2π ωt) V. Assume R_F = 90 kΩ and R_1 = 10 kΩ.

12.19. Determine the overall voltage gain and bandwidth for the circuit in Fig. 12.36. Assume the op amps are AD549Ks with R_F = 10 kΩ and R_1 = 2.5 kΩ.

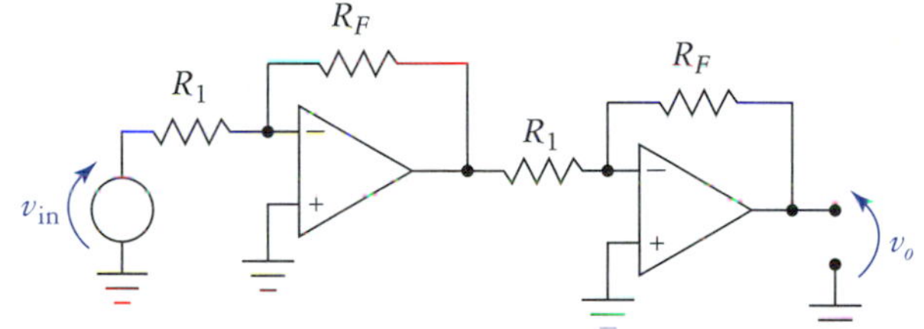

FIGURE 12.36
Circuit for Problems 12.19 and 12.20.

12.20. Determine the overall voltage gain and bandwidth for the circuit in Fig. 12.36. Assume the op amps are LF351s with R_F = 100 kΩ and R_1 = 33.3 kΩ.

12.21. Refer to Fig. 12.35. Determine the value of v_{in} such that v_o = −5 sin (2π 1000t) V. Assume R_F = 12 kΩ and R_1 = 2.2 kΩ.

12.22. Refer to Fig. 12.35. Determine the value of v_{in} such that v_o = 8 sin (2π ft) V. Assume R_F = 6.8 kΩ and R_1 = 1.5 kΩ.

12.23. Refer to Fig. 12.33. The op amp has GBW = 1 MHz, R_F = 9 kΩ, and R_1 = 1 kΩ. Determine the gain magnitude at f = 200 kHz and f = 300 kHz. Assume first-order response characteristics.

12.24. Refer to Fig. 12.33. The op amp has GBW = 4 MHz, R_F = 8 kΩ, and R_1 = 2 kΩ. Determine the gain magnitude at f = 1.6 MHz and f = 3.2 MHz. Assume first-order response characteristics.

12.25. Refer to Fig. 12.34. The op amp has GBW = 1 MHz, R_F = 22 kΩ, and R_1 = 11 kΩ. Determine the gain magnitude at f = 600 kHz and f = 1 MHz. Assume first-order response characteristics.

12.26. Refer to Fig. 12.34. The op amp has GBW = 12 MHz, R_F = 45 kΩ, and R_1 = 15 kΩ. Determine the gain magnitude at f = 6 MHz and f = 7 MHz. Assume first-order response characteristics.

12.27. Refer to Fig. 12.37. Determine the overall gain of the amplifier in decibels. Sketch the asymptotic Bode plot for this circuit assuming 741 op amps are used.

12.28. A certain op amp has A_{OL} = 100 dB at 0 Hz, and f_c = 5 Hz. Determine the numerical and decibel values of A_{OL} at (a) f = 5 Hz, (b) f = 50 Hz, and (c) f = 500 Hz.

12.29. Refer to Fig. 12.33. Assume the op amp has input bias current I_B = 2 nA, I_{io} = 0, R_F = 1 MΩ, and R_1 = 1 MΩ. Determine the output offset voltage V_{oB} that will be produced.

12.30. A certain op amp has SR = 2 V/μs. Determine the power bandwidth for $V_{o(\max)}$ = 12 V.

12.31. A certain op amp has SR = 150 V/μs. Determine the power bandwidth for $V_{o(\max)}$ = 10 V.

12.32. A certain application requires a power bandwidth of 20 kHz with $V_{o(\max)}$ = 10 V. Detemine the minimum required slew rate for the op amp.

12.33. A certain application requires a power bandwidth of 1 MHz with $V_{o(\max)}$ = 5 V. Detemine the minimum required slew rate for the op amp.

12.34. Determine the noise gain at which BW = BW_p for a 741 op amp with $V_{o(\max)}$ = 10 V.

12.35. Determine the noise gain at which BW = BW_p for a LF351 op amp with $V_{o(\max)}$ = 10 V.

12.36. Refer to Fig. 12.38. Op amp U_1 has $V_{io(\max)}$ = 100 μV, R_F = 100 kΩ, R_1 = 4.7 kΩ, and R_{B_2} = 10 Ω. Determine appropriate values for the remaining resistors. Assume V_s = ±15 V.

12.37. Refer to Fig. 12.38. Op amp U_1 has $V_{io(\max)}$ = 1 mV, R_F = 50 kΩ, R_1 = 25 kΩ, and R_{B_2} = 10 Ω. Determine appropriate values for the remaining resistors. Assume V_s = ±15 V.

12.38. Refer to Fig. 12.39. Op amp U_1 has $V_{io(\max)}$ = 5 mV, R_F = 33 kΩ, and R_1 = 3.3 kΩ. Determine appropriate values for the remaining resistors such that the gain error due to the offset compensation circuitry is less than 1%.

12.39. Refer to Fig. 12.39. Op amp U_1 has $V_{io(\max)}$ = 1 mV, R_F = 10 kΩ, and R_1 = 1 kΩ. Determine appropriate values for the remaining resistors such that the gain error due to the offset compensation circuitry is less than 1%.

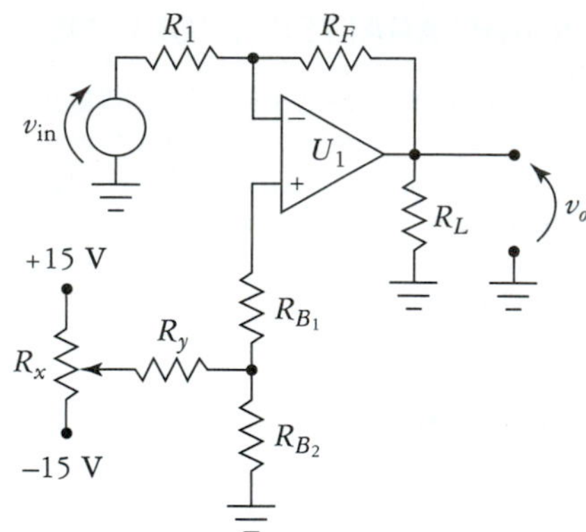

FIGURE 12.38
Circuit for Problems 12.36 and 12.37.

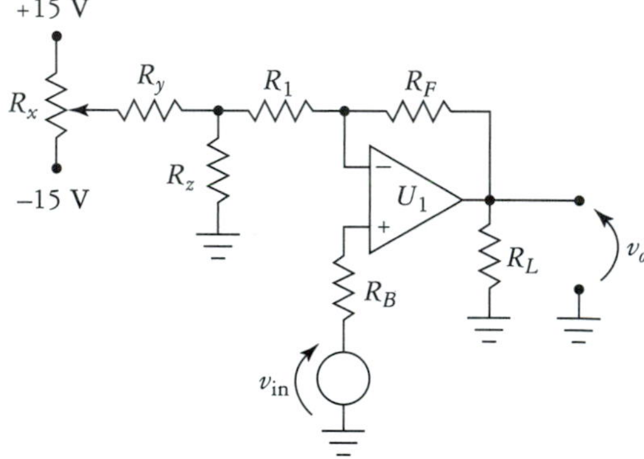

FIGURE 12.39
Circuit for Problems 12.38 to 12.40.

12.40. Refer to Fig. 12.39. Given R_x = 10 kΩ, R_y = 50 kΩ, and R_z = 1 kΩ, determine the maximum and minimum resistance "seen" looking out the left side of R_1.

■ TROUBLESHOOTING PROBLEMS

12.41. A 741 is connected as a noninverting amplifier. Suppose the input voltage source were to become disconnected from the noninverting input terminal. What effect would you expect this to have on the output voltage?

12.42. A voltage follower like that of Fig. 12.3 is driven by a voltage source that has internal resistance of about 10 MΩ. Originally,

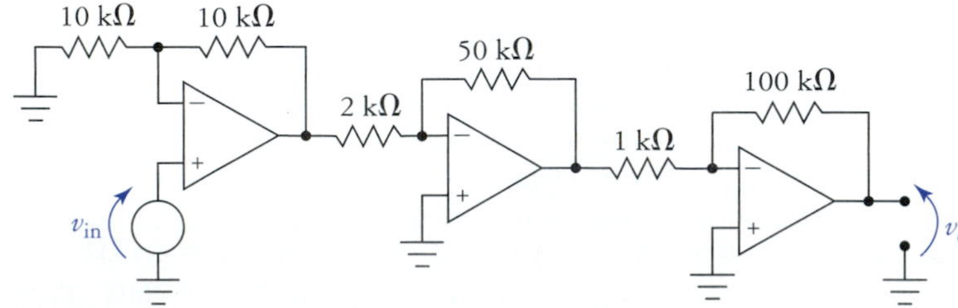

FIGURE 12.37
Circuit for Problem 12.27.

the circuit used an LF351, but this device became defective and someone replaced it with a 741. What problems might this cause in dc signal processing? In ac signal processing?

COMPUTER PROBLEMS

12.43. Use PSpice to generate a plot of R_{in} as a function of frequency for the circuit in Fig. 12.27. (*Hint:* $R = V/I$. Sweep the frequency from 10 Hz to 10 MHz.)

12.44. The circuit in Fig. 12.40 can be used to determine the CMRR of the op amp. The differential gain of this circuit is unity. Set the ac attribute of the signal source to 1 V and sweep the frequency from 10 Hz to 1 MHz. This applies a common-mode voltage that varies in frequency. Figure out a way to produce a Bode plot of the CMRR in decibels.

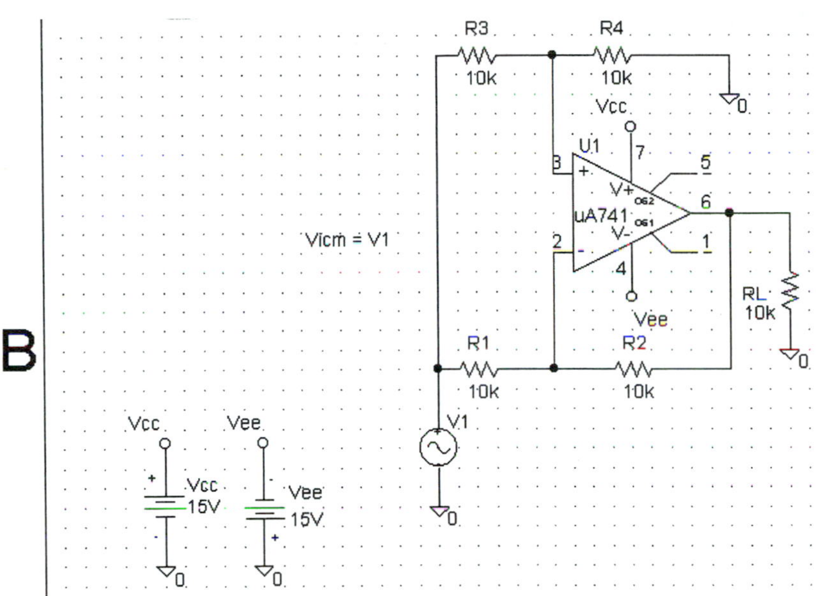

FIGURE 12.40
Circuit for Problem 12.44.

CHAPTER 13

Linear Op Amp Applications

BEHAVIORAL OBJECTIVES

On completion of this chapter, the student should be able to:

- Design and analyze summing amplifiers.
- Determine the bandwidth of a summing amplifier.
- Analyze op-amp-based differential amplifiers.
- Explain the operation and uses of instrumentation amplifiers.
- Analyze op amp VCIS, ICVS, ICIS, and VCVS circuits.
- Describe how to increase op amp output current source/sink capability.
- Explain how op amps can be used to implement adders and subtractors.
- Explain the operation of a differentiator.
- Explain the operation of an integrator.
- Graph a complex waveform using Matlab or a TI-81 graphics calculator.

INTRODUCTION

Op amps are used in tens of thousands of applications. This chapter covers a representative sampling of linear operational amplifier applications. The use of the op amp in summing amplifier, voltage-controlled current source, current-controlled voltage source, and differential amplifier circuits is presented. Instrumentation amplifier circuits and applications are also introduced. Operational amplifiers are still commonly used to generate continuous derivatives and integrals, so these linear applications are also presented. The effects of op amp bandwidth and CMRR limitations in all of these applications are analyzed as well.

13.1 THE SUMMING AMPLIFIER

A summing amplifier is an amplifier whose output is proportional to the sum of the signals applied to its inputs. Summing amplifiers are used in applications where a linear mixing of several signals is required. A common application for a summing amplifier is in music recording, where the signals produced by various musical instruments and voices must be combined and processed to produce a CD.

The block diagram symbol for a summer is shown in Fig. 13.1. Each input may be assigned an arbitrary gain k. The output voltage is given by

$$v_o = k_1 v_1 + k_2 v_2 + \ldots + k_n v_n \tag{13.1}$$

The individual gain k applied to a given input is often called the *weight* of that input. If no weighting is indicated then the gains are assumed to be unity. In principle, there is no limit to the number of inputs that can be summed.

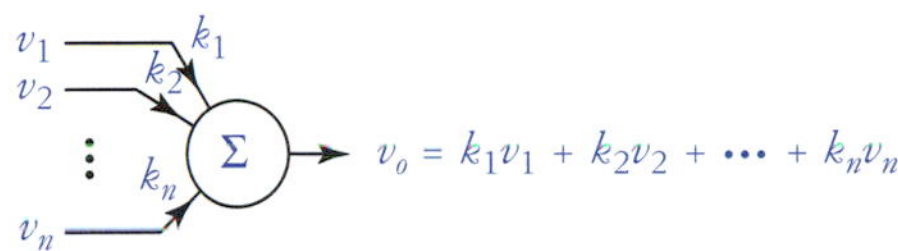

FIGURE 13.1
Block diagram symbol for a summer.

Summing Amplifier Output Voltage Equation

Note
Audio "mixers" are actually variable gain summers, used to combine multiple signal sources such as microphones and guitars.

A two-input op amp summer is shown in Fig. 13.2. Assuming the op amp is ideal, recall that the inverting op amp, with feedback, will drive its output such that $v_{id} = 0$ V. This forces the inverting input of the op amp to become a virtual ground point, which allows us to write

$$i_1 = \frac{v_1}{R_1} \tag{13.2}$$

and

$$i_2 = \frac{v_2}{R_2} \tag{13.3}$$

Since no current enters or leaves the inverting input terminal we can write

$$i_F = i_1 + i_2 \tag{13.4}$$

The output voltage is

$$v_o = -i_F R_F \tag{13.5}$$

Substituting Eqs. (13.1), (13.2), and (13.3) into (13.5) produces

$$v_o = -v_1 \frac{R_F}{R_1} - v_2 \frac{R_F}{R_2} \tag{13.6}$$

If we factor out -1 the summing aspect of the operation becomes more apparent:

$$v_o = -\left(v_1 \frac{R_F}{R_1} + v_2 \frac{R_F}{R_2}\right) \tag{13.7}$$

We can sum additional inputs by adding more input resistors as needed. Because the op amp maintains virtual ground on the inverting input terminal, we simply include an additional term in Eq. (13.7) for each additional input.

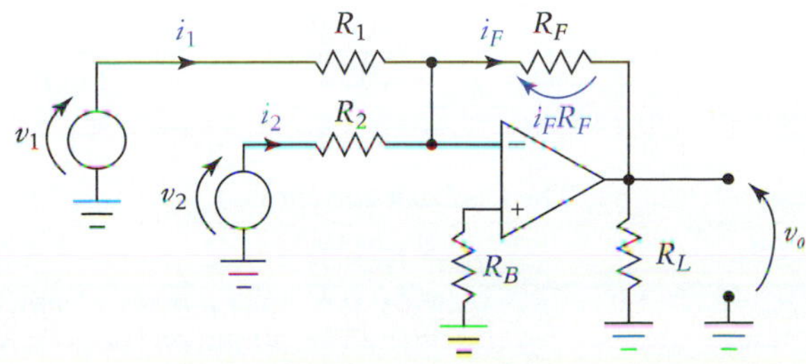

FIGURE 13.2
Two-input op amp summer.

The fact that the op amp maintains virtual ground is the main reason why the inverting configuration works so well as a summer. The presence of the virtual ground prevents a source interaction from occurring, which is a very desirable characteristic. It is a minor inconvenience that the op amp summer inverts the input voltages. However, usually this does not matter, and if it does, a second inverter can be used to change the sign of the output again.

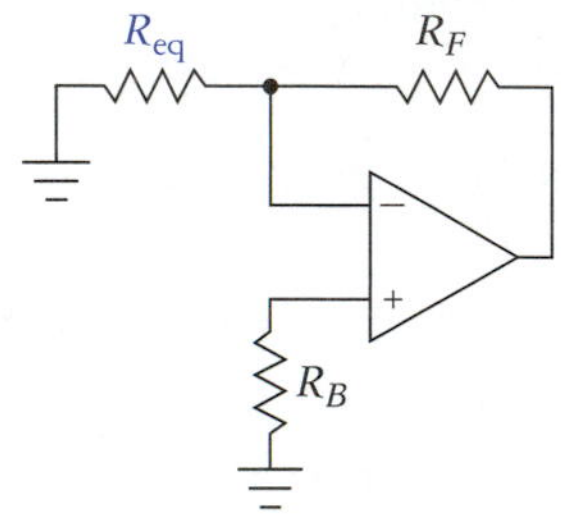

FIGURE 13.3
Summer with simplified input resistor network.

Some applications may require the use of the bias compensation resistor R_B. As before, this resistor is set equal to the equivalent dc resistance seen looking out of the inverting input terminal. In general, for a summing amplifier with n inputs this is

$$R_B = R_F \| R_1 \| R_2 \| \ldots \| R_n \tag{13.8}$$

Bandwidth of the Summing Amplifier

Just like the other amplifier configurations, the bandwidth of the summing amplifier is a function of noise gain and the op amp GBW. The noise gain is determined using the equivalent circuit of Fig. 13.3 where

$$A_N = \frac{R_F + R_{eq}}{R_{eq}} \tag{13.9}$$

The equivalent resistance in this case is

$$R_{eq} = R_1 \| R_2 \| \ldots \| R_n \tag{13.10}$$

A few examples will help tie all of this information together.

EXAMPLE 13.1

Refer to Fig. 13.4. The op amp has GBW = 1 MHz. The input voltages are $v_1 = +50$ mV, $v_2 = -150$ mV, and $v_3 = 200 \sin(2\pi 1000t)$ mV. The resistors are $R_1 = 1$ kΩ, $R_2 = 2$ kΩ, $R_3 = 4$ kΩ, and $R_F = 100$ kΩ. Determine I_1, I_2, i_3, i_F, v_o, BW, and R_B. Sketch the resulting output voltage waveform.

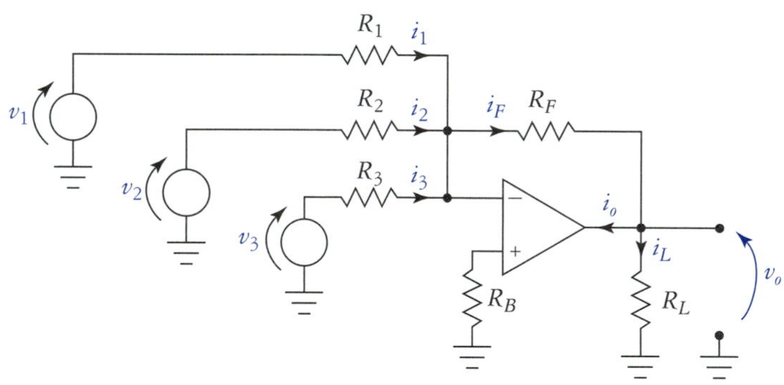

FIGURE 13.4
Circuit for Example 13.1.

Solution The input currents are found as follows:

$$I_1 = \frac{V_1}{R_1} = \frac{50\text{ mV}}{1\text{ k}\Omega} = 50\ \mu\text{A} \qquad I_2 = \frac{V_2}{R_2} = \frac{-150\text{ mV}}{2\text{ k}\Omega} = -75\ \mu\text{A} \qquad i_3 = \frac{v_3}{R_3} = \frac{200}{4\text{ k}\Omega}\sin(2\pi 1000t)\text{ mV} = 50\sin(2\pi 1000t)\ \mu\text{A}$$

The feedback current is

$$\begin{aligned} i_F &= I_1 + I_2 + i_3 \\ &= 50\ \mu\text{A} - 75\ \mu\text{A} + 50\sin(2\pi 1000t)\ \mu\text{A} \\ &= -25 + 50\sin(2\pi 1000t)\ \mu\text{A} \end{aligned}$$

The output voltage is

$$\begin{aligned} v_o &= -i_F R_F \\ &= -[-25 + 50 \sin(2\pi 1000t)\ \mu\text{A}](100\ \text{k}\Omega) \\ &= 2.5 - 5 \sin(2\pi 1000t)\ \text{V} \end{aligned}$$

The voltage waveform is shown in Fig. 13.5. The bias current compensation resistor is given by

$$\begin{aligned} R_B &= R_F \parallel R_1 \parallel R_2 \parallel R_3 \\ &= 100\ \text{k}\Omega \parallel 1\ \text{k}\Omega \parallel 2\ \text{k}\Omega \parallel 4\ \text{k}\Omega \\ &= 568\ \Omega \end{aligned}$$

The noise-gain equivalent resistance is

$$\begin{aligned} R_{\text{eq}} &= R_1 \parallel R_2 \parallel R_3 \\ &= 1\ \text{k}\Omega \parallel 2\ \text{k}\Omega \parallel 4\text{k}\Omega \\ &= 571\ \Omega \end{aligned}$$

and the noise gain is found to be

$$\begin{aligned} A_N &= \frac{R_F + R_{\text{eq}}}{R_{\text{eq}}} \\ &= \frac{100\ \text{k} + 571}{571} \\ &\cong 176 \end{aligned}$$

Finally, the bandwidth is

$$\begin{aligned} \text{BW} &= \frac{\text{GBW}}{A_N} \\ &= \frac{1\ \text{MHz}}{176} \\ &\cong 5.7\ \text{kHz} \end{aligned}$$

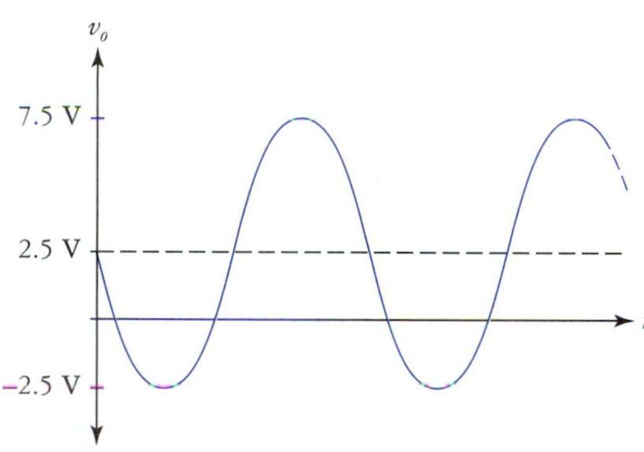

FIGURE 13.5
Output voltage waveform for Example 13.1.

The frequency response analysis of the summing amplifier suggests a method by which the bandwidth of an inverting amplifier can be reduced to any arbitrary value without altering the closed-loop gain. Examine the circuit in Fig. 13.6. The closed-loop voltage gain of the amplifier is $A_v = -R_F/R_1$. Variable resistor R_x has no effect on the closed-loop gain, but it does affect the noise gain. Resistor R_{min} is required to prevent the inverting terminal from being directly grounded. The equivalent noise-gain resistance is given by

$$R_{\text{eq}} = R_1 \parallel (R_x + R_{\text{min}}) \tag{13.11}$$

Why would we want to reduce the bandwidth of an amplifier? Sometimes we do not need an extremely wide bandwidth. If we know that the signal that we wish to amplify is limited to some upper frequency, reducing the bandwidth of the amplifier to the minimum required can reduce interference from higher frequency noise.

Because of the presence of multiple input signals, it is more difficult to quantify the power bandwidth of the summing amplifier in a meaningful way. We can, however, determine the maximum

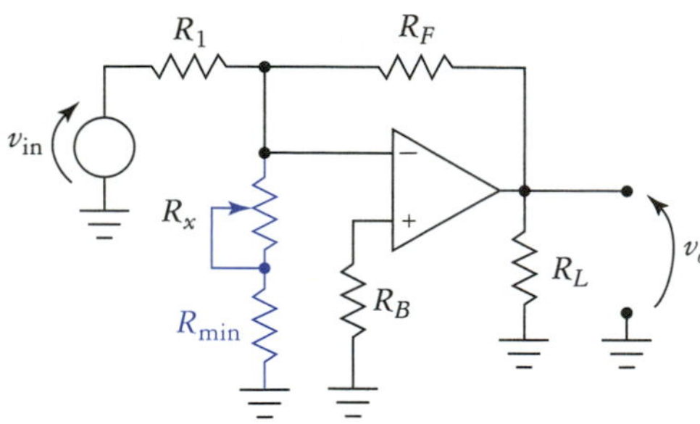

FIGURE 13.6
Changing the bandwidth of the amplifier without affecting closed-loop gain.

rate of change of the output voltage without too much difficulty. Whenever time-varying signals are summed, the net rate of change is the sum of the rates of change of the component signals. That is,

$$\left.\frac{dv_o}{dt}\right|_{\max} = \left.\frac{dv_{o_1}}{dt}\right|_{\max} + \left.\frac{dv_{o_2}}{dt}\right|_{\max} + \ldots + \left.\frac{dv_{o_n}}{dt}\right|_{\max} \tag{13.12}$$

where v_{o_1}, v_{o_2} etc. are the output components associated with the various input sources, as when using superposition. We use Eq. (13.12) to determine whether the maximum rate of change of v_o is greater than SR for the op amp. This is not as difficult as the appearance of Eq. (13.12) might suggest. The next example demonstrates this statement.

EXAMPLE 13.2

Refer to Fig. 13.7. Assume $R_F = 10\ \text{k}\Omega$, $R_1 = 1\ \text{k}\Omega$, $R_2 = 2\ \text{k}\Omega$, $v_1 = 800 \sin(2\pi 1000t)$ mV, and $v_2 = 600 \sin(2\pi 4000t)$ mV. Determine the maximum rate of change of the output voltage.

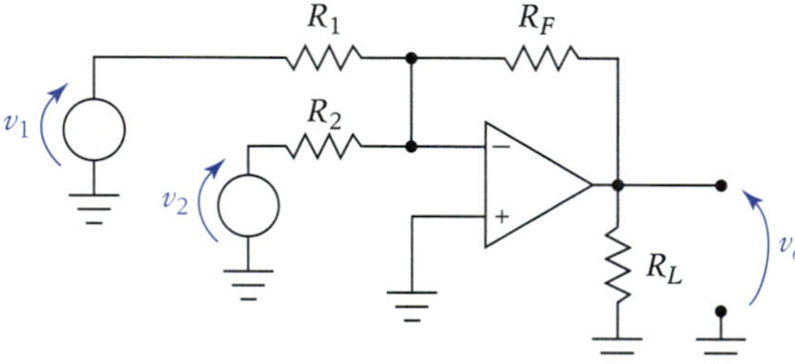

FIGURE 13.7
Circuit for Example 13.2.

Solution Determine the output voltage as follows:

$$\begin{aligned}
v_o &= v_{o_1} + v_{o_2} \\
&= -\left(\frac{R_F}{R_1} v_1 + \frac{R_F}{R_2} v_2\right) \\
&= -\left(\frac{10\ \text{k}\Omega}{1\ \text{k}\Omega} 800 \sin(2\pi 5000t)\ \text{mV} + \frac{10\ \text{k}\Omega}{2\ \text{k}\Omega} 600 \sin(2\pi 8000t)\ \text{mV}\right) \\
&= -8 \sin(2\pi 5000t)\ \text{V} + 3 \sin(2\pi 8000t)\ \text{V}
\end{aligned}$$

This expression cannot be simplified because the signals have different frequencies. The output voltage is simply the instantaneous sum of these two signals, which periodically interfere constructively (producing maximum rate of change) and then destructively. The maximum rates of change are

$$\begin{aligned}
\left.\frac{dv_1}{dt}\right|_{\max} &= \omega_1 V_{p_1} \\
&= (2\pi 5000\ \text{Hz})(8\ \text{V}) \\
&= 16\pi 5000\ \text{V/s} \\
&\cong 0.25\ \text{V}/\mu\text{s}
\end{aligned}
\qquad
\begin{aligned}
\left.\frac{dv_2}{dt}\right|_{\max} &= \omega_2 V_{p_2} \\
&= (2\pi 8000\ \text{Hz})(3\ \text{V}) \\
&= 6\pi 8000\ \text{V/s} \\
&\cong 0.15\ \text{V}/\mu\text{s}
\end{aligned}$$

The net maximum rate of change of the output voltage is

$$\left.\frac{dv_o}{dt}\right|_{\max} = 0.25\ \text{V}/\mu\text{s} + 0.15\ \text{V}/\mu\text{s}$$
$$= 0.4\ \text{V}/\mu\text{s}$$

Averagers

The averager is a special case of the summing amplifier. As you might suspect, the output of an averager is a voltage that is proportional to the average of the input voltages. The arithemetic average of n voltages is

$$v_{\text{AVG}} = \frac{v_1 + v_2 + \cdots + v_n}{n} \tag{13.13}$$

The op amp averager is designed such that the expression for its output voltage is of the same form as Eq. (13.13). Basically, the averager is just a summing amplifier as shown in Fig. 13.8. To design an n-input averager, we start with a summing amplifier, choose some particular R_F, and choose the n input resistor values using

$$R_1 = R_2 = \ldots = R_n = nR_F \tag{13.14}$$

Use of Eq. (13.14) results in an averager with *equally weighted* inputs. Because the op amp summer inverts, the output will have the opposite sign of the true average, but this can be corrected easily if desired.

FIGURE 13.8
An n-input averager.

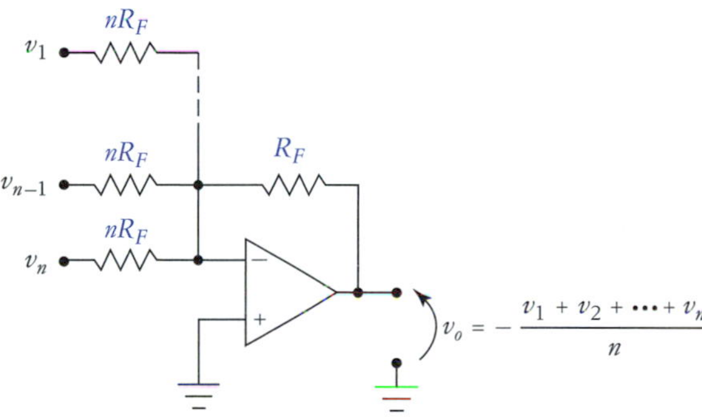

EXAMPLE 13.3

Design a three-input averager given $R_F = 10\ \text{k}\Omega$. Determine V_o given $V_1 = 2.5$ V, $V_2 = -1.5$ V, and $V_3 = 5$ V.

Solution The input resistor values are found using Eq. (13.14):

$$\begin{aligned} R_1 = R_2 = R_3 &= nR_F \\ &= (3)(10\ \text{k}\Omega) \\ &= 30\ \text{k}\Omega \end{aligned}$$

The output voltage can be determined by using standard op amp analysis techniques, or by the use of Eq. (13.13) and multiplying by -1. Since we know that this is an averager, it is easiest to use Eq. (13.13), noting that $V_o = -V_{\text{AVG}}$.

$$\begin{aligned} V_o &= -\frac{V_1 + V_2 + V_3}{3} \\ &= -\frac{2.5\ \text{V} - 1.5\ \text{V} + 5\ \text{V}}{3} \\ &= -2.0\ \text{V} \end{aligned}$$

It is worth the effort to verify that $V_o = -2$ V, using basic op amp analysis techniques.

Occasionally, a weighted average of input variables is required. That is, one or more of the averager's inputs is to have a greater or lesser weight than the other normally weighted inputs. In such a case, the input resistors on the normally weighted inputs are determined using Eq. (13.14). An input that is to have a greater or lesser weighting will have its associated input resistance decreased or increased in inverse proportion to its weighting factor. For example, if in Example 13.3, V_1 were to be weighted twice the normal ($w = 2$), its new value would be given by

$$\begin{aligned} R_w &= nR_F/w \\ &= (3)(10\ \text{k}\Omega)\ /\ 2 \\ &= 15\ \text{k}\Omega \end{aligned} \tag{13.15}$$

If this input were to be weighted at half the normal ($w = 0.5$), its new value would be

$$\begin{aligned} R_w &= nR_F\ /\ w \\ &= (3)(10\ \text{k}\Omega)\ /\ 0.5 \\ &= 60\ \text{k}\Omega \end{aligned}$$

From a circuit analysis standpoint, a weighted averager would require the use of standard op amp analysis methods, because the simple average formula is no longer valid.

Summing Amplifier Variations

It is not uncommon to find signals being applied to summing amplifiers at both the inverting and noninverting input terminals of a given op amp. The analysis of such amplifiers is relatively straightforward and builds on the fundamentals covered thus far. A representative circuit is shown in Fig. 13.9. The output of this circuit can be considered to be the superposition, or sum, of the individual responses as defined below:

$$v_o = v_{o_k} + v_{o_1} + v_{o_2} + \ \dots\ + v_{o_n} \tag{13.16}$$

where v_{o_k} = output due to source v_k, v_{o_1} = output due to source v_1, and so on.

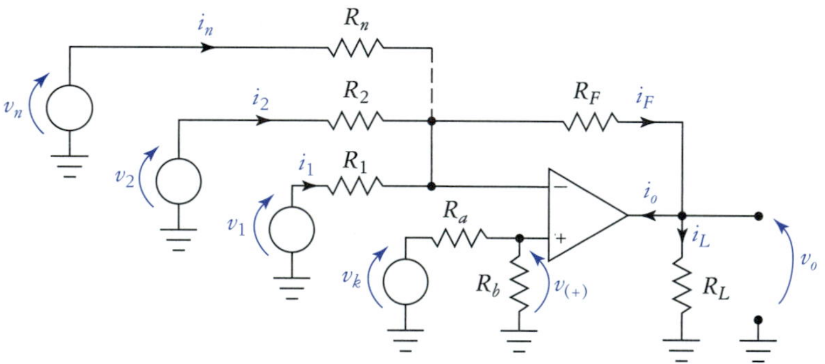

FIGURE 13.9
Applying signals to inverting and noninverting inputs.

The inverting-side responses are derived using superposition and the analysis methods covered previously with source v_k reduced to zero. This is illustrated in Fig. 13.10, where it is clear that the circuit reduces to a simple summing amplifier. We find the resulting output voltage by generalizing Eq. (13.7), producing

$$v_{o(-)} = -\left(v_1 \frac{R_F}{R_1} + v_2 \frac{R_F}{R_2} + \ \dots\ + v_n \frac{R_F}{R_n}\right) \tag{13.17}$$

The response to the signal present at the noninverting side of the amplifier is determined by killing all sources on the inverting side, which produces the equivalent circuit of Fig. 13.11. We apply the usual noninverting gain equation, using the parallel equivalent inverting side resistance and the voltage-divider equation, which yields

$$v_{o(+)} = v_k \left(\frac{R_{\text{eq}} + R_F}{R_{\text{eq}}} \times \frac{R_b}{R_a + R_b}\right) \tag{13.18}$$

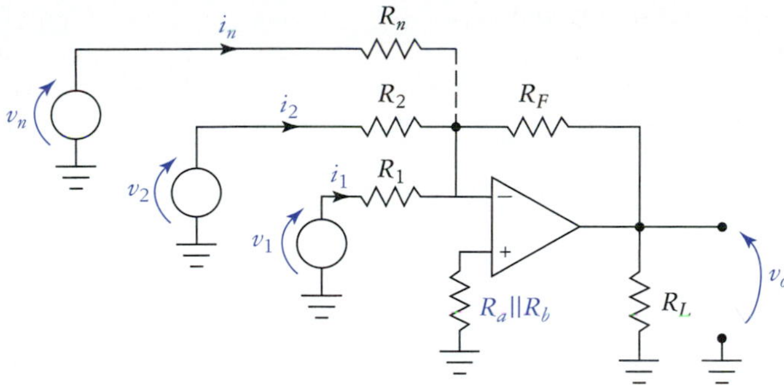

FIGURE 13.10
Analysis of the inverting side.

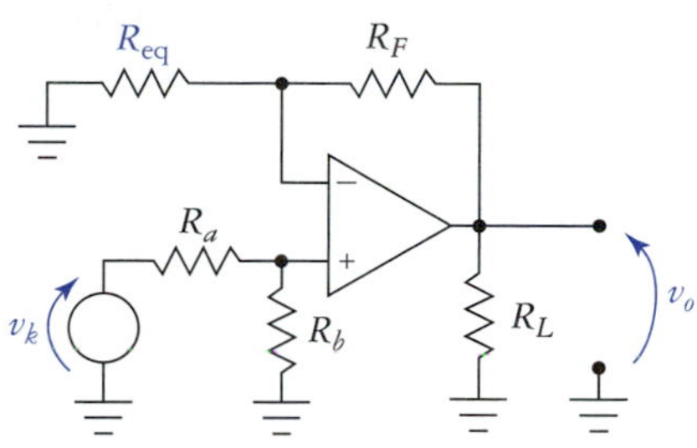

FIGURE 13.11
Analysis of the noninverting side.

where the equivalent resistance is

$$R_{eq} = R_1 \parallel R_2 \parallel \ldots \parallel R_n \tag{13.19}$$

The net output of the amplifier is found by algebraically combining Eqs. (13.17) and (13.18) giving us

$$v_o = v_k\left(\frac{R_{eq} + R_F}{R_{eq}} \times \frac{R_b}{R_a + R_b}\right) - \left(v_1\frac{R_F}{R_1} + v_2\frac{R_F}{R_2} + \ \ldots \ + v_n\frac{R_F}{R_n}\right) \tag{13.20}$$

EXAMPLE 13.4

Refer to Fig. 13.9. Assume that only two inverting-side inputs are used with $R_1 = 2.2\ \text{k}\Omega$, $R_2 = 4.7\ \text{k}\Omega$, $R_a = 10\ \text{k}\Omega$, $R_b = 10\ \text{k}\Omega$, $R_F = 33\ \text{k}\Omega$, $V_1 = -10$ mV, $v_2 = 100 \sin 500t$ mV, and $V_k = +100$ mV. Determine the expressions for i_1, i_2, i_F, and v_o. Sketch the resulting v_o waveform.

Solution The equivalent inverting-side resistance is

$$\begin{aligned} R_{eq} &= R_1 \parallel R_2 \\ &= 2.2\ \text{k}\Omega \parallel 4.7\ \text{k}\Omega \\ &= 1.5\ \text{k}\Omega \end{aligned}$$

and v_o is found using Eq. (13.20):

$$\begin{aligned} v_o &= v_k\left(\frac{R_{eq} + R_F}{R_{eq}} \times \frac{R_b}{R_a + R_b}\right) - \left(v_1\frac{R_F}{R_1} + v_2\frac{R_F}{R_2}\right) \\ &= 100\ \text{mV}\left(\frac{1.5\ \text{k} + 33\ \text{k}}{1.5\ \text{k}} \times \frac{10\ \text{k}}{10\ \text{k} + 10\ \text{k}}\right) \\ &\quad - \left(-10\ \text{mV}\frac{33\ \text{k}}{2.2\ \text{k}} + 100 \sin 500t\ \text{mV}\frac{33\ \text{k}}{4.7\ \text{k}}\right) \\ &= (100\ \text{mV})(23)(0.5) - (-10\ \text{mV})(15) + (100 \sin 500t\ \text{mV})(7) \\ &= 1.15\ \text{V} + 150\ \text{mV} - 700 \sin 500t\ \text{mV} \\ &= 1.3 - 0.7 \sin 500t\ \text{V} \end{aligned}$$

The sketch of v_o is shown in Fig. 13.12. Currents i_1 and i_2 cannot be found by simple application of Eq. (13.2) or (13.3) because the noninverting input is no longer at virtual ground. The voltage at the noninverting input is

$$\begin{aligned} V_{(+)} &= V_k\frac{R_b}{R_a + R_b} \\ &= 100\ \text{mV}\frac{10\ \text{k}\Omega}{20\ \text{k}\Omega} \\ &= 50\ \text{mV} \end{aligned}$$

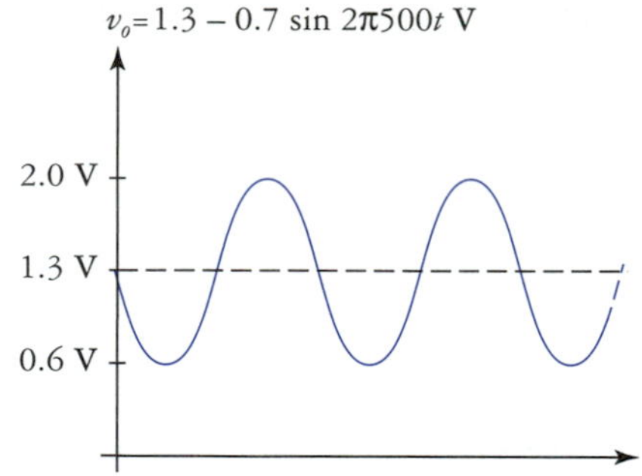

FIGURE 13.12
Waveform for Example 13.4.

The virtual short circuit forces $v_{id} = 0$ V, therefore the voltage at the inverting input is

$$v_{(-)} = v_{(+)}$$

Applying KVL, the currents are

$$\begin{aligned} i_1 &= \frac{v_1 - v_{(-)}}{R_1} \\ &= \frac{-10 \text{ mV} - 50 \text{ mV}}{2.2 \text{ k}\Omega} \\ &= -27.3 \ \mu\text{A} \end{aligned} \qquad \begin{aligned} i_2 &= \frac{v_2 - v_{(-)}}{R_2} \\ &= \frac{100 \sin 500t \text{ mV}}{4.7 \text{ k}\Omega} - \frac{50 \text{ mV}}{4.7 \text{ k}\Omega} \\ &= -10.6 + 21.3 \sin 500t \ \mu\text{A} \end{aligned}$$

The feedback resistor current is

$$\begin{aligned} i_F &= i_1 + i_2 \\ &= -27.3 \ \mu\text{A} + (-10.6 \ \mu\text{A} + 21.3 \sin 500t \ \mu\text{A}) \\ &= -37.9 + 21.3 \sin 500t \ \mu\text{A} \end{aligned}$$

Again, it is a good idea to check the consistency of this result. Be careful though! We cannot simply apply Eq. (13.5) because this equation was derived under virtual ground conditions. We must apply KVL to modify Eq. (13.5) giving us

$$\begin{aligned} v_o &= v_{(-)} - i_F R_F \\ &= 50 \text{ mV} - (-37.9 + 21.3 \sin 500t \ \mu\text{A})(33 \text{ k}\Omega) \\ &= 50 \text{ mV} + 1.25 - 0.7 \sin 500t \text{ V} \\ &= 1.3 - 0.7 \sin 500t \text{ V} \end{aligned}$$

The result checks.

13.2 DIFFERENTIAL AND INSTRUMENTATION AMPLIFIERS

It is reasonable to assume that since the op amp is designed around the differential amplifier, the op amp itself can be used as a differential amplifier. Op amp differential amplifiers generally behave almost ideally and find use in many areas, especially measurement applications.

Basic Differential Amplifier Analysis

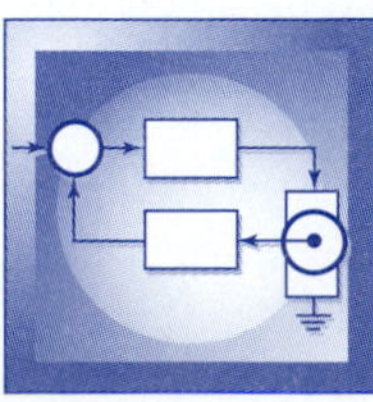

The differential amplifier shown in Fig. 13.13 is really just a special case of the summing amplifier that was shown in Fig. 13.9. To operate correctly as a diff amp, we require $R_1' = R_1$ and $R_F' = R_F$. Making these substitutions into Eq. (13.20)

$$v_o = v_2\left(\frac{R_1' + R_F'}{R_1} \times \frac{R_F'}{R_1' + R_F'}\right) - v_1 \frac{R_F}{R_1} \tag{13.21}$$

Now, since $R_F' = R_F$ and $R_1' = R_1$, this reduces to

$$v_o = v_2 \frac{R_F'}{R_1} - v_1 \frac{R_F}{R_1} \tag{13.22}$$

Factoring this produces the standard form of the diff amp output equation

$$v_o = \frac{R_F}{R_1}(v_2 - v_1) \tag{13.23}$$

Note
Differential amplifiers can be viewed as subtractors, depending on the context of the application.

The output of the differential amplifier is proportional to the difference between the two input voltages. An alternative way of expressing Eq. (13.23) is the familiar equation

$$v_o = A_d v_{id} \tag{13.24}$$

where $A_d = R_F/R_1$ and $v_{id} = v_2 - v_1$.

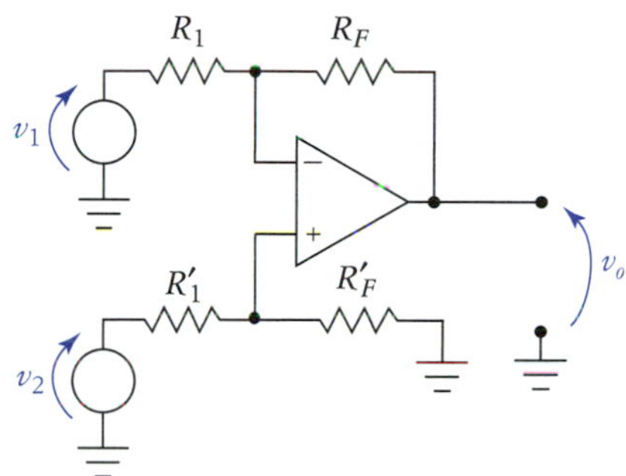

FIGURE 13.13
Op amp closed-loop differential amplifier.

Common-Mode Response

Common-mode response and common-mode rejection are a recurring theme in this book because differential amplifiers are such an important class of circuits. We know that ideally the output of a differential amplifier is zero when equal voltages are present at the input terminals; that is, $v_o = 0$ V when $v_{id} = 0$ V. For a differential amplifier like Fig. 13.13, the only time the common-mode input voltage will be zero is when $v_1 = v_2 = 0$ V, or when $v_1 = -v_2$. A real differential amplifier will produce an output in response to the common-mode input. The response of the amplifier to a common-mode voltage is given by the now familiar equation

$$V_{ocm} = A_{cm} v_{icm}$$

and we still define the common-mode rejection ratio as

$$\text{CMRR} = 20 \log \left| \frac{A_d}{A_{cm}} \right|$$

As far as a circuit like Fig. 13.13 goes, the overall CMRR will be somewhat less than that of the op amp around which it is built. The 741, for example, has a typical CMRR of 90 dB, and so a diff amp built using a 741 would be expected to have CMRR $<$ 90 dB. Other op amps are available with CMRR ratings of $>$110 dB. For diff amps like that of Fig. 13.13, assuming that the op amp itself has a very high CMRR, the main limiting factor, in terms of CMRR is the matching between the $R'_1 - R_1$ and $R'_F - R_F$ pairs. Differential amplifiers are often implemented using hybrid IC techniques where laser-trimmed precision resistors make high CMRR possible.

EXAMPLE 13.5

The circuit in Fig. 13.13 has the following component values: $R'_1 = R_1 = 10\ \text{k}\Omega$ and $R'_F = R_F = 270\ \text{k}\Omega$. Given $V_1 = +2.00$ V and $V_2 = +2.05$ V, determine V_o for the following CMRR values. Assume that the common-mode gain is positive.

(a) CMRR = ∞
(b) CMRR = 100 dB
(c) CMRR = 50 dB

Solution The differential gain is

$$\begin{aligned} A_d &= R_F/R_1 \\ &= 270\ \text{k}\Omega/10\ \text{k}\Omega \\ &= 27 \end{aligned}$$

The differential and common-mode input voltages are

$$V_{id} = V_2 - V_1 = 2.05\text{ V} - 2.00\text{ V} = 50\text{ mV}$$

$$V_{icm} = \frac{V_1 + V_2}{2} = \frac{2.00\text{ V} + 2.05\text{ V}}{2} = 2.025\text{ V}$$

Assuming all other error sources have been nulled, the output voltage is given by

$$V_o = A_d V_{id} + A_{cm} V_{icm}$$

(a) CMRR $= \infty$: This implies $A_{cm} = 0$, therefore

$$\begin{aligned} V_o &= (27)(50\text{ mV}) + (0)(2.035\text{ V}) \\ &= 1.35\text{ V} \end{aligned}$$

(b) CMRR $= 100$ dB: The common-mode gain is

$$\begin{aligned} A_{cm} &= \frac{A_d}{\log^{-1}\left(\frac{\text{CMRR}}{20}\right)} \\ &= \frac{27}{\log^{-1}(5)} \\ &= 260 \times 10^{-6} \end{aligned}$$

The net output voltage is

$$\begin{aligned} V_o &= (27)(50\text{ mV}) + (260 \times 10^{-6})(2.035\text{ V}) \\ &= 1.35\text{ V} + 529\ \mu\text{V} \\ &\cong 1.35\text{ V} \end{aligned}$$

(The voltage is actually $V_o = 1.350529$ V, but since the most accurate input voltage has only three significant figures, this seven-place precision is meaningless.)

(c) CMRR $= 50$ dB: The common-mode gain is $A_{cm} = 0.085$. The output voltage in this case is

$$\begin{aligned} V_o &= (27)(50\text{ mV}) + (0.085)(2.035\text{ V}) \\ &\cong 1.35\text{ V} + 170\text{ mV} \\ &= 1.52\text{ V} \end{aligned}$$

The error in this case is almost 13%, which would most likely be unacceptable.

The preceding example demonstrates the importance of high CMRR in differential amplifier applications. It is not unusual for common-mode voltages to greatly exceed differential signals in practical situations.

Common-mode voltages can also be time varying in nature. The differential amplifier should also reject this type of interference. As pointed out in Chapters 8 and 12, CMRR decreases with frequency. This should be kept in mind when a differential amplifier is used in a "noisy" environment.

Instrumentation Amplifiers

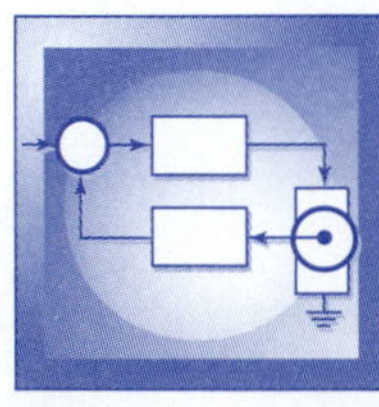

An *instrumentation amplifier* (IA) is basically a buffered-input differential amplifier, like that shown in Fig. 13.14. Op amps A_1 and A_2 are unity-gain buffers, while A_3 is a basic diff amp. The output voltage for this circuit is given by Eq. (13.23). The op amps used in these circuits are designed to have superior CMRR, offset current, offset voltage, bias current, and temperature drift specifications. Instrumentation amplifiers are usually implemented using hybrid IC technology, where accurate laser-trimmed resistors may be employed.

Variable-Gain Instrumentation Amplifiers

In theory, it would be possible to vary the gain of the IA in Fig. 13.14 by simultaneously varying the resistors in pairs: $R_1 - R_1'$ or $R_F - R_F'$. In practice, it is too difficult to maintain accurate matching,

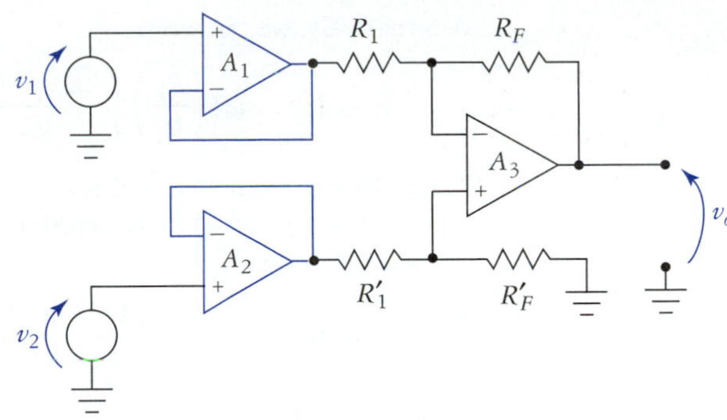

FIGURE 13.14
Buffered diff amp is called an instrumentation amplifier.

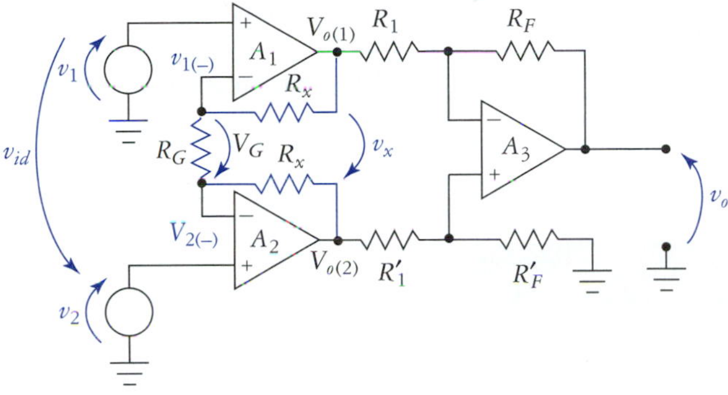

FIGURE 13.15
Variable-gain instrumentation amplifier.

and CMRR decreases dramatically. One of several possible solutions to this problem is the circuit of Fig. 13.15. In this circuit, the buffers have equal-value resistors R_x in their feedback loops, and resistor R_G is varied to adjust the gain. Using this single resistor to provide gain adjustment eliminates tracking problems that reduce CMRR.

Output Voltage Expression Derivation

The virtual short-circuit property allows us to write $v_{1(-)} = v_1$ and $v_{2(-)} = v_2$. Thus, the voltage drop across R_G is

$$v_G = v_2 - v_1 = v_{id} \tag{13.25}$$

The differential input to A_3 is

$$v_x = v_{o(2)} - v_{o(1)}$$

Now, since there is no current flow into or out of the input terminals of an op amp, we can express v_G in terms of $v_{o(1)}$ and $v_{o(2)}$ using the voltage-divider relationship. That is,

$$\begin{aligned} v_G &= (v_{o(2)} - v_{o(1)}) \frac{R_G}{2R_x + R_G} \\ &= v_x \frac{R_G}{2R_x + R_G} \end{aligned}$$

Solving for v_x,

$$v_x = \frac{v_{id}(2R_x + R_G)}{R_G} \tag{13.26}$$

The output of A_3 is

$$v_o = \frac{R_F}{R_1} v_x \tag{13.27}$$

Substituting Eq. (13.26) into (13.27), we obtain

$$v_o = v_{id}\left(\frac{R_F}{R_1}\right)\left(\frac{2R_x + R_G}{R_G}\right) \tag{13.28}$$

Alternatively, we can write

$$v_o = (v_2 - v_1)\left(\frac{R_F}{R_1}\right)\left(\frac{2R_x + R_G}{R_G}\right) \tag{13.29}$$

The LH0036, from National Semiconductor, is a hybrid IC instrumentation amplifier that uses the topology of Fig. 13.15. The LH0036 has $R_F = R_1$ and $R_x = 25$ kΩ. Inserting these values into Eq. (13.29) produces

$$v_o = (v_2 - v_1)\left(\frac{50\text{ k}\Omega + R_G}{R_G}\right)$$

or, equivalently,

$$v_o = (v_2 - v_1)\left(1 + \frac{50\text{ k}\Omega}{R_G}\right) \tag{13.30}$$

which is the equation presented in the LH0036 data sheet.

EXAMPLE 13.6

Refer to Fig. 13.15. Given $R_F' = R_F = 100$ kΩ, $R_1' = R_1 = R_x = 10$ kΩ, and assuming CMRR $\rightarrow \infty$, determine:

a. R_G such that $A_d = 50$.
b. V_o, given $V_1 = 3.250$ V, $V_2 = 3.155$ V.
c. V_o, given $V_1 = 200\angle 30°$ mV, $V_2 = 210\angle 60°$ mV.

Solution (a) The differential gain is v_o/v_{id}, so by Eq. (13.28),

$$A_d = \left(\frac{R_F}{R_1}\right)\left(\frac{2R_x + R_G}{R_G}\right)$$

Solving for R_G we have

$$R_G = \frac{2R_x}{\dfrac{A_d R_1}{R_F} - 1} \tag{13.31}$$

$$= \frac{20\text{ k}\Omega}{\dfrac{(50)(10\text{ k}\Omega)}{100\text{ k}\Omega} - 1}$$

$$= 5\text{ k}\Omega$$

(b) Since we already know that $A_d = 50$ we can simply write

$$\begin{aligned} V_o &= A_d(V_2 - V_1) \\ &= 50(3.155\text{ V} - 3.250\text{ V}) \\ &= -4.75\text{ V} \end{aligned}$$

(c) In this case, we have two polar-form voltages, but the basic differential relationship is still valid. Since we are subtracting, it is most convenient to express the voltages in rectangular form, which is $V_1 + 173 = j100$ mV and $V_2 = 105 + j182$ mV. Now, we have

$$\begin{aligned} V_o &= A_d(V_2 - V_1) \\ &= 50[(105 + j182\text{ mV}) - (173 + j100\text{ mV})] \\ &= 50(68 + j82\text{ mV}) \\ &= 50(107\angle 50.3°\text{ mV}) \\ &= 5.35\angle 50.3°\text{ V} \end{aligned}$$

Notice that even though phasor voltages V_1 and V_2 differ only by a small amount in magnitude (10 mV), the phase difference has caused the magnitude of the differential input to be relatively large (107 mV).

An Instrumentation Amplifier Application

A very common use of the instrumentation amplifier is in the processing of the output of a *bridge circuit.* An example is shown in Fig. 13.16. In this particular case the bridge has two matched *thermistors* located in opposite arms. Recall that thermistors are temperature-dependent resistors. When we use a component such as a thermistor to convert some nonelectrical quantity into a proportional electrical signal, that component is called a *transducer.* The opposite-arm thermistor arrangement doubles the sensitivity of the bridge, compared to using a single thermistor.

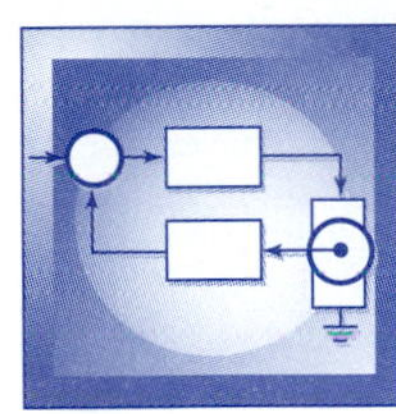

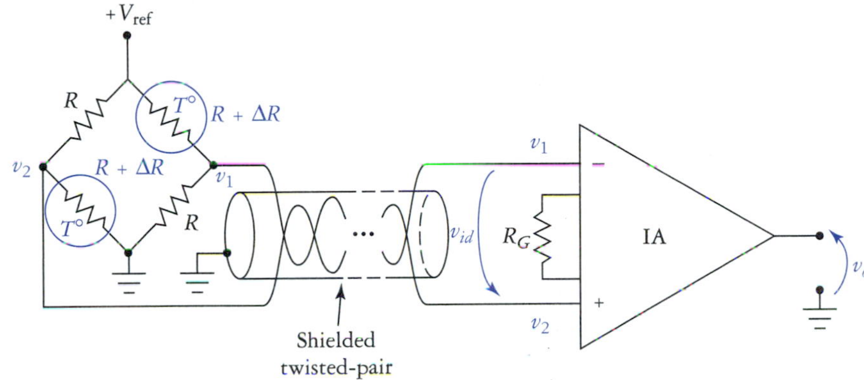

FIGURE 13.16
Using an instrumentation amplifier to process the output of a bridge.

Note
Transducers are devices that convert a non-electrical quantity such as force, temperature, or motion into a voltage or current.

At some particular temperature, say, 25°C, let us assume that the bridge is *in balance.* This means that all resistors have the same value ($\Delta R = 0\ \Omega$). Under these conditions, $V_1 = V_2$, and $V_{id} = 0$ V. Should the temperature change, then ΔR will also change, unbalancing the bridge, causing $V_1 \neq V_2$ and producing an output voltage from the IA.

Circuits like that of Fig. 13.16 are frequently used to provide feedback in process control applications. Because of the extremely high input resistance of the instrumentation amplifier (R_{in} is typically $> 1000\ \mathrm{M\Omega}$), loading of the bridge is essentially nonexistent. When the bridge is in balance $V_1 = V_2 = V_{ref}/2$. This is a common-mode voltage at the input of the amplifier. This is why CMRR is a very important instrumentation amplifier specification. It is interesting to note that when we place matched thermistors (or other transducers) in opposite arms of the bridge that the common-mode voltage remains constant. This can be shown as follows. The voltages at the corners of the bridge are

$$V_1 = V_{ref}\frac{R}{2R + \Delta R} \tag{13.32}$$

$$V_2 = V_{ref}\frac{R + \Delta R}{2R + \Delta R} \tag{13.33}$$

The common-mode input voltage is defined as

$$V_{icm} = \frac{V_1 + V_2}{2} \tag{13.34}$$

Substituting Eqs. (13.32) and (13.33) into (13.34) yields

$$\begin{aligned} V_{icm} &= \frac{V_{ref}\left(\dfrac{R}{2R + \Delta R} + \dfrac{R + \Delta R}{2R + \Delta R}\right)}{2} \\ &= \left(\frac{V_{ref}}{2}\right)\left(\frac{2R + \Delta R}{2R + \Delta R}\right) \\ &= \frac{V_{ref}}{2} \end{aligned}$$

EXAMPLE 13.7

Refer to Fig. 13.16. At 25°C, $R = 10.000\ \text{k}\Omega$ and $\Delta R = 0\Omega$. The bridge uses NTC (negative temperature coefficient) thermistors that vary approximately by TC $= -0.5\%/°\text{C}$. The IA is set for $A_d = 40.32$ and $V_{\text{ref}} = 10.000$ V. Assuming CMRR $\rightarrow \infty$, determine V_o at the following temperatures:

(a) $T = 20°\text{C}$
(b) $T = 25°\text{C}$
(c) $T = 30°\text{C}$

Solution The basic strategy applied to solve this problem is as follows. For each of the temperatures given, voltages V_1 and V_2 can be determined by simply applying the voltage-divider equation to both halves of the bridge (Eqs. 13.31 and 13.32). Before finding these voltages, however, in each case we must first determine the resistance of the thermistor at the temperatures of interest.

$$\begin{aligned}\text{(a) } \Delta T &= 20°\text{C} - 25°\text{C}\\ &= -5°\text{C}\end{aligned}$$

The corresponding change in resistance is

$$\begin{aligned}\Delta R &= (\text{TC})(R)(\Delta T)\\ &= (-0.005)(10.000\ \text{k}\Omega)(-5°\text{C})\\ &= 250\ \Omega\end{aligned}$$

Thus, the voltages are

$$\begin{aligned}V_1 &= V_{ref}\frac{R}{2R + \Delta R}\\ &= 10.000\ V\frac{10.000\ \text{k}\Omega}{20.000\ \text{k}\Omega + 250\ \Omega}\\ &= 4.938\ \text{V}\end{aligned}$$

$$\begin{aligned}V_2 &= V_{ref}\frac{R + \Delta R}{2R + \Delta R}\\ &= 10.000\ \text{V}\frac{10.000\ \text{k}\Omega + 250\ \Omega}{20.000\ \text{k}\Omega + 250\ \Omega}\\ &= 5.062\ \text{V}\end{aligned}$$

The output of the IA is

$$\begin{aligned}V_o &= A_d V_{id}\\ &= A_d(V_2 - V_1)\\ &= 40.32(5.062\ \text{V} - 4.938\ \text{V})\\ &= 4.999\ \text{V}\end{aligned}$$

(b) In this case, $T = T_o = 25°\text{C}$, the bridge is in balance and $V_{id} = 0.000$ V; therefore,

$$V_o = 0.000\ \text{V}.$$

$$\begin{aligned}\text{(c) } \Delta T &= 30°\text{C} - 25°\text{C}\\ &= 5°\text{C}\end{aligned}$$

The corresponding change in resistance is

$$\begin{aligned}\Delta R &= (\text{TC})(R)(\Delta T)\\ &= (-0.005)(10.000\ \text{k}\Omega)(5°\text{C})\\ &= -250\ \Omega\end{aligned}$$

Thus, the voltages are

$$\begin{aligned}V_1 &= V_{\text{ref}}\frac{R}{2R + \Delta R}\\ &= 10.000\ \text{V}\frac{10.000\ \text{k}\Omega}{20.000\ \text{k}\Omega - 250\ \Omega}\\ &= 5.063\ \text{V}\end{aligned}$$

$$V_2 = V_{\text{ref}} \frac{R + \Delta R}{2R + \Delta R}$$

$$= 10.000 \text{ V} \frac{10.000 \text{ k}\Omega - 250 \ \Omega}{20.000 \text{ k}\Omega - 250 \text{ k}\Omega}$$

$$= 4.937 \text{ V}$$

The output of the IA is

$$\begin{aligned} V_o &= A_d V_{id} \\ &= A_d(V_2 - V_1) \\ &= 40.32(4.937 \text{ V} - 5.063 \text{ V}) \\ &= -5.080 \text{ V} \end{aligned}$$

Examining the results of Example 13.7 we find that the output of the IA varies at the very convenient rate of about -1 V/°C. The variation in V_o versus T is approximately linear as long as $\Delta R \ll R$. Also notice that the common-mode input voltage in each case is half of the reference voltage.

Noise Reduction and Guard Drive

Notice in Fig. 13.16 that the bridge is connected to the IA via a shielded, twisted pair of wires. The shield is typically a copper braiding or foil cladding, grounded at one end, that surrounds the twisted pair. This shielding helps prevent pickup of electromagnetic interference, especially in long cable runs. The signal conductors are twisted so that any interference that does penetrate the shielding is picked up equally on each line. This tends to make the interference common-mode noise, which is rejected by the instrumentation amplifier.

A long length of shielded cable can have significant distributed capacitance and resistance in the conductors. These effects can be modeled as shown in Fig. 13.17. If ac common-mode noise is picked up by the twisted pair and $R_{s_1}C_1 \neq R_{s_2}C_2$ then unequal attenuation of this noise occurs in each line. The net effect is the conversion of what was originally common-mode noise into differential-mode noise, which will be amplified greatly by the IA. The solution to this problem is called *guard drive* or *active guarding*, and a common implementation is also shown in Fig. 13.17. The idea here is to force the potential of the shield to track the common-mode noise. This way, there will be no difference in potential, or rate of change of potential, between the shield and the twisted pair (relative to the common-mode noise). Under these conditions there will be no current flow through stray capacitances C_1 and C_2. The common-mode drive voltage is derived from the outputs of the buffer section of the IA. If the stray capacitance of the cable is high, a buffer such as A_4 should be used to prevent loading of the IA.

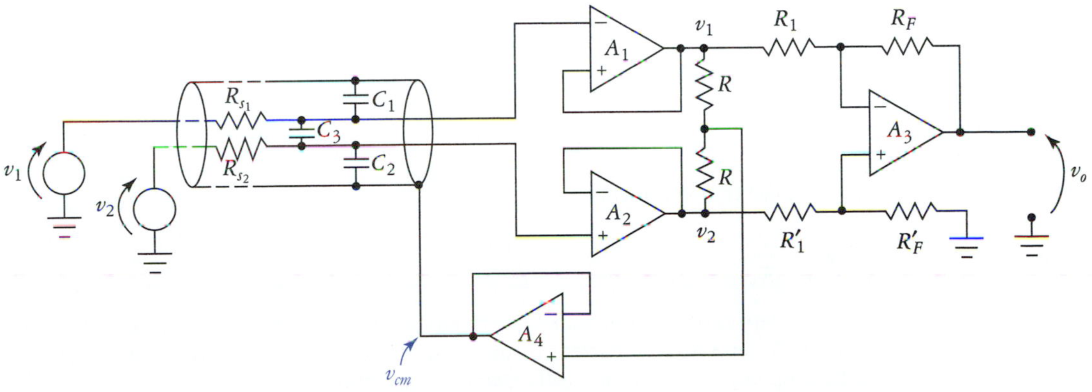

FIGURE 13.17
Active guard drive reduces noise caused by unequal line capacitances.

Stray capacitance C_3 in Fig. 13.17 does not affect the common-mode response of the IA. However, if the differential voltages are the time varying, this capacitance tends to convert the differential-mode signal into a common-mode signal, which is another potential error source.

13.3 VOLTAGE-TO-CURRENT AND CURRENT-TO-VOLTAGE CONVERSION

The circuits presented so far have been voltage amplifiers. Such circuits are also referred to as voltage-controlled voltage sources (VCVS). In the most general of terms, the quantity that relates the input and output of a system (or circuit) is of primary importance to understanding the behavior and possible uses of that circuit. This is what we call the system *transfer function.* The VCVS, or voltage amplifier transfer function, is v_o/v_{in}, the voltage gain.

In terms of transfer function-related classifications, recall that there are three other types of amplifiers: current amplifiers, transconductance amplifiers, and transresistance amplifiers. All four amplifier types and their input–output relationships are listed here:

Transfer Function	Symbol	Dimensions	Amplifier Description
v_o/v_{in}	A_v or μ	V/V	Voltage amplifier, VCVS
i_o/v_{in}	g_m	Siemens (mhos)	Transconductance amplifier, VCIS, I/V converter
v_o/i_{in}	r_m	Ohms	Transresistance amplifier, ICVS, V/I converter
i_o/i_{in}	A_i or β	A/A	Current amplifier, ICIS

The VCIS

The VCIS is used to source or sink current that is proportional to the VCIS input voltage. Recall that the proportionality constant relating output current to input voltage is termed transconductance g_m. Possibly the most descriptive name given to the VCIS is the term voltage-to-current (V/I) converter. Two simple op amp V/I converters are shown in Fig. 13.18. In both cases, the load is in the feedback loop of the op amp.

The analysis of Fig. 13.18(a) is straightforward. In order for the op amp to maintain virtual ground at the inverting input terminal, we know that $i_L = i_1$ and $i_1 = v_{\text{in}}/R_1$. Therefore,

$$i_L = \frac{v_{\text{in}}}{R_1} \tag{13.35}$$

We express the transconductance of this amplifier as

$$g_m = \frac{1}{R_1} \tag{13.36}$$

For the V/I converter of Fig. 13.18(b), because $v_{id} = 0$ V, and the currents flowing into the op amp input terminals are zero, we know that $v_1 = v_{\text{in}}$. Therefore,

$$i_1 = \frac{v_{\text{in}}}{R_1} \tag{13.37}$$

Now, since $i_L = i_1$, we obtain

$$i_L = \frac{v_{\text{in}}}{R_1} \tag{13.38}$$

and, again, the transconductance is

$$g_m = \frac{1}{R_1} \tag{13.39}$$

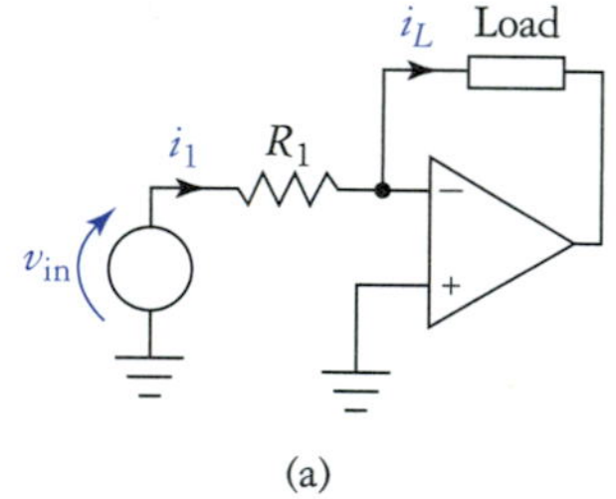

(a)

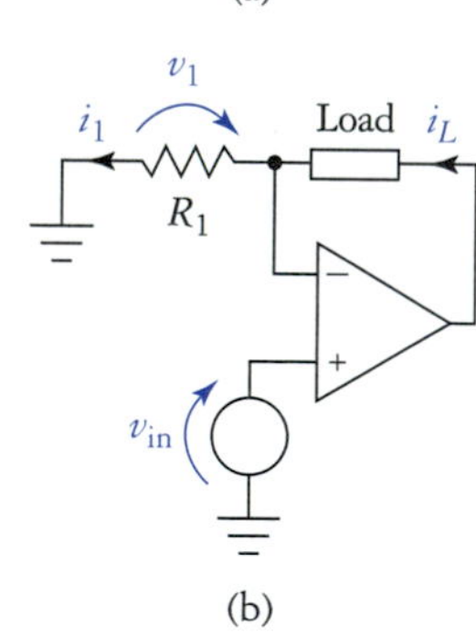

(b)

FIGURE 13.18 *Simple VCIS or V/I converters.*

The main thing to notice about both circuits is that the load current is independent of the load resistance. To force a given current through various load resistances, the voltage produced at the output of the op amp will automatically change as necessary.

The limits of operation for Fig. 13.18(a) or (b) depends on either the current source/sink capability, or the voltage compliance of the op amp, and the load resistance being driven. There is no lower limit on R_L; it is easier to force a given current through a short circuit than a nonzero resistance load.

EXAMPLE 13.8

Refer to Fig. 13.18(a). Assume $R_1 = 1\ \text{k}\Omega$, $V_{\text{in}} = +2$ V, and the op amp has the following limitations: $V_{o(\max)} = \pm 12$ V and $I_{o(\max)} = \pm 10$ mA. Determine g_m and the maximum load resistance that can be used.

Solution The transcondductance is

$$g_m = \frac{1}{R_1}$$
$$= 1 \text{ mS}$$

The load current is

$$I_L = g_m V_{\text{in}}$$
$$= (1 \text{ mS})(2 \text{ V})$$
$$= 2 \text{ mA}$$

The maximum load resistance is determined by

$$R_{L(\text{max})} = V_{o(\text{max})}/I_L$$
$$= 12 \text{ V}/2 \text{ mA}$$
$$= 6 \text{ k}\Omega$$

If $R_L > R_{L(\text{max})}$ then the op amp will saturate.

The Howland Current Source

Floating-load current sources like those in Fig. 13.18 perform quite well; however, in many applications the load must be referred to ground. In these cases, the VCIS of Fig. 13.19 could be used. This circuit is called the *Howland current source*, so named after its inventor. The detailed analysis of this circuit is very complex, so we work only with the final results here. This circuit topology can be used to realize many other useful functions as well, but for now we restrict the discussion to the current-source application. For proper operation, we require the following relationship to hold:

$$R_4/R_3 = R_2/R_1 \tag{13.40}$$

When this is true the transconductance is

$$g_m = \frac{1}{R_1} \tag{13.41}$$

and the load current is

$$i_L = g_m \, v_{\text{in}} \tag{13.42}$$

or, if you prefer,

$$i_L = \frac{v_{\text{in}}}{R_1} \tag{13.43}$$

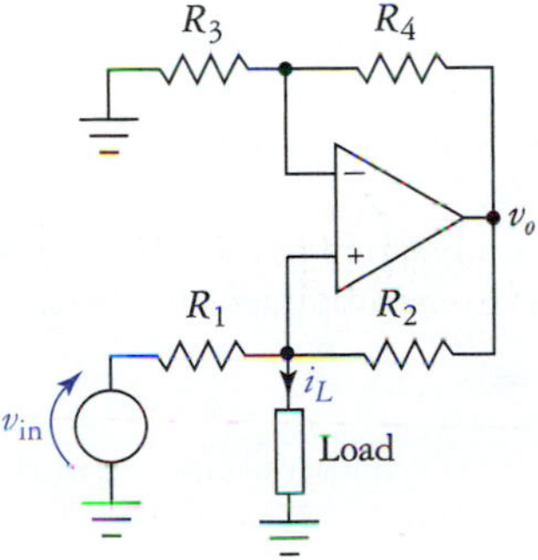

FIGURE 13.19
The Howland current source.

Normally, we just set all of the resistors equal, that is, $R_1 = R_2 = R_3 = R_4$. Under these conditions, the maximum voltage that can be applied across the load is $V_{L(\text{max})} = \pm V_{\text{sat}}/2$.

A V/I Converter Application

A good application for a voltage-to-current converter is in the design of a high-impedance analog voltmeter, as shown in Fig. 13.20. D'Arsonval meter movements are current-dependent devices used to implement analog measuring instruments. Even though digital meters are much more common,

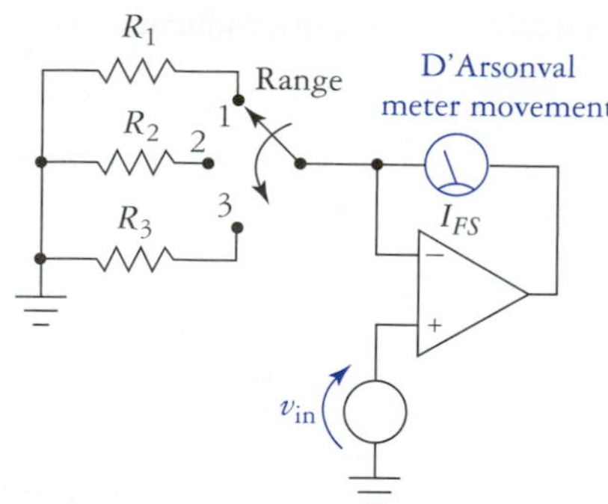

FIGURE 13.20
A high-impedance voltmeter.

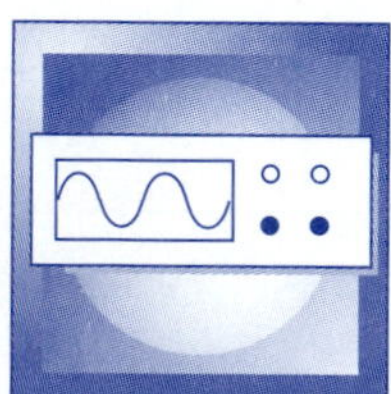

analog meters are still useful when it is desired to observe the behavior of relatively slowly varying signals. In these applications, a digital voltmeter will usually produce an unreadable jumble of numbers. The analog meter will smoothly vary in proportion to the signal.

In a traditional analog voltmeter application, it is necessary to know the internal resistance R_{int} and full-scale current I_{FS} of the meter movement. When the meter movement is driven by a current source as in Fig. 13.20, the internal resistance can be disregarded and we choose range multiplier resistors according to

$$R_n = \frac{V_{FS}}{I_{FS}} \tag{13.44}$$

Applying this information, if the meter in Fig. 13.20 has $I_{FS} = 1$ mA and we wish to choose R_1 such that $V_{FS} = 5.00$ V, then

$$\begin{aligned} R_1 &= 5.00 \text{ V}/1 \text{ mA} \\ &= 5 \text{ k}\Omega \end{aligned}$$

As shown in the schematic, we can use a single-pole n-throw rotary switch to choose different full-scale voltage ranges.

The advantages of using an op amp V/I converter are the extremely high input resistance achieved at the input and the simplicity of the circuit overall. The main disadvantage is the need for a power supply for the op amp. We must also ensure that the input voltage is restricted to

$$-V_{EE} < V_{in} < V_{CC} \tag{13.45}$$

Input voltages that exceed this range will damage the op amp.

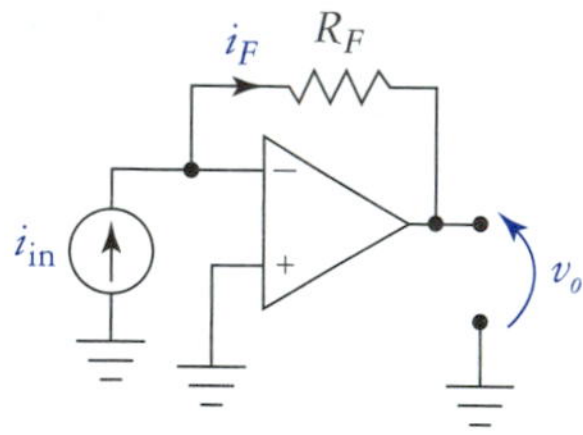

FIGURE 13.21
Basic ICVS or I/V converter.

The ICVS

For low-power applications, the op amp can be used to implement a very effective ICVS, or transresistance amplifier. The main idea here is for the op amp to produce an output voltage proportional to its input current. This is very easy to do, and the basic op amp I/V converter is shown in Fig. 13.21.

The operation of the I/V converter is straightforward. In Fig. 13.21, a current source provides the input to the amplifier. We know that $i_F = i_{in}$ and, therefore,

$$v_o = -i_F R_F \tag{13.46}$$

Apparently the transresistance r_m of the amplifier is

$$r_m = R_F \tag{13.47}$$

Two devices that are typically modeled as current sources are the photodiode and the phototransistor. Figure 13.22(a) illustrates one possible method of converting the diode photocurrent I_{photo} into a voltage. Recall that the photodiode can be approximately modeled as a current source, in which the current is proportional to the intensity of the light impinging on the pn junction. Typical characteristic curves for a photodiode are shown in Fig. 13.22(b). Assuming the photocurrent is linearly related to the light intensity, the transresistance amplifier converts the transfer characteristics of the photodiode as shown in Fig. 13.22(c). Circuits like this are used in fiber optic data communication systems.

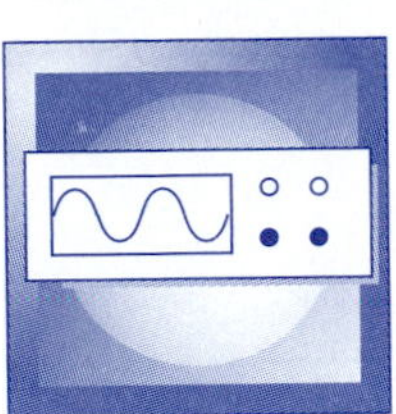

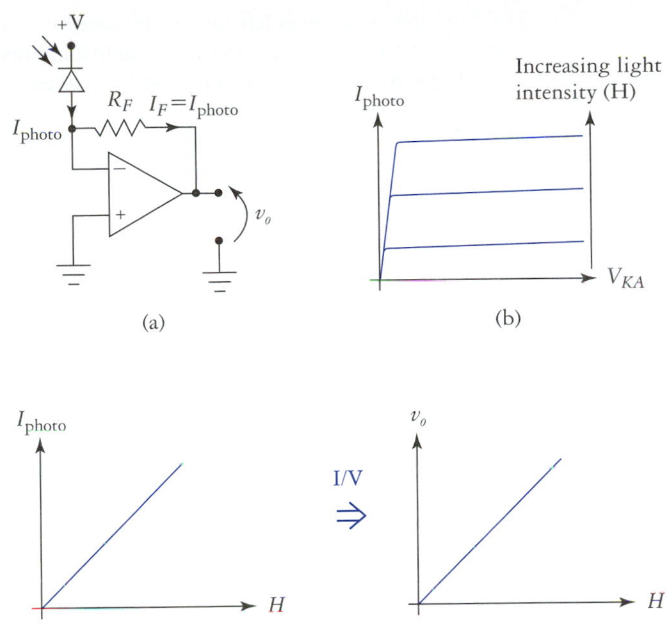

FIGURE 13.22
Using an I/V converter with a photodiode to produce a voltage that is proportional to light intensity.

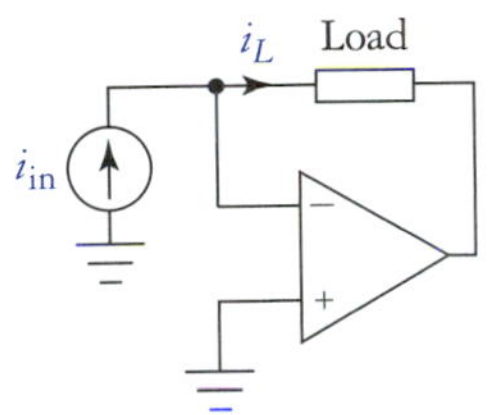

FIGURE 13.23
Simple ICIS.

The ICIS (Current Amplifier)

Current amplifiers are not often implemented using op amps. For the sake of completeness, one possible current amplifier is shown in Fig. 13.23. In this circuit $i_L = i_{in}$. This circuit is analogous to the voltage follower, for which $v_o = v_{in}$.

13.4 DIFFERENTIATORS AND INTEGRATORS

Op amps are used extensively to realize the functions of differentiation and integration, which are the foundations of calculus. The word *calculus* strikes fear into the hearts of many students. Often, this is because calculus is viewed as a difficult, dry, and highly abstract subject and, unfortunately, depending on how it is taught and by whom, this is sometimes true. In any case, some knowledge of integration and differentiation is extremely useful to the electronics technologist. Hopefully, this section will eliminate much anxiety and provide some useful working knowledge and insight into this very important subject.

Differentiation

The concept of the derivative and differentiation have been used many times previously in this text. Technically, *differentiation* is the act of taking the derivative of a function. A *derivative* is the result of the differentiation of a function. In electronics, we call a function a *signal*. As an example a very common signal is a voltage that is a function of time. The derivative of such a signal tells us the rate of change of that signal, with respect to time. The derivative of such a voltage can be written using many different forms of notation; however, we use the following forms:

$$\frac{dv}{dt}, \quad \frac{d}{dt}v(t) \text{ or } \dot{v}(t) \text{ (read "}v\text{ dot }t\text{")}$$

The function notation is left out in the first form to reduce clutter. The main thing to remember is that the derivative is proportional to the instantaneous rate of change of some function.

Capacitance and inductance can be defined in terms of derivatives. Let's restrict the discussion to capacitors, since we deal with them so often in coupling and bypass situations. The instantaneous current $i_c(t)$ through the capacitor in Fig. 13.24 is proportional to the instantaneous rate of change of voltage across the capacitor. That is,

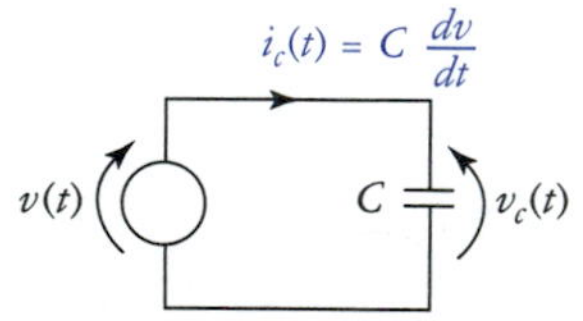

$$i_c = C\frac{dv_c}{dt}$$

FIGURE 13.24
Defining capacitor current in terms of the derivative of capacitor voltage.

where we see that the capacitance C is a proportionality constant. If the source is a constant dc voltage, the rate of change is zero, and no current flows (assume the voltage was applied at $t = -\infty$).

In electronics, a few basic functions appear over and over again and, fortunately, most of them are easy to differentiate, like the constant dc voltage just discussed. We discuss these common signals, using mostly voltages and currents as our dependent variable and time as our independent variable.

Constants

A constant is the simplest time-domain function. It is simply a quantity that never changes value. The derivative of a constant k is zero:

$$\frac{d}{dt}k = 0 \tag{13.48}$$

This is easy to visualize: since the slope of a constant is zero, its rate of change is always zero as well. This is shown in Fig. 13.25. A simple constant is a zero-degree equation because it is a function of $t^\circ = 1$.

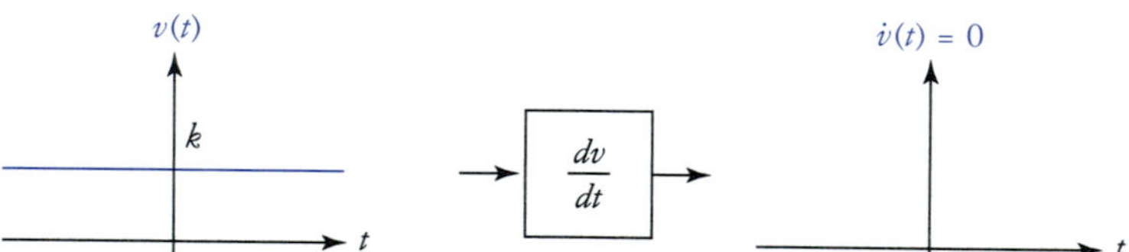

FIGURE 13.25
The derivative (instantaneous rate of change) of a constant is zero.

Linear Ramps

A linear ramp is a function that is always increasing or always decreasing. The rate of change of a linear ramp is the slope of the graph of that function. The simplest example of a linear ramp is shown in Fig. 13.26. Shifting this graph up or down has no effect on the derivative. The shift corresponds to a dc offset that has zero rate of change, and so it produces no output from the differentiator, much like a constant voltage would produce no steady-state current to flow in Fig. 13.24. In equation form

$$\frac{d}{dt}kt = k \tag{13.49}$$

The linear ramp is a first-degree equation. That is, it is a function of t^1.

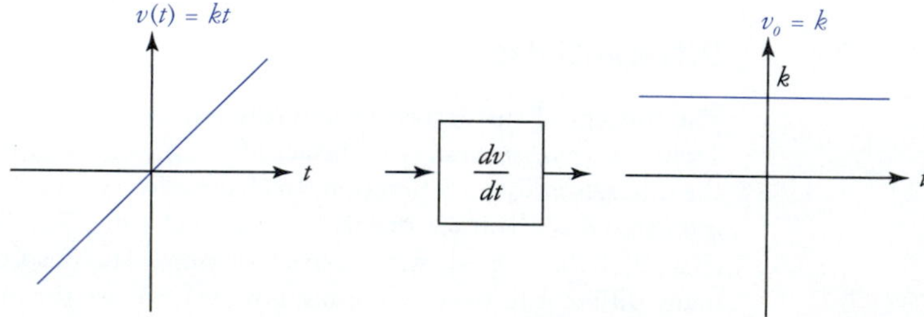

FIGURE 13.26
The derivative of a linear function is a constant.

Quadratic Functions

A quadratic or square-law function and its derivative are shown in Fig. 13.27. It is somewhat difficult to visualize the derivative, but looking at the equation for this function we have

$$\frac{d}{dt}kt^2 = 2kt \tag{13.50}$$

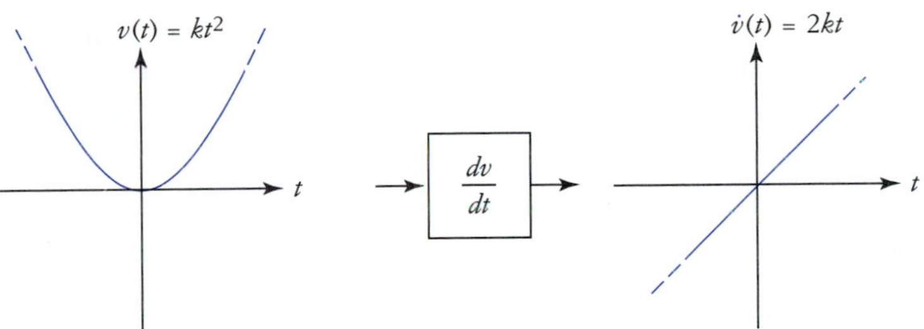

FIGURE 13.27
The derivative of a quadratic function is a linear function.

Each of the three functions covered so far is a *power function.* That is, each is characterized by the constant power to which the independent variable t is raised. The derivative of a power function can be found using the formula

$$\frac{d}{dt}kt^n = nkt^{n-1} \tag{13.51}$$

This formula works for any real value of n. For example,

$$\begin{aligned}\frac{d}{dt}8t^{1/2} &= (1/2)(8)t^{(1/2)-1}\\ &= 4t^{-1/2}\end{aligned}$$

Functions can have more than one term (*polynomials* are a special case, where we have several terms that are power functions with integer exponents). When we have a function that consists of more than a single term, the derivative of the function is the sum of the derivatives of the individual terms. Symbolically, this property of the derivative is written

$$\frac{d}{dt}\left[f_1(t) + f_2(t) + \ \ldots \ + f_n(t)\right] = \frac{d}{dt}f_1(t) + \frac{d}{dt}f_2(t) + \ \ldots \ + \frac{d}{dt}f_n(t) \tag{13.52}$$

This property must be applied to determine the derivative of the signal in Fig. 13.28. The derivative of this voltage is

$$\begin{aligned}\frac{d}{dt}v(t) &= \frac{d}{dt}(2 + 5t)\text{ V}\\ &= \frac{d}{dt}2\text{ V} + \frac{d}{dt}5t\text{ V}\\ &= 0\text{ V} + 5\text{ V}\\ &= 5\text{ V}\end{aligned}$$

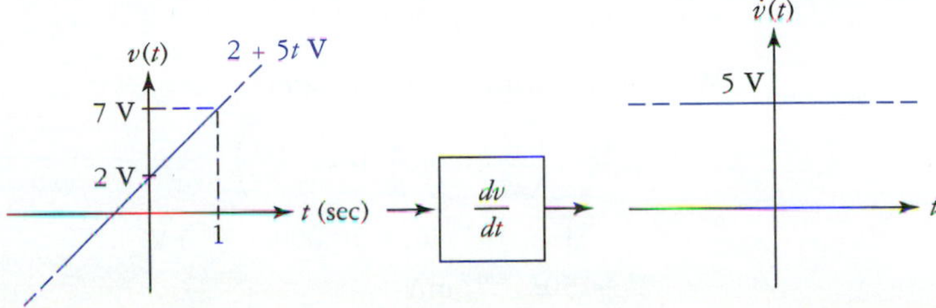

FIGURE 13.28
The input function has two terms. The output is the sum of the derivatives of each input term.

Sinusoidal Functions

Derivatives of sinusoidal functions have been discussed many times previously in this book. Without theoretical justification, the derivative of a sinusoidal voltage is

$$\frac{d}{dt} V_p \sin(\omega t + \theta) \text{ V} = \omega V_p \cos(\omega t + \theta) \text{ V} \qquad (13.53)$$

This equation says that the derivative of a sine is a cosine, scaled by the frequency ω. The constant phase shift θ simply carries through to the result.

Sinusoidal functions (including all phase-shifted versions, such as the cosine) are interesting because we can simplify them using phasor notation. For example, using the polar form

$$V_p \sin(\omega t + \theta) \text{ V} = V_p \angle \theta \text{ V} \qquad (13.54)$$

Note
The cosine function is related to the exponential function via Euler's identity:

$$\cos \theta = \frac{e^{j\theta} + e^{-j\theta}}{2}.$$

If $\theta = 90°$ then we have the special case

$$V_p \sin(\omega t + 90°) \text{ V} = V_p \cos(\omega t) \text{ V} = V_p \angle 90° \text{ V} \qquad (13.55)$$

This gives us some insight into differentiation. It turns out that differentiation in the time domain is equivalent to a 90° phase shift in the phasor domain. This relationship can be used to simplify our work in some cases.

It is also interesting to note that the derivative of a sinusoid is another sinusoid. Unlike power functions, the basic shape of the sinusoidal function is not changed by differentiation. If we repeatedly differentiate a sine function, after four successive derivatives are taken the total phase shift is 360°, and we have the original sine function again.

Exponential Functions

Exponentials are another class of common functions encountered in electronics. Again, we have worked with exponentials previously in this book. Because they are the most common, we deal strictly with decaying exponentials. The basic form of the decaying exponential voltage and its derivative is

$$\frac{d}{dt} V_o e^{-\alpha t} \text{ V} = -\alpha V_o e^{-\alpha t} \text{ V} \qquad (13.56)$$

Like the derivative of a sinusoidal function, the derivative of an exponential function is simply another exponential function. Not that it matters here, but this happens because at more fundamental levels, sinusoidal functions are equivalent to exponential functions with imaginary exponents.

EXAMPLE 13.9

As a 1-μF capacitor is charging, its terminal voltage varies according to

$$v_c = 10(1 - e^{-5000t}) \text{ V}$$

The charging current is given by

$$i_c = C \frac{dv_c}{dt}$$

Determine the expression for this current.

Solution Distributing the constant 10 makes the voltage equation easier to handle, thus

$$v_c = 10 - 10e^{-5000t} \text{ V}$$

We now differentiate term by term:

$$\begin{aligned} i_c &= C\left(\frac{d}{dt} 10 - \frac{d}{dt} 10e^{-5000t}\right) \\ &= 1 \times 10^{-6}(0 + 50{,}000e^{-5000t}) \text{ V} \\ &= 50e^{-5000t} \text{ mA} \end{aligned}$$

Graphs of v_c and i_c are shown in Fig. 13.29.

FIGURE 13.29
Waveforms for Example 13.9.

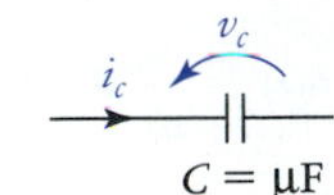

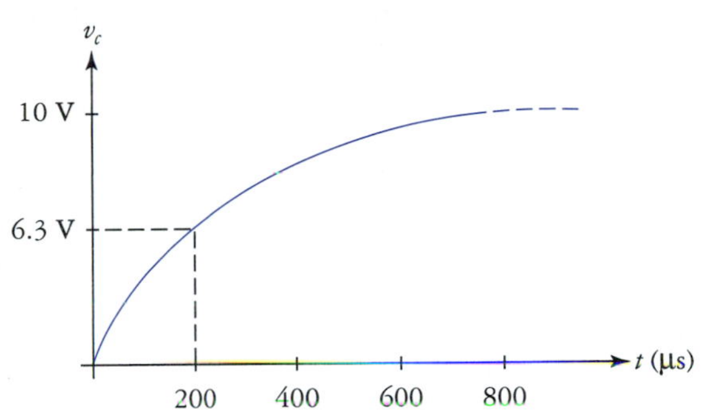

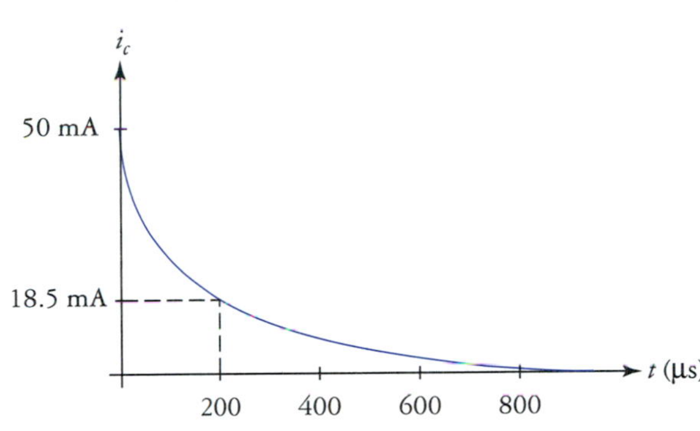

Step Functions

The last function we discuss is the *step function.* The basic step function is the *unit step function u(t)*, which is read "you of tee." The unit step is a discontinuous function defined as

$$u(t) = \begin{cases} 0, & t < 0 \\ 1, & t \geq 0 \end{cases} \tag{13.57}$$

Equation (13.57) is basically the mathematical description of a switch. We use $u(t)$ to switch other functions on and off. For example, in Fig. 13.30 the switch is connected to the battery at $t = 0$ sec, producing the waveform shown. The voltage waveform is written

$$v_x = V_1 u(t) \tag{13.58}$$

or we can express this as in Eq. (13.57) as

$$v_x = \begin{cases} 0, & t < 0 \\ V_1, & t \geq 0 \end{cases} \tag{13.59}$$

If we delay the step function (the time at which the switch turns on) by t_d seconds, as in Fig. 13.31, we write

$$v_y = v_2 u(t - t_d) \tag{13.60}$$

Real-world signals do not actually exist for all time, so the step function is useful for describing signals that are turned on at some particular point in time, usually at $t = 0$. This allows the transient behavior of a network to be analyzed mathematically. There are other uses as well. For example, a square wave may be described as a sequence of time-delayed step functions.

The derivative of a unit step function is called an *impulse*, a *Dirac delta function,* or just a *delta function.* The ideal impulse has zero width, infinite amplitude, and unity area! Mathematically, this is all written as

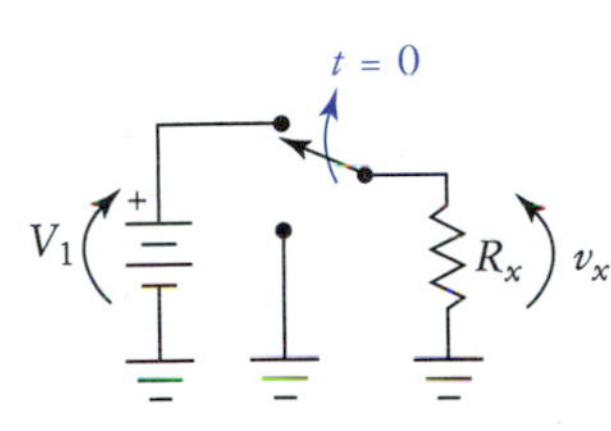

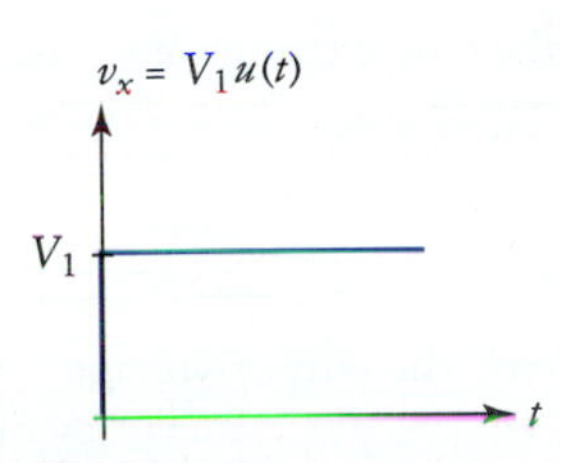

FIGURE 13.30
The closing of an ideal switch produces a step function output.

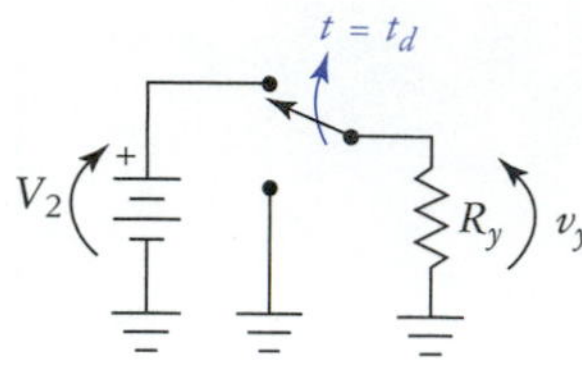

$$\delta(t) = \frac{d}{dt}u(t) = \begin{cases} 0, & t \neq 0 \\ \infty, & t = 0 \end{cases} \tag{13.61}$$

An impulse is represented with a vertical arrow as shown in Fig. 13.32. In this case, the impulse is delayed by t_d seconds so that it can be more easily distinguished from the vertical axis. It can also be shown that the area of this impulse is k volt-seconds.

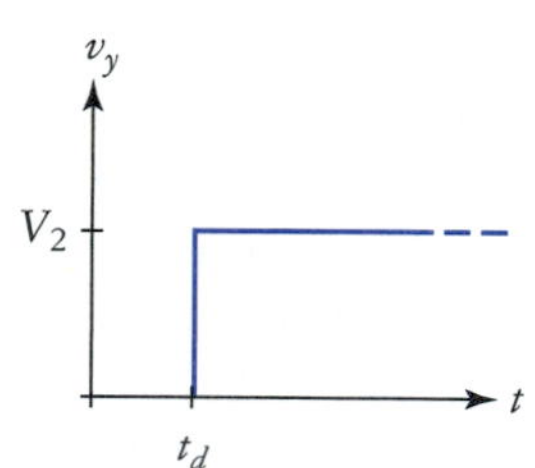

FIGURE 13.31
Delayed step function.

FIGURE 13.32
The derivative of a step function is an impulse or delta function. The impulse has infinite amplitude, zero width, and area equal to the height of the step.

The Differentiator

A simple op amp differentiator is shown in Fig. 13.33. The output of this circuit is the derivative of the input voltage, inverted and scaled by the product R_FC_1. To understand how this circuit performs differentiation, we use the standard op amp analysis techniques. We know that $i_F = i_1$ and the inverting input is at virtual ground, thus

$$i_F = i_1 = C_1\frac{dv_{in}}{dt}$$

and the output voltage is

$$\begin{aligned} v_o &= -i_FR_F \\ &= -R_FC_1\frac{dv_{in}}{dt} \end{aligned} \tag{13.62}$$

If it is required, we set $R_B = R_F$ since the inverting input bias current path is to ground via R_F.

FIGURE 13.33
Basic op amp differentiator.

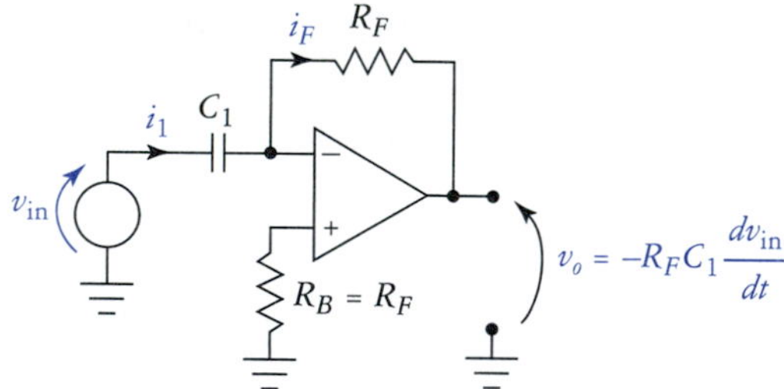

Remember the basic concept behind the differentiator: The greater the rate of change of v_{in}, the larger will be the resulting output voltage.

EXAMPLE 13.10

The differentiator in Fig. 13.33 has $R_F = 10\ \text{k}\Omega$ and $C_1 = 0.1\ \mu\text{F}$. Sketch the output voltage waveform given the triangle wave signal shown in Fig. 13.34.

Solution Let's look at this problem from a qualitative perspective first. The triangle wave is composed of connected linear-ramp segments. The response of a differentiator to a linear ramp is a constant, equal to the slope of the ramp. In this case we see that the slope of successive segments changes sign once every millisecond, so the output of the differentiator should be a square wave with a period of 1 ms as well.

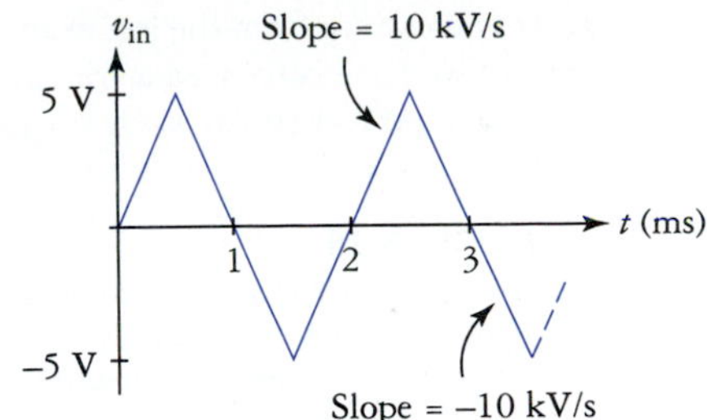

FIGURE 13.34
Input waveform for Example 13.10.

To determine the amplitude of the square wave, we use Eqs. (13.62) and (13.49). This requires us to write the equation for the input signal as well. It is possible to write a formal equation for this signal using delayed step functions, which is*

$$V_{in}(t) = 10{,}000tu(t) - 20{,}000(t - 0.0005)u(t - 0.0005) \\ + 20{,}000(t - 0.001)u(t - 0.001) + \ldots$$

Taking the derivative of this function is not difficult but it is beyond the scope of this book. Fortunately, we can use our qualitative knowledge of this signal and a little bit of calculus. For the first positive-going segment we have $v_{in(+)} = 10{,}000t$ V. Taking advantage of the symmetry of the waveform, we find the next segment is $V_{in(-)} = -10{,}000t$ V. The output during positive-going portions of the input signal is given by

$$v_o = -R_F C_1 \frac{dv_{in}}{dt}$$

$$= -(10\ \text{k}\Omega)(0.1\ \mu\text{F}) \frac{d}{dt} 10{,}000t\ \text{V}$$

$$= -10\ \text{V}$$

During the negative-going portions of the input signal, we get

$$v_o = 10\ \text{V}$$

This sequence repeats indefinitely, producing the response shown in Fig. 13.35.

FIGURE 13.35
Input and output waveforms for Example 13.10.

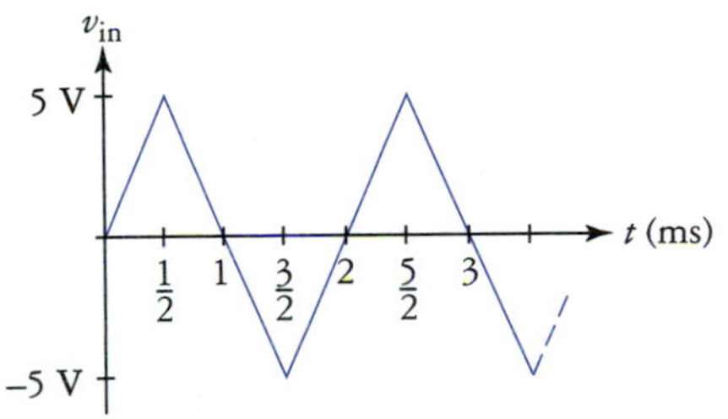

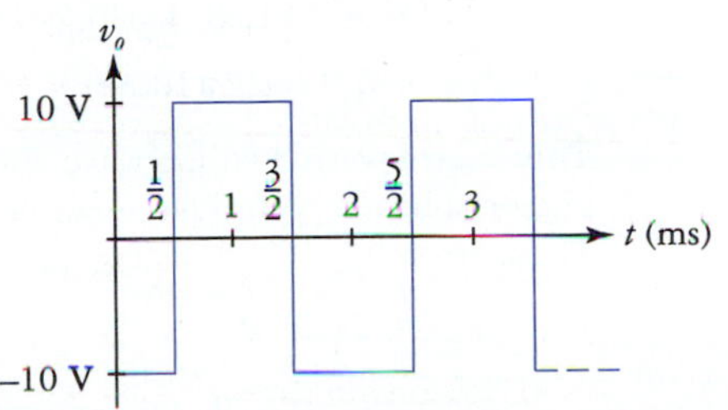

*You may skip this step without loss of continuity. This material is covered in courses on linear circuit analysis and Laplace transform analysis.

As you can see from the last example, differentiators can be used in wave-shaping applications. In this case, we have converted a triangle wave into a square wave. Keep in mind, however, that the output is not a perfect square wave because of the finite slew rate of the op amp.

Note
With a sinusoidal input, the output of the inverting differentiator will always have a +90° phase shift, regardless of frequency.

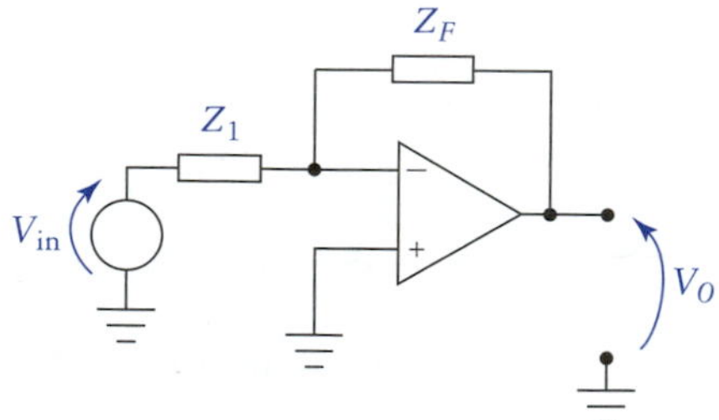

FIGURE 13.36
If the input to the differentiator is a single-frequency sinusoid, phasor analysis can be used to determine the output voltage.

In general, we must use Eq. (13.62) to determine the response of the differentiator to an arbitrary input. However, when the input to the differentiator is a sinusoidal voltage, we can use phasor analysis to determine the steady-state response of the circuit. This is the beauty of phasors: They allow us to perform useful circuit analysis without actually having to resort to calculus. A phasor-domain representation for an op amp with negative feedback is shown in Fig. 13.36. The impedances can be any combination of resistors, capacitors, and inductors, and the output of the circuit is still given by

$$V_o = -\frac{V_{in}Z_F}{Z_1} \tag{13.63}$$

For an ideal differentiator, Z_F is purely capacitive and Z_1 is purely resistive. The next example demonstrates the equivalence of the phasor-domain and time-domain (derivative) approaches.

EXAMPLE 13.11

Determine the output voltage for the circuit in Fig. 13.37 using the derivative and the phasor-domain approaches.

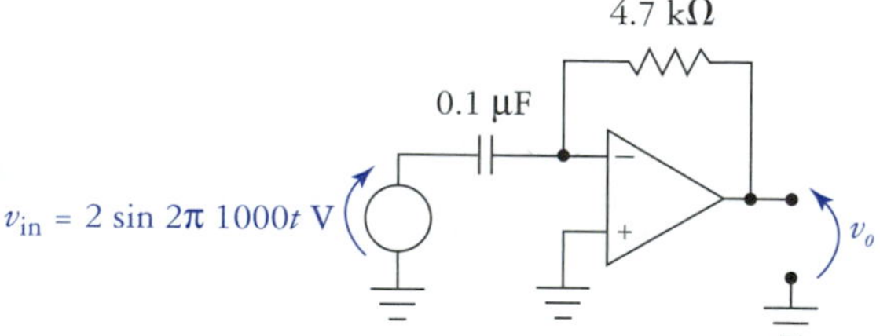

FIGURE 13.37
Circuit for Example 13.11.

Solution Using the general time-domain relationship we have

$$\begin{aligned} v_o &= -(R_FC_1)\frac{d}{dt}2\sin(2\pi1000t)\text{ V} \\ &= -(4.7\text{ k}\Omega)(0.1\ \mu\text{F})(2)(2\pi1000)\cos(2\pi1000t)\text{ V} \\ &= -5.9\cos(2\pi1000t)\text{ V} \end{aligned}$$

This is an inverted cosine wave with an amplitude $V_p = 5.9$ V and a frequency $f = 1$ kHz. The phasor solution technique requires us to determine the phasor equivalent values for the circuit. These are

$$V_{in} = 2\angle 0°\text{ V}$$

$$\begin{aligned} Z_1 = X_C &= \frac{1}{2\pi fC_1}\angle -90°\ \Omega \\ &= \frac{1}{(2\pi1000)(0.1\ \mu\text{F})}\angle -90°\ \Omega \\ &= 1.59\angle -90°\text{ k}\Omega \end{aligned}$$

$$Z_F = R_F = 4.7\angle 0°\text{ k}\Omega$$

The output of the op amp is

$$\begin{aligned} V_o &= -V_{\text{in}} \frac{Z_F}{Z_1} \\ &= -(2\angle 0^\circ \text{ V} \frac{4.7 \angle 0^\circ \text{ k}\Omega}{1.59 \angle -90^\circ \text{ k}\Omega}) \\ &= 2\angle 180^\circ \text{ V} \frac{4.7 \angle 0^\circ \text{ k}\Omega}{1.39 \angle -90^\circ \text{ k}\Omega} \\ &= (2\angle 180^\circ \text{ V})(2.96\angle 90^\circ) \\ &= 5.92\angle 270^\circ \text{ V} \\ &= 5.92\angle -90^\circ \text{ V} \end{aligned}$$

This is equivalent to an inverted cosine wave with $V_p \cong 5.9$ V. The frequency is suppressed in the phasor form, but our calculations are based on $f = 1$ kHz. Thus we see that the two analysis approaches yield identical results.

In the last example, the time-domain analysis (differentiation) was much less work than the phasor analysis approach. Sometimes phasor analysis is simpler than a time-domain analysis. Unfortunately, phasors are only applicable when sinusoidal signals are present.

Practical Differentiator Considerations

Note
Differentiators tend to be susceptible to high-frequency noise.

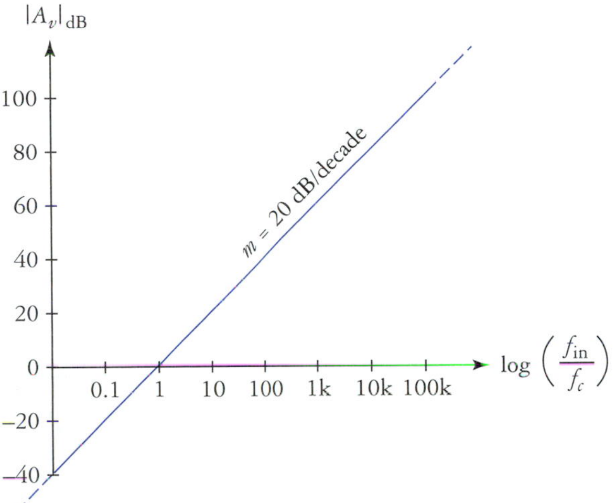

FIGURE 13.38
Frequency response of an ideal differentiator.

An ideal op amp differentiator has a frequency-response plot like that shown in Fig. 13.38. The critical frequency f_c of the differentiator is

$$f_c = \frac{1}{2\pi R_F C_1} \tag{13.64}$$

This is the frequency at which the gain magnitude is unity (0 dB). As you can see in Fig. 13.38, the closed-loop gain of the differentiator increases by 20 dB/decade with frequency. In a real circuit, the closed-loop gain would increase until it equaled the open-loop gain of the op amp, as shown in Fig. 13.39. At this frequency, f_{max}, the differentiator no longer functions correctly.

Because the response of a differentiator increases with frequency, they tend to pick up and greatly amplify noise. To help reduce this tendency, a resistor can be placed in series with R_1 as shown in Fig. 13.40. This causes the high-frequency response of the differentiator to level off at f_H, which is given by

$$f_H = \frac{1}{2\pi R_1 C_1} \tag{13.65}$$

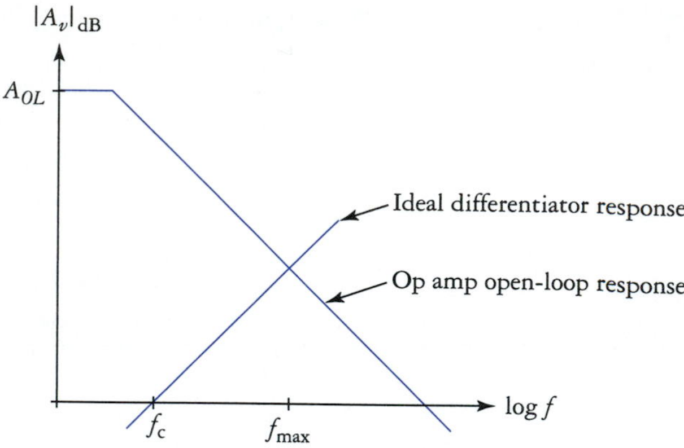

FIGURE 13.39
Op amp frequency response and differentiator response.

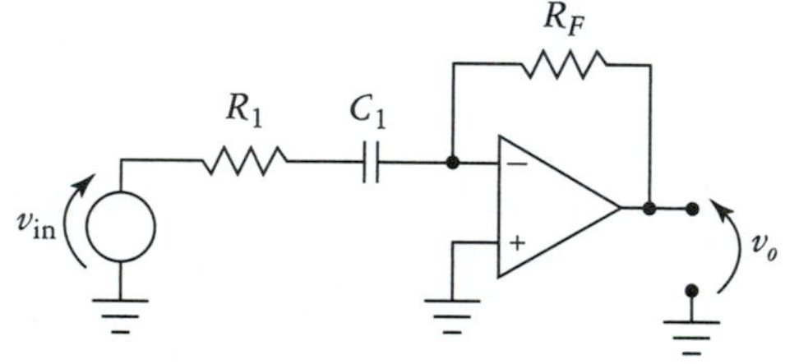

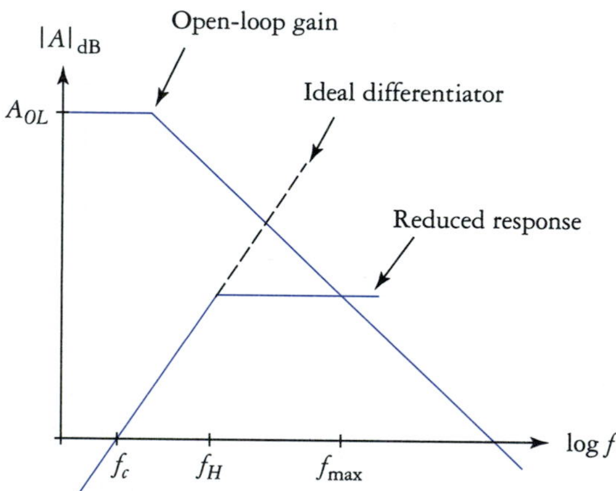

FIGURE 13.40
Modified differentiator and resulting frequency response curves.

We would choose R_1 such that f_H is less than f_{max}, but greater than the highest signal frequency we are going to process, $f_{in(max)}$. In other words

$$f_{in(max)} < f_H < f_{max} \tag{13.66}$$

Often, we can get away with simply choosing $R_1 \leq R_F/100$. Whichever way we choose, the differentiator response is altered as shown in the graph of Fig. 13.40.

Integration

Integration is the complementary operation to differentiation. Symbolically, integration is written

$$\int v(t)\, dt$$

which is read as *the indefinite integral of v(t) with respect to time.* That's not a very satisfying description, however, if you are not familiar with integration. There are several different ways to interpret the meaning of integration. One way is to view integration as "undoing" differentiation. For example, we know that if we differentiate a triangular waveform, a square wave is produced. Thus, if we integrate a square wave, we obtain a triangular waveform as shown in Fig. 13.41.

FIGURE 13.41
Integration is the inverse of differentiation.

Another way of interpreting integration is to view it as a mathematical operation that tells us the area under the graph of a function. For example, in Fig. 13.42, if we wish to know the area under the curve for this signal, from t_1 to t_2, we would evaluate the *definite integral:*

$$A = \int_{t_1}^{t_2} v(t)\,dt$$

A definite integral has limits of integration that tell us over which interval to determine the area under the curve. We now look at a few of the more commonly encountered signals and their integrals.

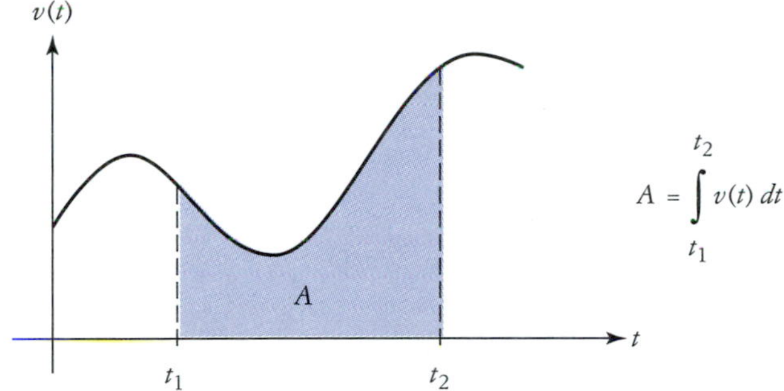

FIGURE 13.42
Integration tells us the area under the graph of a function.

The Integral of a Constant

The integral of a constant is a linear ramp function as shown in Fig. 13.43(a). You can visualize this output being generated by considering the waveforms in Fig. 13.43(b). Here, the input is a 1-V dc level applied at $t = 0$. After 1 sec has passed, the area under the curve is 1 V·1 sec = 1 V·s and after 2 sec have passed, the area is 1 V·2 sec = 2 V·s, and so on. The area increases linearly with time, so the waveform produced by integration is a linear ramp. In equation form

$$\int k\,dt = kt + C \tag{13.67}$$

where C is called the constant of integration. For our purposes, the constant of integration is just the value of the output of the integrator that exists when integration is started. Often, in electronic applications, we reset the output of the integrator to zero, as was apparently done in Fig. 13.43(b).

To better understand the nature of integration, consider the waveforms in Fig. 13.44. The input is two 5-V pulses of 1-sec duration, separated by a 1-sec interval. This sort of signal might be produced by a digital circuit, for example. Discontinuous signals like this must be integrated piece by piece over the continuous portions of the waveform. We would start by evaluating the portion of the signal from $t = 0$ to 1 sec, producing $A_1 = 5$ V·s. During the interval from $t = 1$ to 2, the area under the curve stops increasing and $A_2 = 0$. Now, integrating from $t = 2$ to 3 gives us $A_3 = 5$ V·s. The total area under the curve is

$$\begin{aligned} A_T &= A_1 + A_2 + A_3 \\ &= 5\text{ V·s} + 0\text{ V·s} + 5\text{ V·s} \\ &= 10\text{ V·s} \end{aligned}$$

The output is the accumulated area under the curve. We can consider the result of integration from $t = 0$ to 1 (A_1) as being the constant of integration for the integration from $t = 1$ to 2, and so on.

An important observation to be made here is that the integrator "remembers" past inputs. The integrator's output is the sum of all past responses, unless we force the output to zero at the

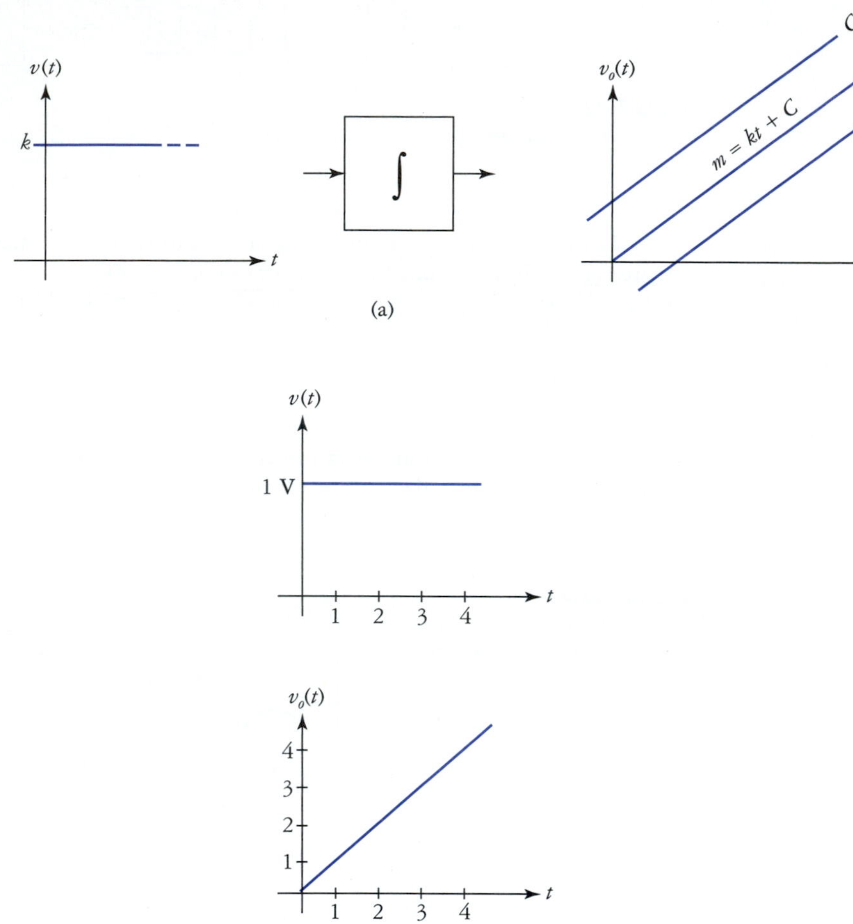

FIGURE 13.43
Integration of a constant. (a) The output is a linear function with an arbitrary vertical translation. (b) Development of the integral as the area under the curve.

FIGURE 13.44
Integration accumulates area over time.

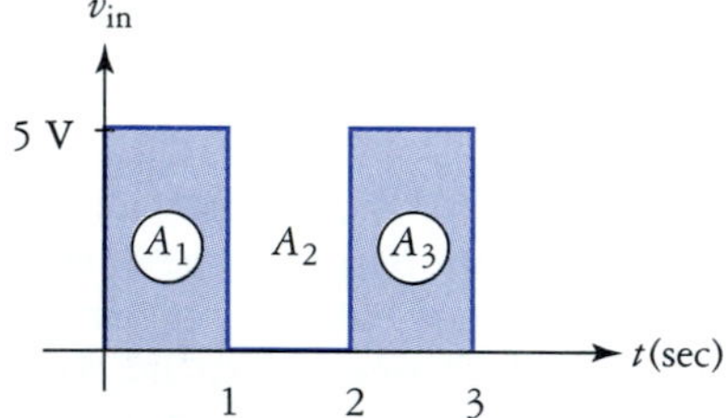

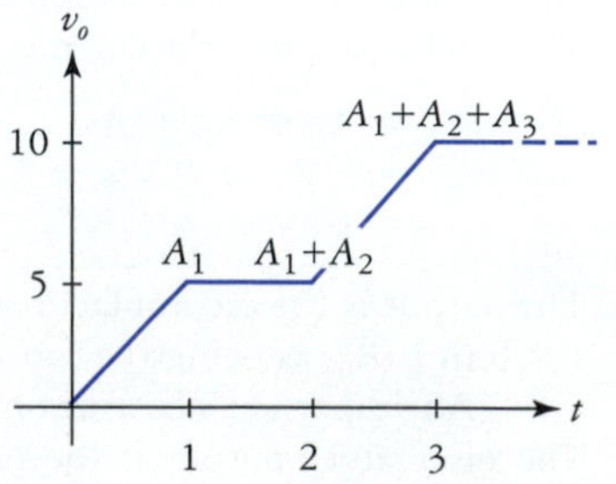

beginning of the integration interval (which is easy to do with op amp integrators). Mathematically for Fig. 13.44, we write this as

$$\int_0^3 v_{\text{in}}\, dt = \int_0^1 v_{\text{in}}\, dt + \int_1^2 v_{\text{in}}\, dt + \int_2^3 v_{\text{in}}\, dt + \ldots \tag{13.68}$$

This generalizes to as many terms as required. Each of the terms causes an associated change in the output of the integrator Δv_o. We can express this related output voltage equation as

$$\Delta v_o = \Delta v_{o(1)} + \Delta v_{o(2)} + \Delta v_{o(3)} + \ldots \tag{13.69}$$

where each change in output voltage is due to the integration over the associated interval. In Fig. 13.44, for example, $\Delta v_{o(1)} = 5$ V, $\Delta v_{o(2)} = 0$ V, and $\Delta v_{o(3)} = 5$ V, hence $\Delta v_o = 10$ V.

The Integral of a Linear Ramp

Recall that when we differentiated a power function signal, the degree of the function was reduced by one (see Eq. 13.51). When we integrate a power function, the degree is increased by one. As we saw in the case of the constant $k = kt^0$, the integral was $kt^{n+1} = kt^1$. In the case of the linear ramp input, the output is a parabolic or second-degree function given by

$$\int kt\, dt = \frac{1}{2}kt^2 + C \tag{13.70}$$

In fact, we can generalize this to form

$$\int kt^n\, dt = \frac{1}{n+1}kt^{n+1} + C \tag{13.71}$$

for all $n \neq -1$.

Integration of Sinusoidal Functions

In this particular case it is easiest to visualize the integrator as the complement of the differentiator. Recall that if we differentiate a sine wave we obtain a cosine wave, which is simply a sine wave shifted by 90°. Similarly, if we integrate a sine wave, we obtain a negative cosine wave; that is a sine wave phase shifted by −90°. The equation is

$$\int V_p \sin(2\pi ft + \theta)\, dt = -\frac{V_p}{2\pi f}\cos(2\pi ft + 0) + C \tag{13.72}$$

As was the case with differentiation, the initial phase shift θ is simply carried through to the output. The integration of a cosine waveform results in a sine wave being produced:

$$\int V_p \cos(2\pi ft + \theta)\, dt = \frac{V_p}{2\pi f}\sin(2\pi ft + \theta) + C \tag{13.73}$$

Integration of Exponential Functions

Integration of an exponential function results in another exponential function. Again, most of the time we are likely to be dealing with decaying exponentials, and for this special case, we have

$$\int ke^{-\alpha t}\, dt = -\frac{k}{\alpha}e^{-\alpha t} + C \tag{13.74}$$

Equations like (13.74) can be difficult to visualize. A graphics calculator like the TI-85 is very useful in these situations. This will be covered in the computer-aided analysis section of this chapter.

The Integrator

The basic op amp integrator is shown in Fig. 13.45. Let's think about the operation of this circuit intuitively for a moment to get a feel for its operation; remember that the basic op amp analysis rules still hold for this circuit. Assume that v_{in} is a positive dc voltage. Because the op amp forces the inverting input terminal to virtual ground, the input current i_1 is constant. Since no current enters the inverting input, $i_F = i_1$. Now, in order for current to flow into (or through) a capacitor, the voltage across the capacitor must be changing. Since the left side of the cap is at virtual ground, the output of the op amp must ramp in a negative-going direction in order to hold i_F constant at the level set by i_1. We know that the integral of a constant is a linear ramp, and this is just what the op amp integrator does, except that it also inverts the signal. Assuming that we integrate over some particular time interval, the output of the op amp integrator is given by

$$V_o = -\frac{1}{R_1 C_1}\int_{t_0}^{t_1} v_{\text{in}}\, dt \qquad (13.75)$$

This equation tells us the value of the output voltage at the end of the integration interval. That is, it simply produces a numerical value that is representative of the area under the input voltage curve. The shape of the output voltage waveform depends on the shape of the input signal. This can be a tricky operation, so we limit ourselves to a few simple but common and useful examples. The next example illustrates how this equation is applied.

EXAMPLE 13.12

The circuit in Fig. 13.45 has $R_1 = 5\ \text{k}\Omega$ and $C_1 = 2\ \mu\text{F}$. The input voltage waveform is shown in Fig. 13.46. Sketch the output voltage waveform that would be produced over the interval from $t = 0$ to 14 ms. Assume the integrator is reset to zero at the beginning of integration.

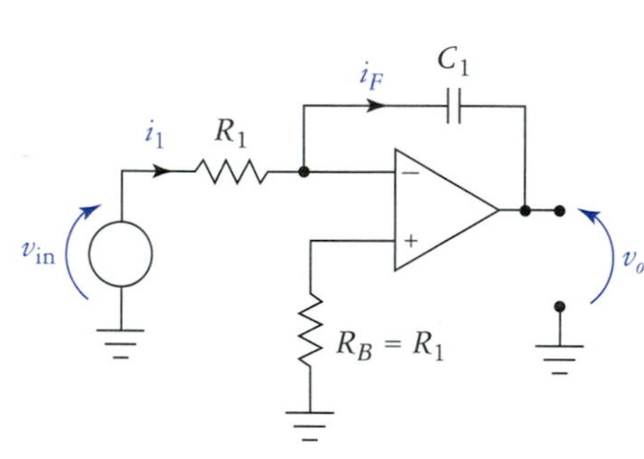

FIGURE 13.45
Basic op amp integrator.

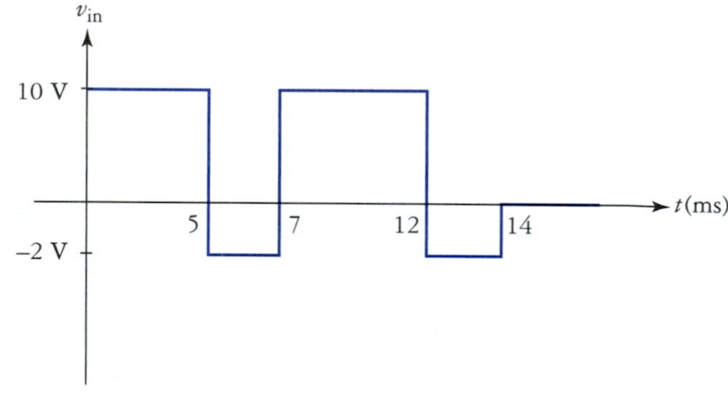

FIGURE 13.46
Input waveform for Example 13.12.

Solution We must evaluate the response of the integrator over each continuous interval of the input signal. This means that we must break the integral into pieces. This is written

$$V_o = -\frac{1}{R_1C_1}\int_0^{14\text{ ms}} v_{\text{in}}\, dt$$

$$= -\frac{1}{R_1C_1}\left[\int_0^{5\text{ms}} v_{\text{in}}\, dt + \int_{5\text{ ms}}^{7\text{ms}} v_{\text{in}}\, dt + \int_{7\text{ ms}}^{12\text{ ms}} v_{\text{in}}\, dt + \int_{12\text{ ms}}^{14\text{ ms}} v_{\text{in}}\, dt\right]$$

or, equivalently,

$$\Delta V_o = \Delta V_{o(1)} + \Delta V_{o(2)} + \Delta V_{o(3)} + \Delta V_{o(4)}$$

This will tell us the change in voltage at the end of each subinterval. Evaluating each term individually we obtain

$$\Delta V_{o(1)} = -\frac{1}{(5\text{ k})(2\mu\text{F})}(10\text{ V})(5\text{ ms})$$
$$= (-100\text{ sec}^{-1})(10\text{ V})(5\text{ ms})$$
$$= -5\text{ V}$$

$$\Delta V_{o(2)} = (-100\ \text{sec}^{-1})(-2\ \text{V})(2\ \text{ms}) \\ = 0.4\ \text{V}$$

$$\Delta V_{o(3)} = (-100\ \text{sec}^{-1})(10\ \text{V})(5\ \text{ms}) \\ = -5\ \text{V}$$

$$\Delta V_{o(4)} = (-100\ \text{sec}^{-1})(-2\ \text{V})(2\ \text{ms}) \\ = 0.4\ \text{V}$$

The net output voltage at $t = 14$ ms is

$$\Delta V_o = \Delta V_{o(1)} + \Delta V_{o(2)} + \Delta V_{o(3)} + \Delta V_{o(4)} \\ = -5\ \text{V} + 0.4\ \text{V} - 5\ \text{V} + 0.4\ \text{V} \\ = -9.2\ \text{V}$$

We use the Δ values for each interval to construct the output waveform as shown in Fig. 13.47.

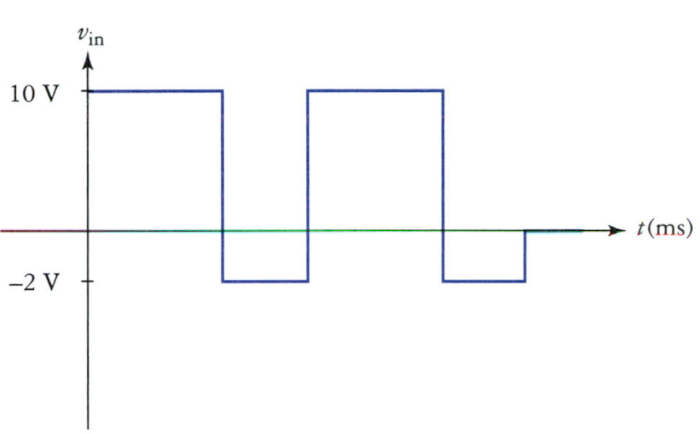

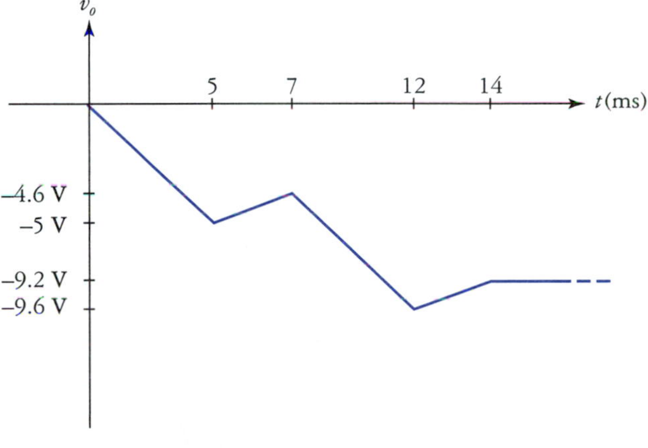

FIGURE 13.47
Input and output waveforms for Example 13.12.

Like the differentiator, if the input to the integrator is a sinusoidal waveform, we can use phasor analysis to determine the resulting output voltage. The next example demonstrates this fact.

EXAMPLE 13.13

Refer to Fig. 13.48. Determine the output voltage using Eq. (13.75) as an indefinite integral and using phasor analysis.

Solution We first use the indefinite form of Eq. (13.75). The definite integral tells us the end result of integration (a final voltage level) over some interval. When we use the indefinite form of an integral, the result is not a number, but rather it is an equation. What this means physically is

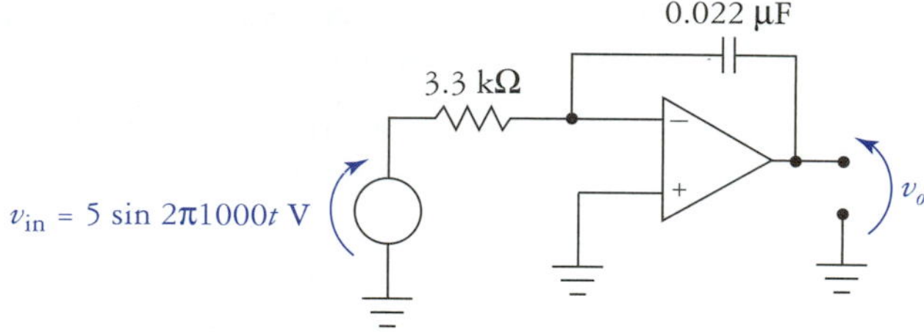

FIGURE 13.48
Circuit for Example 13.13.

that the indefinite integral tells us how the output behaves in general over time. Indefinite integration can only be applied to continuous functions however. But, the steady-state sinusoid of Fig. 13.48 is just such a continuous function. Thus, we have

$$\begin{aligned} v_o &= -\frac{1}{R_1C_1}\int 5\sin(2\pi 1000t)\text{ V} \\ &= (-13{,}774)(-5)(1/2\pi 1000)\cos(2\pi 1000t)\text{ V} \\ &\cong 11\cos(2\pi 1000t)\text{ V} \end{aligned}$$

Using phasor analysis, we can redraw the circuit as shown in Fig. 13.49. The output voltage is

$$\begin{aligned} V_o &= -V_{\text{in}}\frac{Z_F}{Z_1} \\ &= 5\angle 180^\circ\frac{7.2\angle -90^\circ\text{ k}\Omega}{3.3\angle 0^\circ\text{ k}\Omega} \\ &= (5\angle 180^\circ)(2.2\angle -90^\circ) \\ &= 11\angle 90^\circ\text{ V} \end{aligned}$$

This is equivalent to the instantaneous time-domain form obtained using the integral. The input and output voltage waveforms are shown in Fig. 13.50.

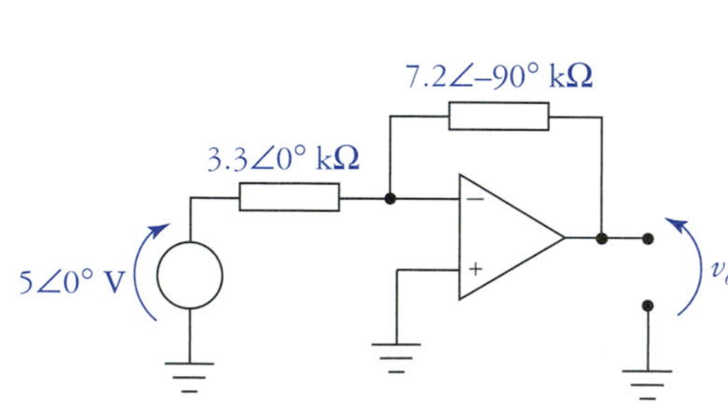

FIGURE 13.49
Phasor equivalent of Fig. 13.48.

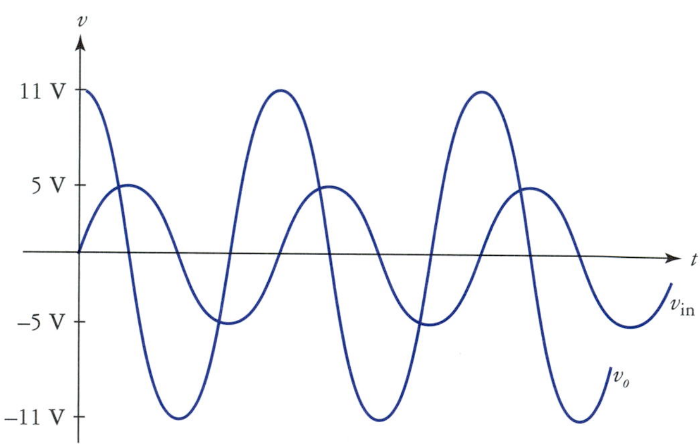

FIGURE 13.50
Input and output waveforms for Example 13.13.

Practical Integrator Considerations

The ideal op amp integrator has a frequency response as shown in Fig. 13.51. The frequency axis has been normalized with the integrator critical frequency f_c, which is given by

$$f_c = \frac{1}{2\pi R_1C_1} \tag{13.76}$$

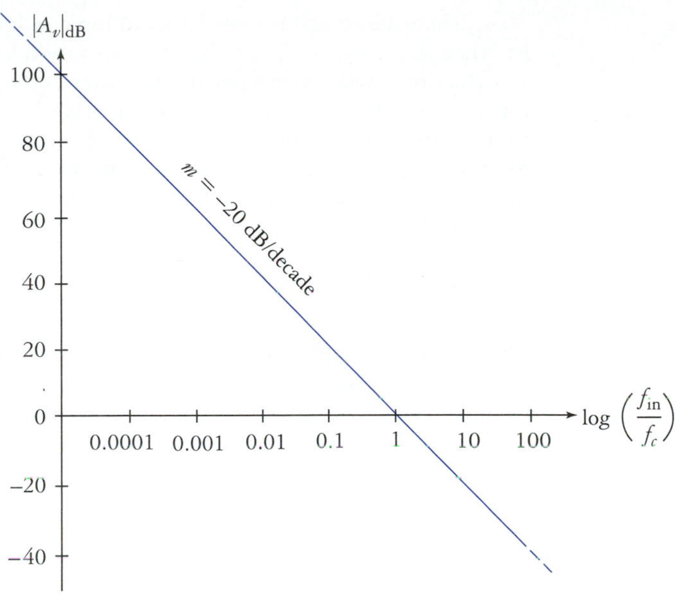

FIGURE 13.51
Ideal integrator frequency response.

Note
Given a sinusoidal input, the output of the inverting integrator will always have a $-90°$ phase shift regardless of frequency.

The ideal integrator response produces infinite gain at dc, rolling off at the first-order rate of -20 dB/decade. This should seem reasonable, because the gain magnitude for sinusoids is

$$|A_v| = \frac{Z_F}{Z_1} = \frac{X_C}{R_1}$$

and at $f = 0$ Hz (dc), we have

$$|A_v| = \frac{\infty}{R_1} = \infty$$

In practice, the integrator gain cannot exceed the open-loop gain of the op amp. A representative plot showing this limitation is presented in Fig. 13.52. Observe here that the integrator must be designed with $f_c < f_{unity}$ for the op amp. In fact, to allow for normal chip-to-chip variation in f_{unity} a reasonable safety margin would be achieved by ensuring that $f_c < f_{unity}/2$.

FIGURE 13.52
Op amp open-loop response and integrator response.

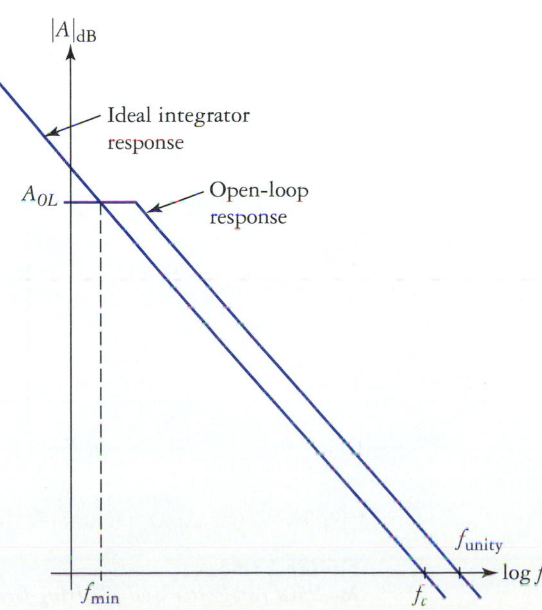

There are several reasons for us to limit to low-frequency gain of the integrator to something less than A_{OL}. First, we know that A_{OL} can vary substantially from unit to unit. For example, for the 741 A_{OL} could easily vary from 100 k to over 200 k, thus making the maximum gain of the integrator unpredictable. Second, and more importantly, if the low-frequency gain of the integrator is extremely high then the small but unavoidable input offset voltage will cause the output of the op amp to drift and eventually saturate. This is one reason why resistor R_B is virtually mandatory in a practical integrator circuit. The low-frequency gain is limited by adding a resistor in parallel with C_1 as shown in Fig. 13.53(a). The low-frequency corner f_L is

$$f_L = \frac{1}{2\pi R_F C_1} \tag{13.77}$$

Note
Integrators tend to drift over time due to a small input offset voltage being present.

If we apply a sinusoidal input that has $f < f_L$, the circuit will not maintain the correct gain and phase relationships to be considered a true integrator. In practical applications, however, this is not usually a problem and a good rule of thumb is to set R_F according to

$$R_F \cong 100R_1 \tag{13.78}$$

This limits the low-frequency gain magnitude to 100, which will reduce output drift significantly.

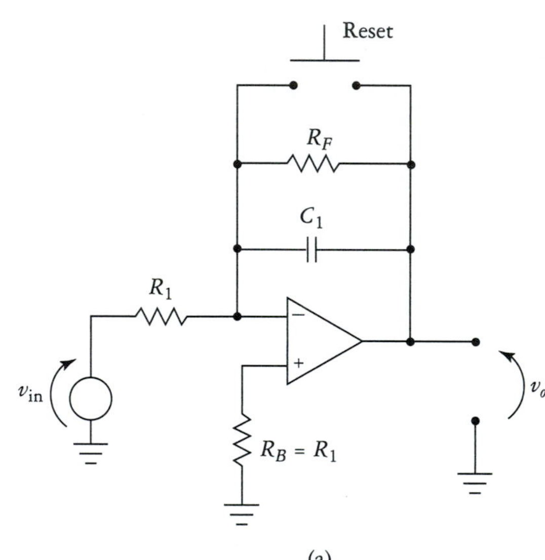

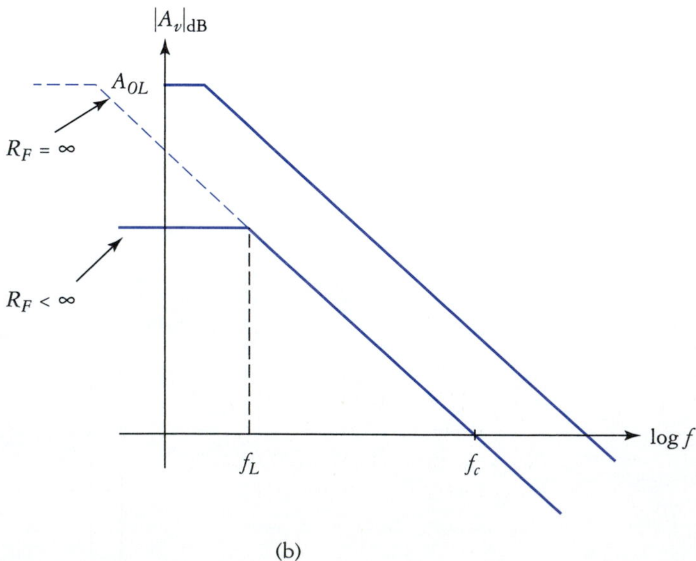

FIGURE 13.53
Modified integrator and resulting frequency response.

Notice that a switch was included in Fig. 13.53. This allows the integrating capacitor to be discharged, resetting the integrator's output to zero. Often this switch would be implemented with an FET or an electronic switch (called an analog switch) in order to allow for more rapid and accurate reset timing.

The Miller Effect

The inverting integrator we have been examining is sometimes called a *Miller integrator.* This name is derived from the Miller effect, which allows us to split the integrating capacitor as shown in Fig. 13.54. The output capacitance has negligible effect on response because $R_o \cong 0\ \Omega$ and $C_o' \cong C_1$. The effective input capacitance is extremely large, however, because A_{OL} is very large. Why is this important? Well, here's what is happening. The Miller input capacitance is typically very large. For $A_{OL} =$ 200 k and $C_1 = 10\ \mu\text{F}$, we have

$$C_{\text{in}}' = (200\text{ k})(10\ \mu\text{F}) = 2 \text{ farads}$$

That's a large amount of capacitance! The input time constant is

$$\tau_{\text{in}} = R_1 C_{\text{in}}' \tag{13.79}$$

Now, assuming a typical value of $R_1 = 1\text{ k}\Omega$, the time constant of the input network is

$$\tau_{\text{in}} = (1\text{ k}\Omega)(2\text{ F}) = 2000 \text{ sec}$$

which is a fairly long time constant, but this is typical of most integrators.

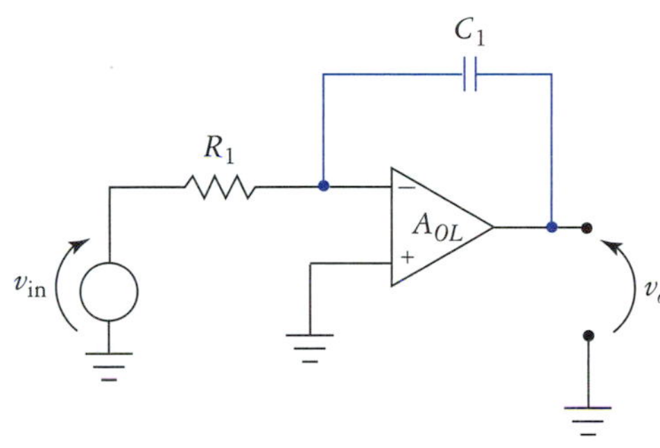

⇓

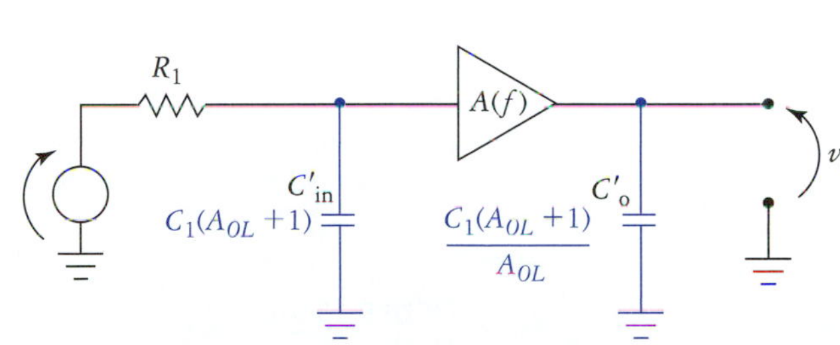

FIGURE 13.54
Miller's theorem applied to the integrator.

The significance of the Miller effect can be understood by considering that when we apply a dc voltage to the input, the Miller input capacitance begins charging exponentially. In order for the circuit to approximate ideal integration, we must ensure that the period of integration T_{int} is much smaller than the input time constant τ_{in}. Another reasonable rule of thumb is

$$T_{\text{int}} < \tau_{\text{in}}/10 \tag{13.80}$$

We know that the application of a dc input voltage should cause the output to produce a linear ramp. Meeting the condition of Eq. (13.80) keeps the integrator operating on the fairly linear portion of the charging curve as shown in Fig. 13.55.

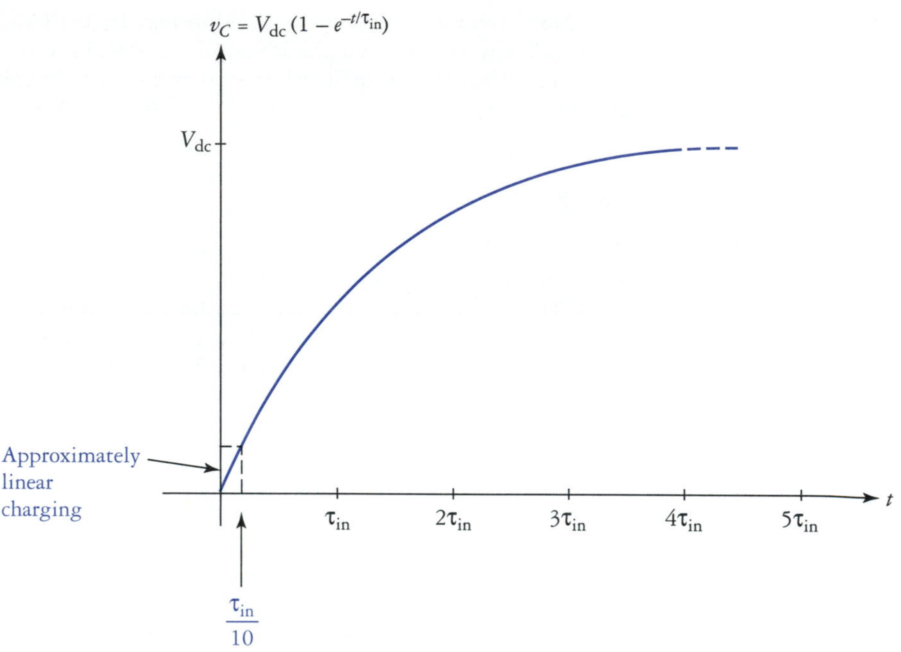

FIGURE 13.55
For time intervals that are short relative to the time constant, the capacitor charging voltage is nearly linear.

A Noninverting Integrator

The instantaneous current flow into a capacitor, as shown in Fig. 13.56(a), is given by

$$i_C = C\frac{dv_C}{dt} \tag{13.81}$$

If we charge a capacitor using a current source, as shown in Fig. 13.56(b), then by definition the capacitor current is forced to equal the input current. That is, $i_C = i_{in}$. We substitute this into Eq. (13.81) and solve for v_C by integrating both sides, which eliminates the derivative, producing

$$\begin{aligned} v_C &= \int \frac{i_{in}}{C}\,dt \\ &= \frac{1}{C}\int i_{in}\,dt \end{aligned}$$

In a practical sense, what makes this relationship useful here is that if we use a V/I converter to supply the input current, the resulting network is an integrator, because the capacitor voltage is then proportional to the integral of the input voltage.

FIGURE 13.56
Defining the capacitor charging voltage as the integral of charging current.

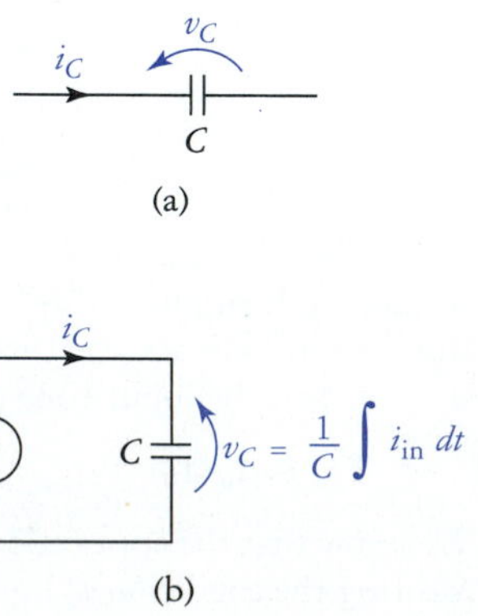

i_C v_C C

(a)

i_C i_{in} C $v_C = \frac{1}{C}\int i_{in}\,dt$

(b)

Using a capacitor as a load on a Howland current source produces a noninverting integrator. This is shown in Fig. 13.57. The Howland current source causes the capacitor current to be proportional to v_{in}. This in turn causes the output voltage to be proportional to the integral of v_{in}. As stated in Section 13.3, we require $R_4/R_3 = R_2/R_1$. If this is true, then the output voltage is

$$v_o = \frac{R_1 + R_2}{R_1^2 C} \int v_{\text{in}}\, dt \tag{13.82}$$

If we set all resistors equal ($R_1 = R_2 = R_3 = R_4$), then this simplifies to

$$v_o = \frac{2}{R_1 C} \int v_{\text{in}}\, dt \tag{13.83}$$

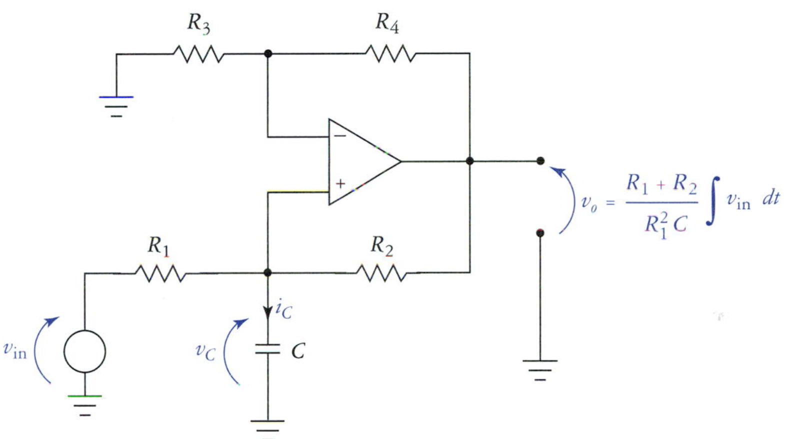

FIGURE 13.57
A noninverting integrator.

13.5 MISCELLANEOUS LINEAR OP AMP APPLICATIONS

There are thousands of linear applications suitable for realization using the circuits presented so far in this chapter. Some of these applications are very important, but they are also so involved that there is not enough room to cover them sufficiently in this chapter. These topics are covered later in their own chapters. However, a few interesting applications and modifications of the basic op amp circuits presented here deserve some mention.

Adders and Subtractors

We have seen that the op amp can be viewed as a real-time mathematical functional block in the section on differentiators and integrators. It is also possible to view the basic inverting, noninverting, summing, and differential amplifiers as adders and subtractors. Consider the following linear equation:

$$v_o = 2x + y - 3z \tag{13.84}$$

where x, y, and z might represent variables such as temperature, pressure, and so on in some industrial process. Voltages that are proportional to these physical quantities could be produced using various transducers. We could then use op amp circuits to combine these signals according to Eq. (13.84) and produce v_o continuously in real time. This output voltage might be used to control some piece of equipment, or perhaps it would simply drive a meter that indicates the status of the process under examination.

There are many ways to realize Eq. (13.84) using op amps. One possibility is shown in Fig. 13.58. The output of amplifier A_1 is found using basic summing amplifier relationships to be

$$v_{o(1)} = -(2x + y) = -2x - y$$

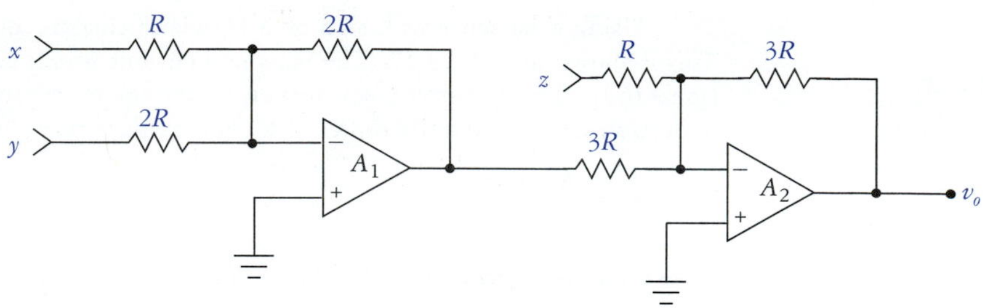

FIGURE 13.58
Op amp realization of a linear equation.

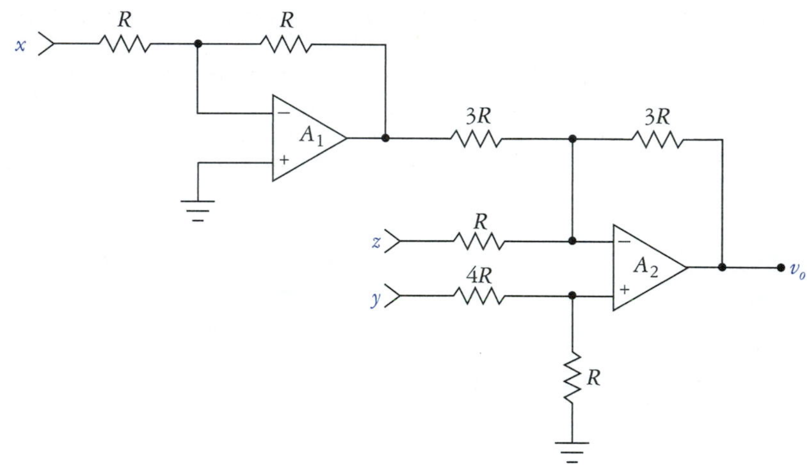

FIGURE 13.59
Alternate form for Fig. 13.58.

The overall output is

$$\begin{aligned} v_o &= -(v_{o(1)} + 3z) \\ &= -(-2x - y + 3z) \\ &= 2x + y - 3z \end{aligned}$$

Another possible implementation of Eq. (13.84) is shown in Fig. 13.59. The analysis of this version is a little more messy, but is interesting to perform. The output of A_1 is

$$v_{o(1)} = -2x$$

The output of A_2 is the superposition of the responses to its three input variables, which is

$$v_o = y\left(\frac{R}{R + 4R}\right)\left(\frac{R_{\text{eq}} + 3R}{R_{\text{eq}}}\right) - \left(-2x\frac{3R}{3R} + z\frac{3R}{R}\right) \tag{13.85}$$

The equivalent resistance R_{eq} is found to be

$$\begin{aligned} R_{\text{eq}} &= \frac{1}{\dfrac{1}{3R} + \dfrac{1}{R}} \\ &= \frac{1}{\dfrac{1}{3R} + \dfrac{3}{3R}} \\ &= \frac{1}{\dfrac{4}{3R}} \\ &= \frac{3R}{4}\ \Omega \end{aligned}$$

Substituting this into Eq. (13.85) yields

$$v_o = y\left(\frac{R}{R+4R}\right)\left(\frac{\frac{3R}{4}+3R}{\frac{3R}{4}}\right) - \left(-2x\frac{3R}{3R} + z\frac{3R}{R}\right)$$

$$= (y)(5)\left(\frac{1}{5}\right) - [(-2x)(1) + (z)(3)]$$

$$= y + 2x - 3z$$

$$= 2x + y - 3z$$

There are many other ways to implement this circuit as well (an infinite number of ways, actually), but the question that might be asked now is which version is preferable? Generally, we wish to use the minimum number of components, particularly op amps. In this case Fig. 13.58 uses six resistors, while Fig. 13.59 uses seven. The difference in cost is probably negligible, but a more subtle consideration favors the circuit in Fig. 13.58. The disadvantage of Fig. 13.59 is that A_2 experiences common-mode error effects. The circuit in Fig. 13.58 is strictly inverting and so has no common-mode error. This characteristic plus the slight cost advantage make Fig. 13.58 the better choice of the two circuits.

EXAMPLE 13.14

Determine the linear equation that is realized by the circuit shown in Fig. 13.60.

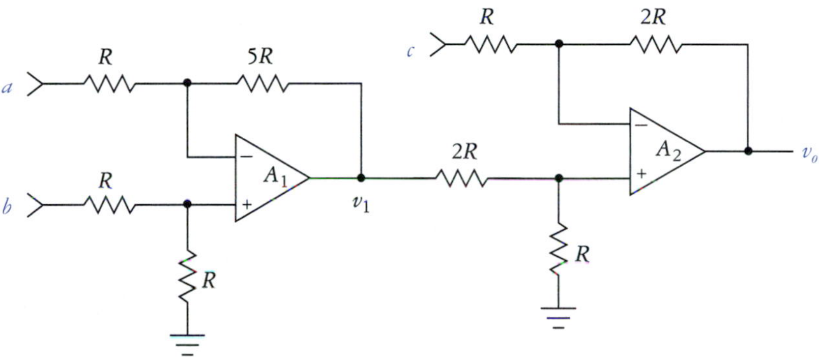

FIGURE 13.60
Circuit for Example 13.14.

Solution The output of A_1 is

$$v_1 = b\left(\frac{R}{2R}\right)\left(\frac{5R+R}{R}\right) - a\left(\frac{5R}{R}\right)$$

$$= 3b - 5a$$

The overall output is found to be

$$v_o = v_1\left(\frac{R}{3R}\right)\left(\frac{R+2R}{R}\right) - c\left(\frac{2R}{R}\right)$$

$$= (3b - 5a)\left(\frac{1}{3}\right)\left(\frac{3}{1}\right) - 2c$$

$$= -5a + 3b - 2c$$

Bridged Amplifiers

Bridging of amplifiers provides an easy way to provide an effective output voltage swing that is twice the normal maximum value. This also increases the maximum power that can be delivered to a load. A simple bridged amplifier is shown in Fig. 13.61. The output voltage is $v_o = v_1 - v_2$. The individual output voltages are found to be $v_1 = -v_{in}$ and $v_2 = v_{in}$, which gives us

$$\begin{aligned} v_o &= -v_{in} - v_{in} \\ &= -2v_{in} \end{aligned}$$

We can set the amplifier up for higher gain magnitude if desired, but in any case, in order to maintain symmetry we require

$$A_{v_1} = -A_{v_2} \tag{13.86}$$

FIGURE 13.61
Bridged amplifiers.

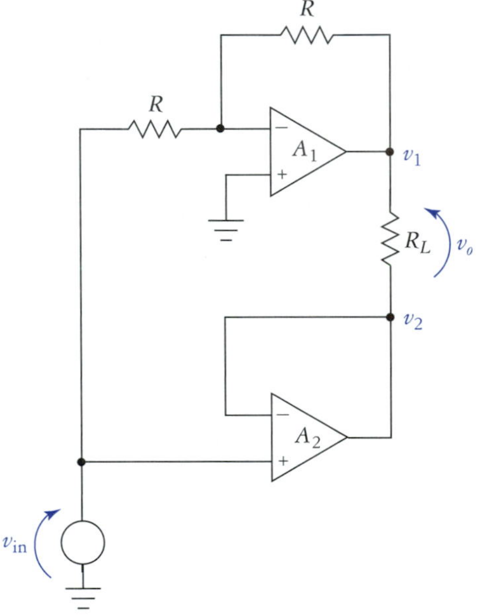

To understand how the output voltage is doubled, refer to Fig. 13.62. Let's assume the signal is just large enough to drive the amplifiers to saturation. For the positive peak of v_{in} we have $v_1 = -V_{sat}$ and $v_2 = +V_{sat}$. The peak output voltage is the difference between these voltages, which is

$$V_{o(pk)} = \pm 2V_{sat} \tag{13.87}$$

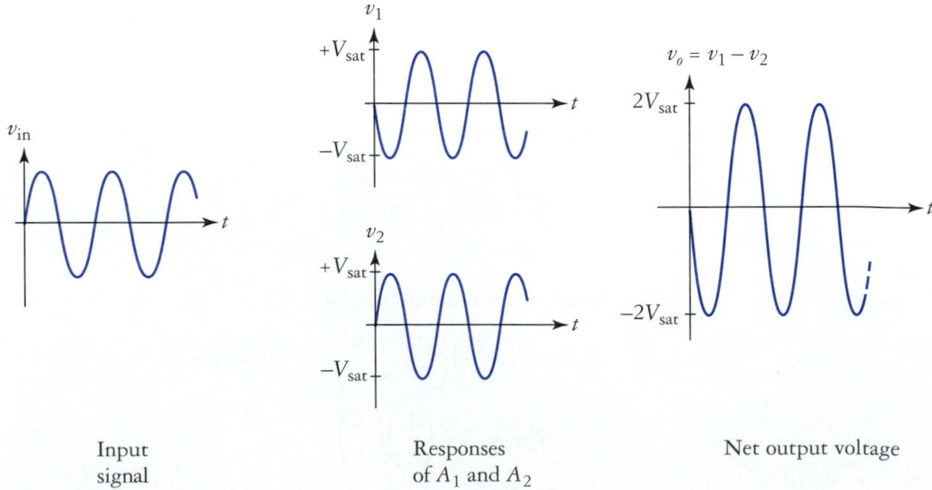

FIGURE 13.62
Development of the bridge amplifier output voltage.

Note

Remember, for a sine wave,

$P_{rms} = \frac{P_{max}}{2}$.

As a practical example, let's assume that in Fig. 13.60 we are using ± 15-V power supplies and 741 op amps. For each 741 we can reasonably expect to have $V_{sat} = \pm 13$ V. Thus, the maximum output voltage is $V_{o(max)} = 52\ V_{p\text{-}p}$ across a suitable load. However, the 741s will still go into current limiting at $I_o \cong \pm 20$ mA.

The increase in maximum output voltage produces an increase in maximum load power as well. Recall form Chapter 10 that the maximum (peak) power that can be delivered to a load is

$$P_{L(max)} = \frac{V_{o(max)}^2}{R_L}$$

and for a sinusoidal output signal, the maximum rms load power is

$$P_{o(rms)} = P_{o(ave)} = \frac{V_{o(max)}^2}{2R_L}$$

EXAMPLE 13.15

Two amplifiers used in an automobile sound system have $V_{o(max)} = \pm 6$ V into a 4-Ω load. Determine the peak and rms power for a single amplifier and for a bridged connection of the two amps.

Solution For a single amplifier, the maximum load power is

$$P_{L(max)} = \frac{V_{o(max)}^2}{R_L} = \frac{6^2}{4} = 9\text{ W}$$

$$P_{rms} = 4.5\text{ W}.$$

For the bridged configuration

$$P_{L(max)} = \frac{12^2}{4} = 36\text{ W}$$

$$P_{rms} = 18\text{ W}.$$

As shown in the preceding example, bridging two amplifiers increases the maximum load power delivery by a factor of 4. This should seem reasonable, since peak voltage doubles and power is proportional to the square of voltage.

The disadvantage of amplifier bridging is that it requires a floating load. If one side of such a load should accidentally be shorted to ground, the amplifier driving that side could be destroyed.

Increasing Op Amp Drive Capability

Some applications require current source/sink capability that is much greater than the typical op amp can deliver. In these situations it may be necessary to add some external circuitry to increase the current handling capability of the op amp.

The addition of an external push-pull output stage is shown in Fig. 13.63. Assuming matched transistors are used ($\beta_1 = \beta_2$), and symmetrical supply voltages, the maximum current drawn from the op amp is

$$I_{o(max)} \cong \pm \frac{V_{CC}}{\beta R_L} \tag{13.88}$$

Thus, the external transistors reduce the load on the op amp approximately by a factor of β.

The external transistors are operating class B, however, there will be little crossover distortion because the B-E junctions of the transistors are in the feedback loop of the op amp. This reduces the effective barrier potential V'_ϕ to

$$V'_\phi = \frac{V_\phi}{1 + \beta A_{OL}} \tag{13.89}$$

In most situations, the crossover distortion will be negligibly small.

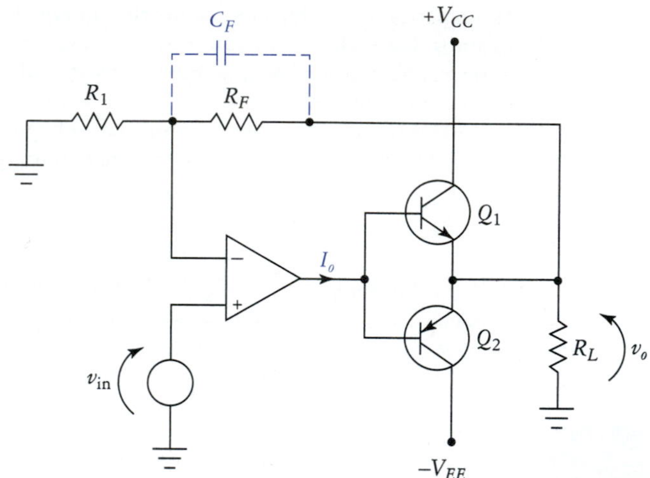

FIGURE 13.63
Buffering the output of an op amp to increase current-handling ability.

You may recall that the emitter-follower configuration is rather susceptible to oscillation. Adding an extra stage to the op amp will in general increase the possibility of oscillation as well. This is especially true at low closed-loop gain settings. In any case, circuit layout is critical, with good supply bypassing required. A small feedback capacitor C_F (around 0.01 μF or so) may also be used to roll off the high frequency gain of the op amp. This also helps to reduce the tendency for oscillation.

A boosted Howland current source is shown in Fig. 13.64. In this circuit, we still conform to the ratios given in Eq. (13.40)($R_4/R_3 = R_2/R_1$). The current source/sink limits of the circuit are generally given by the current-handling capabilitites of the external push-pull transistors.

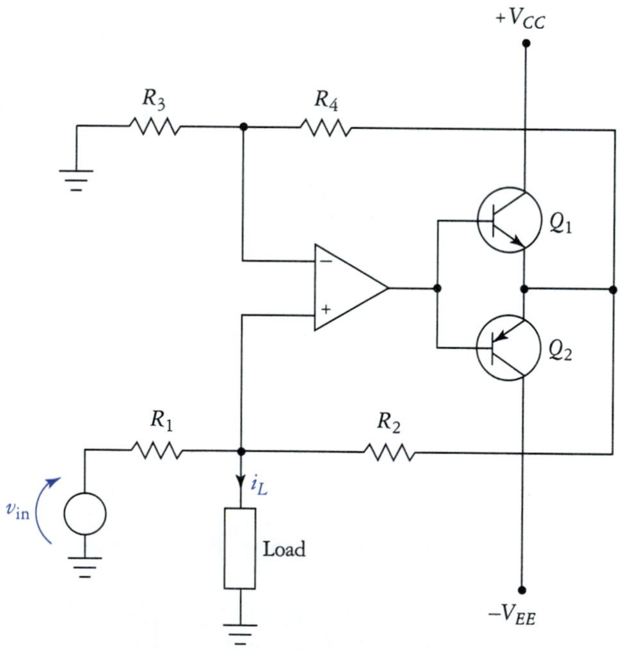

FIGURE 13.64
Buffered Howland current source.

13.6 COMPUTER-AIDED ANALYSIS APPLICATIONS

Although PSpice was designed to be used specifically for circuit analysis and design, other programs are available that are also useful. In this section we look at a few more PSpice examples, the use of a graphics calculator, and a math program called Matlab.

Summing Amplifier Simulation

The circuit used in Example 13.1 is redrawn in the PSpice schematic editor in Fig. 13.65. Notice that the 741 op amp symbol has been rotated and flipped such that the inverting input is at the top left. This also reverses the normal locations of the positive and negative power supply terminals, which can cause some confusion if you're caught off guard.

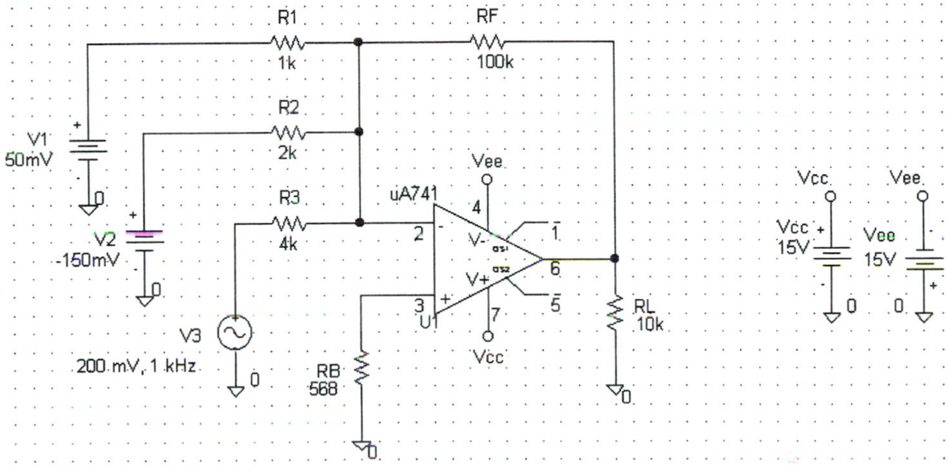

FIGURE 13.65
Summing amplifier.

The simulation setup was configured for a transient analysis from $t = 0$ to 3 ms, and a logarithmic ac frequency sweep from 10 Hz to 1 MHz, with VAC = 1 V. This analysis produced the output voltage waveform shown in Fig. 13.66, which is in very good agreement with the hand-calculated version of Fig. 13.5.

The frequency-domain response, in decibels, is shown in Fig. 13.67. Using the cursor to locate the −3-dB point, we find $f_c = 5.6$ kHz. This is also in good agreement with the value determined in the example (5.7 kHz).

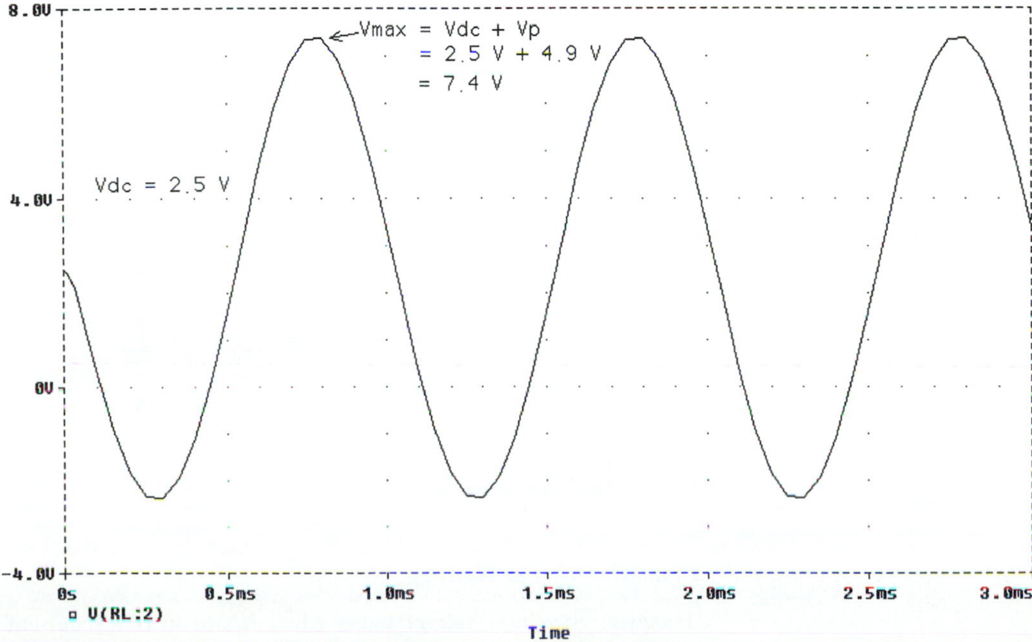

FIGURE 13.66
Output of the summing amp.

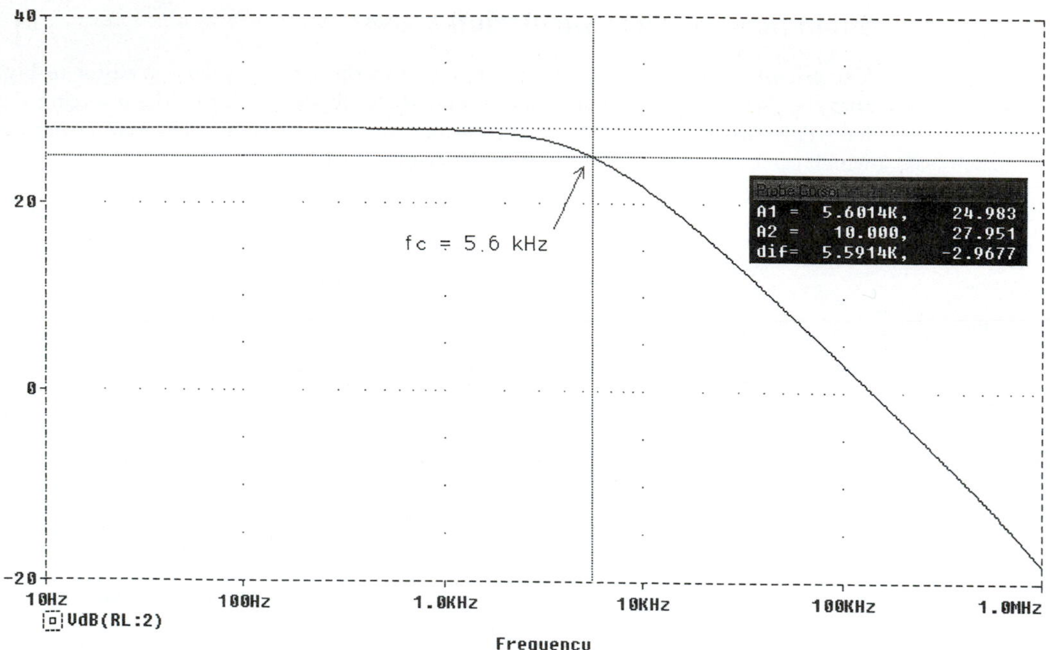

FIGURE 13.67
Frequency response of the summing amplifier.

Integrator and Differentiator Simulation

The schematic in Fig. 13.68 shows a differentiator followed by an integrator. The input to this circuit is supplied by a *piecewise-linear voltage source* (VPWL). We briefly discuss the VPWL source before getting into the details of the actual circuit to be simulated.

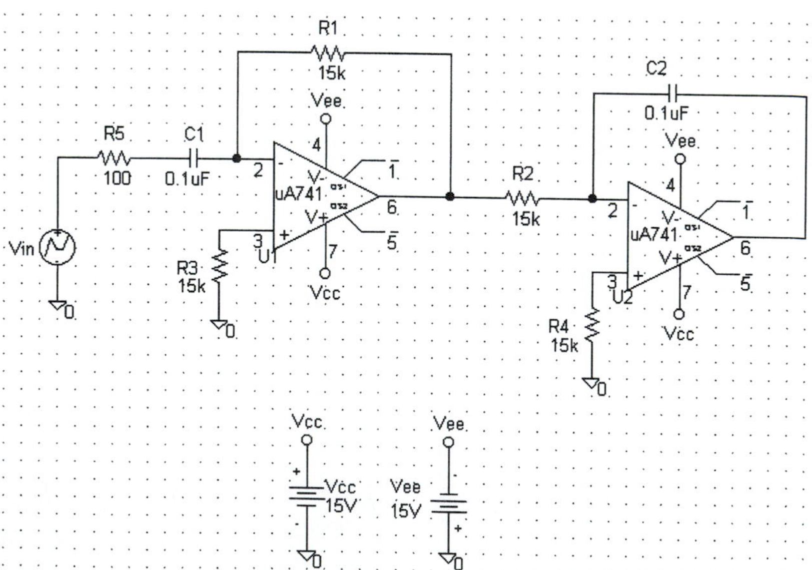

FIGURE 13.68
Cascaded differentiator and integrator.

Piecewise-Linear Sources

The piecewise-linear voltage source (there is also an IPWL current source too) allows us to specify waveforms that consist of linear ramps of arbitrary slope (rate of change) and start/stop times. This sounds more complicated than it really is. For example, if we wish to produce the waveform shown in

Fig. 13.69, we double-click on the VPWL symbol and enter the time and voltage associated with each point on the waveform as shown. Thus, we would enter $T_1 = 0$ ms, $V_1 = 0$ V, $T_2 = 5$ ms, $V_2 = 5$ V, $T_3 = 10$ ms, $V_3 = 5$V, $T_4 = 15$ ms, and $V_4 = 0$ V. If we are not interested in the behavior of the signal after $t = 15$ ms, the remainder of the time/voltage attributes are ignored.

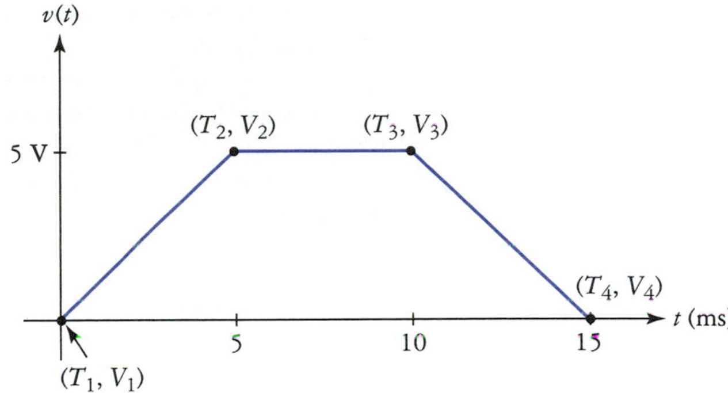

FIGURE 13.69
Defining a piecewise-linear voltage waveform.

Simulation of the Circuit

The circuit was simulated with the input voltage waveform specified as shown in Fig. 13.69. The responses of the differentiator and integrator are shown in Fig. 13.70. During the interval from $t = 0$ to 5 ms, the input voltage is given by $v_{\text{in}} = 1000t$ V. The response of the differentiator is

$$\begin{aligned} v_{o_1} &= -RC\frac{dv_{\text{in}}}{dt} \\ &= (-15\text{ k}\Omega \times 0.1\ \mu\text{F})(1000\text{ V/s}) \\ &= (-1.5\text{ ms})(1000\text{ V/s}) \\ &= -1.5\text{ V} \end{aligned}$$

The responses of the differentiator during the remaining time intervals of interest are similarly found to be

$t = 5$ ms to 10 ms, $v_{o_1} = 0$ V

$t = 10$ ms to 15 ms, $v_{o_1} = -1.5$ V

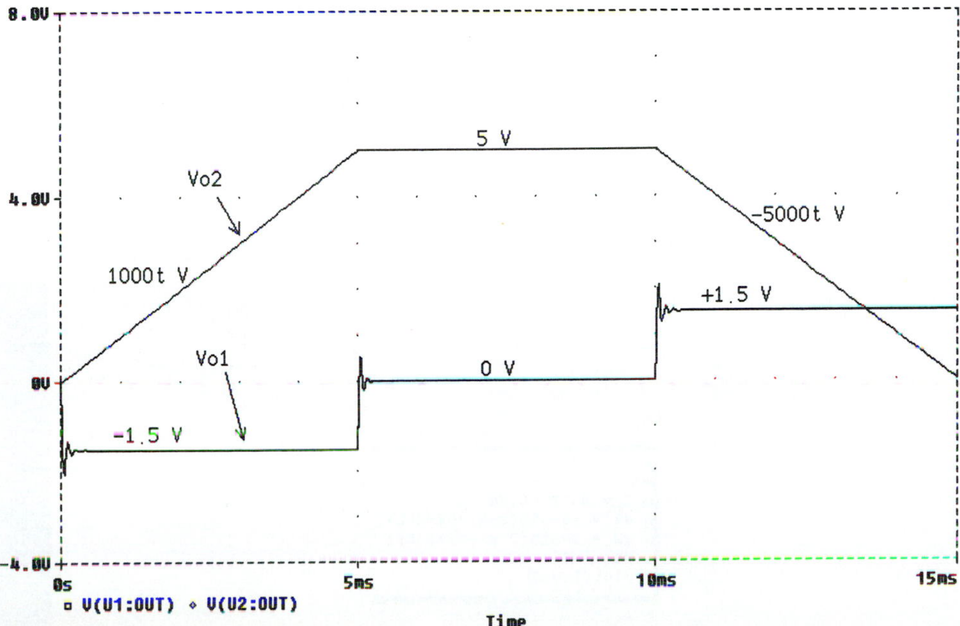

FIGURE 13.70
Response of the circuit to the PWL input.

This produces the stair-step waveform of Fig. 13.70. Notice that there is some ringing present in this signal. To remove this ringing, a 100-Ω resistor was placed in series with the differentiating capacitor C_1.

The integrator is processing the output of the differentiator. Because these are complementary circuits, using the same *RC* time constants, the output of the integrator is (ideally) identical to the original input waveform. This is apparent in Fig. 13.70, where the slopes of the signal are labeled for convenience. Notice that the ringing produced by the differentiator has virtually no effect on the output of the integrator. This is reasonable and we can look at this in two different ways. First, from a time-domain perspective the effect should be small because the net area under the graph of this noise is very small. The second way to view this situation is to realize that the integrator is a form of low-pass filter (see Fig. 13.51). Thus in the frequency domain, the ringing, which is low in amplitude to begin with, is a relatively high-frequency noise component that is attenuated greatly by the integrator.

Waveform Visualization

It can be very difficult to visualize the various signals and responses that occur in circuit analysis. This is especially true when we dealing with circuits like summing amplifiers, that have multiple time-varying sources present. Programs like PSpice provide an excellent environment for graphical representation of circuit behavior. However, other options are available that provide this capability as well. Graphics calculators like the Texas Instruments TI-8x series are useful in this context. Computer programs are also available that can be used to produce waveform graphs. Matlab is an example of a very popular and very powerful general-purpose mathematics program that can be used in electronic circuit analysis applications. Let's take a quick look at these options.

In Example 13.2 it was determined that the output of a summing amplifier was given by

$$v_o = -8\sin(2\pi 5000t)\text{ V} + 3\sin(2\pi 8000t)\text{ V}$$

This sort of signal is difficult to visualize mentally, but it can be graphed easily in Matlab by entering the series of statements shown in Fig. 13.71 in the command window. Matlab is instructed what to do by (usually) short programs like that of Fig. 13.71. Placing a semicolon at the end of a line prevents Matlab from immediately printing out all of the values of the variables on the screen, as each line is entered. Let's take a look at the structure of the program.

The first line steps t from 0 to 2 ms in 10-μs increments. This produces 200 sample points. The second line calculates the value of the 5-kHz signal component at each value of t. The third line does the same for the 8-kHz signal component. Because Matlab calculates trig functions using radians, we must include scaling of the frequency by 2π. Matlab interprets *pi* as a numeric constant, just as the symbol π would. The next line simply sums the two components at each sample point, while the last line produces a plot of v_o as a function of t. Matlab automatically scales the vertical axis to accommodate the amplitude of the signal, while the horizontal axis is set by the start and stop values of the independent variable t. The resulting graph is shown in Fig. 13.72.

If we enter the command *plot(t, v1, t, v2, t, vo)* as shown in Fig. 13.73, then the previous plot is replaced with a plot showing the individual components and the resultant sum, as shown in Fig.

```
>>
>>
>> t = 0:10e-6:2e-3;
>> v1 = -8*sin(2*pi*5000*t);
>> v2 = 3*sin(2*pi*8000*t);
>> vo = v1 + v2;
>> plot(t,vo)
```

FIGURE 13.71
Matlab listing for output voltage in Example 13.2.

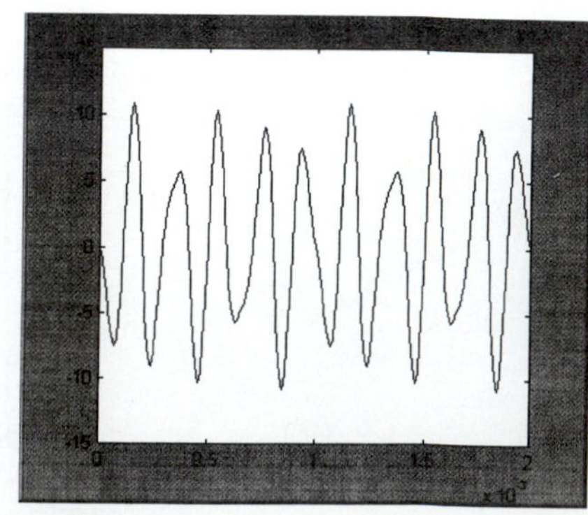

FIGURE 13.72
Matlab plot.

```
>>
>> t = 0:10e-6:2e-3;
>> v1 = -8*sin(2*pi*5000*t);
>> v2 = 3*sin(2*pi*8000*t);
>> vo = v1 + v2;
>> plot(t,vo)
>> plot(t,v1,t,v2,t,vo)
```

FIGURE 13.73
Listing to create plot of all output voltage components.

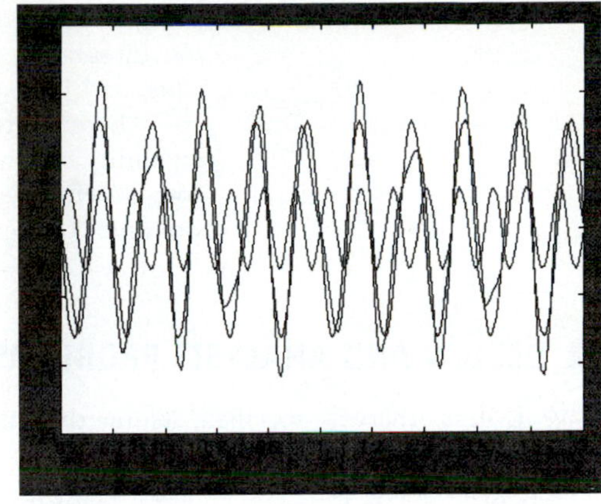

FIGURE 13.74
Resulting plot.

13.74. In this case, Matlab simply used all of the data that was previously calculated and stored in memory to produce this graph. Matlab has many capabilities that cannot be explored here, such as the ability to perform closed-form algebra and calculus. In addition, various extensions called *toolboxes* are available that allow easy analysis of feedback control systems, digital signal processing, and many other electronics-related tasks.

Finally, if we wished to plot the waveform of Fig. 13.72 using a TI-81 graphics calculator, for example, setting the calculator for operation in the radian mode, we would enter

$$Y_1 = -8\sin(2\pi 5000X) + 3\sin(2\pi 8000X)$$

where we substitute X for t as required by the calculator. The range could be set as follows. To match the graph in Fig. 13.72, we set $X_{min} = 0$ and $X_{max} = 2$ E -3. If we wish to use the entire viewing area of the calculator display, the min and max values of Y_1 are set to $\pm(V_1 + V_2)$, thus $Y_{min} = -11$ and $Y_{max} = 11$. This produces a graph much like that obtained using Matlab.

■ CHAPTER REVIEW

Op amps are used as basic linear circuit building blocks. Because it eliminates source interaction, the virtual ground allows the op amp to be used as a nearly ideal summing amplifier. The noise gain and, hence, the bandwidth of the summing amplifier is dependent on the equivalent resistance of all inputs, taken into account simultaneously. The averager is a special case of the summing amplifier.

The differential amplifier configuration allows the op amp to be used to process signals that contain large common-mode noise components. The instrumentation amplifier is a differential amplifier with buffered inputs. Instrumentation amplifiers are available in monolithic and hybrid IC forms. Most instrumentation amplifiers allow gain to be adjusted using a single variable resistor. Instrumentation amplifiers are often used to process the outputs of resistance bridges. A technique called active guard drive is used to further reduce the effects of common-mode input signals.

The basic VCVS, VCIS, ICVS, and ICIS structures can easily be implemented using op amps. The VCVS is the traditional voltage amplifier. The VCIS is used in applications where a load current must be controlled by a voltage signal. Such a load can be placed in the feedback loop of the op amp, or it may be ground referenced if the Howland current source is used.

To increase output voltage compliance, op amps are sometimes bridged. A bridged amplifier requires that the load be connected between the output terminals of two amplifiers. Thus the load driven by a bridged amplifier must be floating. A grounded load cannot be driven by a bridged amplifier.

Op amps are used very often to form differentiators and integrators. A differentiator is a circuit whose output voltage is proportional to the derivative of its input signal. Differentiators are very susceptible to error caused by high-frequency noise. This error can be minimized by placing a resistance in series with the differentiator input capacitor.

An integrator is a circuit for which the output voltage is proportional to the integral of the input signal. The integral mathematically describes the accumulation of the area under the curve of a function. Integrators are most susceptible to drift caused by input offset voltage. A reset switch is often included to allow the integrating capacitor to be discharged prior to integration being performed. The Howland current source can be used to implement a noninverting integrator. Integrators and differentiators are often used in control system designs, analog computer circuits, and in wave-shaping applications.

External transistors can be used in most applications in order to increase the current-handling capability of the op amp. If the B-E junction of such a transistor is within the feedback loop of the op amp, its effective barrier potential will be reduced to a very small value. Care must be taken in these situations to prevent circuit instability and oscillation.

DESIGN AND ANALYSIS PROBLEMS

Note: Unless otherwise specified, assume that the ideal op amp analysis rules apply.

13.1. Refer to Fig. 13-2. Given $R_F = 100$ kΩ, $R_1 = 10$ kΩ, $R_2 = 20$ kΩ, $R_L = 10$ kΩ, $V_1 = 25$ mV, $V_2 = 30$ mV, and GBW = 5 MHz, determine I_1, I_2, I_F, I_L, V_o, and the required value for R_B.

13.2. Refer to Fig. 13-2. Given $R_F = 10$ kΩ, $R_1 = 1$ kΩ, $R_2 = 2.2$ kΩ, $R_L = 10$ kΩ, $V_1 = 500$ mV, $V_2 = 1$ V, and GBW = 2 MHz, determine I_1, I_2, I_F, I_L, V_o, and the required value for R_B.

13.3. Refer to Fig. 13.2. Given $R_F = 50$ kΩ, $R_1 = 10$ kΩ, $R_2 = 20$ kΩ, $R_L = 10$ kΩ, $V_1 = 2.0$ V, $V_2 = -1.0$ V, and GBW = 1 MHz, determine I_1, I_2, I_F, I_L, V_o, and the required value for R_B.

13.4. Refer to Fig. 13.2. Given $R_F = 6.8$ kΩ, $R_1 = 1.5$ kΩ, $R_2 = 1.5$ kΩ, $R_L = 10$ kΩ, $V_1 = 1.20$ V, $v_2 = -1.50$ V, and GBW = 5 MHz, determine I_1, I_2, I_F, I_L, V_o, and the required value for R_B.

13.5. Refer to Fig. 13.2. Given $R_F = 22$ kΩ, $R_1 = 10$ kΩ, $R_2 = 4.7$ kΩ, $R_L = 2$ kΩ, $V_1 = 2.5$ V, $V_2 = 500 \sin(2\pi 1000t)$ mV, and GBW = 5 MHz, determine i_1, i_2, i_F, i_L, i_o, and the required value for R_B. Determine the maximum rate of change of v_o.

13.6. Refer to Fig. 13.2. Given $R_F = 5$ kΩ, $R_1 = 1$ kΩ, $R_2 = 2.5$ kΩ, $R_L = 5$ kΩ, $V_1 = 1.5$ V, $v_2 = -2 \sin(2\pi 5000t)$ V, and GBW = 5 MHz, determine i_1, i_2, i_F, i_L, v_o, and the required value for R_B.

13.7. Refer to Fig. 13.2. Assume that $R_F = 10$ kΩ, $R_1 = 5$ kΩ, $R_2 = 10$ kΩ, $R_L = 10$ kΩ, $V_1 = 2 \sin(2\pi 1000t)$ V, $v_2 = 3 \sin(2\pi 1000t + 30°)$ mV, and GBW = 5 MHz. Determine the values for i_1, i_2, i_F, i_L, and V_o, in instantaneous and polar phasor form. Determine the maximum rate of change of v_o.

13.8. Refer to Fig. 13.2. Assume that $R_F = 33$ kΩ, $R_1 = 3.3$ kΩ, $R_2 = 10$ kΩ, $R_L = 10$ kΩ, $V_1 = 250 \sin(5000t)$ mV, $v_2 = 300 \sin(5000t - 45°)$ mV, and GBW = 5 MHz. Determine the values for i_1, i_2, i_F, i_L, and v_o in instantaneous and polar phasor form. Determine the maximum rate of change of v_o.

13.9. Refer to Fig. 13.4. Assume that $R_F = 10$ kΩ, $R_1 = 5$ kΩ, $R_2 = 10$ kΩ, $R_3 = 10$ kΩ, $R_L = 10$ kΩ, $V_1 = 1.5 \angle 20°$ V, $v_2 = 3 \angle 50°$ V, and $V_3 = 1\angle 180°$ V. Determine the values for I_1, I_2, I_3, I_F, and V_o.

13.10. Refer to Fig. 13.4. Assume that $R_F = 39$ kΩ, $R_1 = 27$ kΩ, $R_2 = 22$ kΩ, $R_3 = 12$ kΩ, $R_L = 10$ kΩ, $V_1 = 1.5\angle 20°$ V, $V_2 = 3 \angle 50°$ V, and $V_3 = 1\angle 180°$ V. Determine the values for I_1, I_2, I_3, I_F, and V_o.

13.11. Refer to Fig. 13.6. Given $R_F = 10$ kΩ, $R_1 = 1$ kΩ, $R_{min} = 100$ Ω, R_x is a 10-kΩ potentiometer, and GBW = 10 MHz. Determine the minimum and maximum small-signal bandwidth of the amplifier.

13.12. Refer to Fig. 13.6. Given $R_F = 10$ kΩ, $R_1 = 10$ kΩ, $R_{min} = 10$ Ω, R_x is a 1-kΩ potentiometer, and GBW = 1 MHz. Determine the minimum and maximum small-signal bandwidth of the amplifier.

13.13. Refer to Fig. 13.6. Given $R_F = 100$ kΩ, $R_1 = 10$ kΩ, and GBW = 1 MHz, determine the required value for $R_x + R_{min}$ such that BW = 10 kHz and BW = 100 Hz.

13.14. Refer to Fig. 13.6. Given $R_F = 20$ kΩ, $R_1 = 5$ kΩ, and GBW = 5 MHz, determine the required value for $R_x + R_{min}$ such that BW = 10 kHz and BW = 100 Hz.

13.15. Draw the schematic diagram for a four-input averager using $R_F = 6.8$ kΩ.

13.16. Draw the schematic for a three-input weighted averager that realizes the following expression:

$$v_o = \frac{v_1 + 2v_2 + v_3}{3}$$

13.17. Determine V_o for the circuit in Fig. 13.75.

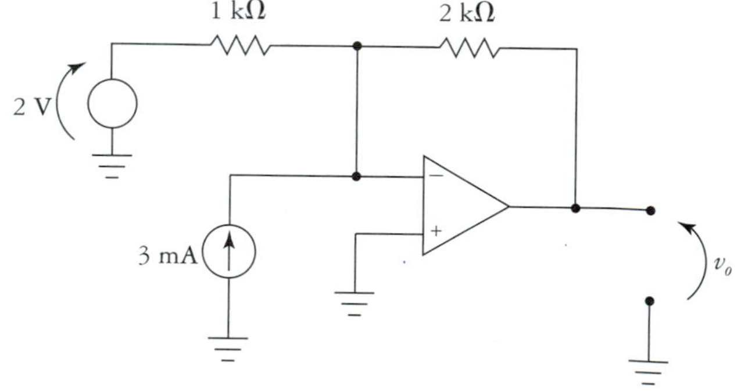

FIGURE 13.75
Circuit for Problem 13.17.

13.18. Refer to Fig. 13.76. Determine I_1, I_2, I_3, I_F, and V_o.

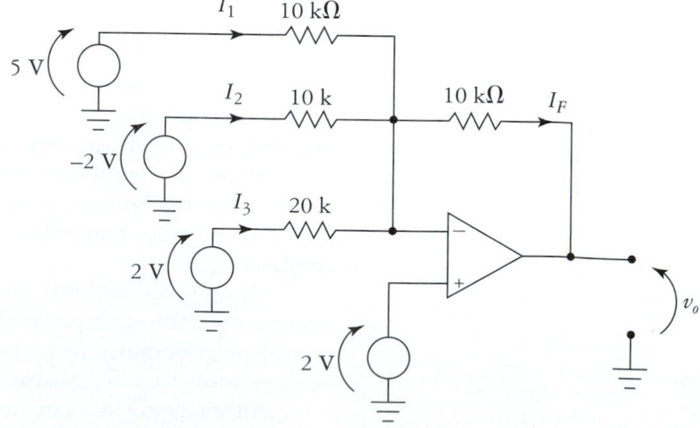

FIGURE 13.76
Circuit for Problem 13.18.

13.19. Refer to Fig. 13.77. Given $V_{in} = 3$ V, determine V_o.

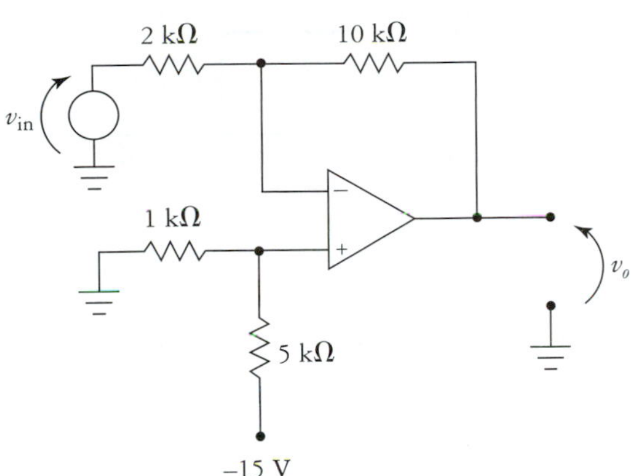

FIGURE 13.77
Circuit for Problems 13.19 and 13.20.

13.20. Refer to Fig. 13.77. Determine the value of V_{in} that will produce $V_o = 0$ V.

13.21. Refer to Fig. 13.78. This circuit is designed to bias an LED at a particular current level. Application of a time-varying input voltage causes a change in LED current, which causes a proportional variation in brightness. If $v_{in} = 2 \sin\omega t$ V, determine the expression for the resulting LED forward current.

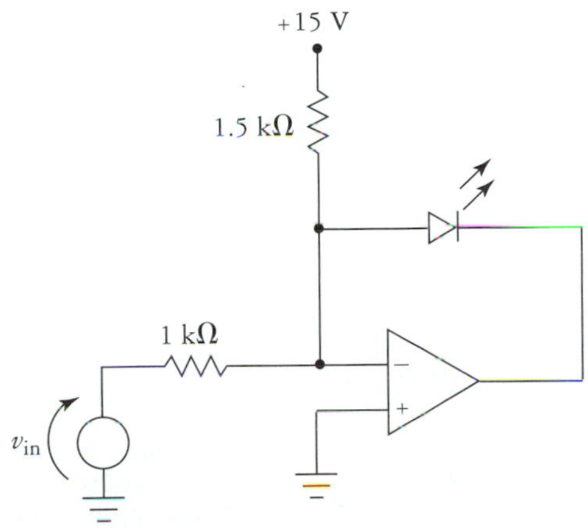

FIGURE 13.78
Circuit for Problem 13.21.

13.22. Refer to Fig. 13.79. Given $v_{in} = 500$ mV and $\beta_{Q_1} = 100$, determine V_o, I_F, I_B, P_L, and P_{DQ_1}.

13.23. Refer to Fig. 13.13. Assume $R_F' = R_F = 50$ kΩ, $R_1' = R_1 = 2$ kΩ, $V_1 = 2.02$ V, and $V_2 = 1.98$ V. Determine V_o for CMRR = ∞ and CMRR = 50 dB.

13.24. Refer to Fig. 13.13. Assume $R_F' = R_F = 100$ kΩ, $R_1' = R_1 = 5$ kΩ, $V_1 = -5.50$ V, and $V_2 = -5.75$ V. Determine V_o for CMRR = ∞ and CMRR = 50 dB.

13.25. Refer to Fig. 13.15. Assume $R_F' = R_F = 50$ kΩ, $R_1' = R_1 = 25$ kΩ, and $R_x = 10$ kΩ. Choose R_G such that $A_d = 10$. The input voltages are $V_1 = 2.02 + 5 \sin\omega t$ V and $V_2 = 1.98 + 5 \sin\omega t$ V. Determine v_{id}, v_{icm}, and v_o for CMRR = ∞ and CMRR = 50 dB.

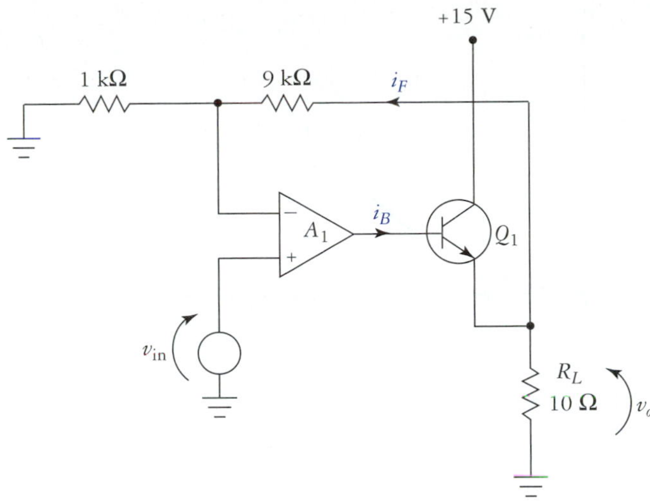

FIGURE 13.79
Circuit for Problem 13.22.

13.26. Refer to Fig. 13.15. Assume $R_F' = R_F = 20$ kΩ, $R_1' = R_1 = 20$ kΩ, and $R_x = 20$ kΩ. Choose R_G such that $A_d = 25$. The input voltages are $V_1 = 2.02 + 5 \sin\omega t$ V and $V_2 = 1.98 + 5 \sin\omega t$ V. Determine v_{id}, v_{icm}, and v_o for CMRR = ∞ and CMRR = 50 dB.

13.27. Refer to Fig. 13.18(a). Assume that the op amp has an output voltage compliance of ±13 V and an output current compliance of ±20 mA. Given $R_1 = 1$ kΩ and $R_L = 100$ Ω, determine the maximum input voltage that can be applied. Which output compliance specification is the limiting factor?

13.28. Refer to Fig. 13.18(b). Assume that the op amp has an output voltage compliance of ±13 V and an output current compliance of ±20 mA. Given $R_1 = 1$ kΩ and $R_L = 1$ kΩ, determine the maximum input voltage that can be applied. Which output compliance specification is the limiting factor?

13.29. Refer to Fig. 13.33. Given: $C_1 = 0.22$ μF and $R_F = 4.7$ kΩ. Where indicated, either sketch v_o or write the equation for v_o for each of the following input voltages:

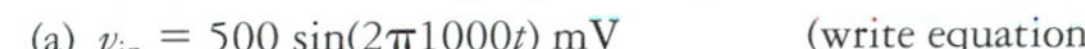

(a) $v_{in} = 500 \sin(2\pi 1000t)$ mV	(write equation)
(b) $v_{in} = 2 \cos(2\pi 250t)$ V	(write equation)
(c) $v_{in} = -5000t$ V	(write equation)
(d) $v_{in} = 1000t^2 + 2000t$ V	(write equation and sketch output)
(e) v_{in} shown in Fig. 13.80	(sketch output)
(f) v_{in} shown in Fig. 13.81	(sketch output)

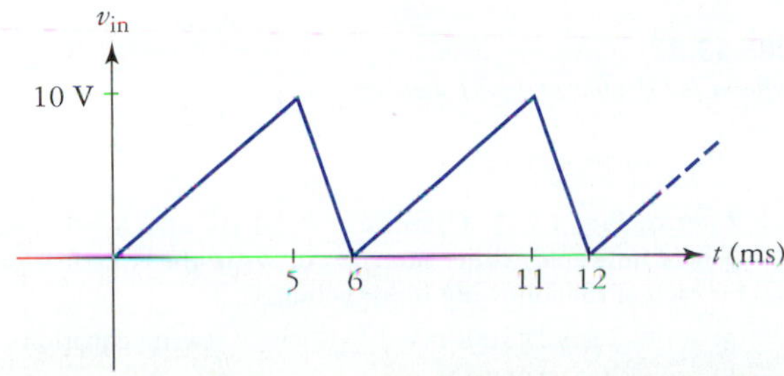

FIGURE 13.80
Waveform for Problems 13.29 and 13.30.

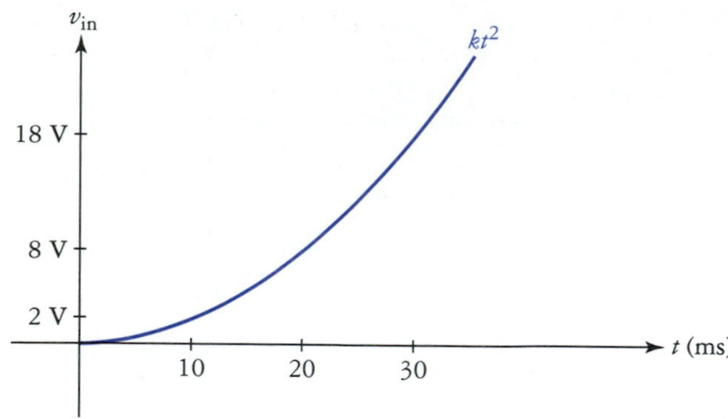

FIGURE 13.81
Waveform for Problems 13.29 and 13.30.

13.30. Refer to Fig. 13.33. Given: $C_1 = 0.1\ \mu F$ and $R_F = 3.3\ k\Omega$. Where indicated, either sketch v_o or write the equation for v_o for each of the following input voltages:

(a) $v_{in} = -250 \sin(2\pi 1000t)$ mV (write equation)
(b) $v_{in} = 2 \sin(2\pi 250t + 30°)$ V (write equation)
(c) $v_{in} = 8000t^{1/2}$ V (write equation)
(d) $v_{in} = 5000t^2 - 6000t$ V (write equation and sketch output)
(e) v_{in} shown in Fig. 13.80 (sketch output)
(f) v_{in} shown in Fig. 13.81 (sketch output)

13.31. Refer to Fig. 13.45. Given: $C_1 = 0.1\ \mu F$ and $R_F = 10\ k\Omega$. Where indicated, either sketch v_o or write the equation for v_o for each of the following input voltages:

(a) $v_{in} = 2 \sin(2\pi 100t)$ mV (write equation)
(b) $v_{in} = 2 \sin(1000t)$ V (write equation)
(c) v_{in} shown in Fig. 13.82 (sketch output)
(d) v_{in} shown in Fig. 13.83 (sketch output)

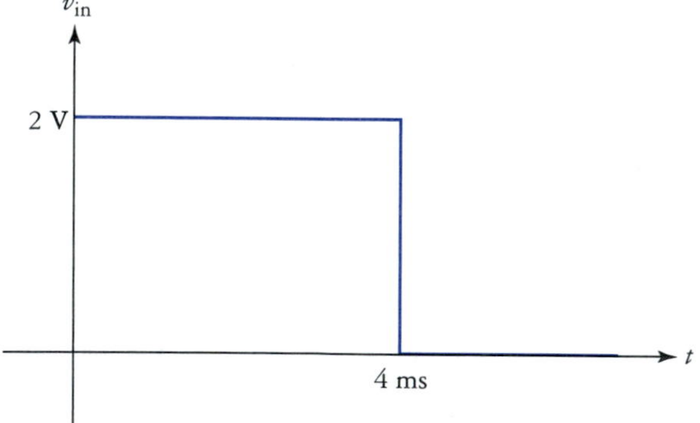

FIGURE 13.82
Waveform for Problems 13.31 and 13.32.

13.32. Refer to Fig. 13.45. Given: $C_1 = 10\ \mu F$ and $R_F = 5\ k\Omega$. Where indicated, either sketch v_o or write the equation for v_o for each of the following input voltages:

(a) $v_{in} = 2 \sin(2\pi 100t)$mV (write equation)
(b) $v_{in} = 2 \sin(1000t)$ V (write equation)
(c) v_{in} shown in Fig. 13.82 (sketch output)
(d) v_{in} shown in Fig. 13.83 (sketch output)

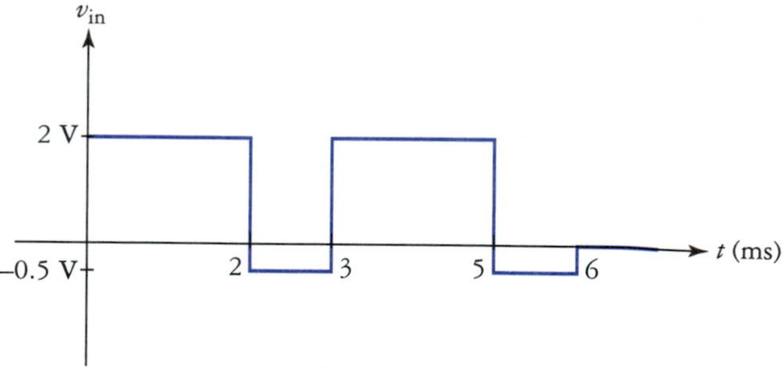

FIGURE 13.83
Waveform for Problems 13.31 and 13.32.

13.33. Determine the expression for v_o in Fig. 13.84(a).

13.34. Determine the expression for v_o in Fig. 13.84(b).

13.35. Refer to the bridge amplifier in Fig. 13.85. Given $R_1 = 5\ k\Omega$, and $R_2 = R_3 = 10\ k\Omega$, determine the required value for R_4 such that symmetrical operation is achieved.

13.36. Refer to the bridge amplifier in Fig. 13.85. Given $R_2 = 10\ k\Omega$, $R_3 = 15\ k\Omega$, and $R_4 = 30\ k\Omega$, determine the required value for R_1 such that symmetrical operation is achieved.

■ TROUBLESHOOTING PROBLEMS

13.37. Refer to Fig. 13.15. Assume that R_G is a 10-kΩ potentiometer. Suppose that this potentiometer has defects that cause the resistance to become an open circuit at some settings of its rotation. In general, what effect would setting the potentiometer to one of these positions have on the differential gain of the circuit?

13.38. Refer to Fig. 13.16. Suppose that the upper-right thermistor became open circuited. What effect would this fault have on v_o?

13.39. A circuit like that in Fig. 13.63 is constructed. When power is applied, both transistors begin to heat up excessively. Assuming a reasonable load is applied and the amplifier is not being overdriven, what is a likely cause for this problem?

■ COMPUTER PROBLEMS

13.40. Refer to the circuit in Fig. 13.86. The inputs to this amplifier are driven by VSIN sources. Use the results of a transient analysis with Voffset_1 = 5.00 V and Voffset_2 = 5.10 V. Generate a logarithmic plot of the common-mode gain (in decibels) of the circuit as a function of frequency from 10 Hz to 1 MHz.

13.41. Simulate the bridge amplifier of Fig. 13.61 using µA741 op amps, operating from a ± 15-V supply. Determine the voltage gain using $V_{in} = 1$ V. Determine the small-signal bandwidth of the circuit.

13.42. Use the VPWL source to simulate the waveform in Fig. 13.86. Drive a 10-kΩ resistor with this source. Plot the resulting load current using PROBE. The trace designation will probably be entered as $-I(R1)$. The negative sign is used to produce the correct polarity, as PSpice uses passive sign convention. PSpice can be used to plot the derivative of a signal by using the prefix *d*. Plot the derivative of the current by using $d(-I(R1))$. Does the waveform appear as you would expect? Plot the derivative of the input voltage. Comment on the amplitude of this signal.

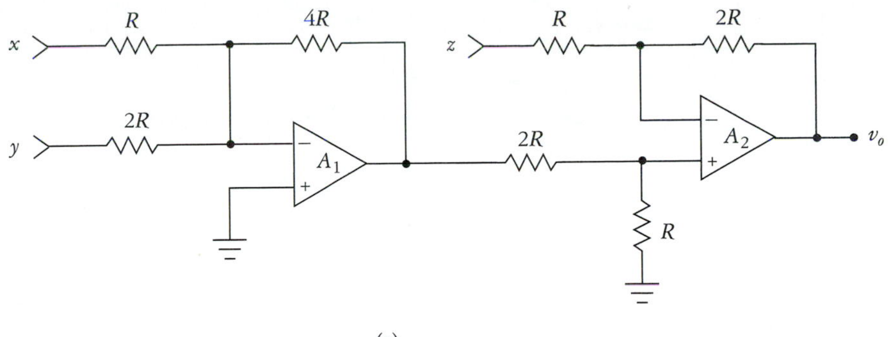

(a)

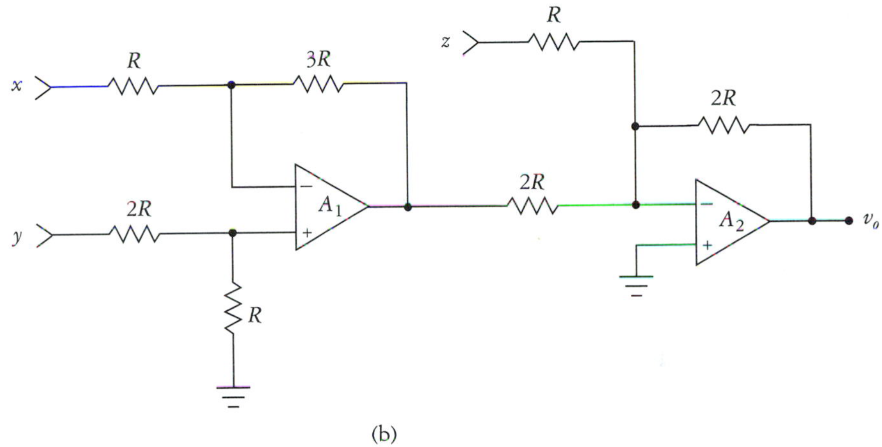

(b)

FIGURE 13.84
Circuits for Problems 13.33 and 13.34.

FIGURE 13.85
Circuit for Problems 13.35 and 13.36.

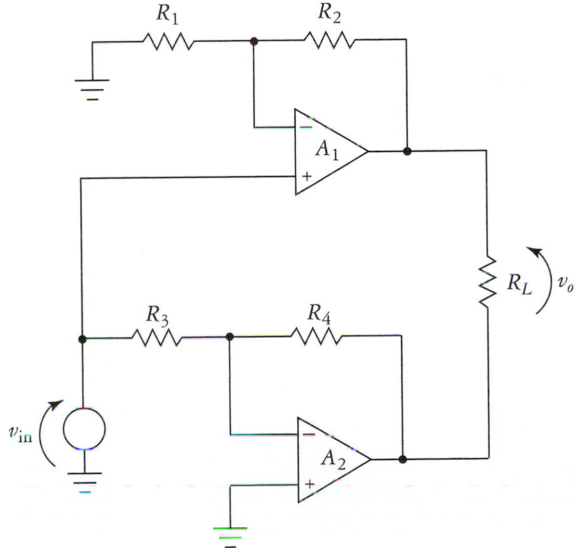

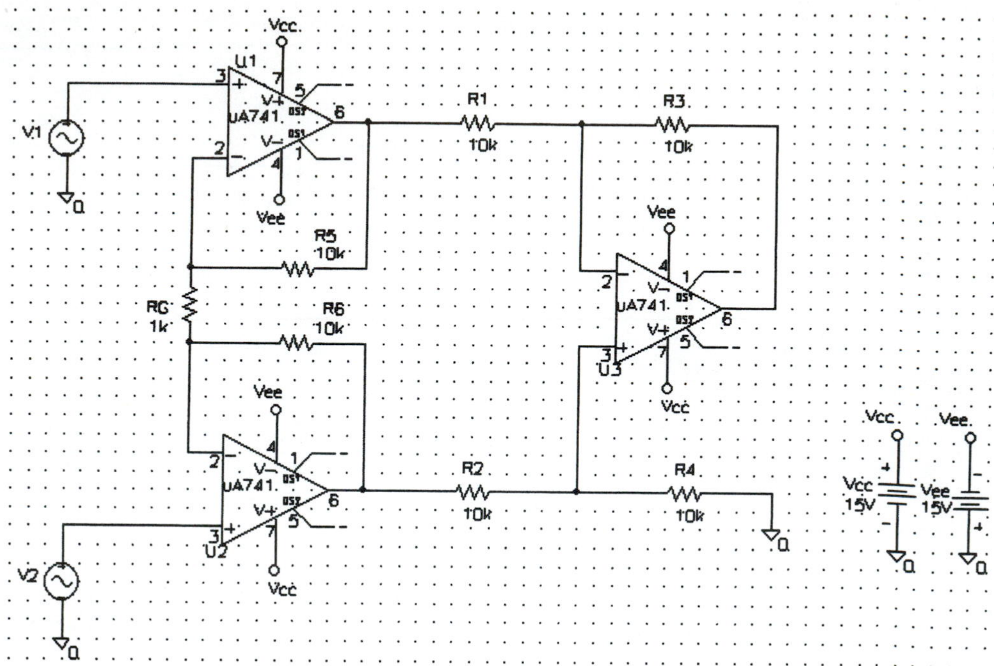

FIGURE 13.86
Circuit for Problems 13.40 and 13.42.

CHAPTER 14

Active Filters

BEHAVIORAL OBJECTIVES

On completion of this chapter, the student should be able to:

- Describe the basic filter response types.
- Explain the effect of damping on filter response.
- Identify the basic active filter circuit structures.
- Describe the advantages of active filters relative to passive filters.
- Describe the characteristics of various filter response types.
- Design equal-component, second-order VCVS active filters.
- Describe the operation of switched-capacitor filters.

INTRODUCTION

A filter can be described simply as a frequency-selective electrical circuit. That is, a filter removes or attenuates signals at some particular frequency or frequencies while passing all others. Traditionally, filters were designed using *RLC* networks of varying complexity. In general, such *RLC* filter networks are called *passive filters.* Active filters are normally inductorless, high-performance filters designed using op amps, resistors, and capacitors.

The subject of active filters is a very extensive one. In addition to the classical continuous-time, linear active filter structures, switched-capacitor active filters are also becoming more widespread. These filters are classified as sampled, discrete-time systems. This chapter presents an overview of some of the more commonly encountered continuous-time and switched-capacitor active filter topologies. Much of the basic terminology and behavior of filters has been covered in previous chapters (especially Chapter 11). We begin with the continuous-time active filters.

14.1 SOME FILTER FUNDAMENTALS

Let's begin by defining some of the notation we will be using. We use low-pass filters in most explanations for the time being. When dealing with filters, it is common to designate the frequency response with H. In the most technically complete sense, we define the frequency response of a filter as

$$H(j\omega) = \frac{V_o(j\omega)}{V_{\text{in}}(j\omega)} \tag{14.1}$$

or equivalently

$$H(jf) = \frac{V_o(jf)}{V_{\text{in}}(jf)} \tag{14.2}$$

where the presence of j indicates that this is a complex-valued function. That is, the response has a magnitude and a phase angle that are functions of frequency. If we are only interested in the magnitude of the response, then we can write

$$H(\omega) = |H(j\omega)| = \left|\frac{V_o(j\omega)}{V_{\text{in}}(j\omega)}\right| \tag{14.3}$$

or

$$H(f) = |H(jf)| = \left|\frac{V_o(jf)}{V_{\text{in}}(jf)}\right| \tag{14.4}$$

Often, to simplify the notation, we will simply use H because the functional relationship to frequency is implied.

Filter Order

Ideally, a filter should completely reject all signal energy in its stopband(s) and pass all signal energy within its passband(s). This behavior would produce what is called a *brick wall response.* In Fig. 14.1(a) we find that the brick wall response is the limiting response as we increase filter order n to infinity. For high-pass and low-pass filters recall that for each increase in filter order, the asymptotic rolloff rate of the filter increases by 20 dB/decade.

There are only two possible ways to implement a first-order low-pass filter; using an *RC* section or an *RL* section as shown in Fig. 14.2(a). We deal strictly with the *RC* version here. In either case, however, a plot of the frequency response will be like that shown in Fig. 14.1(a) with $n = 1$.

Second-order filters can be implemented in many different ways. For example, in Fig. 14.2(b) we have cascaded two first-order low-pass sections. The buffer amplifier simply prevents the second section from loading the first section. This could be considered a simple active filter (it does contain an active device), but other much more versatile circuits are used in active filter designs. This does illustrate, however, that op amps are useful in simple filter applications. The classical passive low-pass *LC* filter approach is shown in Fig. 14.2(c). Most of the time capacitors can be treated as nearly ideal components. However, inductors often behave much more nonideally, exhibiting significant winding resistance. This resistance is shown in Fig. 14.2(c) as r_s.

Damping and Response Shapes

Under no-load conditions, depending on the value of r_s, the shape of the response curve produced by Fig. 14.2(c) could look like any of those shown in Fig. 14.1(b). Basically, this is what's happening. The inductor and capacitor each serve to alternately store and then return energy to the circuit. For ideal components, there is no loss of energy during this process. For real components there will be some energy loss, most of which is in the form of heat via r_s. If r_s is small, then there is little energy loss and we may have a peaked response curve as shown in Fig. 14.2(b). If r_s is relatively large, then we obtain the droopy curve with more gradual initial rolloff, indicated with long dashes. If we have just the right value for r_s, then we obtain the solid curve shown in Fig. 14.2(b). Most often, this is the type of response we will shoot for.

The shape of a filter response curve is determined by the filter's *damping coefficient* α. You may recall that this was discussed briefly in Chapter 10 with tuned class C amplifiers. Depending on the

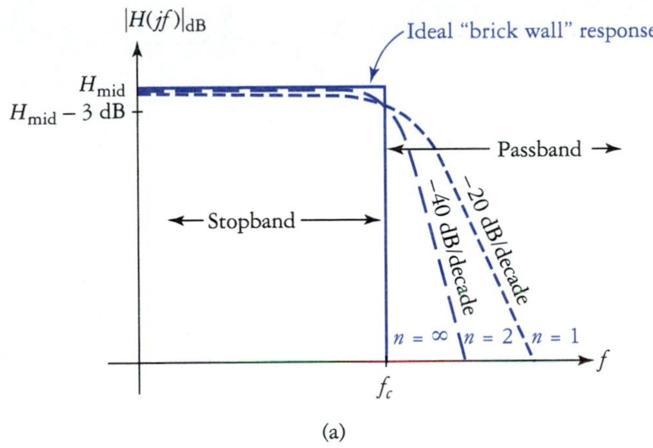

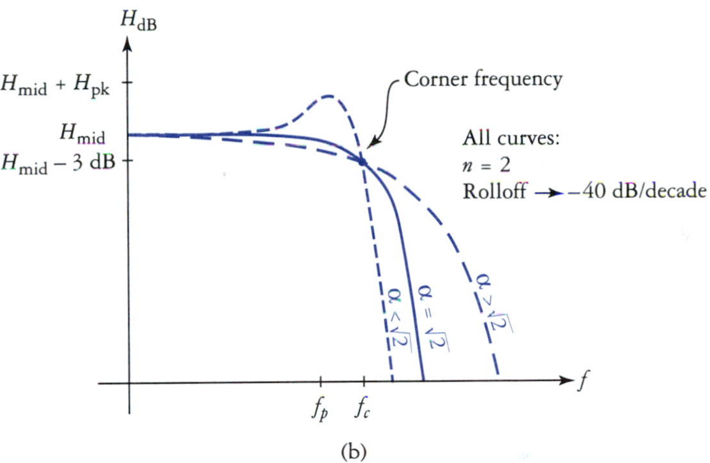

FIGURE 14.1
Low-pass filter response curves. (a) Effect of increasing filter order. (b) Effect of changing damping coefficient on a second-order filter.

type of filter, there are different equations that determine the value of α. For example, for Fig. 14.2(c), the damping coefficient is

$$\alpha = \left.\frac{r_s}{X_L}\right|_{f=f_o} \tag{14.5}$$

As it turns out, if we have $\alpha = \sqrt{2}$ then the corner frequency of the filter coincides with the resonant frequency ($f_o = 1/2\pi\sqrt{LC}$) of the circuit. Any other value of α causes the corner frequency to differ from the resonant frequency.

Critically Damped Response

Note
Critically damped filters are often called *Butterworth filters.* They are very commonly used.

Critical damping exists when we have

$$\alpha = \sqrt{2} \tag{14.6}$$

A critically damped second-order low-pass filter will produce the solid response curve of Fig. 14.1(b). Critical damping results in the flattest possible passband response for a given filter, which is generally a desirable characteristic.

Another name for the critically damped response is the *Butterworth response.* Critically damped filters are sometimes called *Butterworth filters.* The term *Butterworth* comes from the mathematician who studied the characteristics of a certain class of polynomials that are used in the analysis of maximally flat filter responses. Because of the nice mathematical properties they have, Butterworth filters are normally encountered more often than other filter types. Butterworth filters generally provide the

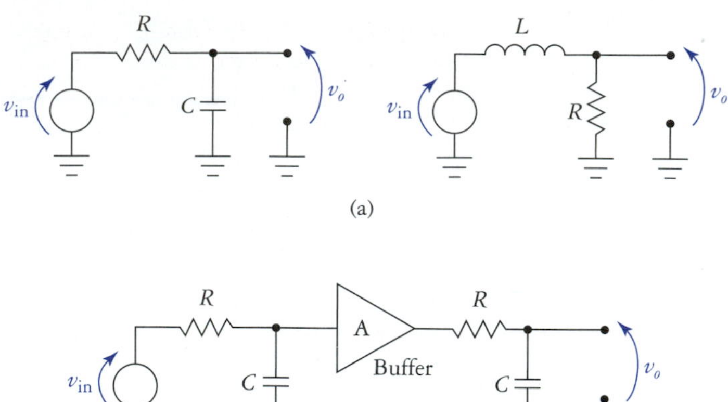

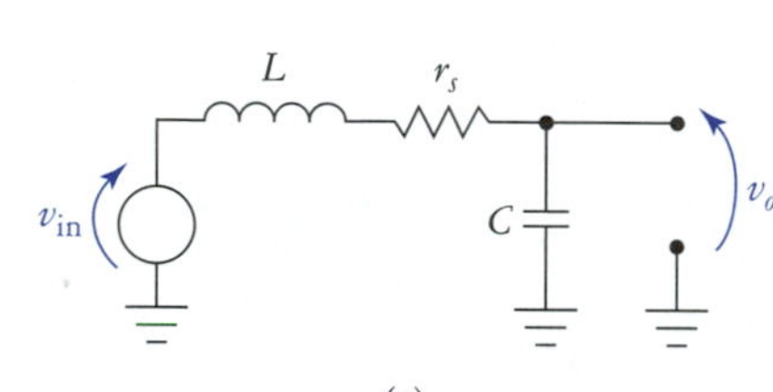

FIGURE 14.2
Low-pass filter structures. (a) First-order low-pass networks. (b) A second-order RC filter with buffering between sections. (c) Second-order low-pass LC filter.

best compromise between rejection of stopband signals, passband flatness, phase distortion near the corner, and overshoot in response to transients.

Underdamped Response

Note
Underdamped filters provide rapid initial rolloff but tend to overshoot and have very nonlinear phase response.

An underdamped filter will exhibit a peaked response as shown by the short-dashed curve in Fig. 14.1(b). An underdamped filter is characterized by

$$\alpha < \sqrt{2} \tag{14.7}$$

The smaller α is, the greater the amplitude of the peaking. If $\alpha = 0$ then the peak becomes infinite and the filter becomes an oscillator. In most practical filter applications, peaking is limited to around 6 dB or so.

Most underdamped filters are classified as *Chebyshev filters,* in honor of the Russian mathematician who studied the properties of certain polynomials that are useful in underdamped filter analysis. Thus, if we had a low-pass filter that exhibited a 3-dB peak in its passband, this filter may be called a 3-dB Chebyshev filter.

Underdamped filters are used when very rapid initial rolloff is required. We know that all second-order filters approach a rolloff rate of 40 dB/decade; however, an underdamped filter will roll off faster than this initially after the peak occurs. The trade-offs for this rapid initial rolloff are passband peaking and severe phase distortion around the peak. Underdamped filters also tend to ring significantly in response to transient inputs (recall the tuned class C amplifier of Chapter 10).

Overdamped Response

The curve shown by long dashes in Fig. 14.1(b) is typical of an overdamped response. Mathematically, a filter is overdamped if

$$\alpha > \sqrt{2} \tag{14.8}$$

Overdamped filters are usually the least commonly encountered. A common overdamped response is the *Bessel response,* so named after Bessel polynomials, which are used in mathematical descriptions of this response shape.

Overdamped filters have a very gradual initial rolloff, which may not provide adequate rejection of out-of-band signals for many applications. On the plus side, however, overdamped filters have a more linear phase response than the other filters. Also, overshoot and ringing are minimized with overdamped filters.

Filter Response Equations

Note

A_0 is the gain of the LP filter for $f_{in} \ll f_c$. A_∞ is the gain of the HP filter for $f_{in} \gg f_c$.

The most common filters will be either first or second order. The expressions for the magnitude response of first-order low-pass (LP) and high-pass (HP) filters were presented in Chapter 11, Eqs. (11.10) and (11.12). For first-order filters, the response phase angle is a rather simple function of frequency, so it is also included here:

$$H(jf) = \frac{A_0}{\sqrt{\left(\frac{f_{in}}{f_c}\right)^2 + 1}} \angle -\tan^{-1}\left(\frac{f_{in}}{f_c}\right) \tag{14.9}$$

and for the high-pass filter

$$H(jf) = \frac{A_\infty}{\sqrt{\left(\frac{f_c}{f_{in}}\right)^2 + 1}} \angle -\tan^{-1}\left(\frac{f_c}{f_{in}}\right) \tag{14.10}$$

In the case of the low-pass filter, the subscript on A_0 indicates that this is the gain of the filter at $f =$ 0 Hz. For the high-pass filter A_∞ represents the gain of the filter at very high frequencies (assuming the response extends to infinity). For passive filters, the numerator coefficient $A = 1$. For active filters, it is generally possible to have any value for A, which represents the passband gain of the filter. If we wish to express the response magnitude in decibels, we would use

$$H(f)_{dB} = 20 \log|H(jf)| \tag{14.11}$$

The generalized equations for the second-order responses are presented next. Filters of order $n > 2$ in general exhibit rather complex behavior. To keep the size of the equations manageable, the phase angle expressions will not be presented here. The second-order low-pass response magnitude is

$$|H(jf)| = \frac{A_0}{\left[\left(\frac{f_{in}}{f_c}\right)^4 + (\alpha^2 - 2)\left(\frac{f_{in}}{f_c}\right)^2 + 1\right]^{1/2}} \tag{14.12}$$

where α is the damping coefficient of the filter. The second-order high-pass response magnitude is

$$|H(jf)| = \frac{A_\infty}{\left[\left(\frac{f_c}{f_{in}}\right)^4 + (\alpha^2 - 2)\left(\frac{f_c}{f_{in}}\right)^2 + 1\right]^{1/2}} \tag{14.13}$$

For practical filter applications we restrict α to the range

$$0 < \alpha < 2 \tag{14.14}$$

If we hold f_c constant and only allow α to vary, Eqs. (14.12) and (14.13) form families of response curves like those shown in Fig. 14.3. For the Butterworth response ($\alpha = \sqrt{2}$) the LP and HP responses produce the same corner frequency. Notice, however, that the corner frequencies for the underdamped and overdamped responses have shifted in opposite directions from the expected corner frequency. If we wish to hold to the traditional definition of the corner frequency (response down 3 dB from the passband value), then we must determine correction factors based on $\alpha \neq \sqrt{2}$. This is another reason why the Butterworth response is used whenever possible. The curves used in Fig. 14.1(b) are based on frequency corrections being applied as α is varied. Just remember that unless some frequency scaling factor is provided, Eqs. (14.12) and (14.13) assume that f_c is the corner frequency of an equivalent Butterworth filter.

The amplitude of the response peak of an underdamped filter (either high pass or low pass) is given by

$$|H|_{pk} = \frac{A}{\alpha} \tag{14.15}$$

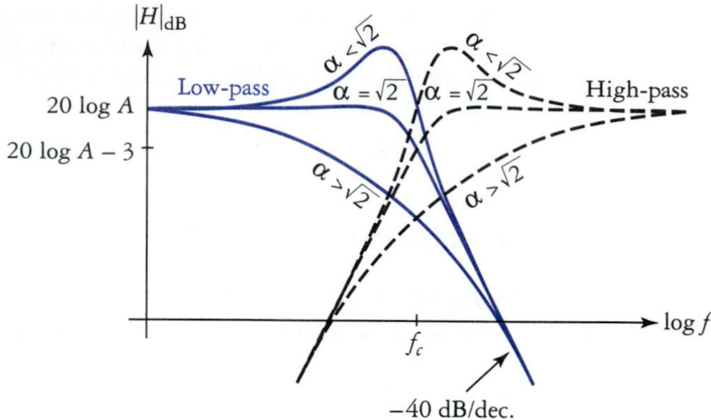

FIGURE 14.3
Second-order LP and HP response curves.

If we wish to know the amplitude of the peak in decibels, as usual we simply take the $\log_{10}$ and multiply by 20.

We usually prefer to work with Butterworth filters for many reasons, and here is another. In the case of the Butterworth response, we know that $\alpha = \sqrt{2}$. When this occurs, the middle term under the radical in Eqs. (14.12) and (14.13) is zero. Another important characteristic is that it can be shown that for Butterworth filters of *any* order, the response equations reduce as follows. For the low-pass filter

$$|H(jf)| = \frac{A_0}{\left[\left(\frac{f_{\text{in}}}{f_c}\right)^{2n} + 1\right]^{1/2}} \tag{14.16}$$

and for the high-pass filter we have

$$|H(jf)| = \frac{A_\infty}{\left[\left(\frac{f_c}{f_{\text{in}}}\right)^{2n} + 1\right]^{1/2}} \tag{14.17}$$

where n is the order of the filter.

EXAMPLE 14.1

A fourth-order high-pass filter has $\alpha = \sqrt{2}$, $A_\infty = 2$, and $f_c = 500$ Hz. Determine the magnitude of the response of the filter given $v_{\text{in}} = 3\sin(2\pi 100t)$ V in linear and decibel forms. Sketch the asymptotic Bode plot of the response and label the input frequency. Determine the rolloff of the filter in decibels per decade. Using the linear response magnitude, determine the amplitude of the resulting output voltage.

Solution The pertinent input signal parameters are $V_p = 3$ V and $f_{\text{in}} = 100$ Hz. Since this is a Butterworth filter ($\alpha = \sqrt{2}$), we can determine the response magnitude by direct application of Eq. (14.17):

$$\begin{aligned} |H(jf)| &= \frac{A_\infty}{\left[\left(\frac{f_c}{f_{\text{in}}}\right)^{2n} + 1\right]^{1/2}} \\ &= \frac{2}{\left[\left(\frac{500}{100}\right)^{8} + 1\right]^{1/2}} \\ &= 3.2 \times 10^{-3} \end{aligned}$$

In decibels, this is

$$\begin{aligned} |H(jf)| &= 20\log(3.2 \times 10^{-3}) \\ &= -50 \text{ dB} \end{aligned}$$

The rolloff rate is

$$\begin{aligned}\text{Rolloff} &= n20 \text{ dB/decade}\\ &= (4)(20 \text{ dB/decade})\\ &= 80 \text{ dB/decade}\end{aligned}$$

The output voltage is $V_o = V_{\text{in}} \times |H(jf)| = 3 \text{ V} \times 3.2 \times 10^{-3} = 9.6 \text{ mV}$.

The asymptotic Bode plot is shown in Fig. 14.4.

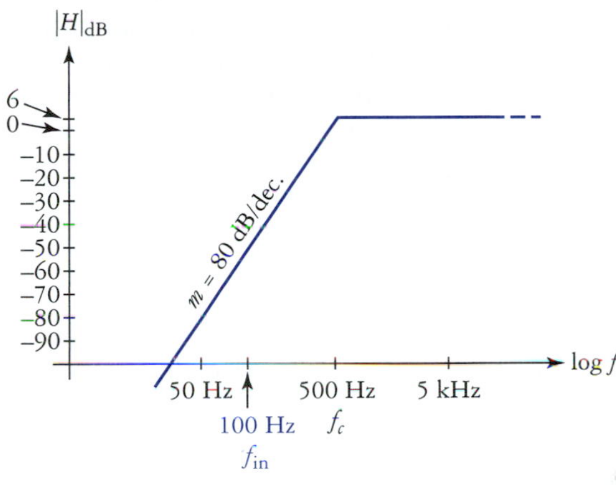

FIGURE 14.4
Bode plot for Example 14.1.

In the next example, we must determine the order of a filter necessary to achieve a desired amount of attenuation of a signal frequency in the stopband.

EXAMPLE 14.2

A certain application requires a low-pass filter with $f_c = 5$ kHz, $A_0 = 1$, with Butterworth response. At $f_{\text{in}} = 15$ kHz the response of the filter is to be down by at least 25 dB. Determine the required filter order.

Solution

Because we are dealing with decibels, we use Eq. (14.16) modified as follows:

$$|H(jf)| = 20 \log \frac{A_0}{\left[\left(\dfrac{f_{\text{in}}}{f_c}\right)^{2n} + 1\right]^{1/2}}$$

We must solve this equation for the filter order n, which yields the following unfriendly looking expression:

$$n = \frac{\log\left[\left(\dfrac{A_0}{\log^{-1}\left(-\dfrac{|H|}{20}\right)}\right)^2 - 1\right]}{2 \log\left(\dfrac{f_{\text{in}}}{f_c}\right)}$$

$$= \frac{\log\left[\left(\dfrac{1}{\log^{-1}\left(\dfrac{-25}{20}\right)}\right)^2 - 1\right]}{2 \log\left(\dfrac{15 \text{ k}}{5 \text{ k}}\right)}$$

$$= 2.6$$

It is not possible to have a fractional order filter. A second-order filter will not produce the required attenuation ($|H|_{f\,=\,15\text{ kHz}} = -19.1$ dB), so we must round up to the next higher order. Therefore, $n = 3$. The response for the third-order filter is $|H|_{f\,=\,15\text{ kHz}} = -28.6$ dB, which does satisfy the stated requirements.

14.2 EQUAL-COMPONENT SALLEN-KEY LP AND HP ACTIVE FILTERS

One of the most popular active filter topologies is the second-order, equal-component, Sallen-Key VCVS. This terminology is derived from the inventor's names and the fact that these filters are usually implemented using the op amp as a VCVS. These are called *equal-component filters* because we use equal-value resistors and capacitors in the frequency-determining portion of the filter. To simplify notation, we simply refer to these filters as VCVS filters. The basic low-pass VCVS is shown in Fig. 14.5. The passband gain A_0 of the filter is determined by R_F and R_1 according to the usual noninverting gain expression

$$A_0 = \frac{R_1 + R_F}{R_1} \tag{14.18}$$

The corner frequency of the filter is determined by the values of R and C according to

$$f_c = \frac{k_f}{2\pi RC} \tag{14.19}$$

where k_f is a frequency correction factor that is determined based on the damping coefficient of the filter, from Table 14.1. The HP and LP corrections are simply reciprocals of one another. Using the low-pass response to visualize these correction factors, we find that the -3-dB corner frequencies of the filters align as shown in Fig. 14.1(b). If these correction factors are not used (we simply set $k_f = 1$ regardless of α), then the response curves will correspond to those of Fig. 14.3.

Note
The gain of the equal-component, Sallen-Key filter is set as required to obtain a desired damping coefficient.

The closed-loop gain of the op amp is related to the damping coefficient of the filter according to the following equation:

$$A_0 = 3 - \alpha \tag{14.20}$$

or equivalently

$$\alpha = 2 - \frac{R_F}{R_1} \tag{14.21}$$

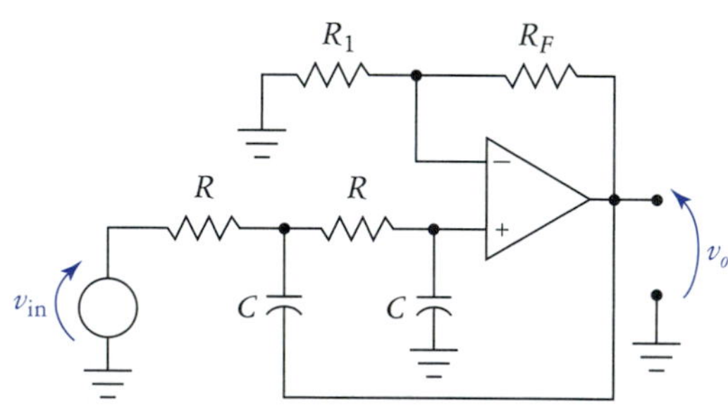

FIGURE 14.5
Equal-component, second-order, Sallen-Key VCVS low-pass filter.

TABLE 14.1

α	k_f(LP)	k_f(HP)	Response
0.500	1.49	0.67	6-dB Chebyshev
0.707	1.43	0.70	3-dB Chebyshev
1.122	1.25	0.80	1-dB Chebyshev
1.414	1.00	1.00	Butterworth (maximally flat passband)
1.500	0.94	1.06	Bessel (low overshoot)
2.000	0.65	1.54	Bessel (most linear phase response)

It is possible to fix the gain of the op amp (to unity for example) but this requires us to use unequal values for the C and R pairs at the filter input in order to set the damping coefficient. The equal-component design is popular because it is generally more convenient to accurately adjust the closed-loop gain than it is to fix the gain and accurately adjust frequency-determining component ratios.

EXAMPLE 14.3

Refer to Fig. 14.5. Given $C = 0.047\ \mu\text{F}$ and $R_1 = 47\ \text{k}\Omega$, determine the required values for R and R_F such that $f_c = 1$ kHz with Butterworth response ($\alpha = \sqrt{2}$).

Solution Because this is a Butterworth filter, we solve Eq. (14.19) for R using $k_f = 1$, which yields

$$
\begin{aligned}
R &= \frac{1}{2\pi f_c C} \\
&= \frac{1}{(2\pi)(1\ \text{kHz})(0.047\ \mu\text{F})} \\
&= 3.39\ \text{k}\Omega
\end{aligned}
$$

We determine R_F by solving Eq. (14.21), which produces

$$
\begin{aligned}
R_F &= R_1(2 - \alpha) \\
&= 47\ \text{k}\Omega(2 - 1.414) \\
&= 27.5\ \text{k}\Omega
\end{aligned}
$$

We can just as easily predict the response shape of an equal-component VCVS using the given component values. This is demonstrated in the next example.

EXAMPLE 14.4

A low-pass filter like that shown in Fig. 14.5 has $R_F = 22.5\ \text{k}\Omega$, $R_1 = 15\ \text{k}\Omega$, $R = 10\ \text{k}\Omega$, and $C = 0.022\ \mu\text{F}$. Determine the closed-loop gain in linear and decibel forms. For convenience let's use A for linear response and H for decibel response. Also determine the damping coefficient α, the response type, and the -3-dB corner frequency. Also determine the total amplitude of the response peak (linear and dB), if such a peak exists.

Solution The passband gain of the filter is found using Eq. (14.18)

$$
\begin{aligned}
A_0 &= \frac{R_1 + R_F}{R_1} \\
&= \frac{15\ \text{k} + 22.5\ \text{k}}{15\ \text{k}} \\
&= 2.5
\end{aligned}
$$

Equivalently, we have

$$
\begin{aligned}
H_0 &= 20 \log A_0 \\
&= 20 \log 2.5 \\
&\cong 8\ \text{dB}
\end{aligned}
$$

The damping coefficient is found using Eq. (14.21), giving us

$$
\begin{aligned}
\alpha &= 2 - \frac{R_F}{R_1} \\
&= 2 - \frac{22.5\ \text{k}}{15\ \text{k}} \\
&= 0.5
\end{aligned}
$$

Referring to Table 14.1, we find that this is a 6-dB Chebyshev filter. That is, the response will have a maximum value 6 dB higher than A_0. This means that the total maximum response is

$$\begin{aligned}|H|_{max} &= H_0 + H_{pk} \\ &= 8 \text{ dB} + 6 \text{ dB} \\ &= 14 \text{ dB}\end{aligned}$$

This corresponds to a linear gain maximum of

$$\begin{aligned}A_{max} &= \log^{-1}(14 \text{ dB}/20) \\ &\cong 5\end{aligned}$$

We could also find this value by multiplying the passband gain by the peak gain:

$$\begin{aligned}A_{max} &= A_0 A_{pk} \\ &= A_0/\alpha \\ &= 2.5/0.5 \\ &= 5\end{aligned}$$

We still need to determine the corner frequency. Consulting Table 14.1 we find for $\alpha = 0.5$, $k_f = 1.49$; therefore,

$$\begin{aligned}f_c &= \frac{1.49}{2\pi RC} \\ &= \frac{1.49}{(2\pi)(10 \text{ k})(0.022\ \mu\text{F})} \\ &= 1078 \text{ Hz}\end{aligned}$$

LP-HP and HP-LP Transformations

If we interchange the positions of R and C in Fig. 14.5, we obtain a high-pass filter. This is shown in Fig. 14.6. The damping coefficient of the filter is independent of whether the filter is HP or LP, so the transformation from one form to another still produces the identical response type (Butterworth, Chebyshev, etc.). If the filter is a Butterworth type, then we can simply switch the R and C positions to obtain a HP (or LP) filter with the same f_c.

Note
Swapping the R and C positions converts HP to LP and vice versa.

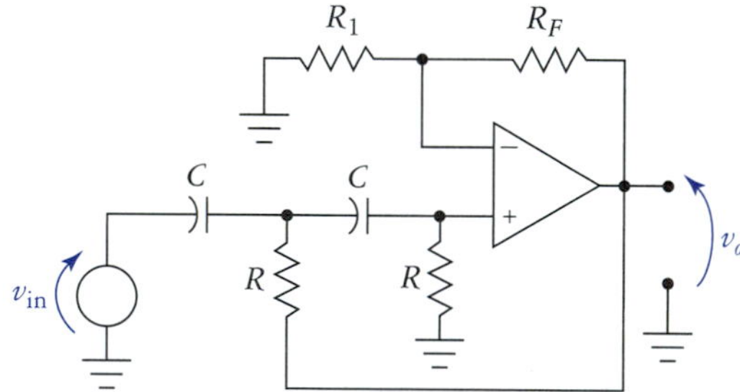

FIGURE 14.6
High-pass second-order VCVS filter.

Unless we are dealing with a Butterworth filter, converting from LP to HP or vice versa requires new frequency-determining resistors and/or capacitors to be chosen to produce the same corner frequency. This is still another reason why we like to use Butterworth filters: HP and LP Butterworth filters have response curves that are symmetrical about f_c.

EXAMPLE 14.5

The LP filter used in Example 14.4 is modified such that a HP response is produced. No component values are changed: The R and C positions are simply interchanged as in Fig. 14.6. The component values remain $R_F = 22.5\text{ k}\Omega$, $R_1 = 15\text{ k}\Omega$, $R = 10\text{ k}\Omega$, and $C = 0.022\ \mu\text{F}$. Determine the new corner frequency of this filter and sketch the response curve, along with the response curve of the original LP filter.

Solution Since the transformation does not affect anything other than f_c, we use the relevant result of Example 14.4, which is $\alpha = 0.5$ (6-dB Chebyshev). The corner frequency of the HP filter is found using Eq. (14.19) and Table 14.1, which yields

$$f_c = \frac{k_f}{2\pi RC}$$

$$= \frac{0.67}{(2\pi)(10\text{ k}\Omega)(0.22\mu\text{F})}$$

$$= 485\text{ Hz}$$

The response curve of the filter is shown in Fig. 14.7.

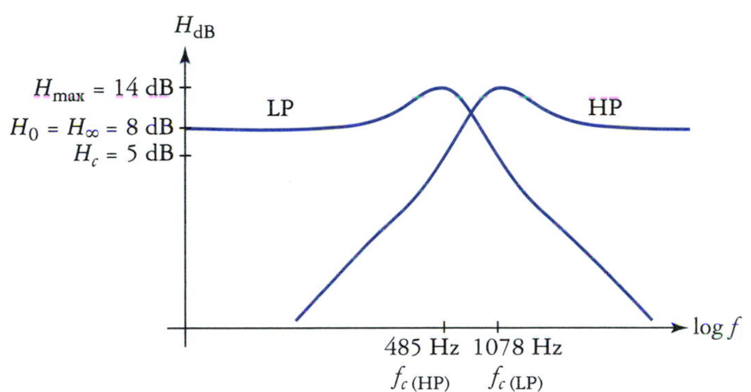

FIGURE 14.7
Response curve for Example 14.5.

14.3 HIGHER ORDER HP AND LP FILTERS

Low- and high-pass active filters of order 3 or greater are formed by cascading first- and second-order sections. The overall order of a filter is the sum of the orders of the individual cascaded sections. This is illustrated in Fig. 14.8.

Note
The overall order of a filter is the sum of the orders of the various sections comprising the filter.

FIGURE 14.8
Higher order filters.

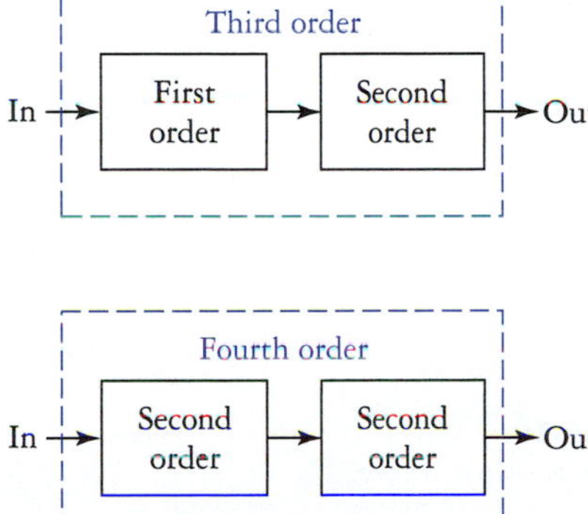

It would be nice if we could simply cascade identical filter sections and obtain the same response of the original filters. Unfortunately, it is not that simple. We would again have to resort to tables of correction factors to set appropriate stage corner frequencies and damping coefficients in order to obtain the desired overall response. Extensive tables of these correction factors are available in more advanced books on active filters, so they are not presented here.

Before moving on, it is worthwhile to take a quick look at why these correction factors are required for higher order filters. Recall from previous discussions of multiple-stage amplifiers that if we cascade sections with identical corner frequencies, the overall corner frequency is lower than that of the individual sections. It also turns out that cascading of filters changes the overall response type as well. For example, let's cascade two identical second-order low-pass Butterworth filters, as shown in

Fig. 14.9. The low-frequency stage gains are $A_{0_1} = A_{0_2} = 1.59$, which is consistent with the equal-component VCVS filters that we have been using. Without further analysis, about all we can say at this point is that overall we have a LP filter with $n = 4$, and $A_0 = 1.59 \times 1.59 = 2.53$.

Let's investigate a little further. The actual corner frequency is found using a form of Eq. (12.31), which is repeated here as

$$\begin{aligned} f_{cT} &= f_{cS}(2^{1/2} - 1)^{1/2} \\ &= (1 \text{ kHz})(0.643) \\ &= 643 \text{ Hz} \end{aligned} \tag{14.22}$$

This is way off, so we know some major frequency scaling is necessary.

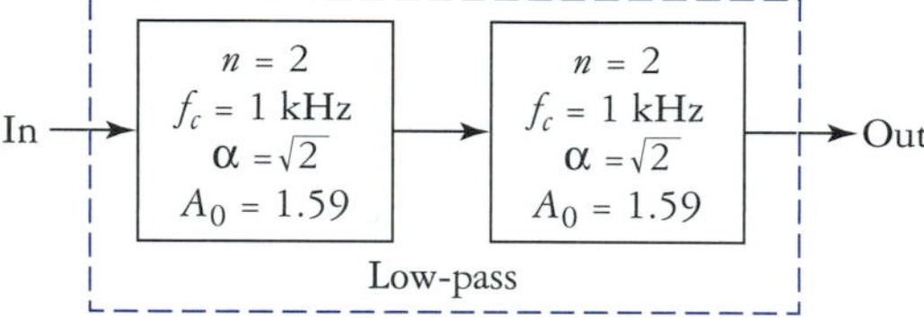

FIGURE 14.9
Cascaded second-order Butterworth filters.

Since both sections are second-order Butterworth filters, we know that the response magnitude of each is given by

$$|H(jf)|_1 = |H(jf)|_2 = \frac{A_0}{\left[\left(\dfrac{f_{\text{in}}}{f_c}\right)^4 + 1\right]^{1/2}}$$

Therefore, the overall response is the product

$$\begin{aligned} |H(jf)|_T &= |H(jf)|_1 \times |H(jf)|_2 \\ &= \left[\frac{A_0}{\left[\left(\dfrac{f_c}{f_{\text{in}}}\right)^4 + 1\right]^{1/2}}\right]^2 \\ &= \frac{A_0^2}{\left(\dfrac{f_{\text{in}}}{f_c}\right)^4 + 1} \end{aligned}$$

This is not the response equation for a fourth-order Butterworth filter. According to Eq. (14.16), such a filter should have a response magnitude given by

$$|H(jf)|_T = \frac{A_0}{\left[\left(\dfrac{f_c}{f_{\text{in}}}\right)^8 + 1\right]^{1/2}}$$

These two responses are similar, but nevertheless they are not identical.

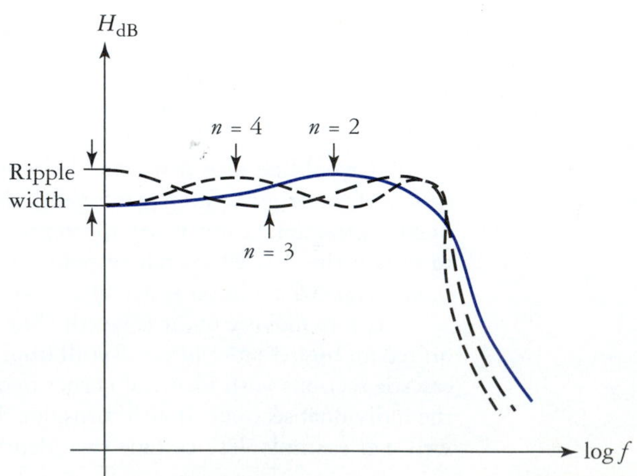

FIGURE 14.10
Chebyshev response curves.

It is also interesting to note that higher order Chebyshev filters produce multiple peaks in the passband. This is shown in Fig. 14.10. These peaks are called *passband ripples,* and they are usually of equal amplitude, which is called the *ripple width.* Passband ripples are the price that is paid for the increasingly greater rolloff gained by increasing the order of a Chebyshev filter.

14.4 BANDPASS FILTERS

Much of the basic bandpass (BP) filter information was covered in Chapter 10 in the context of tuned class C amplifiers. The basic bandpass response curve is shown in Fig. 14.11. Recall the following equations:

$$f_o = \sqrt{f_H f_L} \tag{14.23}$$

$$\text{BW} = f_H - f_L \tag{14.24}$$

$$Q = \frac{f_o}{\text{BW}} \tag{14.25}$$

$$\alpha = \frac{1}{Q} \tag{14.26}$$

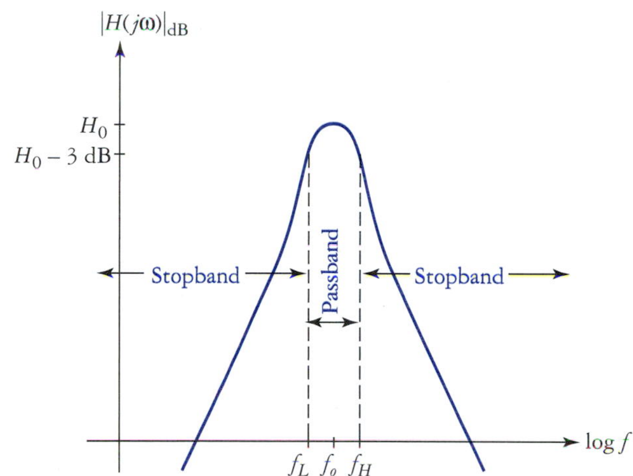

FIGURE 14.11
Bandpass filter response.

Bandpass filters are normally of even order ($n = 2, 4, 6, \ldots$). Also, most bandpass filters have symmetrical response curves. That is, the ultimate (asymptotic) rolloff rate is the same on each side of the center frequency. A second-order LP filter, for example, will have rolloff = −40 dB/decade. A second-order BP filter will split this rolloff between the two sides of the response, producing symmetrical rolloff rates of ±20 dB/decade.

The Q of a BP filter can be interpreted in a couple of ways. For example, it can be thought of as a measure of the energy loss of a resonant circuit. Perhaps a more meaningful interpretation is to view Q as a measure of the sharpness or narrowness of BP response. Figure 14.12 illustrates how the variation of Q affects the shape of a BP response curve. In this case, the order is $n = 2$; therefore, all response curves approach a rolloff of ±20 dB/decade at the asymptote.

Wide-Bandwidth Filters

A filter can be considered to have a wide bandwidth if it has $Q << 1$. For example, suppose we wished to build a BP filter that passed audio frequencies from $f_L = 20$ Hz to $f_H = 20$ kHz. This filter has

$$\begin{aligned}\text{BW} &= 20\text{ kHz} - 20\text{ Hz} \\ &\cong 20\text{ kHz}\end{aligned}$$

$$\begin{aligned}f_o &= \sqrt{f_H f_L} \\ &= \sqrt{400{,}000} \\ &= 632\text{ Hz}\end{aligned}$$

$$\begin{aligned}Q &= f_o/\text{BW} \\ &= 632/20\text{ k} \\ &= 0.0316\end{aligned}$$

FIGURE 14.12
Effect of Q on second-order BP response.

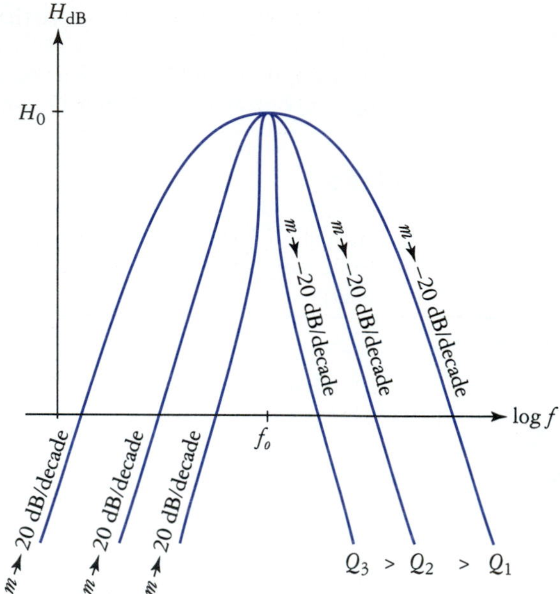

This is a very low Q filter, with a response curve like that of Fig. 14.13(a) where the solid line is the asymptotic response. In a case like this, we would construct the filter from cascaded LP and HP filters, most likely with Butterworth responses to keep the passband flat. The corner of the LP filter is the high corner of the BP filter and the corner of the HP filter is the low corner. That is,

$$f_H = f_{c(\mathrm{LP})} \tag{14.27}$$

$$f_L = f_{c(\mathrm{HP})} \tag{14.28}$$

The HP and LP filters are cascaded as shown in Fig. 14.13(b). The overall passband gain in decibels is

$$H_0 = H_{0(\mathrm{LP})} + H_{\infty(\mathrm{HP})} \tag{14.29}$$

Using A_0 to represent the linear response magnitude (remember, $H_0 = 20 \log A_0$), we have

$$A_0 = A_{0(\mathrm{LP})} \times A_{\infty(\mathrm{HP})} \tag{14.30}$$

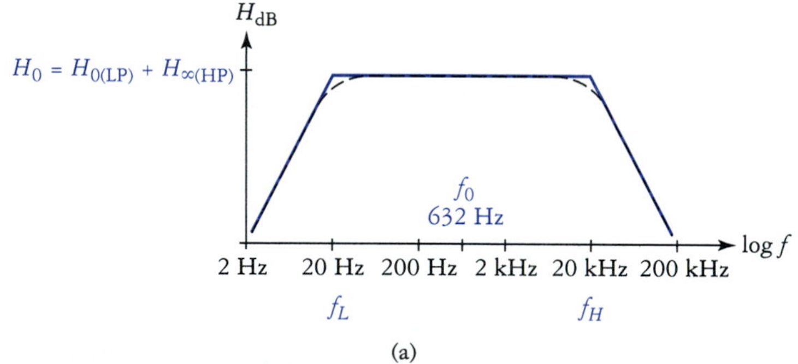

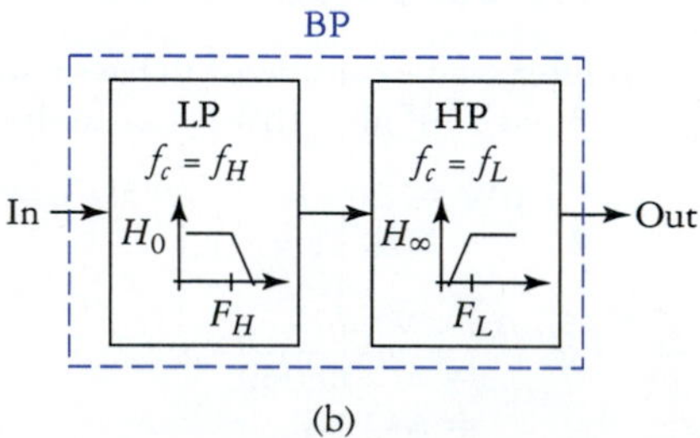

FIGURE 14.13
Wideband filter. (a) Response curve. (b) Cascaded LP-HP realization.

The Infinite-Gain Multiple-Feedback BP Filter

Many designs are used to implement active bandpass filters. One of the most commonly used forms is the *infinite gain multiple-feedback* (IGMF) structure of Fig. 14.14. Although the exact limits of circuit performance depend on the op amp used, the IGMF is generally restricted to $Q \leq 25$.

FIGURE 14.14
IGMF second-order bandpass filter.

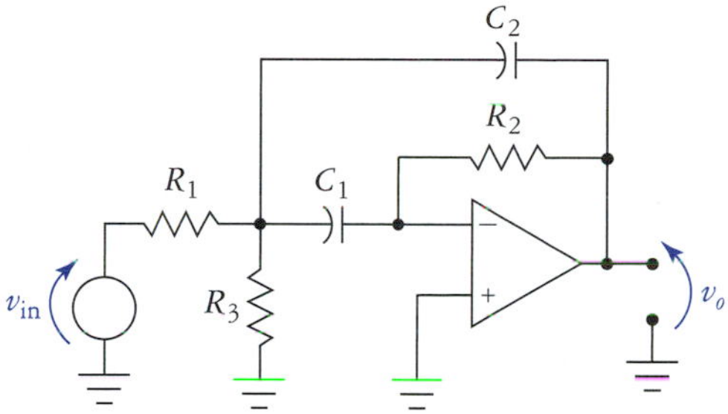

To simplify analysis and design, the capacitors in the IGMF filter are usually equal in value. That is, $C_1 = C_2 = C$. Using this relationship, the analysis and design equations for the IGMF BP filter are as follows:

$$f_o = \frac{\left[\left(\frac{1}{R_2C^2}\right)\left(\frac{1}{R_1} + \frac{1}{R_3}\right)\right]^{1/2}}{2\pi} \tag{14.31}$$

$$R_1 = \frac{Q}{2\pi f_o A_0 C} \tag{14.32}$$

$$R_2 = \frac{Q}{\pi f_o C} \tag{14.33}$$

$$R_3 = \frac{Q}{2\pi f_o C(2Q^2 - A_o)} \tag{14.34}$$

$$A_o = -\frac{R_2}{2R_1} \tag{14.35}$$

Note
You can make a guitar "wah-wah" effect by using a potentiometer for R_3 and placing it in a foot pedal.

Examination of Eq. (14.34) indicates the following restriction on the gain and Q of the circuit:

$$A_v < 2Q^2 \tag{14.36}$$

One of the more useful features of the IGMF filter is that f_o is independently adjustable. Specifically, if we wish to change f_o to a new value f_{new}, the following relationship is used:

$$R_{3(\text{new})} = R_3\left(\frac{f_o}{f_{o(\text{new})}}\right)^2 \tag{14.37}$$

If continuous adjustment of f_o is desired a potentiometer can be used in place of R_3. We must not allow $R_3 = 0\ \Omega$, however, or the circuit will cease operating.

EXAMPLE 14.6

Design an IGMF BP filter with $f_o = 1$ kHz, $A_o = 2$, $Q = 5$, and $C = 0.022\ \mu$F. Determine the bandwidth of the resulting filter.

Solution We simply apply Eqs. (14.32) through (14.34), which produces

$$R_1 = \frac{Q}{2\pi f_o A_o C} = \frac{5}{(2\pi)(1\text{ k})(2)(0.022\ \mu\text{F})} = 18\text{ k}\Omega$$

$$R_2 = \frac{Q}{\pi f_o C}$$
$$= \frac{5}{(\pi)(1\text{ kHz})(0.022\ \mu\text{F})}$$
$$= 72\text{ k}\Omega$$

$$R_3 = \frac{Q}{2\pi f_o C(2Q^2 - A_o)}$$
$$= \frac{5}{(2\pi)(1\text{ k})(0.022\ \mu\text{F})(50 - 2)}$$
$$= 753\ \Omega$$

The bandwidth of the filter is

$$\text{BW} = \frac{f_o}{Q}$$
$$= \frac{1\text{ kHz}}{5}$$
$$= 200\text{ Hz}$$

14.5 NOTCH FILTERS

Notch filters (or *band-reject* or *bandstop* filters) are complementary to bandpass filters. Notch filters are designed to attenuate a specific range of frequencies while passing all others. The notch response is shown in Fig. 14.15. The frequency at which maximum attenuation occurs is called the *null frequency* f_0. The null frequency is the geometric mean of the stopband, which is

$$f_0 = \sqrt{f_H f_L} \tag{14.38}$$

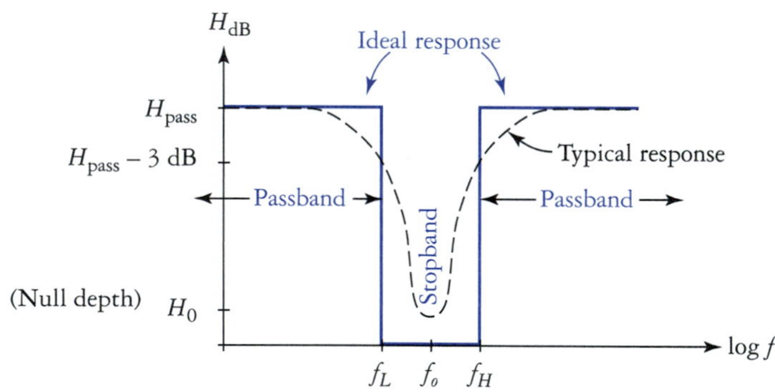

FIGURE 14.15
Bandstop filter response curves.

The maximum attenuation (or minimum response) H_0 is called the *null depth.* Ideally the null depth is $H_0 = -\infty$ dB (or $A_0 = 0$). We define the bandwidth and Q of the notch filter in the same way as for BP filters. That is,

$$\text{BW} = f_H - f_L \tag{14.39}$$

$$Q = \frac{f_0}{\text{BW}} \tag{14.40}$$

A wideband (low-Q) notch filter can be produced by parallel connecting HP and LP filters as shown in Fig. 14.16; however, this is not a common configuration.

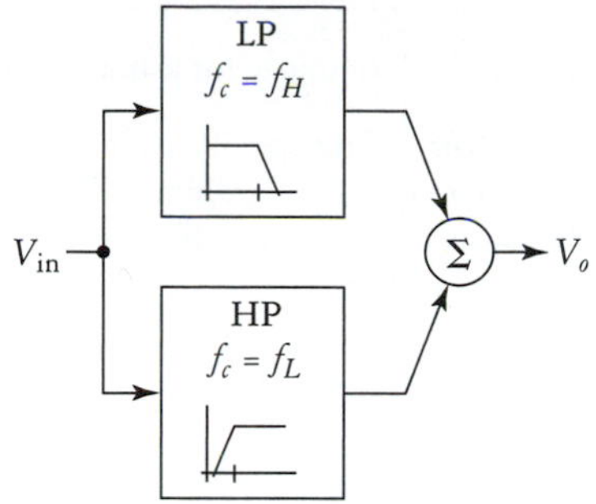

FIGURE 14.16
Realization of bandstop response.

IGMF-Based Notch Filter

A very common notch filter realization is based on the IGMF BP filter. A block diagram representation for the basic IGMF notch filter is shown in Fig. 14.17. Basically, we are summing the output of the BP filter with the original input voltage, scaled by a factor of H_0 (or A_0). The circuit works as a notch filter because the phase of the BP output varies as shown in Fig. 14.17(b). Thus, when the BP filter reaches maximum response, the BP input to the summer is

$$V_1 = (A_0\angle 180°)(V_{in}) = (-A_0)(V_{in})$$

The original input signal is applied to the other input of the summer and scaled by the value of A_0 without being phase shifted. That is,

$$\begin{aligned} V_2 &= (A_0\angle 0°)(V_{in}) \\ &= (A_0)(V_{in}) \end{aligned}$$

Normalizing A_0 to unity, the net output of the summer at f_0 is

$$\begin{aligned} V_o &= V_2 + V_1 \\ &= V_{in} - V_{in} \\ &= 0 \end{aligned}$$

At all other frequencies the phase and amplitudes of the two signals will differ causing a net response like that shown in Fig. 14.15.

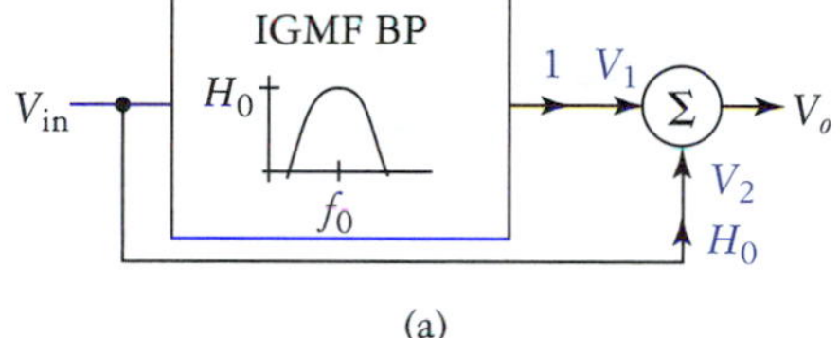

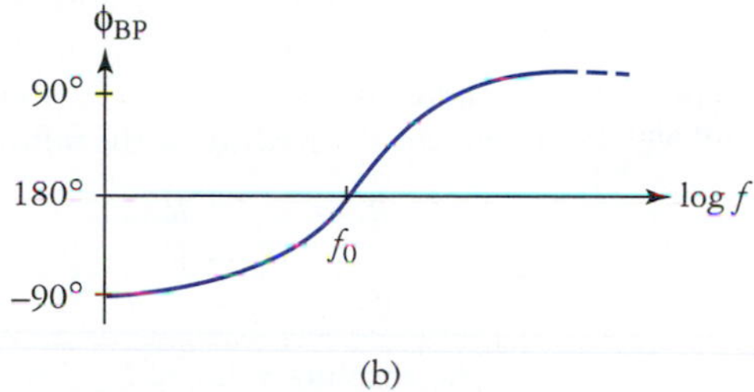

FIGURE 14.17
(a) Realization of a bandstop response using IGMF BP and a summer. (b) Phase response of the BP filter section.

It is impossible to have perfect cancellation; therefore, the maximum attenuation of the notch filter depends on how closely we can match V_1 and V_2. In practice, it is possible to achieve a null depth of about −60 dB with 1% tolerance components. The bandwidth, Q, and passband response of the IGMF notch filter will be the same as for the original BP filter.

EXAMPLE 14.7

Draw the schematic for a notch filter using the IGMF BP filter of Example 14.6.

Solution The specifications of the existing BP filter are $f_o = 1$ kHz, $A_o = 2$, and $Q = 5$. The complete circuit is shown in Fig. 14.18. Note that R_F is chosen somewhat arbitrarily to be 10 kΩ. Because we desire the BP output to be scaled by 1, we use 10 kΩ on that input, whereas 5 kΩ is used to scale v_{in} by 2.

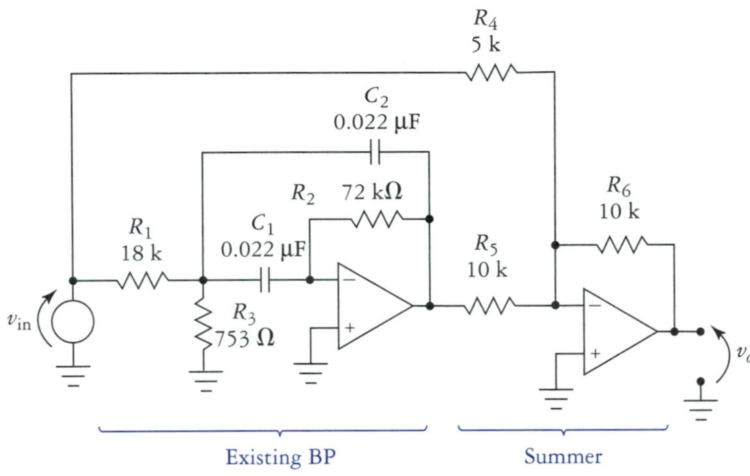

FIGURE 14.18
Circuit for Example 14.7.

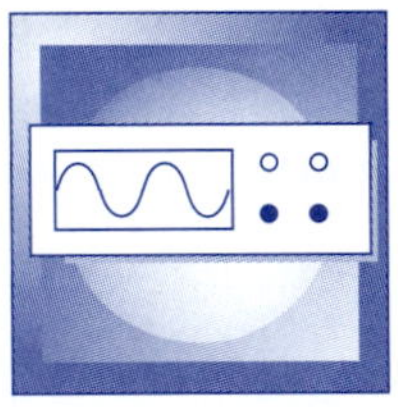

Notch filters are used to remove interference of a specific frequency form a signal. For example, in medical equipment, 60-Hz noise pickup can be a major problem. Consider that the electrical signals being produced by, say, the heart are in the low millivolt range, whereas induced ac line noise is often hundreds of millivolts in amplitude. Rejection of this noise is critical and 60-Hz notch filters are used often in such cases. Of course, we would also take advantage of the common-mode rejection properties of diff amps as well to reduce noise pickup.

14.6 STATE-VARIABLE FILTERS

State-variable filters are high-performance filters that require a minimum of three op amps in their realization. The basic state-variable filter circuit is shown in Fig. 14.19. Notice that the state-variable filter produces simultaneous LP, HP, and BP responses. If we assume that $\alpha = \sqrt{2}$ (Butterworth response), then the responses of the various outputs will be similar to those shown in Fig. 14.19(b). In this case, the corner frequencies of the HP and LP responses match the center of the BP passband.

The state-variable filter basically consists of a summer (A_1) and two integrators (A_2 and A_3). The various corner frequencies are determined by the R_5–C_1 and R_6–C_2 pairs. Mathematically, this circuit is described by a second-order differential equation. To simplify the math, we choose the components according to the following rules:

Note
State variable filters produce simultaneous HP, LP, and BP responses.

$$R_1 = R_2 = R_3$$
$$R_5 = R_6 = R$$
$$C_1 = C_2 = C$$

The absolute value of R_4 is not critical, but it is used with R_α to determine the damping coefficient of the filter according to

$$R_\alpha = R_4\left(\frac{3-\alpha}{\alpha}\right) \tag{14.41}$$

Assuming that the rules above are followed, the center frequency of the BP response is

$$f_o = \frac{1}{2\pi RC} \tag{14.42}$$

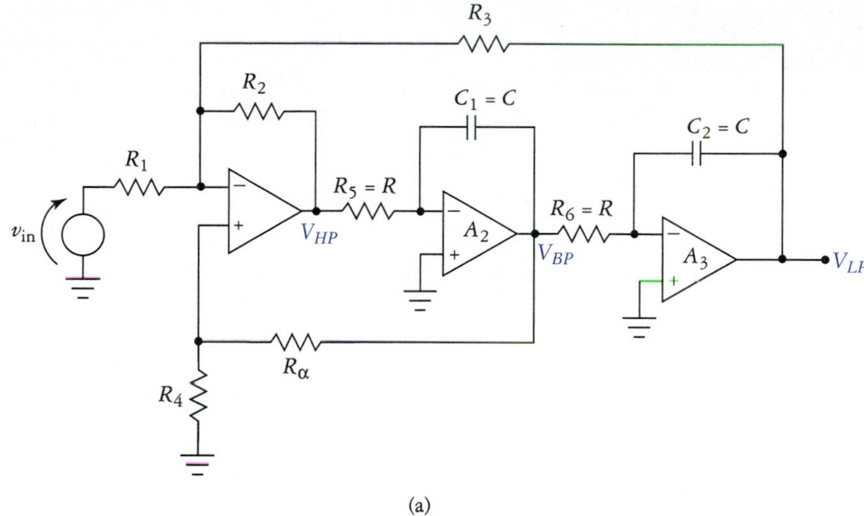

(a)

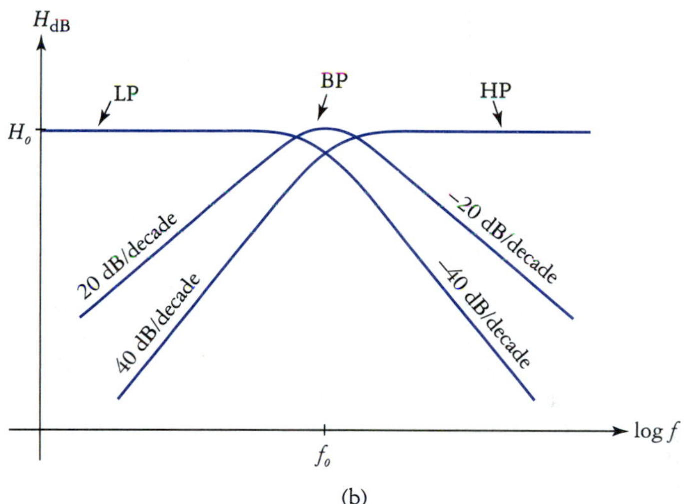

(b)

FIGURE 14.19
State-variable filter. (a) Typical circuit. (b) Response curves.

The passband response magnitude is given by

$$A_0 = 1/\alpha = Q \tag{14.43}$$

And, as stated earlier, if $\alpha = \sqrt{2}$ then the HP and LP corner frequencies are located at f_o as well. If the filter is underdamped then the peaks of the HP and LP response curves are located at f_o. We can use the correction factors of Table 14.1 to determine the actual LP and HP −3-dB frequency locations for the underdamped situation.

To continuously vary f_o, we must vary either pairs of R_5 and R_6 or C_1 and C_2 simultaneously. Usually, the capacitors are switched in pairs to provide a factor of 10 frequency shift, while a ganged potentiometer would be used to smoothly vary the resistance.

The three-op-amp state-variable filter does not allow gain and damping to be adjusted independently. If we add a fourth op amp as shown in Fig. 14.20, then the passband gain magnitude is

$$A_0 = \frac{R_3}{R_1} \tag{14.44}$$

Notice that several new resistors have been defined as repetitions of R_3. This allows R_4 to be used to adjust the damping coefficient according to

$$\alpha = \frac{R_4}{R_3} \tag{14.45}$$

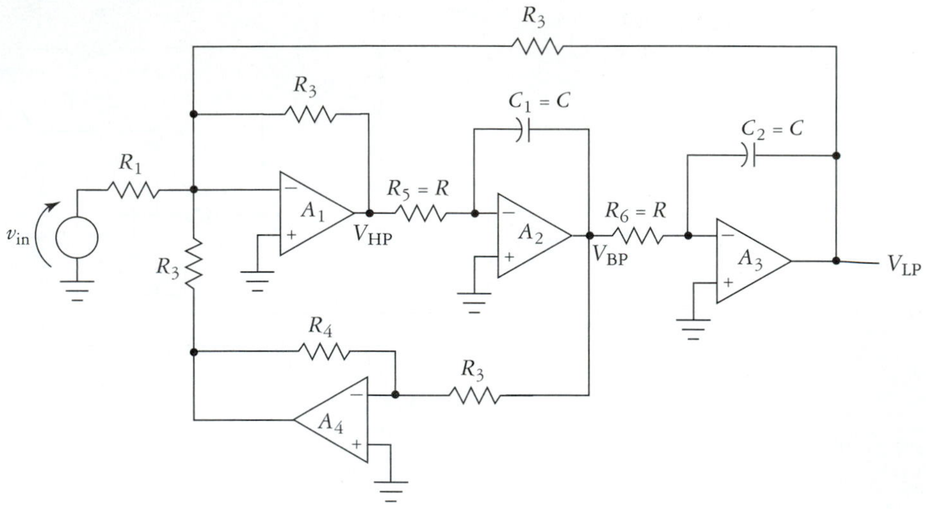

FIGURE 14.20
State-variable filter with independent gain and damping adjustment.

14.7 THE BIQUADRATIC FILTER

A *biquadratic filter* is shown in Fig. 14.21. Sometimes this circuit is called a *Tow-Thomas filter* after its inventors, but usually it is just called a *biquad.* In the version shown, op amp A_1 is called a *lossy integrator,* while A_2 is a *noninverting integrator* formed from a Howland current source. If a standard inverting integrator is used, then a third inverting op amp is required. Biquads produce simultaneous BP and LP response and are capable of realizing Q values as high as 500.

Note
Biquad filters can be used to implement a high -Q BP response.

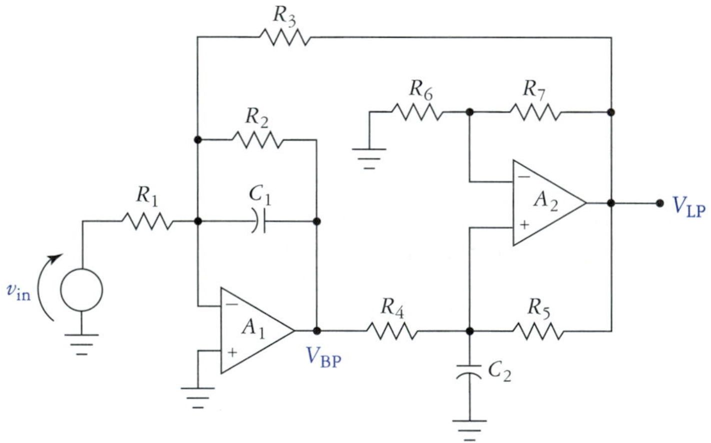

FIGURE 14.21
Biquadratic filter.

For simplicity, we generally prefer to have $R_4 = R_5 = R_6 = R_7$. The analysis equations for the biquad are

$$f_0 = \frac{1}{2\pi\sqrt{R_3R_4C_1C_2}} \tag{14.46}$$

$$Q = \frac{R_2\sqrt{C_1}}{\sqrt{R_3R_4C_2}} \tag{14.47}$$

$$A_{0(LP)} = \frac{R_3}{R_1} \tag{14.48}$$

$$A_{0(BP)} = \frac{R_2}{R_1} \tag{14.49}$$

EXAMPLE 14.8

The biquad filter in Fig. 14.21 has $R_1 = R_4 = R_5 = R_6 = R_7 = 10\ \text{k}\Omega$, and $C_1 = C_2 = 0.1\ \mu\text{F}$. Determine R_3 and R_2 such that $Q = 5$ and $f_0 = 150$ Hz. Determine the resulting filter characteristics.

Solution Examination of the equations reveals that we must begin by solving Eq. (14.46) for R_3 giving us

$$R_3 = \frac{\left(\frac{1}{2\pi f_0}\right)^2}{R_4 C_1 C_2}$$
$$= \frac{\left(\frac{1}{(2\pi)(150)}\right)^2}{(10\ \text{k})(0.1\ \mu\text{F})(0.1\ \mu\text{F})}$$
$$= 11.3\ \text{k}\Omega$$

Because we wish to set the Q, we now solve Eq. (14.47) for R_2 giving us

$$R_2 = \frac{Q\sqrt{R_3 R_4 C_2}}{\sqrt{C_1}}$$
$$= \frac{5\sqrt{(11.3\ \text{k})(10\ \text{k})(0.1\ \mu\text{F})}}{\sqrt{0.1\ \mu\text{F}}}$$
$$= 53.2\ \text{k}\Omega$$

The remaining filter parameters are

$$A_{0(\text{BP})} = \frac{R_2}{R_1}$$
$$= \frac{53.2\ \text{k}}{10\ \text{k}}$$
$$= 5.32$$

which in decibels is

$$H_{o(\text{BP})} = 20 \log A_{o(\text{BP})}$$
$$= 20 \log(5.32)$$
$$= 14.5\ \text{dB}$$

$$A_{0(\text{LP})} = \frac{R_3}{R_1}$$
$$= \frac{11.3\ \text{k}}{10\ \text{k}}$$
$$= 1.13$$

$$H_{0(\text{LP})} = 20 \log A_{0(\text{LP})}$$
$$= 20 \log(1.13)$$
$$= 1.1\ \text{dB}$$

The damping coefficient of the filter is $\alpha = 1/Q = 1/5 = 0.2$. Thus, the LP response is highly underdamped. The LP response maximum amplitude will be

$$A_{\text{LP(max)}} = (A_{\text{pk}})(A_0)$$
$$= Q \times 1.13$$
$$= 5 \times 1.13$$
$$= 5.65$$

In decibels this is

$$H_{\text{LP(max)}} = H_{\text{pk}} + H_0$$
$$= 20 \log Q + 20 \log A_{0(\text{LP})}$$
$$= 14\ \text{dB} + 1.1\ \text{dB}$$
$$= 15.1\ \text{dB}$$

The Bode plot for these responses is shown in Fig. 14.22. A response like this is difficult to sketch, making this an ideal problem for circuit simulation.

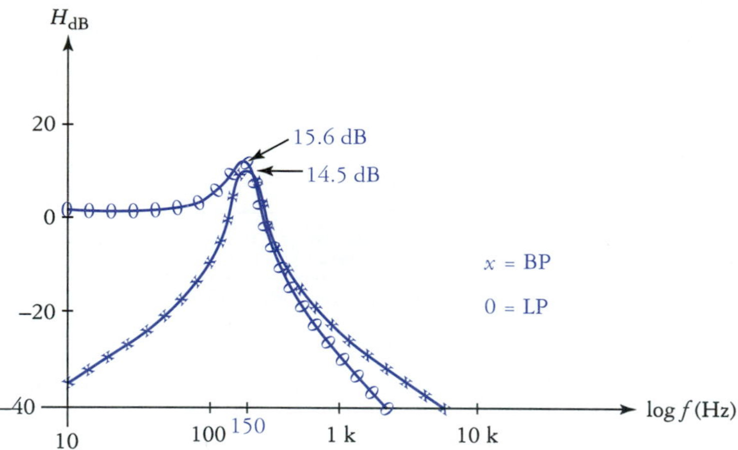

FIGURE 14.22
Biquad BP and LP response curves.

14.8 ALL-PASS FILTERS

The all-pass filter is an interesting circuit. As the name implies, an all-pass filter has a flat passband: ideally, its response is flat from 0 Hz to infinity. A reasonable question to ask is "What good is a filter with a flat response magnitude?" The answer lies in the fact that although the amplitude response is constant, the phase response of the all-pass filter does vary with frequency, which is a useful characteristic.

Note
All-pass filters have a flat amplitude response curve, but the relative phase of the output varies with frequency.

A common all-pass filter and its amplitude and phase response curves are shown in Fig. 14.23. The phase response of this circuit is given by

$$\phi = 2 \tan^{-1} \frac{1}{2\pi f R_1 C_1} \tag{14.50}$$

We can define the center frequency of this filter to be

$$f_0 = \frac{1}{2\pi R_1 C_1} \tag{14.51}$$

The circuit requires the gain-determining resistors to be equal, but there are no restrictions on R_1 other than the standard rules-of-thumb regarding bias current error and so forth; that is, in order to minimize output offset voltage we would set $R_1 = R/2$. This is not likely to be a major concern however.

Derivation of the All-Pass Filter Response*

Although it is a little complicated, it is interesting to look at how the all-pass filter response is derived. Examination of Fig. 14.23 reveals that the op amp is set up as a form of differential amplifier. We can use superposition to determine the output response to the input voltage just like we did in the previous chapter, which certainly seems like a long time ago. Anyway, the output of the op amp is

$$\begin{aligned} V_o &= V_{(+)} \frac{R}{R + X_C} - V_{\text{in}} \frac{R}{R} \\ &= 2V_{(+)} - V_{\text{in}} \end{aligned} \tag{14.52}$$

*This section is optional and can be skipped without loss of continuity.

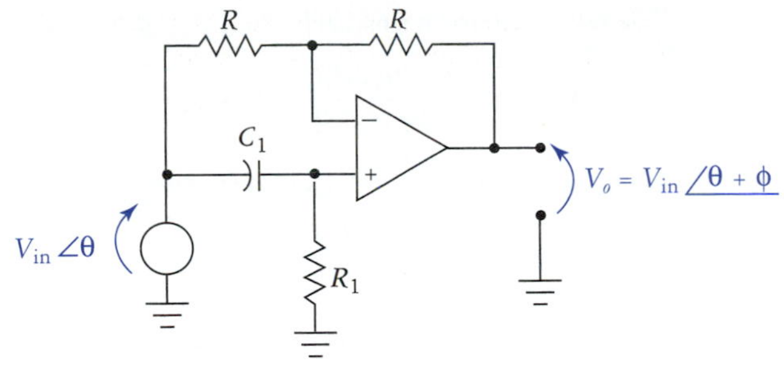

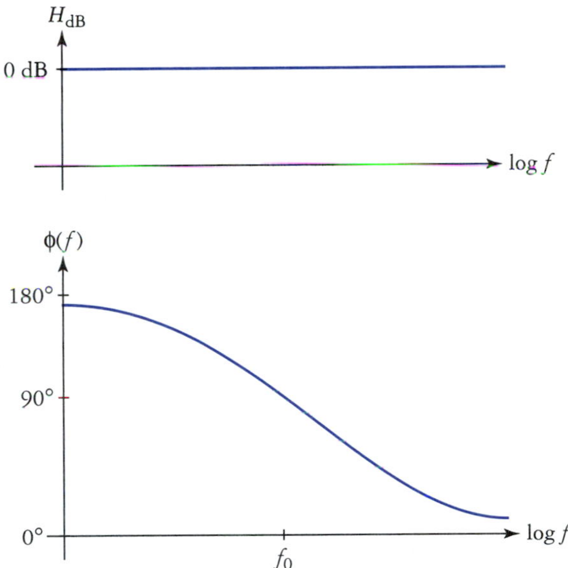

FIGURE 14.23
The all-pass filter.

Now the fun part starts. We must express $V_{(+)}$ as a function of frequency. We will work in radian frequency ω to simplify the notation. Application of the voltage-divider relationship shows the noninverting input to be

$$V_{(+)} = V_{\text{in}} \frac{R_1}{R_1 + X_c}$$

Expressing the capacitive reactance in rectangular form

$$V_{(+)} = V_{\text{in}} \frac{R_1}{R_1 + \dfrac{1}{j\omega C}}$$

Note
Remember, in polar form, $X_C = \dfrac{1}{\omega C} \angle{-90^\circ}\ \Omega$ and in rectangular form, $X_C = \dfrac{1}{j\omega C}$.

Forming a common denominator and simplifying we obtain

$$V_{(+)} = V_{\text{in}} \frac{R_1}{\dfrac{j\omega R_1 C}{j\omega C} + \dfrac{1}{j\omega C}}$$

$$= V_{\text{in}} \frac{R_1}{\dfrac{j\omega R_1 C + 1}{j\omega C}}$$

$$= V_{\text{in}} \frac{j\omega R_1 C}{j\omega R_1 C + 1}$$

We now substitute this back into Eq. (14.52), which produces

$$V_o = V_{in}\left(\frac{2j\omega R_1C}{j\omega R_1C + 1}\right) - V_{in}$$

Dividing both sides by V_{in} produces the frequency response expression:

$$A(j\omega) = \left(\frac{2j\omega R_1C}{j\omega R_1C + 1}\right) - 1$$

Again, we form a common denominator and simplify, which yields

$$A(j\omega) = \left(\frac{2j\omega R_1C}{j\omega R_1C + 1}\right) - \left(\frac{j\omega R_1C + 1}{j\omega R_1C + 1}\right)$$

$$= \frac{j\omega R_1C - 1}{j\omega R_1C + 1}$$

This expression is in rectangular form, which is useful but very difficult to visualize. We must convert to polar form to make physical sense of the function. Applying the usual R-to-P techniques we have

$$A(j\omega) = \frac{\sqrt{(\omega R_1C)^2 + 1}\ \angle 180° - \tan^{-1}\left(\frac{\omega R_1C}{-1}\right)}{\sqrt{(\omega R_1C)^2 + 1}\ \angle \tan^{-1}(\omega R_1C)}$$

Separating the magnitude and angle we find

$$|A(j\omega)| = 1$$

$$\phi(j\omega) = 180° - \tan^{-1}\left(\frac{\omega R_1C_1}{-1}\right) - \tan^{-1}(\omega R_1C_1)$$

which is more conveniently expressed as

$$\phi(j\omega) = 2\tan^{-1}\frac{1}{2\pi fR_1C_1}$$

Thus we find that the amplitude response of the all-pass filter is flat while the relative phase of the output voltage varies as a function of frequency.

Lagging Phase All-Pass Filter

We can easily modify the all-pass filter such that the output voltage lags the input voltage in phase. This modification is shown in Fig. 14.24, along with the phase response curve. As before, the amplitude response is flat, while the phase response is given by

$$\phi = -2\tan^{-1}(2\pi fR_1C_1) \tag{14.53}$$

If you are feeling ambitious you can derive this expression using the basic approach presented in the last subsection.

It is easy to think of applications for the usual HP, LP, BP, and notch filters, but you may be wondering where the all-pass response might be used. Next we examine a few rather interesting applications.

Phase Equalization

One application for all-pass filters is in the correction of phase distortion in signals that are sent over long-distance transmission lines. This is called *phase equalization.* For example, let's consider a signal that consists of two components defined as follows:

$$v_{in} = 1\sin(\omega t) + 1\sin(2\omega t)\ \text{V}$$

This signal is shown by the solid curve in Fig. 14.25, while the constituent components are shown as dashed lines. Now, suppose that we send this signal through a long transmission line and the 2ω component is phase shifted by 90°. This is shown in Fig. 14.26. Notice that the resultant waveform (solid curve) has a different shape than the original input. If this distorted output were processed

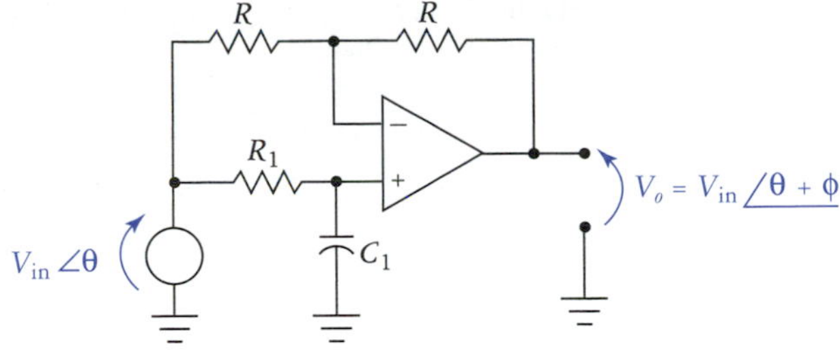

FIGURE 14.24
Alternative all-pass filter.

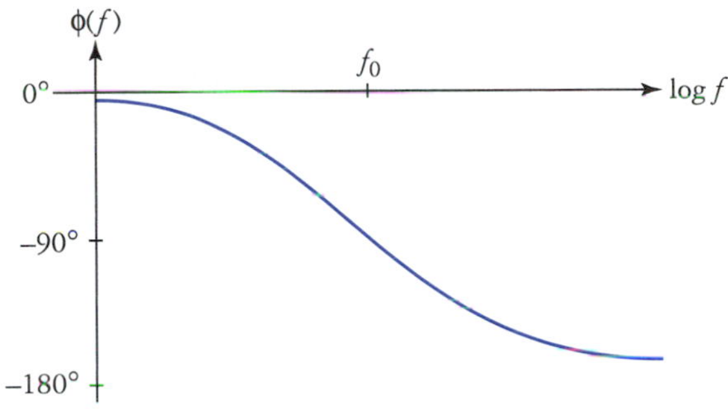

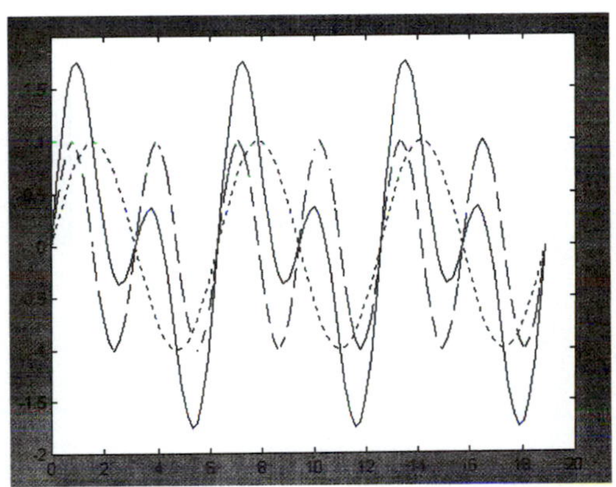

FIGURE 14.25
Input waveforms and the net resultant.

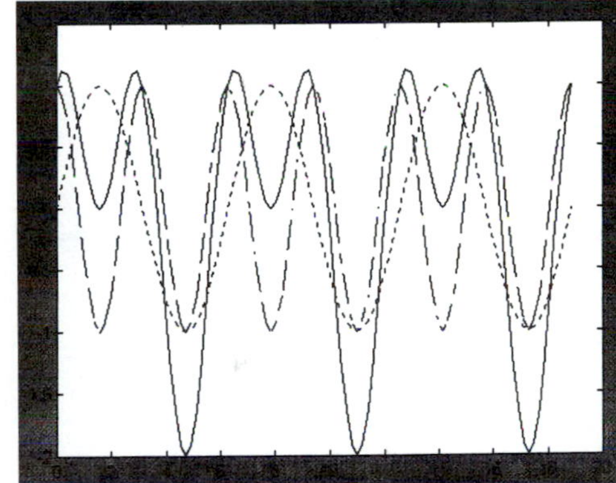

FIGURE 14.26
Phase-shifted input waveforms and resultant.

through an all-pass filter we could correct the phase relationships. We would design the filter such that $\phi_\omega = -\theta$ and $\phi_{2\omega} = -\theta - 90°$. We add the arbitrary (and probably small) phase shift θ because the only time there is no phase shift is when $\omega = 0$. Thus, the important thing is for the relative phase shift between signal components to be correct.

Audio Phase Shifters

An application of all-pass filters that might be familiar to many readers is in the realization of an electronically variable phase shifter such as that used to process the signal from an electric guitar. The effect is a swooshing sound that is difficult to describe but easily recognized when heard. The block diagram for a phase shifter is shown in Fig. 14.27(a). Here's the basic idea behind the circuit: A complex audio signal is applied to the input. The control voltage is varied, which in turn sweeps f_o of the all-pass filter such that the relative phase of the signal frequencies is continuously varied. This phase-altered signal is summed with the original signal. As the various frequency components are swept in phase, they smoothly go through constructive to destructive interference with the original signal producing the classical phase-shift sound. The depth of the phase-shift effect is varied by adjusting a potentiometer. Also, to produce a rich and deep sound, the phase shifter will consist of two or more cascaded all-pass filters.

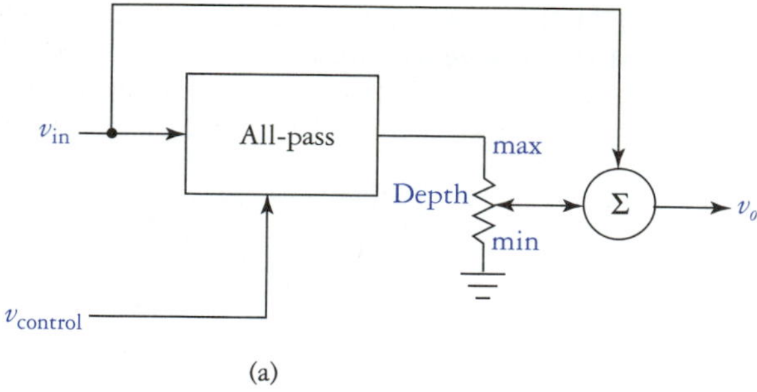

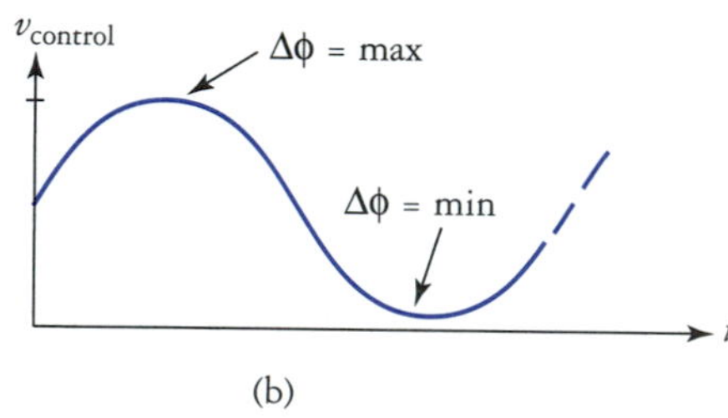

FIGURE 14.27
Audio phase shifter.

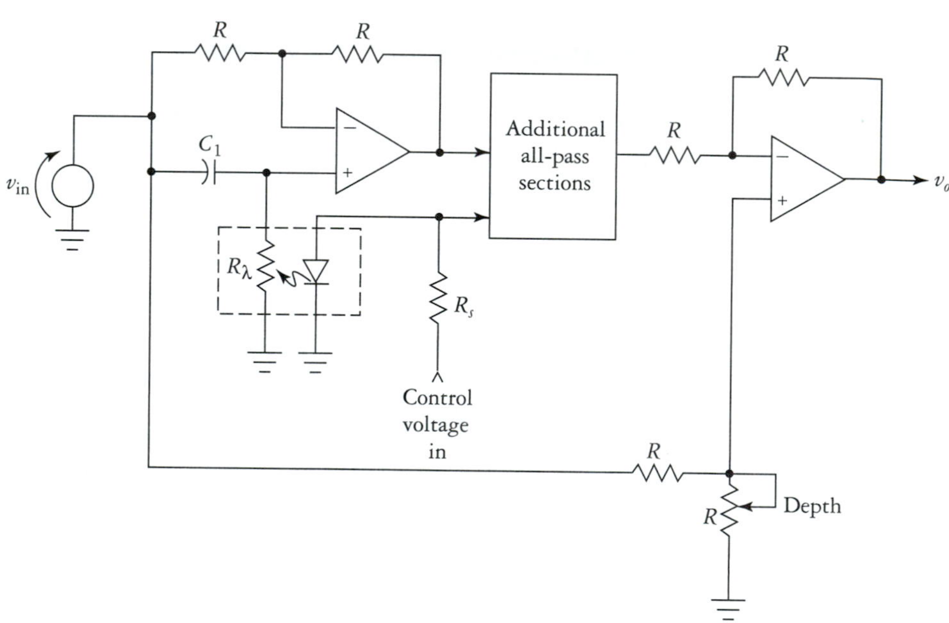

FIGURE 14.28
Phase shifter with electronic control.

The heart of the phase shifter is the electronically variable all-pass filter. One way that such a circuit can be implemented is shown in Fig. 14.28. The device in the dashed box is a *photocoupler,* a light source (an LED or incandescent lamp) enclosed with a light-dependent resistor R_λ. The control voltage varies the brightness of the light source, which in turn varies the phase response of the filter.

14.9 SWITCHED-CAPACITOR FILTERS

The main advantage of active filters is the elimination of inductors, which at low frequencies tend to be large, expensive, heavy, nonideal, and susceptible to noise pickup. The main goal of most circuit designers/manufacturers in general is to produce high-performance circuits that are as small and inexpensive as possible. Ideally we would like to incorporate all of the required components for an active filter into a single IC. There are limits to the level of integration that can be achieved using the circuit topologies covered so far. For example, it is difficult to control the absolute value of resistors in monolithic IC designs. It is also difficult to fabricate high- and low-value resistors as well. If we wish to produce active filters with variable parameters, we must normally use external variable resistors. We also would like to eliminate the need for external capacitors as well. So basically, high-performance, tunable active filters rely on accurate resistors that are easily adjusted. These factors require a different approach to integrated circuit (IC) active filter design.

A few aspects of monolithic IC design can be exploited here. Specifically, using MOS (metal-oxide semiconductor) technology it is possible to produce capacitors that are closely matched in value and temperature characteristics. These capacitors can typically range from 0.1 pF, as used in memory circuits, to around 100 pF, a common size for switched-capacitor filter applications. Of course, it is easy to produce lots of transistors that can be used to implement amplifiers and switches.

Switched-Capacitor Concepts

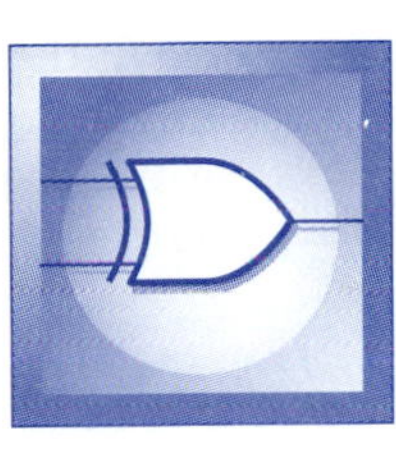

The basic switched-capacitor arrangement is shown in Fig. 14.29. If switches S_1 and S_2 are opened and closed in the proper sequence, the circuit will closely approximate the behavior of an equivalent resistance R_C. Switches S_1 and S_2 are not mechanical switches, but are actually MOSFETs that act as very fast electronic switches. These are usually called *analog switches.* The switches are driven by a two-phase nonoverlapping clock signal as shown in the figure. The clock signals must be nonoverlapping to prevent the input source from being connected directly to the output terminals *a-b*. We are modeling the analog switches as if they have infinite off resistance, while the on-state resistance is labeled R_{on}. All of this circuitry is fabricated on a monolithic IC. Typical component values are R_{on} = 100 Ω and C_s = 100 pF.

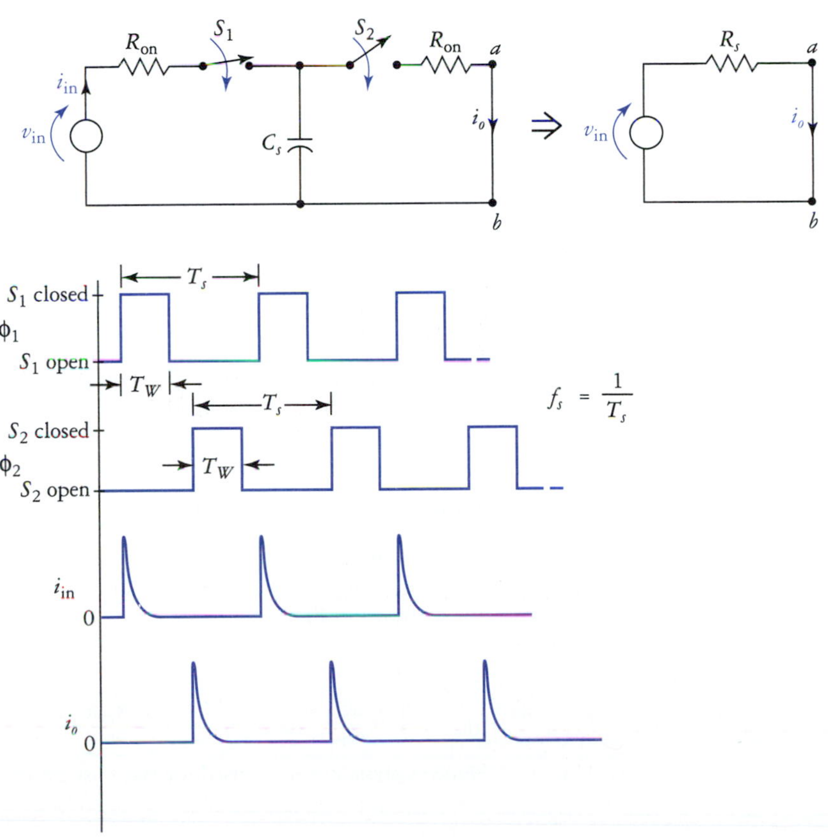

FIGURE 14.29
Switched-capacitor simulation of a resistance.

Here's the main idea behind the operation of the switched capacitor circuit: Assume that S_1 is closed and S_2 is open, and the capacitor is charging through R_{on}. In order for the capacitor to fully charge, we require $T_W \geq 5\tau$. Let's assume that the on resistance of the switches is equal, so the charging and discharging time constants are equal. The time constant of each RC network is

$$\begin{aligned}\tau &= R_{\text{on}}C_s \\ &= (100\ \Omega)(100\ \text{pF}) \\ &= 10\ \text{ns}\end{aligned}$$

This means that the minimum clock pulse width is

$$\begin{aligned}T_W &\geq 5\tau \\ &\geq 50\ \text{ns}\end{aligned}$$

In practice, the maximum clock frequency that can be used with a switched-capacitor filter is about 1 MHz ($T_{\text{clk}} = 1000$ ns), therefore it is easy to obtain a two-phase clock pulse width that is much longer than 5τ, so full charging will be achieved. This charging process produces a narrow current pulse at each rising edge of ϕ_1 as shown in Fig. 14.29. Now assume that S_1 opens and then S_2 closes. If the output is shorted, the capacitor will rapidly discharge, producing the output current pulses shown in the figure. There is a clock-dependent charge transfer taking place.

Here's the key to understanding the circuit. First, realize that reducing the clock frequency will not change the shape or amplitude of the current pulses; they will only spread further apart. In Fig. 14.29 we are looking at the circuit at a very short timescale. Looking at the circuit from a relatively long timescale, relative to the slowest clock frequency, the average current is proportional to the clock frequency f_s. The equation relating the clock frequency to the average current is

$$I_o = I_{\text{in}} = V_{\text{in}}f_sC_s \tag{14.54}$$

One of the first things we learned in electronics is $R = V/I$, so if we rearrange Eq. (14.54) into this form we have the formula for the switched-capacitor resistance R_s:

$$R_s = \frac{1}{C_sf_s} \tag{14.55}$$

From the equation we find that the effective resistance is proportional to the clock frequency. This is very useful because we can form large numbers of switched-capacitor resistors, which, if they all operate from the same clock signal, can be varied simultaneously. This allows for convenient tuning of the active filter. Based on limitations in clock frequency, transistor leakage, and available capacitor sizes, the range for R_s is about 10 kΩ to 100 MΩ.

The relationship in Eq. (14.55) will hold as long as the input voltage changes slowly compared to the minimum clock frequency. Keeping in mind that switched capacitors are used in active filter circuits, the following relationship is usually applied in practical circuits:

$$f_s \geq 50\,f_x \tag{14.56}$$

where $f_x = f_c$ is the corner frequency of a low-pass filter, or $f_x = f_o$ is the center frequency of a bandpass filter.

Switched-Capacitor Integrators

Most switched-capacitor filters use the state-variable topology, which requires the use of two integrators. These integrators are implemented using switched capacitors. A switched-capacitor inventing integrator is shown in Fig. 14.30. Assuming that Eq. (14.56) is satisfied, the output of the integrator is

$$v_o = \frac{1}{R_s\,C_1}\int v_{\text{in}}\,dt$$

This is the same expression that is used for the classical op amp integrator presented in Chapter 13.

The MF4-50

The National Semiconductor MF4-50 is a fourth-order low-pass Butterworth filter that is housed in an 8-in DIP. The pinout and internal block diagram are shown in Fig. 14.31. The MF4-50 is designed to operate with a single-polarity supply ranging from 5 to 14 V, with 10 V being the recommended value. Bipolar operation is also possible from ±2.5 to ±7 V, with ±5 V recommended. The maximum clock frequency is 1 MHz.

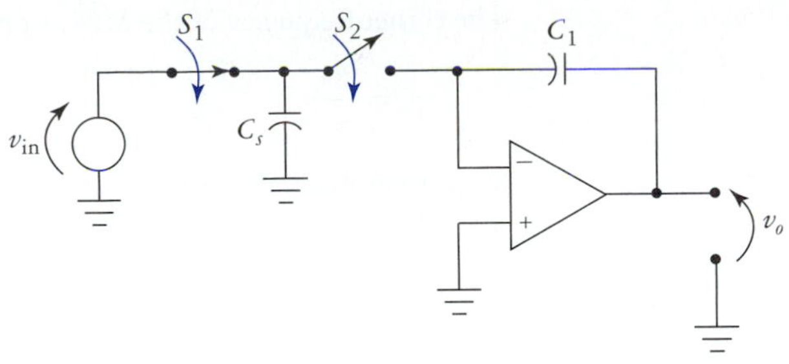

FIGURE 14.30
Switched-capacitor integrator.

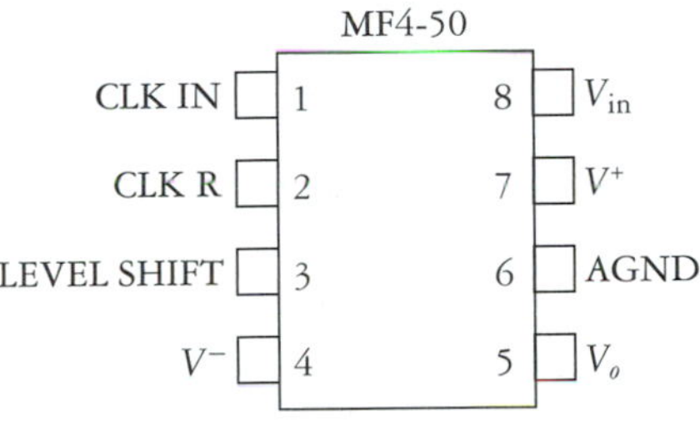

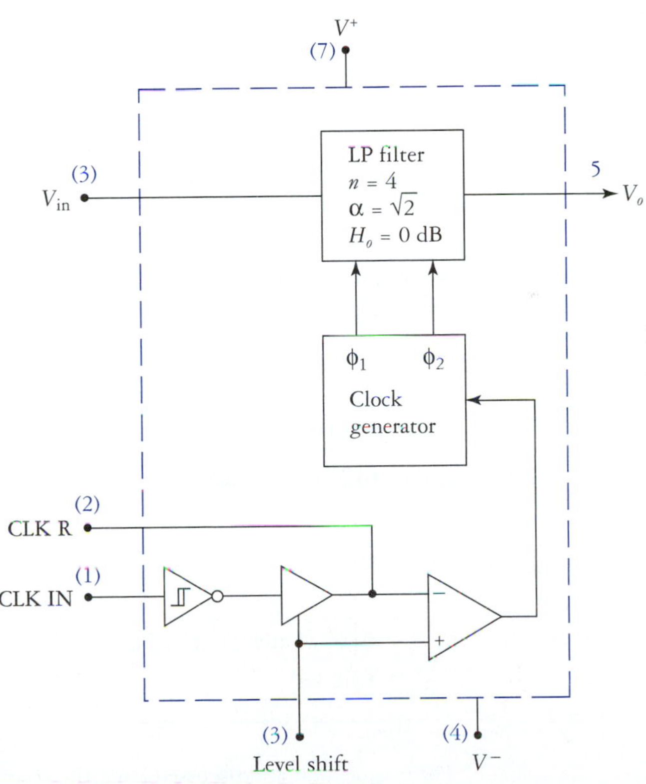

FIGURE 14.31
MF4-50 monolithic switched-capacitor filter.

The MF4 can use an external clock source or may be configured in a self-clocking mode. The single-polarity, self-clocking mode is shown in Fig. 14.32. As you can see, the MF4 requires a minimum number of critical external components for operation. For 10-V (or ±5-V) operation, the clock frequency is

$$f_{clk} \cong \frac{1}{0.69RC} \tag{14.57}$$

The corner frequency of the MF4 is given by

$$f_c = \frac{f_{clk}}{50} \tag{14.58}$$

To prevent distortion, the input signal should be limited to about 2 V less than the rail-to-rail supply voltage.

Note
The MF4 is a fourth-order, low-pass, Butterworth filter.

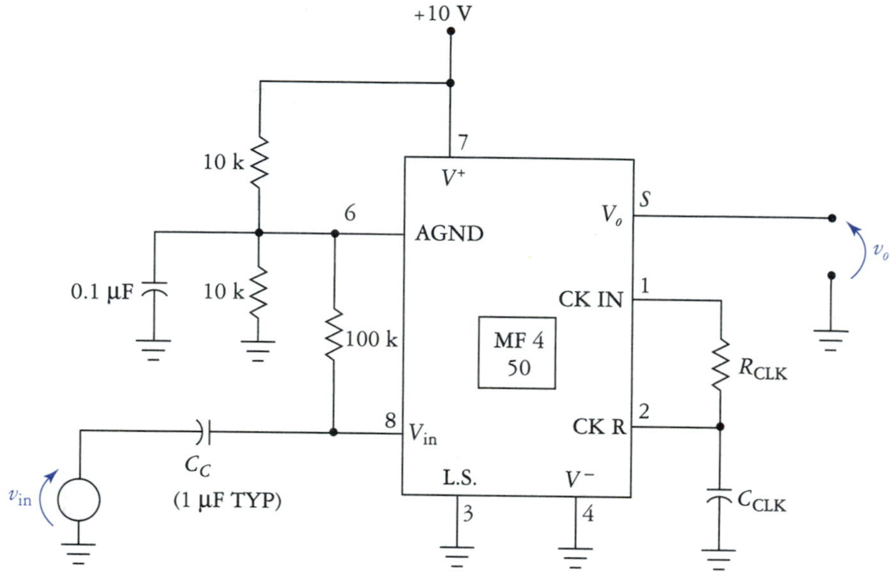

FIGURE 14.32
Typical switched-filter application circuit.

EXAMPLE 14.9

The filter of Fig. 14.32 is required to have f_c = 1 kHz, with C_{clk} = 0.01 μF. Determine the required value for R_{clk}.

Solution The required clock frequency is

$$\begin{aligned} f_{clk} &= 50f_c \\ &= 50 \text{ kHz} \end{aligned}$$

Solving Eq. (14.57), we obtain

$$\begin{aligned} R_{clk} &= \frac{1}{0.69 f_{clk} C} \\ &= \frac{1}{(0.69)(1 \text{ kHz})(0.01\ \mu\text{F})} \\ &= 11.8 \text{ k}\Omega \end{aligned}$$

For reasons that are discussed later, the absolute maximum input frequency that should be applied to a switched-capacitor filter should be limited to

$$f_{in(max)} = f_{clk}/2 \tag{14.59}$$

If the input signal is not band limited to less than $f_{clk}/2$ then an external low-pass passive prefilter should be used to reject out-of-range frequencies.

The MF5 Universal Filter

The MF5 is a monolithic, switched-capacitor, second-order, state-variable filter housed in a 14-pin DIP. Recall that the state-variable topology is very flexible and can be used to produce any filter response. The pinout and internal block diagram for the MF5 are shown in Fig. 14.33. The filter can be

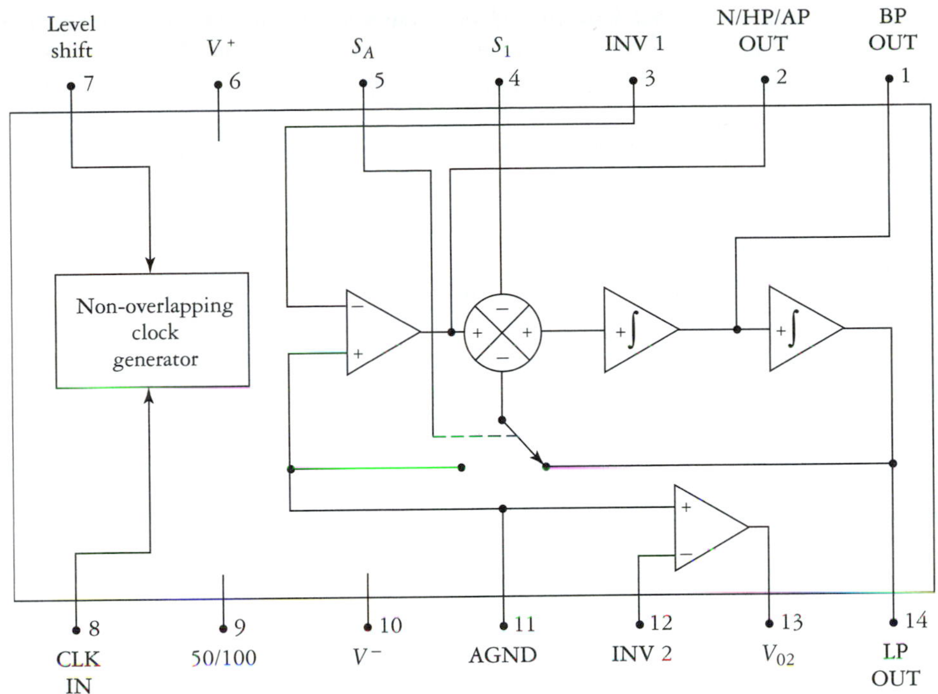

FIGURE 14.33
MF5 universal switched-capacitor filter.

configured to implement various forms of the state-variable filter. Brief descriptions of the various pin functions are as follows (pin numbers are in parentheses):

V^+(6), V^-(10): Positive and negative supply inputs. Total supply voltage (difference between V^+ and V^-) should be between 8 V and 14 V. For single-supply operation, V^- is typically connected to ground.

AGND(11): Analog ground terminal. Connect to system ground for spit-supply operation. Bias to $V^+/2$ for single supply operation. Decouple to system ground with a 0.1-μF capacitor.

LP(14), BP(1), N/HP/AP(2): Second-order low-pass, bandpass, notch, high-pass, and all-pass response output terminals. All outputs have voltage compliance within about the supply rail voltages ±1 V. Current loading should be limited to about ±1 mA.

INV1(3): Inverting input of internal op amp.

S1(4): All-pass input terminal. If unused, S1 should be tied to AGND.

SA(5): Switch control signal: SA = V^-, connects switch to AGND. SA = V^+ connects switch to the LP OUT terminal.

50/100(9): Sets internal clock to f_0 ratio according to 50/100 = +V $f_0 = f_{clk}/50$, if 50/100 = −V $f_0 = f_{clk}/100$.

Level Shift(7): Sets clock input for operation with CMOS or TTL clock signals. Bipolar operation: Connect LS(7) to AGND. This will accommodate both CMOS and TTL clock signals. Single-supply operation: LS(7) = AGND for CMOS clock input, LS(7) = V^- (system ground) for TTL clock signals.

CLK(8): Filter clock input pin. Frequency should be limited to less than 1 MHz, with 50% duty cycle recommended.

INV2(12): Noninverting input of internal op amp.

V_{o2}(13): Uncommitted op amp output. Typically used as an inverting buffer.

The MF5 is very flexible with no less than nine different modes of operation. Because of space limitations, we take a brief look at only *mode 1* operation, which provides simultaneous BP, LP, and notch outputs. Mode 1 is shown in Fig. 14.34, where the uncomitted op amp is used in the inverting configuration to buffer the notch filter output. For the connections shown, the design equations are:

$$f_o = f_{clk}/100 \tag{14.60}$$

$$f_{notch} = f_o \tag{14.61}$$

$$H_{O(LP)} = -\frac{R_2}{R_1} \tag{14.62}$$

$$H_{o(BP)} = -\frac{R_3}{R_1} \tag{14.63}$$

$$H_{O(notch)} = \frac{R_2 R_5}{R_1 R_4} \text{ [as } f_{in} \rightarrow 0 \text{ Hz and } f_{in} \rightarrow f_{max} \text{ (that is, } f_{clk}/2)] \tag{14.64}$$

$$Q = \frac{f_O}{BW} = \frac{R_3}{R_2} \tag{14.65}$$

Note
Eq. 14.64 is the passband gain of the notch response.

Clearly, the MF5 is much easier to set up than an equivalent continuous-time state-variable filter using traditional design techniques.

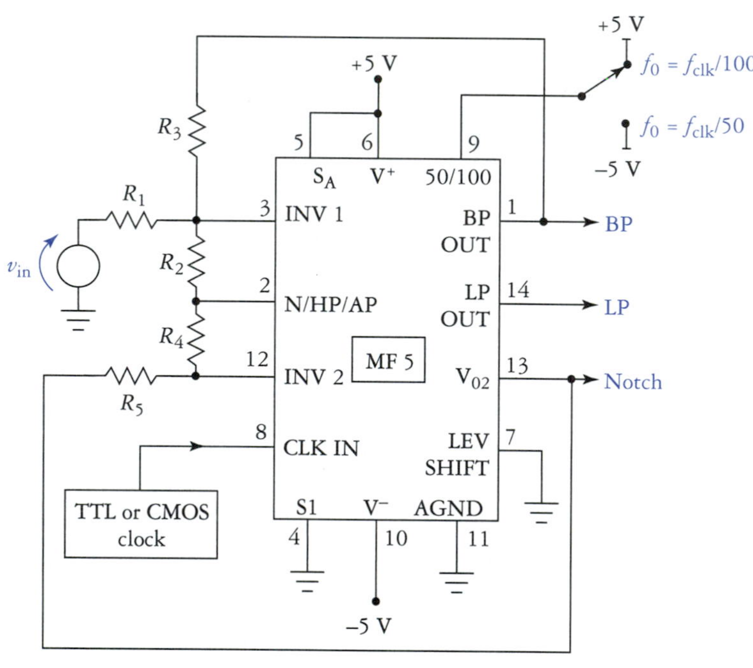

FIGURE 14.34
Typical application circuit.

EXAMPLE 14.10

Refer to Fig. 14.34. The circuit has $f_{clk} = 100$ kHz, $R_1 = 1$ kΩ, $R_2 = 5$ kΩ, and $R_3 = R_4 = R_5 = 10$ kΩ. Determine f_o, f_{notch}, $H_{O(LP)}$, $H_{o(BP)}$, $H_{O(notch)}$, Q, BW, and the peak amplitude of the low-pass response. Reminder: For notational convenience, let's use H to represent response in decibels, and A to represent response as a linear voltage ratio.

Solution We simply apply Eqs. (14.60) through (14.65), giving us

$$f_{O(notch)} = f_O = 1 \text{ kHz}$$

$$H_{O(LP)} = 5 = 14 \text{ dB}$$

$$H_{o(BP)} = 10 = 20 \text{ dB}$$

$$H_{O(notch)} = 5 = 14 \text{ dB}$$

$$Q = 2$$

$$BW = 500 \text{ Hz}$$

To determine the peak response of the low-pass response, we must think about the damping of the filter. The damping coefficient is $\alpha = 1/Q = 0.5$. Thus, we have an underdamped filter. Specifically this is a Chebyshev response filter with

$$\begin{aligned} A_{\text{pk(LP)}} &= A_{0\text{(LP)}}Q \\ &= (5)(2) \\ &= 10 \end{aligned}$$

which in decibels is

$$\begin{aligned} H_{\text{pk(LP)}} &= 20 \log A_{\text{pk(LP)}} \\ &= 20 \log(10) \\ &= 20 \text{ dB} \end{aligned}$$

The maximum response is

$$\begin{aligned} H_{0\max\text{(LP)}} &= H_{0\text{(LP)}} + H_{\text{pk(LP)}} \\ &= 14 \text{ dB} + 20 \text{ dB} \\ &= 34 \text{ dB} \end{aligned}$$

or equivalently

$$A_{0\max\text{(LP)}} = 50$$

Keep in mind that the main advantage of switched-capacitor filters is the elimination of external frequency-determining capacitors, which are generally the most difficult external component to obtain in different values or in adjustable versions.

14.10 COMPUTER-AIDED ANALYSIS APPLICATIONS

As we have seen before, PSpice is very useful for analyzing network frequency response. This is especially true in the case of filter circuits. In this section we look at the time- and frequency-domain characteristics of some of the filters that have been discussed.

Second-Order VCVS Response

In Example 14.4 a second-order low-pass Sallen-Key VCVS Butterworth filter was designed. This circuit is shown in the schematic editor in Fig. 14.35. The analysis was set up for a decade ac sweep from 100 Hz to 10 kHz. A transient analysis was set up to run from 0 to 5 ms. The input signal is created using a pulse source, which allows a step function to be applied (for the transient analysis) as well as a steady-state ac input (for the frequency-response analysis). The pulse source parameters are set as follows:

DC = 0	(No dc offset)
AC = 1	(Phasor voltage magnitude $V_p = 1$ V)
V1 = 0	(Initial voltage of pulse)
V2 = 1	(Final pulse voltage)
TD = 0	(No delay of pulse rising edge)
TR = 1 ns	(Pulse rise time: relatively fast)
TF = 1 ns	(Pulse fall time: does not matter here)
PW = 1 s	(Pulse width: set longer than analysis run)
PER = 1 s	(Pulse period: set ≥ PW)

The circuit is simulated and the ac analysis option is selected for PROBE. The initial plot of the amplitude response appears as expected in Fig. 14.36, with the response down by 3 dB at 1 kHz. A second plot was created in which the phase response of the filter is shown, where we find the phase shift is half the asymptotic value ($-180°/2 = -90°$) at the corner frequency.

The time-domain response of the filter to the step input is shown in Fig. 14.37. As shown, the Butterworth response results in some overshoot (about 4.3%) in response to a pulse input. Recall that an overdamped response such as that provided by a Bessel filter will reduce this overshoot, at the expense of a droopy passband and more gradual rolloff.

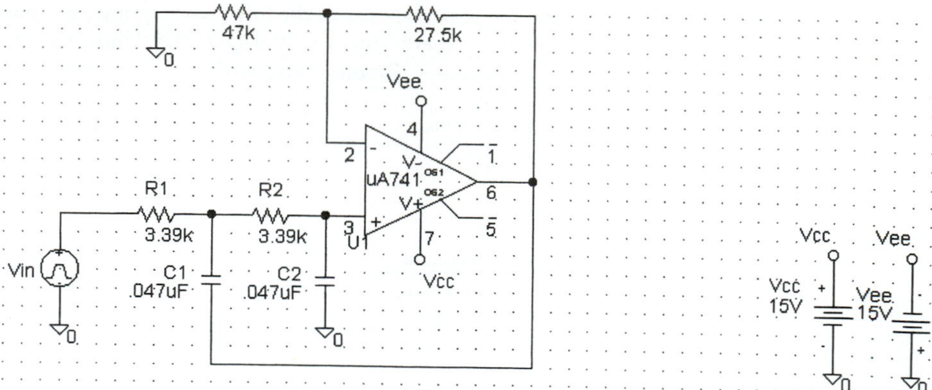

FIGURE 14.35
Second-order VCVS.

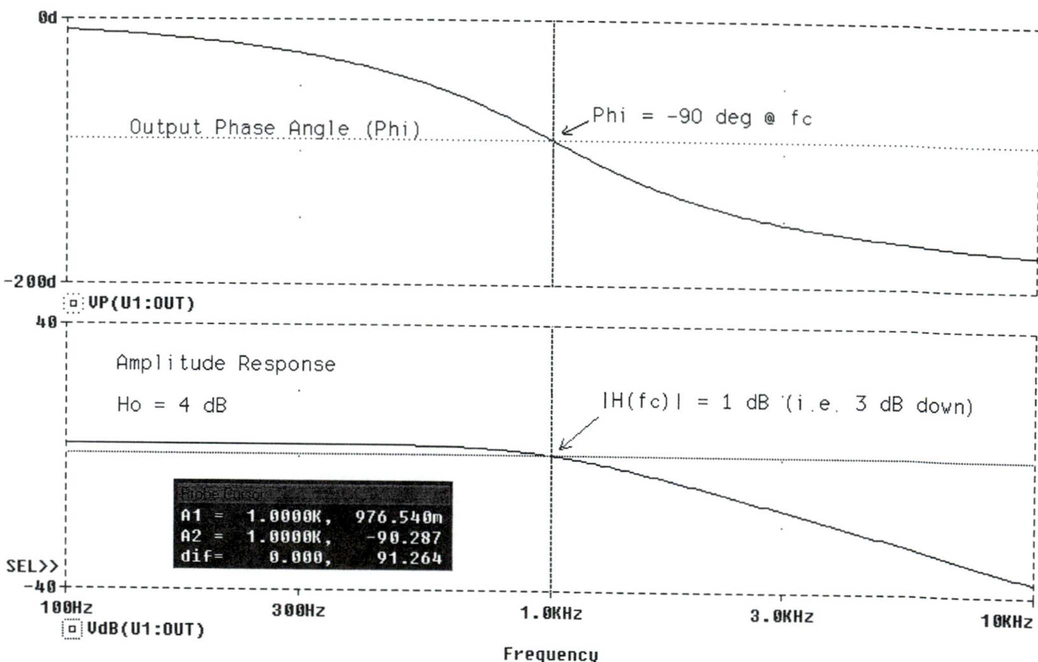

FIGURE 14.36
Filter amplitude and phase response.

The rise time of the output is defined as the time required for v_o to go from 10% to 90% of the final value (assuming slew limiting does not occur), as shown in Fig. 14.38. In equation form, this relationship is written

$$T_r = t_{90\%} - t_{10\%} \tag{14.66}$$

It can also be shown that the corner frequency of the filter (or any system for which we know the rise time) is related to the rise time by the equation

$$f_c = \frac{0.35}{T_r} \tag{14.67}$$

If we substitute the rise time found in Fig. 14.37 into this equation, we find $f_c = 933$ Hz. This is close to the corner frequency found theoretically and via direct frequency-domain simulation.

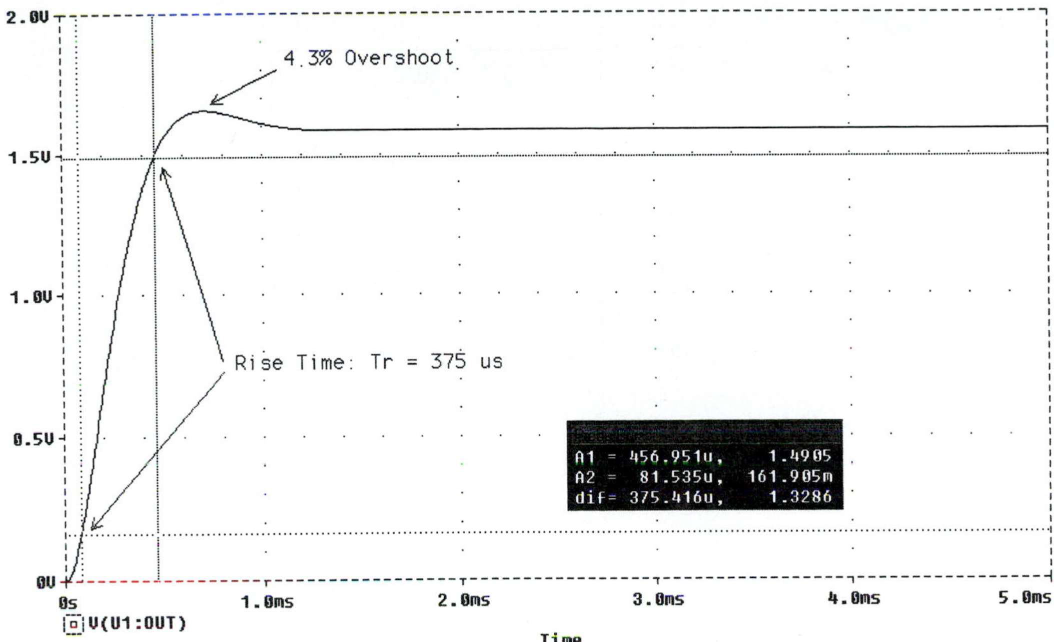

FIGURE 14.37
Filter transient response.

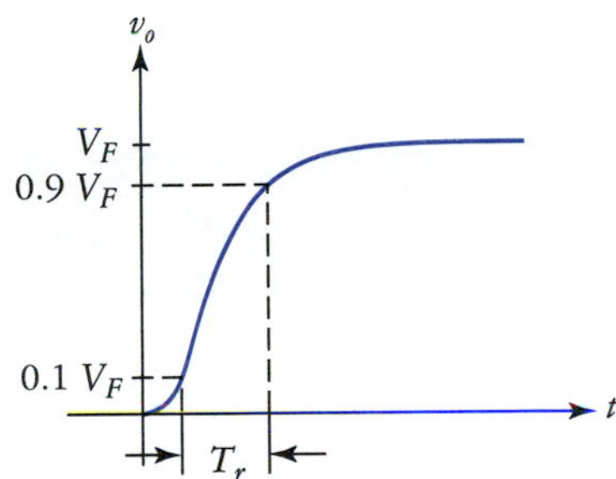

FIGURE 14.38
Specifying the rise time of a signal.

Bandpass Filter Response

We now determine the frequency and pulse response of the IGMF BP filter designed back in Example 14.6. As usual, we will use an ac sweep to obtain the frequency response directly; however, we will also verify much of this information using the transient response results.

The IGMF bandpass filter is shown in Fig. 14.39. The filter was designed such that

$$f_o = 1\text{ kHz}$$

$$Q = 5$$

$$A_o = 2\ (H_o = 6\text{ dB})$$

The pulse source, ac sweep, and transient analysis parameters are set the same as those of the previous analysis given in Fig. 14.35.

The frequency response of the filter is shown in Fig. 14.40. Here we find $f_o = 1$ kHz, BW = 200 Hz, and $Q = 5$, all of which are just as expected. It can also be verified that the response skirts approach rolloff rates of ± 20 dB/decade, which is characteristic of a second-order bandpass filter.

The display of the output step response is shown in Fig. 14.41. The damped oscillations indicate a very underdamped response, which should be expected. The period of the oscillatory response is $T_o = 1$ ms. This relates to the center frequency of the filter simply by the following familiar equation:

$$f_o = 1/T_o = 1\text{ kHz}$$

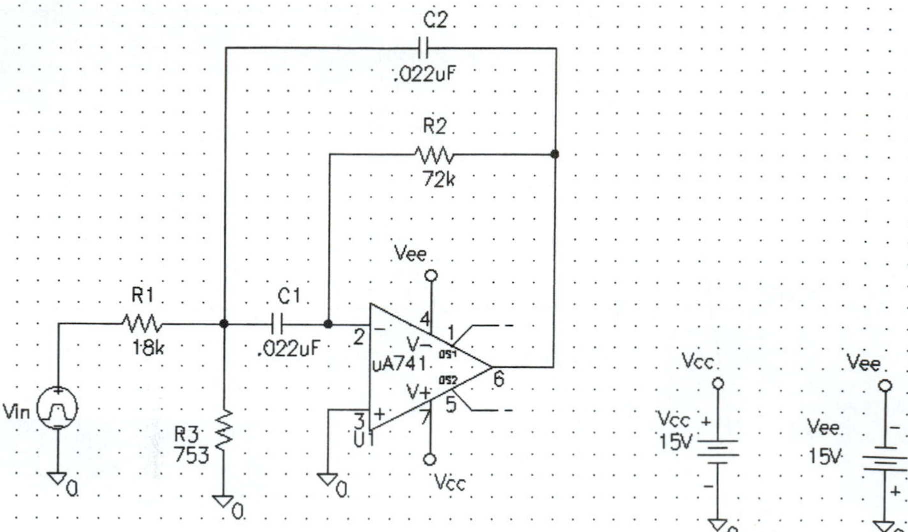

FIGURE 14.39
IGMF BP filter.

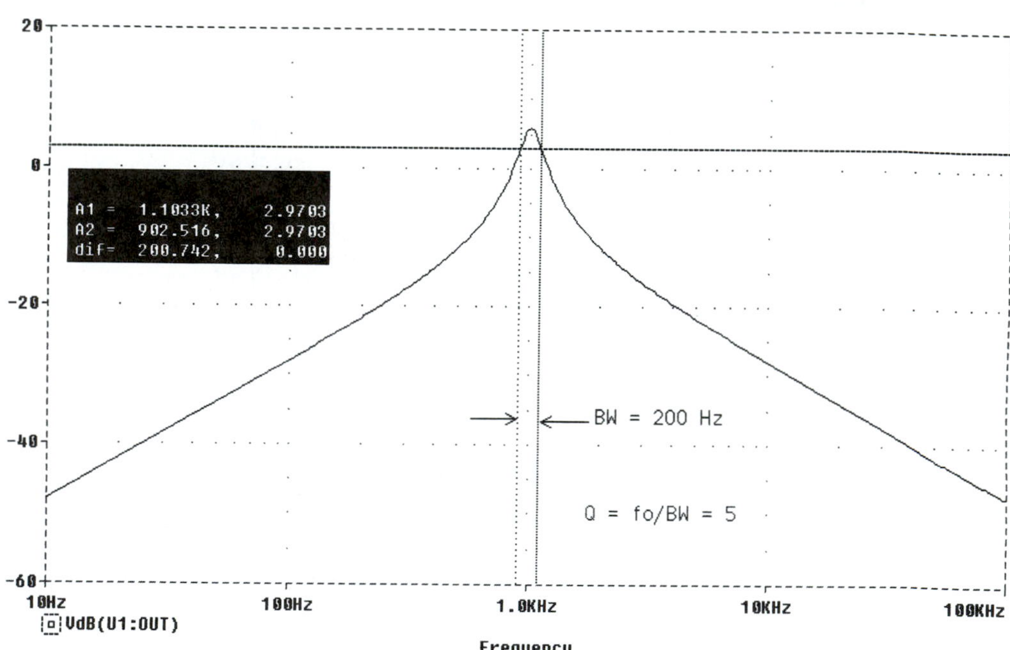

FIGURE 14.40
IGMF amplitude response.

All-Pass Filter Response

We finish this section with the simulation of the lagging-phase all-pass filter shown in Fig. 14.42. The amplitude and phase response plots for this filter are shown in Fig. 14.43.

Ideally the amplitude response of the all-pass filter would be flat for all frequencies, but in this case, the rolloff of the 741's open-loop gain limits the filter to a maximum 20 kHz. Keep in mind that the ac sweep holds the op amp in the small-signal mode, which eliminates the effect of slew limiting, which also limits the useful frequency range of the circuit.

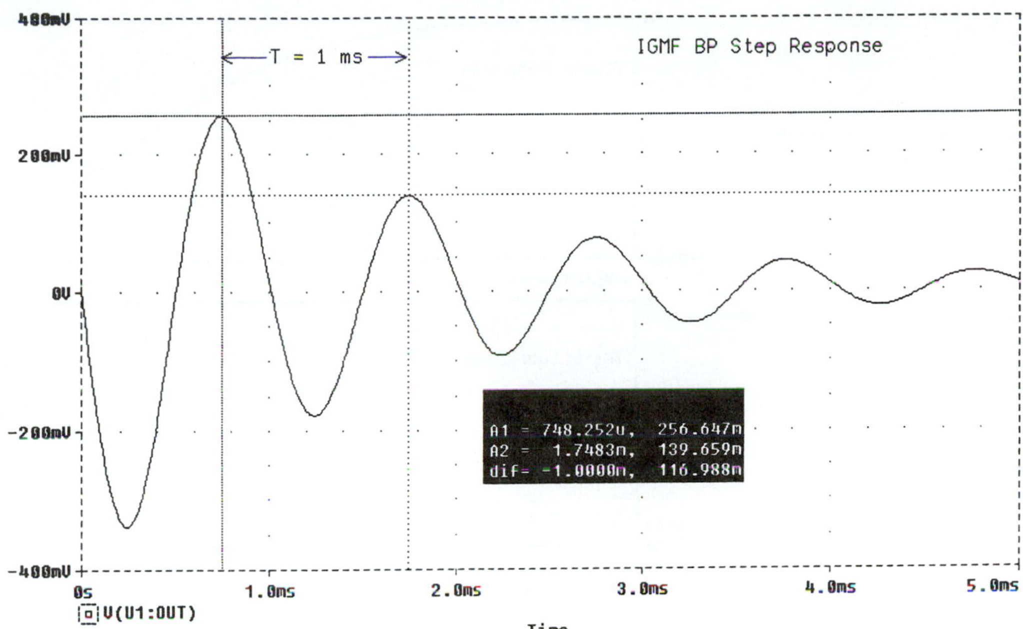

FIGURE 14.41
IGMF step response.

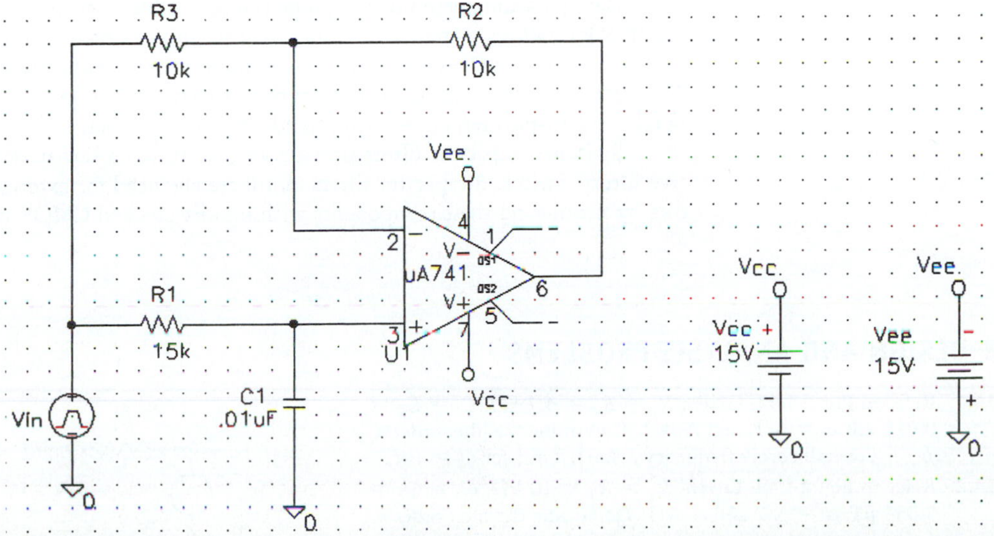

FIGURE 14.42
All-pass filter.

Note
The amplitude response of the all-pass filter drops at higher frequencies because the op amp A_{OL} drops eventually.

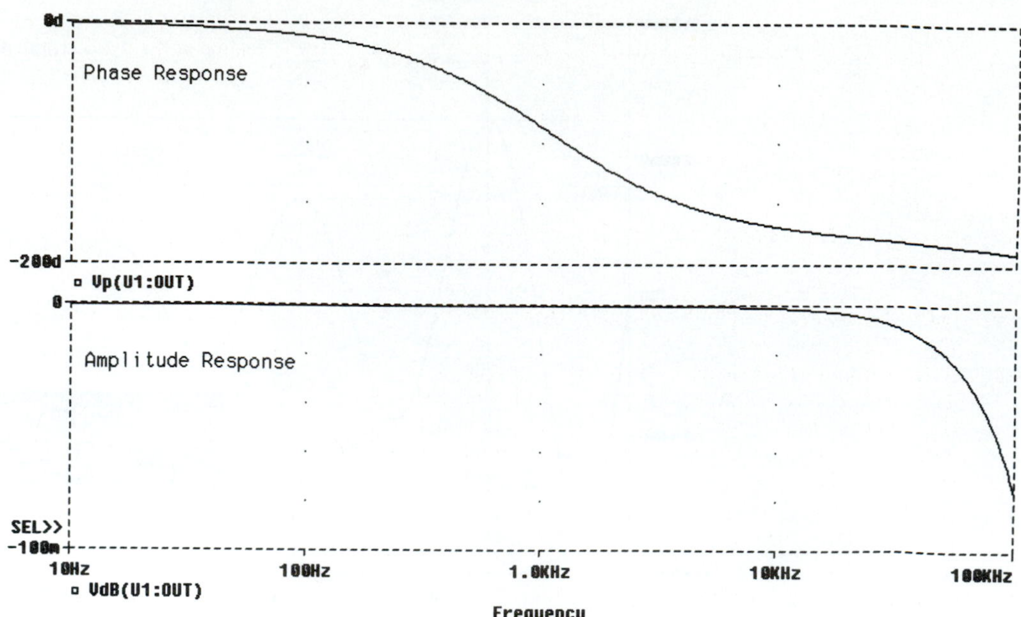

FIGURE 14.43
All-pass filter phase and amplitude response curves.

▪ CHAPTER REVIEW

Active filters have been developed primarily in order to eliminate the need for inductors, which tend to be expensive and behave rather nonideally at lower frequencies. Active filters provide significant advantages over passive filters at frequencies of typically less than a few hundred kilohertz or so, by providing power gain, and convenient frequency and damping adjustment. The advantages of active filters are most significant at very low frequencies, such as those encountered in biomedical instrumentation applications, where classical *RLC* filter designs require very large inductors.

One of the most common active filter topologies is the equal-component Sallen-Key VCVS configuration. The damping coefficient of such a filter is easily adjusted by varying the closed-loop voltage gain of the op amp. LP-to-HP transformations are easily accomplished using the VCVS circuit.

Bandpass and notch filter response can be produced using cascaded HP and LP filters; however, the IGMF topology is more commonly used to implement response *Q* values ranging from around 1 to 25 or so. Higher *Q* responses are usually implemented using state-variable or biquad filter circuits. The state-variable circuit is the most versatile active filter topology, capable of providing LP, HP, BP, notch, and all-pass responses with attendant *Q*s higher than 100.

Switched-capacitor filters provide an easy-to-use alternative to standard continuous-time active filters. Switched-capacitor filters eliminate the need for external frequency-determining capacitors by simulating these components with rapidly clocked CMOS switches.

▪ DESIGN AND ANALYSIS PROBLEMS

14.1. Refer to Fig. 14.44. Given: $R_1 = R_2 = 4.7\ \text{k}\Omega$, $C_1 = C_2 = 0.015\ \text{mF}$, $\alpha = \sqrt{2}$, and A = 1. Determine the filter order *n*, f_{C_1}, f_{C_2}, overall corner frequency f_c and $|H(jf)|$ for $f_{\text{in}} = 10f_c$.

14.2. Refer to Fig. 14.44. Given: $R_1 = R_2 = 10\ \text{k}\Omega$, $C_1 = C_2 = 0.033\ \mu\text{F}$, $\alpha = \sqrt{2}$ and $A = 1$. Determine the filter order *n*, f_{c_1}, f_{c_2}, overall corner frequency f_c, and $|H(jf)|$ for $f_{\text{in}} = 10f_c$.

14.3. Refer to Fig. 14.45. Assume: $L = 10\ \mu\text{H}$, $C = 0.022\ \mu\text{F}$, $r_s = 5\ \Omega$, and $R_x = 0\ \Omega$. Determine f_o, α, *Q*, the response shape (Butterworth, etc.), and the peak amplitude of the output in

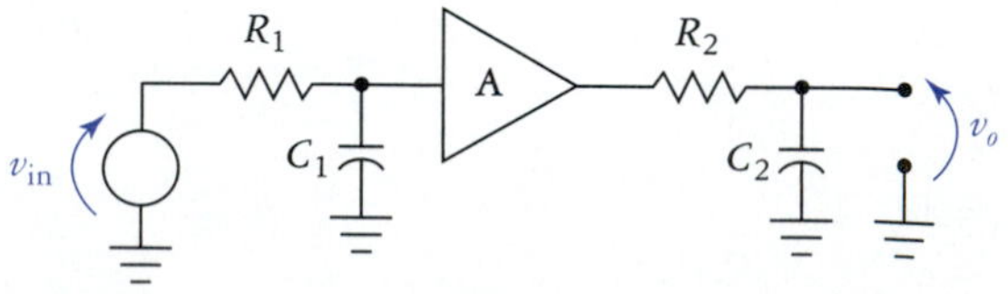

FIGURE 14.44
Circuit for Problems 14.1 and 14.2.

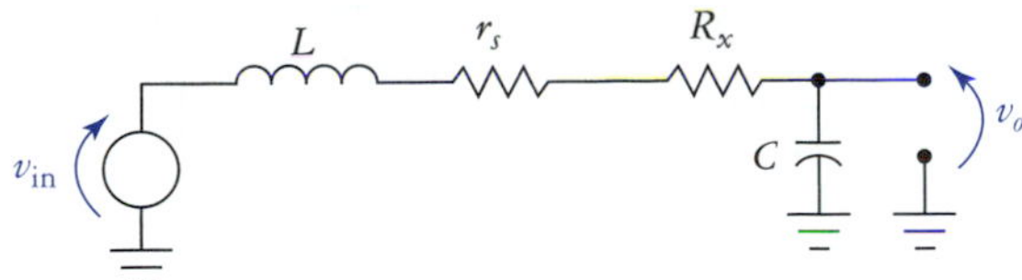

FIGURE 14.45
Circuit for Problems 14.3 and 14.4.

decibels (i.e., H_{pk}). Determine the value required for R_x such that $\alpha = \sqrt{2}$.

14.4. Refer to Fig. 14.45. Assume: $L = 100\ \mu H$, $C = 500$ pF, $r_s = 25\ \Omega$, and $R_x = 0\ \Omega$. Determine f_o, α, Q, the response shape (Butterworth, etc.), and the peak amplitude of the output in decibels (i.e., H_{pk}). Determine the value required for R_x such that $\alpha = \sqrt{2}$.

14.5. A certain first-order LP filter has $f_c = 500$ Hz, $A_0 = 1$, and $\alpha = \sqrt{2}$. Determine $H(jf)$ (magnitude and phase angle) for $f_{in} = 1$ kHz, 2.5 kHz, and 10 kHz.

14.6. A certain first-order HP filter has $f_c = 2$ kHz, $A_\infty = 1$, and $\alpha = \sqrt{2}$. Determine $H(jf)$ (magnitude and phase angle) for $f_{in} = 1$kHz, 500 Hz, and 100 Hz.

14.7. A first-order LP Butterworth filter has $f_c = 1$ kHz and $A_0 = 1$. Find the frequency at which the response magnitude is $|H| = -30$ dB.

14.8. A first-order HP Butterworth filter has $f_c = 1$ kHz and $A_\infty = 1$. Find the frequency at which the response magnitude is $|H| = -25$ dB.

14.9. A first-order LP Butterworth filter has $f_c = 1$ kHz. Find the frequency at which the response phase angle [$\phi = \arg H(jf)$] is $\phi = -50°$.

14.10. A first-order HP Butterworth filter has $f_c = 1$ kHz. Find the frequency at which the response phase angle [$\phi = \arg H(jf)$] is $\phi = 60°$.

14.11. A certain second-order low-pass filter has $f_c = 1$ kHz, $A_0 = 10$, and $\alpha = 0.8$. Determine $|H(jf)|$ at $f_{in} = 3$ kHz and 15 kHz.

14.12. A certain second-order high-pass filter has $f_c = 5$ kHz, $A_0 = 2$, and $\alpha = 1.0$. Determine $|H(jf)|$ at $f_{in} = 1$ kHz and 200 Hz.

14.13. Given a low-pass filter with $\alpha = \sqrt{2}$, $A_0 = 1$, and $f_c = 3$ kHz, determine the required order n such that the response is down by at least 25 dB at $f_{in} = 9$ kHz.

14.14. Given a high-pass filter with $\alpha = \sqrt{2}$, $A_\infty = 1$, and $f_c = 800$ Hz, determine the required order n such that the response is down by at least 60 dB at $f_{in} = 200$ Hz.

14.15. Refer to Fig. 14.46. Given $R_F = 39\ k\Omega$ and $R = 10\ k\Omega$, determine R_1 and C such that a 3-dB Chebyshev response with $f_c = 400$ Hz is produced.

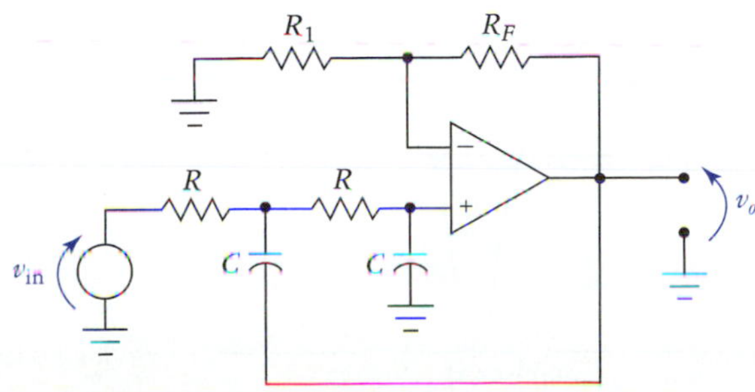

FIGURE 14.46
Circuit for Problem 14.15.

14.16. Refer to Fig. 14.47. Given $R_1 = 10\ k\Omega$ and $C = 0.01\ \mu F$, determine R_F and R such that a Butterworth response with $f_c = 10$ kHz is produced.

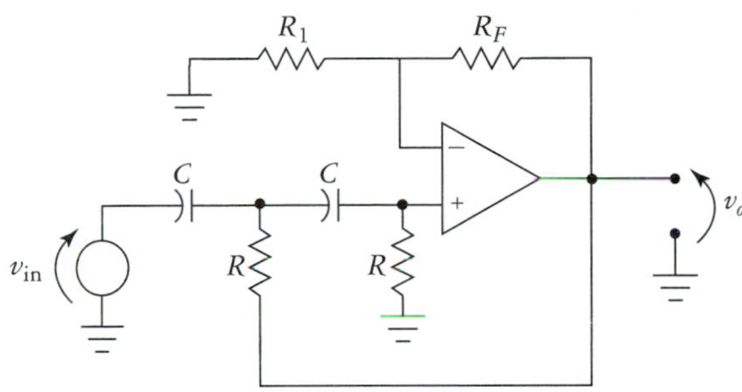

FIGURE 14.47
Circuit for Problem 14.16.

14.17. Refer to Fig. 14.48. Determine A_0, H_0 (in dB), α, A_{pk}, H_{pk} (dB), A_{max}, and H_{max} (dB) and name the response type.

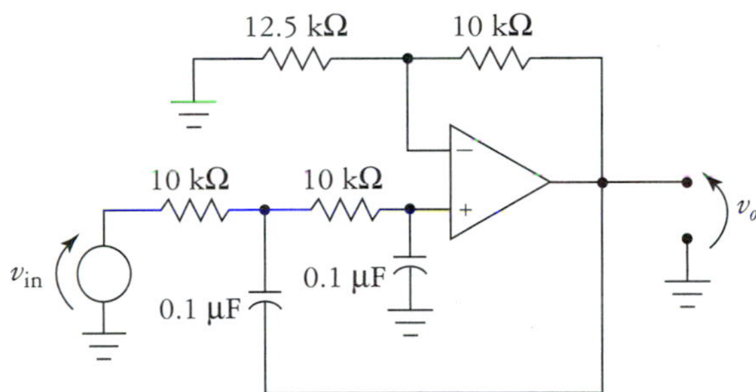

FIGURE 14.48
Circuit for Problem 14.17.

14.18. Refer to Fig. 14.49. Determine A_0, H_0 (in dB), α, A_{pk}, H_{pk} (dB), A_{max}, and H_{max} (dB) and name the response type.

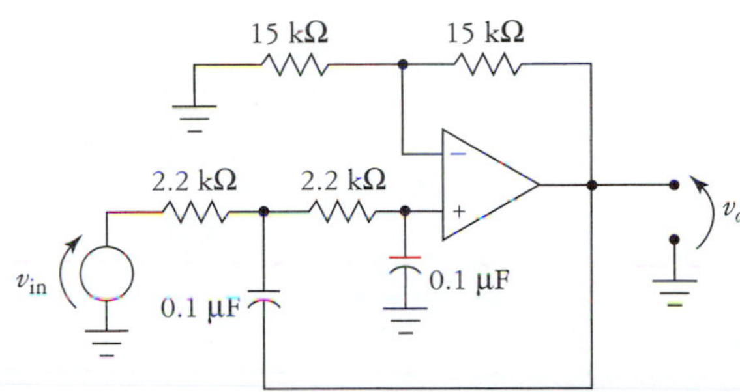

FIGURE 14.49
Circuit for Problem 14.18.

14.19. Refer to Fig. 14.50. Given $R_F = 22\ k\Omega$ and $R = 2.2\ k\Omega$, determine R_1 and C such that a 1-dB Chebyshev response with $f_c = 150$ Hz is produced.

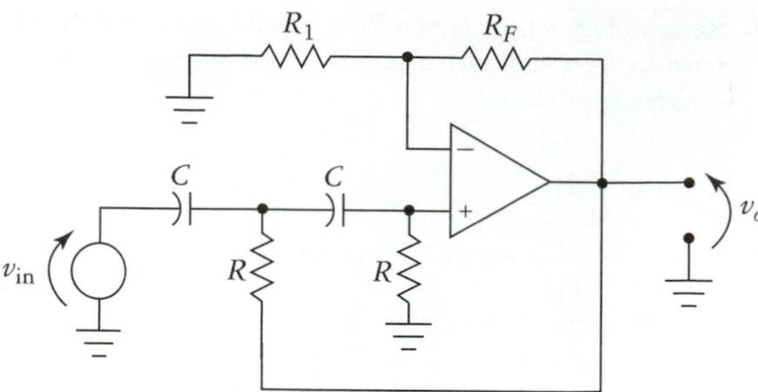

FIGURE 14.50
Circuit for Problems 14.19 and 14.20.

14.20. Refer to Fig. 14.50. Given $R_1 = 12\ \text{k}\Omega$ and $C = 0.0047\ \mu\text{F}$, determine R_F and R such that a Butterworth response with $f_c = 5$ kHz is produced.

14.21. Refer to Fig. 14.51. Determine A_0, H_0 (in dB), α, A_{pk}, H_{pk} (dB), A_{max}, and H_{max} (dB) and name the response type.

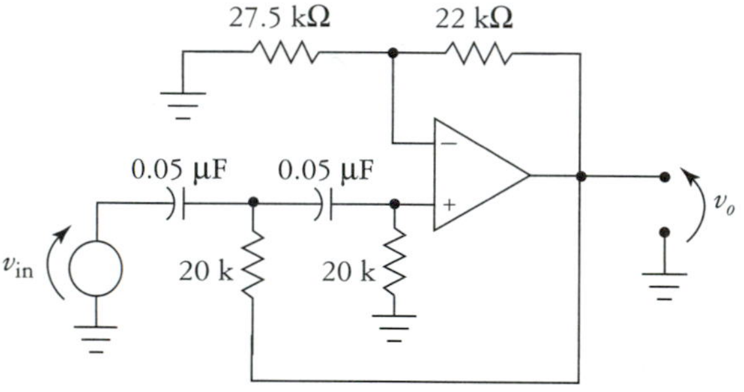

FIGURE 14.51
Circuit for Problem 14.21.

14.22. Refer to Fig. 14.52. Determine A_0, H_0 (in dB), α, A_{pk}, H_{pk} (dB), A_{max}, and H_{max} (dB) and name the response type.

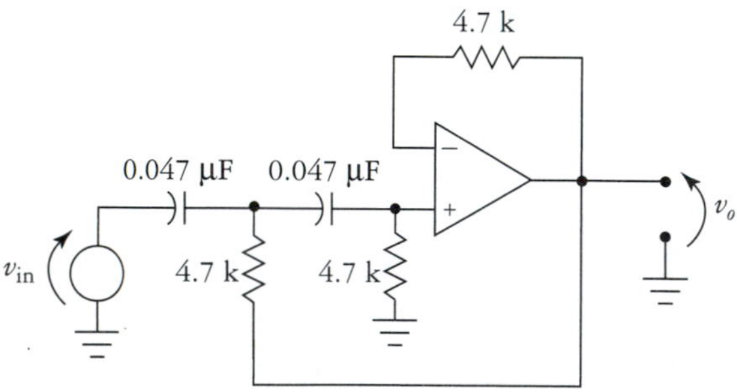

FIGURE 14.52
Circuit for Problem 14.22.

14.23. Refer to the block diagram shown in Fig. 14.53. Write the response magnitude expression and determine the response for $f_{in} = 100$ Hz and 10 kHz.

14.24. Refer to the block diagram shown in Fig. 14.54. Write the response magnitude expression and determine the response for $f_{in} = 100$ Hz and 10 kHz.

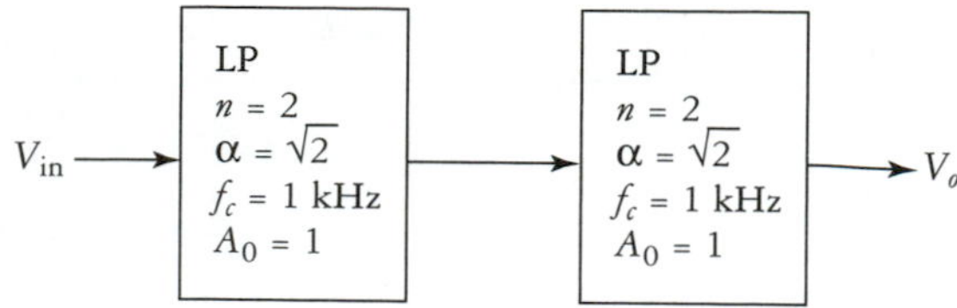

FIGURE 14.53
Circuit for Problem 14.23.

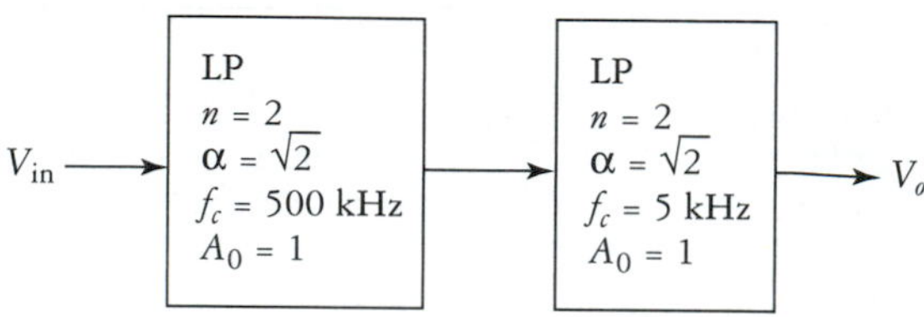

FIGURE 14.54
Circuit for Problem 14.24.

14.25. Refer to the response curve in Fig. 14.55. Determine f_o, BW, Q, and α.

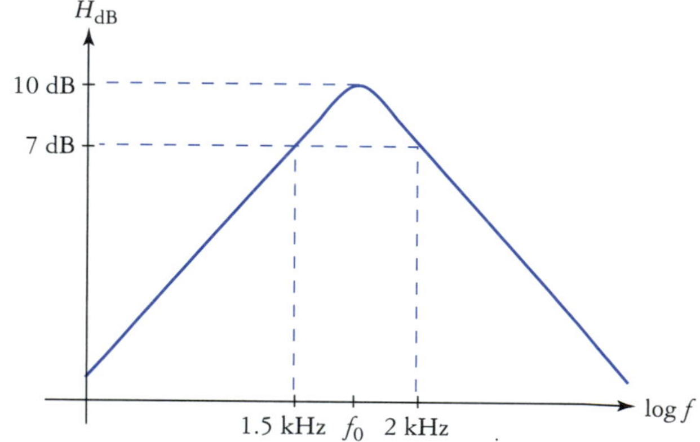

FIGURE 14.55
Circuit for Problem 14.25.

14.26. Refer to the response curve in Fig. 14.56. Determine f_H, BW, Q, and α.

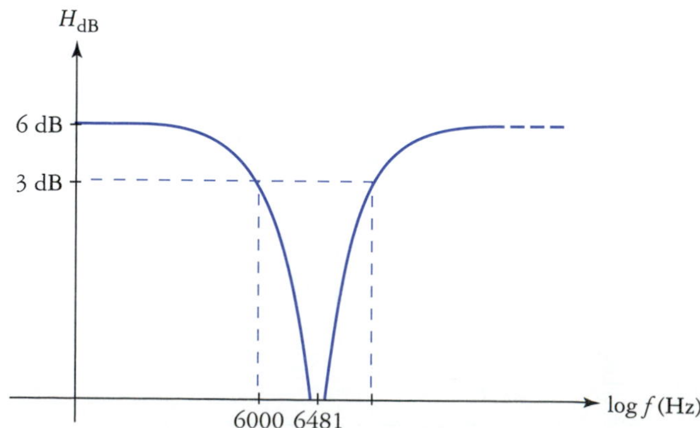

FIGURE 14.56
Circuit for Problem 14.26.

14.27. Design an IGMF BP filter with $f_o = 500$ Hz, $A_v = 1$, and $Q = 10$. Assume that $C = 0.047$ μF.

14.28. Design an IGMF BP filter with $f_o = 2$ kHz, $A_v = 2$, and $Q = 2$. Assume that $C = 0.01$ μF.

14.29. Convert the BP filter of Problem 14.27 into a bandstop filter using a summer. Draw the resulting schematic.

14.30. Convert the BP filter of Problem 14.28 into a bandstop filter using a summer. Draw the resulting schematic.

14.31. Refer to Fig. 14.19. Assume $R_1 = R_2 = R_3 = 10$ kΩ, $R = 15$ kΩ, $C = 0.01$ μF, $R_\alpha = 30.3$ kΩ, and $R_4 = 27$ kΩ. Determine the response type (i.e., Butterworth, etc.). A_o, H_o, (dB), f_o, Q, and α.

14.32. Refer to Fig. 14.19. Assume $R_1 = R_2 = R_3 = 22$ kΩ, $R = 10$ kΩ, $C = 0.33$ μF, $R_\alpha = 5.27$ kΩ, and $R_4 = 4.7$ kΩ. Determine the response type (i.e., Butterworth, etc.), A_o, H_o, (dB), f_o, Q, and α.

14.33. Refer to Fig. 14.21. Assume $R_4 = R_5 = R_6 = R_7 = 10$ kΩ, $R_1 = 10$ kΩ, $R_3 = 10$ kΩ, $R_2 = 22$ kΩ, and $C_1 = C_2 = 0.022$ μF. Determine $A_{LP(max)}$, $A_{BP(max)}$, f_o, Q, and α.

14.34. Refer to Fig. 14.21. Assume $R_4 = R_5 = R_6 = R_7 = 10$ kΩ, $R_1 = 15$ kΩ, $R_3 = 5$ kΩ, $R_2 = 22$ kΩ, and $C_1 = C_2 = 0.01$ μF. Determine $A_{LP(max)}$, $A_{BP(max)}$, f_o, Q, and α.

14.35. Refer to Fig. 14.21. Assume $R_1 = R_4 = R_5 = R_6 = R_7 = 10$ kΩ, and $C_1 = C_2 = 0.068$ μF. Determine R_3 and R_2 such that $Q = 10$ and $f_o = 220$ Hz. Determine the resulting values of $A_{LP(max)}$, A_{BP}, f_o, Q, and α.

14.36. Refer to Fig. 14.21. Assume $R_1 = R_4 = R_5 = R_6 = R_7 = 6.8$ kΩ, and $C_1 = C_2 = 0.022$ μF. Determine R_3 and R_2 such $Q = 5$ and $f_o = 800$ Hz. Determine the resulting values of $A_{LP(max)}$, $A_{BP(max)}$, f_o, Q, and α.

14.37. Refer to Fig. 14.23. Given $R = 10$ kΩ, $C = 0.1$ μF, and $R_1 = 18$ kΩ, determine the output phase shift ϕ for $f_{in} = 100$ Hz, 1 kHz, and 10 kHz.

14.38. Refer to Fig 14.24. Given $R = 47$ kΩ, $C = 0.22$ μF, and $R_1 = 1$ kΩ, determine the output phase shift ϕ for $f_{in} = 100$ Hz, 1 kHz, and 10 kHz.

14.39. Sketch the response curves that should be produced by the filter circuit specified in Example 14.10.

TROUBLESHOOTING PROBLEMS

14.40. An equal-component, second-order, Sallen-Key low-pass filter is used in a certain medical application. The filter is supposed to be critically damped, but is exhibiting peaking in the passband. What needs to be adjusted in order to correct this problem?

14.41. Refer to Fig. 14.18. This circuit is constructed in the lab with resistors R_5 and R_6 replaced with 20-kΩ potentiometers. Which resistor would be varied to obtain maximum null depth?

14.42. An audio-frequency signal is processed by an all-pass filter. Assuming that the filter is functioning properly, if the input and output signals were examined using a spectrum analyzer, what difference, if any, might you expect to find between the two displays?

COMPUTER PROBLEMS

14.43. Using PROBE, we plotted the amplitude and phase response of output of the filters in Section 14.10 by specifying signals using VdB(U1:OUT) and Vp(U1:OUT). This gives us what is equivalent to a polar form representation of the response (magnitude and angle). We can use the rectangular form by specifying plots of the real and imaginary components of the output response using VR(U1:OUT) and VI(U1:OUT). Use these commands to produce frequency response plots of the filters of Section 14.10.

14.44. Determine the step response of the all-pass filter of Fig. 14.42 using PSpice.

CHAPTER 15

Nonlinear Op Amp Applications

BEHAVIORAL OBJECTIVES

On completion of this chapter, the student should be able to:

- Describe the operation of a comparator.
- Describe the effect of positive feedback on the operation of a comparator.
- Determine the trip points for comparators and comparators with hysteresis.
- Explain the purpose of input and output bounding circuits.
- Analyze the operation of a window comparator.
- Explain the operation of a logarithmic amplifier.
- Explain the operation of an antilogarithmic amplifier.
- Describe the operation of an analog multiplier.
- Determine the output of an analog multiplier given various constant and time-varying input signals.
- Analyze the operation of precision rectifier circuits.
- Describe the advantages and disadvantages of precision rectifiers.
- Use the PSpice analog multiplier behavioral model.

INTRODUCTION

Most of the circuits we have covered so far have been linear circuits, or at least we have been able to treat them as linear circuits under certain conditions. It is generally easier to work with linear circuits because they are usually well behaved and can be analyzed using linear mathematical methods. Of course, we have seen many examples of nonlinear behavior as well, such as clipping, saturation, rectification, and so on. Sometimes we try to avoid this behavior, whereas at other times the nonlinear behavior is desirable. In general, however, nonlinear behavior is much more difficult to analyze. The usual approach we have taken in the past was to define the operation of a nonlinear circuit by different linear approximations, depending on the amplitude of the input signal. Most often, this approach is used here as well.

15.1 COMPARATORS

A *comparator* is a circuit that switches output states when its input exceeds a certain level, called the *trip point* or *set point.* The functional block symbol for a comparator is shown in Fig. 15.1(a). The operation of the comparator is illustrated by the waveforms in Fig. 15.1(b). The idea is that as long as v_{in} is less than the trip level V_t, the output is low. When v_{in} exceeds V_t, the output instantly switches high. Mathematically, we could write this as

$$v_o = \begin{cases} V_{max} & \text{for } v_{in} > V_t \\ V_{min} & \text{for } v_{in} < V_t \end{cases} \tag{15.1}$$

By the way, sometimes other terms are applied to the trip-point voltage. For example, V_t is sometimes called the *set-point voltage* or the *reference voltage.* We will use whichever term is most appropriate in a given application.

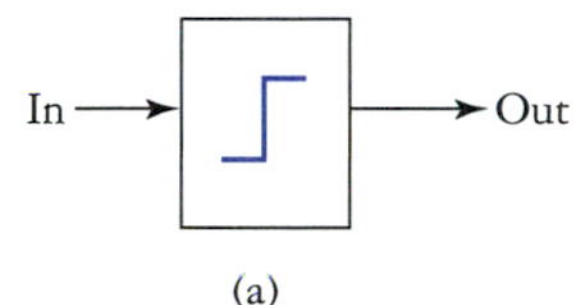

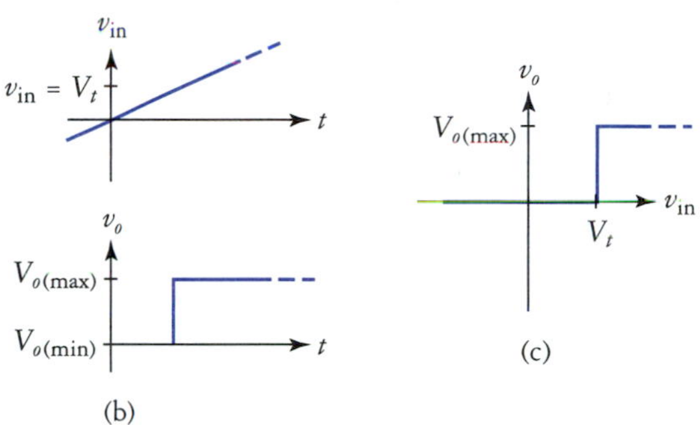

FIGURE 15.1
Comparator: (a) Block-diagram symbol. (b) Input and output waveforms. (c) Transfer characteristic curve.

As we saw in previous study, another useful way to graphically represent the operation of the comparator is with a plot of the transfer characteristics, as shown in Fig. 15.1(c). This is a graphical representation of Eq. (15.1). Using Figs. 15.1(b) and (c) we can tell two things about the ideal comparator: First, Fig. 15.1(b) implies that, ideally, the comparator jumps instantly from min to max or from max to min. This implies high speed or, equivalently, high slew rate. Second, Fig. 15.1(c) implies that the ideal comparator has infinite gain, because the transfer characteristic is vertical at $v_{in} = V_t$. If we implement a comparator with an amplifier that has finite voltage gain, the transfer characteristic curve will have a slope $m = v_o/v_{in} = A_v$ centered at V_t. This is shown in Fig. 15.2. Most often, in a practical circuit A_v is so high that the transfer characteristic is for all practical purposes vertical.

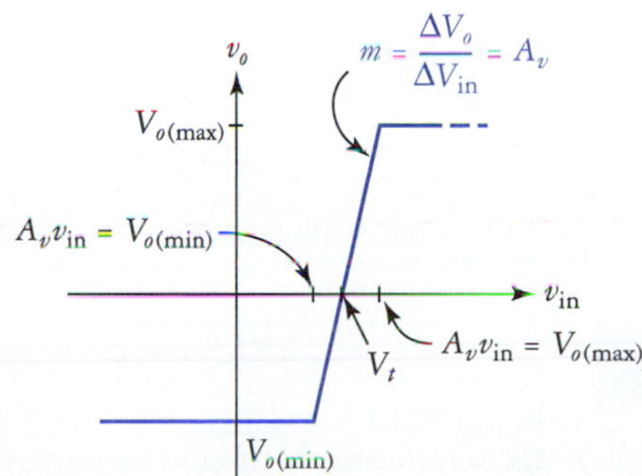

FIGURE 15.2
Effect of finite gain on transfer curve.

It is possible to locate the trip point at any arbitrary location. That is, V_t can be a positive value as shown in Fig. 15.1, or it can be zero or negative as well. In addition, we can design a comparator such that its output has the opposite polarity of the input, using an inverting op amp configuration. This is shown in the following example.

EXAMPLE 15.1

Sketch the transfer characteristic curve that is described by the following equation.

$$V_o = \begin{cases} 10\text{ V} & \text{for } v_{\text{in}} < 5.0\text{ V} \\ -10\text{ V} & \text{for } v_{\text{in}} > 5.0\text{ V} \end{cases}$$

Solution This equation tells us that $V_t = 5.0$ V, $V_{o(\max)} = 10$ V, and $V_{o(\min)} = -10$ V. We also find that when v_{in} is less than 5.0 V, the output is $v_o = 10$ V, and when the input is greater than 5.0 V the output is -10 V. The transfer characteristics are graphed in Fig. 15.3.

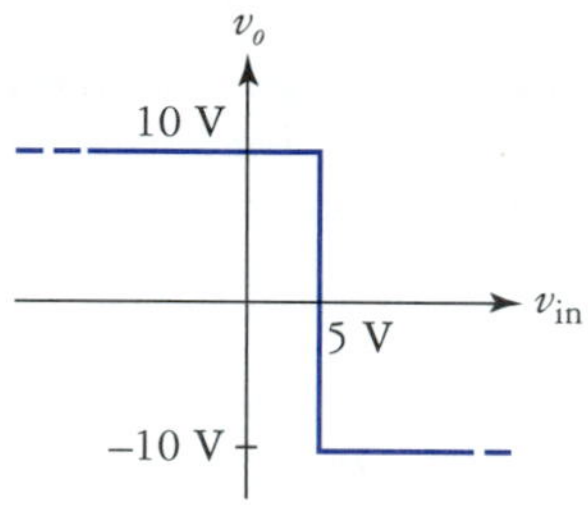

FIGURE 15.3
Transfer curve for Example 15.1.

Op Amp Comparators

Recall that one of the outstanding properties of the typical op amp is high voltage gain. Because of this, in many situations a comparator can be implemented simply by using an op amp that is running in an open-loop mode. For example, the circuit in Fig. 15.4 can be used. In this case, we would most likely use symmetrical bipolar supply voltages ($V_{CC} = V_{EE}$) because the typical op amp is designed to work best under these conditions. The R_1–R_2 voltage divider sets the trip point of the circuit.

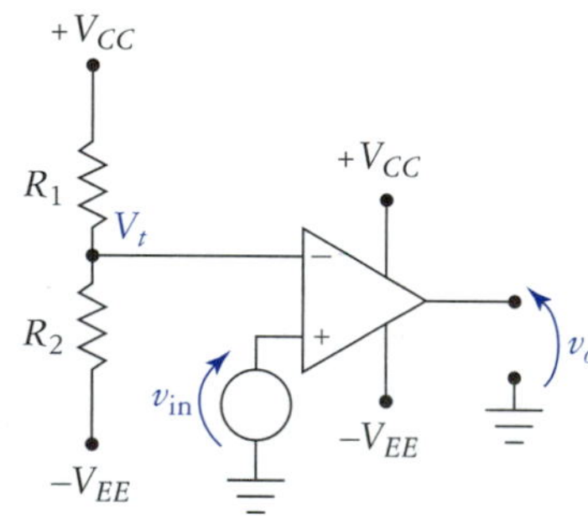

FIGURE 15.4
Typical op amp comparator.

To maintain good op amp performance, we should limit the trip-point voltage to within about a volt of either supply rail. That is,

$$(-V_{EE} + 1\text{ V}) \leq V_t \leq (V_{CC} - 1\text{ V}) \tag{15.2}$$

The input voltage must be limited to within the supply rail values as well. In equation form, this is written

$$-V_{EE} \leq v_{\text{in}} \leq V_{CC} \tag{15.3}$$

If the input voltage exceeds this range the op amp may be permanently damaged.

EXAMPLE 15.2

The circuit in Fig. 15.4 has $V_{EE} = V_{CC} = 15$ V, $R_1 = 5$ kΩ, and $R_2 = 10$ kΩ. The op amp has $V_{o(\max)} = 14$ V and $V_{o(\min)} = -14$ V. Sketch the transfer characteristic curve and the output voltage waveforms that should be produced given the following input voltages:

(a) $$v_{\text{in}} = \begin{cases} 0\text{ V}^* & \text{for } t < 0 \\ 100t\text{ V} & \text{for } t > 0 \end{cases}$$

(b) $$v_{\text{in}} = 6\sin(2\pi 100t)\text{ V}$$

---

* We could also use the unit step function and write this as $v_{\text{in}} = 100tu(t)$ V.

Determine the duty cycle of the output waveform produced by the sinusoidal input of part (b).

Solution First, we determine the trip point for the circuit. One approach using KVL and Ohm's law gives us

$$V_t = V_{CC} - (V_{CC} + V_{EE})\frac{R_1}{R_1 + R_2}$$

$$= 15\text{ V} - (30\text{ V})\frac{5\text{ k}\Omega}{5\text{ k}\Omega + 10\text{ k}\Omega}$$

$$= 15\text{ V} - 10\text{ V}$$

$$= 5\text{ V}$$

The transfer characteristic curve is shown in Fig. 15.5.

FIGURE 15.5
Transfer curve for Example 15.2.

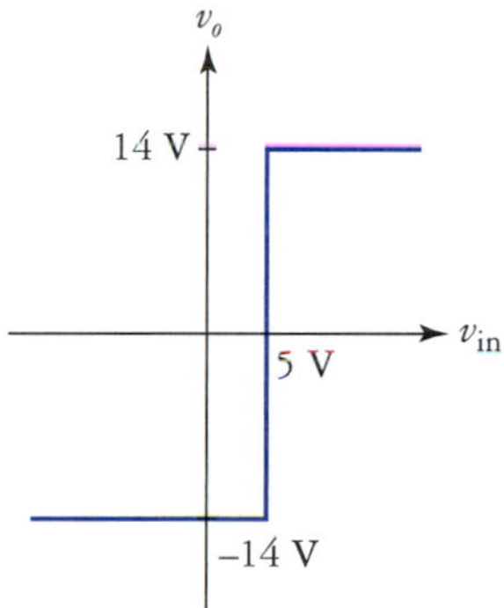

(a) The input voltage is a linear ramp with a slope of 100 V/sec. This is shown in Fig. 15.6. The output of the comparator switches from low to high at the instant v_{in} reaches V_t. To find this time, we solve the time-varying part of the voltage equation for time which in general gives us

$$t = \frac{v}{\text{m}}$$

We substitute the values of interest producing

$$t_{\text{trip}} = \frac{V_t}{100}$$

$$= \frac{5}{100}$$

$$= 50\text{ ms}$$

The resulting output voltage is shown in the lower part of Fig. 15.6.

FIGURE 15.6
(a) Input and (b) output waveforms for Example 15.2.

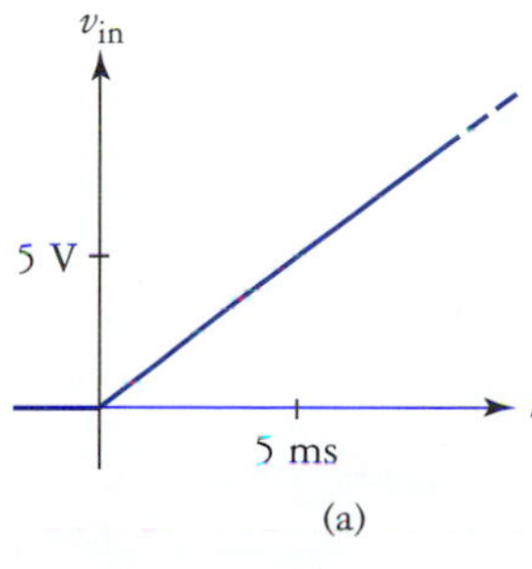

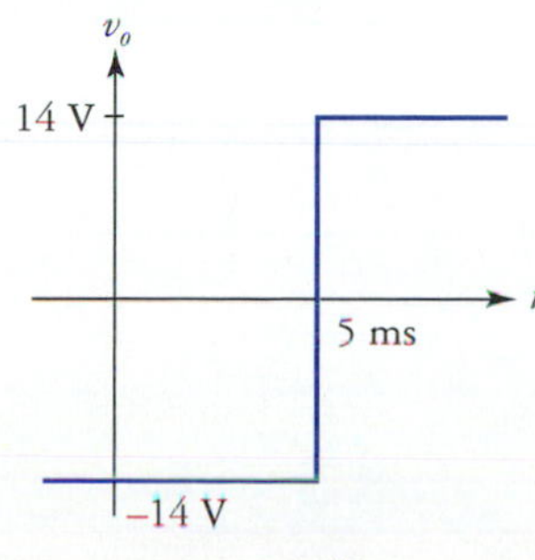

(b) Since the peak amplitude of the input voltage sine wave exceeds the trip voltage, the comparator output will be a pulse train. We must determine the time it takes for the sine wave to reach the trip point. The input voltage equation is $v_{in} = 6 \sin(2\pi 100t)$ V. We substitute the trip voltage, giving us 5 V $= 6 \sin(2\pi 100t)$ V. Solving for the trip-point time t_t we have

$$t_t = \frac{\sin^{-1}\left(\frac{5\text{ V}}{6\text{ V}}\right)}{2\pi 100}$$

$$= \frac{0.985}{628.3}$$

$$= 1.6 \text{ ms}$$

This result tells us that the comparator will be triggered high 1.6 ms after zero crossing of v_{in} in the positive-going direction, and the comparator will drop low again 1.6 ms prior to zero crossing in the negative-going direction. The period of the output pulse train is the same as that of the input signal, which is

$$T = 1/f$$
$$= 1/100 \text{ Hz}$$
$$= 10 \text{ ms}$$

From these values, we find that during a cycle of the sine wave, the output of the comparator will be high for

$$T_H = (T/2) - 2t_t$$
$$= 5 \text{ ms} - 3.2 \text{ ms}$$
$$= 1.8 \text{ ms}$$

These values are shown in Fig. 15.7.

The duty cycle of the output voltage waveform is found using

$$D = \frac{T_H}{T} \tag{15.4}$$

$$= \frac{1.8 \text{ ms}}{10 \text{ ms}}$$

$$= 0.18 = 18\%$$

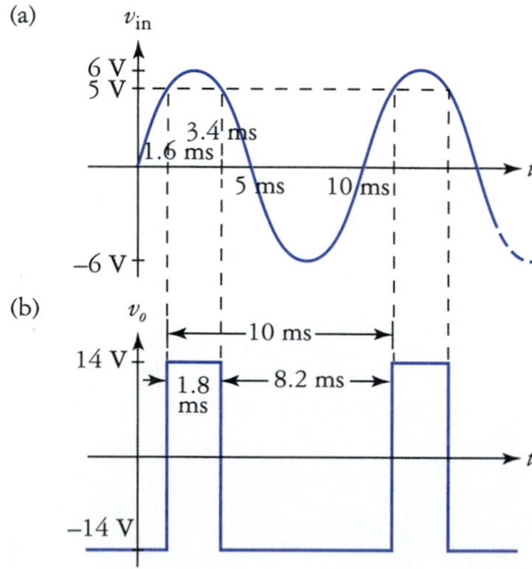

FIGURE 15.7
(a) Input and (b) output waveforms.

The previous example illustrates that a comparator can be used to convert a continuous signal, like a sine wave, into a pulse waveform. If we place a potentiometer in the set-point divider network, then assuming a constant amplitude input voltage, the duty cycle of the output can be easily adjusted from 0% to 100%, continuously.

Zero-Crossing Detector

If we set the trip point to 0 V, we have a special case of the comparator called a *zero-crossing detector.* If we apply the input signal to the noninverting terminal, as shown in Fig. 15.8(a), the zero-crossing detector simply switches states according to

$$V_o = \begin{cases} V_{\text{max}} & \text{for } v_{\text{in}} > 0\text{ V} \\ V_{\text{min}} & \text{for } v_{\text{in}} < 0\text{ V} \end{cases} \tag{15.5}$$

Note
A zero-crossing detector is just a comparator with V_t = 0 V.

If the input signal is applied to the inverting input as shown in Fig. 15.8(b), the output expression is

$$V_o = \begin{cases} V_{\text{min}} & \text{for } v_{\text{in}} > 0\text{ V} \\ V_{\text{max}} & \text{for } v_{\text{in}} < 0\text{ V} \end{cases} \tag{15.6}$$

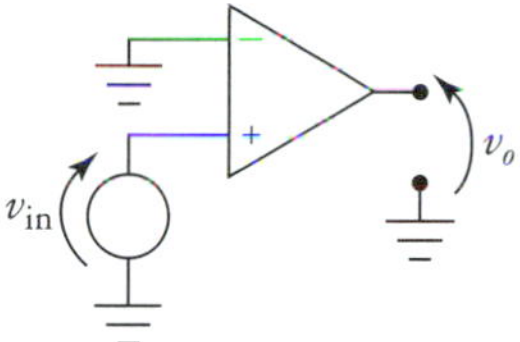

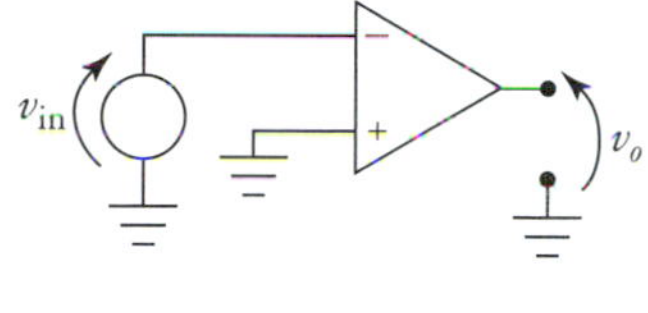

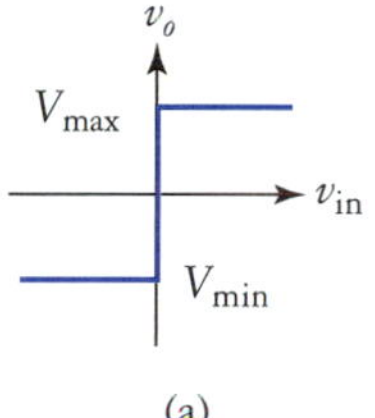

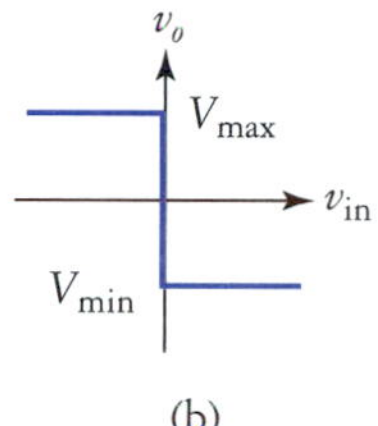

FIGURE 15.8
Simple op amp zero-crossing detectors.
(a) Noninverting.
(b) Inverting.

Bounding Circuits

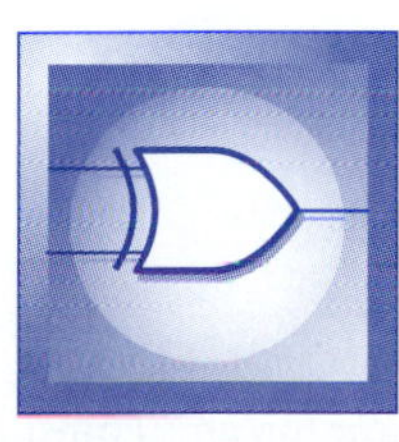

As stated previously, comparators are often used to convert signals like sine waves into pulse waveforms. If we wish to use the output of the circuit in Fig. 15.4 or 15.8 to drive a TTL input, the comparator output voltage should be restricted to TTL-compatible levels $V_H \cong 5$ V and $V_L \cong 0$ V. This can be accomplished in several ways. For example, we could operate the op amp from a single polarity supply with $V_{CC} = 5$ V and $V_{EE} = 0$ V. This would limit the output voltage to the desired range, but it is likely that the op amp performance would be severely degraded. A more suitable solution to this problem is shown in Fig. 15.9. Here, the op amp zero-crossing detector is powered using normal supply voltages, while the output is clamped to $V_{o(\text{max})} \cong 5.7$ V and $V_{o(\text{min})} \cong -0.7$ V. These voltages are slightly outside of the TTL range, but should not be destructive to the gate being driven. A current-limiting resistor is shown connected to the op amp output in Fig. 15.9. Typically, this will be

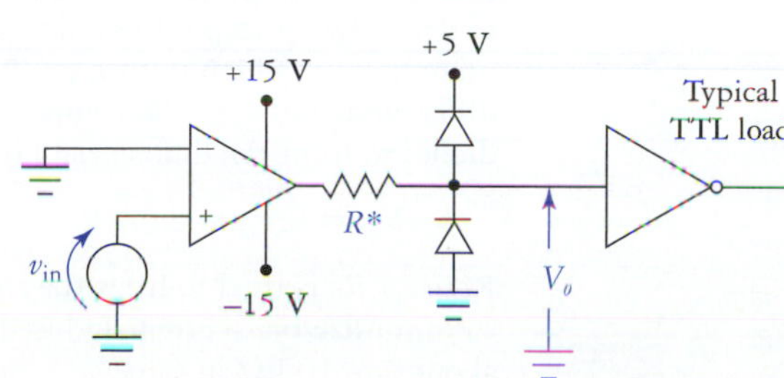

FIGURE 15.9
Interfacing an op amp comparator to a TTL input.

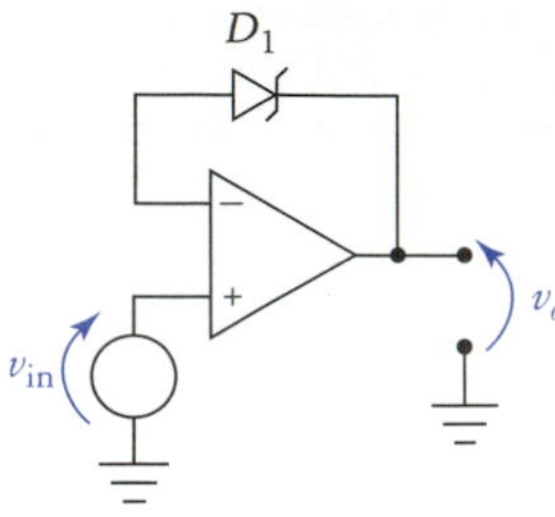

FIGURE 15.10
Output bounding with a zener diode.

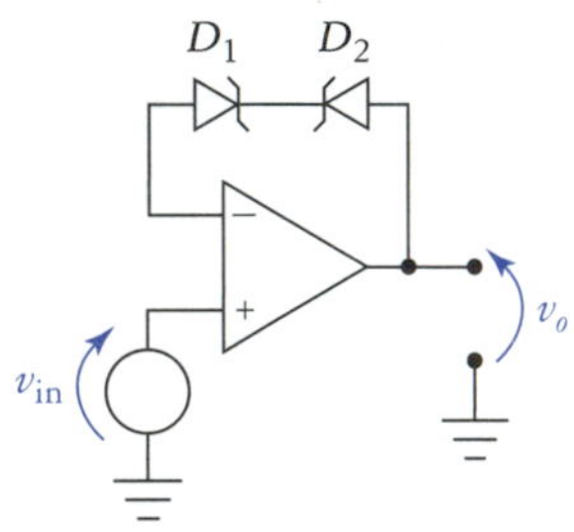

FIGURE 15.11
Alternative output bounding circuit.

between 1 and 10 kΩ. The resistor shown in the circuit may not be required if the op amp is output current limited, like the 741.

The advantage the circuit shown in Fig. 15.9 is that although the output voltage is bounded, the trip voltage can be set to any desired value within the supply rail limits. If we wish to have $V_t = 0$ V (a zero-crossing detector), then we can bound the output by placing an appropriate zener diode in the feedback loop as shown in Fig. 15.10. The idea here is that if $v_{in} < 0$ V, the output of the op amp drives negative until the zener is forward biased. This produces a minimum output voltage of $V_{o(min)} = -V_\phi \cong -0.7$ V. When $v_{in} > 0$ V the output drives positive until the zener breakdown voltage is exceeded. This gives us $V_{o(max)} = V_Z$. What's happening here is that when the diode conducts (forward or in zener breakdown), the feedback loop is closed with a low resistance and the gain of the op amp drops to a relatively low value. When the diode is nonconducting, the feedback loop is basically open and full open-loop gain is present, providing high sensitivity at the trip point.

Another circuit that can be used to provide output bounding both above and below zero volts is shown in Fig. 15.11. Here, two back-to-back connected zener diodes provide bounding at limits given by $V_{o(min)} = -(V_{\phi 1} + V_{Z2})$ and $V_{o(max)} = V_{\phi 2} + V_{Z1}$. We simply choose zener diodes with the appropriate voltage ratings to obtain the desired min and max output voltage levels.

To produce a TTL-compatible output voltage, we could choose a zener diode with $V_Z = 4.3$ V. This type of output bounding normally allows faster switching times. The approach of Fig. 15.9 may provide slower switching because the output of the op amp saturates very heavily compared to the gain-reduction approach of Fig. 15.10. In addition, if the series current-limiting resistor is used, switching speed is compromised further because of the increased time constant associated with the charging of stray load capacitance.

Perhaps the best approach to producing a TTL-compatible output from the comparator is to use a BJT saturated switch as shown in Fig. 15.12, where a zero-crossing detector is used for the sake of simplicity. In this case, when $v_{in} < 0$ V, the output of the op amp drops to about -14 V, which drives Q_1 into cutoff, causing $V_o = V_{o(max)} = +5$ V. Diode D_1 prevents breakdown of the reverse-biased base-emitter junction. When $v_{in} > 0$ V, the output of the op amp drives Q_1 into saturation, where the output is $V_o = V_{o(min)} \cong 0$ V. Because the BJT saturated switch is a common-emitter configuration, the transfer characteristic curve of the comparator will be inverted.

FIGURE 15.12
Using a BJT to provide TTL-compatible output.

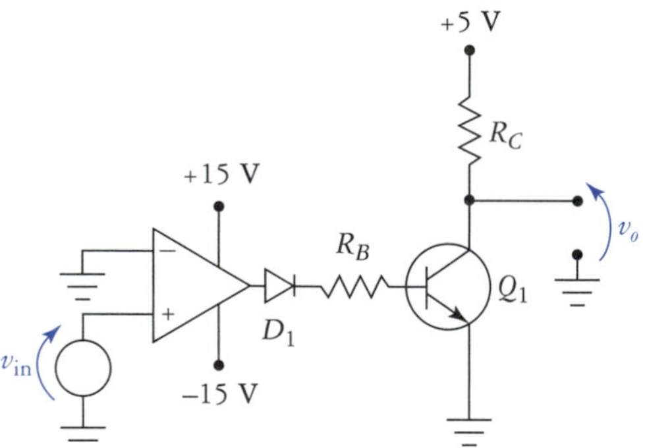

Input Bounding

We already know that the voltage applied to either input of the comparator (op amp) must be within the limits of the supply rails in order to prevent damage to the op amp. It also happens that the comparator will switch states much more quickly if the input differential voltage V_{id} is kept small. Consider that since the gain of the comparator is so high ($A_{OL(typ)} \geq 100$ k), it takes only a few millivolts to drive the op amp into saturation. A large differential input will cause the input transistors to saturate heavily, which in turn results in slower switching time. Thus it is unnecessary to apply a large differential input to the comparator. The circuit in Fig. 15.13 uses two inverse-parallel connected diodes to limit the differential input voltage to

$$V_{id} = \pm V_\phi \cong \pm 0.7 \text{ V} \tag{15.7}$$

Resistor R_2 is used to limit the current drawn from the input source to a safe level, while R_1 is used to minimize bias-current-induced offset. Typically, the resistors are set equal in value ranging from about 1 to 10 kΩ in value.

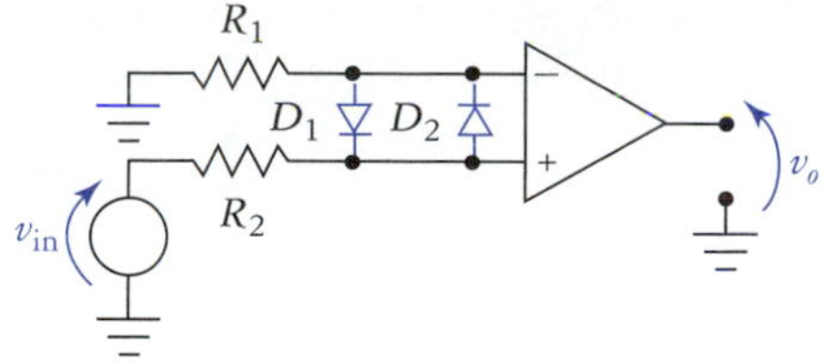

FIGURE 15.13
Input bounding protects the op amp.

Schmitt Triggers

A *Schmitt trigger* is a comparator that employs *positive feedback.* This positive feedback causes the comparator to exhibit *hysteresis.* Consider the circuit in Fig. 15.14. Resistors R_1 and R_2 form a voltage divider from which a feedback voltage v_F is applied to the noninverting input. Assuming symmetrical supply voltages, the feedback voltage is given by

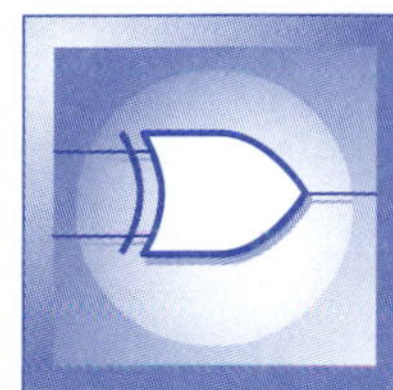

$$v_F = \pm V_{o(\max)} \frac{R_2}{R_1 + R_2} \tag{15.8}$$

Recall that the divider ratio is often called the *feedback factor* β. That is,

$$\beta = \frac{R_2}{R_1 + R_2} \tag{15.9}$$

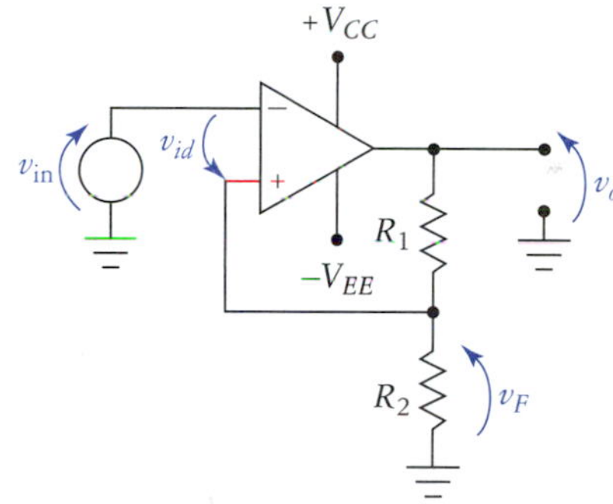

FIGURE 15.14
A Schmitt trigger. Positive feedback creates hysteresis.

To gain an understanding of how hysteresis is produced by this circuit, let's assume that in Fig. 15.14, we have just applied power and $v_{in} = 0$ V. It is physically impossible for the output voltage to be exactly 0 V, thus on power up let us also assume that the output voltage goes slightly positive. A portion of this positive voltage is applied back to the noninverting input, which in turn drives the output further positive. This self-reinforcing positive feedback causes the output to drive into hard saturation at $V_{o(\max)}$. Now, assume that v_{in} is increasing in the positive direction. As soon as v_{in} exceeds v_F, the input differential voltage will cause the output to begin to go negative. The feedback voltage is now negative, which in turn increases the drive on the noninverting input causing the output to drive into hard saturation at $V_{o(\min)}$. This is shown in portions 1 and 2 of Fig. 15.15. The input voltage required to cause the output to drive to $V_{o(\min)}$ is called the *upper trip voltage* V_{UT}, which is given by

Note
Schmitt triggers are often used to "square-up" slowly varying voltages for use in digital logic circuits.

$$V_{UT} = \beta V_{o(\max)} \tag{15.10}$$

If the input voltage begins decreasing, it must decrease to the lower trip voltage in order to cause the comparator to again switch states. This is shown in Fig. 15.15 where the curve is labeled 3 and 4. The lower trip voltage is given by

$$V_{LT} = \beta V_{o(\min)} \tag{15.11}$$

From the previous discussion, we find that hysteresis in effect produces a *dead-band* between the upper and lower trip points. We can set the size of this dead-band by choosing appropriate ratios for R_1 and R_2. In a high noise environment it is desirable to have a wide dead-band to provide increased noise immunity. In Fig. 15.16(a) we have a block diagram showing two comparators driven by the same input signal. The upper comparator is a simple inverting circuit, while the lower comparator is a Schmitt trigger, as indicated by the hysteresis loop shown in the block. If the input signal has a small amount of noise as shown in Fig. 15.16(b), the simple comparator may respond by erroneously switching states near its switching threshold (0 V in this case). This false triggering can cause severe problems if the output of the comparator is used to provide a clock pulse to a digital circuit, for example. How-

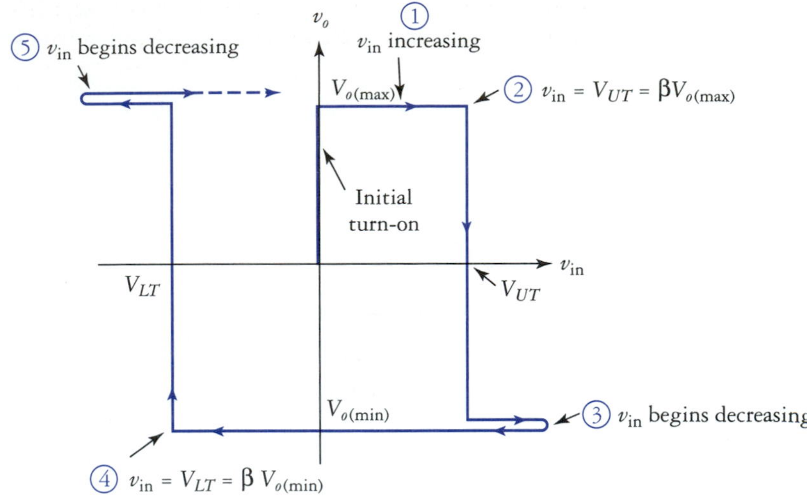

FIGURE 15.15
Schmitt trigger hysteresis loop.

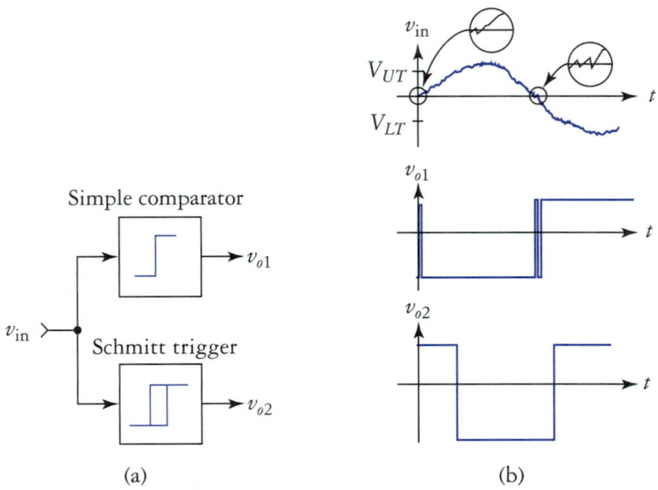

FIGURE 15.16
Comparator responses. (a) Driving simple comparator and Schmitt trigger with a noisy signal. (b) Resulting output waveforms.

ever, since the Schmitt trigger has a wide dead-band, it is immune to the effect of this low-level noise, and we see that a clean output pulse is produced. Finally, as an added benefit, in addition to providing noise immunity, the positive feedback also tends to speed the switching time of the comparator.

EXAMPLE 15.3

Refer to Fig. 15.14. The comparator has $R_1 = R_2 = 10\ \text{k}\Omega$, $V_{o(\min)} = -10$ V, and $V_{o(\max)} = 10$ V. The input voltage is given by $v_{in} = 8 \sin(2\pi 1000t)$ V. Determine the upper and lower trip points, and sketch the input and resulting output voltage waveforms over one cycle. Determine the times at which the comparator will switch states and determine the duty cycle of the output voltage. Assume that the output of the comparator is initially $V_{o(\max)}$.

Solution The circuit will have symmetrical trip points given by

$$V_{LT} = V_{UT} = \pm V_{o(\max)} \frac{R_2}{R_1 + R_2}$$

$$= \pm 10\ \text{V} \frac{10\ \text{k}\Omega}{20\ \text{k}\Omega}$$

$$= \pm 5\ \text{V}$$

As in Example 15.2, we must solve the input voltage equation for t in order to determine the switching times for the comparator. The time it takes for the input to reach the upper trip point is given by

$$t_{UT} = \frac{\sin^{-1}\left(\frac{5}{8}\right)}{2\pi 1000} = \frac{0.675}{6283} = 0.107 \text{ ms}$$

Because of the symmetry of the circuit, the time it takes v_{in} to go from zero crossing in the negative-going direction to reach the lower trip point is the same as the previous time. That is, $t_{LT} = t_{UT} = 0.107$ ms. These time intervals are shown in Fig. 15.17, where we find that disregarding the initial startup delay, the output voltage is a square wave with 50% duty cycle.

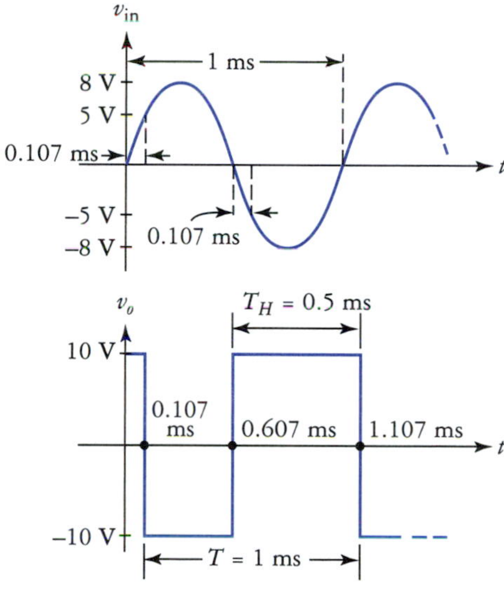

FIGURE 15.17
Waveforms for Example 15.3.

Window Comparators

A window comparator is used to indicate whether or not a voltage is within a range of values that is determined by two reference voltages. An example of a window comparator is shown in Fig. 15.18(a). Divider networks R_1–R_2 and R_3–R_4 are used to establish the upper and lower limits of the "window" according to

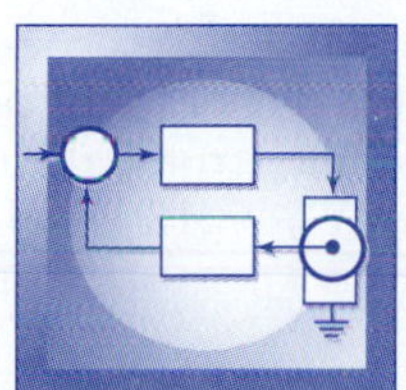

$$V_U = V_2 = \frac{R_4}{R_3 + R_4} V_{CC} \tag{15.12}$$

$$V_L = V_1 = \frac{R_2}{R_1 + R_2} V_{CC} \tag{15.13}$$

In this particular example, the comparator operates from a single-polarity supply; however, a bipolar supply can also be used. In either case we must have

$$V_2 > V_1 \tag{15.14}$$

The response of the window comparator to a linearly increasing input voltage is shown in Figs. 15.18(b) and (c). This waveform is generated as follows. Initially, $v_{\text{in}} = 0$ V, which is less than V_1 or V_2, which causes the output of A_1 to drive high while A_2 drives low. This causes D_1 to be forward biased while D_2 is reverse biased, thus the output voltage will be high. When the input voltage enters the lower limit of the window, A_1 drives low along with A_1, which causes the output to be pulled low

FIGURE 15.18
Window comparator. (a) Circuit Typical (b) input and (c) output waveforms.

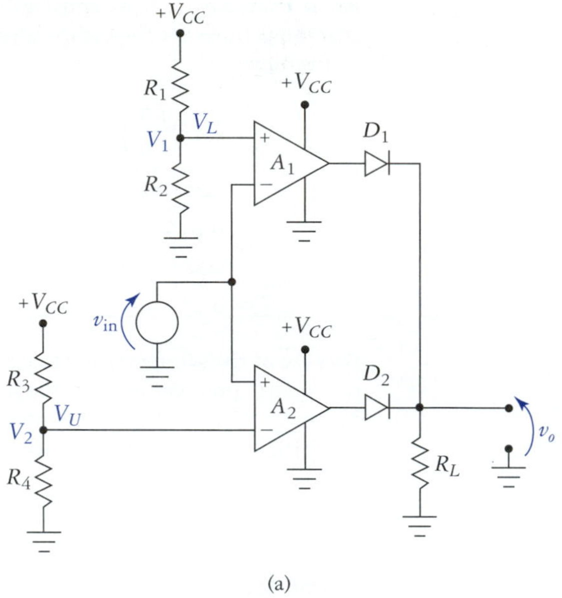

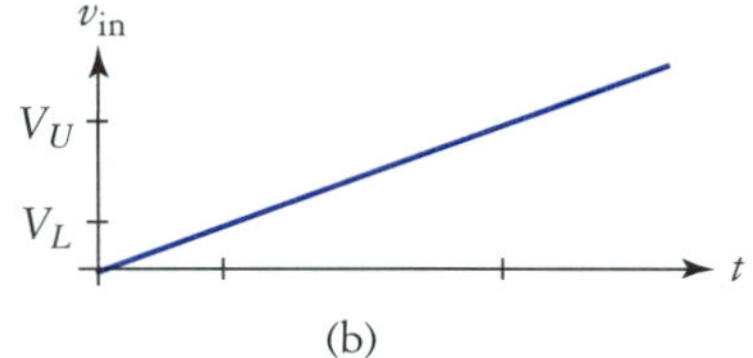

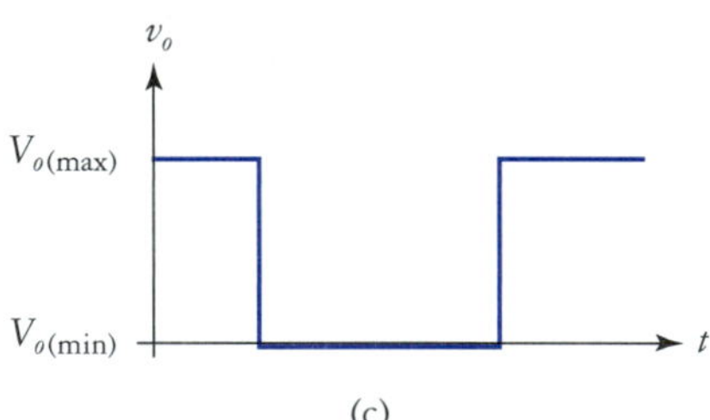

via R_L. When the input voltage exceeds the upper limit of the window, A_2 drives high while A_1 remains low. Thus, we see that as long as the input voltage is within the window, the output of the comparator is low.

EXAMPLE 15.4

Refer to Fig. 15.19. This circuit is part of a system that is used to monitor the temperature of a chemical manufacturing process. The comparator is to be used to trigger an alarm when the temperature drifts outside of a specified range. The resistor and thermistor are chosen such that $R = 10\ \text{k}\Omega$ with $\Delta R = 0\ \Omega$ at $T = 100°\text{C}$. Over the temperature range of interest, the thermistor varies such that $\Delta R = 2\%/°\text{C}$ (a positive temperature coefficient). Given $V_{CC} = 15$ V and $R_1 = R_3 = 10\ \text{k}\Omega$, determine R_2 and R_4 such that the upper temperature limit is $T_U = 105°\text{C}$ and the lower temperature limit is $T_L = 95°\text{C}$.

Solution Over the temperature range of interest, the value of the thermistor will vary symmetrically about the nominal 100°C value:

$$\begin{aligned}\Delta R &= (0.5\%/°\text{C})(\pm\Delta T)\\ &= (0.02)(10\ \text{k})(\pm 5)\\ &= \pm 1000\ \Omega\end{aligned}$$

FIGURE 15.19
Window comparator used as a temperature out-of-range indicator.

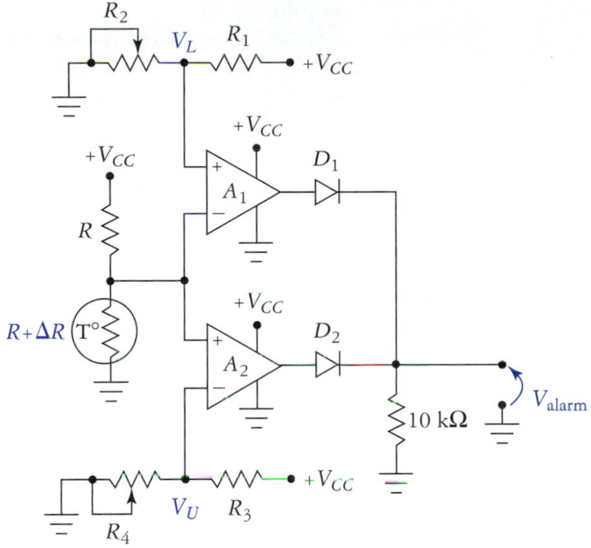

This produces the following upper and lower temperature-related voltages that will be applied to the comparator:

$$V_U = V_{CC}\frac{R + \Delta R}{R + R + \Delta R}$$

$$= 15\text{ V}\frac{10\text{ k} + 1\text{ k}}{10\text{ k} + 10\text{ k} + 1\text{ k}}$$

$$= 7.86\text{ V}$$

$$V_L = \frac{9\text{ k}}{19\text{ k}}15\text{ V}$$

$$= 7.11\text{ V}$$

The required value for R_2 is found by solving the voltage-divider equation, giving us

$$R_2 = R_1\left(\frac{V_{CC}}{V_L} - 1\right)$$

$$= 10\text{ k}\Omega\left(\frac{15\text{ V}}{7.11\text{ V}} - 1\right)$$

$$= 11.1\text{ k}\Omega$$

The value of R_4 is found in a similar manner to be

$$R_4 = 9.08\text{ k}\Omega$$

A short discussion of the previous example will point out some practical considerations. First, the supply voltage that is used to drive the voltage dividers must be relatively stable for good performance. In some cases a very stable and accurate reference voltage, separate from the normal supply voltage, will be employed to derive the trigger voltage levels and for transducer excitation. Resistors R_2 and R_4 would most likely be implemented with multiple-turn potentiometers, which would be adjusted to set the upper and lower trip levels to the desired values. Also, if the transducer is located far from the comparator/alarm circuit, it might be worth the effort to set up a bridge/instrumentation amplifier circuit to take advantage of common-mode rejection.

The LM311 Comparator

Although op amps are often used as comparators, specially designed monolithic IC comparators are available. The popular LM311 IC comparator has been an industry standard for many years. The pinout and simplified internal equivalent circuit are shown in Fig. 15.20. As you can see, the LM311

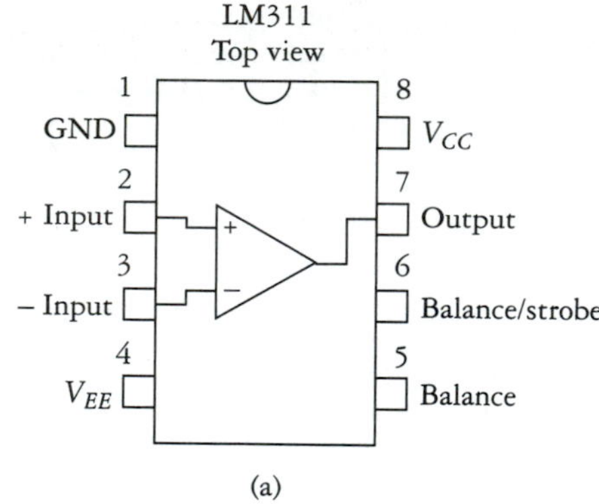

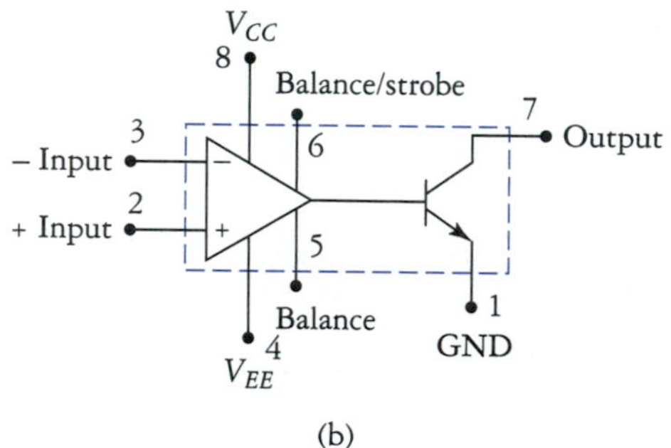

FIGURE 15.20
The LM311 monolithic comparator.

is very similar to the comparator/saturated switch discussed in Fig. 15.12. Because the LM311 was intended to be a comparator from the outset, it is designed for high switching speeds. Recall that op amps are least stable (they may break into oscillation) at low closed-loop gains. Most normal op amps are designed to be relatively stable at low gains, which comes at the expense of increased response time (lower slew rate). The 311 is not normally used with negative feedback anyway, so fast response can be achieved. For the 311, the response time is the time it takes for the output to respond to a step-input differential sufficient to cause output saturation. This will typically be 50% longer than the output rise time.

Note
A strobe input is a digital input used to enable or disable a function.

The 311 may be configured for use as a simple zero-crossing detector as shown in Fig. 15.21. The collector pull-up resistor will typically be around 1 to 2 kΩ. The typical response time using pull-up resistors in this range will be about 400 ns. Reducing the pull-up resistor can decrease the typical response time to about 200 ns. In this example, the balance pins of the 311 are used for input offset compensation.

If input offset compensation is not required, the balance pins (5 and 6) may be left open. Pin 6 may also be used as a *strobe* input. This allows a digital signal to enable or disable the comparator. Additional information about strobing, as well as stability considerations and typical application circuits can be found in the various manufacturers' data books.

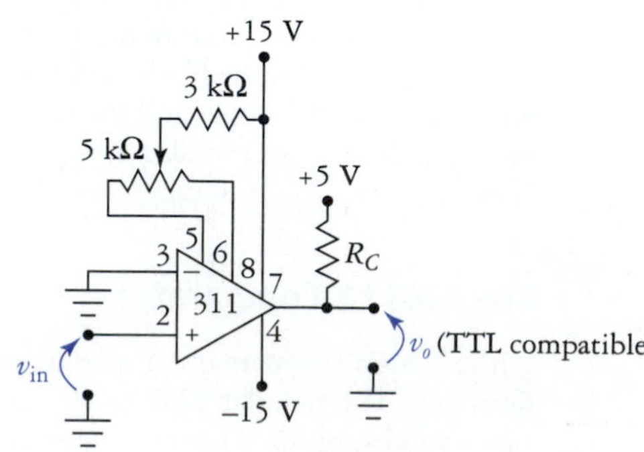

FIGURE 15.21
LM-311 zero-crossing detector.

15.2 LOGARITHMIC AND ANTILOGARITHMIC AMPLIFIERS

With the exception of comparators, up until now we have generally assumed that the output of an amplifier is simply a scaled version of the input signal. In equation form this is

$$v_o = A_v v_{in} \tag{15.15}$$

where A_v is the scaling constant. As long as A_v is constant, Eq. (15.15) is a linear equation. We know that in a real-world amplifier the gain and output phase shift may vary with frequency, but if the amplifier is linear, the output signal will contain only those frequency components that are present in the input signal. A nonlinear amplifier will introduce frequency components into the output signal spectrum that were not present in the input signal.

We have seen amplifiers exhibit nonlinearity many times previously, such as when an amplifier goes into clipping. Another example occurs when due to r_e variations, an unbypassed class A common-emitter amplifier exhibits the characteristic squashing and stretching of its output signal on positive and negative excursions. These are usually undesirable events. However, if we intentionally introduce predictable nonlinearity we can produce some very useful functions.

The Logarithmic Amplifier

The output of a logarithmic amplifier is proportional to the logarithm of the input. In equation form this is written

$$v_o = k \ln(v_{in}) \tag{15.16}$$

In this form, we are using the natural logarithm, however we could just as easily use the common or base-10 logarithm since the two are related by

$$\ln(x) \cong 2.303 \log_{10}(x) \tag{15.17}$$

Thus, the scaling constant k can be modified to conform to any desired base, although bases 10 and e are the most commonly used.

A logarithmic amplifier can be implemented by placing a diode in the feedback loop of an op amp as shown in Fig. 15.22. In these circuits, the diode is called the *logging element.* To emphasize the fact that these circuits require either positive- or negative-polarity inputs, depending on the direction of the diode, variable dc sources have been used to supply the input voltage. Let's examine the circuit in Fig. 15.22(a). Here, the log amp requires $v_{in} \geq 0$ V. That is, the diode must be driven toward forward bias in order for the circuit to work as a log amp. An input voltage that reverse biases the diode will cause the op amp output to saturate.

Qualitative Analysis

Let's get a feel for the operation of the log amp in Fig. 15.22(a) without getting bogged down in a lot of algebra. Recall that at low forward current levels, a diode will exhibit a large dynamic resistance r_D. When V_{in} is small, the diode current is small, which causes r_D to be large. The gain of the op amp is the ratio r_D/R_1, thus for small values of V_{in}, the closed-loop gain of the amplifier can be quite large.

When the input voltage is increased, the diode current increases, which causes a decrease in r_D. Now, since R_1 is fixed and the effective feedback resistance r_D has decreased, the closed-loop gain decreases. This is illustrated in Fig. 15.23, where gain and voltage magnitudes are used to place the graph in the first quadrant. The logarithmic nature of the diode transconductance causes the voltage gain of the amplifier to be logarithmic as well.

Note
Think of a logarithmic amplifier as an amplifier whose gain decreases continuously as the imput voltage increases.

Quantitative Analysis

Now let's determine some analysis equations. Assuming in Fig. 15.22(a) that we have $V_{in} \geq 0$ V and the op amp is ideal, the input current I_1 is

$$I_1 = \frac{V_{in}}{R_1} \tag{15.18}$$

We also know

$$I_F = I_1 \tag{15.19}$$

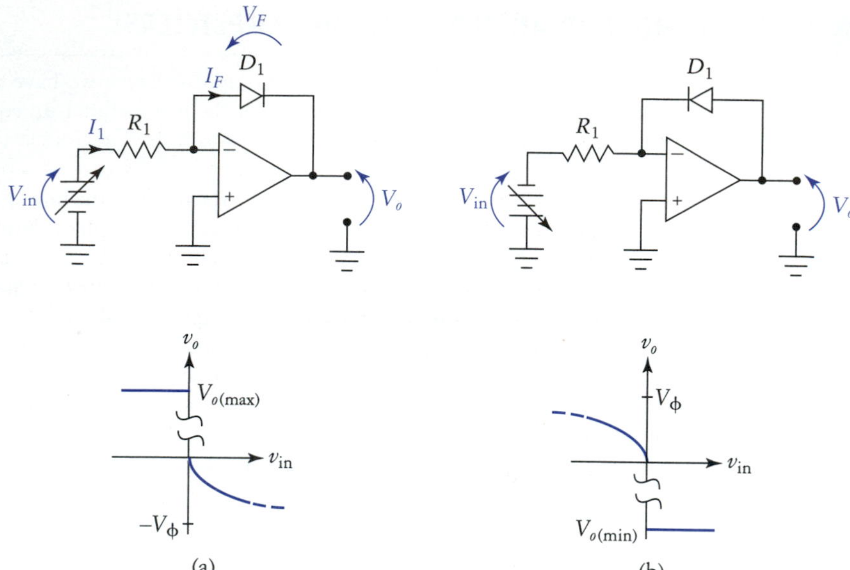

FIGURE 15.22
Logarithmic amplifiers. (a) For positive input voltages. (b) For negative input voltages.

FIGURE 15.23
Logarithmic ampifier gain versus input voltage.

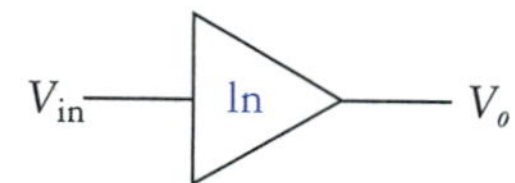

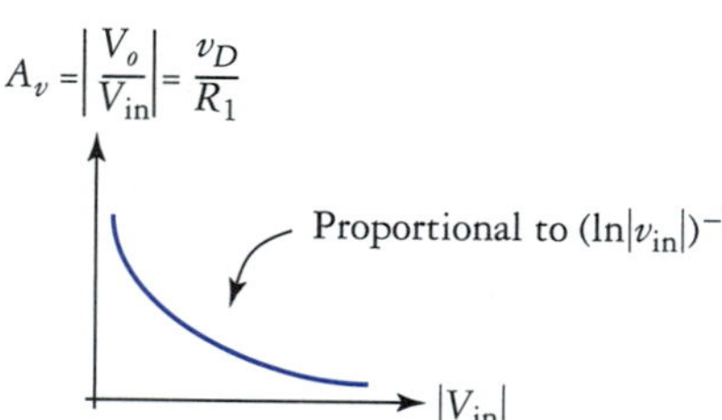

Shockley's equation must now be used. In general, the diode current is given by

$$I_D = I_S(e^{V_D/\eta V_T} - 1) \tag{15.20}$$

where I_S is the diode saturation current, η is the emission coefficient, and V_T is the thermal voltage equivalent ($V_T = kT/q \cong 26$ mV at $T = 25°$C). Because we are only interested in forward-bias operation, and we will operate the diode such that $I_F \gg I_S$, we can write

$$I_F \cong I_S e^{V_F/\eta V_T} \tag{15.21}$$

The voltage drop across the diode for any given forward current is found by solving Eq. (15.21) for V_F, which produces

$$V_F = \eta V_T \ln\left(\frac{I_F}{I_S}\right) \tag{15.22}$$

The output of the op amp is

$$V_o = -V_F \tag{15.23}$$

Thus we can write

$$V_o = -\eta V_T \ln\left(\frac{I_F}{I_S}\right) \tag{15.24}$$

Finally, substituting Eq. (15.18) into (15.24) we have

$$V_o = -\eta V_T \ln\left(\frac{V_{\text{in}}}{R_1 I_S}\right) \tag{15.25}$$

We find that the output of the op amp is indeed proportional to the logarithm of the input voltage.

The analysis of Fig. 15.22(b) is basically the same as that just presented, except that we require $V_{\text{in}} \leq 0$ V, and the output voltage is positive. You might note the similarity between the log amp and the bounded comparators covered in the last section. Whether or not a given circuit is a bounded comparator or a simple log amp will be clear from the application in which the circuit is used. In bounding applications, we generally disregard the exponential behavior of the diode. In log amp applications, the exponential behavior of the diode must be accounted for.

Transdiode Log Amps

Note

The transdiode is usually preferred because $n \cong 1$ over a wider range of feedback current.

To use Eq. (15.25) in a sensible manner, we must know I_S, V_T, and η for the diode being used. Unfortunately, these parameters are rather unpredictable: V_T and I_S are both very temperature dependent, and I_S can vary by an order of magnitude from one diode to another with the same part number. In addition, η (the emission coefficient) varies from 2 at low currents to 1 at higher current levels. We can eliminate the problem of the variable emission coefficient by using a transistor logging element as shown in Fig. 15.24. This configuration is called a *transdiode* logging element. Notice that in order to reverse the polarity of the log amp, we must use either an npn or pnp transistor. This slight inconvenience is more than compensated for by the fact that the transdiode has $\eta = 1$ for all practical current levels (typically 1 nA $\leq |I_E| \leq$ 1 mA).

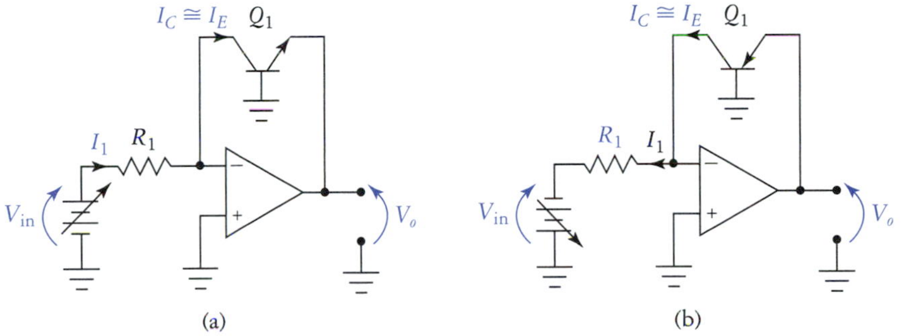

FIGURE 15.24
Transdiode log amplifiers. (a) For positive input voltages. (b) For negative input voltages.

The virtual ground forces the transistor to have $V_{BC} = 0$ V, thus the output voltage produced by the log amp will be

$$V_o = \pm V_{BE} \tag{15.26}$$

where the polarity is determined based on whether an npn or pnp transistor is used. Keep in mind that in this application V_{BE} is not simply approximated as being a constant 0.7 V. We must account for the exponential behavior of the pn junction. Similar to a normal diode, the forward-biased base–emitter junction of the transistor behaves according to

$$I_C = \alpha I_{CES}\, e^{V_{BE}/V_T} \tag{15.27}$$

where α is defined as usual ($\alpha = I_C/I_E \cong 1$) and I_{CES} is the collector to emitter leakage current with $V_{CB} = 0$ V. Following an analysis like that performed for the diode log amp we find that for Fig. 15.24(a)

$$V_o = -V_T \ln\left(\frac{I_C}{I_{CES}}\right) \tag{15.28}$$

Because the emission coefficient for the transdiode $\eta = 1$ and $\alpha \cong 1$, we have simply left them out of Eq. (15.28). This is an improvement, but the strong temperature dependence of I_{CES} and V_T still presents problems. One technique that is sometimes used to reduce temperature dependence is to place the logging element in a temperature-controlled oven. ICs are available with built-in heaters to maintain a constant internal device temperature. These devices are designed for use in log amp applications.

Log-Ratio Amplifier

The main problem that remains with the circuits presented so far is the unit-to-unit variation in device leakage current I_{CES}. A circuit design that eliminates this error source is the *log-ratio amplifier.* A simplified log-ratio amplifier is shown in Fig. 15.25. Temperature compensation is achieved via the use of an on-chip heater that maintains matched transistors Q_1 and Q_2 at a constant temperature. Op amps A_1 and A_2 form matched log amplifiers. Amplifier A_3 is a differential amplifier for which $R_4' = R_4$ and $R_3' = R_3$, as usual. Either V_x or V_y or both may serve as the input voltage(s).

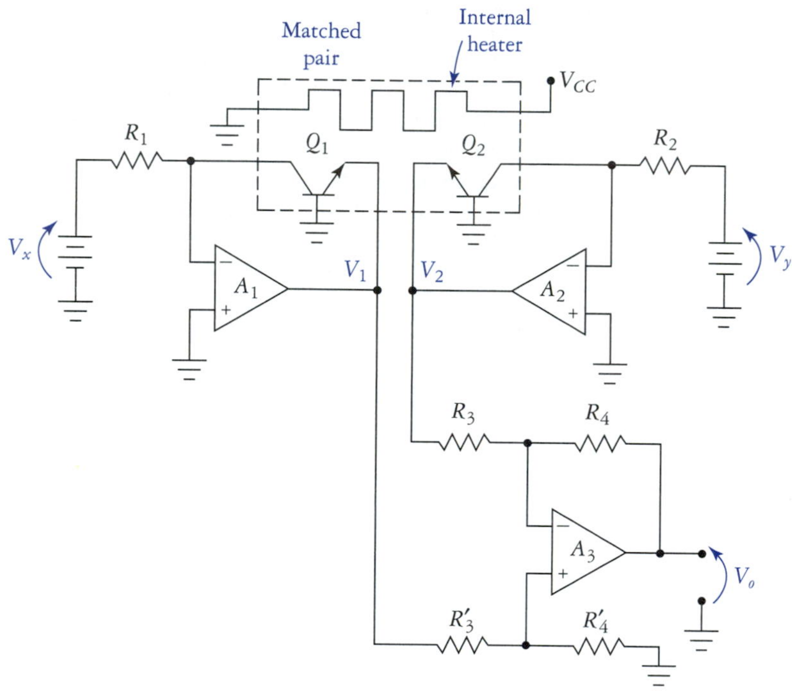

FIGURE 15.25
Log-ratio amplifier.

Because the transistors are matched, let us write

$$I_{CES1} = I_{CES2} = I_{CES}$$

The output voltages produced by A_1 and A_2 are given by

$$V_1 = -V_T \ln\left(\frac{V_x}{R_1 I_{CES}}\right) \quad \text{and} \quad V_2 = -V_T \ln\left(\frac{V_y}{R_2 I_{CES}}\right) \tag{15.29}$$

As stated earlier $R_4' = R_4$ and $R_3' = R_3$, so we can express the output of amplifier A_3 as

$$V_o = \frac{R_4}{R_3}(V_1 - V_2) \tag{15.30}$$

Now, if we substitute Eq. (15.29) into (15.30) we have

$$V_o = -\frac{R_4}{R_3}\left[V_T \ln\left(\frac{V_x}{R_1 I_{CES}}\right) - V_T \ln\left(\frac{V_y}{R_2 I_{CES}}\right)\right]$$

We can factor out V_T, producing

$$V_o = -V_T\left(\frac{R_4}{R_3}\right)\left[\ln\left(\frac{V_x}{R_1 I_{CES}}\right) - \ln\left(\frac{V_y}{R_2 I_{CES}}\right)\right] \tag{15.31}$$

Applying the property of logarithms $\ln(x) - \ln(y) = \ln(x/y)$, we can rewrite Eq. (15.31) as

$$V_o = -V_T\left(\frac{R_4}{R_3}\right)\ln\left[\frac{V_x}{R_1 I_{CES}} \div \frac{V_y}{R_2 I_{CES}}\right] \tag{15.32}$$

Normally, $R_1 = R_2$, so Eq. (15.32) reduces to

$$V_o = -V_T\left(\frac{R_4}{R_3}\right)\ln\left(\frac{V_x}{V_y}\right) \tag{15.33}$$

The output of this amplifier is proportional to the logarithm of the ratio of the two input voltages. We also see that since the transistors are matched that the leakage currents cancel. Resistors R_1 and R_2 are chosen to limit the maximum emitter current to something on the order of 100 μA. The gain of the differential amplifier A_3 scales the output voltage. A variable-gain instrumentation amplifier or a second variable-gain buffer can be used to provide continuous scaling adjustment. The on-chip heater has little effect on the log accuracy because both transistors operate at the same temperature anyway. However, the heater does result in a constant value of V_T, allowing accurate scaling.

The output of the log-ratio amplifier is undefined if both inputs are set to 0 V, therefore at least one input must always be held above ground potential. Also, both inputs must never be driven negative, because this will result in open-loop operation of the op amps and instability.

The graphs of the natural logarithm and base-10 (common) logarithm functions are shown in Fig. 15.26(a). The correlation between the mathematical (y versus x) and physical (V_o versus V_{in}) interpretations is also presented. Figure 15.26(a) is the kind of graph that is usually presented in math books. In this form it is easy to see that the logarithmic functions approach $-\infty$ as x approaches 0.

Note

Properties of logarithms:

$\ln(x) - \ln(y) = \ln\left(\frac{x}{y}\right)$

$\ln(x) + \ln(y) = \ln(xy)$

$x\ln(y) = \ln(y^x)$

$\log_m(x) = \frac{\ln(x)}{\ln(m)}$

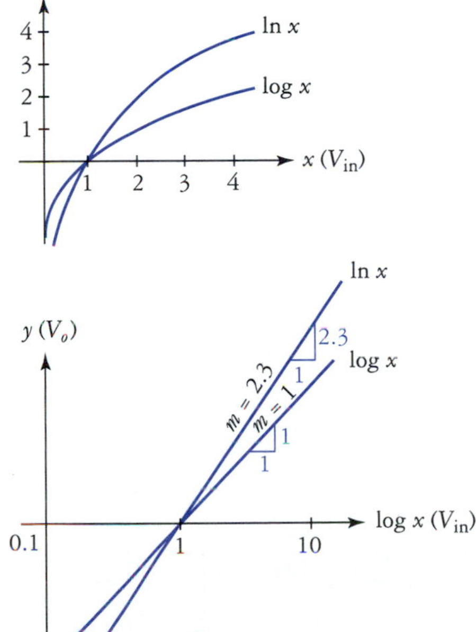

FIGURE 15.26
Comparison of the natural log (ln x) and common log (log x) functions.

If we look at these functions using a logarithmic x axis (using $\log_{10}$ is most common) as shown in Fig. 15.26(b), we find that the graphs of the functions are linear. This is most likely to be the form of the graph that would be used in electronics applications. Very often the "gain" of the logarithmic amplifier is expressed as the slope of this linearized transfer function graph, in volts/decade, or more conveniently mV/dec. The graph of the natural log function [ln(x)] has a slope of 2.3 V/dec, while the common log has a slope of 1 V/dec. The gain of the differential amplifier in Fig. 15.25 determines the slope of the graph. This allows us to scale the amplifier to any convenient base. An offset adjustment can also be included to shift the zero-crossing point of the graph to any convenient input voltage. Keep in mind that because the circuit in Fig. 15.25 is a log-*ratio* amplifier, the input voltage as it relates to Fig. 15.26 is

$$V_{in} = \frac{V_x}{V_y} \tag{15.34}$$

It is possible to logarithmically amplify signals, like sine waves, that may go positive or negative by placing complementary logging elements in parallel. In this manner, only one logging element is active during positive or negative signal excursions. We can prevent problems associated with the log of zero by limiting the maximum gain of the log amp with a feedback resistance placed in parallel with the logging element. Bipolar input signal capability is useful in applications such as signal compression and modulation, which are discussed shortly.

Antilog Amplifiers

As you probably expect, the output voltage produced by an antilog amp is proportional to the antilogarithm, usually $\log^{-1}(V_{in})$ or $\ln^{-1}(V_{in})$, of the input voltage. The antilog amplifier performs the inverse function of the log amplifier much like the differentiator performs the inverse function of the integrator. An antilog amplifier can be realized using a diode logging element; as with the log amplifier, however, the transdiode implementation is usually preferred. The circuit in Fig. 15.27(a) requires $V_{in} \geq 0$ V and produces a negative output voltage. The circuit in Fig. 15.27(b) requires $V_{in} \leq 0$ V and produces a positive output voltage.

Note
The antilog (or inverse log) is equivalent to:
$\log^{-1} x = 10^x$
$\ln^{-1} x = e^x$

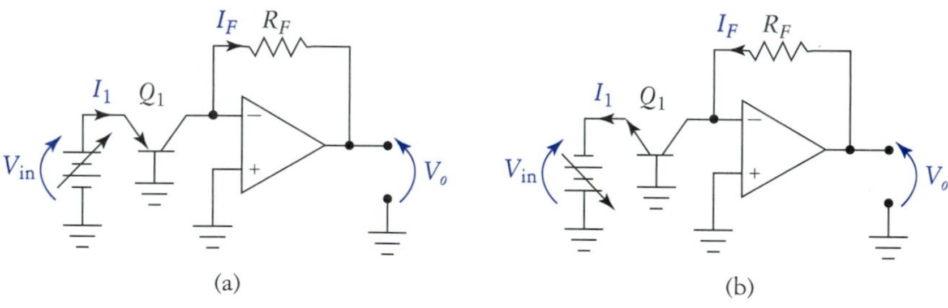

FIGURE 15.27
Antilog amplifiers using the transdiode. (a) For positive inputs. (b) For negative inputs.

The analysis of the antilog amplifier is based on the exponential behavior of the pn junction of the circuit's logging element (a diode or transdiode). Using Fig. 15.27(a) for the derivation example, we begin with the direct application of Shockley's equation (assuming $\eta = 1$) giving us

$$I_1 = I_{CES}\, e^{V_{in}/V_T} \tag{15.35}$$

The output of the op amp is

$$V_o = -I_F R_F \tag{15.36}$$

and since $I_F = I_1$ we can write

$$V_o = -R_F I_{CES}\, e^{V_{in}/V_T} \tag{15.37}$$

Thus we see that the output voltage is indeed proportional to the antilogarithm (or exponential) of the input voltage. The transfer characteristic of the antilog amplifier of Fig. 15.27(a) is shown in Fig. 15.28, where the output voltage magnitude has been used for convenience. The actual numerical val-

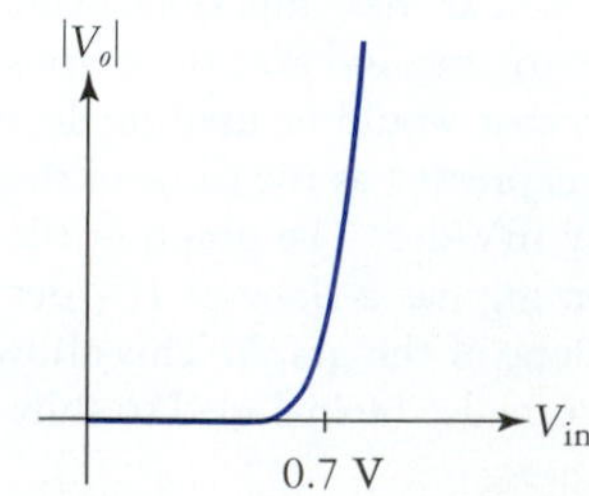

FIGURE 15.28
Typical antilog amp transfer characteristic.

ues associated with the curve will vary from transistor to transistor and with R_1; however, the values shown indicate the typical scale of the graph.

Like the log amplifier, the antilog amplifier has strong temperature dependence and a direct dependence on logging element saturation current. Because of this, the logging element of the antilog amplifier will often be a monolithic IC with self-contained transistor(s) and temperature control circuitry.

Log and Antilog Amplifier Applications

There are many uses for log and antilog amplifiers. This subsection presents a few of the more common applications.

Transducer Linearization

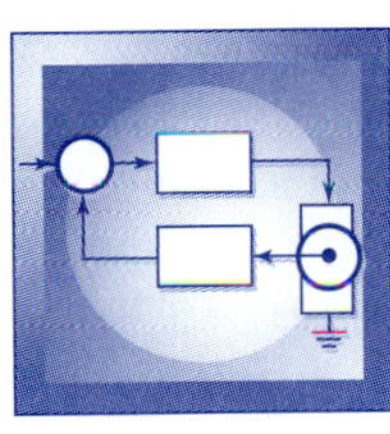

Recall that a transducer is a sensor that converts a physical quantity into a related electrical signal. Examples of transducers include strain gauges, thermistors, thermocouples, photodiodes, and Hall-effect sensors. In many applications the output produced by a transducer may be logarithmically or antilogarithmically related to the quantity being measured. An amplifier that exhibits the inverse characteristics of the transducer response can be used to produce a linearized output. This is illustrated in Fig. 15.29. In this system, the thermistor has an exponential response to temperature. The signal produced by the thermistor is processed by a log amp with an inversely related transfer characteristic. This results in the net response of the system being linear. That is, the output voltage is a linear function of temperature.

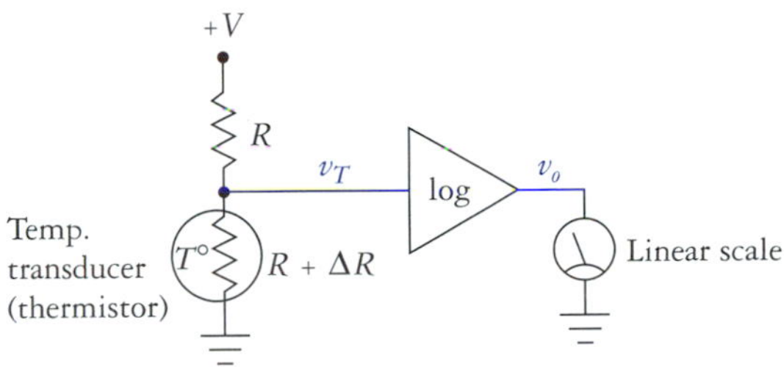

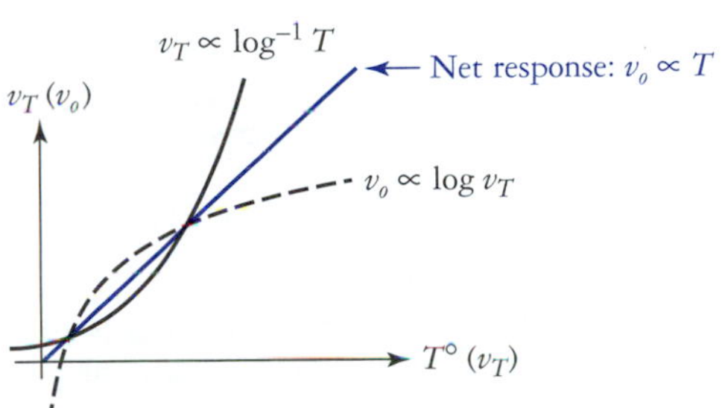

FIGURE 15.29
Using a log amp to linearize a transducer.

Signal Compression

Another use for log and antilog amplifiers is in an application called *signal compression.* The idea behind signal compression is to provide high gain to signals that have low amplitude and low gain to signals of large amplitude. A logarithmic amplifier provides this kind of characteristic, as illustrated in Fig. 15.30. The first two cycles of v_{in} are of low amplitude and the log amp provides significant voltage gain. The second two cycles of v_{in} are much larger, but the gain of the log amp is reduced in response. The net result is that the log amp reduces the *dynamic range* of the signal. Dynamic range is the ratio of the largest signal peak to the smallest. That is,

$$\text{DR} = \frac{V_{p(\max)}}{V_{p(\min)}} \tag{15.38}$$

Normally, dynamic range is expressed in decibels:

$$\text{DR} = 20 \log \frac{V_{p(\max)}}{V_{p(\min)}} \tag{15.39}$$

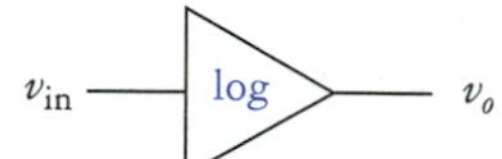

FIGURE 15.30
Using a log amp to compress the dynamic range of a signal.

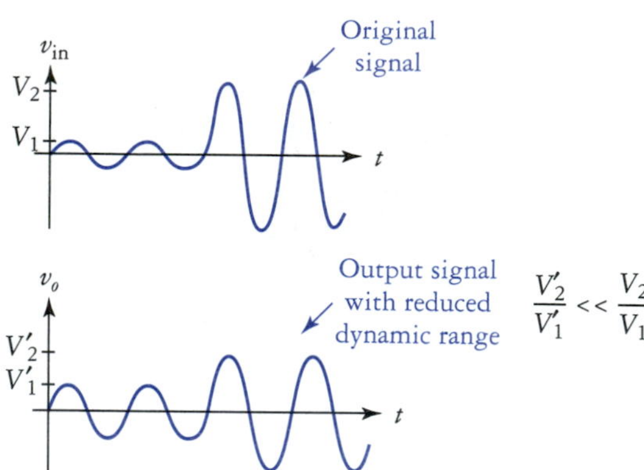

Why would we want to reduce the dynamic range of a signal? There are many applications for which this is useful but perhaps the easiest to explain at this point is in the recording of music on magnetic tape. Such a tape has minimum recording signal amplitude level that is determined by the *noise floor* of the tape. This floor is the point at which the signal is the same amplitude as the inherent noise of the tape (perceived as tape hiss). The maximum input signal level is limited by magnetic saturation of the tape. These limits define the dynamic range of the magnetic tape. We can use the log amp to reduce the dynamic range of the signal being recorded: Quiet passages are amplified well above the noise floor while loud passages are reduced in amplitude, preventing saturation of the tape. If we play back the tape through a normal linear amplifier, the audio sounds very flat and muddy. To restore the original dynamic range to the signal, the playback amplifier must have complementary antilogarithmic characteristics. The typical dynamic range of human hearing is about 90 dB. High-fidelity reproduction of music requires an equal or better dynamic range.

Note
In addition to DR compression, broadcasters now have digital time compression equipment that allows more commercials to be presented in a given time slot!

Dynamic range compression is also used by commercial AM (amplitude modulation) broadcast stations in order to reduce the loudness variation associated with speech. In this manner the average modulation is kept as high as possible. This is one of several reasons why music sounds so terrible on normal AM broadcasts. Television stations generally do not employ compression during programming such as movies so that the audio signal is high fidelity. However, during commercials compression may be used to produce a high average loudness level. That is why commercials are often obnoxiously loud compared to normal programming.

Analog Multipliers

So far we have seen circuits that allow the electronic realization of addition, subtraction, scaling (multiplication by a constant), differentiation, integration, and, of course, logarithms and antilogarithms. The log and antilog amplifiers can be used to provide generalized multiplication. The key to realizing multiplication is related to the rules of logarithms. In particular, we have

$$\ln(x) + \ln(y) = \ln(xy) \tag{15.40}$$

$$\ln(x) - \ln(y) = \ln(x/y) \tag{15.41}$$

A common symbol used to represent an *analog multiplier* is shown in Fig. 15.31(a). The internal circuitry could be arranged as shown in Fig. 15.31(b). Normally, the output of an analog multiplier will be scaled by 1/10 as shown in the illustration.

Analog multipliers are classified according to the polarity restrictions placed on the input voltages. For example, if the log amps used to construct the analog multiplier are like that of Fig. 15.24(a), then the input voltages must always be greater than or equal to zero. Because both inputs must always be positive or zero, the output will also be positive or zero. This is called a *one-quadrant multiplier* because the inputs are restricted to the first quadrant (I) of Fig. 15.32.

If we allow one of the inputs (say, v_x) to be positive or negative while restricting the other to positive values only, then the analog multiplier can operate in either quadrant I or II. This is a two-quadrant multiplier. If neither input is restricted in polarity then the multiplier may operate in quadrants I, II, III, or IV. This is called a four-quadrant multiplier.

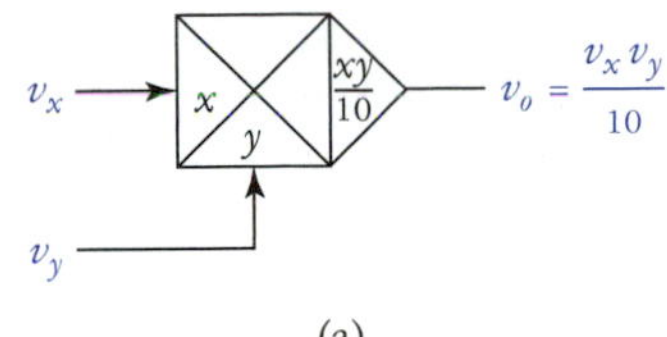

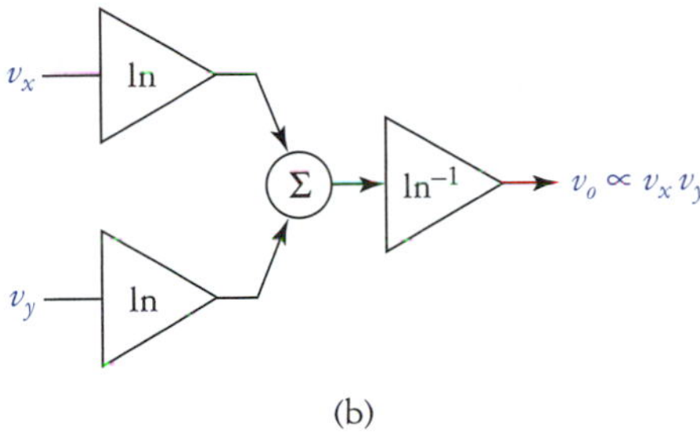

FIGURE 15.31
Analog multipliers. (a) Schematic symbol. (b) Implementation using log amps.

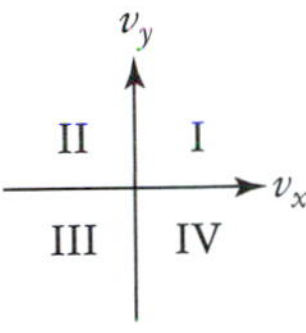

FIGURE 15.32
The four quadrants.

The reason why the output of most analog multipliers is scaled by 1/10 is to produce standard input and output ranges of ±10 V. For example, if both inputs are at their maximum values where $V_x = V_y = 10$ V, then

$$\begin{aligned} V_o &= (V_x V_y)/10 \\ &= 100 \text{ V} / 10 \\ &= 10 \text{ V} \end{aligned}$$

Thus we see that the output of the analog multiplier will always be in the range from -10 to $+10$ V.

EXAMPLE 15.5

Assume that Fig. 15.31 is a four-quadrant multiplier. Determine the equation for v_o given the following input voltages. Sketch the input and output voltages.

a. $V_x = -3$ V, $V_y = 2\sin(2\pi 100t)$ V
b. $v_x = 5\sin(2\pi 800t)$ V, $v_y = 6\sin(2\pi 200t)$ V
c. $v_x = 5\sin(100t)$ V, $v_y = 5\cos(100t)$ V

Solution

a. The output in this case is found by simply multiplying the coefficient of the sine function by the dc value. That is,

$$\begin{aligned} v_o &= V_x V_y \\ &= (-3)(2\sin(2\pi 100t)) \\ &= -6\sin(2\pi 100t) \text{ V} \end{aligned}$$

The input and output voltages are shown in Fig. 15.33(a).

b. In this case we have the product of two sine functions. We must use the following trigonometric identity:

$$A_1\sin(\omega_1 t)A_2\sin(\omega_2 t) = \frac{A_1A_2}{2}\cos[(\omega_1 - \omega_2)t] - \frac{A_1A_2}{2}\cos[(\omega_1 + \omega_2)t] \tag{15.42a}$$

This identity indicates that the output signal contains frequency components that are equal to the sum and difference of the input signal frequencies. Because negative frequency has no physical meaning, we always assign the higher frequency to ω_1. Applying this identity to the given inputs (and remembering to scale the output by 1/10) gives us

$$v_o = 1.5\cos(2\pi 600t) - 1.5\cos(2\pi 1000t) \text{ V}$$

This is a tough one to visualize. A graphics calculator might come in very handy here. The input and output signals are shown in Fig. 15.33(b). Signals like this occur frequently in communication electronics, particularly in the analysis of modulators.

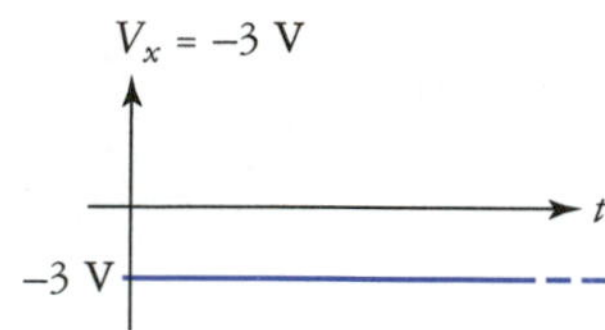

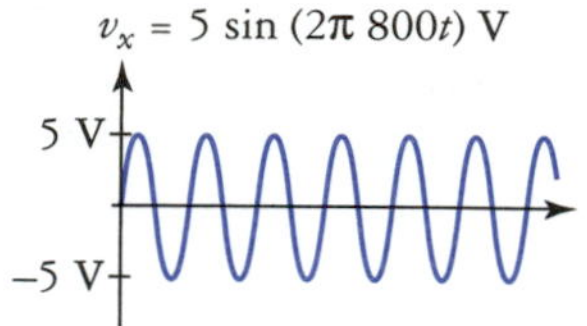

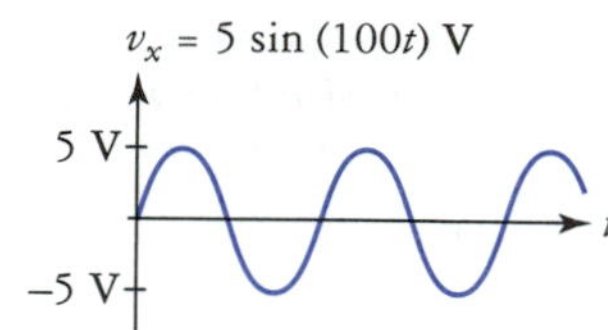

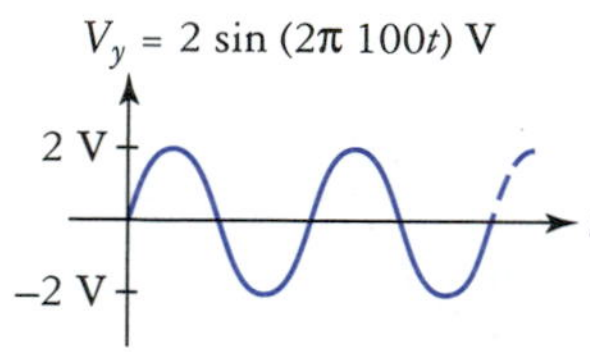

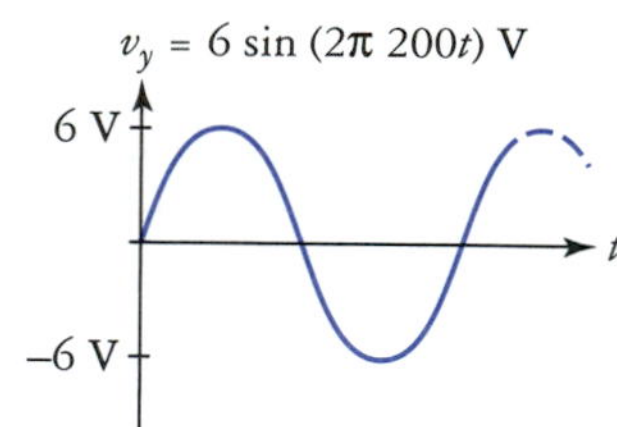

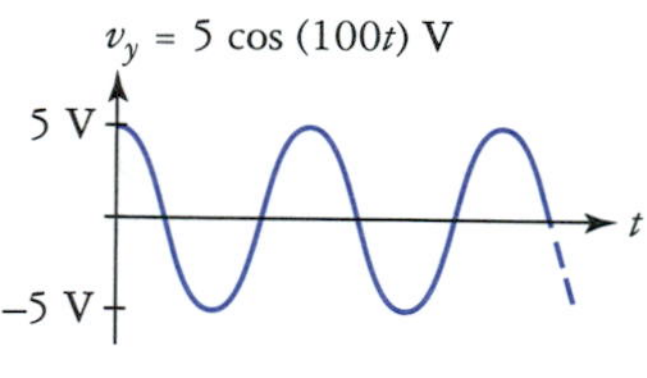

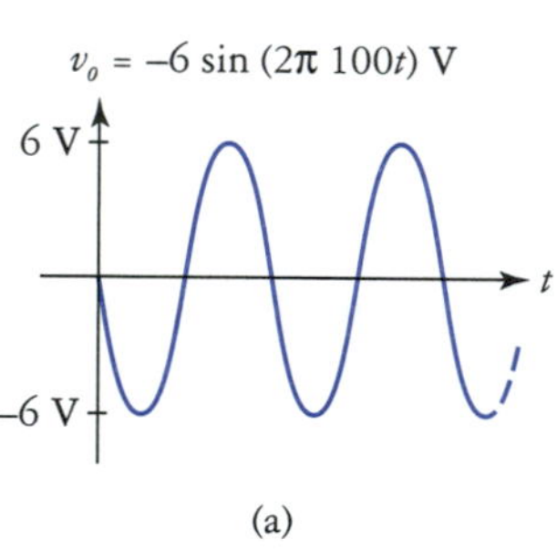

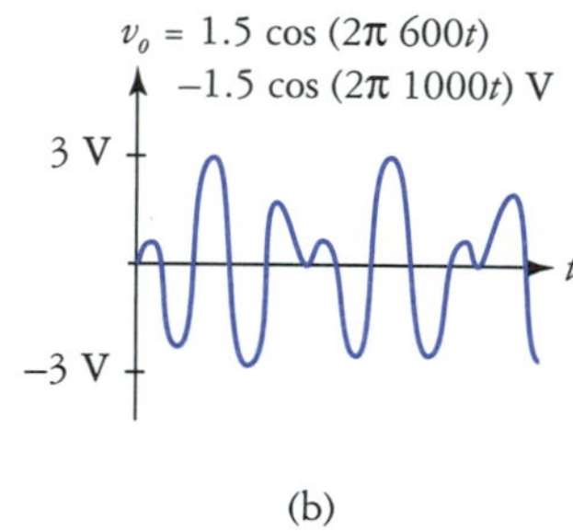

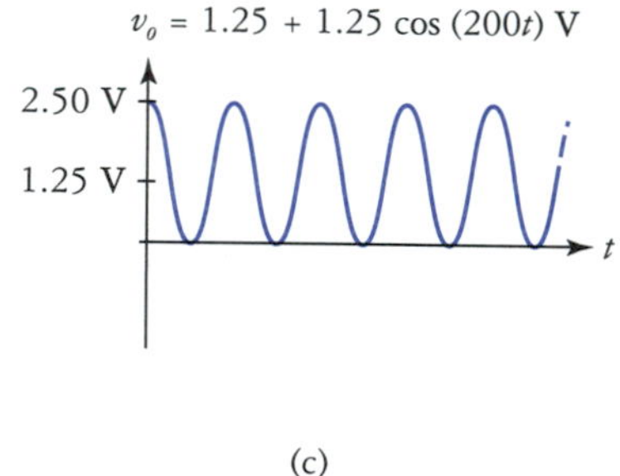

FIGURE 15.33
Waveforms for Example 15.5.

c. We can use the following trigonometric identity in this case:

$$A_1 \cos(\omega_1 t) A_2 \cos(\omega_2 t) = \frac{A_1 A_2}{2} \cos[(\omega_1 - \omega_2)t] + \frac{A_1 A_2}{2} \cos[(\omega_1 + \omega_2)t] \qquad (15.42b)$$

which, when applied to the multiplier, results in

$$\begin{aligned} v_o &= 1.25 \cos(0)t + 1.25 \cos(200t) \text{ V} \\ &= (1.25)(1) + 1.25 \cos(200t) \text{ V} \\ &= 1.25 + 1.25 \cos(200t) \text{ V} \end{aligned}$$

Because the input signals have identical frequency and phase, the output frequency is twice the input frequency. It is interesting to note that the dc offset produced in the output is proportional to the phase shift between the two input voltages. This property could be used to build a phase meter. The input and output signals are shown in Fig. 15.33(c).

Squaring and Square Root Circuits

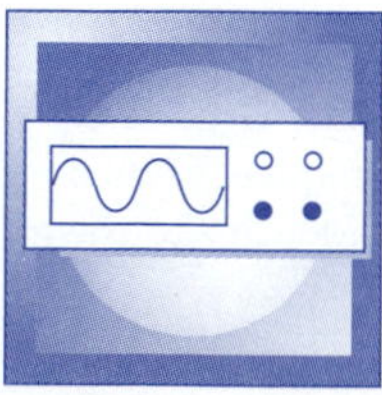

There are many other applications for analog multipliers. For example, if we short the two inputs of the analog multiplier together as shown in Fig. 15.34, we have a squaring circuit. The output of this circuit is

$$v_o = \frac{V_{in}^2}{10} \qquad (15.43)$$

Squarers are used in linearization, instrumentation, frequency multiplication, and modulation applications.

If we place a squarer in the feedback loop of an op amp, we obtain a square root circuit, as shown in Fig. 15.35. Applying the usual op amp analysis techniques we find

$$i_2 = i_1 = \frac{v_{in}}{R_1} = \frac{v_2}{R_2} \qquad (15.44)$$

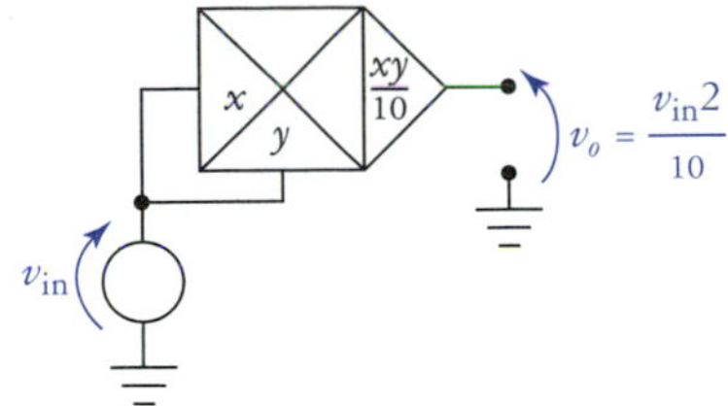

FIGURE 15.34
Using an analog multiplier as a squarer.

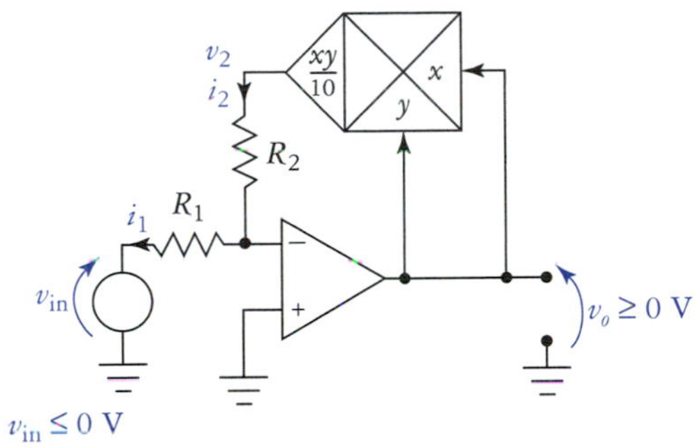

FIGURE 15.35
Using an analog multiplier as a square root circuit.

By inspection we have

$$v_2 = \frac{v_o^2}{10} \tag{15.45}$$

Substituting Eq. (15.45) into (15.44) gives us

$$\frac{v_{in}}{R_1} = \frac{v_o^2}{10\,R_2} \tag{15.46}$$

Solving this equation for v_o produces

$$v_o = \left(-\sqrt{\frac{10R_2}{R_1}}\right)(\sqrt{V_{in}}) \tag{15.47}$$

If we set $R_1 = 10R_2$, the first radical reduces to unity resulting in

$$v_o = -\sqrt{V_{in}} \tag{15.48}$$

Because the op amp inverts, the input to the square root circuit must be less than or equal to 0 V. A positive input voltage would require the op amp to produce an imaginary output voltage, which is an interesting idea, but is not physically realizable.

15.3 PRECISION RECTIFIERS

A half-wave rectifier is shown in Fig. 15.36(a). In the past we have used various approximations to model the behavior of the diode in this kind of application. The transconductance curves for the ideal and real diodes are shown in Fig. 15.36(b). Applying these curves to the half-wave rectifier circuit we obtain the transfer characteristics shown in Fig. 15.36(c). The ideal diode results in the output being exactly equal to the input as long as $v_{in} \geq 0$ V, otherwise $v_o = 0$ V. Mathematically, this is written

$$v_o = \begin{cases} v_{in} & \text{for } v_{in} \geq 0\text{ V} \\ 0 & \text{for } v_{in} < 0\text{ V} \end{cases}$$

In the case of the real diode, the output expression is much more complex and is not discussed further. The main things to notice here are that the ideal diode has zero barrier potential and zero dynamic resistance and is perfectly linear. In this section we examine circuits that use op amps to force normal diodes to behave nearly ideally.

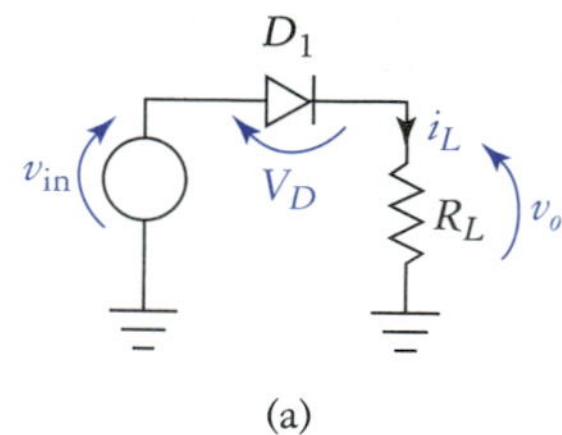

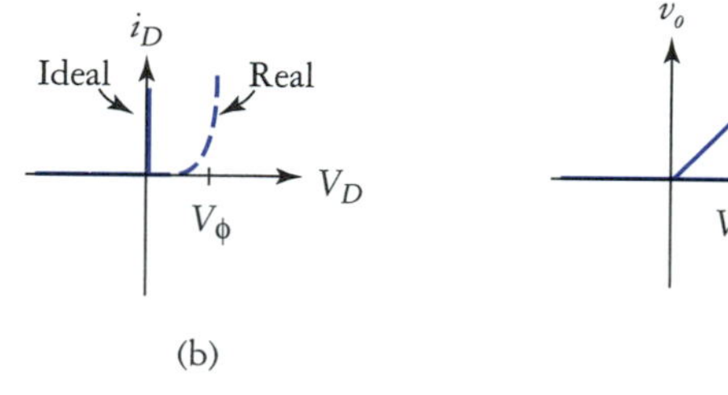

FIGURE 15.36
Half-wave rectification. (a) Simple diode rectifier. (b) Ideal and typical diode transconductance curves. (c) Ideal and typical half-wave transfer characteristics.

Precision Half-Wave Rectifier

The circuit in Fig. 15.37(a) is called a precision half-wave rectifier. Refer to the waveforms in Fig. 15.37(b) and (c) during the following explanation of circuit operation. We begin with sine wave v_{in} crossing zero in the positive-going direction. This causes i_1 to flow in the direction indicated by the arrow. The current is given by

Note
The precision half-wave rectifier is sometimes called a *superdiode*.

$$i_1 = \frac{v_{in}}{R_1} \qquad \text{for } v_{in} > 0 \text{ V}$$

The op amp forces the feedback current i_2 to flow through diode D_1. Notice that the current through R_2 will be zero because the op amp drives its output such that $v_x \cong -0.7$ V. The output voltage will therefore be zero for the entire positive half cycle of the input.

On the negative-going half-cycle of v_{in} the current i_1 is

$$i_1 = -\frac{v_{in}}{R_1} \qquad \text{for } v_{in} < 0 \text{ V}$$

The op amp output drives positively, reverse biasing D_1 and forward biasing D_2. This causes the feedback current to flow through R_2. Because the diode is inside the feedback loop, its barrier potential is effectively reduced to

$$V'_\phi = \frac{V_\phi}{A_{OL}}$$
$$\cong 0 \text{ V}$$

In addition to the reduction in effective barrier potential, the nonlinearity of D_2 is virtually eliminated as well. The output voltage produced during the negative-going half-cycle is inverted and scaled according to

$$v_o = -\frac{R_2}{R_1} \qquad \text{for } v_{in} < 0 \text{ V} \tag{15.49}$$

Putting this all together, the complete expression for the output voltage is

$$v_o = \begin{cases} -v_{in}\dfrac{R_2}{R_1} & \text{for } v_{in} < 0 \text{ V} \\ 0 & \text{for } v_{in} \geq 0 \text{ V} \end{cases} \tag{15.50}$$

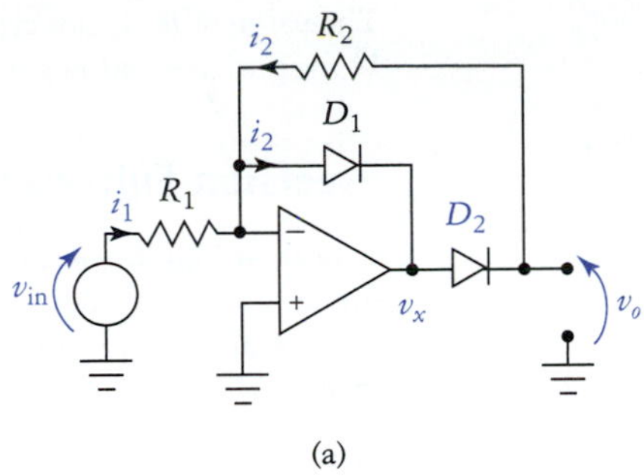

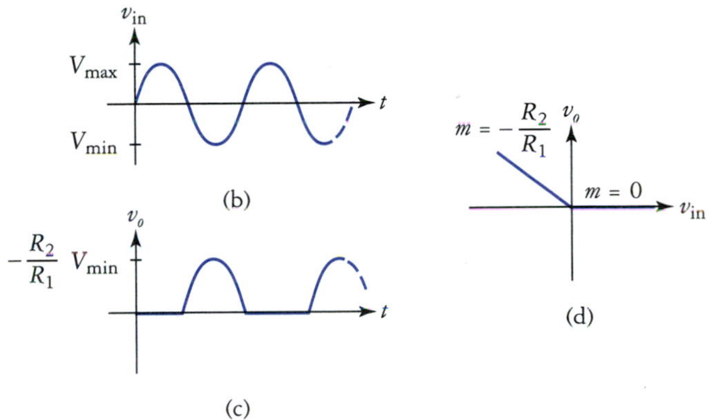

FIGURE 15.37
(a) Precision half-wave rectifier. Typical (b) input and (c) output waveforms. (d) Transfer characteristic.

This equation is shown in graphical form in the transfer characteristics of Fig. 15.37(d). We can also reverse the direction of both diodes in Fig. 15.37(a) if so desired. The analysis of the circuit with this modification is left as an exercise for the reader.

The advantages of the precision half-wave rectifier of Fig. 15.37(a) are elimination of the diode barrier potential, which allows rectification of small signals ($V_{in} \ll V_\phi$), good linearity, and adjustable voltage gain. There are several disadvantages as well. First, the frequency response is limited by the op amp. Typically such rectifiers are useful up to a few hundred kilohertz. Second, the output voltage is limited by the power supply. Third, this is a low-power circuit, useful primarily in signal processing applications. Last, the output resistance of the circuit is relatively high when the input voltage is positive. Specifically, the small-signal output resistance is given by

$$R_o \cong \begin{cases} R_2 & \text{for } v_{in} > 0 \text{ V} \\ 0 & \text{for } v_{in} < 0 \text{ V} \end{cases} \tag{15.51}$$

The basic precision half-wave rectifier can be modified to reduce the small-signal output resistance to nearly zero by adding a second op amp as shown in Fig. 15.38. This circuit has the same gain characteristics as that of Fig. 15.37(a). That is,

$$v_o = \begin{cases} -v_{in}\dfrac{R_2}{R_1} & \text{for } v_{in} < 0 \text{ V} \\ 0 & \text{for } v_{in} \geq 0 \text{ V} \end{cases} \tag{15.52}$$

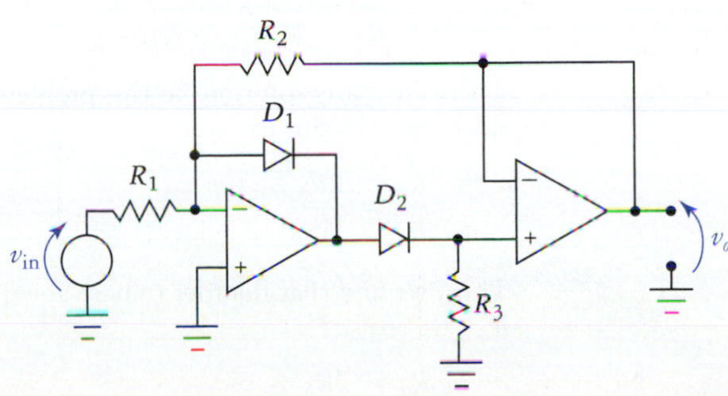

FIGURE 15.38
Modified half-wave rectifier.

The value of R_3 is not critical, with 10 kΩ being typical. The advantage of this circuit is that $R_o \cong 0$ Ω for positive and negative signal swings.

Precision Full-Wave Rectifier

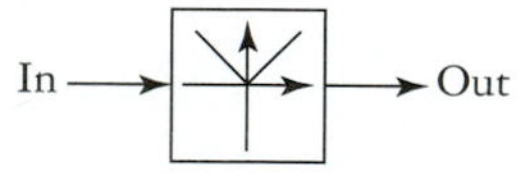

FIGURE 15.39
Absolute value block diagram symbol.

Like their half-wave counterparts, precision full-wave rectifiers are used in low-power signal processing applications. Because of this, they are often given the functional description of *absolute value* circuits or an absolute value block. The block diagram symbol for an absolute value function is shown in Fig. 15.39. Notice that, as with many other block diagram symbols (comparators, Schmitt triggers, and so on), the function performed is described by the transfer characteristic sketched inside the block.

A popular implementation of the precision full-wave rectifier is shown in Fig. 15.40. Rather than attempt to describe the operation of this circuit in its present form, the analysis equations are developed using the positive- and negative-going input signal equivalent circuits.

FIGURE 15.40
Precision full-wave rectifer.

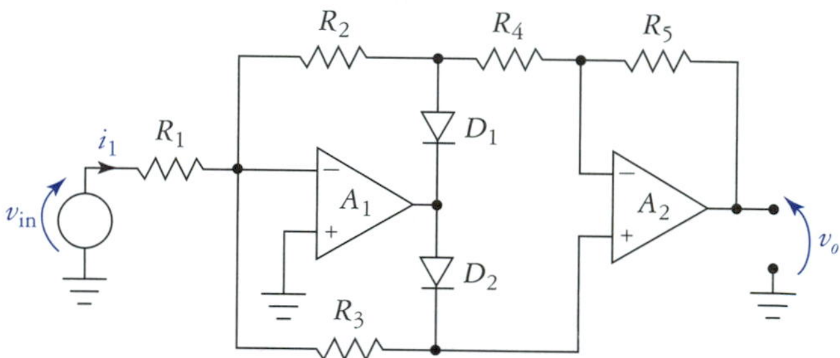

When $v_{in} > 0$ V, D_1 is forward biased and D_2 is reverse biased, producing the equivalent circuit shown in Fig. 15.41. Analysis of this circuit shows that the output voltage for the positive half-cycle of the input signal is

$$v_o \Big|_{v_{in} > 0} = v_{in}\left(\frac{R_2}{R_1}\right)\left(\frac{R_5}{R_4}\right) \tag{15.53}$$

The equivalent circuit of Fig. 15.42 is present when $v_{in} < 0$ V. Analysis of this circuit gives us

$$v_o \Big|_{v_{in} < 0} = -v_{in}\left(\frac{R_3}{R_1}\right)\left(\frac{R_4 + R_5}{R_4}\right) \tag{15.54}$$

If we combine Eqs. (15.53) and (15.54), we obtain the complete general expression for the output of the precision full-wave rectifier:

$$v_o = \begin{cases} v_{in}\dfrac{R_2R_5}{R_1R_4} & \text{for } v_{in} > 0 \text{ V} \\ -v_{in}\left(\dfrac{R_3}{R_1}\right)\left(\dfrac{R_4 + R_5}{R_4}\right) & \text{for } v_{in} < 0 \text{ V} \end{cases} \tag{15.55}$$

Most of the time we will want symmetrical amplification of the input signal. That is, we want equal voltage-gain magnitude for positive- and negative-going half-cycles of the input. This requires us to set Eq. (15.53) equal to Eq. (15.54):

$$\frac{R_2\, R_5}{R_1\, R_4} = \frac{R_3}{R_1} \times \frac{R_4 + R_5}{R_4} \tag{15.56}$$

There are many solutions to this problem; however, to simplify the algebra we set $R_5 = R_4$, which reduces Eq. (15.56) to

$$\frac{R_2}{R_1} = \frac{2R_3}{R_1} \tag{15.57}$$

Thus, we find that in order to have equal gain magnitude for positive and negative inputs (with $R_5 = R_4$), we require

$$R_2 = 2R_3 \tag{15.58}$$

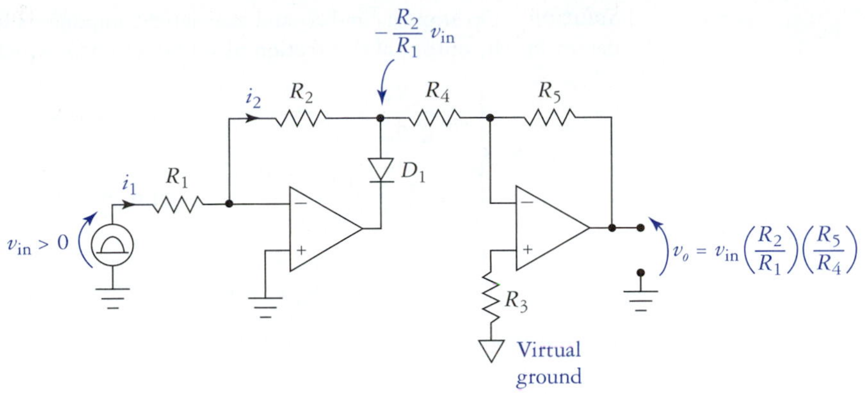

FIGURE 15.41
Equivalent circuit for positive-going input voltage.

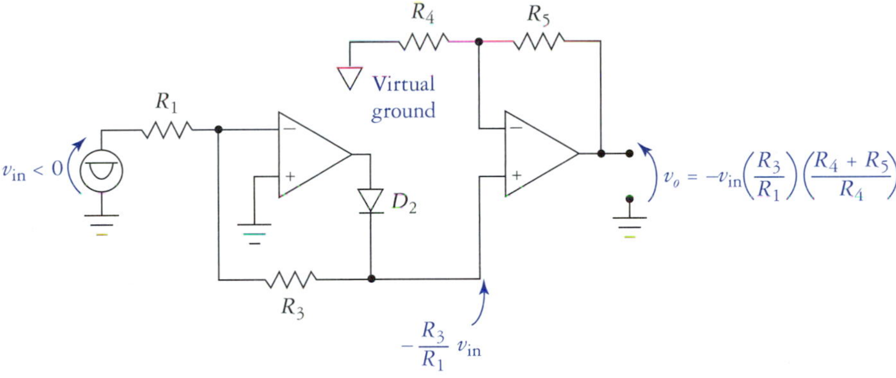

FIGURE 15.42
Equivalent circuit for negative-going input voltage.

If we meet these conditions, then the output of the precision rectifier is given by

$$v_o = |v_{in}| \frac{R_2}{R_1} \tag{15.59}$$

EXAMPLE 15.6

Refer to the circuit in Fig. 15.43. Determine the expression for v_o and sketch the transfer characteristic curve for the circuit.

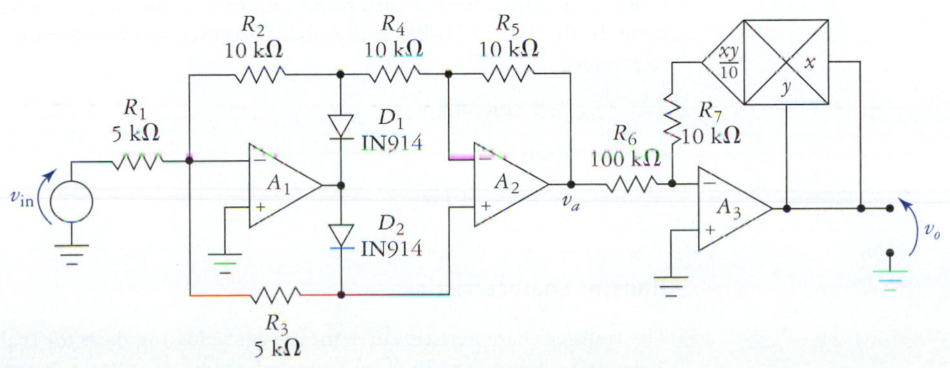

FIGURE 15.43
Circuit for Example 15.6.

Solution Op amps A_1 and A_2 and associated components form an absolute value block. We can determine the output of this section using Eq. (15.55), which yields

$$v_a = \begin{cases} v_{in}\dfrac{R_2R_5}{R_1R_4} & \text{for } v_{in} > 0 \text{ V} \\ -v_{in}\left(\dfrac{R_3}{R_1}\right)\left(\dfrac{R_4 + R_5}{R_4}\right) & \text{for } v_{in} < 0 \text{ V} \end{cases}$$

$$= \begin{cases} 2v_{in} & \text{for } v_{in} > 0 \text{ V} \\ -2v_{in} & \text{for } v_{in} < 0 \text{ V} \end{cases}$$

which is more conveniently written as

$$v_a = 2|v_{in}|$$

This signal drives a square root circuit. Because $R_6 = 10R_7$ the output of this section is given by Eq. (15.48), which is

$$v_o = \sqrt{v_a}$$
$$= \sqrt{2|v_{in}|}$$

The transfer characteristic curve is shown in Fig. 15.44.

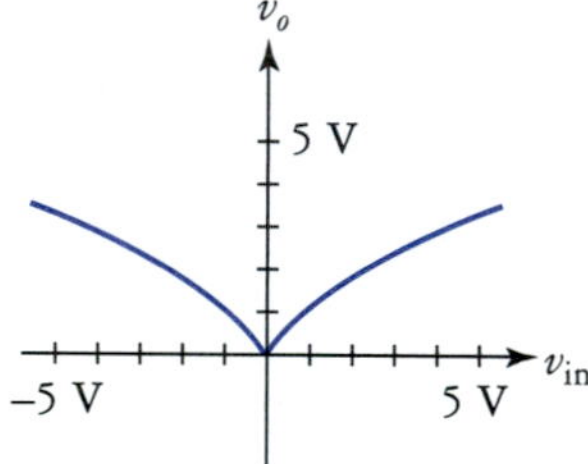

FIGURE 15.44
Transfer characteristic for Example 15.6.

15.4 COMPUTER-AIDED ANALYSIS APPLICATIONS

This section presents PSpice analysis of a representative sample of the circuits covered in this chapter. In addition to the familiar simulation options, the analog behavior model for the analog multiplier is introduced.

Zero-Crossing Detector Simulation

A 741 op amp is used to implement a zero-crossing detector in Fig. 15.45. The output of the op amp must be provided with a path to ground or a simulation error will occur due to the floating output node. In this case, a 10-kΩ load resistor was used to provide this path. The input sine source is set up to produce

$$v_{in} = 1 \sin(2\pi t) \text{ V}$$

In other words,

$$v_{in} = 1 \text{ V}_{pk}, 1 \text{ Hz}$$

Transfer Characteristics

The transfer characteristic curve for the zero-crossing detector is shown in Fig. 15.46. This curve was generated via a dc sweep of v_{in} from -1 to $+1$ V in 0.01-V increments. As expected, the op amp switches between saturation limits very quickly when v_{in} crosses 0 V.

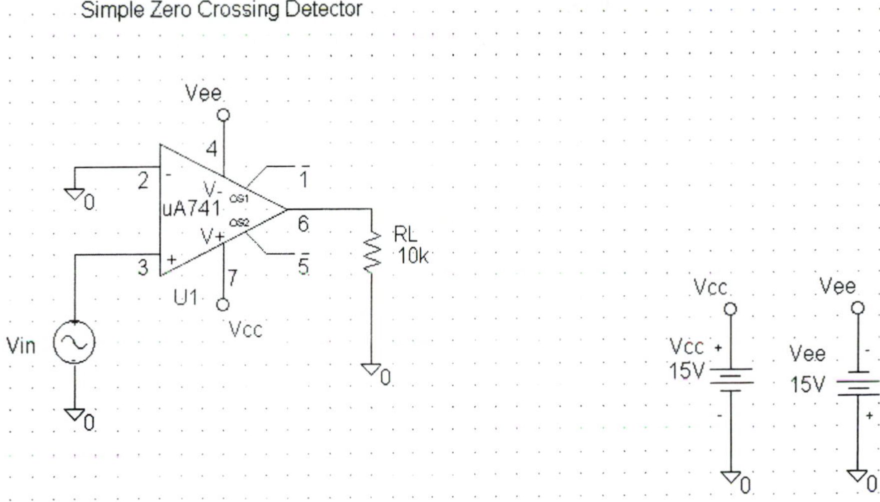

FIGURE 15.45
Simple zero-crossing detector.

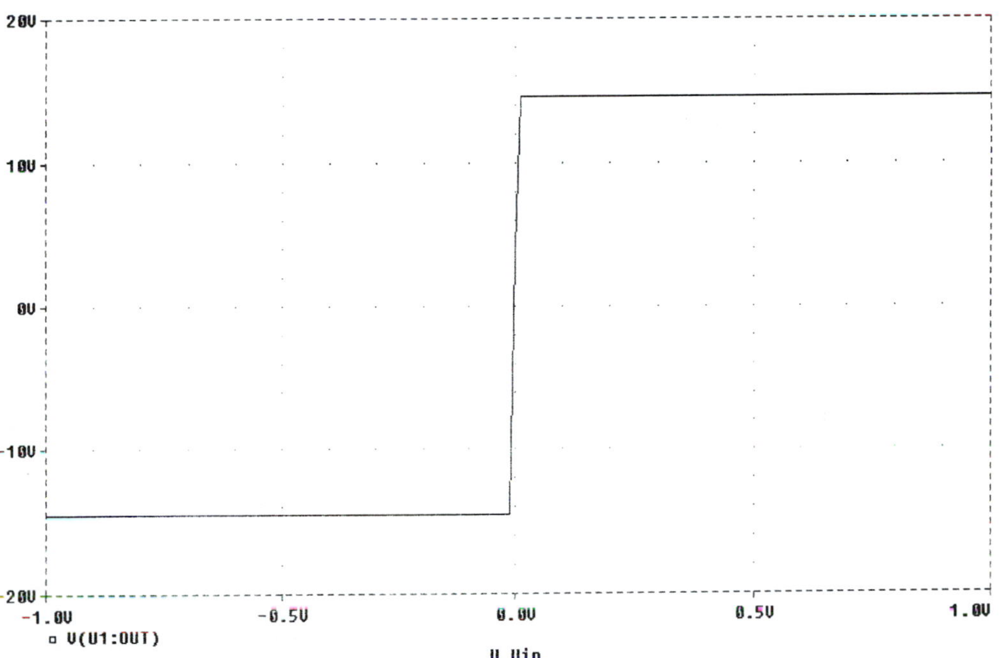

FIGURE 15.46
Zero-crossing detector transfer characteristic.

Transient Analysis

The results of a transient analysis run are shown in Fig. 15.47. In this case, the transient analysis is set to run from $t = 0$ to 3 sec, with a maximum step size of 100 μs. The print step interval was set to 3 sec (the default value is 20 ns) to speed simulation and save disk space. The print step saves data in text format from each step in the simulation to the disk. If you have a lot of data, as would be generated in 20-ns steps from 0 to 3 sec, this can really slow the simulation and waste lots of disk space. Since we do not need these data, it is reasonable to set the print step to the same size as the simulation run time.

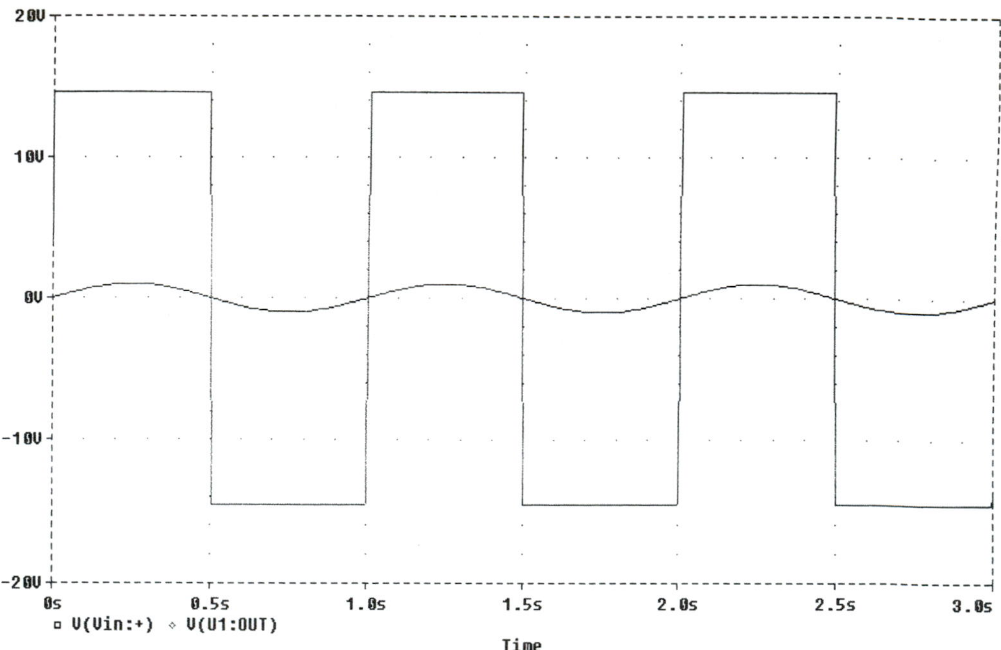

FIGURE 15.47
Response of the zero-crossing detector to a sinusoidal input.

Logarithmic Amplifier Simulation

A log-ratio amplifier is shown in Fig. 15.48. The differential amplifier (U_3 and associated components) has been modified to allow for offset adjustment. The input voltage source is set to an initial value of 0.1 V in order to prevent the circuit from saturating, as would occur if $V_{in} = 0$ V.

Transfer Characteristics

The transfer characteristic curve for the log-ratio amplifier is shown in Fig. 15.49. This curve was obtained by setting the analysis run for a linear sweep of V_{in} from 0.1 to 10 V using 0.01-V increments.

Apparently, the curve of Fig. 15.49 is logarithmic, but the human eye is not a very accurate judge of such curves. We can determine with certainty whether the transfer characteristic is indeed logarithmic by rescaling the x axis logarithmically. The dialog box of Fig. 15.50 is activated by the Plot selection of the PROBE toolbar and configured as shown. This produces the plot shown in Fig. 15.51. The resulting plot is a straight line, which will only occur if the transfer characteristic is logarithmic. Using the cursor, we find that the slope of the characteristic is 615 mV/dec.

If so desired, we could vary the gain of the differential amplifier to obtain any desired transfer characteristic slope. We can also easily vary the offset in order to place the zero crossing of the curve at any arbitrary point on the x axis.

Analog Behavioral Modeling: The Analog Multiplier

We could design and simulate an analog multiplier using op amps, transistors, and so on, but there is an alternative technique that is much simpler. The PSpice library contains a number of idealized functional blocks that are called *analog behavioral models.* This term is used because the blocks are not electronic components in the usual sense but rather are mathematical functions that are placed on the schematic just like normal components. Of course, in computer simulation none of the components we use are "real" but analog behavioral models are a step further into abstraction. As stated earlier, analog behavioral models are ideal. This means that such models will not saturate or clip, and they have no frequency response limitations due to stray capacitance or other factors. This is convenient when you wish to avoid spending lots of time tweaking component values to get a circuit simulation to perform as desired.

B

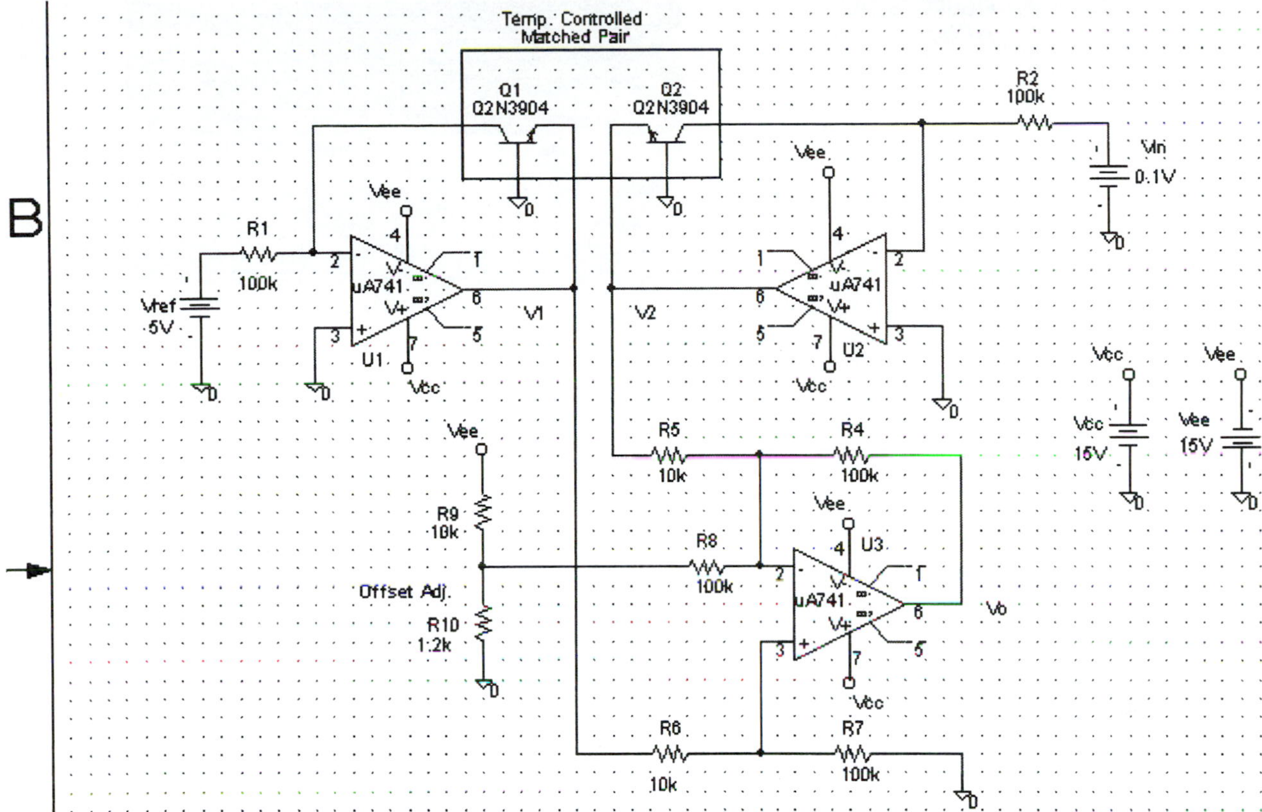

FIGURE 15.48
Log-ratio amplifier.

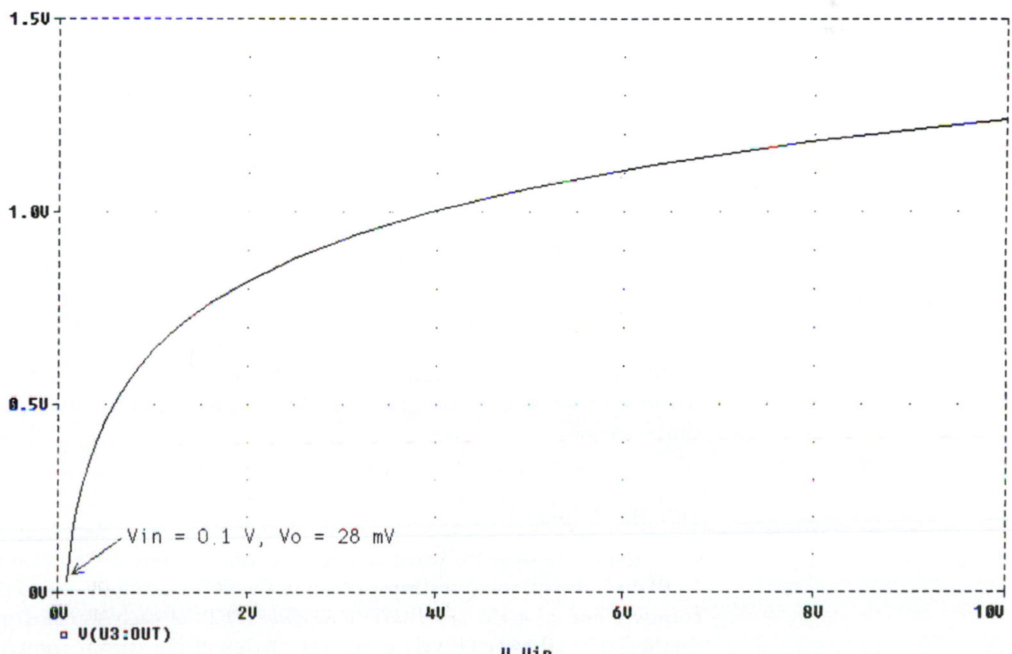

FIGURE 15.49
Transfer characteristic for the log-ratio amplifier.

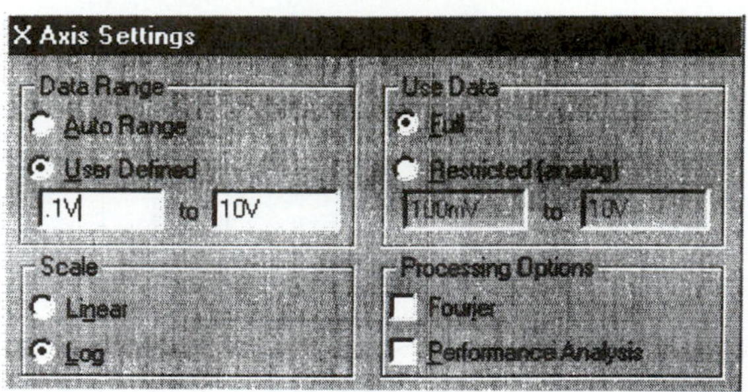

FIGURE 15.50
Setting the x-axis attributes.

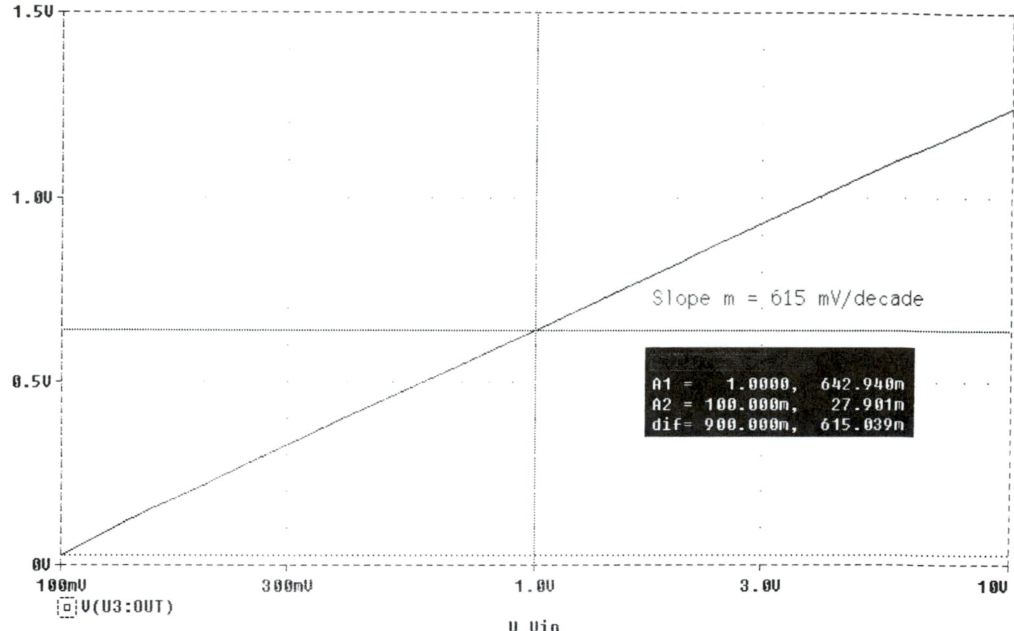

FIGURE 15.51
Transfer characteristics plotted with a logarithmic x axis.

A few of the analog behavioral models available in the PSpice library are the four-quadrant multiplier, log and antilog (exp) amplifiers, summer, transmission lines, trigonometric functions, differentiator, integrator, and various filter models. The four-quadrant analog multiplier is used in Fig. 15.52. The circuit has been configured to correspond to Example 15.5, part (b), where $v_x = 5 \sin(2\pi 800t)$ V and $v_y = 6 \sin(2\pi 200t)$ V.

The analog multiplier used in Example 15.5 has its output scaled by 1/10, as would be the case in a commercially available device such as the AD534 from Analog Devices Inc. A resistive voltage divider is used in the simulation in order to provide this same output scaling. The results of the simulation run are shown in Fig. 15.53. The upper plot shows the input signals while the lower plot shows the product of these signals.

Spectral Analysis

It is interesting to examine the analog multiplier's input and output signals in the frequency domain. In PSpice, this is done by simply clicking the FFT button on the toolbar. This causes the program to compute and plot the *fast Fourier transform* (FFT) of each signal displayed. The FFT is a numerical method that allows relatively quick calculation of the *spectral* components of a signal. The resulting display is shown in Fig. 15.54. Each spike corresponds to energy at a given signal frequency. For the input signals the frequencies are 200 and 800 Hz. The output signal spectrum components are located at 600 Hz and 1 kHz as expected.

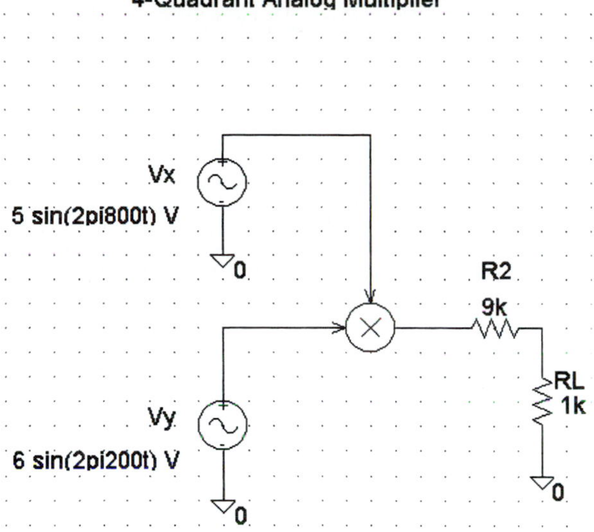

FIGURE 15.52
Four-quadrant multiplier using behavioral model.

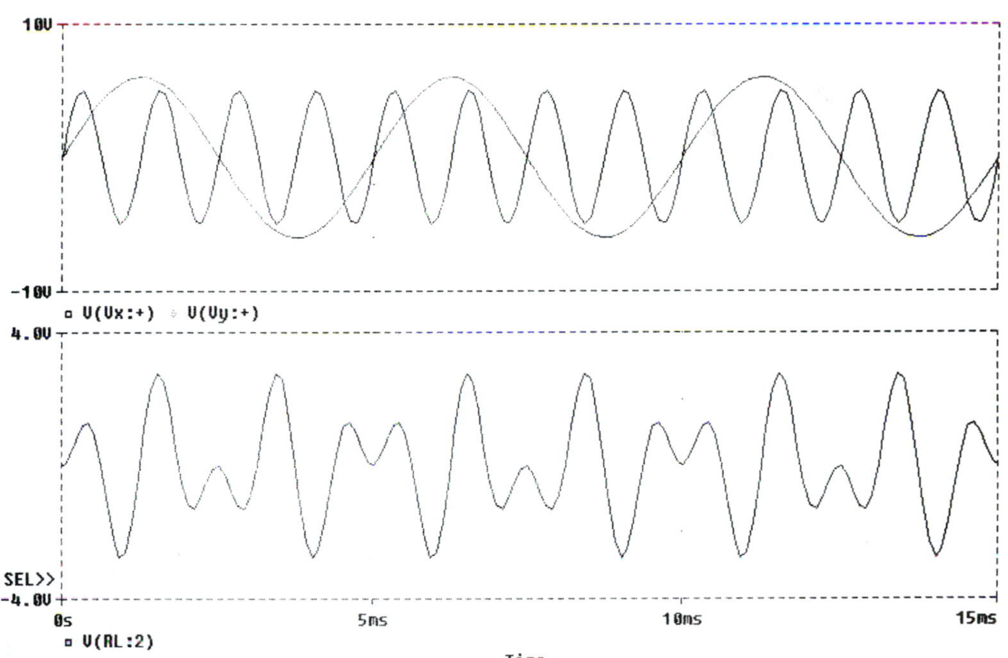

FIGURE 15.53
Input and output voltage waveforms.

Precision Rectifier Simulation

The precision full-wave rectifier (or absolute value) portion of Example 15.6 is shown in Fig. 15.55. The input voltage is a sine save of 1 V_{pk} with $f = 1$ Hz. A signal with such low amplitude would produce almost no output from a normal passive full-wave rectifier. In this case, the precision rectifier compensates for the diode barrier potential and produces a clean, unattenuated output signal as shown in Fig. 15.56.

If we sweep the dc component of v_{in} from -1 to 1 V, the transfer characteristic curve of Fig. 15.57 can be generated. This is the classical absolute value function with the slopes of the curve being ± 1. The absolute value function is also available in the PSpice analog behavioral model library.

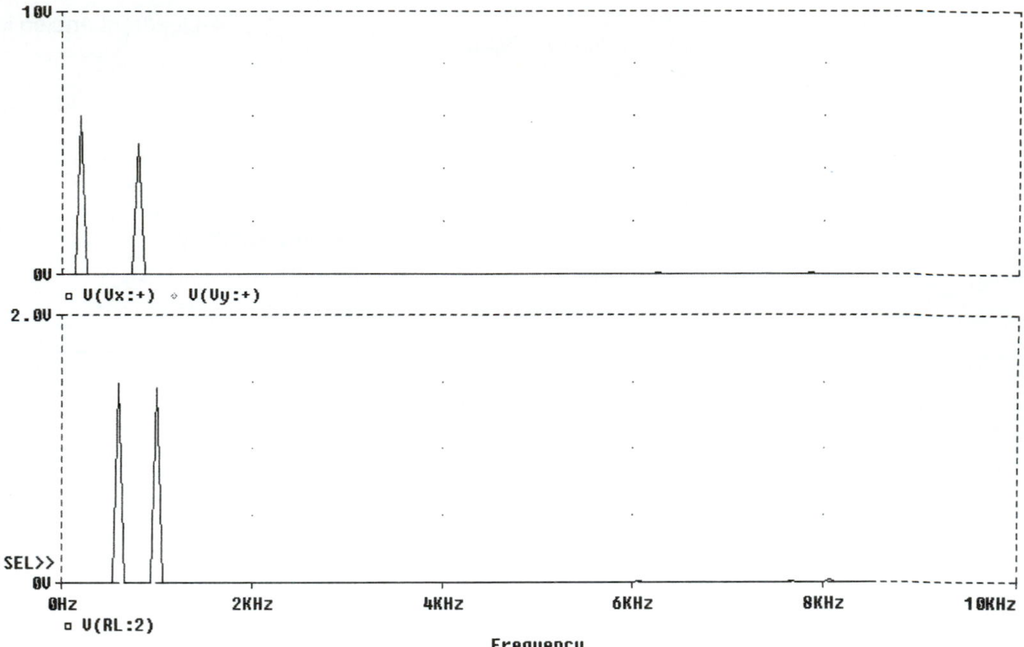

FIGURE 15.54
Input and output voltage frequency-domain representations.

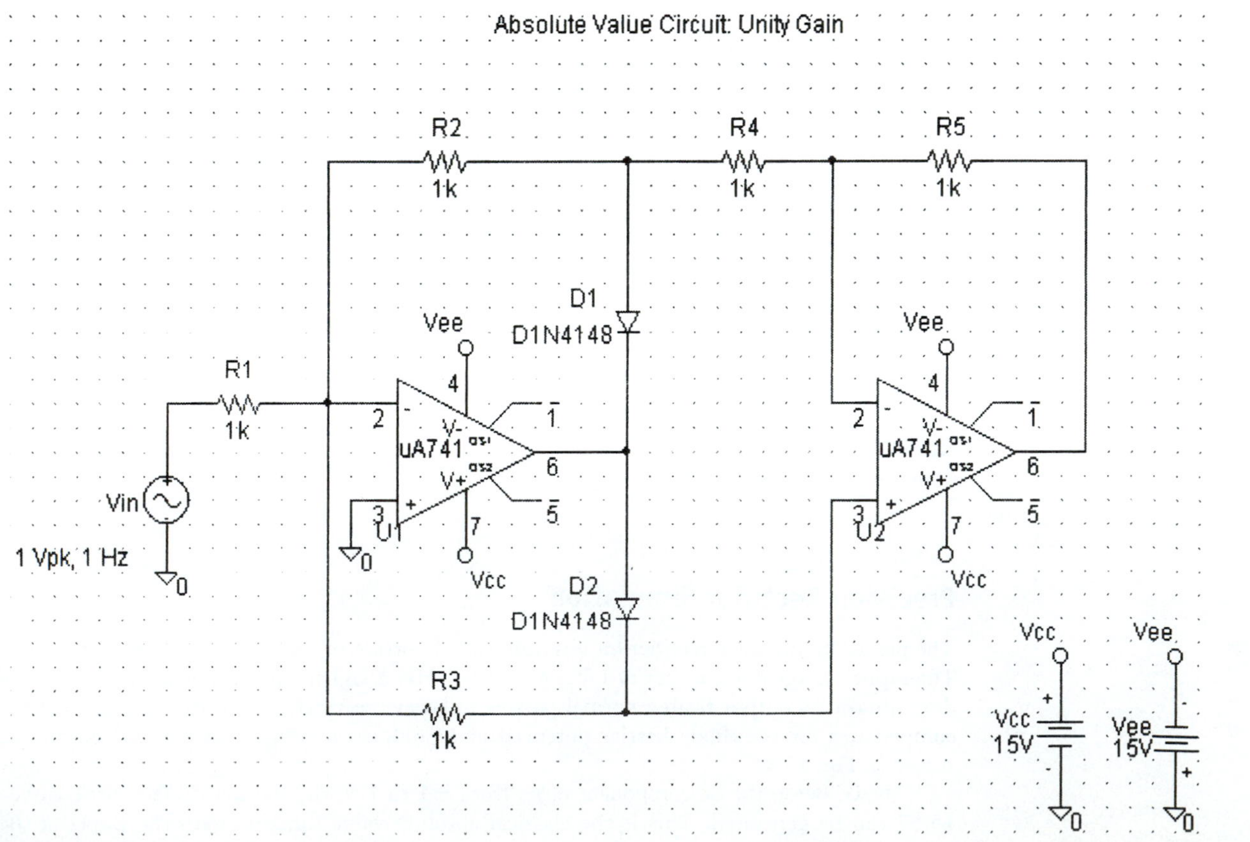

FIGURE 15.55
Absolute value circuit.

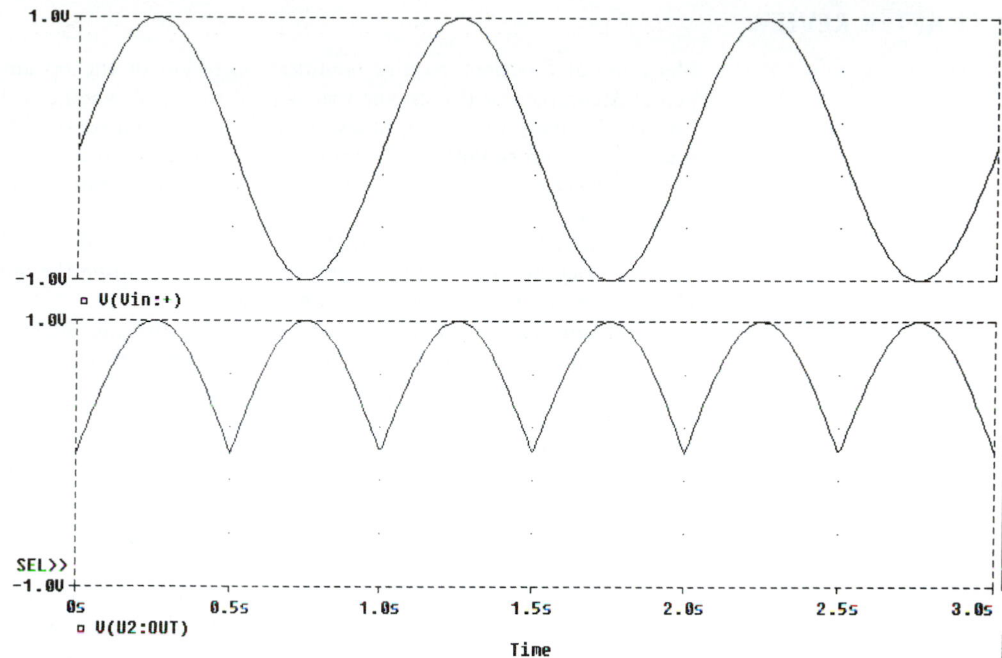

FIGURE 15.56
Input and output waveforms for the absolute value circuit.

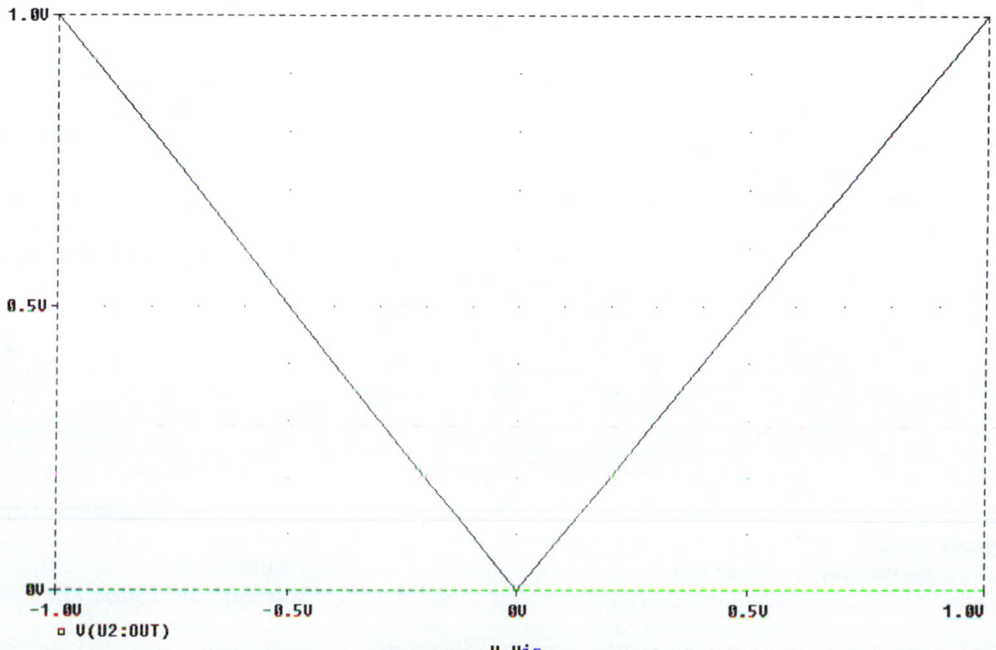

FIGURE 15.57
Transfer characteristic curve for the absolute value circuit.

CHAPTER REVIEW

Many useful functions require nonlinear operation of the op amp and/or its associated components. A comparator is a circuit that is used to signal whether a voltage exceeds some predefined limit(s). In many cases, a comparator can be implemented simply by using an op amp operating under open-loop conditions. Output bounding is used to allow the op amp to be biased up with normal power supply voltages while at the same time limiting the output voltage to lower levels as necessary.

Positive feedback is employed to cause a comparator to exhibit hysteresis. The amount of positive feedback and the output saturation voltage determine the width of the hysteresis dead-band. Hysteresis is used to provide noise immunity.

Op amp circuits can be designed to have logarithmic and antilogarithmic transfer characteristics through the use of diodes and transistors in the feedback network. Such circuits are used to linearize signals, to compress and expand the dynamic range of signals, and to implement analog multipliers. Compression amplifiers are used in audio recording noise reduction circuits and in radio and television broadcasting applications.

Analog multipliers are classified as being one-, two-, or four-quadrant devices, according to the polarity restrictions on their input terminals. Analog multipliers are used to realize mathematical operations in real time and to perform frequency translation functions in modulation applications.

In precision rectifier circuits, the op amp is used to reduce barrier potential and diode nonlinearity to very low levels, allowing nearly ideal rectification to be accomplished. Half-wave and full-wave versions are available. Precision full-wave rectifiers are often called absolute value circuits.

DESIGN AND ANALYSIS PROBLEMS

Note: Unless otherwise noted, assume all op amps are ideal.

15.1. Refer to Fig. 15.4. Given: $V_{CC} = 10$ V, $-V_{EE} = -10$ V, and $R_2 = 10$ kΩ, determine R_1 such that $V_t = -1.0$ V. Determine the duty cycle of the output voltage if $v_{in} = 2 \sin(2\pi 1000t)$ V.

15.2. Refer to Fig. 15.4. Given $V_{CC} = 12$ V, $-V_{EE} = -12$ V, and $R_1 = 22$ kΩ, determine R_2 such that $V_t = 1.5$ V. Determine the duty cycle of the output voltage if $v_{in} = 2 \sin(2\pi 1000t)$ V.

15.3. Refer to Fig. 15.58. Assume that $V_{REF} = 3.0$ V. Sketch the transfer characteristic curve for the circuit. Sketch the input and output waveforms that would be produced given $v_{in} = 4 \sin(2\pi 1000t)$ V.

15.4. Refer to Fig. 15.58. Assume that $V_{REF} = -1.0$ V. Sketch the transfer characteristic curve for the circuit. Sketch the input and output waveforms that would be produced given $v_{in} = 2 \cos(2\pi 1000t)$ V.

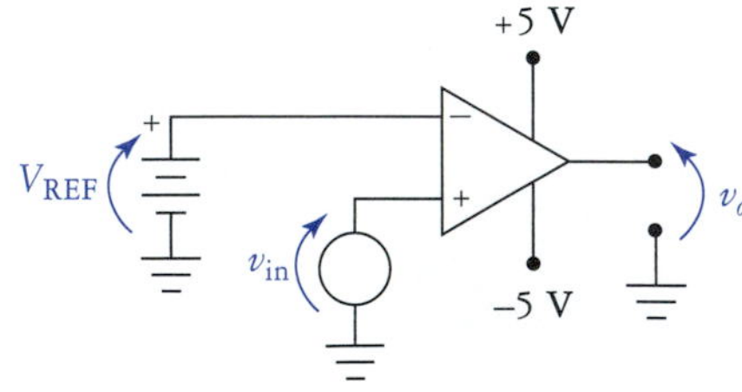

FIGURE 15.58
Circuit for Problems 15.3 and 15.4.

15.5. Refer to Fig. 15.59. Assume that $V_{REF} = 5.0$ V. Sketch the transfer characteristic curve for the circuit. Sketch the input and output waveforms that would be produced given $v_{in} = 6 \sin(2\pi 1000t)$ V.

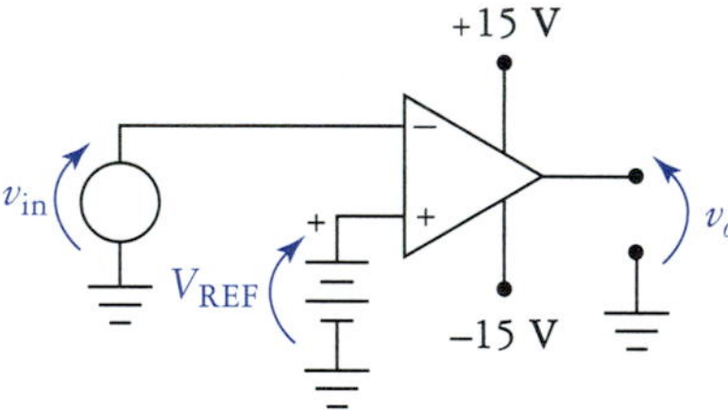

FIGURE 15.59
Circuit for Problems 15.5 and 15.6.

15.6. Refer to Fig. 15.59. Assume that $V_{REF} = 2.0$ V. Sketch the transfer characteristic curve for the circuit. Sketch the input and output waveforms that would be produced given $v_{in} = 10t$ V.

15.7. Design a circuit using a 741 with ±15-V supply rails to approximate the transfer characteristic shown in Fig. 15.60. (There is more than one correct solution.)

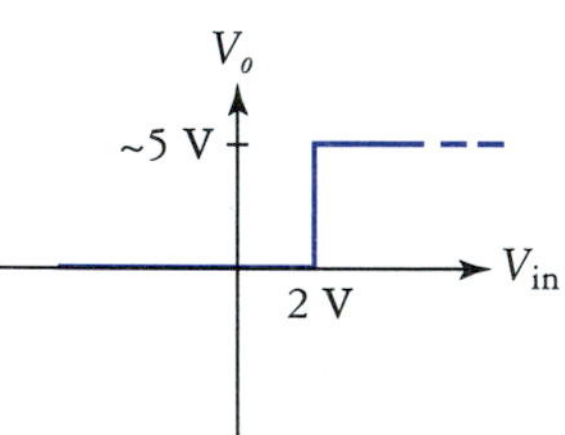

FIGURE 15.60
Waveform for Problem 15.7.

15.8. Design a circuit using a 741 with ±15-V supply rails to approximate the transfer characteristic shown in Fig. 15.61. (There is more than one correct solution.)

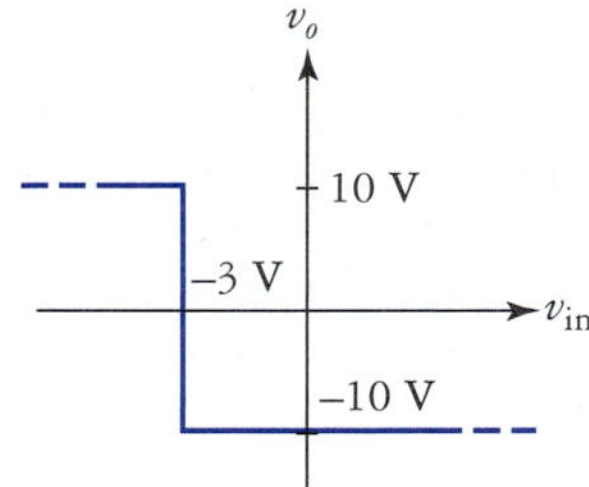

FIGURE 15.61
Waveform for Problem 15.8.

15.9. Refer to Fig. 15.14. Assume $V_{CC} = 15$ V, $-V_{EE} = -15$ V, $V_{o(sat)} = \pm 14$ V, $R_1 = 22$ kΩ, and $R_2 = 2.2$ kΩ. Determine V_{LT} and V_{UT} and sketch the resulting transfer characteristic curve.

15.10. Refer to Fig. 15.14. Assume $V_{CC} = 12$ V, $-V_{EE} = -12$ V, $V_{o(sat)} = \pm 10$ V, $R_1 = 10$ kΩ, and $R_2 = 3.3$ kΩ. Determine V_{LT} and V_{UT} and sketch the resulting transfer characteristic curve.

15.11. Refer to Fig. 15.18. Given $V_{CC} = 10$ V, $R_1 = 9$ kΩ, $R_2 = 1$ kΩ, $R_3 = 10$ kΩ, and $R_4 = 10$ kΩ, sketch the transfer characteristic curve. Assume that $V_{o(max)} = 10$ V and $V_{o(min)} = 0$ V.

15.12. Refer to Fig. 15.18. Given $V_{CC} = 12$ V, $R_1 = 9$ kΩ, $R_2 = 1$ kΩ, $R_3 = 10$ kΩ, and $R_4 = 10$ kΩ, sketch the transfer characteristic curve. Assume that $V_{o(max)} = 12$ V and $V_{o(min)} = 0$ V.

15.13. Refer to Fig. 15.24(a). The transistor has $I_{CES} = 5$ pA, $\eta = 1$, $V_T = 26$ mV, and $\alpha \cong 1$. Given $R_1 = 10$ kΩ, determine V_o for $V_{in} = 1.0$ V and 10.0 V.

15.14. Refer to Fig. 15.24(b). The transistor has $I_{CES} = 20$ pA, $\eta = 1$, $V_T = 26$ mV, and $\alpha \cong 1$. Given $R_1 = 10$ kΩ, determine V_o for $V_{in} = -2.0$ V and -20.0 V.

15.15. Refer to Fig. 15.25. Given $R_1 = R_2 = 100$ kΩ, $R'_3 = R_3 = 10$ kΩ, $R'_4 = R_4 = 50$ kΩ, and $V_T = 26$ mV, determine V_o for (a) $V_x = 1.0$ V, $V_y = 10.0$ V; (b) $V_x = 5.0$ V, $V_y = 2.0$ V.

15.16. Refer to Fig. 15.25. Given $R_1 = R_2 = 100$ kΩ, $R'_3 = R_3 = 10$ kΩ, $R'_4 = R_4 = 100$ kΩ, and $V_T = 26$ mV, determine V_o for (a) $V_x = 1.0$ V, $V_y = 1.0$ V; (b) $V_x = 0.1$ V, $V_y = 10.0$ V.

15.17. Refer to Fig. 15.27(a). Given $R_F = 1$ kΩ, $I_{CES} = 50$ pA, $\eta = 1$, $\alpha = 1$, and $V_T = 26$ mV, determine V_o for $V_{in} = 0.50$ V and $V_{in} = 0.56$ V.

15.18. Refer to Fig. 15.27(b). Given $R_F = 1$ kΩ, $I_{CES} = 0.1$ pA, $\eta = 1$, $\alpha = 1$, and $V_T = 26$ mV, determine V_o for $V_{in} = 0.6$ V and $V_{in} = -0.66$ V.

15.19. Refer to Fig. 15.31(a). Determine the output voltage equation for the following sets of inputs:

(a) $v_x = 5 \sin(2\pi 1000t)$ V $v_y = 2 \sin(2\pi 1500t)$ V
(b) $v_x = 5 \cos(2\pi 100t)$ V $v_y = 5 \sin(2\pi 200t)$ V
(c) $v_x = 2 + 4 \sin(100t)$ V $v_y = 3 \sin(500t)$ V
(d) $v_x = 5t$ V $v_y = 2t$ V

15.20. Refer to Fig. 15.31(a). Determine the output voltage equation for the following sets of inputs:

(a) $v_x = 3 \sin(2\pi 2000t)$ V $v_y = 2 \sin(2\pi 3000t)$ V
(b) $v_x = 5 \cos(800t)$ V $v_y = 5 \cos(400t)$ V
(c) $v_x = 4 \sin(100t)$ V $v_y = 4 \sin(100t)$ V
(d) $v_x = 2t^2$ V $v_y = 4t + 2$ V

15.21. Refer to Fig. 15.35. Given $R_1 = 10$ kΩ and $R_2 = 20$ kΩ, determine the expression for V_o.

15.22. Refer to Fig. 15.35. Given $R_1 = 10$ kΩ, determine R_2 such that $V_o = -5\sqrt{V_{in}}$.

15.23. Refer to Fig. 15.37. Given $R_1 = 10$ kΩ and $R_2 = 50$ kΩ, sketch the output voltage that would be produced for each input signal shown in Fig. 15.62(a).

15.24. Refer to Fig. 15.37. Given $R_1 = 10$ kΩ and $R_2 = 20$ kΩ, sketch the output voltage that would be produced for each input signal shown in Fig. 15.62(b).

15.25. Refer to Fig. 15.40. Assume $R_1 = 10$ kΩ, $R_2 = 20$ kΩ, $R_3 = 10$ kΩ, $R_4 = 10$ kΩ, and $R_5 = 10$ kΩ. Sketch the circuit transfer characteristic curve and express this function in equation form. Sketch the output waveforms that will be produced for the input signals shown in Fig. 15.62(a).

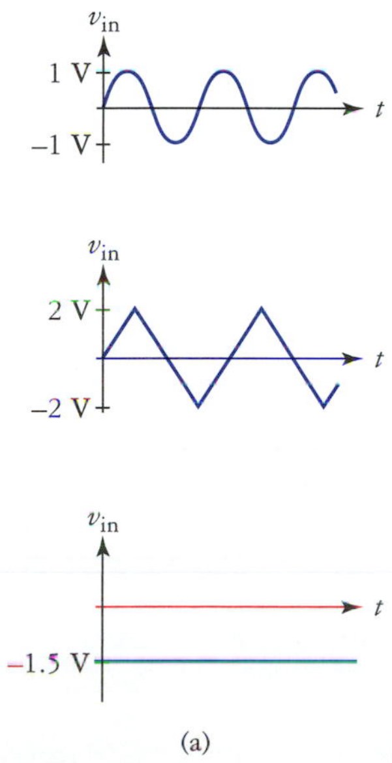

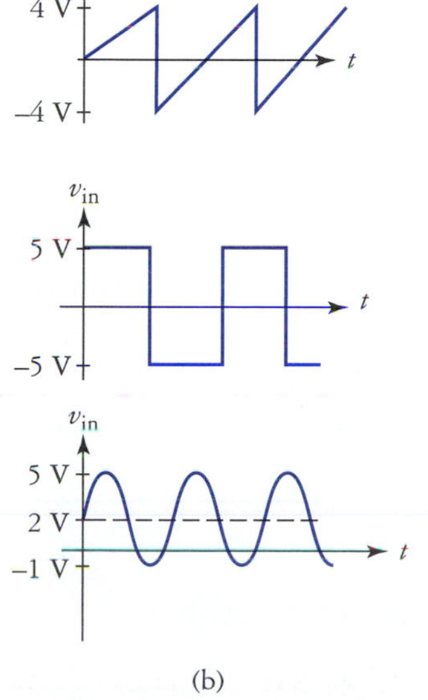

FIGURE 15.62
Waveforms for Problems 15.23 through 15.26.

15.26. Refer to Fig. 15.40. Assume $R_1 = 10\ \text{k}\Omega$, $R_2 = 20\ \text{k}\Omega$, $R_3 = 20\ \text{k}\Omega$, $R_4 = 10\ \text{k}\Omega$, and $R_5 = 10\ \text{k}\Omega$. Sketch the circuit transfer characteristic curve and express this function in equation form. Sketch the output waveforms that will be produced for the input signals shown in Fig. 15.62(b).

■ TROUBLESHOOTING PROBLEMS

15.27. Refer to the comparator circuit in Fig. 15.9. A 1-$V_{p\text{-}p}$ sine wave is applied to the input of the comparator. The input to the TTL inverter is observed on an oscilloscope, as shown in Fig. 15.63. What is the most likely problem with the circuit?

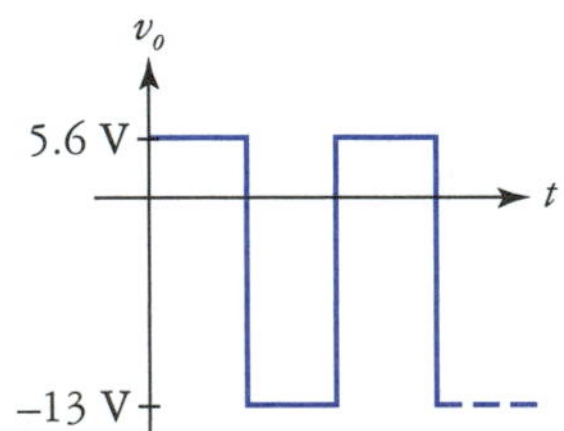

FIGURE 15.63
Waveform for Problem 15.27.

15.28. Refer to Fig. 15.18. Suppose that A_1 became defective and its output terminal became stuck at ground potential. How would this problem affect the waveform shown in Fig. 15.18(b)?

15.29. Refer to Fig. 15.25. Suppose the on-chip heater stopped functioning. In relation to the output voltage transfer characteristic curve, would this affect the offset, the slope, both offset and slope, or neither? Explain your reasoning. [*Hint:* Look at Eq. (15.32).]

15.30. Refer to Fig. 15.40. The circuit is set up for unity gain. Suppose that D_2 became open circuited. How would this affect the output waveform if a sine wave of 1 $V_{p\text{-}p}$ were applied to the input?

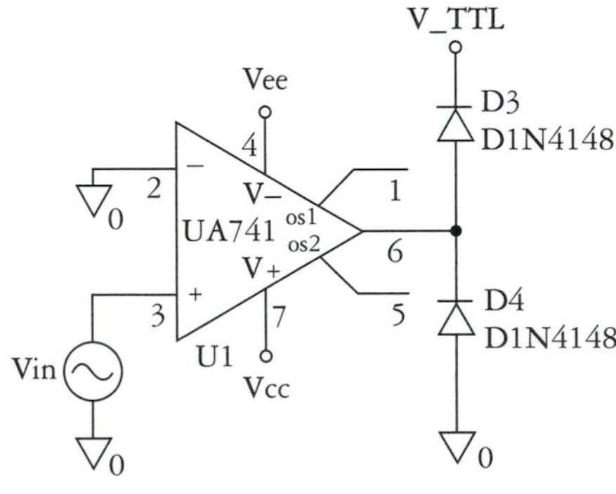

FIGURE 15.64
Circuit for Problem 15.31.

■ COMPUTER PROBLEMS

15.31. Refer to Fig. 15.64. Simulate this circuit. Generate a plot of the transfer function (dc sweep) and generate plots of input and output signals for $v_{in} = 2\sin(2\pi t)$ V.

15.32. If Matlab is available to you, enter the commands listed in Fig. 15.65. This will produce plots of the natural log [entered as log(x)] and the common log [entered as 2.303log(x)] functions for $x = 0$ to 5 in increments of 0.1. The *hold on* command forces Matlab to place multiple plots on the same graph. The command *grid on* causes the scale grid to be drawn.

```
>> x = 0:.1:5;
>> y = log(x);

Warning: Log of zero.
>> f = 2.303*log(x);

Warning: Log of zero.
>> hold on;
>> plot(x, y);
>> plot(x, f);
>> grid on;
>> |
```

FIGURE 15.65
Listing for Problem 15.32.

15.33. Refer to Fig. 15.55. Perform a log ac sweep from 10 Hz to 10 MHz and determine the small-signal corner frequency of the circuit. After determining the corner frequency, set v_{in} such that $V_p = 1$ V, $f = f_c$ and perform a transient analysis over three cycles of v_{in}. Plot the resulting output signal.

CHAPTER 16

Data Conversion Devices

BEHAVIORAL OBJECTIVES

On completion of this chapter, the student should be able to:

- Describe digital-to-analog and analog-to-digital converter errors.
- Analyze op amp-based digital-to-analog converters.
- Describe the operation of a multiplying digital-to-analog converter.
- Explain the meaning of digital-to-analog converter specifications.
- Describe the differences and similarities between sample-and-hold and track-and-hold circuits.
- Describe the operation of weighted-resistor, R-2R ladder, counter, integrating, and successive approximation digital-to-analog converters.
- Explain the consequences of the Nyquist sampling theorem.

INTRODUCTION

A large class of electronic applications is concerned with the measurement and control of various physical processes. Such applications are almost exclusively realized using digital design techniques. Another huge electronics-related area that is going almost exclusively digital is the telecommunications industry. It is a simple fact that this digital revolution is a direct consequence of the incredible power, low cost, and availability of microprocessors and related hardware. But, of course, we all know that none of these devices would be practical without the glue that holds it all together: BJTs and FETs.

Quantum mechanics aside, the world of physical phenomena is an analog place and a computer that cannot interact with the outside world is just an expensive paperweight. To facilitate communication between digital systems and the analog world we must use digital-to-analog converters and analog-to-digital converters. In this chapter, we investigate the operation and applications of these interesting devices.

16.1 SOME BASIC CONCEPTS

Electronic circuits can be classified in many different ways. Using this book as an example, we have so far divided op amp circuits into linear and nonlinear categories. These are general descriptions that relate to the mathematical modeling of a given circuit's behavior. Circuits are also commonly described as being either analog or digital. Usually the distinction is simple: a full adder performs a digital operation, whereas a low-pass *RC* filter is analog. So far, we have concentrated on analog circuits, but here we enter a gray area where sometimes neither description fits perfectly. With this in mind, a few new concepts are now explained.

Note
The x-axis is the domain, or independent variable, while the y-axis is the range or dependent variable.

Continuous and Discrete Signals and Systems

A few examples of analog signals are sine waves, speech, music, and seismic waves, to name just a few. The important thing that all of these signals have in common is that they vary continuously in amplitude and time. Simply put, an analog signal is classified as having a continuous range and a continuous domain. The sine wave shown in Fig. 16.1(a) is a continuous-range, continuous-domain signal. Recall that the range of a signal describes its variation on the vertical y axis. The domain of a signal describes the x-axis independent variable.

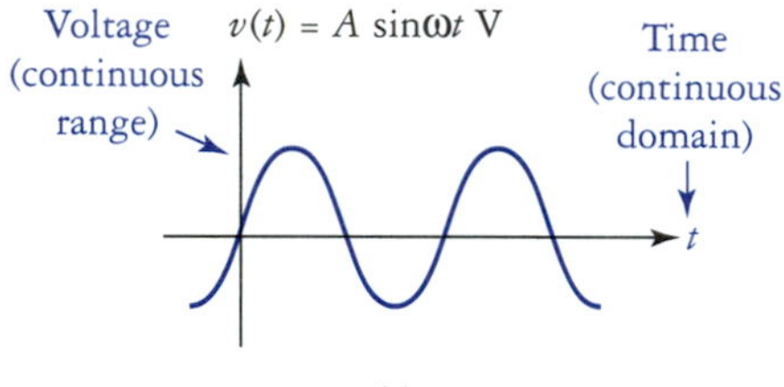

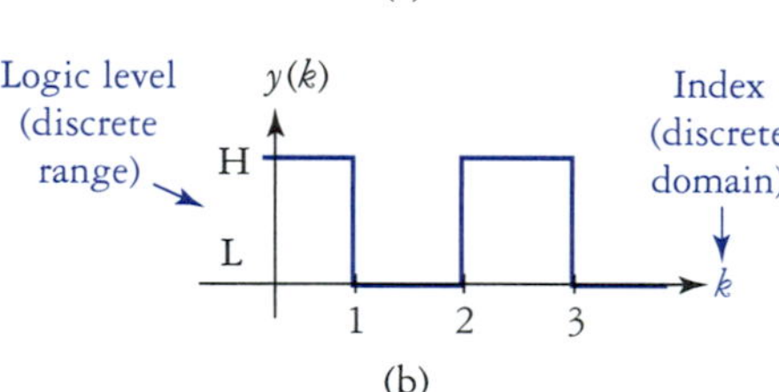

FIGURE 16.1 *Signal classifications. (a) Typical continuous-range, continuous-domain signal. (b) Typical discrete-range, discrete-domain signal.*

A digital signal has a discrete range and a discrete domain. That is, the range of an ideal digital signal consists of two discrete values: high or low. Assuming we are dealing with a synchronous (clocked) digital circuit, the transitions between high and low can only occur at discrete points in time, thus the digital signal has a discrete domain. In Fig. 16.1(b), k is an *index* that defines when the signal may switch states. The x-axis indices are normally uniformly spaced in most practical discrete-time systems. This is equivalent to saying that the clock frequency is constant. With these concepts in mind, we are now prepared to take a look at digital-to-analog converters.

16.2 DIGITAL-TO-ANALOG CONVERSION

A digital-to-analog converter (DAC) is a functional block that accepts a digital code word (usually 8-4-2-1 binary) as an input and produces an output voltage or current that is proportional to the numerical value of that code word. The block diagram symbol used to represent a DAC is shown in Fig. 16.2(a). As shown, most DACs have one output terminal, although occasionally differential outputs are available. In our discussions, we assume that the output of a DAC is a voltage, unless otherwise indicated. In general, a DAC may have any number of inputs n, but most often we find $n = 8, 10, 12$, or 16 bits. The number of bits n determines the number of discrete output voltage levels N that can be produced by the DAC. This relationship is

$$N = 2^n \tag{16.1}$$

Using this equation we find that an 8-bit DAC can produce 256 discrete output voltage levels, and a 16-bit converter can produce 65,536 different voltages. So, there are N possible input and output states. The maximum numerical value M of the binary input word is

$$M = 2^n - 1 \tag{16.2}$$

Thus, the maximum input count is always one less than the number of discrete input states.

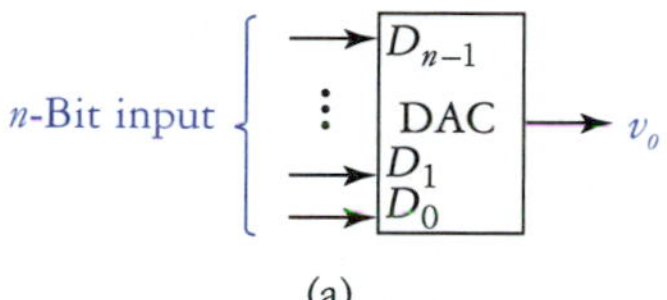

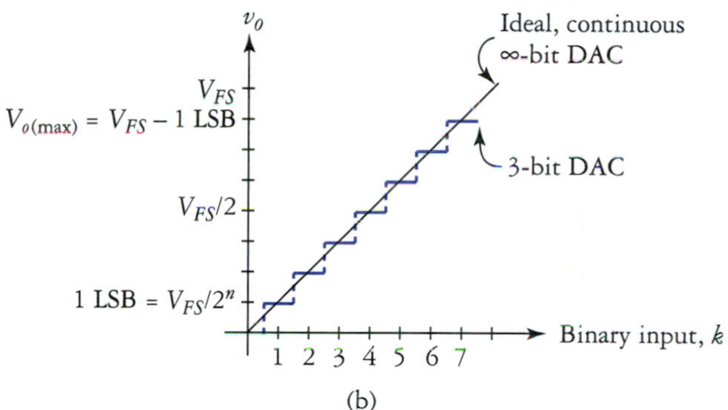

FIGURE 16.2
Digital-to-analog converter.
(a) Block diagram symbol.
(b) Transfer characteristic curve.

DAC Transfer Characteristics

The smallest intentional output change that can occur is caused by toggling (switching states) of the least significant bit (LSB) of the input code word. This change is called an LSB as well.

Note
The maximum output of a DAC approaches V_{FS} in the limit as the number of input bits increases.

Maximum output voltage $V_{o(\max)}$ occurs when all binary inputs to the DAC are high. The maximum output voltage is one LSB less than the *full-scale output voltage* V_{FS}. That is,

$$V_{o(\max)} = V_{FS} - 1 \text{ LSB} \tag{16.3}$$

The full-scale output voltage V_{FS} is the theoretical maximum output that would be produced by the DAC as the number of input bits approaches infinity. It is standard practice to set $V_{FS} = 10.00$ V. To tie these terms together, let's examine Fig. 16.2(b) more closely. This is the transfer characteristic for a 3-bit DAC. There are eight output voltage levels ($2^3 = 8$) represented by the treads of the staircase. Each of the tick marks on the vertical axis represents a 1-LSB change in v_o. The largest possible value of the index is $M = 2^3 - 1 = 7$.

Holding V_{FS} constant, if we increase the number of input bits to 4, the LSB size is cut in half while the number of steps is doubled. This causes the maximum output voltage to move closer to V_{FS}. Increasing the number of bits without limit causes the DAC stairstep transfer characteristic to approach the continuous, straight-line transfer characteristic in which $V_{o(\max)} = V_{FS}$, as shown by the dashed line in Fig. 16.2(b).

Resolution and Accuracy

When dealing with DACs it is important to have a good understanding of the concepts of resolution and accuracy. Accuracy refers to the degree of agreement between an expected and actual measurement. Resolution relates to how precisely the measurement is expressed. In other words, resolution is the same as precision.

Resolution

The resolution of a DAC is determined by the number of input bits available. There are several equivalent ways to express resolution. For example, resolution is often expressed as

$$\text{Resolution} = \frac{1}{2^n} \tag{16.4}$$

which is read "one part in 2^n," where, again, n is the number of bits. Resolution can also be expressed as a percentage:

$$\text{Resolution} = \frac{1}{2^n} \times 100\% \tag{16.5}$$

We can interpret percent resolution as telling us "The smallest possible change in the output voltage that can be caused by a change in the input word is x percent of the full-scale output."

EXAMPLE 16.1

A certain 8-bit DAC has $V_{FS} = 10.00$ V. Determine the LSB size, $V_{o(\max)}$, and the percent resolution of the converter.

Solution

$$\begin{aligned} 1\text{ LSB} &= \frac{V_{FS}}{2^n} \\ &= \frac{10.00\text{ V}}{256} \\ &\cong 39.1\text{ mV} \end{aligned}$$

$$\begin{aligned} V_{o(\max)} &= V_{FS} - 1\text{ LSB} \\ &= 10.00\text{ V} - 39.1\text{ mV} \\ &\cong 9.96\text{ V} \end{aligned}$$

$$\begin{aligned} \text{Resolution} &= \frac{1}{2^n} \times 100\% \\ &= \frac{1}{256} \times 100\% \\ &= 0.39\% \end{aligned}$$

Accuracy

Accuracy and resolution are related, but they are not the same thing. As we have seen, resolution is defined exactly by the number of input bits on the DAC. The resolution of a DAC can be used to define its accuracy. For example, an 8-bit DAC may indeed be accurate to 8 bits, but for various reasons, some 8-bit DACs may only have 7-bit or even 6-bit accuracy. Depending on the application, perhaps only 6- or 7-bit accuracy is required of an 8-bit DAC. A manufacturer will sell the less accurate parts for a lower price.

Like resolution, accuracy can be expressed in several different ways. Here, we are dealing with *absolute accuracy.* In order for an 8-bit DAC to have full 8-bit accuracy, its output voltage must be *at least* within ±1/2 LSB of the ideal, expected value for all input codes. Thus, we can define the accuracy of a DAC by the number of bits for which its output is within ±1/2 LSB of the expected value. A very popular family of DACs, which we cover later in detail, includes the DAC0808, DAC0807, and DAC0806 versions from National Semiconductor. These are all 8-bit DACs with identical pinouts. The last digit in the part number defines the accuracy of the DAC, in bits—that is the only difference between them.

Accuracy can also be expressed in terms of 1 LSB. For example, the DAC0808 has ±½-LSB accuracy. The DAC0807 has ±1-LSB accuracy and the DAC0806 is only accurate to ±4 LSB. Some manufacturers prefer to express overall accuracy in this form.

Some DACs are available that have accuracy of ±¼ LSB. Strictly speaking, it is only necessary for a given n-bit DAC to have ±½-LSB accuracy to have n-bit accuracy. An 8-bit DAC with ±¼-LSB accuracy has 8-bit precision with 9-bit accuracy.

Determination of Output Voltage

The expected output voltage produced by a DAC is found by multiplying the numerical value of the input word k by the value of 1 LSB. In equation form, this is written

$$V_o = k \times 1\ \text{LSB} \tag{16.6}$$

For our convenience, we would normally express k as a base-10 number, although in a digital system the value can be expressed in binary or hexadecimal (base-16) form.

The actual output voltage produced by a DAC should fall within a range determined by the accuracy of the converter, for example, $V_o \pm \frac{1}{2}$ LSB. Unless otherwise specified, we shall assume the output of a DAC is the ideal, expected value.

EXAMPLE 16.2

The DAC813x is a family of 12-bit DACs manufactured by Burr-Brown. The DAC813A is specified accurate to $\pm\frac{1}{4}$ LSB with $V_{FS} = 10$ V. Determine the number of output steps, 1 LSB, $V_{o(\text{max})}$, and the range of V_o given the binary input word $1000\ 0000\ 0000_2$.

Solution The number of output steps is

$$\begin{aligned} N &= 2^n \\ &= 2^{12} \\ &= 4096 \end{aligned}$$

The value of an LSB is

$$\begin{aligned} 1\ \text{LSB} &= V_{FS}/2^n \\ &= 10\ \text{V}/4096 \\ &\cong 2.441\ \text{mV} \end{aligned}$$

The maximum output voltage is

$$\begin{aligned} V_{o(\text{max})} &= V_{FS} - 1\ \text{LSB} \\ &= 10\ \text{V} - 2.441\ \text{mV} \\ &\cong 9.998\ \text{V} \end{aligned}$$

The expected output voltage is found by multiplying the numerical value of the input word by 1 LSB.

$$\begin{aligned} V_o &= k \times 1\ \text{LSB} \\ &= (2048)(2.441\ \text{mV}) \\ &= 5\ \text{V} \end{aligned}$$

The actual output voltage may deviate from the expected value by $\pm\frac{1}{4}$ LSB $\cong 0.61$ mV. Therefore, the actual output voltage should be within the range of $4.9994\ \text{V} \leq V_o \leq 5.00061$ V.

Other DAC Specifications

If there is one thing that can be said about DACs, it is that they certainly have a lot of specifications associated with them. Fortunately, we only have a few more major ones to discuss before we delve into the hardware side of this stuff.

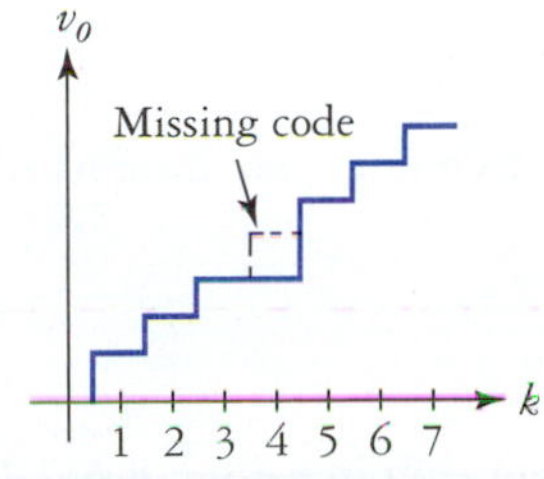

FIGURE 16.3
DAC transfer characteristic with a missing code.

Monotonicity and Missing Codes

A DAC is said to be *monotonic* if its output never decreases with increasing input values. A monotonic DAC will have steps of positive or zero height as the binary input is increased. In Fig. 16.2(b), both the ideal continuous transfer characteristic and the discrete, 3-bit transfer characteristic are monotonic.

A DAC may have *missing codes* and still be monotonic. A missing code is illustrated in Fig. 16.3. Any DAC that has at least $\pm\frac{1}{2}$-LSB accuracy will always be monotonic.

Offset, Gain, and Nonlinearity Error

An offset error is a constant shift of the output in either a positive or negative direction. DAC circuitry is normally designed such that offset adjustment is possible. In Fig. 16.4 we find that offset causes the transfer characteristic to be shifted by a constant amount above or below the expected

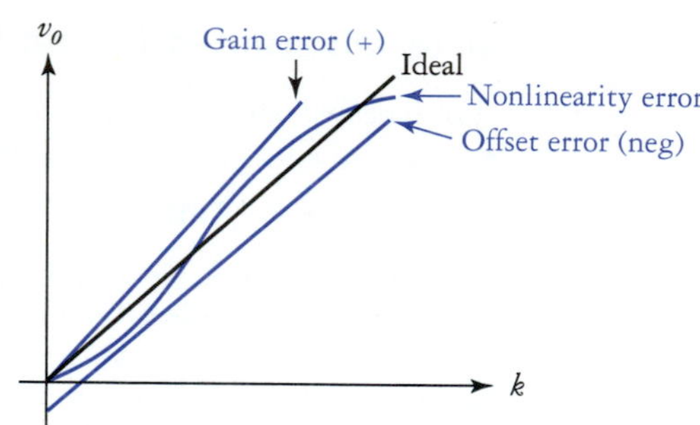

FIGURE 16.4
DAC transfer characteristic errors.

Note
Offset and gain errors are relatively easy to reduce, while nonlinearity errors are difficult to trim out.

curve. When an offset is present, the absolute error is constant, while the relative error is inversely proportional to the input word.

Gain error (sometimes called *scale error*) results in the characteristic curve being steeper or shallower than the expected curve. In other words, if the gain is too high, the steps are too tall. If the gain is low, the output steps will be smaller than expected. This is shown in Fig. 16.4. Gain error causes the absolute error to increase in proportion to the binary input while the relative error remains constant. Provisions for gain adjustment are usually incorporated into DAC designs.

The DAC transfer characteristic is inherently nonlinear, but variations in gain (step size) with input can cause the best-fit continuous curve to deviate from a straight line as shown in Fig. 16.4.

Settling Time

Depending on the application, a given DAC may be required to respond to rapidly changing input code words. It is important then to know the settling time of the DAC. The *settling time* is the time required for the DAC to respond to within a specified range of the final output voltage (usually ±½ LSB) given an input code change from all zeros to all ones or vice versa. As shown in Fig. 16.5, the output may overshoot, depending on the damping of the system.

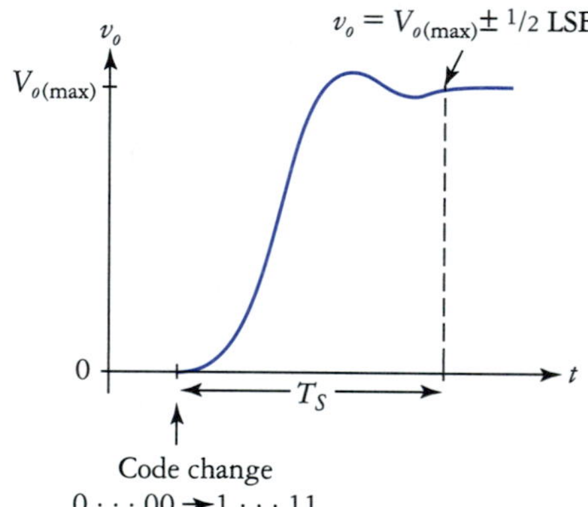

FIGURE 16.5
DAC settling time.

Coding Schemes

Nearly all DACs accept some form of binary code as input. These coding techniques are introduced below.

Natural Binary

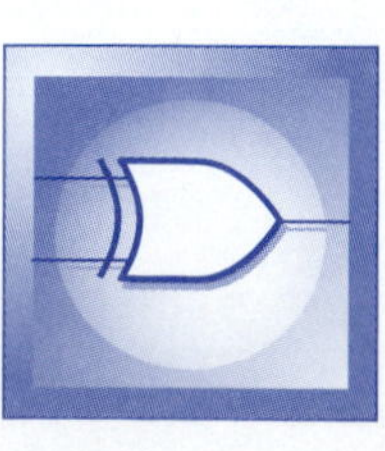

The transfer characteristic in Fig. 16.2 assumes that the DAC accepts input data in natural 8-4-2-1 binary form. This is a very common coding scheme and is used throughout the remainder of the chapter unless otherwise noted.

Offset Binary

Offset binary has the same bit pattern sequence as natural binary, while the output of the DAC is offset to −½ scale as shown in Fig. 16.6. This allows positive and negative output voltages to be produced.

FIGURE 16.6
Offset binary DAC transfer characteristic.

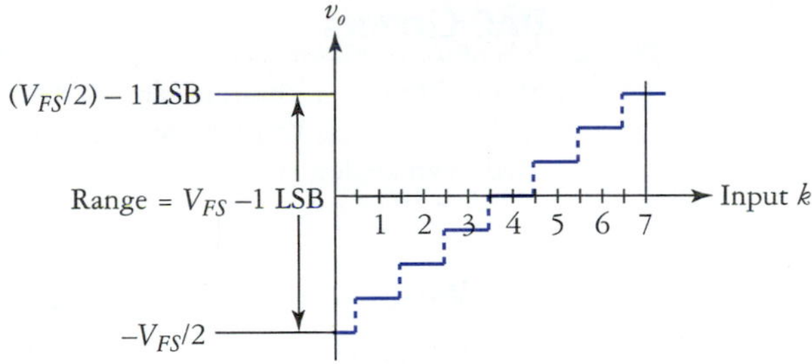

Two's-Complement Binary

Note
Two's complement is the most common system used to represent signed numbers in digital and microprocessor-based systems.

In this coding scheme, the binary code words are applied to the DAC in two's-complement form. In two's-complement form, the MSB is the sign bit. If the MSB is high, the value of the word is negative. If the MSB = 0, the code word is either positive or zero (zero is neither positive or negative). It is most common for the output of the DAC to be shifted to $-\frac{1}{2}\ V_{FS}$ in order to allow a bipolar output voltage to be produced. The transfer characteristic for a 3-bit, two's-complement DAC is shown in Fig. 16.7.

Given an n-bit binary word W, the two's-complement of that code word W^* is given by

$$W^* = \overline{W} + 1 \tag{16.7}$$

Using a 3-bit word as an example, the two's complement representation for −3 is found to be

$$|W| = 3 = 011_2$$

$$\begin{aligned} W^* &= \overline{W} + 1 \\ &= 100 + 1 \\ &= 101 \end{aligned}$$

which agrees with the table in Fig. 16.7. The two's-complement representation is popular because it simplifies assembly language programming and interfacing with microcomputer-based systems.

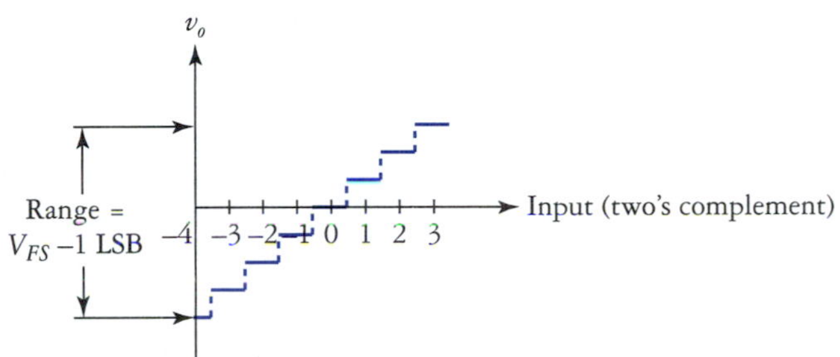

Input Code	Value	Output
100	−4	$-V_{FS}/2$
101	−3	$-3V_{FS}/8$
110	−2	$-V_{FS}/4$
111	−1	$-V_{FS}/8$
000	0	0V
001	1	$V_{FS}/8$
010	2	$V_{FS}/4$
011	3	$3V_{FS}/8$

FIGURE 16.7
Two's-complement binary DAC transfer characteristic.

DAC Circuits

Many different techniques are used to implement DACs. Most implementations use op amps and analog switches. You may wish to refer back to Chapter 6 for a quick review of how a JFET can be used as an analog switch.

The Weighted-Resistor DAC

A simple DAC can be designed using a summing amplifier with binarily weighted input resistors. Two versions of a 4-bit, weighted-resistor DAC are shown in Fig. 16.8. The circuit in Fig. 16.8(a) uses SPST analog switches that are open or closed according to a digital code word applied to $D_3 - D_0$. In this arrangement if $D_k = 1$ means S_k is open, then $D_k = 0$ means S_k is closed.

It can be shown using standard op amp analysis techniques that if the input resistors are weighted according to the weights of the binary bit positions, the output voltage will be similarly scaled. The output voltage is given by

$$V_o = -V_{\text{REF}} R_F \left(\frac{D_3}{R} + \frac{D_2}{2R} + \frac{D_1}{4R} + \frac{D_0}{8R} \right) \tag{16.8}$$

where $D_k = 1$ or 0.

FIGURE 16.8
Weighted-resistor DACs. (a) Using SPST switches. (b) Using SPDT switches.

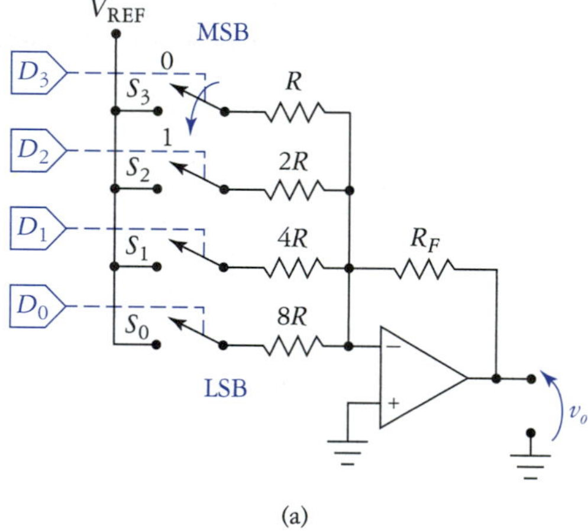

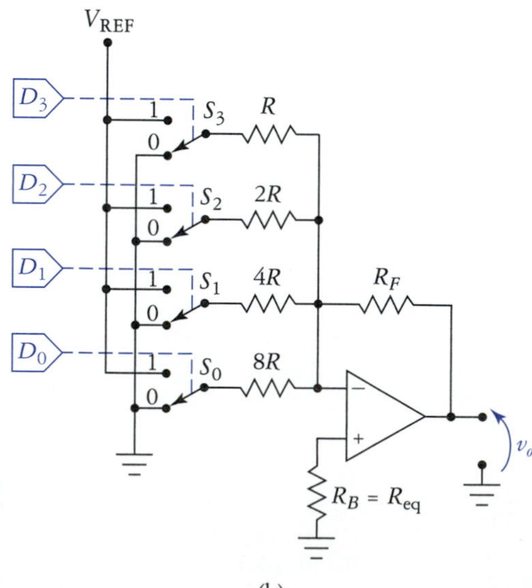

The full-scale output voltage is given by

$$V_{FS} = -2V_{\text{REF}}\frac{R_F}{R} \tag{16.9}$$

If a positive output voltage is required then either V_{REF} must be negative or a second inverting stage must be used. The LSB size is given by

$$1 \text{ LSB} = \frac{V_{FS}}{2^n} \tag{16.10}$$

where n is the number of input bits. The maximum output voltage (assuming positive output) is given by Eq. (16.3): $V_{o(\max)} = V_{FS} - 1$ LSB.

A few characteristics of the weighted-resistor DAC of Fig. 16.8(a) make it impractical for more than 4- or 5-bit input words. Notice that as the analog switches open and close, the equivalent inverting input resistance changes. With all switches open, we have

$$R_{\text{eq}} = R_F \tag{16.11}$$

With all switches closed, we have

$$R_{\text{eq}} = R_F \parallel \frac{1}{\dfrac{1}{R} + \dfrac{1}{2R} + \dfrac{1}{4R} + \dfrac{1}{8R}} \tag{16.12}$$

This resistance change causes a variation in bias-current-induced input offset voltage, which will affect the output voltage. As a compromise, we could use a bias current compensation resistor of $R_B = R_F$, but there is another solution. By using SPDT switches, as shown in Fig. 16.8(b), the equivalent inverting input resistance is constant, given by Eq. (16.12). We would then set $R_B = R_{\text{eq}}$ for bias current offset compensation. The disadvantage to this approach is the added complexity of implementing SPDT analog switches.

Even using the SPDT switches, the weighted-resistor DAC is limited to use with 6- or 7-bit input bits. The reason is that because such a wide range of resistor values is required, monolithic IC fabrication techniques cannot be used for their implementation. Even if hybrid or discrete resistors are used, the range of values is so wide that resistor composition would have to be varied. This in turn would cause temperature tracking characteristics to be very poor, resulting in output drift.

The R-2R Ladder DAC

An R-2R ladder DAC is shown in Fig. 16.9. The ladder-like appearance of the input voltage-divider network is obvious. This DAC structure is very popular because it uses only resistors that vary over a 2-to-1 range. The absolute value of the ladder resistors is not that important, but maintaining the correct 2-to-1 ratio is. Fortunately, such matching is accomplished with high accuracy in both monolithic and hybrid IC technologies. The R-2R ladder DAC can be expanded to accept arbitrarily wide input words with the simple addition of more ladder sections and SPDT analog switches.

FIGURE 16.9
The R-2R ladder DAC.

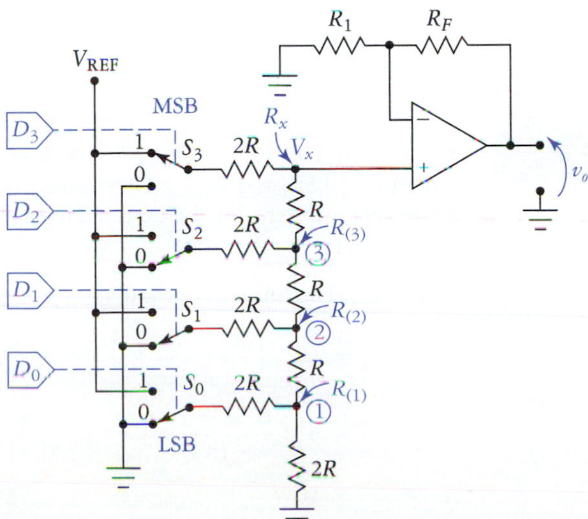

The analysis of the ladder section of the DAC can be tedious, but a sample calculation should serve to illustrate a few interesting points. As shown in Fig. 16.9, the input is 1000_2. Starting at the bottom of the ladder, we find that with S_0 in the ground position, the resistance looking into node (1) is

$$R_{(1)} = 2R \parallel 2R = R$$

Looking down into node (2), with S_1 set to ground, the resistance is

$$R_{(2)} = [(2R \parallel 2R) + R] \parallel 2R = R$$

Similarly, for node (3) we have $R_{(3)} = R$; and, finally, looking from the left into node (x), we have $R_{(x)} = 2R$.

Now, since S_3 is connected to the reference source, the voltage applied to the op amp is

$$V_x = V_{\text{REF}} \frac{2R}{2R + 2R}$$

$$= \frac{V_{\text{REF}}}{2}$$

It can be shown that for the other various combinations of switch settings that V_x varies in proportion to the binary weighting of the input bits. For example, if the input data word is 0100_2 then we find $V_x = V_{\text{REF}}/4$, and so on.

Resistors R_F and R_1 are selected to scale the output for the desired value of V_{FS}. The general output voltage expression is

$$V_o = V_{\text{REF}} \left(\frac{R_1 + R_F}{R_1} \right) \left(\frac{D_3}{2} + \frac{D_2}{4} + \frac{D_1}{8} + \frac{D_0}{16} \right) \tag{16.13}$$

The full-scale output voltage is the first two factors of the output voltage equation. That is,

$$V_{FS} = V_{\text{REF}} \left(\frac{R_1 + R_F}{R_1} \right) \tag{16.14}$$

It is common practice to set the op amp to unity gain, where we have

$$V_{FS} \bigg|_{A_v = 1} = V_{\text{REF}} \tag{16.15}$$

Equation (16.13) was derived specifically for the DAC of Fig. 16.9. If we know the value of an LSB and V_{FS} for a given DAC, the general expression presented in Eq. (16.6) can be used. For convenience, this equation is repeated here:

$$V_o = k \times 1 \text{ LSB}$$

where k is the decimal value of the input code word.

EXAMPLE 16.3

The DAC in Fig. 16.9 has $V_{\text{REF}} = 5$ V, $R = 10$ kΩ, $2R = 20$ kΩ, and $R_F = R_1 = 10$ kΩ. Determine V_{FS}, 1 LSB, and $V_{o(\text{max})}$. Determine the maximum deviation of the output voltage for full 4-bit accuracy. Determine the expected output voltage given the following input words:
(a) 0011_2 (b) 1001_2 (c) 1111_2

Solution The full-scale output voltage is found using Eq. (16.14):

$$V_{FS} = V_{\text{REF}} \left(\frac{R_1 + R_F}{R_1} \right)$$

$$= 5 \text{ V} \frac{10 \text{ k} + 10 \text{ k}}{10 \text{ k}}$$

$$= 10 \text{ V}$$

The LSB size is

$$1\text{ LSB} = \frac{V_{FS}}{2^n} = \frac{10\text{ V}}{16} = 0.625\text{ V}$$

The maximum output voltage is

$$V_{o(\max)} = V_{FS} - 1\text{ LSB} = 10\text{ V} - 0.625\text{ V} = 9.375\text{ V}$$

For full 4-bit accuracy, the output must not deviate more than ±½ LSB, which is $\Delta V_{(\max)} = \pm\, 0.3125$ V. The expected output voltages are given by $k \times 1$ LSB, which gives us

(a) $3 \times 0.625\text{ V} = 1.875\text{ V}$
(b) $9 \times 0.625\text{ V} = 5.625\text{ V}$
(c) $15 \times 0.625 = 9.375\text{ V}$

Monolithic Digital-to-Analog Converters

Many different monolithic DACs are available. A popular 8-bit DAC is the National Semiconductor DAC0808. The DAC0808 pin designations and functional symbol are shown in Fig. 16.10. Supply voltage for the DAC0808 ranges from ±4.5 to ±18 V. Typically, $V_{EE} = -15$ V and V_{CC} is +5 V, which causes the digital inputs to be TTL compatible allowing for easy interfacing to a microprocessor or other digital circuit. Settling time for the DAC is 150 ns. Capacitor C_1 helps prevent oscillation and is typically set to around 0.01 μF. The reference voltage sets the full-scale output current as

$$I_{FS} = \frac{V_{\text{REF}}}{R_1} \tag{16.16}$$

with the limitation of $I_{FS} \leq 1.9$ mA. Often, the reference voltage is taken directly from the +5-V logic supply. To minimize offset error, we set $R_2 = R_1$. The DAC0808's output sinks current in proportion to the value of the applied binary input word. The output current is given by

$$I_o = \frac{V_{\text{REF}}}{R_1}\left(\frac{A_1}{2} + \frac{A_2}{4} + \ldots + \frac{A_8}{256}\right) \tag{16.17}$$

where $A_k = 0$ or 1. Note that National Semiconductor labels the digital input lines A_1 through A_8. This can be confusing when the DAC is interfaced to microprocessor-based equipment where lines labeled A_n are usually part of the address bus. Standard 8-bit data-bus designations are shown external to the DAC in Fig. 16.10, where the LSB has the subscript zero. National Semiconductor's digital input numbering system does, however, allow the output current to be expressed in summation form as

$$I_o = \frac{V_{\text{REF}}}{R_1} \sum_{k=1}^{8} \frac{A_k}{2^k} \tag{16.18}$$

Most often, we prefer the output of a DAC to be a voltage, so a load resistor can be used as shown in Fig. 16.10 to accomplish this conversion. The output voltage expression is

$$V_o = -\frac{V_{\text{REF}} R_L}{R_1}\left(\frac{A_1}{2} + \frac{A_2}{4} + \ldots + \frac{A_8}{256}\right) \tag{16.19}$$

or, equivalently,

$$V_o = -\frac{V_{\text{REF}} R_L}{R_1} \sum_{k=1}^{8} \frac{A_k}{2^k} \tag{16.20}$$

Unfortunately, this produces $V_o \leq 0$ V, while we usually prefer a positive output.

FIGURE 16.10
Typical DAC0808 circuit.

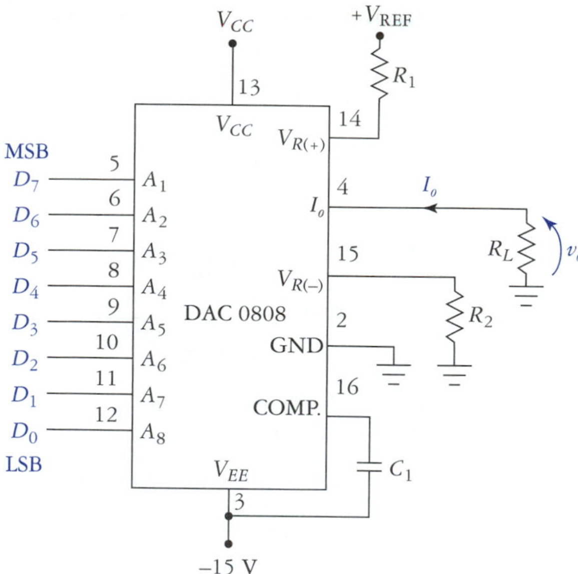

The addition of an op amp I/V converter (transresistance amplifier) allows the output of the DAC0808 to be converted into a positive-going voltage. Such a modification is shown in Fig. 16.11. The output voltage is given by

$$V_o = \frac{V_{REF}R_F}{R_1}\left(\frac{A_1}{2} + \frac{A_2}{4} + \ldots + \frac{A_8}{256}\right) \tag{16.21}$$

Offset compensation is provided by setting $R_B = R_F$ and via adjustment of the 10-kΩ potentiometer.

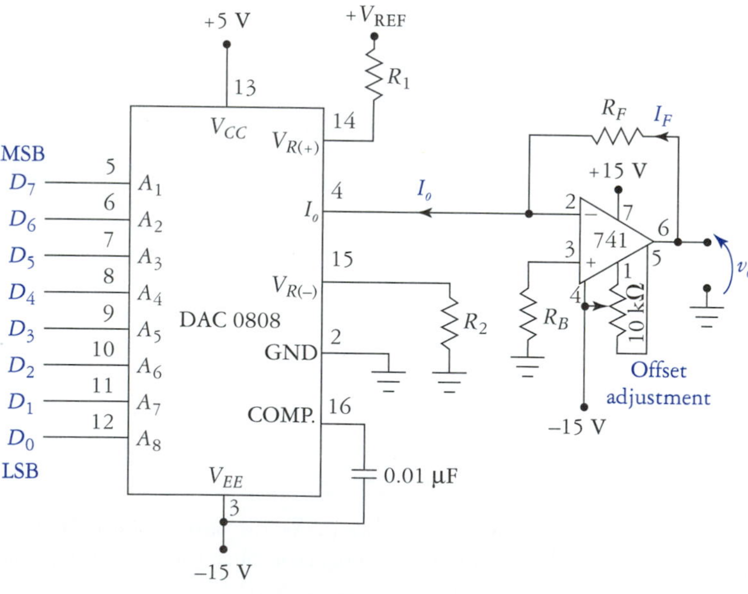

FIGURE 16.11
DAC0808 with I/V converter.

EXAMPLE 16.4

Refer to Fig. 16.11. Given $V_{REF} = 5$ V, determine the required values for R_1, R_2, R_F, and R_B such that $I_{FS} = 1$ mA and $V_{FS} = 10$ V. Determine the maximum rate at which data can be sent to the converter, assuming the 741 has $SR = 0.5$ V/μs.

Solution Starting at the DAC0808, solve Eq. (16.16) for R_1, and assign $R_2 = R_1$, giving us

$$R_2 = R_1 = \frac{V_{\text{REF}}}{I_{FS}} = \frac{5\text{ V}}{1\text{ mA}} = 5\text{ k}\Omega$$

Examination of Eq. (16.21) reveals that the full-scale output voltage is given by

$$V_{FS} = \frac{V_{\text{REF}} R_F}{R_1} = I_{FS} R_F$$

Therefore, the required value for R_F is

$$R_F = \frac{V_{FS}}{I_{FS}} = \frac{10\text{ V}}{1\text{ mA}} = 10\text{ k}\Omega$$

For minimum offset, we have $R_B = R_F = 10\text{ k}\Omega$.

The settling time for the DAC0808 is 150 ns, which is many times faster than the 741 and so contributes negligible delay here. In terms of output time delay, the worst-case situation occurs when the op amp must slew from zero to V_{FS}, or vice versa. Thus, the maximum change in V_o is $\Delta V_{o(\text{max})} = 10$ V. Recall that slew rate is defined as

$$\left.\frac{dv_o}{dt}\right|_{\text{max}} = \left.\frac{\Delta V_o}{\Delta t}\right|_{\text{max}}$$

Solving for $\Delta t_{(\text{max})}$ gives us the worst-case time delay:

$$\Delta t_{(\text{max})} = \frac{\Delta V_o}{SR} = \frac{10\text{ V}}{500\text{ kV/sec}} = 2\ \mu\text{s}$$

This corresponds to a maximum frequency of

$$f_{\text{max}} = 1/\Delta t_{\text{max}} = 1/20\ \mu\text{s} = 50\text{ kHz (i.e., 50,000 words/sec)}$$

This is a theoretical maximum data rate. To be on the safe side, we would probably cut this data rate in half to $f_{\text{max}} = 25$ kHz.

A Few DAC Applications

There are many uses for DACs, and several possibilities are explored here.

Servomechanisms

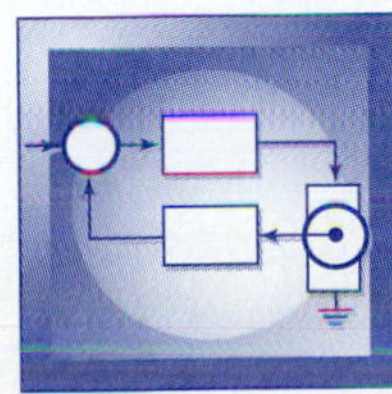

A servomechanism is an electromechanical closed-loop system in which a physical quantity, such as motor speed or position, is held to some preset value determined by a command or reference signal. The reference signal may be generated in a number of different ways, but here we will use a DAC to generate this command. Assuming that a computer or microprocessor development system has a latched output port available, we can control the speed of a dc motor as shown in Fig. 16.12(a). Here, we are simply using the DAC to provide a reference voltage for the servo. The actual closed-loop control system is formed by the differential amplifier (called an *error amplifier* or *error detector* here). The error detector amplifies a difference between the reference voltage and the feedback voltage generated

FIGURE 16.12
Servomechanisms. (a) Using a DAC to set the speed of a motor. (b) Positional control.

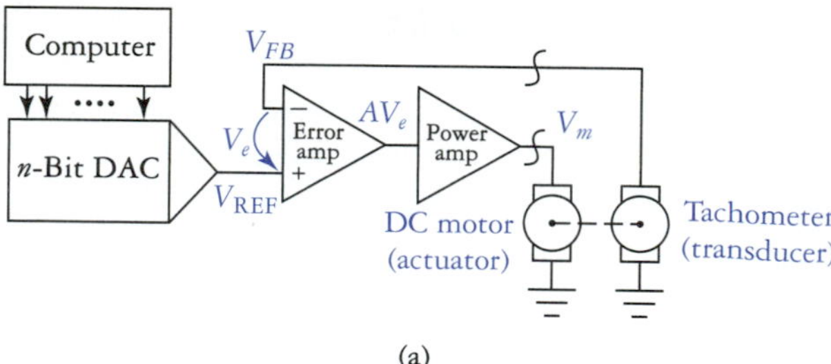

(a)

V_{FB}
V_m
+V
DC motor (actuator)
Potentiometer (transducer)
–V

(b)

Note
Servos are used in applications ranging from CD players to fly-by-wire systems in fighter aircraft such as the F-16 and F/A-18.

by the tachometer, which is the motor speed transducer. The motor is called the *actuator.* The controlled quantity (motor speed) is called the *plant,* in control system terminology. If a load is placed on the motor, it tends to slow down, reducing the feedback voltage below V_{REF}. This difference is amplified and used to increase the motor drive to compensate for the increased load.

If we replace the tachometer with a potentiometer, as shown in Fig. 16.12(b), then the position of the motor is proportional to the reference signal output by the DAC. Normally, the motor is connected to the position-sensing potentiometer via a speed-reducing transmission or gear train. We examine such control systems again in this chapter, but for now realize that the computer is not directly in the feedback loop of the servo as shown. We can place the computer in the loop using analog-to-digital converters, which are covered in the next section.

Arbitrary Function Generators

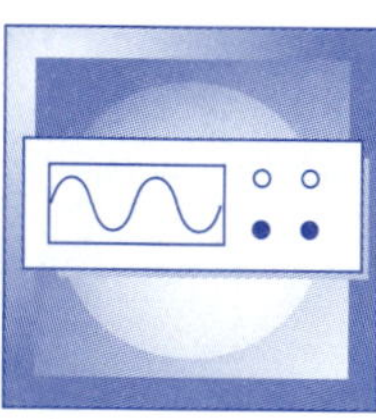

Most readers of this book will be very familiar with the typical laboratory function generator. The basic function generator is a signal source that is usually used to provide sine, triangular, and pulse waveforms of varying amplitude and frequency. The advent of fast digital circuitry and DACs has made a more flexible piece of test equipment called an arbitrary function generator possible.

The block diagram for a hypothetical arbitrary function generator is shown in Fig. 16.13(a). In this example, the microcomputer is used to generate sequences of 8-bit binary numbers that are latched and translated into voltage levels by the DAC. The reconstruction filter smooths the stairstep jumps of the output signal. These small jumps produce unwanted frequency components called *quantization noise.* We would design this LP filter such that the corner frequency is lower than the step (sample) rate, but higher than the frequency of the signal being synthesized. The process is shown pictorially in Fig. 16.13(b) for a simple linearly increasing count sequence. Because this is an 8-bit converter, it will take 256 sample periods for the count to roll over, producing a sawtooth waveform.

The computer can generate the waveform sequence in one of two ways: (1) computation on the fly or (2) lookup table. To synthesize a sine wave for example, we might use the BASIC program shown in Listing 16.1.

LISTING 16.1

```
start:
for theta = 0 to 6.28319 step .02454
    y = sin(theta)
    n = 127*theta + 128
    output(port,n)
next theta
goto start
```

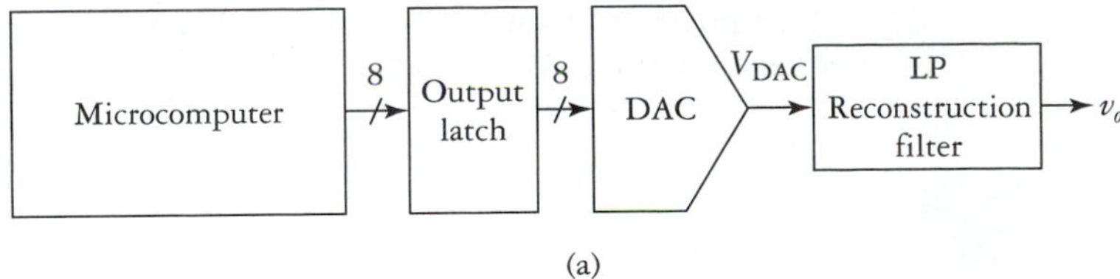

(a)

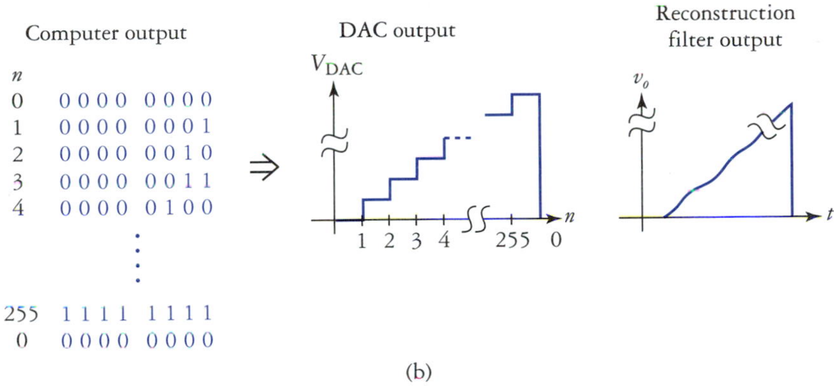

(b)

FIGURE 16.13
Microprocessor-based waveform synthesis. (a) System block diagram. (b) Output code word sequence, DAC response, and filtered DAC response.

This program increments *theta* (in radians) from 0 to 2π in 256 equal steps and calculates sin θ at each step. The offset of 128 is required for a unipolar DAC so that the output is half of full scale at zero crossing of the sine wave. The sine of theta is scaled by 127 to utilize the full range of the 8-bit DAC. The filtered output would appear as shown in Fig. 16.14.

FIGURE 16.14
Sine wave produced by program in Listing 16.1.

We can trade resolution for speed by increasing the step size. For example, doubling the step size roughly doubles the frequency of the sine wave but it also cuts the number of steps used to approximate the waveform in half. This makes filtering out the quantization noise more difficult. The main disadvantage of calculating the waveform values on the fly is that this method is relatively slow.

The speed of execution and hence the frequency of the output waveform can be increased greatly if the program is written in assembly language. The more complex calculations such as trigonometric functions, may still slow us down however. We can really speed the program up by using a *lookup table.* In this approach, the values of the sine function are calculated ahead of time and placed in a table in memory. Now, to generate the correct sequence, we simply write a program that goes down the list, outputting each number to the DAC as it goes. The process is repeated when the end of the list is reached. The beauty of this approach is that any arbitrary sequence of numbers can be stored in a lookup table, which may be as long or short as we please. We can easily switch from waveform to waveform by using different lookup tables as well. The precision of the waveform approximation can be traded for speed by skipping lookup table entries.

A program that will produce the lookup table data for a 16-step approximation of a sine wave is shown in Listing 16.2. This program prints out the hexadecimal numbers for each step.

LISTING 16.2

```
for theta = 0 to 6.28319 step .393
  y = sin(theta)
  n = 127*n + 128
  print hex$(n);
  print " ";
next theta
```

The resulting sequence is 80, B1, DA, F5, FF, F5, DA, B0, 80, 4F, 26, 0B, 01, 0B, 27, 50. Let's assume these numbers are stored in memory in a 6800-based microprocessor trainer at addresses h0100–h010F. Let us also assume that a DAC is located at an output port at address hF000. A program that will generate the sine wave is shown in Listing 16.3. This program includes a delay loop that can be used to reduce the frequency of the resulting waveform.

LISTING 16.3

```
0000 LDX, 0100          ;Set IX to start of lookup table
0003 LDAA, [IX + 00]    ;Load A, indexed-no offset, with table entry
0005 STAA, F000         ;Write byte to DAC
0008 LDAB, xx           ;Start delay loop, xx = delay count (01 − FF)
000A DECB               ;B = B − 1
000B BNE, FD            ;If B≠0 then go back to 000A
000D CPX, 010F          ;IX pointing to last table entry?
0010 BEQ, EE            ;If yes, go back to 0000 and start again
0012 INX                ;Otherwise, IX = IX + 1
0013 BRA, EE            ;Go back to 0003 and get next value from table
```

The unfiltered output that would be produced by this function generator is shown in Fig. 16.15.

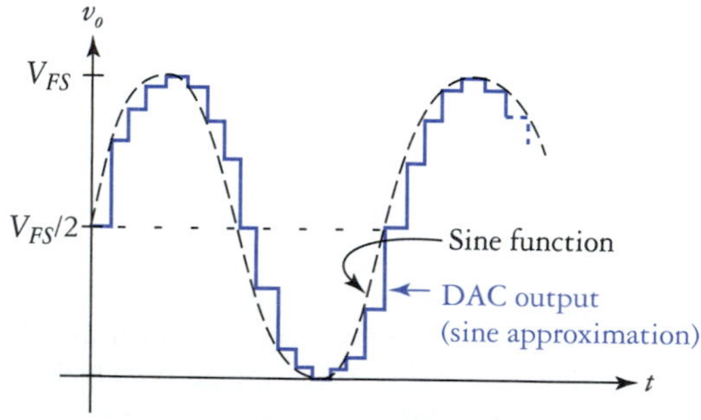

FIGURE 16.15
Sine wave approximation generated by Listing 16.3.

As an alternative to the microprocessor-controlled arbitrary function generator, a circuit like that of Fig. 16.16 could be used. In this design, the lookup table data are stored in an EPROM (erasable, programmable, read-only memory). A variable frequency clock, switchable over three decades, is applied to a 5-bit up-counter comprised of the 74LS193 and 74LS76. The resulting repetitive binary count is used to access sequential address locations in the EPROM. Each lookup table contains 32 bytes of data. Switches S_2 and S_3 are used to select one of four separate tables. More lookup tables can be used by switching EPROM address line A_7–A_9 if desired. The output sequence is applied to the 8-bit DAC, which is then low-pass filtered and ac coupled to a 25-kΩ

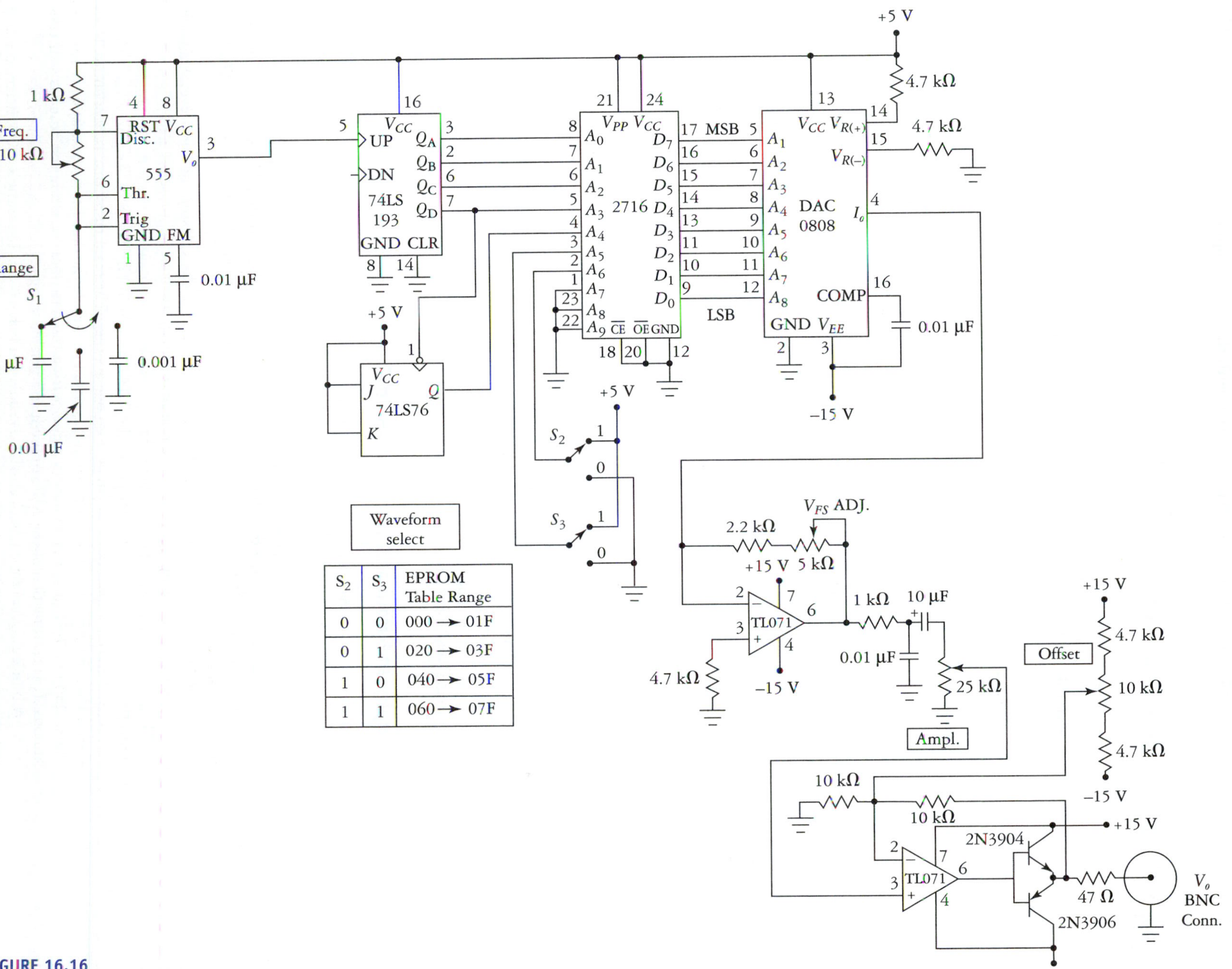

S_2	S_3	EPROM Table Range
0	0	000 → 01F
0	1	020 → 03F
1	0	040 → 05F
1	1	060 → 07F

FIGURE 16.16
Lookup table function generator.

potentiometer, configured as a variable voltage divider. The output of the divider is applied to a noninverting amplifier using discrete transistors to boost output drive capability. Waveform offset is adjustable, though there is some interaction with the signal amplitude.

16.3 ANALOG-TO-DIGITAL CONVERSION

An analog-to-digital converter (ADC) is a functional block that accepts an analog input (usually a voltage) and produces an n-bit binary number that is proportional to that voltage. An ADC is shown symbolically in Fig. 16.17. Typically, the ADC will sample the input signal at a constant rate, called the *sample rate,* which is determined by how often the START command is asserted. The clock signal can be generated externally as shown or internally by the ADC itself. The clock frequency determines how long it takes to perform a conversion. When a conversion is complete, the ADC asserts the EOC (end of conversion) signal. This indicates that the output is a valid binary code word. Once START is asserted, the time required for the ADC to produce a valid output and assert the EOC signal is called the *conversion time* T_C.

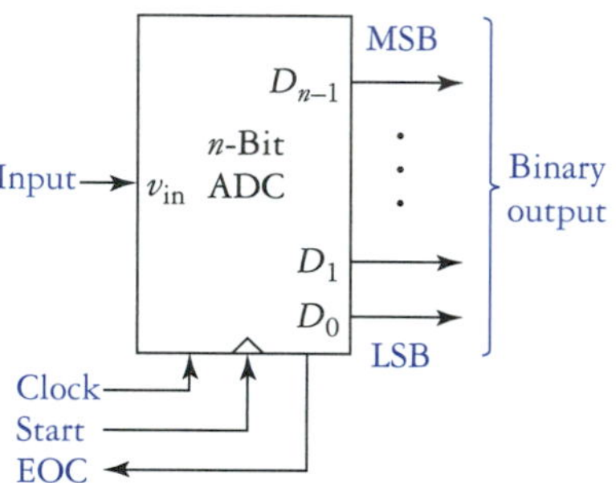

FIGURE 16.17
Block diagram symbol for an analog-to-digital converter.

In general, the input to the ADC is a continuous-domain (time), continuous-range (voltage) analog signal. The range of the signal is restricted to the full-scale range (FSR) of the ADC. Typically, the FSR is from 0 to 10 V for a unipolar ADC, or ±5 V for a bipolar unit. The ADC quantizes the input signal. That is, the ADC produces one of 2^n possible numerical values at its output, in response to a voltage within its FSR. The domain of the input signal is continuous, but the ADC produces a sequence of output codes at discrete time intervals. Thus, the ADC output has a discrete-range (binary integer) and a discrete-domain (sample interval).

Sample-and-Hold and Track-and-Hold Circuits

As stated earlier, an ADC cannot perform a conversion instantaneously. In many situations, we must ensure that the input to the ADC remains constant during the conversion interval. This is accomplished by using either a sample-and-hold (S/H) or a track-and-hold (T/H) circuit to preprocess the input voltage. The basic circuit for a S/H or T/H is shown in Fig. 16.18(a). An analog switch is used to connect the analog signal to buffer A_1. When the switch is closed, the *hold capacitor* C_H is quickly charged to the voltage level at the input. Buffer A_2 prevents the capacitor from being discharged by the input of the ADC.

Whether we have a S/H or a T/H depends on the sample clock. Using the logic levels shown in the illustration, if the sample clock is a train of impulses (pulse width approaches zero), then we have a S/H. The output of the S/H v'_{in} is shown in Fig. 16.18(b). If the sample clock is a rectangular pulse train, then a T/H is formed. The output of the T/H is shown in Fig. 16.18(c). As you can see, the resulting waveforms appear quite different from one another. The main thing that is happening, however, is that the domain of the input signal is continuous, while the domain of the output is discrete. Thus, a S/H or T/H produces an output signal with a discrete domain, while the range is still continuous. It is the function of the ADC to produce a discrete range.

A block diagram for a practical ADC is shown in Fig. 16.19. The sample clock drives the S/H and also asserts the START command to the ADC. We do not need to synchronize the sample clock with the EOC signal if the conversion time is short compared to the sample time. This could be accomplished by setting $f_{ADC} \gg f_{sa}$. However, EOC could be used to latch the output of the ADC so that a valid code is held in between conversions.

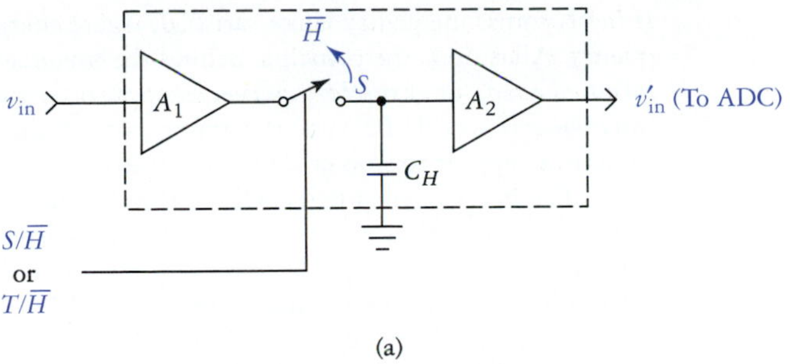

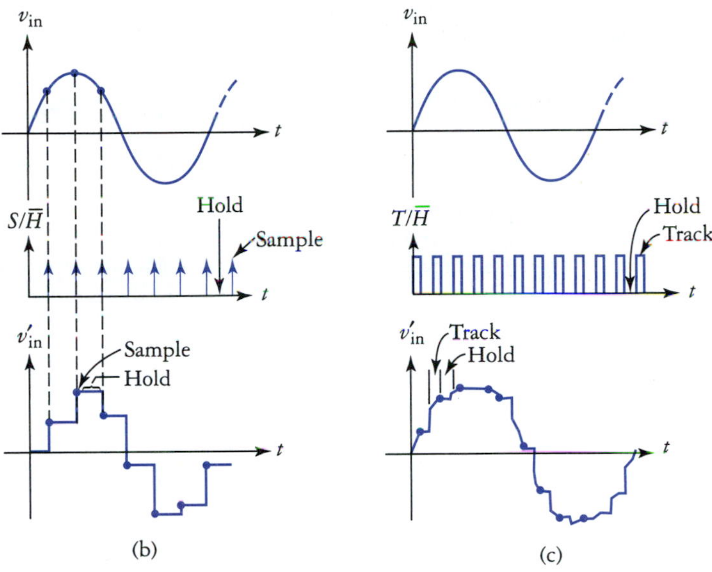

FIGURE 16.18
Sample-and-hold or track-and-hold circuit. (a) Typical circuit. (b) Clocking the switch with a narrow pulse (impulse) produces a T/H response. (c) Clocking with a rectangular pulse train produces a T/H response.

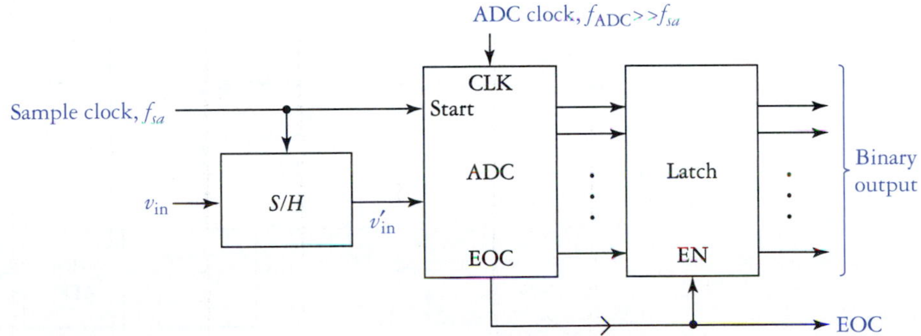

FIGURE 16.19
Practical ADC system.

Sample Rate and Aliasing

When we use an ADC to convert a voltage into a numerical code, we are *quantizing* or *digitizing* the signal. If we wish to digitize a time-varying signal, without losing information about the frequency content of that signal, the sample frequency f_{sa} must be at least twice the highest frequency component present in the input signal spectrum. This is called the *Nyquist sampling theorem,* which is written

$$f_{sa} \geq 2f_{in(max)} \tag{16.22}$$

If the Nyquist inequality is not satisfied, signal energy will be folded back, or *aliased,* to a lower frequency. Aliasing is the principle behind the use of a strobe light to reduce the apparent speed of rotation of a fan, for example. Another example occurs when a motion picture is made of a rapidly spinning wagon wheel. Because the frame rate of the film is not sufficiently high, the wheel may sometimes appear to stop or rotate backward.

The Nyquist inequality tells us the absolute minimum sample frequency required to prevent aliasing. In a practical system, the sample rate is usually set somewhat higher than the Nyquist limit. This is called *oversampling.* The graphs in Fig. 16.20 illustrate these concepts. The left side plots show the result of oversampling a sine wave by $f_{sa} = 10f_{\text{in}}$. The lower plot of the sampled output clearly reflects the characteristics of the original input signal. If we increase the frequency of the input voltage by a factor of 10.1 and maintain the same sample frequency, then we are severely undersampling the signal. This produces the aliased output shown on the right side of Fig. 16.20. The signal has been aliased to a much lower apparent frequency.

Note

Most CD players oversample at slightly higher than twice the maximum audio frequency to be reproduced. That is, $f_{\max} =$ 20 kHz and $f_{\text{sa}} = 44$ kHz.

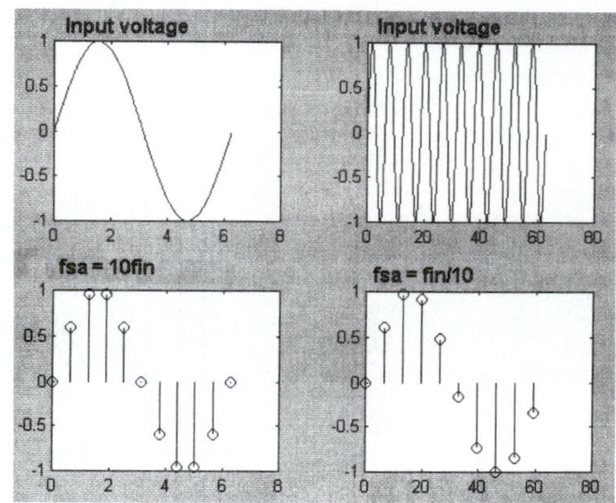

FIGURE 16.20

Effects of sample rate. (a) Sample frequency 10 times signal frequency allows accurate reconstruction. (b) Sample frequency less than the Nyquist limit produces an alias of the original signal.

Once out-of-band signal energy has been aliased to a lower frequency it cannot be distinguished from the desired information signal energy that was originally in the Nyquist range of the system. To help prevent this from occurring, an analog anti-aliasing filter can be placed in front of the sampled data system. This filter *band limits* the input signal. The anti-aliasing filter is designed to provide some minimum amount of attenuation at $f_{sa}/2$ while having minimum effect on the in-band frequency components. The higher the amount of oversampling, the easier it is to implement the anti-aliasing filter. A block diagram of a data acquisition system front end is shown in Fig. 16.21.

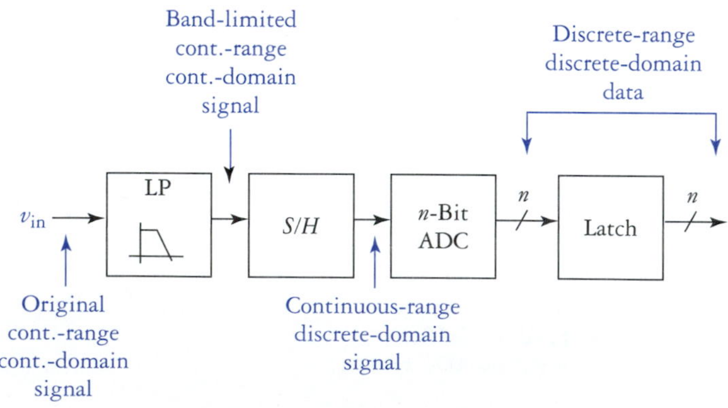

FIGURE 16.21

ADC system with anti-aliasing filter.

EXAMPLE 16.5

A certain data acquisition system has $f_{sa} = 10$ kHz. A LP Butterworth anti-aliasing filter is to be used, with the requirements that $f_c = 2$ kHz, $H_0 = 1$, and the response is down by at least 40 dB at $f_{\text{in}} = f_{sa}/2$. Determine the minimum filter order required.

Solution A quick look back at Example 14.2 presents a similar problem. The expression we must use is found to be

$$n = \frac{\log\left[\left(\dfrac{A_0}{\log^{-1}\left(-\dfrac{|H|}{20}\right)}\right)^2 - 1\right]}{2\log\left(\dfrac{f_{\text{in}}}{f_C}\right)}$$

$$= \frac{\log\left[\left(\dfrac{1}{\log^{-1}\left(-\dfrac{30}{20}\right)}\right)^2 - 1\right]}{2\log\left(\dfrac{5\text{ k}}{2\text{ k}}\right)}$$

$$= 5.02$$

Fractional order filters are not physically possible, so we round to the next higher order where we have $n = 6$.

Dynamic Range

Recall from the log amplifier discussions in Chapter 15 that the dynamic range of a signal is the log of the ratio of the maximum signal level to the minimum. This principle is also applied to ADCs (and DACs) in the following form:

Note
The maximum dynamic range of human hearing is about 120 dB.

$$\text{DR} = 20\log\frac{V_{\max}}{V_{\min}}$$

$$= 20\log\frac{V_{FS} - 1\text{ LSB}}{1\text{ LSB}}$$

In most practical systems, $V_{FS} \cong V_{FS} - 1$ LSB, so we can write

$$\text{DR} \cong 20\log\frac{V_{FS}}{1\text{ LSB}} \tag{16.23}$$

This can also be written as

$$\text{DR} \cong 20\log 2^n \tag{16.24}$$

or

$$\text{DR} \cong 20n\log 2 \tag{16.25}$$

ADC Circuits

There are many different ways to implement an ADC. We examine several of the more common circuits and their characteristics in the following subsections.

Counting ADC

The counting ADC, like most other ADCs, is designed around a DAC. The block diagram for a counting ADC is shown in Fig. 16.22. Assuming that a unipolar DAC is used, the full-scale range of the ADC is

$$\text{FSR} = 0\text{ V to } V_{FS} \tag{16.26}$$

Here's how the circuit works. Assume that V_{in} is some constant value within the FSR of the ADC. $\overline{\text{START}}$ is momentarily asserted, clearing the up-counter. At this point, the output of the DAC is $V_{FB} = 0$ V, which is less than V_{in}. This causes the comparator output to saturate high, gating the clock signal through to the counter. Each falling edge of the clock causes the counter to increment, increasing the output of the DAC by 1 LSB. This process continues until the output of the DAC exceeds the input voltage. When $V_{FB} > V_{\text{in}}$ the output of the comparator goes low, indicating EOC and

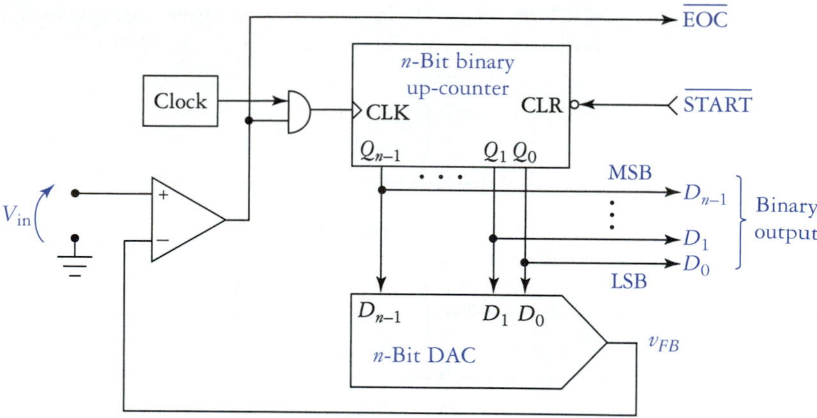

FIGURE 16.22
Counting ADC.

inhibiting the clock. The output of the binary counter is now a binary number that is proportional to the input voltage. The binary number m is given by

$$m = \frac{V_{in}}{1\ \text{LSB}} \tag{16.27}$$

Thus, knowing the count and the LSB size, the unknown input voltage is

$$V_{in} = m \times 1\ \text{LSB} \tag{16.28}$$

The conversion time of the counting converter is variable. For small values of V_{in}, only a few clock pulses are necessary to resolve the final count. For large values of V_{in}, as many as $2^n - 1$ clock cycles are required for a conversion. In general, because the input voltage is unknown, we must assume the worst case, which is

$$T_C = (2^n - 1)T_{clk} \tag{16.29}$$

This equation indicates that the conversion time is proportional to 2^n, which increases very rapidly.

EXAMPLE 16.6

A certain 12-bit ADC operates with FSR = 0 to 10 V, and $f_{clk} = 1$ MHz. Determine the dynamic range of the converter, the conversion time, the conversion rate, and the Nyquist frequency of the input.

Solution The dynamic range is

$$\begin{aligned} \text{DR} &= 20n \log 2 \\ &= (20)(12)(0.301) \\ &= 72.2\ \text{dB} \end{aligned}$$

The clock period is

$$\begin{aligned} T_{clk} &= 1/f_{clk} \\ &= 1/1\ \text{MHz} \\ &= 1\ \mu\text{s} \end{aligned}$$

The conversion time is

$$\begin{aligned} T_C &= (2^n - 1)T_{clk} \\ &= (4095)(1\ \mu\text{s}) \\ &= 4095\ \mu\text{s} \end{aligned}$$

The conversion rate is simply the reciprocal of the conversion time

$$\begin{aligned} f_C &= 1/T_C \\ &= 1/4095\ \mu\text{s} \\ &= 244\ \text{Hz} \end{aligned}$$

The Nyquist frequency $f_{Nyquist}$ is the highest input signal frequency that will not cause aliasing. In other words, $f_{Nyquist} = f_C/2$, and $f_C/2 = 122$ Hz.

The previous example emphasizes the main disadvantage of the counting converter: slow conversion. In this case, even though a relatively high clock frequency of 1 MHz is used, the conversion rate is a leisurely 244 Hz.

Successive Approximation ADC

If we replace the up-counter in the ADC with a *successive approximation register* (SAR) and make a few other slight modifications, the result is a *successive approximation converter.* A functional diagram for a 4-bit SAR converter is shown in Fig. 16.23. The full-scale range of the converter is

$$\text{FSR} = 0 \text{ to } V_{FS} \tag{16.30}$$

Assuming $V_{FS} = 10$ V and $V_{\text{in}} = 7$ V, the circuit works as follows. The START input is asserted, which sets the SAR output to 1000_2, causing the DAC to produce $V_{FB} = V_{FS}/2 = 5$ V. If $V_{FB} < V_{\text{in}}$ the SAR keeps this guess and outputs 1100_2. If $V_{FB} > V_{\text{in}}$ the SAR discards the guess and outputs 0100_2. In this case the SAR keeps the first guess and discards the second guess. The process repeats four times: once for each bit of resolution. The waveforms in Fig. 16.23(b) illustrate the convergence of V_{FB} toward V_{in}. Ideally, a SAR converter will resolve the analog input voltage to within ±1/2 LSB in n clock cycles. That is,

$$T_{C(\text{ideal})} = nT_{\text{clk}} \tag{16.31}$$

The conversion time is independent of V_{in} and directly proportional to the number of bits in the output code word.

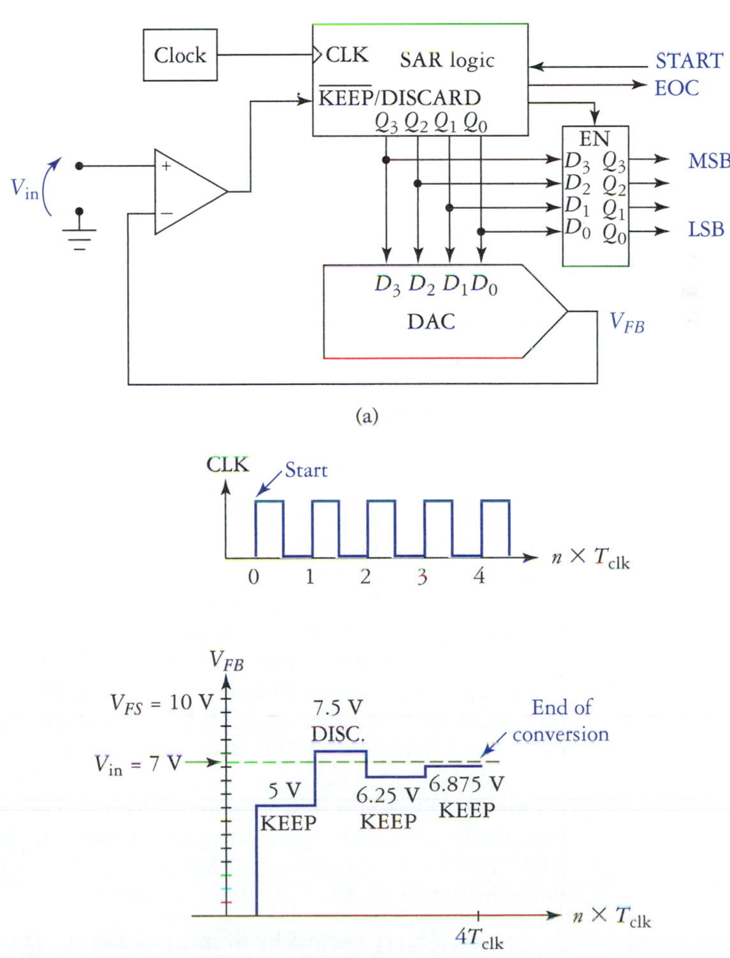

FIGURE 16.23
Successive approximation ADC.

EXAMPLE 16.7

A 12-bit SAR converter has $V_{FS} = 10$ V and $f_{\text{clk}} = 1$ MHz. Determine the ideal conversion time and conversion rate for this circuit.

Solution The clock period is

$$\begin{aligned} T_{\text{clk}} &= 1/f_{\text{clk}} \\ &= 1/1 \text{ MHz} \\ &= 1\ \mu\text{s} \end{aligned}$$

The conversion time is

$$\begin{aligned} T_{C(\text{ideal})} &= nT_{\text{clk}} \\ &= (12)(1\ \mu\text{s}) \\ &= 12\ \mu\text{s} \end{aligned}$$

The conversion rate is

$$\begin{aligned} f_{C(\text{ideal})} &= 1/T_{C(\text{ideal})} \\ &= 1/12\ \mu\text{s} \\ &= 83.3 \text{ kHz} \end{aligned}$$

The converter of this example has the same specifications as the counting ADC of Example 16.6 but the conversion rate is more than 680 times faster! This advantage makes the SAR converter a very popular circuit. The slight disadvantage of the more complex SAR logic is more than offset by the increase in conversion speed. The increased complexity of the SAR logic circuitry decreases the actual conversion speed because several extra clock cycles are needed for internal processing steps and other housekeeping functions. Even so, the conversion rate will usually be hundreds or thousands of times faster than an equivalent counting ADC.

The ADC0801

The ADC0801 is an 8-bit monolithic ADC that is designed for easy interfacing to microprocessor-based systems that are LS TTL compatible. A typical ADC0801 application is presented in Fig. 16.24. The ADC has FSR = 0 to 5 V. An internal clock generator can be used as shown in the diagram, with its frequency given by

$$f_{\text{clk}} = \frac{1.1}{R_1 C_1} \qquad (16.32)$$

Note
The ADC0801 is designed to be easily interfaced with microprocessor-based systems.

The maximum clock frequency is 1 MHz, but full accuracy is guaranteed only for $f_{\text{clk}} \leq 640$ kHz. An external TTL clock may be applied to pin 4, in which case pin 19 is left open.

Notice that two different grounds are specified in the diagram. The digital ground is the reference point for the clock and digital control lines of the ADC. The analog ground is a *quiet ground.* That is, if there is significant noise present due to digital switching transients, the analog ground pins should be connected to a ground point that is filtered and isolated from the digital noise sources. In many cases noise is not a significant problem and the grounds are interchangeable.

The microprocessor interface to the ADC is now explained. The output of the ADC0801 consists of eight, tri-state LS TTL compatible lines. The tri-state capability of the output allows direct connection to the microprocessor data bus. Normally, the ADC is assigned an I/O address via an address decoding circuit. The $\overline{\text{INTR}}$ output of the ADC is asserted when a conversion is complete. This signal can be used to supply an interrupt request to the microprocessor, or it can be polled in order to determine ADC status. If we guarantee that the ADC is given sufficient time to complete a conversion, the $\overline{\text{INTR}}$ signal can be ignored altogether. The following steps would be performed to use the ADC. Assume that initially the $\overline{\text{CS}}$ input is high, resulting in DB_0–DB_7 being placed in the high-impedance state.

1. $\overline{\text{CS}}$ and $\overline{\text{WR}}$ are asserted by the microprocessor. This causes the ADC0801 to drive $\overline{\text{INTR}}$ high.
2. $\overline{\text{CS}}$ and $\overline{\text{WR}}$ are returned high by the microprocessor. The ADC initiates a conversion. Up to eight clock cycles may be required before actual conversion begins.
3. When conversion is complete the ADC asserts $\overline{\text{INTR}}$ low.

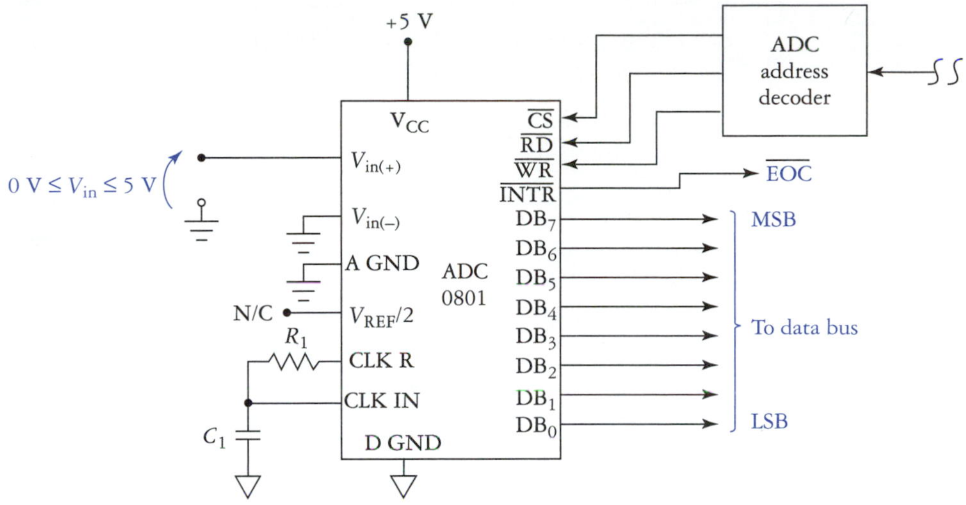

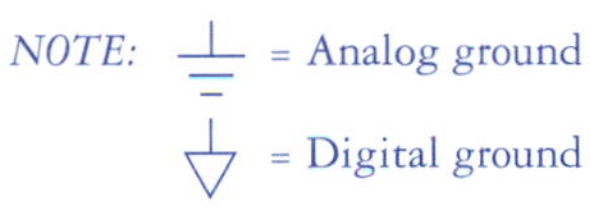

$\overline{CS}$	$\overline{RD}$	$\overline{WR}$	Function
0	0	0	Illegal
0	0	1	Start conversion
0	1	0	Read ADC output byte
0	1	1	No effect (HI-Z)
1	X	X	$DB_0 \rightarrow DB_7$ = High Z

NOTE: Analog ground; Digital ground

FIGURE 16.24
The ADC0801 monolithic 8-bit ADC.

4. $\overline{CS}$ and $\overline{RD}$ are asserted by the microprocessor. The ADC drives the 8-bit conversion result onto DB_7–DB_0 and drives $\overline{INTR}$ high.
5. $\overline{RD}$ is returned high by microprocessor. The ADC places DB_7–DB_0 in the high-impedance state. $\overline{CS}$ may remain asserted if another conversion is to be initiated.

Tracking ADC

A variation of the counting ADC is the *tracking ADC* or *tracking converter.* The block diagram for a tracking ADC is shown in Fig. 16.25(a). Note that there is no $\overline{EOC}$ output, and the counter is a programmable up/down unit. The tracking ADC is a free-running converter. The term *free-running* means that the ADC is allowed to operate continuously. The idea here is that when the converter is started it runs continuously until $\overline{STOP}$ is asserted. As shown in Fig. 16.25(b), the counter increments until $V_{FB} > V_{in}$. If the input remains constant as shown, the converter will continuously toggle the feedback voltage and output count by 1 LSB as shown. Changes of the input voltage are tracked by the converter, hence the name of the circuit.

If the input to a tracking ADC changes too rapidly, the converter will not be able to track the signal, resulting in *slope overload.* The effect is analogous to slew rate limiting of an analog amplifier. The maximum average rate of change of the tracking converter is given by

$$\left.\frac{\Delta V_o}{\Delta t}\right|_{\max} = \frac{1 \text{ LSB}}{T_{\text{clk}}} \text{ (V/s)} \tag{16.33}$$

Dual-Slope ADC

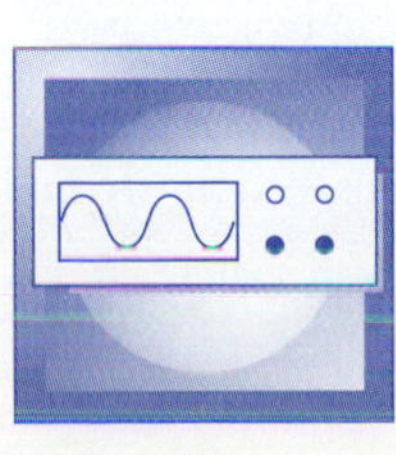

The dual-slope ADC is an example of an *integrating converter.* As shown in Fig. 16.26(a), the dual-slope ADC is built around an integrator, formed by A_1. Because the circuit is relatively complex, a quick overview of operation should give you a good feel for how a conversion is accomplished. Assume a positive input voltage is applied to the circuit.

1. Assertion of $\overline{START}$ clears the counter, resets the integrator for $v_x = 0$ V, and connects the input voltage to the integrator.

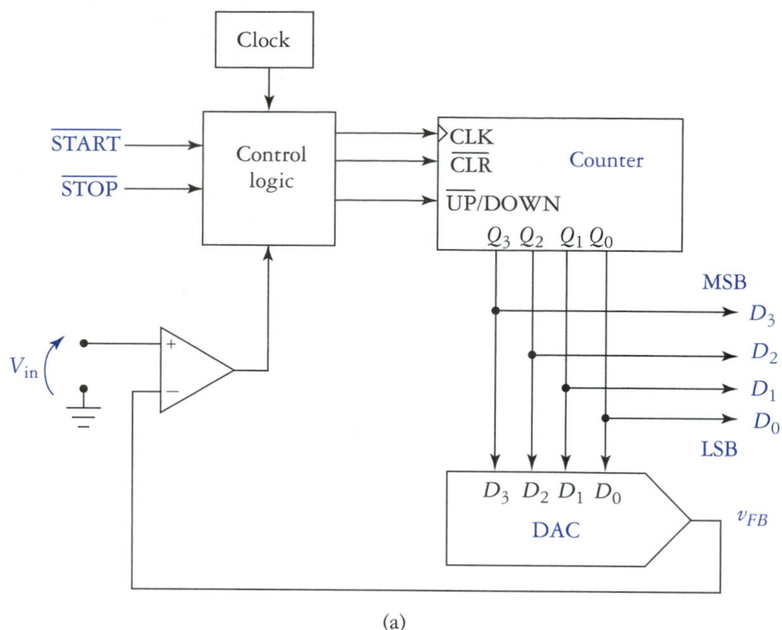

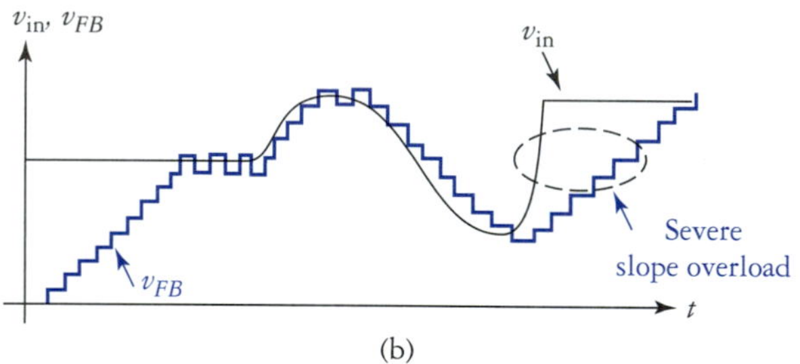

FIGURE 16.25
Tracking ADC. (a) Circuit. (b) Typical circuit waveforms.

2. The integrator ramps negatively for the time required for the counter to reach its maximum count. The control logic detects this occurrence when the counter rolls over to zero. Because the clock frequency is constant, this time interval T_1 is constant. If the input voltage is large, the output of the integrator ramps quickly. A small input voltage results in a slow ramp. This is shown in Fig. 16.26(b). The change of the output of the integrator at the end of this interval is given by

$$\Delta V_{x1} = \frac{-V_{\text{in}} T_1}{R_1 C_1} \tag{16.34}$$

3. Rollover of the counter signals the control logic to connect the integrator to $-V_{\text{REF}}$. This causes the integrator to ramp positively at a constant rate, while at the same time, the counter begins accumulating a count.
4. When the integrator ramps to a value slightly greater than zero, the comparator output goes low. This is detected by the control logic, which stops the counter and transfers the accumulated count to the output latch. The equation for the output voltage change during this time interval is

$$\Delta V_{x2} = \frac{-V_{\text{REF}} T_2}{R_1 C_1} \tag{16.35}$$

Note that the time required for the integrator to ramp up to zero varies, based on the original input voltage level. The larger V_{in} is, the longer the T_2 interval is. By inspection, we find $\Delta V_{x1} = -\Delta V_{x2}$; therefore,

$$\frac{-V_{\text{in}} T_1}{R_1 C_1} = \frac{-V_{\text{REF}} T_2}{R_1 C_1}$$

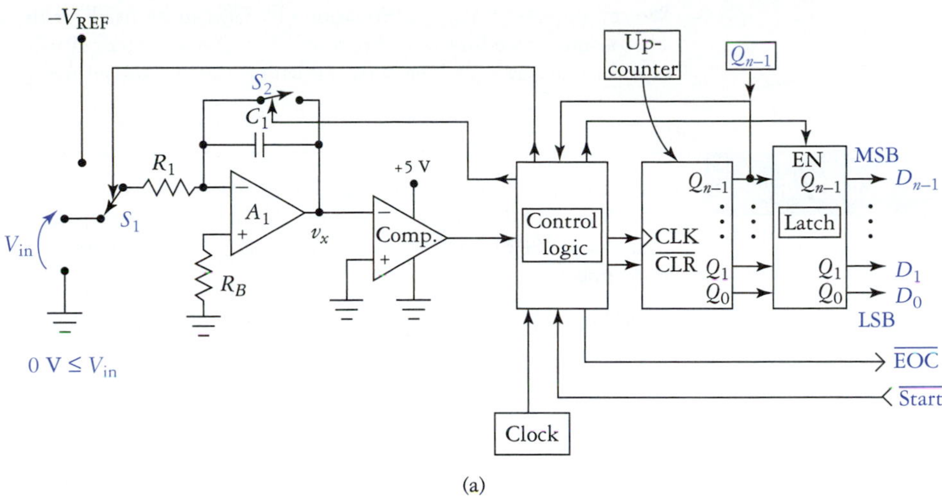

(a)

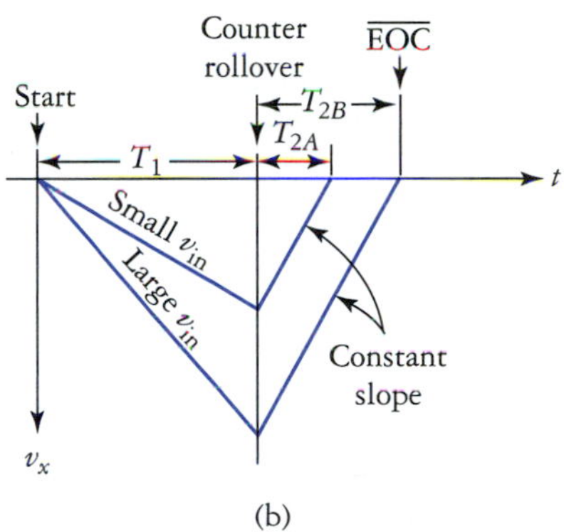

(b)

FIGURE 16.26

Dual-slope integrating ADC. (a) Circuit. (b) Integrator voltage for large and small input voltage levels.

Solving for V_{in}, we obtain

$$V_{in} = -V_{REF}\frac{T_2}{T_1} \tag{16.36}$$

However, the accumulated counts are directly proportional to the time intervals T_1 and T_2, so we can write

$$V_{in} = -V_{REF}\frac{m}{2^n} \tag{16.37}$$

where m is the accumulated count during T_2 and n is the number of bits in the counter.

The full-scale range of the dual-slope ADC is a function of the integrator time constant R_1C_1, the clock frequency f_{clk}, the number of bits of resolution of the ADC n, and the output voltage limits of the integrator $\pm V_{x(max)}$. At first glance, it appears that determining FSR for the dual-slope ADC is complicated, but if we take it a little bit at a time it is not too bad of a problem. First, we know that T_1 is constant, given by

$$T_1 = 2^n T_{clk} \tag{16.38}$$

The output voltage limit of the integrator is set to some convenient value: $V_{x(max)} = \pm 10$ V might be a good choice. Keep in mind, though, that using the design presented here, the output of the integrator will approach the negative supply rail most closely. In any case, the maximum input voltage, and hence the FSR of the ADC, is found by solving Eq. (16.34) for V_{in} giving us

$$V_{in(max)} = \text{FSR} = \frac{R_1C_1\,V_{x(max)}}{\text{T}_1} = \frac{R_1C_1\,V_{x(max)}}{2^n T_{clk}} \tag{16.39}$$

We set $V_{REF} = -V_{in(max)}$. Equation (16.39) can be used for design purposes as well as for analysis. The design procedure would require us to choose practical values for all variables in Eq. (16.39) except for one (say, f_{clk}). The equation would then be solved for the unknown value.

EXAMPLE 16.8

Refer to Fig. 16.26. Assume that $n = 8$, $R_1 = 100\ \text{k}\Omega$, $C_1 = 0.1\ \mu\text{F}$, $f_{clk} = 25.6$ kHz, $V_{REF} = -10$ V, and FSR = 10 V. Determine the input voltage if, after completing a conversion, the final count is $m = 0110\ 0100_2$. Given $V_{in} = 8$ V, determine the final output count m and the conversion time.

Solution Given an output count, we apply Eq. (16.37), where $m = 0110\ 0100_2 = 100_{10}$, producing

$$\begin{aligned} V_{in} &= -V_{REF}\frac{m}{2^n} \\ &= 10\text{ V}\frac{100}{256} \\ &= 3.906\text{ V} \end{aligned}$$

Given an input voltage, the output count can be determined in several different ways, but perhaps the easiest way is to solve Eq. (16.37) for m:

$$\begin{aligned} m &= 2^n\frac{V_{in}}{-V_{REF}} \\ &= 256\frac{8\text{ V}}{10\text{ V}} \\ &= 204.8 \end{aligned}$$

Fractional counts are not possible, but with this design we must always round up to the next higher count, which gives us $m = 205$.

The initial integration time is

$$\begin{aligned} T_1 &= 2^nT_{clk} \\ &= (2^8)(1/25.6\text{ kHz}) \\ &= (256)(39.06\ \mu\text{s}) \\ &= 10\text{ ms} \end{aligned}$$

The time of the second integration interval is found by solving Eq. (16.36), giving

$$\begin{aligned} T_2 &= \frac{V_{in}T_1}{-V_{REF}} \\ &= \frac{(8\text{ V})(10\text{ ms})}{10\text{ V}} \\ &= 8\text{ ms} \end{aligned}$$

The conversion time is

$$\begin{aligned} T_C &= T_1 + T_2 \\ &= 10\text{ ms} + 8\text{ ms} \\ &= 18\text{ ms} \end{aligned}$$

In practice, dual-slope ADCs are relatively slow compared to SAR converters, but they can be more easily designed for high accuracy. For these reasons, dual-slope converters are normally used in measuring equipment such as handheld digital multimeters. In these applications, a conversion (update) rate of 1 or 2 Hz is usually preferred.

The integrator used in the dual-slope ADC must be built using very high-quality components if high accuracy is desired. The op amp should have very low bias current and input offset voltage, and the integrating capacitor should be a very low leakage unit.

Dual-slope ADCs are relatively immune to random short-term fluctuations of the input voltage because of the low-pass filtering effect of the integrator. In this context, *short-term* means fluctuations that occur on a timescale that is short compared to T_1. This noise rejection characteristic allows the dual-slope ADC to be used without a S/H or T/H circuit in many applications. SAR converters are susceptible to very large errors due to short-term input fluctuations and, therefore, almost always require the use of a S/H or T/H circuit.

Parallel ADCs

The fastest ADCs are *parallel* or *flash ADCs.* The schematic diagram for a 2-bit parallel ADC is shown in Fig. 16.27. The input voltage is applied simultaneously to a network of comparators. The number of comparators N required depends on the number of bits of resolution n of the ADC according to

$$N = 2^n - 1 \tag{16.40}$$

The inverting inputs of the comparators are connected to a voltage-divider network consisting of equal-value resistors that divide the reference voltage into $2^n - 1$ equal parts.

FIGURE 16.27
Flash or parallel ADC.

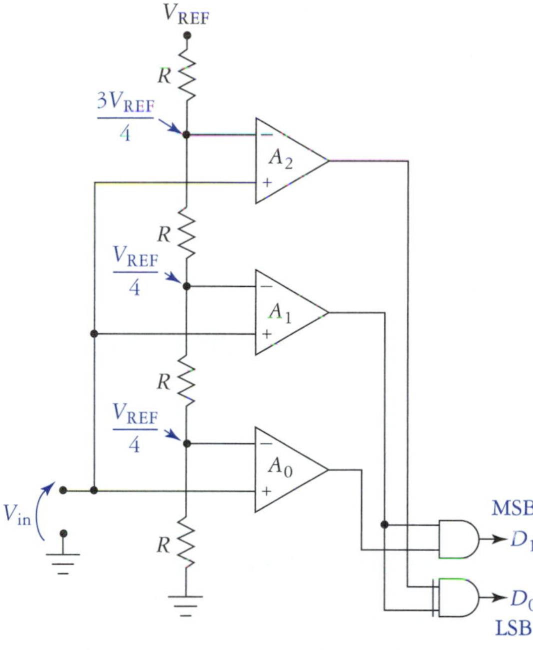

Note
Parallel ADCs are used in applications that require high-speed conversion, such as in digital video systems.

Referring to Fig. 16.28, let us assume that a positively increasing voltage is applied to the input of the ADC. Initially, each comparator output will be low. As the input voltage increases to $V_{REF}/4$, the output of comparator A_0 goes high. As v_{in} increases, each of the higher comparators in turn goes high. Thus, the input voltage is resolved into comparator output patterns. These patterns, shown in the truth table of Fig. 16.28, are encoded into standard binary form by combinational logic. Because the only delays associated with a conversion are comparator state changes and the propagation delay of the encoder, the parallel ADC is capable of very high conversion speeds. Such ADCs have been produced with conversion times on the order of 10 ps (10×10^{-12} sec). For the parallel converter topology of Fig. 16.28, the full-scale range is

$$\text{FSR} = V_{REF} \tag{16.41}$$

The parallel converter of Fig. 16.28 operates asynchronously. That is, this simplified converter changes output states in an unpredictable manner with variations of V_{in}. In a practical application, some form of clock synchronization would probably be implemented in order to ensure that sampling is performed at fixed intervals.

The most significant problem associated with parallel converters is the large number of comparators required for reasonably high resolution. For example, an 8-bit parallel ADC requires 255 comparators, which is at the current limit of linear IC fabrication technology. A 12-bit converter would require 4095 comparators, not to mention a voltage divider consisting of 4096 resistors.

Parallel ADCs are used in applications where conversion speed is the primary concern. Some application examples include digital sampling oscilloscope circuitry and real-time digital signal processing applications, such as audio and video digitization, ultrasound imaging, and sonar data acquisition.

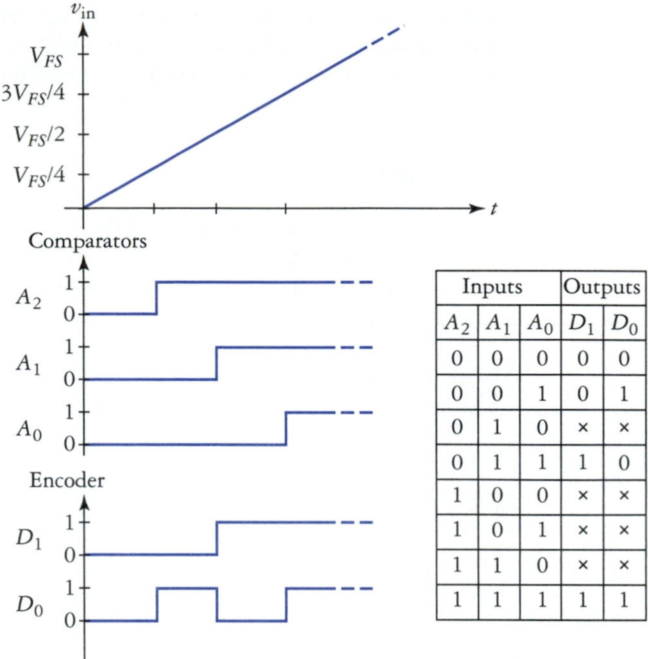

Inputs			Outputs	
A_2	A_1	A_0	D_1	D_0
0	0	0	0	0
0	0	1	0	1
0	1	0	×	×
0	1	1	1	0
1	0	0	×	×
1	0	1	×	×
1	1	0	×	×
1	1	1	1	1

FIGURE 16.28
Response of the parallel ADC to a ramp input voltage.

16.4 COMPUTER-AIDED ANALYSIS APPLICATIONS

Up until now, we have only simulated analog devices and circuits. As you might expect, PSpice has the capability to simulate digital and—most useful to us at this point—mixed-mode circuits. In this section we develop a 4-bit counting ADC and investigate the effects of oversampling and undersampling.

The Ramp Generator

We begin with the design and simulation of the ramp generator section of the counting ADC. The design can be tackled in many different ways, but here we use half of a 74393 dual 4-bit counter and the default 8-bit DAC. This is shown in Fig. 16.29. Notice that supply voltages are not required for these devices: PSpice automatically applies normal supply voltages. The 74393 is found in Eval.lib, while the DAC is found in the breakout sublibrary. To align the MSB of the 74193 with the MSB of

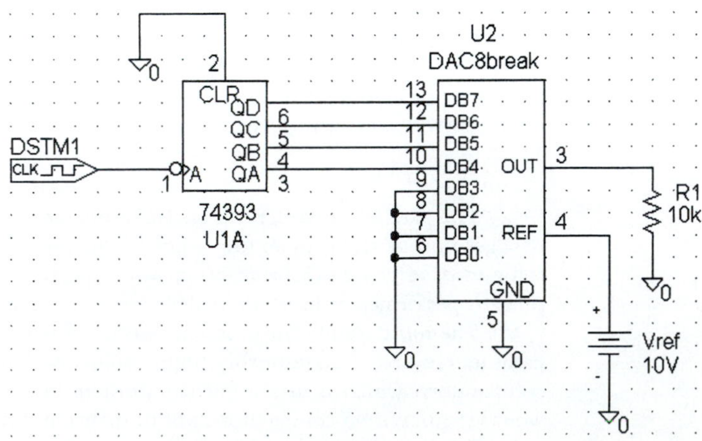

FIGURE 16.29
DAC with binary counter.

the DAC, the counter was flipped once, then rotated until correct orientation was achieved. The lower 4 bits of the DAC are tied to ground. This converts the 8-bit DAC into a 4-bit device, which is done to allow the stairstep appearance of the output to be viewed more easily. The full-scale output voltage of the DAC is $V_{FS} = V_{\text{REF}} = 10$ V.

Digital Stimulus

It is possible to set up an analog pulse source to generate the clock signal, but in this case, the digital stimulus, DigClock, found in Source.lib was used. Double-clicking on this symbol opens its attribute window. The pertinent values assigned here are Delay = 0, Ontime = 500 μs, and Offtime = 500 μs. This simply produces a 1-kHz clock signal with 50% duty cycle ($T_{\text{high}} = T_{\text{low}} = 500$ μs).

Simulation Setup and Results

The simulation was set up for a transient analysis for a final time of 3.5 ms with a print step of 1 ms (to save time and disk space). The digital setup attributes are set as shown in Fig. 16.30. What we are most concerned with here is to ensure that the counter is initialized to all zeros at the start of the simulation. The Timing Mode defaults to Typical, which produces typical propagation delay characteristics for the digital parts. The A/D interface options are used to define various signal strengths interpreted by the digital inputs. We use Level 1, which represents ideal zero–one operation.

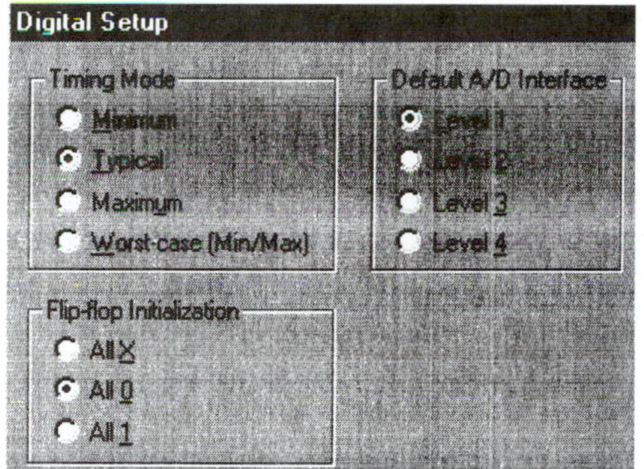

FIGURE 16.30
Setting up the digital simulation parameters.

The results of the analysis run are shown in Fig. 16.31. PSpice automatically places digital and analog waveforms in separate plots, which prevents voltage scaling problems. The circuit performs much as would be expected, except that glitches appear on the output of the DAC at various points in the step sequence. Actually, these glitches are not an artifact of the simulator; rather, they occur because the 74393 is a ripple counter. Propagation delay within the counter causes successively more significant output lines to change states slightly later in time. Notice that the glitch is largest when the count increments from 0111 to 1000. During this time (tens of nanoseconds), the counter actually produces erroneous output counts that cannot be resolved on the plot, but are manifested as short-duration spikes in the output of the DAC. A synchronous counter could be used to prevent these glitches from occurring.

Counting ADC

The ramp generator can be converted into a counting ADC with the addition of a comparator (actually a 741 op amp) and an AND gate, as shown in Fig. 16.32. Diodes D_1 and D_2 are used to limit the voltage applied to the AND gate to safe levels. The input source is a VSIN device with all attributes set to zero except for the offset voltage, which is $V_{\text{offset}} = 7$ V. A dc source could have been used just as easily though.

The circuit is simulated using a transient analysis run from 0 to 1.5 ms, with the resulting Probe output being shown in Fig. 16.33. The upper traces show the output of the counter, which is the output of the ADC. The lower traces are the output of the DAC (V_{FB}) and the output of the comparator (clock inhibit or EOC). The counter increments up from 0000_2. On reaching 1100_2 (12_{10}), the output of the DAC is $V_{FB} = 12 \times 1$ LSB $= 7.5$ V. This causes the output of the comparator to drive low, inhibiting the clock and halting the count.

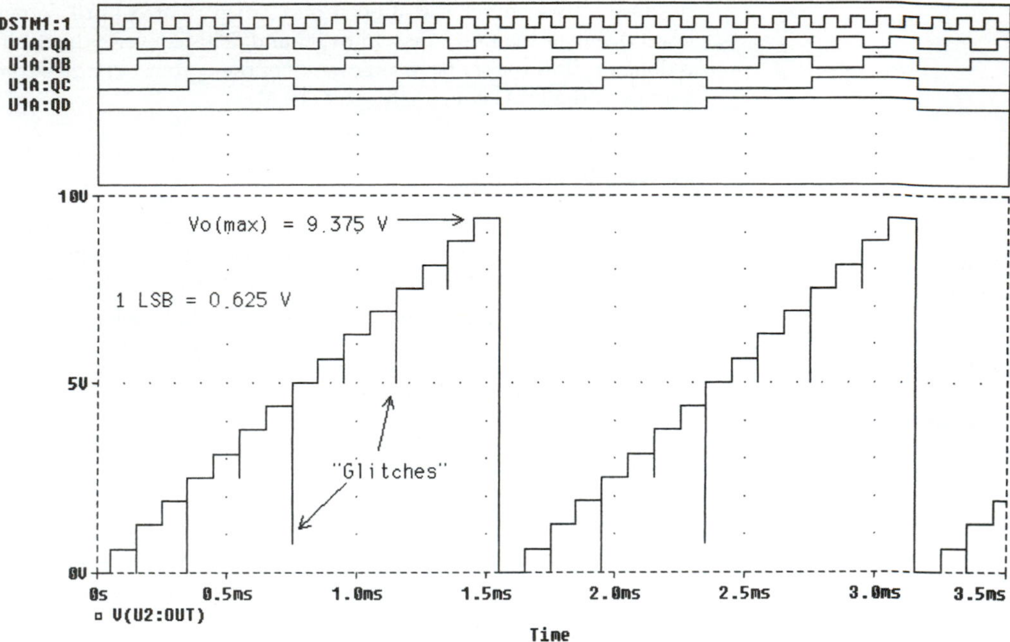

FIGURE 16.31
Counter waveforms and DAC response.

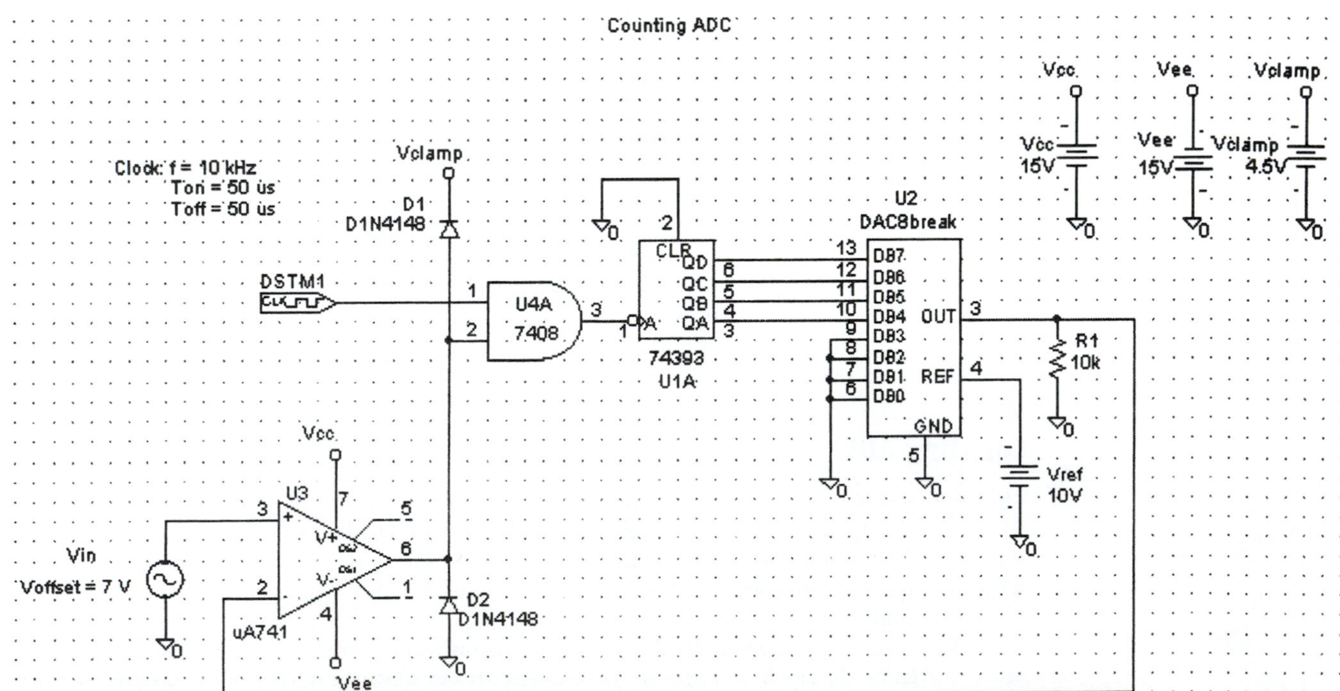

FIGURE 16.32
Counting DAC simulation.

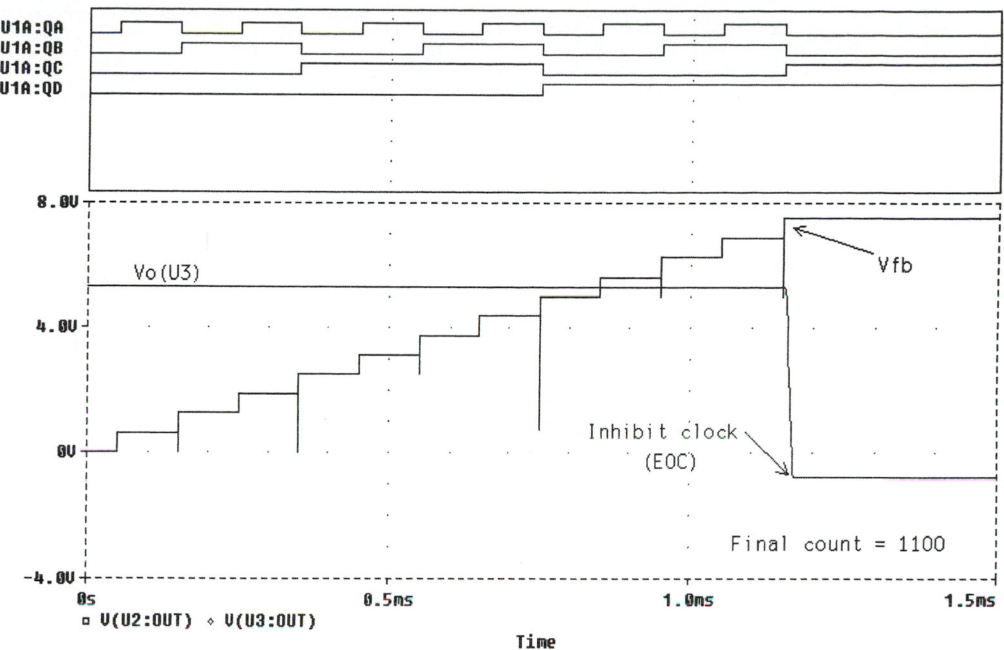

FIGURE 16.33
Response of the counting ADC for $V_{\text{in}} = 7$ V.

Sampled Data Systems

The effects of sampling a continuous signal are very important considerations in many situations. The schematic in Fig. 16.34 is that of a track-and-hold circuit. The active devices used are two GAIN blocks from ABM.slb, a CD4016 analog switch from Eval.slb, and the digital clock stimulus DSTIM1. Capacitor C_1 is the holding cap and R_1 is used to prevent the output gain block from having a floating node. Because the gain blocks are ideal, any capacitor and resistor values could be used, but those shown would provide a reasonable starting place for actual circuit prototyping.

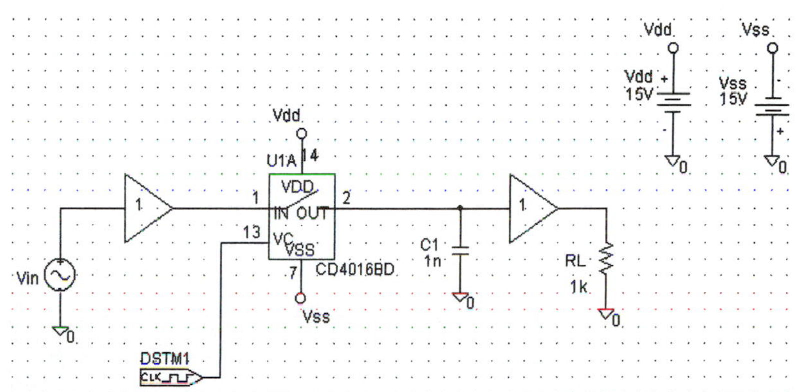

FIGURE 16.34
Track-and-hold circuit simulation using unity-gain blocks and an analog switch.

Oversampling

The clock signal is set for $T_{\text{on}} = T_{\text{off}} = 500$ μs producing a 50% duty cycle sample frequency of $f_{sa} = 1$ kHz for all simulation runs in this discussion. Initially, the analog input voltage is set for 2 $V_{\text{p-p}}$ at 100 Hz. Thus, the input signal is oversampled by a factor of 10. That is, $f_{sa} = 10f_{\text{in}}$. Performing a transient analysis for $t = 0$ to 20 ms and setting up separate plots of v_{in} and v_o produces the results shown in Fig. 16.35. Clearly the output signal captures the character of the input.

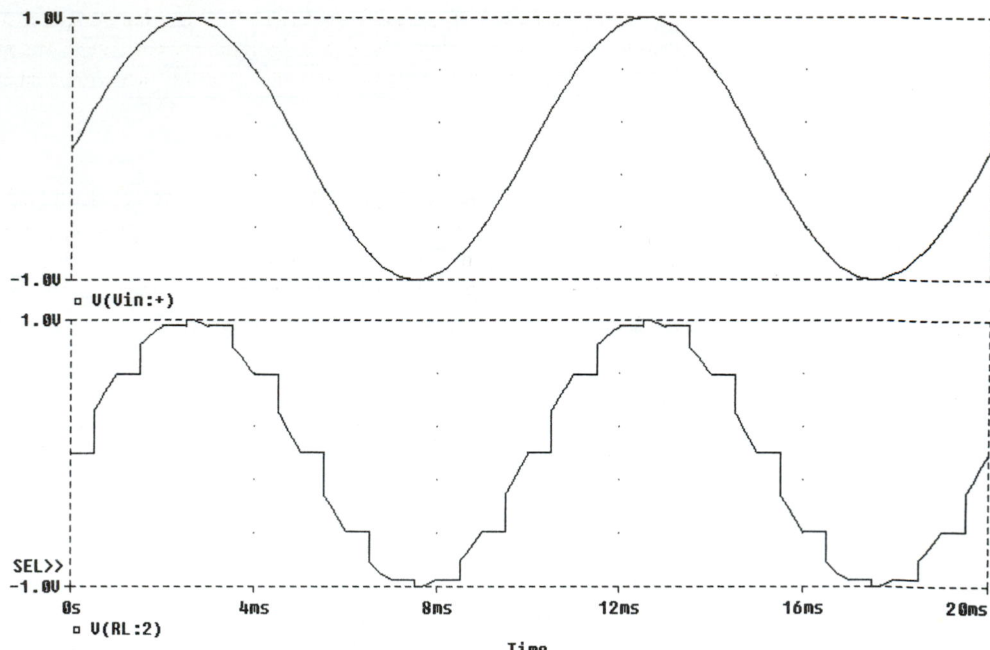

FIGURE 16.35
T/H input and output signals with 10× oversampling.

If we click on the FFT button on the PROBE toolbar, the input and output voltage frequency spectra are produced, as shown in Fig. 16.36. The input contains a single frequency component at 100 Hz with 1-V amplitude. The output contains a large component at 100 Hz (not quite a 1-V amplitude) as we should expect. In addition, higher frequency components of diminishing amplitude centered at about integer multiples of the sample frequency are also present. These spurious frequencies are located at

$$nf_{sa} \pm f_{\text{in}} \qquad (n = 1, 2, \ldots)$$

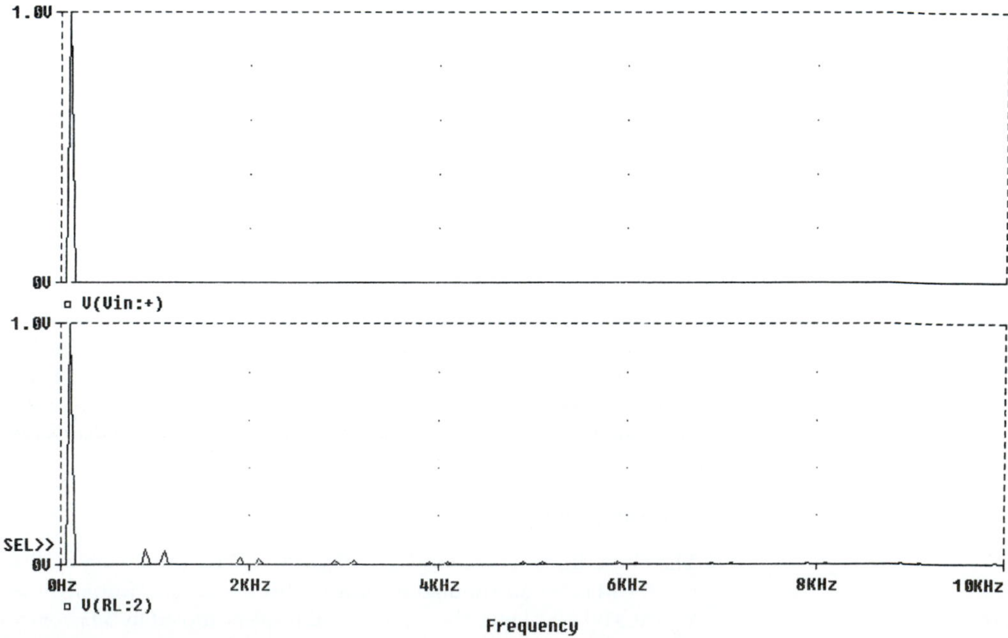

FIGURE 16.36
Signal spectra for the original input (top) and the sampled version (bottom).

The energy lost from the main V_o peak at 100-Hz frequency is distributed among these higher frequencies. What we find here though is that because the signal was highly oversampled, the reduction in the main output peak is quite small and most of the signal energy stayed right where we want it.

Undersampling

Recall that undersampling occurs when $f_{\text{in}} \geq f_{sa}/2$. If we set the frequency of the input signal to $f_{\text{in}} =$ 600 Hz, then with $f_{sa} = 1$ kHz, the Nyquist requirements are not met and we are undersampling by 100 Hz. This is shown in Fig. 16.37. The output has some of the characteristics of the original signal, but it is getting pretty difficult to see the connection.

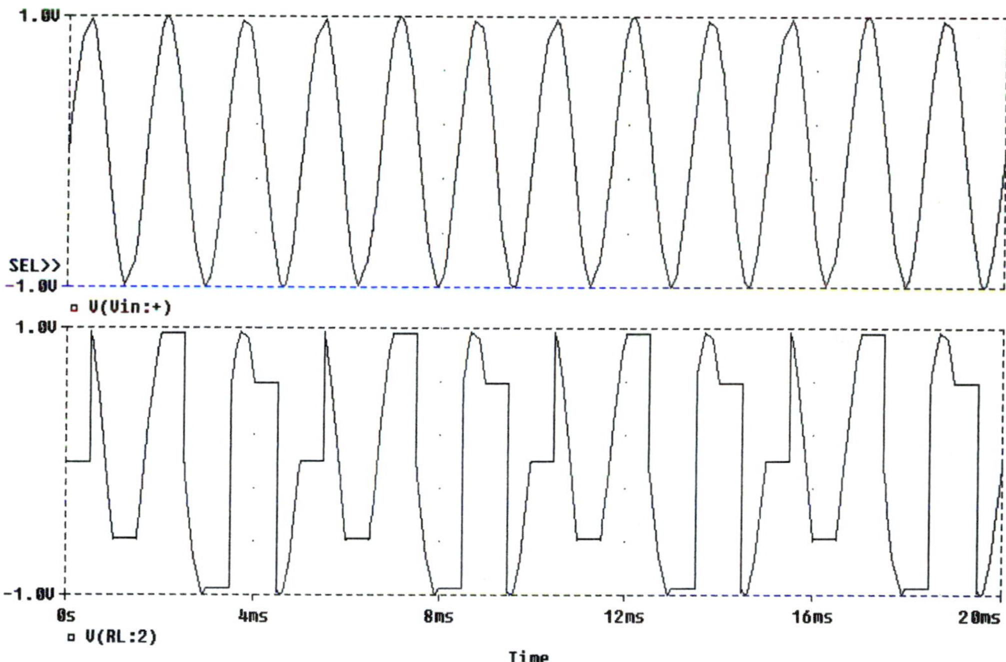

FIGURE 16.37
Effects of undersampling.

Taking the FFT of the signals and setting the x-axis limits to 0–5 kHz we obtain the display of Fig. 16.38. Undersampling has resulted in aliasing of the original signal information. A large 400-Hz frequency component ($f_{sa} - f_{\text{in}}$) has been created. Unless we know all about the input signal *a priori,* which usually only happens in textbooks, this signal energy cannot be distinguished from normal input signal energy.

Taking undersampling a step further, we set $f_{\text{in}} = 1.2$ kHz and obtain the time-domain waveforms of Fig. 16.39. After some practice it is possible to visualize the low-frequency characteristic of such an undersampled output signal, but the FFT is required for confirmation of this observation.

Taking the FFT of the input and output signals produces the plots of Fig. 16.40. This time the undersampling is so severe that two frequency components are created below the desired frequency. The details of how this occurs are beyond the scope of this book, but it is clear that undersampling can have serious consequences.

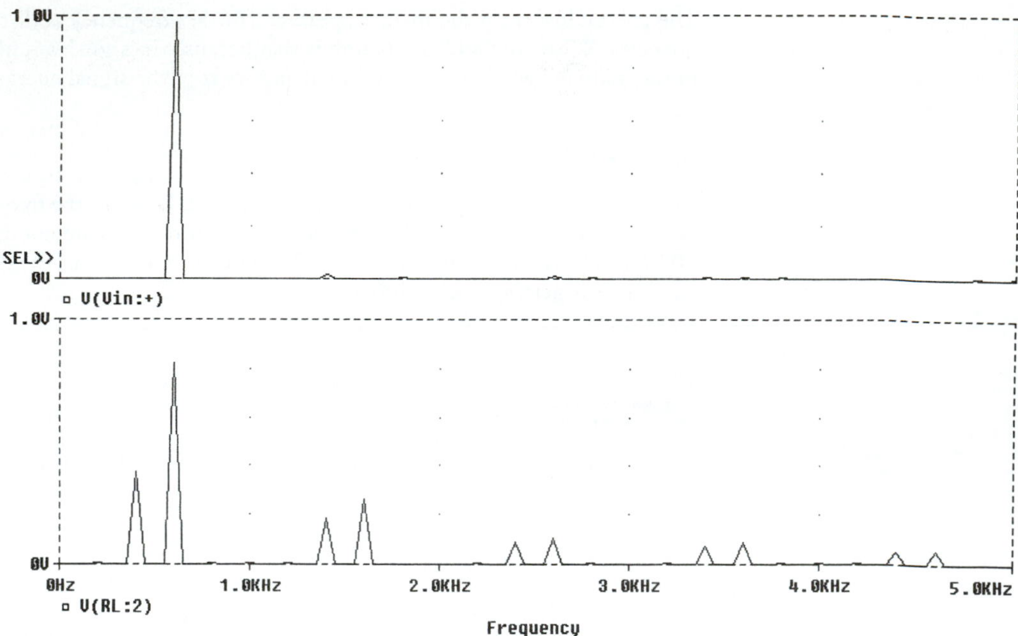

FIGURE 16.38
Spectra for the input signal (top) and the undersampled output signal (bottom). Note aliasing occurs in sampled output spectrum.

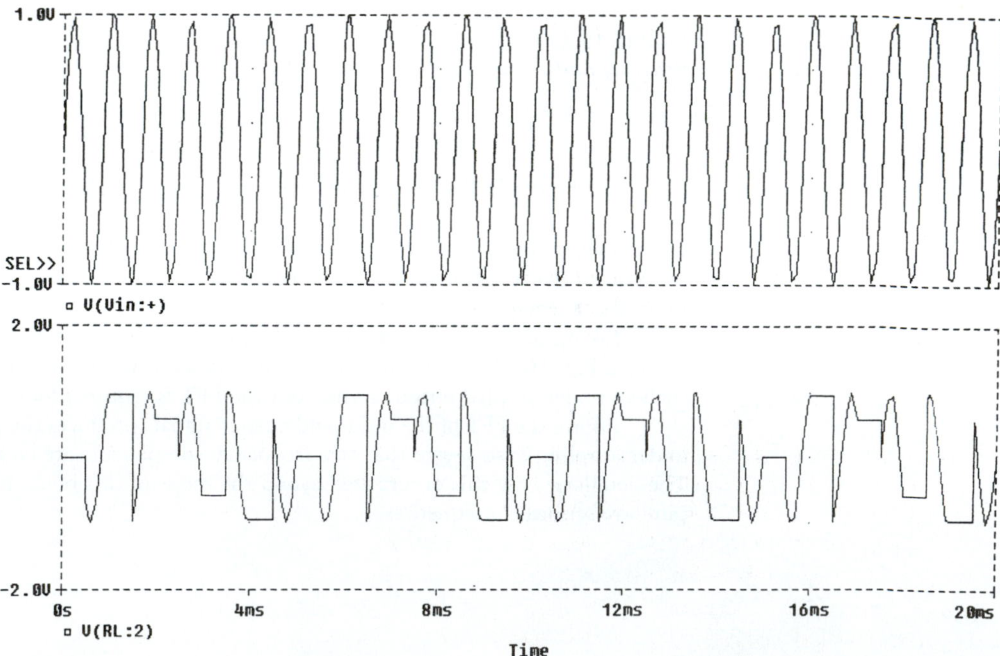

FIGURE 16.39
Severe undersampling ($f_{sa} = f_{in}/10$).

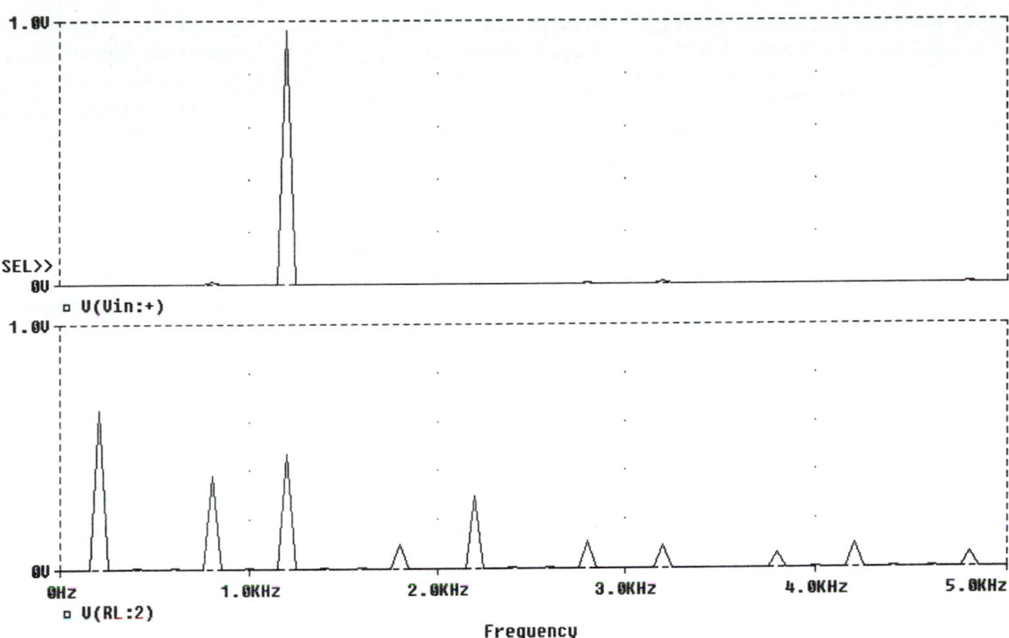

FIGURE 16.40
Frequency spectra of input and 10× undersampled output. Note two significant aliased spectral components.

■ CHAPTER REVIEW

Digital-to-analog and analog-to-digital converters provide an interface between the continuous, analog world and the discrete domain of computers and other digital systems. A DAC accepts a binary input and produces an output voltage that is proportional to the value of that input word. DACs can be implemented using weighted-resistor summers, but the R-2R ladder configuration is more commonly used. The resolution of an n-bit DAC is proportional to $\frac{1}{2}^n$. The accuracy of a DAC is typically expressed in terms of 1 LSB. Most DACs accept input words in normal 8-4-2-1 binary, offset binary, or two's-complement binary forms. Typical uses for DACs include digital measuring equipment, function generators, and control systems.

An ADC samples a continuous analog signal and produces a binary number that is proportional to the value of the input at the time of sampling. This is called *quantization* or *digitization.* Sample-and-hold or track-and-hold circuits are often used to sample a continuous-domain signal to allow sufficient time for the ADC to perform a conversion. According to the Nyquist sampling theorem, in order to prevent aliasing, a time-varying signal must be sampled at a rate of at least twice the highest frequency present in the input. Analog filters are often used to band limit analog signals prior to quantization.

Some of the different types of ADCs available are the counting converter, tracking converter, successive approximation (SAR) converter, integrating converter, and the parallel or flash converter. Parallel converters are the fastest but also the most difficult from which to obtain high resolution. Integrating converters are capable of high accuracy and noise immunity, but are relatively slow. The SAR converter provides a good compromise, with fast conversion speed, good accuracy, and low cost.

■ DESIGN AND ANALYSIS PROBLEMS

Note: Unless otherwise specified, assume all op amps are ideal.

16.1. Refer to Fig. 16.8(a). Given $R = R_F = 10\text{ k}\Omega$ and $V_{REF} = -5$ V, determine V_{FS}, $V_{o(max)}$, and the output voltage for the following input words: 1100_2, 1010_2, 0011_2.

16.2. Refer to Fig. 16.8(a). Given $R = 2\text{ k}\Omega$, $R_F = 4\text{ k}\Omega$, and $V_{REF} = -5$ V, determine V_{FS}, $V_{o(max)}$, and the output voltage for the following input words: 1100_2, 1010_2, 0011_2.

16.3. Refer to Fig. 16.9. Given $R = 10\text{ k}\Omega$, $R_F = 4.7\text{ k}\Omega$, $R_1 = 4.7\text{ k}\Omega$, and $V_{REF} = 5$ V, determine V_{FS}, $V_{o(max)}$, and the output voltage for the following input words: 1010_2, 1011_2, 0111_2.

16.4. Refer to Fig. 16.9. Given $R = 10\text{ k}\Omega$, $R_F = 10\text{ k}\Omega$, $R_1 = 2\text{ k}\Omega$, and $V_{REF} = 3$ V, determine V_{FS}, $V_{o(max)}$, and the output voltage for the following input words: 1010_2, 1011_2, 0111_2.

16.5. Refer to Fig. 16.11. Given $V_{REF} = 5$ V, $R_2 = R_1 = 10\text{ k}\Omega$, and $R_B = R_F = 10\text{ k}\Omega$, determine I_{FS}, V_{FS}, and the output voltage for the following input words: $1010\ 1010_2$, $1111\ 0000_2$, $0111\ 0101_2$.

16.6. Refer to Fig. 16.11. Given $V_{REF} = 10$ V, $R_2 = R_1 = 10$ kΩ, and $R_B = R_F = 5$ kΩ, determine I_{FS}, V_{FS}, and the output voltage for the following input words: $0110\ 0101_2$, $0000\ 1111_2$, $1001\ 1001_2$.

16.7. Refer to Fig. 16.11. Determine the required resistor values such that $I_{FS} = 1$ mA and $V_{FS} = 10$ V, given $V_{REF} = 10$ V.

16.8. Refer to Fig. 16.11. Determine the required resistor values such that $I_{FS} = 1.5$ mA and $V_{FS} = 5$ V, given $V_{REF} = 5$ V.

16.9. Refer to Fig. 16.12(a). Assume the DAC has $n = 4$ and $V_{FS} = 10$ V. The tachometer produces a dc output voltage that is given by $V_o = 2$ V/1000 rev/min. Determine the maximum and minimum motor speed in rev/min (rpm). Determine the motor speed given the following binary input words: 1000_2, 0101_2.

16.10. Refer to Fig. 16.12(a). Assume the DAC has $n = 8$ and $V_{FS} = 10$ V. The tachometer produces a dc output voltage that is given by $V_o = 1$ V/1000 rev/min. Determine the maximum and minimum motor speed in rev/min (rpm). Determine the motor speed given the following binary input words: $1000\ 0000_2$, $0101\ 1100_2$, $1111\ 0000_2$.

16.11. A 4-bit DAC ($n = 4$) is used to generate a 16-step (2^n-step) sine wave approximation with frequency $f_{sine} = 1$ kHz. The quantization noise present in the output signal consists of frequency components located at $m2^n f_{sine} \pm f_{sine}$ as shown in Fig. 16.41. If a second-order LP Butterworth filter with $f_c = 2$ kHz and $H_0 = 1$ is used to smooth the output, determine the response magnitude $|H(f)|$ of this filter at the lowest quantization noise frequency.

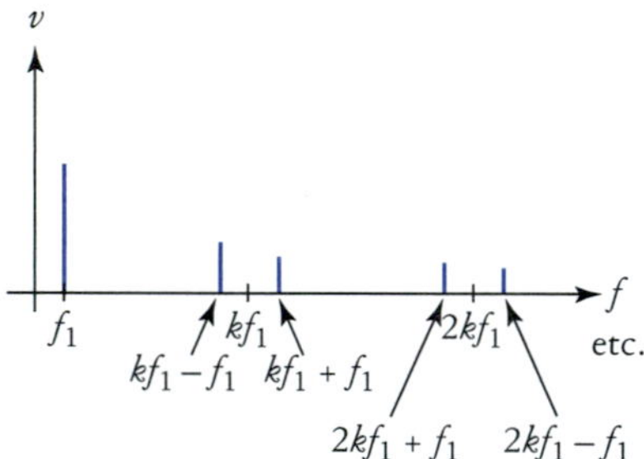

FIGURE 16.41
Signal spectra for Problem 16.11.

16.12. Refer to Fig. 16.13. The microcomputer outputs data at a rate of 10 kbytes/sec to the DAC. Assuming that the data stream represents a 256-step sine wave approximation, determine the frequency of the resulting output sine voltage. Using the information from Problem 16.11, determine the lowest quantization noise frequency that will be present in the output signal.

16.13. Refer to Fig. 16.21. The S/H is clocked at $f_{sa} = 50$ kHz. If the anti-aliasing filter has a fourth-order Butterworth response with $H_0 = 1$ and $f_c = 5$ kHz, determine the response magnitude at the Nyquist frequency. Determine the required filter order such that $|H(f_{Nyquist})| \leq 0.01$.

16.14. Refer to Fig. 16.21. The S/H is clocked at $f_{sa} = 90$ kHz. If the anti-aliasing filter has a fourth-order Butterworth response with $H_0 = 1$ and $f_c = 15$ kHz, determine the response magnitude at the Nyquist frequency. Determine the required filter order such that $|H(f_{Nyquist})| \leq 60$ dB.

16.15. Refer to Fig. 16.22. Assume $n = 8$, $V_{FS} = 10$ V, and $f_{clk} = 25$ kHz. Determine the output count and conversion time for the following input voltages: 3 V, 5 V, and 7.8 V.

16.16. Refer to Fig. 16.22. Assume $n = 12$, $V_{FS} = 10$ V, and $f_{clk} = 100$ kHz. Determine the output count and conversion time for the following input voltages: 3 V, 5 V, and 7.8V.

16.17. A certain 8-bit SAR ADC has $f_{clk} = 500$ kHz. Determine the conversion time assuming that it takes four clock cycles per KEEP/DISCARD operation. Determine the Nyquist frequency of the converter.

16.18. A certain 12-bit SAR ADC has $f_{clk} = 1$ MHz. Determine the conversion time assuming that it takes four clock cycles per KEEP/DISCARD operation. Determine the Nyquist frequency of the converter.

16.19. Refer to the 4-bit tracking converter of Fig. 16.25. Assume that $f_{clk} = 100$ kHz and FSR = 0 to 10 V. Determine the maximum rate of change of v_{in} that can be tracked without causing slope overload. Given a sinusoidal input voltage of 8 $V_{p\text{-}p}$, determine the frequency that will just cause slope overload to begin.

16.20. Refer to the 4-bit tracking converter of Fig. 16.25. Assume that $f_{clk} = 1$ MHz and FSR = 0 to 10 V. Determine the maximum rate of change of v_{in} that can be tracked without causing slope overload. Determine the maximum sinusoidal input frequency that can be applied with $V_{p\text{-}p} = 10$ V.

16.21. Determine the number of bits required of an ADC such that its dynamic range is greater than or equal to 60 dB.

16.22. Determine the number of bits required of an ADC such that its dynamic range is greater than or equal to 90 dB.

16.23. Refer to Fig. 16.26. Assume that $n = 12$, $R_1 = 40.96$ kΩ, $C_1 = 1$ μF, $f_{clk} = 100$ kHz, $V_{REF} = -10$ V, and FSR = 10 V. Determine the input voltage if, after completing a conversion, the final count is $m = 1100\ 0100_2$. Given $V_{in} = 5$ V, determine the final output count m and the conversion time.

16.24. Refer to Fig. 16.26. Assume that $n = 10$, $R_1 = 160$ kΩ, $C_1 = 0.1$ μF, $f_{clk} = 64$ kHz, $V_{REF} = -10$ V, and FSR = 10 V. Determine the input voltage if, after completing a conversion, the final count is $m = 00\ 0111\ 0110_2$. Given $V_{in} = 2$ V, determine the final output count m and the conversion time.

16.25. Determine the number of comparators required to realize a 7-bit parallel ADC.

16.26. Determine the number of comparators required to realize a 12-bit parallel ADC.

16.27. Refer to Fig. 16.42. This is the block diagram for a microcomputer-controlled motor speed control servo. The system works as follows, where X = desired speed byte, Y = actual speed byte, and Z= byte to DAC.

1. $Z = 00_{16}$. (Make sure motor is stopped!)
2. Get user's speed input byte X (00_{16} = Stop, FF_{16} = Max).
3. Start ADC.
4. Delay.
5. Read ADC, Y.
6. Compare Y and X.
7. If $Y = X$ goto 3.
8. If $Y < X$ then $Z = Z + 1$ else $Z = Z - 1$.
9. goto 3.

Assuming the tachometer has $V_T = 1$ V/1000 rpm, $R_1 = 0$ Ω, and $R_2 = 10$ kΩ, determine the maximum motor speed.

16.28. Repeat Problem 16.27 with $R_1 = R_2 = 10$ kΩ.

TROUBLESHOOTING PROBLEMS

16.29. Refer to Fig. 16.16. Suppose the Q output of the JK flip-flop is stuck high. What effect would this fault have on the output waveform if a sine wave is being synthesized?

16.30. Refer to Fig. 16.42. Suppose the load on the motor is increased such that even at maximum output, the DAC is unable to maintain sufficient motor speed. Based on the program description of Problem 16.27, what type of waveform would be observed at the output of the DAC?

COMPUTER PROBLEMS

16.31. Refer to the T/H circuit in Fig. 16.34. This circuit can closely approximate an ideal sample-and-hold circuit if the clock $T_{high} \cong 0$. Simulate the circuit with the digital clock stimulus attributes set for Ontime = 1 μs and Offtime = 999 μs, and the input voltage set for 2 $V_{p\text{-}p}$ at 100 Hz. Plot the output and input voltage waveforms. Plot the FFT of the output signal.

16.32. Figure out a way to simulate a SPDT switch that is controlled by a digital clock stimulus signal, using the CD4016 analog switch in the PSpice Eval.slb.

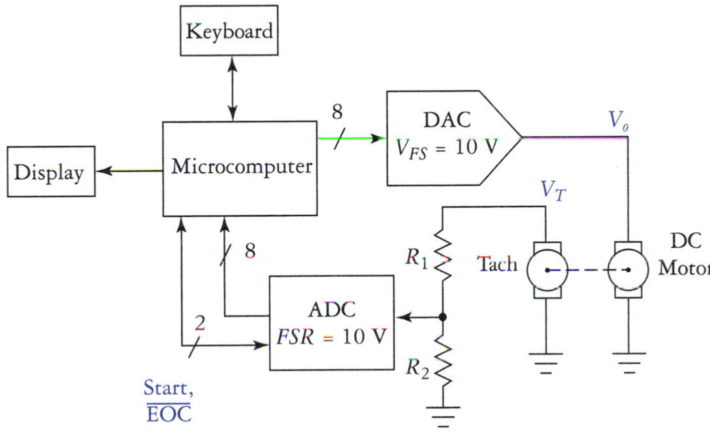

FIGURE 16.42
Circuit for Problems 16.27, 16.28, and 16.30.

CHAPTER 17

Tuned Amplifiers and Oscillators

BEHAVIORAL OBJECTIVES

On completion of this chapter, the student should be able to:

- Determine the dc and ac characteristics of basic linear tuned amplifiers.
- Explain why tuned amplifiers are used in radio-frequency (RF) applications.
- Analyze the operation of ideal transformers and autotransformers in amplifier and oscillator circuits.
- Describe the use of transformer coupling for the purposes of tuning and impedance matching.
- Describe the requirements for sustained oscillation.
- Identify Hartley, Colpitts and Clapp, phase-shift, and Wein-bridge oscillator topologies.

INTRODUCTION

We covered some aspects of tuned amplifier operation in the context of class C amplifiers, back in Chapter 10. In this chapter, we concentrate on linear tuned amplifiers. The techniques used to analyze these tuned amplifiers are directly applicable to the analysis of oscillator circuits. Tuned amplifiers are used extensively in high-frequency applications such as in radio receiver and transmitter circuits. Oscillators are generally used more extensively at both low and high frequencies, but also serve very important functions in communications applications.

In a manner of speaking, we are going to shift gears in this chapter. That is, most of the circuits that are covered here are designed around discrete transistors, as opposed to op amps and other monolithic ICs. The main reason is that except for a relatively few specialized devices, op amps generally exhibit poor performance at high frequencies, relative to discrete devices.

17.1 TRANSFORMER-COUPLED TUNED AMPLIFIERS

A coupling network may serve several purposes. The simple capacitive coupling used so often in previous chapters serves to pass ac signals, while providing isolation of dc voltage levels. Direct coupling is even simpler; it allows both dc and ac signals to pass from one section of a circuit to another. There are applications, however, where a coupling network may be required to provide impedance matching and filtering functions as well. At high frequencies, transformer coupling provides an efficient and convenient way of achieving these objectives, which is the subject of this section.

Transformer Review

The main function performed by a transformer is the magnetic coupling of energy from the primary winding to the secondary. A typical transformer configuration is shown in Fig. 17.1. The pertinent equations for this circuit are

$$V_{\text{sec}} = V_{\text{pri}}\frac{n_s}{n_p} \tag{17.1}$$

and

$$i_{\text{pri}} = i_{\text{sec}}\frac{n_s}{n_p} \tag{17.2}$$

where n_p and n_s are the number of turns in the primary and secondary windings, respectively. The turns ratio is defined as

$$\text{Turns ratio} = \frac{n_p}{n_s} \tag{17.3}$$

When $n_p > n_s$, we have a *step-down* transformer. If $n_s > n_p$ the transformer is a *step-up* unit, and if $n_p = n_s$ it is an *isolation* transformer.

FIGURE 17.1
Transformer with phasing dots.

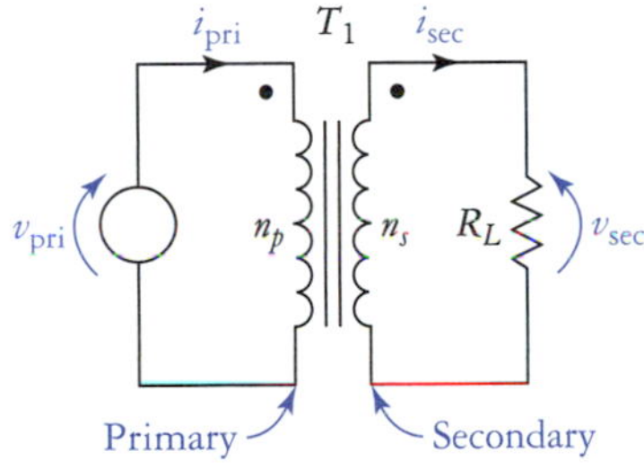

> **Note**
> We are using lowercase k to represent the coupling coefficient, while uppercase K is a proportionality constant.

The dots shown in Fig. 17.1 are called *phasing dots.* Phasing dots are used to indicate the phase relationship between the primary and secondary voltages. The interpretation is that if the dotted end of the primary is driven by a positive-going voltage, then the dotted end of the secondary winding is positive going as well.

The efficiency of energy transfer from primary to secondary is expressed as the transformer coupling coefficient k. For a transformer that is 100% efficient, $k = 1$. Most often we desire near-unity coupling, and unless otherwise noted, this will be assumed in the following discussions.

You may recall from your studies of ac circuits that inductance is directly proportional to the number of turns in a winding. That is,

$$L = Kn$$

where the actual inductance depends on the core material used with the winding and the geometric properties of the core. However, because the primary and secondary are wound on the same core, the proportionality constant K cancels and we may write

$$\frac{n_p}{n_s} = \frac{L_p}{L_s} \tag{17.4}$$

Thus, we can use the inductance ratio to determine the voltage and current relationships in a transformer.

Note
You may recall Eq. 17.5 being used in Ch. 10 as Eq. 10.13.

Impedance Transformation

An interesting characteristic of the transformer is its ability to transform impedance. In reference to Fig. 17.1, the apparent resistance that is being driven by the voltage source is given by

$$R'_L = R_L \left(\frac{n_p}{n_s}\right)^2 \tag{17.5}$$

We call this apparent resistance the *reflected load resistance.*

EXAMPLE 17.1

The transformer in Fig. 17.1 has $n_p/n_s = 10$, $V_{in} = 5\ V_{rms}$, and $R_L = 100\ \Omega$. Determine V_{sec}, I_{sec}, R'_L, and I_{pri}. Also determine the power delivered by the source P_{in}, and the power delivered to the load P_L.

Solution Applying the basic transformer relationships, we have

$$\begin{aligned} V_{sec} &= V_{pri}\frac{n_s}{n_p} \\ &= (5\ \text{V}_{rms})(0.1) \\ &= 0.5\ \text{V}_{rms} \end{aligned}$$

$$\begin{aligned} I_{sec} &= \frac{V_{sec}}{R_L} \\ &= \frac{0.5\ \text{V}_{rms}}{100\ \Omega} \\ &= 5\ \text{mA}_{rms} \end{aligned}$$

$$\begin{aligned} R'_L &= R_L\left(\frac{n_p}{n_s}\right)^2 \\ &= (100\ \Omega)(100) \\ &= 10\ \text{k}\Omega \end{aligned}$$

$$\begin{aligned} I_{pri} &= \frac{n_s}{n_p} I_{sec} \\ &= \frac{5\ \text{mA}_{rms}}{10} \\ &= 500\ \mu\text{A}_{rms} \end{aligned}$$

The power delivered to the primary is found to be

$$\begin{aligned} P_{in} &= V_{pri} I_{pri} \\ &= (5\ \text{V})(500\mu\text{A}) \\ &= 2.5\ \text{mW} \end{aligned}$$

The load power dissipation is

$$\begin{aligned} P_L &= V_{sec} I_{sec} \\ &= (0.5\ \text{V})(5\ \text{mA}) \\ &= 2.5\ \text{mW} \end{aligned}$$

Let's look at the results of Example 17.1. Because the transformer is ideal, we find that the load power dissipation equals the power delivered by the source. Also, note the large increase in the apparent load resistance. This would be useful, for example, if we wish to drive a low resistance load using a common-emitter amplifier, which has rather high output resistance. The transformer prevents the low-resistance load from loading down the CE amplifier, keeping power gain high.

Autotransformers

An autotransformer is basically a tapped inductor, shown schematically in Fig. 17.2. Except that they don't provide electrical isolation between primary and secondary windings, autotransformers are used to perform the same functions as normal transformers. The autotransformer in Fig. 17.2(a) is configured for step-down operation. The number of turns of the primary winding is

$$n_p = n_1 + n_2 \tag{17.6}$$

The number of turns on the secondary winding is

$$n_s = n_2 \tag{17.7}$$

With n_p and n_s thus defined, we can apply Eqs. (17.1) through (17.5) directly to the autotransformer.

Note
Another place where autotransformers are often found is in xenon lamp flash equipment.

FIGURE 17.2
Autotransformers. (a) Step-down. (b) Step-up.

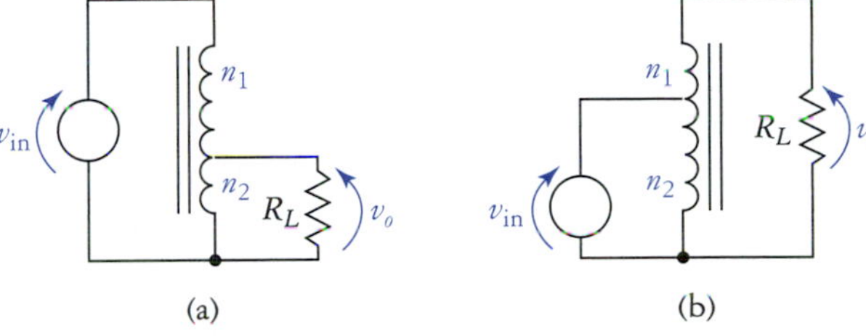

The circuit in Fig. 17.2(b) shows the autotransformer configured for step-up operation. A familiar example of a step-up autotransformer is the ignition coil of an automobile engine. For the step-up configuration, we define the primary winding using

$$n_p = n_2 \tag{17.8}$$

and the secondary winding is

$$n_s = n_1 + n_2 \tag{17.9}$$

Again, Eqs. (17.1) through (17.5) are now directly applicable to this circuit.

Transformer Construction

The core material used in a transformer is indicated on a schematic diagram as shown in Fig. 17.3. Iron cores are typically useful up to around 20 kHz or so. At higher frequencies, hysteresis losses result in decreased coupling efficiency and core heating. Iron core transformers provide very efficient coupling of energy between the primary and secondary windings, and they are relatively inexpensive. Most transformers used in power supply designs, for example, are iron core devices. Actually, high-silicon steel is used as the core material, but the term "iron core" is still widely used.

FIGURE 17.3
Transformer core materials. (a) Iron core. (b) Ferrite core. (c) Air core.

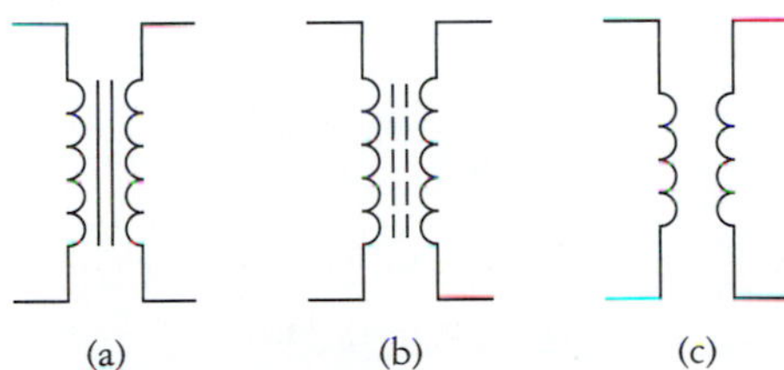

Another popular transformer core material is *ferrite.* Ferrites are powdered magnetic alloys that are *sintered* to form magnetic cores. Sintering is a process whereby the powdered ferrite is heated and compressed into the desired core shape. Ferrites are used over frequencies ranging from less than 100 Hz to hundreds of megahertz.

Air core transformers are typically used at frequencies of 1 MHz and higher. The coupling coefficient of an air core transformer will generally be lower than that of a typical iron or ferrite core transformer, with $k \cong 0.5$ to 0.8 being common. Unless otherwise noted, we assume that all transformers used here have a unity coupling coefficient.

Resonant Circuit Review

Before we explore transformer coupling further, let us review a few characteristics of resonant circuit behavior. A series-resonant circuit and plots of circuit current and input impedance are shown in Fig. 17.4. Recall that in resonance $|X_L| = |X_C|$, which causes the input impedance of the circuit to be real and equal to the series resistance in the circuit. The resonant frequency is given by the very familiar equation

$$f_o = \frac{1}{2\pi\sqrt{LC}} \tag{17.10}$$

The Q of the series resonant circuit is given by

$$Q = \frac{X_L}{R} \quad (\text{at } f = f_o) \tag{17.11}$$

The higher Q is, the sharper the curves in Fig. 17.4 become. If we view the circuit of Fig. 17.4 as a filter with the output taken across the resistor as shown, we find that this is a band-pass filter for which

$$\text{BW} = \frac{f_o}{Q} \tag{17.12}$$

FIGURE 17.4
Series-resonant circuit with current and impedance versus frequency.

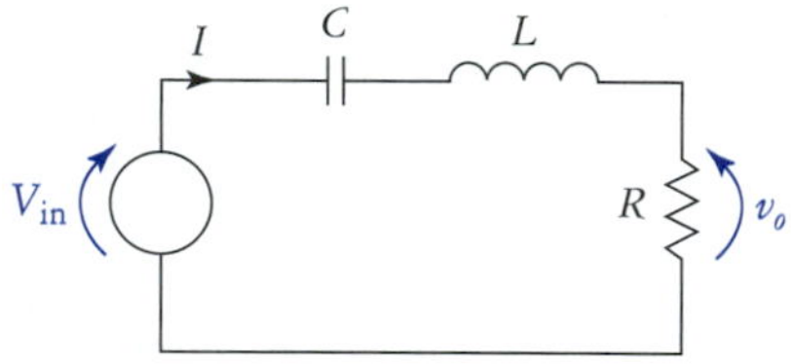

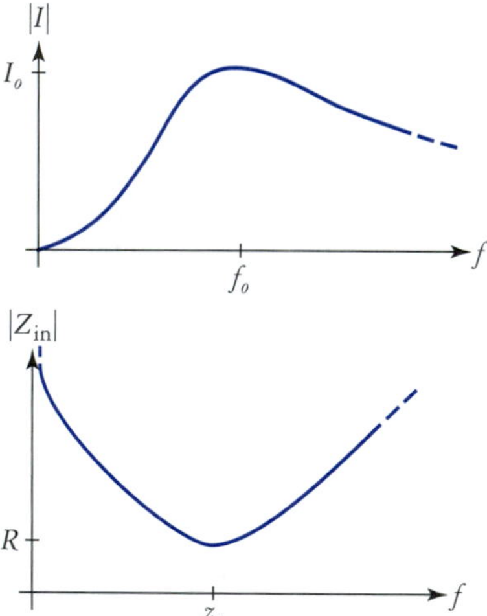

The series-resonant circuit is important, but in most tuned RF amplifier applications, a parallel resonant circuit will be used to modify the frequency response of the amplifier. An ideal parallel resonant circuit, driven by an ac current source, is shown in Fig. 17.5. The associated plots of $|Z_{in}|$ and the voltage across the resonant circuit are shown as well. Notice that in parallel resonance the inductive and capacitive reactances cancel to infinity, causing the input impedance to be resistive, with

$$|Z_{in}| = R_p \quad (\text{at } f = f_o) \tag{17.13}$$

A parallel-resonant circuit with a nonideal inductor is shown on the left side of Fig. 17.6. In high-frequency circuits, low-value capacitors, which have ceramic, mica, or polypropylene dielectrics, for

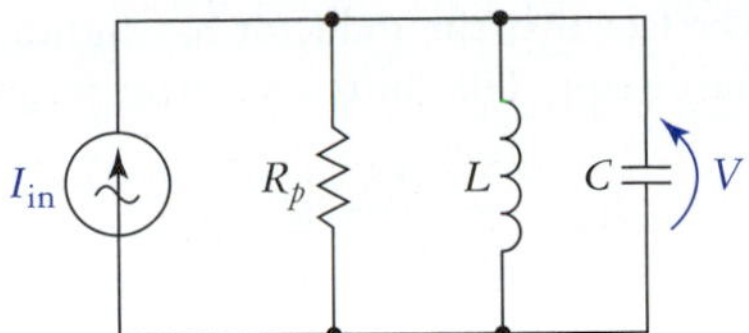

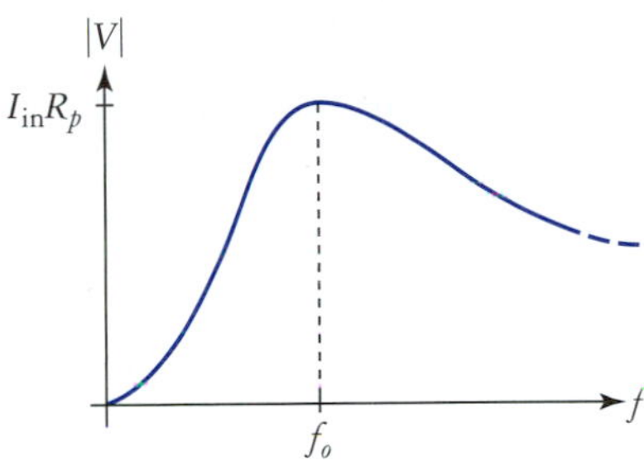

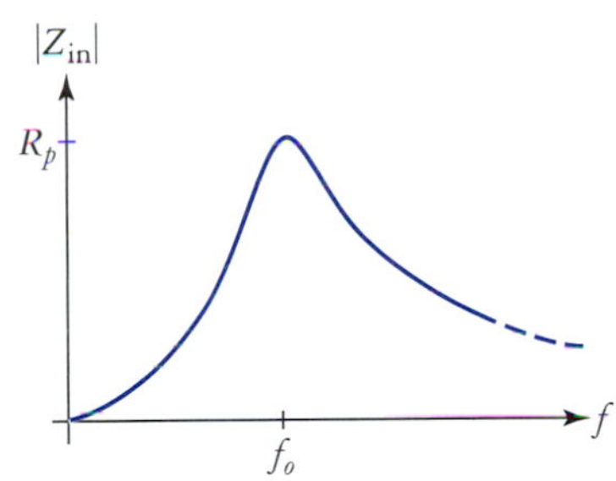

FIGURE 17.5
Parallel-resonant circuit with voltage and impedance versus frequency.

example, behave almost ideally. The series resistance of the inductor, however, must nearly always be accounted for. The Q of this circuit is dominated by that of the inductor, which is

$$Q_L = \frac{|X_L|}{r_s}(f = f_o) \tag{17.14}$$

where we assume that $f_o = 1/2\pi\sqrt{LC}$, as usual.

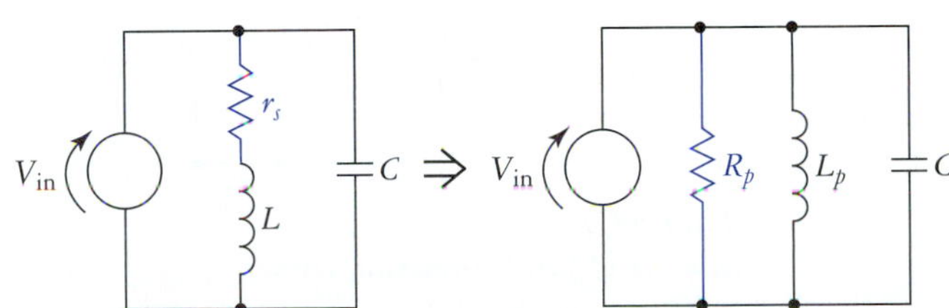

FIGURE 17.6
Nonideal inductor in a parallel-resonant circuit may be transformed to more useful form.

As we shall learn very soon, it turns out that when we perform a tuned RF amplifier analysis, it is most convenient to model the nonideal parallel resonant circuit as shown on the right side of Fig. 17.6. In this form, the series resistance of the inductor r_s is transformed into a parallel equivalent labeled R_p. The effective value of the inductance changes as well. These equivalent circuit values are

$$R_p = r_s(Q_L^2 + 1) \tag{17.15}$$

and

$$L_p = L\left(\frac{Q_L^2 + 1}{Q_L^2}\right) \tag{17.16}$$

The fact that the inductor has significant series resistance causes a shift in the resonant frequency of the circuit. The shifted resonant frequency f'_o is

$$f'_o = \frac{1}{2\pi\sqrt{L_p C}} \tag{17.17}$$

or, equivalently,

$$f'_o = f_o\left(\frac{Q_L^2}{Q_L^2 + 1}\right)^{1/2} \tag{17.18}$$

If $Q_L \geq 10$, then the difference between the ideal and nonideal inductance values is negligible, and we can write

$$L_p \cong L_s \quad (\text{for } Q_L \geq 10) \tag{17.19}$$

and

$$f'_o = f_o \quad (\text{for } Q_L \geq 10) \tag{17.20}$$

In addition to the parallel equivalent resistance of the inductor, the parallel-resonant circuit may also have other resistances shunting the tuned circuit, as shown in Fig. 17.7. In a case like this, the resistances are combined in parallel, as usual, where

$$R_{eq} = R_1 \parallel R_p \tag{17.21}$$

Now, accounting for loading by R_1, the total Q of the circuit is given by

$$Q_t = \frac{R_{eq}}{|X_L|} \quad [\text{at } f = f_o \text{ from Eq. (17.10)}] \tag{17.22}$$

Notice that for the parallel-resonant circuit, the Q_t formula is the inverse of that for the series-resonant circuit. This total Q_t is used to determine the loaded bandwidth of the resonant circuit, which is

$$\text{BW}_t = \frac{f_o}{Q_t} \tag{17.23}$$

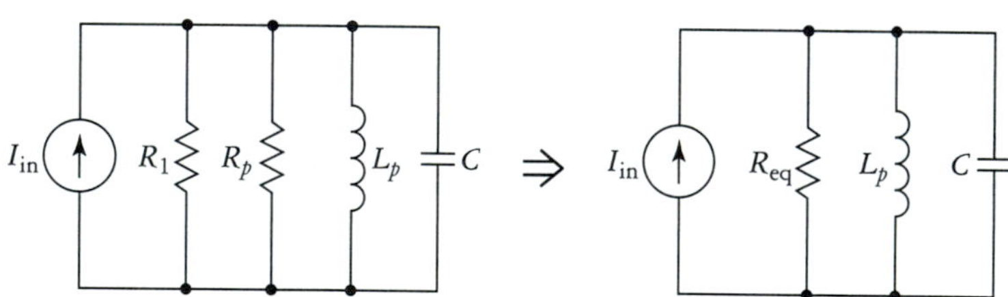

FIGURE 17.7
Reduction of parallel-resonant circuit.

This has been quite a rapid and intense review of the highlights of resonant circuit behavior. An example should help tie up any loose ends that may be present.

EXAMPLE 17.2

Refer to Fig. 17.8. This is the ac-equivalent circuit for a certain common-emitter amplifier. The dependent current source represents the collector of an npn transistor operating in the active region. In this particular case, we are assuming that the current source has infinite internal resistance (the transistor has $r_{CE} = \infty$).

Determine f_o, Q_L, R_p, L_p, f'_o, R'_L, Q_t, and BW.

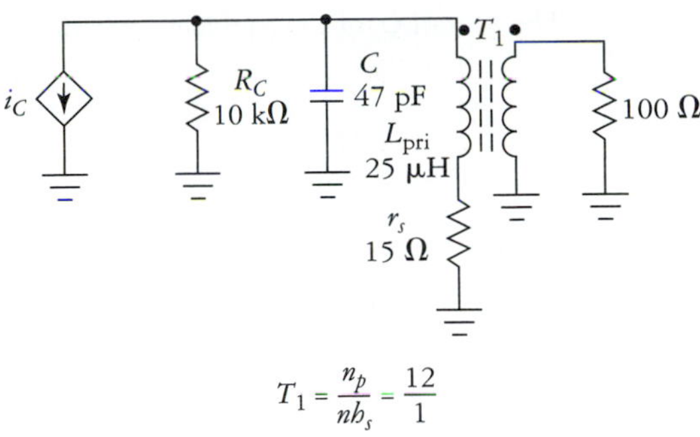

FIGURE 17.8
Circuit for Example 17.2.

Solution The ideal resonant frequency is

$$f_o = \frac{1}{2\pi\sqrt{LC}}$$
$$= \frac{1}{(2\pi)(25\ \mu\text{H} \times 47\ \text{pF})^{1/2}}$$
$$= 4.64\ \text{MHz}$$

The unloaded Q of the inductor is given by

$$Q_L = \left.\frac{|X_L|}{r_s}\right|_{f=f_o}$$
$$= \frac{2\pi f_o L}{r_s}$$
$$= \frac{729\ \Omega}{15\ \Omega}$$
$$= 48.6$$

We can now determine the parallel equivalent of r_s, which is

$$R_p = r_s(Q_L^2 + 1)$$
$$= 15\ \Omega(48.6^2 + 1)$$
$$= 35.4\ \text{k}\Omega$$

The effective inductance of the transformer primary winding is

$$L_p = L_{\text{pri}}\left(\frac{Q_L^2 + 1}{Q_L^2}\right)$$
$$\cong (25\ \mu\text{H})(1)$$
$$= 25\ \mu\text{H}$$

Because $Q_L \gg 10$, the effective inducatance of the primary winding is virtually identical to the ideal value. We find that the same situation occurs for the shifted resonant frequency, which is

$$f_o' = f_o\left(\frac{Q_L^2}{Q_L^2 + 1}\right)^{1/2}$$
$$\cong (4.64\ \text{MHz})(1)$$
$$= 4.64\ \text{MHz}$$

The reflected load resistance is

$$R_L' = R_L\left(\frac{n_p}{n_s}\right)^2$$
$$= (100\ \Omega)(12^2)$$
$$= 14.4\ \text{k}\Omega$$

The overall Q_t of the equivalent circuit is found as follows:

$$\begin{aligned} Q_t &= \frac{R_{eq}}{|X_L|} \\ &= \frac{R_C \| R_p \| R_L'}{|X_L|} \\ &= \frac{10\text{ k} \| 35.4\text{ k} \| 14.4\text{ k}}{729\ \Omega} \\ &= \frac{5058\ \Omega}{729\ \Omega} \\ &= 6.94 \end{aligned}$$

Finally, we determine the bandwidth of the circuit to be

$$\begin{aligned} \text{BW} &= \frac{f_o}{Q_t} \\ &= \frac{4.64\text{ MHz}}{6.94} \\ &= 668.6\text{ kHz} \end{aligned}$$

Transformer-Coupled Tuned Amplifier Analysis

As illustrated in the previous example, in RF amplifier applications, it is common for the inductance of a transformer to serve as part of a parallel-resonant circuit, typically located on the output of that amplifier. A tuned-output, class A, common-emitter amplifier is shown in Fig. 17.9. Don't let the unusual schematic layout throw you. It is common practice for communications circuitry schematics to be drawn with the supply rail at the bottom of the diagram. Let's walk through the analysis of this circuit step-by-step, using previously developed techniques.

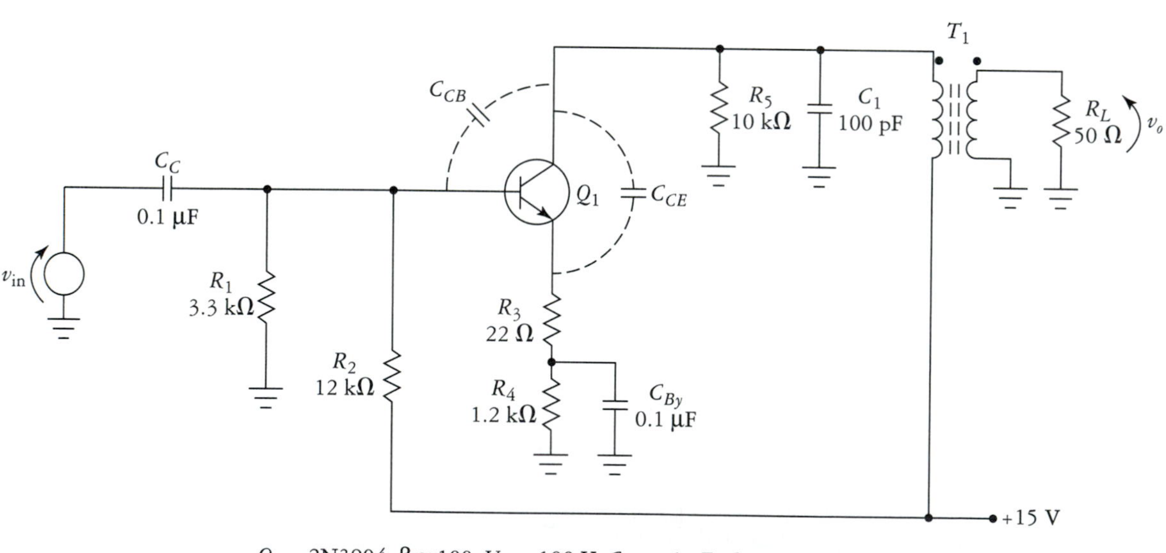

FIGURE 17.9
A transformer-coupled, tuned class A amplifier.

DC Analysis

The dc-equivalent circuit is shown in Fig. 17.10(a). If we redraw the dc equivalent, the familiar form of Fig. 17.10(b) results. Notice that the primary winding of the transformer has such a small winding resistance (r_s = 25 Ω) that we make the simplifying assumption that the collector is connected directly to V_{CC}. The minimum information we require from the dc analysis is V_{CEQ}, I_{CQ}, and r_e, which are found as follows:

$$R_{Th} = R_1 \| R_2 = 3.3\ k\Omega \| 12\ k\Omega = 2.59\ k\Omega$$

$$V_{Th} = V_{CC}\frac{R_1}{R_1 + R_2} = 15\ V\frac{3.3\ k\Omega}{3.3\ k\Omega + 12\ k\Omega} = 3.24\ V$$

$$\begin{aligned} I_{CQ} &= \frac{V_{Th} - V_{BE}}{R_E + R_{Th}/\beta} \\ &= \frac{3.24\ V - 0.7\ V}{1222\ \Omega + 25.9\ \Omega} \\ &\cong 2\ mA \end{aligned}$$

$$\begin{aligned} V_{CEQ} &= V_{CC} - I_{CQ}R_E \\ &= 15\ V - (2\ mA)(1222\ \Omega) \\ &\cong 12.6\ V \end{aligned}$$

$$\begin{aligned} r_e &= \frac{V_T}{I_{CQ}} \\ &= \frac{26\ mV}{2\ mA} \\ &= 13\ \Omega \end{aligned}$$

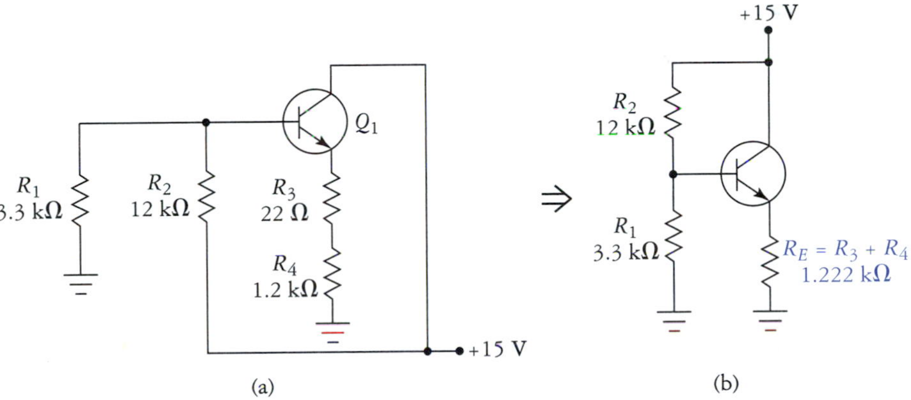

FIGURE 17.10
DC-equivalent circuit. (a) Drawn as shown in the original circuit. (b) Drawn in a more familiar manner.

AC Analysis

The ac-equivalent circuit is shown in Fig. 17.11(a) in unreduced form. The analysis is performed at the resonant frequency of the amplifier, so that is where we begin. The stray capacitances are combined with the tuning capacitor giving us

$$\begin{aligned} C_T &= C_{CB} + C_{CE} + C_1 \\ &= 1\ pF + 0.5\ pF + 100\ pF \\ &= 101.5\ pF \end{aligned}$$

In practice, we would be justified disregarding the stray capacitances because $C_1 \gg C_{CB} + C_{CE}$. It is a matter of judgment; but since we took the time to determine C_T, the accurate value is used this time. The resonant frequency is

$$\begin{aligned} f_o &= \frac{1}{2\pi\sqrt{LC_T}} \\ &= \frac{1}{2\pi\sqrt{50\ \mu H \times 101.5\ pF}} \\ &= 2.23\ MHz \end{aligned}$$

The remaining frequency-dependent circuit parameters are now found:

$$Q_L = \left.\frac{|X_L|}{r_s}\right|_{f=f_o} = \frac{2\pi f_o L}{r_s} = \frac{700\ \Omega}{25\ \Omega} \cong 28$$

$$R_p = r_s(Q_L^2 + 1) = 25\ \Omega(28^2 + 1) = 19.6\ \text{k}\Omega$$

The Q_L is high enough that we can disregard shifts of f_o and L_p caused by r_s. The reflected load resistance is

$$R_L' = R_L\left(\frac{n_p}{n_s}\right)^2 = (50\ \Omega)(100) = 5\ \text{k}\Omega$$

Most of the time we will assume that the internal collector resistance of the active-region transistor is infinite (i.e., $r_o = \infty$ or, equivalently, $r_{CE} = \infty$). Because the Early voltage V_A is given for the transistor, we can easily calculate

$$r_o \cong \frac{V_A}{I_{CQ}} = \frac{100\ \text{V}}{2\ \text{mA}} = 50\ \text{k}\Omega$$

We can finally lump these resistances into the total ac collector resistance R_C', which is

$$R_C' = r_o \parallel R_C \parallel R_p \parallel R_L' = 50\ \text{k}\Omega \parallel 10\ \text{k}\Omega \parallel 19.6\ \text{k}\Omega \parallel 5\text{k}\Omega = 2.7\ \text{k}\Omega$$

This equivalent resistance is seen in Fig. 17-11(b). The total Q_t of the tuned circuit is now determined:

$$Q_t = \frac{R_C'}{|X_L|} = \frac{2.7\ \text{k}\Omega}{700\ \Omega} = 3.85$$

The bandwidth of the amplifier is

$$\text{BW} = \frac{f_o}{Q_t} = \frac{2.23\ \text{MHz}}{3.85} = 579.2\ \text{kHz}$$

The voltage gain of transistor Q_1, which shall be denoted by A_v' in this discussion, is found to be

$$A_v' = -\frac{R_C'}{R_e + r_e} = -\frac{2.7\ \text{k}\Omega}{22\ \Omega + 13\ \Omega} \cong -77$$

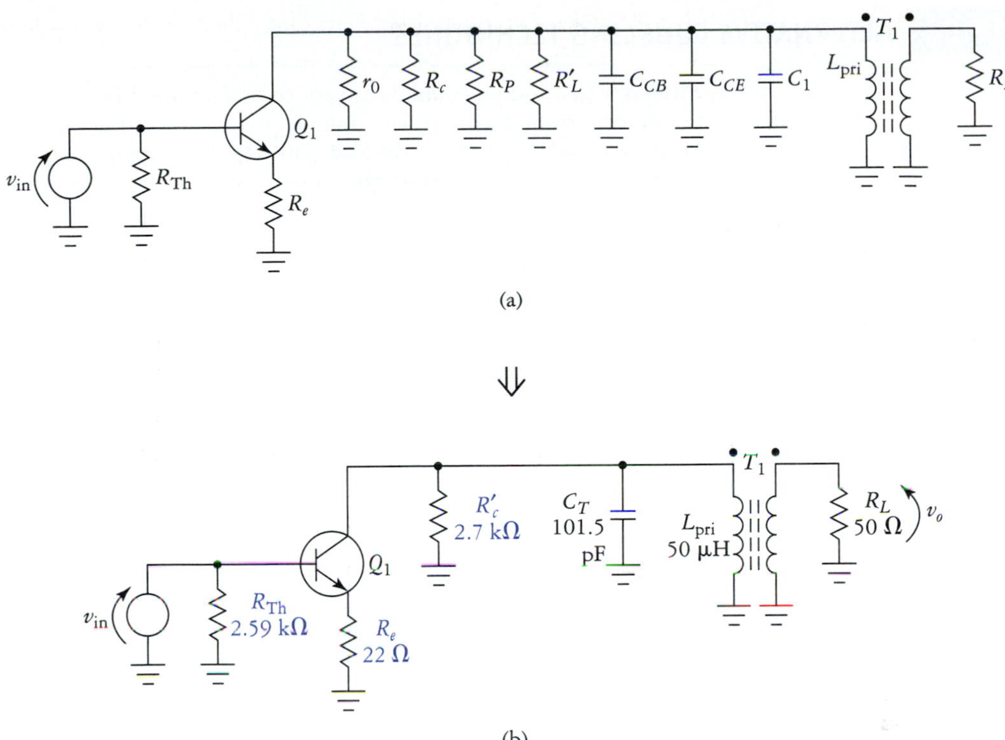

FIGURE 17.11
AC-equivalent circuits. (a) All equivalent resistances shown explicitly. (b) Reduced version.

Note
The presence of the transformer must be accounted for when determining the overall gain of the amplifier.

This is not, however, the voltage gain of the amplifier. Because a step-down transformer is used, the actual overall voltage gain (at resonance) is

$$A_v = A_v' \frac{n_s}{n_p} \tag{17.24}$$

$$= -77 \frac{1}{10}$$

$$= -7.7$$

The negative sign is retained because the phasing dots present on the transformer, in relation to the voltage-sensing arrows, indicate that the transformer preserves the phase of its primary voltage.

We finish this analysis with the calculation of the input resistance, current gain, and power gain for the amplifier. Thus, we find

$$\begin{aligned} R_{\text{in}} &= R_{\text{Th}} \parallel \beta(R_e + r_e) \\ &= 2.59\text{ k}\Omega \parallel 100(22\Omega + 13\ \Omega) \\ &\cong 1.5\text{ k}\Omega \end{aligned}$$

$$\begin{aligned} A_i &= A_v \frac{R_{\text{in}}}{R_L} \\ &= -7.7 \frac{1.5\text{ k}\Omega}{50\ \Omega} \\ &= -231 \end{aligned}$$

$$\begin{aligned} A_p &= A_v A_i \\ &= (-7.7)(-231) \\ &= 1.78\text{ k} \end{aligned}$$

17.2 ALTERNATIVE COUPLING TECHNIQUES

Although transformer coupling is very commonly used in tuned amplifier designs, other coupling techniques are used as well. For example, the circuit in Fig. 17.12 uses a tapped capacitor arrangement to provide coupling to the load. This coupling arrangement is capable of transforming the load resistance value much like the transformer coupling covered previously.

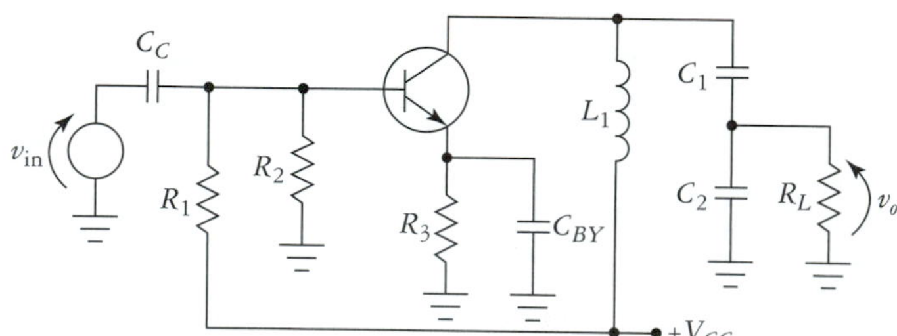

FIGURE 17.12
Tapped capacitor coupling.

The dc analysis of Fig. 17.12 is straightforward and will not be reviewed again at this time. The ac analysis begins with the determination of the equivalent capacitance in the resonant circuit. Because C_1 and C_2 are in series, we have

$$C_{eq} = \frac{1}{\frac{1}{C_1} + \frac{1}{C_2}} \tag{17.25}$$

Of course, stray capacitances may also factor into the equation, if they are of a significant magnitude. For the sake of simplicity, we neglect these stray capacitances here, although C_{CB} and C_{CE} simply add to the total equivalent capacitance. The ideal resonant frequency of the amplifier is

$$f_o = \frac{1}{2\pi\sqrt{L_1 C_{eq}}} \tag{17.26}$$

The split capacitors perform an impedance transformation. Assuming that $|X_{C_2}| \geq 10R_L$, we have

$$R'_L = R_L\left(\frac{C_1 + C_2}{C_1}\right)^2 \tag{17.27}$$

This is the reflected load resistance, as "seen" by the transistor. The ac-equivalent collector resistance, at resonance, is

$$R'_C = r_o \parallel R'_L \parallel R_p \tag{17.28}$$

Recall that R_p is the parallel equivalent of the inductor series resistance r_s, given by Eq. (17.15). As usual, we often choose to disregard r_o (assume the transistor is ideal, with $r_o = \infty$) to simplify the analysis. There may also be other resistances present, to intentionally reduce Q perhaps, which would be factored in as well.

The voltage gain of the CE connected transistor (called A'_v in this discussion) is determined as usual, by

$$A'_v = -\frac{R'_C}{r_e + R_e}$$

However, capacitors C_1 and C_2 form a voltage divider that attenuates the output of the amplifier. The overall voltage gain is found to be

$$A_v = A'_v\frac{C_1}{C_1 + C_2} \tag{17.29}$$

EXAMPLE 17.3

Refer to Fig. 17.12. Assume that the transistor is biased for $I_{CQ} = 1$ mA, $V_{CEQ} = 0.9V_{CC}$ (a reasonable V_{CE} for such a circuit), $r_o = 100$ kΩ, $L_1 = 25$ μH, $r_s = 10$ Ω, $C_1 = 25$ pF, $C_2 = 100$ pF, and $R_L = 1$ kΩ. Determine C_{eq}, f_o, Q_L, R_p, R_L', Q_t, A_v, and BW.

Solution The total equivalent capacitance is

$$C_{\text{eq}} = \frac{1}{\dfrac{1}{C_1} + \dfrac{1}{C_2}}$$
$$= \frac{1}{\dfrac{1}{25\text{ pF}} + \dfrac{1}{100\text{ pF}}}$$
$$= 20\text{ pF}$$

The resonant frequency is found to be

$$f_o = \frac{1}{2\pi\sqrt{L_1 C_{\text{eq}}}}$$
$$= \frac{1}{2\pi\sqrt{25\ \mu\text{H} \times 20\text{ pF}}}$$
$$= 7.11\text{ MHz}$$

The Q_L of the inductor is

$$Q_L = \frac{2\pi f_o L}{r_s}$$
$$= \frac{1.11\text{ k}\Omega}{10\ \Omega}$$
$$= 111$$

This produces a parallel equivalent resistance of

$$R_p = r_s(Q_L^2 + 1)$$
$$= (10\ \Omega)(12.3\text{ k})$$
$$= 123\text{ k}\Omega$$

The reflected load resistance is

$$R_L' = R_L\left(\frac{C_1 + C_2}{C_1}\right)^2$$
$$= (1\text{ k}\Omega)(5^2)$$
$$= 25\text{ k}\Omega$$

Thus, the total ac collector resistance is

$$R_C' = r_o \parallel R_p \parallel R_L'$$
$$= 100\text{ k}\Omega \parallel 123\text{ k}\Omega \parallel 25\text{ k}\Omega$$
$$= 17.2\text{ k}\Omega$$

The voltage gain of the amplifier can now be determined. This is

$$A_v = -\left(\frac{R_C'}{r_e}\right)\left(\frac{C_1}{C_1 + C_2}\right)$$
$$= -\left(\frac{17.2\text{ k}\Omega}{26\ \Omega}\right)\left(\frac{25}{25 + 100}\right)$$
$$= -529$$

The overall Q_t of the circuit is now found:

$$Q_t = \frac{R'_C}{|X_L|}$$
$$= \frac{17.2\text{ k}\Omega}{1.11\text{ k}\Omega}$$
$$= 15.6$$

Finally, the bandwidth of the amplifier is

$$\text{BW} = \frac{f_o}{Q_t}$$
$$= \frac{7.11\text{ MHz}}{15.6}$$
$$= 455.8\text{ kHz}$$

As a final example, let us examine a tuned amplifier that uses an autotransformer to provide load coupling.

EXAMPLE 17.4

Refer to Fig. 17.13. Assume that $I_{CQ} = 2$ mA, $r_o = 50$ kΩ, $L_T = 50$ μH, $r_s = 12$ Ω, $C_1 = 50$ pF, $n_1 = 25$, $n_2 = 5$, and $R_L = 75$ Ω. Determine f_o, Q_L, R_p, R'_L, R'_C, Q_t, A_v, and BW.

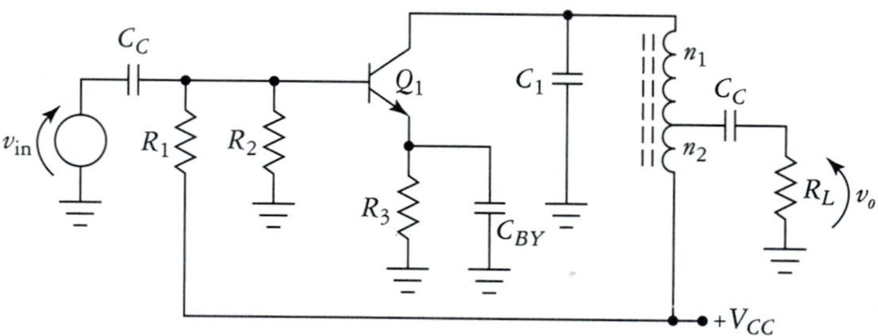

FIGURE 17.13
Autotransformer coupling.

Solution The resonant frequency is

$$f_o = 1/2\pi\sqrt{L_T C_1} = 3.18\text{ MHz}$$

The Q_L of the inductor is

$$Q_L = 2\pi f_o L_T/r_s = 999/12 \cong 83$$

The parallel equivalent of r_s is

$$R_p = r_s(Q_L^2 + 1) = (12\ \Omega)(6.9\text{ k}) = 82.8\text{ k}\Omega$$

The reflected load resistance is found using the basic transformer relationships developed in the previous section:

$$R'_L = R_L\left(\frac{n_p}{n_s}\right)^2$$
$$= R_L\left(\frac{n_1 + n_2}{n_2}\right)^2$$
$$= 75\ \Omega\left(\frac{30}{5}\right)^2$$
$$= 2.7\text{ k}\Omega$$

The ac collector resistance is

$$R'_C = r_o \,\|\, R'_L \,\|\, R_p = 50\text{ k}\Omega \,\|\, 2.7\text{ k}\Omega \,\|\, 82.8\text{ k}\Omega \cong 2.5\text{ k}\Omega$$

The overall Q_T is now found:

$$Q_T = R'_C/|X_L| = 2.5\text{ k}\Omega/999\ \Omega \cong 2.5$$

The bandwidth of the amplifier is

$$\text{BW} = f_o/Q = 3.18\text{ MHz}/2.5 = 1.27\text{ MHz}$$

The voltage gain of the amplifier is found as follows:

$$\begin{aligned} A_v &= -\left(\frac{R'_C}{r_e}\right)\left(\frac{n_s}{n_p}\right) \\ &= -\left(\frac{R'_C}{r_e}\right)\left(\frac{n_2}{n_1 + n_2}\right) \\ &= -\left(\frac{2.5\text{ k}}{13\ \Omega}\right)\left(\frac{5}{30}\right) \\ &= (-192)(0.17) \\ &= -32 \end{aligned}$$

17.3 TRANSISTOR PERFORMANCE CONSIDERATIONS

So far, we have not considered the suitability of a given transistor for use in a given tuned amplifier application. Many factors must be considered, with the main ones being frequency, power dissipation, and noise characteristics.

Frequency Effects

The frequency-related transistor limitations that must be considered in tuned amplifier applications are the same as those presented in Chapter 11. Briefly, depending on the amplifier configuration used, a given transistor parameter may be more or less significant than others. For example, in the common-emitter configuration, the beta-cutoff frequency f_β would be of great importance, while in the case of a common-base amplifier, the alpha-cutoff frequency f_α is most important. Also, recall from Chapter 11 that if good high-frequency performance is desired, then the cascode configuration would be a preferred circuit topology, because it does not suffer from Miller-effect capacitance multiplication. This allows the cascode configuration to provide good current and voltage gain at high frequencies. A typical tuned cascode amplifier is shown in Fig. 17.14. Analysis of this circuit is left as an exercise at the end of this chapter.

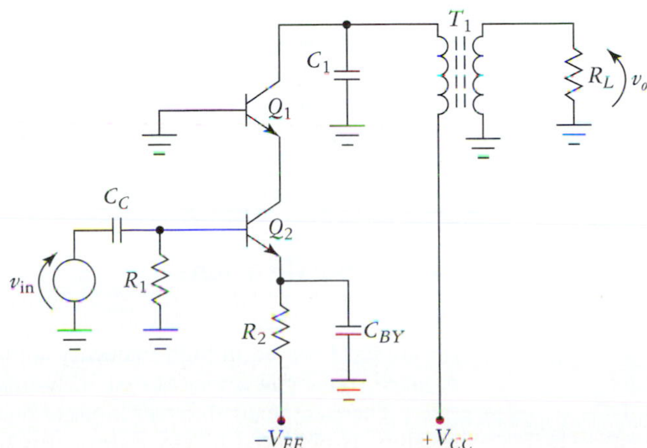

FIGURE 17.14
Tuned cascode amplifier.

Circuit layout becomes very critical at high frequencies. Stray capacitances and inductances make conventional breadboarding techniques very difficult for $f \cong 1$ MHz and nearly impossible for $f \geq 10$ MHz. At still higher frequencies, such as those use in VHF and UHF communication systems (hundreds of megahertz), the use of printed circuit boards is required even for circuit prototyping.

RF Transistors

Specially designed RF transistors are used in high-frequency applications. In the context of this discussion, we may consider high frequency to be anything over 1 MHz, although in modern communication systems we may be dealing with frequencies ranging from 100 MHz to over 1 GHz.

For low-power applications ($P_{D(\text{typ})} \leq 1$ W), these RF transistors are generally packaged the same as general-purpose low-power devices, using such popular case styles as the TO-92 and TO-18. Surface mount packaging, such as the TO-236 (SOT-23) and Micro-X are also very commonly used for low-power RF transistors. These packages are illustrated in Fig. 17.15(a). High-power RF transistors are typically packaged in "pill" or stud forms, which are illustrated in Fig. 17.15(b).

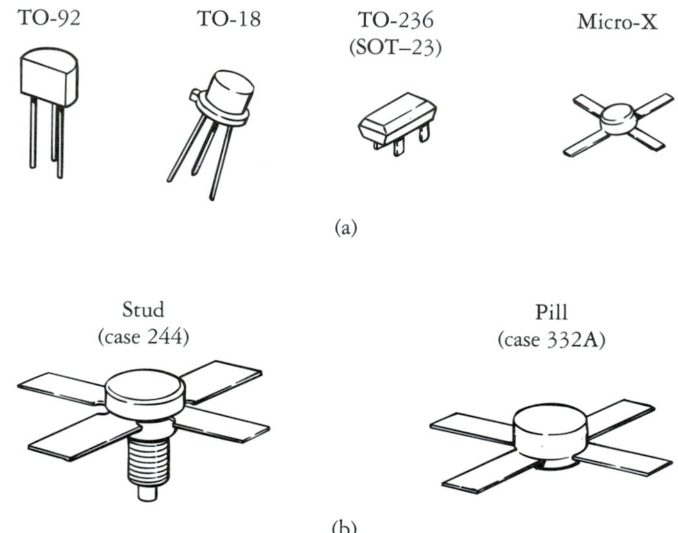

FIGURE 17.15
Common RF transistor packages. (a) Low power. (b) High power.

Some aspects of RF transistor performance are specified in the same manner as for their general-purpose cousins. For example, power dissipation, temperature, voltage, and current limits are fairly standard regardless of device type. Because of the demands placed on RF transistors, however, many of the frequency-related specifications are expressed in the form of *scattering parameters* (often simply called *S parameters*). Although they are beyond the scope of this book, here's a quick explanation of S parameters. Basically, S parameters characterize frequency-dependent transistor behavior by modeling the transistor as a two-port network, much as in hybrid parameter analysis. There are four S parameters: S_{11}, S_{12}, S_{21}, and S_{22}. At very high frequencies, electrical signals must be viewed as traveling waves that are reflected and absorbed by various circuit elements. Scattering parameters are expressed in terms of signal energy absorption, reflection, and transfer from one port to another. S parameters are complex numbers that vary with frequency. Sometimes S parameters are listed in tabular from, or they may be plotted over a specific frequency range on a special graph called a *Smith chart.*

Noise Specifications

Transistor noise specifications may be of great significance in any low-level signal amplification application. However, in high-frequency applications such as radio receiver front-end circuitry, where received signal power may be extremely small, noise is an especially crucial issue.

There are many different kinds of noise. Thermal noise, which is most closely associated with resisters, is produced by any resistor that has a temperature above absolute zero. Thermal noise is *white noise,* and has a flat (relatively constant) average power from near dc up to the gigahertz range. Thermal noise voltage is proportional to temperature, resistance, and bandwidth according to

$$V_{n(\mathrm{rms})} = \sqrt{4kTR\Delta f} \tag{17.30}$$

where k is Boltzmann's constant (1.38×10^{-23} W/K·Hz), T is absolute temperature in kelvin, R is resistance, and Δf is the bandwidth of interest. To reduce thermal noise, designers minimize these parameters as much as possible, especially in low-level amplifiers. In some applications, such as in deep-space communications systems, receiver front ends are cooled with liquid nitrogen in order to reduce thermal noise. The use of narrow bandwidth tuned amplifiers also helps to reduce the effects of noise.

Signal-to-Noise Ratio and Noise Figure

The ratio of signal power to noise power at a given point in a circuit, such as the input to an amplifier, is the signal-to-noise ratio, SNR. If we are dealing with power levels, we shall use p as a subscript and write the equation

$$\mathrm{SNR}_p = \frac{P_{\mathrm{sig}}}{P_n} \tag{17.31}$$

Very often SNR is expressed in decibels:

Note
In low-level RF amplifier designs, transistors are often biased for best performance in terms of device noise figure.

$$\mathrm{SNR}_p = 10 \log\left(\frac{P_{\mathrm{sig}}}{P_n}\right) \tag{17.32}$$

Voltage ratios can also be used in which case we change the subscript accordingly:

$$\mathrm{SNR}_v = 20 \log \frac{V_{\mathrm{sig}}}{V_n} \tag{17.33}$$

The larger the SNR, the lower the relative noise content of the desired signal.

Any amplifier will add some noise to a signal that it processes. Thus, the signal at the output of an amplifier will have a somewhat lower SNR than the input signal. Noise that is generated by a transistor, while somewhat related to temperature, is strongly related to I_{CQ} and to a somewhat lesser extent V_{CEQ}. The most frequently specified transistor noise parameter is *noise figure,* NF. The noise figure of a transistor is given by

$$\mathrm{NF} = 10 \log \frac{\mathrm{SNR}_{p(o)}}{\mathrm{SNR}_{p(\mathrm{in})}} \tag{17.34}$$

where $\mathrm{SNR}_{p(\mathrm{in})}$ and $\mathrm{SNR}_{p(o)}$ are the linear (i.e., not decibel) signal-to-noise power ratios of the input and output signals, respectively.

In addition to being a function of biasing levels I_{CQ} and V_{CEQ}, NF is also strongly dependent on the resistance of the source that is driving the transistor. These relationships are illustrated in the graphs of Fig. 17.16, which are representative of the NF characteristics of a general-purpose transistor such as a 2N3904 or similar device.

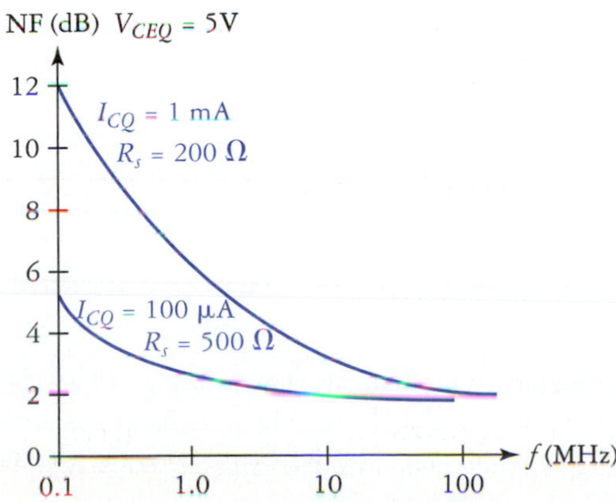

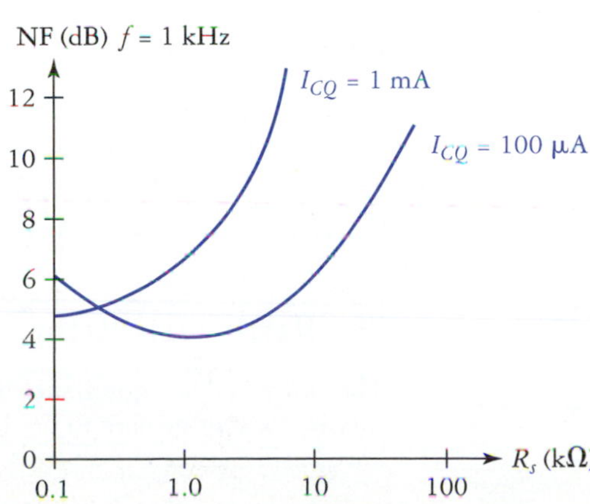

FIGURE 17.16
Various noise figure plots for a typical transistor.

For a single stage of amplification, NF simply tells us how much the signal-to-noise ratio of the output is reduced, relative to the input. For example, if a certain amplifier has $SNR_{(in)} = 12$ dB and NF = 4 dB, then the output signal-to-noise ratio is

$$\begin{aligned} SNR_{(o)} &= SNR_{(in)} - NF \\ &= 12 \text{ dB} - 4 \text{ dB} = 8 \text{ dB} \end{aligned} \tag{17.35}$$

Often the NF graphs for a given transistor are not used for quantitative analysis. Rather, this information is used to help determine a Q-point location that will result in the lowest noise.

17.4 OSCILLATORS

An oscillator is a circuit that produces an output signal in the absence of an input signal. During the course of your laboratory experience you have most likely seen examples of amplifiers that have inadvertently become oscillators. In this section we examine some of the more commonly used sinusoidal oscillator circuit topologies.

The Barkhausen Criteria

Figure 17.17 shows the block diagram for an amplifier with open-loop voltage gain A and a feedback network with gain β (called the feedback factor). This feedback model was used previously at various times (see Section 11.3), where the following closed-loop voltage-gain relationship can be derived*:

$$A_v = \frac{A}{1 - A\beta} \tag{17.36}$$

If negative feedback is employed, then the *loop gain* $A\beta \leq 0$ and we have a normal amplifier. If, however, the feedback signal is made to be in phase with the input disturbance that caused the output to be created, then we have positive feedback where the product $A\beta > 0$. This does not necessarily mean that the circuit will oscillate, however. To guarantee oscillation, several requirements must be met. First, the following inequality must be satisfied:

$$A\beta \geq 1 \tag{17.37}$$

In addition, in order for sustained oscillation to occur, the feedback network must be frequency selective. We have already seen the effects of positive feedback without frequency selectivity: the Schmitt trigger. A Schmitt trigger will not normally oscillate, but rather it will *latch up* to $\pm V_{o(max)}$ when power is applied. This is a direct result of positive feedback. Thus, to guarantee oscillation, we also require the following phasor relationship to be satisfied:

$$V_{fb} = V_o \angle n360° \qquad (n = 0, 1, \ldots) \tag{17.38}$$

Because we are using phasors, a single preferred frequency is assumed to be in existence. The feedback network is designed such that only one frequency will satisfy both Eqs. (17.37) and (17.38). These requirements are called the *Barkhausen criteria.*

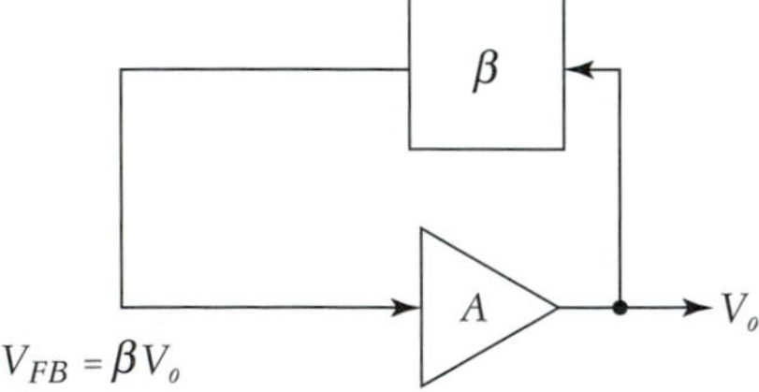

FIGURE 17.17
Amplifier with feedback.

The Hartley Oscillator

The simplified ac-equivalent circuit for a Hartley oscillator is shown in Fig. 17.18. Hartley oscillators are most often used to produce radio-frequency signals; therefore, such circuits are commonly realized using discrete transistors. The key to identifying a Hartley oscillator is the presence of a tapped

*In Section 11.3 the form $A/(1 + A\beta)$ was used. It is usually more convenient to use the form given in Eq. (17.36) when dealing with oscillator circuits.

inductor in the feedback loop. In this case L_1 and L_2 may be windings on a transformer, if the output is taken from v_{o_2}. If transformer coupling is not required, the output may be taken from the collector of the transistor as indicated by v_{o_1}.

FIGURE 17.18
Simplified transformer-coupled Hartley oscillator.

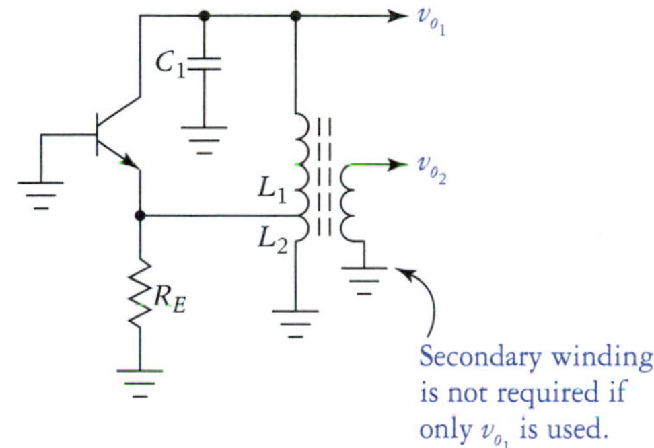

Here's the basic idea behind how the oscillator shown in Fig. 17.18 works. When the circuit is initially powered up, random noise is generated over a wide range of frequencies. Capacitor C_1 and inductors L_1 and L_2 form a parallel resonant circuit, which at f_o is purely resistive. The collector signal energy at f_o is fed to the emitter via the voltage divider of L_1 and L_2. Here we have an amplifier for which the input is the emitter and the output is the collector, that is, a common-base amplifier. In this case, however, the amplifier provides its own input signal. Recall that the common-base amplifier provides high voltage gain without phase shift. Thus, the tuned circuit filters out all but the f_o signal energy, which in turn drives the emitter with the required phase. If the loop gain is sufficiently high, then sustained oscillation will occur. The oscillation frequency is the same as the resonant frequency of the tuned circuit. That is,

$$f_o = \frac{1}{2\pi\sqrt{(L_1 + L_2)\,C_1}} \tag{17.39}$$

As far as the actual component-level design of a given oscillator circuit is concerned, hundreds of variations may be possible. Therefore, it is not possible to present a set of formulas that will work in all situations. Several analysis examples, however, are presented. The techniques used to analyze these circuits are simply those that we have been developing all along in the course of this book. As is the case with most other electronic circuits, it is important to be able to recognize a given oscillator topology on sight. With this ability we are able to get a general idea of how the circuit should behave, and then we apply quantitative analysis techniques as necessary.

EXAMPLE 17.5

Refer to Fig. 17.19. Assume the coupling and bypass capacitors are ideal and the transistor has $\beta \cong 100$, $\alpha \cong 1$, $V_\phi = 0.7$ V, and $r_o \to \infty$. Also, $C_1 = 630$ pF, $R_1 = 10$ kΩ, $R_2 = 1.2$ kΩ, $R_3 = 470$ Ω, $R_L = 50$ Ω, $V_{CC} = 15$ V, and the transformer T_1 has $n_1 = 15$ turns ($L_1 = 30$ μH, $r_{s_1} = 6$ Ω), $n_2 = 5$ turns ($L_2 = 10$ μH, $r_{s_2} = 2$ Ω), and $n_3 = 4$. Determine the Q point of the transistor, the oscillation frequency f_o, and the loop gain $A\beta$. Assuming sustained oscillation does occur, determine the approximate peak-to-peak output voltage.

FIGURE 17.19
Circuit for Example 17.5.

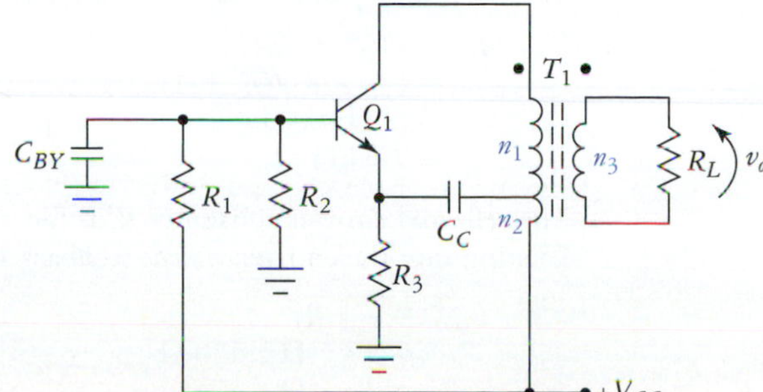

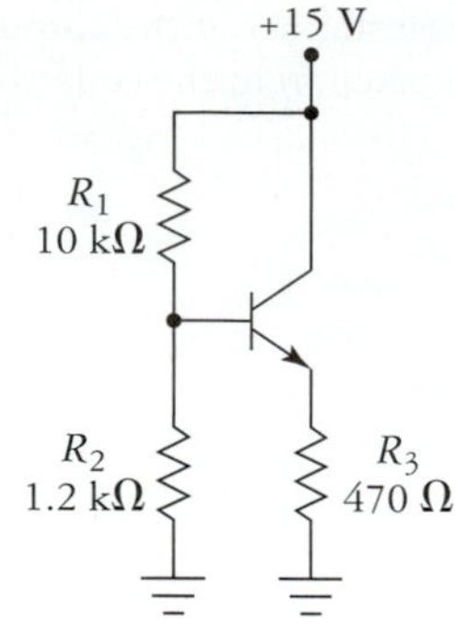

FIGURE 17.20
DC-equivalent circuit.

Solution The dc-equivalent circuit appears as shown in Fig. 17.20. Applying basic Q-point analysis techniques, we find

$$I_{CQ} = 1.9 \text{ mA}$$
$$V_{CEQ} = 14.1 \text{ V}$$
$$r_e = 13.7 \ \Omega$$

The ac-equivalent circuit shown in Fig. 17.21 requires some explanation. Let's start with the easy part; the resonant frequency is

$$f_o = \frac{1}{2\pi\sqrt{(L_1 + L_2)\,C_1}}$$
$$= \frac{1}{2\pi\sqrt{40 \ \mu\text{H} \times 630 \text{ pF}}}$$
$$\cong 1 \text{ MHz}$$

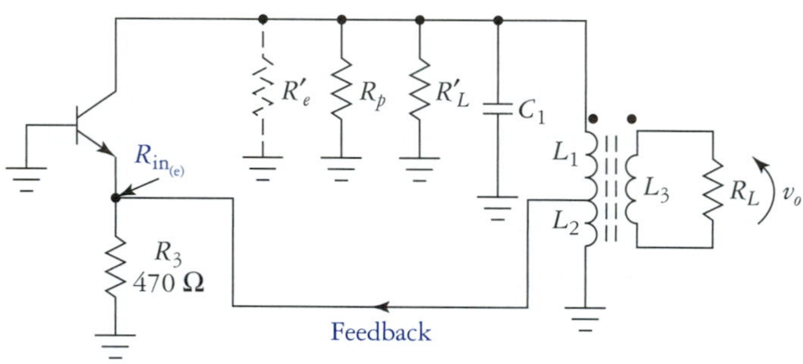

FIGURE 17.21
AC-equivalent circuit.

The reflected load resistance is

$$R'_L = R_L\left(\frac{n_p}{n_s}\right)^2$$
$$= R_L\left(\frac{n_1 + n_2}{n_3}\right)^2$$
$$= (50 \ \Omega)(5^2)$$
$$= 1.25 \text{ k}\Omega$$

A useful parameter to determine now is the Q of the primary winding of the transformer.

$$Q_L = \frac{2\pi f_o(L_1 + L_2)}{r_{s_1} + r_{s_2}}$$
$$= \frac{251 \ \Omega}{8 \ \Omega}$$
$$= 31.3$$

The equivalent parallel resistance of the transformer's primary is

$$R_p = r_s(Q_L^2 + 1)$$
$$= (r_{s_1} + r_{s_2})(Q_L^2 + 1)$$
$$\cong (8 \ \Omega)(981)$$
$$\cong 7.9 \text{ k}\Omega$$

The reflected emitter resistance R'_e is found by treating the left side of T_1 as an autotransformer. Looking into the emitter via the feedback loop, we find

$$R_{in(e)} = r_e \parallel R_3$$
$$= 13.7 \ \Omega \parallel 470 \ \Omega$$
$$= 13.3 \ \Omega$$

This resistance is reflected to the collector via the autotransformer as follows:

$$R'_e = R_{\text{in}(e)}\left(\frac{n_p}{n_s}\right)^2$$
$$= R_{\text{in}(e)}\left(\frac{n_1 + n_2}{n_2}\right)^2$$
$$= (13.3\ \Omega)(4^2)$$
$$= 212.8\ \Omega$$

These resistances are combined in parallel to determine the ac-equivalent collector resistance. This is

$$R'_C = R'_e \parallel R_p \parallel R'_L$$
$$= 212.8\ \Omega \parallel 7.9\ \text{k}\Omega \parallel 1.25\ \text{k}\Omega$$
$$\cong 178\ \Omega$$

Using this value, we can now determine the open-loop gain of the common-base amplifier. Since the base is bypassed to ground, the voltage gain is

$$A_v = \frac{R'_C}{r_e}$$
$$= \frac{178\ \Omega}{13.7\ \Omega}$$
$$= 13$$

To determine the loop gain, we must now determine the feedback factor β. This is found using the autotransformer relationships for the left side of T_1, or if you prefer, we can treat L_1 and L_2 as a voltage divider. In either case, the result is the same, which is

$$\beta = \frac{n_2}{n_1 + n_2}$$
$$= \frac{5}{20}$$
$$= 0.25$$

The loop gain is now found to be

$$A\beta = (13)(0.25)$$
$$= 3.25$$

Since $A\beta > 1$, oscillation is guaranteed. In fact, since the inequality is exceeded by a factor of 3.25, we can be fairly certain that the circuit will oscillate even under rather wide component parameter variations. The output voltage is found as follows:

$$V_{0(p-p)} \cong 2V_{CEQ}\left(\frac{n_s}{n_p}\right)$$
$$\cong 2V_{CEQ}\left(\frac{5}{20}\right)$$
$$\cong (28.2\ \text{V})(0.2)$$
$$= 5.64\ \text{V}_{p-p}$$

Another variation of the Hartley oscillator is examined in the next example.

EXAMPLE 17.6

Refer to Fig. 17.22. The CA3028A is a monolithic differential/cascode amplifier. In this example, the CA3028A is configured as a single-ended-input, single-ended-output amplifier. This configuration ideally produces no phase shift, and when the tuned circuit is in resonance the loop phase shift is zero degrees as well, resulting in positive feedback. All transistors have $\beta \cong 100$, $\alpha \cong 1$,

$V_{\phi} = 0.7$ V, and $r_o \to \infty$. Given $V_{CC} = V_{EE} = \pm 15$ V, $R_4 = R_5 = 1.5$ kΩ, $R_L = 1$ kΩ, $L_1 = 22$ μH ($r_{s_1} = 12$ Ω), $L_2 = 6.8$ μH ($r_{s_2} = 3.7$ Ω), and $C_1 = 100$ pF, determine f_o and the loop gain $A\beta$.

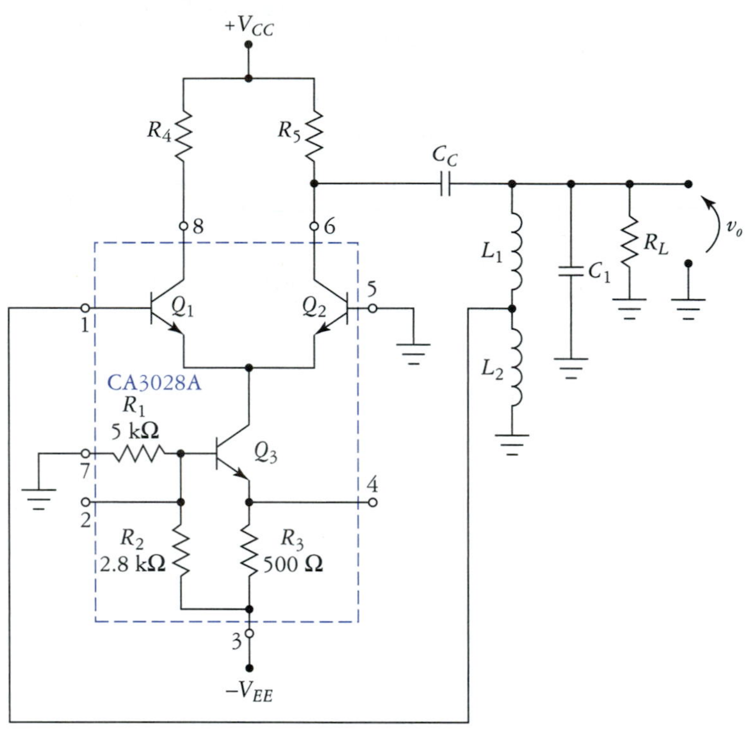

FIGURE 17.22
Circuit for Example 17.6.

Solution The dc-equivalent circuit is shown in Fig. 17.23. Routine analysis gives us

$$I_{CQ_1} = I_{CQ_2} \cong 4.5 \text{ mA}$$
$$V_{CEQ_1} = V_{CEQ_2} \cong 9 \text{ V}$$
$$r_{e_1} = r_{e_2} = 5.8 \ \Omega$$

It will also be useful to determine the small-signal input resistance of the diff amp, which is $R_{\text{in}} = 1.16$ kΩ. As used here, the CA3028A can be modeled as shown in Fig. 17.24. The oscillation frequency is

$$f_o = \frac{1}{2\pi\sqrt{(L_1 + L_2)\,C_1}}$$
$$\cong 3 \text{ MHz}$$

The feedback factor is now determined to be

$$\beta = \frac{L_2}{L_1 + L_2}$$
$$= 0.24$$

To determine the loaded voltage gain A of the diff amp, we must calculate the reflected input resistance R'_{in} and the parallel equivalent resistance of the inductors R_p. Starting with R_p we have

$$R_p = (r_{s_1} + r_{s_2})(Q_L^2 + 1)$$
$$= (12\ \Omega + 3.7\ \Omega)\{\left(\frac{(2\pi \times 3 \text{ MHz} \times 28.8\ \mu\text{H})}{12\ \Omega + 3.7\ \Omega}\right) + 1\}$$
$$= (15.7\ \Omega)(34.6^2 + 1)$$
$$= 18.8 \text{ k}$$

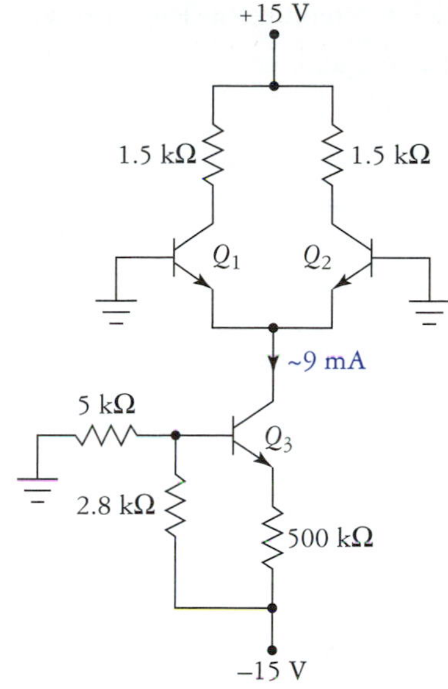

FIGURE 17.23
DC-equivalent circuit.

The reflected input resistance is

$$
\begin{aligned}
R'_{\text{in}} &= R_{\text{in}}\left(\frac{L_1 + L_2}{L_2}\right)^2 \\
&= 1.16\ \text{k}\Omega\left(\frac{28.8}{6.8}\right)^2 \\
&= (1.16\ \text{k}\Omega)(17.9) \\
&= 20.8\ \text{k}\Omega
\end{aligned}
$$

The total collector resistance is

$$
\begin{aligned}
R'_C &= R_C \parallel R'_{\text{in}} \parallel R_L \parallel R_p \\
&= 1.5\ \text{k}\Omega \parallel 20.8\ \text{k}\Omega \parallel 1\ \text{k}\Omega \parallel 18.8\ \text{k}\Omega \\
&= 566\ \Omega
\end{aligned}
$$

We can now determine the loaded voltage gain A of the diff amp, which is

$$
\begin{aligned}
A_v &= \frac{R'_C}{2r_e} \\
&= \frac{566\ \Omega}{11.6\ \Omega} \\
&= 48.8
\end{aligned}
$$

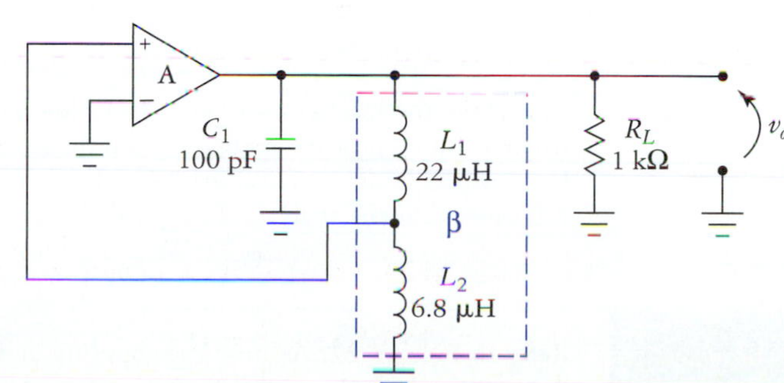

FIGURE 17.24
AC-equivalent circuit.

At last, we can determine the loop gain, which we find to be

$$A\beta = (48.8)(0.24) = 11.7$$

Since $A\beta > 1$, oscillation is guaranteed.

It is very convenient to implement a variable-frequency oscillator using the Hartley topology, simply by means of a variable capacitor as shown in Fig. 17.25. Such variable-frequency oscillators are commonly used in radio receiver tuning circuits.

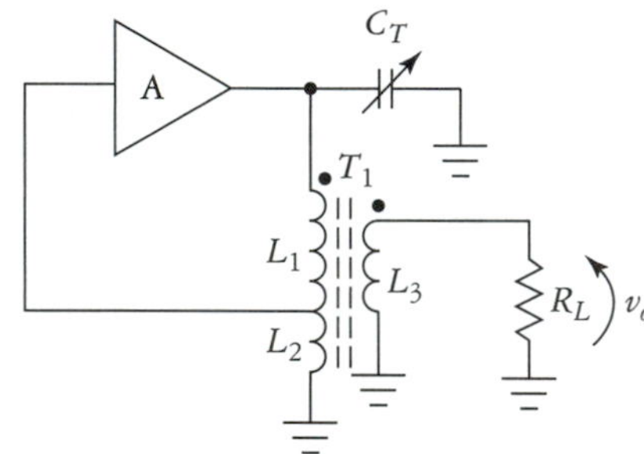

FIGURE 17.25
Variable-frequency Hartley oscillator.

The Colpitts Oscillator

The Colpitts oscillator topology is identified by the presence of a tapped capacitor in the feedback loop/resonant circuit, as shown in the simplified circuit of Fig. 17.26. Keep in mind that a transformer need not be used if the output is taken directly from the collector of the transistor. However, transformer coupling is most commonly used because it reduces the sensitivity of the oscillator to varying load conditions. As usual, the oscillation frequency is given by

$$f_o = \frac{1}{2\pi\sqrt{L_1 C_{eq}}} \tag{17.40}$$

where the equivalent capacitance is

$$C_{eq} = \frac{1}{\dfrac{1}{C_1} + \dfrac{1}{C_2}} \tag{17.41}$$

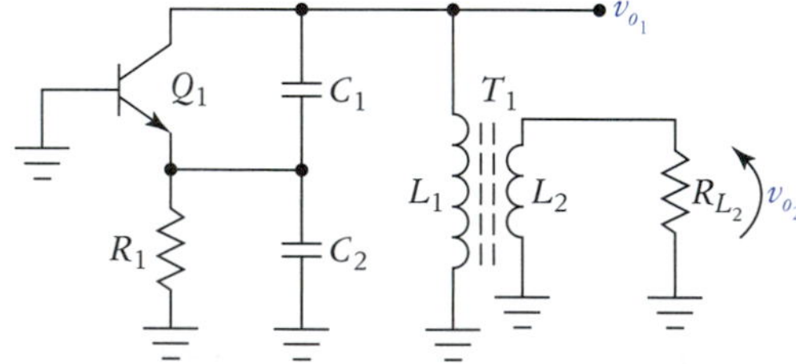

FIGURE 17.26
Simplified Colpitts oscillator.

Like the Hartley oscillator, the Colpitts oscillator of Fig. 17.26 is built around a CB amplifier, which exhibits positive feedback when the tuned circuit (L_1, C_1, and C_2) is in resonance. Also like the Hartley oscillator, many variations of the Colpitts oscillator are possible. We cannot cover all of these variations, but the analysis of a typical circuit is very similar to that of the oscillators covered thus far.

EXAMPLE 17.7

Refer to Fig. 17.27. Assume the coupling and bypass capacitors are ideal and the transistor has $\beta \cong 100$, $\alpha \cong 1$, $V_\phi = 0.7$ V, and $r_o \to \infty$. Also, $C_1 = 150$ pF, $C_2 = 470$ pF, $R_1 = 15$ kΩ, $R_2 =$ 3.9 kΩ, $R_3 = 560$ Ω, $R_4 = 100$ Ω, $R_L = 100$ Ω, $V_{CC} = 15$ V, and the transformer T_1 has $n_1 =$

25 turns ($L_1 = 50\ \mu H$, $r_{s_1} = 10\ \Omega$), $n_2 = 5$ turns ($r_{s_2} = 2\ \Omega$). Determine the Q point of the transistor, the oscillation frequency f_o, and the loop gain $A\beta$. Assuming sustained oscillation does occur, determine the approximate peak-to-peak output voltage.

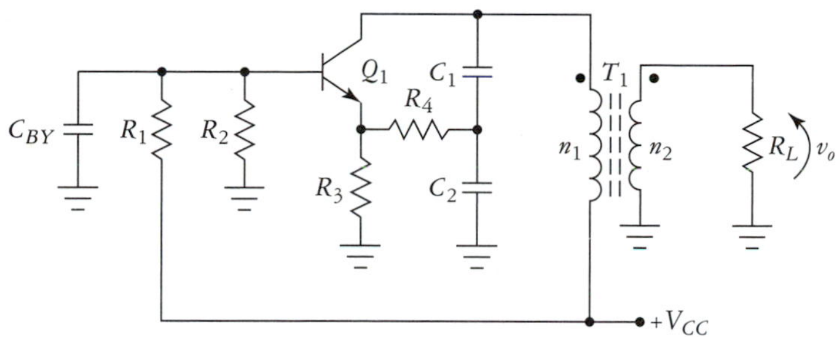

FIGURE 17.27
Circuit for Example 17.7.

Solution Analysis of the dc-equivalent circuit shown in Fig. 17.28 gives the following Q-point related information:

$$I_{CQ} \cong 4\ \text{mA}$$

$$V_{CEQ} \cong 12.8\ \text{V}$$

$$r_e = 6.5\ \Omega$$

FIGURE 17.28
DC-equivalent circuit.

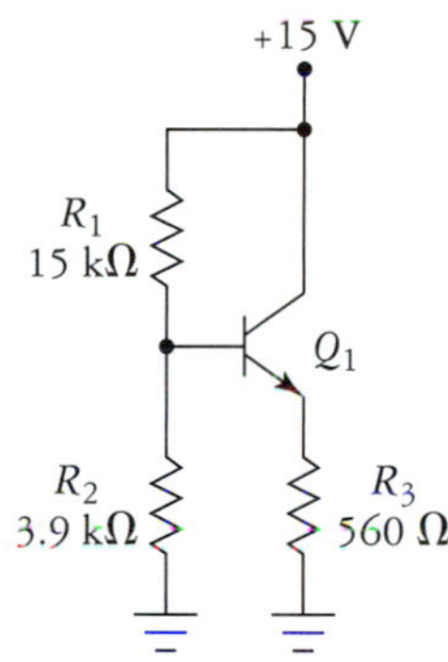

The equivalent circuit of Fig. 17.29 is used in the ac analysis. The equivalent capacitance is

$$C_{eq} = \frac{1}{\dfrac{1}{C_1} + \dfrac{1}{C_2}}$$

$$= \frac{1}{\dfrac{1}{150\ \text{pF}} + \dfrac{1}{470\ \text{pF}}}$$

$$= 114\ \text{pF}$$

This gives us an oscillation frequency of

$$f_o = \frac{1}{2\pi\sqrt{L_1 C_{eq}}}$$

$$= \frac{1}{2\pi\sqrt{50\ \mu\text{H} \times 114\ \text{pF}}}$$

$$= 2.1\ \text{MHz}$$

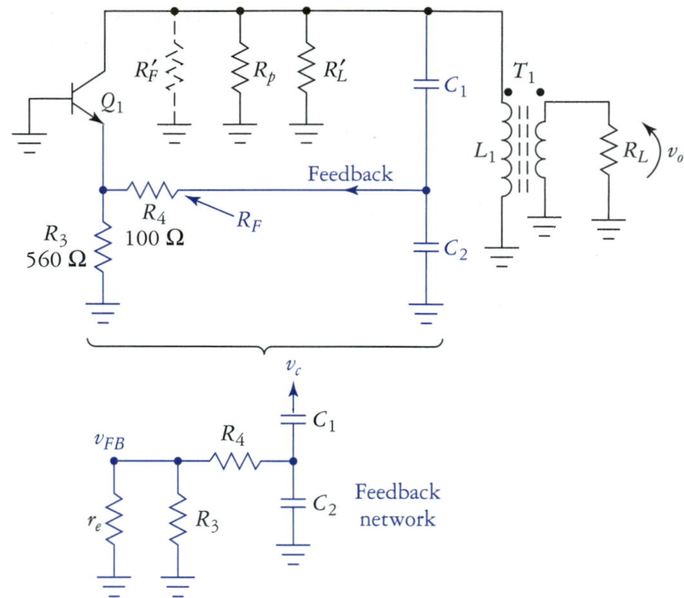

FIGURE 17.29
AC-equivalent circuit.

The reflected load resistance is

$$
\begin{aligned}
R'_L &= R_L\left(\frac{n_p}{n_s}\right)^2 \\
&= 100\ \Omega\left(\frac{25}{5}\right)^2 \\
&= 2.5\ \text{k}\Omega
\end{aligned}
$$

The Q of the primary winding is

$$
\begin{aligned}
Q_L &= 2\pi f_o L/r_s \\
&= 660/10 \\
&= 66
\end{aligned}
$$

The parallel equivalent of the primary winding resistance is

$$
\begin{aligned}
R_p &= r_{s_1}(Q_L^2 + 1) \\
&= 10(66^2 + 1) \\
&= 43.6\ \text{k}\Omega
\end{aligned}
$$

The resistance R_F seen looking from the tuned circuit toward the emitter must now be determined. This is found using

$$
\begin{aligned}
R_F &= (r_e \parallel R_3) + R_4 \\
&= (6.5\ \Omega \parallel 560\ \Omega) + 100\ \Omega \\
&\cong 106.5\ \Omega
\end{aligned}
$$

We use this information, and Eq. (17.28) to determine the reflected feedback resistance R'_F:

$$
\begin{aligned}
R'_F &= R_F\left(\frac{C_1 + C_2}{C_1}\right)^2 \\
&= 106.5\ \Omega\left(\frac{620}{150}\right)^2 \\
&\cong 1.8\ \text{k}\Omega
\end{aligned}
$$

The collector equivalent resistance is

$$
\begin{aligned}
R'_C &= R'_F \parallel R_p \parallel R'_L \\
&= 1.8\ \text{k}\Omega \parallel 43.6\ \text{k}\Omega \parallel 2.5\ \text{k}\Omega \\
&\cong 1\ \text{k}\Omega
\end{aligned}
$$

The voltage gain of the amplifier is now found to be

$$A_v = \frac{R'_C}{r_e}$$
$$= \frac{1\text{ k}\Omega}{6.5\ \Omega}$$
$$\cong 154$$

The feedback factor is now determined. For the sake of simplicity, we assume that there is no loading effect between the split capacitors and the r_e, R_3, R_4 resistor network. This equivalent feedback network is shown in the lower part of Fig. 17.29. Since $r_e \ll R_3$, the feedback factor is approximated by

$$\beta = \left(\frac{r_e}{r_e + R_4}\right)\left(\frac{C_1}{C_1 + C_2}\right)$$
$$= \left(\frac{6.5}{106.5}\right)\left(\frac{150}{620}\right)$$
$$= (0.061)(0.24)$$
$$= 0.015$$

The loop gain of the oscillator is

$$A\beta = (154)(0.015)$$
$$= 2.31$$

Since $A\beta > 1$, oscillation will occur. The peak-to-peak output voltage is approximately

$$V_o \cong 2\,V_{CEQ}\left(\frac{n_s}{n_p}\right)$$
$$= (2)(12.8\text{ V})(0.2)$$
$$= 5.12\ \text{V}_{\text{P-P}}$$

Colpitts oscillators are used often in applications such as communication circuits and test equipment, generally to produce frequencies that range from hundreds of kilohertz up. To produce very high frequencies, it is necessary to use very small values for L_1, C_1, and C_2. Unfortunately, under these conditions variations in the transistor junction capacitances tend to cause frequency instability and feedback factor instability as well. The solution to this problem is the Clapp oscillator.

The Clapp Oscillator

The simplified schematic for a Clapp oscillator is shown in Fig. 17.30. The similarity between this circuit and the Colpitts oscillator is apparent. In this design, however, capacitors C_1 and C_2 are used primarily to set the feedback factor β. They are also chosen such that the following inequalities are satisfied:

$$C_1, C_2 \gg C_{CB} \tag{17.42}$$

$$C_1, C_2 \gg C_3 \tag{17.43}$$

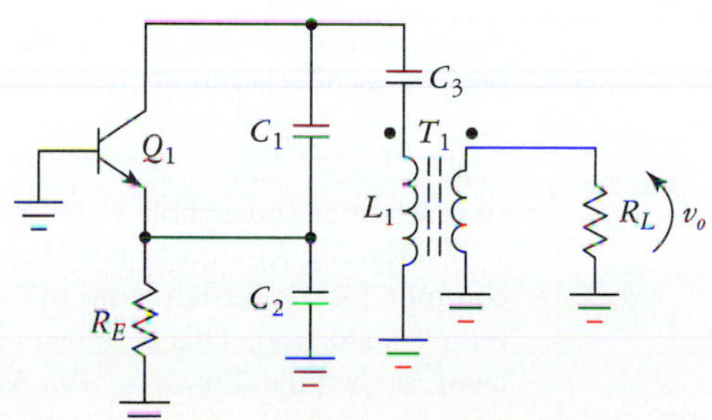

FIGURE 17.30
Simplified Clapp oscillator circuit.

Usually, a ten to one, or higher ratio will satisfy a *much greater than* inequality. Inequality (17.42) ensures that transistor junction capacitances will have negligible effect on the oscillation frequency or feedback factor. Inequality (17.43) allows the oscillation frequency to be closely approximated by

$$f_o \cong \frac{1}{2\pi\sqrt{L_1C_3}} \tag{17.44}$$

The Clapp oscillator has the added feature that it can be made to have adjustable frequency by using a variable capacitor for C_3. This is not generally possible in the Colpitts design, because two capacitors must be varied simultaneously while maintaining a constant ratio as well, so as not to alter β.

The Phase-Shift Oscillator

The Hartley, Colpitts, and Clapp circuits are classic RF oscillator designs. There are, however, many other designs in use. The phase-shift oscillator shown in Fig. 17.31 is one such circuit.

In Fig. 17.31, R_F and R set the voltage gain of the op amp. The inverting configuration provides gain with a phase shift of 180° while the RC network provides an additional phase shift of 180° to the feedback signal at a frequency given by

$$f_o = \frac{1}{2\pi RC\sqrt{6}} \tag{17.45}$$

for equal capacitors C and resistors R. It can be shown that to produce oscillation, we must have $|A_v| = R_F/R \geq 29$. Therefore, we require

$$R_F \geq 29R \tag{17.46}$$

Phase-shift oscillators are not as stable or easily tuned as the resonant oscillators covered previously. Because of this, they are not used in RF applications, but rather are used in less critical audio frequency applications.

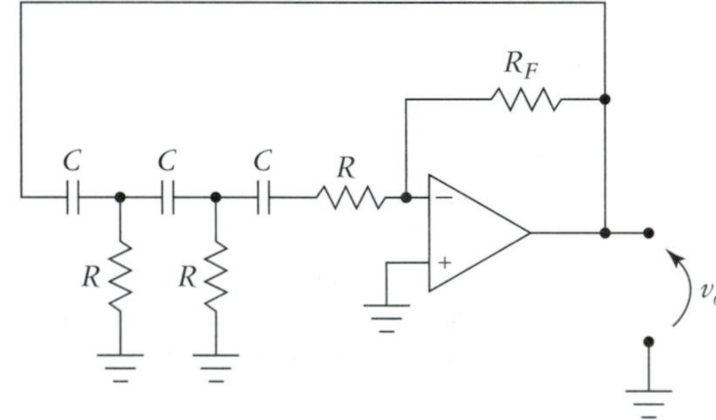

FIGURE 17.31
Phase-shift oscillator.

The Wein Bridge Oscillator

An op-amp-based implementation of a Wein bridge (pronounced "Veen" bridge) oscillator is shown in Fig. 17.32(a). The RC network in the positive feedback loop is called a *lead-lag network.* The lead-lag network functions such that at one particular frequency f_o the phase of the feedback voltage is zero degrees, as shown in the phase plot of Fig. 17.32(b). As shown in the voltage magnitude plot of Fig. 17.32, the amplitude of the positive feedback voltage V_{fb} is one-third of the output voltage at f_o. Thus, in resonance we have

$$\beta = 1/3 \angle 0°$$

To satisfy the Barkhausen criteria we choose R_F and R_1 such that $A \geq 3$. Keeping in mind that the positive feedback voltage is applied to the noninverting input, in equation form, we have

$$R_F \geq 2R_1 \tag{17.47}$$

In practice, it is desirable to set R_F just slightly greater than $2R_1$ to ensure oscillation.

Making R_F significantly larger than $2R_1$ results in distortion of the output signal. It is not uncommon for a tungsten lamp to be placed in series with R_1 in order to ensure oscillation startup and reduce distortion. This is shown in Fig. 17.33. The tungsten lamp has a positive temperature coefficient, as shown in Fig. 17.33(b). At startup, $I_{\text{Lamp}} = 0$ and R_{Lamp} is low. Resistors R_F and R_1 are chosen

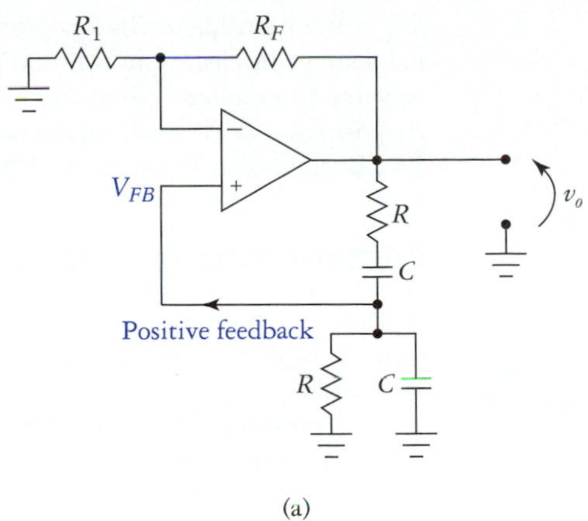

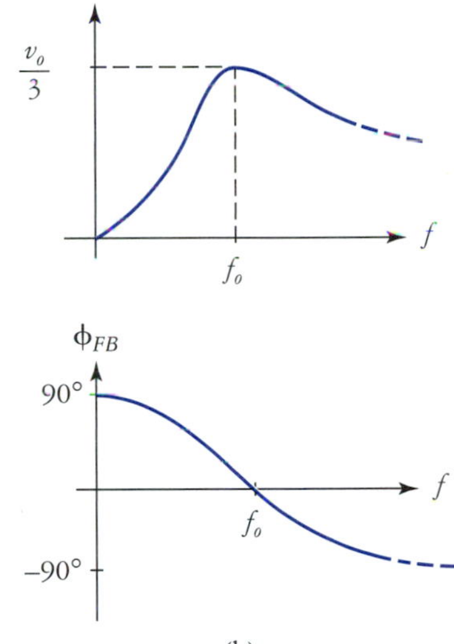

FIGURE 17.32
Wein bridge oscillator. (a) Circuit. (b) Feedback network amplitude and phase response.

such that at startup the op amp has $A > 3$, guaranteeing oscillation. When the output voltage increases, the lamp current increases, which in turn increases R_{Lamp}, until equilibrium is reached at $A = 3$. Thus, the loop gain of the oscillator stabilizes at $A\beta = 1$.

With equal value capacitor and resistor pairs in the lead-lag network, the oscillation frequency of the Wein bridge oscillator is given by

$$f_o = \frac{1}{2\pi RC} \tag{17.48}$$

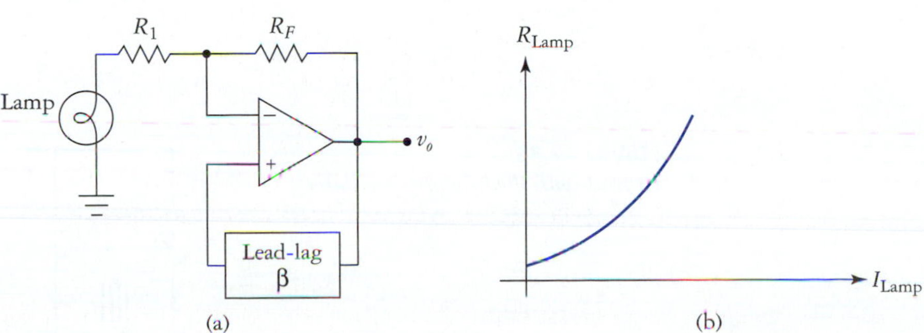

FIGURE 17.33
Stabilized Wein bridge oscillator. (a) Circuit using nonlinear resistance in feedback network. (b) Lamp resistance versus current.

Wein bridge oscillators produce very low distortion sine waves and are used extensively as signal sources in electronic test equipment from sub-audiofrequencies up to around 10 MHz. By the way, for an excellent discussion of oscillators (and lots of other cool stuff), see *Analog Circuit Design: Art, Science, and Personalities,* Section 7, *Max Wein, Mr. Hewlett, and a Rainy Afternoon* (Jim Williams, Ed., Butterworth-Heinemann, 1991).

Crystal-Controlled Oscillators

Note
A common crystal frequency is 3.5795 MHz, which is called a *colorburst* crystal. These are used in color television receivers.

Modern communication systems and digital test equipment require the use of oscillators of extremely high accuracy and stability. These applications often require the use of crystal-controlled oscillators.

A crystal is a thin slice of piezoelectric material, such as quartz, housed in a metal can as shown in Fig. 17.34(a). The schematic symbol for a crystal is shown in Fig. 17.34(b). Most often, crystals are designated with an uppercase Y on schematic diagrams, although sometimes X or XTAL may be encountered.

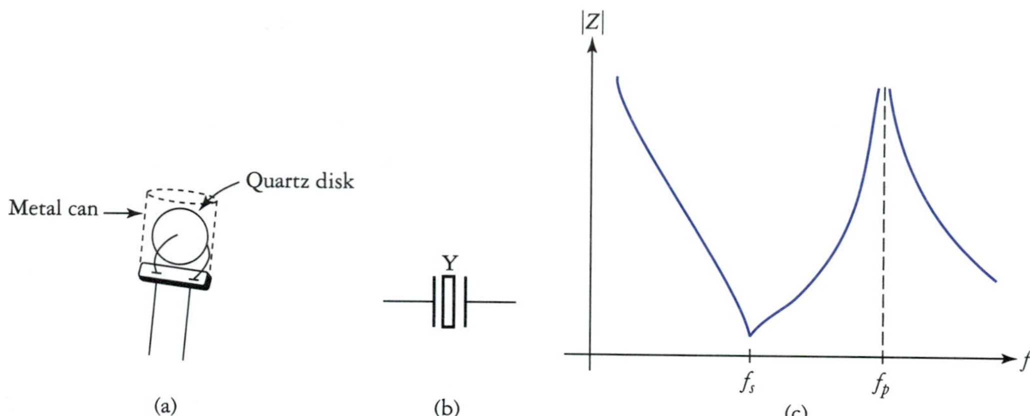

FIGURE 17.34
Quartz crystal. (a) Typical physical structure. (b) Schematic symbol. (c) Crystal impedance versus frequency.

Electrically, a crystal behaves as a resonant circuit with two modes of operation: series resonant or parallel resonant. The impedance versus frequency plot of Fig. 17.34(c) for a typical crystal shows the relationship between series and parallel resonant frequencies f_s and f_p, respectively. Most often, a given crystal will be designed to operate most efficiently in one mode or the other.

A crystal can be integrated into the design of any of the *LC* oscillator circuits that were covered previously in this chapter. The circuit of Fig. 17.35 shows a Colpitts oscillator that uses crystal Y_1 to maintain frequency stability. To provide the maximum level of positive feedback, the crystal must operate in the series resonant mode; therefore,

$$f_o = f_s \tag{17.49}$$

FIGURE 17.35
Crystal-controlled Colpitts oscillator.

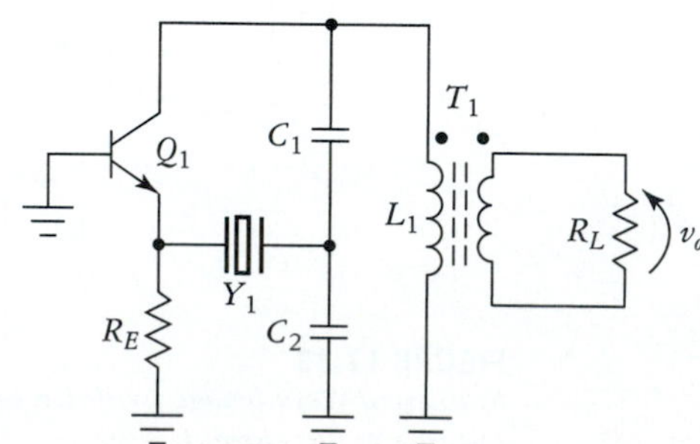

The parallel resonant circuit is designed such that its resonant frequency [given by Eq. (17.40)] is approximately equal to f_s to ensure startup. At f_s, capacitors C_1 and C_2 set the feedback factor β to

$$\beta = \frac{C_1}{C_1 + C_2} \tag{17.50}$$

This assumes that the series impedance of the crystal is negligibly small at f_s. Very slight adjustment of frequency ($\Delta f \ll 1\%$ of f_o) can be effected by using an adjustable core on L_1, or by using a small trimmer capacitor across C_1 or C_2.

A crystal can have a specified accuracy as good as 10 ppm (parts per million) with Q as high as 10,000. A crystal oscillator can be made even more stable if the crystal is enclosed in a temperature-controlled oven.

17.5 COMPUTER-AIDED ANALYSIS APPLICATIONS

The analysis of the tuned amplifiers and oscillators presented in this chapter is performed using the same basic techniques used in previous chapters. We have not yet used the PSpice transformer model, however. Therefore, in this section we simulate examples of transformer coupled circuits.

Transformer Coupled Tuned Amplifier Simulation

The circuit of Fig. 17.9 was analyzed in Section 17.1. The pertinent results of that analysis are repeated here:

$$I_{CQ} = 2 \text{ mA}$$

$$V_{CEQ} = 12.6 \text{ V}$$

$$f_o = 2.23 \text{ MHz}$$

$$\text{BW} = 579.2 \text{ kHz}$$

$$|A_v| = 7.7 = 17.7 \text{ dB}$$

The PSpice schematic editor version of this amplifier is shown in Fig. 17.36. The linear transformer is used, with attributes set as shown in Fig. 17.37. The coupling coefficient is set to one, in accordance with our previous assumptions. Rather than allowing the number of turns to be specified, PSpice requires us to specify the inductance ratio of the transformer. The specifications for this circuit call for

$$n_p/n_s = 10, \qquad L_{\text{pri}} = 50 \ \mu\text{H}$$

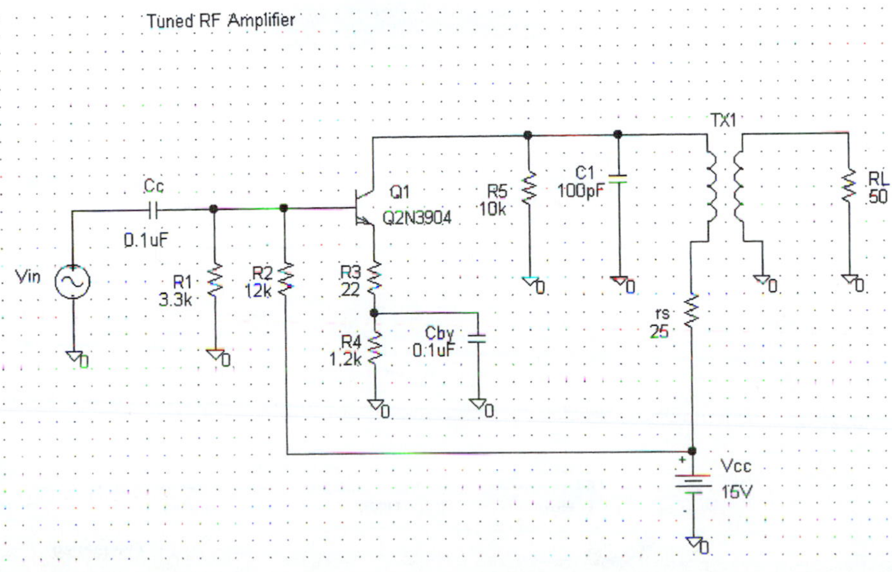

FIGURE 17.36
Tuned RF amplifier simulation.

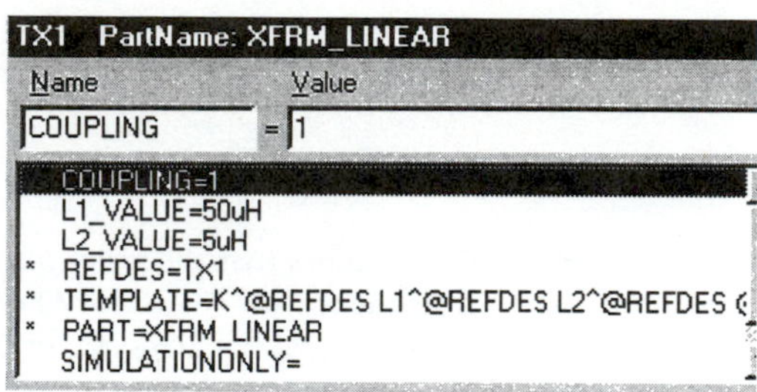

FIGURE 17.37
Setting the transformer parameters.

Because the inductance ratio is equal to the turns ratio ($n_p/n_s = L_{pri}/L_{sec}$), we simply assign the secondary inductance as $L_2 = L_{pri}/10 = 5\ \mu H$. This produces the desired inductance and turns ratio values. The series resistance of the primary winding is included as a separate resistor in the schematic.

The results of a logarithmic frequency response run are shown in Fig. 17.38. The input voltage ac attribute was set for 1 V. The ac sweep was set to run from 1 kHz to 100 MHz with 101 points per decade. The upper plot shows V_o (linear y axis) versus log f. Using the cursors, we can determine the following:

$$f_o = 2.2 \text{ MHz}$$

$$A_o = 4.2 \cong 12.5 \text{ dB}$$

$$\text{BW} = 3.2 \text{ MHz}$$

Clearly, the resonant frequency is in close agreement with our pencil-and-paper approximation. However, the simulation voltage gain and bandwidth are significantly different. The main reasons for these discrepancies are the simplifying assumptions that we make in order to allow the circuit to be analyzed with a reasonable amount of effort. We could achieve closer agreement with the simulator if second-order transistor parameters are taken into account. All things considered though, the agreement between PSpice and manual analysis is not too bad. Often, we use manual calculations just to obtain order-of-magnitude estimates, especially in more complex circuits.

It is interesting to examine the decibel response of the circuit, which is shown in the lower plot of Fig. 17.38. Here, we find the classical constant-slope response skirts that are characteristic of log–log plots. Notice that there is a change in the slope of the response curve at around 50 kHz or so. Examination of the response curve shows a rolloff rate of about 20 dB/decade between 100 kHz and

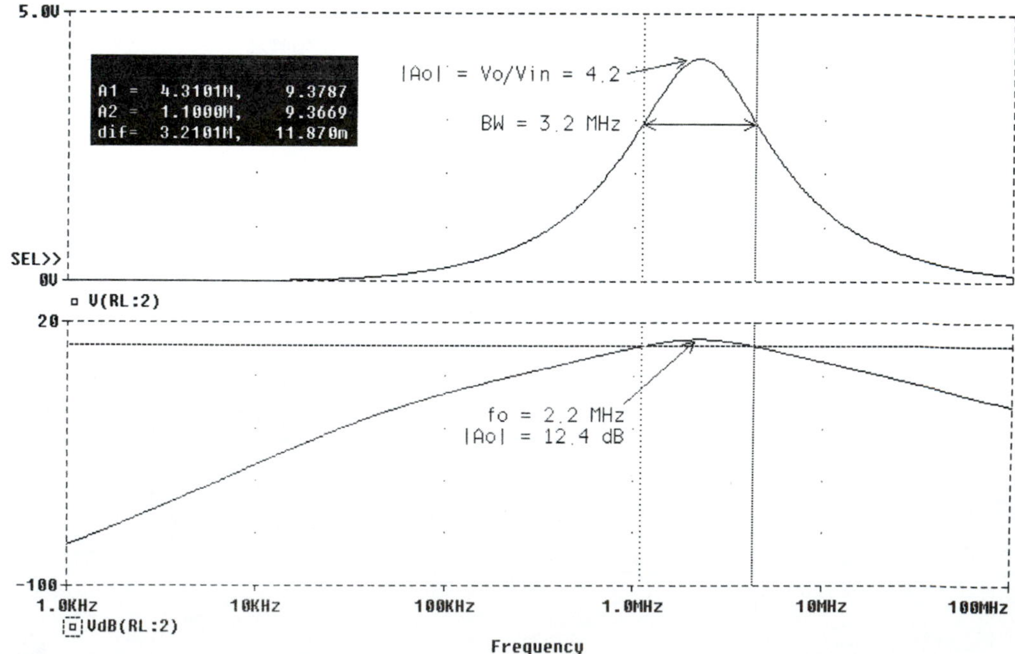

FIGURE 17.38
Amplifier frequency response with linear y axis (top) and decibel y axis (bottom).

1 MHz, which is a first-order HP response. For frequencies below about 50 kHz, the rolloff rate is about 40 dB/decade, which is a second-order HP response. This additional HP break frequency is caused by the emitter bypass network. The emitter network is redrawn in Fig. 17.39. Examination of the reduced equivalent circuit yields

$$\begin{aligned} f_c &= \frac{1}{2\pi RC} \\ &= \frac{1}{(2\pi)(35\ \Omega)(0.1\ \mu\text{F})} \\ &= 45.5\text{ kHz} \end{aligned}$$

which agrees well with the simulation results.

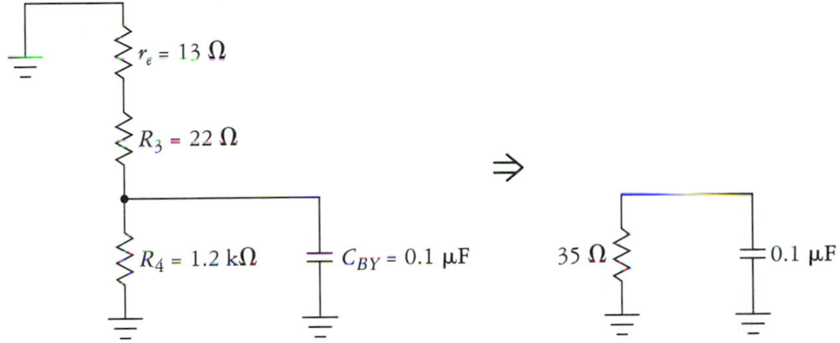

FIGURE 17.39
Examination of the emitter-bypass circuitry.

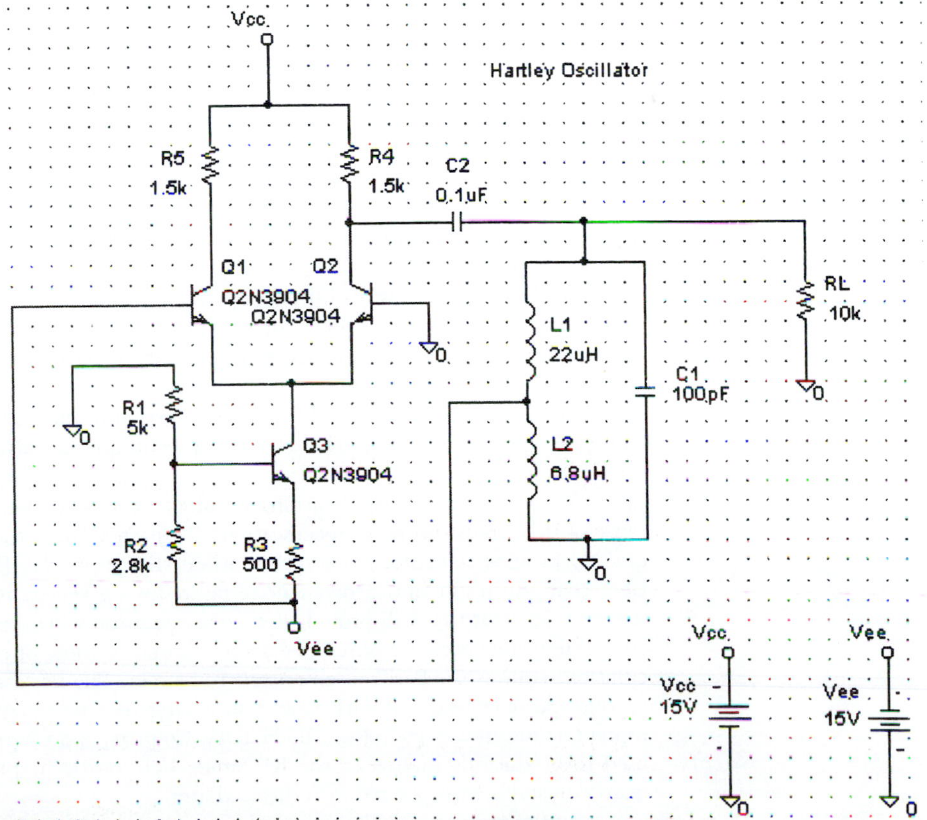

FIGURE 17.40
Diff amp Hartley oscillator.

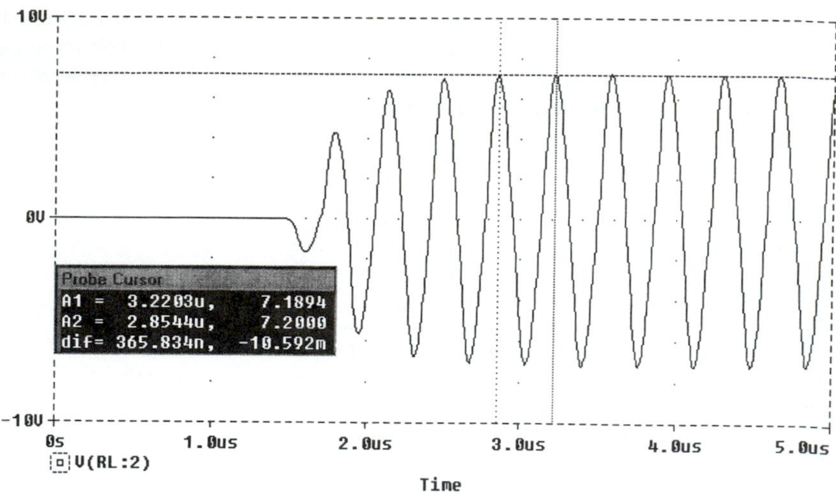

FIGURE 17.41
Oscillator output voltage waveform.

Hartley Oscillator Simulation

The Hartley oscillator of Example 17.6 is shown in Fig. 17.40. Prior pencil-and-paper analysis resulted in $f_o \cong 3$ MHz. Display of the transient analysis run produced the output waveform shown in Fig. 17.41. Once steady-state oscillation has begun, the cursors are used to find the period of the signal, which is $T_{osc} \cong 366$ ns. This corresponds to a frequency of $f = 2.73$ MHz, which agrees reasonably well with previous results.

■ CHAPTER REVIEW

Tuned amplifiers and oscillators are used extensively in communications applications. Tuned amplifiers use resonant circuits to obtain a desired frequency response, which is typically a bandpass response. Tuned amplifiers most often operate at relatively high signal frequencies. Discrete transistor circuits are still widely used in these applications because of their superior high-frequency performance. Special RF transistors are available that have been designed specifically to operate at very high frequencies.

High-frequency, tuned-response operation makes transformer coupling very practical. In addition, transformer coupling allows for very convenient impedance matching as well. Both standard transformers with separate primary and secondary winding and autotransformers are used in tuned amplifier and oscillator designs.

In addition to transformer coupling, tuned amplifiers can also incorporate direct, capacitive, and split capacitor coupling. These coupling techniques are also used to provide feedback in oscillator circuits.

In communication applications we often must work with extremely low power signals. Because of this, transistor noise specifications are often an important consideration. Manufacturers frequently provide transistor noise specifications that allow the designer to determine the bias conditions that will result in the lowest noise figure for a given transistor. These data can also be used to select a transistor that will provide low noise operation when driven by a given source resistance.

Sinusoidal oscillator circuits are very similar to those of tuned linear amplifiers. Sustained oscillation requires the Barkhausen criteria to be met. The Barkhausen criteria are the conditions that produce positive feedback with loop gain greater than or equal to unity. Many sinusoidal oscillator topologies are in use, including the Hartley, Colpitts, and Clapp configurations. These are normally used to produce oscillation in the RF range. Very stable frequency operation can be obtained if a quartz crystal is incorporated into the oscillator.

Audiofrequency oscillators more often use the phase-shift or Wein bridge topologies. These circuits are inductorless, which is an advantage at lower frequencies, where high-value inductors tend to be rather large and expensive.

DESIGN AND ANALYSIS PROBLEMS

Note: Unless otherwise noted, assume that all coupling and bypass capacitors are ideal, and all transistors have negligible stray capacitance, $\beta = 100$, $V_\phi = 0.7$ V, and $V_A = \infty$.

17.1. Refer to Fig. 17.42. Assume that $R_1 = 22$ kΩ, $R_2 = 6.8$ kΩ, $R_3 = 33$ Ω, $R_4 = 470$ Ω, $R_L = 75$ Ω, $V_{CC} = 12$ V, $C_1 = 47$ pF, and T_1 has $L_p = 50$ µH, $n_p/n_s = 12$, $r_{s(\text{pri})} = 15$ Ω. Determine I_{CQ}, V_{CEQ}, r_e, f_o, the reflected load resistance R_L', the ac-equivalent collector resistance R, the ac-equivalent external emitter resistance R_e, the unloaded Q of the inductor Q_L, the parallel resonant equivalent of the inductor winding resistance R_P, the overall Q_t of the circuit, the voltage gain of Q_1 referred to the collector A_v', the bandwidth of the amplifier BW, and the voltage gain referred to the load resistor A_v. Determine the maximum unclipped output voltage. Determine the required value for C_1 such that $f_o = 4$ MHz.

17.2. Refer to Fig. 17.42. Assume that $R_1 = 10$ kΩ, $R_2 = 2.7$ kΩ, $R_3 = 22$ Ω, $R_4 = 1.2$ kΩ, $R_L = 150$ Ω, $V_{CC} = 15$ V, $C_1 = 22$ pF, and T_1 has $L_p = 12$ µH, $n_p/n_s = 8$, $r_{s(\text{pri})} = 20$ Ω. Determine I_{CQ}, V_{CEQ}, r_e, f_o, the reflected load resistance R_L', the ac-equivalent collector resistance R_C', the ac-equivalent external emitter resistance R_e, the unloaded Q of the inductor Q_L, the parallel resonant equivalent of the inductor winding resistance R_P, the overall Q_t of the circuit, the voltage gain of Q_1 referred to the collector A_v', the bandwidth of the amplifier BW, and the voltage gain referred to the load resistor A_v. Determine the maximum unclipped output voltage. Determine the required primary inductance such that $f_o = 8$ MHz.

17.3. Refer to Fig. 17.43. Assume that $R_1 = 10$ kΩ, $R_2 = 2.7$ kΩ, $R_3 = 12$ Ω, $R_L = 300$ Ω, $V_{CC} = 15$ V, $-V_{EE} = -15$ V, $C_1 = 100$ pF, and T_1 has $L_1 = 25$ µH, $L_2 = 5$ µH, and $r_s = 25$ Ω. Determine I_{CQ}, V_{CEQ}, r_e, f_o, the reflected load resistance R_L', the ac-equivalent collector resistance R_C', the ac-equivalent external emitter resistance R_e, the unloaded Q of the inductor Q_L, the parallel resonant equivalent of the inductor winding resistance R_p, the overall Q_t of the circuit, the voltage gain of Q_1 referred to the collector A_v', the bandwidth of the amplifier BW, and the voltage gain referred to the load resistor A_v. Determine the maximum unclipped output voltage.

17.4. Refer to Fig. 17.43. Assume that $R_1 = 10$ kΩ, $R_2 = 2.2$ kΩ, $R_3 = 18$ Ω, $R_L = 200$ Ω, $V_{CC} = 15$ V, $-V_{EE} = -15$ V, $C_1 = 25$ pF, and T_1 has $L_1 = 15$ µH, $n_1 = 30$ turns, $L_2 = 3$ µH, $n_2 = 6$ turns, and $r_s = 25$ Ω. Determine I_{CQ}, V_{CEQ}, r_e, f_o, the reflected load resistance R_L', the ac-equivalent collector resistance R_C', the ac-equivalent external emitter resistance R_e, the unloaded Q of the inductor Q_L, the parallel resonant equivalent of the inductor winding resistance R_P, the overall Q_t of the circuit, the voltage gain of Q_1 referred to the collector A_v', the bandwidth of the amplifier BW, and the voltage gain referred to the load resistor A_v. Determine the maximum unclipped output voltage.

17.5. Refer to Fig. 17.44. This is the ac-equivalent circuit for a tuned amplifier. Assuming that $I_{CQ} = 2$ mA, $C_1 = 33$ pF, $C_2 = 220$ pF, $L_1 = 30$ µH and $r_s = 5$ Ω, determine f_o, R_p, R_L', R_C', A_v, Q_t, and BW.

17.6. Refer to Fig. 17.44. This is the ac-equivalent circuit for a tuned amplifier. Assuming that $I_{CQ} = 5$ mA, $C_1 = 22$ pF, $C_2 = 120$ pF, $L_1 = 15$ µH, and $r_s = 8$ Ω, determine f_o, R_p, R_L', R_C', A_v, Q_t, and BW.

17.7. Refer to Fig. 17.45. Given: $V_{CC} = 24$ V, $R_1 = 22$ kΩ, $R_2 = 2.7$ kΩ, $R_3 = 920$ Ω, $R_L = 200$ Ω, and T_1 has $n_1 = 25$ turns,

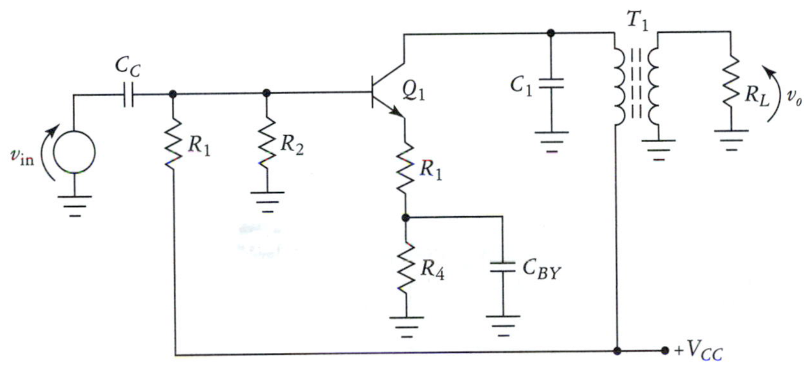

FIGURE 17.42
Circuit for Problems 17.1 and 17.2.

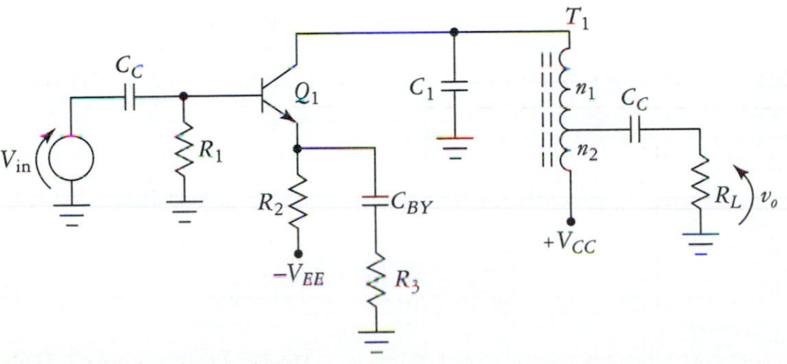

FIGURE 17.43
Circuit for Problems 17.3 and 17.4.

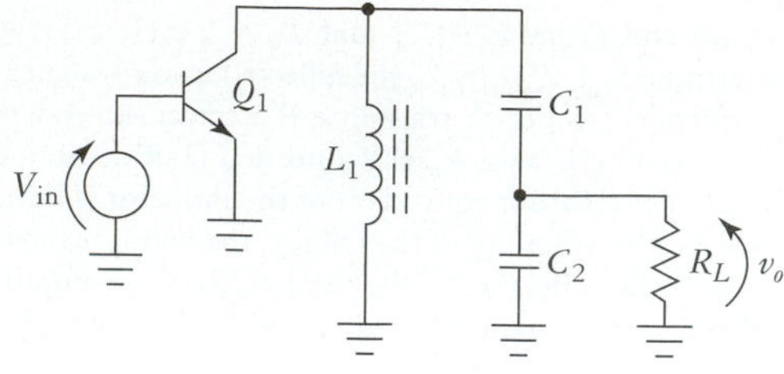

FIGURE 17.44
Circuit for Problems 17.5 and 17.6.

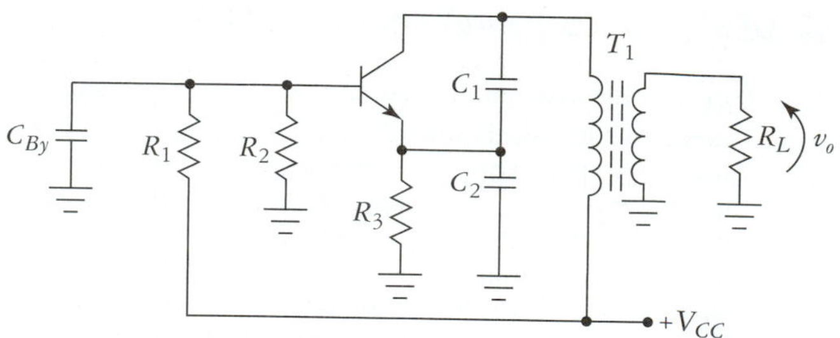

FIGURE 17.46
Circuit for Problems 17.9 and 17.10.

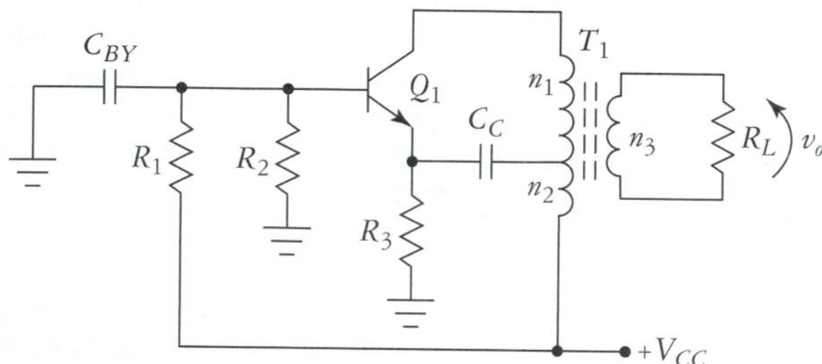

FIGURE 17.45
Circuit for Problems 17.7 and 17.8.

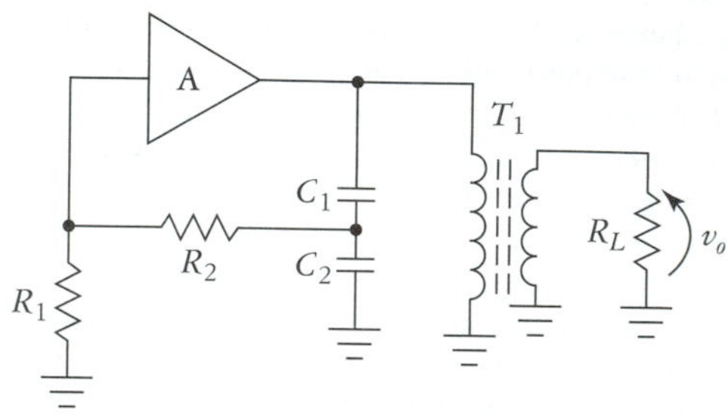

FIGURE 17.47
Circuit for Problems 17.11 and 17.12.

$n_2 = 5$ turns, $n_3 = 6$ turns, $L_1 = 12\ \mu\text{H}$, $L_2 = 2.5\ \mu\text{H}$, $C_1 = 100\text{pF}$, and $r_{s(\text{pri})} = 10\ \Omega$. Determine the oscillation frequency f_o, the reflected load resistance R'_L, the reflected emitter resistance R'_e, the feedback factor β, the open-loop gain of Q_1, the loop gain $A\beta$, and the approximate peak-to-peak output voltage, assuming oscillation will be sustained.

17.8. Refer to Fig. 17.45. Given: $V_{CC} = 15$ V, $R_1 = 22\ \text{k}\Omega$, $R_2 = 3.3\ \text{k}\Omega$, $R_3 = 1.2\ \text{k}\Omega$, $R_L = 500\ \Omega$, $C_1 = 100\text{pF}$ and T_1 has $n_1 = 20$ turns, $n_2 = 4$ turns, $n_3 = 6$ turns, $L_1 = 40\ \mu\text{H}$, $L_2 = 8\ \mu\text{H}$, and $r_{s(\text{pri})} = 15\ \Omega$. Determine the oscillation frequency f_o, the reflected load resistance R'_L, the reflected emitter resistance R'_e, the feedback factor β, the open-loop gain A, the loop gain $A\beta$, and the approximate peak-to-peak output voltage, assuming oscillation will be sustained.

17.9. Refer to Fig. 17.46. Assume that $R_1 = 15\ \text{k}\Omega$, $R_2 = 2.2\ \text{k}\Omega$, $R_3 = 560\ \Omega$, $R_L = 250\ \text{k}\Omega$, $V_{CC} = 9$ V, $C_1 = 470$ pF, $C_2 = 330$ pF, and T_1 has $L_p = 100\ \mu\text{H}$, $L_s = 10\ \mu\text{H}$, and $r_{s(\text{pri})} = 50\ \Omega$. Determine I_{CQ}, V_{CEQ}, r_e, f_o, the reflected load resistance R'_L, the reflected emitter resistance R'_e, the ac-equivalent collector resistance R'_C, the unloaded Q of the primary Q_L, the parallel resonant equivalent of the inductor winding resistance R_P, the overall Q_t of the circuit, the open-loop voltage gain A, and the feedback factor β. Determine the peak-to-peak output voltage, assuming oscillation is sustained.

17.10. Refer to Fig. 17.46. Assume that $R_1 = 15\ \text{k}\Omega$, $R_2 = 3.9\ \text{k}\Omega$, $R_3 = 560\ \Omega$, $R_L = 1\ \text{k}\Omega$, $V_{CC} = 9$ V, $C_1 = 47$ pF, $C_2 = 470$ pF, and T_1 has $L_1 = 150\ \mu\text{H}$, $L_2 = 30\ \mu\text{H}$, and $r_{s(\text{pri})} = 50\ \Omega$. Determine I_{CQ}, V_{CEQ}, r_e, f_o, the reflected load resistance R'_L, the reflected emitter resistance R'_e, the ac-equivalent collector resistance R'_C, the unloaded Q of the primary Q_L, the parallel resonant equivalent of the transformer primary winding resistance R_P, the overall Q of the circuit, the open-loop voltage gain A, the feedback factor β, and the loop gain $A\beta$. Determine the peak-to-peak output voltage, assuming oscillation is sustained.

17.11. Refer to Fig. 17.47. Assume that the amplifier has $R_{\text{in}} \to \infty$ and that the resistive voltage divider has negligible loading effect on the capacitive divider (that is, $R_1 + R_2 >> X_{C_2}$ at f_o). Given: $A = 100$, $R_1 = 1\ \text{k}\Omega$, $R_2 = 9\ \text{k}\Omega$, $R_L = 500\ \Omega$, $C_1 = 0.001\ \mu\text{F}$, $C_2 = 0.01\ \mu\text{F}$, and T_1 has $n_p/n_s = 5$, $L_p = 200\ \mu\text{H}$, and $r_{s(\text{pri})} = 12\ \Omega$. Determine f_o, β, and $A\beta$. Will sustained oscillation occur? Determine the minimum value for A that will result in sustained oscillation.

17.12. Refer to Fig. 17.47. Assume that the amplifier has $R_{\text{in}} \to \infty$ and that the resistive voltage divider has negligible loading effect on the capacitive divider (that is, $R_1 + R_2 >> X_{C_2}$ at f_o). Given: $A = 100$, $R_1 = 10\ \text{k}\Omega$, $R_2 = 10\ \text{k}\Omega$, $R_L = 200\ \Omega$, $C_1 = 0.0022\ \mu\text{F}$, $C_2 = 0.01\ \mu\text{F}$, and T_1 has $n_p/n_s = 10$, $L_p = 100\ \mu\text{H}$, and $r_{s(\text{pri})} = 10\Omega$. Determine f_o, β, and $A\beta$. Will sustained oscillation occur? Determine the minimum value for A that will result in sustained oscillation.

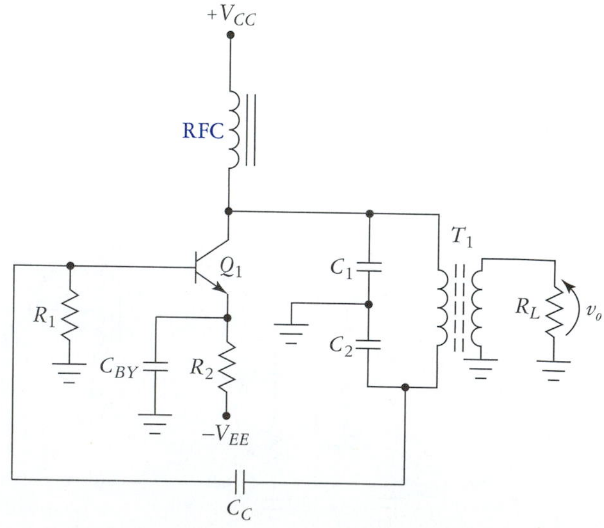

FIGURE 17.48
Circuit for Problem 17.13.

17.13. The circuit in Fig. 17.48 is a version of the Colpitts oscillator that uses feedback from the collector to the base. The inductor labeled RFC is a radio-frequency choke. The choke has low resistance at dc, while having very high reactance at the oscillation frequency, allowing the CE gain of the amplifier to be high while also preventing signal energy from feeding through to the power supply. Equation (17.27) is used to determine the reflected base resistance by substituting R_{in} for R_L, and the feedback factor is still $\beta = C_1/(C_1 + C_2)$. Assume the circuit has V_{CC}, $V_{EE} = \pm 15$ V, $R_1 = 10$ kΩ, $R_2 = 2.7$ kΩ, $R_L = 50$ Ω, $X_{L(RFC)} \rightarrow \infty$, $C_1 = 100$ pF, $C_2 = 470$ pF, and T_1 has $L_P = 100$ μH, $r_{s(pri)} = 9$ Ω, and $n_p/n_s = 12$. Determine I_{CQ}, V_{CEQ}, f_o, the open-loop gain A, the feedback factor β, and the loop gain $A\beta$.

17.14. The circuit in Fig. 17.49 is an example of an *Armstrong oscillator.* In this circuit, positive feedback is obtained through the use of traditional transformer coupling (as opposed to coupling via an autotransformer). The feedback factor is given by $\beta = n_3/n_1$. The oscillation frequency is given by $f_o = 1/2\pi\sqrt{L_1C_1}$. Given: V_{CC}, $V_{EE} = \pm 12$ V, $R_1 = 10$ kΩ, $R_2 = 1.5$ kΩ, $R_L = 300$ Ω, $L_1 = 50$ μH, $r_{s(pri)} = 15$ Ω, $n_1 = 100$ turns, $n_2 = 25$ turns, $n_3 = 10$ turns, and $C_1 = 150$ pF. Determine I_{CQ}, V_{CEQ}, f_o, the open-loop gain A, the feedback factor β, and the loop gain $A\beta$.

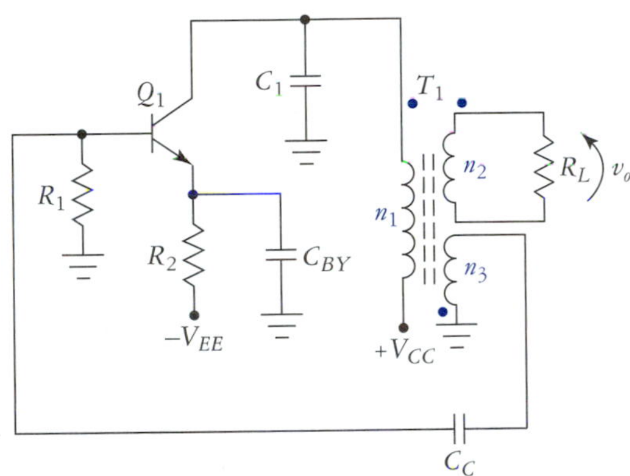

FIGURE 17.49
Circuit for Problem 17.14.

17.15. The ac equivalent for a Clapp oscillator is shown in Fig. 17.50. Assume that the transistor is operating at $I_{CQ} = 5$ mA and that $r_e << R_E$. Given $C_1 = 0.001$ μF, $C_2 = 0.01$ μF, $C_3 = 22$ pF, and $L_1 = 10$ μH with $r_s = 13$ Ω, determine f_o, the open-loop gain A, the feedback factor β, and the loop gain $A\beta$.

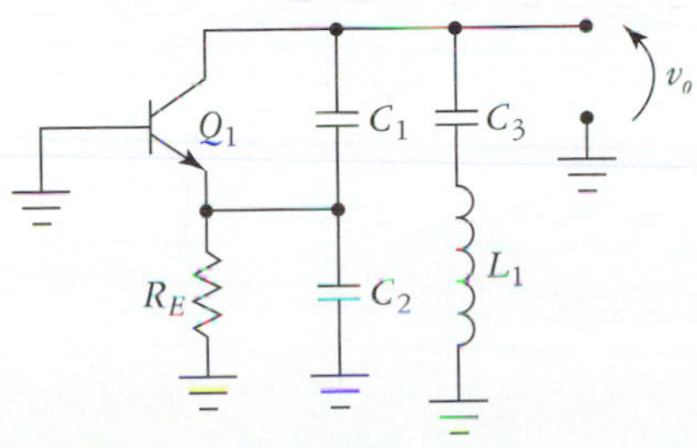

FIGURE 17.50
Circuit for Problems 17.15 and 17.16.

17.16. Modify the circuit of Fig. 17.50 such that a series mode crystal could be used to stabilize the oscillation frequency.

17.17. Refer to Fig. 17.51. This is a form of the Hartley oscillator in which a varactor diode is used to electronically vary the oscillation frequency. Such a circuit is usually called a *voltage-controlled oscillator,* or simply a VCO. The total capacitance is given by $C_t = 1/(1/C_1 + 1/C_2)$. Resistor R_1 is chosen such that the biasing voltage source V_B does not load down the resonant circuit appreciably. That is, for all practical purposes $R_1 \rightarrow \infty$. Assume that the varactor diode capacitance varies from 5 to 30 pF, $C_1 = 50$pF, and that $L_1 = 50$ μH, $r_s = 20$ Ω, $n_1 = 15$, $n_2 = 5$, and $A = 10$. Determine the minimum and maximum oscillation frequencies $f_{o(min)}$ and $f_{o(max)}$ and the loop gain $A\beta$.

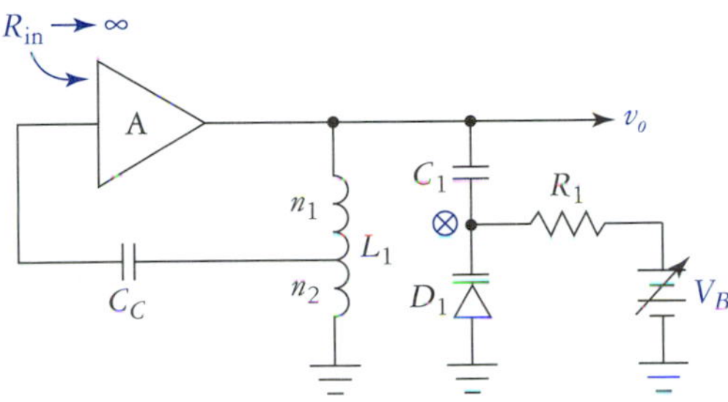

FIGURE 17.51
Circuit for Problems 17.17 and 17.18.

17.18. Refer to Fig. 17.51. Suppose that the feedback was taken from node *x*, instead of from the inductor. Explain how this change would affect the feedback factor.

17.19. Refer to Fig. 17.52. This phase-shift oscillator is constructed around a collector-feedback-biased CE amplifier. The quiescent collector current is found using Eq. (4.19). The small-signal analysis equations are Eqs. (5.29) through (5.31), for which we assume $R'_C \cong R_C \| R_L \| R$. Recall that we require $|A_v| > 29$ for sustained oscillation. Given $V_{CC} = 10$ V, $R_C = 1$ kΩ, $R_F = 100$ kΩ, $R_L = 10$ kΩ, $R = 1$ kΩ, and $C = 0.01$ μF, determine I_{CQ}, V_{CEQ}, f_o, R'_C, and A_v. Will oscillation be sustained?

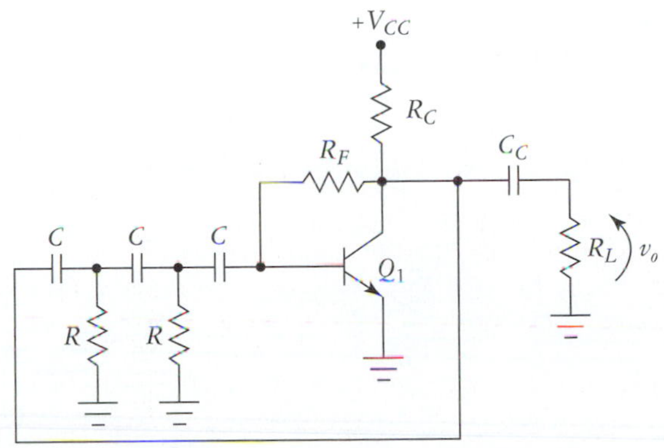

FIGURE 17.52
Circuit for Problems 17.19 and 17.20.

17.20. Refer to the phase-shift oscillator of Fig. 17.52. At the oscillation frequency, assume that each section of the phase shift network contributes 60° of phase shift (ignoring loading effects).

This produces a right triangle like that shown in Fig. 17.53, where R is normalized to 1. Write the equation for X in terms of R and M.

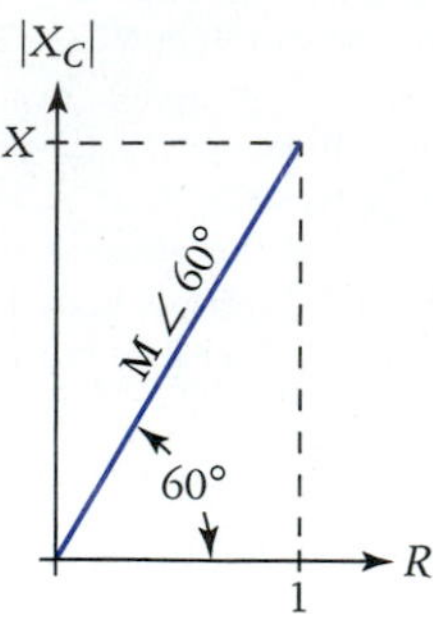

FIGURE 17.53
Phase shift oscillator feedback network phase relationships.

17.21. Refer to Fig. 17.54. Given $R_F = 10\ \text{k}\Omega$, $C = 0.022\ \mu\text{F}$, and R varies from 1 to 50 kΩ, determine the maximum value of R_1 that will result in sustained oscillation and the minimum and maximum oscillation frequencies.

17.22. Refer to Fig. 17.54. Explain the meaning of the dashed line that connects the variable resistors in this circuit.

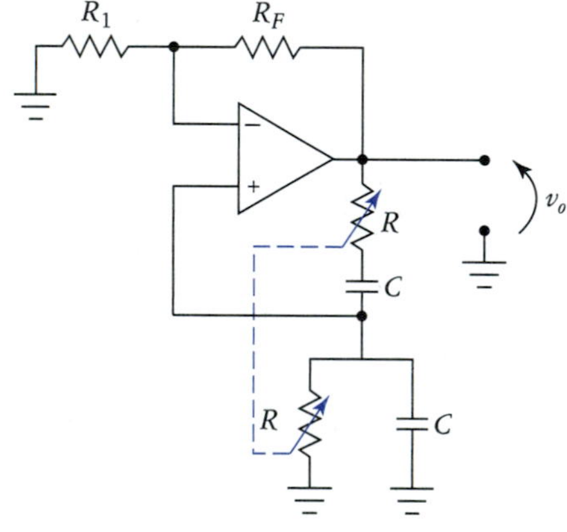

FIGURE 17.54
***Circuit for Problems** 17.21 **and** 17.22.*

■ TROUBLESHOOTING PROBLEMS

17.23. Refer to Fig. 17.9. Suppose that transformer T_1 was replaced with a unit that had a higher turns ratio. Would the overall voltage gain of the amplifier increase or decrease?

17.24. Refer to Fig. 17.9. Suppose that the original transistor was replaced with a device that had much higher current-handling and power dissipation capability. What effect, if any would you expect this change to have on the center frequency of the amplifier?

17.25. Refer to Fig. 17.27. Suppose that the secondary of the transformer was connected to the load such that the dotted end was connected to ground. Would this affect the operation of the circuit?

■ COMPUTER PROBLEMS

17.26. Refer to the tuned amplifier of Fig. 17.36. Simulate the circuit using transformer coupling coefficients of 0.75 and 0.5. Describe the effect of the coupling coefficient on bandwidth and overall voltage gain.

17.27. Refer to the transient analysis of Fig. 17.41. Display the oscillator output signal in the frequency domain (use FFT) in decibels. Determine the relationship between the unwanted signal energy frequencies and the fundamental oscillation frequency.

CHAPTER 18

A Sampling of Linear Integrated Circuits and Their Applications

BEHAVIORAL OBJECTIVES

- On completion of this chapter, the student should be able to:
- Describe the operation of a 555 timer in the astable mode.
- Describe the operation of a 555 timer in the monostable mode.
- Analyze a linear, series voltage regulator.
- Describe the operation of a switching voltage regulator.
- Determine the operating point and gain of a 3080 OTA.
- Explain how an OTA may be used as an AM modulator.
- Describe the operation of a diode AM demodulator.
- Describe the operation and applications of a balanced modulator.
- Describe the operation and applications of a phase-locked loop.

INTRODUCTION

Literally thousands of linear integrated circuits (ICs) are available today—far too many to be covered in even 10 books of this size. Certainly, many of these devices are different varieties of op amps. However, many other general purpose and specialized devices also warrant coverage. With this thought in mind, this chapter covers a representative sampling of some common linear ICs. We begin with a look at the very popular 555 timer. This device is analyzed in depth because it provides much insight into the operation of many related circuits such as voltage-to-frequency converters and pulse modulation circuits.

The operation of linear voltage regulators is covered in this chapter. As with many other linear ICs, the operation of these regulators is easily understood using the basic op amp theory covered previously. Switching regulators are used extensively in microcomputer power supply applications, so their operation is discussed as well.

Electronic communications is another area for which many specialized linear ICs are available. The operational transconductance amplifier is a device that is useful as a general-purpose amplifier, but also very useful in the design of amplitude modulator circuits. This chapter also presents the basic operating principles of phase-locked loops and balanced modulators. These are fundamentally important communications building blocks, but their operating principles are very general and can be applied to many other systems and applications.

18.1 THE 555 TIMER

The 555 timer is one of the most common ICs used to generate pulse and square waveforms. The internal block diagram and pin diagram for the 555 are shown in Figs. 18.1(a) and (b). The 555 can be connected in either of two basic configurations: astable and monostable. In the astable configuration, the 555 functions as an astable multivibrator. *Astable multivibrator* is the technical term for a rectangular waveform or pulse-train generator. This is the most common 555 configuration. The 555 can also be configured as monostable multivibrator or a *one shot.* These configurations are discussed in detail shortly.

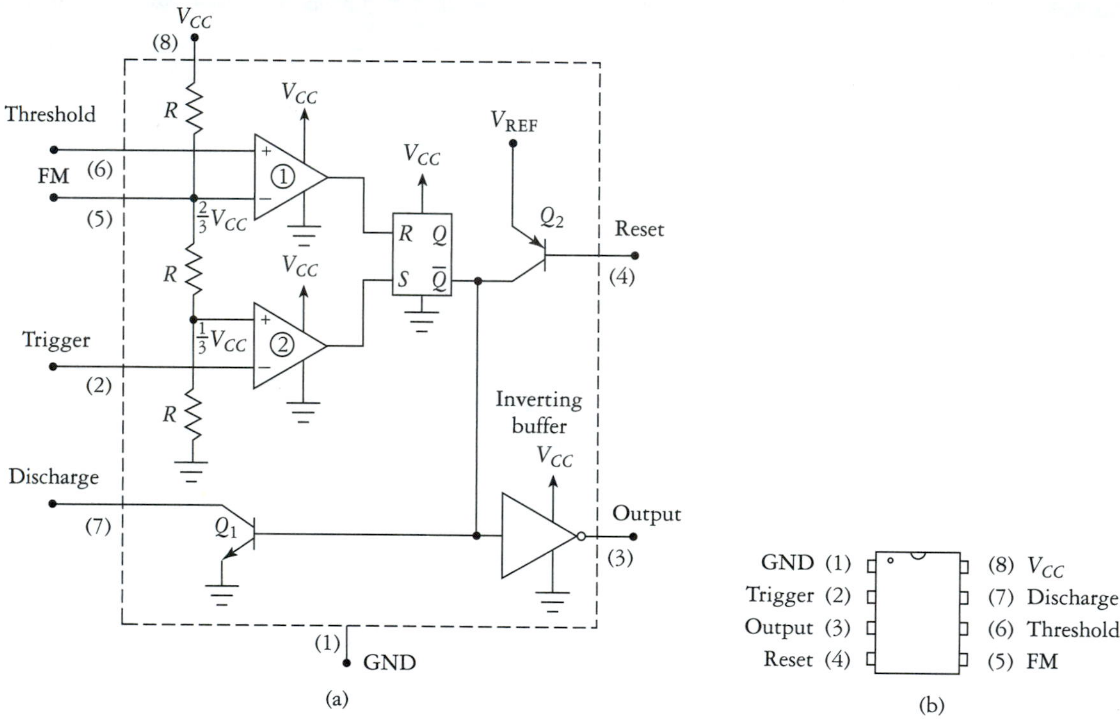

FIGURE 18.1
The 555 timer. (a) Internal equivalent circuit. (b) Pin designations.

The 555 timer is designed to operate over a supply voltage range from 4.5 to 18 V, and it is capable of sinking or sourcing ±200 mA. The 555 timer is limited to operation below 1 MHz, with no lower frequency limit.

Astable Operation

The 555 timer is connected as shown in Fig. 18.2 for operation in the astable mode. The load can be referred to ground or V_{CC}, as shown in dashed line form in the schematic. The v_o waveform of Fig. 18.2(b) is derived for the grounded-load configuration. Examination of the waveforms in Fig. 18.2(b) reveals that

$$T_H = T_2$$
$$T_2 = T_3 - T_1$$
$$T_L = T_4$$
$$T = T_L + T_H$$

We begin with an analysis of the astable circuit. Initially, C_1 charges through resistors R_1 and R_2. The charging voltage across C_1 is sampled by the threshold and trigger inputs of the 555. Examination of Figs. 18.1 and 18.2(b) reveals that as the capacitor charges toward $2V_{CC}/3$, the output of the 555 is high. During this interval, transistor Q_1 is in cutoff, placing the discharge pin in an open-circuit condition. The equation for v_C as the capacitor charges is

$$v_C = V_{CC}(1 - e^{-T_x/\tau_{chg}}) \tag{18.1}$$

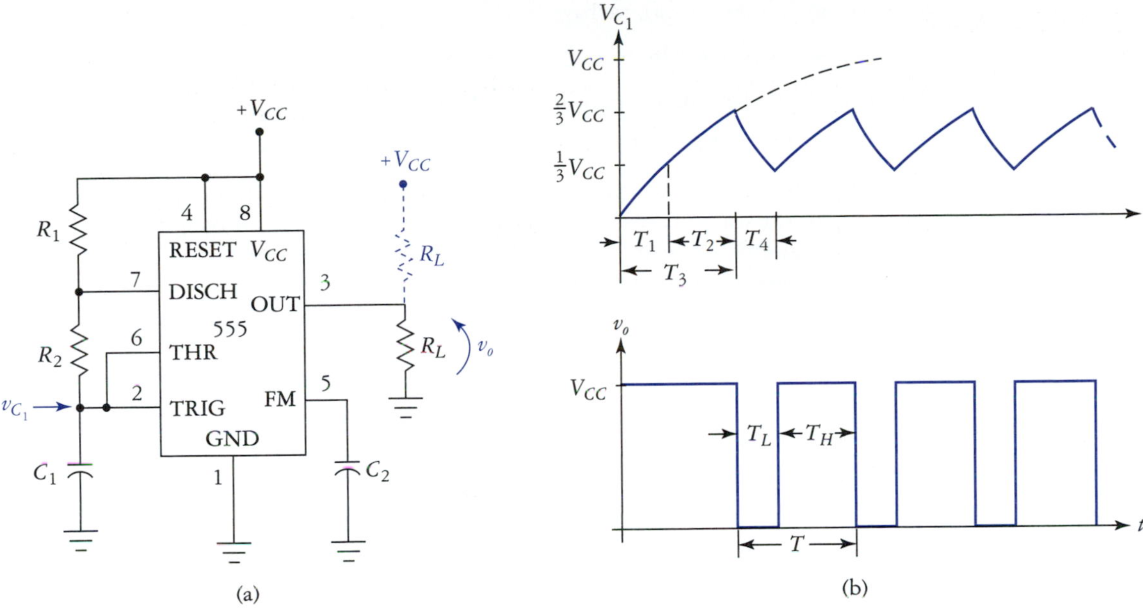

FIGURE 18.2
The 555 timer astable configuration. (a) The circuit. (b) Timing capacitor voltage and output voltage waveforms.

We determine time interval T_1 by substituting $v_C = V_{CC}/3$, $T_n = T_1$, and solving Eq. (18.1) for T_1. This gives us

$$T_1 = -\tau_{\text{chg}} \ln\left(\frac{V_{CC}/3}{V_{CC}}\right) - 1$$

Simplifying the argument of the natural logarithm, we find that V_{CC} cancels, leaving us with

$$T_1 = -\tau_{\text{chg}} \ln \frac{2}{3} \tag{18.2}$$

In a similar manner, we find the time interval T_3 to be

$$T_3 = -\tau_{\text{chg}} \ln\left(1 - \frac{2V_{CC}/3}{V_{CC}}\right)$$

This equation simplifies to

$$T_3 = -\tau_{\text{chg}} \ln \frac{1}{3} \tag{18.3}$$

The output high time interval is $T_H = T_2 = T_3 - T_1$; thus,

$$\begin{aligned} T_H &= -\tau_{\text{chg}} \ln \frac{1}{3} - \left(-\tau_{\text{chg}} \ln \frac{2}{3}\right) \\ &= \tau_{\text{chg}}\left(\ln \frac{2}{3} - \ln \frac{1}{3}\right) \end{aligned}$$

Recall that $\ln x - \ln y = \ln(x/y)$, which allows us to write

$$T_H = \tau_{\text{chg}} \ln \frac{2/3}{1/3}$$

This finally simplifies to

$$T_H = \tau_{\text{chg}} \ln 2 \tag{18.4}$$

Since the capacitor charges through both resistors, the charging time constant is

$$\tau_{\text{chg}} = (R_1 + R_2)C_1 \tag{18.5}$$

Substituting into Eq. (18.4), we have

$$T_H = C_1(R_1 + R_2)\ln 2 \tag{18.6a}$$

We can evaluate the natural logarithm, which produces

$$T_H \cong 0.693C_1(R_1 + R_2) \tag{18.6b}$$

We now know T_H, which is how long it takes to charge C_1 from $V_{CC}/3$ to $2V_{CC}/3$. When the capacitor voltage reaches $2V_{CC}/3$, comparator 1 resets the flip-flop, driving the 555's output low and also driving Q_1 into saturation. This causes the discharge pin to drop to ground potential, causing C_1 to begin discharging through R_2. The discharge equation is

$$v_C = V_0 e^{-T_{\text{dis}}/\tau_{\text{dis}}}$$

Substituting $2V_{CC}/3 = V_0$, $V_{CC}/3 = v_C$, and $T_L = T_{\text{dis}}$, and solving for T_L gives us

$$T_L = -\tau \ln\frac{1}{2}$$

Since the capacitor discharges through R_2, we have

$$T_L = -R_2C_1 \ln\frac{1}{2} \tag{18.7}$$

Evaluating the logarithm, we obtain

$$T_L \cong 0.693R_2C_1 \tag{18.8}$$

Recall that the period of the output waveform is $T = T_H + T_L$; thus,

Note
The oscillation frequency of the 555 is independent of the supply voltage.

$$T = 0.693C_1(R_1 + 2R_2) \tag{18.9}$$

The oscillation frequency is simply the reciprocal of the period. This is normally written as

$$f = \frac{1.44}{C_1(R_1 + 2R_2)} \tag{18.10}$$

Although some of the other equations derived in this discussion will be useful to us later, Eq. (18.10) is more generally useful. Also in the case of pulse- and rectangular-type waveforms, the oscillation frequency is often called the *pulse repetition rate* (PRR) or the *pulse repetition frequency* (PRF).

It is interesting to note that because the v_C trigger points of the 555 are based on constant proportions of V_{CC}, the oscillation frequency is independent of the power supply voltage. This is a very desirable characteristic, which helps to keep the oscillation frequency quite stable.

In many applications, it is important to know the duty cycle of the 555's output signal. Recall that the duty cycle of a rectangular waveform, like that of Fig. 18.2(b) is

$$D = \frac{T_H}{T} \tag{18.11}$$

This is often expressed as a percent:

$$\%D = \frac{T_H}{T} \times 100\% \tag{18.12}$$

If we substitute Eqs. 18.8 and 18.9 into Eq. 18.11, we obtain

$$D = \frac{R_1 + R_2}{R_1 + 2R_2} \tag{18.13}$$

As we shall see in a moment, the value of R_1 cannot be reduced to zero, therefore the duty cycle of the circuit in Fig. 18.2(a) will always be greater than 50%. If a 50% duty cycle (or less) is required, then the circuit must be modified.

EXAMPLE 18.1

Refer to Fig. 18.2(a). Assume that the grounded-load configuration is used. Given $R_1 = 6.8\ \text{k}\Omega$, $R_2 = 4.7\ \text{k}\Omega$, $R_L = 1\ \text{k}\Omega$, $C_1 = 0.1\ \mu\text{F}$, and $V_{CC} = 10$ V, determine f, D, and the average current flow I_L through the load.

Solution

$$f = \frac{1.44}{C_1(R_1 + 2R_2)}$$
$$= \frac{1.44}{0.1\ \mu\text{F}(6.8\text{ k} + 9.4\text{ k})}$$
$$= 889\text{ Hz}$$

$$D = \frac{R_1 + R_2}{R_1 + 2R_2}$$
$$= \frac{6.8\text{ k} + 4.7\text{ k}}{6.8\text{ k} + 9.4\text{ k}}$$
$$= 0.71 = 71\%$$

To determine the average load current, we first calculate the peak load current:

$$I_{L(\text{pk})} = \frac{V_{CC}}{R_L}$$
$$= \frac{5\text{ V}}{1\text{ k}\Omega}$$
$$= 5\text{ mA}$$

For rectangular waveforms, the average value is the product of the peak amplitude and the duty cycle. Thus,

$$I_L = DI_{L(\text{pk})}$$
$$= (0.71)(5\text{ mA})$$
$$= 3.55\text{ mA}$$

Practical Design Considerations

Many operating and component value limits are easily obtained from the data sheets for a given device. In other cases, some constraints may not be given, but are instead deduced through examination of the IC internal circuitry, or through experience and trial and error. We now look at a few of these restrictions on the 555.

Restrictions on C_1

Generally, there is no upper limit on the value of C_1 that can be used with the 555. High-value capacitors ($C \geq 10\ \mu$F) can be used to obtain low frequencies, but these capacitors will usually be either tantalum or aluminum electrolytic units. Aluminum electrolytic capacitors are relatively unstable and tend to have high leakage currents, whereas, tantalum caps are much more expensive. If accurate timing is not critical, aluminum electrolytic caps are usually chosen.

A practical lower limit on C_1 is 0.001 μF. This prevents stray capacitance from significantly affecting the 555's oscillation frequency. In equation form, this is

$$C_1 \geq 0.001\ \mu\text{F}$$

Restrictions on R_1

Recall that we cannot replace R_1 with a short circuit. To understand this, refer to the schematic of Fig. 18.2(a). When the output is low internal transistor Q_1 pulls the discharge terminal (pin 2) to ground. If R_1 is too low in value, excessive current will be drawn from the power supply. This will damage or destroy the 555 timer. As a general rule, R_1 should be limited to

$$1\text{ k}\Omega \leq R_1 \leq 1\text{ M}\Omega$$

The upper limit of 1 MΩ helps reduce temperature and comparator bias current sensitivity.

Restrictions on R_2

The lower limit on R_2 is a function of the supply voltage and capacitor C_1. A large capacitor can store a substantial amount of energy. Thus, if R_2 is too low in value, the 555's discharge transistor could be damaged. A good rule of thumb is to limit the value of R_2 such that

$$1\ \text{k}\Omega \le R_2 \le 1\ \text{M}\Omega$$

As with R_1, a maximum value of 1 MΩ helps reduce temperature and bias current sensitivity.

Unused Inputs

The reset input (pin 4) of the 555 is active low and if unused it should be tied to V_{CC}. The FM input (pin 5) is used to electronically vary the 555's oscillation frequency. Normally this input is not used and it should be bypassed to ground with a 0.01-μF capacitor.

EXAMPLE 18.2

Refer to Fig. 18.2(a). Assume that the grounded-load configuration is used. Given $R_1 = 10\ \text{k}\Omega$ and $C_1 = 0.1\ \mu\text{F}$, determine R_2 such that $D = 80\%$ and determine the resulting oscillation frequency. Determine the required value of R_2 such that $f = 10$ kHz.

Solution We solve Eq. (18.13) for R_2, which results in

$$R_2 = \frac{R_1 - DR_1}{2D - 1} \tag{18.14}$$

$$= \frac{10\ \text{k} - (0.8)(10\text{k})}{(2)(0.8) - 1}$$

$$= 3.33\ \text{k}\Omega$$

The resulting oscillation frequency is

$$f = \frac{1.44}{0.1\ \mu\text{F}(10\ \text{k} + 6.66\ \text{k})}$$

$$= 864\ \text{Hz}$$

To set $f = 10$ kHz, we determine the required value for R_2 by solving Eq. (18.10), producing

$$R_2 = \frac{1.44}{2fC_1} - \frac{R_1}{2} \tag{18.15}$$

$$= \frac{1.44}{(2)(10\ \text{kHz})(0.1\ \mu\text{F})} - \frac{10\ \text{k}}{2}$$

$$= -4280\ \Omega$$

It is not possible to have a negative resistance (at least in this context); therefore, using the given values of R_1 and C_1 it is not possible to obtain $f = 10$ kHz.

Note
Recall that there are three types of multivibrator: Astable (oscillator), Bistable (latch), and Monostable (one-shot).

The previous example serves to show that we cannot simply take the results of our calculations (or the results of some computer program) for granted. This is another example of why critical thinking and problem-solving skills are so important. Anyone can blindly plug numbers into a calculator or a computer, but it takes a higher level of understanding to determine whether the resulting answers are realistic or ridiculous.

Monostable Operation

Figure 18.3 shows a 555 timer configured for operation in the monostable, or one-shot, mode. Assuming C_1 is initially discharged and the trigger input is high ($V_{trig} = V_{CC}$), when power is applied the output will be low and the discharge pin is held at ground potential. When the trigger input drops below $V_{CC}/3$, the output of the 555 goes high and the discharge pin becomes an open circuit. This allows C_1 to begin charging through R_1. Charging continues until $v_{C_1} = 2V_{CC}/3$, at which time

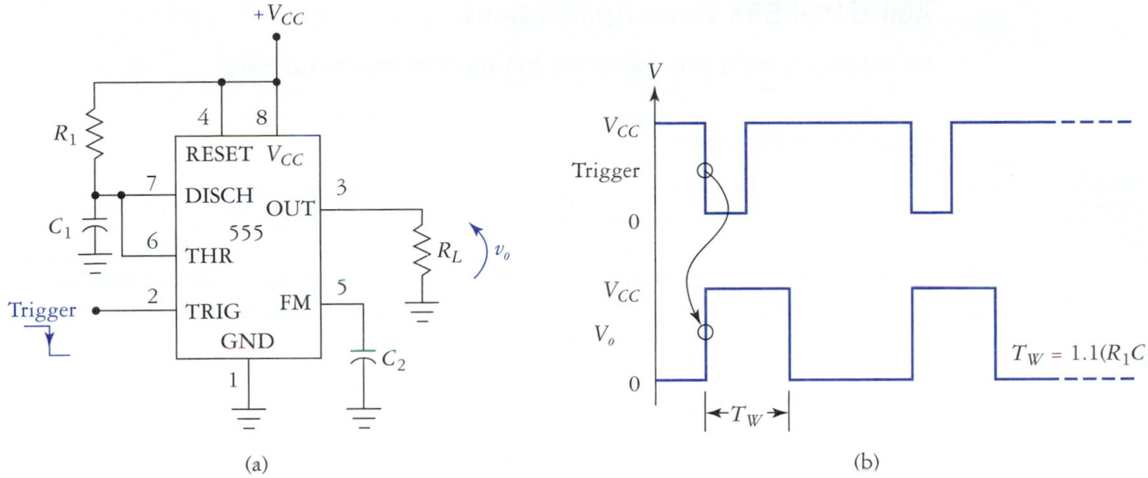

FIGURE 18.3
The 555 in a one-shot configuration. (a) Basic circuit. (b) Typical trigger and output voltage waveforms.

the discharge pin and the output go low. The capacitor is discharged rapidly and the 555 is ready for another trigger pulse. The output pulse width is given by

$$T_W \cong 1.1\, R_1 C_1 \tag{18.16}$$

It is difficult to describe the operation of the 555 one-shot device because it has characteristics of both edge-triggered and level-triggered devices. As long as we ensure that the trigger input is high when T_W expires, the 555 one-shot behaves as a non-retriggerable, edge-triggered one-shot. Thus, once a trigger occurs, multiple pulses within the T_W interval are ignored. This is shown in Fig. 18.4. However, if the trigger input is held low continuously, the output will remain in the high state. National Semiconductor suggests that the trigger input be placed in the high state before T_W expires.

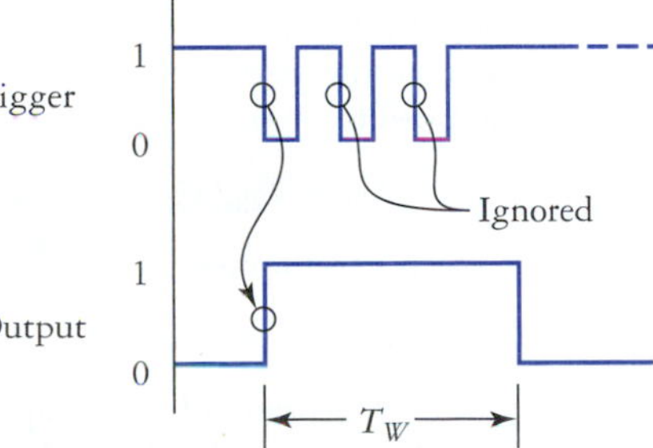

FIGURE 18.4
Multiple trigger events are ignored, but trigger input should be high prior to output reset.

EXAMPLE 18.3

Refer to Fig. 18.3. The circuit has $R_1 = 3.3\text{ k}\Omega$. Determine the required value for C_1 that will produce $T_W = 1$ ms.

Solution Solving Eq. (18.16), we have

$$C_1 = \frac{T_W}{1.1\,R}$$

$$= \frac{1\text{ ms}}{(1.1)(3.3\text{ k}\Omega)}$$

$$= 0.28\ \mu\text{F}$$

Additional 555 Timer Applications

Several interesting applications of the 555 timer are worth examining.

Pulse-Width Modulator

In electronic communication applications, it is common to vary a signal parameter (amplitude or frequency, for example) in proportion to an information source. This process is called *modulation.* The signal that represents the information source is called the *modulating signal,* while the signal that is altered by the modulation is called the *carrier.* A familiar example is commercial amplitude modulation (AM) broadcasting, in which the amplitude of a high-power, high-frequency carrier voltage is caused to vary in proportion to an audio signal.

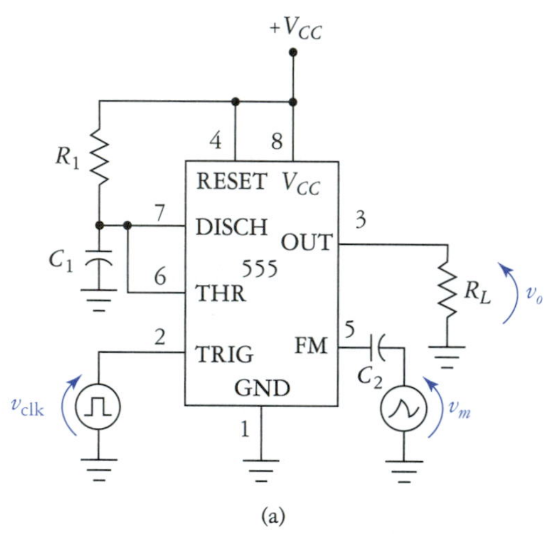

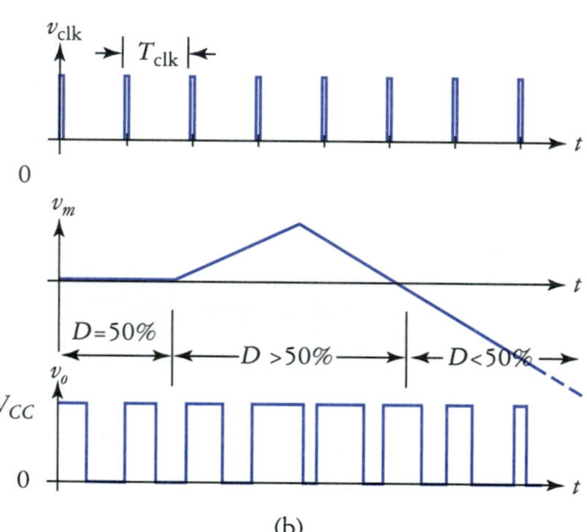

FIGURE 18.5
Pulse-width modulator circuit.

If a 555 is configured as a one-shot device, and an information signal is coupled to the FM input pin, a pulse width modulator is formed. Such a circuit is shown in Fig. 18.5. The clock signal is a narrow (low duty cycle) pulse train, and the 555 is designed such that with no modulation signal applied ($m = 0$) we have

$$T_W = T_{clk}/2$$

which also implies

$$D = 50\%$$

When modulation is applied, the modulating signal is summed with the 555's internal voltage divider, causing the threshold voltage to vary in proportion to the modulating voltage. Normally, the threshold voltage is $2V_{CC}/3$. As v_m increases, so does the threshold voltage. Because the threshold voltage is greater, it takes longer for C_1 to charge to this level. This increases the pulse width T_W. Since the clock frequency is constant, the duty cycle of v_o increases in proportion to the increase in v_m. This is shown in Fig. 18.5. When the modulating voltage decreases, the threshold voltage decreases, decreasing T_W and hence the duty cycle of the output.

Pulse modulation is used extensively in fiber optic data communications systems. In this application, the PWM generator is used to drive an LED or laser diode. The width of the light pulses varies in proportion to the modulating signal. Because the light source is either fully on or off, pulse modulation techniques are very efficient and immune to noise.

A PWM fiber optic communication system is shown in Fig. 18.6. A representative modulating signal is shown in the waveform. The PWM signal is converted to light pulses by the LED and transmitted to a photodiode via an optical fiber. The signal derived from the photodiode may exhibit some distortion due to attenuation and pulse dispersion in the fiber. The waveform is "squared up" by a Schmitt trigger. At this point, the signal can be processed digitally (not shown), but in this system we use an analog technique to extract the information from the signal. The cleaned-up pulse train is summed with a bias voltage that effectively removes the received dc signal component. The

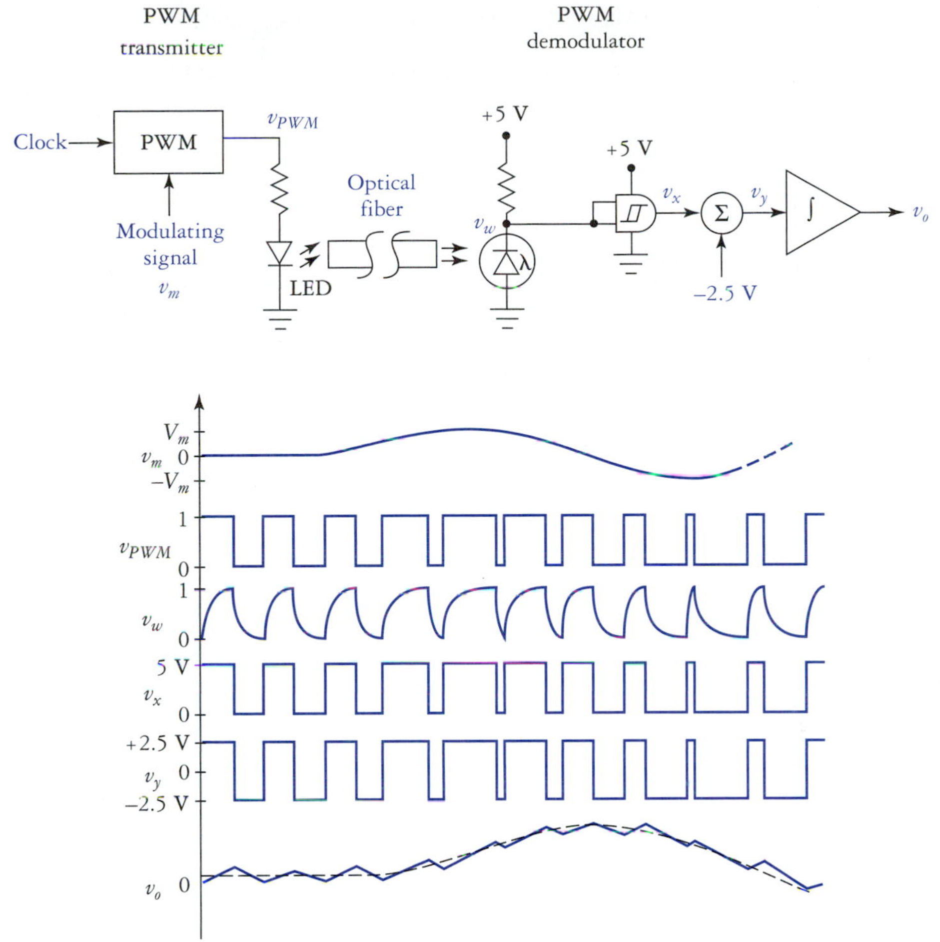

FIGURE 18.6
PWM fiber optic data communication system with typical waveforms.

bipolar pulse signal is integrated, producing a replica of the original modulating signal, as shown in the bottom of the figure. Some additional low-pass filtering can be used to further smooth the signal. The process whereby the original information signal is extracted from the transmitted signal is called *demodulation.*

In a practical pulse modulation system, the clock frequency is normally much higher than that of the modulating signal. The greater the difference, the easier it is to filter the resulting demodulated signal. To prevent overdriving of the 555 timer, the modulating signal should be limited such that

$$-\frac{V_{CC}}{3} \leq v_m \leq \frac{V_{CC}}{3} \tag{18.17}$$

The coupling capacitor C_2 is chosen such that $|X_C| \ll 5\ \text{k}\Omega$ at the lowest modulating frequency to be used. A good rule to use here is

$$|X_C|\bigg|_{f=f_{m(\min)}} \leq 500\ \Omega \tag{18.18}$$

If an electrolytic capacitor is used, the positive terminal should be connected to the FM input.

Pulse Position Modulator

The 555 timer can be used to implement a pulse-position modulator (PPM). As seen in Fig. 18.7(a), the 555 is configured in the astable mode in which the modulating signal is coupled to the FM input. The modulating signal is summed with the internal threshold reference voltage. Thus, when v_m increases, the threshold voltage increases as well, which causes C_1 to take longer to charge to the

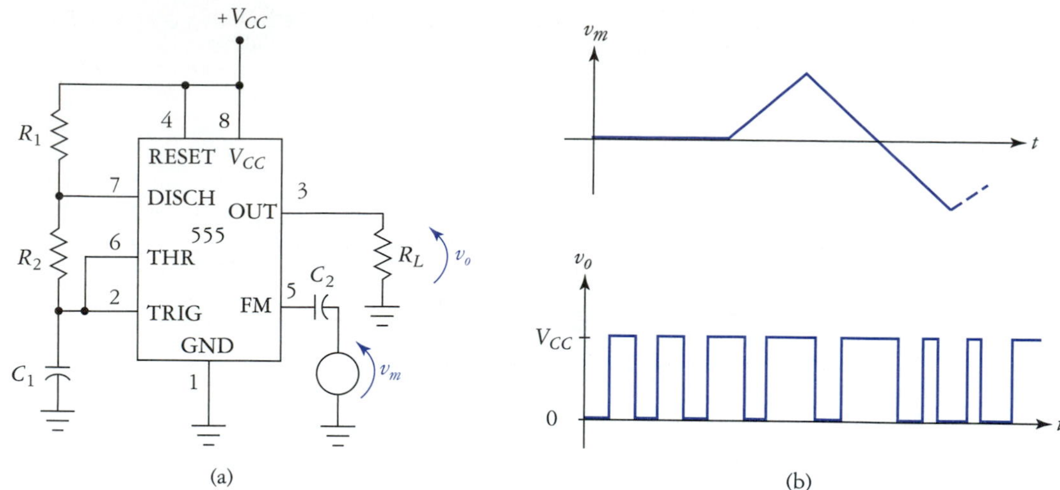

FIGURE 18.7
Pulse-position modulator. (a) The circuit. (b) Typical input and output waveforms.

threshold voltage. In this manner, a positive-going modulating signal decreases the oscillation frequency of the timer. Similarly, when the modulating signal is negative going, the frequency of oscillation increases. Typical waveforms are shown in Fig. 18.7(b).

As with the 555 PWM circuit, the PPM modulating input signal should be restricted as defined by Eq. (18.17), and the modulation coupling capacitor is chosen according to Eq. (18.18). The PPM signal is not as easy to demodulate as that of the PWM. PPM is most effectively demodulated using a phase-locked loop. Phase-locked loops are discussed later in this chapter.

Fifty Percent Duty Cycle Timer

It is possible to achieve nearly 50% duty cycle from the 555 astable using the circuit shown in Fig. 18.8. The circuit works as follows: Capacitor C_1 charges through R_1 and D_1 until $v_C = 2V_{CC}/3$. Once this voltage is reached, pin 7 is grounded, and C_2 discharges through R_2. The oscillation frequency is given by

Note
We can also obtain exactly 50% duty cycle by using a toggle-made flip-flop on the output of the 555.

$$f \cong \frac{1.44}{C_1(R_1 + R_2)} \tag{18.19}$$

The duty cycle is given by

$$D \cong \frac{R_1}{R_1 + R_2} \tag{18.20}$$

FIGURE 18.8
Modified astable 555 timer.

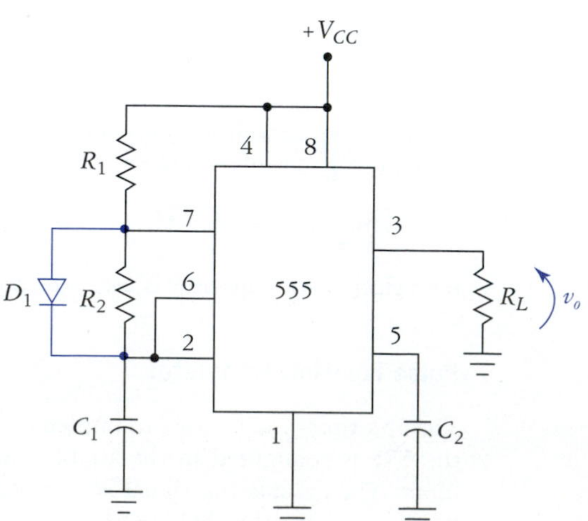

These equations are approximations because the effect of the barrier potential of D_1 is not accounted for. If we set $R_1 = R_2$, then Eq. (18.20) indicates $D = 50\%$.

18.2 VOLTAGE REGULATORS

Voltage regulation is a virtual necessity in most electronic circuits. Voltage regulators can be divided into two basic categories: linear and switching regulators. Switching regulators are more efficient than linear regulators. For low-power applications, say, $P \leq 10$ W or so, linear regulators are generally simpler and less expensive than switching regulators. As power goes up, switching power supply regulators become far less expensive, more efficient, and much smaller than linear regulators. Still, linear regulators are still widely used and important to understand.

Linear Voltage Regulators

Most linear voltage regulators are series regulators. A classic positive regulator circuit is shown in Fig. 18.9. Resistor R_1 and zener diode D_1 produce a stable reference voltage $V_{REF} = V_Z$, that is applied to the noninverting input of the op amp. Transistor Q_1 is called a *series pass transistor.* This is an emitter-follower that boosts the current handling capability of the op amp. Resistors R_1 and R_2 form a voltage divider that samples the output voltage providing a feedback voltage V_F to the inverting input of the op amp. Because this is a negative feedback loop, the op amp ideally forces its differential input voltage to zero. Should the input voltage decrease, or load current increase, the op amp compensates by driving the pass transistor closer to saturation. If the input voltage increases, or the load current decreases, the op amp applies less drive to Q_1.

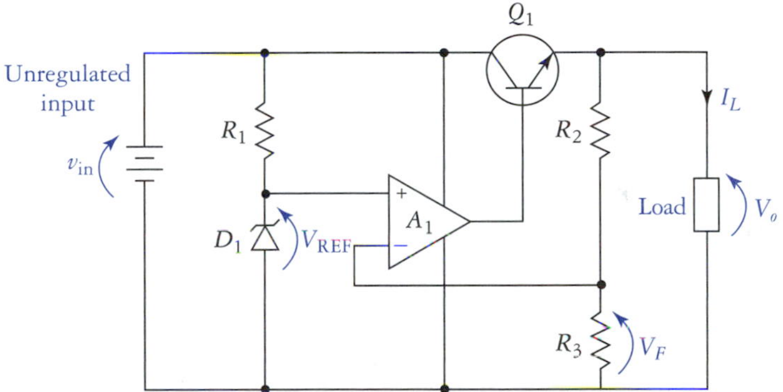

FIGURE 18.9
Basic series voltage-regulator circuit.

We now derive a few useful analysis and design equations for the series regulator. Again, because the op amp forces $V_{id} \rightarrow 0$ V we have

$$V_F = V_{REF} \tag{18.21}$$

Applying the voltage-divider equation, we also have

$$V_F = V_o \frac{R_3}{R_2 + R_3} \tag{18.22}$$

Substituting Eq. (18.21) into (18.22) and solving for V_o we obtain

$$V_o = V_{REF} \frac{R_2 + R_3}{R_3} \tag{18.23}$$

This is the noninverting op amp gain equation that we derived in Chapter 12. The alternative form is

$$V_o = V_{REF}\left(1 + \frac{R_2}{R_3}\right) \tag{18.24}$$

The pass transistor in a series regulator may be subject to high-power dissipation. Because of this, the transistor will normally be mounted on a heat sink. Recall that the power dissipation of any transistor is given by

$$P_{DQ} = I_{CQ}V_{CEQ} \tag{18.25}$$

Using KVL and Eq. (18.25), we obtain the following equation for the power dissipation of Q_1:

$$P_{DQ} = I_L(V_{\text{in}} - V_o) \tag{18.26}$$

Let's apply these relationships to an analysis example.

EXAMPLE 18.4

Refer to Fig. 18.9. Assume that $V_{\text{in}} = 15$ V, $V_Z = 5.2$ V, $R_1 = 680\ \Omega$, $R_2 = 1\ \text{k}\Omega$, and $R_3 = 2\ \text{k}\Omega$. Determine the zener current I_Z and V_o. Determine I_L, P_L, and P_{DQ} given $R_L = 50\ \Omega$ and $R_L = 10\ \Omega$.

Solution The zener current is found as follows:

$$I_Z = \frac{V_{\text{in}} - V_Z}{R_1} = \frac{15\text{V} - 5.2\text{ V}}{680\ \Omega} = 14.4\text{ mA}$$

The output voltage is now determined:

$$V_o = V_{\text{REF}}\frac{R_2 + R_3}{R_3} = 5.2\text{ V}\left(\frac{1\text{ k}\Omega + 2\text{ k}\Omega}{2\text{ k}\Omega}\right) = 7.8\text{ V}$$

For $R_L = 50\ \Omega$, we have

$$I_L = \frac{V_o}{R_L} = \frac{7.8\text{ V}}{50\ \Omega} = 156\text{ mA}$$

$$P_L = V_oI_L = (7.8\text{ V})(156\text{ mA}) = 1.22\text{ W}$$

$$P_{DQ} = I_C(V_{\text{in}} - V_o) = (156\text{ mA})(7.2\text{ V}) = 1.12\text{ W}$$

In the case of $R_L = 10\ \Omega$, we obtain

$$I_L = \frac{V_o}{R_L} = \frac{7.8\text{ V}}{10\ \Omega} = 780\text{ mA}$$

$$P_L = V_oI_L = (7.8\text{ V})(780\text{ mA}) = 6.80\text{ W}$$

$$P_{DQ} = I_C(V_{\text{in}} - V_o) = (780\text{ mA})(7.2\text{ V}) = 5.62\text{ W}$$

Notice in the example that the pass transistor dissipates nearly as much power as the load. In some situations, the pass transistor will dissipate far more power than the load. This is the main problem with linear power supplies.

EXAMPLE 18.5

Using the results of Example 18.4, with $R_L = 10\ \Omega$, assume a 2N3055 is used as a pass transistor. The 2N3055 has a TO-3 case with junction to ambient thermal resistance $\theta_{JA} = 34.5$°C/W. Determine the junction temperature T_J of the pass transistor if the ambient temperature is $T_A = 25$°C.

Solution A quick review of Chapter 10 provides us with the necessary equation, Eq. (10.42), which gives us

$$\begin{aligned} T_J &= T_A + P_{DQ}\,\theta_{JA} \\ &= 25°\text{C} + (5.62\ \text{W})(34.5°\text{C/W}) \\ &= 219°\text{C} \end{aligned}$$

This exceeds even the maximum allowable storage temperature for the 2N3055, which is 200°C! Unless we use a substantial heat sink, this transistor is toast.

The lesson to be learned from the preceding example is that although a transistor may have a very high power rating, we must be careful not to overheat the device. The 2N3055 is rated at 115 W, yet in the example where $P_{DQ} = 5.62$ W, the transistor is well out of its safe operating area (SOA). We do not dwell on this subject now, but it is a good thing to keep in mind.

In order for the regulator in Fig. 18.9 to operate correctly, the input voltage must not drop below some minimum value. Of course, we require $V_{\text{in}} > V_o$, but we also require sufficient input voltage to keep the op amp working correctly and to keep the zener diode biased properly as well. The minimum output voltage for the circuit of Fig. 18.9 is

$$V_{o(\text{min})} = V_{\text{REF}} \tag{18.27}$$

Zener diodes have a minimum value of about $V_Z = 4$ V. If a lower output voltage than this must be obtained, it is possible to modify the reference source as shown in Fig. 18.10. The reference voltage is

$$V_{\text{REF}} = V_Z \frac{R_3}{R_2 + R_3} \tag{18.28}$$

We must choose R_2 and R_3 such that the zener current is greater than the zener knee current at the minimum allowable input voltage. Generally, $R_1 < 1\ \text{k}\Omega$, and if we set $R_1 + R_2 \geq 10\ \text{k}\Omega$, the zener will be unaffected by this modification.

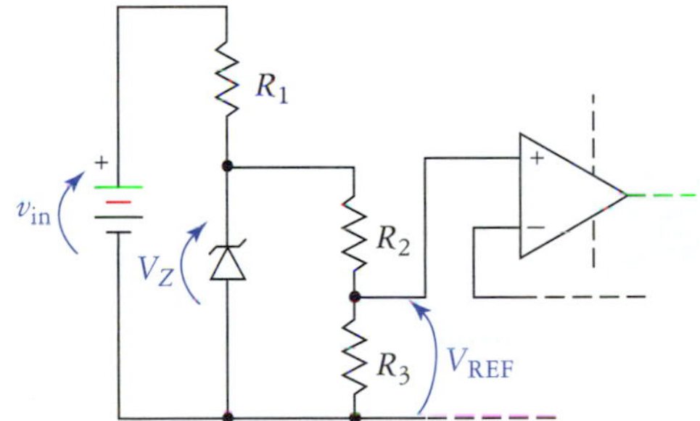

FIGURE 18.10
Reducing the reference voltage.

Adjustable Regulators

The output of the regulator can be adjusted by the addition of a potentiometer as shown in Fig. 18.11. If we neglect transistor saturation voltage, emitter resistor voltage drops, and other small effects, the minimum and maximum output voltages are given by

$$V_{o(\text{min})} = V_{\text{REF}}\left(1 + \frac{R_2}{R_3 + R_4}\right) \tag{18.29}$$

and

$$V_{o(\text{max})} = V_{\text{REF}}\left(1 + \frac{R_2 + R_3}{R_4}\right) \tag{18.30}$$

In order for the regulator to work properly, we also require

$$V_{o(\text{min})} > V_{\text{REF}} \tag{18.31}$$

$$V_{o(\text{max})} < V_{\text{in}} \tag{18.32}$$

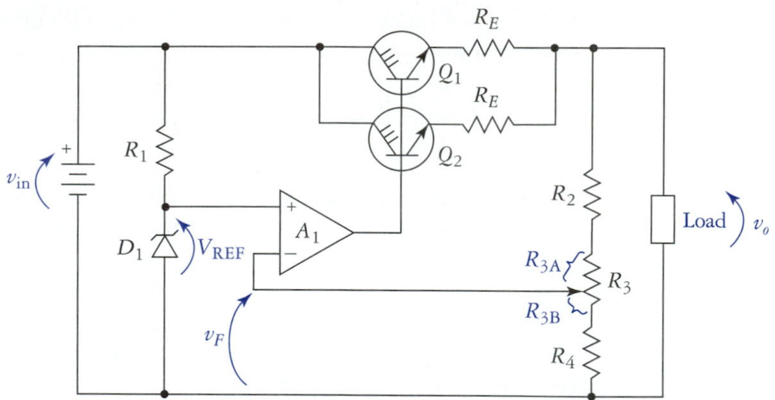

FIGURE 18.11
Modified series pass linear regulator.

The regulator in Fig. 18.11 also uses two parallel-connected pass transistors. This cuts the per transistor power dissipation in half, reducing thermal stress on these devices. The short lines on the collector leads indicate that the transistors are mounted on a heat sink. It is desirable to match the pass transistors as closely as possible, however, emitter swamping resistors are required to prevent current hogging. Typically, $R_E \leq 1\ \Omega$. Parallel-connected transistors also have the undesirable characteristic that the effective total beta β_T of the group is reduced proportionate to the number of devices. Assuming n matched transistors ($\beta_1 = \beta_2 = \cdots = \beta_n$) connected in parallel we can write

$$\beta_T = \frac{\beta_n}{n} \tag{18.33}$$

Low β_T increases the load on the op amp. For example, consider that the 741 op amp has a typical $I_{o(\max)} = \pm 15$ mA. If we use a 741 in Fig. 18.11 and $\beta_T = 50$, then the maximum load current will be

$$\begin{aligned} I_{L(\max)} &= \beta_T I_{o(\max)} \\ &= (50)(15\text{ mA}) \\ &= 750\text{ mA} \end{aligned}$$

We could use a Darlington configuration to reduce the load on the op amp.

EXAMPLE 18.6

Refer to Fig. 18.11. The circuit has $V_{in} = 18$ V, $V_Z = 4.2$ V, $R_2 = 1\ \text{k}\Omega$, $R_3 = 2\ \text{k}\Omega$ pot, and $R_4 = 1\ \text{k}\Omega$. Determine the minimum and maximum output voltages.

Solution

$$\begin{aligned} V_{o(\min)} &= V_{REF}\left(1 + \frac{R_2}{R_3 + R_4}\right) \\ &= 4.2\text{ V}\left(1 + \frac{1\text{ k}}{2\text{k} + 1\text{k}}\right) \\ &= 5.6\text{ V} \end{aligned}$$

$$\begin{aligned} V_{o(\max)} &= V_{REF}\left(1 + \frac{R_2 + R_3}{R_4}\right) \\ &= 4.2\text{ V}\left(1 + \frac{1\text{ k} + 2\text{ k}}{1\text{k}}\right) \\ &= 16.8\text{ V} \end{aligned}$$

Overcurrent Protection

Should the output of a series regulator become short circuited, the pass transistor and other expensive components could be destroyed. Current limiting prevents this from happening.

Simple Current Limiting

Simple current limiting was discussed as it relates to power amplifiers in Section 10.5 of Chapter 10. The same principle can be used to provide current limiting in series regulators. Such a circuit is shown in Fig. 18.12(a). As the load current increases, the voltage drop across R_E approaches the turn-on voltage for Q_2. This is about 0.7 V, as usual, for silicon devices. Once Q_2 is biased on, it robs the pass transistor of base current. Now, as the op amp attempts to force Q_1 further into conduction, Q_2 shunts this drive current to the load. Eventually the op amp output will saturate, but this is a relatively light load for Q_2, because the typical op amp can source current in the tens of milliamperes range. The load current (and pass transistor collector current) is limited to

$$I_{L(\max)} \cong \frac{V_{BE}}{R_E} \tag{18.34}$$

The graph in Fig. 18.12(b) shows how current is limited as the load resistance decreases toward zero.

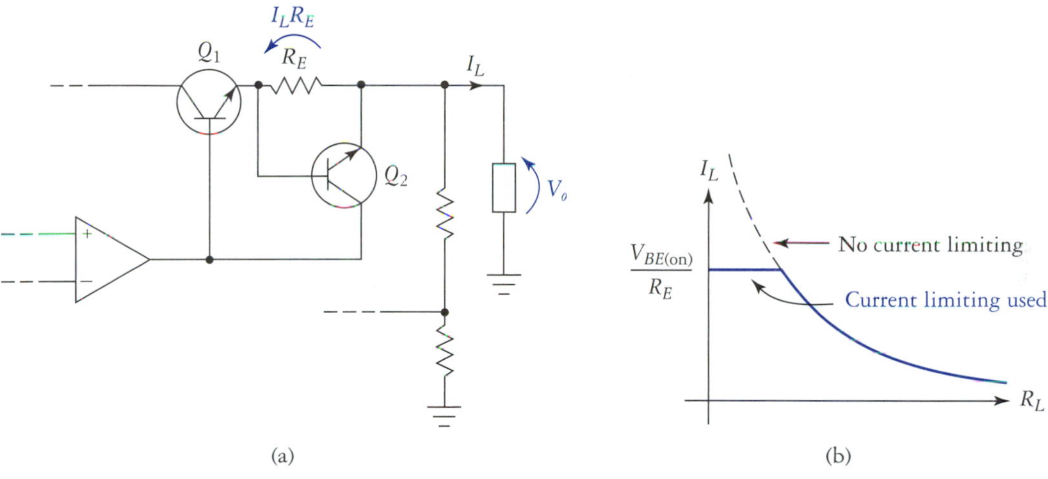

FIGURE 18.12
Overcurrent protection. (a) Circuit modification. (b) Load current versus load resistance.

EXAMPLE 18.7

Determine the required value of R_E in Fig. 18.12, such that the load current is limited to 500 mA. Determine the power dissipation of R_E under overload conditions.

Solution Solve Eq. (18.34) for R_E, which produces

$$R_E = \frac{V_{BE}}{I_{L(\max)}} = \frac{0.7\text{ V}}{500\text{ mA}} = 1.4\ \Omega$$

The maximum emitter resistor power dissipation is

$$P_{RE(\max)} = I_{L(\max)}^2 R_E = (500\text{ mA})^2 \times 1.4\ \Omega = 350\text{ mW}$$

One big disadvantage of simple current limiting is that the lower the output voltage is when limiting occurs, the greater the power dissipation of the pass transistor will be. This is so because at low output voltages the pass transistor has higher V_{CE}, and $P_{DQ} = V_{CEQ}I_{CQ}$. For a given short-circuit current, the power dissipation of the pass transistor increases as the output voltage decreases. This means that the current limiter must be designed to protect the transistor under worst-case conditions, where $V_o = V_{o(\min)}$. We do not go any further into the details of this situation; instead we move on to a protection scheme that solves this problem.

Foldback Current Limiting

Foldback limiting provides much greater protection of the regulator than simple current limiting. A foldback limiter is shown in Fig. 18.13(a). This circuit operates such that the short-circuit current decreases as output voltage decreases. In this way the regulator maintains approximately constant pass transistor power dissipation under all overload conditions. The graph in Fig. 18.13(b) illustrates the foldback limiting characteristic curve.

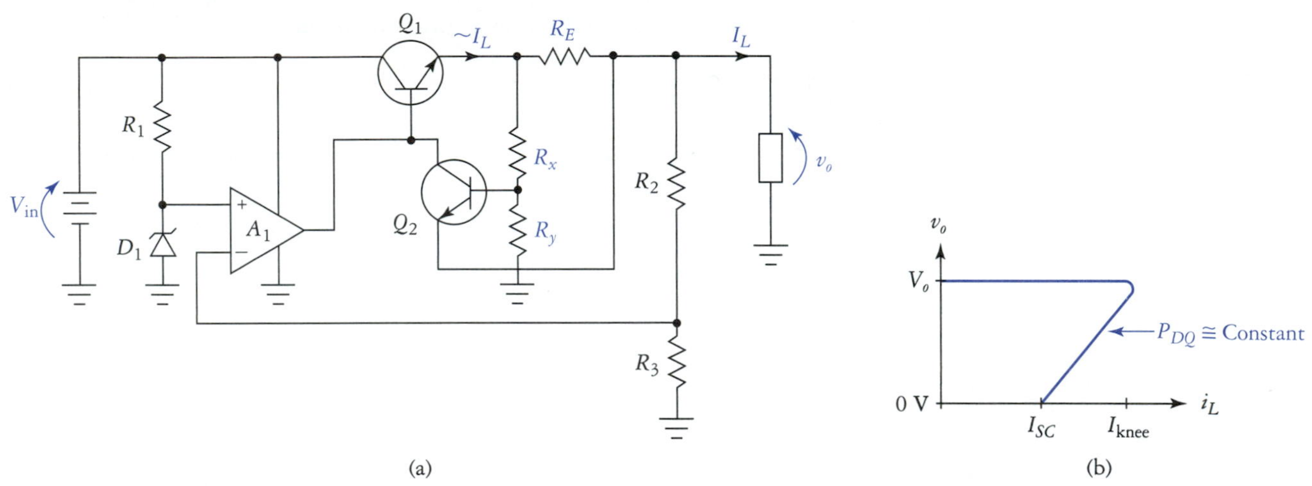

FIGURE 18.13
Regulator with foldback current limiting.

A foldback-limited regulator is designed to limit the maximum *power dissipation* of the pass transistor rather than the maximum current. Referring to Fig. 18.13(b), the short-circuit current is given by

$$I_{SC} \cong \frac{V_{BE}}{R_E}\left(\frac{R_x + R_y}{R_y}\right) \tag{18.35}$$

The knee current is

$$I_{knee} \cong \frac{V_o R_x + V_{BE}(R_x + R_y)}{R_E R_y} \tag{18.36}$$

EXAMPLE 18.8

Refer to Fig. 18.13. Given $V_{in} = 12$ V, $R_2 = 3.3$ kΩ, $R_3 = 10$ kΩ, $R_E = 0.6$ Ω, $R_x = 150$ Ω, $R_y = 520$ Ω, $V_{in} = 12$ V, and $V_Z = 6$ V, determine V_o, I_{knee}, I_{SC} and $P_{DQ(max)}$.

Solution First, we determine the output voltage:

$$V_o = V_{REF}\left(1 + \frac{R_2}{R_3}\right)$$
$$= 6\text{ V}\left(1 + \frac{3.3\text{ k}}{10\text{ k}}\right)$$
$$\cong 8\text{ V}$$

The knee current is

$$I_{knee} = \frac{V_o R_x + V_{BE}(R_x + R_y)}{R_E R_y}$$
$$= \frac{(8\text{V})(150\ \Omega) + 0.7\text{ V}(150\ \Omega + 520\ \Omega)}{(0.6\ \Omega)(520\ \Omega)}$$
$$= 5.35\text{ A}$$

The short-circuit current is

$$I_{SC} = \frac{V_{BE}}{R_E}\left(\frac{R_x + R_y}{R_y}\right)$$
$$= \frac{0.7\text{ V}}{0.6\ \Omega}\left(\frac{150\ \Omega + 520\ \Omega}{520\ \Omega}\right)$$
$$= 1.5\text{ A}$$

The power dissipation of the pass transistor under short-circuit conditions ($R_L = 0\ \Omega$, $V_o = 0$ V) is

$$P_{DQ} = I_{SC}V_{CEQ}$$
$$\cong I_{SC}\,(V_{\text{in}} - V_o)$$
$$= 1.5\text{ A}(12\text{ V} - 0\text{ V})$$
$$= 18\text{ W}$$

Under normal output voltage conditions the power dissipation of the pass transistor is

$$P_{DQ} = I_{\text{knee}}\,V_{CEQ}$$
$$\cong I_{\text{knee}}(V_{\text{in}} - V_o)$$
$$= 5.35\text{ A}(12\text{ V} - 8\text{ V})$$
$$= 21.4\text{ W}$$

Ideally the power dissipation of the pass transistor would be identical for both P_{DQ} conditions in the previous example. The difference is due to design approximations that are beyond the scope of this discussion. The main point to notice though is that the foldback limiting circuit does keep the pass transistor power dissipation fairly constant over all overload conditions.

18.3 MONOLITHIC VOLTAGE REGULATORS

While it is certainly important to understand the fundamental principles of linear voltage regulators, most often an off-the-shelf monolithic IC will be used to perform this function.

The LM317 and LM337 Regulators

An example of a regulated bipolar power supply is shown in Fig. 18.14. Capacitors C_1 and C_2 are the main filter caps, typically high-value electrolytic units. The positive side of the supply is regulated with an LM317 three-terminal adjustable regulator, while the negative side uses a complementary LM337. These regulators feature internal overcurrent, overtemperature, and SOA protection. The outputs are adjustable from 1.2 to 37 V (−1.2 to −37 V). Both regulators are available in TO-3 (LM317K, 337K) and TO-220 (LM317T, 337T) case styles, with current ratings of 1.5 A. Assuming adequate heat sinking is available, the TO-3 case has lower thermal resistance and so at a given power dissipation will run cooler. The capacitors marked with asterisks help prevent oscillation and should be mounted close to the regulators. The LM317 may be damaged if the output should become more positive than its input pin. The LM337 may be damaged if its output terminal is driven more negative than its input. Diodes D_5 and D_6 clamp these reverse voltages to about 0.7 V, preventing this damage from occurring.

The LM317 and LM337 have typical line regulation specifications of 0.01%/V. Here's what this means. Say the supply is designed to operate from 120 V with $V_o = 10$ V. Should the ac line voltage decrease by 20 V, for example, the output will typically decrease by just 2 mV.

The LM317 and LM337 load regulation specifications apply over the range of 10 mA $\leq I_L \leq I_{\text{max}}$. For $V_o \leq 5$ V, $\text{Reg}_{\text{load}} = 5$ mV. That is, the output voltage will not change by more than 5 mV under normal load conditions. For $V_o \geq 5$ V, $\text{Reg}_{\text{load}} = 0.1\%\ V_o$. Thus, for example, if $V_o = 10$ V, the output will not change by more than 0.01 V over all normal loading conditions. The output voltage of the LM317 (multiply by −1 for the LM337) is

$$V_o = 1.25\text{ V}\left(\frac{R_1 + R_2}{R_1}\right) + R_2(50\ \mu\text{A}) \qquad (18.37)$$

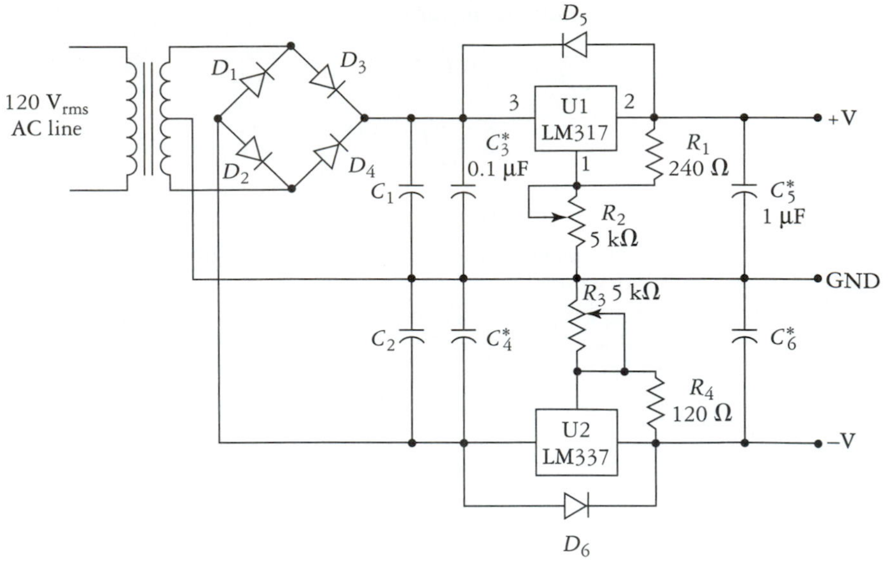

FIGURE 18.14
Adjustable bipolar regulated power supply.

The 50-μA term is caused by the adjustment terminal bias current. Often this term is small enough to neglect.

18.4 SWITCHING REGULATORS

Most modern high-current power supplies employ switching regulators. This can be understood by comparing a linear regulated supply with an equivalent switching supply. A common personal computer supply is rated at 250 W. If the supply delivers 5 V to the computer then $I = 250/5 = 50$ A. Such a power supply would likely measure about 6″ × 6″ × 6″ and would weigh about 3 pounds. These switching supplies are also available for less than $20 as well. An equivalent linear regulated supply would be much larger and would probably weigh close to 100 pounds. In addition, the linear regulated supply would cost (at least) hundreds of dollars.

Note
MOSFETs are commonly used in switching regulator designs because of their high switching speeds and inherent thermal stability.

The three basic switching regulator configurations are step-down, step-up, and inverting. These basic configurations are shown in Fig. 18.15. Each regulator employs an inductor as its primary energy storage element, and a diode for current steering. Although a variable frequency oscillator such as a VCO could be used, most switching regulators rely on the use of a pulse-width modulator to control the duty cycle of the MOSFET pass transistor. Typical PWM frequencies range from around 20 to about 100 kHz. Because such high switching frequencies are used, the filter capacitance can be small (compared to an equivalent line-operated linear supply) and still achieve good ripple rejection. The circuit operates such that if varying load or line conditions cause the output voltage to decrease, the duty cycle of the PWM is increased to compensate. Conversely, if for some reason the output voltage begins to increase, the PWM duty cycle will decrease. MOSFETs are used often because they have superior switching characteristics compared to BJTs. Also, recall that multiple FETs can be connected in parallel without the use of efficiency reducing swamping resistors.

The Step-Down Configuration

The circuit in Fig. 18.15(a) illustrates the step-down configuration. In this case, the inductor and capacitor form an averaging filter. When the FET is conducting, current flows through the inductor, which stores energy in its magnetic field. The diode provides a low-impedance path for inductor current when the FET is biased off, preventing a high-voltage inductive kick from occurring. The output voltage is proportional to the average of the input voltage, the load current, and the duty cycle of the PWM. That is,

$$V_o = xDV_{\text{in}} \tag{18.38}$$

where x depends on the values of L, C, and I_L. Given that L and C are constant, x varies with I_L. Thus, the regulator varies the duty cycle D to compensate.

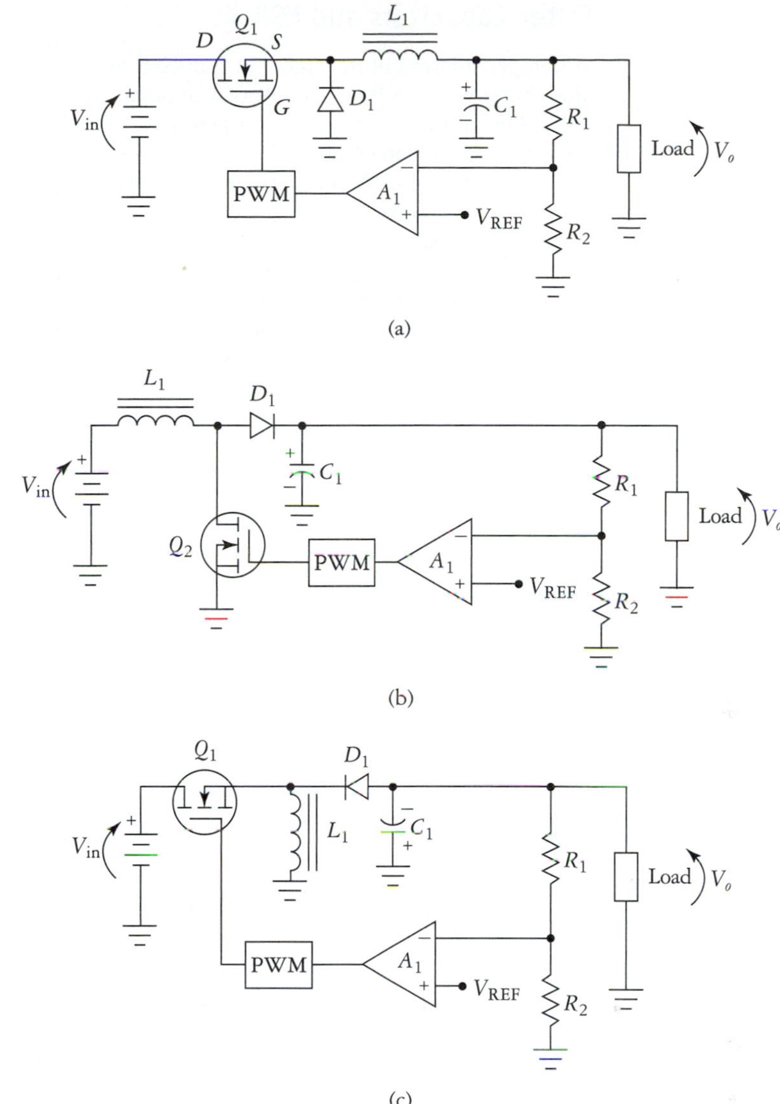

FIGURE 18.15
Switching regulators. (a) Step-down. (b) Step-up. (c) Inverting.

The Step-Up Configuration

The step-up switching regulator is shown in Fig. 18.15(b). In this configuration, when the FET is turned on, the inductor is energized. When the FET is turned off, the inductor's magnetic field collapses, inducing a voltage that is series aiding the input voltage. This causes C_1 to charge to a value greater than the input voltage. The diode prevents the capacitor from being discharged through the FET when it is driven into conduction. As before, the duty cycle of the PWM automatically varies to compensate for load and line variations. The output voltage is given by Eq. (18.38).

The Inverting Configuration

The inverting configuration is shown in Fig. 18.15(c). In this case, when the FET is on, the inductor is energized and D_1 is reverse biased. When the FET is switched off, the collapsing magnetic field induces a voltage that drives the top of the inductor negative with respect to ground. This forward biases the diode, allowing the capacitor to charge with the polarity indicated in the schematic.

It is interesting to note that the inverting regulator is capable of producing $|V_o| > V_{in}$ as well as $|V_o| < V_{in}$. Linear regulators are not capable of voltage inversion or step-up operation.

Filter Capacitors and ESR

A detailed analysis of the choice of inductor and capacitor values is beyond the scope of this discussion, but a word on filter capacitors is in order.

One of the reasons why high-power linear regulated supplies are so large is the requirement for large filter capacitors. A 25-amp, 5-V, line-operated linear power supply would typically require a filter capacitance of $C > 200{,}000\ \mu\text{F}$. An equivalent switching supply might require around 10,000 μF or less. If you have ever examined a switching power supply you may have noticed that a rather large number of relatively small (100 to 1000 μF) aluminum electrolytic or tantalum capacitors are used. Rather than using a single large filter capacitor, most switching supply designs employ lower value capacitors connected in parallel to obtain the necessary total capacitance. The reason is that a single large capacitor will have an unacceptably large *equivalent series inductance* (ESL) and to a lesser extent *equivalent series resistance* (ESR), as shown in Fig. 18.16(a). Connecting several smaller capacitors in parallel increases the total capacitance while decreasing the overall ESL and ESR. This is shown in Fig. 18.16(b). Low ESR is required for efficient switching regulator operation.

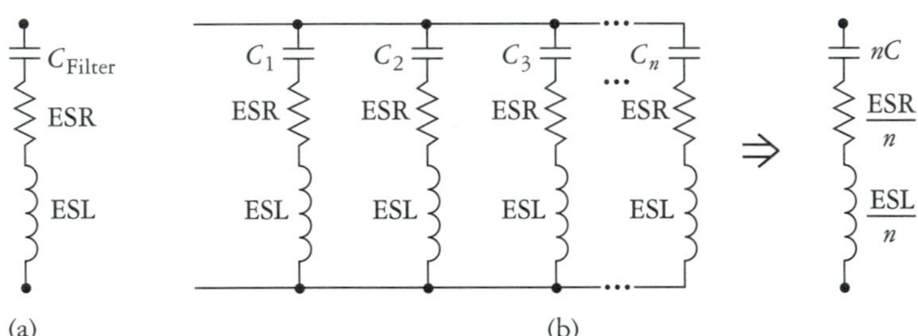

FIGURE 18.16
Effective series resistance of a capacitor.

18.5 OPERATIONAL TRANSCONDUCTANCE AMPLIFIERS

An operational transconductance amplifier (OTA) is a differential amplifier that has a high-impedance current-source output. The simplified internal circuit for an LM3080 OTA is shown in Fig. 18.17(a). The pin designations are shown in Fig. 18.17(b). Because the output variable is a current and the input is a voltage, the transfer parameter is out/in $= i_o/v_{id} = g_m$, thus the gain of the OTA has the dimensions of Siemens (S) or equivalently, mhos (℧)

As can be seen in the schematic diagram of Fig. 18.17(a), the input of the OTA is a differential pair, while the output stage is a complementary-symmetry *common-emitter* amplifier, which accounts for the high output resistance. Standard op amps generally use an emitter-follower output stage, which provides low output resistance. An external resistor must be used to bias up the current mirror. This current is called the *amplifier bias current* I_{ABC}, and it establishes the Q point and voltage gain of the differential input stage.

OTA Biasing and Operation

Note
To review the current mirror, see p. 148 in Chapter 4.

The basic inverting and noninverting OTA configurations are presented in Figs. 18.18(a) and (b), respectively. Notice that an external feedback loop is not used. The gain of the OTA is set by adjusting I_{ABC} and R_L to suit a given application. The following analysis equations apply to both inverting and noninverting configurations. The amplifier bias current is given by the familiar current-mirror equation

$$I_{\text{ABC}} = \frac{V_{EE} - V_{BE}}{R_{\text{ABC}}} \tag{18.39}$$

The transconductance is given by

$$g_m = kI_{\text{ABC}} \tag{18.40}$$

where $k \cong 19.2\ \text{V}^{-1}$ at 27°C. The load resistance develops a voltage drop (output voltage V_o), which is given by

$$V_o = \pm v_{\text{in}} g_m R_L \tag{18.41}$$

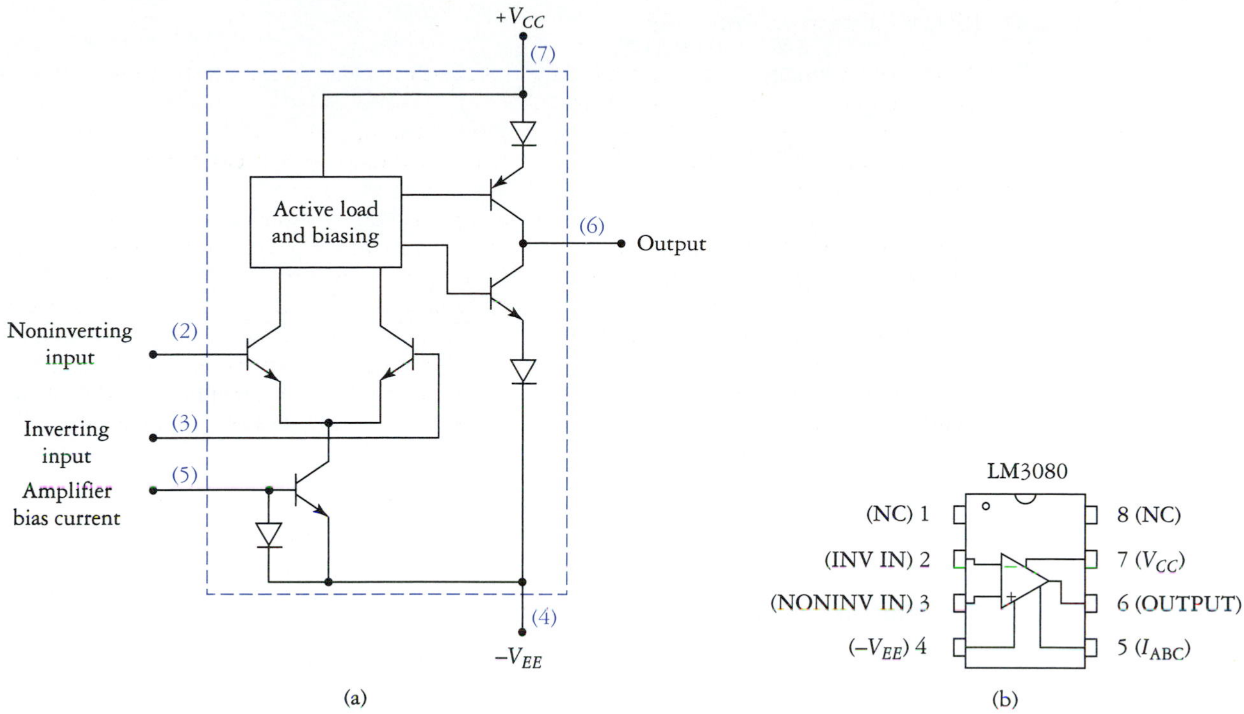

FIGURE 18.17
Operational transconductance amplifier. (a) Equivalent internal circuitry. (b) Pin designations.

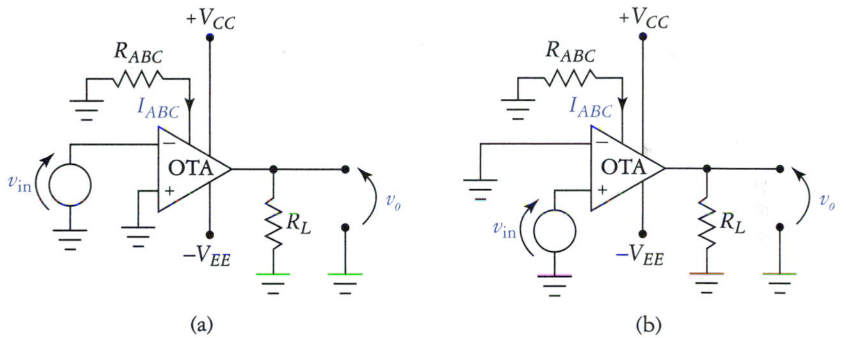

FIGURE 18.18
OTA amplifier configurations. (a) Inverting. (b) Noninverting.

where the negative sign is used with the inverting configuration. The voltage gain is therefore

$$A_v = \pm g_m R_L \tag{18.42}$$

EXAMPLE 18.9

Refer to Fig. 18.18(a). Given $V_{CC} = V_{EE} = 15$ V, $R_{ABC} = 47$ kΩ, and $R_L = 10$ kΩ, determine I_{ABC}, g_m, A_v, and v_o for $v_{in} = 25 \sin \omega t$ mV.

Solution

$$I_{ABC} = \frac{V_{EE} - V_{BE}}{R_{ABC}}$$

$$= \frac{15\text{ V} - 0.7\text{ V}}{47\text{ k}\Omega}$$

$$= 304\ \mu\text{A}$$

$$g_m = kI_{ABC} = (19.2\ \text{V}^{-1})(304\ \mu\text{A}) = 5.84\ \text{mS}$$

$$A_v = -g_m R_L = (-5.84\ \text{mS})(10\ \text{k}\Omega) = -58.4$$

$$v_o = A_v v_{in} = (-58.4)(25 \sin \omega t\ \text{mV}) = -1.46 \sin \omega t\ \text{V}$$

To determine whether the 3080 will operate correctly in a given application, we need to be familiar with a few of its specifications. The data sheet for the LM3080 provides the following information:

Supply Voltage Range

$V_{CC}, V_{EE} = \pm 5$ V to ± 18 V

Output Current

$I_o = \pm 300\ \mu$A (guaranteed over operating temperature range, $R_L = 0\ \Omega$)

Amplifier Bias Current

$0.1\ \mu\text{A} \leq I_{ABC} \leq 1$ mA

Slew Rate

$\text{SR}_{typ} = 50\ \text{V}/\mu\text{s}$

Open-Loop Bandwidth

BW = 2 MHz

It can be shown that the input resistance R_{in} of the 3080 varies with I_{ABC} according to

$$R_{in} \cong \frac{2\beta V_T}{I_{ABC}} \tag{18.43a}$$

Assuming that $\beta \cong 100$ and $V_T \cong 26$ mV, we have

$$R_{in} \cong \frac{5.2}{I_{ABC}} \tag{18.43b}$$

The output resistance of the OTA is related to I_{ABC} as well. That is,

$$R_o \cong \frac{V_A}{I_{ABC}} \tag{18.44}$$

where the Early voltage is approximately $V_A \cong 100$ V.

The output voltage compliance of the circuits in Fig. 18.18 may be limited by either the OTA output current drive capability or the supply rails, whichever causes limiting first. Assuming symmetrical supply voltages ($V_{CC} = V_{EE} = V_S$) we can write

$$V_{o(max)} = I_{o(max)} R_L < (V_S - V_{BE}) \tag{18.45}$$

where we find that the maximum output voltage is one barrier potential drop (about 0.7 V) less than the supply voltage.

EXAMPLE 18.10

Refer to Fig. 18.18(a). Given the conditions of Example 18.9 ($V_{CC}, V_{EE} = \pm 15$ V, $R_{ABC} = 47$ kΩ, and $R_L = 10$ kΩ), determine the maximum output voltage that can be produced by the OTA. Is the output voltage range limited by the supply rails or by the output current compliance of the OTA? Determine the value of R_L that will allow $V_{o(max)} \cong \pm 15$ V.

Solution The LM3080 is guaranteed to deliver ±300 μA. Some units may exceed this specification, but we cannot rely on chance. Using this specification, we have

$$\begin{aligned} V_{o(\max)} &= I_{o(\max)}R_L \\ &= (300\ \mu\text{A})(10\ \text{k}\Omega) \\ &= 3\ \text{V} \end{aligned}$$

Using the values specified in Example 18.9, the amplifier is limited by OTA output current compliance. To achieve $V_{o(\max)}$ we use

$$\begin{aligned} R_L &= \frac{V_{o(\max)}}{I_{o(\max)}} \\ &= \frac{15\ \text{V} - 0.7\ \text{V}}{300\ \mu\text{A}} \\ &= 47.7\ \text{k}\Omega \end{aligned}$$

Keep in mind that if we increase the output voltage compliance by increasing R_L, the voltage gain of the amplifier will increase as well.

An OTA AM Modulator

It is easy to electronically vary the gain of the OTA by varying I_{ABC}. This allows the output of the OTA to be *amplitude modulated.* Amplitude modulation is used to shift or translate a low-frequency information signal (such as music or speech) to a higher frequency that will propagate efficiently through space. A block diagram representation for an AM modulator and some representative input and output signals are shown in Fig. 18.19. The signals are shown in the time domain on the left, while frequency-domain representations are shown at right.

The modulation index m quantifies the amount of modulation that is applied to the carrier. Using the time-domain output voltage waveform, as would be seen on an oscilloscope, we find m using

$$m = \frac{V_A - V_B}{V_A + V_B} \tag{18.46}$$

Using the frequency-domain representation, we determine m using

$$m = \frac{2V_{SF}}{V_{oc}} \tag{18.47}$$

The maximum modulation index is $m_{\max} = 1$. Overmodulation occurs if $m > 1$. Often, the modulation index is expressed as a percentage. That is, $\%m = 100m$.

An OTA amplitude modulator is shown in Fig. 18.20. The low-frequency *modulating signal* v_m causes a modulation current i_{abc} to be superimposed on the dc bias current I_{ABC}. The total instantaneous amplifier bias current $i_{ABC} = I_{ABC} + i_{abc}$ causes the OTA gain to vary in direct proportion to the instantaneous amplitude of the modulating signal. We next develop an expression for the output voltage of this circuit. First, let's define the input signals as

$$v_c = V_c \sin \omega_c t\ \text{V}$$

$$v_m = V_m \sin \omega_m t\ \text{V}$$

The transconductance of the OTA under no-modulation conditions ($m = 0$) is

$$g_{mQ} = kI_{ABC} \tag{18.48}$$

The instantaneous transconductance of the OTA is given by

$$g_m = g_{mQ} + ki_{abc} \tag{18.49}$$

where i_{abc} is given by

$$i_{abc} = \frac{V_m}{R_{abc}} \sin \omega_m t\ \text{V} \tag{18.50}$$

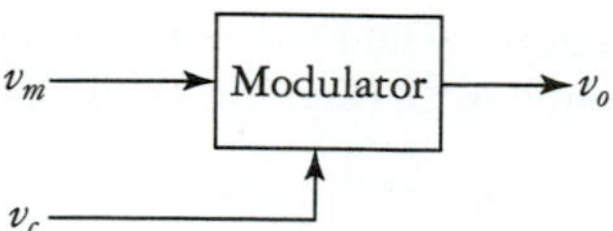

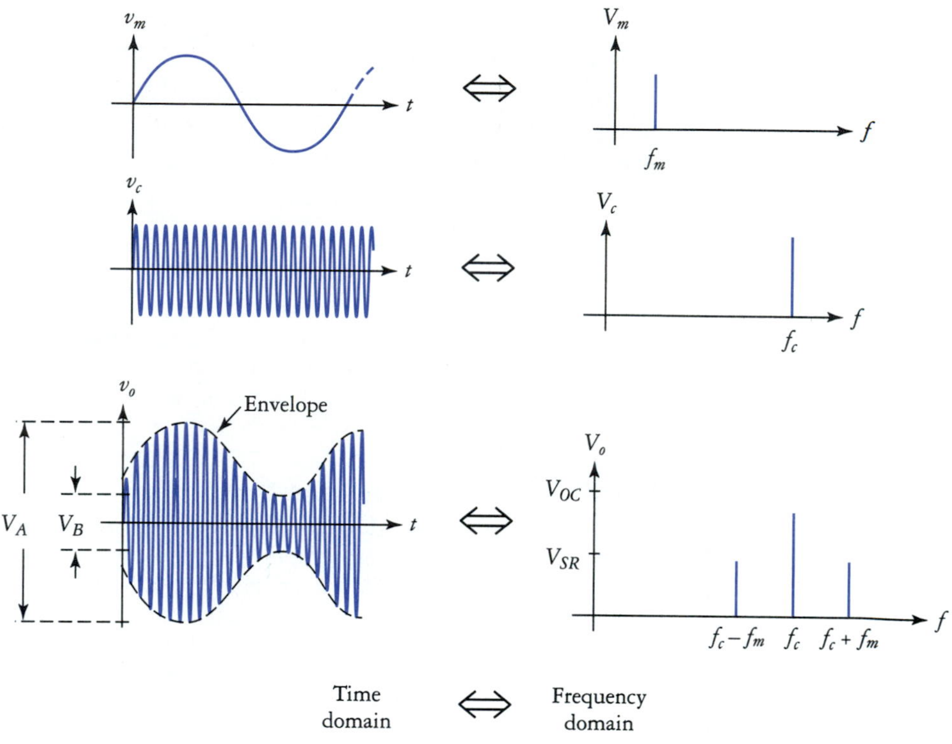

FIGURE 18.19
Amplitude modulation waveforms and frequency spectra.

FIGURE 18.20
An OTA AM modulator.

Substituting Eq. (18.50) into (18.49) gives us

$$g_m = g_{mQ} + \frac{kV_m}{R_{\text{abc}}} \sin \omega_m t \text{ V} \tag{18.51}$$

Multiplying by R_L we have the voltage gain of the amplifier, referred to the carrier input, which is

$$A_v = R_L g_{mQ} + \frac{kV_m R_L}{R_{\text{abc}}} \sin \omega_m t \tag{18.52}$$

The output voltage is $v_o = A_v v_{in}$, so we have

$$v_o = v_c R_L g_{mQ} + v_c \frac{kV_m R_L}{R_{abc}} \sin \omega_m t$$

$$= V_c R_L g_{mQ} \sin \omega_c t + V_c \sin \omega_c t \frac{kV_m R_L}{R_{abc}} \sin \omega_m t \tag{18.53}$$

The following trigonometric identity is useful in this situation, where we choose $\omega_1 \geq \omega_2$ to avoid negative frequencies, which are physically unrealizable:

$$(A_1 \sin \omega_1 t)(A_2 \sin \omega_1 t) = \frac{A_1 A_2}{2} \cos(\omega_1 - \omega_2)t - \frac{A_1 A_2}{2} \cos(\omega_1 - \omega_2)t \tag{18.54}$$

Applying Eq. (18.54) to (18.53) we have

$$v_o = V_c R_L g_{mQ} \sin \omega_c t + \frac{kV_m V_C R_L}{2R_{abc}} \cos(\omega_c - \omega_m)t$$

$$- \frac{kV_m V_C R_L}{2R_{abc}} \cos(\omega_c + \omega_m)t \tag{18.55}$$

The first term of Eq. (18.55) is the carrier component of the output. The second term has a frequency of $\omega_c - \omega_m$ and is called the *lower side frequency* (LSF). The third term has $\omega_c + \omega_m$ and is called the *upper side frequency* (USF). If a complex modulating signal were used then *upper* and *lower side bands* (USB, LSB) would be produced.

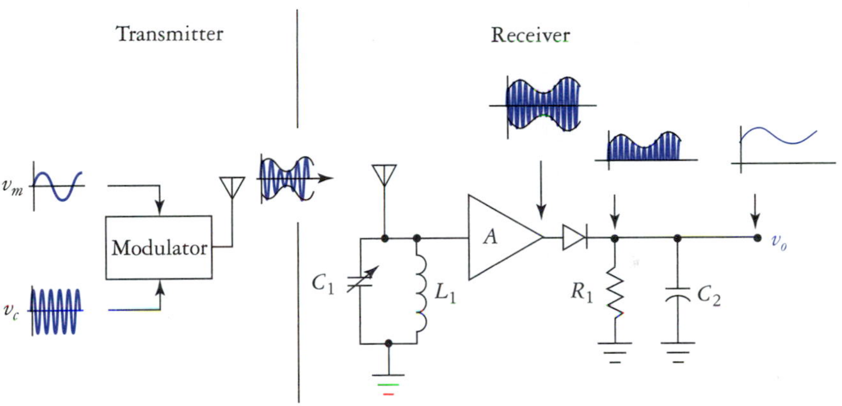

FIGURE 18.21
Transmission and reception of AM signals.

> **Note**
> The tank circuit in Fig. 18.21 acts as a bandpass filter, centered at f_c of the station we wish to receive.

As stated earlier, if the carrier frequency is high enough, the output of the modulator can be easily transmitted through space to a receiver. This is the basis for commercial AM broadcast transmitters, CB transceivers, and walkie-talkies, for example. A simple half-wave rectifier can be used to demodulate a received AM signal. This is illustrated in Fig. 18.21. The tank circuit is tuned such that $f_o = f_c$. This causes reception of the desired signal while others are rejected. The received signal is amplified and half wave rectified. The rectifier removes the lower halves of the modulation envelope and the carrier signal. The remaining high-frequency carrier energy is removed by a low-pass filter resulting in an output signal that is a replica of the original modulating signal with a dc offset.

EXAMPLE 18.11

The modulator in Fig. 18.20 has $V_{CC} = V_{EE} = \pm 15$ V, $R_{ABC} = 47$ kΩ, $R_{abc} = 10$ kΩ, $R_L = 47$ kΩ, $v_c = 20 \sin(2\pi 10{,}000t)$ mV, and $v_m = 3 \sin(2\pi 1000t)$ V. Determine v_o and m, and sketch the output signal spectrum (frequency domain) and the output modulation envelope (time domain). Also, determine the maximum rate of change of the output voltage.

Solution The dc bias current is

$$I_{ABC} = \frac{V_{EE} - V_{BE}}{R_{ABC}} = \frac{15\text{ V} - 0.7\text{ V}}{47\text{ k}\Omega} = 304\ \mu\text{A}$$

The quiescent (unmodulated) transconductance is

$$g_{mQ} = kI_{ABC} = (19.2)(304)\ \mu\text{A}) = 8.54\text{ mS}$$

The output is found via direct application of Eq. (18.55):

$$\begin{aligned} v_o &= V_c R_L g_{mQ} \sin \omega_c t \ \frac{kV_m V_c R_L}{2R_{\text{abc}}} \cos(\omega_c - \omega_m)t \\ &\quad + \frac{kV_m V_c R_L}{2R_{\text{abc}}} \cos(\omega_c + \omega_m)t \\ &= (0.02)(47\text{ k})(8.54\text{ mS}) \sin \omega_c t \\ &\quad + \frac{(19.2)(3)(0.02)(47\text{ k})}{(2)(10\text{ k})} \cos(2\pi 10{,}000 - 2\pi 1000)t \\ &\quad - \frac{(19.2)(3)(0.02)(47\text{ k})}{(2)(10\text{ k})} \cos(2\pi 10{,}000 + 2\pi 1000)t \\ &= 5.5 \sin(2\pi 10{,}000t) + 2.7 \cos(2\pi 9{,}000t) - 2.7 \cos(2\pi 11{,}000t)\text{ V} \end{aligned}$$

The modulation index can be found directly from the instantaneous output voltage equation and Eq. (18.47):

$$m = \frac{2V_{SF}}{V_c} = \frac{(2)(2.7)}{5.5} = 0.98\ (98\%)$$

The frequency-domain representation of v_o is shown in Fig. 18.22(a). The amplitudes of the frequency components are simply the coefficients of the terms in the output voltage equation. The time-domain representation is shown in Fig. 18.22(b). The peak amplitude of the upper half of the envelope is $V_{oc} + 2V_{SF}$, while the minimum of the envelope is $V_{oc} - 2V_{SF}$. The lower half of the envelope is a mirror image of the top.

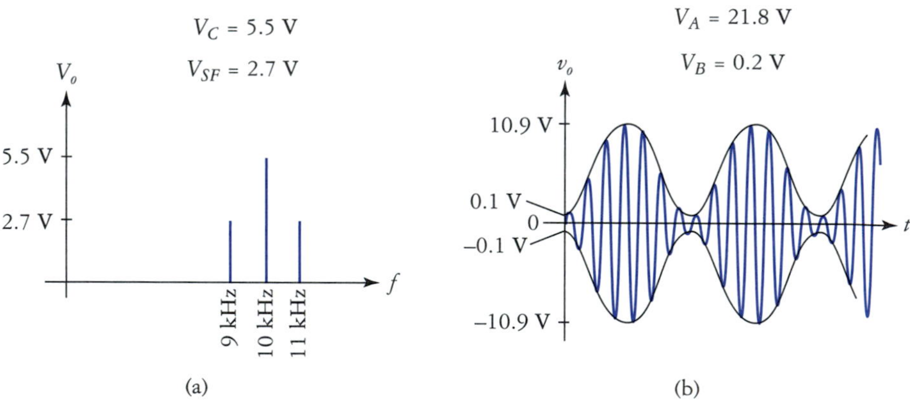

FIGURE 18.22
Waveforms for Example 18.11.

The maximum rate of change (ROC or $dv_o/dt|_{\max}$) of the output voltage occurs periodically when all three output-frequency components cross zero in the same direction at the same time. Thus, the net max ROC is found by adding the max ROCs of each component:

$$\left.\frac{dV_c}{dt}\right|_{\max} = \left.\frac{d}{dt}\right|_{\max} 5.5 \sin(2\pi 10{,}000t) = 0.346\text{ V}/\mu\text{s}$$

$$\left.\frac{dV_{\text{VSF}}}{dt}\right|_{\max} = \left.\frac{d}{dt}\right|_{\max} 2.7 \sin(2\pi 11{,}000t) = 0.187\text{ V}/\mu\text{s}$$

$$\left.\frac{dV_{\text{LSF}}}{dt}\right|_{\max} = \left.\frac{d}{dt}\right|_{\max} 2.7 \sin(2\pi 9{,}000t) = 0.153 \text{ V}/\mu\text{s}$$

$$\left.\frac{dv_o}{dt}\right|_{\max} = 0.346 \text{ V}/\mu\text{s} + 0.187 \text{ V}/\mu\text{s} + 0.153 \text{ V}/\mu\text{s}$$

$$= 0.686 \text{ V}/\mu\text{s}$$

Any amplifier that is nonlinear can be used as a modulator, although for good operation, the nonlinearity should be quadratic or higher-degree. A bypassed CE amplifier for example will exhibit much nonlinearity if the input signal is very large. In cases where this nonlinearity is used to produce modulation, a tuned circuit is employed to filter out unwanted frequency components. The advantage of using a modulator like the OTA presented here is that no tuned circuits are required. Because of this, the OTA modulator can easily be used over a wide range of carrier frequencies.

The OTA Integrator

As in the case of a standard op amp, an OTA can be used to form an integrator. The OTA integrator, however, can easily be configured as either an inverting or noninverting integrator, or even as a differential integrator. These three circuits are shown in Fig. 18.23(a), (b), and (c), respectively. Notice that the only difference between the circuits is how the input voltage source(s) is connected!

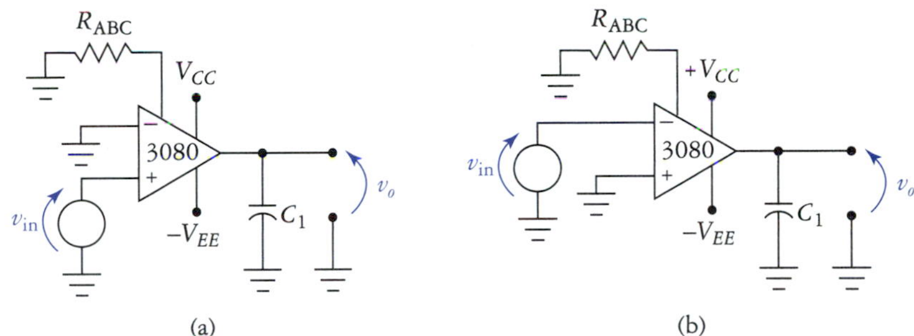

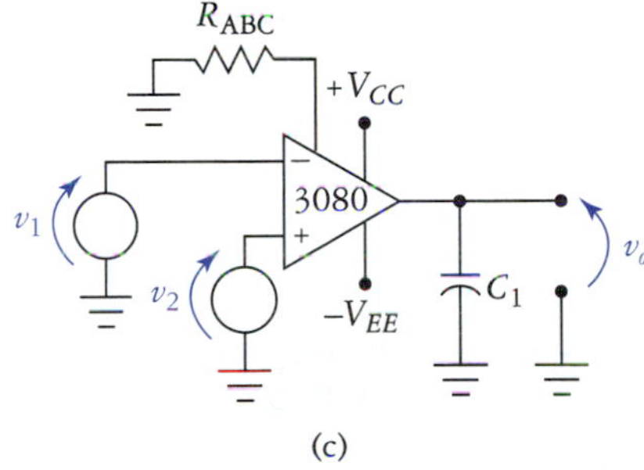

FIGURE 18.23
OTA integrator circuits. (a) Noninverting. (b) Inverting. (c) Differential.

The output voltage expression for the OTA integrator is easily developed as follows. The voltage across capacitor C_1 is given by

$$v_o = v_{C_1} = \frac{1}{C_1}\int i_o \, dt \qquad (18.56)$$

The output current is

$$i_o = g_m v_{\text{in}} \qquad (18.57)$$

Using these equations, the output current expressions for each of the OTA integrator configurations in Fig. 18.23 are as follows, where the constant g_m is taken outside of the integral:

$$v_o = \frac{g_m}{C_1}\int v_{\text{in}}\,dt \qquad \text{(noninverting integrator)} \tag{18.58}$$

$$v_o = \frac{-g_m}{C_1}\int v_{\text{in}}\,dt \qquad \text{(inverting integrator)} \tag{18.59}$$

$$v_o = \frac{g_m}{C_1}\int (v_2 - v_1)\,dt \qquad \text{(differential integrator)} \tag{18.60}$$

In this application we are taking advantage of the fact that the output of the OTA behaves as a dependent current source. The current-source output of the OTA forces the capacitor current to be directly proportional to v_{in}. In this way, the fundamental behavior of the capacitor forces the output voltage to be proportional to the integral of the input voltage.

EXAMPLE 18.12

Refer to the differential integrator of Fig. 18.23(c). Assume $V_{CC} = V_{EE} = \pm 15$ V, $R_{\text{ABC}} = 47$ kΩ, $C_1 = 10$ μF, $v_1 = 0.5$ V, and $v_2 = 3 \sin 1000t$ V. Determine the output voltage equation.

Solution Analysis of the OTA reveals $g_m = 8.54$ mS. The output voltage is

$$\begin{aligned} v_o &= \frac{g_m}{C_1}\int (v_2 - v_1)\,dt \\ &= \frac{g_m}{C_1}\int v_2\,dt - \frac{g_m}{C_1}\int v_1\,dt \\ &= 854\int -3\sin 1000t\,dt - 854\int 0.5\,dt \\ &= \frac{(854)(-3)}{1000}\cos 1000t - (0.5)(854)t \text{ V} \\ &= -2.56\cos 1000t - 427\,t \text{ V} \end{aligned}$$

18.6 BALANCED MODULATORS

As you can probably guess, a balanced modulator is normally used to produce some sort of amplitude modulation. The block diagram symbol for a balanced modulator is shown in Fig. 18.24. The output voltage is given by

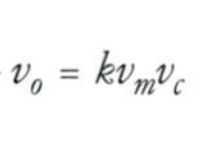

FIGURE 18.24 *Block diagram symbol for a balanced modulator.*

$$v_o = kv_mv_c \tag{18.61}$$

The constant k depends on the particular components used in the circuit. Balanced modulators are closely related to analog multipliers. That is, the output of a balanced modulator is proportional to the product of its input signals. Functionally, the main difference between analog multipliers and balanced modulators is that analog multipliers are designed for high accuracy with limited frequency capability, whereas balanced modulators are designed to operate at high frequencies with less absolute accuracy.

Double-Sideband, Suppressed-Carrier Modulator

Let's assume we have a high-frequency carrier and a modulating signal given by

$$v_c = V_c \sin \omega_c t \text{ V}$$

$$v_m = V_m \sin \omega_m t \text{ V}$$

If these signals are applied to the balanced modulator of Fig. 18.24, we can apply the trigonometric identity of Eq. (18.54) to obtain

$$v_o = \frac{mV_cV_m}{2}\cos(\omega_c - \omega_m)t - \frac{mV_cV_m}{2}\cos(\omega_c + \omega_m)t \text{ V} \tag{18.62}$$

Notice that the output signal contains the sum and difference frequencies, but the original carrier and modulating signals are absent. This is called a *double-sideband suppressed-carrier* (DSB-SC) modulator. Typical modulating and carrier signals and the resulting output voltage are shown in Fig. 18.25.

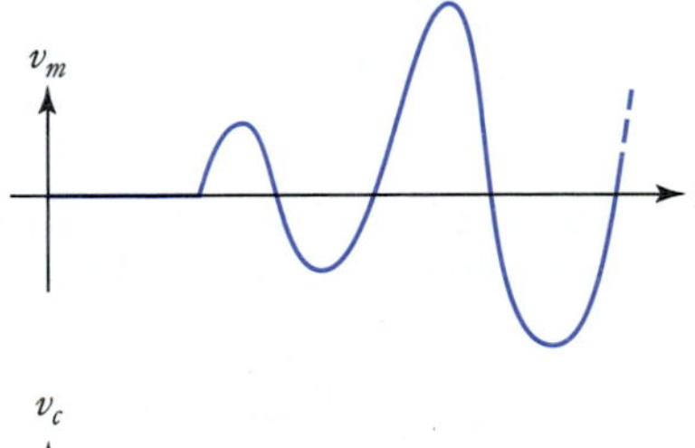

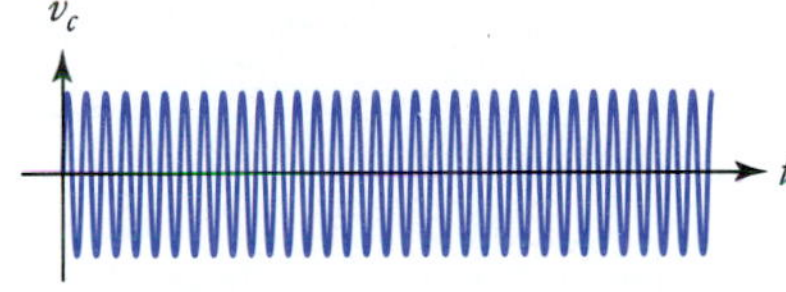

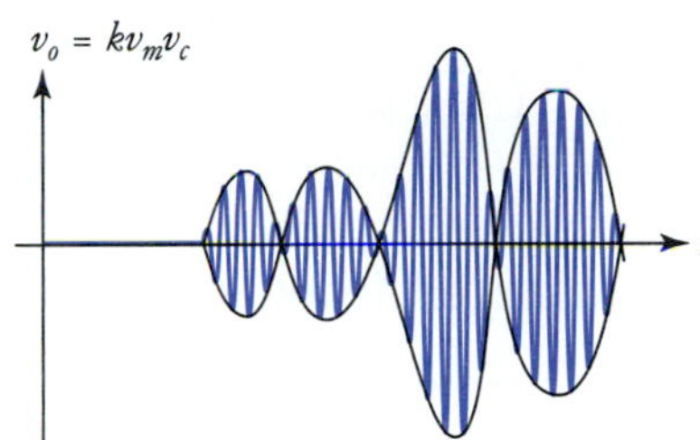

FIGURE 18.25
Typical balanced modulator input and output signals.

Single-Sideband Modulator

The block diagram in Fig. 18.26 is that of a single-sideband, suppressed-carrier (SSB-SC) modulator. The carrier and modulating signals are applied directly to the upper modulator, while they are phase shifted by 90° before being applied to the lower modulator. The resulting signals v_x and v_y are summed to produce the output signal.

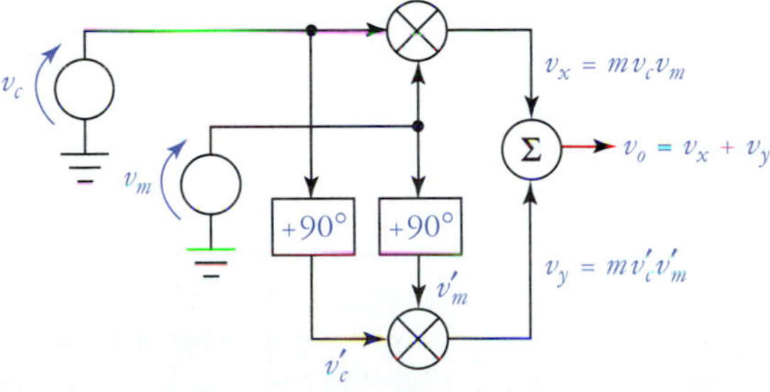

FIGURE 18.26
Single-sideband, suppressed-carrier (SSB-SC) generator.

Note
A network that provides ±90° phase shift, with a flat amplitude vs frequency characteristic is sometimes called a *Hilbert transformer*.

As usual, we assume that the input signals are given by

$$v_c = V_c \sin \omega_c t \text{ V}$$
$$v_m = V_m \sin \omega_m t \text{ V}$$

The phase-shifted versions of these signals are

$$v_c' = V_c \cos \omega_c t \text{ V}$$
$$v_m' = V_m \cos \omega_m t \text{ V}$$

We can apply Eq. (18.62) to find the output of the upper modulator, which is

$$v_x = \frac{mV_cV_m}{2}\cos(\omega_c - \omega_m)t - \frac{mV_cV_m}{2}\cos(\omega_c + \omega_m)t \text{ V}$$

The output of the bottom modulator is found using the following trigonometric identity:

$$(A_1 \cos \omega_1 t)(A_2 \cos \omega_2 t) = \frac{A_1 A_2}{2} \cos(\omega_1 - \omega_2)t + \frac{A_1 A_2}{2} \cos(\omega_1 + \omega_2)t$$

Application of this identity yields

$$v_y = \frac{mV_c V_m}{2} \cos(\omega_c - \omega_m)t + \frac{mV_m V_c}{2} \cos(\omega_c + \omega_m)t$$

The output voltage is now determined by summing v_x and v_y, where we find that the first terms of each voltage are in phase (they add constructively), and the last terms cancel, thus:

$$\begin{aligned} v_o &= v_x + v_y \\ &= mV_c V_m \cos(\omega_c - \omega_m)t \text{ V} \end{aligned} \tag{18.63}$$

The output of this modulator consists of only the lower side frequency (LSF). If a more complex signal such as speech were used for modulation, the lower sideband (LSB) alone would be created.

If we reverse the phase shift of the signals to $-90°$, the upper sideband would be produced. The advantage of single-sideband modulation is that since the bandwidth of the output signal is half that of DSB modulation, precious space in the frequency spectrum is conserved.

Sometimes the carrier is summed back into the output of the system producing a *single-sideband, full-carrier* (SSB-FC) modulator. This allows demodulation to be accomplished more easily. A simple half-wave rectifier cannot be used to demodulate either SSB-SC or SSB-FC signals. A more complex demodulating scheme is required that is beyond the scope of this discussion.

Balanced Modulator Phase Detector

An interesting application for the balanced modulator is as a *phase detector* (sometimes also called a *phase comparator*). This is shown in Fig. 18.27(a). From a circuit standpoint, a balanced modulator phase detector is identical to a DSB-SC modulator cascaded with a low-pass filter. If the two input signals are equal in frequency, then the output is a dc voltage that is proportional to the phase difference between the inputs. Let's assume that the two input signals are given by

$$v_{\text{in}} = V_1 \sin \omega t \text{ V}$$

$$v'_{\text{in}} = V_2 \sin(\omega t + \phi) \text{ V}$$

Referring back to the trigonometric identity of Eq. (18.54), we find that since the frequencies of the inputs are equal, the output of the balanced modulator consists of a dc offset voltage $[\cos(\omega - \omega) = \cos 0 = 1]$ and an ac component at 2ω. The low-pass filter removes the 2ω component, leaving only the dc offset. It can be shown that the dc level is given by

Note
Recall that A_0 is the passband gain of the low-pass filter. For a passive filter, $A_0 = 1$.

$$V_o = V_{\text{dc}} = \frac{kA_0 V_1 V_2}{2} \cos \Delta\phi \tag{18.64}$$

Thus we see that the output voltage is proportional to the phase difference between the input signals. The transfer characteristic curve is shown in Fig. 18.27(b). Notice that the circuit does not distinguish between positive and negative phase differences. This is so because either input can serve as the reference, therefore the sign of the phase difference is arbitrary. It is also useful to notice that the "gain" of the phase detector is also greatest (the slope of the transfer curve is steepest) for $\Delta\phi \approx \pm 90°$. The gain is very linear in this region as well. Recall that we define gain by out÷in. It can be shown that the gain $A\phi$ of the detector, for phase changes in the range of $60° \leq |\Delta\phi| \leq 120°$, is constant within about $\pm 5\%$ of

$$A_\phi = \frac{\Delta V_o}{\Delta\phi} = A_0 k \text{ V/deg} \tag{18.65}$$

where A_0 is the dc gain of the LP filter and k is the gain of the balanced modulator. A_ϕ is sometimes referred to as the *conversion gain* of the phase detector.

We can use the phase detector to build a *phase meter* if we connect a dc voltmeter to the output of the circuit. This would be useful for monitoring the synchronization of ac electrical power generating machinery, for example. This is shown in Fig. 18.28. The alternators must be in phase ($\Delta\phi = 0$) in order to be connected in parallel as shown. Thus, when the alternators are in phase, the dc voltmeter will read maximum. Should the phase change ($\Delta\phi \neq 0$) due to a speed difference between the alternators, the dc voltage will decrease. This simple phase detector arrangement cannot tell us which alternator has changed speed, however, so a more sophisticated system would be required to be practical.

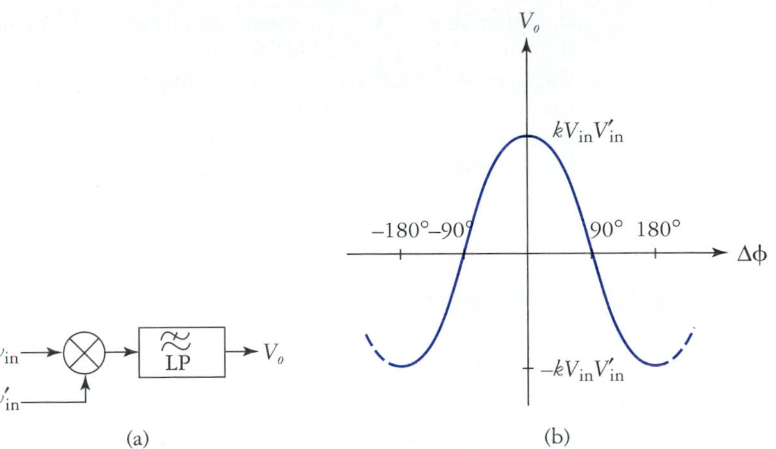

FIGURE 18.27
Balanced modulator phase detector.

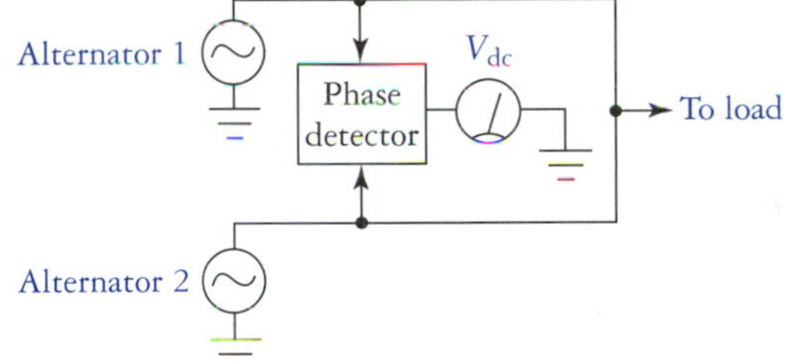

FIGURE 18.28
Using a phase detector to synchronize ac sources.

Phase detectors are also used in the design of phase-locked loops, which are covered in the next section.

The LM1496 Monolithic Balanced Modulator

The LM1496 is a very easy to use and popular balanced modulator IC. The LM1496 is available in standard 14-pin DIP and metal can packages. An LM1496 is configured as a DSB-SC modulator in Fig. 18.29. This circuit is useful with carrier frequencies as high as 100 MHz. The carrier signal

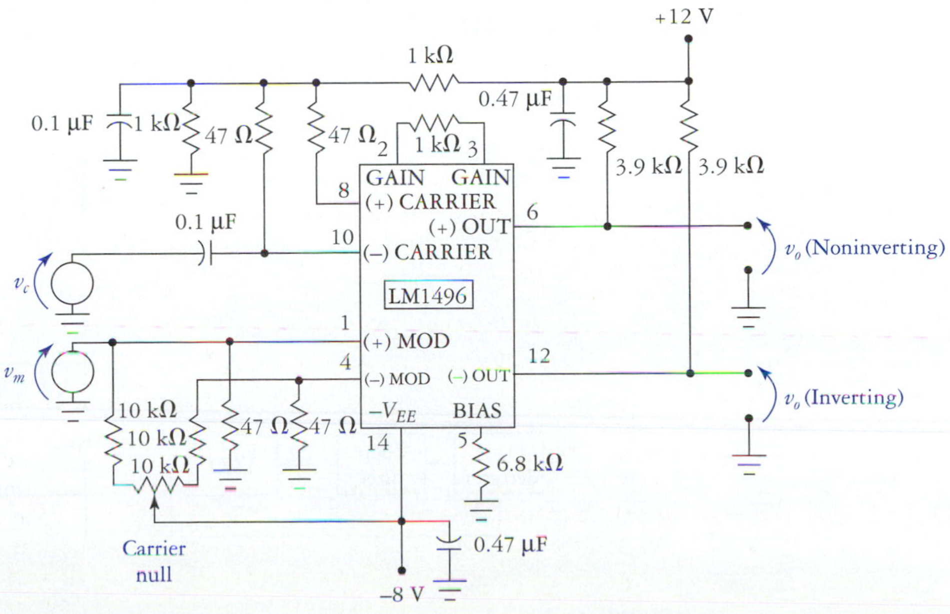

FIGURE 18.29
Typical LM1496 balanced modulator circuit.

should be about 1 $V_{p\text{-}p}$ for correct operation, while the modulating signal amplitude may range from 0 to about 1.5 $V_{p\text{-}p}$. The LM1497 has inverting (−) and noninverting (+) inputs, and can be used in single-ended or differential output configurations. The carrier null potentiometer is used to trim out the carrier from the DSB-SC output signal. This works by acting as an offset null for the modulation input. If the potentiometer is adjusted toward either extreme, a dc offset is summed with the modulating signal, resulting in a large carrier frequency component being added to the output signal. This is useful if we wish to produce DSB-FC (normal AM) modulation.

18.7 THE PHASE-LOCKED LOOP

A phase-locked loop (PLL) is a closed-loop system in which an oscillator is forced to track a reference signal in frequency. This can be best understood by referring to the block diagram of Fig. 18.30. An oscillator is used to generate a reference signal V_{ref}. This signal and the output of a voltage-controlled oscillator (VCO) are applied to the inputs of a phase detector. Upon power up, the VCO operates at its free-running frequency f_O. If the reference signal frequency is close to the free-running frequency of the VCO, the PLL will lock onto the reference signal and force the VCO to track such that the output of the phase comparator is zero volts. This means that the VCO output will be forced to either lead or lag the reference signal by $|\theta| = 90°$, with exactly the same frequency ($f_{VCO} = f_{ref}$). Should the reference frequency increase or decrease, a phase difference will occur ($|\theta| \neq 90°$). This phase difference produces a dc voltage at the output of the phase comparator, which forces the VCO to match the reference frequency change. The loop filter, which may be the low-pass filter associated with the phase detector, is used to control the response speed of the loop. This is illustrated by the waveforms of Fig. 18.30(b), where the PLL is initially locked onto the reference signal. For convenience we have assumed that $f_o = f_{ref}$ so that the VCO bias voltage is zero. The frequency of the reference signal suddenly increases, which causes the VCO bias voltage to increase, forcing the VCO frequency to increase and match the reference. As shown in the diagram, depending on various loop parameters, the PLL may respond slowly (overdamped response) or quickly, possibly with some overshoot (underdamped or critically damped response). The loop filter can be modified to obtain the desired response.

We have previously discussed the concept of conversion gain in terms of the phase detector, which is characterized as A_ϕ (V/deg.). The VCO also has a conversion gain that is defined as

$$A_f = \frac{\Delta f_o}{\Delta V_{bias}} \qquad \text{(Hz/V)} \tag{18.66}$$

Capture and Lock Range

The incoming reference frequency must be within a certain distance (in hertz) of the VCO free-running frequency in order for the PLL to acquire lock. This span of frequencies is called the *capture range* of the PLL.

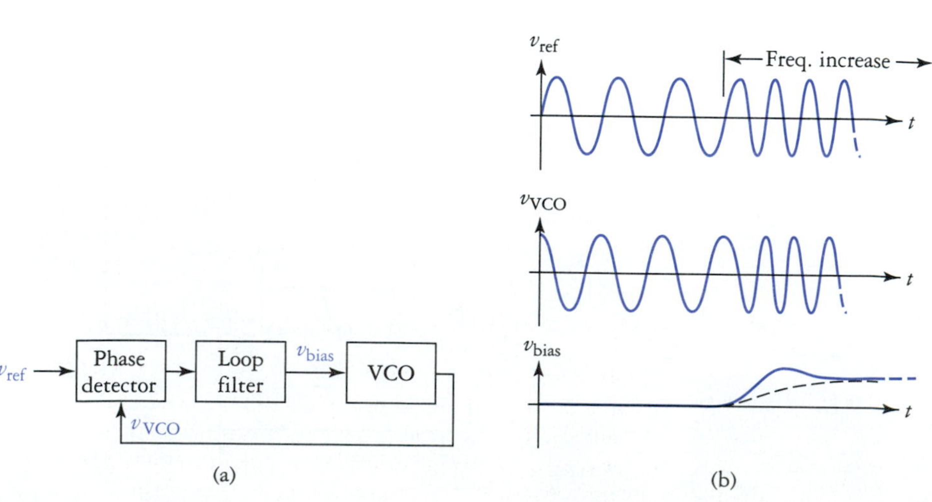

FIGURE 18.30
A phase-locked loop. (a) Block diagram. (b) Typical system waveforms.

Once the PLL has acquired lock, it will track changes in the reference frequency that do not exceed a span of frequencies called the *lock range* or sometimes the *tracking range.* It is common for the lock range to exceed the capture range of the PLL. This is illustrated in Fig. 18.31.

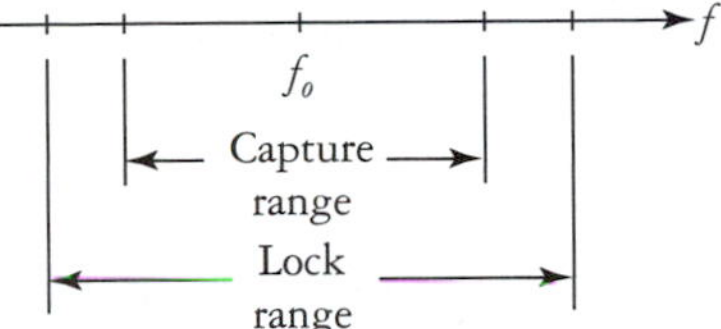

FIGURE 18.31
Graphical representation of the PLL capture and lock ranges.

PLLs can be used in many applications. The following subsections describe just a small sampling.

FM Demodulation

Frequency modulation (FM) is produced when an information-bearing signal is used to proportionally shift the frequency of an oscillator, such as a VCO. The 555 timer (see Section 18.1), when used in the PPM configuration generates a frequency-modulated output signal. The original modulating (information) signal can be extracted from the FM signal with a PLL as shown in Fig. 18.32.

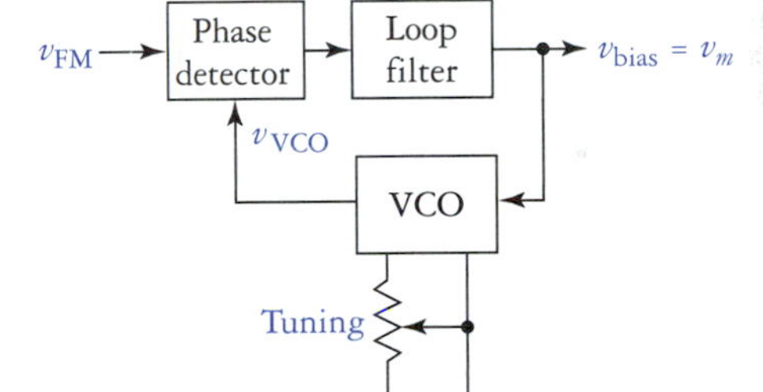

FIGURE 18.32
Using a PLL as an FM demodulator.

The FM demodulator works as follows. Assume that an FM signal v_{FM} is present, with a center frequency f_{FM} that is outside of the capture range of the PLL. As the free-running frequency of the VCO is adjusted, the capture range shifts as well, eventually enclosing f_{FM} and causing the PLL to lock onto the FM signal. Now, as the modulating signal v_m causes frequency shifts $f_o \pm \Delta f$, the PLL must reproduce an identical signal $v_{bias} = v_m$, in order to force the VCO to track the frequency changes. This demodulation technique is used extensively in modern FM receivers.

FSK Demodulator

A popular digital data encoding shceme is called *frequency-shift keying* (FSK). This is basically a form of FM where the frequency of a VCO is shifted between two frequencies in accordance with a digital signal. The frequency used to represent logic zero is called the *space* frequency, while the logic-high frequency is called the *mark* frequency. FSK may be used to transmit digital data over the telephone system. A PLL may be used at the receiving end of the system to regenerate the original data stream. A simplified representation of this process is shown in Fig. 18.33. This demodulator is very similar to the FM demodulator of Fig. 18.32, except for the addition of a Schmitt trigger, which is used to square up and buffer the output of the loop filter.

PLL Motor Speed Control Servomechanism

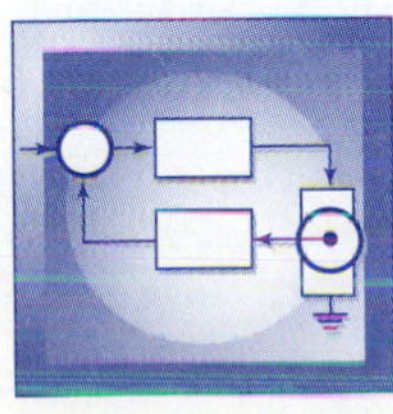

A servomechanism (or just a servo) is a closed-loop electromechanical control system in which some parameter such as motor speed or position is controlled. A typical application for a servo might be in the automatic control of a valve in some industrial process. Another application would be in the flight control system of an airplane. Recall that in such systems, the mechanical device that effects a physical change in the system is called the *actuator.* Typical actuators include motors and hydraulic pistons. The physical quantity being controlled (liquid flow rate, aircraft pitch, automobile speed, etc.) is called the *plant.* The sensor that is used to monitor the physical parameter being controlled is

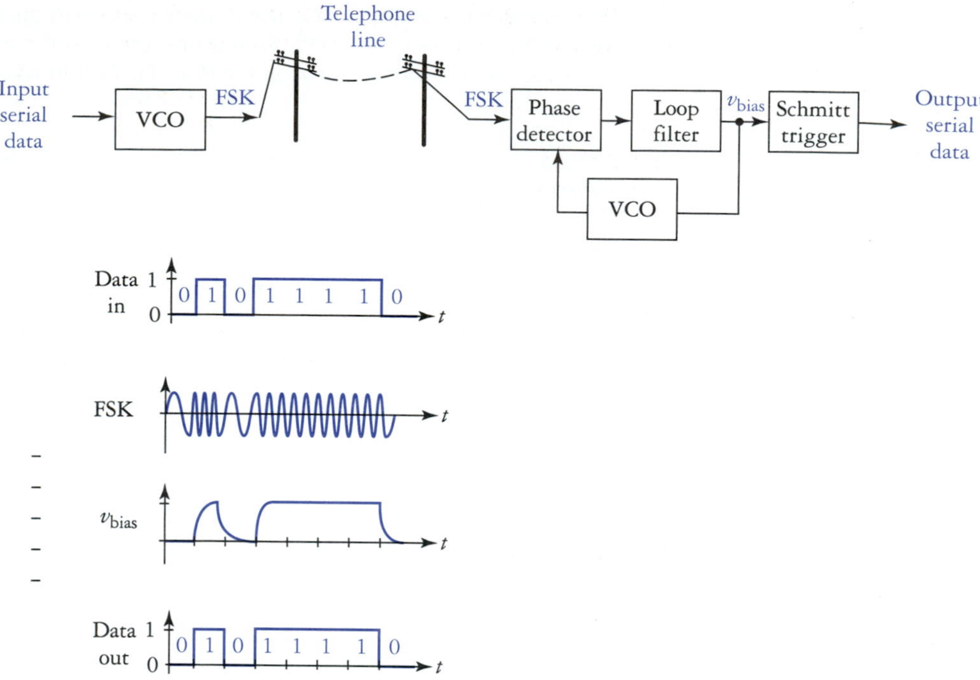

FIGURE 18.33
FSK data transmission. (a) Possible system. (b) Typical system waveforms.

called a *transducer*. The transducer converts changes in the physical system (flow rate, etc.) into a representative electrical signal. This feedback signal is compared to a reference signal that is typically determined by the operator of the system (plant operator, pilot, etc.). This reference signal is usually called the *command* input. The feedback signal is compared to the command input and if there is a difference, an error signal is produced. This error signal is amplified and used to drive the actuator such that the disturbance that caused the error is compensated for.

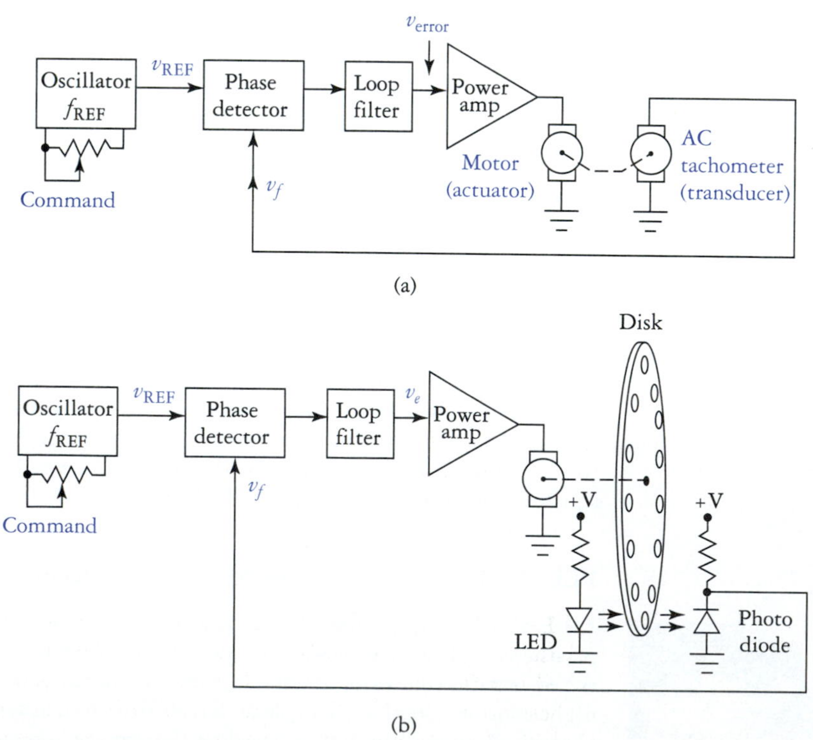

FIGURE 18.34
Servomechanisms using PLLs. (a) Motor speed control using a tachometer. (b) Motor speed control using an optical interrupter.

Two versions of a simple motor speed control servo are shown in the block diagrams of Fig. 18.34. In 18.34(a), the speed of the motor is proportional to the frequency of an oscillator that is set by a human operator. Assuming the motor is initially running at the correct speed, the frequency of the tachometer is equal to that of the reference oscillator. Should loading of the motor cause it to slow down, the feedback frequency decreases, causing an error signal to be produced. This signal causes the motor torque to increase, which will hold the motor speed steady. The greater the difference between the command and feedback signals, the harder the servo drives the actuator to compensate. Such systems are called *proportional control systems.*

The system in Fig. 18.34(b) produces a feedback signal that is also proportional to motor speed. As the disk rotates, the light produced by an LED is alternately passed and blocked, producing a signal whose frequency is proportional to motor speed and the number of holes in the disk. This approach has the advantage that the feedback signal amplitude is constant, regardless of motor speed. The output voltage produced by the tachometer will vary with motor speed, requiring additional signal processing.

EXAMPLE 18.13

Refer to Fig. 18.34(b). The optical interrupter disk has 100 equally spaced holes drilled around its circumference. Assume the reference oscillator operates at $f_{REF} = 1$ kHz. Determine the speed of the motor in rev/min (rpm).

Solution The following equation relates the reference frequency f_{REF} and the number of holes in the disk n to the speed of the motor S in rpm:

$$S = \frac{1 \text{ rev.}}{n \cancel{\text{cycles}}} \times \frac{f_{REF} \cancel{\text{cycles}}}{\cancel{\text{sec}}.} \times \frac{60 \cancel{\text{sec}}}{\text{min}}$$

Note that cancellation produces a result that is dimensionally correct. In a more simple form we have

$$S = \frac{60 f_{REF}}{n} \tag{18.67}$$

$$= \frac{(60)(1000)}{100}$$

$$= 600 \text{ rpm}$$

PLL Frequency Multipliers

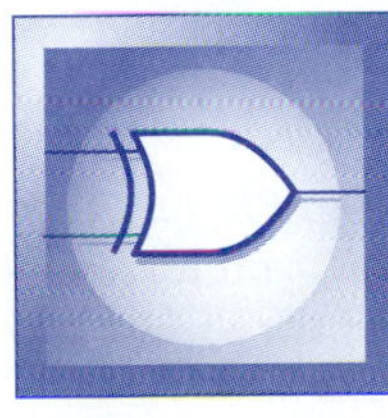

The frequency of a signal can be changed in many different ways. In digital electronics, we use flip-flops and counters to divide frequency. Counters are often used in conjunction with PLLs to perform frequency multiplication. The block diagram of Fig. 18.35(a) shows the PLL configured as a frequency multiplier. In order for the PLL to be in phase lock, the VCO frequency must be N times the frequency of the reference oscillator, where N is the modulus of the divider (counter) in the feedback loop. A few typical dividers are shown in Fig. 18.35(b). The 7490 is a TTL decade counter that is connected to provide a 50% duty cycle output. Because both dividers are digital circuits, either a TTL-compatible VCO must be used (something like a 555 timer, for example), or the sinusoidal output must be squared up, perhaps with a Schmitt trigger. As we shall see, PLLs are available that have been designed specifically for use in digital systems.

The LM565 PLL

The LM565 is a monolithic PLL, available in a 14-pin DIP or 10-pin metal can package. This device consists of three main sections: a VCO, a balanced modulator phase detector, and an amplifier. A simplified internal diagram is shown in Fig. 18.36. The LM565 is designed for operation at VCO frequencies of less than 500 kHz, although PLL ICs that can operate into the hundreds of megahertz are available. The 565 is designed to operate with supply voltages from ±5 to ±12 V. The phase detector has three inputs, all of which must have a dc path to ground (for transistor bias currents). The inputs at pins 2 and 3 are interchangeable in most applications, and they are TTL compatible if $V_{CC} = 5$ V. The pin 5 input is to be used to close the feedback loop from the VCO.

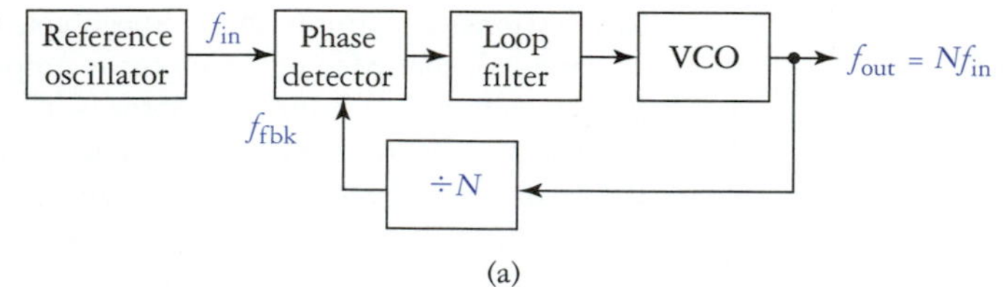

(a)

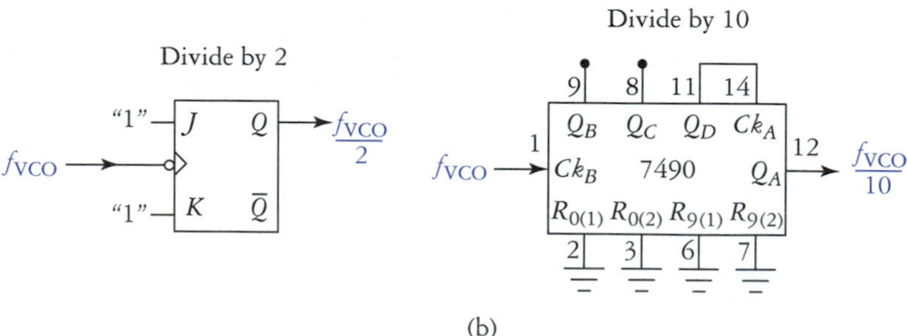

(b)

FIGURE 18.35
PLL-based frequency multiplier. (a) Block diagram. (b) Toggle-mode flip-flop. (c) 7490 decade counter.

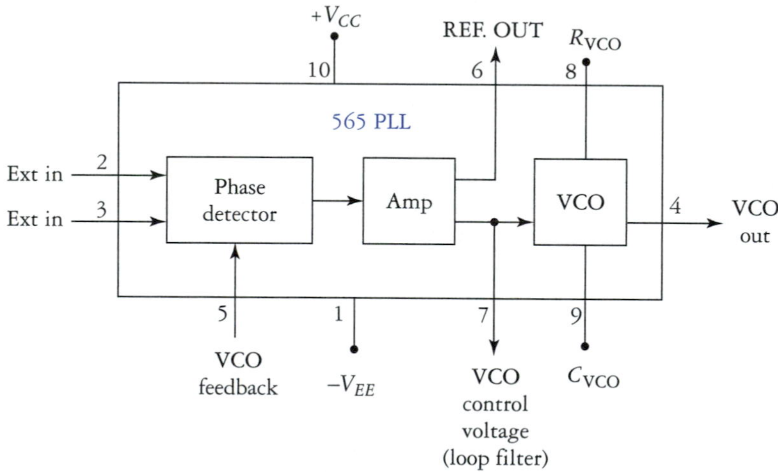

FIGURE 18.36
Internal block diagram for the 565 PLL.

The circuit in Fig. 18.37 shows the 565 configured to demodulate a PPM (or FSK) signal, as would be produced by the 555-based modulator in Fig. 18.7. Resistors R_1 and R_2 provide dc bias current paths, and are typically about 1 kΩ. The free-running frequency of the VCO is given by

$$f_o \cong \frac{1}{3.7\, R_{VCO}\, C_{VCO}} \tag{18.68}$$

To demodulate the PPM signal, the free-running frequency of the VCO would be set approximately the same as the zero-modulation frequency of the 555 timer.

The tracking range of the 565 is dependent on the total power supply voltage, the free-running frequency of the VCO, and other second-order effects. If pin 6 is left open circuited, the tracking range is approximately

$$f_{track} \cong f_o \pm 0.4 f_o \tag{18.69}$$

We can also express the tracking range as a percentage of the free-running frequency. Thus, we can write

$$f_{track} \cong f_o \pm 40\% \tag{18.70}$$

If the PPM signal frequency change (deviation) exceeds this range, the PLL will lose its lock.

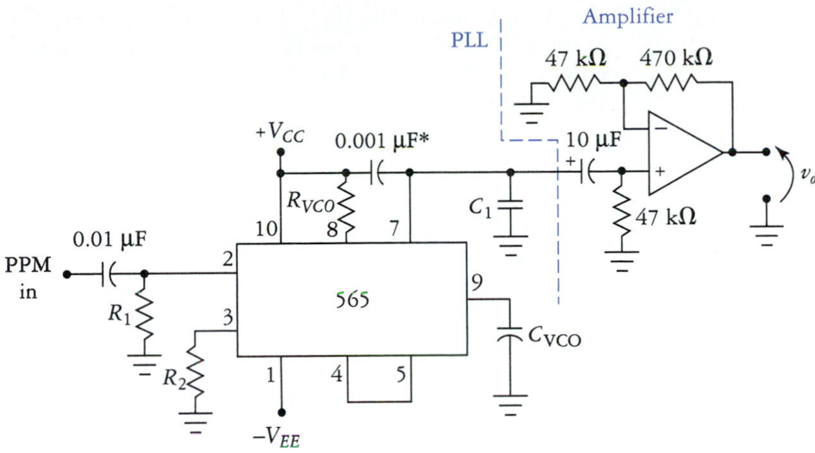

FIGURE 18.37
Using the 565 as a PPM demodulator.

Capacitor C_1 and the output resistance of the internal amplifier ($R_o \cong 3.6\ \text{k}\Omega$) form a simple low-pass loop filter. Choose loop-filter capacitor C_1 using

$$C_1 \cong \frac{10}{2\pi R_o f_o} \tag{18.71}$$

EXAMPLE 18.14

The PPM demodulator in Fig. 18.37 is to be used as a data communication system in which $f_o = 50$ kHz. Assuming $V_s = \pm 5$ V and $R_{\text{VCO}} = 4.7\ \text{k}\Omega$, determine the required values for C_{VCO} and C_1. If the PPM signal varies by ± 5 kHz, will the PLL maintain phase lock on the signal?

Solution Solve Eq. (18.68) for C_{VCO}, which gives us

$$\begin{aligned} C_{\text{VCO}} &\cong \frac{1}{3.7\ R_{\text{VCO}} f_o} \\ &= \frac{1}{(3.7)(4.7\ \text{k})(50\ \text{kHz})} \\ &= 0.001\ \mu\text{F} \end{aligned}$$

The loop-filter capacitor is now determined:

$$\begin{aligned} C_1 &\cong \frac{10}{2\pi\ R_o f_o} \\ &= \frac{10}{(2\pi)(3.6\ \text{k})(50\ \text{k})} \\ &= 0.009\ \mu\text{F} \end{aligned}$$

The tracking range of the PLL is

$$\begin{aligned} f_{\text{track}} &\cong f_o \pm 0.4 f_o \\ &= 50 \pm 20\ \text{kHz} \end{aligned}$$

Because the input PPM frequency deviates ± 10 kHz, the PLL will remain in lock.

The LM567 Tone Decoder

A tone decoder is a PLL that is designed specifically to signal the presence of a specific frequency. While a bandpass filter could also do this job, the advantage of a PLL is that once locked, the PLL will track changes in the detected frequency, even in the presence of noise. A simplified block diagram for the 567 is shown in Fig. 18.38. The I-phase detector compares the phase of the input signal with that of the VCO. If the 567 is not in lock, the Q-phase detector will cause the comparator output to be in a high-impedance state. A pullup resistor is required to interface this output to a logic gate. When the 567 is in phase lock, the output of the comparator is driven low.

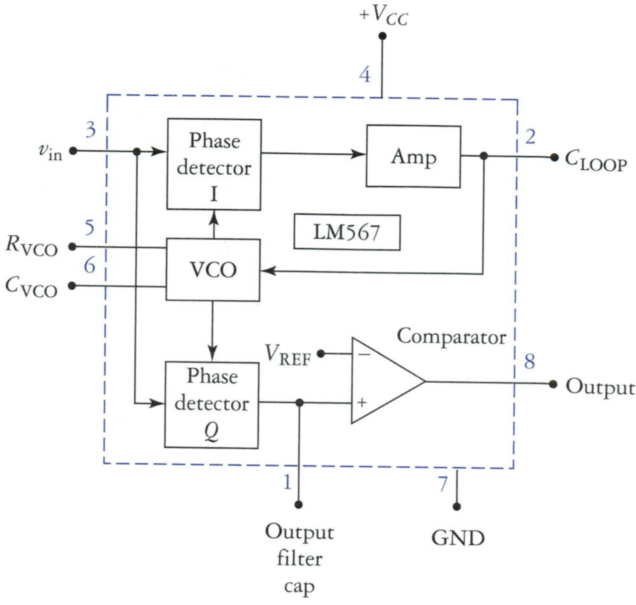

FIGURE 18.38
Internal block diagram for the 567 tone decoder.

The LM567 is designed for operation below 500 kHz, and is typically used in applications such as telephone touch-tone decoders and FSK demodulation. A typical 567 configuration is shown in Fig. 18.39. The input source must be capacitively coupled to the tone decoder, typically with a 0.47-μF cap. For best operation, the input to the 567 should be limited to about 200 mV_{rms}. The VCO free-running frequency is set by R_1 and C_1 according to

$$f_o \cong \frac{1}{1.1\, R_1 C_1} \tag{18.72}$$

Capacitor C_2 forms the loop filter, which determines the tracking range of the PLL. Expressing the tracking range as a percentage of f_o, we have

$$\text{BW}_{\text{track}} = 1070 \left(\frac{V_{\text{in}}}{f_o C_2} \right)^{1/2} \% \tag{18.73}$$

where $V_{in} \leq 200\ mV_{rms}$ and C_2 is in microfarads.

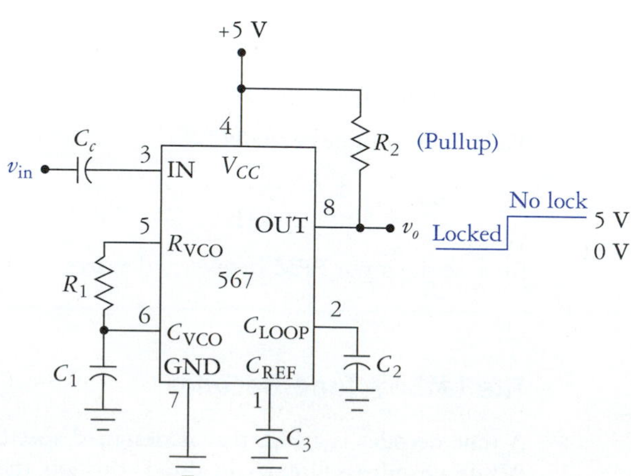

FIGURE 18.39
Typical 567 circuit.

EXAMPLE 18.15

Refer to Fig. 18.39. The circuit has $C_c = 0.47\ \mu F$, $R_1 = 10\ k\Omega$, $R_2 = 10\ k\Omega$, $C_1 = 0.022\ \mu F$, $C_2 = 0.47\ \mu F$, $C_3 = 4.7\ \mu F$, and $V_{CC} = 5$ V. Assuming that $V_{in} = 200\ mV_{rms}$, determine the free-running frequency and the lock range of the circuit. What range of input frequencies can be tracked under these conditions?

Solution

$$f_o \cong \frac{1}{1.1\ R_1 C_1}$$

$$= \frac{1}{(1.1)(10\ k)(0.022\ \mu F}$$

$$= 4132\ Hz$$

$$BW_{track} = 1070\left(\frac{V_{in}}{f_o C_2}\right)^{1/2}\%$$

$$= 1070\left(\frac{200\ mV}{(4132\ Hz)(0.47)}\right)^{1/2}\%$$

$$= 10.9\%$$

The 567 can track frequencies that range from 3682 Hz to 4582 Hz.

18.8 COMPUTER-AIDED ANALYSIS APPLICATIONS

In this section we verify a few of the results presented in this chapter using PSpice.

555 Timer Simulation

The circuit of Example 18.1 is shown in the schematic editor in Fig. 18.40. The 555 timer modeled here is denoted as X1. It is found in the Eval.lib sublibrary. It is worth noting that PSpice does not allow the Control (FM) pin to be left open, thus, we are required to use the noise-suppression capacitor C_2 in this simulation. Often, in an actual circuit, this pin is simply left open.

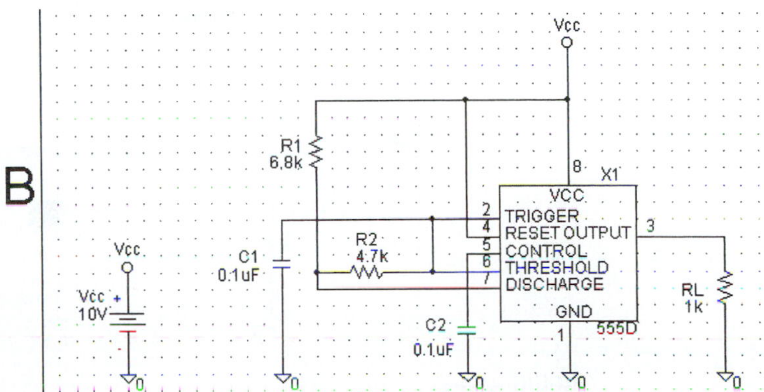

FIGURE 18.40
Simulation of the 555 timer.

A transient analysis run from 0 to 5 ms results in the waveforms shown in Fig. 18.41. Use of the cursors reveals $T_H = 782\ \mu s$, $T_L = 324\ \mu s$, and $T = 1106\ \mu s$. These correspond to $f = 904$ Hz and $D = 70.7\%$. This is in good agreement with the manual analysis results, which were $f = 889$ Hz and $D = 71\%$.

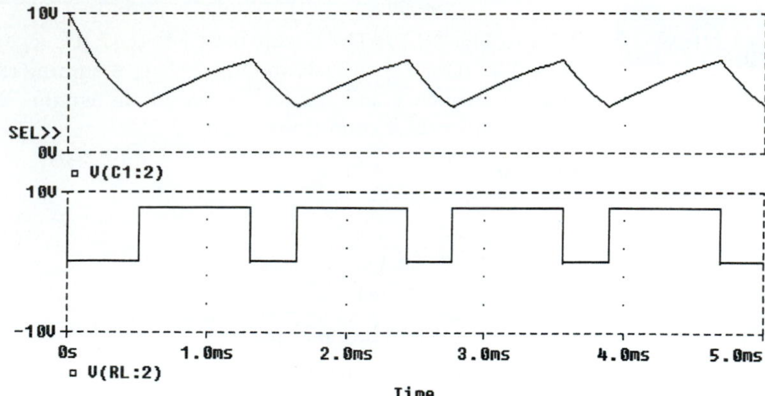

FIGURE 18.41
Timing capacitor and output voltage waveforms.

Linear Regulator Simulation

A linear voltage regulator is shown in Fig. 18.42. The dc bias-point voltages are shown on the schematic. We can verify the results by using Eq. (18.24), which produces

$$\begin{aligned} V_o &= V_{\text{REF}}\left(1 + \frac{R_2}{R_3}\right) \\ &= 4.685\ \text{V}\left(1 + \frac{1}{2}\right) \\ &= 7.028\ \text{V} \end{aligned}$$

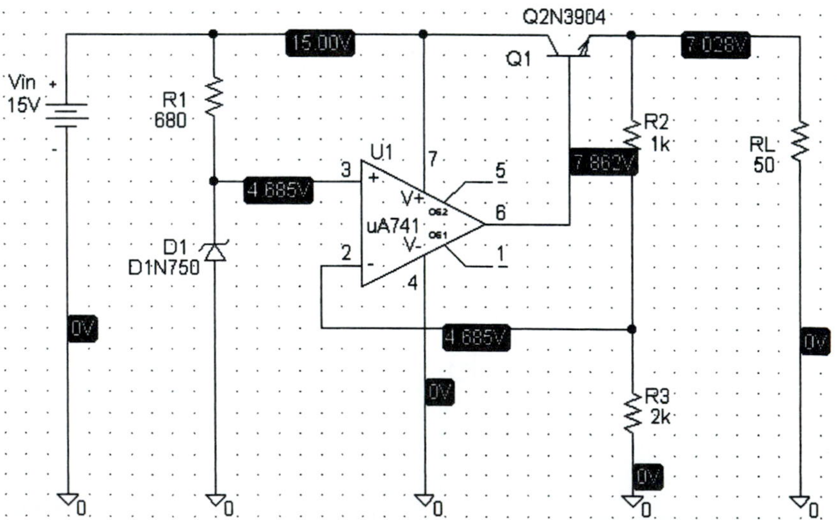

FIGURE 18.42
Linear series regulator simulation.

We can get a good idea of how well the regualtor works by performing a dc sweep of the V_{in} source, say, from 10 to 20 V. Performing this sweep using linear, 2-V increments, and plotting V_o [that is, V(RL:2) in this case] results in the plot of Fig. 18.43. The relationship is fairly linear over this wide range of input voltages. Based on a nominal output voltage of 7.028 V, the percent change of V_o over this range is approximately $\Delta V_o = 0.44\%/\text{V}$.

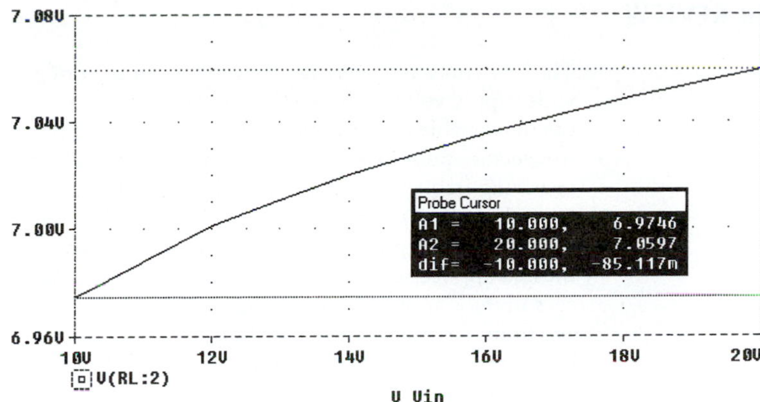

FIGURE 18.43
Output voltage versus input voltage.

Balanced Modulator Simulation

A balanced modulater is formed using the MULT (multiplier) block found in the ABM (Analog Behavioral Model) sublibrary. This is a general-purpose time-domain multiplication block that is useful in many system-level simulations. The circuit in Fig. 18.44 uses the multiplier and two sine sources to produce SSB-SC modulation. The output signal is shown in Fig. 18.45. Generation of the frequency-domain representation of the output is left as an exercise for the reader.

FIGURE 18.44
Balanced modulator simulation.

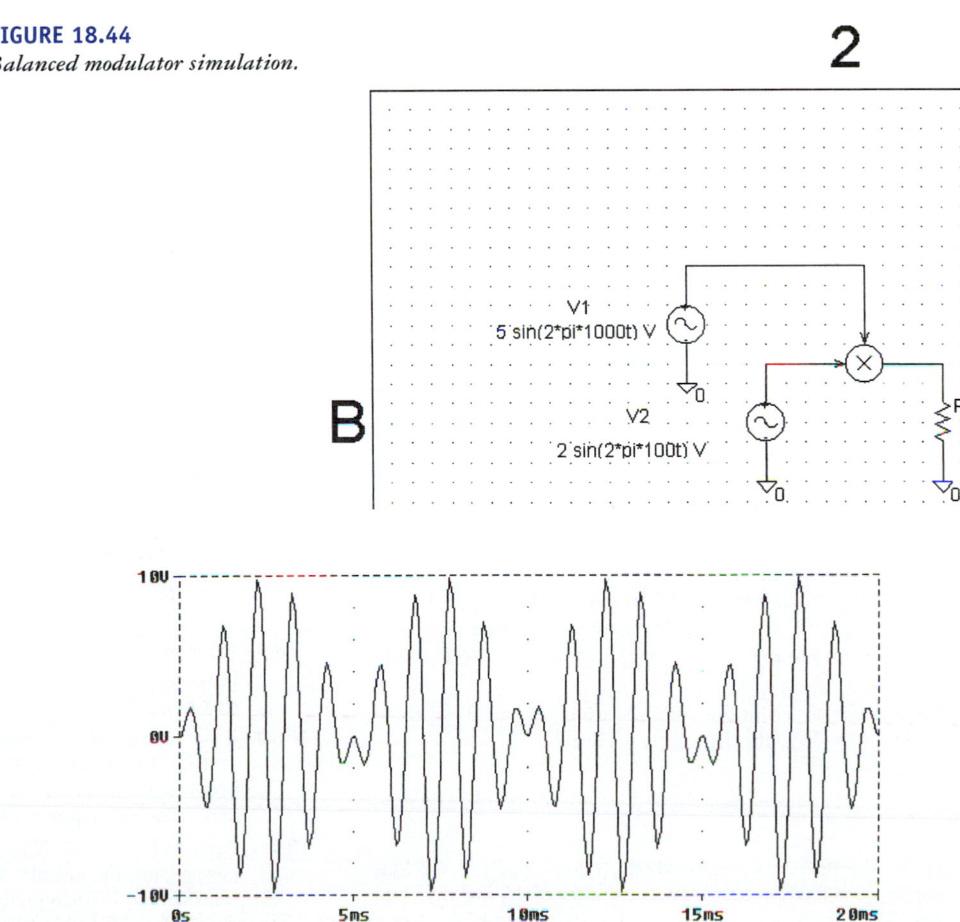

FIGURE 18.45
Output of the balanced modulator.

CHAPTER REVIEW

The 555 timer is a versatile device that can be configured for astable or monostable operation over a wide supply voltage range. The oscillation frequency of the astable configuration, and pulse width of the monostable configurations are independent of power supply voltage. The 555 timer may be used to generate pulse-position modulation (PPM) and pulse-width modulation (PWM). The maximum operating frequency of the 555 is about 1 MHz.

Voltage regulators are available in a wide variety of forms. Most linear voltage regulators are of the series pass type. An error amplifier senses changes in output voltage and compensates by varying the bias applied to the series pass element. Zener diodes are often used to provide stable reference voltages in these applications. Voltage regulators must usually be mounted on heat sinks to prevent damage from overheating. Current limiting is also used to protect the power supply. Foldback limiting limits maximum regulator power dissipation to a constant value. A wide variety of three-terminal monolithic voltage regulators is available, making low- and medium-power supply design quite simple.

Switching regulators are used very often today because of their high efficiency, relative to linear regulators. Switching regulators can be configured for step-down, step-up, or inverting operation. Switching regulators are generally much more complex than linear regulators and they can generate large amounts of electrical noise, requiring careful attention to shielding and filtering.

Operational transconductance amplifiers (OTAs) are amplifiers with a voltage-sensing, differential input stage, and a high-impedance current-source output stage. The gain of the OTA is determined by an external resistor that sets the amplifier bias current. If the bias current is varied in proportion to some signal, the gain of the OTA varies as well, forming an amplitude modulator. Because the output of the OTA is a current source, it is very simple to implement both inverting and noninverting integrators.

Balanced modulators are frequently used in the design of communication circuits such as amplitude modulators and demodulators. The output of the balanced modulator is proportional to the product of its input voltages. Balanced modulators are similar to analog multipliers in operation, however, analog multipliers are optimized for accuracy at the expense of speed whereas balanced modulators are designed for high-frequency operation. Balanced modulators are also used to implement phase detectors.

Phase-locked loops are used in many different applications, but are primarily found in digital and analog communication systems. A phase-locked loop will lock onto and track an input signal that falls within its capture range. Phase-locked loops are used in FM demodulation, frequency synthesizers, and servomechanisms. A tone decoder is a specialized PLL that is used to signal the presence of a particular signal frequency. Tone decoders are typically used in touch-tone decoders and data communication systems.

DESIGN AND ANALYSIS PROBLEMS

18.1. Refer to the circuit in Fig. 18.2. Given $V_{CC} = 10$ V, $R_1 = 10$ kΩ, $R_2 = 3.3$ kΩ, $C_1 = 0.01$ μf, and $R_L = 1$ kΩ, determine f_{osc} and the duty cycle, assuming the load resistor is connected to ground. Determine the average load voltage and current.

18.2. Refer to the circuit in Fig. 18.2. Given $V_{CC} = 15$ V, $R_1 = 2.2$ kΩ, $R_2 = 4.7$ kΩ, $C_1 = 0.022$ μF, and $R_L = 5$ kΩ, determine f_{osc} and the duty cycle, assuming the load resistor is connected to ground. Determine the average load voltage and current.

18.3. Refer to the circuit in Fig. 18.2. Assuming the load resistor is connected to V_{CC}, the duty cycle is given by

$$D = 1 - \frac{R_1 + R_2}{R_1 + 2R_2}$$

Given $V_{CC} = 5$ V, $R_L = 1$k Ω, $R_1 = 3.3$ kΩ, $R_2 = 3.3$ kΩ, and $C_1 = 0.1$ μF, determine f_{osc}, the duty cycle, and the average load current.

18.4. Refer to Fig. 18.2. Given $V_{CC} = 12$ V, $R_L = 250$ Ω, $R_1 = 15$ kΩ, $R_2 = 10$ kΩ, and $C_1 = 10$ μF, determine f_{osc}, the duty cycle, and average load current, assuming the load resistor is connected to V_{CC}.

18.5. Refer to Fig. 18.2. Given $R_1 = 1.5$ kΩ and $R_2 = 3.3$ kΩ, determine the required value for C_1 such that $f_{osc} \cong 5.4$ kHz.

18.6. Refer to Fig. 18.2. Given $R_1 = 4.7$ kΩ and $C_1 = 0.022$ μF, determine the required value for R_2 such that $f_{osc} \cong 3.6$ kHz.

18.7. Refer to Fig. 18.2. Given $R_2 = 10$ kΩ and $C_1 = 2.2$ μF, determine the required value for R_2 such that $f_{osc} \cong 19$ Hz.

18.8. Refer to Fig. 18.2. Assume the load is connected to ground. Given $R_1 = 10$ kΩ, determine the required value for R_2 such that $D = 75\%$.

18.9. Refer to Fig. 18.2. Assume the load is connected to ground. Given $R_2 = 1.5$ kΩ, determine the required value for R_1 such that $D = 80\%$.

18.10. Refer to Fig. 18.46. The LED is to be pulsed at a 2-Hz rate. The LED specs are $I_F = 25$ mA @ $V_F = 1.5$ V (for rated brightness). Given $V_{CC} = 5$ V, $R_1 = 22$ kΩ, and $R_2 = 8.2$ kΩ, determine the required values for C_1 and R_3.

18.11. Refer to Fig. 18.47. This circuit is to be used to flash a high-power lamp on and off. The relay has $R_{coil} = 200$ Ω, while transistor Q_1 is silicon with $\beta = 100$. Determine an appropriate value for R_3 (that will ensure saturation of the transistor), and determine the frequency of oscillation and the duty cycle of the lamp (T_{on}/T).

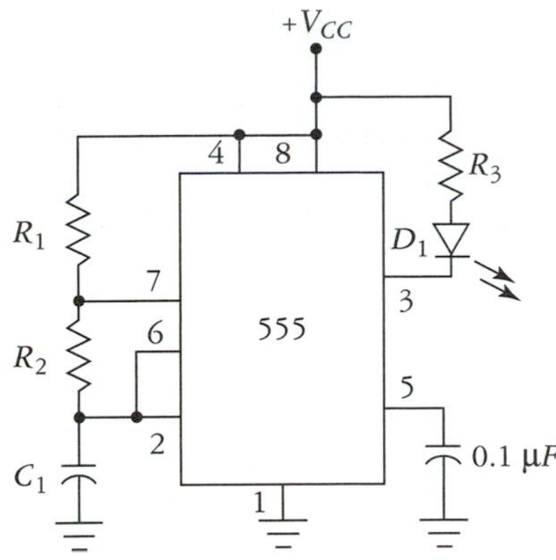

FIGURE 18.46
Circuit for Problem 18.10.

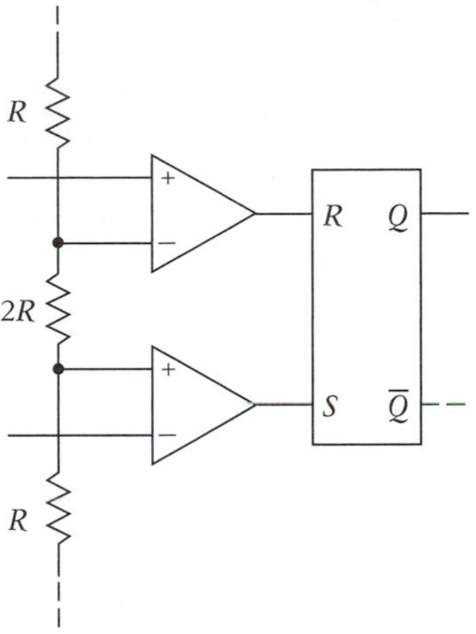

FIGURE 18.48
Circuit for Problem 18.12.

18.12. Refer to Fig. 18.1. Assume that the voltage-divider resistors are modified as shown in Fig. 18.48. Derive the new oscillation frequency equation.

18.13. Refer to Fig. 18.3. Given $R_1 = 5.6\ \text{k}\Omega$, determine C_1 such that $T_W = 0.5$ sec.

18.14. Refer to Fig. 18.3. Given $C_1 = 0.47\ \mu\text{F}$, determine R_1 such that $T_W = 5$ ms.

18.15. Refer to Fig. 18.49. Given $R_1 = R_2 = 10\ \text{k}\Omega$, and $C_1 = C_2 = 0.1\ \mu\text{F}$, sketch the waveforms for v_x and v_y given the trigger signal from Fig. 18.49(b).

18.16. Refer to Fig. 18.5. Assume that $V_{CC} = 10$ V. Based on the internal functional diagram of Fig. 18.1, determine the maximum and minimum amplitude limits on v_m.

18.17. Refer to Fig. 18.9. Given $V_{in} = 24$ V, $R_1 = 680\ \Omega$, $V_Z = 5.6$ V, $R_2 = 7.9\ \text{k}\Omega$, $R_3 = 10\ \text{k}\Omega$, and the load draws $I_L = 2.0$ A, determine I_Z, V_o, P_L, and P_{DQ}.

18.18. Refer to Fig. 18.9. Given $V_{in} = 24$ V, $R_1 = 680\ \Omega$, $D_Z = 6.2$ V, $R_2 = 10\ \text{k}\Omega$, $R_3 = 10\ \text{k}\Omega$, and the load draws $I_L = 2.0$ A, determine V_o, P_L, and P_{DQ}.

18.19. Refer to Fig. 18.11. Given $V_{in} = 24$ V, $R_1 = 680\ \Omega$, $D_Z = 5.6$ V, $R_E = 2\ \Omega$, $R_2 = 2\ \text{k}\Omega$, $R_3 = 2$-kΩ pot, and $R_4 = 5\ \text{k}\Omega$, determine the maximum and minimum output voltages. Assuming that $\beta_1 = \beta_2 = 50$, $V_o = 10.0$ V and $I_L = 2$ A, determine P_L, $P_{DQ_1} = P_{DQ_2}$, $P_{R_{E(1)}} = P_{R_{E(2)}}$, and the current that must be sourced by op amp A_1. Assuming that the ambient temperature is $T_A = 25°\text{C}$ and the transistors are mounted on a heat sink such that $\theta_{JA} = 8°\text{C/W}$, determine the transistor junction temperatures.

18.20. Refer to Fig. 18.11. Given $V_{in} = 20$ V, $R_1 = 470\ \Omega$, $D_Z = 6.2$ V, $R_E = 1\ \Omega$, $R_2 = 1\ \text{k}\Omega$, $R_3 = 2\ \text{k}\Omega$ pot, and $R_4 = 2\ \text{k}\Omega$, determine the maximum and minimum output voltages. Assuming that $\beta_1 = \beta_2 = 50$, $V_o = 10.0$ V, and $I_L = 1$ A, determine P_L, $P_{DQ_1} = P_{DQ_2}$, $P_{RE_1} = P_{RE_2}$, and the current that must be sourced by op amp A_1. Assuming that the ambient temperature is $T_A = 25°\text{C}$ and the transistors are mounted on a heat sink such that $\theta_{JA} = 12°\text{C/W}$, determine the transistor junction temperatures.

18.21. Refer to Fig. 18.12. Assuming that Q_2 is a silicon transistor and $R_E = 0.5\ \Omega$, determine the approximate maximum collector current for Q_1. Determine the required value for R_E such that the maximum collector current for Q_1 is 500 mA.

18.22. Refer to Fig. 18.12. Assuming that Q_2 is a silicon transistor and $R_E = 0.2\ \Omega$, determine the approximate maximum collector current for Q_1. Determine the required value for R_E such that the maximum collector current for Q_1 is 2 A.

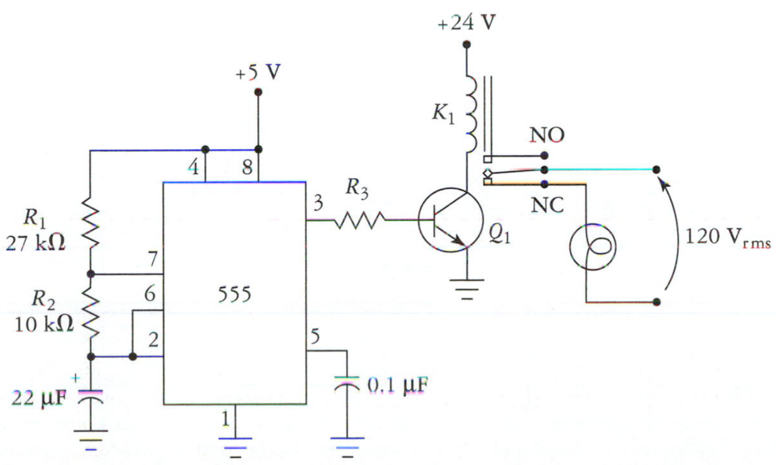

FIGURE 18.47
Circuit for Problem 18.11.

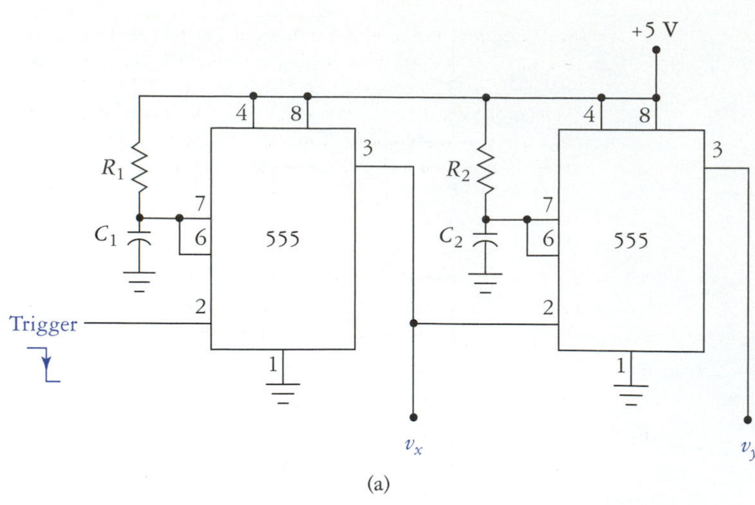

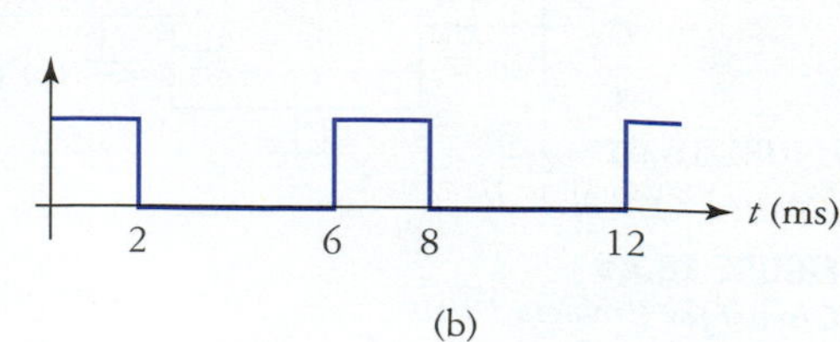

FIGURE 18.49
Circuit for Problem 18.15.

18.23. Refer to Fig. 18.13. Given $V_{\text{in}} = 15$ V, $V_Z = 6.7$ V, $R_2 = 5$ kΩ, $R_3 = 10$ kΩ, $R_E = 0.47$ Ω, $R_x = 220$ Ω, and $R_y = 470$ Ω, determine V_o, I_{SC}, and I_{knee}.

18.24. Refer to Fig. 18.13. Given $V_{\text{in}} = 16$ V, $V_Z = 5.0$ V, $R_2 = 5$ kΩ, $R_3 = 5$ kΩ, $R_E = 0.68$ Ω, $R_x = 330$ Ω, and $R_y = 680$ Ω, determine V_o, I_{SC}, and I_{knee}.

18.25. Refer to Fig. 18.50. Assume that $V_{\text{in}} = 10$ V, $V_Z = 5.0$ V, $R_1 = 470$ Ω, $R_2 = 4.7$ kΩ, $R_3 = 4.7$ kΩ, $R_4 = R_5 = 1$ Ω, $R_6 = 180$ Ω, $R_7 = 470$ Ω $R_8 = 3.3$ kΩ, $R_9 = 1$ kΩ pot, and $R_{10} = 3.3$ kΩ. Determine I_Z, $V_{o(\max)}$, $V_{o(\min)}$, $I_{SC}\big|_{V_o=\max}$, $I_{SC}\big|_{V_o=\min}$, $I_{\text{knee}}\big|_{V_o=\max}$, and $I_{\text{knee}}\big|_{V_o=\min}$.

18.26. Refer to Fig. 18.50. Assume that $V_{\text{in}} = 12$ $V_Z = 6.2$ V, $R_1 = 560$ Ω, $R_2 = 2.2$ kΩ, $R_3 = 1$ kΩ, $R_4 = R_5 = 0.82$ Ω, $R_6 = 270$ Ω, $R_7 = 820$ Ω, $R_8 = 5$ kΩ, $R_9 = 10$ kΩ pot, and $R_{10} = 5$ kΩ. Determine I_z, $V_{o(\max)}$, $V_{o(\min)}$, $I_{sc}\big|_{V_o=\max}$, $I_{sc}\big|_{V_o=\min}$, $I_{\text{knee}}\big|_{V_o=\max}$, and $I_{\text{knee}}\big|_{V_o=\min}$.

18.27. State two advantages of switching regulators over linear voltage regulators. State two disadvantages.

18.28. Explain why MOSFETs are usually preferred to BJTs in switching regulator designs.

18.29. Refer to Fig. 18.18(a). Given: $V_{CC} = V_{EE} = 15$ V, $R_{\text{ABC}} = 68$ kΩ, and $R_L = 27$ kΩ. Determine I_{ABC}, g_m, A_v, R_{in}, and v_o given $v_{\text{in}} = 20 \sin \omega t$ mV. Assuming that $I_{o(\max)} = \pm 300$ μA and $V_{o(\max)} = \pm 14$ V, determine the maximum output voltage compliance ($V_{\text{p-p}}$).

18.30. Refer to Fig. 18.18(b). Given: $V_{CC} = V_{EE} = 15$ V, $R_{\text{ABC}} = 100$ kΩ, and $R_L = 33$ kΩ. Determine I_{ABC}, g_m, A_v, R_{in}, R_o, and v_o given $v_{\text{in}} = 50 \sin \omega t$ mV. Assuming that $I_{o(\max)} = \pm 300$ μA and $V_{o(\max)} = \pm 14$ V, determine the maximum output voltage compliance ($V_{\text{p-p}}$).

18.31. Refer to Fig. 18.51. Given $V_{CC} = V_{EE} = 12$ V, $R_{\text{ABC}} = 100$ kΩ, and $R_L = 47$ kΩ, determine I_{ABC}, g_m, and A_v.

18.32. Refer to Fig. 18.52. Given $V_{CC} = V_{EE} = 15$ V, $R_{\text{ABC}} = 33$ kΩ, and $R_F = 10$ kΩ, determine I_{ABC}, g_m, and A_v. Describe the function of the 741 op amp in this circuit.

18.33. Refer to Fig. 18.20. Given $V_{CC} = V_{EE} = 15$ V, $R_{\text{ABC}} = 68$ kΩ, $R_{\text{abc}} = 22$ kΩ, and $R_L = 33$ kΩ, determine I_{ABC} and

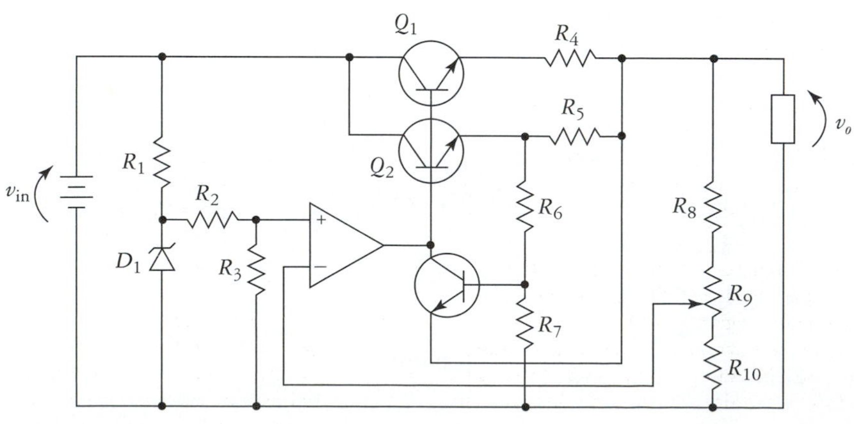

FIGURE 18.50
Circuit for Problems 18.25 and 18.26.

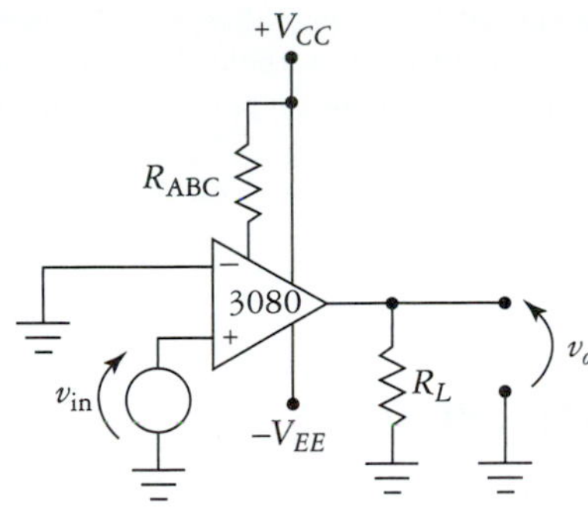

FIGURE 18.51
Circuit for Problem 18.31.

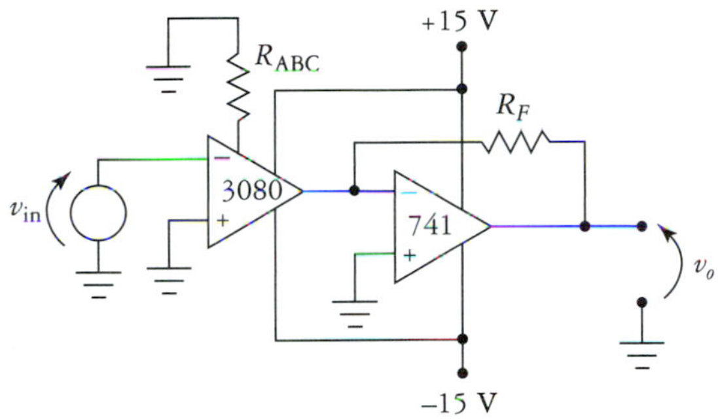

FIGURE 18.52
Circuit for Problem 18.32.

g_{mQ}. Assume $v_c = 30 \sin 10{,}000t$ mV and $v_m = 4 \sin 1000t$ V. Determine the expression for v_o and the modulation index.

18.34. Refer to Fig. 18.20. Given $V_{CC} = V_{EE} = 15$ V, $R_{ABC} = 100$ kΩ, $R_{abc} = 15$ kΩ, and $R_L = 47$ kΩ, determine I_{ABC} and g_{mQ}. Assume $v_c = 50 \sin 25{,}000t$ mV and $v_m = 0.5 \sin 5000t$ V. Determine the expression for v_o and the modulation index. Sketch the time- and frequency-domain representations for the output voltage.

18.35. Refer to Figs. 18.23(a) and (b). Given: $V_{CC} = V_{EE} = 15$ V, $R_{ABC} = 150$ kΩ, and $C_1 = 22$ μF, determine the output voltage expression for each OTA given $v_{in} = 2 \sin 2\pi 100t$ V and $v_{in} = 2e^{-250t}$ V. Assume that the integrators are reset to zero before input signals are applied. *Note:* Setting arbitrary constants to zero, $\int A \sin \omega t \, dt = (-1/\omega A) \cos \omega t$ and $\int ke^{-\alpha t} dt = (-k/\alpha)e^{-\alpha t}$.

18.36. Refer to Figs. 18.23(a) and (b). Given: $V_{CC} = V_{EE} = 12$ V, $R_{ABC} = 100$ kΩ, and $C_1 = 47$ μF. Determine the output voltage expression for each OTA given $v_{in} = 5 \sin 800t$ V and v_{in}, $= 0.5e^{-120t}$ V. Assume that the integrators are reset to zero before input signals are applied. Sketch the input and output waveforms.

18.37. Refer to Fig. 18.23(c). Given $V_{CC} = V_{EE} = 10$ V, $R_{ABC} = 120$ kΩ, and $C_1 = 4.7$ μF. Determine the output voltage expression for the differential integrator given $v_1 = 2$ V and $v_2 = 5$ V. Assume that the integrators are reset to zero before input signals are applied. Assuming that the output of the OTA saturates at ±9 V, determine the time it takes for the output of the OTA to saturate.

18.38. Refer to Fig. 18.23(c). Given: $V_{CC} = V_{EE} = 10$ V, $R_{ABC} = 120$ kΩ, and $C_1 = 4.7$ μF, determine the output voltage expression for the differential integrator given $v_1 = 3 \sin 1200t$ V and $v_2 = 5 \cos 1200t$ V. Assume that the integrators are reset to zero before input signals are applied. [*Hints:* The difference of v_2 and v_1 can be expressed as a phase-shifted sinusoid using a trigonometric identity or by phasor algebra. The integral of a phase-shifted sinusoid is $\int A \sin(\omega t + \phi)\, dt = (-A/\omega)\cos(\omega t + \phi) + K$. We assume the constant $K = 0$ because the integrator is reset prior to operation.]

18.39. Refer to the balanced modulator in Fig. 18.24. Assume $k = 1$, $A_0 = 1$. Determine the expression for v_o given:
(a) $v_m = 3 \sin 2\pi 500t$ V, $v_c = 6 \sin 2\pi 2000t$ V
(b) $v_m = 4 \sin 250t$ V, $v_c = 5 \cos 1000t$ V
(c) $v_m = 2$ V, $v_c = 5 \cos 1000t$ V
(d) $v_m = 5 \sin 400t$ V, $v_c = 5 \sin 400t$ V

18.40. Refer to the balanced modulator in Fig. 18.24. Assume $k = 1$, $A_0 = 1$. Determine the expression for v_o given:
(a) $v_m = 5 \sin 100t$ V, $v_c = 4 \sin 800t$ V
(b) $v_m = 2 \cos 25t$ V, $v_c = 3 \sin 100t$ V
(c) $v_m = 3 + 4 \sin 500t$ V, $v_c = 5 \sin 1000t$ V
(d) $v_m = 5 \cos 1000t$ V, $v_c = 5 \sin 1000t$ V

18.41. Refer to the phase detector in Fig. 18.27. Determine the dc output voltage given the following inputs (assume the balanced modulator has $k = 1$ and the low-pass filter has $A_0 = 1$):
(a) $v_{in} = 5 \sin 2500t$ V, $v'_{in} = 30 \sin(2500t + 30°)$ V
(b) $v_{in} = 2 \cos 500t$V, $v'_{in} = 5 \cos(500t - 45°)$ V
(c) $v_{in} = 3 \cos 2\pi 1000t$ V, $v'_{in} = 4 \sin 2\pi 1000t$ V
(d) $V_{in} = 0.5\angle 0°$ V, $V'_{in} = 0.8\angle 60°$ V
(e) $V_{in} = 2 + 2j$ V, $V'_{in} = 3 + 0j$V

18.42. Refer to Fig. 18.27. The input voltages are equal amplitude sinusoids with $V_p = 4$ V, the balanced modulator has $k = 1$, and the low-pass filter has $A_0 = 1$. The output of the phase comparator is 1.414 V. Determine the amplitudes of the input voltages and the phase difference.

18.43. Refer to Fig. 18.32. The PLL has $f_o = 1$ MHz, capture range $f_{capture} = f_o \pm 10$ kHz and lock range $f_{lock} = f_o \pm 20$ kHz. Express the capture and lock ranges as a percent of f_o.

18.44. A certain PLL has $f_o = 3$ MHz. The capture range is $f_{capture} = f_o \pm 5\%$ and $f_{capture} = f_o \pm 15\%$. Determine the capture and track ranges in hertz.

18.45. Refer to Fig. 18.34(b). The reference oscillator is set for $f_o = 10$ kHz and the optical interrupter disk has 100 holes drilled around its edge. Determine the motor rotation speed S in rpm.

18.46. Refer to Fig. 18.34(b). Assuming the disk has 250 holes, what reference oscillator frequency is required to produce $S = 500$ rpm?

18.47. Refer to Fig. 18.35(a). Assume $N = 100$ and $f = 25$ kHz. Determine the resulting output frequency.

18.48. The 565-based PPM demodulator in Fig. 18.37 is to be used as a data communication system in which $f_o = 150$ kHz. Assuming $V_s = \pm 5$ V and $R_{VCO} = 2.7$ kΩ, determine the required values for C_{VCO} and C_1. If the PPM signal varies by ±10 kHz, will the PLL maintain phase lock on the signal?

18.49. Refer to Fig. 18.39. The circuit has $C_c = 0.47$ μF, $R_1 = 33$ kΩ, $R_2 = 10$ kΩ, $C_1 = 0.001$ μF, $C_2 = 0.33$ μF, $C_3 = 4.7$ μF, and $V_{CC} = 5$ V. Assuming that $V_{in} = 200$ mV_{rms}, determine the free-running frequency. What range of input frequencies can be tracked under these conditions?

18.50. Determine the relationship between the output frequency and the input frequency for the PLL in Fig. 18.53.

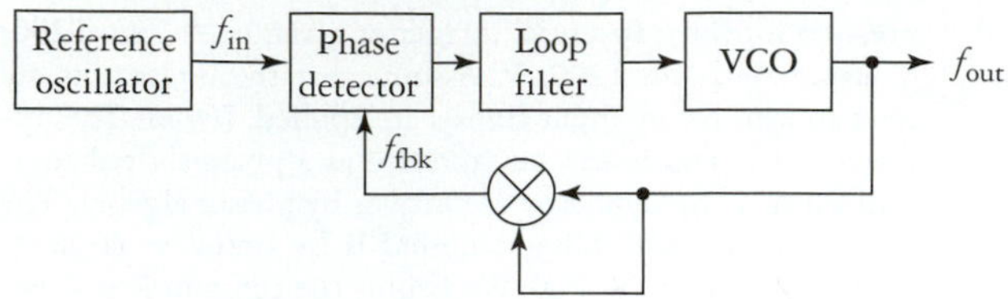

FIGURE 18.53
Circuit for Problem 18.50.

■ TROUBLESHOOTING PROBLEMS

18.51. Refer to the PWM system of Fig. 18.6. Suppose that the −2.5-V reference was to become defective and was measured to be 0 V. What effect would this have on the output of the integrator?

18.52. Suppose the diode in Fig. 18.8 was replaced and accidentally connected in the wrong direction. What effect would this have on the frequency and the duty cycle of v_o?

18.53. Refer to Fig. 18.9. Determine the effect on the output voltage under each fault condition listed.

(a) D_1 open circuited
(b) D_1 short circuited
(c) R_3 short circuited
(d) R_3 open circuited

18.54. Refer to Fig. 18.11. Suppose that the B-E junction of Q_1 became open circuited. How could this condition be detected without opening the power supply and taking voltage or current measurements?

18.55. Refer to Fig. 18.34(a). Suppose that the tachometer was to become disconnected from the motor. What effect would this have on the speed of the motor?

■ COMPUTER PROBLEMS

18.56. Refer to Fig. 18.40. Modify the circuit by placing a diode across R_2 (anode to pins 6 and 2 of the 555). Determine the resulting oscillation frequency and duty cycle.

18.57. Use the FFT option of PROBE to determine the frequency spectrum of the output of the circuit in Fig. 18.44.

CHAPTER 19

Thyristors and Other Miscellaneous Devices

BEHAVIORAL OBJECTIVES

On completion of this chapter, the student should be able to:

- Describe the operation of a four-layer diode.
- Describe the operation of a silicon-controlled rectifier.
- Explain how thyristors are used to control the power delivered to a load.
- Analyze simple thyristor-based relaxation oscillator circuits.
- Explain the use of snubber circuits.
- Explain the operation of the unijunction transistor (UJT).
- Analyze simple UJT-based relaxation oscillator operation.
- Describe the uses of a tunnel diode.

INTRODUCTION

We have come to the last chapter of the book. It's been a long journey and it is now time to clean up a few loose ends. In this chapter we examine the operation and uses of thyristors, which include silicon-controlled rectifiers (SCRs), triacs, diacs, and Shockley diodes, just to name a few. Thyristors are most often used in the control of high-power equipment such as motors, heaters, and lamps.

We also examine the unijunction transistor (UJT). UJTs are frequently used in conjunction with thyristors, particularly to generate pulse trains, which are used to control the thyristors.

The final device covered here is the tunnel diode. Although not commonly encountered, tunnel diodes are interesting because they exhibit negative resistance. This allows the tunnel diode to act as a two-terminal amplifier.

19.1 SILICON-CONTROLLED RECTIFIERS AND FOUR-LAYER DIODES

A silicon-controlled rectifier (usually called an SCR) is a four-layer npnp, three-terminal device. The schematic symbol and internal structure for the SCR are depicted in Fig. 19.1. The three terminals are the anode (A), the cathode (K), and the gate (G).

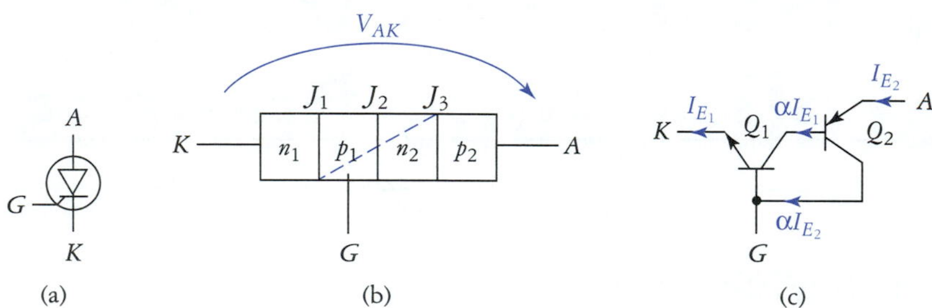

FIGURE 19.1
The silicon-controlled rectifier. (a) Schematic symbol. (b) Internal junction structure. (c) Equivalent internal circuit.

SCR Behavior

As seen in Fig. 19.1(b), the four-layer construction of the SCR results in at least one reverse-biased pn junction regardless of the polarity of V_{AK}. For example, if $V_{AK} > 0$ V, then J_2 is reverse biased. When $V_{AK} < 0$ V junctions J_1 and J_3 are reverse biased. Thus it appears that the SCR will not conduct in either direction. This is true, to an extent, but as we shall see, the SCR can be made to conduct very well when $V_{AK} > 0$ V.

If we visualize the SCR as being sliced along the dashed line in Fig. 19.1(b), the equivalent bipolar transistor circuit of Fig. 19.1(c) results. Examination of this circuit reveals that when $V_{AK} > 0$ V the base–emitter junctions of both the npn and pnp equivalent transistors tend toward forward bias, whereas the collector–base junctions are definitely reverse biased. Recall that these are the conditions required for active-region operation. Under this biasing condition it is possible to cause the SCR to switch into a conductive state. We examine this condition first.

The circuit shown in Fig. 19.2(a) can be used to determine the terminal characteristics of the SCR. Notice that the V_1 bias voltage orientation causes $V_{AK} \geq 0$ V. If we initially set the gate current source for $I_G = 0$ A, and slowly increase the bias voltage, V_{AK} will initially increase as shown in Fig. 19.1(b). Here's what's happening inside the SCR: As V_{AK} increases, the only current that flows is due to device leakage. Two things are happening here. First, although leakage currents are relatively constant, they do increase slightly with voltage, thus, the internal base currents will increase with V_{AK}. Second, as V_{AK} increases, the C-B depletion regions widen, which in turn increases the forward beta (and alpha) of the transistors. Because of these factors, a voltage is eventually reached at which the SCR suddenly goes into conduction (fires). In general, this is called the *breakover voltage* V_{BO}. Since

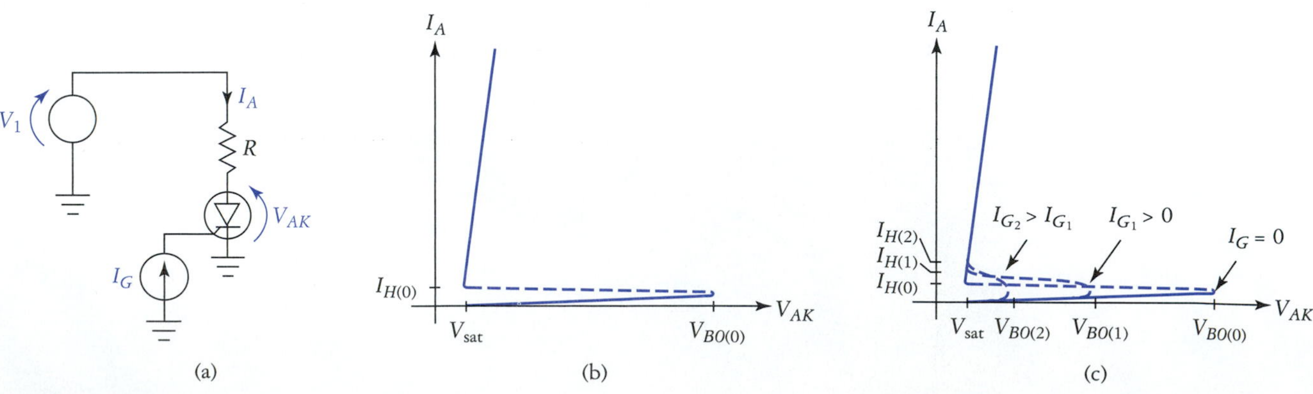

FIGURE 19.2
SCR terminal characteristics. (a) Test circuit. (b) Transconductance curve with zero gate current. (c) Curves obtained under varying gate current conditions.

$I_G = 0$, this particular voltage is called $V_{BO(0)}$. Once breakover has occurred, the equivalent internal transistors drive each other into heavy saturation in a positive-feedback loop, causing a rapid decrease in V_{AK} (often to as low as a few tenths of a volt) and an increase in I_A, which is limited by the external resistance. The SCR will remain latched in conduction as long as the anode current is greater than a minimum value called the *holding current* $I_{H(0)}$.

Injection of an external gate current causes the breakover voltage to decrease, as shown in Fig. 19.2(c). In this manner it is possible to reduce the breakover voltage of the SCR to a small fraction of $V_{BO(0)}$. Typically, the gate current required to accomplish this is very small (in the milliampere range) compared to the rated anode current of the SCR (in the ampere range). The minimum gate current required to trigger the SCR (typically, for $V_{AK} \cong 5$ V or so) is called I_{GT}. The corresponding gate turn-on voltage is called V_{GT}. Notice in Fig. 19.2(c) that the holding current increases slightly with I_G as well. Usually, the holding current is quite low relative to the average anode operating current (on the order of $<1\%$), so the small change in I_H can be neglected. The saturation voltage for a typical SCR is low, ranging from around 0.8 to 2 V, depending on the anode current.

SCRs are available with breakover voltages that range from around 30 V to more than 1000 V. Similarly, current-handling specifications typically vary from 1 A to more than 1000 A. A few examples of typical SCR packages are shown in Fig. 19.3. The larger hockey-puck style devices are capable of carrying over 1500 A. An example of a typical rather low-power SCR is the 2N5064. The 2N5064 comes in a small TO-226 epoxy package, which at $T_A = 25°C$, has $V_{BO(0)} = 200$ V, $I_{D(rms)} = 0.8$ A, $I_{H(max)} = 5$ mA, $I_{GT} = 200$ μA, and $V_{GT} = 0.8$ V.

Commutation

Once the SCR is latched, or triggered on, it will remain so until either I_A drops below I_H, or a negative pulse is applied to the gate terminal. Very often SCRs are used to control the average current flow through a load connected to the ac line. In such cases, the line zero crossing will turn off the SCR. This is called *natural commutation.*

It is possible to force an SCR into the blocking state by shunting the load current around the SCR with a switch, as shown in Fig. 19.4(a). When the switch is closed, current is diverted around the SCR and $I_A \cong 0$. If no gate trigger is applied, then the SCR will remain off when the switch is opened. This is called *forced commutation.* Forced commutation can also be achieved with the application of a negative gate current pulse. Some SCRs can be damaged by negative gate triggering; however, some SCRs are designed to operate in this manner. These SCRs are called *gate turn-off SCRs* (GTOs), or sometimes *gate-controlled switches* (GCS).

Reverse Blocking

If an SCR is biased such that $V_{AK} < 0$ V, then the equivalent internal transistors of Fig. 19.1(c) will have reverse-biased emitter-base junctions, while the collector–base junctions are forced toward forward bias. You may recall from Chapter 3 that this results in operation in the inverse active region. Transistor gain is very low in this region, and so sufficient positive feedback, required to latch the SCR on, cannot occur. The SCR simply blocks reverse current until reverse breakdown (V_{RM} or V_{BR}, depending on the manufacturer) occurs. As with a normal rectifier diode, reverse breakdown is generally destructive and should be avoided. The magnitude of the reverse breakdown voltage is the same as the maximum forward breakover voltage.

SCRs have many uses, but before discussing them, we first take a look at a close relative, the four-layer diode.

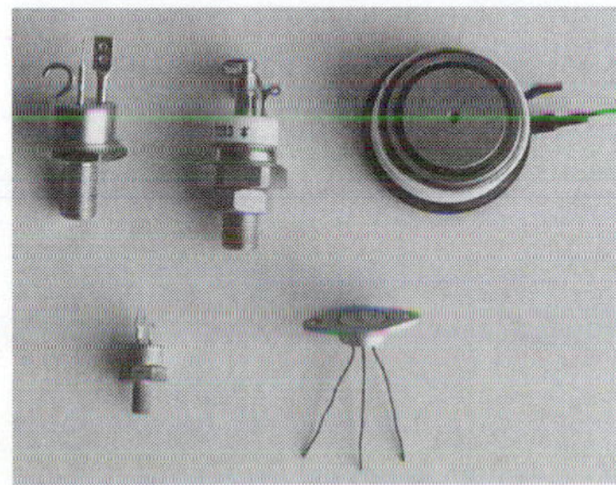

FIGURE 19.3
Common SCR packages.

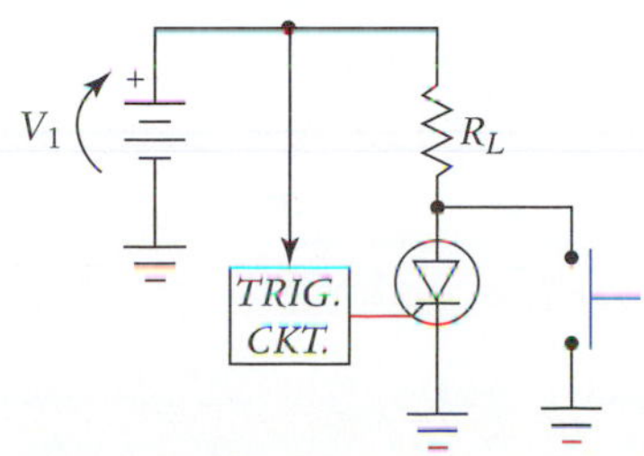

FIGURE 19.4
Forced commutation of an SCR.

Four-Layer Diodes

Structurally, a *four-layer diode* (sometimes called a *Shockley diode*) is very similar to an SCR without a gate lead. The schematic symbol for a four-layer diode and its terminal characteristic curve are shown in Fig. 19.5. The voltage at which the four-layer diode fires is called the *breakover voltage* V_{BO}. Once fired, the four-layer diode remains latched in conduction until either $I_A < I_H$, or $V_{AK} < V_{sat}$. Four-layer diodes are typically available with $25\text{ V} \le V_{BO} \le 150\text{ V}$, and corresponding saturation voltage values are in the neighborhood of $5\text{ V} \le V_{sat} \le 50\text{ V}$.

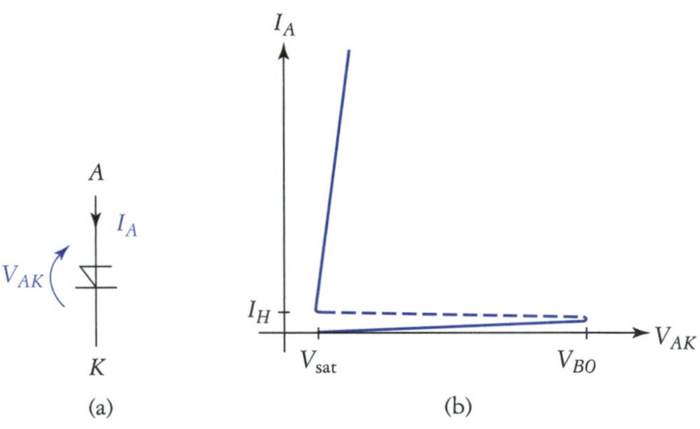

FIGURE 19.5
Four-layer (Shockley) diode. (a) Schematic symbol. (b) Transfer characteristic.

Four-Layer Diode Relaxation Oscillator

A classical application for the four-layer diode is in the design of a relaxation oscillator. Such a circuit is shown in Fig. 19.6(a). In this circuit, the capacitor charges until $v_C = V_{BO}$, at which time the four-layer diode triggers. This causes the capacitor to discharge rapidly down to the saturation voltage of the diode. The process then repeats, as shown in Fig. 19.6(b).

We can determine the oscillation frequency formula much as was done for the 555 timer in the previous chapter. Neglecting the fall time of the output voltage waveform, the period of the output voltage signal is given by

$$T = R_1 C_1 \ln\left(\frac{V_{in} - V_{sat}}{V_{in} - V_{BO}}\right) \tag{19.1}$$

FIGURE 19.6
Four-layer diode relaxation oscillator.

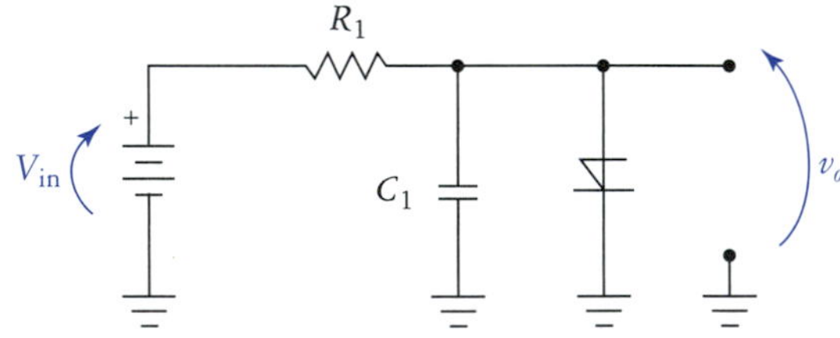

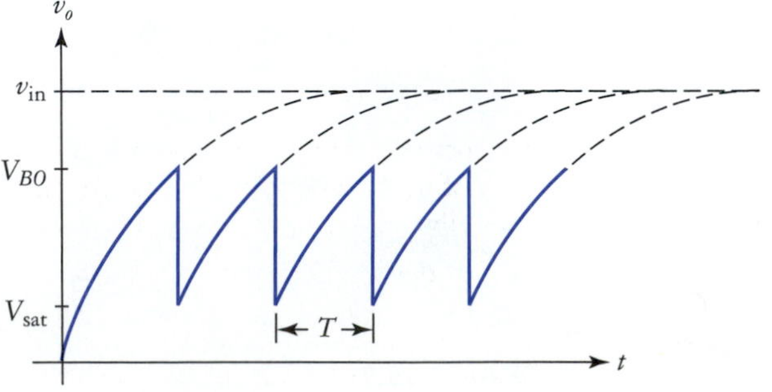

EXAMPLE 19.1

The circuit in Fig. 19.6 has $V_{in} = 50$ V, $R_1 = 47$ kΩ, $C_1 = 0.1$ μF, $V_{BO} = 30$ V, and $V_{sat} = 10$ V. Determine the oscillation period and frequency. Determine the value for R_1 that will result in $f = 1$ kHz.

Solution Direct application of Eq. (19.1) yields

$$
\begin{aligned}
T &= R_1C_1 \ln\left(\frac{V_{in} - V_{sat}}{V_{in} - V_{BO}}\right) \\
&= (47\ \text{k}\Omega)(0.1\ \mu\text{F}) \ln\left(\frac{50\ \text{V} - 10\ \text{V}}{50\ \text{V} - 30\ \text{V}}\right) \\
&= (4.7\ \text{ms})(0.693) \\
&= 3.3\ \text{ms} \\
f &= 1/T \\
&= 1/3.3\ \text{ms} \\
&= 303\ \text{Hz}
\end{aligned}
$$

To set the frequency to 1 kHz, we solve Eq. (19.1) for R_1 (this is left as an exercise for the reader), which gives us $R_1 = 14.4$ kΩ.

SCR Phase Control

SCRs are often used to control the average power delivered to a load. When the ac line is the power source, this is accomplished by delaying the firing of the SCR until late in a given cycle. Consider the circuit in Fig. 19.7(a). Resistors R_1 and R_2 form a voltage divider that controls the voltage v_x applied to the four-layer diode. By changing the ratio of these resistors, we can vary the firing time delay of the SCR. In other words, the voltage divider causes a delay that prevents the SCR from firing until $v_x > V_{BO}$ of the four-layer diode. This is shown in the waveforms of Fig. 19.7(b). The four-layer diode is used to provide a very fast gate triggering pulse. Notice that natural commutation occurs each time the ac line voltage crosses zero (negative going). Also notice that the SCR is a half-wave control device.

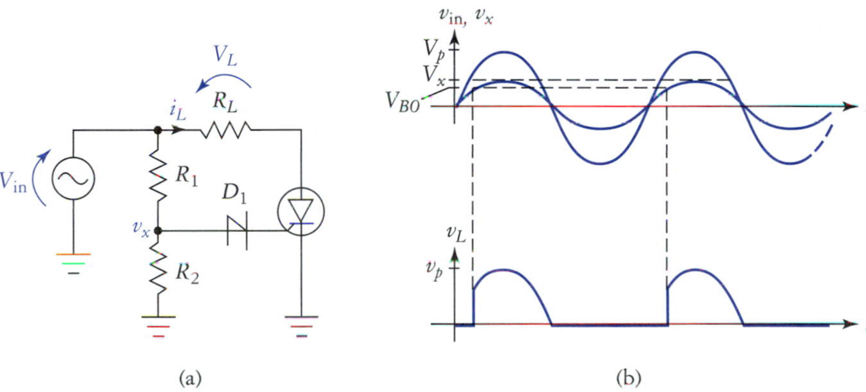

FIGURE 19.7
Phase control of an SCR.

Conduction Angle

The circuit in Fig. 19.7 controls the *conduction angle* θ of the SCR. The conduction angle is the number of degrees per cycle for which the SCR is latched on. Using a four-layer diode and resistive voltage divider as shown in Fig. 19.7, the conduction angle will be within the following range:

$$90° < \theta < 180° \tag{19.2}$$

The conduction angle cannot exceed 180° because the SCR can only be triggered on during positive half-cycles of the ac line voltage. The conduction angle will be exactly 90° when $V_x = V_{BO}$ for the four-layer diode. If $V_{BO} > V_x$ then the SCR will never fire, so technically the conduction angle can be set to 0°, but it is not continuously adjustable below 90°.

EXAMPLE 19.2

Refer to Fig. 19.7. The input voltage is 120 V_{rms}, 60 Hz. Given $R_1 = 10\ k\Omega$, $R_2 = 2.2\ k\Omega$, $R_L = 10\ \Omega$, and the four-layer diode has $V_{BO} = 25$ V, determine the following:

(a) The time delay from positive zero crossing until the SCR is fired.
(b) The conduction angle of the SCR.
(c) The value of v_{in} at the instant the SCR fires.

Solution First, we must determine the peak input voltage. This is

$$\begin{aligned} V_p &= V_{rms}\sqrt{2} \\ &= (120\ \text{V})(1.414) \\ &= 170\ \text{V} \end{aligned}$$

The line frequency is 60 Hz, therefore, the instantaneous input voltage is given by $v_{in} = 170 \sin 2\pi 60t$ V. The voltage divider attenuates the input voltage such that

$$\begin{aligned} v_x &= v_{in}\frac{R_2}{R_1 + R_2} \\ &= (170 \sin 2\pi 60t\ \text{V})(0.18) \\ &= 30.7 \sin 2\pi 60t\ \text{V} \end{aligned}$$

The SCR fires when $v_x = V_{BO} = 25$ V, so substituting this gives us $25 = 30.7 \sin 2\pi 60t$ V. Solving for t gives us the time delay Δt, which is

$$\Delta t = \frac{\sin^{-1}\left[\dfrac{V_{BO}}{V_{in}}\right]}{2\pi f} \tag{19.3}$$

$$\begin{aligned} &= \frac{0.95}{377} \\ &= 2.52\ \text{ms} \end{aligned}$$

Note that the arcsine in the numerator of Eq. (19.3) is evaluated on a calculator in radian mode. However, if we evaluate this numerator in the degree mode, we can obtain the delay angle ϕ, which is

$$\phi = \sin^{-1}\left(\frac{V_{BO}}{V_{in}}\right) \tag{19.4}$$

$$\begin{aligned} &= \sin^{-1}\left(\frac{25}{30.7}\right) \\ &= 54.5^\circ \end{aligned}$$

We can use this to determine the conduction angle θ, which is given by

$$\begin{aligned} \theta &= 180^\circ - \phi \\ &= 180^\circ - 54.5^\circ \\ &= 125.5^\circ \end{aligned} \tag{19.5}$$

The instantaneous value of v_{in} when the SCR fires is found using

$$\begin{aligned} V_{fire} &= V_{in} \sin[2\pi 60(\Delta t)] \\ &= 170 \sin[(2\pi)(60)(2.52\ \text{ms})] \\ &= 162\ \text{V} \end{aligned} \tag{19.6}$$

Improved Phase Control

The problem with the resistive divider firing control is that the conduction angle control is limited to a rather narrow range, as defined in Eq. (19.2). The simple modification shown in Fig. 19.8 allows us much wider phase control. In this circuit, the resistive divider is replaced with an RC network that

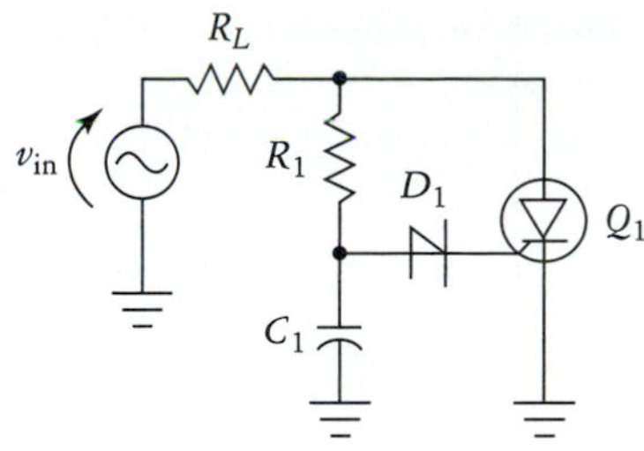

FIGURE 19.8
Phase control using a lag network.

causes v_x to lag in phase as well as providing attenuation. Assuming v_{in} is a sine wave, and disregarding loading effects on the lag network (which are extremely nonlinear, but negligible prior to firing of the four-layer diode), the peak amplitude of v_x is given by

$$V_x \cong \frac{|X_C|}{\sqrt{X_C^2 + R^2}} V_{in} \tag{19.7}$$

The phase lag α of v_x is found using

$$\alpha = -\tan^{-1}\left(\frac{R}{|X_C|}\right) \tag{19.8}$$

Because of the combined phase lag and attenuation, the conduction angle can be continuously adjusted over a much wider range, given by

$$0° \geq \theta < 180° \tag{19.9}$$

EXAMPLE 19.3

Refer to Fig. 19.8. Given an input voltage of 120 V_{rms}, 60 Hz, $R_1 = 6.8\ k\Omega$, $C_1 = 0.68\ \mu F$, and the four-layer diode has $V_{BO} = 30$ V, determine the following:

(a) The expression for the instantaneous value of v_x.
(b) The SCR firing time delay
(c) The conduction angle.

Solution We first determine $|X_C|$:

$$|X_C| = \frac{1}{2\pi fc}$$
$$= \frac{1}{(2\pi)(60)(0.68\ \mu F)}$$
$$= 3.9\ k\Omega$$

The amplitude and phase angle of v_x are found to be

$$V_x = V_{in}\frac{|X_C|}{\sqrt{X_C^2 + R_2}} \qquad \alpha = -\tan^{-1}\left(\frac{R}{|X_C|}\right)$$
$$= 170\frac{3.9}{\sqrt{3.9^2 + 6.8^2}} \qquad = -\tan^{-1}\left(\frac{6.8}{3.9}\right)$$
$$= 85\ V \qquad = -60°$$

Using this information, the instantaneous value of v_x is given by $v_x = 85 \sin(2\pi 60t - 60°)$ V.

We find the firing time delay by setting $v_x = V_{BO}$ for the diode and solving for Δt. Note: 1 rad $\cong$ 57.3°.

$$\Delta t = \frac{\sin^{-1}\left(\frac{V_{BO}}{V_X}\right) + |\alpha|}{2\pi f} \tag{19.10}$$
$$= \frac{\sin^{-1}\left(\frac{30}{85}\right) + 1.05\ \text{rad}}{(2\pi)(60)}$$
$$= 3.74\ \text{ms}$$

Similar to the previous example, the numerator of Eq. (19.10) is the delay angle ϕ, which is

$$\phi = \sin^{-1}\left(\frac{V_{BO}}{V_X}\right) + |\alpha| \tag{19.11}$$

$$= \sin^{-1}\left(\frac{30}{85}\right) + 60°$$

$$= 20.7° + 60°$$

$$= 80.7°$$

The conduction angle is now found to be

$$\theta = 180° - \phi$$
$$= 180° - 80.7°$$
$$= 99.3°$$

19.2 TRIACS AND DIACS

One disadvantage associated with half-wave SCR phase control is that the maximum conduction angle is only 180°. If we connect two SCRs in inverse parallel, as shown in Fig. 19.9(a), it is possible to use Q_1 to control conduction during the positive half-cycle of v_{in} and Q_2 to control conduction during the negative half-cycle. In this configuration, a positive gate pulse triggers Q_1 while a negative gate pulse triggers Q_2. We can also place two four-layer diodes in inverse parallel as in Fig. 19.9(b) to obtain a bidirectional triggering diode.

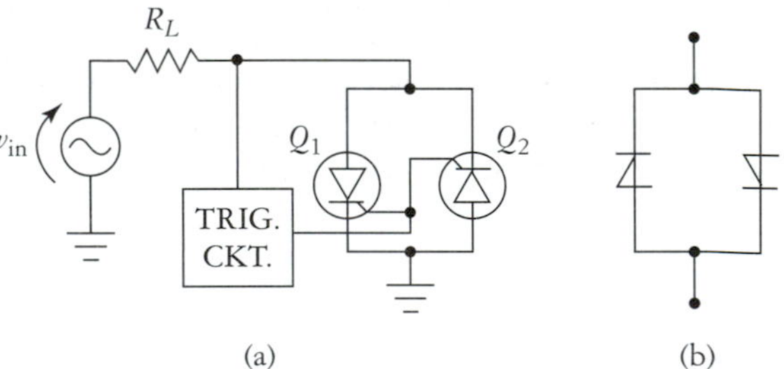

FIGURE 19.9
(a) Inverse-parallel connected SCRs.
(b) Four-layer diodes.

A triac is a three-terminal thyristor that is functionally equivalent to two inverse-parallel SCRs. The schematic symbol for a triac and a typical characteristic curve are shown in Figs. 19.10(a) and (b). The device that is functionally equivalent to two inverse-parallel four-layer diodes is called a *diac.* Two commonly used symbols for the diac and a representative characteristic curve are shown in Figs. 19.10(c) and (d). Diacs are also sometimes referred to as bidirectional trigger diodes.

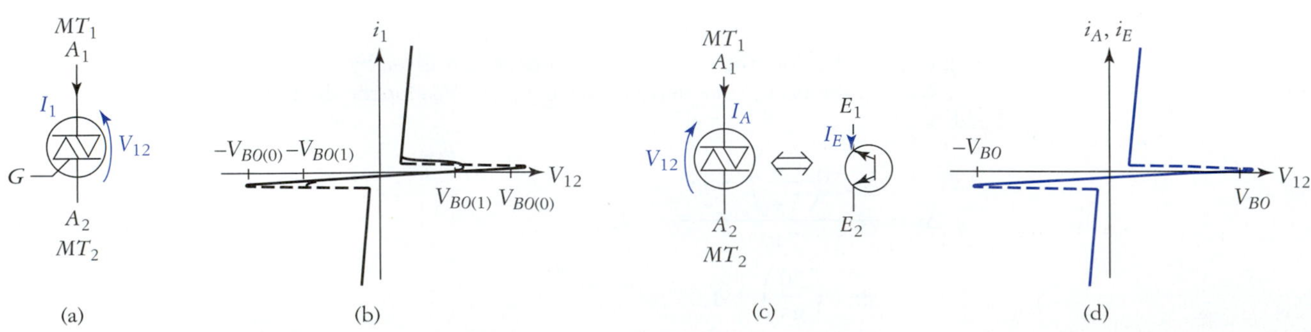

FIGURE 19.10
Triacs and diacs. (a) Triac symbol. (b) Triac transfer curves. (c) Diac symbols. (d) Diac transfer curves.

Triac Phase Control

Triacs and diacs are commonly used in applications similar to those of SCRs and four-layer diodes. A typical simple phase control circuit is shown in Fig. 19.11(a). The waveforms in Figs. 19.11(b), (c), and (d) represent those that might be present under low, medium, and high conduction angle conditions. Because the triac is a bidirectional (or bilateral), the conduction angle of the circuit is adjustable over the range

$$0° \leq \theta < 360° \tag{19.12}$$

Notice that the conduction angle is defined as the number of degrees of load current conduction over one cycle of the line voltage. Thus, for $\theta = 180°$, for example, we have 90° of conduction during the positive half cycle and 90° during the negative-going half of a given cycle of the ac line voltage. This is shown in waveform (c) of Fig. 19.11.

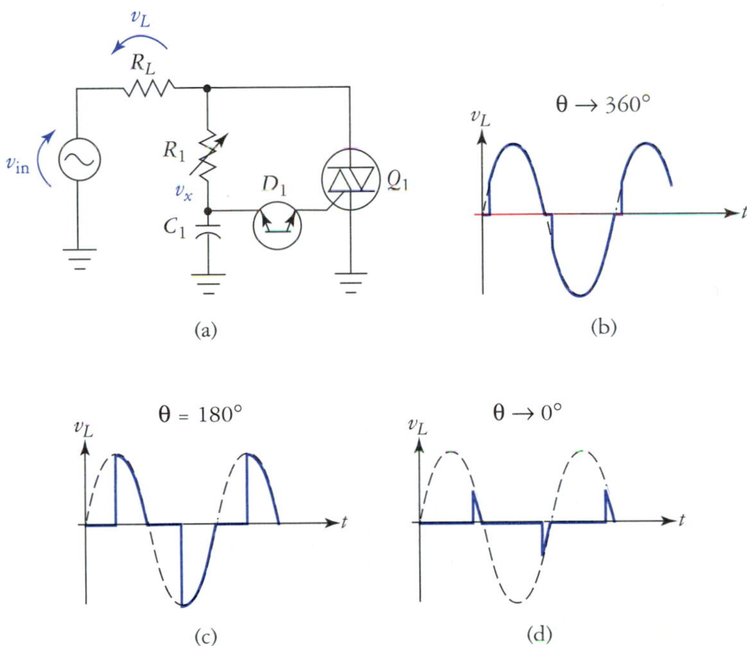

FIGURE 19.11
Simple triac phase control and representative load voltage waveforms.

The analysis of the triac phase control circuit is almost identical to the SCR circuit of Fig. 19.8. The next example illustrates this fact.

EXAMPLE 19.4

Refer to Fig. 19.11. Given an input voltage of 120 V_{rms}, 60 Hz, $R_1 = 6.8\ k\Omega$, $C_1 = 0.68\ \mu F$, and the diac has $V_{BO} = \pm 30$ V, determine the following:
(a) The expression for the instantaneous value of v_x.
(b) The triac firing time delay on each half-cycle.
(c) The conduction angle.

Solution Again, we first determine $|X_C|$, which is 3.9 kΩ. The amplitude and phase angle of v_x are found to be

$$v_x = V_{in}\frac{|X_C|}{\sqrt{X_C^2 + R_2}} \qquad \alpha = -\tan^{-1}\left(\frac{R}{X_C}\right)$$

$$= 170\frac{3.9}{\sqrt{3.9^2 + 6.8^2}} \qquad = -\tan^{-1}\left(\frac{6.8}{3.9}\right)$$

$$= 85 \text{ V} \qquad = -60° = 1.05 \text{ rad}$$

from which the instantaneous value of v_x is $v_x = 85\sin(2\pi 60t - 60°)$ V.

We find the firing time delay by using Eq. (19.10), setting $v_x = V_{BO}$ for the diac and solving for Δt:

$$\Delta t = \frac{\sin^{-1}\left(\frac{V_{BO}}{V_x}\right) + |\alpha|}{2\pi f}$$

$$= \frac{\sin^{-1}\left(\frac{30}{85}\right) + 1.05}{(2\pi)(60)}$$

$$= 3.74 \text{ ms}$$

Because we are using a triac, there is a delay of 3.74 ms after both positive- and negative-going zero crossings of the line voltage. The resulting load voltage is shown in Fig. 19.12.

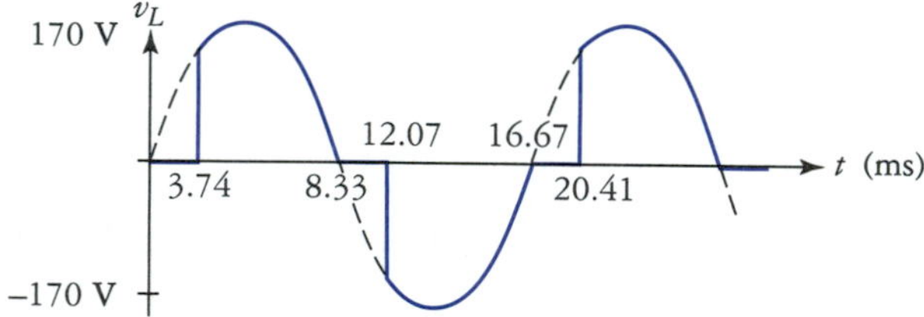

FIGURE 19.12
Waveform for Example 19.4.

We can still define the delay angle ϕ with Eq. (19.11), which yields

$$\phi = \sin^{-1}\left(\frac{V_{BO}}{V_x}\right) + |\alpha|$$

$$= \sin^{-1}\left(\frac{30}{85}\right) + 60°$$

$$= 20.7° + 60° = 80.7°$$

Now, because the triac conducts on both half-cycles, the conduction angle θ is found using

$$\theta = 360° - \phi$$
$$= 360° - 80.7° = 279.3°$$

Filters and Snubbers

The switching action of a thyristor may generate a large amount of electrical noise well into the radio-frequency range. This noise will often contain a large amount of energy because thyristors are usually used to control high-power loads. One way to help reduce the emission of noise from circuits that employ thyristors is to use shielded enclosures. In addition, the thyristor control circuit may include a filter such as that shown in Fig. 19.13. Such filtering helps to prevent switching noise from being transferred onto the ac line.

Because thyristors are often designed to handle large currents, they also tend to have a large junction area, which in turn results in high internal parasitic capacitances. In an industrial setting, the switching of heavy loads such as large motors can cause transients or rapid changes in line voltage. Such transients can cause significant charging currents to flow in thyristor interjunction capacitances, causing false triggering. For the Motorola 2N5060 through 2N5064 SCRs, the maximum allowable rate of change of the applied voltage is

$$\left.\frac{dV_{AK}}{dt}\right|_{\max} = 30 \text{ V}/\mu\text{s}$$

A line transient that exceeds this rate of change will cause the SCR to trigger.

To help prevent this *dv/dt* triggering, an *RC* filter, called a *snubber,* can be connected across the SCR as shown in Fig. 19.14. The snubber provides a relatively low-impedance path for high *dv/dt* transients, preventing false triggering of the thyristor.

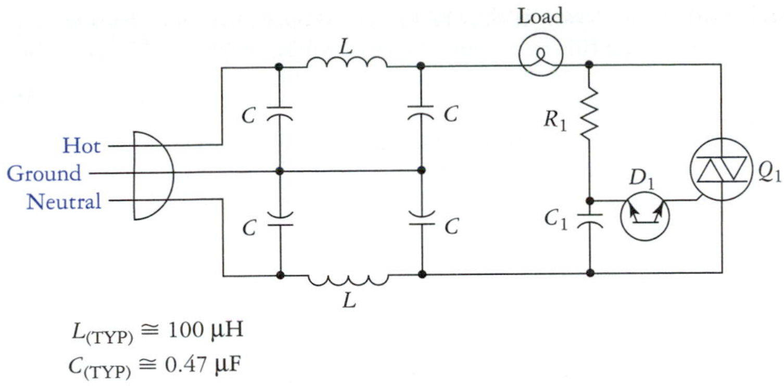

$L_{(TYP)} \cong 100\ \mu H$
$C_{(TYP)} \cong 0.47\ \mu F$

FIGURE 19.13
Filter prevents SCR switching noise from feeding into ac line.

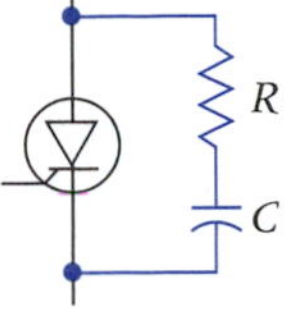

FIGURE 19.14
SCR snubber network.

19.3 UNIJUNCTION TRANSISTORS

Unijunction transistors (UJTs) are specialized three-terminal devices that are most often used to implement relaxation oscillator circuits. Such UJT oscillators are frequently used to provide triggering pulses in thyristor power control circuits.

The internal structure, schematic symbol, emitter characteristic curve, and equivalent internal model are shown in Fig. 19.15. As you can see, the term *unijunction* derives from the fact that there is a single pn junction in the device. Examination of the characteristic curve in Fig. 19.15(c) indicates that the interbase resistance r_{BB_2} decreases when the emitter current exceeds the peak current I_P, which corresponds with the peak voltage V_P. Once I_E exceeds I_P, interbase resistance r_{BB_1} decreases with further increases in emitter current until $i_E > I_V$, the valley current.

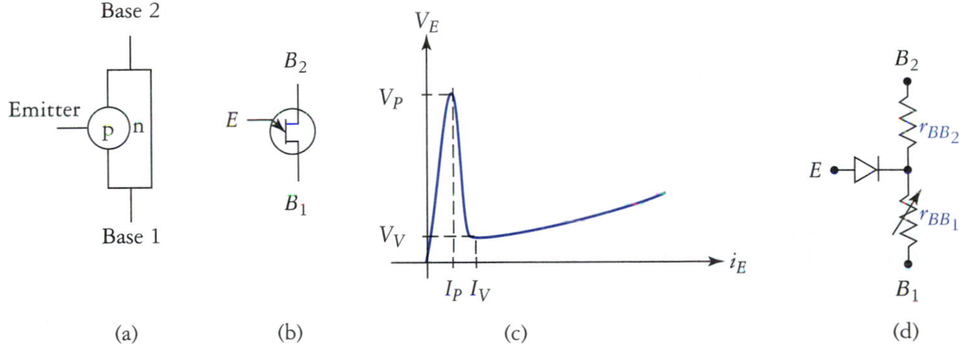

FIGURE 19.15
Unijunction transistor. (a) Internal structure. (b) Symbol. (c) Transfer characteristic. (d) Equivalent internal model.

For the interval $0 < I_E < I_P$, the UJT is considered to be off, and the total interbase resistance $r_{BB(\text{off})}$ will typically be around 10 kΩ. In general, we can express interbase resistance as

$$r_{BB} = r_{BB_1} + r_{BB_2} \tag{19.13}$$

The interbase on-resistance $r_{BB(\text{on})}$ typically drops to a few thousand ohms at the valley point, when the UJT is turned on. It turns out that the ratio of r_{BB_1} to r_{BB} is a constant. This is called the *intrinsic standoff ratio* η. In equation form this is written

$$\eta = \frac{r_{BB_1}}{r_{BB}} \tag{19.14}$$

Typically, $0.4 < \eta < 0.8$, $1\text{ V} \le V_V \le 3\text{ V}$, and $4\text{ mA} \le I_V \le 8\text{ mA}$.

A typical UJT relaxation oscillator is shown in Fig. 19.16. For proper operation, resistors R_2 and R_3 are small, relative to $r_{BB(off)}$. Assuming this to be true, the peak voltage of the UJT is given by

$$V_P = V_\phi + \eta V_{BB} \tag{19.15}$$

where V_ϕ is the barrier potential of the pn junction (usually silicon). As can be seen in the emitter voltage waveform, once capacitor C_1 charges to V_P, r_{BB_1} decreases, allowing the capacitor to rapidly discharge to V_V. This causes the circuit to produce the waveforms shown in the figure. The minimum value of the V_{B_2} waveform and the maximum value of the V_{B_1} waveform are dependent on many factors that are beyond the scope of this discussion.

FIGURE 19.16
UJT relaxation oscillator.

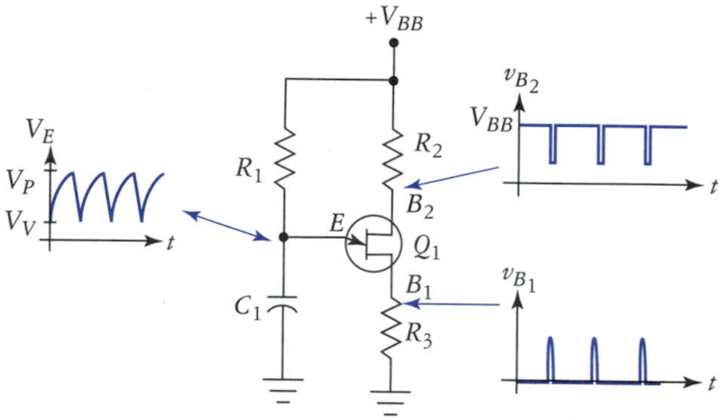

An interesting property of Eq. 19.15 is that for $\eta V_{BB} \gg V_\phi$, which is common, the UJT peak voltage is the same proportion of the supply voltage V_{BB}. This means that unlike the four-layer diode relaxation oscillator, the UJT relaxation oscillator has a frequency that is nearly independent of supply voltage variation. The oscillation frequency of the UJT relaxation oscillator is approximated by

$$f \cong \frac{1}{R_1 C_1 \ln\left(\frac{1}{1-\eta}\right)} \tag{19.16}$$

EXAMPLE 19.5

The circuit in Fig. 19.16 has $R_1 = 10\text{ k}\Omega$, $C_1 = 0.22\ \mu\text{F}$, $R_2 = 680\ \Omega$, $R_3 = 47\Omega$, $V_{BB} = 12$ V, and a silicon UJT with $\eta = 0.65$ and $V_V = 2$ V. Determine the oscillation frequency of the circuit, the peak voltage of the UJT, and sketch the emitter voltage waveform.

Solution

$$f \cong \frac{1}{R_1 C_1 \ln\left(\frac{1}{1-\eta}\right)}$$

$$= \frac{1}{(10\text{ k}\Omega)(0.22\ \mu\text{F})\left(\frac{1}{1-0.65}\right)}$$

$$= 433\text{ Hz}$$

The peak voltage of the UJT is

$$\begin{aligned} V_P &= V_\phi + \eta V_{BB} \\ &= 0.7\text{ V} + (0.65)(12\text{ V}) \\ &= 8.5\text{ V} \end{aligned}$$

The emitter voltage waveform is shown in Fig. 19.17.

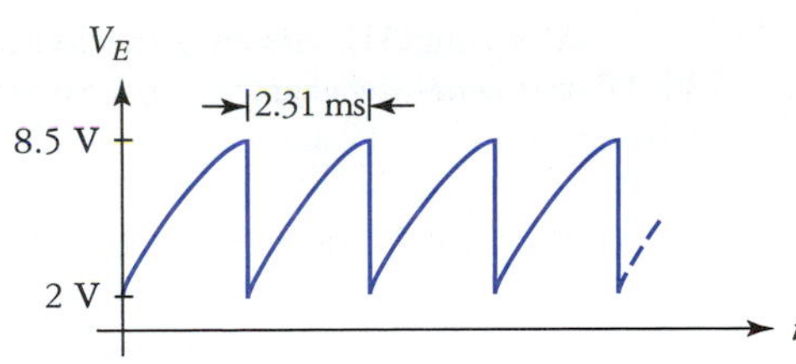

FIGURE 19.17
Waveform for Example 19.5.

As stated earlier, UJTs are often used to trigger thyristors. A typical circuit is shown in Fig. 19.18. The full-wave rectifier provides pulsating dc, which is clipped at 20 V by the zener diode. The 10-kΩ resistor and 33-μF capacitor provide a constant 20-V_{dc} level for the UJT timing network R_1 and C_1. The UJT fires at a point in a given half-cycle that depends on the setting of R_1. This induces a pulse that is coupled through T_1, firing the triac. The pulsating dc voltage applied to the UJT causes the oscillation to be synchronized with the ac line. If this was not the case, say, if the output of the bridge was filtered, the UJT would oscillate asynchronously, producing triggering of the triac at random times.

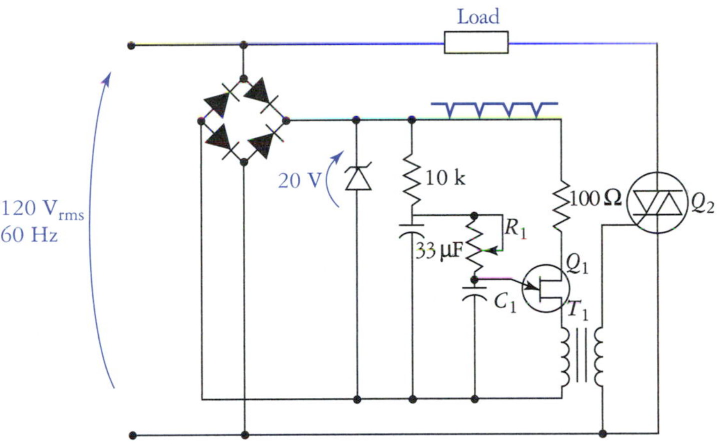

FIGURE 19.18
Using a UJT to control triac triggering.

19.4 TUNNEL DIODES

Tunnel diodes are pn junction diodes that rely on extremely heavy doping in order to achieve negative resistance characteristics. The schematic symbol and typical transconductance curve for a tunnel diode are shown in Fig. 19.19. Negative resistance is a result of the fact that the slope of the transconductance curve is negative for $V_P < V_{AK} < V_V$. This is not really a new phenomenon. All of the devices covered in this chapter exhibit negative resistance. In fact, even bipolar transistors can exhibit negative resistance characteristics when they go into C-E breakdown. The difference here is that tunnel diodes behave in this manner because of quantum mechanical tunnelling. See Chapter 2 for a brief discussion of this phenomenon.

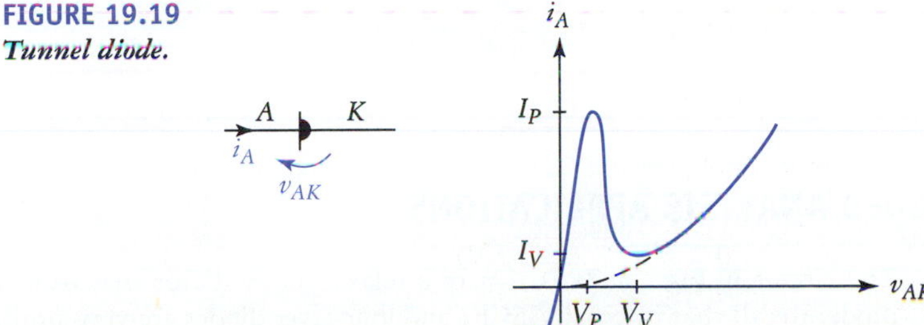

FIGURE 19.19
Tunnel diode.

Most tunnel diodes are either germanium or gallium-arsenide devices. The dashed line in Fig. 19.19 represents the characteristic curve of a normal germanium diode. Typical values for the points labeled in Fig. 19.19 are

$$V_P = 0.1\text{V}$$
$$V_V = 0.35\text{ V}$$
$$I_P = 2\text{ mA}$$
$$I_V = 0.12\text{ mA}$$

The negative resistance of the typical tunnel diode will be around −100 to −500 ohms.

Tunnel diodes are most often used to implement low-power, high-frequency sinusoidal oscillators, such as that shown in Fig. 19.20. Frequencies of more than 10 GHz are attainable using tunnel diode oscillators. The basic idea behind the tunnel diode oscillator is to use the negative resistance of the tunnel diode to cancel the positive resistance associated with the resonant circuit, thereby reducing the damping coefficient of the circuit to zero. This is equivalent to a circuit with infinite Q. The oscillation frequency is found using the familiar equation

$$f_o = \frac{1}{2\pi\sqrt{R_1C_1}} \tag{19.17}$$

Resistors R_1 and R_2 are adjusted such that the Q point of the diode is located in the negative-resistance region. That is,

$$V_P < V_{AKQ} < V_V \tag{19.18}$$

Once the Q point is established, via adjustment of R_2, electrical noise that is invariably present will cause the circuit to begin oscillation. Tunnel diode oscillators are inherently low-power circuits, with power output on the order of 1 mW or less.

FIGURE 19.20
Simple tunnel diode oscillator.

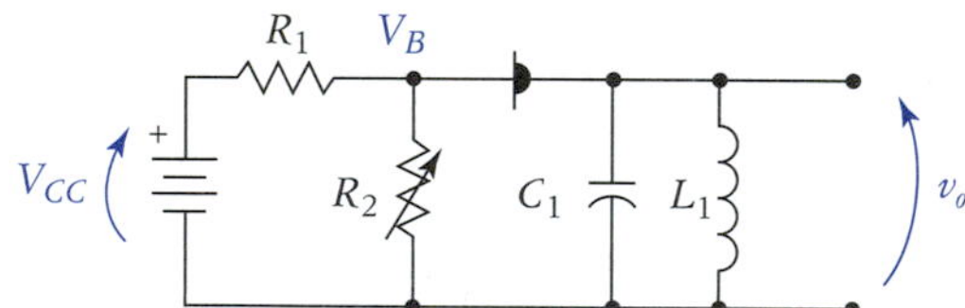

EXAMPLE 19.6

Refer to Fig. 19.20. Given $L_1 = 0.5\ \mu\text{H}$ and $C_1 = 1\text{ pF}$, determine the oscillation frequency of the circuit.

Solution

$$f_o = \frac{1}{2\pi\sqrt{R_1C_1}}$$
$$= \frac{1}{2\pi\sqrt{0.5\ \mu\text{H} \times 1\text{ pF}}}$$
$$= 225.1\text{ MHz}$$

19.5 COMPUTER-AIDED ANALYSIS APPLICATIONS

The circuit in Fig. 19.21 is that of a relaxation oscillator that uses an SCR in place of a four-layer diode. Recall that internally, SCRs and four-layer diodes are very similar and the devices have similar transconductance characteristics. Thus, if we short the gate of the SCR to the cathode, we effectively have a four-layer diode.

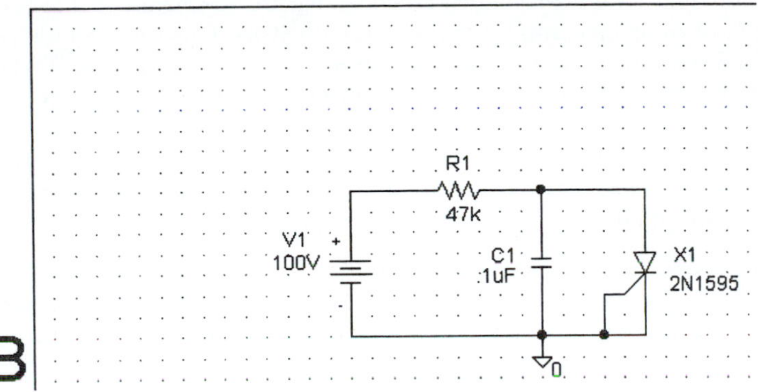

FIGURE 19.21
Using an SCR as a four-layer diode.

```
.SUBCKT 2N1595-X          anode gate cathode

* "Typical" parameters
X1 anode gate cathode Scr PARAMS:
+ Vdrm=50v  Vrrm=50v   Ih=5ma      Vtm=1.1v    Itm=1
+ dVdt=1e9  Igt=2ma    Vgt=.7v     Ton=0.8u    Toff=10u
+ Idrm=10u
* 90-5-18   Morotola    DL137, Rev 2, 3/89
.ENDS
*$
```

FIGURE 19.22
The SCR parameters.

The attributes for the 2N1595 are shown in Fig. 19.22. Here, we find that $V_{BO} = 50$ V. Most SCRs have very low on-state saturation voltage, so we may assume that the effective valley voltage for this device will be $V_V \cong 0$ V. Applying this information to Eq. (19.1) we find

$$
\begin{aligned}
T &= R_1C_1 \ln\left(\frac{V_{\text{in}} - V_{\text{sat}}}{V_{\text{in}} - V_{BO}}\right) \\
&= (4.7\ \text{k}\Omega)(0.1\ \mu\text{F}) \ln\left(\frac{100}{100 - 50}\right) \\
&= (4.7\ \text{ms})\,(0.693) \\
&= 3.3\ \text{ms} \\
f &= 1/T \\
&= 1/3.3\ \text{ms} \\
&= 303\ \text{Hz}
\end{aligned}
$$

The output voltage resulting from the simulation is shown in Fig. 19.23. The period of the signal is found to be about 3.5 ms, which agrees well with the predicted value.

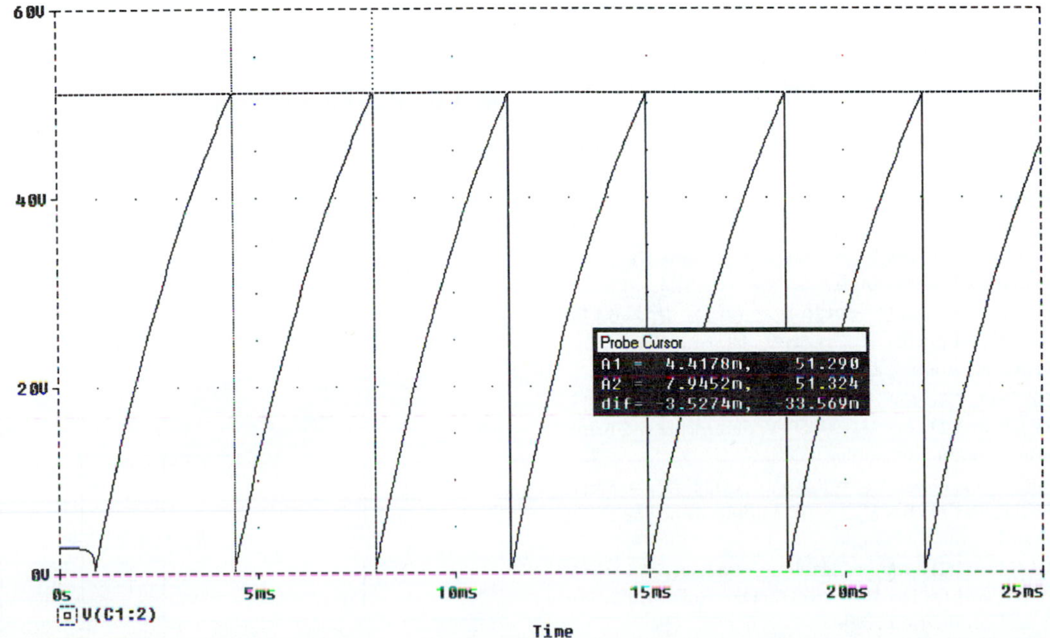

FIGURE 19.23
Relaxation oscillator output voltage waveform.

CHAPTER REVIEW

Thyristors are a family of nonlinear, four-layer semiconductor devices that are commonly used to perform switching functions. The most common examples of thyristors are SCRs, four-layer (Shockley) diodes, triacs, and diacs. SCRs and triacs are most often used to control the power dissipation of relatively large loads such as heaters, motors, and lamps. Diacs and triacs are most often used in SCR and triac triggering circuits.

Unijunction transistors are nonlinear three-terminal devices that are used to implement relaxation oscillators and to produce trigger pulses for thyristor circuits. Tunnel diodes are negative-resistance devices that are most often used to produce low-power, high-frequency sinusoidal oscillators.

DESIGN AND ANALYSIS PROBLEMS

19.1. Refer to Fig. 19.2(a). The SCR has $I_G = 100\ \mu A$, which results in $V_{BO} = 25$ V. If the anode supply voltage is varied such that $v_1 = 500t\ u(t)$ V, how long will it take for the SCR to fire? Remember that multiplication by $u(t)$ simply means that the input voltage is

$$v_1 = \begin{cases} 0\text{ V} & \text{for } t < 0 \\ 500t\text{ V} & \text{for } t \geq 0 \end{cases}$$

19.2. Refer to Fig. 19.2(a). The SCR has $I_G = 200\ \mu A$, which results in $V_{BO} = 20$ V. The anode supply voltage is $v_1 = (30 \sin 377t)\ u(t)$ V. Determine the SCR firing time delay. Again, multiplication by $u(t)$ simply means that the input voltage is

$$v_1 = \begin{cases} 0\text{ V} & \text{for } t < 0 \\ 30 \sin 377t\text{ V} & \text{for } t \geq 0 \end{cases}$$

19.3. Refer to Fig. 19.6. The circuit had $V_{in} = 50$ V, $R_1 = 15$ kΩ, $C_1 = 0.033\ \mu F$, $V_{BO} = 25$ V, and $V_{sat} = 10$ V. Determine the oscillation frequency.

19.4. Refer to Fig. 19.6. The circuit has $V_{in} = 35$ V, $R_1 = 5$ kΩ, $C_1 = 0.1\ \mu F$, $V_{BO} = 20$ V, and $V_{sat} = 5$ V. Determine the oscillation frequency.

19.5. Solve Eq. (19.1) for V_{in}.

19.6. Solve Eq. (19.1) for V_{BO}.

19.7. Refer to Fig. 19.24. Given $V_{in} = 100$ V, $R_1 = 10$ kΩ, $R_2 = 10$ kΩ, $C_1 = 0.01\ \mu F$, $V_{BO} = 20$ V, and $V_{sat} = 10$ V, determine the oscillation frequency.

19.8. Refer to Fig. 19.24. Given $V_{in} = 100$ V, $R_1 = 20$ kΩ, $R_2 = 30$ kΩ, $C_1 = 10\ \mu F$, $V_{BO} = 38$ V, and $V_{sat} = 20$ V, determine the oscillation frequency.

19.9. Refer to Fig. 19.7. Given $V_{in} = 120\ V_{rms}$, 60 Hz, $R_1 = 5$ kΩ, $R_2 = 1$ kΩ, and D_1 has $V_{BO} = 25$ V, determine the firing delay time and conduction angle.

19.10. Refer to Fig. 19.7. Given $V_{in} = 120\ V_{rms}$, 60 Hz, $R_1 = 10$ kΩ, $R_2 = 4.7$ kΩ, and D_1 has $V_{BO} = 30$ V, determine the firing delay time and conduction angle.

19.11. Refer to Fig. 19.8. Given $V_{in} = 120\ V_{rms}$, 60 Hz, $R_1 = 5$ kΩ, $C_1 = 0.33\ \mu F$, and D_1 has $V_{BO} = 50$ V, determine the firing delay time and conduction angle.

19.12. Refer to Fig. 19.8. Given $V_{in} = 120\ V_{rms}$, 60 Hz, $R_1 = 3.9$ kΩ, $C_1 = 0.68\ \mu F$, and D_1 has $V_{BO} = 60$ V, determine the firing delay time and conduction angle.

19.13. Refer to Fig. 19.11. Given $V_{in} = 120\ V_{rms}$, 60 Hz, $R_1 = 5$ kΩ, $C_1 = 0.47\ \mu F$, and D_1 has $V_{BO} = 50$ V, determine the firing delay time and conduction angle.

19.14. Refer to Fig. 19.11. Given $V_{in} = 120\ V_{rms}$, 60 Hz, $R_1 = 3.9$ kΩ, $C_1 = 0.68\ \mu F$, and D_1 has $V_{BO} = 60$ V. Determine the firing delay time and conduction angle.

19.15. Refer to Fig. 19.16. The circuit has $R_1 = 22$ kΩ, $C_1 = 0.022\ \mu F$, $\eta = 0.5$, and $V_{BB} = 15$ V. Determine V_P and the oscillation frequency of the circuit.

19.16. Refer to Fig. 19.16. The circuit has $R_1 = 10$ kΩ, $C_1 = 0.1\ \mu F$, $\eta = 0.7$, and $V_{BB} = 15$ V. Determine V_P and the oscillation frequency of the circuit.

19.17. Solve Eq. (19.16) for η.

19.18. Determine the value of η that reduces Eq. (19.16) to

$$f = \frac{1}{R_1 C_1}$$

19.19. Refer to Fig. 19.20. Given $C_1 = 1$ pF, determine L_1 such that $f = 100$ MHz.

19.20. Refer to Fig. 19.20. Given $L_1 = 2\ \mu H$, determine C_1 such that $f = 120$ MHz.

TROUBLESHOOTING PROBLEMS

19.21. A relaxation oscillator like that in Fig. 19.6 has stopped working. The main symptom is that the output voltage simply rises to V_{in} when power is applied. What is most likely to be wrong with the four-layer diode?

19.22. In a certain SCR control circuit, the SCR triggers into conduction occasionally when it shouldn't. The SCR is tested and found to be good, and even if a replacement SCR is used, the same problem occurs. What might be a cause for this problem and how might it be solved?

19.23. Refer to Fig. 19.18. Explain why it could be dangerous to attempt to view the voltage waveform across the 33-μF capacitor with a normal line-operated oscilloscope.

COMPUTER PROBLEMS

19.24. Use the FFT function to plot the spectrum of the waveform in Fig. 19.23.

19.25. Examine the SCR parameters in Fig. 19.22. Determine the maximum rate of change of V_{AK} and the holding current for this device.

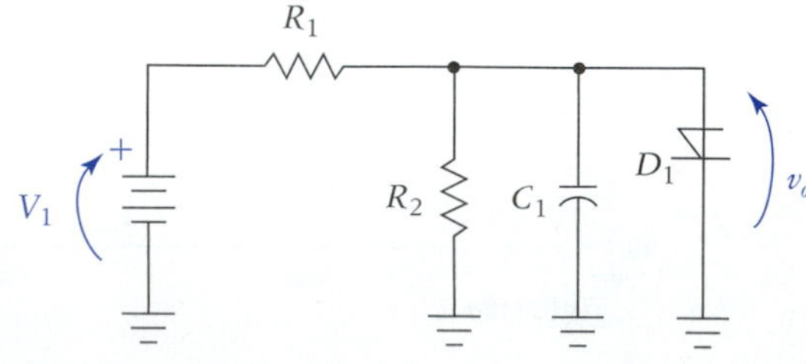

FIGURE 19.24
Circuit for Problems 19.7 and 19.8.

APPENDIX 1

Solutions to Odd-Numbered Problems

Chapter 1

1.1. $R_S = 1.43\ \text{k}\Omega$

1.3. $\eta = 1{:}I_F = 14.4$ A, $\eta = 2$, $I_F = 54\ \mu\text{A}$

1.5. Approximately 0.72 V

1.7. Approximately 0.72 V

1.9. $I_D = 6.25$ mA

1.11. $I_D = 6.2$ mA

1.13. Assume $V_\phi = 0.7$ V, $I_D = 19$ mA

1.15. $I_{DQ} = 4.06$ mA, $V_{DQ} = 0.7$ V, $r_D = 6.4\ \Omega$, $i_D = 4.06 + 0.039 \sin \omega t$ mA, $V_D = 700 + 0.25 \sin \omega t$ mV

1.17. $R_S = 3.3\ \text{k}\Omega$

1.19. $I = 0.9\ I_{Zm}$: $R_S = 108\ \Omega$, $I = 0.5\ I_{Zm}$: $R_S = 186\ \Omega$

1.21. $V_Z = 5$ V, $I_{Zm} = 160$ mA, $I_{Zk} = 1$ mA, $R_L = 150\ \Omega$

1.23. $R = 82\ \Omega$, $P_{R_1} = 820$ mW, $I_S = 147.1$ mA, $P_Z = 569$ mW

1.25. $V_o = 10$ V

1.27. $C_{j(\text{max})} = 15$ pF, $f_{o(\text{min})} = 6.04$ MHz, $C_{j(\text{min})} = 2.7$ pF, $f_{o(\text{max})} = 14.13$ MHz

1.29.

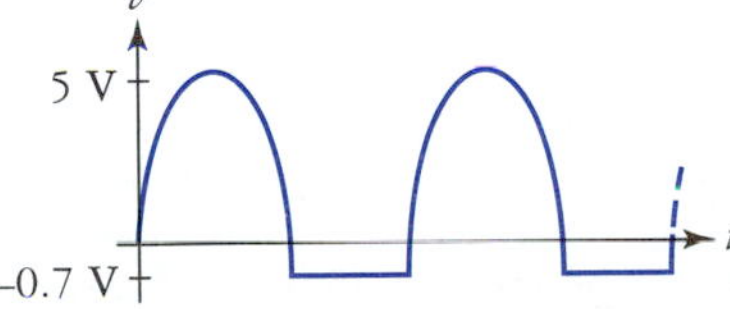

1.31.

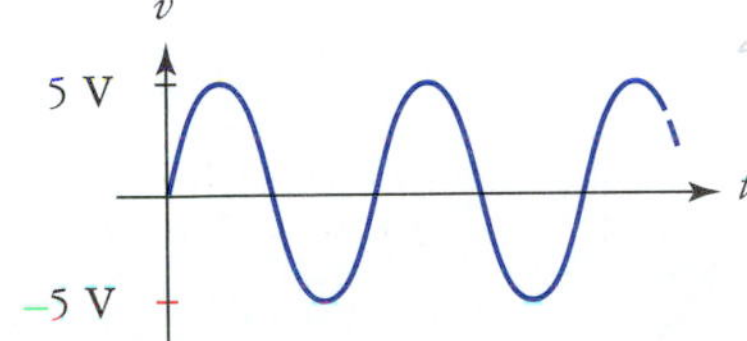

1.33. No. A good Si diode should have reverse resistance $\gg 20$ MΩ.

Chapter 2

2.1. $V_{o(\text{av})} = 1.37$ V, $I_{o(\text{av})} = 13.7$ mA

2.3. $n_S/n_p = 8.7$

2.5. $V_{S(\text{rms})} = 6$ V, PIV = 8.5 V

2.7. $f_{\text{rip}} = 120$ Hz

2.9. $V_{r(\text{p-p})} = 226$ mV, $r = 0.52\%$, PIV = 28.3 V

2.11. $R_L = 477\ \Omega$

2.13. $C_1 \geq 315\ \mu\text{F}$, $v_o = 9.3 + 10 \sin 500t$ V, $i_L = 23.3 + 25 \sin 500t$ mA

2.15.

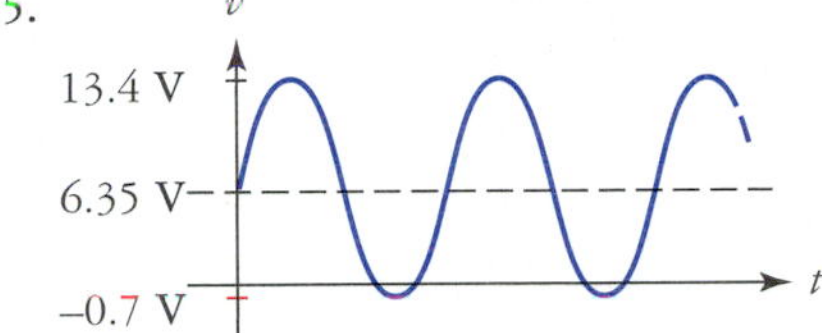

2.17. $I_{\text{surge}} = 12.75$ A

2.19. $V_{r(\text{p-p})} = 32$ mV, $r = 0.3\%$

2.21. $V_{L(\text{dc})} = V_{2(\text{dc})} = 9.8$ V, $V_{r_1(\text{p-p})} = 0.38$ V, $V_{r_2(\text{p-p})} = 152\ \mu\text{V}$, $r = 0.0016\%$

2.23. PIV $\cong V_o = 140$ V

2.25. $f_{\min} = 455$ kHz

2.27.

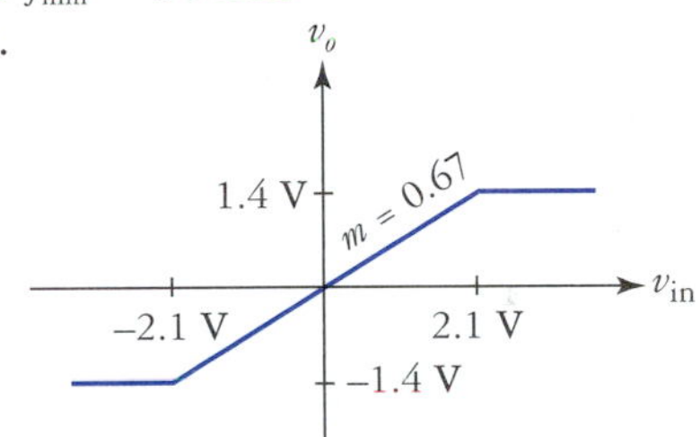

2.29.

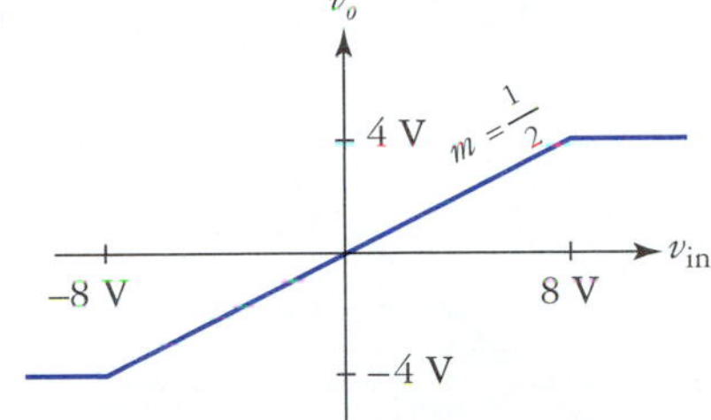

2.31. $R_S = 55.6\ \Omega$, $R_{L(\min)} = 113.6\ \Omega$, $P_{R_S} = 0.45$ W, $r = 0.036\%$, VR = 7.5%

2.33. See *Solutions Manual*.

2.35.

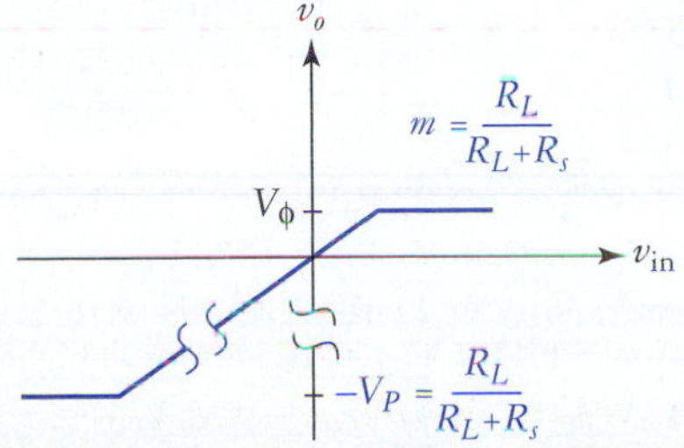

2.37. Approx. 5.2 V

2.39. The ground or the connection to the center tap is open.

2.41. Most likely, the zener diode is in backward.

Chapter 3

3.1. (a) Active (b) Active (c) Saturation (d) Cutoff (e) Active (f) Active

3.3. $I_C = 7.96$ mA, I_B 36.2 μA

3.5. $I_{CEO} = 3$ μA, $I_C = 753$ μA

3.7. $i_B = 100$ μA, $i_C = 1.9$ mA, $r_e = 13\ \Omega$

3.9. $g_m = 0.5$ S

3.11. $V_A = 30$ V, $r_o = 30$ kΩ, $\beta = 110$, $\alpha = 0.991$

3.13. $r_\pi = 663\ \Omega$

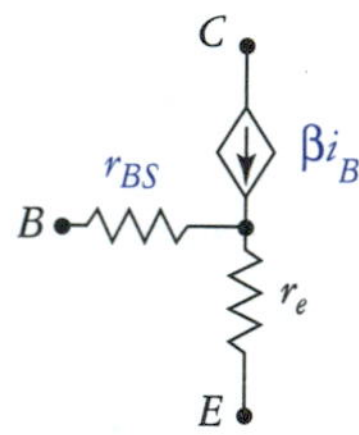

3.15. For $V_{CE} = 10$ V:
$10\ \mu A \le I_B \le 20\ \mu A$: $\beta_{ac} = 240$,
$20\ \mu A \le I_B \le 30\ \mu A$: $\beta_{ac} = 220$
$30\ \mu A \le I_B \le 40\ \mu A$: $\beta_{ac} = 213$
$40\ \mu A \le I_B \le 50\ \mu A$: $\beta_{ac} = 215$

3.17. $r_o \cong 50$ kΩ

3.19. Actual β is twice the estimated value.

3.21. (1) High (2) Low (3) Low (4) Low

Chapter 4

4.1. $V_{BB} = 4.45$ V

4.3. $R_C = 1.67$ kΩ, $R_B = 45$ kΩ

4.5. $R_B = 86$ kΩ

4.7. $V_{CEQ} = -4.1$ V, $I_{CQ} = 2.4$ mA, $I_{C(sat)} = 3.6$ mA, $V_{CE(cut)} = -12$ V, $V_{CQ} = 7.9$ V, $V_{BQ} = 11.7$ V, $V_{EQ} = 12$ V

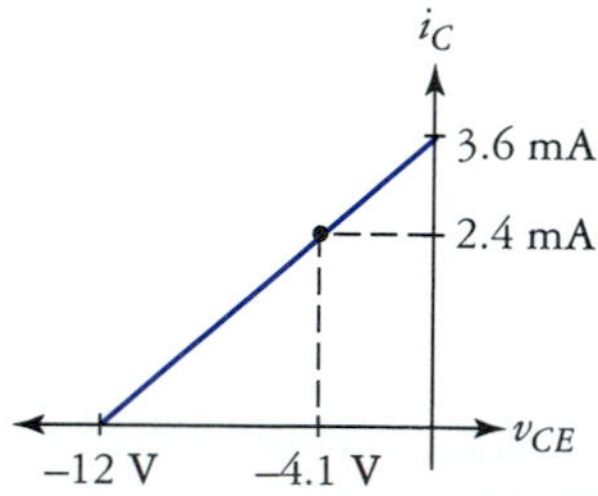

4.9. $I_{CQ} = 3.8$ mA, $V_{CEQ} = -8.4$ V, $V_{CQ} = -9.4$ V, $V_{CB} = -7.7$ V

4.11. $V_{CC} \cong 13.13$ V

4.13. $R_C \cong 1.8$ kΩ

4.15. $R_F = 188.3$ kΩ

4.17. $I_{CQ} = 973$ μA, $V_{CEQ} = -5.27$ V, $V_{CQ} = 9.73$ V, $V_{BQ} = 14.3$ V

4.19. $I_{CQ} = 4$ mA, $V_{CEQ} = 8.56$ V, $V_{CQ} = 12$ V, $V_{BQ} = 4.14$ V

4.21. $I_{CQ} = 2.25$ mA, $V_{CEQ} = 11.18$ V, $V_{CQ} = 9.75$ V, $V_{BQ} = -0.73$ V, $V_{EQ} = -1.43$ V, $V_{CB} = 10.48$ V

4.23. $I_{CQ} = 2$ mA, $V_{CEQ} = 13$ V, $V_{CQ} = 0$ V, $V_{BQ} = -12.3$ V, $V_{EQ} = -33$ V, $V_{CB} = 12.3$ V

4.25. $I_{CQ} \cong V_{Th}/[(R_{Th}/\beta) + R_E]$, where $V_{Th} = V_{CC}\,[R_2/(R_1 + R_2)]$

4.27. $R_1 = 4.66$ kΩ, $R_2 = 2.4$ kΩ

4.29. See *Solutions Manual.*

4.31. With the zener diode open, $I_C = 0$ A. Since $I_E = I_C + I_B$, $I_E = 50$ μA.

4.33. The transistor is in the active region, but very close to saturation.

4.35. Toward saturation.

Chapter 5

5.1. $R_{in} = 1.6$ kΩ, $A_v = -17.2$, $A_i = -27.5$, $A_p = 473$

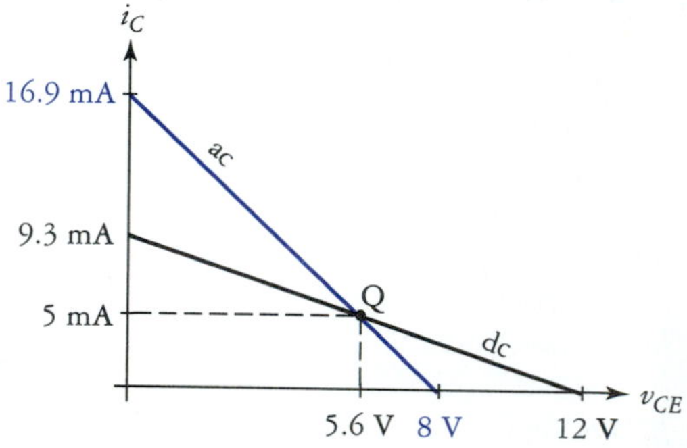

5.3. $i_{in} = 15.6 \sin \omega t$ μA, $i_C = 5 - 0.953 \sin \omega t$ mA, $i_L = -0.43 \sin \omega t$ mA

5.5.

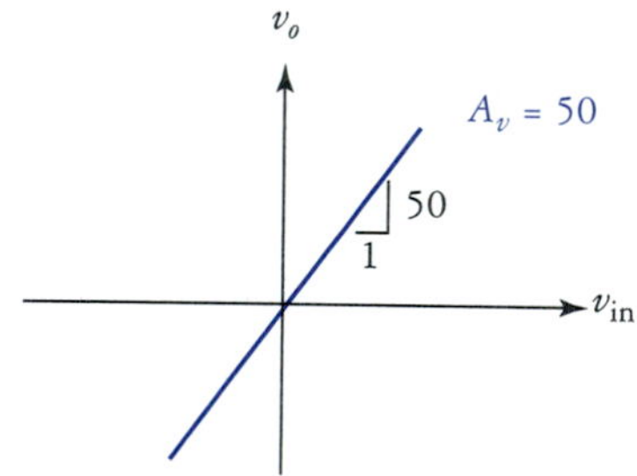

5.7. $V_{o(max)} = 3.3$ V, $V_{o(min)} = -4$ V

5.9. $I_{CQ} = 7.4$ mA, $V_{CEQ} = 10.2$ V, $R_{in} = 1.3$ kΩ

5.11. $I_{CQ} = 2$ mA, $V_{CEQ} = 9.8$ V, $I_{C(sat)} = 3$ mA, $V_{CE(cut)} = 30$ V, $R_{in} = 151$ kΩ, $A_v = -34$, $A_i = -1556$, $A_p = 52.9$ k

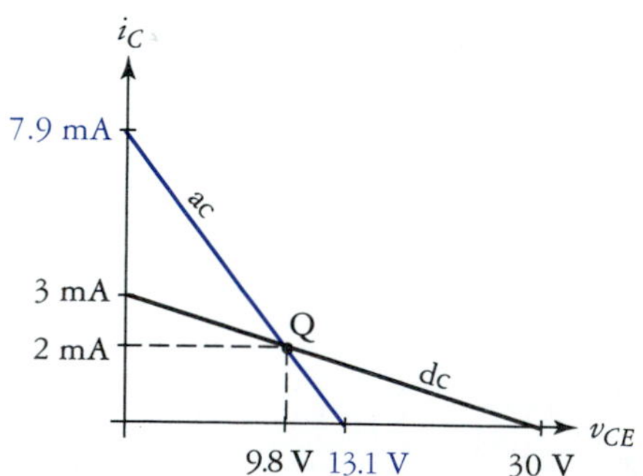

5.13. $I_{CQ} = 1.1$ mA, $V_{CEQ} = 12$ V, $R_{in} = 1.3$ kΩ, $A_v = -119$, $A_i = -55$, $A_p = 6545$, $V_{o(max)} = 3.1$ V, $V_{o(min)} = -12$ V

5.15. $I_{CQ} = 3.9$ mA, $V_{CEQ} = 6.4$, $R_{in} = 459\ \Omega$, $A_v = -102$, $A_i = -47$, $A_p = 4794$

5.17. $I_{CQ} = 2.7$ mA, $V_{CEQ} = 6.1$, $R_{in} = 951\ \Omega$, $A_v = -115$, $A_i = -50$, $A_p = 5750$

5.19. $0.8 \sin \omega t$ V

5.21. $I_{CQ} = 1.1$ mA, $V_{CEQ} = 18.4$ V, $R_{in} = 1.5$ kΩ, $A_v = 0.8$, $A_i = 12$, $A_p = 9.6$, $R_o = 23\ \Omega$

5.23. $I_{CQ} = 2.9$ mA, $V_{CEQ} = 18$ V, $R_{in} = 52.8$ kΩ, $A_v = 0.9$, $A_i = 216$, $A_p = 194$, $R_0 = 17.5$ Ω

5.25. $I_{CQ} = 5.1$ mA, $V_{CEQ} = 9.3$ V, $R_{in} = 5.6$ Ω, $A_v = 83.6$, $A_i = 0.3$, $A_p = 25$, $V_{CQ} = 8.5$ V

5.27. See *Solutions Manual.*

5.29. The transistor would saturate.

Chapter 6

6.1. $I_{DQ} = 7.1$ mA, $V_{DSQ} = 8$ V, $V_{GSQ} = -0.85$ V, $V_{DQ} = 8.9$ V

6.3. $R_S = 62$ Ω, $R_D = 1.6$ kΩ

6.5. $I_{GSS} = 5$ nA

6.7. $I_{DQ} \cong 12$ mA, $V_{DSQ} = 12.7$ V, $V_{GSQ} = -0.35$ V, $V_{DQ} = 15.4$ V, $V_{SQ} = 2.6$ V

6.9. $R_S = 760$ Ω, $R_D = 2.2$ kΩ

6.11. $I_{DQ} \cong 1.5$ mA, $V_{DSQ} = 16.4$ V, $V_{GSQ} = -1.8$ V, $V_{DQ} = 8.2$ V, $V_{SQ} = -8.2$ V

6.13. $I_{DQ} \cong 9.9$ mA, $V_{DSQ} = -8.1$ V, $V_{GSQ} = 1.2$ V, $V_{DQ} = 6.7$ V, $V_{SQ} = 14.8$ V

6.15. $I_T = 8$ mA

6.17. $I_{DQ} = I_{DSS} = 10$ mA, $V_{DSQ} = 10$ V, $V_{GSQ} = 0$ V

6.19. $R_D = 1$ kΩ

6.21. $I_{DQ} \cong 6.2$ mA, $V_{DSQ} = 7.1$ V, $V_{DQ} = 8.8$ V, $V_{SQ} = 1.7$ V

6.23. $I_{DQ} \cong 3.6$ mA, $V_{DSQ} = 11.1$ V, $V_{DQ} = 11.9$ V, $V_{SQ} = 0.8$ V

6.25. $I_{DQ} \cong 3.4$ mA, $V_{DSQ} = 5.5$ V, $V_{DQ} = 5.8$ V, $V_{SQ} = 0.34$ V

6.27. $I_{DQ} \cong 15.1$ mA, $V_{DSQ} = 7.45$ V, $V_{GSQ} = 4.47$ V

6.29. $I_{DQ} = I_{CQ} = 5$ mA, $V_{DSQ} = 9.12$ V, $V_{GSQ} = -0.88$ V, $V_{CEQ} = 15.88$ V, $V_{DQ} = 10$ V

6.31. $V_{GS} \cong -0.6$ V

6.33. $I_{GSS} = 180$ pA

Chapter 7

7.1. $I_{DQ} \cong 6.5$ mA, $V_{DSQ} = 9.6$ V, $g_{m0} = 4$ mS, $V_{GSQ} = -1$ V, $g_{mQ} = 3.2$ mS, $A_v = -1.3$, $R_{in} = 1$ MΩ, $A_i = -1300$, $A_p = 1690$, $R_o \cong 680$ Ω

7.3. $v_o = -130 \sin \omega t$ mV, $i_D = 6.5 - 0.32 \sin \omega t$ mA, $i_L = -0.13 \sin \omega t$ mA, $\omega = 2\pi 1000$ rad/s

7.5. $V_P = 3.2$ V, $I_{DQ} \cong 6.7$ mA, $V_{DSQ} = 7.4$ V, $V_{GSQ} = -0.85$ V, $g_{mQ} = 4.4$ mS, $A_v = -1.97$, $R_{in} = 1$ MΩ, $A_i = -1970$, $A_p = 3881$, $R_{in} = 1$ MΩ

7.7. $g_{m0} = 6$ mS, $I_{DQ} \cong 9.9$ mA, $V_{DSQ} = 7.7$ V, $V_{GSQ} = -0.98$ V, $V_{GQ} = 1.2$ V, $g_{mQ} = 4.8$ mS, $A_v = -3.4$, $R_{in} = 2$ MΩ, $A_i = -1447$, $A_p = 4920$, $R_o = 820$ Ω

7.9. $v_o/v_s = -3.3$

7.11. $g_{m0(Q1)} = 12$ mS, $I_{D(Q1)} = I_{D(Q2)} = I_{DSS(2)} = 9$ mA, $g_{m0(Q2)} = 7.2$ mS, $V_{GS(Q1)} = -0.9$ V, $g_{m(Q1)} = 8.4$ mS, $R_{in} = 4.7$ MΩ, $R_o = 560$ Ω, $A_v = -3$, $A_i = 14{,}100$, $A_p = 42{,}300$

7.13. $g_{m0} = 8.3$ mS, $I_{DQ} \cong 14.5$ mA, $V_{GSQ} = -1.5$ V, $g_{mQ} = 6.2$ mS, $A_v = 0.24$, $R_{in} = 1$ MΩ, $A_i = 10{,}000$, $A_p = 2400$, $R_o = 62$ Ω

7.15 $V_p = 5$ V, $I_{DQ} \cong 8.1$, $V_{GSQ} = -1.5$ V, $V_{DSQ} = 9.7$ V, $g_{mQ} = 4.2$ mS, $A_v = 1$, $R_{in} = 103$ Ω, $A_i = 0.21$, $A_p = 0.21$, $R_o = 470$ Ω

7.17. $I_{DQ} \cong 20$, $V_{DSQ} = 7.6$ V, $g_{mQ} = 9.5$ mS, $A_v = -1.7$, $R_{in} = 1$ MΩ, $A_i = -1700$, $A_p = 2890$, $R_o = 220$ Ω

7.19. $I_{DQ} \cong 11.5$, $V_{GSQ} = -0.8$ V, $g_{m0} = 9.5$ mS, $g_{mQ} = 7.5$ mS CS: $A_v = -6.8$, $R_{in} = 1$ MΩ, $A_i = -680$, $A_p = 4624$ SF: $A_v = 0.18$, $R_{in} = 1$ MΩ, $A_i = 3600$, $A_p = 648$

7.21 $V_{GSQ} = 6.19$ V, $k = 2.5 \times 10^{-3}$, $I_{DQ} = 25.4$ mA, $V_{DSQ} = 9.9$ V, $g_{mQ} = 16$ mS, $R_{in} = 1.9$ MΩ, $R_o = 200$ Ω, $A_v = -2.3$, $A_i = -8740$, $A_p = 20{,}102$

7.23. $k = 3.75 \times 10^{-3}$. For $I_{DQ} = 19$ mA, $V_{GSQ} = 4.75$ V; therefore, $R_2 = 3.3$ MΩ (std. value). For $V_{DSQ} = 11$ V, $R_3 = 579$ Ω (use 560 Ω). $R'_D = 256$ Ω, $g_{mQ} = 16.9$ mS, $A_v = -4.3$, $R_{in} = 2.6$ MΩ, $A_i = -23{,}787$, $A_p = 102{,}284$, $R_o = 560$ Ω.

7.25. $k = 5.6 \times 10^{-3}$, $V_{GQ} = 4.8$ V, $I_{DQ} = 3$ mA, $V_{DSQ} = 14.6$ V, $V_{GSQ} = 4.4$ V, $g_{mQ} = 6.7$ mS, $R_{in} = 3.2$ MΩ, $R_o = 67$ Ω, $A_v = 0.29$, $A_i = -7{,}733$, $A_p = 2243$

7.27. $k = 1.9 \times 10^{-3}$, $V_{GSQ} = 5.8$ V, $I_{DQ} = 6.2$ mA, $V_{DSQ} = 5.8$ V, $g_{mQ} = 6.8$ mS, $R_{in} = 227$ kΩ, $A_v = -3.4$, $A_i = -772$, $A_p = 2625$

7.29. $k = 4.63 \times 10^{-3}$, $I_{DQ} = I_{CQ} = 4.8$ mA, $V_{DSQ} = 11.5$ V, $V_{GSQ} = 5.25$ V, $g_{mQ} = 9.7$ mS, $V_{CEQ} = 3.75$ V, $A_v = -5.8$, $A_i = -4833$, $A_p = 28{,}031$

7.31. A bad FET or capacitor C_1 is shorted.

7.33. With source bypassing there should be no signal voltage dropped across R_S. Apparently, C_3 is open or disconnected, causing a large reduction in voltage gain.

Chapter 8

8.1. $I_{EE} = 3$ mA, $I_{C_1} = I_{C_2} = 1.5$ mA, $V_{CE_1} = V_{CE_2} = 10.8$ V, $A_d = 190$, $A_v = 95$, $R_{in} = 347$ Ω, $A_{cm} = -0.35$, CMRR = 54.7 dB, $V_{od} = -9.5$V

8.3. $\Delta V_{C_1} = -1.75$ V

8.5. $I_{EE} = 2.7$ mA, $I_{C_1} = I_{C_2} = 1.35$ mA, $V_{CE_1} = V_{CE_2} = 7.6$ V, $V_{CE_3} = 11.8$ V, $A_d = 311$, $A_v = 155$, $R_{in} = 3.9$ kΩ, $A_{cm} = -0.08$, CMRR = 71.8 dB, $r_{o(Q_3)} = 37$ kΩ, $V_{od} = 3.11$ V

8.7. $R_3 = 1.13R_4 = (1.13)(4.7$ kΩ$) = 5.3$ kΩ: $R_3 = 5.3$ k $\therefore R_4 = 4.7$ kΩ

8.11. $I_{EE} = 500$ μA, $I_{C_1} = I_{C_2} = 250$ μA, $V_{CE_1} = 10.7$ V, $V_{CE_2} = 5.7$ V, $V_{CE_3} = 9.3$, $V_{oQ} = 5$ V, $A_v = 96$, $R_{in} = 21.8$ kΩ, $A_{cm} = -0.05$, $V_o = 3.85$

8.13. $V_o = 4.8$ V

8.15. $I_{EE} = 4.5$ mA, $I_{C_1} = I_{C_2} = 2.25$ mA, $V_{CE_1} = V_{CE_2} = 0.89$ V, $V_{CE_3} = 8.27$, $V_{icm} = -0.23$ V, $I_z = 11.5$ mA, $A_d = 431$, $A_v = 216$, $R_{in} = 2.24$ kΩ, $A_{cm} = -0.025$

8.17 $R_6 = 683$ Ω, $R_5 = 850$ Ω, $R_3 = R_4 = 2.9$ kΩ

8.19. $I_{EE} = 2$ mA, $I_{C_1} = I_{C_2} = 1$ mA, $V_{CE_1} = V_{CE_2} = 8.7$ V, $V_{CE_3} = 10.5$ V, $A_d = 31.7$, $A_v = 15.9$, $R_{in} = 25.2$ kΩ

8.21. $I_{EE} = 200$ μA, $I_{C_1} = I_{C_2} = 100$ μA, $V_{CE_1} = V_{CE_2} = 5.7$ V, $V_{CE_3} = 14.25$ V, $A_d = 385$, $A_v = 193$, $R_{in} = 52$ kΩ, $A_{cm} = 0.1$, CMRR = 71.7 dB

8.23. $R_3 = 11.3$ kΩ, $R_4 = 508.6$ Ω

8.25. $I_{EE} = 2$ mA, $I_{C_1} = I_{C_2} = 1$ mA, $V_{CE_1} = V_{CE_2} = 10.4$ V, $V_{CE_3} = 13.6$ V, $A_d = 115$, $A_v = 58$, $R_{in} = 1.04$ MΩ, $V_2 = -29.8$ mV

8.27. $I_{SS} = 5$ mA, $I_{D_1} = I_{D_2} = 2.5$ mA, $V_{DS_1} = V_{DS_2} = 8.8$ V, $V_{DS_3} = 15.7$ V, $g_{mQ} = 2.8$ mS

8.29. See *Solutions Manual.*

8.31. $I_L = 3$ mA

8.33. $I_{EE} = 4$ mA, $I_{C_1} = I_{C_2} = 2$ mA, $V_{CE_1} = V_{CE_2} = -9.7$ V, $V_{CE_3} = -14.3$ V, $V_{C_1} = V_{C_2} = -9$ V, $A_d = 231$, $A_v = 116$, $R_{in} = 2.6$ kΩ

8.35. Opening Q_4 would cause saturation of Q_1, Q_2 and Q_3, thus $V_{C_1} = V_{C_2} \cong 0.2$ V.

Chapter 9

9.1. $I_{C_1} = 947\ \mu A$, $V_{CE_1} = 2.83$ V, $I_{C_2} = 2.1$ mA, $V_{CE_2} = 4.2$ V, $A_{v_2} = -121$, $A_{v_1} = -6.8$, $A_v = 823 = 58.3$ dB, $R_{in} = 3.7\ k\Omega$, $R_o = 2.7\ k\Omega$, $A_i = 923$, $A_p = 760\ k\Omega$

9.3. $v_o = 1.65 \sin(2\pi 1000t)$ V.

9.5. $I_{C_1} = 4.7$ mA, $V_{CE_1} = 8.9$ V, $I_{C_2} = 8.5$ mA, $V_{CE_2} = 10.7$ V, $A_{v_2} = 0.99$, $A_{v_1} = -86.2$, $A_v = -85.3 = 38.6$ dB, $A_i = -44.4$, $A_p = 3827$, $R_{in} = 521\ \Omega$, $R_o = 7.8\ \Omega$

9.7. $I_{C_1} = 5$ mA, $V_{CE_1} = 4.9$ V, $I_{C_2} = 4.1$ mA, $V_{CE_2} = 5.1$ V, $A_{v_2} = -286$, $A_{v_1} = -3.2$, $R_{in} = 27\ k\Omega$, $A_v = 915 = 59.2$ dB, $A_i = 2471$, $A_p = 2.26 \times 10^6$

9.11. $I_{C_1} = 1.3$ mA, $I_{C_2} = 1.9$ mA, $I_{C_3} = 3.6$ mA, $I_{C_4} = 9.5$ mA, $V_{CE_1} = 4.4$ V, $V_{CE_2} = 6.3$ V, $V_{CE_3} = 6.1$ V, $V_{CE_4} = 9.1$ V, $A_{v_1} = -44$, $A_{v_2} = -6.7$, $A_{v_3} = -11.9$, $A_{v_4} = 0.75$, $R_{in} = 101\ k\Omega$, $R_o = 19\ \Omega$, $A_V = -2631 = -68.4$ dB, $A_i = -33.2 \times 10^6$, $A_p = 87.3 \times 10^9 = 109.4$ dB

Note: The difference between decibel power gains is due to the assumption that $R_L = R_{in}$ for $A_{v(dB)} = 20 \log A_v$.

9.13. $I_{C_1} = 1.9$ mA, $V_{CE_1} = 7.1$ V, $I_{C_2} = 3.2$ mA, $V_{CE_2} = 2.6$ V, $V_{B_1} = -0.19$ V, $V_{oQ} = 8.16$ V

9.15. $I_{C_1} = I_{C_2} = 2.1$ mA, $V_{CE_2} = 13.4$ V, $V_{CE_1} = 15.9$ V, $I_{C_3} = 5.2$ mA, $A_v = -0.37$

9.17. $I_{C_4} = 500\ \mu A$, $I_{C_1} = I_{C_2} = 250\ \mu A$, $V_{CE_1} = V_{CE_2} = 7.7$ V, $V_{CE_3} = 13.6$ V, $I_{C_4} = I_{C_5} = 1.25$ mA, $V_{CE_4} = 8.7$ V, $V_{CE_5} = 4.7$ V, $V_{CE_6} = 21.3$ V, $I_{C_8} = I_{C_9} = 4.8$ mA, $R_7 = 2.15 k\Omega$, $A_d = 2722$, $= 69$ dB, $R_{in} = 20.8\ k\Omega$, $R_o = 2.19\ k\Omega$

9.19. See *Solutions Manual.*

9.21. $I_{C_3} = 2.8$ mA, $I_{C_1} = I_{C_2} = 1.4$ mA, $V_{CE_1} = V_{CE_2} = 11.1$ V, $V_{CE_3} = 12.1$ V, $V_1 \cong 0$ V, $V_2 = 2.1$ V, $V_3 = -15$ V, $V_4 = -13.6$ V, $V_5 = 0$ V, $V_6 = 10.4$ V, $V_7 = 2.1$ V, $V_8 = 10.4$ V, $v_o/v_s = 0.88$

9.23. $I_{C_3} = 1$ mA, $I_{C_1} = I_{C_2} = 500\ \mu A$, $V_{CE_1} = V_{CE_2} = 5.7$ V, $V_{CE_3} = 12$ V, $I_{C_6} = 2$ mA, $V_{CE_6} = 15.3$ V, $I_{C_4} = I_{C_5} = 1$ mA, $V_{CE_4} = 10.7$ V, $V_{CE_5} = 7$ V, $I_{C_7} = I_{C_8} = 3$ mA, $V_{CE_7} = 4.5$ V, $V_{CE_8} = 11.7$ V, $V_{oQ} = 0$ V, $A_d = 5688$, $R_{in} = 10.4\ k\Omega$, $R_o = 3.5\ k\Omega$

9.25. V_{CE_1} would decrease, which in turn increases V_{B_2}, causing Q_2 to saturate.

9.27. Low R_5 would cause Q_1 to move closer to cutoff, which in turn would drive Q_2 closer to saturation.

Chapter 10

10.1. $I_{CQ} = 8.3$ mA, $V_{CEQ} = 4.8$ V, $P_{DQ} = 40$ mW, $i_{C(sat)} = 21$ mA, $v_{CE(cut)} = 7.9$ V, $V_{o(max)} = 3.1$ V, $V_{o(min)} = -7.9$ V, $A_v = -375$, $P_{o(max)} = 19.2$ mW, $P_{o(ave)} = 9.6$ mW, $\eta = 12\%$

10.3. $\Delta T_J = 3°C$

10.5. $V_{CEQ} = 1.7$ V, $R_E = 626\ \Omega$, $R_C = 2.7\ k\Omega$, $R_x = 507\ \Omega$, $A_V = -1.4$, $R_{in} = 3.7\ k\Omega$, $A_i = -11$, $A_p = 15.4$, $V_{o(max)} = 1$ V, $V_{o(min)} = -1$ V, $P_{DQ} = 4.25$ mW, $P_{o(max)} = 2.1$ mW, $P_{o(ave)} = 1.05$ mW, $\eta = 25\%$

10.7. $\theta_{SA} \leq 10.5°C/W$

10.9. $R_1 = 4\ k\Omega$, $R_2 = 1\ k\Omega$, $R_3 = 1075\ \Omega$

10.11. $R_{3(min)} = 20.6\ \Omega$

10.13. $R_I = 1.89\ k\Omega$

10.15. $I_{CQ} = 4.6$ mA, $V_{CEQ} = 9.4$ V, $I_{C(sat)} = 21.4$ mA, $V_{CE(cut)} = 12$ V, $R'_C = 1.25\ k\Omega$, $R_{in} = 424\ \Omega$, $A_v = 50$, $A_i = 424$, $A_p = 21.2$ k, $i_{C(sat)} = 12.1$ mA, $v_{CE(cut)} = 15.2$ V, $V_{o(max)} = 1.2$ V, $V_{o(min)} = -1.9$ V, $P_{o(max)} = 72.2$ mW, $P_{o(ave)} = 36.1$ mW

10.17. $I_{CQ} = 5.4$ mA, $I_{C(sat)} = 1.5$ A, $R_{in} = 307\ \Omega$, $R_o = 5.2\ \Omega$, $A_v = 0.94$, $A_i = 36$, $A_p = 33.8$, $P_{o(max)} = 18$ W, $P_{o(ave)} = 9$ W

10.19. $I_{CQ} \cong 16$ mA, $I_{C(sat)} \cong 2.4$ A, $R_{in} = 257\ \Omega$, $R_o = 4\ \Omega$, $A_v = 0.89$, $A_i = 28.6$, $A_p = 25.4$, $P_{o(max)} = 46$ W, $P_{o(ave)} = 23$ W

10.21. $I_{CQ} \cong 29.6$ mA, $I_{Sc} = 3.18$ A ($I_{C(sat)} = 3$ A, with $R_L = 8\Omega$), $R_{in} = 211\ \Omega$, $R_o = 2.3\ \Omega$, $A_v = 0.92$, $A_i = 24$, $A_p = 22$, $P_{o(max)} = 72$ W, $P_{o(ave)} = 36$ W, $f_{C(in)} = 51$ Hz, $f_{C(out)} = 4.7$ Hz

10.23. $I_{C_1} = 8.6$ mA

10.25. $I_{C_3} = 1$ mA, $I_{C_1} = I_{C_2} = 500\ \mu A$, $V_{CE_1} = V_{CE_2} = 7.7$ V, $V_{CE_3} = 13.6$ V, $I_{C_4} = I_{C_5} = 2$ mA, $V_{CE_4} = 8.4$ V, $V_{CE_5} = 14.6$ V, $I_{C_6} = I_{C_7} = 2.9$ mA, $A_d = 154$, $R_{in} = 10.4\ k\Omega$, $R_o = 19.2\ \Omega$, V_1 = noninverting, V_2 = inverting

10.27. $C = 0.8\ \mu F$

10.29. $R_1 \geq 909.1\ \Omega$

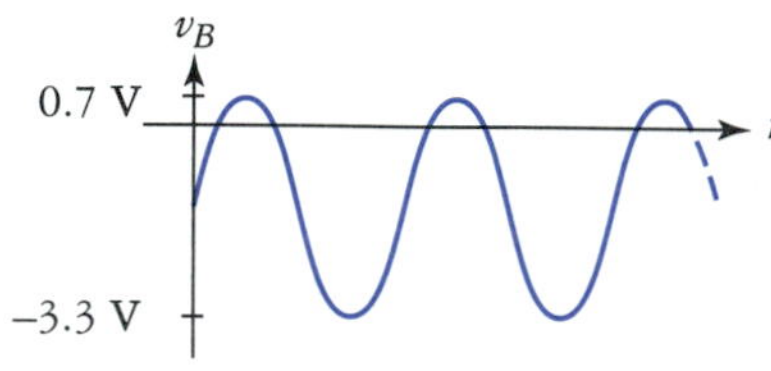

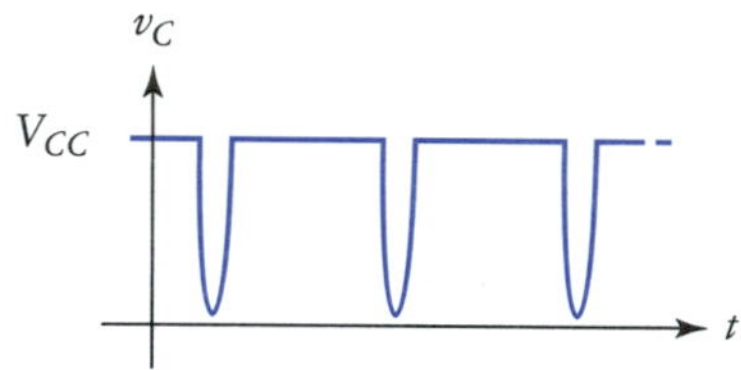

10.31. $P_{ave} = 187.5$ mW

10.33. $f_o = 969$ kHz, $Q_L = 51$, $Q = 3.4$

10.35. $C_2 = 862$ pF, $R_L = 9.3\ k\Omega$

10.37. No effect on I_{CQ} or V_{CEQ}.

10.39. There would be unwanted crossover distortion.

Chapter 11

11.1. BW = 999 kHz, $f_o = 31.62$ kHz

11.3. $A|_{f=50\ MHz} = 40$ dB, $A|_{f=500\ MHz} = 20$ dB

11.5. $A|_{f=400\ Hz} = 74$ dB, $A|_{f=800\ Hz} = 68$ dB, $f \cong 409.6$ kHz

11.7. High-pass, $f_c = 6.3$ Hz

11.9. $C \geq 6\ \mu F$

11.11. Low-pass, $f_c = 244$ kHz

11.13. $C_T = 8.8\ \mu F$

11.15. $I_{CQ} = 2$ mA, $V_{CEQ} = 7.2$ V, $A_v = -104$, $R_{in} = 867\ \Omega$, $f_{c_1} = 83.4$ Hz, $f_{c_2} = 260$ Hz, $f_{c_3} = 118$ Hz

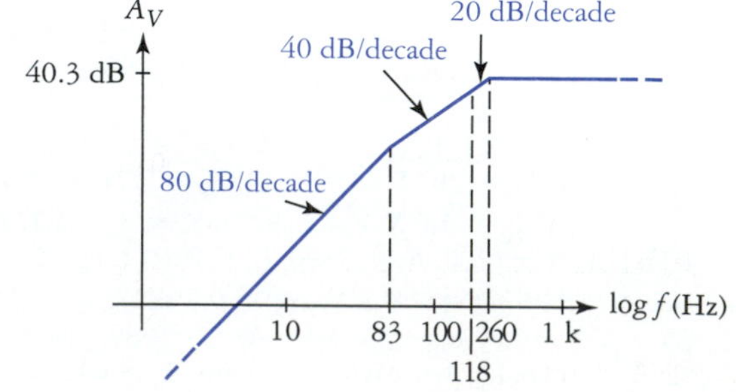

11.17. I_{DQ} = 6.5 mA, V_{DSQ} = 7.9 V, V_{GSQ} = −0.65 V, g_{m0} =6.7 mS, g_{mQ} = 5.2 mS, A_v = −2.6 = 8.3 dB, R_{in} = 1 MΩ, f_{c_1} = 1.6 Hz, f_{c_2} = 242 Hz, f_{c_3} = 318 Hz

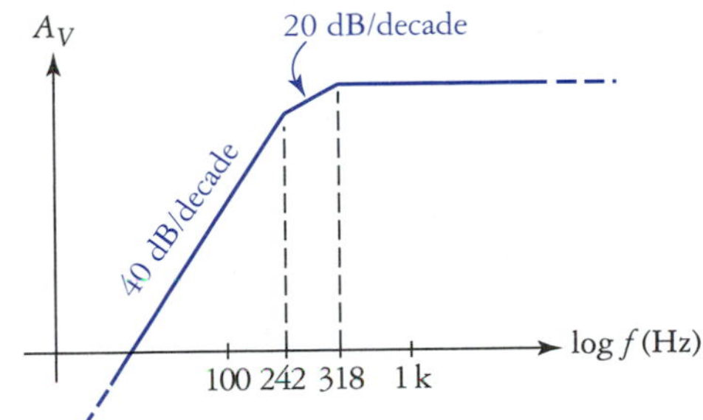

11.19. C'_{in} = 0.002 μF, C'_o = 10 pF

11.21. C'_{in} = 318 pF, C'_o = 5 pF, f_c = 5.6 MHz

11.23. f_c = 17.3 kHz

11.25. I_{CQ} = 902 μA, V_{CEQ} = 28 V, A_v = 0.77, R_{in} = 1.6 kΩ, R'_e = 96 Ω, C'_{in} = 16 pF, C'_o = 17 pF, $f_c = f_{c(out)}$ = 97.5 MHz

11.27. I_{CQ} = 2.1 mA, V_{CEQ} = 7.5 V, A_v = 226, R_{in} ≅ 12.4 Ω, R'_C = 2.8 kΩ, $f_{c(in)}$ = 7.4 GHz, $f_{c(out)}$ = 11.4 MHz, $f_c = f_{c(out)}$ = 11.4 MHz

11.29. A = 8

11.31. A_d = 16100, A_{v_1} = 48, A_{v_2} = 10

11.33. This would remove the emitter HP break frequency, most likely reducing f_L.

Chapter 12

12.1. A_v = 196

12.3. β = 0.0125

12.5. β = 0.167, A_v = 5.95, R_{inF} = 3.36 MΩ, R_o = 0.12 Ω

12.7. β = 0.1, A_v = −9.52, R_{in} = 1.5 kΩ

12.9. A_v = 6.45, i_{R_1} = 909.1 sin(2π1000t) μA, i_{RF} = 909.1 sin(2π1000t) μA, v_o = 12.9 sin(2π1000t) V, i_L = 12.9 sin(2π1000t) mA

12.11. A_v = −4.7, i_{R_1} = 2 sin(2π1000t) mA, i_F = 2 sin(2π1000t) μA, v_o = −sin(2π1000t) V, i_L = 12.9 sin(2π1000t) mA

12.13. BW = 644.4 kHz

12.15. v_{in} = 50 sin(2π1000t) mV

12.17. A_v = 25, BW_T = 32.2 kHz

12.19. v_{in} = −192 sin(2π1000t) mV

12.21. $|A|_{f=200\text{ kHz}}$ = 4.47 = 13 dB, $|A|_{f=300\text{ kHz}}$ = 3.16 = 10 dB

12.23. $|A|_{f=600\text{ kHz}}$ = 0.97 = −0.25 dB, $|A|_{f=1\text{ MHz}}$ = 0.63 = −3.98 dB

12.25. A_v = 74 dB

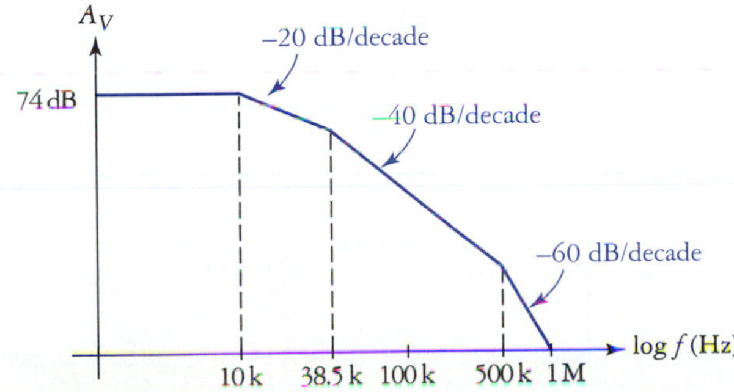

12.27. V_{oB} = −2 mV

12.29. BW_p = 2.39 MHz

12.31. SR = 31.4 V/μs

12.33. A_N = 19.3

12.35. R_{B_1} = 16.7 kΩ, R_y = 75 kΩ, R_x ≤ 18.7 kΩ

12.37. R_B = 909 Ω, R_z = 9 Ω, R_y = 67.8 kΩ, R_x ≤ 16.9 kΩ

12.39. The op amp will saturate because there is no dc return for input bias current.

Chapter 13

13.1. I_1 = 2.5 μA, I_2 = 1.5 μA, I_F = 4 μA, V_o = −0.4 V, R_B = 6.25 kΩ, I_L = 40 μA

13.3. I_1 = 200 μA, I_2 = −50 μA, I_F = 150 μA, V_o = −7.5 V, R_B = 5.9 kΩ, I_L = 750 μA

13.5. I_1 = 250 μA, i_2 = 106.4 sinωt μA, i_F = 250 + 106.4 sinωt μA, v_o = −5.5 − 2.34 sinωt V, i_L = −2.75 − 1.17 sinωt mA, i_o = −3 − 1.28 sinωt mA, R_B = 2.8 kΩ, $\left.\frac{dv_o}{dt}\right|_{max}$ = 8042 V/s

13.7. i_1 = 400 sin(2π1000t) μA = 400 ∠0° μA, i_2 = 300 sin(2π1000t + 30°) μA = 300∠30° μA, i_F = 677 sin(2π1000t + 13°) μA = 677 ∠13° μA, v_o = −6.77 sin(2π1000t + 13°) V = 6.77∠193°V, i_L = −677 sin(2π1000t + 13°) μA = 677 ∠193° μA, $\left.\frac{dv_o}{dt}\right|_{max}$ = 0.043 V/μs

13.9. I_1 = 300∠20° μA, I_2 = 300∠50° μA, I_3 = 100∠180° μA, I_F = 502∠42°, V_o = 5.02∠42°V

13.11. BW_{min} = 90.09 kHz, BW_{max} = 833.3 kHz

13.13. R_{max} = 1010 Ω, R_{min} = 10 Ω

13.15. 27.2 kΩ (×4)

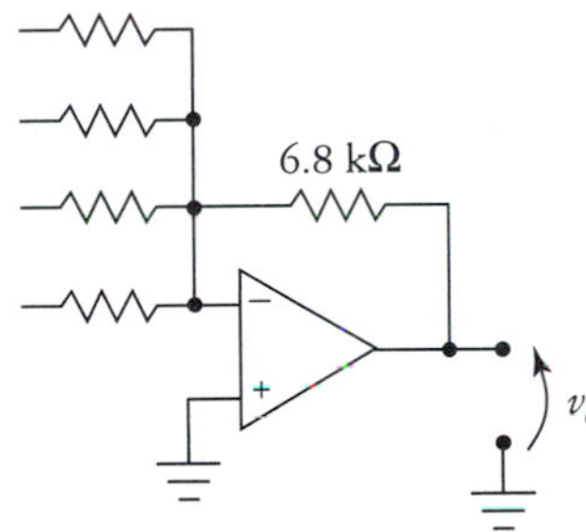

13.17. V_o = −10 V

13.19. v_o = −27.5 V

13.21. i_F = 10 + 2 sin ωt mA

13.23. $V_o|_{CMRR=\infty}$ = −1 V, $V_o|_{CMRR=50\text{ dB}}$ = −0.842 V

13.25. R_G = 5 kΩ, v_{id} = −0.04 V, v_{icm} = 2 + 5 sinωt V, $V_o|_{CMRR=\infty}$ = −0.4 V, $V_o|_{CMRR=50\text{ dB}}$ = −0.34 + 0.16 sinωt V

13.27. $V_{in(max)}$ = ±20 V, output voltage compliance is the limiting factor.

13.29. (a) −3.25 cos(2π1000t) V
(b) 3.25 sin(2π1000t) V
(c) 5.17 V
(d) −2.07t − 2.07 V
(e)

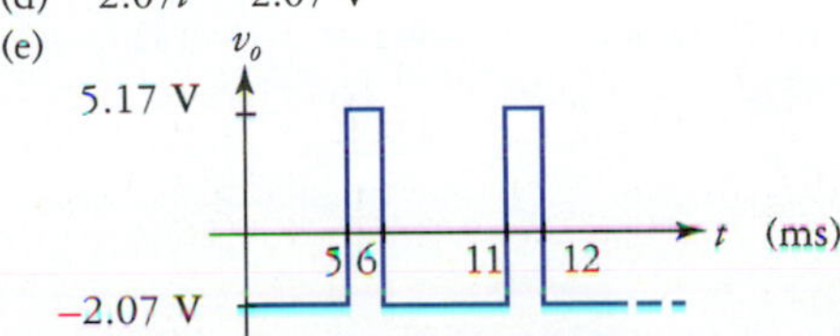

(f)

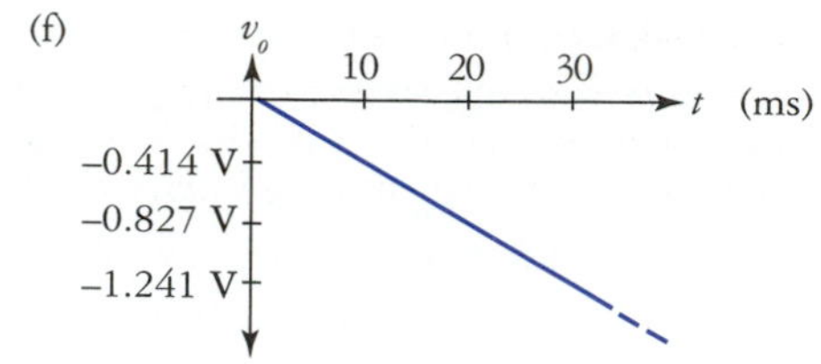

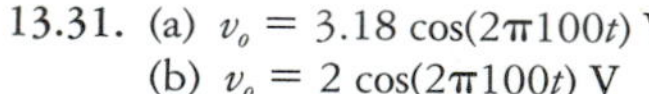

13.31. (a) $v_o = 3.18 \cos(2\pi 100t)$ V
(b) $v_o = 2 \cos(2\pi 100t)$ V
(c)

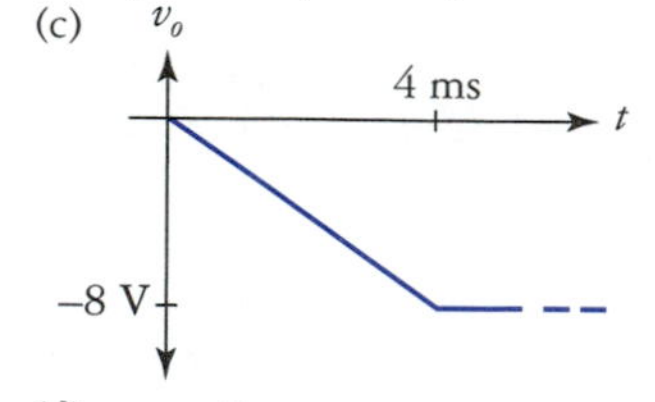

(d)

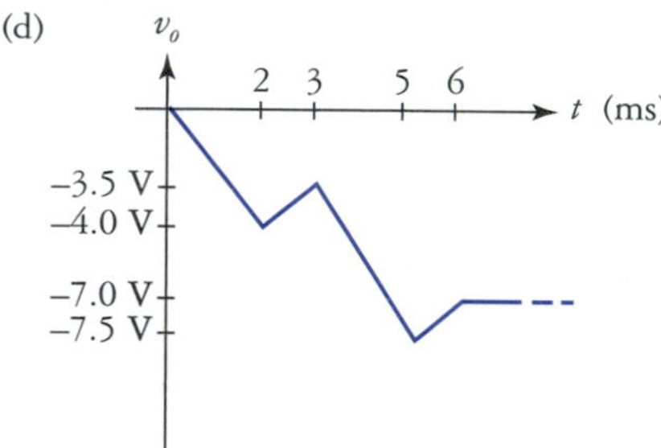

13.33. $v_o = -4x - 2y - 2z$

13.35. $R_4 = 30\ \text{k}\Omega$

13.37. A_d would jump to R_F/R_1 at these points.

13.39. Such heating is very often caused by oscillation.

Chapter 14

14.1. $n = 2$, $f_{c_1} = f_{c_2} = 4.52$ kHz, $f_c = 2.91$ kHz, $|H(jf)|_{f=10f_c} = 0.01 = -40$ dB

14.3. $f = 333.3$ kHz, $Q = 4.3$, $\alpha = 0.23$, Chebyshev response, $H_{pk} = 4.3 = 12.7$ dB, $R_x = 25.1\ \Omega$

14.5. $|H(jf)|_{f=1\text{ kHz}} = 0.45\angle -63°$, $|H(jf)|_{f=2.5\text{ kHz}} = 0.2\angle -79°$, $|H(jf)|_{f=10\text{ k}\Omega} = 0.05\angle -87°$

14.7. $f = 31.6$ kHz

14.9. $f = 1.19$ kHz

14.11. $|H(jf)|_{f=1\text{ kHz}} = 1.2$, $|H(jf)|_{f=15\text{ kHz}} = 0.045$

14.13. $n = 3$

14.15. $R_1 = 30.2\ \text{k}\Omega$, $C = 0.057\ \mu\text{F}$

14.17. $A_o = 1.8$, $H_0 = 5.1$ dB, $\alpha = 1.2$, Chebyshev response, $A_{pk} = 1.2$, $H_{pk} = 1.6$ dB, $H_{max} = 6.7$ dB, $A_{max} = 2.2$

14.19. $R_1 = 25\ \text{k}\Omega$, $C = 0.39\ \mu\text{F}$

14.21. $A_o = 1.8$, $H_0 = 5.1$ dB, $\alpha = 1.2$, Chebyshev response, $A_{pk} = 1.2$, $H_{pk} = 1.6$ dB, $H_{max} = 6.7$ dB, $A_{max} = 2.2$

14.23. $A = \dfrac{1}{\left(\dfrac{f_{in}}{f_c}\right)^4 + 1}$, $A\big|_{f=100\text{ Hz}} \cong 1 = 0$ dB, $A\big|_{f=10\text{ kHz}} \cong 0.001 = -80$ dB

14.25. BW = 500 Hz, $f_o = 1.73$ kHz, $Q = 3.46$, $\alpha = 0.29$

14.27. $R_1 = 67.7\ \text{k}\Omega$, $R_2 = 135.5\ \text{k}\Omega$, $R_3 = 340\ \Omega$

14.29.

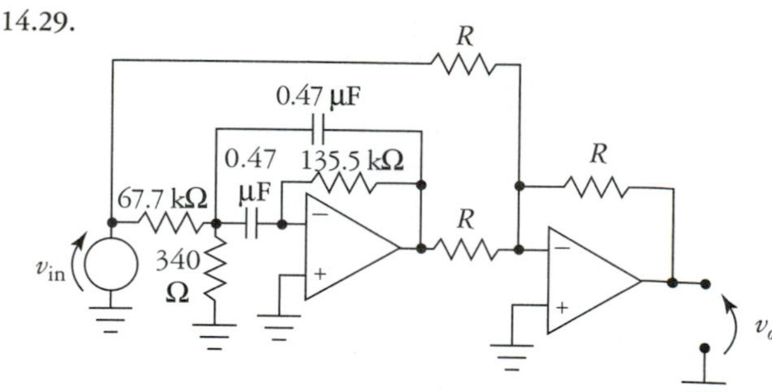

14.31. $A_o = Q = 0.707$, $\alpha = 1.414$, Butterworth response, $H_0 = -3$ dB, $f_o = 1.6$ kHz

14.33. $Q = 2.2$, $A_{LP(\max)} = 2.2 = 6.8$ dB, $A_{o(BP)} = 2.2$, $f_o = 723$ Hz, $\alpha = 0.45$

14.35. $\alpha = 0.1$, $R_3 = 11.3\ \text{k}\Omega$, $R_2 = 106.3\ \text{k}\Omega$, $A_{LP(\max)} = 11.3 = 21.1$ dB, $A_{o(BP)} = 10.6 = 20.5$ dB

14.37. $\phi\big|_{f=100\text{ Hz}} = 116°$, $\phi\big|_{f=1\text{ kHz}} = 18°$, $\phi\big|_{f=10\text{ kHz}} = 1.8°$

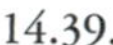

14.39.

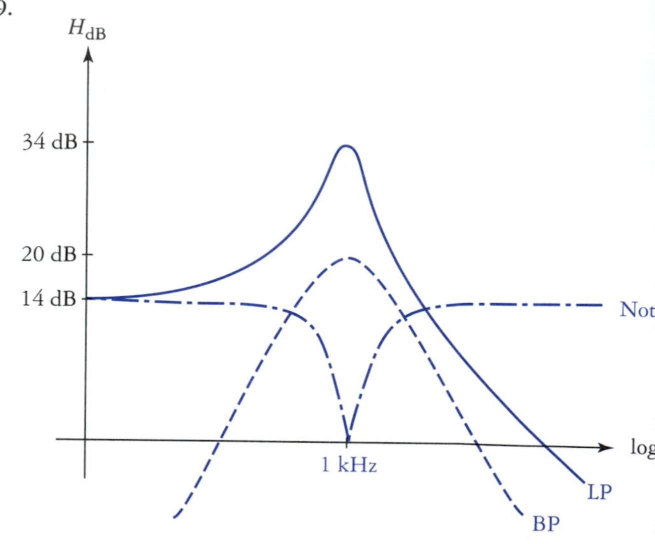

Chapter 15

15.1. $R_1 = 12.2\ \text{k}\Omega$, $D = 66.7\%$

15.3.

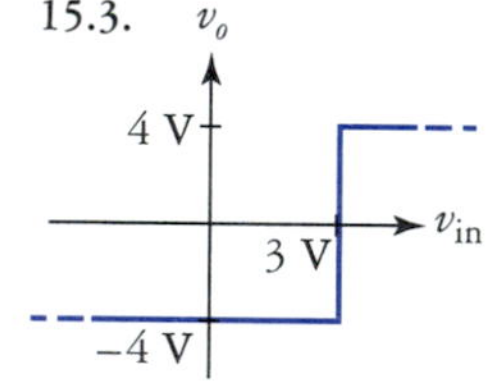

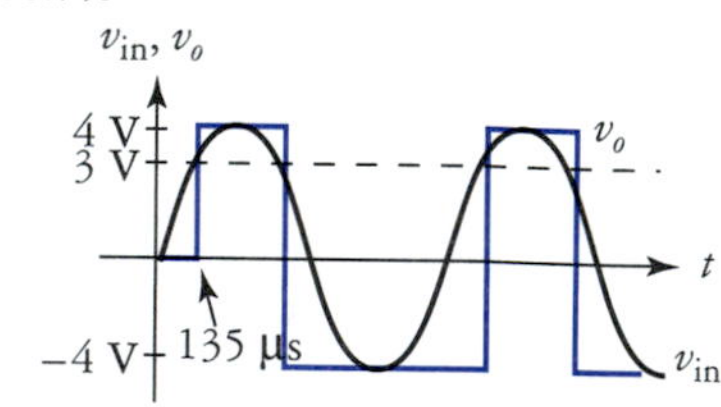

15.5.

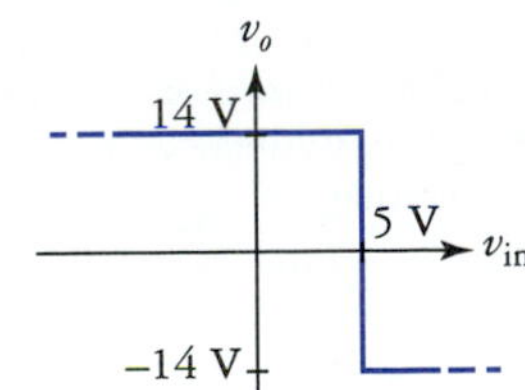

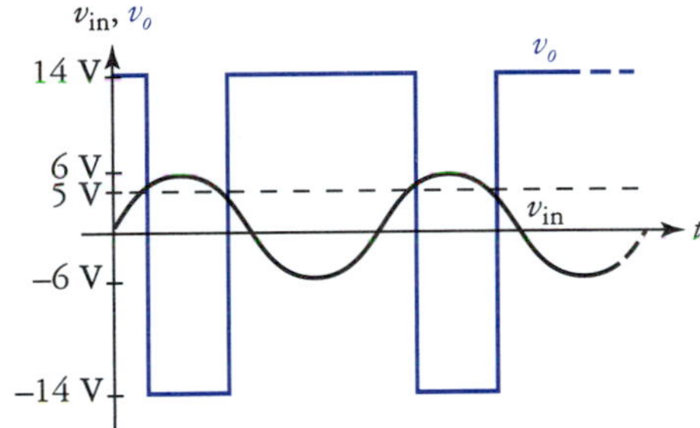

15.7.

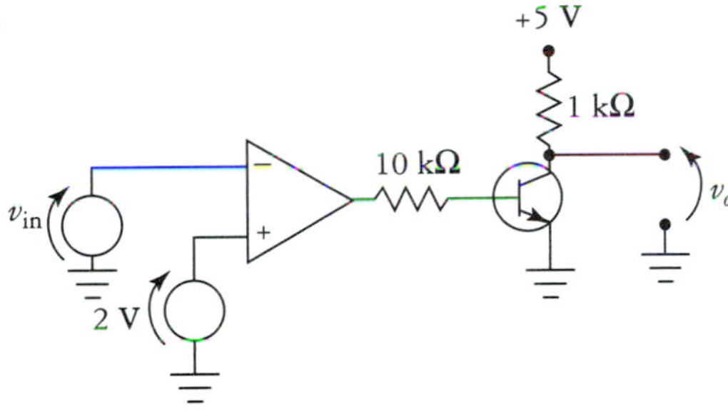

15.9. $V_{LT} = -1.3$ V, $V_{UT} = 1.3$ V

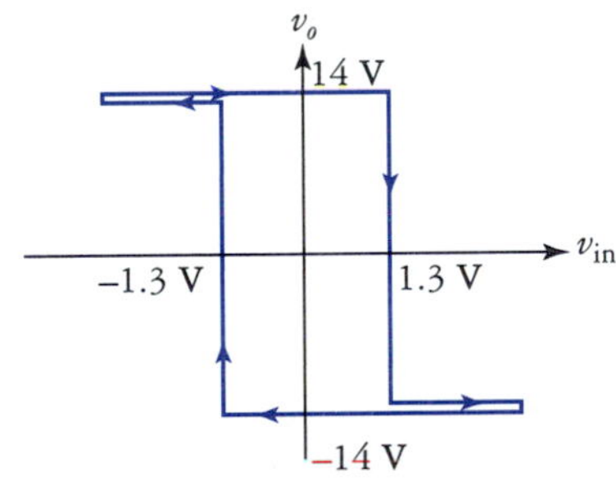

15.11. $V_U = 5$ V, $V_L = 1$ V

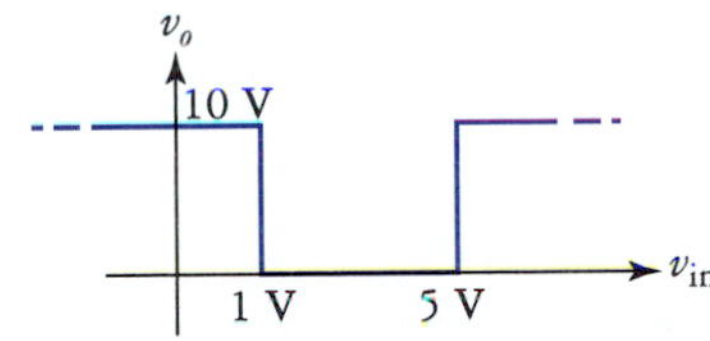

15.13. $V_o\Big|_{v_{in}=1\text{ V}} = -437$ mV, $V_o\Big|_{v_{in}=10\text{ V}} = -497$ mV

15.15. (a) $V_o = -299.3$ mV (b) $V_o = 119$ mV

15.17. $V_o\Big|_{v_{in}=-0.5\text{ V}} = -11.2$ V, $V_o\Big|_{v_{in}=-0.56\text{ V}} = -113$ V

15.19. (a) $v_o = 0.5\cos(2\pi 500t) - 0.5\cos(2\pi 2000t)$ V
(b) $v_o = 1.25\sin 300t + 1.25\sin 100t$ V
(c) $v_o = 0.6\sin 500t + 0.6\sin 400t - 0.6\cos 600t$ V
(d) $v_o = t^2$ V

15.21. $v_o = \sqrt{20v_{in}}$

15.23. See *Solutions Manual.*

15.25. See *Solutions Manual.*

15.27. The lower diode is open circuited.

15.29. Because I_{ES} changes equally for both Q_1 and Q_2, the offset will not drift; however, because changes in V_T do not cancel, the slope (gain) of the circuit will drift with temperature.

Chapter 16

16.1. $V_{FS} = 10.0$ V, $V_{o(\max)} = 9.375$ V, $N = 1100_2$, $V_o = 7.5$ V
$N = 1010_2$, $V_o = 6.25$ V
$N = 0011_2$, $V_o = 1.875$ V

16.3. $V_{FS} = 10.0$ V, $V_{o(\max)} = 9.375$ V, $N = 1010_2$, $V_o = 6.25$ V
$N = 1011_2$, $V_o = 6.875$ V
$N = 0001_2$, $V_o = 0.625$ V

16.5. $I_{FS} = 1$ mA, $V_{FS} = 10.0$ V, $N = 1010\ 1010_2$; $V_o = 6.641$ V
$N = 1111\ 0000_2$; $V_o = 9.375$ V
$N = 0111\ 0101_2$; $V_o = 4.570$ V

16.7. $R_1 = 10$ kΩ, $R_F = 10$ kΩ

16.9. $S_{\max} = 4687.5$ rpm, $N = 1000_2$; $S = 2500$ rpm, $N = 0001_2$; $S = 312.5$ rpm

16.11. See *Solutions Manual.*

16.13. $f_{\text{Nyquist}} = 25$ kHz, $|H(f_{\text{Nyquist}})| = 0.0016 = -56$ dB, $n = 3$

16.15. $V_{in} = 3$ V, count $= 0100\ 1101_2 = 77_{10}$, $T_c = 3.08$ ms
$V_{in} = 5$ V, count $= 1000\ 0001_2 = 129_{10}$, $T_c = 5.16$ ms
$V_{in} = 7.8$ V, count $= 1100\ 1000_2 = 200_{10}$, $T_c = 8$ ms

16.17. $T_c = 64$ μs, $f_{\text{Nyquist}} = 7.813$ kHz

16.19. $\dfrac{\Delta V_o}{\Delta V_{in}}\Bigg|_{\max} = 62{,}500$ V/s, $V_{in} = 8$ $V_{p\text{-}p}$, $f_{\max} = 2.49$ kHz

16.21. $n = 10$ bits

16.23. $V_{in} = 0.4875$ V; for $V_{in} = 5$ V, $m = 2048_{10} = 1000\ 0000\ 0000_2$, $T_c = 61.44$ ms

16.25. $N = 127$

16.27. $S_{\max} = 9960$ rpm

16.29. Only the second half of the 32-byte lookup table data would be accessed. This would produce the lower half of the sine wave as shown in the accompanying figure.

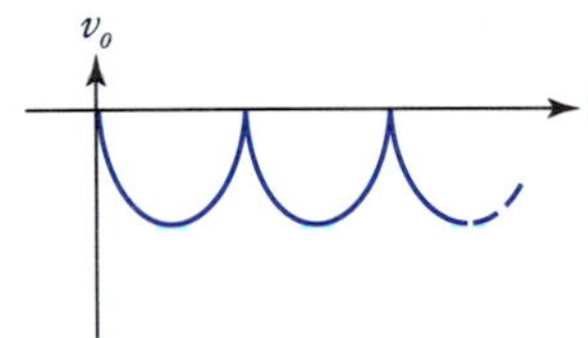

Chapter 17

17.1. $I_{CQ} = 3.8$ mA, $V_{CEQ} = 10$ V, $r_e = 6.8$ Ω, $f_o = 3.3$ MHz, $R'_L = 10.8$ kΩ, $R_e = 33$ Ω, $Q_L = 69$, $R_p = 9.4$ kΩ, $Q_T = 9.1$ $A'_v = -326$, BW = 363.6 kHz, $A_v = 19.7$, $V_{o(p\text{-}p)} = 1.7$ $V_{p\text{-}p}$, $C_1 = 46.5$ pF

17.3. $I_{CQ} = 6.2$ mA, $V_{CEQ} = 16.4$ V, $r_e = 4.2$ Ω, $f_o = 2.9$ MHz, $R'_L = 10.8$ kΩ, $R'_C = 5.7$ kΩ, $R_e \cong 12$ Ω, $Q_L = 22$, $R_p = 12.1$ kΩ, $Q_T = 10.4$, $A'_v = -352$, BW = 279 kHz, $A_v = 59$, $V_o = 5.5$ $V_{p\text{-}p}$

17.5. $f_o = 5.4$ MHz, $R_p = 207$ kΩ, $R'_L = 58.8$ kΩ, $R'_C = 45.8$ kΩ, $A_v = 460$, $Q_t = 45$, BW = 120 kHz

17.7. $f_o = 4.2$ MHz, $R'_L = 5$ kΩ, $R'_C = 410$ Ω, $R'_E = 461$ Ω, $R_p = 14.7$ kΩ, $A_v = 31.5$, $\beta = 0.167$, $A\beta = 5.3$, $V_o = 8.8$ $V_{p\text{-}p}$

17.9. $I_{CQ} = 2.1$ mA, $V_{CEQ} = 13.8$ V, $r_e = 12.4$ Ω, $f_o = 2.5$ MHz, $R'_L = 2.5$ kΩ, $R_p = 51.3$ kΩ, $R'_E = 772$ Ω, $Q_L = 32$, $R'_C = 583$ Ω, $A_v = 47$, $\beta = 0.125$, $A\beta = 5.9$, $V_o = 2.8$ $V_{p\text{-}p}$

17.11. $f_o = 118$ kHz, $\beta = 0.0091$, $A\beta = 0.9$ (oscillation not sustained), $V_o = 8.8\ V_{\text{p-p}}$, $A_{\min} = 109.9$

17.13. $I_{CQ} = 5.1$ mA, $V_{CEQ} = 16.2$ V, $f_o = 1.8$ MHz, $A = 961$, $\beta = 0.18$, $A\beta = 173$

17.15. $f_o = 10.7$ MHz, $A = 119$, $\beta = 0.091$, $A\beta = 10.8$

17.17. $f_{o(\min)} = 5.2$ MHz, $f_{o(\max)} = 10.6$ MHz, $A\beta = 2.5$

17.19. $I_{CQ} = 4.7$ mA, $V_{CEQ} = 5.3$ V, $f_o = 6.5$ kHz, $R'_C = 476\ \Omega$, $A_v = 86$, oscillation is sustained

17.21. $f_{\min} = 145$ Hz, $f_{\max} = 7.2$ kHz, $R_{1(\max)} = 5\ \text{k}\Omega$

17.23. Decreased A_v.

Chapter 18

18.1. $f_{\text{osc}} = 8.67$ kHz, $D = 0.8 = 80\%$, $V_{\text{av}} = 8$ V, $I_{\text{av}} = 8$ mA

18.3. $f_{\text{osc}} = 1.45$ kHz, $D = 0.33 = 33\%$, $I_{\text{av}} = 1.67$ mA

18.5. $C_1 = 0.033\ \mu$F

18.7. $R_2 = 12.2\ \text{k}\Omega$

18.9. $R_1 = 4.5\ \text{k}\Omega$

18.11. $R_3 \le 3.6\ \text{k}\Omega$, $f_o = 1.39$ Hz, $D = 0.787 = 78.7\%$

18.13. $C_1 = 81.1\ \mu$F

18.15.

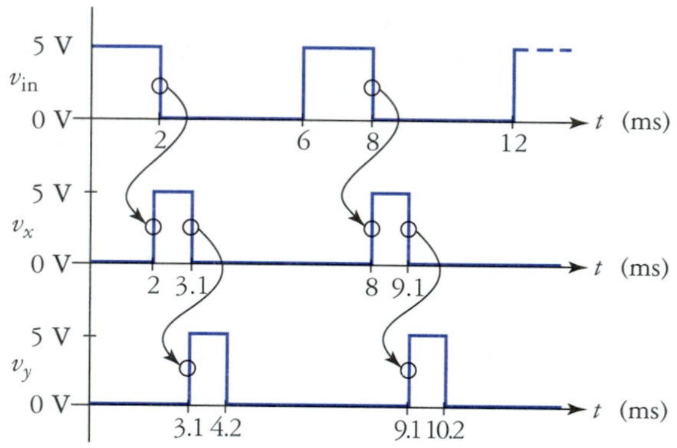

18.17. $I_Z = 27.1$ mA, $V_o = 10$ V, $P_L = 20$ W, $P_{DQ} = 28$ W

18.19. $V_{o(\max)} = 10.08$ V, $V_{o(\min)} = 7.2$ V, $P_L = 20$ W, $V_{CE_1} = V_{CE_2} = 10$ V, $P_{RE_1} = P_{RE_2} = 8$ W, $I_{\text{opamp}} = 80$ mA, $T_{J_1} = T_{J\text{Hooker}} = 105°$C

18.21. $I_{C(\max)} \cong 1.4$ A, $R_E = 1.4\ \Omega$

18.23. $V_o = 10.05$ V, $I_{Sc} = 2.19$ A, $I_{\text{knee}} = 12.2$ A

18.25. $I_Z = 10.6$ mA, $V_{o(\max)} = 5.76$ V, $V_{o(\min)} = 4.42$ V, $I_{sc}\Big|_{v_o=\max} = I_{sc}\Big|_{v_o=\min} = 1.94$ A, $I_{\text{knee}}\Big|_{v_o=\max} = 6.35$ A, $I_{\text{knee}}\Big|_{v_o=\min} = 5.32$ A

18.27. Switching regulator advantages: higher efficiency, smaller, lighter weight. Disadvantages: More complex, high-frequency noise generation.

18.29. $I_{\text{ABC}} = 210\ \mu$A, $g_m = 4$ mS, $A_v = 108$, $R_{\text{in}} \cong 49.5\ \text{k}\Omega$, $v_o = 2.16 \sin \omega t$ V, output compliance $= \pm 8.1$ V

18.31. $I_{\text{ABC}} = 113\ \mu$A, $g_m = 2.17$ mS, $A_v = 102$

18.33. $I_{\text{ABC}} = 210\ \mu$A, $g_m = 4$ mS, $v_o = 3.96 \sin(10{,}000t) + 1.73 \cos(9000t) - 1.73 \cos(11{,}000t)$ V, $m = 87\%$

18.35. Noninverting: $v_o = -0.27 \cos(2\pi 100t)$ V, $v_o = -0.67e^{-250t}$ V
Inverting: $v_o = 0.27 \cos(2\pi 100t)$ V, $v_o = 0.67\ e^{-250t}$ V

18.37. $I_{\text{ABC}} = 77.5\ \mu$A, $g_m = 1.5$ mS, $v_o = 957t^2$ V, $T_{\text{sat}} = 97$ ms

18.39. (a) $9 \cos(2\pi 1500t) - 9 \cos(2\pi 2500t)$ V
(b) $10 \sin(2\pi 1250t) + 10 \sin(2\pi 750t)$ V
(c) $10 \cos(1000t)$ V
(d) $-12.5 \cos(800t)$ V

18.41. (a) $V_o = 6.5$ V (b) $V_o = 3.54$ V (c) $V_o = 0$ V (d) $V_o = 0.1$ V (e) $V_o = 3$ V

18.43. $f_{\text{capture}} = f_o \pm 1\%$, $f_{\text{lock}} = f_o \pm 2\%$

18.45. $S = 6000$ rpm

18.47. $f_{\text{out}} = 2.5$ MHz

18.51. With $v_y < 0$ V, the output of the integrator will eventually drift to $+V_{\text{sat}}$.

18.53. (a) $V_o \cong V_{\text{in}}$ (b) $V_o = 0$ V (c) $V_o = 0$ V (d) $V_o = V_{\text{ref}}$

Chapter 19

19.1. $t = 50$ ms

19.3. $f = 4.3$ kHz

19.5. $$v_{in} = \frac{V_{BO} \ln^{-1}\left(\dfrac{T}{R_1C_1}\right) - V_{\text{sat}}}{\ln^{-1}\left(\dfrac{T}{R_1C_1}\right) - 1}$$

19.7. $f = 69.5$ kHz

19.9. $\Delta t = 2.9$ ms, $\phi = 62°$, $\theta = 118°$

19.11. $\Delta t = 2.4$ ms, $\phi = 32°$, $\theta = 148°$

19.13. $\Delta t = 3$ ms, $\phi = 65°$, $\theta = 295°$

19.15. $V_p = 8.2$ V, $f_o = 2.98$ kHz

19.17. $$\eta = 1 - \left[\ln^{-1}\left(\frac{1}{fR_1C_1}\right)\right]^{-1}$$

19.19. $L = 2.53\ \mu$H

19.21. The four-layer diode is open circuited.

19.23. Connecting the ground lead of the scope probe to the negative side of the capacitor will short the bridge to ground, with possibly spectacular (and dangerous) results.

APPENDIX 2 Data Sheets

MOTOROLA
SEMICONDUCTOR TECHNICAL DATA

Order this document by 1N4001/D

Axial Lead Standard Recovery Rectifiers

1N4001 thru 1N4007

1N4004 and 1N4007 are Motorola Preferred Devices

LEAD MOUNTED RECTIFIERS 50–1000 VOLTS DIFFUSED JUNCTION

This data sheet provides information on subminiature size, axial lead mounted rectifiers for general–purpose low–power applications.

Mechanical Characteristics

- Case: Epoxy, Molded
- Weight: 0.4 gram (approximately)
- Finish: All External Surfaces Corrosion Resistant and Terminal Leads are Readily Solderable
- Lead and Mounting Surface Temperature for Soldering Purposes: 220°C Max. for 10 Seconds, 1/16″ from case
- Shipped in plastic bags, 1000 per bag.
- Available Tape and Reeled, 5000 per reel, by adding a "RL" suffix to the part number
- Polarity: Cathode Indicated by Polarity Band
- Marking: 1N4001, 1N4002, 1N4003, 1N4004, 1N4005, 1N4006, 1N4007

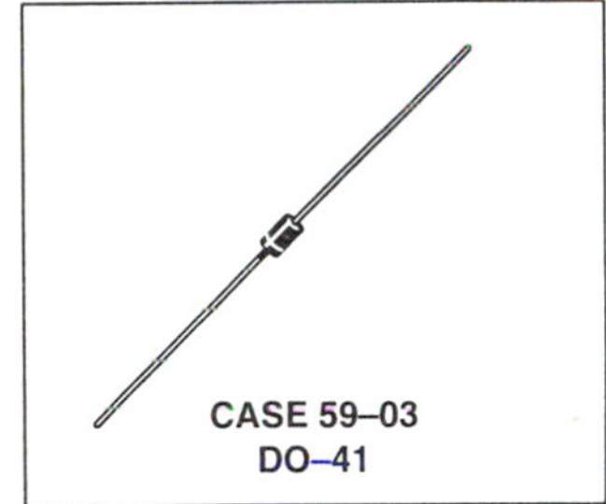

CASE 59–03
DO–41

MAXIMUM RATINGS

Rating	Symbol	1N4001	1N4002	1N4003	1N4004	1N4005	1N4006	1N4007	Unit
*Peak Repetitive Reverse Voltage Working Peak Reverse Voltage DC Blocking Voltage	V_{RRM} V_{RWM} V_R	50	100	200	400	600	800	1000	Volts
*Non–Repetitive Peak Reverse Voltage (halfwave, single phase, 60 Hz)	V_{RSM}	60	120	240	480	720	1000	1200	Volts
*RMS Reverse Voltage	$V_{R(RMS)}$	35	70	140	280	420	560	700	Volts
*Average Rectified Forward Current (single phase, resistive load, 60 Hz, see Figure 8, T_A = 75°C)	I_O	1.0							Amp
*Non–Repetitive Peak Surge Current (surge applied at rated load conditions, see Figure 2)	I_{FSM}	30 (for 1 cycle)							Amp
Operating and Storage Junction Temperature Range	T_J T_{stg}	– 65 to +175							°C

ELECTRICAL CHARACTERISTICS*

Rating	Symbol	Typ	Max	Unit
Maximum Instantaneous Forward Voltage Drop (i_F = 1.0 Amp, T_J = 25°C) Figure 1	v_F	0.93	1.1	Volts
Maximum Full–Cycle Average Forward Voltage Drop (I_O = 1.0 Amp, T_L = 75°C, 1 inch leads)	$V_{F(AV)}$	—	0.8	Volts
Maximum Reverse Current (rated dc voltage) (T_J = 25°C) (T_J = 100°C)	I_R	 0.05 1.0	 10 50	μA
Maximum Full–Cycle Average Reverse Current (I_O = 1.0 Amp, T_L = 75°C, 1 inch leads)	$I_{R(AV)}$	—	30	μA

*Indicates JEDEC Registered Data

Preferred devices are Motorola recommended choices for future use and best overall value.

1N4001 thru 1N4007

PACKAGE DIMENSIONS

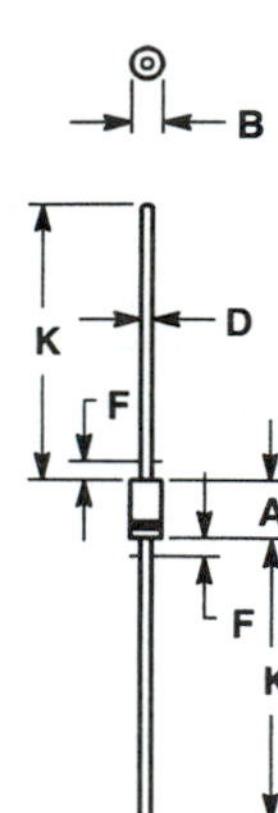

NOTES:
1. ALL RULES AND NOTES ASSOCIATED WITH JEDEC DO–41 OUTLINE SHALL APPLY.
2. POLARITY DENOTED BY CATHODE BAND.
3. LEAD DIAMETER NOT CONTROLLED WITHIN F DIMENSION.

	MILLIMETERS		INCHES	
DIM	MIN	MAX	MIN	MAX
A	4.07	5.20	0.160	0.205
B	2.04	2.71	0.080	0.107
D	0.71	0.86	0.028	0.034
F	—	1.27	—	0.050
K	27.94	—	1.100	—

CASE 59–03
(DO–41)
ISSUE M

Motorola reserves the right to make changes without further notice to any products herein. Motorola makes no warranty, representation or guarantee regarding the suitability of its products for any particular purpose, nor does Motorola assume any liability arising out of the application or use of any product or circuit, and specifically disclaims any and all liability, including without limitation consequential or incidental damages. "Typical" parameters which may be provided in Motorola data sheets and/or specifications can and do vary in different applications and actual performance may vary over time. All operating parameters, including "Typicals" must be validated for each customer application by customer's technical experts. Motorola does not convey any license under its patent rights nor the rights of others. Motorola products are not designed, intended, or authorized for use as components in systems intended for surgical implant into the body, or other applications intended to support or sustain life, or for any other application in which the failure of the Motorola product could create a situation where personal injury or death may occur. Should Buyer purchase or use Motorola products for any such unintended or unauthorized application, Buyer shall indemnify and hold Motorola and its officers, employees, subsidiaries, affiliates, and distributors harmless against all claims, costs, damages, and expenses, and reasonable attorney fees arising out of, directly or indirectly, any claim of personal injury or death associated with such unintended or unauthorized use, even if such claim alleges that Motorola was negligent regarding the design or manufacture of the part. Motorola and Ⓜ are registered trademarks of Motorola, Inc. Motorola, Inc. is an Equal Opportunity/Affirmative Action Employer.

Mfax is a trademark of Motorola, Inc.

How to reach us:
USA/EUROPE/Locations Not Listed: Motorola Literature Distribution; P.O. Box 5405, Denver, Colorado 80217. 303–675–2140 or 1–800–441–2447

Mfax™: RMFAX0@email.sps.mot.com – TOUCHTONE 602–244–6609
– US & Canada ONLY 1–800–774–1848

INTERNET: http://motorola.com/sps

JAPAN: Nippon Motorola Ltd.: SPD, Strategic Planning Office, 4–32–1, Nishi–Gotanda, Shinagawa–ku, Tokyo 141, Japan. 81–3–5487–8488

ASIA/PACIFIC: Motorola Semiconductors H.K. Ltd.; 8B Tai Ping Industrial Park, 51 Ting Kok Road, Tai Po, N.T., Hong Kong. 852–26629298

MOTOROLA
SEMICONDUCTOR TECHNICAL DATA

Order this document
by MMBT3904LT1/D

General Purpose Transistor

NPN Silicon

MMBT3904LT1

Motorola Preferred Device

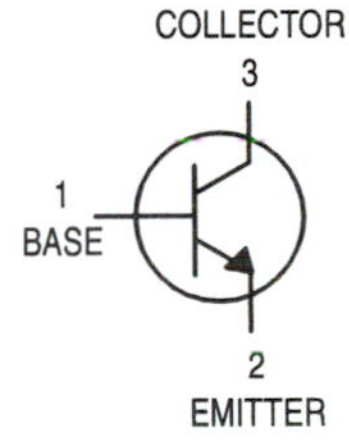

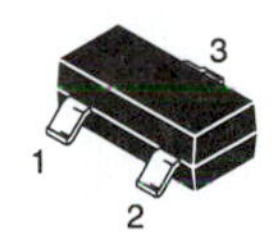

CASE 318–08, STYLE 6
SOT–23 (TO–236AB)

MAXIMUM RATINGS

Rating	Symbol	Value	Unit
Collector–Emitter Voltage	V_{CEO}	40	Vdc
Collector–Base Voltage	V_{CBO}	60	Vdc
Emitter–Base Voltage	V_{EBO}	6.0	Vdc
Collector Current — Continuous	I_C	200	mAdc

THERMAL CHARACTERISTICS

Characteristic	Symbol	Max	Unit
Total Device Dissipation FR–5 Board(1) $T_A = 25°C$	P_D	225	mW
Derate above 25°C		1.8	mW/°C
Thermal Resistance Junction to Ambient	$R_{\theta JA}$	556	°C/W
Total Device Dissipation Alumina Substrate,(2) $T_A = 25°C$	P_D	300	mW
Derate above 25°C		2.4	mW/°C
Thermal Resistance Junction to Ambient	$R_{\theta JA}$	417	°C/W
Junction and Storage Temperature	T_J, T_{stg}	–55 to +150	°C

DEVICE MARKING

MMBT3904LT1 = 1AM

ELECTRICAL CHARACTERISTICS ($T_A = 25°C$ unless otherwise noted)

Characteristic	Symbol	Min	Max	Unit
OFF CHARACTERISTICS				
Collector–Emitter Breakdown Voltage (3) ($I_C = 1.0$ mAdc, $I_B = 0$)	$V_{(BR)CEO}$	40	—	Vdc
Collector–Base Breakdown Voltage ($I_C = 10$ µAdc, $I_E = 0$)	$V_{(BR)CBO}$	60	—	Vdc
Emitter–Base Breakdown Voltage ($I_E = 10$ µAdc, $I_C = 0$)	$V_{(BR)EBO}$	6.0	—	Vdc
Base Cutoff Current ($V_{CE} = 30$ Vdc, $V_{EB} = 3.0$ Vdc)	I_{BL}	—	50	nAdc
Collector Cutoff Current ($V_{CE} = 30$ Vdc, $V_{EB} = 3.0$ Vdc)	I_{CEX}	—	50	nAdc

1. FR–5 = 1.0 × 0.75 × 0.062 in.
2. Alumina = 0.4 × 0.3 × 0.024 in. 99.5% alumina.
3. Pulse Test: Pulse Width ≤ 300 µs, Duty Cycle ≤ 2.0%.

Thermal Clad is a registered trademark of the Berquist Company.

Preferred devices are Motorola recommended choices for future use and best overall value.

REV 1

© Motorola, Inc. 1996

MMBT3904LT1

ELECTRICAL CHARACTERISTICS (T_A = 25°C unless otherwise noted) (Continued)

Characteristic	Symbol	Min	Max	Unit
ON CHARACTERISTICS(3)				
DC Current Gain (1)	H_{FE}			—
(I_C = 0.1 mAdc, V_{CE} = 1.0 Vdc)		40	—	
(I_C = 1.0 mAdc, V_{CE} = 1.0 Vdc)		70	—	
(I_C = 10 mAdc, V_{CE} = 1.0 Vdc)		100	300	
(I_C = 50 mAdc, V_{CE} = 1.0 Vdc)		60	—	
(I_C = 100 mAdc, V_{CE} = 1.0 Vdc)		30	—	
Collector–Emitter Saturation Voltage (3)	$V_{CE(sat)}$			Vdc
(I_C = 10 mAdc, I_B = 1.0 mAdc)		—	0.2	
(I_C = 50 mAdc, I_B = 5.0 mAdc)		—	0.3	
Base–Emitter Saturation Voltage (3)	$V_{BE(sat)}$			Vdc
(I_C = 10 mAdc, I_B = 1.0 mAdc)		0.65	0.85	
(I_C = 50 mAdc, I_B = 5.0 mAdc)		—	0.95	
SMALL–SIGNAL CHARACTERISTICS				
Current–Gain — Bandwidth Product (I_C = 10 mAdc, V_{CE} = 20 Vdc, f = 100 MHz)	f_T	300	—	MHz
Output Capacitance (V_{CB} = 5.0 Vdc, I_E = 0, f = 1.0 MHz)	C_{obo}	—	4.0	pF
Input Capacitance (V_{EB} = 0.5 Vdc, I_C = 0, f = 1.0 MHz)	C_{ibo}	—	8.0	pF
Input Impedance (V_{CE} = 10 Vdc, I_C = 1.0 mAdc, f = 1.0 kHz)	h_{ie}	1.0	10	k ohms
Voltage Feedback Ratio (V_{CE} = 10 Vdc, I_C = 1.0 mAdc, f = 1.0 kHz)	h_{re}	0.5	8.0	X 10^{-4}
Small–Signal Current Gain (V_{CE} = 10 Vdc, I_C = 1.0 mAdc, f = 1.0 kHz)	h_{fe}	100	400	—
Output Admittance (V_{CE} = 10 Vdc, I_C = 1.0 mAdc, f = 1.0 kHz)	h_{oe}	1.0	40	μmhos
Noise Figure (V_{CE} = 5.0 Vdc, I_C = 100 μAdc, R_S = 1.0 k ohms, f = 1.0 kHz)	NF	—	5.0	dB

SWITCHING CHARACTERISTICS

Characteristic	Conditions	Symbol	Min	Max	Unit
Delay Time	(V_{CC} = 3.0 Vdc, V_{BE} = −0.5 Vdc, I_C = 10 mAdc, I_{B1} = 1.0 mAdc)	t_d	—	35	ns
Rise Time		t_r	—	35	
Storage Time	(V_{CC} = 3.0 Vdc, I_C = 10 mAdc, I_{B1} = I_{B2} = 1.0 mAdc)	t_s	—	200	ns
Fall Time		t_f	—	50	

3. Pulse Test: Pulse Width ≤ 300 μs, Duty Cycle ≤ 2.0%.

MMBT3904LT1

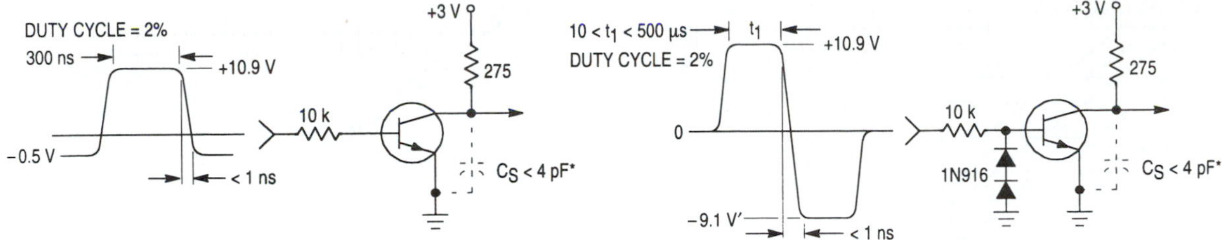

* Total shunt capacitance of test jig and connectors

Figure 1. Delay and Rise Time Equivalent Test Circuit

Figure 2. Storage and Fall Time Equivalent Test Circuit

TYPICAL TRANSIENT CHARACTERISTICS

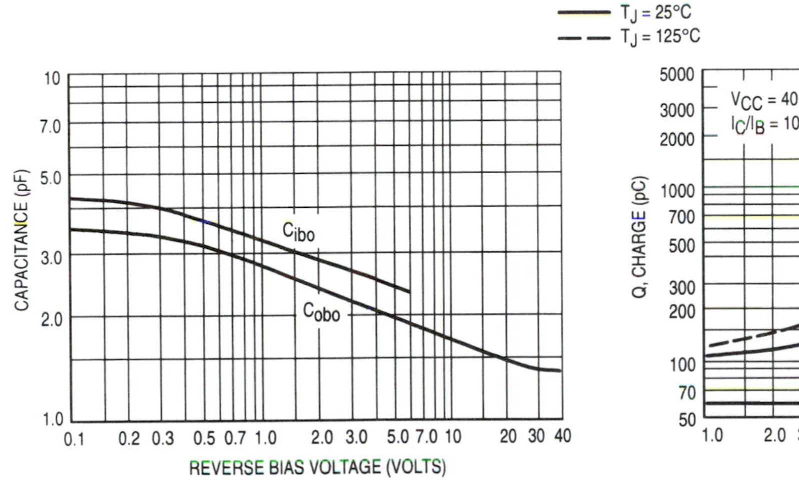

Figure 3. Capacitance

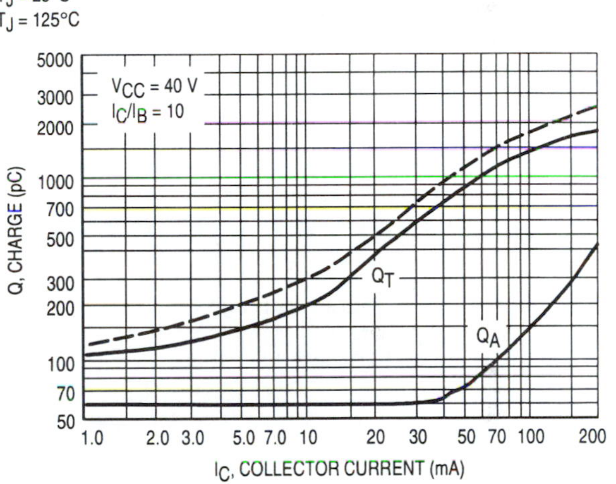

Figure 4. Charge Data

MMBT3904LT1

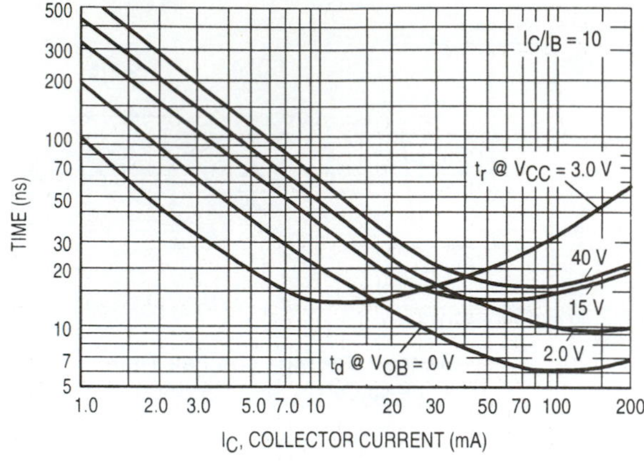

Figure 5. Turn–On Time

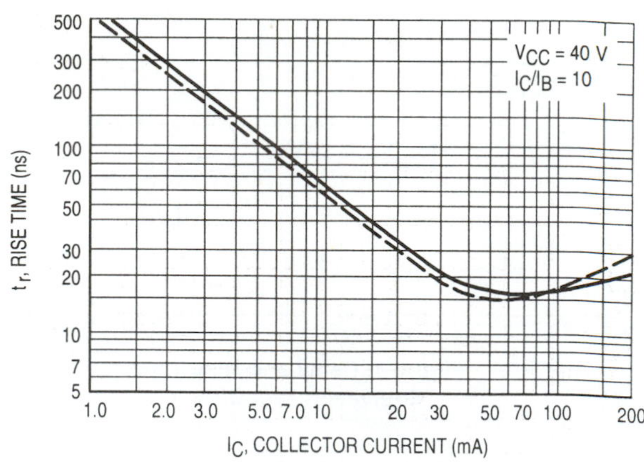

Figure 6. Rise Time

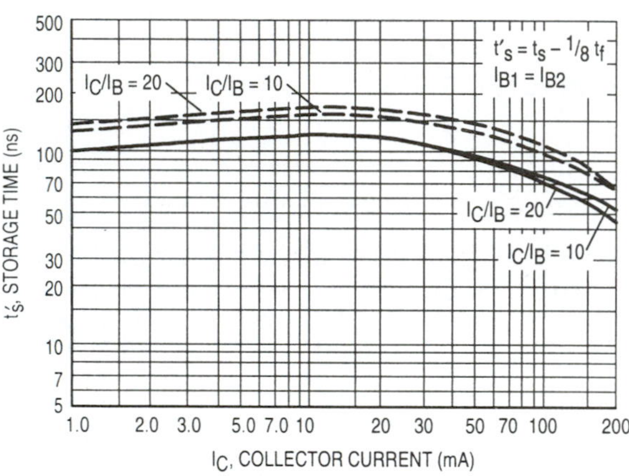

Figure 7. Storage Time

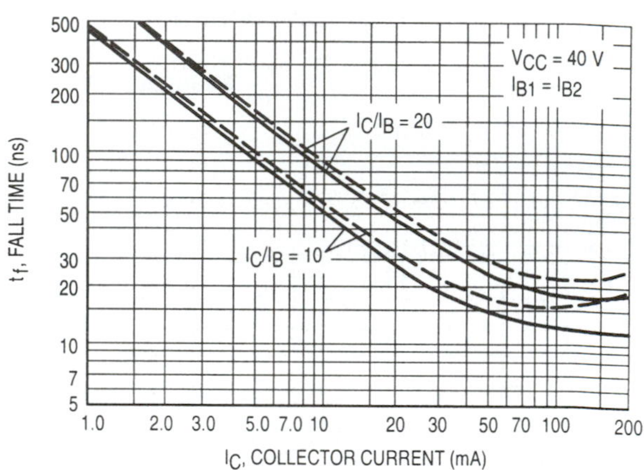

Figure 8. Fall Time

TYPICAL AUDIO SMALL–SIGNAL CHARACTERISTICS
NOISE FIGURE VARIATIONS

(V_{CE} = 5.0 Vdc, T_A = 25°C, Bandwidth = 1.0 Hz)

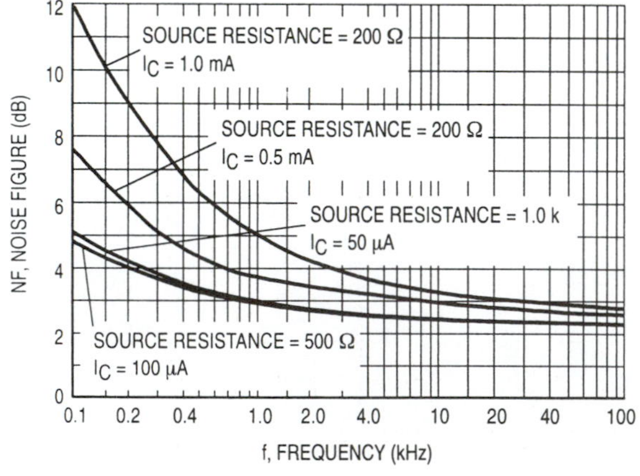

Figure 9.

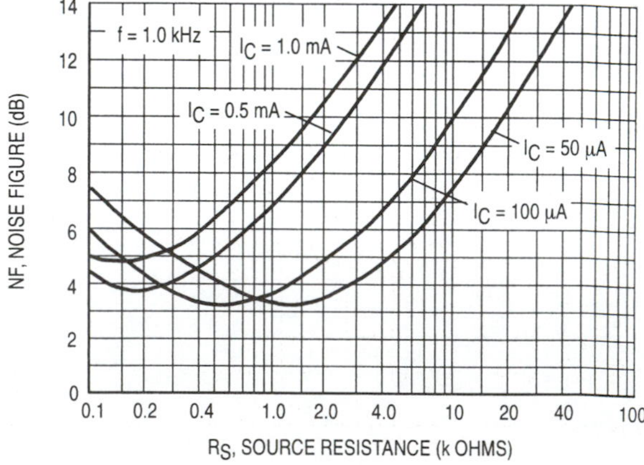

Figure 10.

MMBT3904LT1

h PARAMETERS

(V_{CE} = 10 Vdc, f = 1.0 kHz, T_A = 25°C)

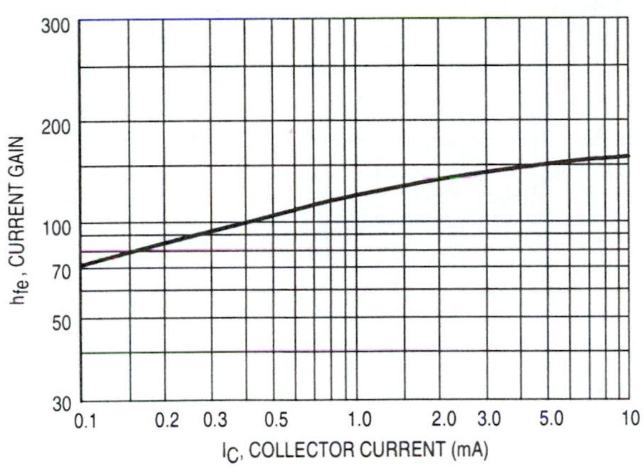

Figure 11. Current Gain

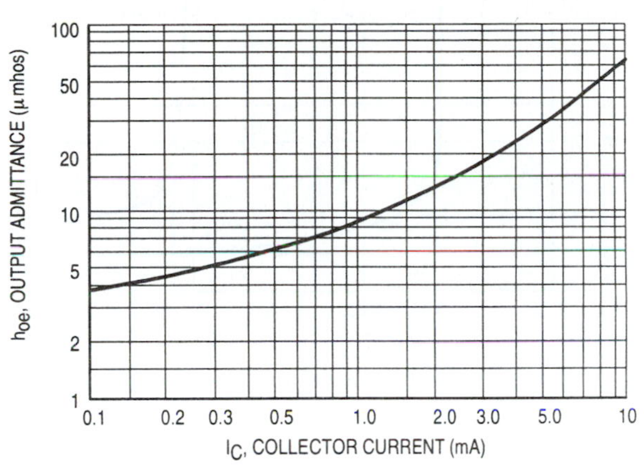

Figure 12. Output Admittance

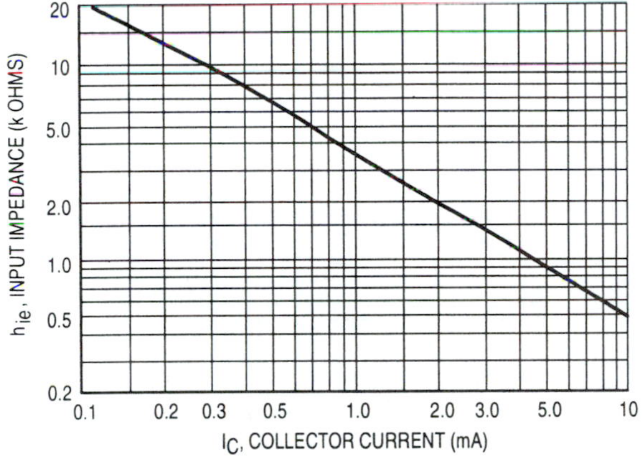

Figure 13. Input Impedance

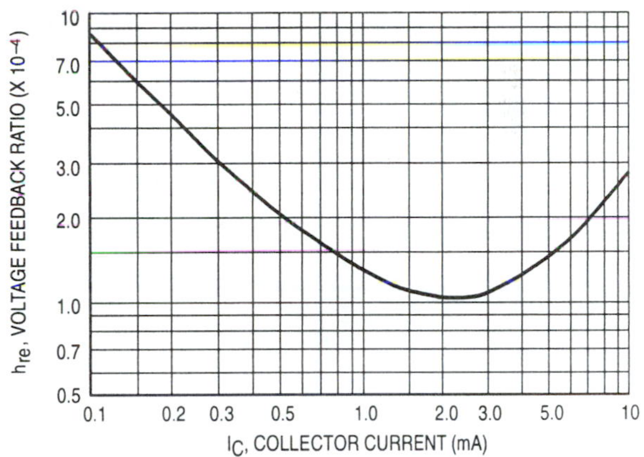

Figure 14. Voltage Feedback Ratio

TYPICAL STATIC CHARACTERISTICS

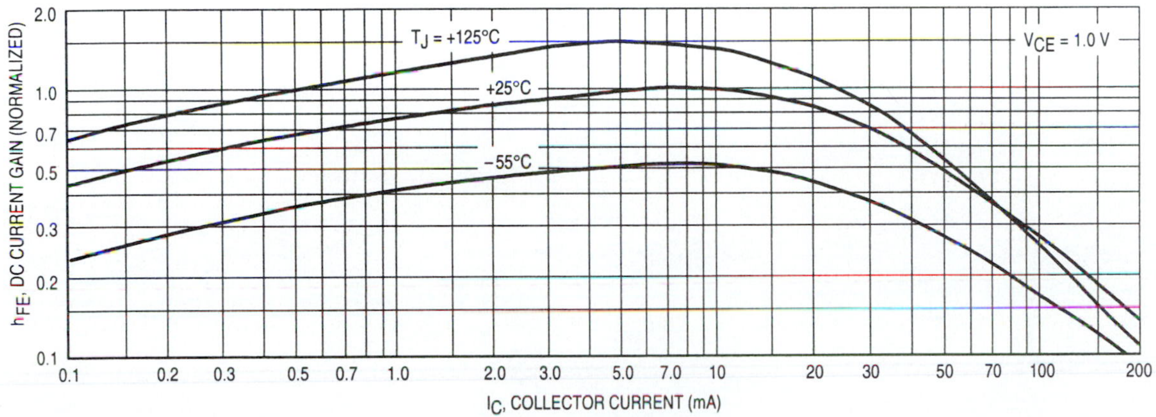

Figure 15. DC Current Gain

MMBT3904LT1

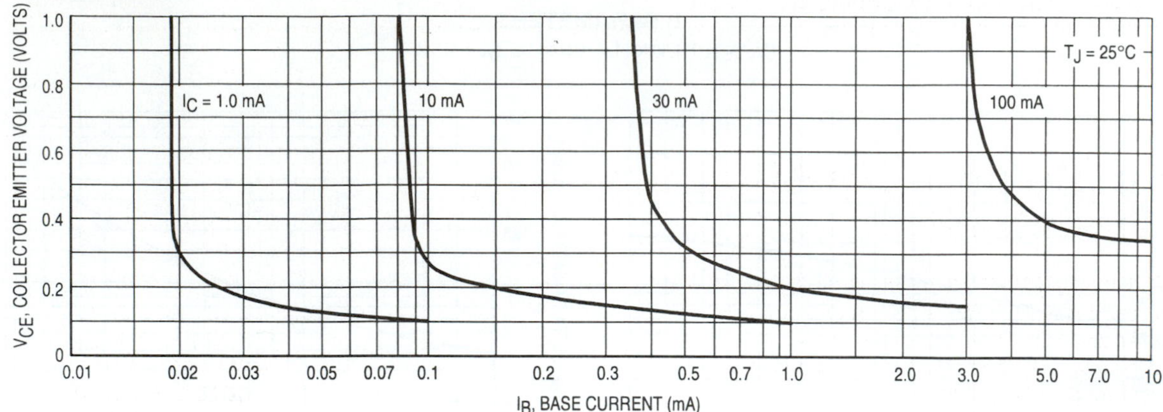

Figure 16. Collector Saturation Region

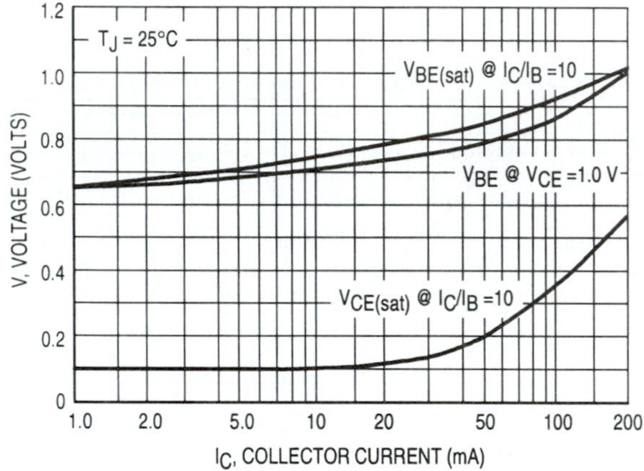

Figure 17. "ON" Voltages

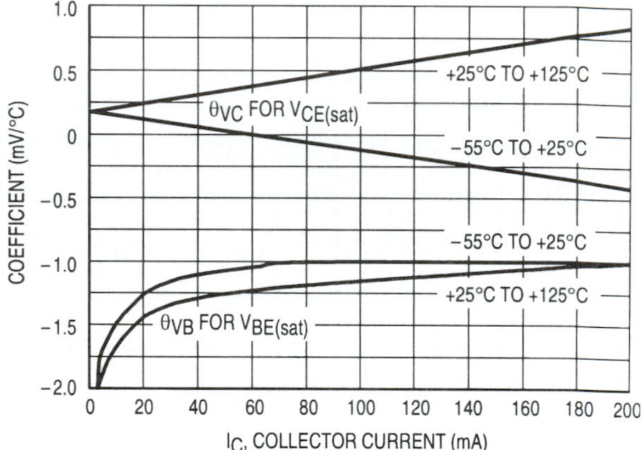

Figure 18. Temperature Coefficients

MOTOROLA
SEMICONDUCTOR TECHNICAL DATA

Order this document
by 2N3055/D

Complementary Silicon Power Transistors

. . . designed for general–purpose switching and amplifier applications.

- DC Current Gain — h_{FE} = 20–70 @ I_C = 4 Adc
- Collector–Emitter Saturation Voltage — $V_{CE(sat)}$ = 1.1 Vdc (Max) @ I_C = 4 Adc
- Excellent Safe Operating Area

NPN
2N3055*
PNP
MJ2955*

*Motorola Preferred Device

15 AMPERE
POWER TRANSISTORS
COMPLEMENTARY
SILICON
60 VOLTS
115 WATTS

CASE 1–07
TO–204AA
(TO–3)

MAXIMUM RATINGS

Rating	Symbol	Value	Unit
Collector–Emitter Voltage	V_{CEO}	60	Vdc
Collector–Emitter Voltage	V_{CER}	70	Vdc
Collector–Base Voltage	V_{CB}	100	Vdc
Emitter–Base Voltage	V_{EB}	7	Vdc
Collector Current — Continuous	I_C	15	Adc
Base Current	I_B	7	Adc
Total Power Dissipation @ T_C = 25°C Derate above 25°C	P_D	115 0.657	Watts W/°C
Operating and Storage Junction Temperature Range	T_J, T_{stg}	–65 to +200	°C

THERMAL CHARACTERISTICS

Characteristic	Symbol	Max	Unit
Thermal Resistance, Junction to Case	$R_{θJC}$	1.52	°C/W

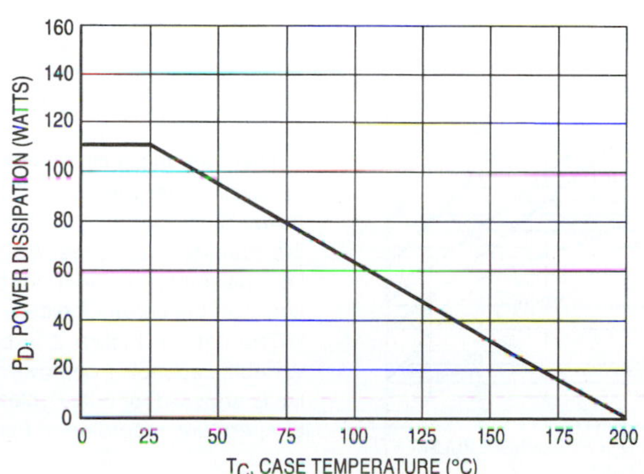

Figure 1. Power Derating

Preferred devices are Motorola recommended choices for future use and best overall value.

© Motorola, Inc. 1995

2N3055 MJ2955

ELECTRICAL CHARACTERISTICS (T_C = 25°C unless otherwise noted)

Characteristic	Symbol	Min	Max	Unit
***OFF CHARACTERISTICS**				
Collector–Emitter Sustaining Voltage (1) (I_C = 200 mAdc, I_B = 0)	$V_{CEO(sus)}$	60	—	Vdc
Collector–Emitter Sustaining Voltage (1) (I_C = 200 mAdc, R_{BE} = 100 Ohms)	$V_{CER(sus)}$	70	—	Vdc
Collector Cutoff Current (V_{CE} = 30 Vdc, I_B = 0)	I_{CEO}	—	0.7	mAdc
Collector Cutoff Current (V_{CE} = 100 Vdc, $V_{BE(off)}$ = 1.5 Vdc) (V_{CE} = 100 Vdc, $V_{BE(off)}$ = 1.5 Vdc, T_C = 150°C)	I_{CEX}	 — —	 1.0 5.0	mAdc
Emitter Cutoff Current (V_{BE} = 7.0 Vdc, I_C = 0)	I_{EBO}	—	5.0	mAdc
***ON CHARACTERISTICS (1)**				
DC Current Gain (I_C = 4.0 Adc, V_{CE} = 4.0 Vdc) (I_C = 10 Adc, V_{CE} = 4.0 Vdc)	h_{FE}	 20 5.0	 70 —	—
Collector–Emitter Saturation Voltage (I_C = 4.0 Adc, I_B = 400 mAdc) (I_C = 10 Adc, I_B = 3.3 Adc)	$V_{CE(sat)}$	 —	 1.1 3.0	Vdc
Base–Emitter On Voltage (I_C = 4.0 Adc, V_{CE} = 4.0 Vdc)	$V_{BE(on)}$	—	1.5	Vdc
SECOND BREAKDOWN				
Second Breakdown Collector Current with Base Forward Biased (V_{CE} = 40 Vdc, t = 1.0 s, Nonrepetitive)	$I_{s/b}$	2.87	—	Adc
DYNAMIC CHARACTERISTICS				
Current Gain — Bandwidth Product (I_C = 0.5 Adc, V_{CE} = 10 Vdc, f = 1.0 MHz)	f_T	2.5	—	MHz
*Small–Signal Current Gain (I_C = 1.0 Adc, V_{CE} = 4.0 Vdc, f = 1.0 kHz)	h_{fe}	15	120	—
*Small–Signal Current Gain Cutoff Frequency (V_{CE} = 4.0 Vdc, I_C = 1.0 Adc, f = 1.0 kHz)	f_{hfe}	10	—	kHz

* Indicates Within JEDEC Registration. (2N3055)
(1) Pulse Test: Pulse Width ≤ 300 μs, Duty Cycle ≤ 2.0%.

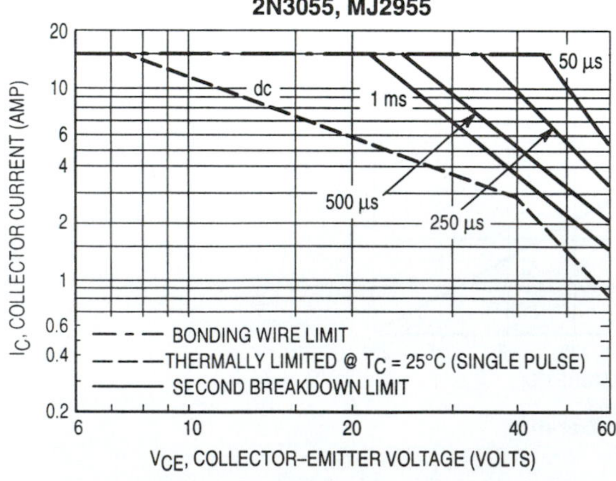

Figure 2. Active Region Safe Operating Area

There are two limitations on the power handling ability of a transistor: average junction temperature and second breakdown. Safe operating area curves indicate $I_C - V_{CE}$ limits of the transistor that must be observed for reliable operation; i.e., the transistor must not be subjected to greater dissipation than the curves indicate.

The data of Figure 2 is based on T_C = 25°C; $T_{J(pk)}$ is variable depending on power level. Second breakdown pulse limits are valid for duty cycles to 10% but must be derated for temperature according to Figure 1.

2N3055 MJ2955

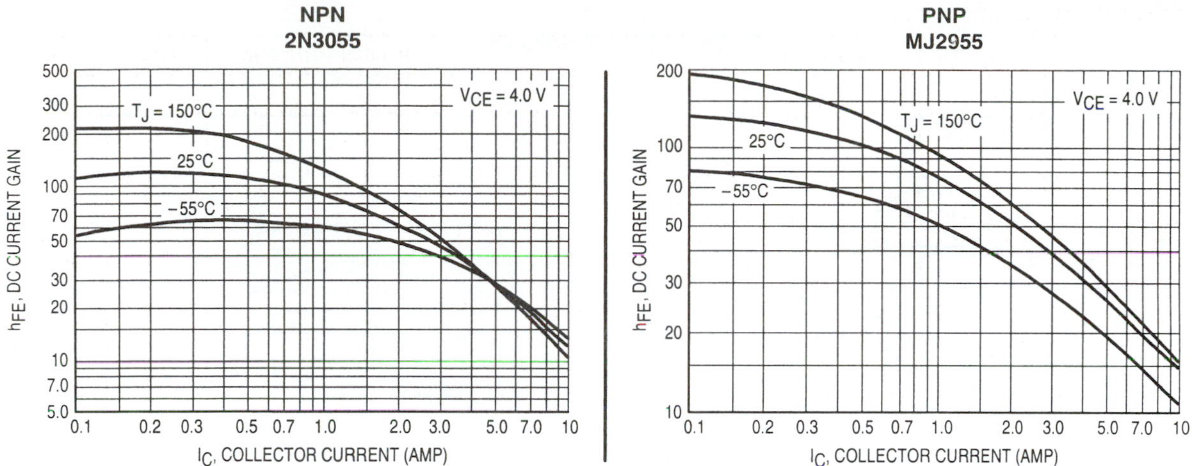

Figure 3. DC Current Gain

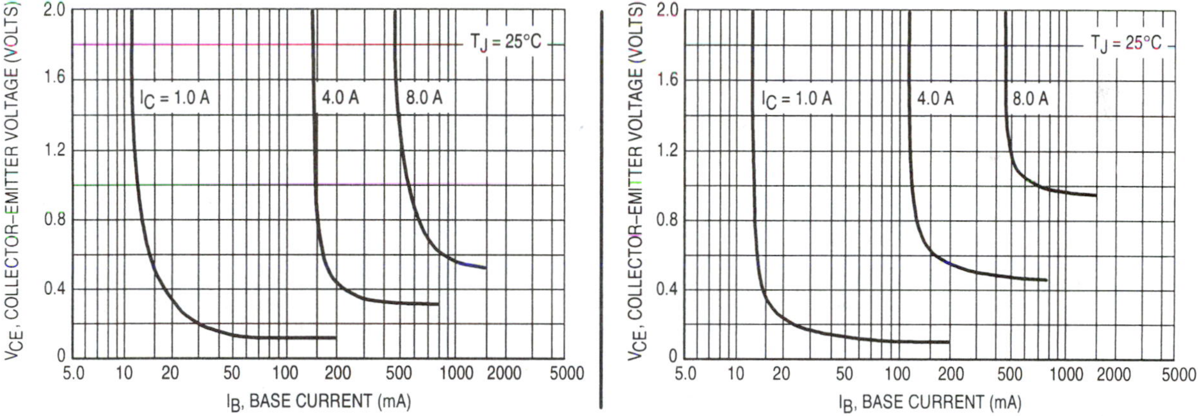

Figure 4. Collector Saturation Region

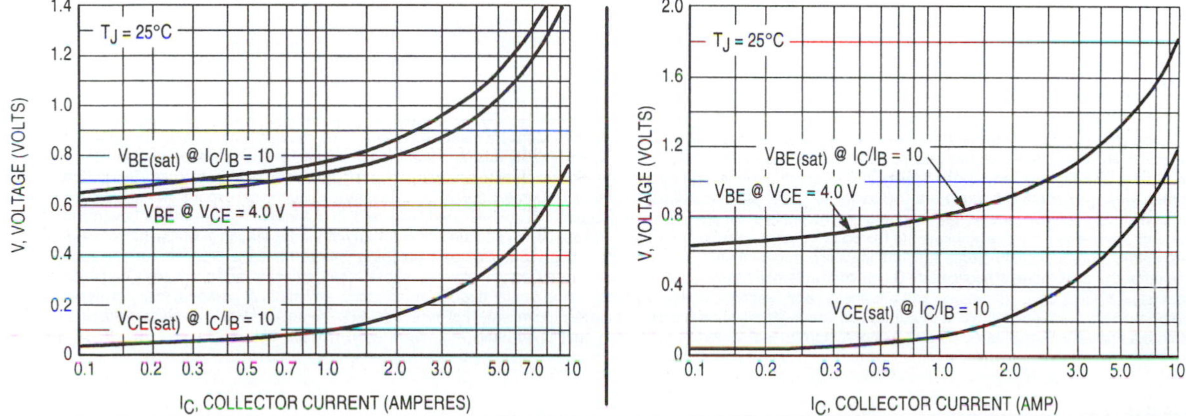

Figure 5. "On" Voltages

2N3055 MJ2955

PACKAGE DIMENSIONS

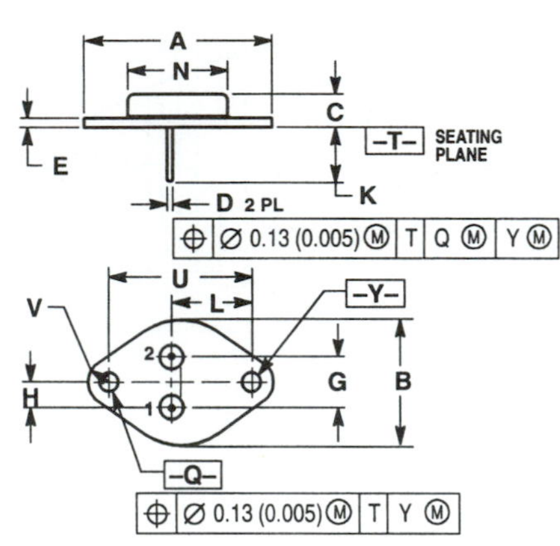

NOTES:
1. DIMENSIONING AND TOLERANCING PER ANSI Y14.5M, 1982.
2. CONTROLLING DIMENSION: INCH.
3. ALL RULES AND NOTES ASSOCIATED WITH REFERENCED TO–204AA OUTLINE SHALL APPLY.

DIM	INCHES MIN	INCHES MAX	MILLIMETERS MIN	MILLIMETERS MAX
A	1.550 REF		39.37 REF	
B	—	1.050	—	26.67
C	0.250	0.335	6.35	8.51
D	0.038	0.043	0.97	1.09
E	0.055	0.070	1.40	1.77
G	0.430 BSC		10.92 BSC	
H	0.215 BSC		5.46 BSC	
K	0.440	0.480	11.18	12.19
L	0.665 BSC		16.89 BSC	
N	—	0.830	—	21.08
Q	0.151	0.165	3.84	4.19
U	1.187 BSC		30.15 BSC	
V	0.131	0.188	3.33	4.77

STYLE 1:
PIN 1. BASE
2. EMITTER
CASE: COLLECTOR

CASE 1–07
TO–204AA (TO–3)
ISSUE Z

Motorola reserves the right to make changes without further notice to any products herein. Motorola makes no warranty, representation or guarantee regarding the suitability of its products for any particular purpose, nor does Motorola assume any liability arising out of the application or use of any product or circuit, and specifically disclaims any and all liability, including without limitation consequential or incidental damages. "Typical" parameters can and do vary in different applications. All operating parameters, including "Typicals" must be validated for each customer application by customer's technical experts. Motorola does not convey any license under its patent rights nor the rights of others. Motorola products are not designed, intended, or authorized for use as components in systems intended for surgical implant into the body, or other applications intended to support or sustain life, or for any other application in which the failure of the Motorola product could create a situation where personal injury or death may occur. Should Buyer purchase or use Motorola products for any such unintended or unauthorized application, Buyer shall indemnify and hold Motorola and its officers, employees, subsidiaries, affiliates, and distributors harmless against all claims, costs, damages, and expenses, and reasonable attorney fees arising out of, directly or indirectly, any claim of personal injury or death associated with such unintended or unauthorized use, even if such claim alleges that Motorola was negligent regarding the design or manufacture of the part. Motorola and Ⓜ are registered trademarks of Motorola, Inc. Motorola, Inc. is an Equal Opportunity/Affirmative Action Employer.

How to reach us:
USA / EUROPE: Motorola Literature Distribution;
P.O. Box 20912; Phoenix, Arizona 85036. 1–800–441–2447

MFAX: RMFAX0@email.sps.mot.com – TOUCHTONE (602) 244–6609
INTERNET: http://Design–NET.com

JAPAN: Nippon Motorola Ltd.; Tatsumi–SPD–JLDC, Toshikatsu Otsuki,
6F Seibu–Butsuryu–Center, 3–14–2 Tatsumi Koto–Ku, Tokyo 135, Japan. 03–3521–8315

HONG KONG: Motorola Semiconductors H.K. Ltd.; 8B Tai Ping Industrial Park,
51 Ting Kok Road, Tai Po, N.T., Hong Kong. 852–26629298

MOTOROLA
SEMICONDUCTOR TECHNICAL DATA

Order this document by MPF102/D

JFET VHF Amplifier
N–Channel — Depletion

MPF102

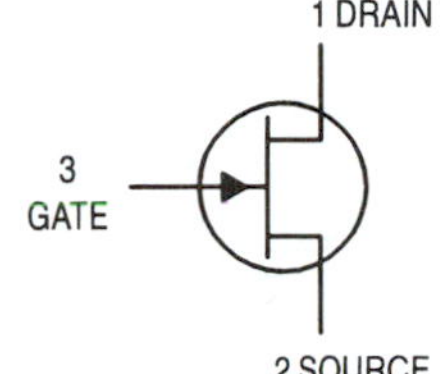

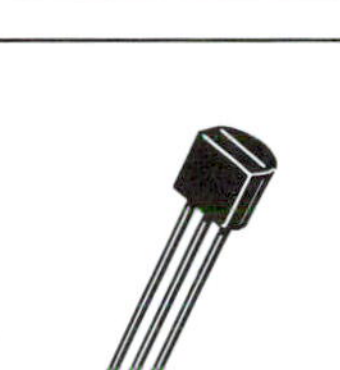

CASE 29–04, STYLE 5
TO–92 (TO–226AA)

MAXIMUM RATINGS

Rating	Symbol	Value	Unit
Drain–Source Voltage	V_{DS}	25	Vdc
Drain–Gate Voltage	V_{DG}	25	Vdc
Gate–Source Voltage	V_{GS}	–25	Vdc
Gate Current	I_G	10	mAdc
Total Device Dissipation @ T_A = 25°C Derate above 25°C	P_D	350 2.8	mW mW/°C
Junction Temperature Range	T_J	125	°C
Storage Temperature Range	T_{stg}	–65 to +150	°C

ELECTRICAL CHARACTERISTICS (T_A = 25°C unless otherwise noted)

Characteristic	Symbol	Min	Max	Unit
OFF CHARACTERISTICS				
Gate–Source Breakdown Voltage (I_G = –10 μAdc, V_{DS} = 0)	$V_{(BR)GSS}$	–25	—	Vdc
Gate Reverse Current (V_{GS} = –15 Vdc, V_{DS} = 0) (V_{GS} = –15 Vdc, V_{DS} = 0, T_A = 100°C)	I_{GSS}	 — —	 –2.0 –2.0	 nAdc μAdc
Gate–Source Cutoff Voltage (V_{DS} = 15 Vdc, I_D = 2.0 nAdc)	$V_{GS(off)}$	—	–8.0	Vdc
Gate–Source Voltage (V_{DS} = 15 Vdc, I_D = 0.2 mAdc)	V_{GS}	–0.5	–7.5	Vdc
ON CHARACTERISTICS				
Zero–Gate–Voltage Drain Current(1) (V_{DS} = 15 Vdc, V_{GS} = 0 Vdc)	I_{DSS}	2.0	20	mAdc
SMALL–SIGNAL CHARACTERISTICS				
Forward Transfer Admittance(1) (V_{DS} = 15 Vdc, V_{GS} = 0, f = 1.0 kHz) (V_{DS} = 15 Vdc, V_{GS} = 0, f = 100 MHz)	$\|y_{fs}\|$	 2000 1600	 7500 —	μmhos
Input Admittance (V_{DS} = 15 Vdc, V_{GS} = 0, f = 100 MHz)	$Re(y_{is})$	—	800	μmhos
Output Conductance (V_{DS} = 15 Vdc, V_{GS} = 0, f = 100 MHz)	$Re(y_{os})$	—	200	μmhos
Input Capacitance (V_{DS} = 15 Vdc, V_{GS} = 0, f = 1.0 MHz)	C_{iss}	—	7.0	pF
Reverse Transfer Capacitance (V_{DS} = 15 Vdc, V_{GS} = 0, f = 1.0 MHz)	C_{rss}	—	3.0	pF

1. Pulse Test; Pulse Width ≤ 630 ms, Duty Cycle ≤ 10%.

© Motorola, Inc. 1997

MOTOROLA

MPF102

POWER GAIN

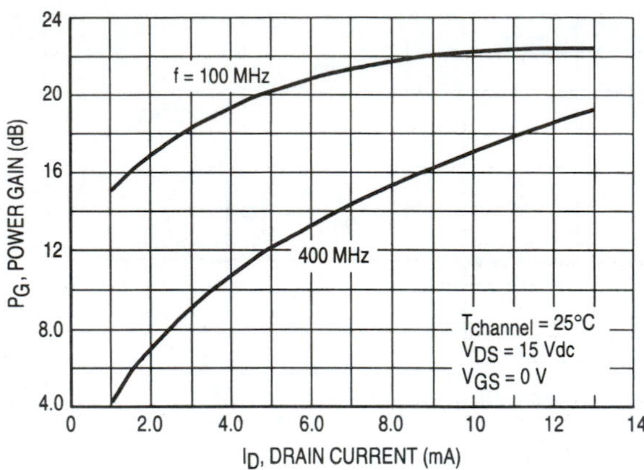

Figure 1. Effects of Drain Current

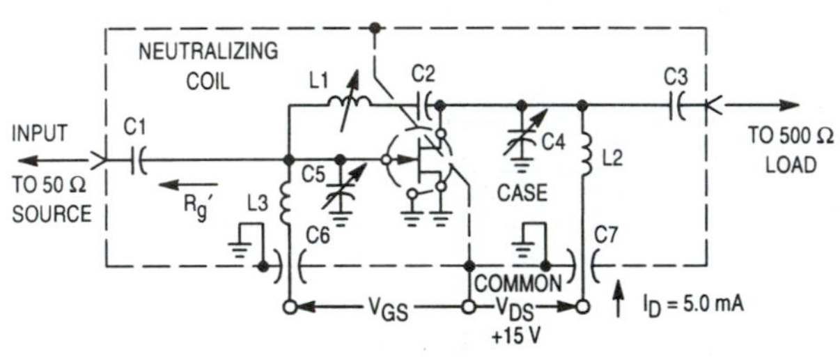

Adjust V_{GS} for
I_D = 50 mA
V_{GS} < 0 Volts

NOTE: The noise source is a hot–cold body (AIL type 70 or equivalent) with a test receiver (AIL type 136 or equivalent).

Reference Designation	VALUE	
	100 MHz	400 MHz
C1	7.0 pF	1.8 pF
C2	1000 pF	17 pF
C3	3.0 pF	1.0 pF
C4	1–12 pF	0.8–8.0 pF
C5	1–12 pF	0.8–8.0 pF
C6	0.0015 μF	0.001 μF
C7	0.0015 μF	0.001 μF
L1	3.0 μH*	0.2 μH**
L2	0.15 μH*	0.03 μH**
L3	0.14 μH*	0.022 μH**

*L1 17 turns, (approx. — depends upon circuit layout) AWG #28 enameled copper wire, close wound on 9/32″ ceramic coil form. Tuning provided by a powdered iron slug.

*L2 4–1/2 turns, AWG #18 enameled copper wire, 5/16″ long, 3/8″ I.D. (AIR CORE).

*L3 3–1/2 turns, AWG #18 enameled copper wire, 1/4″ long, 3/8″ I.D. (AIR CORE).

**L1 6 turns, (approx. — depends upon circuit layout) AWG #24 enameled copper wire, close wound on 7/32″ ceramic coil form. Tuning provided by an aluminum slug.

**L2 1 turn, AWG #16 enameled copper wire, 3/8″ I.D. (AIR CORE).

**L3 1/2 turn, AWG #16 enameled copper wire, 1/4″ I.D. (AIR CORE).

Figure 2. 100 MHz and 400 MHz Neutralized Test Circuit

MPF102

NOISE FIGURE
($T_{channel}$ = 25°C)

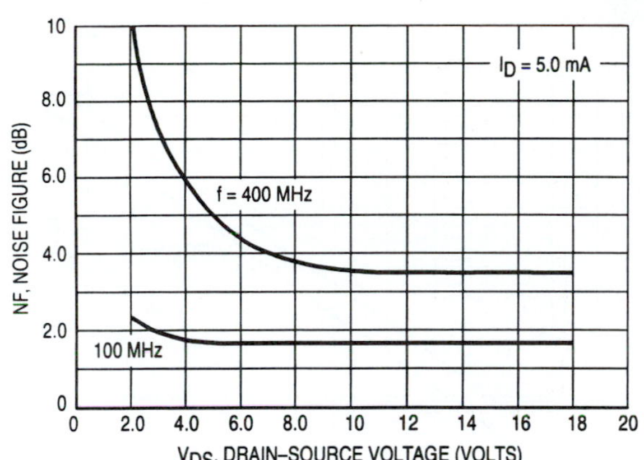

Figure 3. Effects of Drain–Source Voltage

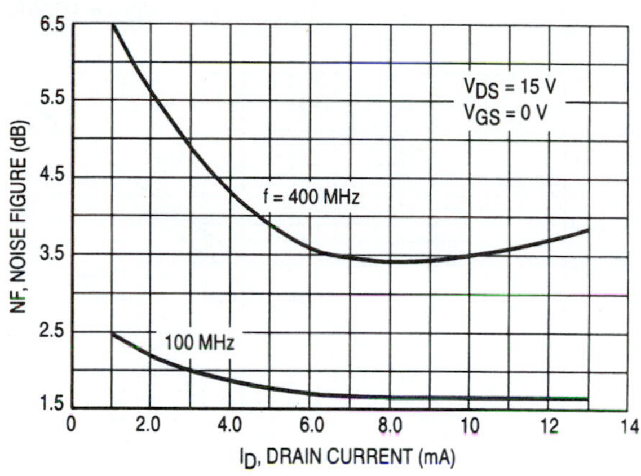

Figure 4. Effects of Drain Current

INTERMODULATION CHARACTERISTICS

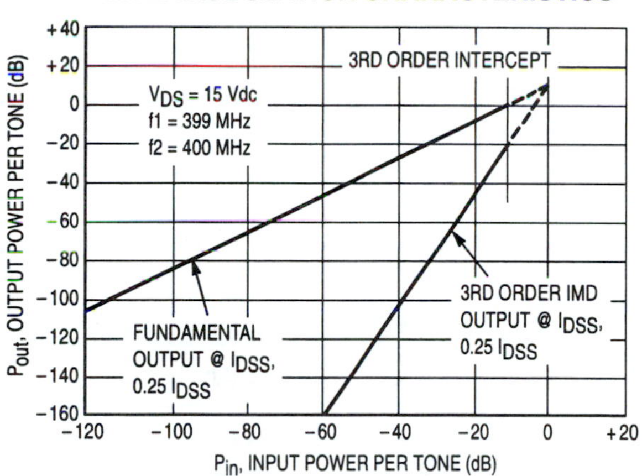

Figure 5. Third Order Intermodulation Distortion

MPF102

COMMON SOURCE CHARACTERISTICS
ADMITTANCE PARAMETERS
(V_{DS} = 15 Vdc, $T_{channel}$ = 25°C)

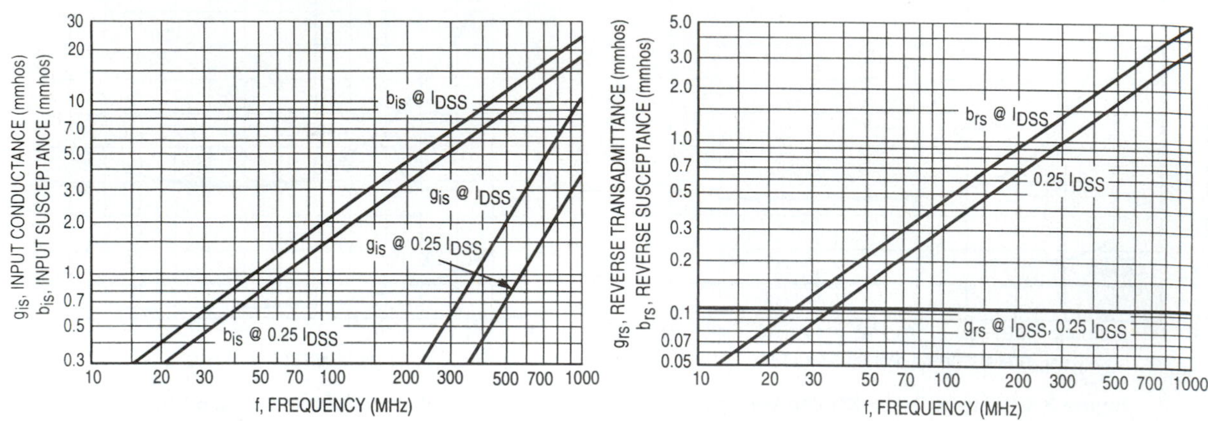

Figure 6. Input Admittance (y_{is})

Figure 7. Reverse Transfer Admittance (y_{rs})

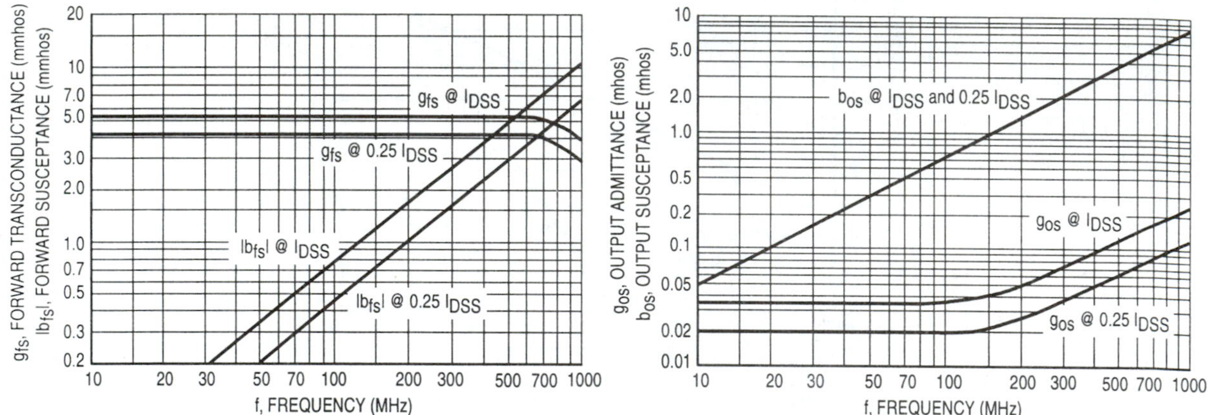

Figure 8. Forward Transadmittance (y_{fs})

Figure 9. Output Admittance (y_{os})

MPF102

COMMON SOURCE CHARACTERISTICS
S–PARAMETERS

(V_{DS} = 15 Vdc, $T_{channel}$ = 25°C, Data Points in MHz)

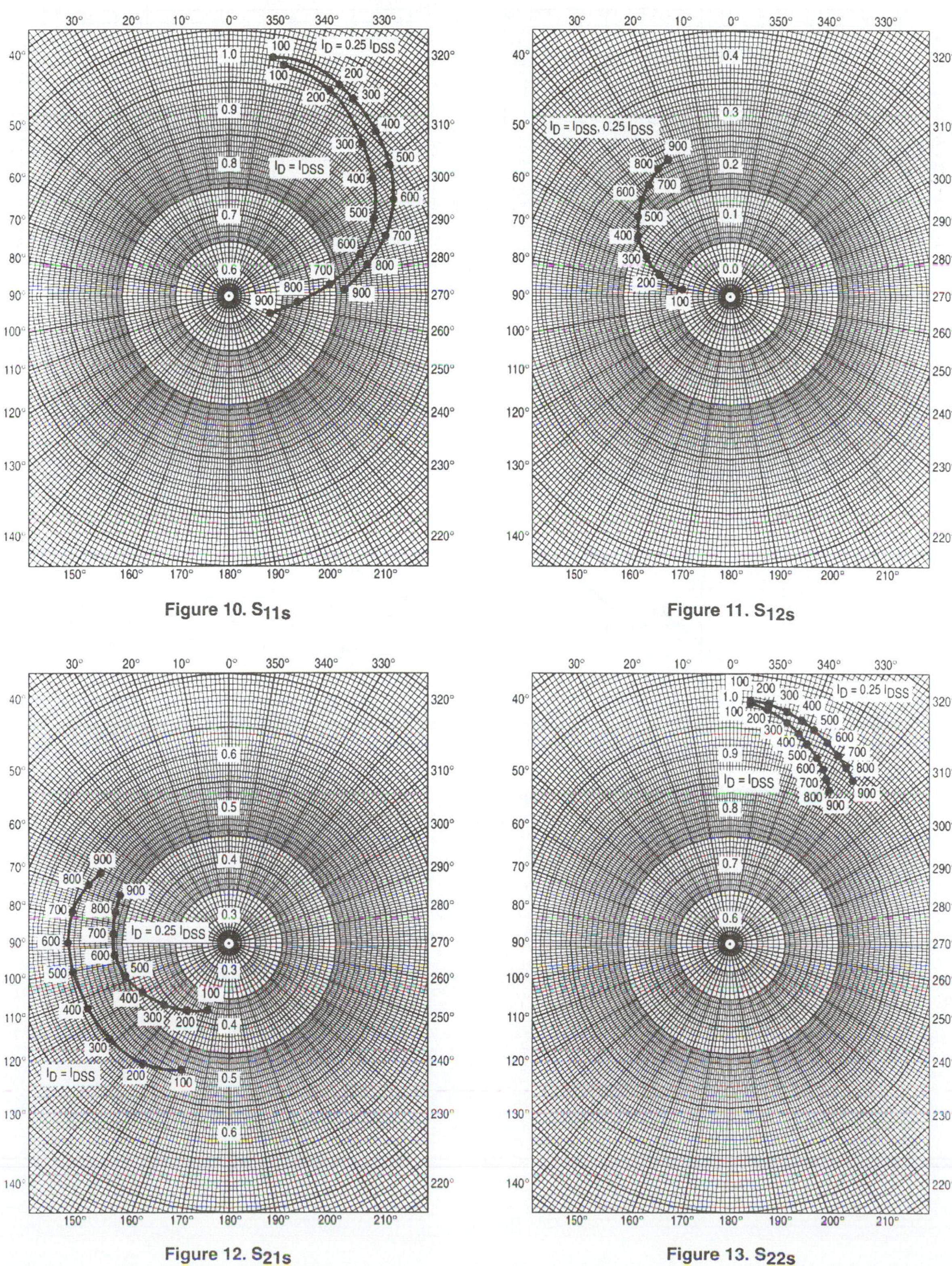

Figure 10. S_{11s}

Figure 11. S_{12s}

Figure 12. S_{21s}

Figure 13. S_{22s}

MPF102

COMMON GATE CHARACTERISTICS

ADMITTANCE PARAMETERS

(V_{DG} = 15 Vdc, $T_{channel}$ = 25°C)

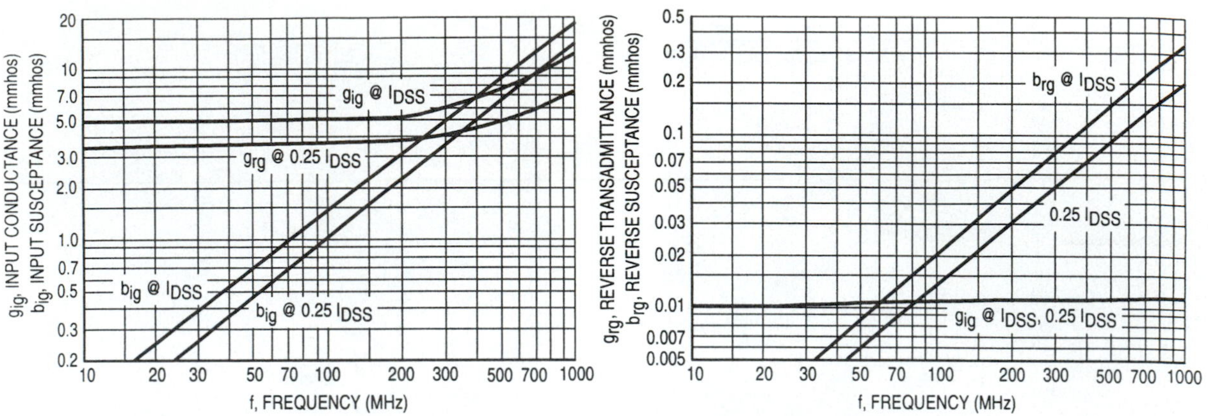

Figure 14. Input Admittance (y_{ig})

Figure 15. Reverse Transfer Admittance (y_{rg})

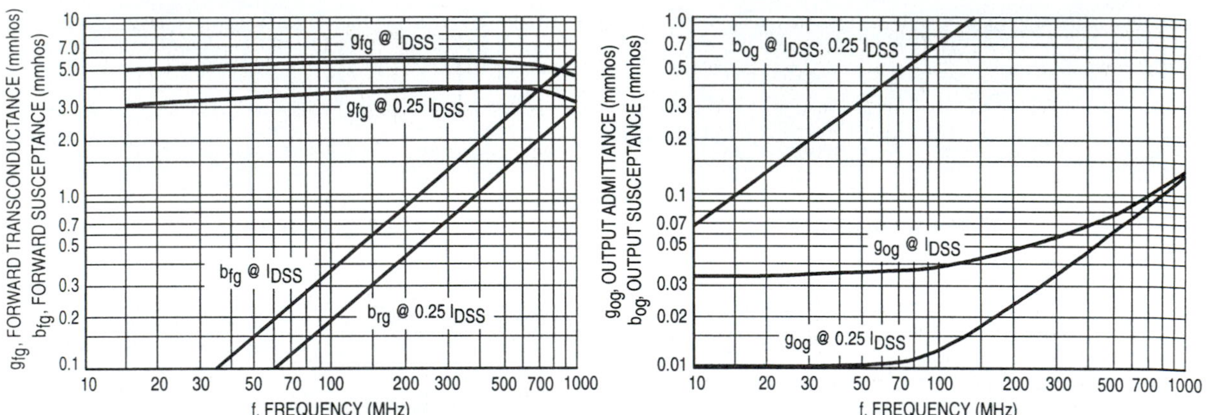

Figure 16. Forward Transfer Admittance (y_{fg})

Figure 17. Output Admittance (y_{og})

COMMON GATE CHARACTERISTICS
S–PARAMETERS
(V_{DS} = 15 Vdc, $T_{channel}$ = 25°C, Data Points in MHz)

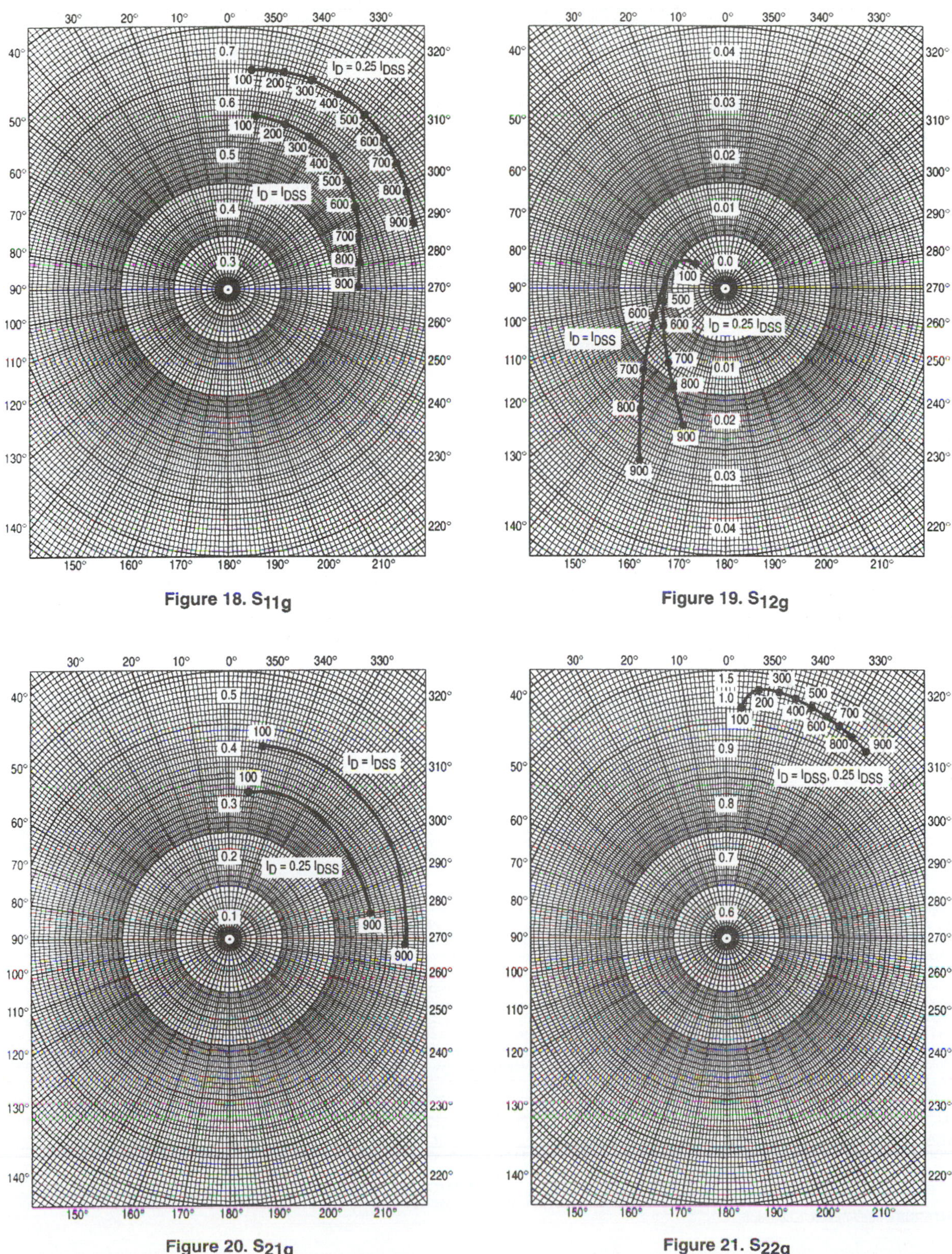

Figure 18. S_{11g}

Figure 19. S_{12g}

Figure 20. S_{21g}

Figure 21. S_{22g}

MPF102

PACKAGE DIMENSIONS

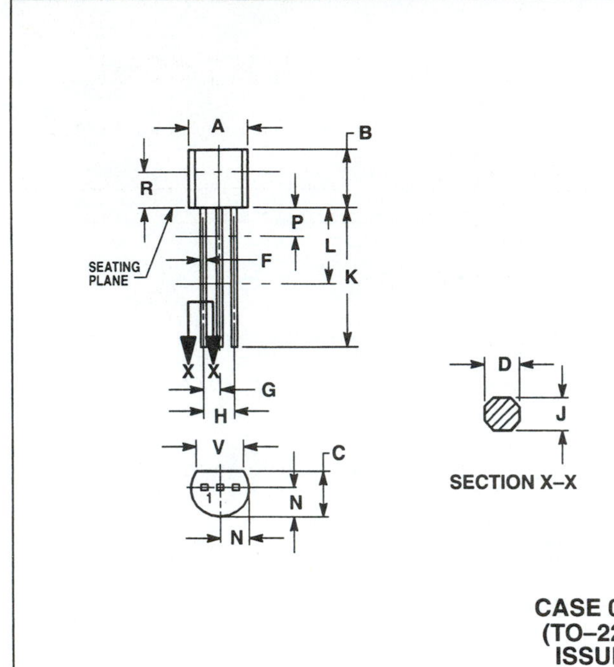

NOTES:
1. DIMENSIONING AND TOLERANCING PER ANSI Y14.5M, 1982.
2. CONTROLLING DIMENSION: INCH.
3. CONTOUR OF PACKAGE BEYOND DIMENSION R IS UNCONTROLLED.
4. DIMENSION F APPLIES BETWEEN P AND L. DIMENSION D AND J APPLY BETWEEN L AND K MINIMUM. LEAD DIMENSION IS UNCONTROLLED IN P AND BEYOND DIMENSION K MINIMUM.

	INCHES		MILLIMETERS	
DIM	MIN	MAX	MIN	MAX
A	0.175	0.205	4.45	5.20
B	0.170	0.210	4.32	5.33
C	0.125	0.165	3.18	4.19
D	0.016	0.022	0.41	0.55
F	0.016	0.019	0.41	0.48
G	0.045	0.055	1.15	1.39
H	0.095	0.105	2.42	2.66
J	0.015	0.020	0.39	0.50
K	0.500	—	12.70	—
L	0.250	—	6.35	—
N	0.080	0.105	2.04	2.66
P	—	0.100	—	2.54
R	0.115	—	2.93	—
V	0.135	—	3.43	—

CASE 029–04
(TO–226AA)
ISSUE AD

STYLE 5:
PIN 1. DRAIN
2. SOURCE
3. GATE

Motorola reserves the right to make changes without further notice to any products herein. Motorola makes no warranty, representation or guarantee regarding the suitability of its products for any particular purpose, nor does Motorola assume any liability arising out of the application or use of any product or circuit, and specifically disclaims any and all liability, including without limitation consequential or incidental damages. "Typical" parameters which may be provided in Motorola data sheets and/or specifications can and do vary in different applications and actual performance may vary over time. All operating parameters, including "Typicals" must be validated for each customer application by customer's technical experts. Motorola does not convey any license under its patent rights nor the rights of others. Motorola products are not designed, intended, or authorized for use as components in systems intended for surgical implant into the body, or other applications intended to support or sustain life, or for any other application in which the failure of the Motorola product could create a situation where personal injury or death may occur. Should Buyer purchase or use Motorola products for any such unintended or unauthorized application, Buyer shall indemnify and hold Motorola and its officers, employees, subsidiaries, affiliates, and distributors harmless against all claims, costs, damages, and expenses, and reasonable attorney fees arising out of, directly or indirectly, any claim of personal injury or death associated with such unintended or unauthorized use, even if such claim alleges that Motorola was negligent regarding the design or manufacture of the part. Motorola and Ⓜ are registered trademarks of Motorola, Inc. Motorola, Inc. is an Equal Opportunity/Affirmative Action Employer.

Mfax is a trademark of Motorola, Inc.

How to reach us:
USA/EUROPE/Locations Not Listed: Motorola Literature Distribution; P.O. Box 5405, Denver, Colorado 80217. 303–675–2140 or 1–800–441–2447

Mfax™: RMFAX0@email.sps.mot.com – TOUCHTONE 602–244–6609
– US & Canada ONLY 1–800–774–1848

INTERNET: http://motorola.com/sps

JAPAN: Nippon Motorola Ltd.: SPD, Strategic Planning Office, 4–32–1, Nishi–Gotanda, Shinagawa–ku, Tokyo 141, Japan. 81–3–5487–8488

ASIA/PACIFIC: Motorola Semiconductors H.K. Ltd.; 8B Tai Ping Industrial Park, 51 Ting Kok Road, Tai Po, N.T., Hong Kong. 852–26629298

MOTOROLA
SEMICONDUCTOR TECHNICAL DATA

Order this document by VN2406L/D

TMOS FET Transistor

N–Channel — Enhancement

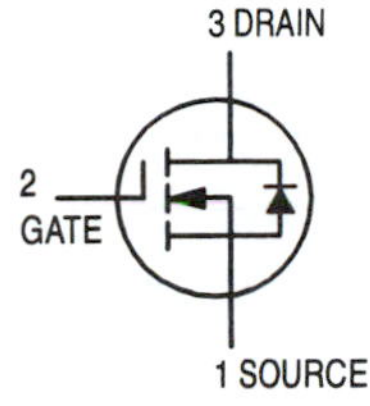

VN2406L

Motorola Preferred Device

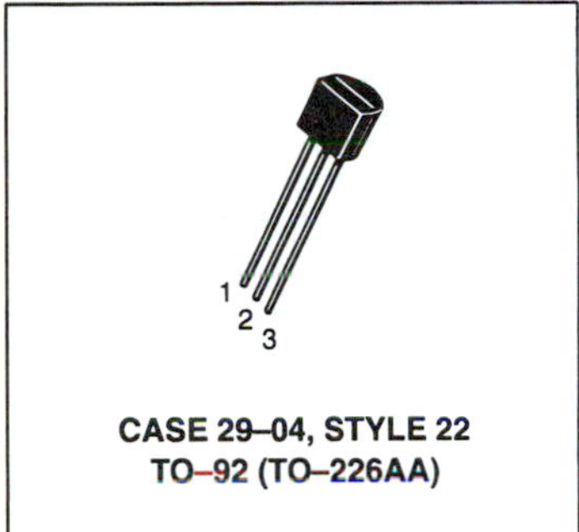

CASE 29–04, STYLE 22
TO–92 (TO–226AA)

MAXIMUM RATINGS

Rating	Symbol	Value	Unit
Drain–Source Voltage	V_{DSS}	240	Vdc
Drain–Gate Voltage	V_{DGR}	60	Vdc
Gate–Source Voltage – Continuous – Non–repetitive ($t_p \leq 50\ \mu s$)	 V_{GS} V_{GSM}	 ± 20 ± 40	 Vdc Vpk
Continuous Drain Current	I_D	200	mAdc
Pulsed Drain Current	I_{DM}	500	mAdc
Power Dissipation @ $T_C = 25°C$ Derate above 25°C	P_D	350 2.8	mW mW/°C
Operating and Storage Temperature	T_J, T_{stg}	—	°C

THERMAL CHARACTERISTICS

Characteristic	Symbol	Max	Unit
Thermal Resistance, Junction to Ambient	$R_{\theta JA}$	312.5	°C/W
Maximum Lead Temperature for Soldering Purposes, 1/16″ from case for 10 seconds	T_L	300	°C

ELECTRICAL CHARACTERISTICS ($T_A = 25°C$ unless otherwise noted)

Characteristic	Symbol	Min	Max	Unit
STATIC CHARACTERISTICS				
Drain–Source Breakdown Voltage ($V_{GS} = 0$, $I_D = 100\ \mu A$)	$V_{(BR)DSS}$	240	—	Vdc
Zero Gate Voltage Drain Current ($V_{DS} = 120$ Vdc, $V_{GS} = 0$) ($V_{DS} = 120$ Vdc, $V_{GS} = 0$, $T_A = 125°C$)	I_{DSS}	 — —	 10 500	µAdc
Gate– Body Leakage ($V_{DS} = 0$, $V_{GS} = \pm 15$ V)	I_{GSS}	—	±100	nAdc
Gate Threshold Voltage ($V_{DS} = V_{GS}$, $I_D = 1.0$ mA)	$V_{GS(th)}$	0.8	2.0	Vdc
On–State Drain Current(1) ($V_{GS} = 10$ V, $V_{DS} \geq 2.0\ V_{DS(on)}$)	$I_{D(on)}$	1.0	—	Adc
Drain–Source On Resistance(1) ($V_{GS} = 2.5$ V, $I_D = 0.1$ A) ($V_{GS} = 10$ V, $I_D = 0.5$ A)	$r_{DS(on)}$	 — —	 10 6.0	Ω
Forward Transconductance(1) ($V_{DS} = 10$ V, $I_D = 0.5$ A)	g_{fs}	300	—	mS

1. Pulse Test; Pulse Width < 300 µs, Duty Cycle ≤ 2.0%.

TMOS is a registered trademark of Motorola, Inc.

Preferred devices are Motorola recommended choices for future use and best overall value.

REV 1

© Motorola, Inc. 1997

VN2406L

ELECTRICAL CHARACTERISTICS (T_A = 25°C unless otherwise noted) (Continued)

Characteristic		Symbol	Min	Max	Unit
DYNAMIC CHARACTERISTICS					
Input Capacitance	(V_{DS} = 25 Vdc, V_{GS} = 0, f = 1.0 MHz)	C_{iss}	—	125	pF
Output Capacitance		C_{oss}	—	50	pF
Reverse Transfer Capacitance		C_{rss}	—	20	pF
SWITCHING CHARACTERISTICS					
Turn–On Time	(V_{DD} = 60 Vdc, I_D = 0.4 A, R_L = 150 Ω, R_G = 25 Ω)	$t_{(on)}$	—	8.0	ns
		$t_{(r)}$	—	8.0	ns
Turn–Off Time		$t_{(off)}$	—	23	ns
		$t_{(f)}$	—	34	ns

Order this document by TL071C/D

MOTOROLA

TL071C,AC
TL072C,AC
TL074C,AC

Low Noise, JFET Input Operational Amplifiers

These low noise JFET input operational amplifiers combine two state–of–the–art analog technologies on a single monolithic integrated circuit. Each internally compensated operational amplifier has well matched high voltage JFET input device for low input offset voltage. The BIFET technology provides wide bandwidths and fast slew rates with low input bias currents, input offset currents, and supply currents. Moreover, the devices exhibit low noise and low harmonic distortion, making them ideal for use in high fidelity audio amplifier applications.

These devices are available in single, dual and quad operational amplifiers which are pin–compatible with the industry standard MC1741, MC1458, and the MC3403/LM324 bipolar products.

- Low Input Noise Voltage: 18 $nV/\sqrt{Hz}$ Typ
- Low Harmonic Distortion: 0.01% Typ
- Low Input Bias and Offset Currents
- High Input Impedance: 10^{12} Ω Typ
- High Slew Rate: 13 V/μs Typ
- Wide Gain Bandwidth: 4.0 MHz Typ
- Low Supply Current: 1.4 mA per Amp

LOW NOISE, JFET INPUT OPERATIONAL AMPLIFIERS

SEMICONDUCTOR TECHNICAL DATA

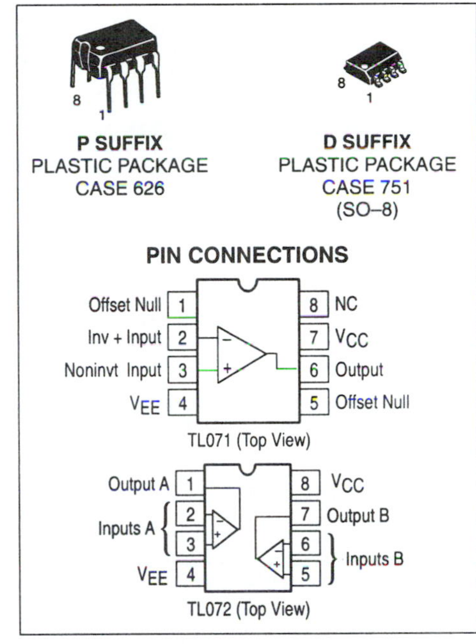

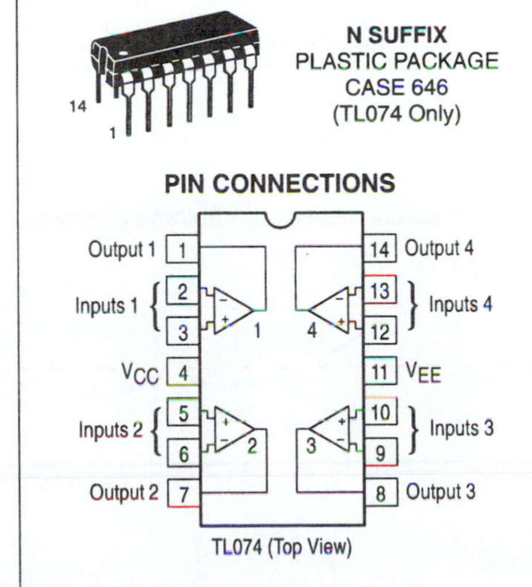

ORDERING INFORMATION

Op Amp Function	Device	Operating Temperature Range	Package
Single	TL071CD	T_A = 0° to +70°C	SO–8
	TL071ACP		Plastic DIP
Dual	TL072CD	T_A = 0° to +70°C	SO–8
	TL072ACP		Plastic DIP
Quad	TL074CN, ACN	T_A = 0° to +70°C	Plastic DIP

© Motorola, Inc. 1997 Rev 1

TL071C,AC TL072C,AC TL074C,AC

MAXIMUM RATINGS

Rating	Symbol	Value	Unit
Supply Voltage	V_{CC} V_{EE}	18 –18	V
Differential Input Voltage	V_{ID}	±30	V
Input Voltage Range (Note 1)	V_{IDR}	±15	V
Output Short Circuit Duration (Note 2)	t_{SC}	Continuous	
Power Dissipation Plastic Package (N, P) Derate above T_A = 47°C	 P_D $1.0/\theta_{JA}$	 680 10	 mW mW/°C
Operating Ambient Temperature Range	T_A	0 to +70	°C
Storage Temperature Range	T_{stg}	–65 to +150	°C

NOTES: 1. The magnitude of the input voltage must not exceed the magnitude of the supply voltage or 15 V, whichever is less.
2. The output may be shorted to ground or either supply. Temperature and/or supply voltages must be limited to ensure that power dissipation ratings are not exceeded.
3. ESD data available upon request.

ELECTRICAL CHARACTERISTICS (V_{CC} = 15 V, V_{EE} = –15 V, T_A = T_{high} to T_{low} [Note 1])

Characteristics	Symbol	Min	Typ	Max	Unit
Input Offset Voltage ($R_S \leq$ 10 k, V_{CM} = 0)	V_{IO}				mV
TL071C, TL072C		–	–	13	
TL074C		–	–	13	
TL07_AC		–	–	7.5	
Input Offset Current (V_{CM} = 0) (Note 2)	I_{IO}				nA
TL07_C		–	–	2.0	
TL07_AC		–	–	2.0	
Input Bias Current (V_{CM} = 0) (Note 2)	I_{IB}				nA
TL07_C		–	–	7.0	
TL07_AC		–	–	7.0	
Large–Signal Voltage Gain (V_O = ±10 V, $R_L \geq$ 2.0 k)	A_{VOL}				V/mV
TL07_C		15	–	–	
TL07_AC		25	–	–	
Output Voltage Swing (Peak–to–Peak)	V_O				V
($R_L \geq$ 10 k)		24	–	–	
($R_L \geq$ 2.0 k)		20	–	–	

NOTES: 1. T_{low} = 0°C for TL071C,AC T_{high} = 70°C for TL071C,AC
0°C for TL072C,AC T_{high} = 70°C for TL072C,AC
0°C for TL074C,AC T_{high} = 70°C for TL074C,AC
2. Input Bias currents of JFET input op amps approximately double for every 10°C rise in junction temperature as shown in Figure 3. To maintain junction temperature as close to ambient temperature as possible, pulse techniques must be used during testing.

Figure 1. Unity Gain Voltage Follower

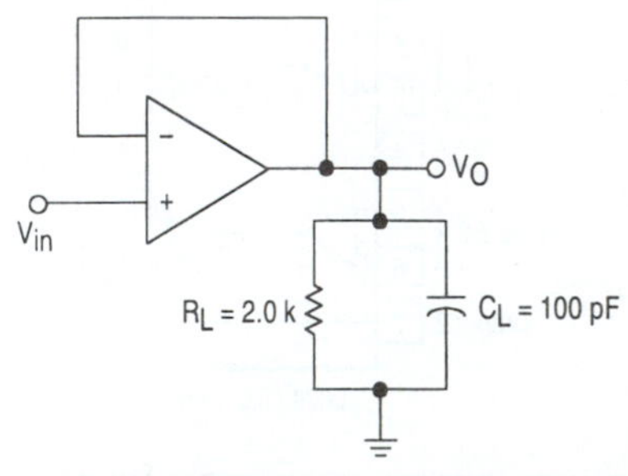

Figure 2. Inverting Gain of 10 Amplifier

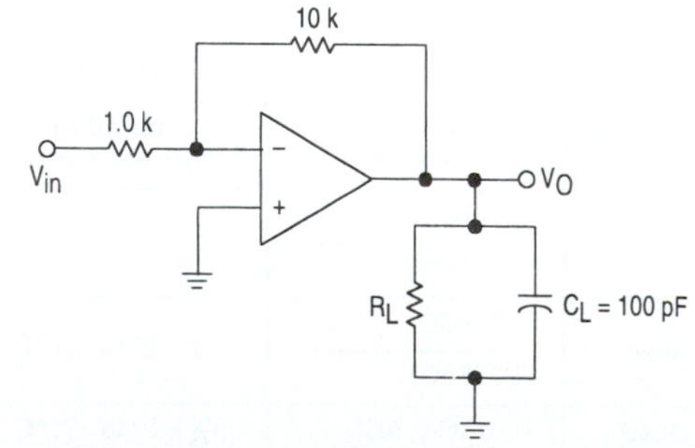

TL071C,AC TL072C,AC TL074C,AC

ELECTRICAL CHARACTERISTICS (V_{CC} = 15 V, V_{EE} = –15 V, T_A = 25°C, unless otherwise noted.)

Characteristics	Symbol	Min	Typ	Max	Unit
Input Offset Voltage (R_S ≤ 10 k, V_{CM} = 0)	V_{IO}				mV
TL071C, TL072C		–	3.0	10	
TL074C		–	3.0	10	
TL07_AC		–	3.0	6.0	
Average Temperature Coefficient of Input Offset Voltage R_S = 50 Ω, T_A = T_{low} to T_{high} (Note 1)	$\Delta V_{IO}/\Delta T$	–	10	–	µV/°C
Input Offset Current (V_{CM} = 0) (Note 2)	I_{IO}				pA
TL07_C		–	5.0	50	
TL07_AC		–	5.0	50	
Input Bias Current (V_{CM} = 0) (Note 2)	I_{IB}				pA
TL07_C		–	30	200	
TL07_AC		–	30	200	
Input Resistance	r_i	–	10^{12}	–	Ω
Common Mode Input Voltage Range	V_{ICR}				V
TL07_C		±10	15, –12	–	
TL07_AC		±11	15, –12	–	
Large–Signal Voltage Gain (V_O = ±10 V, R_L ≥ 2.0 k)	A_{VOL}				V/mV
TL07_C		25	150	–	
TL07_AC		50	150	–	
Output Voltage Swing (Peak–to–Peak) (R_L = 10 k)	V_O	24	28	–	V
Common Mode Rejection Ratio (R_S ≤ 10 k)	CMRR				dB
TL07_C		70	100	–	
TL07_AC		80	100	–	
Supply Voltage Rejection Ratio (R_S ≤ 10 k)	PSRR				dB
TL07_C		70	100	–	
TL07_AC		80	100	–	
Supply Current (Each Amplifier)	I_D	–	1.4	2.5	mA
Unity Gain Bandwidth	BW	–	4.0	–	MHz
Slew Rate (See Figure 1) V_{in} = 10 V, R_L = 2.0 k, C_L = 100 pF	SR	–	13	–	v/µs
Rise Time (See Figure 1)	t_r	–	0.1	–	µs
Overshoot (V_{in} = 20 mV, R_L = 2.0 k, C_L = 100 pF)	OS	–	10	–	%
Equivalent Input Noise Voltage R_S = 100 Ω, f = 1000 Hz	e_n	–	18	–	nV/√Hz
Equivalent Input Noise Current R_S = 100 Ω, f = 1000 Hz	i_n	–	0.01	–	pA/√Hz
Total Harmonic Distortion V_O (RMS) = 10 V, R_S ≤ 1.0 k, R_L ≥ 2.0 k, f = 1000 Hz	THD	–	0.01	–	%
Channel Separation A_V = 100	CS	–	120	–	dB

NOTES: 1. T_{low} = 0°C for TL071C,AC T_{high} = 70°C for TL071C,AC
0°C for TL072C,AC T_{high} = 70°C for TL072C,AC
0°C for TL074C,AC T_{high} = 70°C for TL074C,AC

2. Input Bias currents of JFET input op amps approximately double for every 10°C rise in junction temperature as shown in Figure 3. To maintain junction temperature as close to ambient temperature as possible, pulse techniques must be used during testing.

TL071C,AC TL072C,AC TL074C,AC

Figure 3. Input Bias Current versus Temperature

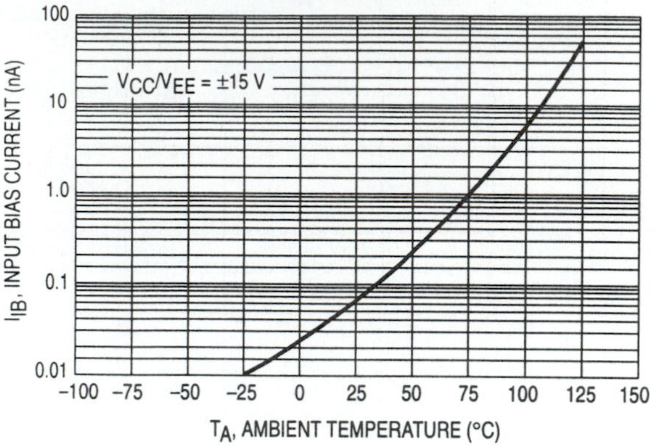

$V_{CC}/V_{EE} = \pm15$ V

I_{IB}, INPUT BIAS CURRENT (nA)

T_A, AMBIENT TEMPERATURE (°C)

Figure 4. Output Voltage Swing versus Frequency

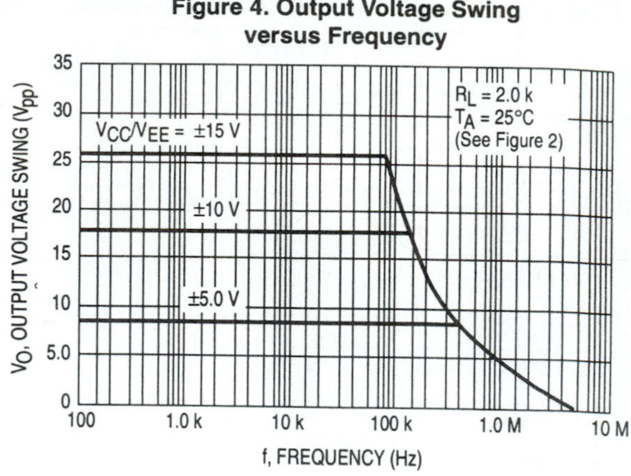

Figure 5. Output Voltage Swing versus Load Resistance

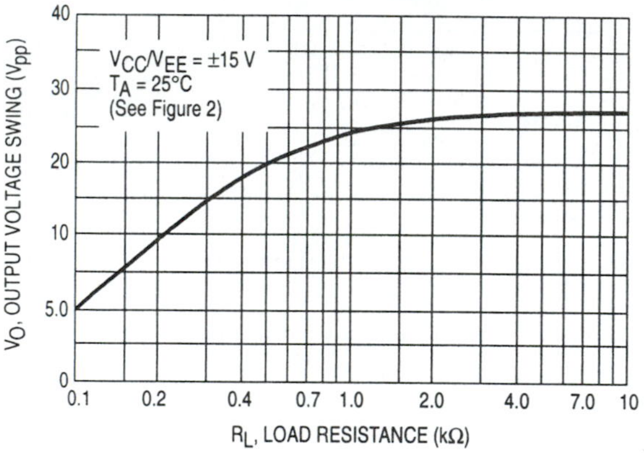

$V_{CC}/V_{EE} = \pm15$ V
$T_A = 25°C$
(See Figure 2)

V_O, OUTPUT VOLTAGE SWING (V_{pp})

R_L, LOAD RESISTANCE (kΩ)

Figure 6. Output Voltage Swing versus Supply Voltage

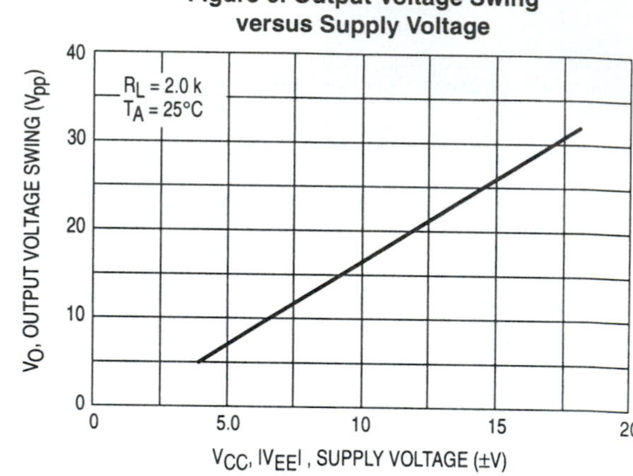

Figure 7. Output Voltage Swing versus Temperature

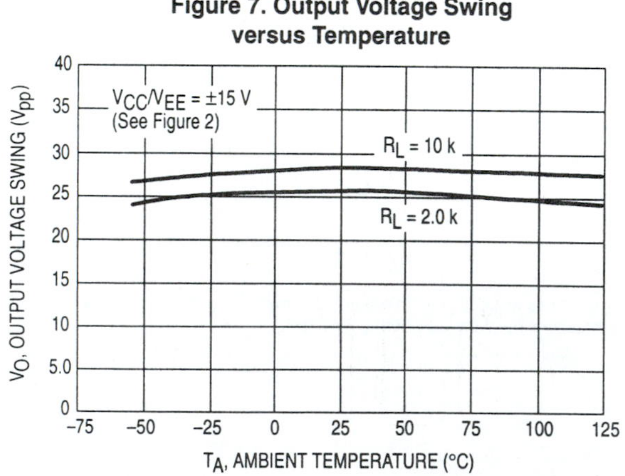

Figure 8. Supply Current per Amplifier versus Temperature

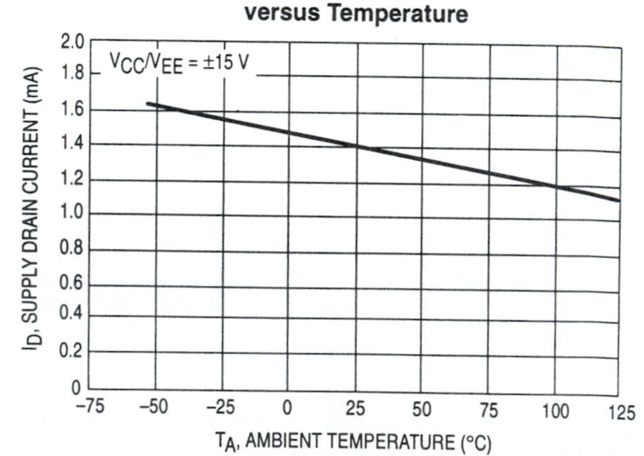

TL071C,AC TL072C,AC TL074C,AC

Figure 9. Large Signal Voltage Gain and Phase Shift versus Frequency

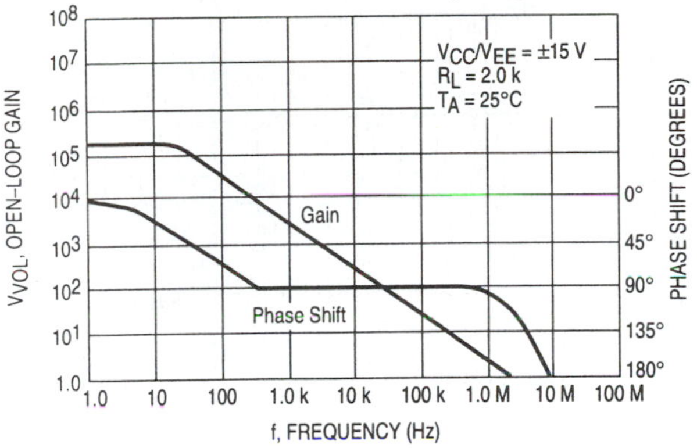

V_{CC}/V_{EE} = ±15 V
R_L = 2.0 k
T_A = 25°C
Gain
Phase Shift
V_{VOL}, OPEN-LOOP GAIN
PHASE SHIFT (DEGREES)
f, FREQUENCY (Hz)

Figure 10. Large Signal Voltage Gain versus Temperature

V_{CC}/V_{EE} = ±15 V
V_O = ±10 V
R_L = 2.0 k
V_{VOL}, VOLTAGE GAIN (V/mV)
T_A, AMBIENT TEMPERATURE (°C)

Figure 11. Normalized Slew Rate versus Temperature

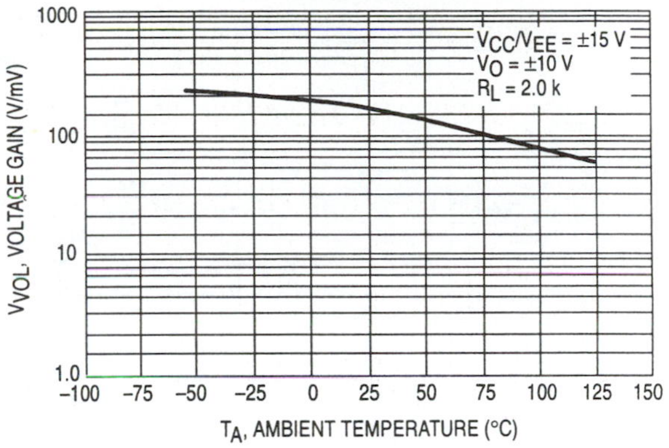

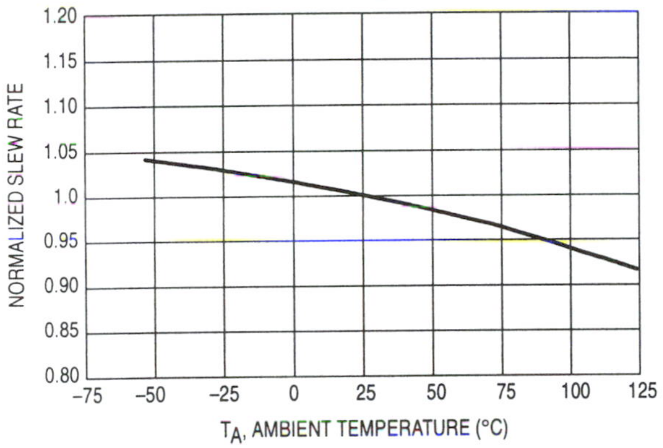

NORMALIZED SLEW RATE
T_A, AMBIENT TEMPERATURE (°C)

Figure 12. Equivalent Input Noise Voltage versus Frequency

V_{CC}/V_{EE} = ±15 Vdc
A_V = 10
R_S = 100 Ω
T_A = 25°C
e_n, EQUIVALENT INPUT NOISE VOLTAGE (nV/√Hz)
f, FREQUENCY (Hz)

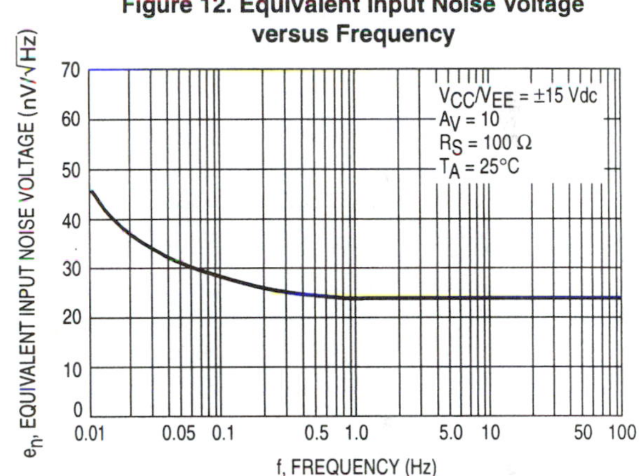

Figure 13. Total Harmonic Distortion versus Frequency

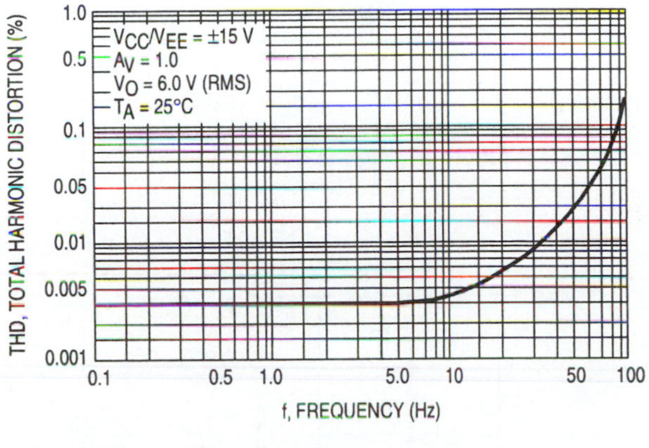

V_{CC}/V_{EE} = ±15 V
A_V = 1.0
V_O = 6.0 V (RMS)
T_A = 25°C
THD, TOTAL HARMONIC DISTORTION (%)
f, FREQUENCY (Hz)

TL071C,AC TL072C,AC TL074C,AC

Representative Schematic Diagram
(Each Amplifier)

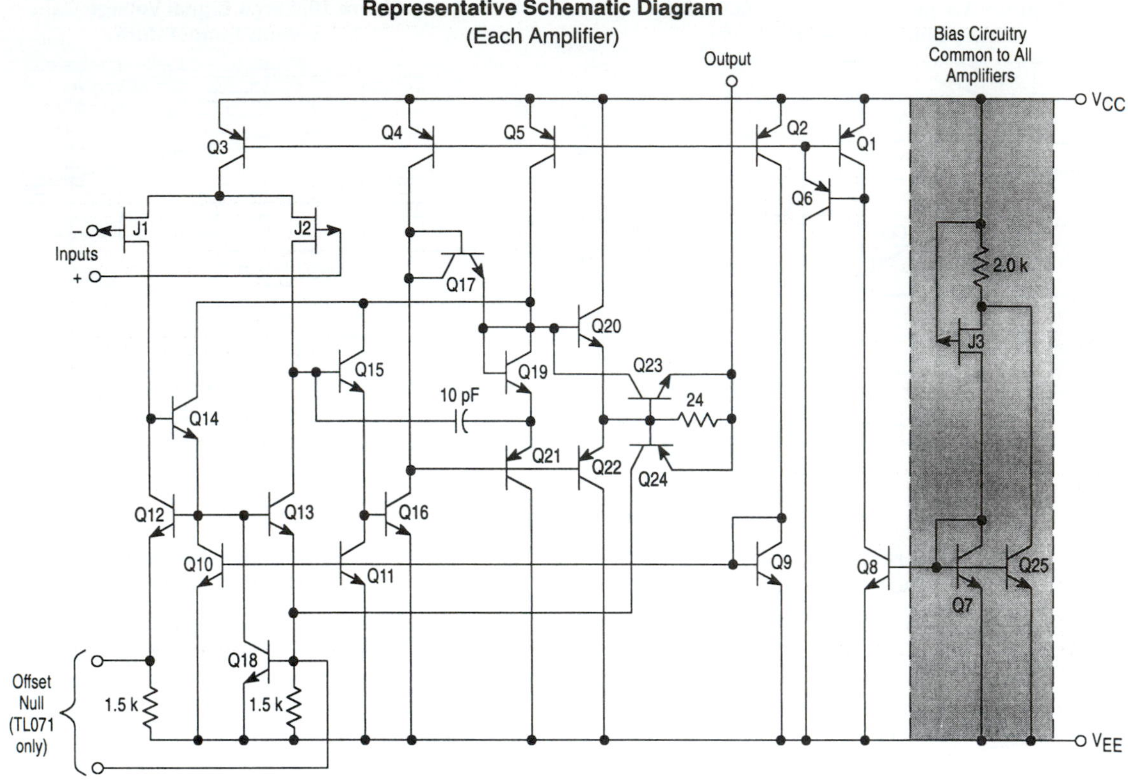

Figure 14. Audio Tone Control Amplifier

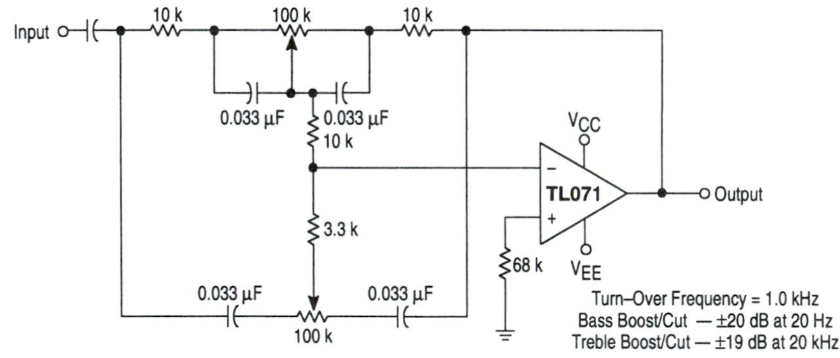

Figure 15. High Q Notch Filter

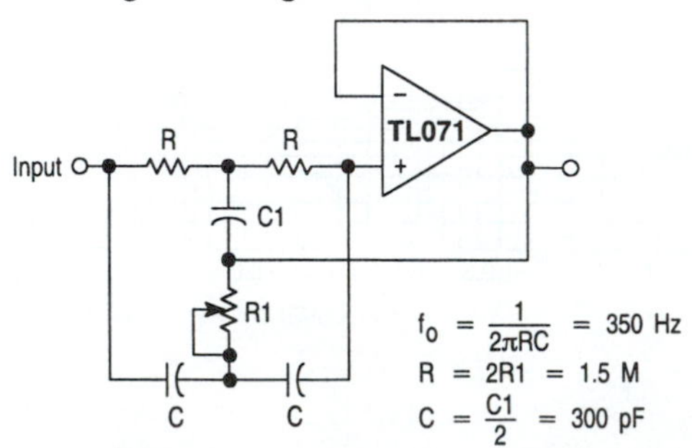

$$f_o = \frac{1}{2\pi RC} = 350\ \text{Hz}$$

$$R = 2R1 = 1.5\ \text{M}$$

$$C = \frac{C1}{2} = 300\ \text{pF}$$

TL071C,AC TL072C,AC TL074C,AC

OUTLINE DIMENSIONS

P SUFFIX
PLASTIC PACKAGE
CASE 626–05
ISSUE K

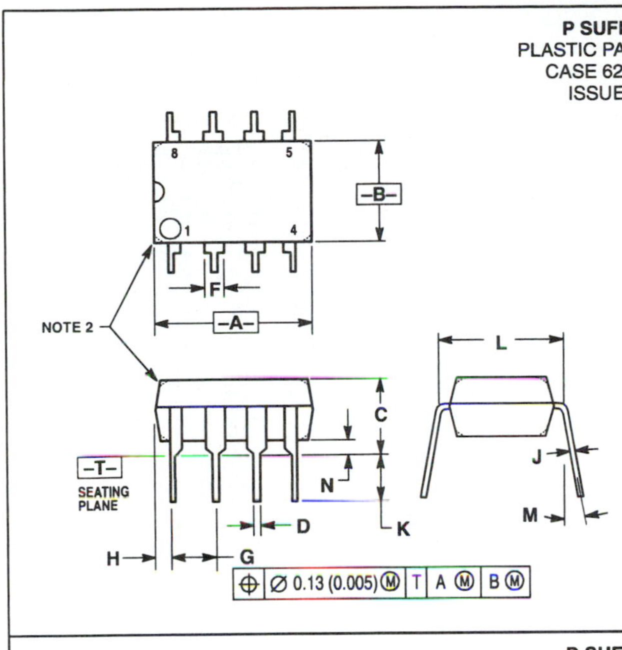

NOTES:
1. DIMENSION L TO CENTER OF LEAD WHEN FORMED PARALLEL.
2. PACKAGE CONTOUR OPTIONAL (ROUND OR SQUARE CORNERS).
3. DIMENSIONING AND TOLERANCING PER ANSI Y14.5M, 1982.

DIM	MILLIMETERS		INCHES	
	MIN	MAX	MIN	MAX
A	9.40	10.16	0.370	0.400
B	6.10	6.60	0.240	0.260
C	3.94	4.45	0.155	0.175
D	0.38	0.51	0.015	0.020
F	1.02	1.78	0.040	0.070
G	2.54 BSC		0.100 BSC	
H	0.76	1.27	0.030	0.050
J	0.20	0.30	0.008	0.012
K	2.92	3.43	0.115	0.135
L	7.62 BSC		0.300 BSC	
M	—	10°	—	10°
N	0.76	1.01	0.030	0.040

D SUFFIX
PLASTIC PACKAGE
CASE 751–05
(SO–8)
ISSUE S

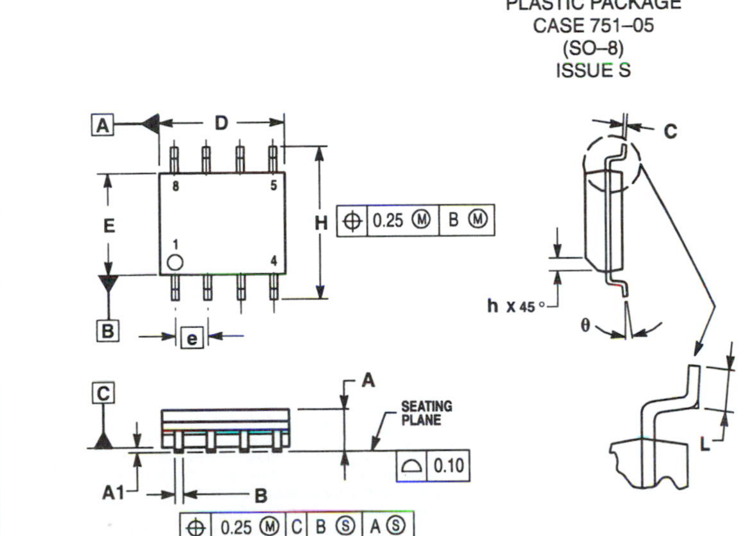

NOTES:
1. DIMENSIONING AND TOLERANCING PER ASME Y14.5M, 1994.
2. DIMENSIONS ARE IN MILLIMETERS.
3. DIMENSION D AND E DO NOT INCLUDE MOLD PROTRUSION.
4. MAXIMUM MOLD PROTRUSION 0.15 PER SIDE.
5. DIMENSION B DOES NOT INCLUDE MOLD PROTRUSION. ALLOWABLE DAMBAR PROTRUSION SHALL BE 0.127 TOTAL IN EXCESS OF THE B DIMENSION AT MAXIMUM MATERIAL CONDITION.

DIM	MILLIMETERS	
	MIN	MAX
A	1.35	1.75
A1	0.10	0.25
B	0.35	0.49
C	0.18	0.25
D	4.80	5.00
E	3.80	4.00
e	1.27 BSC	
H	5.80	6.20
h	0.25	0.50
L	0.40	1.25
θ	0°	7°

TL071C,AC TL072C,AC TL074C,AC

OUTLINE DIMENSIONS

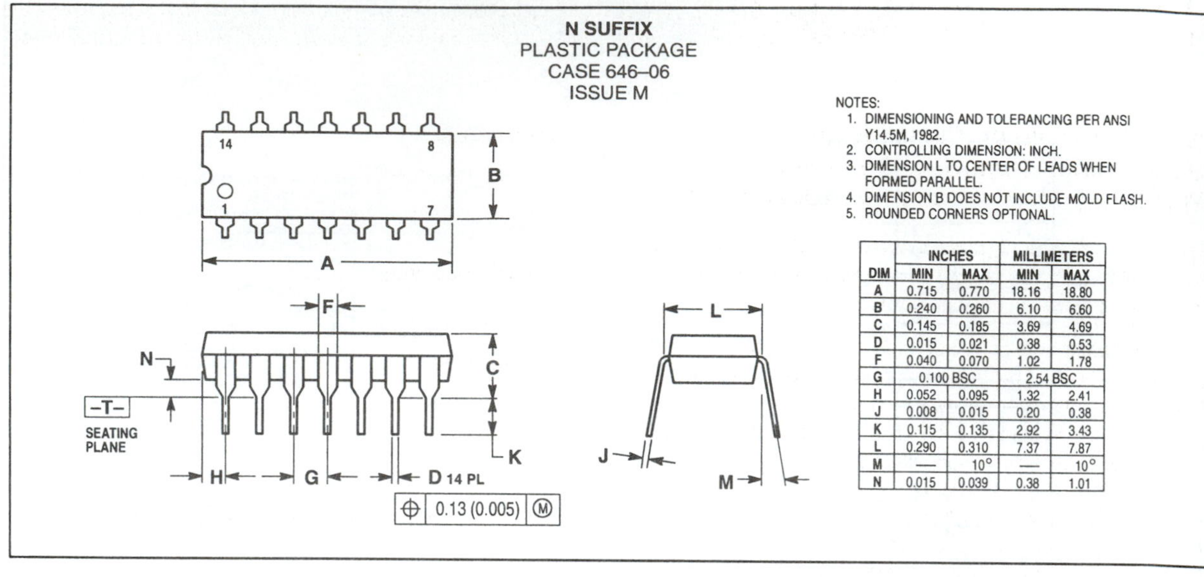

NOTES:
1. DIMENSIONING AND TOLERANCING PER ANSI Y14.5M, 1982.
2. CONTROLLING DIMENSION: INCH.
3. DIMENSION L TO CENTER OF LEADS WHEN FORMED PARALLEL.
4. DIMENSION B DOES NOT INCLUDE MOLD FLASH.
5. ROUNDED CORNERS OPTIONAL.

	INCHES		MILLIMETERS	
DIM	MIN	MAX	MIN	MAX
A	0.715	0.770	18.16	18.80
B	0.240	0.260	6.10	6.60
C	0.145	0.185	3.69	4.69
D	0.015	0.021	0.38	0.53
F	0.040	0.070	1.02	1.78
G	0.100 BSC		2.54 BSC	
H	0.052	0.095	1.32	2.41
J	0.008	0.015	0.20	0.38
K	0.115	0.135	2.92	3.43
L	0.290	0.310	7.37	7.87
M	—	10°	—	10°
N	0.015	0.039	0.38	1.01

Motorola reserves the right to make changes without further notice to any products herein. Motorola makes no warranty, representation or guarantee regarding the suitability of its products for any particular purpose, nor does Motorola assume any liability arising out of the application or use of any product or circuit, and specifically disclaims any and all liability, including without limitation consequential or incidental damages. "Typical" parameters which may be provided in Motorola data sheets and/or specifications can and do vary in different applications and actual performance may vary over time. All operating parameters, including "Typicals" must be validated for each customer application by customer's technical experts. Motorola does not convey any license under its patent rights nor the rights of others. Motorola products are not designed, intended, or authorized for use as components in systems intended for surgical implant into the body, or other applications intended to support or sustain life, or for any other application in which the failure of the Motorola product could create a situation where personal injury or death may occur. Should Buyer purchase or use Motorola products for any such unintended or unauthorized application, Buyer shall indemnify and hold Motorola and its officers, employees, subsidiaries, affiliates, and distributors harmless against all claims, costs, damages, and expenses, and reasonable attorney fees arising out of, directly or indirectly, any claim of personal injury or death associated with such unintended or unauthorized use, even if such claim alleges that Motorola was negligent regarding the design or manufacture of the part. Motorola and Ⓜ are registered trademarks of Motorola, Inc. Motorola, Inc. is an Equal Opportunity/Affirmative Action Employer.

Mfax is a trademark of Motorola, Inc.

How to reach us:
USA/EUROPE/Locations Not Listed: Motorola Literature Distribution; P.O. Box 5405, Denver, Colorado 80217. 1–303–675–2140 or 1–800–441–2447

JAPAN: Nippon Motorola Ltd.: SPD, Strategic Planning Office, 4–32–1, Nishi–Gotanda, Shinagawa–ku, Tokyo 141, Japan. 81–3–5487–8488

Customer Focus Center: 1–800–521–6274

Mfax™: RMFAX0@email.sps.mot.com – TOUCHTONE 1–602–244–6609
Motorola Fax Back System – US & Canada ONLY 1–800–774–1848
– http://sps.motorola.com/mfax/

ASIA/PACIFIC: Motorola Semiconductors H.K. Ltd.; 8B Tai Ping Industrial Park, 51 Ting Kok Road, Tai Po, N.T., Hong Kong. 852–26629298

HOME PAGE: http://motorola.com/sps/

intersil

CA741, CA741C, CA1458, CA1558, LM741, LM741C, LM1458

Data Sheet | September 1998 | File Number 531.4

0.9MHz Single and Dual, High Gain Operational Amplifiers for Military, Industrial and Commercial Applications

The CA1458, CA1558 (dual types); CA741C, CA741 (single types); high-gain operational amplifiers for use in military, industrial, and commercial applications.

These monolithic silicon integrated circuit devices provide output short circuit protection and latch-free operation. These types also feature wide common mode and differential mode signal ranges and have low offset voltage nulling capability when used with an appropriately valued potentiometer. A 10k potentiometer is used for offset nulling types CA741C, CA741 (see Figure 1). Types CA1458, CA1558 have no specific terminals for offset nulling. Each type consists of a differential input amplifier that effectively drives a gain and level shifting stage having a complementary emitter follower output.

The manufacturing process make it possible to produce IC operational amplifiers with low burst "popcorn" noise characteristics.

Technical Data on LM Branded types is identical to the corresponding CA Branded types.

Features

- Input Bias Current . . . . 500nA (Max)
- Input Offset Current . . . . 200nA (Max)

Applications

- Comparator
- Multivibrator
- DC Amplifier
- Summing Amplifier
- Integrator or Differentiator
- Narrow Band or Band Pass Filter

Ordering Information

PART NUMBER	TEMP. RANGE (°C)	PACKAGE	PKG. NO.
CA0741E	-55 to 125	8 Ld PDIP	E8.3
CA0741CE	0 to 70	8 Ld PDIP	E8.3
CA1458E	0 to 70	8 Ld PDIP	E8.3
CA1558E	-55 to 125	8 Ld PDIP	E8.3
CA0741T	-55 to 125	8 Pin Metal Can	T8.C
CA0741CT	0 to 70	8 Pin Metal Can	T8.C
CA1558T	-55 to 125	8 Pin Metal Can	T8.C
LM741N	-55 to 125	8 Ld PDIP	E8.3
LM741CN	0 to 70	8 Ld PDIP	E8.3
LM1458N	0 to 70	8 Ld PDIP	E8.3

Pinouts

CA741, CA741C (CAN)
TOP VIEW

NC 8; OFFSET NULL 1; V+ 7; INV. INPUT 2; OUT 6; NON-INV. INPUT 3; OFFSET NULL 5; V- 4

CA741, CA741C, LM741, LM741C (PDIP)
TOP VIEW

OFFSET NULL 1; INV. INPUT 2; NON-INV. INPUT 3; V- 4; NC 8; V+ 7; OUTPUT 6; OFFSET NULL 5

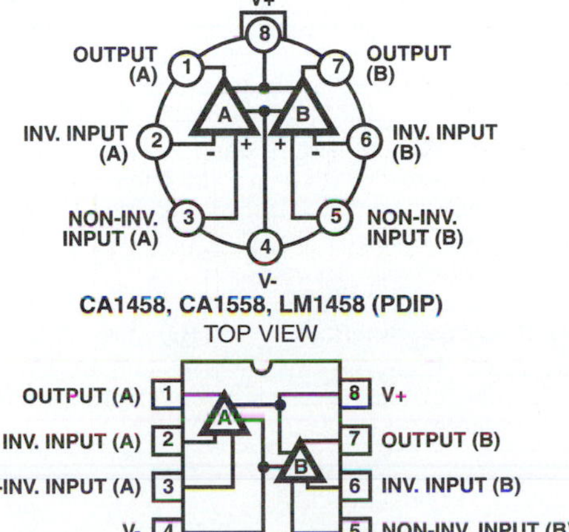

CAUTION: These devices are sensitive to electrostatic discharge; follow proper IC Handling Procedures.
1-888-INTERSIL or 321-724-7143 | Copyright © Intersil Corporation 1999

CA741, CA741C, CA1458, CA1558, LM741, LM741C, LM1458

Absolute Maximum Ratings

Supply Voltage
- CA741C, CA1458, LM741C, LM1458 (Note 1) 36V
- CA741, CA1558, LM741 (Note 1) 44V

Differential Input Voltage 30V
Input Voltage . . . . V_{SUPPLY}
Offset Terminal to V- Terminal Voltage (CA741C, CA741) 0.5V
Output Short Circuit Duration. . . . . Indefinite

Thermal Information

Thermal Resistance (Typical, Note 3)	JA (°C/W)	JC (°C/W)
PDIP Package	130	N/A
Can Package	155	67

Maximum Junction Temperature (Can Package). . . . . 175°C
Maximum Junction Temperature (Plastic Package) . . . . 150°C
Maximum Storage Temperature Range . . . . -65°C to 150°C
Maximum Lead Temperature (Soldering 10s) . . . . 300°C

Operating Conditions

Temperature Range
- CA741, CA1558, LM741 . . . . -55°C to 125°C
- CA741C, CA1458, LM741C, LM1458 (Note 2) . . . . 0°C to 70°C

CAUTION: Stresses above those listed in "Absolute Maximum Ratings" may cause permanent damage to the device. This is a stress only rating and operation of the device at these or any other conditions above those indicated in the operational sections of this specification is not implied.

NOTES:
1. Values apply for each section of the dual amplifiers.
2. All types in any package style can be operated over the temperature range of -55°C to 125°C, although the published limits for certain electrical specification apply only over the temperature range of 0°C to 70°C.
3. JA is measured with the component mounted on an evaluation PC board in free air.

Electrical Specifications

Typical Values Intended Only for Design Guidance, V_{SUPPLY} = 15V

PARAMETER	SYMBOL	TEST CONDITIONS	TYPICAL VALUE (ALL TYPES)	UNITS
Input Capacitance	C_I		1.4	pF
Offset Voltage Adjustment Range			15	mV
Output Resistance	R_O		75	
Output Short Circuit Current			25	mA
Transient Response		Unity Gain, V_I = 20mV, R_L = 2k C_L 100pF		
Rise Time	t_r		0.3	s
Overshoot	O.S.		5.0	%
Slew Rate (Closed Loop)	SR	R_L 2k	0.5	V/ s
Gain Bandwidth Product	GBWP	R_L = 12k	0.9	MHz

Electrical Specifications

For Equipment Design, V_{SUPPLY} = 15V

PARAMETER	TEST CONDITIONS	TEMP (°C)	(NOTE 4) CA741, CA1558, LM741			(NOTE 4) CA741C, CA1458, LM741C, LM1458			UNITS
			MIN	TYP	MAX	MIN	TYP	MAX	
Input Offset Voltage	R_S 10k	25	-	1	5	-	2	6	mV
		Full	-	1	6	-	-	7.5	mV
Input Common Mode Voltage Range		25	-	-	-	12	13	-	V
		Full	12	13	-	-	-	-	V
Common Mode Rejection Ratio	R_S 10k	25	-	-	-	70	90	-	dB
		Full	70	90	-	-	-	-	dB
Power Supply Rejection Ratio	R_S 10k	25	-	-	-	-	30	150	V/V
		Full	-	30	150	-	-	-	V/V
Input Resistance		25	0.3	2	-	0.3	2	-	M

CA741, CA741C, CA1458, CA1558, LM741, LM741C, LM1458

Electrical Specifications For Equipment Design, V_{SUPPLY} = 15V **(Continued)**

PARAMETER	TEST CONDITIONS	TEMP (°C)	(NOTE 4) **CA741, CA1558, LM741**			(NOTE 4) **CA741C, CA1458, LM741C, LM1458**			UNITS
			MIN	**TYP**	**MAX**	**MIN**	**TYP**	**MAX**	
Input Bias Current		25	-	80	500	-	80	500	nA
		Full	-	-	-	-	-	800	nA
		-55	-	300	1500	-	-	-	nA
		125	-	30	500	-	-	-	nA
Input Offset Current		25	-	20	200	-	20	200	nA
		Full	-	-	-	-	-	300	nA
		-55	-	85	500	-	-	-	nA
		125	-	7	200	-	-	-	nA
Large Signal Voltage Gain	R_L 2k , V_O = 10V	25	50,000	200,000	-	20,000	200,000	-	V/V
		Full	25,000	-	-	15,000	-	-	V/V
Output Voltage Swing	R_L 10k	25	-	-	-	12	14	-	V
		Full	12	14	-	-	-	-	V
	R_L 2k	25	-	-	-	10	13	-	V
		Full	10	13	-	10	13	-	V
Supply Current		25	-	1.7	2.8	-	1.7	2.8	mA
		-55	-	2	3.3	-	-	-	mA
		125	-	1.5	2.5	-	-	-	mA
Device Power Dissipation		25	-	50	85	-	50	85	mW
		-55	-	60	100	-	-	-	mW
		125	-	45	75	-	-	-	mW

NOTE:

4. Values apply for each section of the dual amplifiers.

Test Circuits

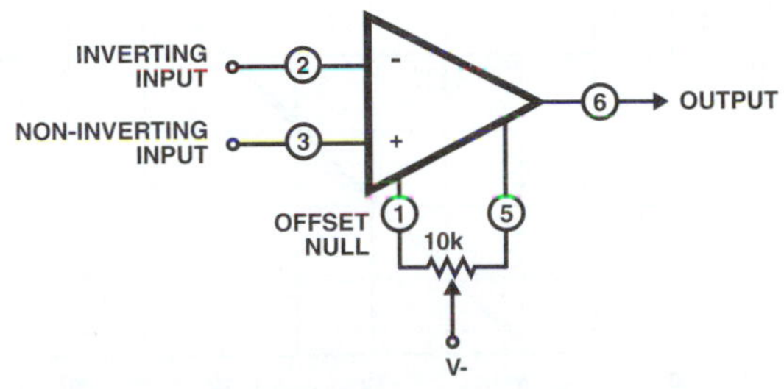

FIGURE 1. OFFSET VOLTAGE NULL CIRCUIT FOR CA741C, CA741, LM741C, AND LM741

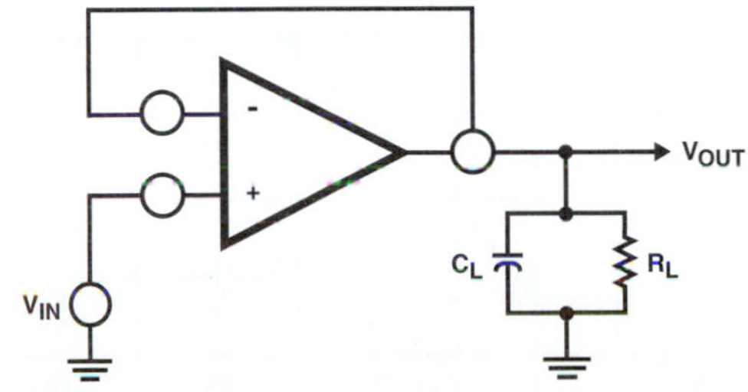

FIGURE 2. TRANSIENT RESPONSE TEST CIRCUIT FOR ALL TYPES

CA741, CA741C, CA1458, CA1558, LM741, LM741C, LM1458

Schematic Diagram (Notes 5, 6)

CA741C, CA741, LM741C, LM741 AND FOR EACH AMPLIFIER OF THE CA1458, CA1558, AND LM1458

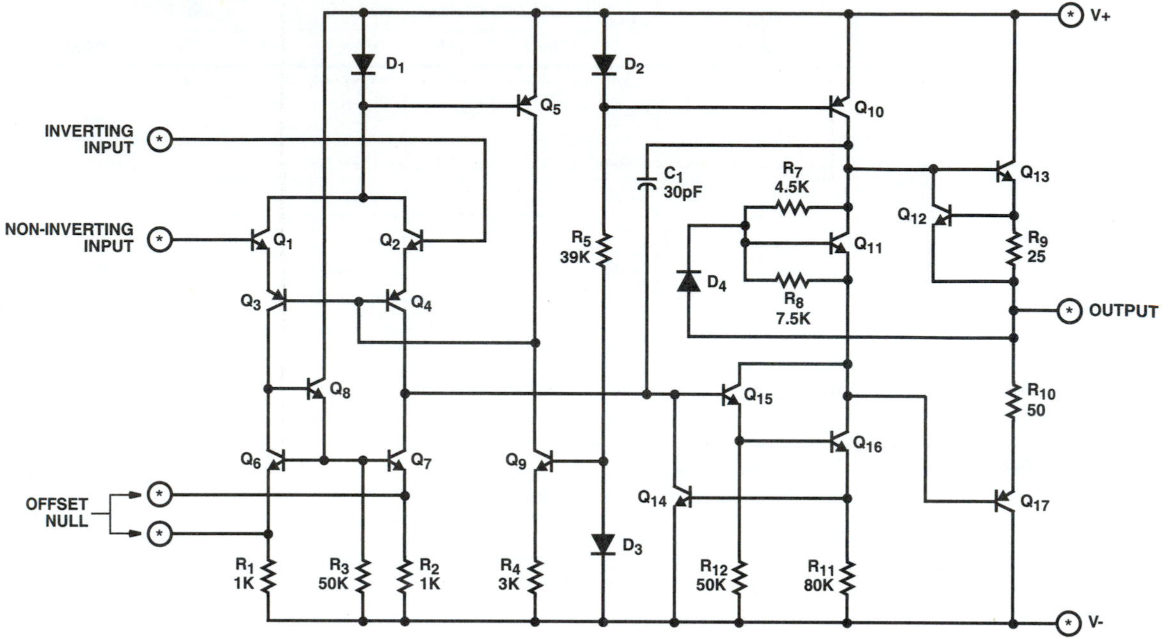

NOTES:

5. See Pinouts for Terminal Numbers of Respective Types.
6. All Resistance Values are in Ohms.

Typical Performance Curves

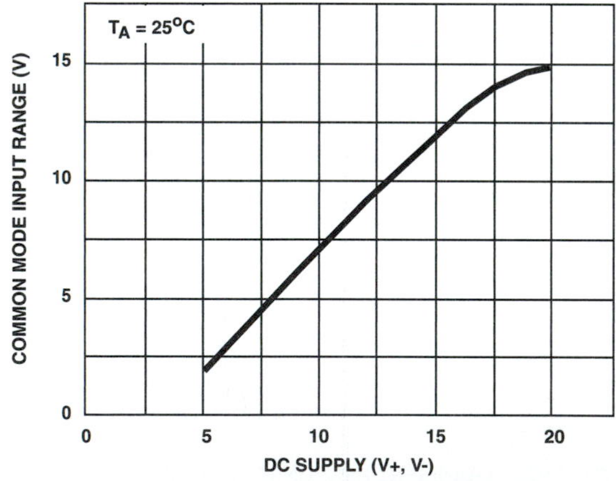

FIGURE 3. COMMON MODE INPUT VOLTAGE RANGE vs SUPPLY VOLTAGE FOR ALL TYPES

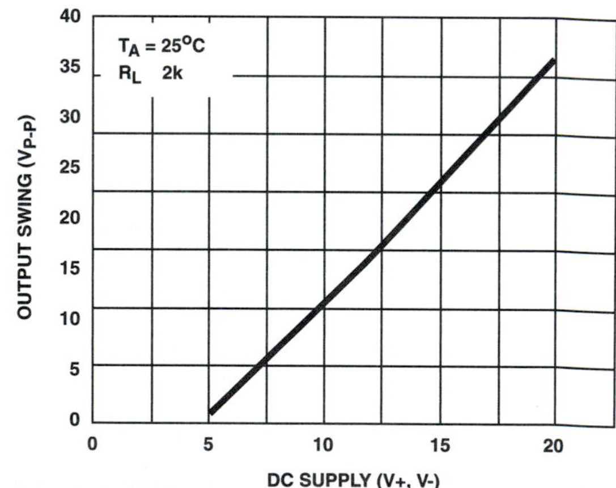

FIGURE 4. OUTPUT VOLTAGE vs SUPPLY VOLTAGE FOR ALL TYPES

CA741, CA741C, CA1458, CA1558, LM741, LM741C, LM1458

Typical Performance Curves (Continued)

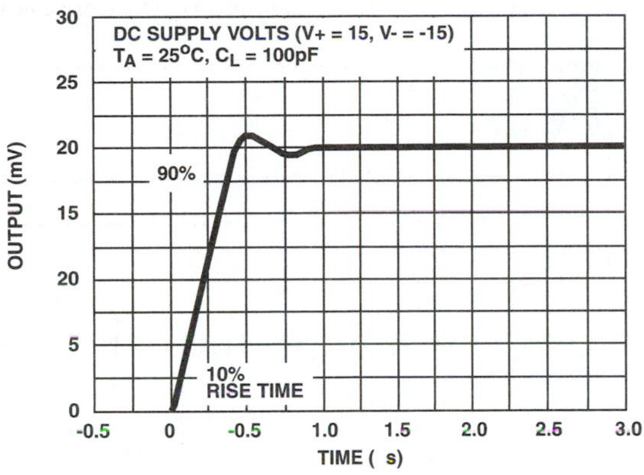

FIGURE 5. TRANSIENT RESPONSE FOR CA741C AND CA741

Metallization Mask Layout

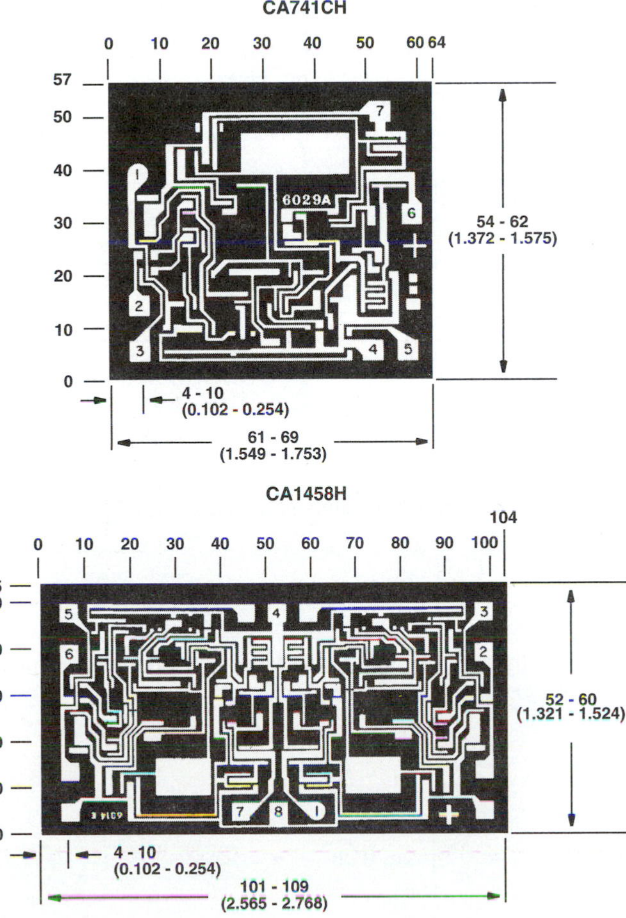

NOTE: Dimensions in parentheses are in millimeters and are derived from the basic inch dimensions as indicated. Grid graduations are in mils (10^{-3} inch).

APPENDIX 3

DERIVATIONS

DERIVATION OF THE AVERAGE VALUE OF THE FULL-WAVE AND HALF-WAVE RECTIFIED SINE WAVE

A full-wave (FW) rectified sine wave is shown in Fig. A3.1. The horizontal axis is scaled in radians to make the derivation simpler. The signal is a period of π radians, as compared to 2π radians for a normal sine wave. The average value of this function is the height of the rectangle that has the same area, over one period, as that of the original FW rectified signal.

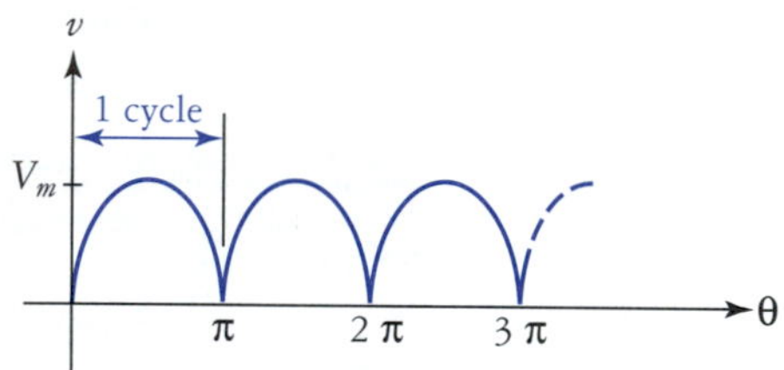

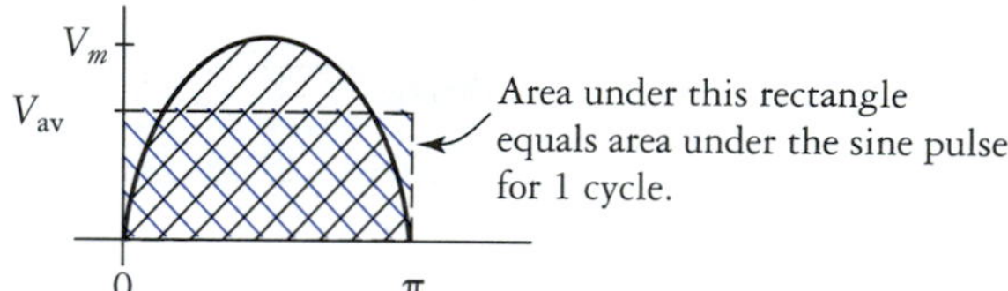

FIGURE A3.1
Full-wave rectified signal.

The average, or mean value, of a function (or signal) over any interval is given by

$$V_{av} = \frac{1}{b-a}\int_a^b f(t)\,dt$$

We set the limits of the integration for $a = 0$ and $b = \pi$, which is one cycle of the signal, and evaluate the integral:

$$\begin{aligned}
V_{av(FW)} &= \frac{1}{\pi}\int_0^{\pi} V_m \sin\theta\, d\theta \\
&= \frac{V_m}{\pi}\int_0^{\pi} \sin\theta\, d\theta \\
&= \frac{V_m}{\pi}(-\cos\theta)\Big|_0^{\pi} \\
&= \frac{V_m}{\pi}[(-\cos\pi) - (-\cos 0)] \\
&= \frac{V_m}{\pi}[(-1) - (-1)] \\
&= \frac{2\,V_m}{\pi}
\end{aligned}$$

For the half-wave (HW) rectified sine wave, the process is similar except that, as shown in Fig. A3.2, the signal has a period of 2π radians (relative to a normal sine function) and has a value of zero for $\pi < \theta < 2\pi$. Because of this, we can use proportionality to infer that the average value is half of that for the full-wave rectified signal. That is,

$$V_{\text{av(HW)}} = \frac{V_m}{\pi}$$

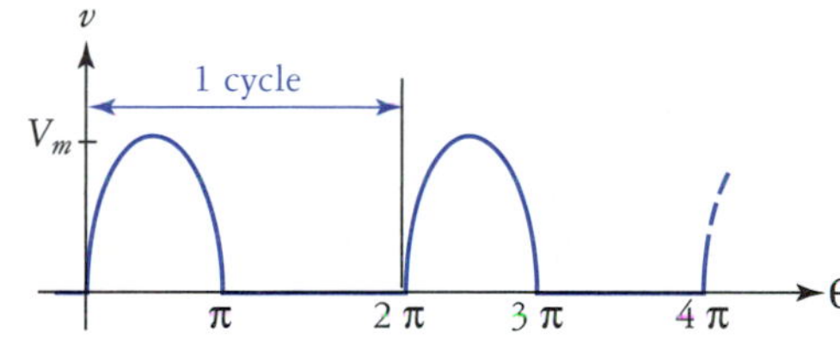

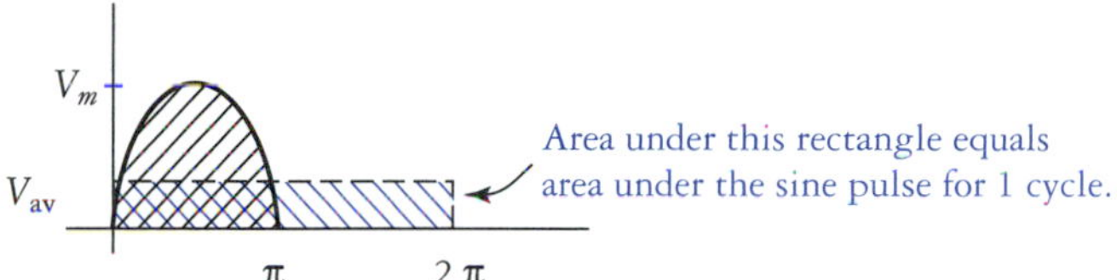

FIGURE A3.2
Half-wave rectified signal.

DERIVATION OF THE DYNAMIC RESISTANCE OF THE FORWARD-BIASED PN JUNCTION

Shockley's equation for a forward-biased pn junction is approximately

$$I_F = I_S e^{V_F\left(\frac{1}{\eta V_T}\right)}$$

Taking the derivative of I_F with respect to V_F gives us the conductance of the forward-biased junction. That is,

$$g_F = \frac{dI_F}{dV_F}$$

$$= \frac{1}{\eta V_T} I_S e^{V_F\left(\frac{1}{\eta V_T}\right)}$$

However, the factor $I_S e^{V_F\left(\frac{1}{\eta V_T}\right)}$ is simply the quiescent current flow through the junction (I_{DQ} for a diode, and I_{CQ} for a transistor). In other words,

$$I_{DQ} = I_S e^{V_F\left(\frac{1}{\eta V_T}\right)}$$

Thus, we can write

$$g_F = \frac{I_{CQ}}{\eta V_T}$$

When we are dealing with a bipolar transistor, this parameter is called *transconductance,* which is often denoted as g_m. Since resistance is simply the reciprocal of conductance, the dynamic resistance of a forward-biased pn junction is

$$r_F = \frac{1}{g_F} = \frac{\eta V_T}{I_{CQ}}$$

Most of the time we assume $\eta = 1$. Using the BJT dynamic emitter resistance notation gives us

$$r_e = \frac{V_T}{I_{CQ}} \cong \frac{26\text{ mV}}{I_{CQ}} \quad \text{(at room temperature)}$$

DERIVATION OF g_{mQ} FOR THE JFET

The drain current of the JFET is given by

$$I_D = I_{DSS}\left(1 - \frac{V_{GS}}{V_{GS(\text{off})}}\right)^2$$

The transconductance of the FET is the derivative of I_D with respect to V_{GS}. To make it easier to differentiate, we can expand the drain current equation and rewrite it in standard polynomial form as

$$I_D = -\frac{I_{DSS}}{V_{GS(\text{off})}^2} V_{GS^2} - \frac{I_{DSS}}{V_{GS(\text{off})}} V_{GS} + I_{DSS}$$

Taking the derivative term by term and performing some algebra yields

$$\begin{aligned} g_{mQ} &= \frac{dI_D}{dV_{GS}} \\ &= -\frac{2\,I_{DSS}}{V_{GS(\text{off})}^2} V_{GS} - \frac{I_{DSS}}{V_{GS(\text{off})}} \\ &= -\frac{2\,I_{DSS}}{V_{GS(\text{off})}}\left(\frac{V_{GS}}{V_{GS(\text{off})}} - 1\right) \\ &= \frac{2\,I_{DSS}}{V_{GS(\text{off})}}\left(1 - \frac{V_{GS}}{V_{GS(\text{off})}}\right) \end{aligned}$$

or, equivalently,

$$g_{mQ} = g_{m0}\left(1 - \frac{V_{GS}}{V_{GS(\text{off})}}\right)$$

NONLINEARITY OF THE BIPOLAR JUNCTION TRANSISTOR

The ac equivalent of a common-emitter BJT amplifier is shown in Fig. A3.3. We assume that a stable Q point has been established by some suitable biasing arrangement. Application of a signal results in an instantaneous base-emitter voltage of

$$v_{BE} = V_{BEQ} + v_{\text{in}}$$

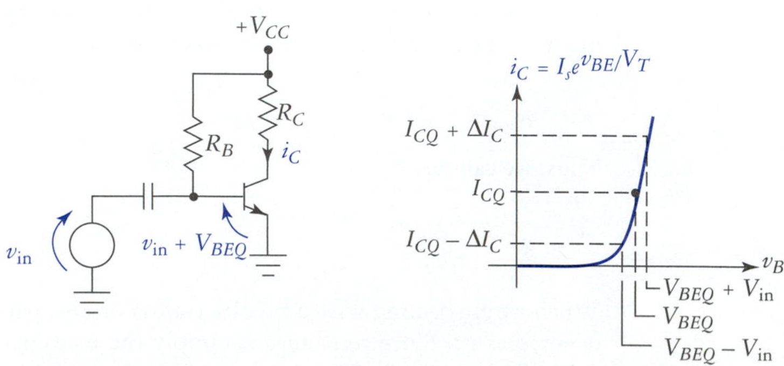

FIGURE A3.3
Simple BJT CE amplifier and device transconductance curve.

That is, the total B-E voltage is the superposition of the quiescent B-E voltage and the input voltage. Substituting this into Shockley's equation (with $\eta = 1$, to reduce the clutter) produces

$$i_C = I_S\, e^{(V_{BEQ} + v_{\text{in}})/V_T}$$

The exponential can be factored giving us

$$i_C = I_S\, e^{V_{BEQ}/V_T} e^{v_{\text{in}}/V_T}$$

The two left-hand factors represent the quiescent collector current, so we can write

$$i_C = I_{CQ}\, e^{v_{\text{in}}/V_T}$$

The exponential can be expanded into an infinite series of terms, called a *power series.* For x between 0 and 1, this expansion is

$$e^x = x^0 \frac{I_{CQ}}{0!} + x^1 \frac{I_{CQ}}{1!} + x^2 \frac{I_{CQ}}{2!} + x^3 \frac{I_{CQ}}{6!} + \ \ldots$$
$$= 1 + x + x^2 \frac{1}{2} + x^3 \frac{1}{6} + \ \ldots$$

We substitute v_{in}/V_T for x, and use this series in Shockley's equation, which gives us

$$i_C = I_{CQ} + v_{\text{in}} \frac{1}{V_T} + v_{\text{in}}^2 \frac{1}{2\,V_T} + v_{\text{in}}^3 \frac{1}{3\,V_T} + \ \ldots$$

Successively higher order terms decrease rapidly when $v_{in} \ll V_T$. Thus, under small-signal conditions, only the first two terms are significant, resulting in linear amplification. In this way, we can define a small signal in terms of the thermal voltage V_T.

As the amplitude of the input becomes larger, the quadratic term becomes significant, then the cubic term, and so on. Assuming that the linear and quadratic factors are significant, if we apply a signal that consists of two or more frequencies, the output will contain the original signal frequencies plus their sums and differences. This characteristic allows the BJT to be used in modulator applications.

APPENDIX 4

OrCAD PSPICE, RELEASE 9

The majority of the computer-aided circuit analysis examples in this book were entered using PSpice, release 7 or release 8. Just as this book was being finished, Microsim, the original producer of PSpice merged with OrCAD. Soon after, PSpice, release 9, was made available. Although there are many similarities between release 9 and earlier versions, the differences are significant. In fact, based on the author's conversations with many colleagues, many users of PSpice have chosen to stick with version 8 for the time being. However, this appendix provides a quick introduction for those who wish to use release 9. As of this writing, evaluation versions of both PSpice release 8 and release 9 are available from OrCAD. The OrCAD website is www.orcad.com.

STARTING THE PROGRAM

Click **Start, Programs, OrCAD Demo,** then select **Capture CIS Demo.** The OrCAD Capture screen should now be on the screen.

STARTING A NEW PROJECT

Click **File, New,** then select **Project.** The New Project window should pop onto the screen, as shown in Fig. A4.1

Enter an appropriate name for your project and select the **Analog, Mixed-Signal Circuit Wizard** button, then click **OK.**

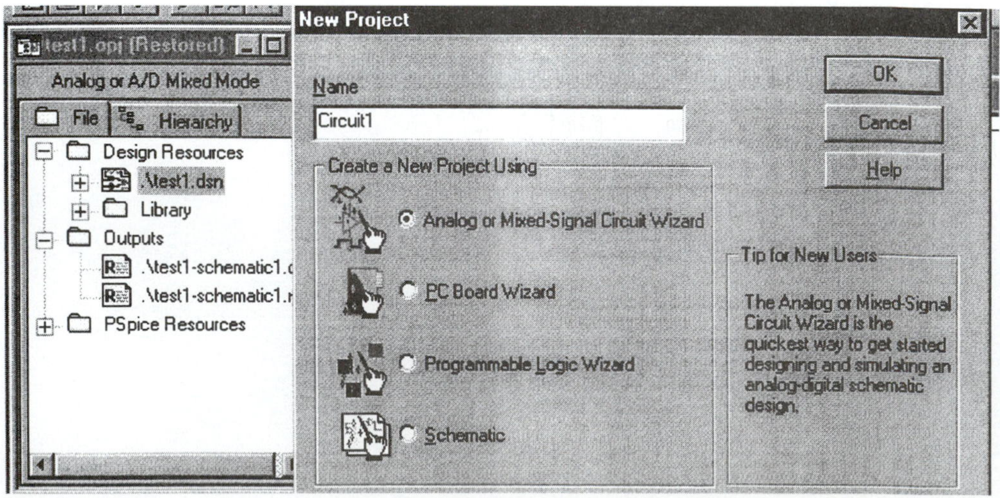

FIGURE A4.1
Starting a new project using Capture.

SELECTING COMPONENT LIBRARIES

At this point, a library selection window opens. Here, you can select which component libraries your project will have access to. It is generally best to select all of the libraries. That is, left click on each library in the left side window and then click **Add.** Click **Finish** when all libraries are in the right-hand window. At this point, you should have a display like that of Fig. A4.2.

The small box at the left of Fig. A4.2 is called the Project Manager. Clicking on the + signs expands each folder. Double clicking on the **PAGE 1** icon brings up the blank schematic page, if it is not already present, and activates the schematic editing selections on the toolbar.

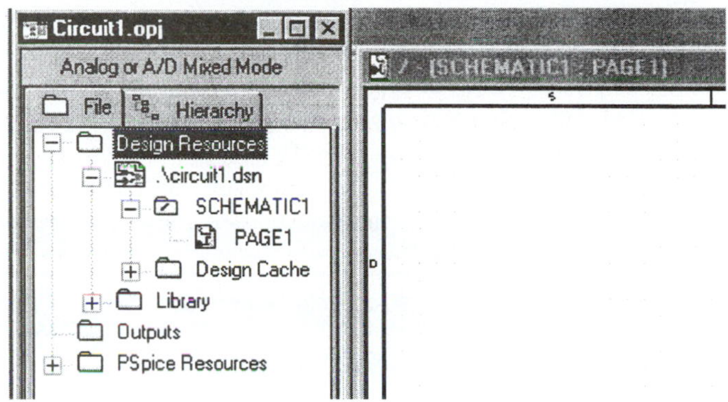

FIGURE A4.2
Opening a new schematic page.

ENTERING A SCHEMATIC

We enter a voltage-divider-biased BJT amplifier. Click **Place, Part,** and select Eval.lib. Choose the 2N3904 (look under Q2N3904) and click **OK.** Place the transistor on the schematic by left clicking once. To exit this part-dropping mode, right click and select **End Mode.**

Enter the remaining parts, as shown in Fig. A4.3 Add the V_{CC} bubbles by selecting **Place, Power, VCC_CIRCLE/CAPSYM.** Place a VSIN source for the input signal and a VDC source for the power supply voltage. Note that components are rotated by pressing **Ctrl-R,** as in earlier versions of PSpice.

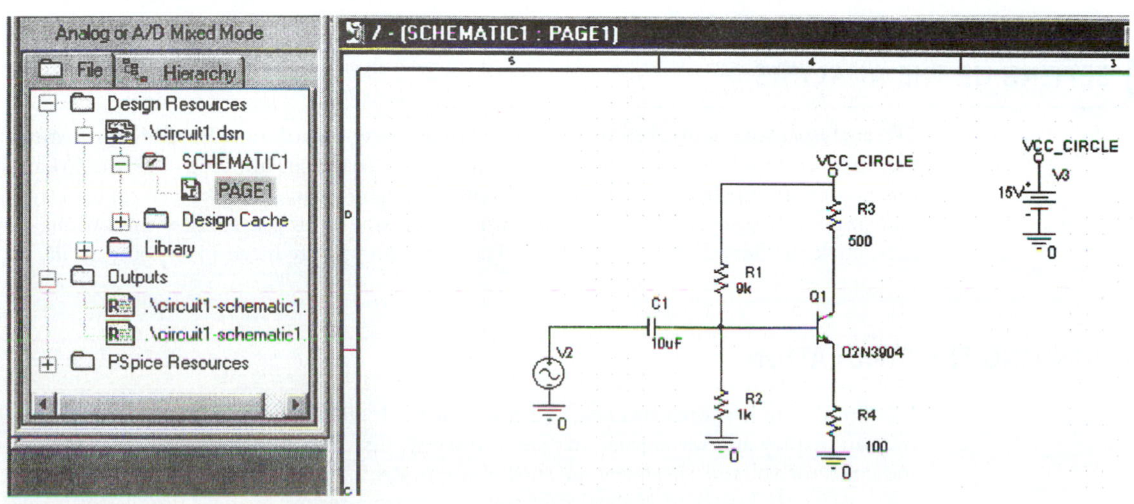

FIGURE A4.3
Sample schematic.

Use the button to enter the wire-placement mode. Right click and select **End Mode** to exit this mode.

All PSpice circuits require a ground reference point. In release 9, the ground is chosen by clicking **Place, Ground,** then selecting **0/SOURCE.**

EDITING COMPONENT PROPERTIES

Double clicking on a part pulls up its Property Editor window. For example, using V2, the sine source, we obtain the display shown in Fig. A4.4. Here, we can change the appropriate device properties, such as the resistance of a resistor, source voltage, and so on. Simply clicking the X to close the Property Editor window saves any changes you made. Set the input source for VDC = 0 V, VOFF = 0 V, VAC = 1 V, VAMPL = 1 V, and FREQ = 1k.

You can also edit the value of a component such as a resistor or capacitor by double clicking on the value of that part. This is the simplest way to change the value.

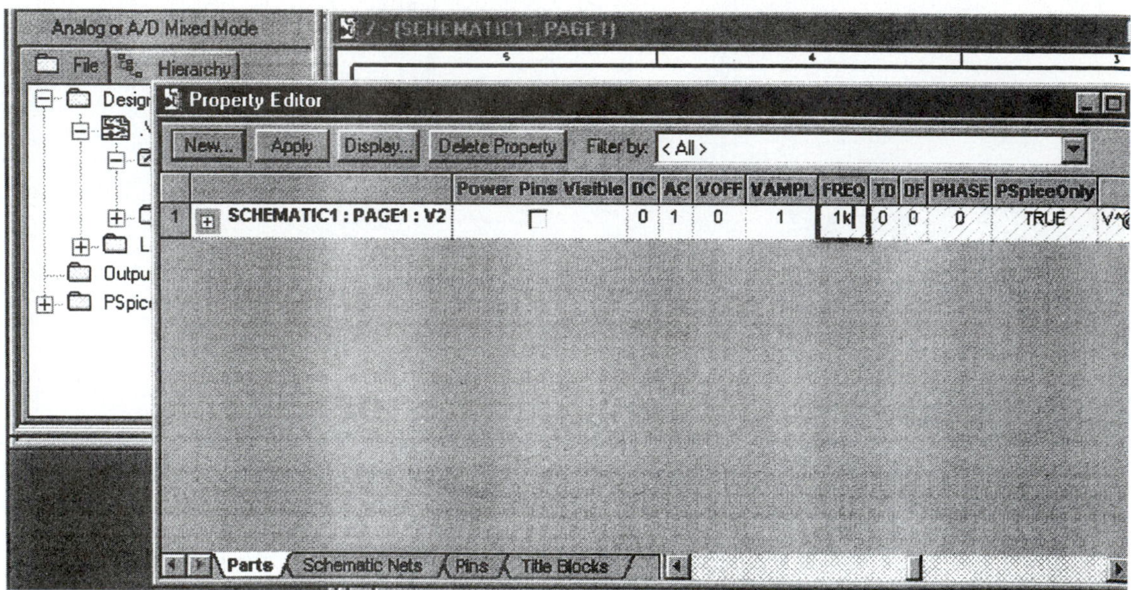

FIGURE A4.4
Editing component properties.

SETTING UP THE ANALYSIS

The analysis setup is invoked by clicking **PSpice, New Simulation Profile** on the menu bar. Enter a simulation profile name, such as *Profile 1* in the box that pops up, then click the **Create** button. At this point you can choose the simulation type(s) that you desire. First we will set up a transient analysis using a run time of 5 ms with a maximum step size of 10 μs. To view the various dc node voltages, click the **Save Bias Point** button. The node voltages are listed in the output file.

RUNNING THE SIMULATION

The simulation is started by clicking **PSpice, Run.** If all is well, the PROBE window will open and you can display any waveforms that are of interest, just like in earlier versions of PSpice. The input and collector voltage waveforms are shown in Fig. A4.5.

After the transient analysis run, you can set up a second analysis profile, say, *Profile 2,* during which a frequency response analysis is performed. Again, after this simulation is performed, PROBE allows the viewing of various circuit waveforms.

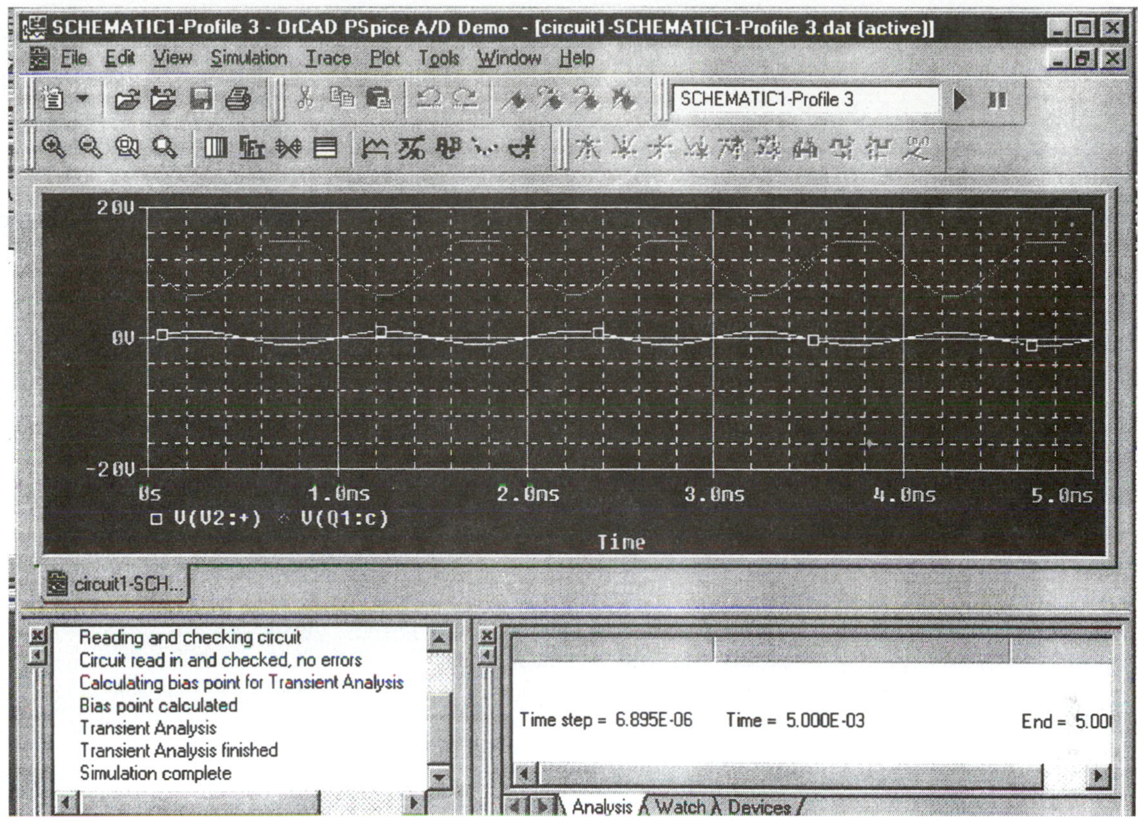

FIGURE A4.5
Viewing circuit waveforms.

INDEX